1st International Conference of Combinatorics, Graph Theory, and Network Topology (ICCGANT 2017)

Journal of Physics: Conference Series Volume 1008

Jember, Indonesia
25 - 26 November 2017

Editor:

Dafik

ISBN: 978-1-5108-6240-1
ISSN: 1742-6588

Printed from e-media with permission by:

Curran Associates, Inc.
57 Morehouse Lane
Red Hook, NY 12571

proceedings.com

Some format issues inherent in the e-media version may also appear in this print version.

Printed by Curran Associates, Inc. (2018)

For permission requests, please contact the Institute of Physics at the address below.

Institute of Physics
Dirac House, Temple Back
Bristol BS1 6BE UK

Phone: 44 1 17 929 7481
Fax: 44 1 17 920 0979

techtracking@iop.org

Additional copies of this publication are available from:

Curran Associates, Inc.
57 Morehouse Lane
Red Hook, NY 12571 USA
Phone: 845-758-0400
Fax: 845-758-2633
Email: curran@proceedings.com
Web: www.proceedings.com

Table of contents

Volume 1008

The 1st International Conference of Combinatorics, Graph Theory, and Network Topology

25–26 November 2017, The University of Jember, East Java, Indonesia

Accepted papers received: 9 April 2018
Published online: 27 April 2018

Preface

Papers

Applied Mathematics

Combinatorics

Mathematics Education

IOP Conf. Series: Journal of Physics: Conf. Series **1008** (2018) 011001 doi:10.1088/1742-6596/1008/1/011001

The First International Conference on Combinatorics, Graph Theory and Network Topology (ICCGANT)

Dafik

Editor in Chief of ICCGANTs Publication, University of Jember, Jember, Indonesia

E-mail: d.dafik@unej.ac.id

Preface

It is with my great pleasure and honor to organize the First International Conference on Combinatorics, Graph Theory and Network Topology which is held from 25-26 November 2017 in the University of Jember, East Java, Indonesia and present a conference proceeding index by Scopus. It is the first international conference organized by CGANT Research Group University of Jember in cooperation with Indonesian Combinatorics Society (INACOBMS). The conference is held to welcome participants from many countries, with broad and diverse research interests of mathematics especially combinatorical study. The mission is to become an annual international forum in the future, where, civil society organization and representative, research students, academics and researchers, scholars, scientist, teachers and practitioners from all over the world could meet in and exchange an idea to share and to discuss theoretical and practical knowledge about mathematics and its applications. The aim of the first conference is to present and discuss the latest research that contributes to the sharing of new theoretical, methodological and empirical knowledge and a better understanding in the area mathematics, application of mathematics as well as mathematics education.

The themes of this conference are as follows: (1) Connection of distance to other graph properties, (2) Degree/diameter problem, (3) Distance-transitive and distance-regular graphs, (4) Metric dimension and related parameters, (5) Cages and eccentric graphs, (6) Cycles and factors in graphs, (7) Large graphs and digraphs, (8) Spectral Techniques in graph theory, (9) Ramsey numbers, (10) Dimensions of graphs, (11) Communication networks, (12) Coding theory, (13) Cryptography, (14) Rainbow connection, (15) Graph labelings and coloring, (16). Applications of graph theory

The topics are not limited to the above themes but they also include the mathematical application research of interest in general including mathematics education, such as:(1) Applied Mathematics and Modelling, (2) Applied Physics: Mathematical Physics, Biological Physics, Chemistry Physics,(3) Applied Engineering: Mathematical Engineering, Mechanical engineering, Informatics Engineering, Civil Engineering,(4) Statistics and Its Application,(5) Pure Mathematics (Analysis, Algebra and Geometry),(6) Mathematics Education, (7) Literacy of Mathematics,(8) The Use of ICT Based Media In Mathematics Teaching and Learning,(9) Technological, Pedagogical, Content Knowledge for Teaching Mathematics, (10) Students Higher Order Thinking Skill of Mathematics, (11) Contextual Teaching and Realistic Mathematics, (12) Science, Technology, Engineering, and Mathematics Approach, (13) Local Wisdom Based

IOP Conf. Series: Journal of Physics: Conf. Series **1008** (2018) 011001 doi:10.1088/1742-6596/1008/1/011001

Education: Ethnomathematics, (14) Showcase of Teaching and Learning of Mathematics, (16) The 21st Century Skills: The Integration of 4C Skill in Teaching Math.

The participants of this ICCGANT 2017 conference were 200 people consisting research students, academics and researchers, scholars, scientist, teachers and practitioners from many countries. The selected papers to be publish of Journal of Physics: Conference Series are 80 papers. On behalf of the organizing committee, finally we gratefully acknowledge the support from the University of Jember of this conference. We would also like to extend our thanks to all lovely participants who are joining this unforgettable and valuable event.

Prof. Drs. Dafik, M.Sc., Ph.D.

IOP Conf. Series: Journal of Physics: Conf. Series **1008** (2018) 011002 doi:10.1088/1742-6596/1008/1/011002

The Committees of The First International Conference on Combinatorics, Graph Theory and Network Topology (ICCGANT)

Dafik

Editor in Chief of ICCGANTs Publication, University of Jember, Jember, Indonesia
Professor of Combinatorics and Graph Theory

E-mail: d.dafik@unej.ac.id

IOP Conf. Series: Journal of Physics: Conf. Series **1008** (2018) 011002 doi:10.1088/1742-6596/1008/1/011002

Scientific Committee and Reviewers

The committees of the First International Conference on Combinatorics, Graph Theory and Network Topology would like to express gratitude to all Committees for the volunteering support and contribution in the editing and reviewing process.

IOP Conf. Series: Journal of Physics: Conf. Series **1008** (2018) 011003 doi:10.1088/1742-6596/1008/1/011003

Peer review statement

All papers published in this volume of *Journal of Physics: Conference Series* have been peer reviewed through processes administered by the proceedings Editors. Reviews were conducted by expert referees to the professional and scientific standards expected of a proceedings journal published by IOP Publishing.

IOP Conf. Series: Journal of Physics: Conf. Series **1008** (2018) 012001 doi:10.1088/1742-6596/1008/1/012001

The effect of heat generation on mixed convection flow in nano fluids over a horizontal circular cylinder

Bagus Juliyanto[1], Basuki Widodo[1], and Chairul Imron[1]

[1]Department of Mathematics, Institut Teknologi Sepuluh Nopember, Arif Rahman Hakim Street, Surabaya, Indonesian.

E-mail: bagus14@mhs.matematika.its.ac.id; b_widodo@matematika.its.ac.id; imron-its@matematika.its.ac.id

Abstract. The purpose of this research is to study the effect of heat generation on mixed convection flow on Nano fluids over a horizontal circular cylinder of a heated in two dimension form. A stream of fluids are steady and incompressible, a stream flowing vertically upwards for circular cylinder and the boundary layer at the stagnation point. Three different types of nanoparticles considered are Cu, Al_2O_3, and TiO_2. Mixed convection flow in Nano fluids on the surface of a circular cylinder will cause the boundary layer. The governing boundary layer equations are transformed into a non-dimensional form, and then the non-dimensional forms are transformed into a similar boundary equations by using stream function. Furthermore, an implicit finite-difference scheme known as the Keller-box method is applied to solve numerically the resulting similar boundary layer equations. The result of the research by varying the non-dimensional parameters are mixed convection, Prandtl number, nanoparticle volume fraction, heat generation, and radius of a cylinder are as follows. First, the velocity profile increase and temperature profile decrease when mixed convection parameter increase. Second, the velocity and temperature profiles decrease when Prandtl number parameter increase. Third, the velocity profile with the variation of nanoparticle volume fraction (χ) is increased when the value of χ is $0,1 \leq \chi \leq 0,15$ and the velocity profile decreases when the value of χ is $0,19 < \chi \leq 0,5$ while the temperature profile is increasing when the value of χ is $0,1 \leq \chi \leq 0,5$. Fourth, the velocity and temperature profiles increase when heat generation and the radius of the cylinder increase. The last, *Cu*, Al_2O_3, and TiO_2 nanoparticles produce the same velocity and temperature profiles, but the three types of nanoparticles are different at the velocity and temperature values.

1. Introduction

Convection is the heat transfer due to movement / flow / fluid mixing from the hot part to the cold part. Free convection is a mixing motion caused by the difference in density, whereas when the mixing motion is caused by a particular tool from the outside it is referred to as forced convection. The combination of free convection and forced convection is known as mixed convection. Several studies on the use of boundary layers in a mixed convection stream which is in [1] said that the flow of mixed convection pass over the circular cylinder on the viscoelastic fluid. [3] examined the problem of mixed convection flow on a pseudo-plastic fluid pass over a horizontal circular cylinder with uniform surface heat flux. On the other hand, [8] examined the flow of mixed convection over a horizontal circular cylinder with Newtonian heating.

There are some characteristics of fluids that accept shape changes and flow's ability. Those characteristics are usually caused by inability's fluid function to hold a shear stress in static equilibrium. The consequence of this trait is Pascal's law which emphasizes the importance of pressure in characterizing fluid forms. Fluids can be characterized as Newtonian Fluids and Non-Newtonian Fluids. Newtonian fluid is a fluid whose behavior is consistent with Newton's law, for example, water. A Non-Newtonian fluid is a fluid whose behavior diverges against Newton's law [11]. The examples of non-Newtonian fluids are a pseudoplastic fluid, shear-thinning fluid, thixotropic fluid, viscoelastic fluid, and Nano fluids. In this study, the effect of mixed convection on Nano fluids is given. The term Nano fluids was first used by [4] defined as the dilution of nanometer-sized particles (smaller than 100 nm) in a fluid [5].

Studies on Nano fluids are getting an increase since the fluid is widely used in the fields of defense industry, oil drilling, food, and paper industry. Suppose that in oil drilling, hot Nano fluids flows through the surface of a medium and causes heat transfer process from the fluid to media. [10] in a study entitled mixed convection boundary layer flow from a horizontal circular cylinder in Nano fluids discussed the flow moving up and down perpendicular to the surface of the horizontal circular cylinder in two cases of heated and cooled cylinders. [6] studied the mixed convection flow of Nano fluids in the boundary layer with a porous conical shaped. [7] examined heat transfer problem in the mixed convection flow of Nano fluids that pass through a horizontal circular cylinder.

2. Methode Research

This research is conducted by the following steps.

1). Develop the Mathematical Model by the following steps:

a. Derive the mass conservation law, the second Newton law, and the first Thermodynamic law to obtain the governing equation.

b. The governing equation is simplified by using the boussinesq approach and boundary layer theory to obtain a dimensionless governing equation.

c. Transform the dimensionless governing equation into non-dimensional form of equation by using the non-dimensional parameters, which are λ (mixed convection), *Pr* (Prandtl number), *Gr* (Grashof number), and *Re* (Reynolds number).

d. The non-dimensional equations are transformed into similarity equations by using the flow function.

2). Solving the Mathematical Model

In this step, the obtained model is solved numerically by the Keller-Box method as the explained procedures. First, discretize the model by central difference method. Second, Linearization the obtained equations by Newton method and it is formed into vector matrix. Third, the result of step 2 is solved by the elimination tridiagonal block-matrix technique.

3). Numerical Simulation

The numerical simulation has been conducted to determine the effect of parameters of nanoparticle volume fraction (χ), mixed convection (λ), Prandtl number (*Pr*), and heat source (Q) the velocity profile (f') and the temperature profile (T).

3. Results and Discussion

3.1 Mathematics Models

IOP Conf. Series: Journal of Physics: Conf. Series **1008** (2018) 012001 doi:10.1088/1742-6596/1008/1/012001

The mathematical model was developed by deriving the law of conservation of mass, the second law of Newton, and the first law of Thermodynamics. Furthermore, it has been obtained the continuity equation, the momentum equation, and the energy equation from those above as follows:

$$\frac{\partial u}{\partial x}+\frac{\partial v}{\partial y}=0 \quad (1)$$

$$\left(u\frac{\partial u}{\partial x}+v\frac{\partial u}{\partial y}\right)=-\frac{1}{\rho_{nf}}\frac{\partial p}{\partial x}+\frac{\mu_{nf}}{\rho_{nf}}\left(\frac{\partial^2 u}{\partial x^2}+\frac{\partial^2 u}{\partial y^2}\right)+\frac{\beta_{nf}}{\rho_{nf}}(T-T_\infty)g_x \quad (2)$$

$$\left(u\frac{\partial v}{\partial x}+v\frac{\partial v}{\partial y}\right)=-\frac{1}{\rho_{nf}}\frac{\partial p}{\partial y}+\frac{\mu_{nf}}{\rho_{nf}}\left(\frac{\partial^2 v}{\partial x^2}+\frac{\partial^2 v}{\partial y^2}\right)+\frac{\beta_{nf}}{\rho_{nf}}(T-T_\infty)g_y \quad (3)$$

$$u\frac{\partial T}{\partial x}+v\frac{\partial T}{\partial y}=\alpha_{nf}\frac{\partial^2 T}{\partial y^2}+\frac{Q_0}{(\rho C_p)_{nf}}(T-T_\infty) \quad (4)$$

where:

$$\rho_{nf}=(1-\chi)\rho_f+\chi\rho_s\,;\ \beta_{nf}=\chi\rho_s\beta_s+(1-\chi)\rho_f\beta_f$$

$$\mu_{nf}=\mu_f\frac{1}{(1-\chi)^{2.5}};\ \alpha_{nf}=\frac{k_{nf}}{(\rho C_p)_{nf}}$$

$$(\rho C_p)_{nf}=(1-\chi)(\rho C_p)_f+\chi(\rho C_p)_s$$

$$k_{nf}=\frac{k_s+2k_f+2\chi(k_s-k_f)}{k_s+2k_f-\chi(k_s-k_f)}k_f$$

Non-dimensional variables [2]:

$$x=\frac{\bar{x}}{a}\,;\ y=Re^{1/2}\left(\frac{\bar{y}}{a}\right)$$

$$u=\frac{\bar{u}}{U_\infty}\,;v=Re^{1/2}\left(\frac{\bar{v}}{U_\infty}\right)$$

$$T=\frac{\bar{T}-T_\infty}{T_w-T_\infty}\,;\ p=\frac{\bar{p}}{\rho_{nf}U_\infty^2}$$

$$u_e=\frac{\bar{u}_e(x)}{U_\infty}$$

where $Re=\frac{U_\infty a\rho_f}{\mu_f}$ and $u_e(x)=\sin(x)$.

Non-dimensional parameters [12] :

$$Pr=\frac{1}{\alpha_f}\frac{\mu_f}{\rho_f};\ \lambda=\frac{Gr}{Re^2};\ Gr=\frac{g\beta_f(T_w-T_\infty)a^3\rho_f^2}{\mu_f^2}$$

The Governing Non-dimensional equations as follows :

$$\frac{\partial u}{\partial x}+\frac{\partial v}{\partial y}=0 \quad (5)$$

$$u\frac{\partial u}{\partial x}+v\frac{\partial u}{\partial y}=u_e\frac{du_e}{dx}+\frac{\mu_{nf}\rho_f}{\rho_{nf}\mu_f}\frac{\partial^2 u}{\partial y^2}+\frac{\beta_{nf}}{\beta_f\rho_{nf}}\lambda T\sin(x) \quad (6)$$

$$u\frac{\partial T}{\partial x}+v\frac{\partial T}{\partial y}=\frac{1}{Pr}\frac{\alpha_{nf}}{\alpha_f}\frac{\partial^2 T}{\partial y^2}+QT \quad (7)$$

where:

$$Q=\frac{a^{1/2}Q_0}{(g(T_w-T_\infty)\beta_f)^{\frac{1}{2}}(\rho C_p)_{nf}}$$

with boundary conditions:

$$u = v = 0, \qquad T = 1, \ \text{when } y = 0$$
$$u = u_e(x), \qquad T = 0, \ \text{when } y \to \infty$$

where λ is mixed convection parameter, Pr is Prandtl number parameter, and Q is heat source parameter.

Stream function (ψ) is a function used to simplify the governing equations into equations that depends only on one variable. The flow function of the fluid which flows passing through the horizontal cylindrical surface is expressed by:

$$u = \frac{\partial \psi}{\partial y} \text{ dan } v = -\frac{\partial \psi}{\partial x} \tag{8}$$

then, non-dimensional equation on stream function as below:

$$\frac{\partial \psi}{\partial y}\frac{\partial^2 \psi}{\partial x \partial y} - \frac{\partial \psi}{\partial x}\frac{\partial^2 \psi}{\partial y^2} = u_e \frac{du_e}{dx} + \frac{\mu_{nf}\rho_f}{\rho_{nf}\mu_f}\frac{\partial^3 \psi}{\partial y^3} + \frac{\beta_{nf}}{\beta_f \rho_{nf}}\lambda T \sin(x)$$

$$\left(\frac{\partial \psi}{\partial y}\right)\frac{\partial T}{\partial x} - \left(\frac{\partial \psi}{\partial x}\right)\frac{\partial T}{\partial y} = \frac{1}{Pr}\frac{\alpha_{nf}}{\alpha_f}\frac{\partial^2 T}{\partial y^2} + QT$$

Relationship between stream function (ψ) and similarity equation as follows [9]:

$$\psi = xf(x,y) \text{ and } T = \theta(x,y)$$

Then, the similarity equation as below:

$$0 = 1 - (f')^2 + ff'' + \frac{\mu_{nf}}{\mu_f}\frac{\rho_f}{\rho_{nf}}(f''') + \frac{\beta_{nf}}{\beta_f \rho_{nf}} \lambda T$$

and

$$0 = fT' + \frac{1}{Pr}\frac{\alpha_{nf}}{\alpha_f}T'' + QT \tag{9}$$

where the boundary condition:

$$f = f' = 0 \text{ and } T = 1 \text{ when } y = 0 \; ; f' \to 1 \text{ and } T \to 0 \text{ when } y \to \infty$$

3.2 Numerical Simulation

Numerical simulation was conducted using MATLAB R2012a software. This simulation is held with $\Delta\eta$ = 0,01; partition η as much as 500 and variation of the non-dimensional parameters of mixed convection parameters (λ), Prandtl number (Pr), nanoparticle volume fraction (χ), and heat source (Q).

Variation of Mixed Convection Parameters (λ)

In this simulation, we used nanoparticles *Cu* with parameter variation λ = 0,5; 2; 4; 6, fixed parameters Pr = 6,2; χ = 0.1; Q = 0.0027; and a = 100. The selected mixed convection parameter (λ) is $0 < \lambda \leq 6$. The velocity and temperature profiles with variations of the mixed convection parameters are shown in Figures 1 and 2.

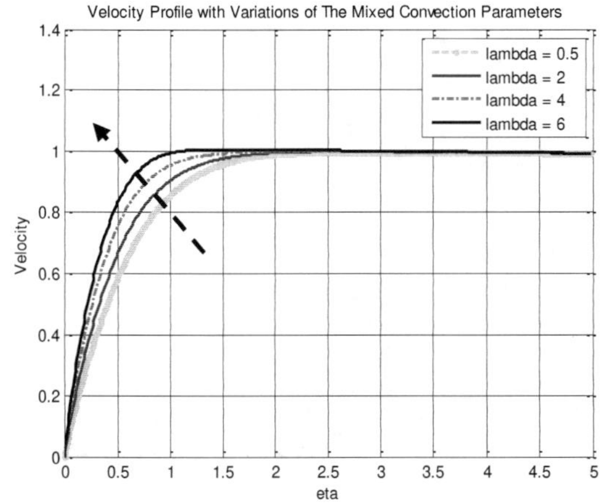

Figure 1: Velocity profile with variations of the mixed convection parameters

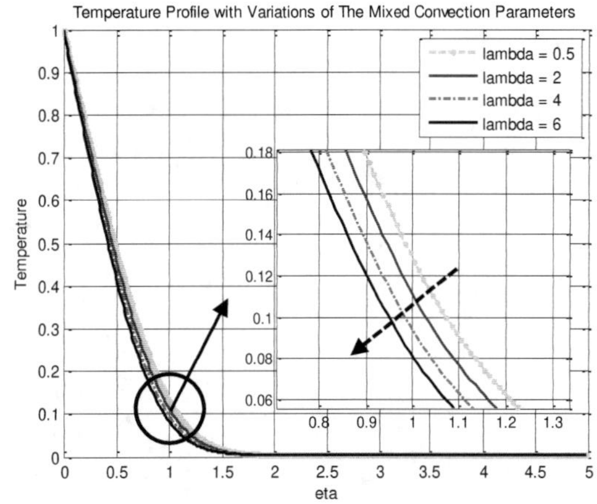

Figure 2: Temperature profile with variations of the mixed convection parameters

In Figure 1, it can be shown that the value of fluid flow velocity increases from $f' = 0$ to $f' \approx 1$. If the value of the mixed convection parameter is increased then the velocity increases. This is because a mixed convection is occured when the buoyancy on free convection becomes significant. Thus, if the value of the mixed convection parameter is magnified then the buoyancy will increase. As the buoyancy force grows, then the flow velocity increases.

Figure 2 shows that the temperature of the fluid flow decreases from T = 1 to $T \approx 0$ as the value of eta (η) increases. If the value of the mixed convection parameter is magnified then the temperature will decrease. In this study, the result that mixed convection parameters (λ) is inversely proportional to the parameter of heat source (Q) which is mathematically expressed by $Q = \frac{aQ_0}{U_\infty (\rho C_p)_{nf}\sqrt{\lambda}}$. The greater of the mixed convection value (λ) causes the heat source (Q) decreases. With the smaller heat source (Q), the heat generated by fraction between the nano particles decreases so that the temperature of the nano fluid gets smaller.

Variation Prandtl Number Parameter (Pr)

In this simulation we apply Cu nanoparticles with variation of Prandtl number parameter Pr = 1,2; 3.2; 6.2; 13.2 with fixed parameters $\lambda = 1$; $\chi = 0.1$; $Q = 0.0027$; and $a = 100$. The selected Prandtl number is $0.7 \leq Pr \leq 100$. The velocity and temperature profiles with variations of Prandtl number parameters are shown in Figures 3 and 4.

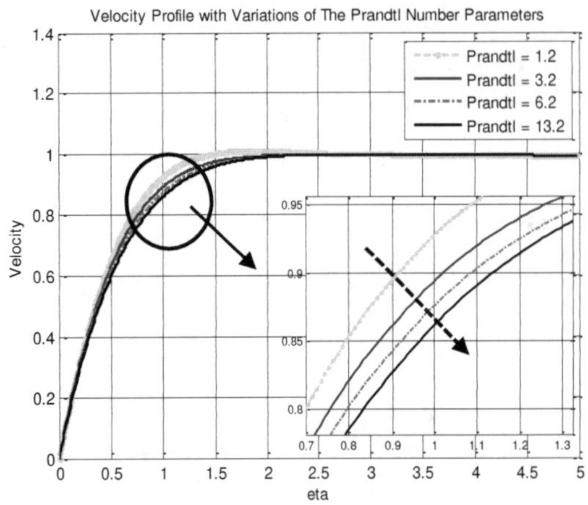

Figure 3: Velocity profile with variations of the Prandtl number parameters

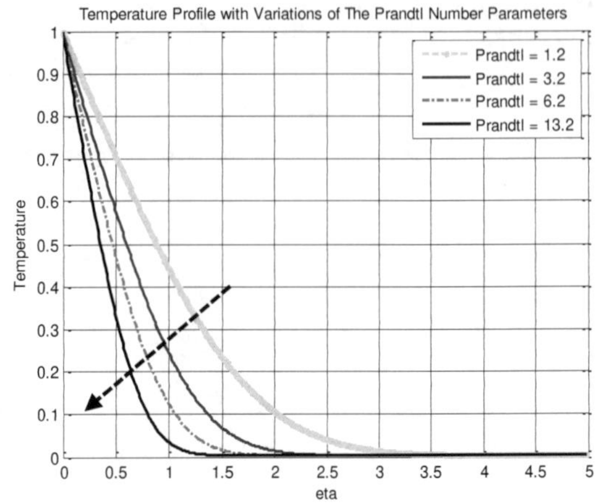

Figure 4: Temperature profile with variations of the Prandtl number parameters

In Figure 3 it can be seen that the fluid flow velocity decreases as the parameter value of Prandtl number increases. Mathematically, Prandtl number is the ratio between kinematic viscosity and fluid heat diffusivity$\left(Pr = \frac{\nu_f}{\alpha_f}\right)$. As the Prandtl number increases, the kinematic viscosity increases. As the kinematic viscosity increases, the fluid increases so the fluid flow rate decreases.

In Figure 4 it can be seen that the greater the Prandtl number the faster the temperature drops. This happens because of the increasing of Prandtl numbers, then the thermal conductivity becomes descending so that the cylinder surface becomes faster heat than its fluid.

Variation of Nanoparticle Volume Fraction Parameter (χ)

In this simulation, *Cu* nanoparticles with variation parameter χ = 0,1; 0,12; 0,135; 0,15, fixed parameters $\lambda = 1$; $Pr = 6.2$; $Q = 0.0027$; and $a = 100$. The velocity and temperature profiles with variations of the nanoparticle volume fraction parameters are shown in Figures 5 and 6.

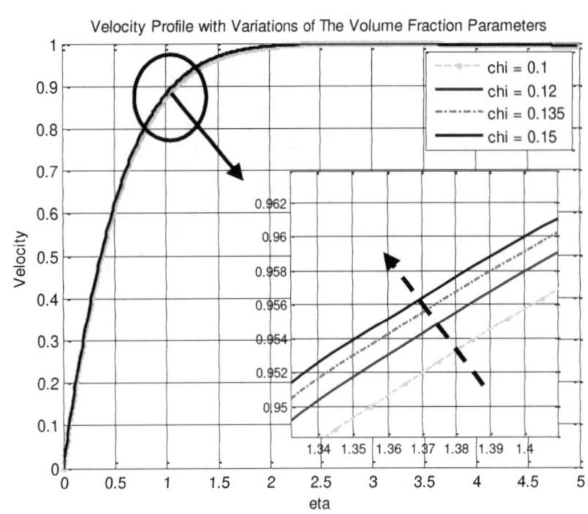

Figure 5: Velocity profile with variations of the *volume fraction* (χ)

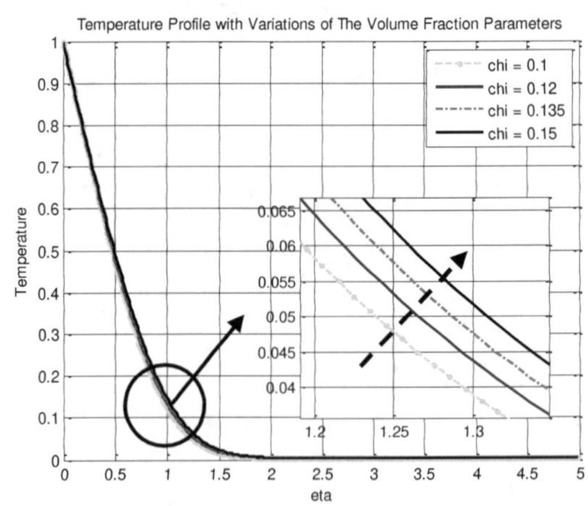

Figure 6: Temperature profile with variations of the *volume fraction* (χ)

Figure 5 shows that the velocity profile of nanoparticle volume fraction (χ) increases when the value of χ is $0,1 \leq \chi \leq 0,15$ and it decreases when the value of χ is $0,19 < \chi \leq 0,5$. This differences velocity profile are due to the viscosity influence of the nano fluid. The greater the value of χ the greater the viscosity of the fluid. The greater the viscosity of the fluid, the friction among the particles in the fluid will be greater, it cause the flow velocity increasing when the value of χ is $0{,}1 \leq \chi \leq 0{,}15$ and the velocity decreases when the value χ is $0{,}19 < \chi \leq 0{,}5$.

In Figure 6 it can be seen that the temperature decreases from T = 1 to $T \approx 0$ as the value of eta (η) increases. If the parameter value of fraction volume increases then the temperature increases. This is due to the increasing of concentrate of nano fluid with a larger value of χ so that the friction between the nanoparticles inside, then the fluid gives rise of heat. As a result the temperature of the fluid becomes increasing.

Variation of Heat Source Parameters (Q)

In this simulation, we use Cu nanoparticles with variation of parameter is Q = 0,0027; 0,1076; 0,1884; 0,3229 with fixed parameter λ = 1; Pr = 6.2; χ = 0.1; and a = 100. The preferred heat source (Q) parameter value is $5{,}3823 \times 10^{-6} \leq Q \leq 0{,}3229$. The velocity and temperature profiles with variations of the heat source parameters are shown in Figures 7 and 8.

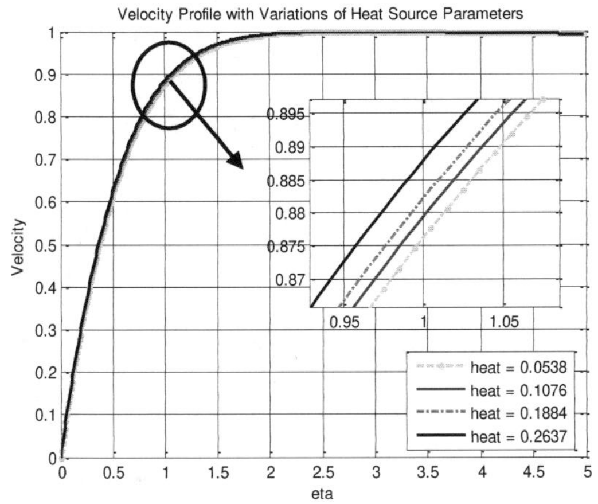

Figure 7: Velocity profile with variation of heat source (Q)

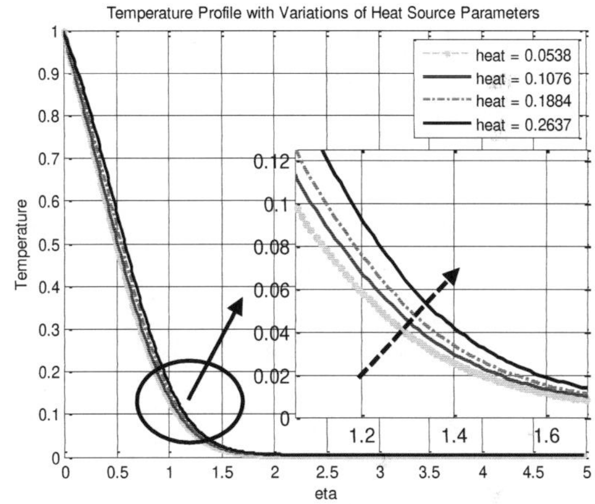

Figure 8: Temperature profile with variation of heat source (Q)

In Figure 7, judging from the variation of Q value, the value of Q increases, the fluid flow velocity increases. If the heat source (Q) is added into the system, the friction among the nanoparticles gets faster. With the faster friction between nanoparticles, the velocity of nano fluid flow through the cylindrical surface becomes increasing.

In Figure 8 it can be seen that the temperature increases with increasing Q value. This is due to the increase of the initial heat source Q_0 as shown mathematically by $Q = \frac{Q_0\sqrt{a}}{(\rho C_p)_{nf}\sqrt{g\beta_f(T_w - T_{inf})}}$. With the increasing value of Q_0, the heat generated in the nano fluid is also greater, so the temperature of the fluid increases.

3.3 Case Study

Nano fluid is formed from basic fluid (water) and nano particles that are 1-100 nm in size. In this study, the nanoparticles are Cu, Al_2O_3, and TiO_2. Each nanoparticle has a value of heat capacity (Cp), density (ρ), thermal conductivity (k), and thermal expansion coefficient (β) as shown in Table 1 below.

IOP Conf. Series: Journal of Physics: Conf. Series **1008** (2018) 012001 doi:10.1088/1742-6596/1008/1/012001

Table 1: Nano fluid thermophysical materials

Coefisien in Physic	**Base Fluid (water)**	**Nanoparticle *Cu***	**Nanoparticle Al_2O_3**	**Nanoparticle TiO_2**
$C_p\ (kg/m^3)$	4179	385	765	686,2
$\rho\ (kg/m^3)$	997,1	8933	3970	4250
$k\ (W/mK)$	0,613	400	40	8,9538
$\beta \times 10^{-5}\ (1/K)$	21	1,67	0,85	0,9

(Source : Tham *et.al*, 2010)

The data in Table 1 is used as input in the physical case simulation to determine the differences of *Cu*, Al_2O_3, and TiO_2 nanoparticles to the velocity and temperature profile of the conventional mixture of the nano-fluid mixture that passes through the surface of the horizontal circular cylinder with the effect of heat generation. The variation of the value of the heat source parameter Q is 0.0538, 0.1076, 0.1884, 0.2637 with fixed parameters ie mixed convection, Prandtl number, volume fraction, and cylinder radius with $\lambda = 1$; $Pr = 6.2$; $\chi = 0.1$; and $a = 100$. The velocity and temperature profiles with variations of the heat source parameters with *Cu*, Al_2O_3, and TiO_2 nanoparticles are shown in Figure 7 to Figure 12.

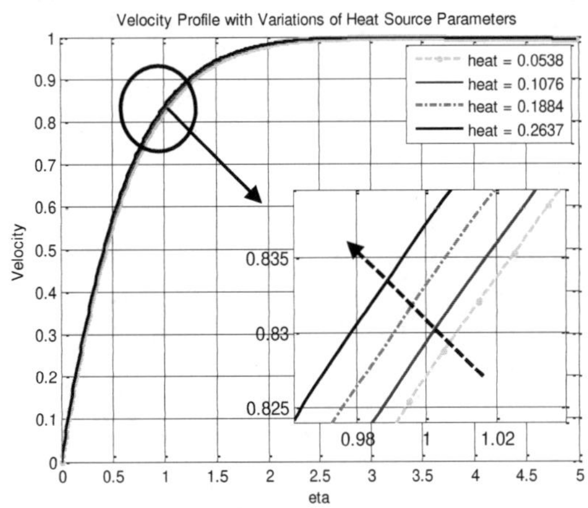

Figure 9: Velocity profile with variation heat source (Q) with nanoparticle Al_2O_3

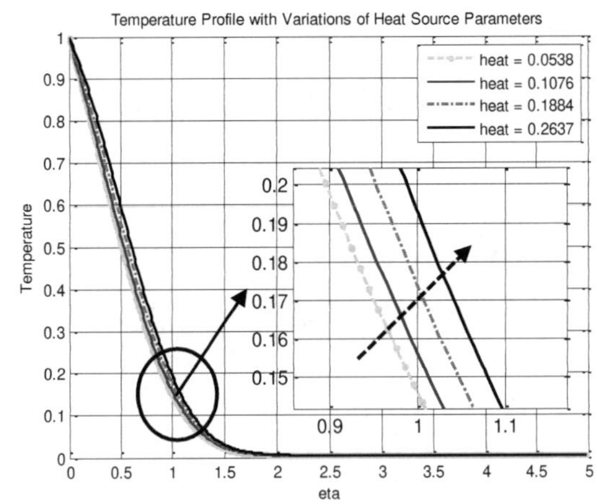

Figure 10: Temperature profile with variation heat source (Q) with nanoparticle Al_2O_3

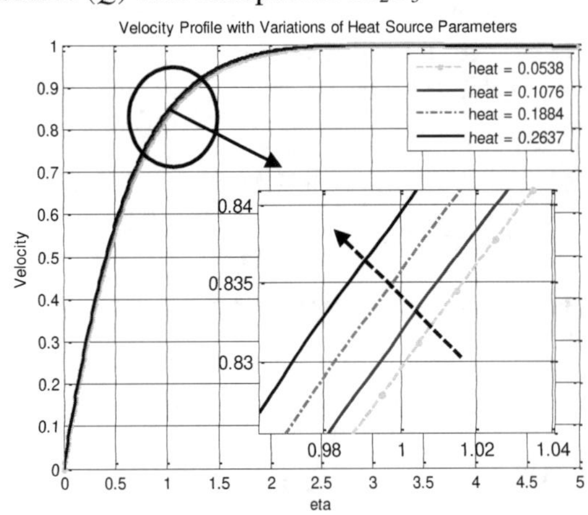

Figure 11: Velocity profile with variation heat source (Q) with nanoparticle TiO_2

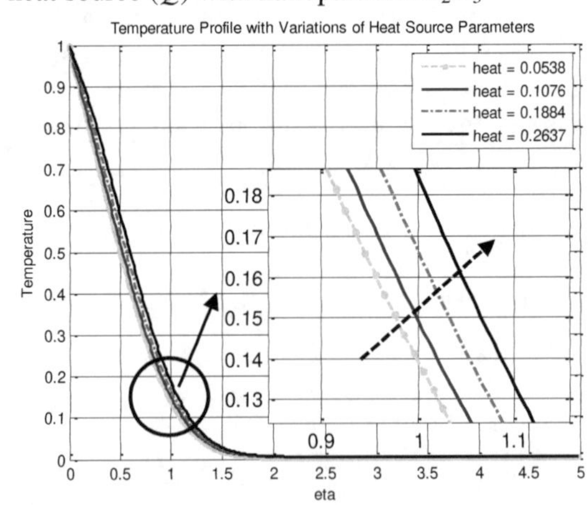

Figure 12: Temperature profile with variation heat source (Q) with nanoparticle TiO_2

In Figure 7, 9, and 11, it can be seen that the increasing value of Q increases the fluid flow rate. The effect of variation of heat source (Q) with TiO_2 nanoparticles has been found to be the same result as Cu and Al_2O_3 nanoparticles to the velocity profile. The difference among the Cu, Al_2O_3, TiO_2 nanoparticles to the velocity profile with the variation of the heat source parameter (Q) is sited in the value of the velocity f' in each η with $f'_{Cu}(\eta) > f'_{Tio2}(\eta) > f'_{Al2o3}(\eta)$.

In Figure 8, 10 and 12, it can be seen that the temperature will increase as weel as the increasing of Q. This results that in the effect of the variation of the heat source Q parameters with the nanoparticles of TiO_2 is found to be the same as Cu and Al_2O_3 nanoparticles to the temperature profile. The difference among Cu, Al_2O_3, and TiO_2 nanoparticles against the temperature profile with the variation of the heat source parameter Q lies in the value of temperature (T) in each η with $T_{Cu}(\eta) < T_{Tio2}(\eta) < T_{Al2o3}(\eta)$.

4. Conclusions

The conclusions of the research are as follows. The velocity profile increase and temperature profile decrease when mixed convection parameter increase. The velocity and temperature profiles decrease when Prandtl number parameter increase. The velocity profile with the variation of nanoparticle volume fraction (χ) is increased when the value of χ is $\mathbf{0,1 \leq \chi \leq 0,15}$ and the velocity profile decreases when the value of χ is $\mathbf{0,19} < \chi \leq 0{,}5$ while the temperature profile is increase when nanoparticle nanofluid parameter increase. The velocity and temperature profiles increase when heat generation and the radius of the cylinder increase. The last, Cu, Al_2O_3, and TiO_2 nanoparticles produce the same velocity and temperature profiles, but the three types of nanoparticles are different at the velocity value (f') with $\boldsymbol{f'_{Cu}(\eta) > f'_{Tio2}(\eta) > f'_{Al2o3}(\eta)}$ and the temperature value (T) with $\boldsymbol{T_{Cu}(\eta) < T_{Tio2}(\eta) < T_{Al2o3}(\eta)}$.

Acknowledgements

The first author acknowledges support of the Indonesian National Council of the Ministry of Research, Technology and Higher Education by BPPDN.

References

[1] Anwar I, Amin N, and Prop I 2008 Mixed Convection Boundary Layer Flow of a Viscoelastic Fluid Over a Horizontal Circular Cylinder *J. Non-Linear Mechanics* **43** 9 p814-821.
[2] Alkasasbeh H T 2015 *Numerical Solution for Convection Boundary Layer Flow Over A Solid Sphere of Newtonian and Non-Newtonian Fluid* (Pahang : Universiti Malaysia)
[3] Bhowmick S, Molla Md M, Mia M, and Saha S C 2014 Non-Newtonian Mixed Convection Flow from a Horizontal Circular Cylinder with Uniform Surface Heat Flux *Procedia Engineering* **90** 510-516
[4] Choi S 1995 Enhancing Thermal Conductivity of Fluids with Nanoparticles *ASME International Mechanical Engineering Congress and Exposition* **66** 99-105
[5] Das S K, Choi S U S, Yu W, and Pradet T 2007 *Nanofluids : Science and Tecnology*, (NJ Wiley)
[6] Fauzi E L H, Ahmad S, and Pop I 2012 MixedConvection Boundary Layer Flow from a Vertical Cone in a Porous Medium Filled with a Nanofluid *J. Mathematical, Computational, Physical, Electrical and Computer Engineering* **6** 10
[7] Rabeti M 2014 Mixed Convection Heat Transfer of Nanofluids about a Horizontal Circular Cylinder in Porous Media *SOP Transaction on Nano Technology* **1** 1.
[8] Salleh M Z, Nazar R, and Pop I 2010 Mixed Convection Boundary Layer Flow Over a Horizontal Circular Cylinder with Newtonian Heating (DOI 10.1007/s00231-010-0662-y) *J.*

IOP Conf. Series: Journal of Physics: Conf. Series **1008** (2018) 012001 doi:10.1088/1742-6596/1008/1/012001

Heat Mass Transfer August 2010.

[9] Sarif N M, Salleh M Z, Tahar R M 2013 *Numerikal Solution Of The Free Convection Boundary Layer Flow Over A Horizontal Circular Cylinder With Convective Boundary Conditions* (Pahang : Universiti Malaysia)

[10] Tham, L, Nazar R, and Pop I 2012 Mixed Convection Boundary Layer Flow Over a Horizontal Circular Cylinder in a Nanofluid *J. Numerical Methods for Heat & Fluid Flow* **22** 5 p576 606.

[11] Widodo B 2012 *Mathematical Models* (Surabaya: ITS Press)

[12] Widodo B, Khalimah D A, Zainal F D S, and Imron C 2015 Numerical Solution of Heat Transfer Unsteady Boundary Layer Magnetohydrodynamics in Micropolar Fluid Past a Sphere *J. Far East J.Mathematical Sciences* (India : Puspha Publishing House)

Performance comparison analysis library communication cluster system using merge sort

D A R Wulandari[1], M E Ramadhan[2]
[1]Information System of Department, University of Jember, Jember, Indonesia
[2]Mechanical engineering of Department, University of Jember, Jember, Indonesia

E-mail: edowarditsdesain@gmail.com; diah.retnaniw@gmail.com

Abstract. Begins by using a single processor, to increase the speed of computing time, the use of multi-processor was introduced. The second paradigm is known as parallel computing, example cluster. The cluster must have the communication potocol for processing, one of it is message passing Interface (MPI). MPI have many library, both of them OPENMPI and MPICH2. Performance of the cluster machine depend on suitable between performance characters of library communication and characters of the problem so this study aims to analyze the comparative performances libraries in handling parallel computing process. The case study in this research are MPICH2 and OpenMPI. This case research execute sorting's problem to know the performance of cluster system. The sorting problem use mergesort method. The research method is by implementing OpenMPI and MPICH2 on a Linux-based cluster by using five computer virtual then analyze the performance of the system by different scenario tests and three parameters for to know the performance of MPICH2 and OpenMPI. These performances are execution time, speedup and efficiency. The results of this study showed that the addition of each data size makes OpenMPI and MPICH2 have an average speed-up and efficiency tend to increase but at a large data size decreases. increased data size doesn't necessarily increased speed up and efficiency but only execution time example in 100000 data size. OpenMPI has a execution time greater than MPICH2 example in 1000 data size average execution time with MPICH2 is 0,009721 and OpenMPI is 0,003895 OpenMPI can customize communication needs.

1. Introduction

The technology grows rapidly and new computation paradigms are also being developed. Started by computation process with single processor in the past, recently the usage of multi-processor computers for parallel computing is very common, in order to speed up the computation period. To carry out various kinds of parallel computing, then there is a need for parallel machines infrastructures that consist of a number of computers connected by network able to solve certain problem in parallel.

There are two types of parallel computers, which are Shared Memory Multiprocessor and Distributed Memory Multicomputer. Cluster system belongs to Distributed Memory Multicomputer, which is two or more computers or nodes that are connected into a single integrated system. In a previous study found that the distributed computing make many motivation for implementation because can solve many large problem from large data sets using more than one computers that to be a single integrated resource system. The main point in cluster system is how to the resource sharing all members by efficiently for handling the large problem or data. Each cluster members have each fungtion. there are master and slave. a master node manages the resources needed for the job, divides the job into parts of the job then assigns the job to each slave. and slave node do task from master node. when the resource is available the task from master gets executed. master give a job to a vailable of members. when there is no available members, a task goes into a work queue. [1]. The most important part of cluster computer is middleware application that is capable to integrate all computer members within the cluster so that they can work together. The main task of middleware is to communicate and synchronize computers within that particular cluster. One famous cluster's middleware is MPI (Message Passing Interface), which is an independent language of communication protocol in parallel programming. Many MPI libraries have strengths and weaknesses in each implementations [2]. There are numerous libraries in MPI, some of them are GridMPI, OpenMPI, MPICH-Madeleine, MPICH-G2, MPICH-VMI, MPICH2, LAM/MPI [2] [3]. all library of MPI have different process management but the purpose is integreted more than one computers or processors to be a single system with parallel process and establishing a portable, efficient, and flexible standard for message passing that used for writing message passing programs from node to node[4] [5].

There are several available libraries that facilitate the implementation of MPI, such as MPICH2 and openMPI. OpenMPI is proven to have the ability to adjust communication management efficiently, based on the characteristics of existing infrastructure [6]. The testing of cluster system performance can be conducted by implementing certain computer program, and in this research was parallel mergesort. Parallel mergesort is a modification of the old sequential mergesort, in which it supports parallel sorting. Parallel mergesort algorithm is relatively easy to set up since it has divide and conquer method, which can work in parallel [7].

In a previous study found that the distributed computing especially cluster make many motivation for implementation. Implementation using PC-cluster. In this research study about performance analysis communication middleware. Communication middleware in cluster system is very important because the cluster system needs communication tools to work together between many computers to be single integrated system resource. This researce compare between MPICH2 and OpenMPI. In a previous study speed-up and efficiency can be parameters of performance when both were solving similar computation process was conducted so this research use speed-up and efficiency to know the parameters both MPICH2 and OpenMPI[8]. In this case, a sorting problem was used. Mergesort was implemented in this research for solving sorting problem. It was used since they are the fastest existing algorithms and is capable of working in parallel. Cluster system was established using virtualization so that it could undertake parallel computing process more efficiently. MPICH2 and OpenMPI were chosen among many other available algorithms for their advantages among their competitors, in which both of them are more well-known and are more commonly utilized.

2. Methods

The analysis of computation process within this research was conducted by comparing the performance of MPICH2 and OpenMPI in solving a sorting problem. In order to carry out the research, the following steps should be conducted:

1. Study related papers about OpenMPI and MPICH2 clustering system, sorting program using mergesort and quicksort methods. and measurement analysis. This literature study was undertaken by utilizing various resources, such as books, journal papers, and electronic documents on the Internet.
2. Design and implement cluster processes, which were conducted in the following orders:
   a. All virtual computers were installed with *distro Linux Ubuntu desktop v.10.04 – Lucid Lynx* operating system. This research use 5 virtual computer for master and slave (slave is member names of this cluster members). Virtual computer from IT as a service cloud computing was used because the cost-efficient of infrastructure [9].
   b. All of computers has conFigured in network connection without coFigured biecause using virtual computere from IAAS cloud system, Figure 1.

c. identification all node by defining the ip address address and hostname name of each node. Master must know abaout ip address all of member in system for giving their task and each member must know ip address their master for getting their task from master.
d. Configuring SSH can communicate and conduct data exchange to all nodes without having to provide authentication process. In the implementation master node generates a public key which is a random number. This public key will then be sent to the SSH folder on the slave node. Once the public key of the master node is sent into the SSH folder, the public key is converted into an authorized key so that when the master node performs an SSH connection on the slave node, SSH on the slave node will check whether the computer that will access it has been registered in the authorized keys slave node. If indeed the computer that will access it is listed, then the master node can access without having to enter the slave node password first.
e. Configuring NFS, In this cluster system, the directory to be shared is the data directory, in this directory is used to store the files used for the test sample. NFS serves to perform a shared directory that will be used for file sharing.
f. MPICH2 and openMPI were installed and conFigured in five virtual computers. Hence, there are one master computer and four slave computers.
g. Installation and configuration of Build Essential for running and excuting parallel sorting program using C language program .

3. Implementing cluster system and testing analysis.

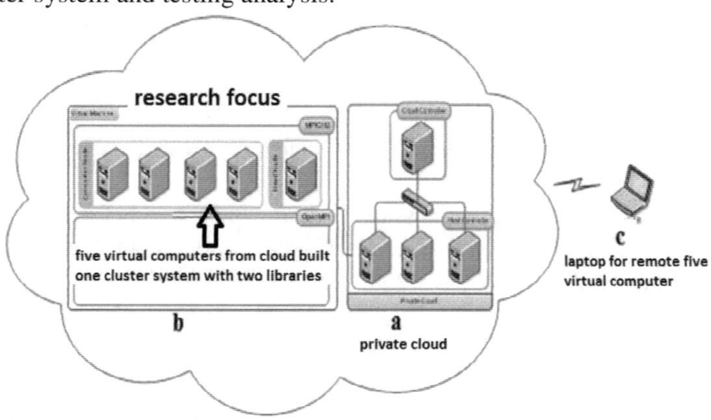

Figure 1. Cluster system design; (a) private cloud computing architecture, (b) focus of this research on cluster system design, (c) notebook as remote controller

The cluster system consisted of five homogeneous virtual computers provided by cloud service. One of those five virtual computers was dedicated as master node and the rest four computers were used as slave nodes. Each node was installed with open source operating system Linux Ubuntu 10.04 LTS (Lucid Lynx). From existing five virtual nodes, it was then possible to establish two cluster system, which were Cluster A implementing MPICH2 library and Cluster B implementing openMPI library. Figure 3.1 describes the architectural design of the cluster system implemented in this research. openMPI library was chosen among other MPI libraries because it is one of most famous libraries in parallel computing and because of its good performance. Both libraries has its own unique characteristics, syntaxes, and advantages that are worth comparing.

4. Cluster system performance test was conducted by executing computation program on cluster system with the following scenario:

The purpose of cluster system performance testing was to understand the performance comparison between MPICH2 and openMPI in terms of speed-up and efficiency. Random number sorting problem was implemented in C programming language to test system performance. Two sorting methods used in this research was mergesort algorithms. Number of tested data was conFigured to be varied, ranging from 100, 1000, 10.000 to 100.000 data size. The data size is number random.

5. Resource monitoring system using HTOP tool.
6. Comparing the result of parallel computing using MPICH2 and that of openMPI, and then deriving conclusion.

Merge short is used for testing the performance of this cluster system. Mergesort is a sorting algorithm that uses the divide and conguer approach. In a parallel process, the first uncompleted list is divided into two sublists by a single processor, then the sublists are sent to another processor. Each processor handles one sublist. Each sublist is subdivided into two smaller sections, the subdist division is done until the sublist is unbreakable or subdivided, until there is only a number in a sublist. Then each sublist is merged into a new sublist with two numbers then reassembled into each sublist of four numbers until all the numbers on the sublist are combined into one ordered list, Figure 2.

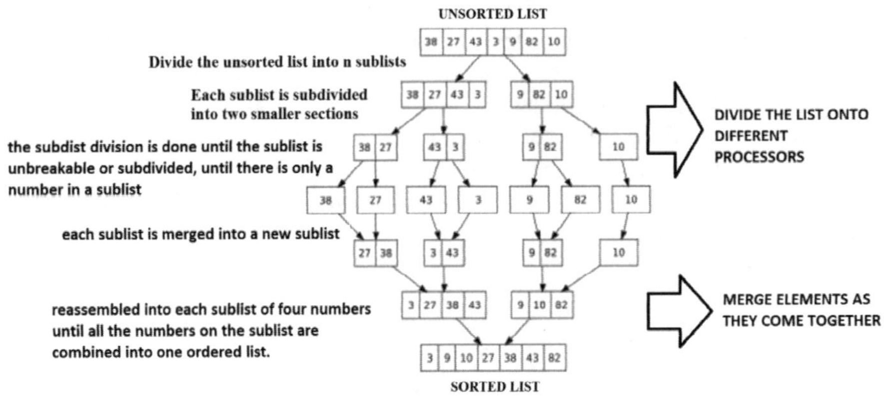

Figure 2. How To Work Parallel Merge Sort

In this research use three parameters. They are time execution, speed-up and efficiency for testing performance analysis [10]. Speed-up of a program (s_p) is defined as the time it takes a program to execute in serial using one processor or computer (t_s) divided by the time it takes to execute in parallel using many processors or many computer (t_p). The formula for speedup is in equation (1).

$$S_p = \frac{t_s}{t_p} \qquad (1)$$

Another metric to measure the performance of a parallel algorithm is efficiency. Efficiency is speed-up devide processors that this system use. The formula for efficiency is in equation (2).

$$E = \frac{S_p}{p} \qquad (2)$$

3. Result and Discussion

Cluster system capability testing in computing process on both cluster systems using execution time, speed-up parameters, and efficiency. Cluster testing is done by running some sample array sorting program which contains integer numbers in four scenarios. The mergesort method is chosen because it uses the divide and conquer method approach, including the fastest and stable sorting method. In another research Divide and Conquer has been designed to work in machines with multiple processors. This test is by sorting random numbers of integer types with intervals of 100 to 100000 numbers or data size. This test will be done using several scenarios. That is using one (sequential) until five processors or computer (multiprocessors). The selection of data between 100 to 100000 random numbers is intended to show the significant difference in execution time. This scenario aims to see the effect of changing the number of nodes with the large amount of data on the execution time, and know what computing program is appropriate so that the performance of both libraries can work optimally. This research have four scenario. The scenario are executing the random number with interval 100,

1000, 10000, and 100000 data size by using sequential and parallel between two until five nodes. The purpose of this testing is performance of Open MPI and MPICH2 with three parameters. That is speed-up efficiency and execution time. The result of this testing are:

1. Execution time base on number of processor

Figure 3. Graphical comparison OpenMPI and MPICH2 on average execution time for mergesort algorithm

Comparison OpenMPI and MPICH2 on average execution time for merge sort at fig 3, show that The sistem cluster have good performance in data size more than 10000 random number because when the increasing processor execution time more faster then sequential, but have bad performance when the data size less than 1000 random number. When the data size less than 1000 random number the sequential performance more better then many processor. In sequential computing applications the resulting process time is computational time, which is the time needed to calculate the steps of computing. In parallel computing applications, in addition to determining computational time also needs to take into account the time required for communication in sending messages to the parallel application. increasing processor can retarded execution time when the data size less than 1000 random number. This incident caused all of processor have communication time for solve sorting number. Communication time be used for sharing data, divison of task each processor. OpenMPI has a execution time greater than MPICH2 example in 1000 data size average execution time WITH MPICH2 is 0,009721 and OpenMPI is 0,003895 OpenMPI can customize communication needs. OpenMPI faster then MPICH2 at 100 until 1000 data size but at 10000 until 100000 data size OpenMPI and MPICH2 have the same performance. This incident caused merge sort method have many communication time for divide and merge list of random number. This is in accordance with previous research that OpenMPI is able to adjust the communication settings efficiently in accordance with the characteristics of existing infrastructure [6], because in the openMPI architecture there is an MCA as a layer component that provides management services for all other layers [11].

2. Speed-up and efficiency

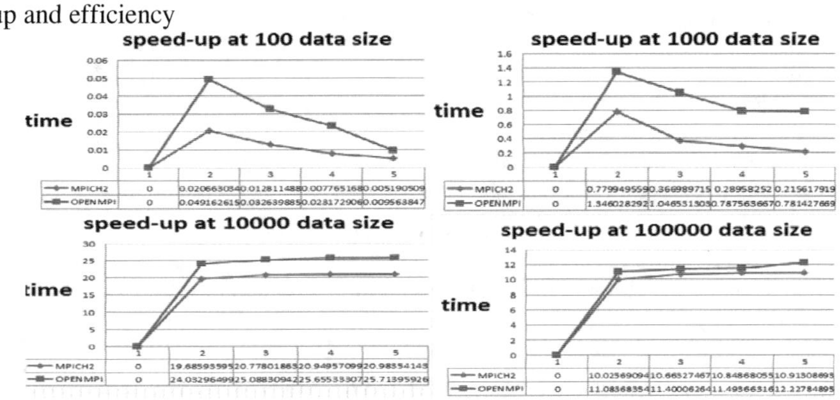

Figure 4. Graphical comparison OpenMPI and MPICH2 on average execution time for mergesort algorithm

comparison OpenMPI and MPICH2 on speed-up for merge sort at fig 4, show that OpenMPI and MPICH2 have almost the same performance at more than 10000 data size. the speed-up value of MPICH2 is slightly smaller than that of openMPI. However, as the number of random number increased as large as 100,000, the speed-up value of MPICH2 significantly improved, compared to that of openMPI. The speed-up values on both libraries decreased as there were addition of more processors.

3. efficiency

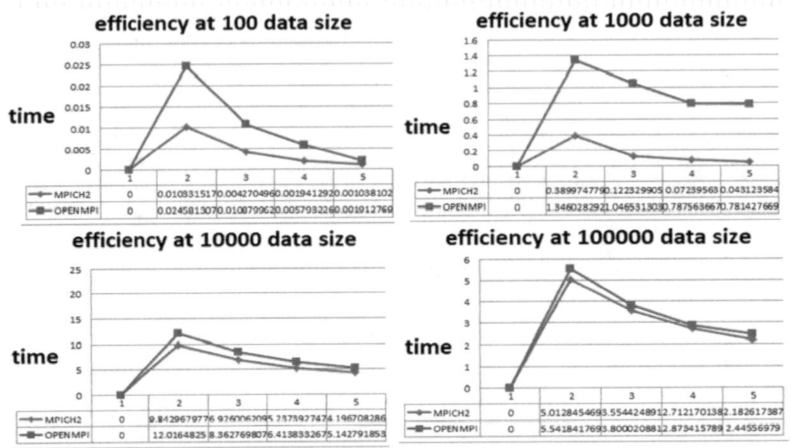

Figure 5. Graphical comparison OpenMPI and MPICH2 on average execution time for mergesort algorithm

Figure 5 described efficiency comparison on both libraries based on the number of nodes. Firstly, the average efficiency rate on program was calculated when using openMPI library. Then, similar process conducted under MPICH2 was also done. The results of those to processes were then compared. Figure 5 showed that efficiency of MPICH2 was less than that of openMPI. Efficiency rates were also decreasing on both libraries as the number of processors increased. The unideal speed-up and efficiency was due to the existence of overhead in parallel system. For example, additional computation which was only required to parallel computing, communication among processors, and synchronization process. Such phenomenon worked on all parallel systems and the speed-up and efficiency trends followed Amdahl's Law. However, speed-up and efficiency would also increase as the data size also getting larger, and that was in parallel with Gustafson's Law. While the addition of processors would decrease efficiency rate, the increase of data size would improve the efficiency [12].

According to Figure four and five, it was clearly seen that as the number of data ranged from 100 up to 10,000, the speed-up and efficiency rate increased as the data grew larger. However, when the random number were as large as 100,000 data, both speed-up and efficiency rare got smaller. Therefore, it was then known that the maximum speed-up and efficiency rate were on 10,000 data size, for all possible test cases scenario, from using two nodes up to five nodes. After conducting this research, it was proven that increasing the size of problem by executing larger data was not necessarily improving speed-up and efficiency. According to Ahmdal's Law, after some certain limit, speed-up and efficiency would decrease because the existence of larger communication time overhead compared to that of computation time on parallel algorithm. It was also possible due to the unpredicted computation task, imbalance of tasks distribution among processors, the existence of excessive sequential code, or the existence of pure sequential code. In some algorithms, the existence of might reduce the average speed-up time. As the data grew, the execution time both on serial and parallel processing would also increase. As the data grew as large as 100.000, the speed-up decreased because execution time on serial and parallel were getting slower, and therefore the difference between the two were smaller.

4. Conclusion

In this paper discuss about performance between two library communication of MPI to handly task in cluster system. There are MPICH2 AND OpenMPI. The task is sorting number 100 until 100000 data size using mergesort methode. Performence parameters use speed-up, efficiency and execution time. The result are when the small data size execute in sequential processing, the performance is better than executing in cluster system using more than one processors / computers. When the data size are large the performance of cluster system more faster than sequential process using one computer or one processor. Execution time of openMPI tends to be faster than that of MPICH2 for small size data, because programs conduct more communication process. However, execution time of MPICH2 tends to be faster as the data gets larger in size, since the increase of computation process reduces the communication effect. The values of openMPI's efficiency rate and speed-up tend to be greater than that of MPICH2 on small size data, ranging from 100 to 10.000 data. In contrary, as the number of data increases, such as 100.000, the values of efficiency rate and speed-up of MPICH2 are better than that of openMPI. Within such scenario, the speed-up difference is 3,1429 and the efficiency rate difference is 0,754. From the result found that implementation library communication of middleware communication have affect in performance of cluster system. Many library communication of middleware communication is in cluster system, but we must to know about suitability between characters of the task and characters of library comunication from communication time parameters for optimation of cluster system . cluster system need communication time to share the task in all members but do not spend long time to communicate between all members because it make decreasing performance, the execution time will be slower, so we need discuss about that in further research

Acknowledgments

This paper is a part of my research when i was in student of master degree that I developed again. I would like to express my gratitude to my advisors and my team of cluster system for supporting this research. This paper could not be written to its fullest without them, who served as well as one who challenged and encouraged me throughout my time spent studying. they would have never accepted anything less than my best efforts, and for that, I thank them for all support and togetherness they give when we did this research.

Reference

[1] Kahanwal B and Singh T P 2012 The Distributed Computing Paradigms: P2P, Grid, Cluster, Cloud, and Jungle *International Journal of Latest Research in Science and Technology* **1**(2) 183-187

[2] Hoefler T, Squyres J M, Mehlan T, Mietke F and Rehm W 2005 Implementing a Hardware-based Barrier in Open MPI *In Proceedings of 2005 KiCC Workshop, Chemnitzer Informatik Berichte*

[3] Hablot L, Glück O, Mignot J C, Genaud S and Primet P V B 2007 Comparison and tuning of MPI implementations in a grid context Université de Lyon INRIA LIP (Laboratoire de l'Informatique du Parallélisme)

[4] Desai N, Lusk A, Bradshaw R and Lusk E 2005 MPISH: A parallel shell for MPI programs. In Proceedings of the 1st Workshop on System Management Tools for Large-Scale Parallel Systems (IPDPS '05) Denver Colorado USA

[5] Gaurav D, Arora S and Gupta P Effect of parallelization, execution time and inter process communication on sorting techniques using Message Passing Interface *International Journal of Computer Applications* **101** (5) 0975 – 8887

[6] Graham R L, Shipman G M, Barrett B W, Castain R H, Bosilca G and Lumsdaine A 2006 Open MPI: A High-Performance Heterogeneous MPI (USA: University of Tennessee)

[7] Marszałek Z 2017 Parallelization of Modified Merge Sort Algorithm *Symmetry* **9**(9) 1-18

[8] Eager D L, Zoharjan J and Lazowska E D 1989 Speed-up Versus Efficiency in Parallel Sistems *IEEE Transactions On Komputers* **38** (3)

[9] Shawish A and Salama M 2014 Cloud Computing: Paradigms and Technologies *Studies in Computational Intelligence* **495** 39-67

[10] ALECU F 2007 performance analysis of parallel algorithms *journal of applied quantitative methods* **2** (1)

[11] Graham R L, Woodall T S and Squyres J M 2005 Open MPI: A Flexible High Performance MPI *Proceedings 6th Annual International Conference on Parallel Processing and Applied Mathematics*

[12] Mathew J, Vijayakumar R 2011 The Performance of Parallel Algorithms by Amdahl's Law, Gustafson's Trend *International Journal of Computer Science and Information Technologies* **2** (6) 2796-2799

IOP Conf. Series: Journal of Physics: Conf. Series **1008** (2018) 012003 doi:10.1088/1742-6596/1008/1/012003

The Development of Web-based Graphical User Interface for Unified Modeling Data with Multi (Correlated) Responses

I Made Tirta[1], Dian Anggraeni[1]

[1] Mathematics of Department, University of Jember, Jember, Indonesia

E-mail: itirta.fmipa@unej.ac.id

Abstract. Statistical models have been developed rapidly into various directions to accommodate various types of data. Data collected from longitudinal, repeated measured, clustered data (either continuous, binary, count, or ordinal), are more likely to be correlated. Therefore statistical model for independent responses, such as Generalized Linear Model (GLM), Generalized Additive Model (GAM) are not appropriate. There are several models available to apply for correlated responses including GEEs (Generalized Estimating Equations), for marginal model and various mixed effect model such as GLMM (Generalized Linear Mixed Models) and HGLM (Hierarchical Generalized Linear Models) for subject spesific models. These models are available on free open source software R, but they can only be accessed through command line interface (using scrit). On the othe hand, most practical researchers very much rely on menu based or Graphical User Interface (GUI). We develop, using Shiny framework, standard pull down menu Web-GUI that unifies most models for correlated responses. The Web-GUI has accomodated almost all needed features. It enables users to do and compare various modeling for repeated measure data (GEE, GLMM, HGLM, GEE for nominal responses) much more easily trough online menus. This paper discusses the features of the Web-GUI and illustrates the use of them. In General we find that GEE, GLMM, HGLM gave very closed results.

1 Introduction

Regresion are among statistical methods which are frequently employed to model dependences between responses and several explanatory variables. Regresion or Statistical models have been developed into various directions to handle various type and complex relationship of data. Data collected from multilevel, repeated meassure or longitudinal studies are more likely to be correlated, hence various models for independent responses such as GLM, GAM, GAMLSS [1] are not appropriate. Rich variety of model for handling correlated or clustered data are available on the open source software R. They can be classified into 2 main types, namely marginal model for modeling population average, such as GEE (Generalized Estimating Equation), and the other is conditional model for modeling individual, such as GLMM (Generalized Linear Mixed Model) and HGLM (Hierarchical Generalized Linear Model). The models are much more complex than those for independent rensponses, since we must take into account the correlation or the clustering. Moreover, these advanced statistical models, can only be accessed using script or command line interface. Therefore it is essential to build users friendly interface that unify the various models and can be more easily accessed by nonstatistician researchers. The kind of interface aims to "closing the gap between methodologist and end user of R" [2]

GEE was first introduced in 1986 [3], and later is also called GEE order 1 (GEE1). This GEE can be regarded as a multivariate extension of GLM (Generalized Linear Model). GLM and GEE have fsimmilar class of family distributios. In addition to GLM, GEE also offer several correlation structures to account for correlation among the responses. Therefore the variance-covariance matrices are generally not diagonal matrices. The first GEE (GEE1) can be writen as

$$S(\boldsymbol{\beta}) = \sum_{i=1}^{n} \mathbf{X}_i^T \left(\frac{\partial \boldsymbol{\mu}_i}{\partial \boldsymbol{\eta}_i}\right) \mathrm{Var}(\mathbf{Y}_i)^{-1} (\mathbf{Y}_i - \boldsymbol{\mu}_i) \text{ with } g(\mu) = \eta = \mathbf{x}_i \beta \tag{1}$$

for link function g. Later, not only the mean parameters, are modelled, but also the scale parameter and the correlation[4]. This type of GEE is later popular as GEE order 2 (GEE2). With GEE 2, in addition to link function (g_1) for the mean, there are also other two link functions g_2 and g_3 related to the scale and correlation parameters. The link functions are:

$$g_2(\phi_i) = \eta_2 = \mathbf{x}_{2i}\gamma \text{ and } g_3(\rho_i) = \eta_3 = \mathbf{x}_{3i}\alpha \tag{2}$$

where $\mathbf{X}_2$ and $\mathbf{X}_3$ are respectively vectors of predictors for the scale and correlation parameters. The GEE model has been implemented into R through `geepack` package [5]. GEE has also been extended for multinomial and ordinal responses (such as level of sickness of patient during time of study, for example weekly for 1 month) through `multgee` package [6].

The Generalized Linear Mixed Models handle the multi-response by accomodating random effects, and the model are called mixed models since they consists of both fixed effects (related to mean) and the random effects. The family distribution for the matrix responses **Y** follows the family for GLM, however the distribution of random effects are limited to Gaussian distribution. The model for the random effects **u** can be written as

$$g(\boldsymbol{\mu}_i) = \boldsymbol{\eta}_i = \mathbf{X}_i\beta + \mathbf{Z}_i\mathbf{u}_i \tag{3}$$

The GLMM has been implemented into R through `lme4` package that can also handle multiple random effects [7].

The later model for random effect is HGLM where the distribution of random effect is not limited to Gaussian distribution but also other distributions, either conjugate or nonconjugate distributions (including Gaussian, Gamma, Beta, Inverse-Gamma). HGLM was introduced in 1996 [8] and has been implemented into R through `hglm` package [9]. The package already accomodate multiple random effects, and model for dispersion, however random effect only applied on modeling the mean. The relationship among the models are as follows: (i) GEE2 with fixed scale parameter and identity correlation link is equivalent to GEE1, while (ii) HGLM with constant dispersion parameter and Gaussian random effects is equivalent to GLMM.

2 Methods

The interface is mainly built by following shiny framework [13] and in this interface focus of pulldown-menu type (utilizing two files `ui.r` and `server.r`). This framework has been succesfully applied for developing unified regression for independent (uni) response[1]. The web-GUI cocsists of 3 main parts: (i) The input and exploration of data which is mainly supported by graphics visualization; (ii) the building and fitting of model, (iii) checking goodness of fit of the model in order to get the most appropriate model (mainly using AIC, BIC, QAIC, QLKL). The interface is also enriched with various high quality of graphics to visualize data exploration. R has many graphics packages which can be utilized to visualize data exploration such as correlations (`corrplot` [10], `pair.plot` [11]), spline smoothers (`ggplot2` [12]). The development involves several steps including (i) identifying variety of input-output options, (ii) identifying related (needed packages), (iii) defining (writing) user interface script, (iv) defining (writing) related server script, (v) testing the whole (integrated) interface. The detail of components and related packages are given in Table 1.

Table 1. Summary of Features of the Web-GUI

No	Components/ Model	Input Option	Output Option	Relate packages

IOP Conf. Series: Journal of Physics: Conf. Series **1008** (2018) 012003 doi:10.1088/1742-6596/1008/1/012003

No	Components/ Model	Input Option	Output Option	Relate packages
	Components/Parts			
1	Input Data	• Internal database • Import data (.csv, .txt)	• Summary of data • List of data	
2	Exploratory Data	• General exploration • Correlation among cluster within responses, • Correlation among numeric variables	• Correlations matrix • Correlation Diagram • Scatter plot matrix	• `corrplot` • `psych` • `car`
	Models Fitting			
3	GEE1	• Xs and Y • Correlation structure • id for cluster identity • Family distributions and link functiins	• Estimates (with robust or sandwich and naive standard error and p-values), • QIC	• `gee` • `MuMIn`
4	GEE2	• as GEE1 plus • predictor and link for scale parameter • link for correlation	• Estmates (with sandwich standard error and p values)	• `geepack`
5	GEE for ordinal responses	• as GEE1 except Y on ordinal scale • correlation structure	• sandwich parameters estimates for each category • p values	• `multgee`
5	GLMM	• Y and predictors for the mean (fixed effect parameters) • family distributons for the response • predictors for random effect • alternative formula script for complex formula (See Fig 1)	• Estimates (with p-values), • prediction for random effects • AIC, BIC • Graphics	• `lme4`
6	HGLM	• as GLMM plus • Family (link):for random effect • alternative formula script for complex formula	• Estimates for parameters related to the mean (with p-values), • lrt (log ratio test)	• `hglm`

GEEs do not originally provide goodness of fit such as AIC or BIC. However, the quasi version, QIC can be computed through `MuMIn` package [14]

3 Results and Discussions

3.1 General Features of the Web GUI interface

The main aim of the Web-GUI is to enable user do GEES, and Mixed Models via menus offered via web (but as complete as if they are done via scripting). At this stage, we have developed online statistical model fitting for independent responses, covering several models described previously with general features as follows (see Figure 1).

1) Data Input: two main options available: (i) select some internally provided data (for practical purposes), (ii) import users' own data with csv or text format (for real data analyses).
2) Data exploration: several graphical representations on relationship among variables (various correlation diagram and scatter plot diagram, some with smoother).
3) Model options: GEE1, GEE2, GEE multinom, GLMM, HGLM
4) Input-output options: as previously given in Table 1. For complex model (GLMM and HGLM), there are also option to combine menu with additional formula model through script input (Figure 1).
5) Goodness Of Fit (GOF): GEE1 and GEE2 fit are acompanied (QIC), GLMM and HGLM are accompanied by conditional-AIC, BIC.

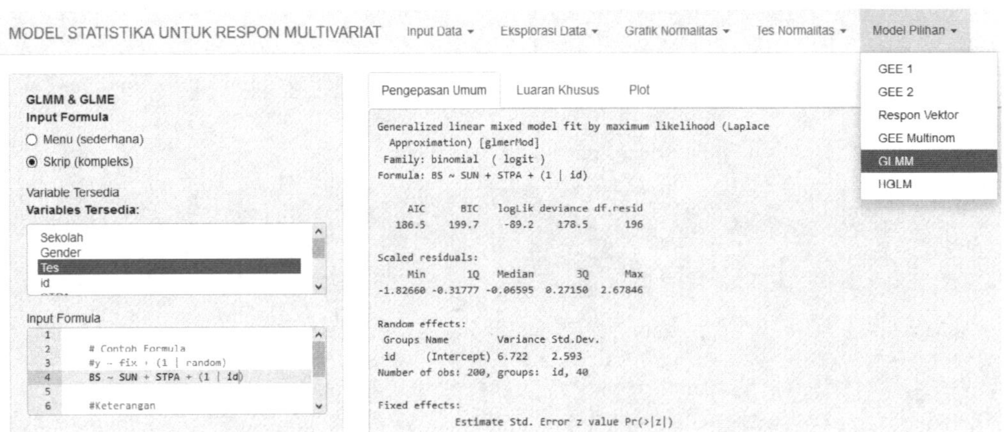

Figure 1. Web Appearrance and The Main Menu and GLMM menu. In addition to menu, there is also space for short formula script.

3.2 Numerical Illustrations

The purposes of the numerical illustration are: (i) to ensure that the main components of the interface are integrated and work well, (ii) to show, in term of the numerical calculation of similar model (between GEEs and between Mixed models), if they are comparable. The numerical illustration uses simulated data with time constant covariates (during repeated measures of Test scores *Y* (SHB), the observed data of Xs remain the same). The data are presummably taken from 40 subjects (id=individu) with 3 factors, 2 numerical predictor. The response is on 5 types of Test (each individu were tested/ measured 5 times). The summary of the statistics are given below.

```
Sekolah Gender  Tes          id              STPA              SUN
 A:50    L:100   T1:40   Min.   : 1.00   Min.   : 33.66   Min.   : 16.46
 B:50    P:100   T2:40   1st Qu.:10.75   1st Qu.: 63.26   1st Qu.: 63.74
 C:50            T3:40   Median :20.50   Median : 70.13   Median : 73.81
 D:50            T4:40   Mean   :20.50   Mean   : 73.07   Mean   : 72.48
                 T5:40   3rd Qu.:30.25   3rd Qu.: 82.24   3rd Qu.: 87.25
                         Max.   :40.00   Max.   :104.18   Max.   :116.94
      SHB                BS
```

```
Min.   :  3.571   Min.    :0.000
1st Qu.: 44.404   1st Qu.:0.000
Median : 66.101   Median :1.000
Mean   : 63.697   Mean    :0.535
3rd Qu.: 79.276   3rd Qu.:1.000
Max.   :131.377   Max.    :1.000
```

3.2.1 Data Exploration

We proceed with general exploration to have ideas of the correlation and distribution of the responses, then we do alternative modellings using avaliable models (except for multinomial) and compare the results. In addition to exploring data convensionally to get numerical information of summary statistics. In this stage users can also explore and visualize correlation, both correlation among all numeric variables (Figure 2), and correlation among repeated mesures respons (Figure 3). Various visualization of the correlation among repeated observations help users to choose reasonable type of candidate for correlation structures.

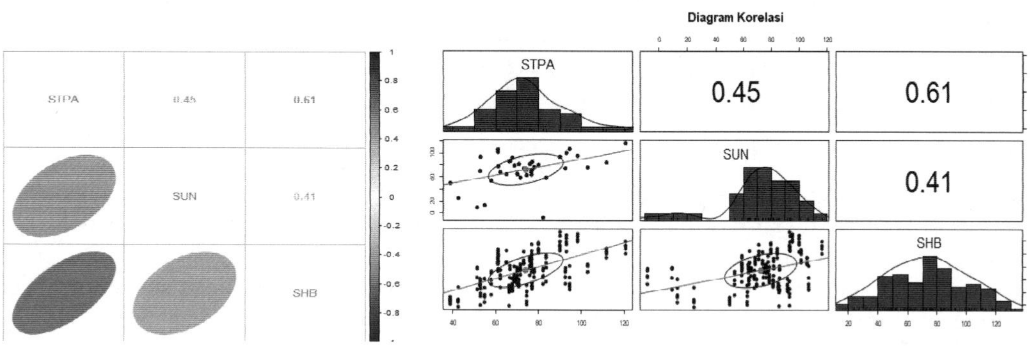

Figure 2. Correlation among 3 numeric variables both graphics describe the same things, exept on the right hand side presents combination of correlation, scatter plot matrix and density of the variables

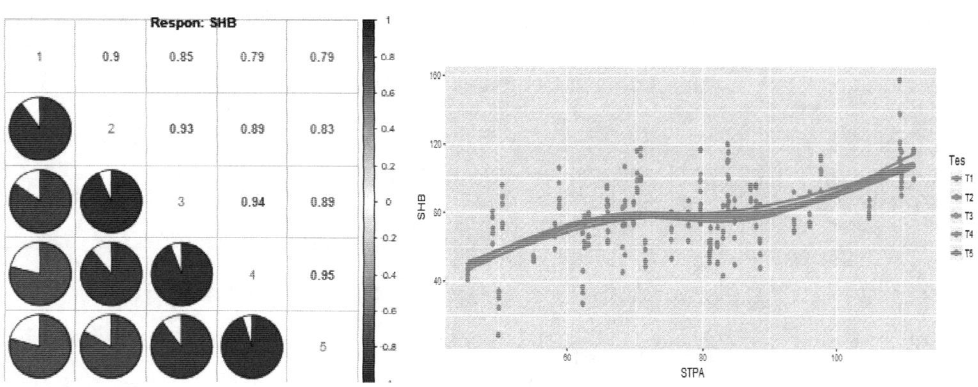

a b

Figure 3. Correlation among 5 repeated measures responses (a) and natural spline between one predictor (STPA) and 5 responses SHB (b)

IOP Conf. Series: Journal of Physics: Conf. Series **1008** (2018) 012003 doi:10.1088/1742-6596/1008/1/012003

3.2.2 Alternatives Model Fittings

We apply 4 models (GEE1, GEE2, GLMM, and HGLM). The main results are given in Table 2.

In general, the web GUI has already accomodate mainly needed features, for data exploration, input and output (estimation and GOF), for the 4 modeling GEE1, GEE2, multgee, GLMM and HGLM. With the unified or integrated web-GUI, users can be more easily fitting and comparing several models and choose the most appropriate one. QAIC has taken into account the correlation structures of GEE. This will help practical researchers alot, since these types of modelings are not available on well known RGUI (RCommander [15]), neither on its original, nor on its extended (pug.in) version. More over, when it is provided on the server, the users even do not need to install anyting on their computers, and they can access the modeling by using various gadgets such as tablets and smartphones. In addition, for modeling with GLMM and HGLM, complex model formula can also be inputed as a short script (feature taken from `shinyAce` package [16]). The menu tree is given in Appendix A.

In term of numerical illustration, after checking various combination of predictors, correlation structures and family distributions, we find that the best GEE1 model is M4, one with STPA as the only predictor and AR-1 correlation structure (with correlation parameter α=0.9369637 and QIC=1247.155). QIC for GEE has not taken into account the dispersion modeling in GEE2. However, although, the QIC values does not change, the fixed parameters estimates (as well as their standard errors), actually change slightly as we change the dispersion model with GEE2.

Choosing the id (individu) as random effects, we find that the values of the fixed parameters estimates (even compare to those of GEEs) are in general comparable (not noticeably different). The estimates of GLMM are very similar (as expected) to those of HGLM with Gaussian random effects. This mixed model has cAIC = 1500.123, still slighty better than applying random effect with Gamma distribution (cAIC = 1504.259). Therefore in general, the estimates of the 4 types of modeling are comparable (similar). The comparison of the GOF are given in Table 2 and the more detail output are given in Appendix B). While GEEs can be considered as average models, the mixed model can be considered as subject spesific models. That is, in this case, every individu has each own intercept.

Tabel 2. Comparison of models

No	**Type of Model**	**Family**	**Predictors**	**Corr.Struct.**	**GOF Type**	**GOF value**
M1	GEE1	Gaussian	Sekolah, Tes, Gender, STPA, SUN	Exchangeable	QIC	1275.89
M2	GEE1	Gaussian	STPA, SUN	Exchangeable	QIC	1256.35
M3	GEE1	Gaussian	STPA, SUN	AR1	QIC	1251.723

No	Type of Model	Family	Predictors	Corr.Struct.	GOF Type	GOF value
				(0.9385780)		
M4	**GEE1**	**Gaussian**	**STPA**	**AR1 0.9369637**	**QIC**	**1247.155**
M5	GEE2 (Dispersion ~ SUN)	Gaussian	STPA	AR1 0.888199	QIC	1247.155
M6	(Dispersion ~STPA)	Gaussian	STPA	0.8944808	QIC	1247.155
				-		
M7	GLMM (id=RE)	Gaussian	STPA	-	cAIC	1585.787
M8	HGLM (RE Gauss)	Gaussian	STPA	-	cAIC	1500.123
M9	HGLM (RE Gamma)	Gaussian	STPA		cAIC	1504.259

3.3 Future Development
The developed web-GUI has covered almost all parametric modelings for correlated responses. However, it is natural to think if these parametric models can be combined with nonparametric or additive model (GAM). The simplest way to include the spline smoother is by including either natural spine or b-spline especially in the fixed model (see [17]). A more comprehensive approach has been done by combining Aditive and Mixed model and known as GAMM (Generalized Aditive Mixed Model) and has been implemented in gamm4 package [18]. However, yet at the moment, GAMM has not been included in the current interface, but it will be considered in the next improvement, so that the users have more modeling choices, for similar data.

4 Conclussion
Web-based Graphical User Inerface for modelling data with multiresponses using R based on R-shiny toolkit framework has been developed. The interface has accomodated main features of GEE1, GEE2, GEE for repeated Ordinal/multinomial responses, GLMM and HGLM. For complex models in GLMM and HGLM model formula can also be alternatively inputed via short script. This interface will help end R users alot, that they can now access the most advanced statistical modelings more easily. However, the interface still can be extended to include GAMM

References

[1]. Tirta I M , Anggraeni D, Pandutama M 2017 Online Statistical Modeling (Regression Analysis) for Independent Responses *IOP Conf. Series: Journal of Physics: Conf Series* **855** 012054

[2]. Wallace B C, Dahabreh I J, Trikalinos T A, Lau J, Trow P and Schmid C H 2012 Closing the Gap between Methodologists and End-Users: R as a Computational Back-End *Journal of Statistics Softwares* **49** (5)

[3]. Liang K Y, Zeger S L 1986 Longitudinal data analysis using generalized linear models *Biometrika* **73**(1) 13–22

[4]. Yan J and Fine J P 2004 Estimating Equations for Association Structures *Statistics in Medicine* **23** 859-880

[5]. HÃjsgaard S, Halekoh U and Yan J 2006 The R Package geepack for Generalized Estimating Equations *Journal of Statistical Software* **15**(2) 1-11

[6]. Touloumis A 2015 R Package multgee: A Generalized Estimating Equations Solver for Multinomial Responses *Journal of Statistical Software* **64**(8) 1-14

[7]. Bates D, Maechler M, Bolker B and Walker S 2015 Fitting Linear Mixed-Effects Models Using lme4 *Journal of Statistical Software* **67**(1) 1-48

[8]. Lee Y and Nelder J A 1996 Hierarchical generalised linear models (with discussion). *Journal of the Royal Statistical Society B* **58** 619-656

[9]. Ronnegard L, Shen X and Alam M 2010 hglm: A Package for Fitting Hierarchical Generalized Linear Models. *The R Journal*, 2(2): 20-28. URL http://journal.r-project.org/archive/2010-2/RJournal_2010-2_Roennegaard~et~al.pdf.

[10]. Taiyun Wei and Viliam Simko (2016). *corrplot: Visualization of a Correlation Matrix*. R package version 0.77. https://CRAN.R-project.org/package=corrplot

[11]. Revelle W 2016 *psych: Procedures for Personality and Psychological Research*, Northwestern University, Evanston, Illinois, USA, https://CRAN.R-project.org/package=psych Version = 1.6.9.

[12]. Wickham H 2009 *ggplot2: Elegant Graphics for Data Analysis* (Springer-Verlag New York)

[13]. Chang W, Cheng J, Allaire J J, Xie J and McPherson J 2017 *shiny: Web Application Framework for R*. R package version 1.0.5. https://CRAN.R-project.org/package=shiny

[14]. Barton K 2017 *MuMIn: Multi-Model Inference*. R package version 1.40.0. https://CRAN.R-project.org/package=MuMIn

[15]. Fox J 2017 *Using the R Commander: A Point-and-Click Interface or R* Boca Raton FL: Chapman and Hall/CRC Press.

[16]. Trestle Technology LLC 2016 *shinyAce: Ace Editor Bindings for Shiny* R package version 0.2.1. https://CRAN.R-project.org/package=shinyAce

[17]. Tirta I M, Anggraini D and Octaviani L C 2016 Online and Interactive Web For Fitting GEES With Natural Splines For Longitudinal Data. *IBSC:International Basic Scince Conference*. FMIPA Universitas Jember. 26-27 September 2016. URL: http://statslab-rshiny.fmipa.unej.ac.id/JORS/GEE/ (revised and submitted for Pertanika)

[18]. Wood S and Scheipl F 2017 *gamm4: Generalized Additive Mixed Models using 'mgcv' and 'lme4'*. R package version 0.2-5. https://CRAN.R-project.org/package=gamm4

Mammogram classification scheme using 2D-discrete wavelet and local binary pattern for detection of breast cancer

Januar Adi Putra[1]
[1]Department of Information Technology, Faculty of Computer Science, University of Jember, Indonesia

Email:_januaradi.putra@gmail.com

Abstract. In this paper, we propose a new mammogram classification scheme to classify the breast tissues as normal or abnormal. Feature matrix is generated using Local Binary Pattern to all the detailed coefficients from 2D-DWT of the region of interest (ROI) of a mammogram. Feature selection is done by selecting the relevant features that affect the classification. Feature selection is used to reduce the dimensionality of data and features that are not relevant, in this paper the F-test and Ttest will be performed to the results of the feature extraction dataset to reduce and select the relevant feature. The best features are used in a Neural Network classifier for classification. In this research we use MIAS and DDSM database. In addition to the suggested scheme, the competent schemes are also simulated for comparative analysis. It is observed that the proposed scheme has a better say with respect to accuracy, specificity and sensitivity. Based on experiments, the performance of the proposed scheme can produce high accuracy that is 92.71%, while the lowest accuracy obtained is 77.08%.

1. Introduction

One of the most effective methods to detect and identify breast cancer is through mammography examination using X-ray, images generated from the X-ray called a mammogram. An analysis of the mammography images that were previously done manually by the radiologist can be replaced with renewable technologies by utilizing digital image processing, so that the results obtained is not subjective and to overcome the radiologist's feel that not confident with the results of the analysis [1]. The main purpose of image processing is to gather information, screening or invertigation, diagnose, therapy and control, as well as monitoring and evaluation[2]. There are several studies that have been done by utilizing digital image processing, Dhawan et al in [3] using wavelet decomposition on a gray-level image to classify the mammograms and obtained the accuracy values is 81%. Liu et al in [4] using a wavelet-based statistical feature and the binary tree as a classifier obtain classification accuracy of 84.2% on a mammogram. Rashed et al in [5] use some type of daubechies wavelets on the classification mamm-ogram, obtained a classification accuracy of 87.06%. Buciu et al in [6] using a filter gabor based on wavelet for feature extraction, PCA for feature selection (dimension reduction) and support vector machine as classifier to classify images of mammograms as a normal-abnormal (performance classification 79%) and benign-malignant (performance classification 78 %).

The literature survey reveals about the existing classification schemes for digital mammogram images. However, most of them are not able to provide a high accuracy. It has been seen that the dimension of extracted feature space is so high due to large varieties of normal and abnormal tissues present in the breast. The use of high dimensional feature space may degrade the performance of the classification scheme. From a large feature space, only some of the features are effective and significant for the mammogram classification. Therefore, in addition to feature extraction, feature

IOP Conf. Series: Journal of Physics: Conf. Series **1008** (2018) 012004 doi:10.1088/1742-6596/1008/1/012004

selection is also the key step in mammogram classification, which selects only the significant features from available feature space. So there is a need to develop some new feature extraction and selection algorithms to increase the accuracy of classification rate. In this paper, we have proposed an effective feature extraction algorithm using two dimensional discrete wavelet transform (2D-DWT) based multiresolution analysis along with gray-level co-occurrence matrix (GLCM) to compute texture features for mammographic images. A feature selection algorithm has been applied using two statistical feature selection methods such as two-sample t and F-test to select significant features from extracted features. Utilizing these significant features, a back propagation neural network (BPNN) has been used as classifier to predict the mammogram, whether it is a normal or abnormal.

2. Methods

In this chapter elaborated on the research methodology used in this study as well as the contribution of the proposed. Illustration groove research methodology can be seen in Figure 1.

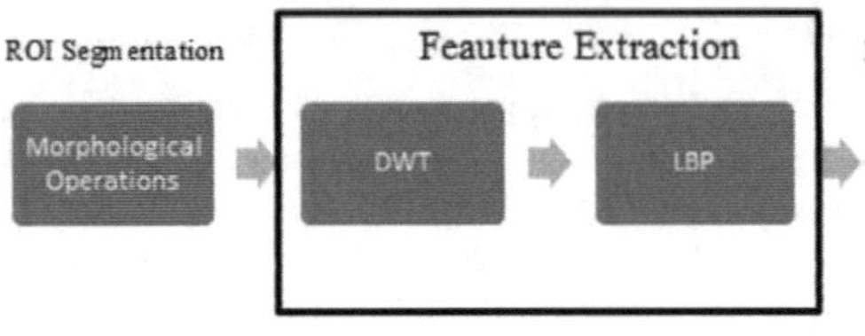

Figure 1. Research Methodology.

On a mammogram, a region that represents cancer is a region that containing a mass, so that before the classification is done then the mammogram image will be preprocessing to detect the location of the mass. Areas that are important in an image or referred to as the region of interest (ROI)[7,8]. Examples of ROI on the mammogram image is shown in Figure 2. There are several stages in the detection and segmentation ROI on a mammogram image. The purpose of this phase is to get the point of cancer on a mammogram image (mass) which may indicates the someone has breast cancer or not. Stages of detection and segmentation of the mass can be seen in Figure 3.

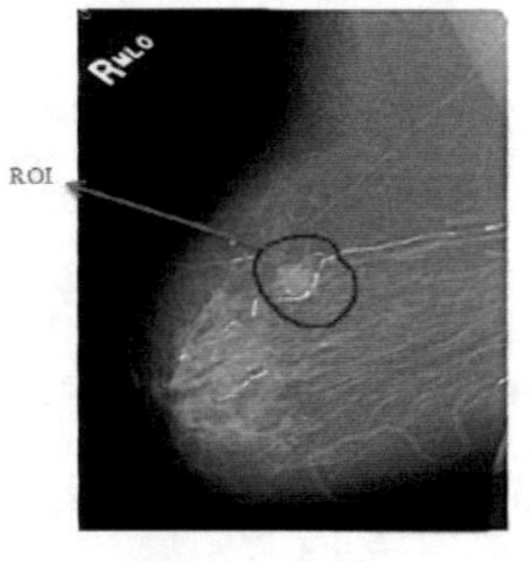

Figure 2. ROI in mammogram.

Figure 3. Stage detection and cropping ROI.

2.1 Discrete Wavelet Transform

Extraction of the image features a mammogram done with the method of combining 2D-DWT and LBP. DWT can be used for image transformation, wavelet transformation process is done in a pretty simple concept. Transformed the original image is divided (decomposed) into four sub-new image to replace him. Each sub-image of ¼ times the size of the original image. Sub-image at the top right, bottom left and bottom right will seem like a crude version of the original image because it

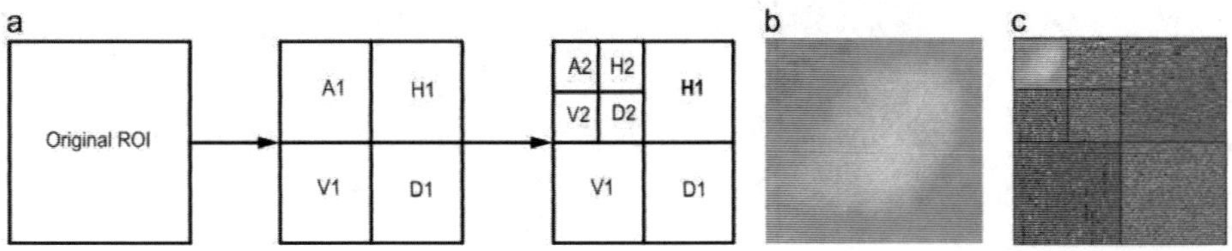

Figure 4. a) decomposition process b) Original ROI c) Result.

contains high frequency components of the original image. As for the first sub-image on the left looks like the original image and looks smoother (smooth) because it contains low frequency components of the original image. Discrete wavelete transform performed in this study is a level 2 which will produce 7 new sub-image. Stages of DWT in this study can be seen in Figure 4.

2.2 Local Binary Pattern

The original LBP operator labels the pixels of an image with decimal numbers, called Local Binary Patterns or LBP codes, which encode the local structure around each pixel[9, 10]. It proceeds thus, as illustrated in Fig.1: Each pixel is compared with its eight neighbors in a 3x3 neighborhood by subtracting the center pixel value; The resulting strictly negative values are encoded with 0 and the others with 1; A binary number is obtained by concatenating all these binary codes in a clockwise direction starting from the top-left one and its corresponding decimal value is used for labeling. The derived binary numbers are referred to as Local Binary Patterns or LBP codes.

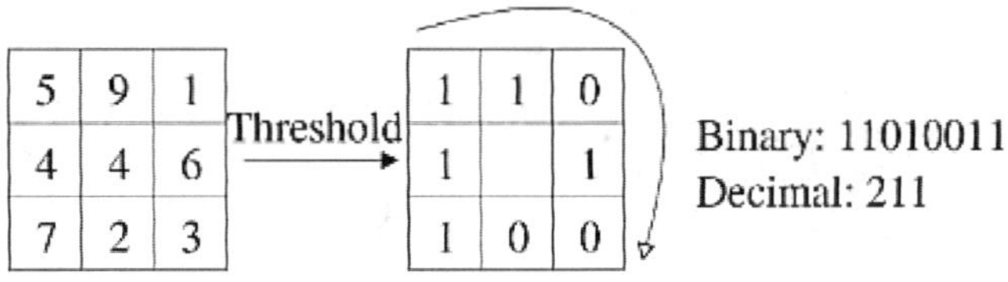

Figure 5. An example of the basic LBP operator

2.3 Feature Selection

Selection feature is one of the stages before the classification process. Feature selection is done by selecting the relevant features that affect the classification. Selection feature is used to reduce the dimensionality of data and features that are not relevant. Selection feature is also used to enhance the effectiveness and efficiency of the performance of classification algorithms. In this research feature selection using statistical methods that is two sample T-test and F-test. T-test known as partial test, which is to examine how the influence of each independent variable individually against the dependent variable. This test can be done with compare T arithmetic with T-table or view column on the significance of each T. F-test known as simultaneous test or tests Model / Anova test, a test to see how the influence of all independent variables together against the dependent variable. T-test or F-test functions to generate hypotheses whether a feature that a significant class of the same features in other classes. If the model is significant then the model can not be used for classification, otherwise if non / not significant then the regression model can be used for the classification process. For algorithm T-

```
R: Total number of selected features
Functions ttest() and vartest() compute the null hypothesis
values of two vectors at different values of significance
level, by two-sample t and F-test respectively
Create two empty vectors v1 and v2
Initialize α, 0 < α < 1
for i ← 1 to M do
   Clear contents of vector v1 and vector v2
   for j ← 1 to N do
      if target_class[j] = 1 then
         Append feature_matrix[i,j] to v1
      else
         Append feature_matrix[i,j] to v2
      end if
   end for
   h1[i] ← ttest(v1, v2, α)
   h2[i] ← vartest(v1, v2, α)
   for l ← 1 to 2 do
      if hk[i] = 1 then
         Append feature_matrix[i, 1 : N] to selected_featurek
      end if
   end for
end for
```

Figure 6. F-test and T-test Algorithm

test and F-test can be seen in Figure 6.

3. Result and Discussion

The data used in this study are the images obtained from the database mammographic Image Analysis Society (MIAS) and DDSM. Data obtained through the website http://peipa.essex.ac.uk-/pix/mias, and DDSM from http://marathon.csee.usf.edu/. With the amount of data mammogram image is 317, with 208 images is normal and 109 is abnormal.

3.1 Selection Featuren Experiment

Selection feature is used to reduce the dimensionality of data and features that are not relevant. Selection feature is also used to enhance the effectiveness and efficiency of the performance of classification algorithms. Feature selection method used is F-test and T-test and parameters that affect the two methods are the significant level. The trial also performed on stage feature selection in a way to alter the parameter values significant level, it is intended to determine the relationship level with a significant variable selection results of existing features. Values that becoming significant level test parameters are 0.1, 0.3, 0.5, 0.7, 0.9.

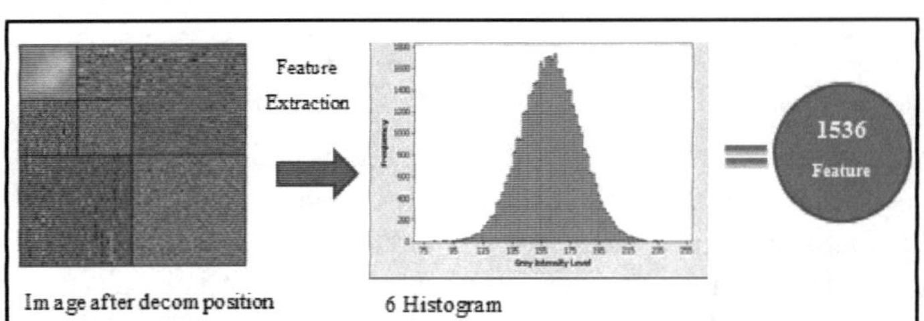

Figure 7. Simulation of feature extraction and selection.

The F-test and T-test will be performed to the results of the feature extraction dataset. Every single mammogram image will produce 7 sub image after decomposed using discrete wavelet transform, but in this research, image that to be used for the extraction feature process are only 6 sub image, that is sub image of a horizontal decomposition level 1 and 2, vertical decomposition level 1 and 2, and the diagonal decomposition level 1 and 2. One sub image of mammogram will produce one histogram after feature extraction process using LBP, while the one histogram is composed of 256 features so 6 sub image mammogram will generate 1536 feature. Simulation of feature extractiom and selection mammogram image untill be a dataset features can be seen in Figure 7 and the results of feature selection can be seen in Table 1.

Table 1. Feature Selection Result.

Significant level	Number of Feature LBP (R)	
	T-test	F-test
0.1	758	453
0.3	1010	775
0.5	1173	1021
0.7	1330	1226
0.9	1478	1443

Can be seen in Table 1, the largest number of feature is 1478 that produced when using T-Test as selection features method, LBP as feature extraction method and the value of significant level is 0.9. Based on the results obtained in Table 1 also can be concluded that higher the value of significant level make the feature will be less to reducted so the number of selection feature result will be same with the initial feature before the feature selection phase, and vice versa the lower the value of significant level make the feature will be more to reducted and produce less independent features. The cause of significant level is inversely proportional to the results of feature selection is due to the significant level used as a benchmark to determine a feature considered independent or not, so if the value of the significant level higher then the chances of a feature considered to be independent will be smaller.

3.2 Classification Experiment

Testing is done by comparing the traditional LBP method using discrete wavelet transform and without discrete wavelet transform. The performance of the method can be seen in Table 2 and Table 3.

Can be seen in Table 2 the performance of a combination of discrete wavelet transform and local binary pattern generating high accuracy that is 92.70% of feature selection F-test with a significant level value of 0.9 and 0.7. At trial using feature selection T-test combinations of discrete wavelet transform and a full neighbor local binary pattern get the highest accuracy is 85.42% on the value of the significant level is 0.1. For the sensitivity and specificity of this combination produces the highest value on a test using the F-test with a significant level of 0.9 and 0.7. Lowest accuracy value for a combination of discrete wavelet transform and a full neighbor local binary pattern obtained at feature selection methods are used T-test with a significant level value of 0.5 which is 77.08% while for feature selection methods F-test lows accuracy obtained at significant level of 0.5. From Table 2 also obtained the fact that the specificity and sensitivity is directly proportional to the accuracy.

Table 2. Performance of DWT+lbp.

Method	Significant Level	R	TP	TN	FP	FN	Sensitivity	Specifisity	Accuration (%)
T-test	0.9	213	21	57	12	6	0.78	0.83	81.25
	0.7	206	22	56	11	7	0.76	**0.84**	81.25
	0.5	197	16	62	17	1	0.94	0.78	81.25
	0.3	188	17	60	16	3	0.85	0.79	80.20
	0.1	175	22	51	11	12	0.64	0.83	76.04
F-test	0.9	216	21	57	12	6	0.78	0.83	81.25
	0.7	211	19	59	14	4	0.83	0.81	81.25
	0.5	207	22	58	11	5	0.81	**0.84**	**83.33**
	0.3	192	22	57	11	6	0.79	**0.84**	82.29
	0.1	174	1	63	32	0	**1.00**	0.67	66.67

Table 3. Performance of LBP.

Method	Significant Level	R	TP	TN	FP	FN	Sensitivity	Specifisity	Accuration (%)
T-test	0.9	213	26	55	7	8	0.76	0.89	84.37
	0.7	206	25	56	8	7	0.78	0.88	84.37
	0.5	197	20	54	13	9	0.69	0.81	77.08
	0.3	188	24	54	9	9	0.73	0.86	81.25
	0.1	175	26	56	7	7	0.79	0.89	85.42
F-test	0.9	216	29	60	4	3	**0.91**	**0.94**	**92.71**
	0.7	211	29	60	4	3	**0.91**	**0.94**	**92.71**
	0.5	207	26	50	7	13	0.67	0.88	79.17
	0.3	192	25	57	8	6	0.81	0.88	85.42
	0.1	174	26	54	7	9	0.74	0.89	83.33

For the traditional local binary pattern the performance can be seen in Table 3 to produce the highest accuracy is 83.33% on feature selection F-test with a significant level value of 0.5. At trial using feature selection using the T-test traditional local binary pattern methods the highest value only on the value of significant levels of 0.9, 0.7, and 0.5 that the accuracy is 81.25%. The sensitivity of this combination produces a perfect score when use feature selection F-test with a value of significant levels of 0.1, while for specificity highest value obtained when accuracy also got the highest score that is the feature selection F-test with a value of significant levels of 0.5 and 0.3 with a value specificity is 0.84, and feature selection T-test with a significant level of 07. The lowest accuracy on local binary pattern obtained when feature selection F-test used with a significant level value of 0.7 which is 66.67%. From Table 4 and Table 5 obtained the fact that the proposed method keeps getting higher accuracy than traditional local binary pattern although without performing discrete wavelet transform decomposition on mammogram image.

4. Conclusions

Based on the method of application has been made and the results obtained from a series of trials that have been conducted on the proposed scheme 2D-discrete wavelet transform and local binary pattern (LBP) traditional, it can be concluded on this study that the proposed scheme is better in an average accuracy than other scheme. Based on experiments, the performance of the proposed scheme can produce high accuracy that is 92.71%, while the lowest accuracy obtained is 77.08%. In the histogram result when using LBP the trend of high bit value histogram is 0 and 255, it indicates that proposed method (dwt+lbp) is sensitive against noise. Higher The value of significant level make the feature will be less to reducted so the number of selection feature result will be same with the initial feature before the feature selection phase, and the lower value of significant level make the feature will be more to reducted and produce less independent features.

Acknowledments

We gratefully acknowledge the support from Faculty of Computer Science & CGANT - University of Jember 2018

References

[1] Uyun S and Harjoko A 2011 Detection of Abnormality in Mammogram Images using Gray-Level Co-Occurrence Matrix
[2] Adollah R, Mashor M Y, Nasir N F M, Rosline H, Mahsin H and Adilah H 2008 Blood Cell Image Segmentation: A Review *Biomed, Proceedings* **21** 141–144
[3] Dhawan A 1996 Analysis of mammographic microcalcifications using greylevel image structure features *IEEE Trans Med. Imaging* **15** (3) 246–259
[4] Liu S, Babbs C F and Delp E J 2001 Multiresolution detection of spiculated lesions in digital mammograms *IEEE Trans Image Process* **10** (6) 874–884
[5] Rashed E A, Ismail I A, Zaki S I 2007 Multiresolution mammogram analysis in multilevel decomposition *Pattern Recognit Lett* **28** (2) 286–292
[6] Buciu, Gacsadi A 2011 Directional features for automatic tumor classification of mammogram images *Biomed Signal Process Control* **6** (4) 370–378
[7] Shen S, Sandham W, Granat M and Sterr A 2005 MRI Fuzzy Segmentation of Brain Tissue Using Neighborhood Attraction With Neural-Network Optimization *IEEE transactions on information technology in biomedicine* **9** (3)
[8] Mancas M and Gosselin B 2003 Fuzzy Tumor Segmentation based on Iterative Watersheds
[9] Ahonen T, Hadid A and Pietikainen M 2004 Face Recognition with Local Binary Patterns *Machine Vision Group Infotech Oulu*
[10] KayaY, Ertugrul O F and Taekin R 2015 Two Novel Local Binary Pattern Descriptor for Texture Analysis *Applied Soft Computing* **34** 728-735

IOP Conf. Series: Journal of Physics: Conf. Series **1008** (2018) 012005 doi:10.1088/1742-6596/1008/1/012005

Continuous connection of two adjacent pipe parts defined by line, bézier and hermit center curves

Kusno[1], Antonius Cahyo Prihandoko[1]
[1]Department of Mathematics, University of Jember, Jember, Indonesia

E-mail: kusno.fmipa@unej.ac.id

Abstract. A shape of pipe part consist of the longitudinal and cross section boundary curve. Also, it can be defined by only its center curve of the pipe. To build a complete pipe, we can connect continually the pipe parts. This paper discuss about continuous connection of two adjacent pipe parts that are defined by line, Bézier and Hermit curves. The method is as follows, we join continually two adjacent center curves of the pipe parts and then by using the same formulae, we define its coincidence cross section boundary curves of the pipes. Finally, the joining of the longitudinal section boundary curves of the pipe parts can be done. The results of this study show that we can evaluate the joint continuity between two adjacent pipe parts defined by line, Bézier and Hermit curves, using the various forms of its center and cross-longitudinal section boundary curves.

1. Introduction

There are many introduced methods for modeling a shape of pipe. We can use a single-valued tubular patches and a cyclide to model radius and pipe design of tubular geometry [7,8]. By using the geometric continuity condition for algebraic surface we can unify some pieces of pipes at common vertex [1]. By joining two cylinders of revolution with axes in a common plane and different radii we will find a circular surfaces [4]. Also, the shape of pipe can be evaluated by geometric properties of canal surfaces in E^3 [6]. The physical model of transitional pipeline parts can be made of materials that cannot be wrinkled or stretched [5]. Then, a modular Pipe-Z parametric design system will give the results for a trefoil, a gure-eight knot and a pentafoil [9]. Different from the earlier methods, this paper discusses about constructing the whole pipe using small parts connection of the pipes. In this discussion, we calculate the continuous connection of two adjacent pipe parts defined by line, Bézier and Hermit curves.

This paper is organized in the following sections. In the first section, we talk about the formulation of pipe parts that depend on its cross, longitudinal and center curves. In the second, we evaluate the continuous connection of two adjacent pipe parts of the line, Bézier and Hermit center curves. Finally, the results will be summarized in the conclusion section.

2. Pipe Parts Formulation

Consider a regular curve $\Gamma(u)$ continuous on the interval $0 \leq u \leq 1$ that can be differentiated twice. Also it can be expressed as a function of the natural parameter $\Gamma(s)$. The tangent unit vector **t** and the normal unit vector **n** are orthogonal in the form (Figure 1a)

$$\mathbf{t} = \frac{d\mathbf{\Gamma}}{ds} = \dot{\mathbf{\Gamma}} = \frac{\mathbf{\Gamma}'}{|\mathbf{\Gamma}'|} \tag{1a}$$

$$\mathbf{n} = -\frac{\mathbf{k}}{|\mathbf{k}|} \tag{1b}$$

and

$$\mathbf{k} = \frac{d\mathbf{t}}{ds} = \dot{\mathbf{t}} = \frac{\mathbf{t}'}{|\mathbf{\Gamma}'|}. \tag{1c}$$

Using the cross product operation of the both vectors **t** and **n**, we can define the unit binormal vector $\mathbf{b} = \mathbf{t} \wedge \mathbf{n}$ such that the triplet [**t**, **n**, **b**] form the Frenet frame of curve [3].

Consider along the direction of parameter u each points of the curve $\Gamma(u)$ as the center points of the defined circles of radius $\Upsilon(u,v)$ in the normal plane [**b**,**n**] of the curve $\Gamma(u)$ that are orthogonal to unit tangent vector **t**. Using the parametric tubular surface formulae of the curve $\Gamma(u)$, we can define a part of tubular pipe in the form [2]

$$\mathbf{T}(u,v) = \Gamma(u) + \Upsilon(u,v).[cos\,(\varphi)\,\mathbf{b} + sin\,(\varphi)\,\mathbf{n}] \qquad (2)$$

with the real function $\Upsilon(u,v) = \rho(u).r(v)$ expresses the radius of the pipe, $\varphi = 2\pi v$ and $0 \leq u,v \leq 1$. In this case, the real function $\rho(u)$ characterizes the inflate-deflate surfaces form of the pipe patches along its center curve, meanwhile $r(v)$ defines the cross-section curve form of the pipe. To facilitate the creation of pipe part, we determine the real function $\rho(u)$ and $r(v)$ from the Bézier, Hermit and trigonometric curves as follows

$$\rho_1(u) = P_o(1-u)^3 + 3P_1(1-u)^2u + 3P_2(1-u)u^2 + P_3u^3 \qquad (3a)$$

$$\rho_2(u) = \rho_2(0)H_1(u) + \rho_2(1)H_2(u) + \rho_2^u(0)H_3(u) + \rho_2^u(1)H_4(u) \qquad (3b)$$

$$\rho_3(u) = a + b\,cos\,(u) + c\,sin\,(u) \qquad (3c)$$

$$r_1(v) = a.cos\,(n.\varphi) \pm b.sin\,(n.\varphi) \qquad (4a)$$

where

$$H_1(u) = 2u^3 - 3u^2 + 1; \quad H_2(u) = 2u^3 + 3u^2;$$
$$H_3(u) = u^3 - 2u^2 + u; \quad H_4(u) = u^3 - u^2$$

with a, b, c real constants, $\varphi = 2\pi.v$ and n is the number of defined rose leafs, $0 \leq u \leq 1$ and $0 \leq v \leq 1$. In addition, we define the cross section curve of pipe using unify n curves of the different circle parts to the same origin O in the form

$$r_2(v) = r_o.\left[cos\left((2i+1).\tfrac{\pi}{n} - \tfrac{\varphi}{n}\right)\right] \pm \sqrt{\left[r_o^2.cos^2\left((2i+1).\tfrac{\pi}{n} - \tfrac{\varphi}{n}\right) - (r_o^2 - \tau^2)\right]} \qquad (4b)$$

for $i = 0,1, ..., n-1$ and r_o as a ray of polar form, $\varphi = 2\,\pi v$ with $0 \leq v \leq 1$. The first, we determine $\Gamma(u)$ as a line curve of equation

$$\mathbf{L}(u) = \mathbf{c} + \lambda\, u.\mathbf{l} \qquad (5)$$

with **c** constant vector, **l** unit direction vector, λ positive real constant and $0 \leq u \leq 1$. Because of the $\mathbf{L}(u)$ line, we must change the circles of radius $\Upsilon(u,v)$ in the normal plane [**b**,**n**] in equation (2) become in plane $[\mathbf{v}_1,\mathbf{v}_2]$ with $\mathbf{v}_1$ and $\mathbf{v}_2$ any unit constant vectors such that $\mathbf{l} = \mathbf{v}_1 \wedge \mathbf{v}_2$. The second, we define the center curve $\Gamma(u)$ in the form cube Bézier curve $\mathbf{B}(u)$ and Hermit curve $\mathbf{H}(u)$ with

$$\mathbf{B}(u) = \mathbf{P}_o\,(1-u)^3 + 3\,\mathbf{P}_1\,(1-u)^2\,u + \mathbf{P}_2\,(1-u)\,u^2 + \mathbf{P}_3\,u^3 \qquad (6)$$
$$\mathbf{H}(u) = \mathbf{H}_o H_1(u) + \mathbf{H}_1 H_2(u) + \mathbf{H}_o^u H_3(u) + \mathbf{H}_1^u H_4(u). \qquad (7)$$

The vectors $\mathbf{P}_o$, $\mathbf{P}_1$, $\mathbf{P}_2$ and $\mathbf{P}_3$ are the control points of the Bézier curve $\mathbf{B}(u)$, $\mathbf{H}_o = \mathbf{H}(0)$, $\mathbf{H}_1 = \mathbf{H}(1)$, $\mathbf{H}_o^u = \mathbf{H}'(0)$, $\mathbf{H}_1^u = \mathbf{H}'(1)$ and $0 \le u \le 1$.

The implementation of formulae (2-7) can be demonstrated as follows. Let $\mathbf{v}_1 = \langle 0,1,0 \rangle$, $\mathbf{v}_2 = \langle 0,0,1 \rangle$ and $\mathbf{L}(u) = \langle -15,0,0 \rangle + 25.u. \langle 1,0,0 \rangle$. When we decide $\rho(u) = 2 + cos\,(2\pi u) - sin\,(2\pi u)$ and $r(v) = 1$ such that $\Upsilon(u,v) = 2 + cos\,(2\pi u) - sin\,(2\pi u)$, equation (2) will show Figure 1a. If we determine $\rho(u) = 1.5$; $r(v) = -2 - \cos(4\pi v) - \cos(8\pi v)$ and $\rho(u) = 2\,(1-u)^3 + 12\,(1-u)^2.\,u + 3\,(1-u).u^2 + 2u^3$; $r(v) = -2 - \cos(4\pi v) - \cos(8\pi v)$, we will find Figure 1b and 1c, respectively. Figure 1d and 1e show the graph of equation (2) and (6) with control points [$\mathbf{P}_o = \langle -22,-18,15 \rangle$, $\mathbf{P}_1 = \langle 8,-2,-50 \rangle$, $\mathbf{P}_2 = \langle 18,3,16 \rangle$, $\mathbf{P}_3 = \langle 25,-14,10 \rangle$], $\rho(u) = 2\,(1-u)^3 + 6\,(1-u)^2.\,u + 3\,(1-u).u^2 + 2u^3$; $r(v) = 4$ and $\rho(u) = 2\,(2u^3-3u^2+1) + 2\,(2u^3+3u^2) + 8\,(= u^3 - 2u^2 + u) + 10\,(u^3 - u^2)$; $r(v) = 2$. Base on the equations (1-7), then we discuss about the continuous connection of two adjacent pipe parts defined by line, Bézier and Hermit curves

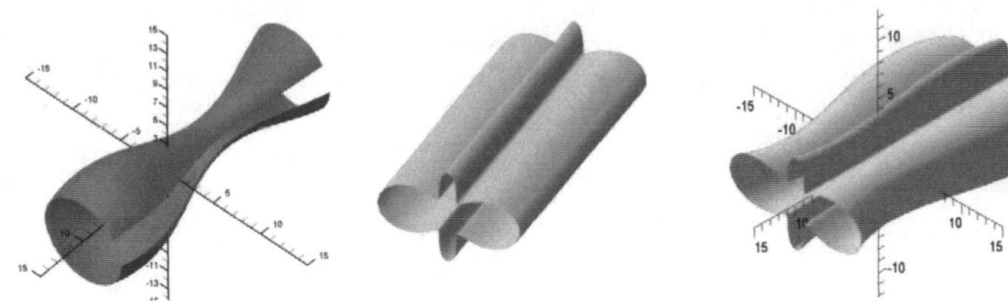

(a) (b) (c)

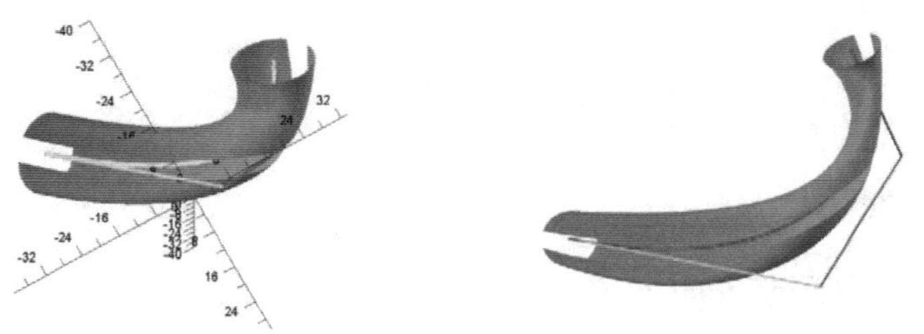

(d) (e)

Figure 1. Pipe parts of the line center curve (1a,b,c) and the cubic Bézier center curve (1d,e)

3. Connection of Two Adjacent Pipe Parts

Consider two parametric tubular surface pieces $\mathbf{T}_1(u,v)$ and $\mathbf{T}_2(u,v)$ of equation (2). Its center curve and its cross-longitudinal boundary curves are respectively [$\mathbf{L}_1(u)$,$\mathbf{L}_2(u)$] of line curve (5) and [$\Upsilon_1(u,v)$, $\Upsilon_2(u,v)$] of equations (3-4) in the same direction condition. Because of the center curves of degree one, we can join the both tubular surfaces using tangential continuity, that is $\mathbf{T}_{L1}(1,v) = \mathbf{T}_{L2}(0,v)$ and $\mathbf{T}_{L1}{}^{u}(1,v) = \mathbf{T}_{L1}{}^{u}(0,v)$ along the interval $0 \le v \le 1$. These mean that, the first, at $\mathbf{T}_{L1}(1,v)$ and $\mathbf{T}_{L2}(0,v)$ it must be

$$[\mathbf{c} + \lambda.\mathbf{l}]_{L1} = [\mathbf{c}]_{L2} \tag{8a}$$

$$\Upsilon_{L1}(1,v) = \Upsilon_{L2}(0,v) \tag{8b}$$

in plane [$\mathbf{v}_1$,$\mathbf{v}_2$]. The second, along the interval $0 \le v \le 1$, the tangent vectors $\mathbf{L}_1{}^{u}(u)$ and $\mathbf{L}_2{}^{u}(u)$ of its center curves and the first derivation of the tube radius $\Upsilon_{L1}{}^{u}(u,v)$ and $\Upsilon_{L2}{}^{u}(u,v)$ have to

$$\mathbf{L}_1^{u}(1) = \sigma\,\mathbf{L}_2^{u}(0) \quad \text{or} \quad \mathbf{l}_1 = \sigma\,\mathbf{l}_2 \tag{9a}$$
$$\Upsilon_{L1}^{u}(1,v) = \Upsilon_{L2}^{u}(0,v) \tag{9b}$$

with σ positive real scalar. When the center curve of pipe $\mathbf{T}_2(u,v)$ is a cubic Bézier curve of equation (6), then $\mathbf{B}'(u) = 3[(\mathbf{P}_1\text{-}\mathbf{P}_o)(1\text{-}u)^2 + 2(\mathbf{P}_2\text{-}\mathbf{P}_1)(1\text{-}u).u+(\mathbf{P}_3\text{-}\mathbf{P}_2).u^2]$, $\mathbf{B}'(0) = 3[(\mathbf{P}_1\text{-}\mathbf{P}_o)]$ and the tangential continuity condition must be

$$[\mathbf{c} + \lambda.\mathbf{l}]_{L1} = [\mathbf{P}_o]_{L2} \tag{10a}$$
$$\Upsilon_{L1}(1,v) = \Upsilon_{L2}(0,v) \tag{10b}$$

$$\mathbf{l}_1 = 3\,\tau(\mathbf{P}_1\text{-}\mathbf{P}_o) \tag{10c}$$
$$\Upsilon_{L1}^{u}(1,v) = \Upsilon_{L2}^{u}(0,v) \tag{10d}$$

with τ positive real scalar.

Two tubular surface pieces $\mathbf{T}_1(u,v)$ and $\mathbf{T}_2(u,v)$ of the formulae (2) with its center curves $\mathbf{\Gamma}_1(u)$ and $\mathbf{\Gamma}_2(u)$ of equation (6) or (7) in the same direction can be joined in the tangential continuity, if along the interval $0 \le v \le 1$, they verify the conditions

$$\mathbf{T}_1(1,v) = \mathbf{T}_2(0,v) \tag{11a}$$
$$\mathbf{t}_1(1) = \alpha.\mathbf{t}_2(0) \tag{11b}$$
$$\Upsilon_1^{u}(1,v) = \Upsilon_2^{u}(0,v) \tag{11c}$$

with α positive real scalar.

To Justify the implementation of equations (8-11), we show in the following connection of pipes. Let two parametric tubular surface pieces $\mathbf{T}_1(u,v)$ with its center curve $\mathbf{L}(u) = \langle -15,0,0\rangle + 20.u.\ \langle 1,0,0\rangle$ and $\mathbf{T}_2(u,v)$ with its center curve of equation (6) where $[\mathbf{P}_o = \langle 5,0,0\rangle,\ \mathbf{P}_1 = \langle 15,0,0\rangle,\ \mathbf{P}_2 = \langle 25,3,16\rangle,\ \mathbf{P}_3 = \langle 35,14,10\rangle]$. Meanwhile $\Upsilon_1(u,v) = \Upsilon_2(u,v) = 2 + \cos(2\pi u) - \sin(2\pi u)$. It is clear that $\mathbf{T}_1(u,v)$ and $\mathbf{T}_2(u,v)$ satisfy all conditions of equations (10) and the result is shown in Figure 2a. Other tangential connections of pipes can be showed in Figures 2b, 2c and 2d. Base on this calculation method, in the next section we discuss about the Frenet connection between two tubular pieces Bézier and Hermit of the equation (6) and (7) respectively.

The tangential continuity conditions in equation (11) of two pipe parts $\mathbf{T}_1(u,v)$ and $\mathbf{T}_2(u,v)$ are also in the moving trihedron continuity, if the triplet $[\mathbf{t}_1(1),\mathbf{n}_1(1),\mathbf{b}_1(1)]$ and $[\mathbf{t}_2(0),\mathbf{n}_2(0),\mathbf{b}_2(0)]$ of the center curves $\mathbf{\Gamma}_1$(u) and $\mathbf{\Gamma}_2$(u) coincide each other respectively and the second derivation of the direction u of radius $\Upsilon_1(u,v)$ and $\Upsilon_2(u,v)$ at $\mathbf{T}_1(1,v)$ and $\mathbf{T}_2(0,v)$ are equal respectively. So, it must be

$$\mathbf{n}_1(1) = \beta.\mathbf{n}_2(0) \tag{12a}$$
$$\mathbf{b}_1(1) = \gamma.\mathbf{b}_2(0) \tag{12b}$$
$$\Upsilon_1^{uu}(1,\mathrm{v}) = \Upsilon_2^{uu}(0,\mathrm{v}) \tag{12c}$$

with β and γ positive real scalars. The calculation of the unit vectors **t**, **n** and **b** is as follows.
$\mathbf{B}'(u) = \langle R_x(u), R_y(u), R_z(u)\rangle$
with

$$R_x(u) = 3[(P_{1x} - P_{ox}).(1\text{-}u)^2 + 2(P_{2x} - P_{1x})(1\text{-}u).u + (P_{3x} - P_{2x}).u^2]$$
$$R_y(u) = 3[(P_{1y} - P_{oy}).(1\text{-}u)^2 + 2(P_{2y} - P_{1y})(1\text{-}u).u + (P_{3y} - P_{2y}).u^2]$$
$$R_z(u) = 3[(P_{1z} - P_{oz}).(1\text{-}u)^2 + 2(P_{2z} - P_{1z})(1\text{-}u).u + (P_{3z} - P_{2z}).u^2].$$

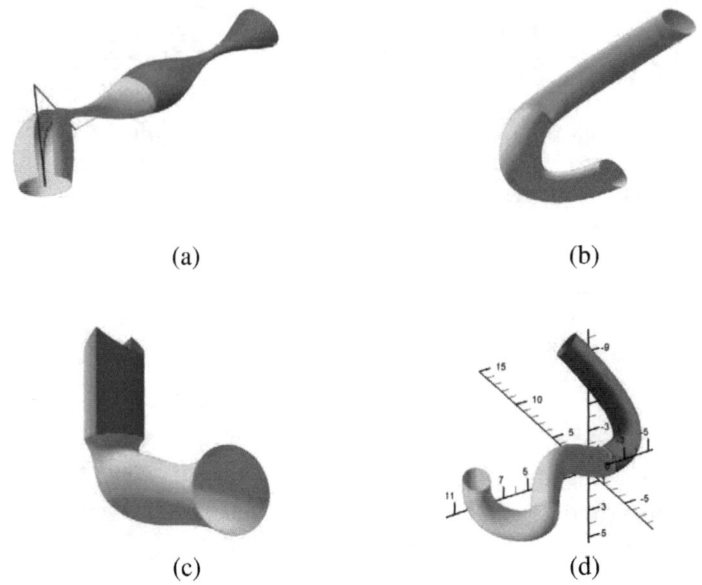

Figure 2. Connection 2 pipes with its centers of line, Bézier and Hermit curves

$\mathbf{B}''(u) = < W_x(u), W_y(u), W_z(u) >$
with

$$W_x(u) = 6[(P_{2x}\text{-}2P_{1x} + P_{ox}).(1\text{-}u) + (P_{3x}\text{-}2P_{2x} + P_{1x}).u]$$
$$W_y(u) = 6[(P_{2y}\text{-}2P_{1y} + P_{oy}).(1\text{-}u) + (P_{3y}\text{-}2P_{2y} + P_{1y}).u]$$
$$W_z(u) = 6[(P_{2z}\text{-}2P_{1z} + P_{oz}).(1\text{-}u) + (P_{3z}\text{-}2P_{2z} + P_{1z}).u].$$

$\mathbf{H}'(u) = < N_x(u),N_y(u),N_z(u)>$
with

$$N_x(u) = H_{ox}(6u^2-6u)+H_{1x}(-6u^2+6u)+H^u_{ox}(3u^2-4u+1)+H^u_{1x}(3u^2-2u)$$
$$N_y(u) = H_{oy}(6u^2-6u)+H_{1y}(-6u^2+6u)+H^u_{oy}(3u^2-4u+1)+H^u_{1y}(3u^2-2u)$$
$$N_z(u) = H_{oz}(6u^2-6u)+H_{1z}(-6u^2+6u)+H^u_{oz}(3u^2-4u+1)+H^u_{1z}(3u^2-2u).$$

$\mathbf{H}''(u) = < Z_x(u), Z_y(u), Z_z(u)>$
with

$$Z_x(u) = H_{ox}(12u-6)+H_{1x}(-12u+6)+H^u_{ox}(6u-4)+H^u_{1x}(6u-2)$$
$$Z_y(u) = H_{oy}(12u-6)+H_{1y}(-12u+6)+H^u_{oy}(6u-4)+H^u_{1y}(6u-2)$$
$$Z_z(u) = H_{oz}(12u-6)+H_{1z}(-12u+6)+H^u_{oz}(6u-4)+H^u_{1z}(6u-2).$$

$\mathbf{k}_B = < M_x(u), M_y(u), M_z(u) >$
with

$$M_x = [s^2.W_x(u)\text{-}[R_x^2(u).W_x(u) + R_x(u).R_y(u).W_y(u) + R_x(u).R_z(u).W_z(u)]]/s^4$$
$$M_y = [s^2.W_y(u)\text{-}[R_y(u).R_x.W_x(u) + R_y^2(u).W_y(u) + R_y(u).R_z(u).W_z(u)]]/s^4$$
$$M_z = [s^2.W_z(u)\text{-}[R_z(u).R_x(u).W_x(u) + R_z(u).R_y(u).W_y(u) + R_z^2(u).W_z(u)]]/s^4$$
$$s = [R_x^2(u) + R_y^2(u) + R_z^2(u)]^{1/2}.$$

$\mathbf{k}_H = < S_x(u); S_y(u); S_z(u) >$

with

$S_x = [n^2.Z_x(u) - [N_x^2(u).Z_x(u) + N_x(u).N_y(u).Z_y(u) + N_x(u).N_z(u).Z_z(u)]]/n^4$
$S_y = [n^2.Z_y(u) - [N_y(u).N_x.Z_x(u) + N_y^2(u).Z_y(u) + N_y(u).N_z(u).Z_z(u)]]/n^4$
$S_z = [n^2.Z_z(u) - [N_z(u).N_x(u).Z_x(u) + N_z(u).N_y(u).Z_y(u) + N_z^2(u).Z_z(u)]]/n^4$
$n = [N_x^2(u) + N_y^2(u) + N_z^2(u)]^{1/2}$.

So, the unit vector tangent, normal and binormal of the Bezier center curve is

$$\mathbf{t}_B = (1/s) < R_x(u), R_y(u), R_z(u) > \quad (13a)$$
$$\mathbf{n}_B = <(M_x(u)/s_o), (M_y(u)/s_o), (M_z(u)/s_o)> \quad (13b)$$
$$\mathbf{b}_B = <[R_y(u).M_z(u) - M_y(u).R_z(u)]/(s.s_o), [M_x(u).R_z(u) - R_x(u).M_z(u)]/(s.s_o), [R_x(u):M_y(u) -M_x(u):R_y(u)]/(s.s_o)> \quad (13c)$$

with $s_o = [M_x^2(u) + M_y^2(u) + M_z^2(u)]^{1/2}$. On the other hand, we have the unit vector tangent, normal and binormal of the Hermit center curve

$$\mathbf{t}_H = 1/n < N_x(u); N_y(u); N_z(u) > \quad (14a)$$
$$\mathbf{n}_H = < S_x(u)/n_o, S_y(u)/n_o, S_z(u)/n_o > \quad (14b)$$
$$\mathbf{b}_H = <[N_y(u).S_z(u) - S_y(u).N_z(u)]/(n.n_o), [S_x(u).N_z(u) - N_x(u).S_z(u)]/(n.n_o), [N_x(u).S_y(u) - S_x(u).N_y(u)]/(n.n_o)> \quad (14c)$$

with $n_o = [S_x^2(u) + S_y^2(u) + S_z^2(u)]^{1/2}$.

4. Conclusions

We have formulated the continuous connection of various pipe shapes which its center curves $\mathbf{\Gamma}(u)$ are defined by line, Bézier and Hermit curves. We can conclude that two parametric tubular surface pieces $\mathbf{T}_1(u,v)$ and $\mathbf{T}_2(u,v)$ connect continuously order-1, if they verify the conditions: first, $\mathbf{T}_1(1,v) = \mathbf{T}_2(0,v)$. Second, its tangent vectors of center curve $\mathbf{\Gamma}_1(u)$ and $\mathbf{\Gamma}_2(u)$ at the values $\mathbf{T}_1(1,v)$ and $\mathbf{T}_2(0,v)$ and its tangent vectors of longitudinal boundary curve along the cross section boundary curve are equal. They are also in the moving trihedron continuity, if its triplet [**t**,**n**,**b**] of the center curves at $\mathbf{\Gamma}_1(1)$ and $\mathbf{\Gamma}_2(0)$ coincide each other respectively and the second derivation of the direction u of radius $\gamma_1(u,v)$ and $\gamma_2(u,v)$ respectively at $\mathbf{T}_1(1,v)$ and $\mathbf{T}_2(0,v)$ are equal.

The continuous connection of various pipe shapes have been introduced. The interesting thing to discuss ahead is how to build a complete pipe using some parts of pipes in various thickness. Also, the connection have to be varied and continuous.

References

[1] Chen F L, Chen, C S and Deng J S 2000 Blending Pipe Surfaces with Picewise Algebraic Surfaces *Chinese Journal of Computers* **9**
[2] Kusno 2017 On the Modeling of the Various Inflate-Deflate Pipes Patches Based on a Curve Shape *Journal of Computational Design and Engineering* Submitted
[3] Lipschultz M 1969 *Theory and Problems of Differential Geometry* (Schaum's Outline Series McGraw-Hill New York).
[4] Malaček K and Šibrava Z 2006 Blending Circular Pipes With a Cyclic Surface *Journal for Geometry and Graphics* **10** 99–107
[5] Obradovic R, Popkonstantinovic B and Beljin B 2012 Algorithm for Approximation Transitional Developable Surfaces Between Two Polygons *Technics Technologies*

Education Management **7** 1907 1913
[6] Öztürk G, Byram B K and Arslan K 2010 On Canal Surfaces in E3 *Selcuk Journal of Applied Mathematics* **11** 103-108
[7] Sanchez-Reyes J 1994 Single-valued Tubular Patches *Computer Aided Geometric Design* **11** 565-592
[8] Srinivas Y L and Debashish D 1995 Cyclides in Geometric Modeling: Computational Tools for an Algorithmic in Infrasructure *Journal of Mechanical Design* **117** 363-373
[9] Zawidzki M and Nishinari K 2013 Modular Pipe-Z System for Three-Dimensional Knots *Journal for Geometry and Graphics* **17** 81-87

IOP Conf. Series: Journal of Physics: Conf. Series **1008** (2018) 012006 doi:10.1088/1742-6596/1008/1/012006

The development rainfall forecasting using kalman filter

Mohammad Zulfi[1], **Moh.Hasan**[1], **Kosala Dwidja Purnomo**[1]

[1]Department of Mathematics, University of Jember, Indonesia

E-mail: mohammadzulfi94@yahoo.com, hasan@unej.ac.id, kosaladp.fmipa@unej.ac.id

Abstract. Rainfall forecasting is very interesting for agricultural planing. Rainfall information is useful to make decisions about the plan planting certain commodities. In this studies, the rainfall forecasting by ARIMA and Kalman Filter method. Kalman Filter method is used to declare a time series model of which is shown in the form of linear state space to determine the future forecast. This method used a recursive solution to minimize error. The rainfall data in this research clustered by K-means clustering. Implementation of Kalman Filter method is for modelling and forecasting rainfall in each cluster. We used ARIMA (p,d,q) to construct a state space for KalmanFilter model. So, we have four group of the data and one model in each group. In conclusions, Kalman Filter method is better than ARIMA model for rainfall forecasting in each group. It can be showed from error of Kalman Filter method that smaller than error of ARIMA model.

1. Introduction

Rainfall information play an important role in agricultural planning. Therefore, the success of an agricultural plan depends on the availability of good rainfall forecasting. Rainfall data can be analyzed using time series analysis. Time series forecasting is an important area of forecasting in which past observations of the same variable are collected and analyzed to develop a model describing the underlying relationship [5]. Kalman filter is one of several methods that can be used for time series forecasting.

The Kalman filter is essentially a set of mathematical equations that implement a predictor-corrector type estimator that is optimal in the sense that it minimizes the estimated error covariance when some presumed conditions are met[4].The Kalman Filter method is used to represent a time series model that is displayed in linear state space[1]. Kalman Filter is an estimator that serves to estimate the state of output. The advantages of the Kalman Filter method are easy to apply in various sciences because they are recursive [4]. While the disanvantages of the Kalman Filter method is the success in getting optimal predictive results depending on the accuracy in estimating the initial state on the latest observation data [5]. In this research we used ARIMA model to construct a state space model in Kalman Filter.

ARIMA forecasting model is a forecasting technique that does not require specific assumptions about historical data, but uses an iterative model to determine the best model. The selected model will be evaluated with historical data whether it has accurately described the data. The best models will produce relatively small residuals, independent and have random distributions.

This research aim to know ARIMA model and state space model in rainfall data. then we will see how the comparison of forecasting between Kalman Filter and ARIMA.

2. Basic Concept of Kalman Filter

Kalman Filter is a recursive procedure used to forecast state vectors. This method uses a recursive technique to integrate the latest observational data into a model to correct previous predictions and predict subsequently optimally based on past data information as well as on current data information[3]. The Kalman Filter concept consists of two stages. The first stage is the forecasting stage. The second stage is the renewal stage. In the forecasting stage, an estimated value is generated for the current state and the value of the error covariance used as the initial estimation information for the next step. The renewal stage serves as a corrector [4]. The correction value is determined by how well the initial guess predicted the new observation [2].

The general equation of the Kalman Filter model is given $\bar{X}_t$ denoted as the vector of size (n x 1) variables observed at time t, the unobserved dynamic model for $\bar{X}_t$ can be explained $\bar{Z}_t$ on the vector of size (r x 1). The space state representation for y is given in the following equation [2]:

State transition equation : $\bar{Z}_{t+1} = F\bar{Z}_t + G\bar{a}_{t+1}$ (1)

output equation : $\bar{X}_t = H\bar{Z}_t + \bar{b}_t$ (2)

where:

F = (3) and (4)

$\bar{Z}$: state vector size (r x 1) in state equation
F : transition matrix size (r x r)
G : input matrix
H : matrix parameters size (n x r) in observation equation
$\bar{a}$: noise vector size (r x 1) in state equation
b : noise vector size (n x 1) in observation equation
$\bar{X}$: vector of the observed variables size (n x 1)

Forecasting uses Kalman filter through two stages. The first stage is the prediction stage (time update) and the next stage is measurement update. All the steps can be seen in table 1

Table 1. Kalman Filter Stages

construction of state space	1) state equation $\bar{Z}_{t+1} = F\bar{Z}_t + G\bar{a}_{t+1}$ 2) output equation $\bar{X}_t = H\bar{Z}_t + \bar{b}_t$	
prediction stage	1) estimation of state equations $\bar{Z}_{t+1} = F\bar{Z}_t + G\bar{a}_{t+1}$ 2) covarian error $R_{t+1} = F\Gamma_t F' + G\Sigma G'$	(5) (6)
Measurement update	1) determine kalman gain matrix $R_{t+1} = F\Gamma_t F' + G\Sigma G'$ 2) update estimation $\hat{Z}_{t+1} = \bar{Z}_{t+1} + K_{t+1}(\bar{X}_t - H\bar{Z}_{t+1})$ 3) covarian error $\Gamma_{t+1} = (I - K_{t+1}H)R_{t+1}$	(7) (8) (9)

3. Methods

In this research we used rainfall data of Kabupaten Jember which have been grouped based on K-mean clustering. In general, this data is divided into two parts, namely data insampel that began from January 2005 to December 2015 and data outsampel which starts from January 2016 until December 2016. The insampel data is the data used to construct the ARIMA model in each cluster, while the outsampel data is used to test how well the estimation results obtained. The stages in this research are as follows:

1. Plotting Data
   Plotting data is an important part of the analysis. In addition to viewing data patterns, plotting also provides an overview of stationary data in variance. The basic assumption that must be met is the stationary data in mean and variance. If the time series data plot does not show any apparent change in variance over time, it can be said that the time series data is stationary in its variance.
2. Augmented Dickey Fuller (ADF) test
   Augmented Dickey Fuller (ADF) test is used to see the stationarity of the data in the mean. If the time series data has been stationary in its mean and variance, then the data can be used to construct the ARIMA model. But if the data is not stationary then need to be done differencing transformation process.
3. ARIMA Forecasting Model
   After passing the stationary test, the insampel data can be used to construct the ARIMA forecasting model. Since the data is divided into four clusters, we will construct a model for each cluster. In the end we have four ARIMA models that may all be different.
4. Forecasting using Kalman Filter
   Calculation of Kalman Filter recursive equation includes:
   a. time update
   b. measurement update
5. Comparison of Forecasting Kalman Filter and ARIMA
   After the forecasting results are produced, the forecasting results are compared with the outsample data. From this process we will get the RMSE value on each forecasting model. Small RMSE values indicate that the model is better than the other model.

4. Results

4.1 Plotting Rainfall Data

Figure 1 is a graph of monthly rainfall time series data in 4 groups. The plot shows the amount of rainfall from January 2005 to December 2016. The amount of rainfall in the form of graph shows that there is an indication of season time series model with time series pattern that repeats at certain time and seen the pattern of up and down regularly formed from plot of rainfall data. In addition, the graph indicates a correlation that is time-sequence associated with the number of observations per season period. Therefore, before estimating the model, the data is tested for the mean and variance. To determine the stationarity of the data on the variance can be seen descriptively plot time series in Figure 1 which shows that the data over time have fixed or constant data fluctuations and are constantly changing on all clusters, so that the data on each cluster has been stationary variance.

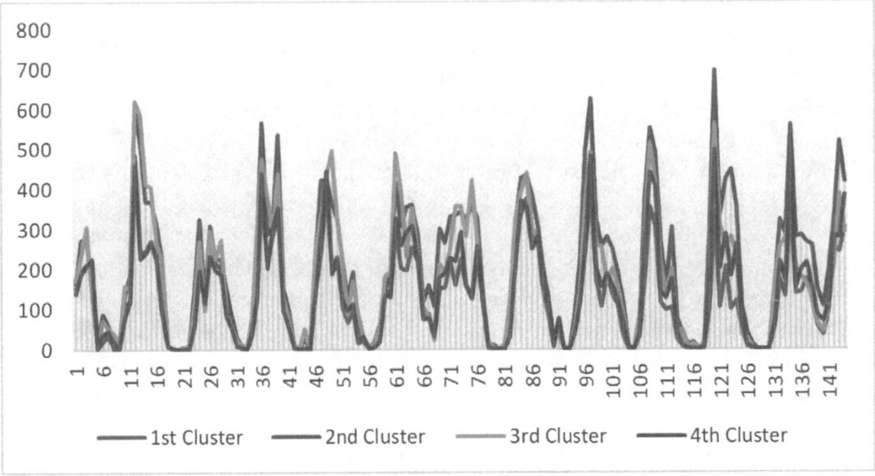

Figure 1 graph of monthly rainfall time series data

Figure 1 is a graph of monthly rainfall time series data in 4 groups. The plot shows the amount of rainfall from January 2005 to December 2016. The amount of rainfall in the form of graph shows that there is an indication of season time series model with time series pattern that repeats at certain time and seen the pattern of up and down regularly formed from plot of rainfall data. In addition, the graph indicates a correlation that is time-sequence associated with the number of observations per season period. Therefore, before estimating the model, the data is tested for the mean and variance. To determine the stationarity of the data on the variance can be seen descriptively plot time series in Figure 1 which shows that the data over time have fixed or constant data fluctuations and are constantly changing on all clusters, so that the data on each cluster has been stationary variance.

4.2 Augmented Dickey Fuller (ADF) Test

The stationarity in mean test is done by using Augmented Dickey Fuller (ADF) test. Based on the results of the ADF test (table 2), it can be seen that the four locations have been stationary without going through differencing process, which is indicated by the p −value is less than 0.05.

Table 2. ADF- Test results

Cluster	ADF-Test	p - value
1	-7.6688	0.01
2	-7.3469	0.01
3	-7.2546	0.01
4	-8.0086	0.01

4.3 ARIMA Forecasting Model

In this step we will construct the ARIMA model for each cluster. The ARIMA model is required to construct the state space model before making a forecasting using Kalman Filter. $\hat{Y}_{C1}$, $\hat{Y}_{C2}$, $\hat{Y}_{C3}$ and $\hat{Y}_{C4}$ respectively are ARIMA models in the first cluster, second cluster, third cluster and fourth cluster. The ARIMA equation is as follows:

$$Y_{C1} = 0.917Y_{t-1} - 0.145Y_{t-2} - 0.12Y_{t-3} - 0.046Y_{t-4} - 0.201Y_{t-5} + a_t + 0.502a_{t-1} \quad (10)$$
$$Y_{C2} = 1.5538Y_{t-1} - 0.7702Y_{t-2} + a_t + 0.8567a_{t-1} \quad (11)$$
$$Y_{C3} = 1.102Y_{t-1} - 0.377Y_{t-2} - 0.308Y_{t-3} - 0.028Y_{t-4} - 0.195Y_{t-5} + a_t + 0.56a_{t-1} \quad (12)$$
$$Y_{C1} = 0.99Y_{t-1} - 0.279Y_{t-2} - 0.080Y_{t-3} + 0.0197Y_{t-4} - 0.252Y_{t-5} + a_t + 0.61a_{t-1} \quad (13)$$

In this section we used auto.arima to get the best arima model on every cluster. Auto.arima can be used if the data has been stationary in mean and variance. Equation (10), (12) and (13) shows that there is a similarity of orders between models in the first cluster, third cluster and fourth cluster. On the other hand, in the second cluster obtained the best model of ARIMA (2.0,1). The equation (10), (11), (12) and (13) will be used to construct the state space model in Kalman Filter. However, this equation is actually a model forecasting of time series data for each cluster. Then we have two forecasts on each cluster that is Kalman Filter forecasting and ARIMA forecasting.

4.4 Forecasting using Kalman Filter
Forecasting using Kalman Filter and ARIMA method is shown in table 3 below.

Table 3. Forecasting Results in 2016

Time	First Cluster		Second Cluster		Third Cluster		Fourth Cluster	
	KF	ARIMA	KF	ARIMA	KF	ARIMA	KF	ARIMA
Jan-16	174.25	218.37454	323	369.4883	256	327.73332	218.37454	264.07617
Feb-16	230.076	242.66848	312.536	368.8745	398.0193	355.44275	242.66848	279.33621
Mar-16	199.46937	241.58537	342.427	332.3858	320.5746	351.48885	241.58537	270.94149
Apr-16	112.456	215.38891	290.3753	276.1615	248.9465	299.27477	215.38891	241.23351
May-16	181.3746	162.20481	232.475	216.902	216.018	222.75925	162.20481	174.63679
Jun-16	139.2501	107.33191	214.572	168.1265	148.5362	143.50329	107.33191	106.1033
Jul-16	78.46288	63.026	185.273	137.9788	79.3547	81.34673	63.026	54.62466
Aug-16	36.142	38.1777	174.8274	128.7009	28.599	47.40059	38.1777	29.26174
Sep-16	51.284	36.15553	290.3753	137.5041	74.9736	48.31659	36.15553	30.03783
Oct-16	136.0947	56.49887	237.309	158.3284	103.647	81.31062	56.49887	57.52409
Nov-16	190.2093	91.55585	473.5847	183.9056	259.4637	135.67046	91.55585	103.10807
Dec-16	285.394	131.08579	383.4096	207.6094	222.301	196.23412	131.08579	153.40549

Forecasting with the Kalman Filter method is obtained after we construct the state space model as in (1). For example we use ARIMA model in second cluster. Based on the equation (11) then obtained parameters of the state space model as follows:

$$\bar{Z}_{t+1} = \begin{bmatrix} 1 & 1 & 0 \\ 1.5538 & 0 & 1 \\ -0.7702 & 0 & 0 \end{bmatrix} \begin{bmatrix} Z_{t-2} \\ Z_{t-1} \\ Z_t \end{bmatrix} + \begin{bmatrix} 1 & -0.8567 & 0 \end{bmatrix} \begin{bmatrix} a_{t+1} \\ a_t \\ a_{t-1} \end{bmatrix} \quad (14)$$

$$\bar{X}_t = \begin{bmatrix} 1 & 0 & 0 \\ 0 & 1 & 0 \\ 0 & 0 & 1 \end{bmatrix} \begin{bmatrix} Z_{t-2} \\ Z_{t-1} \\ Z_t \end{bmatrix} + \begin{bmatrix} b_{t-2} \\ b_{t-1} \\ b_t \end{bmatrix} \quad (15)$$

This process also applies to all clusters. The state space model depends on the ARIMA order size and the value of each parameter.

4.5 Comparison of Forecasting Kalman Filter and ARIMA
To see the accuracy of forecasting using Kalman Filter and ARIMA, then we compare the forecast result in table 3 with outsampel data. One of the indicators we use is Root Mean Square Error (RMSE) which is expressed by the following equation.

$$RMSE = \sqrt{\frac{\sum_{i=1}^{n}(Y_i - \hat{Y}_i)^2}{n}} \quad (16)$$

Based on the forecast result on each cluster, we get the RMSE value shown in table 4.

Table 4. RMSE of ARIMA and Kalman Filter

Number Of Cluster	METHOD	RMSE
1	ARIMA	106.9148
	KF	62.38718
2	ARIMA	144.1151
	KF	89.61118
3	ARIMA	107.6135
	KF	70.55476
4	ARIMA	128.6514
	KF	86.30452

The RMSE value forecasting using the Kalman Filter method is smaller than the ARIMA model. This is shown in table 4. In general, forecasting using the Kalman Filter method is better than ARIMA. Table 4 shows the relationship between RMSE value in Kalman Filter and RMSE value in ARIMA. the RMSE value in Kalman Filter depends on the RMSE value of the ARIMA model. Forecasting using Kalman Filter in the first cluster produces a smaller RMSE value than the other clusters. Similarly, forecasting using ARIMA. Forecasting using ARIMA model also applies the same thing.

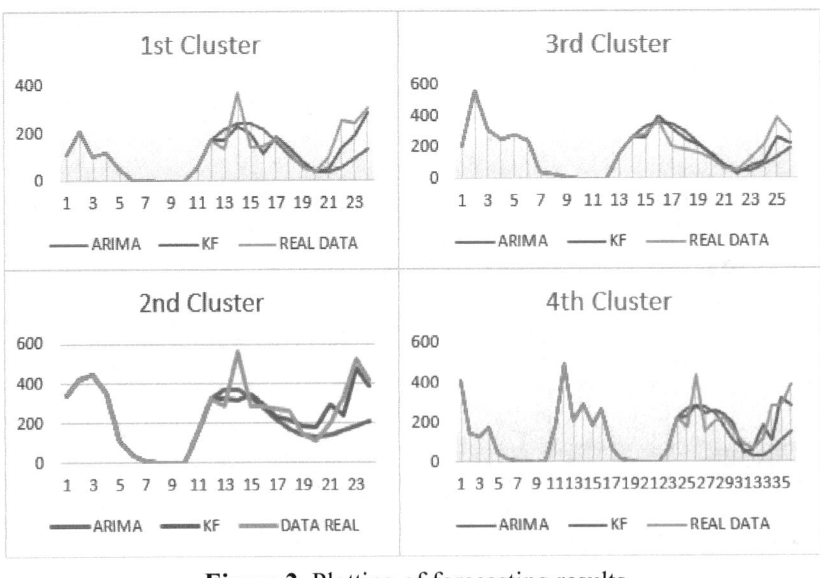

Figure 2. Plotting of forecasting results

5. Conclusion

In general, this study concluded that:

1. The Kalman Filter method gives better results compared to the ARIMA model. This is indicated by the RMSE value in Kalman Filter is smaller than ARIMA model. Based on Figure 2 we can conclude that the error of forecasting using Kalman Filter is relatively

IOP Conf. Series: Journal of Physics: Conf. Series **1008** (2018) 012006 doi:10.1088/1742-6596/1008/1/012006

unchanged significantly at any time. This is due to the measurement update process at each time of forecasting. On the other hand, the error of ARIMA model is getting bigger.

2. Forecasting results using the Kalman Filter method depends on the ARIMA model used as the state space model. The result is an error of the Kalman Filter method depending on the ARIMA model.

References

[1] Brocwell P J and Davis R A 1991 *Time Series: Theory and Methods Second Edition* (New York: Springer-Verlage, Inc)

[2] Meinhold R J and Singpurwala N D 1983 Understanding The Kalman Filter *American Statistical Association* **37** (2) 123-127

[3] Wei W W S 1989 *Time Series Analysis: Univariate and Multivariate Methods* (Canada. Addison Wesley Publishing Company)

[4] Welch G and Bishop G 2006 *An Introduction to the Kalman Filte*r (ACM In)

[5] Zhang G P 2001 Time Series using Hybrid ARIMA and Neural Network Model *Neurocomputing*

IOP Conf. Series: Journal of Physics: Conf. Series **1008** (2018) 012007 doi:10.1088/1742-6596/1008/1/012007

Comparison of exact, efron and breslow parameter approach method on hazard ratio and stratified cox regression model

Mohamat Fatekurohman[1], **Nita Nurmala**[1], **Dian Anggraeni**[1]

[1]Mathematics of Department, University of Jember, Indonesia

E-mail: mfatekurohman.fmipa@unej.ac.id , nitanurmala8@gmail.com , dian_a.fmipa@unej.ac.id

Abstract. Lungs are the most important organ, in the case of respiratory system. Problems related to disorder of the lungs are various, i.e. pneumonia, emphysema, tuberculosis and lung cancer. Comparing all those problems, lung cancer is the most harmful. Considering about that, the aim of this research applies survival analysis and factors affecting the endurance of the lung cancer patient using comparison of exact, Efron and Breslow parameter approach method on hazard ratio and stratified cox regression model. The data applied are based on the medical records of lung cancer patients in Jember *Paru-paru* hospital on 2016, east java, Indonesia. The factors affecting the endurance of the lung cancer patients can be classified into several criteria, i.e. sex, age, hemoglobin, leukocytes, erythrocytes, sedimentation rate of blood, therapy status, general condition, body weight. The result shows that exact method of stratified cox regression model is better than other. On the other hand, the endurance of the patients is affected by their age and the general conditions.

1. Introduction

Health is one of the most important thing in our daily. Considering about that, one of the most essential organ to be concerned is lungs. Lungs is the most vital organ, in the case of respiratory system, if lungs get a problem, the respiratory process will get a problem too. That condition brings other problems to other organ, which further end up with mortality. Considering to the fatal effect of lungs disorder, lung cancer is the most harmful since it can contribute to the mortality. Life style, e.g. smoking, is one of the factors increasing the risk of lung cancer. According to National Cancer Institute, a total of 1.685.370 new diagnoses of lung cancer and 589,430 cancer deathsare projected to occur in United States in 2015 [1] . The patients of lung cancer usually survive for short period. The endurance of the patient is affected by many factors in perspective of statistics, the endurance of patient can be predicated using survival analysis.

Survival analysis is a statistical method used to learn the endurance of a patient, which has correlation with time including time origin and end point [2]. Survival function defined as $S(t)$ is described as the following

$$S(t) = P \text{ (probability of survival with time period)}$$

from definition of cumulative distribution function $F(t)$ from T, then

$$S(t) = 1 - P \text{ (individual who survives with time period} < t)$$
$$= 1 - F(t) \quad (1)$$

The survival function $S(t)$ is a nonincreasing function with respect to t (time. Each of the survival function has the correlation to each other, namely :

hazard function defined in the form of cumulative distribution function $F(t)$ and density as below

$$h(t) = \frac{f(t)}{1-F(t)} = \frac{f(t)}{s(t)} \quad (2)$$

the probability density function defined as the differential of cumulative function

$$f(t) = \frac{d}{dt}\big(1 - s(t)\big) = -s'(t)\;) \tag{3}$$

integral from 0 to t and $S(0) = 1$

$$-\int H\,(x)dx = \log S\;(t) \tag{4}$$

$$H\,(t) = -\log S\,(t)$$

(5) $S\,(t) \;= \exp[-H(t)]$

(6)

$$= \exp\,[-\int h(x)\;dx] \tag{7}$$

we get,

$$f(t) = h(t)\;\exp\,(-H(t)) \tag{8}$$

according to Collect (1992), hazard proportional model for the unsuccessful survival, denote i for every time t, can be written as below:

$$h_i(t) = h_0(t)\exp\,\big((\beta_1 x_{1i} + \beta_2 x_{2i} + \cdots + \beta_p x_{pi}\big) \tag{9}$$

The advantage of Cox method is the flexibility of the presence of density function from parametric distribution. Cox method is applied since each independent variable is proportional with respect to time. The proportional assumption of the model can be indicated by plot $log\,[-log\,(s(t)]$ with respect to survival time t for every time category of descriptive variable p, which form a parallel pattern on different level or uncrossing. The result of plot $log\,[-log\,(s(t)]$ will always increase (exponential function). Furthermore, using plot $log\,[-log\,(s(t)]$ which given information does the variable will proportional in every time [3]. The model $log\,[-log\,(s(t)]$ that is

$$log\,[-lo\,g(s(t)] = \textstyle\sum_{i=1}^{p} \beta_i x_i + log\,[-log\,(S_0((t))] \tag{10}$$

The equation (13) shows that the function of $log\,[-log\,(s(t)]$ is not depended on the time. So, $log\,[-log\,(s(t)]$ function is applicable if it is portrayed toward survival time, then the curve will shape parallel. In the plot of $log\,[-log\,(s(t)]$, the data grouped depend on the level or category on the each variable, then they are needed to be category variable. When the plot of $log\,[-log\,(s(t)]$ shows parallel curve, the proportional hazard assumption is fulfilled. The weakness of $log\,[-log\,(s(t)]$ plot is subjective character, the parallel or not of the curves depend on the method of researcher rate [4]. Hazard ratio value on the regression model cox proportional hazard can be written

$$\widehat{HR} = \frac{\hat{h}(t,x')}{\hat{h}(t,x)} \tag{11}$$

$$= \frac{\hat{h}_0(t)\,\exp\,[\sum_{i=1}^{p} \beta_i x'_i]}{\hat{h}_0(t)\,\exp\,[\sum_{i=1}^{p} \beta_i x_i]} \tag{12}$$

$$= \exp\,[\textstyle\sum_{i=1}^{p} \beta_i (x'_i - x_i)] \tag{13}$$

The object comparison is equal in the all times or not are depended on the time that indicate proportional hazard assumption are not fulfilled (nonproportional hazard) that is:

$$\widehat{HR} = \frac{\hat{h}(t,x')}{\hat{h}(t,x)} \tag{14}$$

$$= \frac{h_0(t)\,\exp[\sum_{i=1}^{p_i} \beta_i x'_i + \sum_{j=1}^{p_j} \delta_j x'_j g_j(t)]}{h_0(t)\,\exp[\sum_{i=1}^{p_i} \beta_i x_i + \sum_{j=1}^{p_j} \delta_j x'_j g_j(t)]} \tag{15}$$

$$= \exp[\textstyle\sum_{i=1}^{p} \beta_i (x'_i - x_i) + \sum_{j=1}^{p_j} \delta_j (x'_j(t) - x_j(t))] \tag{16}$$

the model can be written as:

IOP Conf. Series: Journal of Physics: Conf. Series **1008** (2018) 012007 doi:10.1088/1742-6596/1008/1/012007

$$h(t, x(t)) = h_0(t)\exp\left[\sum_{i=1}^{p_i} \beta_i x_i + \sum_{j=1}^{p_j} \delta_j x'_j g_j(t)\right] \quad (17)$$

while the interaction model of stratified cox regression as:

$$h_g(t) = h_{0_g}(t)\exp[\beta_{1_g} X_1 + \cdots + \beta_{p_g} X_p] \quad (18)$$

the difference to the no-interaction model is that the coefficients β depend on the strata.

On the survival analysis is possible happened to or more individuals experience occurrence in the same time. Generally, there are 3 methods of parameter approach to complete occurrence in the same time, they are exact, Efron and Breslow parameter approach methods. The estimation equal parameter model as:

Exact partial likelihood estimator

$$L \approx \prod_{j=1}^{K} \frac{\exp(S_j\beta)}{\sum_{l\subset S(j,d_j)} \exp(S_{jl}\beta)} \quad (19)$$

Efron partial likelihood estimator

$$L(\beta) \approx \prod_{j=1}^{r} \frac{\exp(\beta'^{s_i})}{\prod_{j=1}^{d_i}[\sum_{k\epsilon R(t_i)} \exp(\beta'^{x_k}) - \frac{j-1}{di}\sum_{k\epsilon d_i} \exp(\beta'^{x_k})} \quad (20)$$

By:

d_i : the group of all individuals who died at t_i

x_k : the variable of individual that still survive and it is the element of $R(t_i)$ and d_i

Breslow partial likelihood estimator

$$L(\beta) \approx \prod_{i=1}^{r} \frac{\exp(\beta'^{s_i})}{\left[\sum_{j\epsilon R(t_i)} \exp(\beta'^{x_j})\right]^{di}} \quad (21)$$

By:

d_i : the amount of occurrence in the same time on the beginning of t_i

$R(t_i)$: the group of individual that has a risk on time t_i which consists of individual who survives on the time t_i

In this research, the result of three parameter approaches is compared and modelled by using hazard ratio model and stratified cox regression model, therefore choosing the best model from two model above by looking in the smallest AIC value and using the best parameter approach method.

The method that can be used for choosing the best model is the AIC value. It can be seen in the equation as:

$$AIC = -2\, In\ (\text{maximum likelihood}) + 2m \quad (22)$$

Where m is the number of the estimated parameters and n is the number of observations. The best model is the model which has minimum AIC value [5].

2. Methods

The data of this research are secondary data which are taken from the *paru-paru* hospital of Jember in 2016 in the form lung cancer patients. There are 64 patients. In this case, the variables used are: independent variable (y) is the life time (in a day) of lung cancer patient in the hospital of Jember, gender (x_1), age (x_2), hemoglobin (x_3), leukocytes (x_4), erythrocytes (x_5), sedimentation rate of blood (x_6), therapy status (x_7), disease history (x_8), general (x_9), body weight (x_{10}).

The AIC (Akaike Information Criteria) value is used to understand the better method for modeling the smallest value. The steps used are:

1 log-rank test
2 proportional test toward ten variables through *log* [-*log* *S*(*t*)]
3 comparing the result of parameter estimate with exact, Efron and Breslow method on Hazard Ratio model

IOP Conf. Series: Journal of Physics: Conf. Series **1008** (2018) 012007 doi:10.1088/1742-6596/1008/1/012007

4 comparing the result of parameter estimate with exact, Efron and Breslow method on Stratified Cox Regression model
5 Choosing the best model and the best estimate approach by looking in the smallest AIC value

3. Result

The occurrence which is analysed in this research is died occurrence when the patients are in the Jember *paru-paru* hospital. For the patients who are out from another hospital then include in the censored data. There are three types of censor in this research. The patient who includes in the research in the different time of January to May 2016, by "1" is the patient who are not censored and "0" is the patients who are censored. There are 10 variables for shaping Hazard ratio and stratified cox regression modal from lung cancer data, namely gender, age, haemoglobin, leukocytes, erythrocytes, sedimentation rate of blood, therapy status, disease history, general condition and body weight. The result of R and SAS program for every variable which is analysed is

Table 1. Log-rank test for ten variables

	Test statistics	$(\chi^2)_{table}$	significant	decision
Gender	2.9	3.84	0.09	Accept H_0
Age	14	7.82	0	reject H_0
Hemoglobin	0.1	5.99	0.96	Accept H_0
Leukocytes	0.6	5.99	0.75	Accept H_0
Erythrocytes	1.3	5.99	0.53	Accept H_0
Sedimentation rate of blood	3.3	3.84	0.07	Accept H_0
Therapy status	1.4	3.84	0.24	Accept H_0
Disease history	0.2	3.41	0.68	Accept H_0
General condition	6	5.99	0.05	reject H_0
Body weight	5.1	9.49	0.28	Accept H_0

Testing the assumption of proportional hazards modeling to ten explanatory variables through the plot $log\ [-log\ S\ (t)]$ obtained 10 variables is

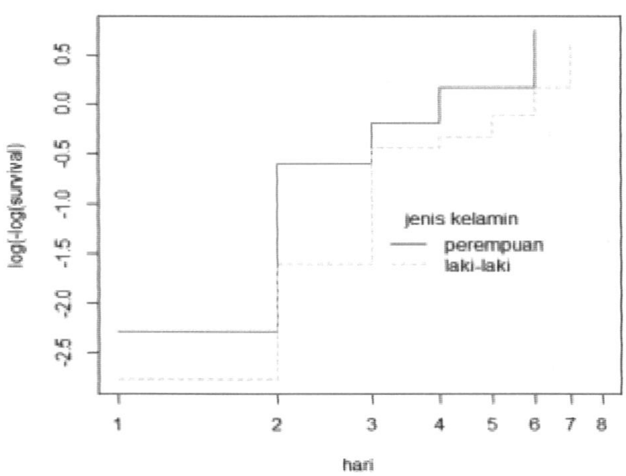

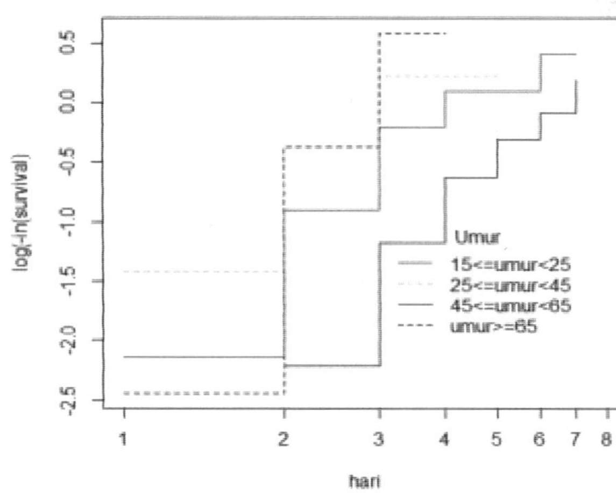

IOP Conf. Series: Journal of Physics: Conf. Series **1008** (2018) 012007 doi:10.1088/1742-6596/1008/1/012007

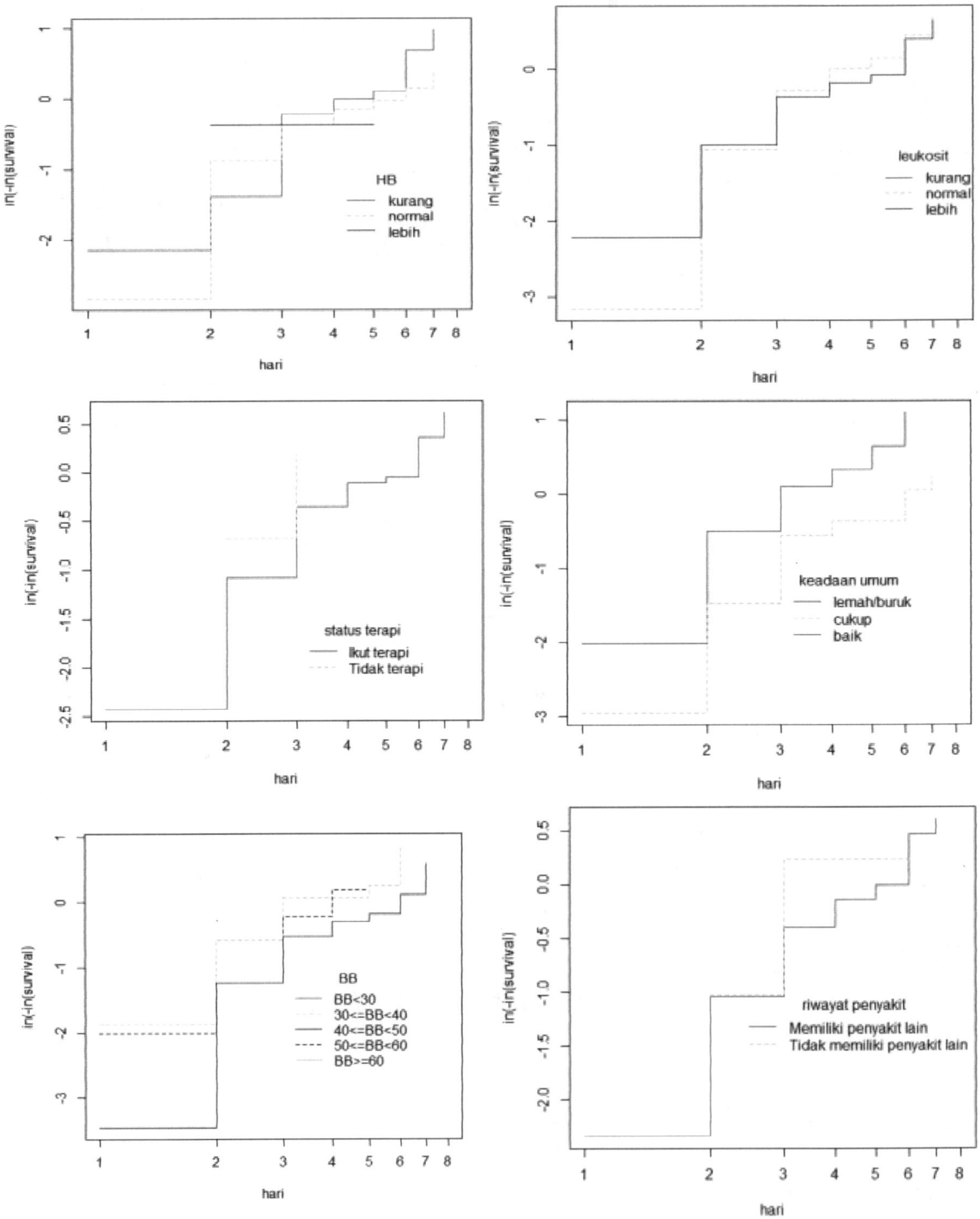

Figure 1. Plot against time $log\ [-log\ (S(t)]$

The results of the comparison exact, Efron and Breslow parameter approach method on a model hazard ratio from the combination of 10 variables got the best models from each of the parameter estimation approach are :

hazard ratio model using exact parameter approach of AIC = 178.938

$$h(t,X) = h_0(t)\exp\left[-0{,}15503x_1g_1(t) + 0{,}04754x_5g_5(t) - 0{,}28071x_6g_6(t) + 0{,}2951x_7g_7(t) - 0{,}31782x_9g_9(t)\right]$$

hazard ratio model using Efron parameter approach AIC = 307.86

$$h(t,X) = h_0(t)\exp\left[-0{,}10907x_1g_1(t) + 0{,}02795x_5g_5(t) - 0{,}27743x_6g_6(t) + 0{,}29232x_7g_7(t) - 0{,}27633x_9g_9(t)\right]$$

hazard ratio model using Breslow parameter approach with AIC = 325.86

$$h(t,X) = h_0(t)\exp\left[-0{,}09034x_1g_1(t) + 0{,}02608x_5g_5(t) - 0{,}23051x_6g_6(t) + 0{,}26914x_7g_7(t) - 0{,}2467x_9g_9(t)\right]$$

The results of the comparison exact, Efron and Breslow parameter approach method on a stratified cox regression models are

stratified cox regression model using exact parameter approach of AIC = 99.091

$$h_g(t) = h_{0_g}(t)\exp\left[-0{,}17829x_1g_1(t) + 0{,}0904x_5g_5(t) - 0{,}27545x_6g_6(t) + 0{,}20410x_7g_7(t) - 0{,}48958x_9g_9(t)\right]$$

by stating strata, $g = 1,2,\dots,12$

stratified cox regression model using Efron parameter approach AIC = 141.94

$$h_g(t) = h_{0_g}(t)\exp\left[-0{,}06363x_1g_1(t) + 0{,}07273x_5g_5(t) - 0{,}26984x_6g_6(t) + 0{,}22026x_7g_7(t) - 0{,}3684g_9(t)\right]$$

by stating strata, $g = 1,2,\dots,12$

stratified cox regression model using Breslow parameter approach with AIC = 157.698

$$h_g(t) = h_{0_g}(t)\exp\left[-0{,}05366x_1g_1(t) + 0{,}06394x_5g_5(t) - 0{,}25124x_6g_6(t) + 0{,}18644x_7g_7(t) - 0{,}32879g_9(t)\right]$$

by stating strata, $g = 1,2,\dots,12$

Of the three approaches parameter estimation showed that exact method is the best parameter estimation approach for modeling the hazard ratio and stratified Cox regression model with the best are

hazard ratio model with the exact parameter approach of AIC = 178.938

$$h(t,X) = h_0(t)\exp\left[-0{,}15503x_1g_1(t) + 0{,}04754x_5g_5(t) - 0{,}28071x_6g_6(t) + 0{,}2951x_7g_7(t) - 0{,}31782x_9g_9(t)\right]$$

stratified cox regression model with the exact parameter approach of AIC = 99.091

$$h_g(t) = h_{0_g}(t)\exp\left[-0{,}17829x_1g_1(t) + 0{,}0904x_5g_5(t) - 0{,}27545x_6g_6(t) + 0{,}20410x_7g_7(t) - 0{,}48958x_9g_9(t)\right]$$

by stating strata, $g = 1,2,\dots,12$

4. Discussion

Log-rank test is often used to see the difference of life endurance in the group with related group. From the log-rank test showed that :

1 gender, HB, leukocytes, erythrocytes, LED, therapy status, history of disease and body weight does not cause any difference in the function of survival in patients with lung cancer

2 differences in age and general condition causes the difference in the function of survival in patients with lung cancer

According testing the assumption of proportional hazards modelling to ten explanatory variables through the plot log [-log S (t)] obtained 10 variables are not proportional for forming a graph that is not parallel. Nonproportional hazard because the study time in the model it uses a hazard ratio and stratified cox regression model. Then from the best model hazard ratio and stratified cox regression model to estimate the exact parameter approach, obtained the exact parameter estimation approach will provide the best model with the smallest AIC by using stratified cox regression model.

5. Conclusion

Best parameter estimation approach is to use the exact method by using a stratified cox regression model, namely

$$h_g(t) = h_{0_g}(t)\exp\ [-0{,}17829x_1g_1(t) + 0{,}0904x_5g_5(t) - 0{,}27545x_6g_6(t) + 0{,}20410x_7g_7(t) - 0{,}48958x_9g_9(t)]$$

by stating strata, $g = 1,2,\dots,12$

Acknowledgement

Thanks to the Director General of Research and Community Service of the Ministry of Research and Technology Higher Education for funding support trouhgt Post Graduate Doctoral Research 2017.

References

[1] Siegel R L, Miller K D and Jemal A 2015 *Cancer Statistics* **65** 5-29
[2] Kleinbaum D G & Klein M 2011 *Survival Analysis a Self-Learning Text Third Edition* (New York : Springer)
[3] Ata N and Tekin M T 2007 Cox Regression Model with Nonproportional Hazard Applied to Lung Cancer Survival Data *Hacettepe Journal of Mathematics and Statistics* **2** 157-167
[4] Collect D 2003 *Modelling Survival Data in Medical Research* (London : Chapman and Hall)
[5] Wang Yand Liu Q 2006 Comparison of Akaike Information Criterion (AIC) and Bayesin Information Criterion (BIC) in Selection of Stock-recruitment Relationships *Fisheries Research of China Ocean University* **77** 220-225

Fractional kalman filter to estimate the concentration of air pollution

Yessy Vita Oktaviana[1], Erna Apriliani[1], Didik Khusnul Arif[1]

[1]Department of Mathematics, Faculty of Mathematics,Computing, and Data Science
Sepuluh Nopember Institute of Technology, Indonesia

Email: yessyvitaoktaviana@gmail.com

Abstract. Air pollution problem gives important effect in quality environment and quality of human's life. Air pollution can be caused by nature sources or human activities. Pollutant for example Ozone, a harmful gas formed by NO_x and volatile organic compounds $(VOCs)$ emitted from various sources. The air pollution problem can be modeled by TAPM-CTM (The Air Pollution Model with Chemical Transport Model). The model shows concentration of pollutant in the air. Therefore, it is important to estimate concentration of air pollutant. Estimation method can be used for forecast pollutant concentration in future and keep stability of air quality. In this research, an algorithm is developed, based on Fractional Kalman Filter to solve the model of air pollution's problem. The model will be discretized first and then it will be estimated by the method. The result shows that estimation of Fractional Kalman Filter has better accuracy than estimation of Kalman Filter. The accuracy was tested by applying RMSE (Root Mean Square Error).

1. Introduction

Nowadays, many cities and their surrounding areas are facing serious air pollution problem due to the increasing of human's population, motor vehicles, and industries. As the result, the number of cities with poor environment has grown. Morever, air pollution problem which gives important effect on the quality of environment, further affects the process of human's life.

Air pollution is the presence of one or more physical, chemical, or biological substances in the atmosphere in quantities that could harmful for human, animal and plant health and impair aesthetics and comfort, or damage property. Air pollution can be caused by natural sources or human activities. Source pollutants that pollute the air in the form of gas and smoke. The gas and smoke are derived from the incomplete combustion process of fuel produced by factory machinery, power plants and motor vehicles.

Many researchers observed pollutant gases such as Ozone (O_3), Nitrogen Oxide (NO), Nitrogen Dioxide (NO_2), Sulfur Dioxide (SO_2), Particulate Matter $(PM_{2.5}$ & $PM_{10})$ to determine the air quality in an area as well controlling the air quality stability. The research process is carried out by forming air quality modeling and subsequently formed a control device to control or reduce the levels of pollutant substances in the air. Because of the high cost of the control device and the amount of maintenance costs it can not be placed as much as possible a tool to measure the concentration of pollutants. Therefore, estimation of pollutant concentration is necessary in addition to predicting future pollutant concentrations as well as to estimate pollutant concentrations in areas not including the radius of the measuring instrument.

In 2010, Erna Apriliani has conducted research on "The Square Root Ensemble Kalman Filter to Estimate the Concentration of Air Pollution". In that research, Square Ensemble Kalman Filter can be used to maintain the stability of computing time [2]. Furthermore, in 2013, Santanu Metia et al conducted an air pollution modeling study entitled "Environmental Time Series Analysis and Estimation with Extended Kalman Filtering". In this study discussed estimation of air pollutant profile by using Extended Kalman Filter algorithm. The Kalman Filter Extended algorithm is used to estimate missing temporary data. It was found that the estimation result of Extended Kalman Filter method contributed significantly [7]. In 2014, Santanu developed his research on "Air Pollution Prediction Using Matern Function Based Extended Fractional Kalman Filtering". In this study, the estimation results show an increase in air quality predictions in the observed region [8]. Morever In 2014, Didik Khusnul Arif has also developed a research on Kalman Filtering entitled "Construction of the Kalman Filter Algorithm on the Model Reduction". This research discussed about a construction of Kalman Filter algorithm on the reduced model. It aimed to obtain accurate estimation with short computing time on the reduced model [3]. In 2017, Risa Fitria and Didik Khsnul Arif had conducted research on "State Variable Estimation of Nonisothermal Continuous Stirred Tank Reactor using Fuzzy Kalman Filter". Fuzzy Kalman Filter is a modification of Kalman Filter that combines with fuzzy theory. This method presented to estimate the state variable of Non-Isothermal CSTR. The estimation results shows that estimation using Fuzzy Kalman Filter had better accuracy than Ensemble Kalman Filter [4]. On the other side, in 2017 Prima Aditya and Didik Khusnul Arif also conducted a research about applied of Kalman Filter entitled "Modelling of Three-Dimensional Radar Tracking System and Its Estimation using Extended Kalman Filter". In this study, Kalman Filter is developed into Extended Kalman Filter to estimate the measurement of nonlinier model of three-dimensional radar tracking system. The results of the simulation show that the allegedly proposed formulation is very efective in the measurement calculation that is not linear [1].

Based on some previous research about Kalman Filter, the authors examine the problem of air pollution. The assessment was done by applying the modification of Kalman Filter namely Fractional Kalman Filter method using several different parameter values. So that we will get the estimated value approaching real system.

2. The Air Pollution Model with Chemical Transport Model (TAPM-CTM)

The TAPM-CTM air pollution model developed by Commonwealth Scientific and Industrial Research Organization (CSIRO). This model is a prognostic 3D equation to determine concentration of pollutant concentrations in the air [5].

$$\frac{\partial C}{\partial t} = \frac{\partial}{\partial x}\left(K_x \frac{\partial C}{\partial x}\right) + \frac{\partial}{\partial y}\left(K_y \frac{\partial C}{\partial y}\right) - \left(\frac{\partial \sigma}{\Delta l}\right)\frac{\partial}{\partial \sigma}(\overline{w'C'}) + S_C + R_C \tag{1}$$

with

$$\overline{w'C'} = -K_\sigma \frac{\partial C}{\partial \sigma}\frac{\partial \sigma}{\Delta l}$$

where C is the concentration of pollutants, $K_x, K_y,$ and K_σ is diffusion coefficient in $x, y,$ and σ direction, respectively, S_C is pollutant emission, and R_C is the effect of chemical reaction.

Suppose we assume the diffusion coefficient are constant, so that we can be rewritten the Eq. (1) as follows

$$\frac{\partial C}{\partial t} = K_x \frac{\partial^2 C}{\partial x^2} + K_y \frac{\partial^2 C}{\partial y^2} + K_\sigma \left(\frac{\partial \sigma}{\partial z}\right)\frac{\partial}{\partial \sigma}\left(\frac{\partial C}{\partial \sigma}\frac{\partial \sigma}{\partial z}\right) + S_C + R_C$$

$$\frac{\partial C}{\partial t} = K_x \frac{\partial^2 C}{\partial x^2} + K_y \frac{\partial^2 C}{\partial y^2} + K_\sigma \left(\frac{\partial \sigma}{\partial z}\right)\left(\frac{\partial^2 C}{(\partial \sigma)^2}\frac{\partial \sigma}{\partial z} + \frac{\partial C}{\partial \sigma}\frac{\partial \sigma}{\partial z}\right) + S_C + R_C \tag{2}$$

IOP Conf. Series: Journal of Physics: Conf. Series **1008** (2018) 012008 doi:10.1088/1742-6596/1008/1/012008

3. Discretization of Model

Before we applied Fractional Kalman Filter to estimate the concentration of air pollution, we discretize Eq. (2) respect to space x, y, σ and time t and then we write in the state space form.

The mathematical model of the pollutant concentration model above is a dynamic system, so the time is continuous, while Fractional Kalman Filter method can only be implemented for the system with discrete time, so it must be discretized with change of state variable to time with finite difference method. To simplify, it is often written with index notation. The first and second subscript indexes as the third space and subscript variables as time variables. So, it can be written:

$$c(x,y,\sigma,t) \sim c_{i,j,k}^{n}$$

$$c(x,y,\sigma,t) \sim c(i\Delta x, j\Delta y, k\Delta\sigma, n\Delta t) \sim c_{i,j,k}^{n} \text{ where } t \approx n\Delta t \text{ and } n = 0,1,2,3,\ldots$$

so the discretization in i, j, k

$$c_t(x,y,\sigma,t) = \frac{c_{i,j,k}^{n+1} - c_{i,j}^{n}}{\Delta t}$$

Forward difference is implemented to $c_\sigma(x,y,\sigma,t)$ in i,j,k (at time n) is

$$c_\sigma(x,y,\sigma,t) = \frac{c_{i,j,k+1}^{n} - c_{i,j,k}^{n}}{\Delta\sigma}$$

On the same way, Center difference is implemented to c_{xx}, c_{yy}, and $c_{\sigma\sigma}$ as follow :

$$c_{xx}(x,y,\sigma,t) = \frac{c_{i+1,j,k}^{n} - 2c_{i,j,k}^{n} + c_{i-1,j,k}^{n}}{\Delta x^2}$$

$$c_{yy}(x,y,\sigma,t) = \frac{c_{i,j+1,k}^{n} - 2c_{i,j,k}^{n} + c_{i,j-1,k}^{n}}{\Delta y^2}$$

$$c_{\sigma\sigma}(x,y,\sigma,t) = \frac{c_{i,j,k+1}^{n} - 2c_{i,j,k}^{n} + c_{i,j,k-1}^{n}}{\Delta\sigma^2}$$

Then the discretization will be implemented to the model (2), it can be :

$$\begin{aligned} c_{i,j,k}^{n+1} &= c_{i,j}^{n} + K_x \frac{\Delta t}{\Delta x^2}\left(c_{i+1,j,k}^{n} - 2c_{i,j,k}^{n} + c_{i-1,j,k}^{n}\right) + K_y \frac{\Delta t}{\Delta y^2}\left(c_{i,j+1,k}^{n} - 2c_{i,j,k}^{n} + c_{i,j-1,k}^{n}\right) + \\ & K_\sigma \frac{\Delta t}{\Delta z^2}\left(c_{i,j,k+1}^{n} - 2c_{i,j,k}^{n} + c_{i,j,k-1}^{n}\right) + K_\sigma \Delta t \frac{\Delta\sigma}{\Delta z}\left(c_{i,j,k+1}^{n} - c_{i,j,k}^{n}\right) + \Delta t S_C + \Delta t R_C \end{aligned} \quad (3)$$

The discretization in Eq. (3) can be written by using tridiagonal matrix as follow :

$$\begin{bmatrix} c_{i,j,1}^{n+1} \\ c_{i,j,2}^{n+1} \\ \vdots \\ c_{i+1,j+1,1}^{n+1} \\ c_{i+1,j+1,2}^{n+1} \\ \vdots \\ c_{u-2,j-2,k-2}^{n+1} \\ c_{u-1,j-1,k-1}^{n+1} \end{bmatrix} = \begin{bmatrix} F & B & D+E & 0 & A1 & 0 & \ldots & 0 & 0 & \ldots \\ B & F & 0 & D+E & 0 & A1 & \ldots & 0 & 0 & \ldots \\ D & 0 & F & B & D+E & 0 & \ldots & 0 & 0 & \ldots \\ 0 & D & B & F & 0 & D+E & \ddots & \vdots & \vdots & \ddots \\ A1 & 0 & D & 0 & F & B & \ldots & 0 & 0 & \ldots \\ 0 & A1 & 0 & 0 & B & F & \ldots & D+E & 0 & \ldots \\ \vdots & \vdots & \vdots & \ddots & \vdots & \vdots & \ddots & \vdots & \vdots & \ddots \\ \vdots & \vdots & \vdots & \ldots & \vdots & \vdots & \ldots & \vdots & \vdots & \ldots \end{bmatrix} \begin{bmatrix} c_{i,j,1}^{n} \\ c_{i,j,2}^{n} \\ \vdots \\ c_{i+1,j+1,1}^{n} \\ c_{i+1,j+1,2}^{n} \\ \vdots \\ c_{u-2,j-2k-2}^{n} \\ c_{u-1,j-1,k-1}^{n} \end{bmatrix} + \Delta t\, S_C + \Delta t\, R_C \quad (4)$$

Generally, The mathematic model in discrete time in Eq. (4) can be written as the discrete state space system of air pollution modelling as follow :

$$\boldsymbol{x_{n+1}} = \mathrm{A}\boldsymbol{x_n} + \mathrm{B}\boldsymbol{u_n}$$

The TAPM-CTM model on air pollution problems is used to determine the concentration of a pollutant in the air pollution process. But in the process of spreading often found problems that contain elements of uncertainty or known probabilistic. The model is not exactly the same with the real system, we take some assumption that cannot be written in the model so that we write the stochastic system as

$$\boldsymbol{x_{n+1}} = \mathrm{A}\boldsymbol{x_n} + \mathrm{B}\boldsymbol{u_n} + \boldsymbol{w_n} \quad (5)$$

where $\boldsymbol{x_{n+1}}$ is the state variable in time $\boldsymbol{n+1}$, $\boldsymbol{x_n}$ is the state variable in time $\boldsymbol{n}$, $\boldsymbol{u_n}$ is system input, A is matrix whose its element is the coefficient of variable state $\boldsymbol{x_n}$ in model system, B is matrix whose its element is the coefficient of variable state $\boldsymbol{u_n}$, $\boldsymbol{w_n}$ in this case is system noise namely Gaussian white noise, respectively, $\boldsymbol{w_n} \sim \boldsymbol{N}(0, \boldsymbol{Q_n})$.

To make correlation between the state which we will estimate and the measurement data, we define the measurement equation as follow :

$$z_n = Hx_n + v_n \tag{6}$$

where z_n is measurement data, H is measurement matrix, x_n state variable in time n, and v_n measurement Gaussian white noise where $v_n \sim N(0, R_n)$.

Noise system and noise measurements that are assumed normally distributed and zero mean values are generated through computer aids. In general the system noise variance is expressed by Q_n and the measurement noise variance is expressed by R_n. Both depend on time and value assumed to be constant.

4. Fractional Kalman Filter Method

Suppose, we has a dynamic stochastic system Eq. (5) and measurement Eq. (6). By using Kalman filter we estimate the state variables of Eq. (1) based on the data measurement Eq. (6). Kalman filter is one of data assimilation method, because in Kalman filter, we combine the model of system Eq. (5) and the data measurement Eq. (6). Fractional Kalman filter is one of Kalman filter modifications.

Kalman Filter is used in network systems for fusion data or error detection in the system. But this method has limitations in estimating unknown state variables to find out the accuracy at a higher level. On the other hand the Fractional Kalman Filter algorithm has been used to estimate unknown state variables in complex systems where fractional derivatives can be used to describe good accuracy.

In the Kalman filter algorithm, we give just one vector for initial estimation x_0, and the covariance of prediction step is determined from the equation [6].

But in the Fractional Kalman Filter, we use a definition of fractional discrete derivative, Grünwald-Letnikov definition to get fractional order difference [9].

Definition 1. The fractional order Grünwald-Letnikov difference is given by the following equation

$$\Delta^r x_n = \frac{1}{h^r}\sum_{s=0}^{n}(-1)^s \binom{r}{s} x_{n-s} \tag{7}$$

where r is a fractional order and h is a sampling time later equal to 1, k is a number of sample for which the derivative is calculated. The factor $\binom{r}{s}$ can be obtained from

$$\binom{r}{s} = \begin{cases} 1 & for\ s = 0 \\ \dfrac{r(r-1)\ldots(r-s+1)}{s!} & for\ s > 0 \end{cases}$$

According to this definition, it is possible to obtained a discrete equivalent of the derivative (when r positive), a discrete equivalent of integration (when r negative), or when r equals 0, the origin function [9].

Now the generalization of the discrete state space model for fractional first order can be obtained by using Definition 1.

$$\Delta^1 x_{n+1} = x_{n+1} - x_n$$

Using formula stochastic system in Eq.(5) and Eq. (6), we can get the linier fractional first order stochastic discrete state-space system as follow :

$$\Delta^1 x_{n+1} = A_d x_n + Bu_n + w_n$$
$$x_{n+1} = \Delta^1 x_{n+1} + x_n$$
$$z_n = Hx_n + v_n$$

where $\Delta^1 x_{n+1}$ is the first order difference for state x_{n+1}, u_n is a system input, w_n is a system noise, $A_d = A - I$ and I is an identity matrix.

Variables $w_n \sim N(0, Q)$ and $v_k \sim N(0, R)$ are assumed white noise (normally distributed with mean 0), not correlated to each other nor with the initial estimation value $\bar{x}_0$.

The algorithm of the Fractional Kalman Filter is built from Definition 1.

First Step: Initialization

Initial value :

$$\hat{x}_0 = \bar{x}_0$$

covariance

$$P_0 = E[(\hat{x}_0 - x_0)(\hat{x}_0 - x_0)^T]$$

Second Step: Prediction
We get prediction value from $\boldsymbol{x_n}$ from the previous step by adding noise system $\boldsymbol{w_n}$

$$\hat{\boldsymbol{x}}_{n+1}^{-} = A_d\hat{\boldsymbol{x}}_n^{-} + B\boldsymbol{u_n}$$

with error covariance

$$\boldsymbol{P}_{n+1}^{-} = A_d\boldsymbol{P}_n A_d^T + \boldsymbol{Q}_n$$

Third Step: Correction
Kalman Gain:

$$\boldsymbol{K}_{n+1} = \boldsymbol{P}_{n+1}^{-}H^T\left(H_{n+1}\boldsymbol{P}_{n+1}^{-1}H^T + \boldsymbol{R}_{n+1}\right)^{-1}$$

Estimation:

$$\hat{\boldsymbol{x}}_{n+1} = \hat{\boldsymbol{x}}_{n+1}^{-} + \boldsymbol{K}_{n+1}(\boldsymbol{z}_{n+1} - H\hat{\boldsymbol{x}}_{n+1}^{-})$$

Error covariance:

$$\boldsymbol{P}_{n+1} = (I - \boldsymbol{K}_{n+1}H)\boldsymbol{P}_{n+1}^{-}[I - K_{n+1}H]^T + \boldsymbol{K}_{n+1}\boldsymbol{R}_{n+1}\boldsymbol{K}_{n+1}^T$$

5. Main Result

Actually, it is difficult to get real data, the concentration of pollution, because the measurement tools are limited, therefore, we concern on the algorithm Kalman Filter, and the Fractional Kalman filter, so that the measurement data is generated from MATLAB program that represents Eq (6).

The simulation results will be evaluated by performing several simulations with many different iteration. And we can know the computing time of the program that has been run. The parameter value used diffusion coefficient which assumed constant with $x_0 = 0.02$, $K = 0.5$ and $K_x = K_y = K_\sigma = \min(10, K)$ and $\Delta t = 0{,}01$.

The simulation is done several times by changing the number of iterations to see the movement condition of the pollution dispersion graph. Noise system $(\boldsymbol{w_k})$ and noise measurement $(\boldsymbol{v_k})$ are searched for the most appropriate.

Here we compare the accuracy of estimation by using Kalman Filter, with by using Fractional Kalman Filter. The simulation results of the pollutant concentration estimate on the pollution process are shown in the following figure:

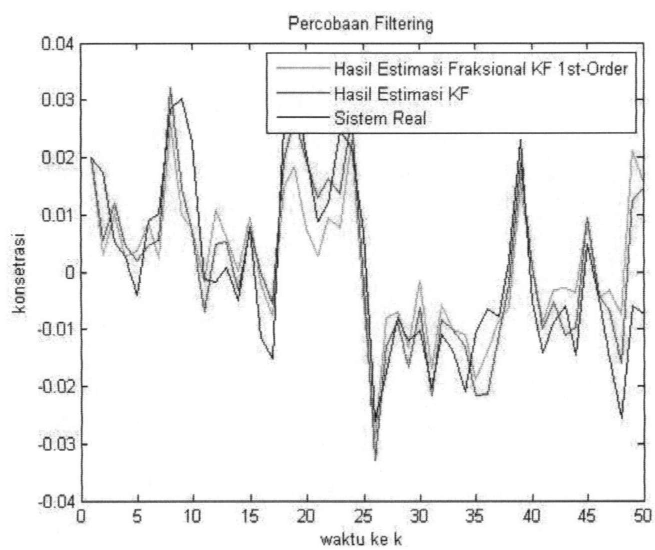

Figure 1. The concentration of pollutant when 50 iteration.

Figure 1 shows real and estimation graphic of concentration pollutant with 50 iteration using Fractional Kalman filter first order. It shows the fluctuation of concentarion pollutant. It shows that the concentration polutant is smaller by the time, so it can give influence to the quality of the air. The graphic shows that the highest concentration happens at 10 iteration with the concentration value is about $0.03 - 0.035\frac{kg}{m^3}$.

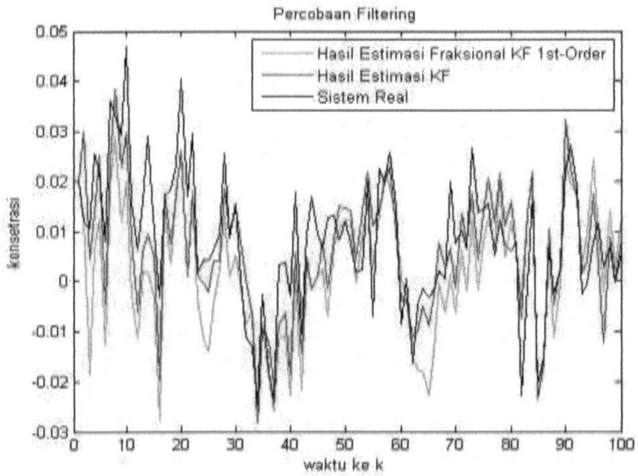

Figure 2. The concentration of pollutant when 100 iteration.

Figure 2 shows real and estimation graphic of concentration pollutant with 100 iteration. It shows the fluctuation of concentration pollutant. The graphic shows that distance between real value and estimation value is adiquate small. The highest concentration happens at 10 iteration with the concentration value is about $0.04 - 0.048\frac{kg}{m^3}$

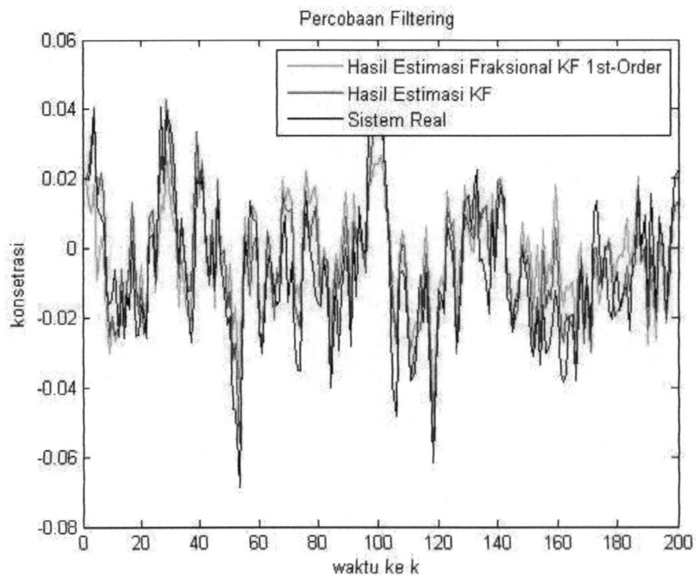

Figure 3. The concentration of pollutant when 200 iteration.

Figure 3 shows real and estimation graphic of concentration pollutant with 200 iteration using Fractional Kalman filter method. The graphic shows density because the iteration is more than before and the distance between real value and estimation value is adiquate small.

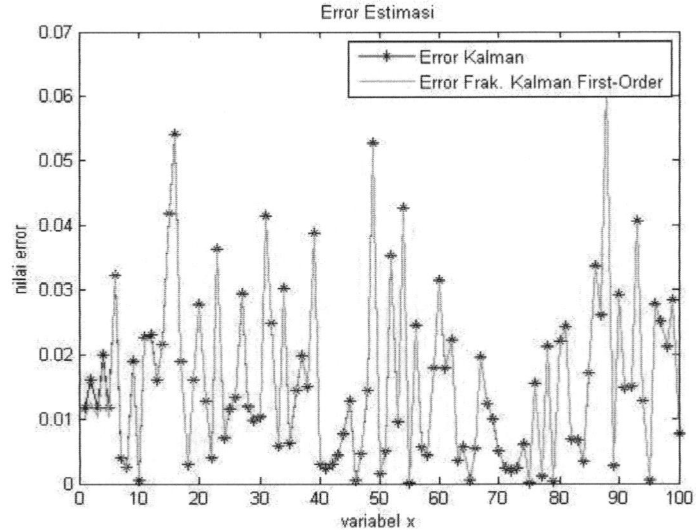

Figure 4. The Error estimation.

Figure 4 shows the error value of concentration pollutant using Kalman Filter and Fractional Kalman Filter is small. But from many iteration we can know that error value of Fractional Kalman Filter is smaller than error value of Kalman Filter.

The accuracy of the method of each experiment is presented in the table as follows:

Table 1. Error estimation and Computational time

	Average from 10 times simulation with many different iteration (N)					
RMSE	N=50		N=100		N=200	
	KF	Fractional KF	KF	Fractional KF	KF	Fractional KF
	0,017289	0,0167837	0,0171377	0,164392	0,0170269	0,0161744
Computational Time	0,839224	1,2665177	0,8826203	1,3403171	0,9632163	1,3830469

It can be seen from Table 1 that the simulation is done with predetermined initial value parameters, and performed with different iterations to see the error values of each different iteration time. Table 1 is the RMSE average of 10 experiments. From the average of RMSE value, it can be seen that Fractional Kalman Filter first order is better than Kalman Filter where the average of RMSE value is smaller than Kalman Filter. For more iteration time, it shows that the average of error value generated smaller. But the computation time of Fractional Kalman Filter first order method tends longer than Kalman Filter method. This is because the Fractional Kalman Filter has a slightly longer step than the Kalman Filter.

6. Conclusions

Based on the analysis and discussion that has been done then can be drawn conclusion as follows:

1. The Fractional Kalman Filter Method can be applied in the estimation of state variables on the model of air pollution process.
2. Based on the estimation results, it is shown that the estimation using Fractional Kalman Filter has better accuracy than Kalman Filter.
3. The more iterations are done to see the error value of pollutant and computational time. It shows that for more iteration time, the average of error value is adequate smaller.

References

[1] Aleksandrovich KA, Mohan SH and Trujillo Juan J 2006 Theory and Applications of Fractional Differential Equations *Elsevier* **204**

[2] Aditya P and Arif D K 2017 Modelling of Three-Dimensional Radar Tracking System and Its Estimation using Extended Kalman Filter *Proceedings of the International Conference on Instrumentation, Control, and Automation*

[3] Apriliani E, Arif D K and Sanjoyo B A 2010 The Square Root Ensemble Kalman Filter to Estimate the Concentration of Air Pollution *International Conference on Mathematical Applications in Engineering* (*ICMAE'10*), Kuala Lumpur Malaysia

[4] Arif D K 2014 Construction of the Kalman Filter Algorithm on the Model Reduction *International Journal Control and Automation* **7** 9 257-270

[5] Fitria R and Arif D K 2017 State Variable Estimation of Nonisothermal Continuous Stirred Tank Reactor using Fuzzy Kalman Filter *International Journal of Computing Science and Applied Mathematics* **3** 1 16-20

[6] Hurley PJ, Physick WL and Luhar AK 2005 TAPM: a practical approach to prognostic meteorological and air pollution modelling. *Environmental Modelling & Software* **20** 6737-752

[7] Lewis F L Optimal estimation: with an introduction to stochastic control theory New York et al: Wiley; 1986 Apr

[8] Lewis JM, Lakshmivarahan S and Dhall S 2006 *Dynamic data assimilation: a least squares approach* Cambridge University Press

[9] Metia S, Oduro SD and Ha QP 2013 Environmental time series analysis and estimation with extended kalman filtering *IEEE First International Conference on Artificial Intelligence, Modelling & Simulation* 202–207

[10] Metia S, Oduro S D and Ha QP 2014 Air Pollution Prediction Using Matern Function Based Extended Fractional Kalman Filtering *International Conferences and Control* 758-763

[11] Sierociuk D and Dzielinski A 2006 Fractional Kalman Filter Algorithm for The States, Parameters, And Order of Fractional System Estimation *International Journal Applied Mathematics Computation Science* **16** 1 129-140

Fire spread estimation on forest wildfire using ensemble kalman filter

Wardatus Syarifah[1] and Erna Apriliani[1]

[1]Department of Mathematics, Sepuluh Nopember Institute of Technology, Surabaya, Indonesia

Email: syarifah13.ws@gmail.com

Abstract. Wildfire is one of the most frequent disasters in the world, for example forest wildfire, causing population of forest decrease. Forest wildfire, whether naturally occurring or prescribed, are potential risks for ecosystems and human settlements. These risks can be managed by monitoring the weather, prescribing fires to limit available fuel, and creating firebreaks. With computer simulations we can predict and explore how fires may spread. The model of fire spread on forest wildfire was established to determine the fire properties. The fire spread model is prepared based on the equation of the diffusion reaction model. There are many methods to estimate the spread of fire. The Kalman Filter Ensemble Method is a modified estimation method of the Kalman Filter algorithm that can be used to estimate linear and non-linear system models.In this research will apply Ensemble Kalman Filter (EnKF) method to estimate the spread of fire on forest wildfire. Before applying the EnKF method, the fire spread model will be discreted using finite difference method. At the end, the analysis obtained illustrated by numerical simulation using software. The simulation results show that the Ensemble Kalman Filter method is closer to the system model when the ensemble value is greater, while the covariance value of the system model and the smaller the measurement.

1. Introduction

Fires are one of methodthe most common disasters in the world, for example in densely populated and forested housing. This has resulted in declining forest populations worldwide, especially in Indonesia. Fires not only cause material losses but also cause loss of life and injuries that can not be assessed how great their losses are when compared to material losses[1].

Recently forest fires have become an international concern as environmental and economic issues, particularly after the ENSO 1997/98 disaster that devastated 25 million hectares of forest land worldwide[2]. Fires are considered a potential threat to sustainable development because of their direct effect on the ecosystem, the contribution of carbon emissions and their impact on biodiversity. According to the government, 2.6 million hectares of land and forests have been burning between June and October 2015, equivalent to four and a half times the size of Bali Island. As of October 2015, each of the eight provinces recorded fires with an area exceeding 100.000 hectares[3].

The development of science, especially in the field of mathematics provides an important role in overcoming the spread of fire. Many mathematical models were created to analyze the nature of the fire so as to investigate the spreading area of fire. In research Renggana discussed mathematical model of spreading fire between laboratory scale buildings. Experiments conducted using the scale down model so that the cost and the level of danger can be reduced. This study mambahas value of heat flux and temperature as the critical value of the spread of fire at a predetermined distance from the results of a survey conducted in 2008.

Andrianus Radipta examines the modeling of fire wildfire distribution. This research designs a Wireless Sensor Network (WSN) system to monitor the occurrence of forest fires wirelessly. The fire spreading model used follows the model of fire spray generated by burners based on the classification of burners in the forest. This study also discusses the calculation of rate of spread and flame length of forest fire causes[4].

In this research writer use fire spread model which have been made by previous researcher by using data assimilation technique. The method applied to this mathematical model uses the Ensemble Kalman Filter method to estimate the spread of fire. Next at the end will be done simulation using software.

2. Reaction-diffusion equation model

The equation of the diffusion reaction is that equationmathematics describing one or more distributed substances in space is altered by the influence of two processes of local chemical reactions in which the substance is converted to another and the diffusion that causes the substance to spread in space. The equation can be represented in a general form[5]:

$$\frac{\partial T}{\partial t} = \mathrm{k}\frac{\partial^2 \mathrm{T}}{\partial x^2} + f(T)$$

where k is the coefficient associated with the diffusion properties of *T*, and *f (T)* is a function that describes how T increases or decreases depending on the current concentration. A reaction often refers to the interaction between two or more objects. Consider a model consisting of a layer of fuel with concentration $F(x,y,t)>0(kg/m^2)$, $k(m^2\, s^{-1})$ is the thermal diffusivity,and temperature $T(x,y,t)(K)$where *x* and *y* are horizontal coordinates *(m)* on the Earth's surface and *t(s)* is the time. The fuel is assumed to burn at the relative rate *r(T)(1/s)*, dependent only on the temperature. In the diffusion reaction model the fire spread in forest fires uses the following model[6]:

$$c\frac{\partial T}{\partial t} = \nabla(k\nabla T) - c(\vec{v} + \gamma\nabla\vec{z})\cdot\nabla T + f(T)$$
$$= \nabla(k\nabla T) - c(\vec{v} + \gamma\nabla\vec{z})\cdot\nabla T + AFr(T) - \Phi$$
$$= \nabla\cdot(k\nabla T) - c(\vec{v} + \gamma\nabla\vec{z})\cdot\nabla T + \hat{A}F\hat{D}e^{-\hat{B}/(T-T_a)} - \hat{\mathrm{C}}(\mathrm{T} - \mathrm{T_a})$$
(1)

$\hat{A}(J/kg)$ is the temperature rise per second at maximum burn rate with full initial fuel load no cooling, c (J/m^2 K) is the 2D thermal capacity of the surface layer at constant pressure, $\gamma\nabla\vec{z}$ is the correction for the gradient of the height of the area $\vec{z}$, $\hat{D}(K)$and$\hat{B}(K)$ are the coefficient of proportionality in the modified Arrhenius law, $\hat{C}(\mathrm{W/m^2\,K})$ is the scale coefficient of heat transfer to the environment, $\mathrm{T_a}$(K) is the ambient temperature, $\vec{v}(m/s)$ is wind speed, $\nabla = [(\partial/\partial \mathrm{x}), (\partial/\partial \mathrm{y})]$ is a gradient, $c\,\partial T/\partial t\,(\mathrm{W/m^2})$ is a heat flow model that is absorbed in the fuel and changes the temperature of the fuel layer, $\nabla(k\nabla T)$ is a diffusion model of heat transfer by short-range radiation and air mixture which causes the fire to spread between adjacent fuel particles such as branches and tree roots, $\vec{v}\nabla T$ is the heat flow due to wind, $f(T)(\mathrm{W/m^2})$ is an external heat flow comprising heat flow $\hat{A}Fr(T)$ which is generated by burning fuel, reduced by heat flow $\hat{\mathrm{C}}(\mathrm{T} - \mathrm{T_a})$ which passes into the atmosphere at ambient temperature $\mathrm{T_a}$ according to Newton's law of cooling which says that the rate of heat loss of a body is directly proportional to the difference in the temperaturesbetweenthebody and its surroundings provided the temperature difference is small and the nature of radiating surface remains same.

Equation (1) denotes the spread of combustion waves, at least for coefficients in a given range. At the beginning edge of the combustion wave, the term diffusion diffuses heat to the unfitted fuel and rises in temperature. The heat generated by the reaction causes the temperature to rise rapidly and fuel fused. The fuel then burns out and cools, creating a line of fire propagating on the combustion wave. As a result, combustion waves move on unburned fuels and leave burning fuel for quantities that are no longer enough to sustain the reaction.

IOP Conf. Series: Journal of Physics: Conf. Series **1008** (2018) 012009 doi:10.1088/1742-6596/1008/1/012009

However, taking the coefficients from the fuel properties and expecting to get the correct wave speed of combustion is not easy. The reason is the unknown homogeneous shape coefficient of micro data, as well as many important physical processes that are not modeled.

3. Dicretization of model

Ensemble kalman filter algorithm can be implemented only for discrete system. Equation (1) is discriminated using a different method to an explicit (Forward Time Center Space) scheme because the time variable $n + 1$ is calculated based on the variable at time n. For degrees with degrees one is discriminated using different to advance because it will estimate for one step ahead. As for the derivative with the second degree is discriminated using a different method to the center. For the three variable functions x, y, and t, the Taylor series is partially derived (partial) to the variable. The model can be written as follows:

$$c\frac{\partial T}{\partial t} = \nabla\cdot(k\nabla T) - c(\vec{v}+\gamma\nabla\vec{z})\cdot\nabla T + \hat{A}F\hat{D}e^{-\hat{B}/(T-T_a)} - \hat{C}(\mathrm{T}-\mathrm{T_a})$$
$$= k\frac{\partial^2 T}{\partial x^2} + k\frac{\partial^2 T}{\partial y^2} - c\vec{v}\frac{\partial T}{\partial x} - c\vec{v}\frac{\partial T}{\partial y} - c\gamma\frac{\partial \vec{z}}{\partial x}\left(\frac{\partial T}{\partial x}\right) - c\gamma\frac{\partial \vec{z}}{\partial y}\left(\frac{\partial T}{\partial y}\right) + \hat{A}F\hat{D}e^{-\hat{B}/(T-T_a)} - \hat{C}(\mathrm{T}-\mathrm{T_a}) \quad (2)$$

That model discretized using forward finite difference method[7]:

$$\frac{\partial T(x_i,y_j,t_n)}{\partial x} \approx \frac{T(x_{i+1},y_j,t_n)-T(x_i,y_j,t_n)}{\Delta x}, \quad \frac{\partial T(x_i,y_j,t_n)}{\partial y} \approx \frac{T(x_i,y_{j+1},t_n)-T(x_i,y_j,t_n)}{\Delta y}$$
$$\frac{\partial T(x_i,y_j,t_n)}{\partial t} \approx \frac{T(x_i,y_j,t_{n+1})-T(x_i,y_j,t_n)}{\Delta t}, \quad \frac{\partial z(x_i,y_j,t_n)}{\partial x} \approx \frac{z(x_{i+1},y_j,t_n)-z(x_i,y_j,t_n)}{\Delta x}$$
$$\frac{\partial z(x_i,y_j,t_n)}{\partial y} \approx \frac{z(x_i,y_{j+1},t_n)-z(x_i,y_j,t_n)}{\Delta y}$$

and using centered finite difference method:

$$\frac{\partial^2 T(x_i,y_j,t_n)}{\partial x^2} \approx \frac{T(x_{i+1},y_j,t_n)-2T(x_i,y_j,t_n)+T(x_{i-1},y_j,t_n)}{\Delta x^2}$$

$$\frac{\partial^2 T(x_i,y_j,t_n)}{\partial y^2} \approx \frac{T(x_i,y_{j+1},t_n)-2T(x_i,y_j,t_n)+T(x_i,y_{j-1},t_n)}{\Delta y^2}$$

So the Equation (2) becomes:

$$T_{i,j}^{n+1} = \frac{k\Delta t}{c\Delta x^2}T_{i-1,j}^n + \frac{k\Delta t}{c\Delta y^2}T_{i,j-1}^n + \left(1 - 2\frac{k\Delta t}{c\Delta x^2} - 2\frac{k\Delta t}{c\Delta y^2} + \frac{\vec{v}\Delta t}{\Delta x} + \frac{\vec{v}\Delta t}{\Delta y} + \frac{\gamma\Delta t}{\Delta x^2}\left(z_{i+1,j}^n - z_{i,j}^n\right) + \frac{\gamma\Delta t}{\Delta y^2}\left(\vec{z}_{i,j+1}^n - \vec{z}_{i,j}^n\right)\right)T_{i,j}^n + \left(\frac{k\Delta t}{c\Delta y^2} - \frac{\vec{v}\Delta t}{\Delta y} - \frac{\gamma\Delta t}{\Delta y^2}\left(\vec{z}_{i,j+1}^n - \vec{z}_{i,j}^n\right)\right)T_{i,j+1}^n + \left(\frac{k\Delta t}{c\Delta x^2} - \frac{\vec{v}\Delta t}{\Delta x} - \frac{\gamma\Delta t}{\Delta x^2}\left(\vec{z}_{i+1,j}^n - \vec{z}_{i,j}^n\right)\right)T_{i+1,j}^n \quad (3)$$

for example $\vec{A} = 1 - 2\frac{k\Delta t}{c\Delta x^2} - 2\frac{k\Delta t}{c\Delta y^2} + \frac{\vec{v}\Delta t}{\Delta x} + \frac{\vec{v}\Delta t}{\Delta y} + \frac{\gamma\Delta t}{\Delta x^2}(\vec{z}) + \frac{\gamma\Delta t}{\Delta y^2}(\vec{z})$, $\vec{B} = \frac{k\Delta t}{c\Delta y^2} - \frac{\vec{v}\Delta t}{\Delta y} - \frac{\gamma\Delta t}{\Delta y^2}(\vec{z})$, $\vec{C} = \frac{k\Delta t}{c\Delta x^2} - \frac{\vec{v}\Delta t}{\Delta x} - \frac{\gamma\Delta t}{\Delta x^2}(\vec{z})$, $\vec{D} = \frac{k\Delta t}{c\Delta y^2}$, $\vec{E} = \frac{k\Delta t}{c\Delta x^2}$, the Equation (3) can be written as follows

$$\begin{bmatrix} T_{1,1} \\ T_{1,2} \\ T_{1,3} \\ T_{1,4} \\ T_{1,5} \\ \vdots \\ T_{10,6} \\ T_{10,7} \\ T_{10,8} \\ T_{10,9} \\ T_{10,10} \end{bmatrix}^{n+1} = \begin{bmatrix} \vec{A} & \vec{B} & 0 & 0 & 0 & \cdots & 0 & 0 & 0 & 0 & 0 \\ \vec{D} & \vec{A} & \vec{B} & 0 & 0 & \cdots & 0 & 0 & 0 & 0 & 0 \\ 0 & \vec{D} & \vec{A} & \vec{B} & 0 & \cdots & 0 & 0 & 0 & 0 & 0 \\ 0 & 0 & \vec{D} & \vec{A} & \vec{B} & \cdots & 0 & 0 & 0 & 0 & 0 \\ 0 & 0 & 0 & \vec{D} & \vec{A} & \cdots & 0 & 0 & 0 & 0 & 0 \\ \vdots & \vdots & \vdots & \vdots & \vdots & \vdots & \vdots & \vdots & \vdots & \vdots & \vdots \\ 0 & 0 & 0 & 0 & 0 & \cdots & \vec{A} & \vec{B} & 0 & 0 & 0 \\ 0 & 0 & 0 & 0 & 0 & \cdots & \vec{D} & \vec{A} & \vec{B} & 0 & 0 \\ 0 & 0 & 0 & 0 & 0 & \cdots & 0 & \vec{D} & \vec{A} & \vec{B} & 0 \\ 0 & 0 & 0 & 0 & 0 & \cdots & 0 & 0 & \vec{D} & \vec{A} & \vec{B} \\ 0 & 0 & 0 & 0 & 0 & \cdots & 0 & 0 & 0 & \vec{D} & \vec{A} \end{bmatrix} \begin{bmatrix} T_{1,1} \\ T_{1,2} \\ T_{1,3} \\ T_{1,4} \\ T_{1,5} \\ \vdots \\ T_{10,6} \\ T_{10,7} \\ T_{10,8} \\ T_{10,9} \\ T_{10,10} \end{bmatrix}^{n} + \begin{bmatrix} \vec{E}T^n_{0,1} + \vec{D}T^n_{1,0} \\ \vec{E}T^n_{0,2} \\ \vec{E}T^n_{0,3} \\ \vec{E}T^n_{0,4} \\ \vec{E}T^n_{0,5} \\ \vdots \\ 0 \\ 0 \\ 0 \\ 0 \\ 0 \end{bmatrix}$$

(4)

The model in Equations (1) used to understand the balance of heat in the fuel layer on forest wildfire. In fact, there is temperature, fuel, and amount of forest wildfire that didn't match with the model, which was called by noise. Noise caused deterministic model turn into stochastic model. The Equation (4) can be written in the form

$$x_{k+1} = f(x_k, u_k) + w_k$$

$$z_k = H_k x_k + v_k$$

The system noise, w_k, and noise measurements, v_k, are generated through the computer and are generally taken normally distributed and have mean value zero, the system noise variation is expressed by $Q_k$and the measurement noise variation is expressed by R_k. Both depend on time and the value assumed to be constant.

The matrix of square-shaped system noise variations corresponds to the size of the error covariance of the estimates. While the variance matrix size of the measurement noise is square-shaped in accordance with the number of rows from the observed H-vector.

The measurement vector H is determined from the state variable that is used as the measurement variable. In this case, it is assessed by the measurement variable that there are state variables T and R so that the measurement vector (H) is

$$H = \begin{bmatrix} 0 & 0 & 0 & 0 & 0 & 0 & 0 & 0 & \cdots & 0 & 0 & 0 & 0 & 0 \\ 0 & 0 & 0 & 0 & 0 & 0 & 0 & 0 & \cdots & 0 & 0 & 0 & 0 & 0 \\ 0 & 0 & 1 & 0 & 0 & 0 & 0 & 0 & \cdots & 0 & 0 & 0 & 0 & 0 \\ 0 & 0 & 0 & 0 & 0 & 0 & 0 & 0 & \cdots & 0 & 0 & 0 & 0 & 0 \\ 0 & 0 & 0 & 0 & 0 & 0 & 0 & 0 & \cdots & 0 & 0 & 0 & 0 & 0 \\ 0 & 0 & 0 & 0 & 0 & 0 & 0 & 0 & \cdots & 0 & 0 & 0 & 0 & 0 \\ 0 & 0 & 0 & 0 & 0 & 0 & 0 & 0 & \cdots & 0 & 0 & 0 & 0 & 0 \\ 0 & 0 & 0 & 0 & 0 & 0 & 0 & 1 & \cdots & 0 & 0 & 0 & 0 & 0 \\ \vdots & \vdots & \vdots & \vdots & \vdots & \vdots & \vdots & \vdots & \vdots & \vdots & \vdots & \vdots & \vdots & \vdots \\ 0 & 0 & 0 & 0 & 0 & 0 & 0 & 0 & \cdots & 0 & 0 & 1 & 0 & 0 \\ 0 & 0 & 0 & 0 & 0 & 0 & 0 & 0 & \cdots & 0 & 0 & 0 & 0 & 0 \\ 0 & 0 & 0 & 0 & 0 & 0 & 0 & 0 & \cdots & 0 & 0 & 0 & 0 & 0 \end{bmatrix}$$

4. Assimilation method

In the Ensemble Kalman Filter method, the estimation process is done by three stages: initialization stage by generating a number of N_e ensembles, the time of the update, and the measurement update ensemble[8].

a) Initialization stage

Resurrected a number N_e ensemble with mean 0 and covariance P according to the initial guess $\bar{x}_0$ with N_e is the number of ensembles that are raised. Let's say $l = N_e$ ensemble the resurrected is:

IOP Conf. Series: Journal of Physics: Conf. Series **1008** (2018) 012009 doi:10.1088/1742-6596/1008/1/012009

$x_{0,l} = [x_{0,1} \quad x_{0,2} \quad x_{0,3} \quad \cdots \quad x_{0,N_e-1} \quad x_{0,N_e}]$

the resulting ensemble is randomly distributed normally. Next set the initial value

$$\hat{x}_0 = \frac{1}{N_e}\sum_{l=1}^{N_e} x_{0,l}$$

b) Prediction stage

At the pediction stage, the estimation of the ensemble, mean and covariance estimation error is determined. Further ensemble estimation is determined as follows:

$\hat{x}_{n+1,l}^{-} = \mathrm{A}\hat{x}_{n,l} + \mathrm{B}u_{n,l} + w_{n,l}$

with $w_{n,l} \sim N(0, Q)$

So from the estimation equation N_e The ensemble, we get the estimation equation with $l = N_e$ ensemble and n iteration as follows:

$\hat{x}_{n+1,l}^{-} = [\mathrm{A}\hat{x}_n + \mathrm{B}u_n + w_{n,1} \quad \mathrm{A}\hat{x}_n + \mathrm{B}u_n + w_{n,2} \quad \cdots \quad \mathrm{A}\hat{x}_n + \mathrm{B}u_n + w_{n,N_e}]$

So the state of estimation becomes:

$$\hat{x}_{n+1}^{-} = \frac{1}{N_e}\sum_{l=1}^{N_e} \hat{x}_{n+1,l}^{-}$$

From the above equation obtained the average of ensemble number N_e and $n = 1$ as follows:

$$\hat{x}_1^{-} = \frac{1}{N_e}\sum_{l=1}^{N_e} \hat{x}_{1,l}^{-}$$

with the following error covariance:

$$P_1^{-} = \frac{1}{N_e - 1}\sum_{l=1}^{N_e} (\hat{x}_{1,i}^{-} - \hat{x}_1^{-})(\hat{x}_{1,l}^{-} - \hat{x}_1^{-})^T$$

Furthermore, the estimated value and the covariance error that has been obtained will be corrected for accuracy in estimating the temperature change and parameter values.

c) Correction stage

In the correction phase begins by replicating the measurement data into a sized matrix $m \times N_e$ with m is the number of measurement variables.

$z_{n,l} = z_n + v_{n,l}$

$v_{n,l} \sim N(0, R)$

for $n = 1$ obtained

$z_{1,l} = z_1 + v_{1,l}$

for $l = 1,2,3, \cdots, N_e$

Then we get the Kalman Gain matrix as follows:

$K_1 = P_1^{-} H^T (H P_1^{-} H^T + R)^{-1}$

The Kalman Gain matrix is used to determine the ensemble estimate as follows:

$\hat{x}_{1,l} = \hat{x}_{1,l}^{-} + K_1(z_{1,l} - H\hat{x}_{1,l}^{-})$

From the equation, the mean estimation is obtained:

$$\hat{x}_1 = \frac{1}{N_e}\sum_{l=1}^{N_e} \hat{x}_{1,l}$$

with the covariant error is:

$P_1 = [I - K_1 H]P_1^{-}$

5. Simulation of the model

In this section, the model is simulated with given noise value in state space (Q) and noise measurement in state space (R). With the parameter values used are as follows:

$\Delta t = 0{,}001 \qquad \Delta x = 0{,}1 \qquad \Delta y = 0{,}1$

IOP Conf. Series: Journal of Physics: Conf. Series **1008** (2018) 012009 doi:10.1088/1742-6596/1008/1/012009

The concentration of fuel is 0,1 kg/m^2 , is the thermal diffusivity is 0,2 m^2 s^{-1}, the temperature of fuel is 250°C, wind speed is 10 m/s[9].

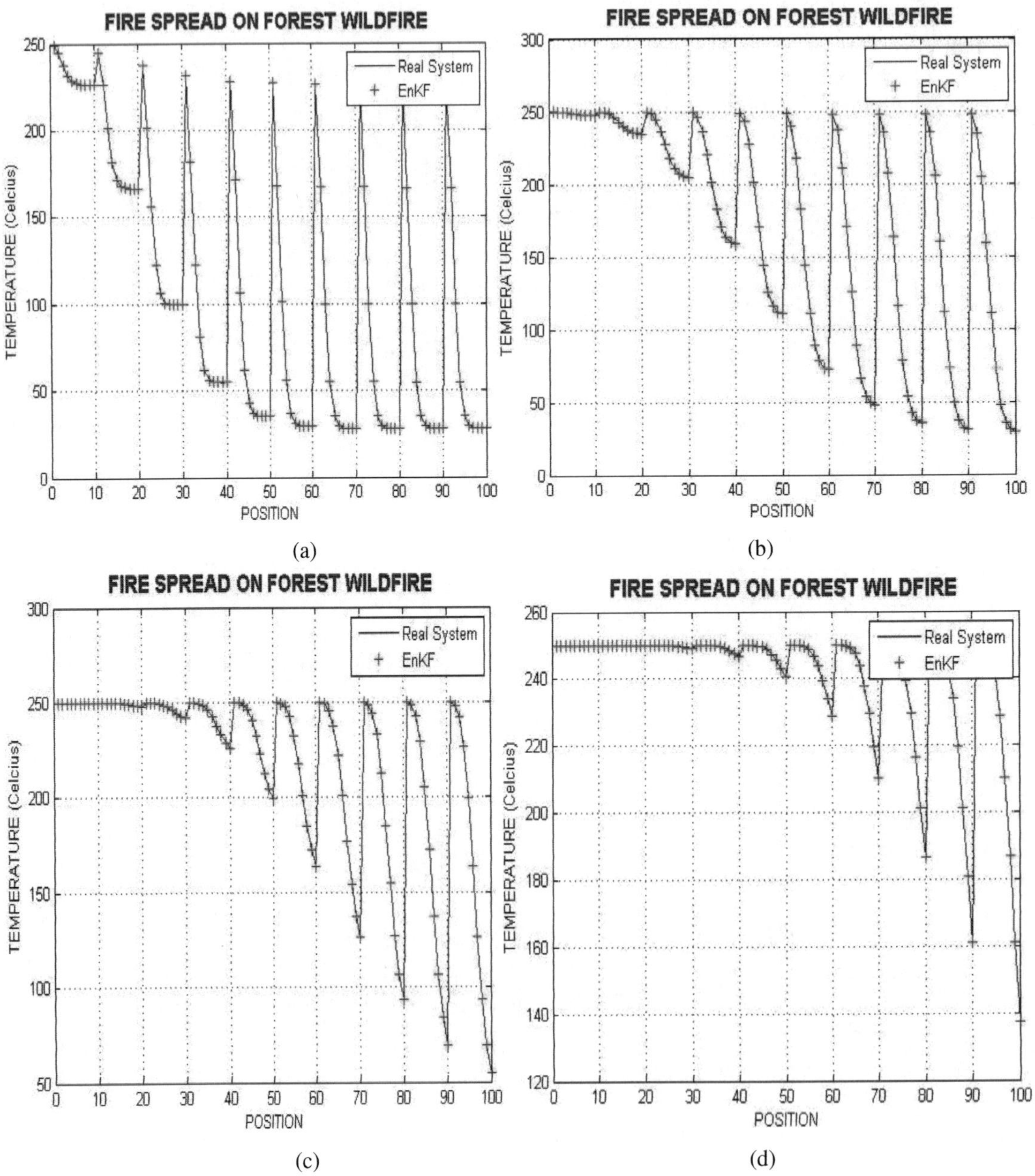

Figure 1. The first simulation using the value of N is 100 and the value of Q and R are 0,1. Simulation results of system model and Ensemble Kalman Filter (a), Error value Ensemble Kalman Filter (b), Simulation results of system model in the contour form (c), and Ensemble Kalman Filter simulation results in contour form (d).

IOP Conf. Series: Journal of Physics: Conf. Series **1008** (2018) 012009 doi:10.1088/1742-6596/1008/1/012009

In the first simulation, figure 1 with the number of iteration times 25 at the last iteration date at point 1,1 is the hotspot with the initial temperature of 28°C then after a rise in temperature around 250°C. The high fire temperature occurs at the closest points of the point 1,1 is called point 1,2, point 1,3, point 2,1, andpoint 3,1. The error value is obtained from the difference between the real system and the estimated results of the Kalman Filter Ensemble Kalman Filter method. The highest error value is 0,47. The fire spread on the real system can be seen after the 25th iteration time at the starting point ie at index i,j=1,1 fire temperature between 245°C -250°C. High fire temperature at points around point 1,1, example point 1,2 is 249,98°C, point 1,3 is 249,97°C, point 2,1 is 249,99°C, and point 3,1 is 249,98°C. The fire temperature that occurs at point 10,10 is 35,69°C. From this explanation it can be said that the fire point at point 1,1 spreads to the points to the midpoint of 10,10 at the time of the sixteenth iteration with the final temperature at the 25th iteration time of 35,69°C. The fire spread on Ensemble Kalman Filter considerations can be seen after the 25th iteration time at the starting point ie at index i,j=1,1 fire temperature ranges between 249,97°C. High fire temperature at points around point 1,1, such as point 1,2 is 249,96°C, point 1,3 is 249,95°C, point 2,1 is 249,98°C, and point 3,1 is 249,97°C. The smallest fire temperature that occurs at point 10,10 is 35,67°C. From this explanation it can be said that the fire point at point 1,1 spreads to the points to the midpoint of 10,10 at the time of the sixteenth iteration with the final temperature at the 25th iteration time of 35,67°C.

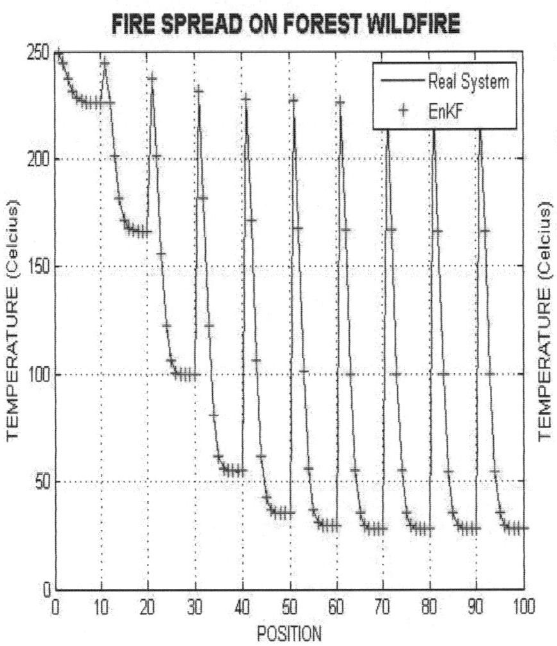

(a)

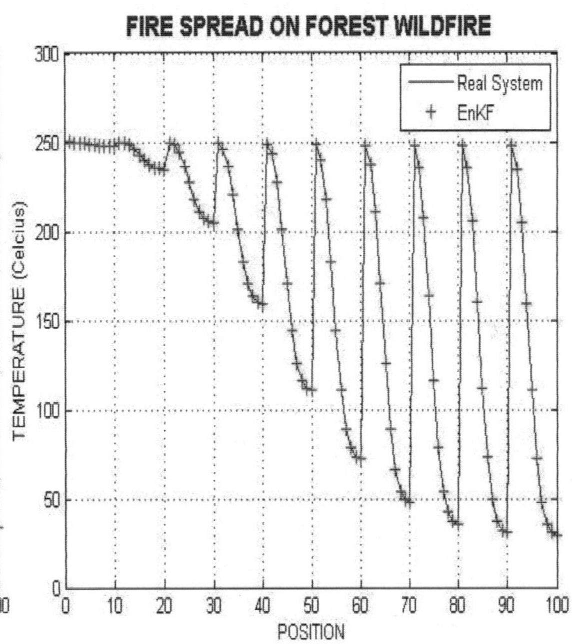

(b)

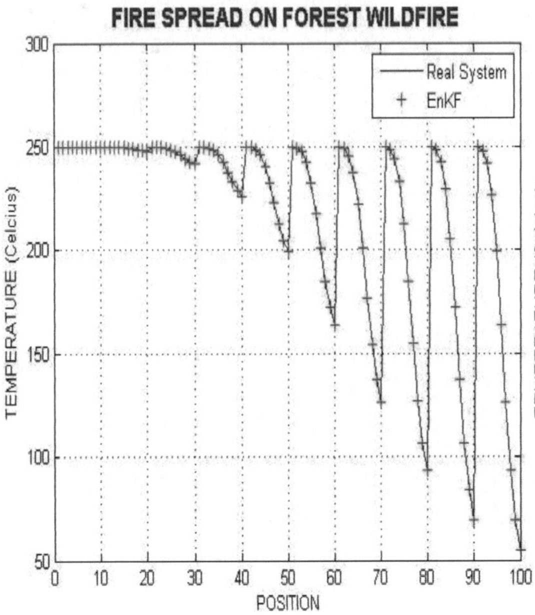

(c)

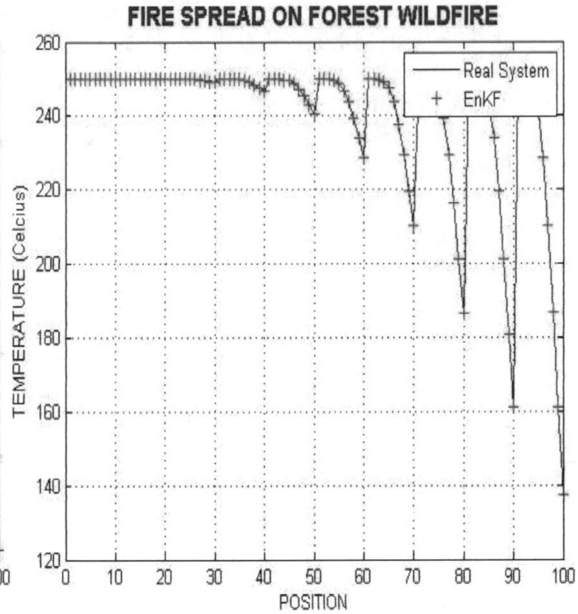

(d)

Figure 2. The second simulation using the value of N is 200 and the value of Q and R are 0,1. Simulation results of system model and Ensemble Kalman Filter (a), Error value Ensemble Kalman Filter (b), Simulation results of system model in the contour form (c), and Ensemble Kalman Filter simulation results in contour form (d).

In the first simulation, figure 2 with the number of iteration times 25 at the last iteration date at point 1,1 is the hotspot with the initial temperature of 28°C then after a rise in temperature around 248°C -250°C. The high fire temperature occurs at the closest points of the point 1,1 is called point 1,2, point 1,3, point 2,1, andpoint 3,1. The error value is obtained from the difference between the real system and the estimated results of the Kalman Filter Ensemble Kalman Filter method. The highest error value is 0,44. The fire spread on the real system can be seen after the 25th iteration time at the starting point ie at index i,j=1,1 fire temperature between 250°C -250,05°C. High fire temperature at points around point 1,1, example point 1,2 is 249,97°C, point 1,3 is 249,97°C, point 2,1 is 250,01°C, and point 3,1 is 249,98°C. The fire temperature that occurs at point 10,10 is 35,695°C. From this explanation it can be said that the fire point at point 1,1 spreads to the points to the midpoint of 10,10 at the time of the sixteenth iteration with the final temperature at the 25th iteration time of 35,695°C. The fire spread on Ensemble Kalman Filter considerations can be seen after the 25th iteration time at the starting point ie at index i,j=1,1 fire temperature ranges between 249,98°C. High fire temperature at points around point 1,1, such as point 1,2 is 250°C, point 1,3 is 249,98°C, point 2,1 is 250°C, and point 3,1 is 249,975°C. The smallest fire temperature that occurs at point 10,10 is 35,69°C. From this explanation it can be said that the fire point at point 1,1 spreads to the points to the midpoint of 10,10 at the time of the sixteenth iteration with the final temperature at the 25th iteration time of 35,69°C.

IOP Conf. Series: Journal of Physics: Conf. Series **1008** (2018) 012009 doi:10.1088/1742-6596/1008/1/012009

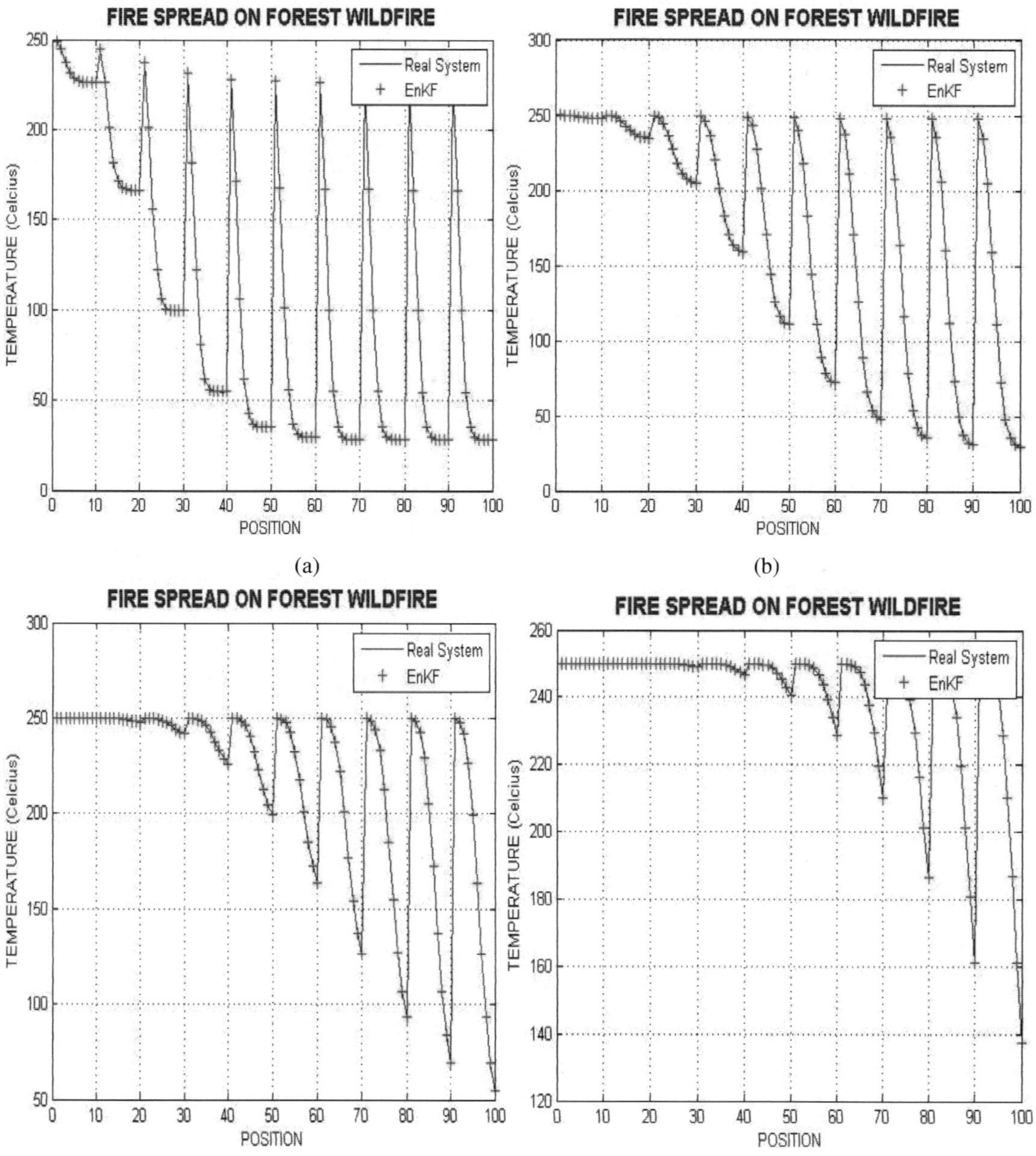

Figure 3. The fifth simulation using the value of N is 300 and the value of Q and R are 0,1. Simulation results of system model and Ensemble Kalman Filter (a), Error value Ensemble Kalman Filter (b), Simulation results of system model in the contour form (c), and Ensemble Kalman Filter simulation results in contour form (d).

In the first simulation, Figure 3 with the number of iteration times 25 at the last iteration date at point 1,1 is the hotspot with the initial temperature of 28°C then after a rise in temperature around 250°C. The high fire temperature occurs at the closest points of the point 1,1 is called point 1,2, point 1,3, point 2,1, andpoint 3,1. The error value is obtained from the difference between the real system and the estimated results of the Kalman Filter Ensemble Kalman Filter method. The highest error value is 0,129. The fire spread on the real system can be seen after the 25th iteration time at the starting point ie at index i,j=1,1 fire temperature between 250°C -250,3°C. High fire temperature at points around point 1,1, example point 1,2 is 250°C, point 1,3 is 249,98°C, point 2,1 is 249,98°C, and point 3,1 is 249,97°C. The fire temperature that occurs at point 10,10 is 35,69°C. From this explanation it can be said that the fire point at point 1,1 spreads to the points to the midpoint of 10,10 at the time of the sixteenth iteration with the final temperature at the 25th iteration time of 35,69°C. The fire spread on Ensemble Kalman Filter considerations can be seen after the 25th iteration time at the starting point ie at index i,j=1,1 fire temperature ranges between 250,5°C. High fire temperature at points around point 1,1, such as point 1,2 is 250°C, point 1,3 is 249,98°C, point 2,1 is 249,99°C, and point 3,1 is 249,98°C. The smallest fire temperature that occurs at point 10,10 is 35,692°C. From this explanation it can be said that the fire point at point 1,1 spreads to the points to the midpoint of 10,10 at the time of the sixteenth iteration with the final temperature at the 25th iteration time of 35,692°C.

In this simulation experiment was also conducted as much as 10 times to see the average error value of Ensemble Kalman Filter method. As in Table 1 discussed the average RMSE value of 10 experiments that have been done.

Table 1. Average RMSE Ensemble Kalman Filter Comparison

Simulation	100 of ensemble	200 of ensemble	300 0f ensemble
1	0,5629	0,53311	0,55559
2	0,5555	0,49214	0,47845
3	0,54713	0,54637	0,50312
4	0,57728	0,52836	0,50014
5	0,53152	0,54032	0,47642
6	0,70025	0,47933	0,41237
7	0,63195	0,38991	0,52197
8	0,54526	0,5812	0,40445
9	0,55473	0,54805	0,48721
10	0,51832	0,46225	0,53872
Mean	0,572484	0,510104	0,487844

It can be seen from Table 1 that after performing ten simulations, the RMSE value of the Ensemble Kalman Filter will be smaller if the ensemble value is greater. In addition, the covariance value of the system model and the measurement model also affects the size of the RMSE value. This causes the Ensemble Kalman Filter method to approach the system model to estimate the occurrence of fire spread.

6. Conclusion

The result of simulation analysis of Ensemble Kalman Filter method will be closer to real system when real system covariance value and covariance value of measurement system is smaller while ensemble value is higher. The average RMSE value in the Ensemble Kalman Filter method simulation results will be smaller when the real system covariance value and the covariance value of the measurement system are smaller while the ensemble value is greater.

References

[1] Renggana E A 2009 *Penyebaran Api Antar Bangunan: Pemodelan dan Eksperimen Skala Laboratorium* (Depok: Universitas Indonesia)

[2] Tacconi L 2003 *Kebakaran Hutan di Indonesia: Penyebab, Biaya dan Implikasi Kebijakan* (Bogor: *Center for International Forest Research*)
[3] Glauber A J, Moyer S, and Andriani, M 2015 *Analisa Dampak Ekonomi dari Krisis Kebakaran Tahun 2015* (Jakarta: *WorldBank Group*)
[4] Radipta A, Kristalina P, and Susetyoko R 2011 *Pemodelan Sebaran Api Kebakaran Hutan Berdasarkan Jenis dan Lokasi Bahan Pembakar* (Surabaya: Politeknik Elektronika Negeri Surabaya)
[5] Kimura Y T 2014 *The Mathematics of Patterns: The modelling and analysis of reaction-diffusion equations.* (New Jersey: Princeton University)
[6] Mandel J, Bezzley J D, and Coen J L 2009 *Data Assimilation for Wildland Fires.* IEEE Control System Magazine.
[7] Davis M 2010 *Finite Difference Method* (London: Imperial College London)
[8] Purnomo K D and Apriliani E 2008 *Estimasi Populasi Plankton dengan Ensemble Kalman Filter* Jurnal Ilmu Dasar. **9** 4
[9] Mandel J, Bezzley J D, and Coen J L 2008 *A Wildland Fire Model with Data Assimilation.* (United States: Colorado Denver University)

IOP Conf. Series: Journal of Physics: Conf. Series **1008** (2018) 012010 doi:10.1088/1742-6596/1008/1/012010

Determination system for solar cell layout in traffic light network using dominating set

Windi Eka Yulia Retnani[1], Brelyanes Z. Fambudi[1], Slamin[1]
[1]Faculty of Computer Science, University of Jember, Jember, Indonesia

Email: windi.ilkom@unej.ac.id, slamin@unej.ac.id

Abstract. Graph Theory is one of the fields in Mathematics that solves discrete problems. In daily life, the applications of Graph Theory are used to solve various problems. One of the topics in the Graph Theory that is used to solve the problem is the dominating set. The concept of dominating set is used, for example, to locate some objects systematically. In this study, the dominating set are used to determine the dominating points for solar panels, where the vertex represents the traffic light point and the edge represents the connection between the points of the traffic light. To search the dominating points for solar panels using the greedy algorithm. This algorithm is used to determine the location of solar panel. This research produced applications that can determine the location of solar panels with optimal results, that is, the minimum dominating points.

1. Introduction

Energy plays a very important role in human life. Any activity from homework to office work requires energy resources. The utilization of energy resources in infrastructure development and facilities is needed. The energy sources that are mostly used are non-renewable energy sources (conventional). In fact, that energy sources are starting to decrease and not sustainable. To overcome this problem, the utilization of new energy and renewable energy sources is an alternative that must be developed and improved as a source of energy in the future.

Law of the Republic of Indonesia Number 30 Year 2007 on Energy Article 20 Verse (4) states that the provision of new energy and renewable energy shall be enhanced by the Government and the regional government according to their respective authority. Furthermore, Verse (5) states that any business entity, permanent business establishment and individual that provide energy from new energy sources and renewable energy sources may obtain facilities and/or incentives from the Government and/or the regional government according to their respective authority for a certain period until the economic value is reached [1].

The energy source that can be utilized is solar light by using solar panels as a power plant. Solar panels have the advantage of being a practical energy source, because they can be installed modularly in every location as required. Ones that can utilize this energy source are traffic lights. The determination of solar panels at traffic light points requires the right strategy. In this case it should pay attention to several aspects such as, the efficiency of the number of solar panels used, so that any solar panels placed at one traffic point can distribute electrical power to at least one nearest red light point. A mathematical approach that can be used to minimize the number of solar panels is by Graph Theory. The essence of applying graph is how to model the problem, and then defines which will be the vertices of the graph. Once the vertices have been obtained, it will be easy to build a graph by providing edges to connect the vertices accordingly [2].

The efficiency factor is used in the determination of the location of solar panels, so that any solar panels placed at one traffic point can supply power to at least one traffic light point. The distribution of electrical power from the solar panels to the traffic light points using the power cord. Solar panels have the advantage of being a practical and eco-friendly energy source considering it does not require transmission like a conventional power grid, as it can be installed modularly in any location that requires [3]. There are various types of solar panels, some are called Poly-crystal, Mono-crystal, Morpheus and Compound [4]. But now there are two kinds of popular solar panels, namely poly-crystalline and mono-crystalline. Poly-crystalline is a solar panel that has a random crystal arrangement. This type requires a larger surface area compared with the mono-crystal type to produce the same electrical power, but can generate electricity during clouding. Mono-crystalline is the most efficient panels, producing the highest electrical power unity. It has an efficiency of up to 15%. The disadvantage of this type of panel is that it will not work well in areas with less sunlight (shade) so that the efficiency will drop dramatically in cloudy weather.

The problem is that how to determine the location of solar panels on the traffic light network. These problems can be solved using the dominating set of a graph. A dominating set provides a way of vertices determination on a graph as a dominating set that can reach all vertices in the graph [2]. A dominating set in a graph G is a set S of vertices of G such that every vertex in $V(G) \setminus S$ is adjacent to at least one vertex in S. The domination number of G is the minimum cardinality of a dominating set of G. For more details about the notion of domination in graphs, see [5].

In this paper, we develop an information system to determine the location of solar panels that will be placed on traffic light network using dominating set. To implement the system, we choose the traffic light networks in the city area of Banyuwangi as a case study although it can be used for any traffic light networks providing the graph as its representation.

2. Methods

This research used *System Development Life Cycle* (SDLC) methods with *Waterfall* model. The Waterfall model is a systematic and linear sequential design approach through some phases in SDLC to develop a system or an application. This model is suitable to develop a medium scale application that involves limited human resources. The illustration of phases in Waterfall model is shown in Figure 1.

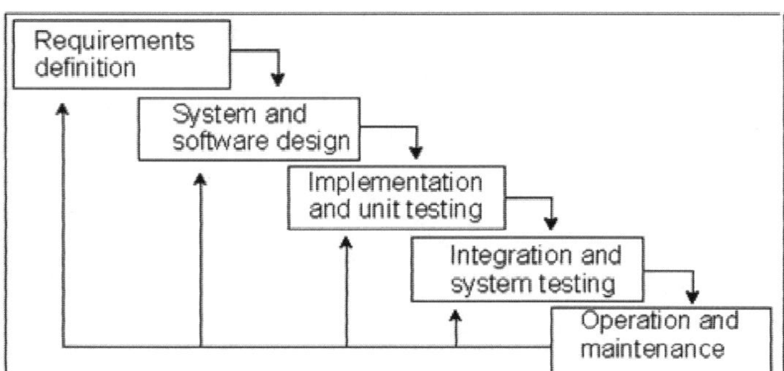

Figure 1. Phases in Waterfall model resource: Sommerville, 2010

3. Result and Discussion

The determination of the location of solar panels that will be placed on traffic light network makes use the concept of dominating set of Graph Theory. Consequently, we require an algorithm to find an optimal solution on determining the dominating set of a graph that represents the traffic light network.

There are two algorithms that can be selected to support the concept of the dominating set. The first algorithm is the Greedy algorithm. The Greedy algorithm is based on the transfer of edges per edge and at every step taken no thought to the future consequences. Greedy algorithm does not operate thoroughly against all alternative solutions and some problems do not always succeed in providing a truly optimum but definitely provide a solution that approaches the optimum value [6]. The second algorithm is the Breadth-First Search algorithm. The advantages of the Breadth-First Search algorithm

IOP Conf. Series: Journal of Physics: Conf. Series **1008** (2018) 012010 doi:10.1088/1742-6596/1008/1/012010

is that it will not find a dead end, because all nodes will be checked thoroughly at each level and if there is one solution then Breadth-First Search will find it, and if there is more than one solution then the minimum solution will be found. However, Breadth-First Search takes much longer execution time than Greedy algorithm [7]. Thus, the Greedy algorithm is used to find an optimal solution in this research.

This study begins by representing the traffic light networks as vertices and edges in a graph. The vertices in the represented graph are sorted by the greatest degree. The next stage is to determine the dominating vertex based on the list of vertices owned. If the vertex list has not been dominated by another vertex, then that vertex becomes the dominating vertex. Mark the vertex with red which is a vertex that has been covered by another vertex that becomes dominator. At the next stage, the remaining search is done. The result of the selection of the remaining vertex can be seen in Table 1. The vertex with yellow color is a vertex that has not been covered by the dominating vertex. The next stage is to find the dominating vertex for the remaining vertices that have not been dominated. The result of selection of dominating vertices can be seen in Table 2. It can be seen that the dominating vertex is traffic light 2. Traffic lights 1, 3, 4 and 5 are the vertices that has been covered by traffic light 2. The next step is to combine the vertices that becomes dominating vertices into the domination set. The results obtained from the search dominating vertices using the Greedy algorithm is 3 vertices. The result of domination set can be seen in Figure 2.

The domination set of traffic lights which become the sources of solar cell consists of 3 traffic lights, namely, 2, 7, and 14. This result is optimal because it satisfies the lower and upper bounds, that is, $2 \leq 3 \leq 8$ [8]. This means that the determination of solar panel point by using Greedy algorithm has been optimized because approaching the minimum domination set.

Table 1. Calculation of the third stage of dominating sets.

Traffic Light	Number of Degree	List	Information
7	6	4,6,8,9,10,11	Dominator
4	5	1,2,3,5,7	Not Dominator
2	4	1,3,4,5	Not Dominator
5	4	1,4,5,8	Not Dominator
10	4	7,9,11,13	Not Dominator
1	3	2,4,5	Not Dominator
3	3	2,4,6	Not Dominator
6	3	3,7,9	Not Dominator
8	3	5,7,12	Not Dominator
9	3	6,7,10	Not Dominator
11	3	7,10,12	Not Dominator
12	3	8,11,14	Not Dominator
13	2	10,14	Not Dominator
14	2	12,13	Dominator

Table 2. Calculation of the fourth stage of dominating sets.

Traffic Light	Number of Degree	List	Information
2	3	1,3,4,5	Dominator
5	2	1,4,5,8	Not Dominator
1	2	2,4,5	Not Dominator
3	1	2,4,6	Not Dominator

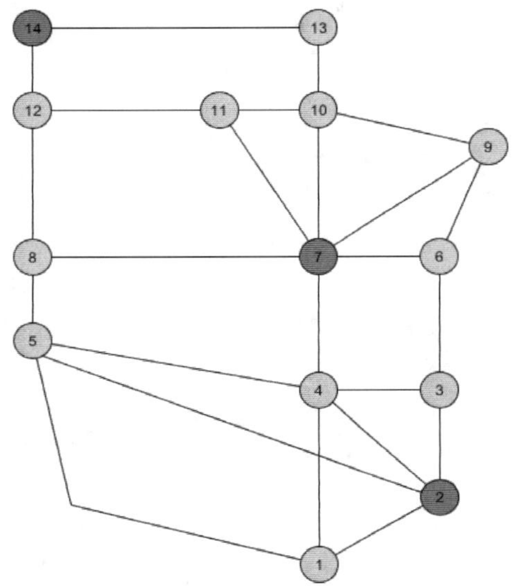

Figure 2. The determination of solar cell location at the traffic light network

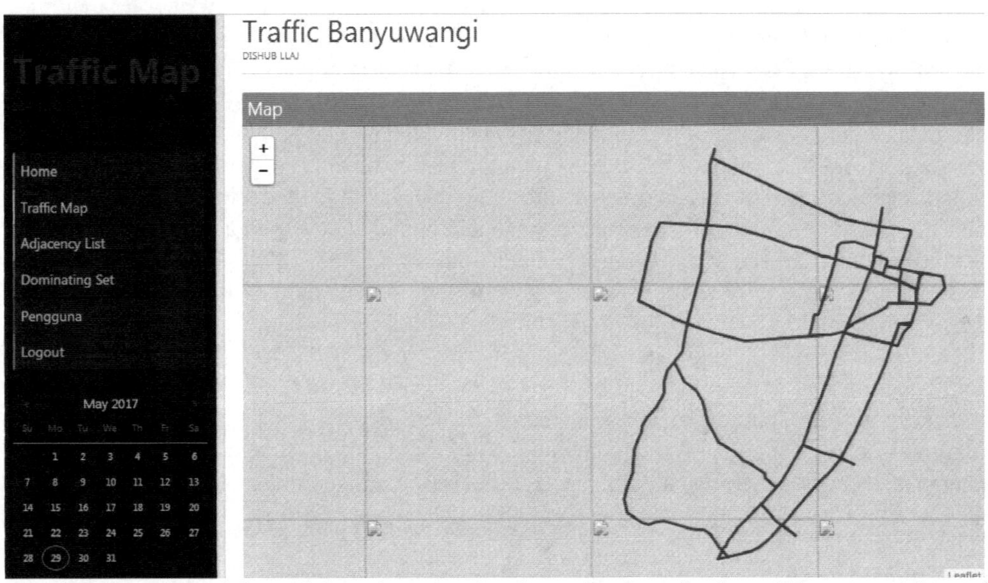

Figure 3. Homepage of the web based system for determination of solar cell location at the traffic light network using dominating set

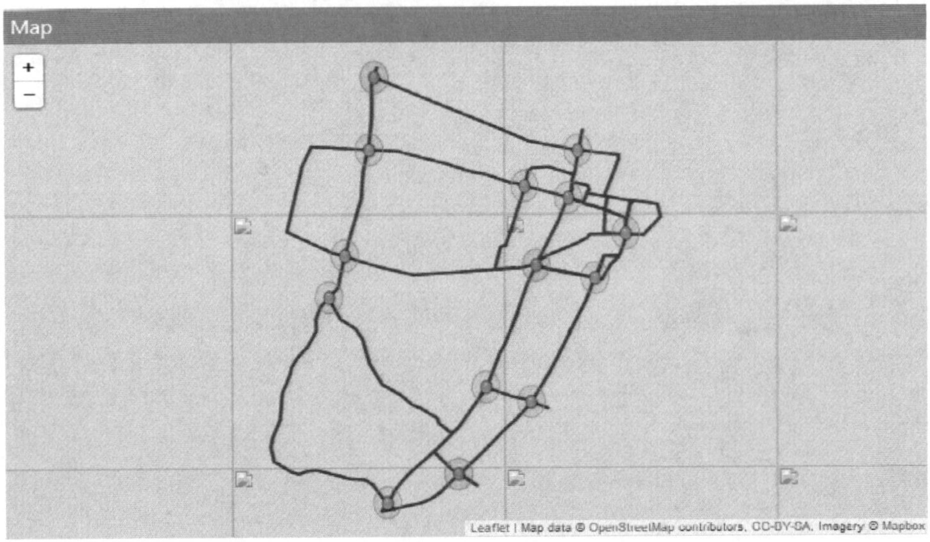

Figure 4. The results of solar cells determination in the system

Figure 3 shows the homepage of the web based system for determination of solar cell location at the traffic light network using dominating set. When this application is running, the system will look for dominating vertices by using Greedy algorithm. If the desired result is optimized based on the concept of dominating set, then the application will display the result as shown in Figure 4.

The data used to compare cost efficiency between conventional and renewable energy sources is which obtained from the results of interviews with the Office of Transportation (DLLAJ) Banyuwangi and manual calculations. The results obtained from the interviews are as follows:

1. each traffic light point that utilizes the energy source of PLN is charged a monthly electrical charge of Rp 300,000 and the required power of 860 wh; and
2. procurement of solar panels for power source of any traffic light point of Rp 80,000,000 and generate power of 1500 wh.

The calculation results from the simulation of the above electrical load usage shows that PLN's electricity load for the fourteen traffic lights for 10 years is Rp 504,000,000. In order to meet the needs of the traffic lights, the required solar panels are three sets of solar panels with the amount of power generated at 4500W. The cost of procurement for the three sets of solar panels is Rp 240,000,000. Calculation of the cost of electricity for 10 years, solar panels far more efficient than conventional energy. This simulation does not include repair and battery replacement.

4. Conclusions

The information system is developed to provide the determination of solar cell location by using Greedy algorithm. The determination of solar cell location using Greedy algorithm in this research is conducted in 5 stages. The first step is to sort the data of traffic light points by the number of degrees from big to small. The second stage is to find the dominating point by considering the list point, in this process produces 2 dominator point. The third stage is to find the point that has not been covered, at this stage produces 4 remaining points. The fourth stage is to find the dominating vertex from the rest of vertices which produces 1 dominating vertex. The last stage is to combine the whole dominating vertices which produces 3 dominating vertices. The solar cell determination result of this system increases the efficiency of electricity costs compared to the use of conventional power within the next 10 years.

References

[1] President of the Republic of Indonesia 2007 *Law of the Republic of Indonesia Number 30 Year 2007 on Energy*

[2] Sari N Y 2015 Pengembangan Himpunan Dominasi dan Himpunan Total Dominasi pada Hasil Operasi Graf *Repos. UNEJ*

[3] Roifah M and Dafik 2014 Kajian Himpunan Dominasi pada Graf Khusus dan Operasinya *Prosiding Seminar Nasional Matematika* 191– 6

[4] Bagher A M, Vahid M M A, and Mohsen M 2015 Types of Solar Cells and Application *American Journal of Optics and Photonics* **3** 5 94 – 113

[5] Chartrand G 2011 *Graph and Digraph* New York: Taylor and Francis Group

[6] Akutsu T, Miyano S and Kuhara S A 2003 Simple Greedy Algorithm for Finding Functional Relations: Efficient Implementation and Average Case Analysis *Theorical Computer Science* **292** 2 481–95

[7] Alanko S D 2011 Computing the Domination Number of Grid Graphs *The Electronic Journal of Combinatorics* **18**

[8] Thai M T, Wang F, Liu D, Zhu S and Du D 2007 Connected Dominating Sets in Wireless Networks with Different Transmission Ranges *IEEE Transactions on Mobile Computing* **6** 7 7 21–30

IOP Conf. Series: Journal of Physics: Conf. Series **1008** (2018) 012011 doi:10.1088/1742-6596/1008/1/012011

Sentiment analysis system for movie review in Bahasa Indonesia using naive bayes classifier method

Yanuar Nurdiansyah[1], Saiful Bukhori[1], Rahmad Hidayat[1]
[1]Faculty of Computer Science, University of Jember, Indonesia

E-Mail : yanuar_pssi@unej.ac.id, saiful.ilkom@unej.ac.id

Abstract—There are many ways of implementing the use of sentiments often found in documents; one of which is the sentiments found on the product or service reviews. It is so important to be able to process and extract textual data from the documents. Therefore, we propose a system that is able to classify sentiments from review documents into two classes: positive sentiment and negative sentiment. We use Naive Bayes Classifier method in this document classification system that we build. We choose Movienthusiast, a movie reviews in Bahasa Indonesia website as the source of our review documents. From there, we were able to collect 1201 movie reviews: 783 positive reviews and 418 negative reviews that we use as the dataset for this machine learning classifier. The classifying accuracy yields an average of 88.37% from five times of accuracy measuring attempts using aforementioned dataset.

1. Introduction

Information presented in cyberspace is now more diverse and the media used means in the process of information diffusion are growing. One of the main media used in the process diffusing information in cyber media is text or document media. The ability in order to extract information from documents is absolutely necessary. The method of extracting information from data in the form of documents is known as text mining. Over 80% of information is currently stored in the form of text, so that text mining is believed to have high commercial value potential [1]. describe the steps taken in text mining: tokenizing, filtering, stemming, tagging, and analyzing [2].

Sentiment analysis is one of the new branches in the domain of text mining or data extraction in the form of text, consisting of processing and extracting textual data automatically in order to obtain information [3, 4, 5, 6, 7]. Sentiment analysis can be utilized as a tool in seeing the public response of a particular event, either positive or negative response, so that the next strategic steps can be undergone immediately. An example of sentiment data in the form of document is movie review from various sites on the internet. Reviews obtained from movie reviews sites can be used a reference for movie fans to know recommended and also a medium for movie producers to know the public responses towards the movie released. Movie review can be divided into a number of categories based on the sentiments contained in the document.

There are some difficulties in developing sentimental analysis study on movie review using Bahasa Indonesia partly due to the lack of studies on movie review documents using Bahasa Indonesia and the lack of corpus or textual literature for movie review using Bahasa Indonesia as the original language of the document.

Based on the description above, the researchers are interested in creating a system which is capable of categorizing movie review in Bahasa Indonesia into two categories of positive and negative sentiments using the method of Naïve Bayes Classifier [8, 9]. Naïve Bayes Classifier method is one of the supervised learning models based on statistics and probability with high level of accuracy [10].

2. Result and Discussion

The steps in this study include the step of literature review, data collection, and system development. The method employed in system development is the method of Naive Bayes Classifier [11]. The flow diagram of the study steps is illustrated in Figure 1.

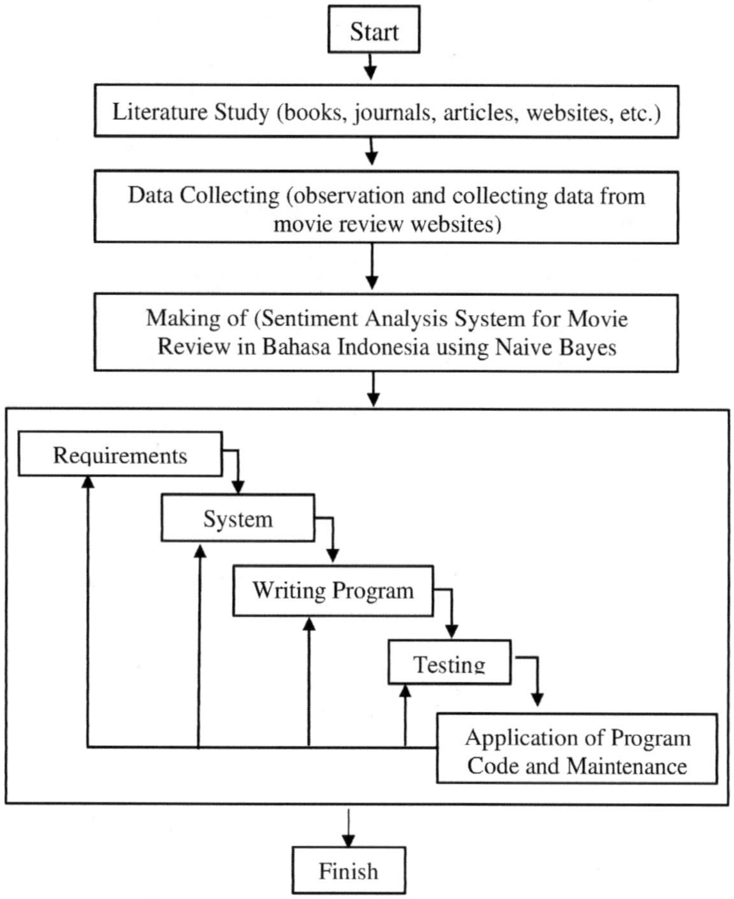

Figure 1. Flow Chart of Research Phases

2.1 Data Collection

The data collection is performed in order to obtain the information needed to determine the system requirements. The data used in the study is divided into two categories based on the data source: primary data and secondary data. The primary data required in the development of the Sentiment Analysis System for Movie Review is a dataset of movie review. The movie review dataset is made by the researcher using a movie review obtained from the movie review web page of Movienthusiast. Secondary data required in this study is the root words in the Great Dictionary of the Indonesia Language (KBBI) and stop words dictionary or set of words which are considered less important and is able to be removed from the movie review text [12, 13, 14]. The dictionary of stop words used comes from the study conducted by Tala.

2.2 The Naive Bayes Classifier Method Implementation

The Sentiment Analysis System for Movie Review which is developed implements the Naive Bayes Classifier method in classifying the movie review documents into two classes: positive sentiment class and negative sentiment class. The flowchart regarding the overview of the Sentiment Analysis System for Movie Review is illustrated in Figure 2.

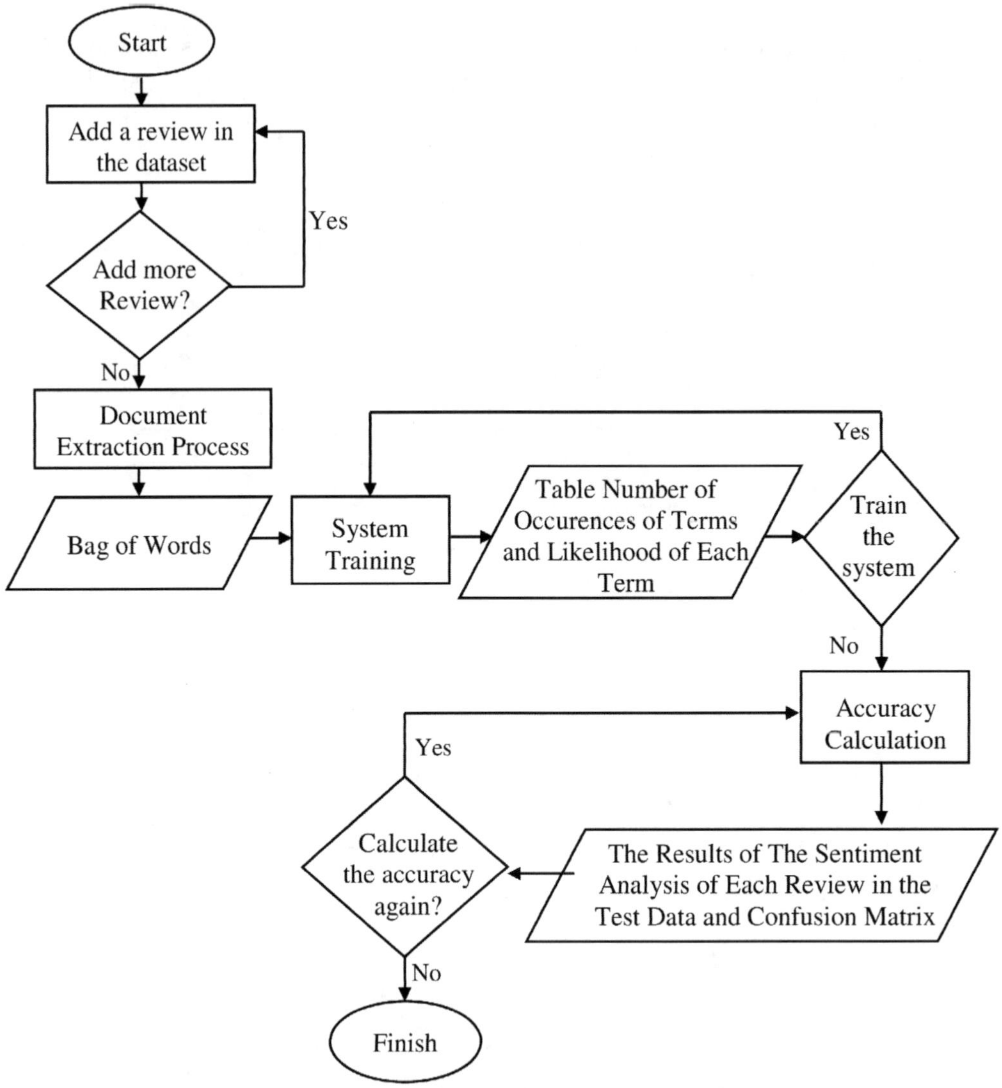

Figure 2. Overview of Film Review Sentiment Analysis System

The explanation on Figure 2. above is: firstly, the derived from the movie review is submitted by the user one by one. The submitted review may be categorized into training data or test data based on the user's wishes. The system, then, executes the document extraction process in order to generate a collection of terms from each review text and so-called bag of words. The next step is to train the system. At this step, the system constructs a model of the training data by calculating the occurrences of each term in the training data and likelihood value [15]. The result of the system training process is a table of feature sets containing the number of occurrences and likelihood values of each term in the training data. The likelihood value is, then, incorporated into the next step which is the calculation of the system accuracy [16].

The system performs the classification process using the Naive Bayes Classifier method in order to obtain the best class of sentiment of each review in the category of data train. The results of the accuracy calculation are the sentiments of the classification of each test data review and the confusion matrix table obtained from the original sentiment comparison of the test data review and the sentiment of the analysis result. System accuracy can be concluded from the confusion matrix. The planning steps which are implemented based on the waterfall method [17].

2.3 Requirement Analysis

System requirement is the ability of the system to meet the condition desired by the users. System requirement analysis is performed by grouping the needs into functional requirements and non-functional requirements.

2.3.1 Functional Requirements

The functional requirements of the Sentiment Analysis System for Moview Review are as follows:

1. The system can manage (add, read, update, and delete) movie review datasets,
2. The system can manage (add, read, update, and delete) the dictionary of root words in Bahasa Indonesia,
3. The system can manage (add, read, update, and delete) the dictionary of stop words,
4. The system can perform the classifier training process and display the model in the form of feature sets of the term data from the training data,
5. The system can display the test data result and display confusion matrix generated from the classifier testing,
6. The system can display a set of movie review dataset terms derived from tokenizing, filtering, and stemming processes, and
7. The system can display sentiment analysis result derived from reviews submitted by users.

2.3.2 Non-Functional Requirements

Meanwhile, the non-functional requirements of the Sentiment Analysis System for Movie Review are as follows:

1. The system uses an authentication through login process in order to differentiate user level.
2. The system can run in various web browsers which support the system environment,
3. The system gives a fast response, and
4. The system has a user-friendly interface design.

2.4 System Design

The design of the Sentiment Analysis System for Movie Review uses Unified-Modeling Language (UML) as the modeling language and object-oriented programming concept. The documentation of the created system includes:

2.4.1 Business Process

Business process diagram describes a series of inputs, needs, and outputs processed by the system in order to produce the desired destination by the user. Description of business process system can be seen in Figure 3.

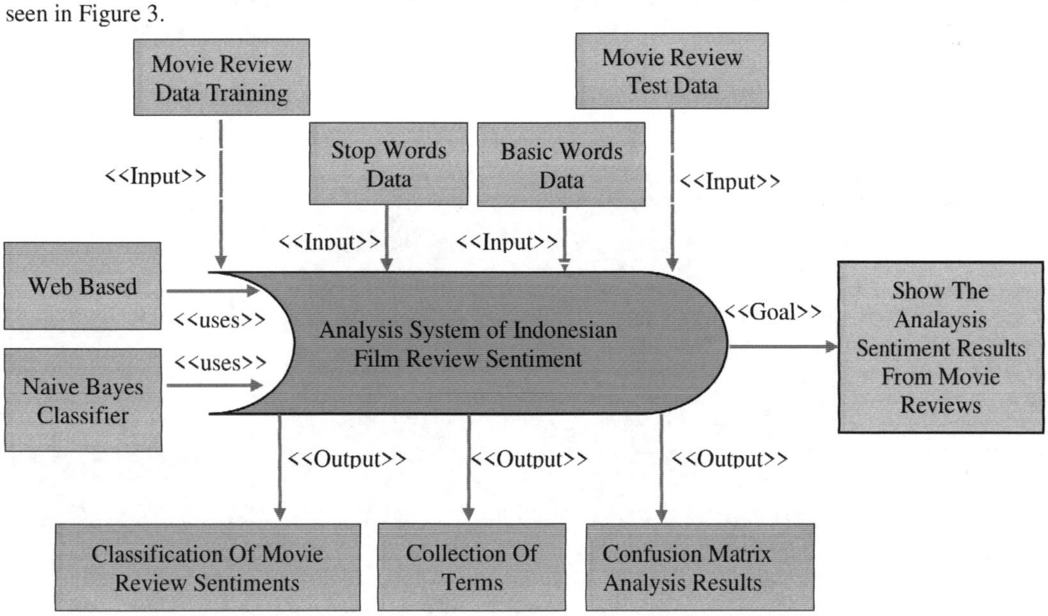

Figure 3. Bussiness Process Movie Sentiment Analysis System Review

IOP Conf. Series: Journal of Physics: Conf. Series **1008** (2018) 012011 doi:10.1088/1742-6596/1008/1/012011

2.4.2 Use case Diagram

Use case diagram describes the system's functional needs in the form of diagram. Use case is needed to describe typical interactions between users and system. Use case diagram of the Sentiment Analysis System for Movie Review is illustrated in Figure 4.

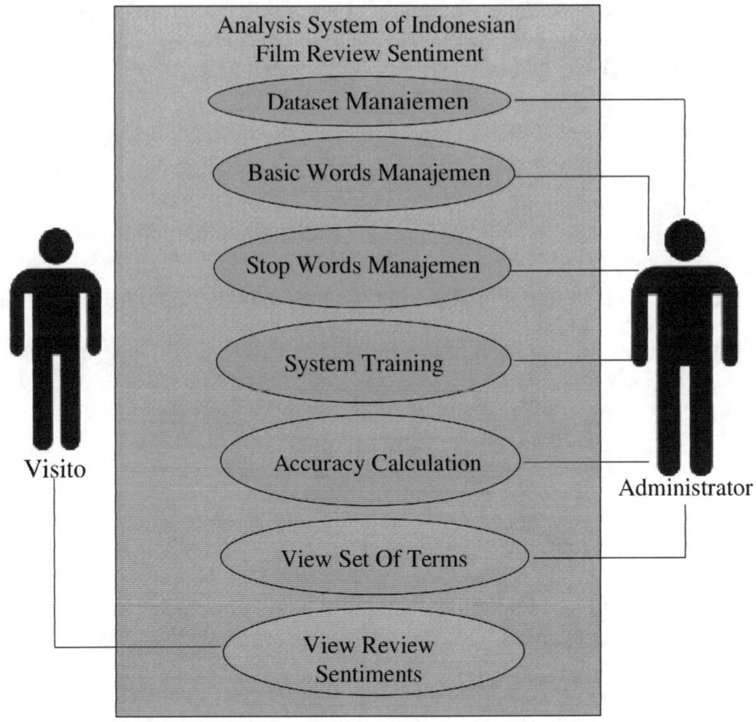

Figure 4. Use Case Diagram Sentiment Analysis Movie Review System

2.4.3 Deployment Diagram

Deployment diagram is a diagram illustrating the physical layout of the system. This diagram is included in order to describe the hardware and software required for the system to run according to the system needs. Deployment diagram of the Sentiment Analysis System for Movie Review can be seen in Figure 5.

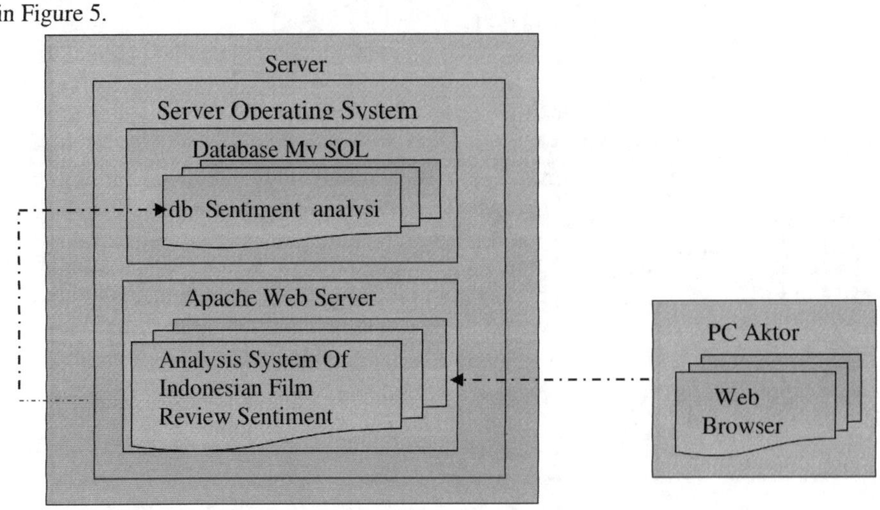

Figure 5. Deployment diagram of the Sentiment Analysis System for Movie

Sentiment Analysis System for Movie Review is developed as a web-based application. The webpage of Sentiment Analysis System for Movie Review is divided into two levels based on the user level: private page for the administrators and public page for the visitors. The private page contains the management dataset menu of the movie review, the stop words management menu, the training system menu, the accuracy calculation menu, and menu containing the set of terms from each movie review on the dataset. The public page contains an interface for visitors wanting to obtain sentiment analysis result from the movie reviews within the system. The main features displayed in the discussion are the dashboard page, system training page, system accuracy calculation page for the visitors.

2.5 System Interface Implementation`

The dashboard page is the first accessed page once the administrator has successfully logged in to the system. The dashboard contains brief information on datasets, such as the number of review for each category, the ratio of training and test data as well as the ratio of positive and negative training data. The information is displayed in the form of numbers and graphs of donut charts. System training page contains a table on the occurrences and likelihood values of each term in all train data. The process of system training in order to form feature sets is performed by the administrators through the interface page by selecting the button of "Latih".

The accuracy calculation page contains a list of reviews which become test data within the dataset, probability of the positive sentiment and negative sentiment for each test data, the result of the class analysis with the highest probability, and the accuracy value of the analysis result towards the original sentiment of the reviews. In addition to the review table of the test data, there are accuracy values of the system's last accuracy test result which have been converted into percentage format and confusion matrix table of the last accuracy test result. Accuracy calculation process is also initiated by the administrators through the interface page by selecting the "Hitung (calculate)" button.

The visitor page is the page which is for the visitors. The visitors can access this page in order to enter the review data they want to know the sentiment of. After selecting the "Lihat Sentiment (view sentiment)" button, the system classifies using the Naive Bayes method in order to obtain positive and negative sentiment probabilities as well as the sentiment result obtained based on the probability ratio of the two sentiment classes.

2.6 Accuracy Calculation Result

The testing process is conducted five times. The first test was conducted by selecting 600 reviews as the training data and 200 reviews as the random test data. Further tests were performed by adding 100 movie reviews as the test data and test data of 20 movie reviews into the dataset utilized for the previous test. Details of the composition of each training and test data for each test process can be seen in Table 1.

Table 1. Composition each training and test data for each test process

Testing no.	Number of Training Data	Number of Testing Data	Number of Positive Training Data	Number of Negative Training Data	Number of Positive Testing Data	Accuracy (%)
1	600	120	390	210	78	87,50%
2	700	140	454	246	91	87,86%
3	800	160	520	280	104	88,13%
4	900	180	585	315	117	88,33%
5	1000	201	650	350	132	90,05%

Based on the five times accuracy calculation processes of the Sentiment Analysis System for Movie Review, the obtained accuracy data for each process is available in the following table 2.

Table 2. Obtained accuracy data for each process

Testing no.	Number of Training Data	Number of Testing Data	Accuracy (%)
1	600	120	87,50%
2	700	140	87,86%

3	800	160	88,13%
4	900	180	88,33%
5	1000	201	90,05%
	Average Value:		88,37%

3. Conclusions

The Sentiment Analysis System for Movie Reviews implements the method of Naive Bayes Classifier in order to obtain the highest posterior probability value of the two review classes of sentiment. The posterior probability value is obtained from the total sum of log prior probability and log likelihood of each term in each review of the training data for each sentiment class. The average accuracy produced by the Sentiment Analysis System for Movie Review is 88.74% out of five time testing processes. The first test was conducted by training 600 movie reviews as the training data and classifying 120 movie reviews as the random test data. For each test, the dataset from the previous conducted test is added with 100 movie reviews as the training data and 20 movie reviews as the test data. The highest accuracy is obtained when the dataset used consists of 1,000 review of training data and 201 test data review and the percentage is 90.05%. The accuracy value is directly proportional to the number of reviews used as the dataset.

References

[1] Grimes S 2008 *Unstructured Data and 80 Percent Rule* Carabridge Bridgepoints

[2] Feldman R and Sanger J 2007 *The Text Mining Handbook* Advanced Approaches in Analyzing Unstructure Data Cambridge: Cambridge University Press.

[3] Pang B and Lee L 2002 Sentiment Classification using Machine Learning Techniques *Proceeding of ACL-02 Conference on Empirical Methods in Natural Language Processing* **10** 79 - 86

[4] Kalaivani P Sentiment Classification of Movie Reviews by supervised machine learning approaches et.al *Indian Journal of Computer Science and Engineering* (*IJCSE*) **4** 4

[5] Rambocas M and Gama J 2013 *Marketing research: The role of sentiment analysis* 489 Universidade do Porto Faculdade de Economia do Porto

[6] Vinodhini, G. and Chandrasekaran, R.M., 2012. Sentiment analysis and opinion mining: a survey. *International Journal*, *2*(6), pp.282-292.

[7] Prabowo R and Thelwall M 2009 Sentiment Analysis: A Combined Approach *Journal of Informatics* **3** 1 143–157

[8] Zulkifli A 2005 *Manajemen Sistem Informasi* Jakarta: Gramedia Pustaka Utama

[9] Blitzer J, Dredze M, Pereira F 2007 Biographies, bollywood, boom-boxes and blenders: domain adaptation for sentiment classification *In: Proceedings of ACL* **7** 440 - 447

[10] Kusrini and Lutfi E T 2009 *Algoritma Data Mining* Yogyakarta: Penerbit Andi.

[11] Pakpahan D and Widyastuti H 2014 Aplikasi Opinion Mining dengan Algoritma Naïve Bayes untuk Menilai Berita Online *Jurnal Integrasi Program Studi Teknik Informatika Politeknik Negeri Batam* **6** 1-10

[12] Candra T 2009 *Metode Pembobotan Statistical Concept Based untuk Klastering dan Kategorisasi Dokumen Berbahasa Indonesia* Tugas Akhir. Bandung: Departemen Teknik Informatika Institut Teknologi Telkom Bandung

[13] Tala F Z 2003 *A Study of Stemming Effects on Information Retrieval in Bahasa Indonesia* Tesis Amsterdam: Institute for Logic, Language and Computation Universiteit van Amsterdam

[14] Agusta L 2009 Perbandingan Algoritma Stemming Porter dengan Algoritma Nazief & Adriani untuk Stemming Dokumen Teks Bahasa Indonesia *Konferensi Nasional Sistem dan Informatika 2009* 196 – 201

[15] Dhande L L and Patnaik G K 2014 Analyzing sentiment of movie review data using Naive Bayes neural classifier *International Journal of Emerging Trends & Technology in Computer Science (IJETTCS)* **3** 4

[16] Kohavi R and R. Provost. 1998. Glossary of Terms. *Machine Learning* **30** 271-274

[17] Rizky S 2011 *Konsep Dasar Rekayasa Perangkat Lunak (Software Engineering)* Jakarta: Prestasi Pustaka

IOP Conf. Series: Journal of Physics: Conf. Series **1008** (2018) 012012 doi:10.1088/1742-6596/1008/1/012012

Tunneling effect on double potential barriers GaAs and PbS

S H B Prastowo[1], B Supriadi[1], Z R Ridlo[1], T Prihandono[1]

[1]Department of Physics Education, University of Jember, Jember, Indonesia

Email : srihandono.fkip@unej.ac.id

Abstract. A simple model of transport phenomenon tunnelling effect through double barrier structure was developed.In this research we concentrate on the variation of electron energy which entering double potential barriers to transmission coefficient. The barriers using semiconductor materials GaAs (Galium Arsenide) with band-gap energy 1.424 eV, distance of lattice 0.565 nm, and PbS (Lead Sulphide) with band gap energy 0.41 eV distance of lattice is 18 nm. The Analysisof tunnelling effect on double potentials GaAs and PbS using Schrodinger's equation, continuity, and matrix propagation to get transmission coefficient. The maximum energy of electron that we use is 1.0 eV, and observable from 0.0025 eV- 1.0 eV.The shows the highest transmission coefficient is0.9982 from electron energy 0.5123eV means electron can pass the barriers with probability 99.82%. Semiconductor from materials GaAs and PbS is one of selected material to design semiconductor device because of transmission coefficient directly proportional to bias the voltage of semiconductor device. Application of the theoretical analysis of resonant tunnelling effect on double barriers was used to design and develop new structure and combination of materials for semiconductor device (diode, transistor, and integrated circuit).

1. Introduction

A familiar fact of quantum Physics is about microscopic system on small size scale, the interaction between atoms and charge. This certainly holds on nanometres-size of a Bohr radius ~ 0.5 Å (0.05 nm).The concept of tunnelling effect is quantum mechanical phenomenon which focus on particle can pass through potentials barriers well even when classically don'thave the energy to do $\boldsymbol{E} < \boldsymbol{V_0}$. In Tunnelling effect we consider the wave nature of the electron to describe its interaction with series of potential Barriers [3]. In tunnelling double barriers occurs resonant condition. Resonant tunnelling of nonequilibrium double barriers as obtained by injection with a ballistic launcher [1]. The application of tunnelling effect is responsible for many important research areas such as development of semiconductor device, photovoltaic cell and imaging (scanning tunnelling microscopy). General application of tunnelling effect is the design of TFET (tunnel field-effect transistor) which has been recognized as feasible choice since its subshreshold swing at room temperature is able to be less than 60mV/decade which is a physical limit of conventional MOSFET(metal oxide semiconductor filed effect transistor) [2]. Quantum confined structures on the nanometer scale, such as quantum wells, wires or dots, have been increasingly used as active components in novel optical and electrical devices, including lasers, photodetectors, and photovoltaics [7]. Materials that exhibit quantum confinement tend to have remarkable thermal, optical and electronic properties useful in many nanotechnology-based applications [6]. In this research we combine two kinds semiconductor device and simulated to give information about tunnelling pheneomenon. Atomistic and quantum simulation are necessary to study carrier transport at nano scale [10].

With continuous scaling of semiconductor devices, it is valuable to study material electronic properties and device transport characteristics at the atomic level [9]. The theoretical analysis application of tunneling effect isto obtain information about process in order toprove the formulation and investigating tunneling effect of GaAs (Gallium Arsenide) and PbS (lead sulphide) on double potentials (resonant tunnelling). Analysis of tunneling effect start from defining equation for each condition before strike the barrier, in barrier, and after pass barrier[4]. Second, find the *k vector* by comparingthe electron energy and "height" of barrier. Third, get the wave equations for each condition using continuity of state. The last is arranging complete equation to solve the Transmission and Reflection of wave function. On Single Potential is easy to find another characteristic of Tunneling effect, we just finish three step above to find it.On the contrary, indouble potential barriers of asymmetric potential, this step is difficult because we have five wave function from double potentials in this condition. Another solution is easier to solve the problems in multiple potentials using matrix propagation to solve equation and get transmission coefficient. Transmission coefficient determine the probability of the particle going through the barriers. Transmission coefficient also contribute to resistance of semiconductor materials. Because all semiconductor device are free from the resistance mismatch problem [5].

In this case we consider the analysis of Tunneling effect on double barriers, which consist of GaAs (Galium Arsenide) and Lead Sulphide (PbS).Characteristics of GaAs consist two kinds of atoms inclung Galium and Arsenide with band gapenergy of 1.424 eVand distance of Lattice is 0.565 nm as width of barriers. Device engineers are interested in GaAs because it is an example of III-V compound semiconductor that used to make laser diodes and high-speed transistors [7].PbS (lead Sulphide) consist of two kind atoms Lead (Pb) and Sulphide (S). Lead from group IV-A and Sulphide from VI-A has small band gap ~ 0.41 eV [11]. Bohr Radius excitation 18nm, and small effective mass from GaAs and lead Sulphide is $0.075m_0$[12],A few studies have been performed on films thinner than 20 nm as combination from semiconductor device[6]. In this research the film thinner about~ 19.565 nm from total thickness of GaAs and PbS,itis an attractive material for quantum confined structures. Combination between GaAs and PbS make a new tunnelling effect analysis and characterization to predict the value of transmission coefficient. Method of matrix propagation is one of the best method to solve tunnelling effect on another types of barrier for example: single barrier, symmetric multiple barrier, asymmetric multiple barrier, periodical barrier, sinusoid barrier, or combination of barrier [8].Transmission coefficient of electrons occurs on resonant tunnelling heterostructure double barriers has attracted much attention in recent years due the negative differential conductance behaviour [1].

2. Methods

Analysis of tunneling effect on GaAs andPbS begin by giving a potential shape definition. There are kinds of potentials shapes: single, double, multiple, and series potentials [8]. GaAs and PbS has different band gap energy, the shape of potential is asymmetric.We can calculate the propagation of matrix to find $\hat{p}_{step}$ consist of transmission and reflection coefficient. Both of them represent energy for each electron in different positions [8]. On multiple standardpotentials we can obtain the potential notation by using x_n for first potential, x_{n+1}, for second potential and according to figure 1, distance of first potential and second potentials are denoted by $L_n = (x_{n+1} - x_n)$.

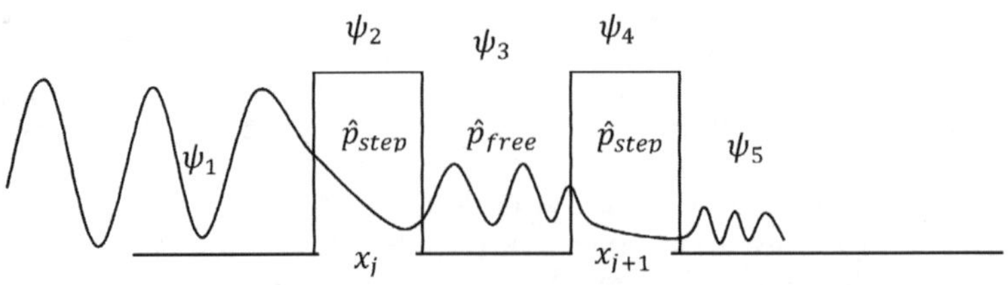

Figure 1. Tunneling in Double Potential Barriers [3]

Analysis and calculation propagation matrix n_{th} region can be solved by multiplying each region matrix $\hat{p}_{step}$ and $\hat{p}_{free}$become $\hat{p}_j = \hat{p}_{step} \cdot \hat{p}_{free}$. Two or more wave function must be continuous, this means that at potential step occur at the boundary between region$j$ and j_{th}. Energy of electron moving circular on atom can be describe as.

$$E = \frac{\hbar^2 k^2}{2m} \tag{1}$$

IOP Conf. Series: Journal of Physics: Conf. Series **1008** (2018) 012012 doi:10.1088/1742-6596/1008/1/012012

Equation (1) is energy formulation with relation of k number [3],wave vector from particle (electron) before and after strike the potentials can be written as function of energy.

$$k = \frac{2m\,(E)^{1/2}}{\hbar} \; and \; k' = \frac{2m\,(E-V)^{\frac{1}{2}}}{\hbar} \tag{2}$$

The electron motion in microscopic condition can be described as wave function for each region on tunnelling effect can be define on equation (3) and (4) [11],ψ_j used forthe first potentials barriers' wave function and ψ_{j+1} for next barriers. There is no limit to use the barriers on analyse of tunnelling effect phenomena but in this research only use double barriers with different energy height.

$$\psi_j = A_j e^{ik_n x} + B_j e^{-ik_n x} (3)$$

$$\psi_{j+1} = C_{j+1} e^{ik_{n+1} x} + D_{j+1} e^{-ik_{n+1} x} \tag{4}$$

Two wave function on each regionmust be continuous on tunnelling process.The analysis of continuity using first order derivation of wave function on equation (5).

$$\psi_n = \psi_{n+1}$$

$$e^{ik_n x} + B_n e^{-ik_n x} = C_{n+1} e^{ik_{n+1} x} + D_{n+1} e^{-ik_{n+1} x}$$

$$\frac{d\psi_n}{dx} = \frac{d\psi_{n+1}}{dx}$$

$$(ik_n)\, A_j e^{ik_n x} + (-\, ik_n) B_n e^{-ik_n x} = (ik_{n+1}) C_{n+1} e^{ik_{n+1} x} + (-\, ik_{n+1}) D_{n+1} e^{-ik_{n+1} x} \tag{5}$$

Equation (5) is the result of wave function derivation for each region and with simple process we can get equation (6) for grouping factor k_n and k_{n+1} (wave vector for each wave) [8].

$$(k_n A_j) e^{ik_n x} - k_n B_n e^{-ik_n x} = (k_{n+1}) C_{n+1} e^{ik_{n+1} x} + (-\, k_{n+1}) D_{n+1} e^{-ik_{n+1} x} \tag{6}$$

The Equation (7) divided by wave vectork_n, and we get eq (10) comparison between k_{n+1} *and* k_n.

$$A_n e^{ik_n x} - B_n e^{-ik_n x} = \left(\frac{k_{n+1}}{k_n}\right) C_{n+1} e^{ik_{n+1} x} - \left(\frac{k_{n+1}}{k_n}\right) D_{n+1} e^{-ik_{n+1} x} \tag{7}$$

Effective electron mass for simulation can describe as comparison between wave vector k_{n+1} and k_n

$$\left(\frac{k_{n+1}}{k_n}\right) = \left(\frac{m_n (k_{n+1})}{m_{n+1}\, (k_n)}\right)$$

The equation (6) and (7) can be combined and re-write into square-matrix approximation.

$$\begin{bmatrix} 1 & 1 \\ 1 & -1 \end{bmatrix} \begin{bmatrix} A_n \\ B_n \end{bmatrix} = \begin{bmatrix} 1 & 1 \\ \left(\frac{k_{n+1}}{k_n}\right) & \left(\frac{k_{n+1}}{k_n}\right) \end{bmatrix} \begin{bmatrix} (C_{n+1}) \\ (D_{n+1}) \end{bmatrix} \tag{8}$$

The Value of A_n *and* B_n from equation (8) can solve by using invers method of matrices to obtain matrices $\hat{p}_{step}$.

$$\begin{bmatrix} A_n \\ B_n \end{bmatrix} = \underbrace{\frac{1}{2} \begin{bmatrix} 1 & 1 \\ 1 & -1 \end{bmatrix} \begin{bmatrix} 1 & 1 \\ \left(\frac{k_{n+1}}{k_n}\right) & \left(\frac{k_{n+1}}{k_n}\right) \end{bmatrix}} \begin{bmatrix} (C_{n+1}) \\ (D_{n+1}) \end{bmatrix}$$

$$\begin{bmatrix} A_n \\ B_n \end{bmatrix} = \hat{p}_{step} \begin{bmatrix} (C_{n+1}) \\ (D_{n+1}) \end{bmatrix}$$

$$\widehat{p}_{step} = \frac{1}{2}\begin{bmatrix}\left(1+\frac{k_{n+1}}{k_n}\right) & \left(1-\frac{k_{n+1}}{k_n}\right)\\ \left(1-\frac{k_{n+1}}{k_n}\right) & \left(1+\frac{k_{n+1}}{k_n}\right)\end{bmatrix} \quad (9)$$

Matrix propagation on potential step can express on equation (9). The same steps to findmatrices of$\hat{p}_{free}$can be obtained by using elimination factor of reflection wave from eq (10) and (11).

$$\psi\, A_n e^{ik_n x} = \psi_{cn} \quad (10)$$

$$\psi\, B_n e^{-ik_n x} = \psi_{Dn} \quad (11)$$

The result of matrix propagation operation identity appears and with invers matrix we can write on equation .

$$\begin{bmatrix} e^{ik_n x} & 0 \\ 0 & e^{-ik_n x}\end{bmatrix}\begin{bmatrix}A_n\\ B_n\end{bmatrix} = \begin{bmatrix}(C_{n+1})\\ (D_{n+1})\end{bmatrix} \quad (12)$$

Simple expression of value A_n and $B_n$as amplitude from reflection wave express with equation (12)

$$\begin{bmatrix}A_n\\ B_n\end{bmatrix} = \begin{bmatrix} e^{-ik_n x} & 0 \\ 0 & e^{ik_n x}\end{bmatrix}\begin{bmatrix}(C_{n+1})\\ (D_{n+1})\end{bmatrix}$$

$$\begin{bmatrix}A_n\\ B_n\end{bmatrix} = \hat{p}_{free}\begin{bmatrix}(C_{n+1})\\ (D_{n+1})\end{bmatrix} (16)$$

$$\widehat{p}_{free} = \begin{bmatrix} e^{-ik_n x} & 0 \\ 0 & e^{ik_n x}\end{bmatrix} \quad (13)$$

Matrix propagation of $\hat{p}_{free}$from particle in outer of barriers can be written as equation (13). Scalar product for $\hat{p}_{free} \cdot \hat{p}_{step}$ can be written as $\hat{p}_{n-th\ region}$ as matrix propagation for amount of potential barriers.

$$\hat{p}_{free} \cdot \hat{p}_{step} = \hat{p}_{n-th\ region}$$

$$\widehat{p}_{n-th\ region} = \frac{1}{2}\begin{bmatrix}\left(1+\frac{k_{n+1}}{k_n}\right)e^{-ik_n x} & \left(1-\frac{k_{n+1}}{k_n}\right)e^{-ik_n x}\\ \left(1-\frac{k_{n+1}}{k_n}\right)e^{ik_n x} & \left(1+\frac{k_{n+1}}{k_n}\right)e^{ik_n x}\end{bmatrix} = \begin{bmatrix}P_{11} & P_{12}\\ P_{12}^{*} & P_{11}^{*}\end{bmatrix} \quad (14)$$

For multiple potential in general we can find$q$ potential step. Propagation matrix can be written as:

$$\hat{P} = \prod_{n=1}^{n=q} \hat{P}_n$$

Transmission of probability can be calculated by Determinant of matrix $\widehat{p}_{n-th\ region}$

$$T = \left|\frac{1}{P_{11}}\right|^2$$

$$T = \left(1 + \frac{1}{4}\frac{V_0}{E(E-V_0)} sin^2(k'L)\right) \quad (15)$$

The expression of transmission coefficient in other style can be written as function energy of particle E and potential barrier energy V_0 express on equation (19).

3. Result and Discussion

The simulation of tunneling effect on double potentials barriers GaAs and PbS shows by computer program. The aim of this research is analyzingthe value of transmission coefficient and the relation of semiconductor materials application.The value of transmission coefficient indicated probability of electron to pass the barriers higher, transmission coefficient making semiconductor materials easier to increase bias voltageon electronical device, because more electron on conduction band can moving easily. Simulation on computer programming of tunneling effect using maximum electron energy 1 eV and observable on computer simulation from 0.0025 eV - 1.0 eV because energy 0.0025 eV is the minimum value that can recorded on computer programming according to width of first potentials and second potentials. Effective electron mass is 0.075 m_0 from reduction electron mass. Energy for first and second barriers using bandgap energy 1.424 eV for GaAs, and 0.41 eV for PbS, the space between barriers using double Bohr radius (1.0 nm) to approach the general size of semiconductor layer~ 20 nm [6].

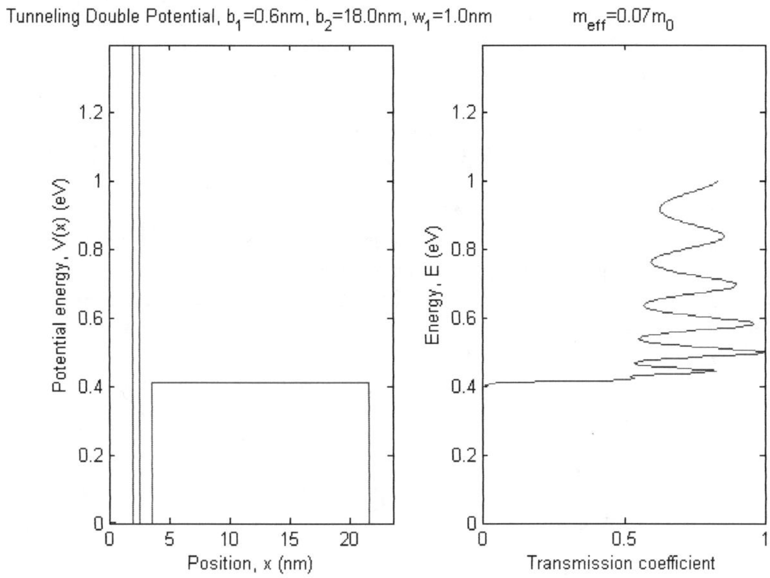

Figure 2. Result from ComputerSimulation (left side is model of double barriers and right side is relation between electron energy to transmission coefficient)

The result of simulation shows on figure 2withtwo pictures, left side isfirst potential of GaAs with band gap energy 1.424 eV barrier width 0.565 nm and the second barriers PbS band gap energy 0.41 eV and barrier width 18 nm.Figure 2 also explain about the relation between electron energy and transmission coefficient, according to equation (15) there are relations between electron energy to transmission coefficient, energy of electron on *(y-axis)* and transmission coefficient on *(x-axis)*. Analysis from figure 2 left side and right side explain the value of transmission coefficient make sinusoid graphic to electron energy, if electron strike double barriers with energy $E < V_0$, electron will pass the barriers even the classical mechanics don't do this, but in this condition electron that moving on microscopic scale has duality character asparticle and wave. In this case electron moving as wave, so if electron pass the barriers electron can penetrate barriers although electron energy less than barriers energy. Numerical result of tunnelling effect shows on table 1 and nine sample data were chosen from picture 2 on relative maximum transmission coefficient to approach maximum transmission coefficient.

Table 1. Energy of electron and Transmission coefisient

Data	Electron Energy (eV)	Transmission Coefficient
1	0.0025	$3.45 \cdot 10^{-17}$
2	0.3975	0.0035
3	0.4259	0.5229
4	0.4453	0.8249
5	0.5123	0.9982
6	0.5886	0.9616
7	0.6975	0.8993
8	0.8425	0.8524
9	1.0	0.8291

Data simulation from Table 1 as first peak graphic given electron energy 0.0025 eV with transmission coefficient is $3.45 \cdot 10^{-17}$, in this condition electron energy less than energy of barriers. Small energy electron on data number 1 try to penetrate double barriers because electron characteristic as wave can pass barriers in small transmission coefficient, because strike double potential with high potentials first and second potential is wide. Wide potential also give contribution on transmission coefficient, the wider of barriers will make small transmission coefficient [8].Second peak observable from data number 2 on table 1 with energy of electron is 0.3975 eV and transmission coefficient is 0.0035 means only 0.35% total of electron can pass the barriers, this data shows increasing of electron energy and transmission coefficient from data before. On simulation data number 1 until number 4, electron pass double barriers in resonant condition tunneling with double penetrate barriers. Data number 4 with electron energy 0.4453 eV and transmission of coefficient indicates that 0.8249 or 82.49% amount of electron can pass the barriers. Electron which pass the barriers on the limit of second barriers height 0.41 eV givenvalue transmission coefficient increase significantly from 0.5229 (data number 3) to 0.8249 (data number 4), this condition depend on type of potentials barriers shape, if energy of electron approach on the limitof double barrier on system barriers, electron more difficult to pass the barriers on wide potential size, because the probability of electron wave function decrease rapidly on second potentialsaccording to first potential.Datanumber 5 is the maximum value of transmission coefficient, proportion electron energy directly l to transmission coefficient, but on data number 5 and data number 6 although electron energy increase transmission coefficient is not directly proportional because wave factor*(k and k')* making important contribution to the transmission coefficient value.

Simulation from data number 6 until number 9, electron energy increase rapidly but transmission coefficient decrease. This condition happen because theoretical background of quantum mechanics, electron energy is bigger than second barriers but smaller than first barrier.This condition makes electrononly pass one barriers, but the second barrier also contributing to perturbation energy that influence of transmission coefficient. Maximum electron energy 1 eV on last data make the transmission coefficient value 0.8291 or 82.91% probability of electron can pass the barrier. The relation between energy and transmission coefficient also shows on figure 3 that explain nine dots on blue thick colour express nine transmission coefficient. Generally, transmission coefficient directly proportional withelectron energy from the curve on picture 3, but on special condition,transmission coefficient also depends on another variable of potential barriers (energy barrier, width of barrier, and distance between barriers).

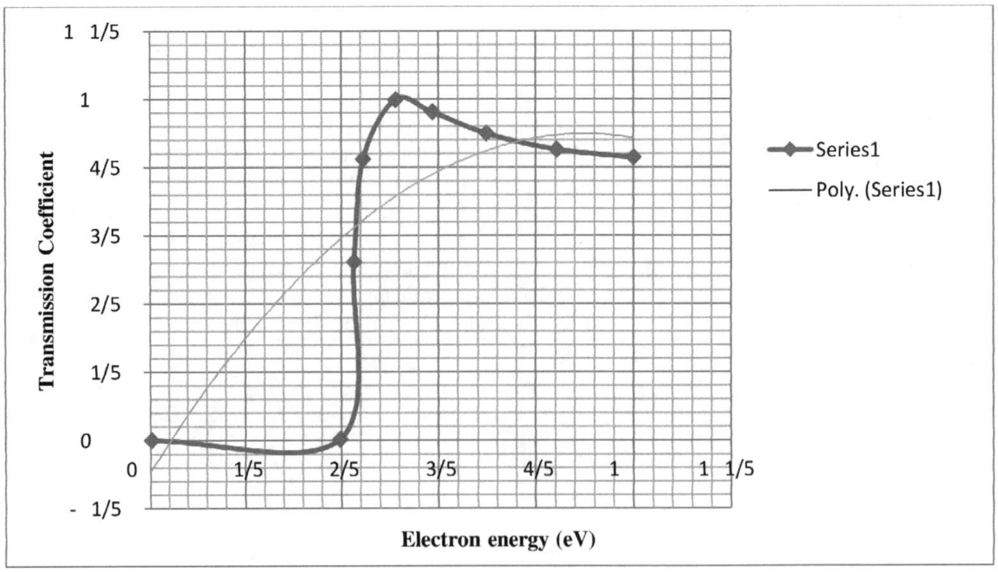

Figure 3. Relation energy with Transmission Coefficient

4. Conclusion

From the analysis result and tunnelling effect simulationon double barriers GaAs and PbS, themaximumnumber of transmission coefficient is 0.9982 or 99.82% on energy of electron0.5123 eVwhich indicates ideal condition, because maximum transmission coefficient reaches relative small energy related to small power consumption for electrical component based on semiconductor device.This research give information about combination of semiconductor materials (GaAs and PbS) in application tosemiconductor layerto design and develop diode, transistor and another semiconductor device like IC (integrated circuit) and computer processor. Analysis of tunnelling effect GaAs (gallium arsenide) and PbS (lead sulphide) ascombination of materials on semiconductor with transmission coefficient 99.82 % and small power consumption to move electron in order to pass the barriers and make electricity is suggested material for semiconductor device and also can give new information about theoretical analysis on Quantum Mechanics especially in tunnelling effect. On further research about tunnelling effect, we can expand analysis of tunnelling effect not only for double barriers but also on series barriers or shape of barriers on parabolic potentials.

Acknowledgment

We gratefully acknowledge the support of 3th Research group Physics Education from FKIP – University of Jember of year 2018.

References

[1] Arvind S, Vengurlekar, Capasso F 1990 *Superfattices and Microstructures* **7** 2

[2] Chien N D, Anh D T K , Shih C H *Vietnam Journal of Science and Technology.* **55** 316

[3] Crankshaw S 2001 A microscopic model of resonant double-barrier tunneling in a quantum system. *Dep Physic, Univ Florida.*

[4] Ellingson R J, Beard M C, Johnson J C, Yu P, Micic O I, Nozik A J, Shabaev A, and Efros A L. 2005 *Nano Lett.* **5** 865

[5] Kim N 2014 *Applied Science and Convergence Technology.* **23** 128

[6] King D M, Johnson S I, Li J, Du X, Liang X, and Weimer, A W 2009 *Nanotechnology* **20** 1

[7] Lee W, Dasgupta, Jung H J, Lee J R, Sinclair R, and Printz F B 2010 *Nanotechnology* **21** 1

[8] Levi A F J 2003 *Applied Quantum Mechanics* (Cambridge: Cambridge University Press)

[9] Liu F, Shi Q, Wang J, Guo, H. 2016. *Atomistic Simulations of Tunneling FETs* (Switzerland :Springer International Publishing Switzerland)

[10] Nanni L, 2017 Superluminal Tunneling of a Relativistic Half-Integer Spin Particle through a Potential Barrier *Preprint*

[11] Phillips A C, 2003 *Introduction to Quantum Mechanics* (New York: John Wiley and Sons)

[12] Schaller R D, Agranovich V M, Klimov V.2005. I. *Nat. Phys.* **1** 189.

[13] Torimoto T, Takabayashi S, Mori H, Kuwabata S.2002. *J. Electroanal. Chem.* **33** 522.

[14] Yu T, Hoyt J L, Antoniadis D A. 2016. Tunneling FET Fabrication and Characterization. In *Tunneling Field Effect Transistor Technology*. (Switzerland: Springer International Publishing Switzerland)

IOP Conf. Series: Journal of Physics: Conf. Series **1008** (2018) 012013 doi:10.1088/1742-6596/1008/1/012013

The stark effect on the spectrum energy of tritium in first excited state with relativistic condition

S H B Prastowo[1], B Supriadi, S Bahri[1] and Z R Ridlo[1]
[1] Physics Education Departement, University of Jember, Jember, Indonesia

E-mail: srihandono.fkip@unej.ac.id

Abstract. This research discussed about the correction of Stark Effect on Tritium atoms in the first excited state with relativistic conditions. The approach used to solve this Stark Effect correction was the perturbation theory which was from time independent degenerate perturbation theory to second-order correction. The Stark Effect on the excited state made the spectrum energy polarization of Tritium which was included in the isotope of hydrogen with an electron moving around the nucleus with high velocity. Hence, the relativistic correction affected the spectrum energy shift. Tritium was a radioactive material having half-time 12,3 years and relatively safe. The Tritium application was a material for the manufacture of nuclear battery. The most effective external electric field that should give to Tritium was 10^8 V/mith the total correction energy that was 0,97398557 × 10^{-21} *Joule*. Therefore, its effect reduced the binding energy between electron and nucleus, and increased the power of Tritium Betavoltaics Battery.

1. Introduction

The microscopic phenomena can be explained by using quantum theory. For more than 100 years, the quantum theory has existed and influenced various aspects of physics such as in Mechanics, Electrodynamics and Thermodynamics[1]. Quantum mechanics can be explained the microscopic phenomena by using theory of particle wave dualism. In 1926, Erwin Schrodinger made a great equation that can give the definition of a particle wave function. The name of this equation was Schrodinger Equation [2,3] which was formed by using the energy conservation law. It was obedient to the hypothesis of de Broglie, so it produced a complex analytical solution in the form of wave functions of single value, continuous, and finite. The Schrodinger equation consisted of kinetic energy operators and potential energy V [4].In the atomic case, the potential energy is caused by charges interaction between proton and electron.

The solution of Schrodinger equation for the real system is usually complicated. The exact solution is very difficult to find, or the solution does not exist; for example, Hydrogen atom on external field or in strong magnetic field. Therefore, to solve this problem, we need to rely on approximation method. The simple and powerful approximation method is perturbation theory. We usually call this method with Rayleigh-Schrodinger Perturbation Theory [5]. The research about external electric field on Atom was first noticed in 1913 by Stark who observed the separation (polarization) of Balmer lines by using external electric field 100.000 V/cm[6]. The Stark effect correction will be found until second order correction.

By using Bohr formula, we can find the velocity of electron, and the value is about $2{,}6\times10^6\, m/s$. Hence, to get the real condition of the hydrogenic atom, we should calculate the relativity effect of the electrons movement. The relativistic correction used in this study was only in the first order because in the second order correction, it gives very small value and hardly any effect. The relativistic correction gives an energy shift that is depended by principal quantum number and orbital quantum number.

Tritium is one of unique types of atom which includes radioactive isotope of Hydrogen with 2 neutrons and a proton in nucleus [7]. Tritium emits Beta ray radiation energy with 0 – 18,6 keV (5,7 keV in average) and 12,323 years of a half-life [8]. The kinetic energy of Beta radiation is used as a source of energy for Betavoltaic battery in semiconductors [9]. This battery has the advantages of being able to survive in a long time and in extreme conditions [10]. The problem of Tritium Betavoltaics Battery is the power of this battery is too small. The principle of the Betavoltaics battery is similar to the Photovoltaics[11].The isotopes of Hydrogen atoms are likely to be easily disturbed like electric field or external magnetic field. In this research, The data were analyzed theoretically about the effect of perturbation caused by external electric field to reduce the binding energy between nucleus and electron. The effect was electron would be easier to release from its state and would increase the beta radiation of Tritium. The larger beta radiation could increase the power of Tritium Betavoltaics battery.

2. Method

2.1 Wave function and Energy Tritium in the first excitation condition.

Tritium has a lot of similar with Hydrogen, like an energy and wave function. The different between Tritium and Hydrogen is just in reduction mass. The mass used in Tritium is reduction mass (μ_t) that occurs due to the interaction between the mass of electron (m_e) and the mass of nucleus (m_N). The formula of reduction mass is shown in eq. (1),

$$\frac{1}{\mu_t} = \frac{1}{m_e} + \frac{1}{m_N} \tag{1}$$

The wave functions of Tritium in the first excited condition can be found by using Schrodinger equation (using spherical coordinates).
The wave functions are shown in table 1.

Table 1. Wave Function of Tritium in the First Excitation State

States	Wave Functions
$\vert 2s\rangle$	$\frac{1}{4\sqrt{2\pi a_{1T}^3}}\left(2-\frac{r}{a_0}\right)e^{-r/2a1T}$
$\vert 2p_z\rangle$	$\frac{r}{4a_{1T}}\frac{1}{\sqrt{2\pi a_{1T}^3}}(\cos\theta)e^{-r/2a_{1T}}$
$\vert 2p_y\rangle$	$\frac{r}{4a_{1T}}\frac{1}{\sqrt{2\pi a_{1T}^3}}(\sin\theta)(\sin\phi)e^{-r/2a_{1T}}$
$\vert 2p_x\rangle$	$\frac{r}{4a_{1T}}\frac{1}{\sqrt{2\pi a_{1T}^3}}(\sin\theta)(\cos\phi)e^{-r/2a_{1T}}$

The energy of Tritium can we found by using Bohr's formula [6],

$$E_n = -\left[\frac{\mu_T}{2\eta^2}\left(\frac{e^2}{4\pi\varepsilon_0}\right)^2\right]\frac{1}{n^2} \tag{2}$$

2.2 Perturbation Theory

This research used perturbation theory to solve the problem. Perturbation theory divides into two types: time-independent perturbation theory and time-dependent perturbation theory. In this research, we used time-independent perturbation theory because the Hamiltonian from Tritium was time-independent.

The Hamiltonian H consists of two parts: imperturbation Hamiltonian H_0 and perturbation Hamiltonian αV . α means the expansion parameter for the perturbation correction order.

$$H = H_0 + \alpha V \tag{3}$$

$$H_0\left|l^0\right\rangle = E_0\left|l^0\right\rangle \tag{4}$$

Because the perturbation V assumed to be very small, it should be possible to expand $\left|l^n\right\rangle$ and E_n as a power series in V. The expansion of energy and wave function are,

$$E_n = E^0 + \alpha E^1 + \alpha^2 E^2 + \ldots \tag{5}$$

$$\left|l^n\right\rangle = \left|l^0\right\rangle + \alpha\left|l^1\right\rangle + \alpha^2\left|l^2\right\rangle + \ldots \tag{6}$$

The first order correction of stark effect is degenerate case because the unperturbed energy eigen kets are degenerate. The first order correction solves with matrix method like eq (5).

$$\begin{pmatrix} V_{11} & V_{12} & V_{13} & V_{14} \\ V_{21} & V_{22} & V_{23} & V_{24} \\ V_{31} & V_{32} & V_{33} & V_{34} \\ V_{41} & V_{42} & V_{43} & V_{44} \end{pmatrix}\begin{pmatrix} a \\ b \\ c \\ d \end{pmatrix} = E_i^1 \begin{pmatrix} a \\ b \\ c \\ d \end{pmatrix} \tag{7}$$

The second order correction of stark effect on Tritium is non degenerate case. By using eq. (3) and eq. (4), we could find the second order correction of Energy.

$$E^2 = \sum_{l \neq m} \frac{\left|\left\langle m^0 \right| V \left| l^0 \right\rangle\right|^2}{E_l^{(0)} - E_m^{(0)}} \tag{8}$$

2.3 Relativistic Correction

The Hamiltonian of Tritium consists of two parts: potential energy and kinetic energy. The relativistic formula of kinetic energy is,

$$K = \frac{mc^2}{\sqrt{1 - \left(v/c\right)^2}} - mc^2 \tag{9}$$

Momentum (p) expression on Tritium atom is,

$$K = \sqrt{p^2c^2 + \mu_t{}^2c^4} - \mu_t c^2 \tag{10}$$

The eq. (10) can solve to get easier expressions by using Taylor expansions.

$$\sqrt{p^2c^2 + \mu_t{}^2c^4} - \mu_t c^2 \cong \frac{p^2}{2\mu_T} - \frac{p^4}{8\mu_T^3 c^2} + .. \tag{11}$$

The first order in eq.(11) is kinetic energy like classical theory and the second one is the perturbation Hamiltonian because of the relativistic condition. By using perturbation theory, we could find the energy correction.

$$ER_{nl}^{(1)} = \left\langle nlm \left| \frac{p^4}{8\mu_t^3 c^2} \right| nlm \right\rangle \tag{12}$$

By using Schrodinger equation, we could get the solution of quadratic momentum of electron to solve the eq. (12). Quadratic momentum is derived from Schrodinger equation given by eq (13).

$$p^2 = 2\mu_t\left(E_n - V(r)\right) \tag{13}$$

Then substitution of equation (13) into equation (12) functions to obtain the first order energy correction of relativity effect.

IOP Conf. Series: Journal of Physics: Conf. Series **1008** (2018) 012013 doi:10.1088/1742-6596/1008/1/012013

$$ER_{nl}^{(1)} = \frac{1}{8\mu_T c^2}\left\langle nlm \left| E_n^2 - \frac{2E_n e^2}{4\pi\varepsilon}\frac{1}{r} + \frac{e^4}{(4\pi\varepsilon)^2}\frac{1}{r^2} \right| nlm \right\rangle$$

$$ER_{nl}^{(1)} = \frac{1}{8\mu_T c^2}\left(E_n^2 - \frac{2E_n e^2}{4\pi\varepsilon}\left\langle\frac{1}{r}\right\rangle + \frac{e^4}{(4\pi\varepsilon)^2}\left\langle\frac{1}{r^2}\right\rangle \right) \tag{14}$$

By using special theorem, we could describe the expectation of $\frac{1}{r}$ and $\frac{1}{r^2}$,

$$\left\langle\frac{1}{r}\right\rangle_{nl} = \frac{1}{a_0 n^2} \tag{15}$$

$$\left\langle\frac{1}{r^2}\right\rangle_{nl} = \frac{1}{a_0^3 n^3\left(l + \frac{1}{2}\right)} \tag{16}$$

Then substitution of equation (15) and equation (16) to equation (14) resulted that a first-order tritium energy correction solution would be obtained,

$$ER_{nl}^{(1)} = -\frac{\beta^2 E_n}{4n^4}\left(\frac{8n}{2l+1} - 3\right) \tag{17}$$

β is fine structure constant, and $ER_{nl}^{(1)}$ is the first-order relativistic correction of Energy Tritium in various state.

3. Result and Discussion

3.1 Energy Tritium in Relativistic Condition

By using equation (1), we could calculate the result of reduction mass of Tritium (μ_T) and it would be compared with the reduction mass of Hydrogen (μ_H) and mass of electron (m_e).

$$\mu_T = \frac{m_e(m_p + 2m_n)}{(m_p + 2m_n) + m_e} = 9{,}1067819 \times 10^{-31} Kg \tag{18}$$

$$\mu_H = \frac{m_e . m_p}{m_p + m_e} = 9{,}10443157 \times 10^{-31} Kg \tag{19}$$

$m_e = 9{,}109390 \times 10^{-31} Kg$, $m_p = 1{,}672623 \times 10^{-27} Kg$, and $m_n = 1{,}674929 \times 10^{-27} Kg$. Jika Now, we could compare among the reduction mass of Tritium, Hydrogen, and electron. When the nucleus mass was getting higher, the reduction mass was getting higher too. This reduction mass affected the radius of the electron's orbit. If the number of neutrons in the nucleus got larger, the binding force of the nucleus also got larger. This phenomenon caused the radius of the electron's orbit getting smaller. It could be proven from the equations of the radius of Bohr atoms in the ground state as below,

$$a_{1H} = \frac{4\pi\varepsilon_0 \eta^2}{\mu_H e^2} = 5{,}294541195 \times 10^{-11} m \tag{20}$$

$$a_{1T} = \frac{4\pi\varepsilon_0\eta^2}{\mu_T e^2} = 5{,}29317475 \times 10^{-11} m \tag{21}$$

By using equation (2), we could find the energy of Tritium atom for the principal quantum number 1 to principal quantum number 3.

$$E_1^{(0)} = -21{,}7947692 \times 10^{-19} Joule$$
$$E_2^{(0)} = -5{,}4486923 \times 10^{-19} Joule$$
$$E_3^{(0)} = -2{,}42164102 \times 10^{-19} Joule$$

The relativistic correction was depended by principal quantum number (*n*) and orbital quantum number (*l*). It was very different from classical theory of energy. In classical theory, energy of atom was just depended by principal quantum number (*n*). Energy Tritium for every state in relativistic condition was,

a. $n = 1$ and $l = 0\,(1s)$

$$ER_{1s}^{(1)} = -\frac{5\beta^2 E_1^{(0)}}{4} = 1{,}45075 \times 10^{-22} Joule$$

The total energy in the state of *1s* under relativistic conditions is,

$$ER_{1s} = E_1^{(0)} + ER_{1s}^{(1)}$$
$$ER_{1s} = -21{,}79331845 \times 10^{-19} Joule$$

b. $n = 2$ and $l = 0\,(2s)$

$$ER_{2s}^{(1)} = -\frac{13\beta^2 E_2^{(0)}}{16} = 2{,}35747 \times 10^{-23} Joule$$

The total energy in the state *2s* in relativistic conditions is,

$$ER_{2s} = E_2^{(0)} + ER_{2s}^{(1)}$$
$$ER_{2s} = -5{,}448456553 \times 10^{-19} Jouole$$

c. $n = 2$ and $l = 1\,(2p)$

$$ER_{2p}^{(1)} = -\frac{7\beta^2 E_2^{(0)}}{48} = 4{,}231357 \times 10^{-24} Joule$$

The total energy in the state *2p* in relativistic conditions is,

$$ER_{2p} = E_2^{(0)} + ER_{2p}^{(1)}$$
$$ER_{2p} = -5{,}448649986 \times 10^{-19} Joule$$

d. $n = 3$ and $l = 0\,(3s)$

$$ER_{3s}^{(1)} = -\frac{21\beta^2 E_3^{(0)}}{36} = 7{,}5224 \times 10^{-24} Joule$$

The total energy in the state *3s* in relativistic conditions is,

$$ER_{3s} = E_3^{(0)} + ER_{3s}^{(1)}$$
$$ER_{3s} = -2{,}421565796 \times 10^{-19} Joule$$

e. $n = 3$ and $l = 1\,(3p)$

$$ER_{3p}^{(1)} = -\frac{5\beta^2 E_3^{(0)}}{36} = 1{,}79105 \times 10^{-24} Joule$$

The total energy in the state *3p* in relativistic conditions is,

$$ER_{3p} = E_3^{(0)} + ER_{3p}^{(1)}$$

$$ER_{3p} = -2{,}42162311 \times 10^{-19}\,Joule$$

f. $n = 3$ and $l = 2\,(3d)$

$$ER_{3d}^{(1)} = -\frac{9\beta^2 E_3^{(0)}}{180} = 6{,}44778 \times 10^{-25}\,Joule$$

The total energy in the state *3d* in relativistic conditions is,

$$ER_{3d} = E_3^{(0)} + ER_{3d}^{(1)}$$

$$ER_{3d} = -2{,}421634572 \times 10^{-19}\,Joule$$

The correction of relativity gave a shift in the energy value of Tritium at every level[13,14]. The higher of the atomic state, the smaller shift of energy would be. It happened because the speed of electron orbit would also be reduced.The principal quantum numbers described the atoms' energy levels, whereas the azimuth quantum numbers described the shape of the electrons' movement around the nucleus. At the same principal quantum number, the energy correction would decrease as the quantum azimuth number increased.

3.2 The Stark Effect on Energy of Tritium in the First Excitation State with Relativistic Condition

The correction result of external electric field affected on Tritium atom for first order was shown in equation (25),

$$\begin{vmatrix} -E_2^1 & \langle 2p_z|V|2s\rangle \\ \langle 2s|V|2p_z\rangle & -E_2^1 \end{vmatrix} = \left(E_2^1\right)^2 - \left|\langle 2s|V|2p_z\rangle\right|^2 = 0 \tag{22}$$

$$E_2^1 = \pm\left|\langle 2s|V|2p_z\rangle\right| \tag{23}$$

Where,

$$\langle 2s|V|2p_z\rangle = e\xi\langle 2s|z|2p_z\rangle = -3e\xi a_{1T} \tag{24}$$

So, the first order correction was.

$$E_2^1 = 3ea_{1T}\xi$$
$$E_2^1 = 25{,}4418016 \times 10^{-30}\,\xi \tag{25}$$

While the calculation result of external electric field correction on Tritium atom for second order correction was shown in equation (26),

$$E_2^2 = \sum_{l\neq m} \frac{\left|\langle m^0|V|l^0\rangle\right|^2}{E_l^{(0)} - E_m^{(0)}} \tag{26}$$

$|m^0\rangle$ was the wave function of the second excitations state of Tritium, and $|l^0\rangle$ was the wave function of first excitations state of Tritium. Therefore, the value of second order correction became,

$$E_2^2 = \frac{(25{,}36243e\xi a_{1T})^2}{(ER_{2s} - ER_{3pz})} + \frac{(0{,}54178e\xi a_{1T})^2}{(ER_{2pz} - ER_{3s})} + \frac{(2{,}45185e\xi a_{1T})^2}{(ER_{2pz} - ER_{3dzz})} + \frac{(1{,}06168e\xi a_{1T})^2}{(ER_{2pz} - ER_{3dxy})}$$

$$+\frac{(1{,}06168e\xi a_{1T})^2}{(ER_{2pz} - ER_{3dx^2-y^2})} + \frac{(2{,}12336e\xi a_{1T})^2}{(ER_{2py} - ER_{3dyz})} + \frac{(2{,}12336e\xi a_{1T})^2}{(ER_{2px} - ER_{3dxz})} \tag{27}$$

$$E_2^2 = -2{,}183233874 \times 10^{21} e^2 a_{1T}^2 \xi^2$$
$$E_2^2 = -1{,}57019459 \times 10^{-37}\,\xi^2 \tag{28}$$

So the total Energy of Tritium in the first excitated state with *l=0* was,

$$E_{2s} = ER_{2s} + E_2^1 + E_2^2$$
$$E_{2s} = \left(-5{,}448456553 \times 10^{-19} + 2{,}54418016 \times 10^{-29}\xi - 1{,}57019459 \times 10^{-37}\xi^2\right) joule \quad (29)$$

And the total Energy of Tritium in the first excitated state with *l=1* was,

$$E_{2p} = ER_{2p} + E_2^1 + E_2^2$$
$$E_{2p} = \left(-5{,}448649986 \times 10^{-19} + 2{,}54418016 \times 10^{-29}\xi - 1{,}57019459 \times 10^{-37}\xi^2\right) Joule \quad (30)$$

Table 2. The results of correction energy with various external electric field

External Electric Field $\xi(V/m)$	The first order correction of Energy (Joule)	The second order Correction of Energy (Joule)	Total Energy Correction (Joule)
10^5	$2{,}54418016 \times 10^{-24}$	$-1{,}57019459 \times 10^{-27}$	$2{,}54260997 \times 10^{-24}$
10^6	$2{,}54418016 \times 10^{-23}$	$-1{,}57019459 \times 10^{-25}$	$2{,}52847821 \times 10^{-23}$
10^7	$2{,}54418016 \times 10^{-22}$	$-1{,}57019459 \times 10^{-23}$	$2{,}3871607 \times 10^{-22}$
10^8	$2{,}54418016 \times 10^{-21}$	$-1{,}57019459 \times 10^{-21}$	$0{,}97398557 \times 10^{-21}$
10^9	$2{,}54418016 \times 10^{-20}$	$-1{,}57019459 \times 10^{-19}$	$-1{,}31577657 \times 10^{-19}$
10^{10}	$2{,}54418016 \times 10^{-19}$	$-1{,}57019459 \times 10^{-17}$	$-1{,}54475279 \times 10^{-17}$

In this research, the disturbance was the external electric field (ξ) that imposed on Tritium atom in the *z*-axis. The external electric field given to Tritium atom would affect to the level of Hamiltonian energy of Tritium[15]. In the first order correction, the perturbation theory was degenerate case because the energy value in the first excited state was identical. On the other hand, in the second order correction, the perturbation theory was non-degeneration case because the energy value in the first excited state energy was different from the second excited state. From equation (25), the first order correction would reduce the binding energy between electron and nucleus, but it would increase the Beta radiation from Tritium Atom. However, in equation (28), the second order correction would increase the binding energy between electron and nucleus, so the Beta radiation would reduce. From Table 2, the binding energy would reduce when the external electric field given to Tritium was $10^5 V/m - 10^8 V/m$. It was because the first order correction was bigger than the second order correction. The binding energy between electron and nucleus increased when the external electric field given to Tritium was higher than $10^9 V/m$. The effective external electric field that should give to Tritium atom to increase the Beta radiation and the power of Betavoltaic Battery was $10^8 V/m$ with the total correction energy that was $0{,}97398557 \times 10^{-21} Joule$.

4. Conclusion

The results of energy correction of Tritium until the second order caused by external electric field in relativistic condition indicated that in the relativistic conditions, the energy value of the Tritium atom depended on the principal quantum number (*n*) and the quantum azimuth number (*l*). The energy value correction caused by relativity would decrease as the value of *n* got higher. At the same *n* value, the energy correction value caused by relativity would decrease as well as the greater of the *l* value. Tritium Energy correction at the first order caused by external electric field used degenerate case, while in second order correction used non-degenerate case. The effective external electric field that should give to Tritium atom was $10^8 V/m$.For the future research, it is expected to use other forms of perturbation such as external magnetic field, the influence of spin coupling, and the others.

Acknowledgment

We gratefully acknowledge the support of 3th Research group Physics Education from FKIP – University of Jember of year 2018.

References

[1] Krumm M, Barnum H, Barret J and Muller M P 2017 *New J.Phys.* **19** 043025
[2] Zettili N 2009 *Quantum mechanics Concepts and Application* (USA: John Wiley & Sons Ltd) 489
[3] Hosseini H, Kumar D, Kaplan M, and Bejarbaneh E Y 2017 *Commun. Theor. Phys.* **68** 761-767
[4] Chen J P, Molchanov S and Teplyaev A 2015 *J. Phys. A: Math. Theor.* **48** 395203 27
[5] Schwabl F 2007 *Quantum mechanics 4th edition* (Munchen : Springer) 206
[6] Libof R L 1980 *Introductory Quantum Mechanics* (Canada: Addison-Wesley) 568
[7] Krasheninnikov S I, Marenkov E D, Smirnov R D and Pisarev A A 2014 *Phys. Scr.* **T159** 014060 5
[8] Lee S K, Son S H, Kim K S, Park J W, Lim H, Lee J M and Chung E S 2009 *Applied Radiation and Isotopes* **67** 1234-1238
[9] Alam T R, Pierson M A and Prelas M A 2017 *Applied Radiation and Isotopes* **130** 80-89
[10] Gui G, Zhang K, Blanchard J P and Ma Z 2017 *Applied Radiation and Isotopes* **130** 272-277
[11] Sachenko A V, Shkrebtii A I, Korkishko R M, Kostylyov V P, Kulish M R, Sokolovskyi I O and Evstigneev M 2015 *J. Phys. D: Appl. Phys.* **48** 455101 8
[12] Krane K 2012 *Modern Physics Third Edition* (United State of America: John Willey & Sons, Inc) 102
[13] Yokozuka T, Ido K, Clark R, Takeda K and Tokura Y 2014 *Japanese Journal of Applied Physics* **53** 031801
[14] Sakho I 2012 Relativistic theory of one-and two electron system: valley of stability in the Helium-like ions *Journal of Atomic and Molecul Science* **3** 23-40
[15] Hogan S D, Seiler Ch and Merkt F 2013 *J. Phys. B: At. Mol. Opt. Phys.* **46** 045303 8

IOP Conf. Series: Journal of Physics: Conf. Series **1008** (2018) 012014 doi:10.1088/1742-6596/1008/1/012014

Water hyacinth cellulose-based membrane for adsorption of liquid waste dyes and chromium

Cintia Agtasia Putri[1], Ian Yulianti[1], Ika Desianna[1], Anisa Sholihah[2] and Sujarwata[1]

[1]Physics Education Postgraduate Program Universitas Negeri Semarang, Semarang, Indonesia

[2]Department of Physics Education, Universitas Negeri Semarang, 50229 Semarang, Central Java, Indonesia

E-mail: cintiaagtasiaputri@gmail.com

Abstract. Water hyacinth (*Eichornia crassipes*) is a weed in aquatic area whose trunk contains a lot of cellulose. Cellulose contained can be used as dyes adsorbent in a form of composite membrane. This study aims to investigate the capacity of water hyacinth cellulose-based membrane to adsorb dye and Chromium (Cr) contained in liquid. The process of membrane fabrication begins with isolation of water hyacinth cellulose. The isolated cellulose powder was used to make the membrane by mixing it with polyvinyl alcohol-polyethylene glycol (PVA-PEG) with various compositions. The morphology of membrane surface was analyzed using CCD microscope. The analysis using Ultraviolet Visible Spectroscopy (UV-Vis) and Atomic Absorption Spectroscopy (AAS) indicate that the membrane with composition ratio of cellulose: PVA: PEG of 6.5: 2.5: 1 adsorb Cr up to 38.75%.

1. Introduction

Indigosol with IR code is one of dye used in production process in fabric manufacturer. To enhance the dye's capacity to adhere to the fabric, indigosol dyes are first activated using acid and nitrite solutions. The activation process results in Cr residue in manufacturing waste. Therefore, effort should be made to reduce the dyes and Cr from the waste before it is disposed to the environment. The common method in reducing dyes and heavy metal ions in liquid and waste water treatment is using membrane-bioreactor. However, membrane-bioreactor is expensive and requires complicated process. Various types of membranes have been investigated to give alternative solution for waste water treatment such as pressure-driven membrane [1], nanoadsorbents based on conducting polymer nanocomposites [2], composite graphene oxide membrane [3], thin film nanocomposite (TFN) membrane [4] nanofiltration [5] and hydrophilic nanoporous ion-exchange barrier membranes [6].

Other alternative for dyes removal is using organic membrane. Various organic membranes have been fabricated using pineapple leaf [7], banana stem [8], pandan duri leaves [9], and polymeric membrane cellulose-chitosan [10]. As stated in other researches mentioned above, the main material of organic membrane is cellulose. Therefore, organic material with high cellulose content is desired to be used as organic membrane. Priya et al [11] showed that water hyacinth (*Eichornia crassipes*) contain 30-50% of cellulose. Water hyacinth has been used for various application such is Carboxymethyl Cellulose (CMC) [12] and Cellulose Acetate Membrane-PVC [13]. In this work, water hyacinth was used as cellulose source for organic membrane for dyes and Cr removal application.

IOP Conf. Series: Journal of Physics: Conf. Series **1008** (2018) 012014 doi:10.1088/1742-6596/1008/1/012014

2. Research Methods

The equipment used in this research are a set of extraction tools, filter paper, beaker glass, oven, petri dish, CCD balance, magnetic stirrer, 0.1 mm^2 sized filter, FTIR, UV-Vis and AAS. Meanwhile, the materials used are $NaClO_2$, NaOH, toluene, 96% alcohol, PVA, PEG, distilled water, water hyacinth sticks, and dyes liquid waste.

2.1 Delignification of Water Hyacinth

The fabrication of composite membranes begins with the fabrication of cellulose powders. First, water hyacinth was cleaned, dried and powdered with a 0.1 mm^2 sieve. To investigate the chemical composition of the powder, the cellulose powder was characterized using *Fourier Transform Infra Red* (FTIR) Spectroscopy method. The chemical compound was determined by comparing the peak of the obtained FTIR spectrum with FTIR table.

The content of wax on the water hyacinth was removed by extraction process. The extraction process uses a mixture of 96% toluene and alcohol with a volume ratio of 2:1. Lignin was removed by bleaching process using 1% $NaClO_2$ solution at temperature of 80 °C for one hour. Meanwhile, the hemicellulose was removed by using 17.5% w/v NaOH solution for one hour at room temperature. After hemicellulose was removed, the obtained cellulose was then washed with aquades and 96% alcohol. The obtained cellulose powder was also characterized for its chemical compound composition using FTIR.

2.2 Preparation of Composite Membrane

The membrane was fabricated by mixing the water hyacinth cellulose powder with PVA and PEG with various compositions as shown in Table 1. The mixture was put on a glass plate forming a thin layer and then dried in a room temperature for an hour. After dry, the membranes were then washed with 1% NaOH and were dried in a room temperature to remove residual moisture content.

Table 1. Variation of cellulose membrane composition

Membrane	Cellulose (%b/b)	PVA (%b/b)	PEG (%b/b)
1	5.5	3.5	1
2	6	3	1
3	6.5	2.5	1

2.2.1 Morphological Characterization of Composite Membrane

The morphological (surface) membrane structure was characterized using a CCD (Charge Coupled Device) MS-804 Microscope. CCD microscope produces a morphological image of the membrane surface. The morphology of the membrane surface shows the pore size of the membrane that has been made.

2.2.2 Characterization of Membrane Adsorption to Dyes and Cromium

The membrane adsorption characterization was done by filtering the liquid waste containing Indigosol IR code-dyes using the fabricated membrane. The filtered liquid was then investigated using UV-Vis. For comparison purposes, the un-filtered liquid was also characterized.

The decrease of Cr content in waste water was determined through AAS test. The result of initial AAS test before the adsorption process showed Cr metal content of 5.349 mg / L. This Cr content of metal will certainly pollute the environment if it is discharged directly prior to adsorption.

3. Results and Discussion

3.1 Characterization of Membrane Adsorption to Dyes and Cromium

The result of FTIR spectroscopy of the water hyacinth powder and the cellulose powder are shown in Figure 1 and Figure 2, respectively.

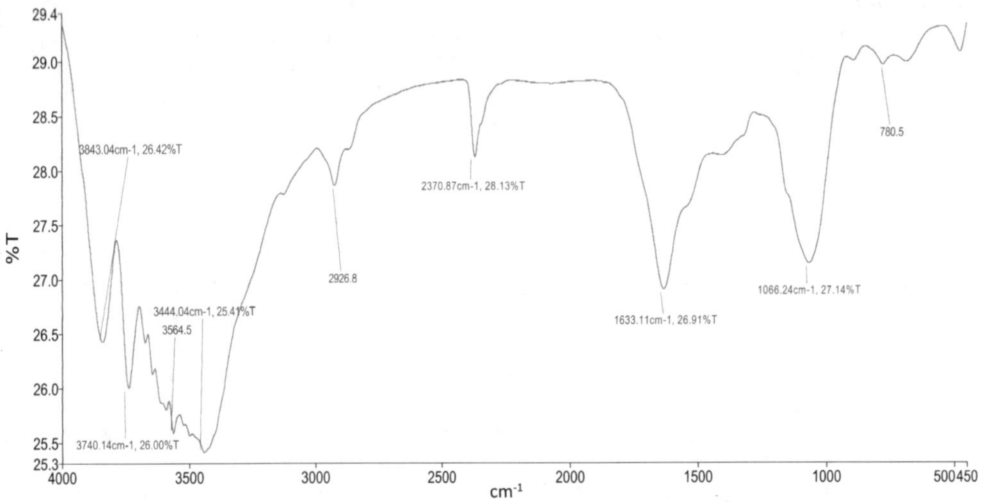

Figure 1. FTIR spectra from water hyacinth powder before the cellulose isolation process.

The main cellulose functional are represented by the bound of O-H, C-H and C-O glycosidic on a material [13]. Figure 1 shows that the peak at 2348.03 cm^{-1} is indicating the C-H, the peak at 3700 - 3800 cm^{-1} is indicating the formation of several O-H in the hydrogen bounding of hydroxyl and oxygen atoms of other hydroxyl groups on the cellulose polymer chain. The peak at 1651.79 cm^{-1} indicates of a water-absorption fiber-OH and the peak at 1064.51 cm^{-1} showing the C-O strain. Meanwhile, the peak at 806.10 cm^{-1} shows the absorption peak due to the C-H vibration. Therefore, it can be concluded that the water hyacinth powder contains cellulose as shown by the FTIR spectrum. However, there are also exist other chemical compounds which are indicated by the peaks at wave number of 4000-3500.

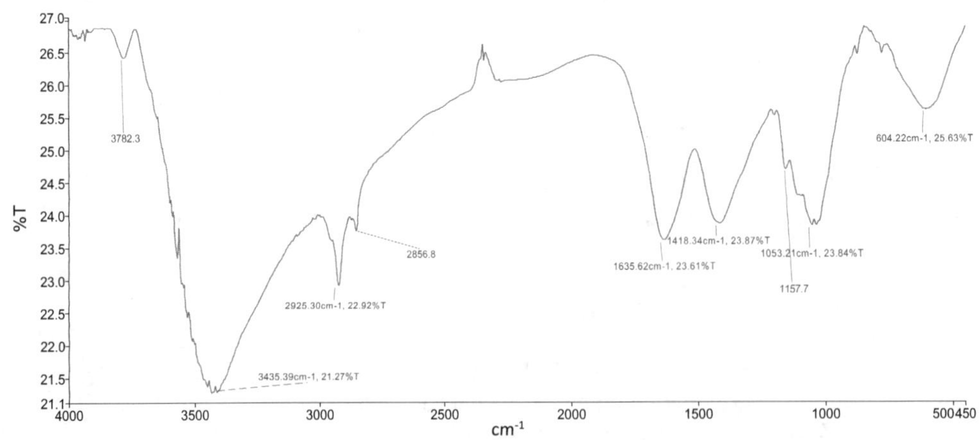

Figure 2. FTIR spectra from cellulose powder after the cellulose isolation process.

Figure 2 shows that the spectrum of the cellulose powder has peaks at 1418.34 cm^{-1}, 1635.62 cm^{-1}, 2925.3 cm^{-1}, 3444.04 cm^{-1} and 3435.39 cm^{-1} which indicates –CH, C-O, C-H, OH, and –O-H, respectively. It is also shown that the peaks at wave number of 4000-3500 no longer appear. It means that the powder only consist of cellulose.

3.2 Surface Morphology of Membrane Cellulose Water Hyacinth
The surface morphology of the membranes are shown in Figure 3. It is shown that as PVA-PEG composition increases, the surface is more covered by PVA-PEG.

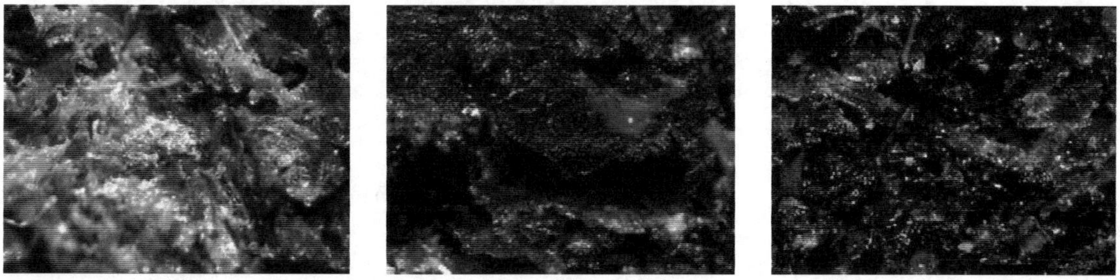

(a) (b) (c)

Figure 3. Membrane surface with composition variation a) membrane 1, b) membrane 2, and c) membrane 3.

3.3 Characterization of Membrane Adsorption to Dyes and Cromium
Figure 4 shows the UV-Vis results of unfiltered liquid (membrane 0), filtered liquid using membrane 1, membrane 2 and membrane 3, respectively.

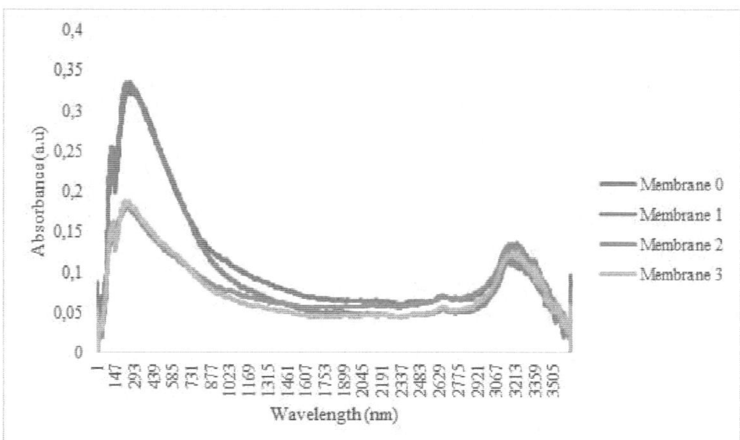

Figure 4. UV-Vis results of filtered liquid through membrane 1, 2 and 3 and unfiltered liquid.

Based on Figure 4 it can be seen that the dye concentration decrease after filtration using cellulose membrane as indicated by the decrease of the absorbance peak. The decreased of the dye levels occurs since the pores size of the membrane is lower than the indigosol dye particles size. Therefore, the dye was maintained on the surface of the membrane. It is also shown that the membrane with highest cellulose composition, which is membrane 3, has the highest adsorption capacity.

The Cr content in the filtered liquid waste was obtained from the AAS analysis as shown in Table 2. It can be seen that the Cr concentration is decreased after the liquid filtered using the membrane. It is also shown that membrane 3 has the highest capacity to reduce the Cr concentration. Compared to other composite membranes, such as polyethylene glycol-polystyrene-cellulose acetate from pineapple [7], banana stem [8] and pandan duri leaves [9], the fabricated water hyacinth membrane has lower adsorption capacity. However, the proposed water hyacinth membrane provides short fabrication time and simple fabrication process without the need for further cellulose treatment such as acetylation [7].

Table 2. AAS results of the liquid waste before and after filtration using water hyacinth membrane

Membrane	Cr Metal Metals Before Filtration	Cr Metal Content After Filtration	Percentage of Waste Dye Decrease
Membrane 1	5.439 mg/L	3.901 mg/L	28.27%
Membrane 2	5.439 mg/L	3.610 mg/L	33.62%
Membrane 3	5.439 mg/L	3.331 mg/L	38.75%

4. Conclusion

Based on the results, it can be concluded that the water hyacinth membrane has a good adsorption to Indigosol IR and Cr. The optimum composition of the membrane is consisting of 6.5: 2.5: 1 of cellulose: PVA: PEG with Cr adsorption of 38.75%. Further improvement of Cr adsorption can be done by combining water hyacinth cellulose with cellulose from other plants such as pineapple and banana stems.

Acknowledgements
This work was supported by Postgraduate of Physics Education and Physics Education Universitas Negeri Semarang.

References

[1] Bruggen BVD, Vandecasteele C, Gestel TV, Doyen W and Leysen R 2003 A review of pressure-driven membrane processes in wastewater treatment and drinking water production *Environ. Prog.* **22** 46–56
[2] Zare EN and Sillanpää M 2018 Nano adsorbents based on conducting polymer nano composites with main focus on polyaniline and its derivatives for removal of heavy metal ions/dyes: A review *Environmental Research* **162** 173-195
[3] Zhang P, Gong JL, Zeng GM, Deng CH, Yang HC, Liu HY, Huan HY 2017 Cross-linking to prepare composite graphene oxide-framework membranes with high-flux for dyes and heavy metal ions removal *Chemical Engineering Journal* **322**(15) 657-666
[4] Yin J, Zhu G, and Deng B 2016 Graphene oxide (GO) enhanced polyamide (PA) thin-film nanocomposite (TFN) membrane for water purification *Desalination* **379** 93–101
[5] Sukma FM, Emecen PZÇ 2018 Cellulose membranes for organic solvent nanofiltration *Journal of Membrane Science* **545** 329–336
[6] Song J, Niu X, Li XM and He T 2018 Selective separation of copper and nickel by membrane extraction using hydrophilic nanoporous ion-exchange barrier membranes *Process Safety and Environmental Protection* **113** 1–9
[7] Delsy EVY, Irmanto and Kazanah FN 2017 Utilization of composite membrane polyethyleneglycol-polystyrene-cellulose acetate from pineapple leaf fibers in lowering levels of methyl orange batik waste *IOP Conference Series: Materials Science and Engineering* **172** 1-6
[8] Anirudhan TS, Fernandez NB and Mullassery MD 2012 Removal of Cd (II) ions from aqueous solution using a cation exchanger derived from banana stem. *J Chem Technol Biotechnol* **87** 714-722
[9] Laksono EW, Marfuatun and Aji D 2016 Conductivity of cellulose acetate membranes from pandan duri leaves (*pandanus tectorius*) for li-ion battery *MATEC Web of Conferences* **64** 04001
[10] Istirokhatun T, Rachmawaty R, Meriyani M, Rokhati N, and Susanto H 2017 Polymeric membrane made of cellulose isolated from tropical water hyacinth blended with chitosan *IOP Conf. Ser.: Earth Environ.* Sci. **55** 1-7
[11] Priya ES and Selvan PS 2017 Water hyacinth (*Eichhornia crassipes*)–An efficient and economic adsorbent for textile effluent treatment – A review *Arabian Journal of Chemistry* **10** S3548-S3558
[12] Gendi AE, Abdallah H, Amin A, and Amin SK 2017 Investigation of polyvinylchloride and cellulose acetate blend membranes for desalination *Journal of Molecular Structure* **1146** 14-22

[13] Thaiyibah N, Alimuddin, and Panggabean AS 2016 Synthesis and characterization of cellulose acetate membrane-PVC from water hyacinth (*Eichhornia crassipes*) for Copper (II) Adsorption *Jurnal Kimia Mulawarman* **14** 1 29-35

Wireless SAW passive tag temperature measurement in the collision case

A. Sorokin[1], A. Shepeta[1], M. Wattimena[1]

[1]Major Problem-Oriented Computer Complexes Department, State University of Aerospace Instrumentation (SUAI), St. Petersburg, Russia

E-mail: mauritswattimena@yahoo.com, ultramagnus88@gmail.com

Abstract. This paper describes temperature measurement in the multisensor systems based on the radio-frequency identification SAW passive tags which are currently applied in the electric power systems and the switchgears. Different approaches of temperature measurement in the collision case are shown here. The study is based on the tag model with specific topology, which allows us to determine temperature through the response signal with time-frequency information. This research considers the collision case for several passive tags as the temperature sensors which are placed in the switchgear. This research proposal is to analyze the possibility of using several SAW passive sensors in the collision case. We consider the using of the different typical elements for passive surface acoustic wave tag which applies as an anticollision passive sensor. These wireless sensors based on the surface acoustic waves tags contain specifically coded structures. This topology makes possible the reliability of increasing tag identification and the temperature measurement in the collision case. As the results for this case we illustrate simultaneous measurement of at least six sensors.

1. Introduction

Modern solutions in the field of monitored parameters should meet the requirements of increased noise immunity. In addition, they should have high speed, resistance to external climatic influences, high reliability and durability in operation. Modern monitoring based on radio frequency systems should be passive (without any energy source). Such requirements are met by radio frequency identification (RFID) technology, based on the use of passive acoustoelectronic devices on surface acoustic waves (SAW) [1], [2].

For temperature monitoring, RFID reader interrogates several passive SAW tags which are placed on the monitored object. However, in the simultaneous monitoring of several temperature sensors is limited to the problem of collision for passive acoustoelectronic RFID devices. In the case when several sensors are interrogated simultaneously, their response signals can overlap each other in the time domain, making it impossible to read the code of each device and obtain information about the temperature.

The last analysis shows that solutions cannot satisfy the actual tasks of monitoring and control objects with a wide temperature range and certain operating conditions in the collision case. This problem is the most evident in multisensory systems which are using the electric power systems. The temperature is the most important indicator of the electrical equipment current status. It makes possible to provide information about the emergency and electrical fire caused by short circuit or overheating of the equipment in the switchgear. Monitoring systems which are applied for temperature measurement would increase reliability and safety of high-voltage equipment.

The monitoring of contact connections status can be carried out with the thermal imager. However, this method has a number of disadvantages which are related to the lack of automation.

Figure 1. Contacts overheat in switchgears. Left picture illustrates temperature of three contacts. Overheating impacts are demonstrateв in the right picture

2. Basic approaches to temperature measurement using passive SAW tags

As a rule, for temperature measurement time impulse coded passive SAW tad are applied. Increase in the number of the tags which are simultaneous interrogated by RFID reader causes collision for one or more group of code. It makes impossible to identify each tag. For this reason several approaches are applied:

1. Signal separation using matched filtering is the basic method for accomplishing the required signal based anti-collision. processing methods.
2. Time domain separation.
3. Spatial separation.
4. Processing technique based on the fractional Fourier transform (FrFT).

All common anticollision techniques and schemes are proposed in [3]. But all approaches are restricted by the information capacity which defines total number of tags which could be interrogated by RFID reader simultaneously. In order to enlarge the number of tags we propose specific topology of RFID SAW tag which aims to expand possibilities for anticollision processing signal approaches.

3. Temperature measurement of passive SAW RFID tag

The RFID SAW temperature sensors application helps to achieve an increase in the level of automation of control over the status of the main parts of the electrical installations and other electric power units, as well as to increase the level of efficiency in obtaining information on the mode of operation of power facilities and preventing supernumerary situations [4], [5].

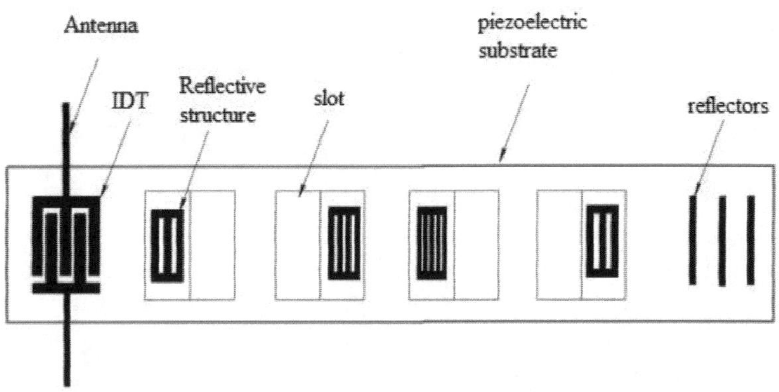

Figure 2. Simplified illustration of the SAW tag as the temperature sensor

IOP Conf. Series: Journal of Physics: Conf. Series **1008** (2018) 012015 doi:10.1088/1742-6596/1008/1/012015

Passive SAW anticollision temperature sensor with frequency-time code difference is presented in Figure 2 [6]. It consists of a piezoelectric substrate, where are consistently located a unidirectional broadband interdigital transducer (IDT) with antenna and four orthogonally frequency-coded reflecting structures which are located in the time slots. The OFC structures position defines the identification code of the temperature sensor [7], [8], [9]. The three consecutive reflectors make possible the obtaining information about temperature.

Slots are located at a certain distance *R*, and they have n variants of OFC structure position. The number of slots on the entire label is defined as *M*. In each slot, only one orthogonally frequency-coded structure can be placed. The reflection coefficients of the structures are determined by different frequencies that are not repeating for subsequent structures. Reflectors 6 are located at a distance *X* between themselves.

To obtain information about the temperature of the sensor, the reader emits an interrogation signal which is represented by expression:

$$S_{in}(t)=\left[\left(h\left(t-T_1^{'}\right)-h\left(t-T_2^{'}\right)\right)\cdot\left(A\cdot\left(\sum_{k=1}^{M}\sin\left(2\pi\cdot w_k t-\varphi_k\tau_k\right)\right)+A\cdot\sin\left(2\pi f_c+\varphi_0\right)\right)\right] \quad (1)$$

Where T_1' and T_2', the initial and final time of the interrogation signal, *A*- is the amplitude of the signal, w_k - frequency of *k*-th pulse, φ_k – phase of *k*-th pulse, *N*- is the number of pulses during the period of the interrogation signal duration, f_c – is the carrier frequency the last pulse in the interrogation signal, φ_0 – is the initial phase of the frequency f_c for the last pulse.

The incoming interrogated electromagnetic signal is converted by the IDT into the surface acoustic waves, which completely repeats the shape of the interrogation signal. Acoustic pulses propagate along the surface of the piezoelectric substrate. In the case where the carrier frequency of the *k*-th pulse coincides with the frequency that determines the reflection coefficient of the *k*-th reflecting structure - this pulse is reflected back to the IDT. The remaining pulses continue their propagation until they reach the reflecting structures in which the frequency that determines the reflection coefficient coincides with the carrier frequency of the acoustic pulse. The reflected pulses come back to the IDT, and thus the response of the identification tag is formed in the form of consecutive pulses.

A pulse with a carrier frequency f_c that does not coincide with any of the orthogonally coded structures 3 continues to propagate until it encounters an inhomogeneity in its path in the form of reflectors. The part of the acoustic wave is reflected back into the IDT 2, and the part continues to propagate to the next reflector 6, where part of the wave is reflected, and the part passes to the last reflector, reflected from it.

The response signal from the anticollision passive temperature sensor on surface acoustic waves can be represented as:

$$S_{out}(t)=\left(\sum_{k=1}^{M}\left[h(t-T'_{1k})-h(t-T'_{2k})\right]\cdot A_k\cdot\sin(2\cdot\pi\cdot w_k\cdot t-\varphi_k\cdot\tau_k)\right)+x(t) \quad (2)$$

Where: T'_{1k} and T'_{2k} are the initial and final time point of each pulse duration. Each pulse reflects from the orthogonally frequency-coded structure (OFC), A_k is the amplitude of the reflected *k*-th pulse of the OFC structure, w_k is the frequency of the *k*-th pulse of the OFC structure, φ_k is the *k*-th reflected pulse from the OFC structure, *M* is the number of response pulses from the surface acoustic wave of the tag, *x(t)* is the signal reflected from the reflectors 5.

The frequency values for the reflecting structures are selected from wk to $w_k = M$. Thus, from the IDT, a signal is sent to the RFID reader in the form of a sequence of delayed pulses. Each pulse has own carrier frequency. The last three response pulses, reflected from the reflectors, have the same carrier frequency.

The pulses which come from the reflecting structures 3 are delayed relative to each other in time proportional to the distance determined by the tag topology. Thus, the topology of the passive SAW tag allows obtain a time and frequency information for further time-frequency coding of each tag and its identification in the collision tags case. The signal reflected from the tag can be expressed as:

$$x(t)=\mathrm{Re}\left[\sum_{i=1}^{N} A_i h\left(t-\tau_c-\tau_{r_i}\right)\exp\left(-j\left(2\pi f_c\left(\tau_c+\tau_{r_i}\right)+\theta_{r_i}\right)\right)\exp\left(j\left(2\pi f_c+\varphi_0\right)\right)\right] \tag{3}$$

Where A_i is the attenuation coefficient of the amplitude due to propagation loss of acoustic waves, τc and τ_{ri} are the propagation delay of the electromagnetic wave and acoustic wave in free space, respectively, θri is the phase shift of the central frequency related to the reflection characteristics from the *i*-th reflector, f_c is the carrier frequency, φ_0 is the initial phase of the carrier frequency, *h(t)* is the envelope function.

The propagation delay of the acoustic wave τ_i and the phase delay φ_i for the *i*-th reflector can be represented as [10]:

$$\tau_i = \tau_c + \tau_{r_i} + \frac{\theta_{r_i}}{2\pi f_c} \tag{4}$$

$$\varphi_i = 2\pi f_c(\tau_c + \tau_{r_i}) + \theta_r \tag{5}$$

At different ambient temperatures, the propagation delay of the acoustic wave τ_i will change due to a change in the distance between the reflectors. The ambient temperature T is determined by means of reflectors 6 as:

$$T = \frac{\Delta\phi_{ij}}{TCD\phi_{ij,0}} + T_0 \tag{6}$$

Where $\Delta\phi_{ij}$ – is the phase difference, defined as: $2\pi f_c(\tau_j + \tau_i)$, TCD – temperature coefficient of delay, φ_{ij} is the difference between the phase delays between the reflections reflected from the *i*-th and *j*-th reflector of pulses corresponding to the *i*-th and j-th reflector, φ_{ij}, 0 design value at ambient temperature, T_0 - reference temperature [10].

4. Anticolision encoding of passive SAW RFID temperature sensors

The RFID SAW temperature sensors application helps to achieve an increase in the level of automation of control over the status of the main parts of the electrical installations and other electric power units, as well as to increase the level of efficiency in obtaining information on the mode of response pulse.

Let's consider the solution described in [5] where radio-frequency reader interrogates tags using the signal

$$S(t) = \left[h(t-T_1^{'}) - h(t-T_2^{'})\right]\cdot A\cdot\sum_{i=1}^{N}\sin\left(2\cdot\pi\cdot F_i\cdot t-\phi_i\right) \tag{7}$$

here $T_1^{'}$ and $T_2^{'}$ are the initial and finite times of the interrogation pulse, A is the pulse amplitude, F_i is the carrier frequency of the pulse, φi is the phase of the pulse, and N is the number of pulses for one period of the interrogation signal.

The tag's response is formed in the IDT in the form of a pulse sequence:

$$S(t) = \sum_{i=1}^{N}\left[h(t-T_{1i}^{'}) - h(t-T_{2i}^{'})\right]\cdot A_i\cdot\sin\left(2\cdot\pi\cdot F_i\cdot t-\phi_i\right) \tag{8}$$

where T_{1i}' and T_{2i}' are the initial and finite times of the response pulse.

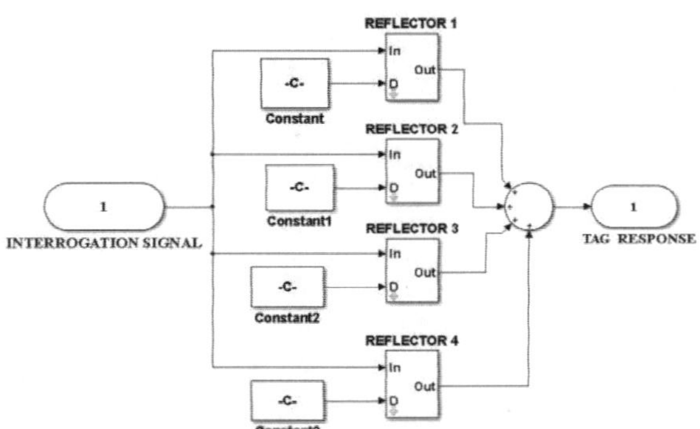

Figure 3. SAW RFID model as a time delay in the Simulink

IOP Conf. Series: Journal of Physics: Conf. Series **1008** (2018) 012015 doi:10.1088/1742-6596/1008/1/012015

This research takes a simple model with four slots and one reference reflective structure. This model uses time slots T_1–T_{12}, as shown in Figure. 5.

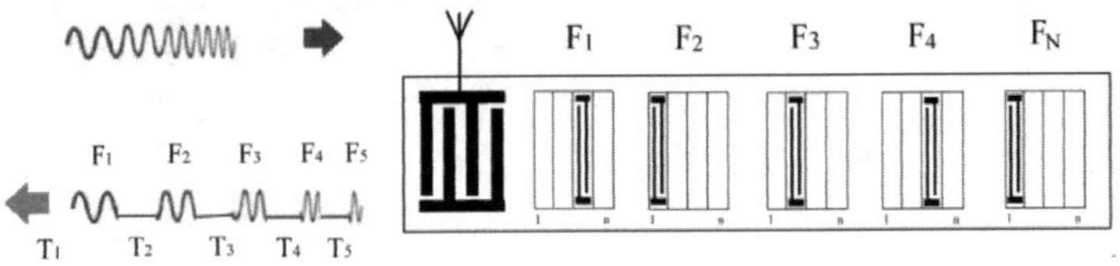

Figure 4. Passive tag with reflective structures in the slots

In this model, the position of the reflective structure in the time slot is matched with a certain frequency; it determines the binary code. The first slot has durations from T_1 to T_3, second from T_4 to T_6, third from T_7 to T_9, fourth from T_{10} to T_{12}. Time slots $T_{i1} = T_4 - T_2$, $T_{i2} = T_7 - T_6$, and $T_{i3} = T_{10} - T_9$, are set between slots containing reflective structures [11-15]. The time–frequency attributes allow build a time–frequency matrix for tag A, which is shown in Figure. 5.

	T_2-T_1	T_3-T_2	T_4-T_3	T_5-T_4	T_6-T_5	T_7-T_6	T_8-T_7	T_9-T_8	$T_{10}-T_9$	$T_{11}-T_{10}$	$T_{12}-T_{11}$
F_1	0	1	0	0	0	0	0	0	0	0	0
F_2	0	0	0	0	1	0	0	0	0	0	0
F_3	0	0	0	0	0	0	0	0	0	0	0
F_4	0	0	0	0	0	0	0	0	0	0	0

Figure 5. Time–frequency matrix for tag A

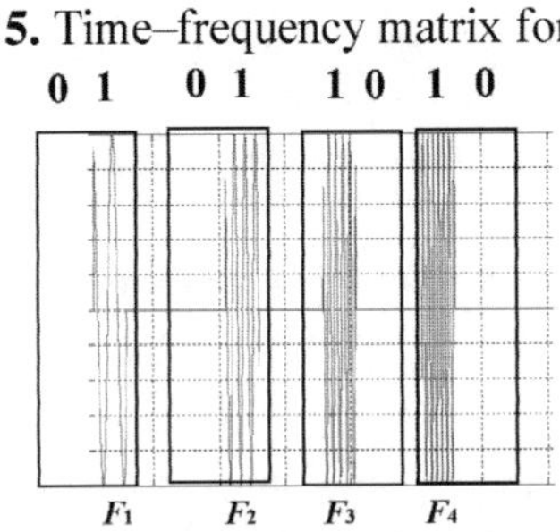

Figure 6. Response signal of the passive SAW tag A

Thus, using information about time and frequency of the first tag, the RFID reader forms the binary code A2 = 01001000000. Zeros fill spacing between slots. Placement of the reflective structure in a certain position denotes either "0" or "1." The reflected signal from tag A in the time domain is shown in Figure. 6.

Here, T_d is the pulse length, T_{slot} is the time slot boundary, T_s is the shift time of the reflective structure. Using time intervals and frequencies of pulses at the collision situation moment allows building the time–frequency matrix. Then, shifting both matrices in the time domain with a given step allows them to be separated.

	T_2-T_1	T_3-T_2	T_4-T_3	T_5-T_4	T_6-T_5	T_7-T_6	T_8-T_7	T_9-T_8	$T_{10}-T_9$	$T_{11}-T_{10}$	$T_{12}-T_{11}$	$T_{12}-T_{13}$
F_1	0	1	1	0	0	0	0	0	0	0	0	0
F_2	0	0	0	0	1	1	0	0	0	0	0	0
F_3	0	0	0	0	0	0	0	0	1	0	0	0
F_4	0	0	0	0	0	0	0	0	0	0	1	1

Figure 7. Time–frequency matrix of tags A and B

Similarly, we obtained the code of tag B_2 = 01001001001. Forming two time–frequency arrays *A* and *B* allows us to create a model of the tags' collision.

A shift of matrices *A* and *B* relative to each other with the pitch T_{slot} creates the *A* and *B* tag codes.

The code bits that trapped the separation intervals $T_{i1} = T_4 - T_3$, $T_{i2} = T_7 - T_6$, $T_{i3} = T_{10} - T_9$, $T_{i4} = T_{12} - T_{13}$, determine the place of the reflective structure of tag B. The presence of a reference pulse makes it possible to separate tags in the given frequency range from F_1 to F_2.

Specific passive SAW tag topology allows us to get time–frequency information for coding, which is represented in Equation 8. It provides the possibility to build a time–frequency matrix in the reader and measure temperature of each tag. The reader processes the response signals and executes the anticollision algorithm.

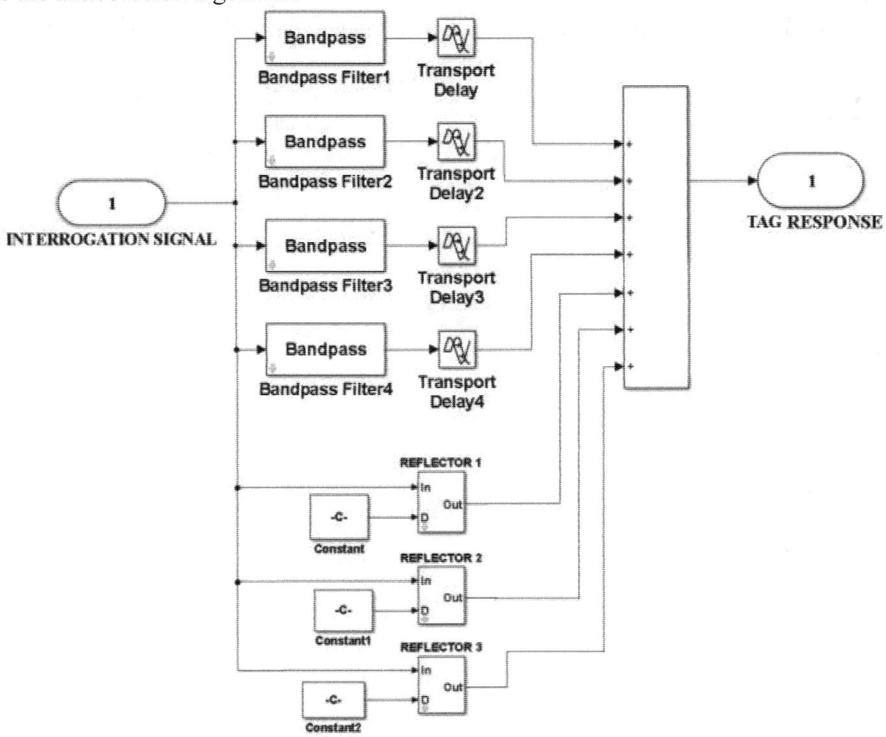

Figure 8. A simplified model of the passive SAW RFID temperaturey sensor with time–frequency information

For this case we consider the collision case between two passive sensors A and B, with the same number of reflecting structures and different identification codes.

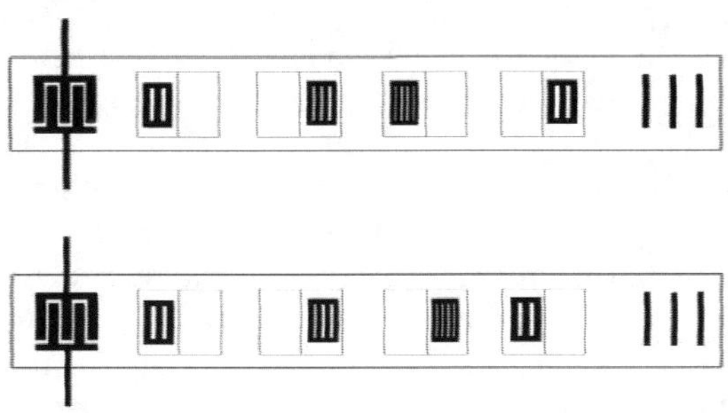

Figure 9. Collision case for two SAW temperature sensors.

As is shown in Figure. 9 [6], the first and the second structures of sensors A and B are placed in the same time slot places. In this way, these structures have different central frequencies but similar time delays. The reader receives reflected signals from the passive sensor, which contain information about frequencies and time delays. As a result, the reader creates matrix as is similar to the matrix which is illustrated in the Figure. 7.

The reader uses this information and makes a shift of columns in the time domain. This shift is equal to the $T_{12} - T_1$ interval. Thus, information about frequency and time delays allows highlighting the identification code of each tag in the collision case. For sensor *A*, the code is 01 10 10 10. For second, sensor B, the code is 01 01 10 01 [16, 17].

Thus, such an approach could be possible for consider in the temperature encoding of passive SAW *RFID* temperature sensors in collision case.

Reader measures temperature through time delay measurement of last three reflectors as it is shown in the expression (6).

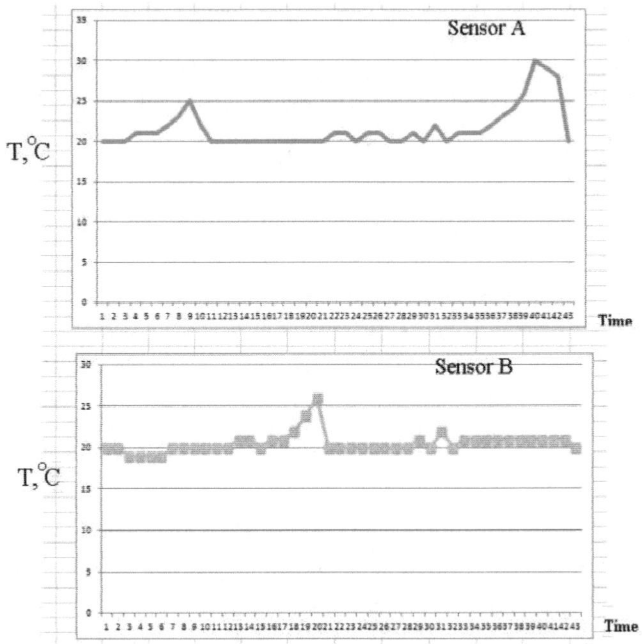

Figure 10. Temperature measurement of two passive SAW RFID sensors

5. Results

For results illustration we review of application anticollision identification approach for temperature measurement in the switchgears and other high-voltage devices. SAW RFID based on devices allows us to carry out monitoring of temperature. Temperature alters correlation properties of the tag response. As a rule, response signal modulation allows us to get a temperature value from each tag. As an example, we consider the innovative technology of wireless passive temperature sensors, which have already been used for a real-time online monitoring system. Passive wireless temperature sensors are installed in the contact point of the distribution equipment to measure the contact temperature. An antenna is installed in the line of sight from the sensors to provide wireless communication with the temperature sensor. The temperature reader is placed in the compartment of the automatic devices. If the sensor temperature exceeds the specified threshold level, the reader indicates the event on the LED panel and sends a message to the automation system by the commutation relays. The reader transmits the temperature data to the industrial computer by an *RS*-485 interface. The industrial computer is connected to a local network. In this case, the operator can view and analyze the temperature data by web access.

Let's consider the solution described in [5] where radio-frequency reader interrogates tags using the signal from the passive SAW tag comes to the reader.

After shifting the columns of the matrix, we get identification codes (Figure 11). Thus, we can separate the labels from each other using time-frequency information. As a result of the separation of the identification codes, we use the information on the modulation of the signals of each tag, that is, obtain the temperature value of each tag [5]. Therefore, for the collision case, this solution can be implemented in monitoring systems and control the temperature of high-voltage parts of electric power equipment.

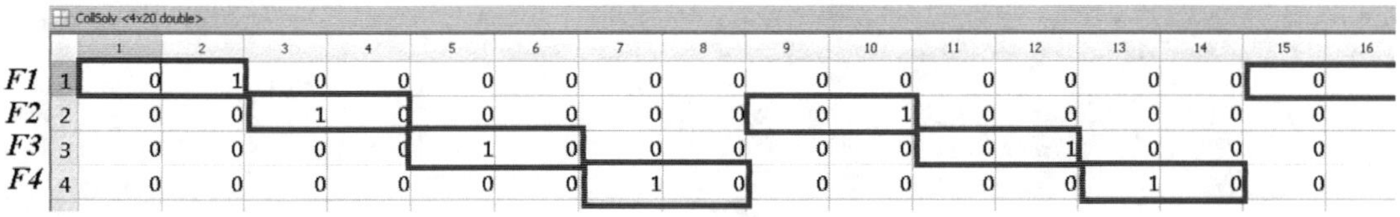
ColSolv <4x20 double>

		1	2	3	4	5	6	7	8	9	10	11	12	13	14	15	16
F1	1	0	1	0	0	0	0	0	0	0	0	0	0	0	0	0	
F2	2	0	0	1	0	0	0	0	0	0	1	0	0	0	0	0	
F3	3	0	0	0	0	1	0	0	0	0	0	0	1	0	0	0	
F4	4	0	0	0	0	0	0	1	0	0	0	0	0	1	0	0	

Figure11. Matrices with IDs of two RFID SAW sensors. Due to specific topology RFID reader gets ID of each sensor and determines temperature in collison case

Based on the modern existing solutions in the field of RFID SAW technology we consider using intelligent sensors for monitoring the temperature of the switchgear. It makes possible to increase the number of simultaneously interrogated sensors. Our anticollision approach increases the reliability and accuracy of the system, that makes it attractiveness for large-scale application in the electricity industry.

The innovative technology of wireless temperature sensors is already used today in the real-time monitoring systems [5, 18, and 19]. Passive temperature sensors are installed at the contact point of the distribution equipment for measuring the contact temperature.

The reader's antenna is installed at a distance of up to 10 meters from the sensors to provide wireless communication. The RFID SAW temperature sensor reader is placed in the switchgear cell. If the temperature sensor exceeds the specified threshold level, the reader indicates an event on the LED panel and sends a message to the automation system using switching relays. The reader transmits temperature data to the industrial computer via the RS-485 interface. The industrial computer is connected to the local network. In this case, the operator sees and analyzes the temperature data using web access (Figure. 12).

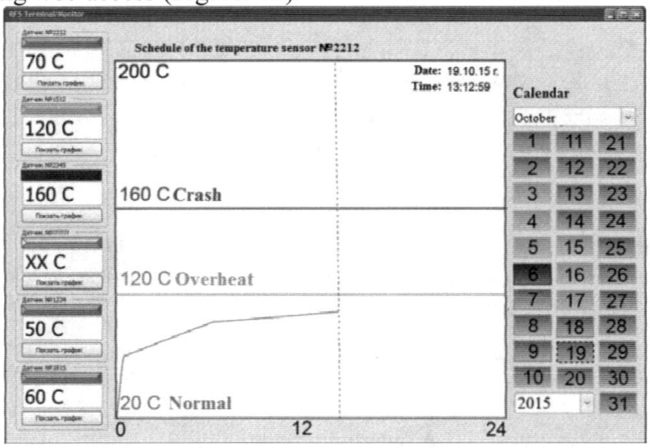

Figure 12. PC software interface which illustrates temperature measurement for several sensors

Considered system is used for wireless monitoring and temperature monitoring. The offered approach allows solving a tag collision problem at the reading of codes and temperature values in multisensory systems, which are applied for the control of temperature in the electric equipment.

The collision problem is illustrated by the fact that in one cell of the switchgear there can be more than a dozen sensors and their temperature varies from -40 ° C to 200 ° C (Figure 12, 13, 14).

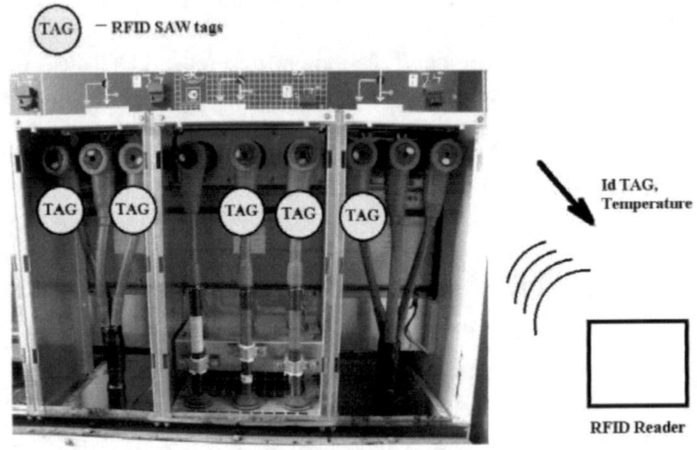

Figure 13. The example of SAW RFID temperature sensors placement in the switchgear

Figure 14. SAW sensors which are placed in the switchgear

The proposed technology, which has the industrial name "*RFSens*", increases the number of simultaneously sampled sensors, the temperature range and the accuracy of this parameter. This becomes especially relevant for equipment with a large number of bends. Sensors for SAW technology are absolutely passive and consist of a metal antenna and a chip. The resulting solution is very important for the energy market.

In general, such systems use only one sensor for wireless monitoring. Our solution expands the possibilities for RFID wireless temperature monitoring systems based on the passive SAW sensors [5]. SAW RFID tags with time–frequency information provide the possibility of increasing number of passive temperature sensors for one switchgear. In this case, we could potentially use at least seven tags.

With compare general anticollison approaches like signal separation using correlation methods and time-spatial separation we assume that suggested anticollison algorithm significant increases information capacity each tag and enlarge the number of possible interrogated tags.

Thus, the temperature measurement example explains real-world measurement results. In addition, it illustrates the basic idea of how to realize our solution in practice.

6. Conclusion

In this paper, we propose a model that allows solving the collision problem using the frequency–time separation of passive SAW temperature sensor. Specific reflective OFC structures, which are placed in time slots on the surface of the piezoelectric substrate, make it possible to get a unique identification code, which contains time–frequency information. Reflectivity structure positioning and its placement in the time slots determine the total number of codes and decrease the possibility of incorrect identification.

Our approach is based on a combined coding method including orthogonal-frequency coding and encoding of the pulse time position. The combined coding method allows us to increase the number of temperature that are interrogated simultaneously in the electric power system and switchgears.

Anti-collision algorithm increases the number of any passive SAW RFID wireless smart sensors due to increasing of the number of variants for placing reflective structures in time slots, the number of slots, as well as the technological capabilities of producing SAW devices and the dimensions of the piezoelectric material substrate determine the number of identified sensors.

Our research covers the application of the SAW RFID temperature sensors in temperature measurement in the electrical industry.

Acknowledgment
The authors wish to acknowledge continuing support from RF SAW R&D Center in St. Petersburg, Russia, which operates with SAW RFID technology in the smart sensors domain. This company is developing smart wireless passive temperature sensors based on SAW RFID technology.

The authors express thanks to all the current and past colleagues for their varied contributions to this research.

References

[1] Jalal A S 2015 Passive RFID Tags *Wulfenia Journal* **22** 12 415-35
[2] Plessky V P and Reindl L M 2010 Review on SAW RFID tags *Proc.IEEE Trans Ultrason Ferroelectr Freq Control* **57** 3 654-68
[3] Karmakar NC, Koswatta R, Kalansuriya P and Rubayet E 2013 Chipless RFID reader architecture *Artech House*
[4] Malocha D C, Gallagher M, Fisher B, Humphries J, Gallagher D and Kozlovski N Passive 2013 Wireless Multi-Sensor SAW Technology Device and System Perspective *Sensors* **13** 1 1-27
[5] Sorokin AV and Shepeta AP Anti-collision radio-frequency identification system using passive SAW tags *InSmart Sensors, Actuators, and MEMS VIII* **10246** 1024613 International Society for Optics and Photonics
[6] Podoplekin Y F, Sorokin A V and Shepeta A P 2017 Anticollision passive radio frequency tag with time–frequency information *Marine Radio Electronics* **1** 59 41
[7] Sharma M and Singh S K Orthogonality Measurent of OFDM Signal *Indonesian Journal of Electrical Engineering and Computer Science* **9** 3 595-598
[8] Rafiqul I M, Rafiq S, Yasmin MS and Habaebi MH 2017 A 2X2 MIMO Patch Antenna for Multi-Band Applications *Indonesian Journal of Electrical Engineering and Informatics (IJEEI)* **5** 4 383-389
[9] Habaebi M, Rosli R and Islam M R 2017 RSSI-based Human Presence Detection System for Energy Saving Automation *Indonesian Journal of Electrical Engineering and Informatics (IJEEI)* **5** 4 339-350
[10] Kang A, Zhang C, Ji X, Han T, Li R and Li X 2013 SAW-RFID enabled temperature sensor. *Sensors and Actuators A: Physical* **201** 105-113
[11] Harma, S., Plessky, V. P. "Surface Acoustic Wave RFID Tags," Development and Implementation of RFID Technology 1(1), 145-158 (2009).
[12] Plessky VP and Reindl LM 2010 Review on SAW RFID tags *IEEE transactions on ultrasonics, ferroelectrics, and frequency control*, **57** 3 654 - 68
[13] Harma S, Arthur W G, Hartman C S, Maev R G and Plessky V P 2008 Inline SAW RFID Tag Using Time Position and Phase Encoding *Proc. IEEE Transactions on Ultrasonics, Ferroelectrics and Frequency Control* **55** 8 145-158
[14] Binder G 2011 SAW Transponder–RFID for Extreme Conditions *Deploying RFID – Challenges, Solutions and Open Issues* **1** 1 303-318
[15] Stelzer A, Scheiblhofer S, Schuster S and Brandl M 2008 Multi-Reader/Multi-Tag SAW RFID Systems combining Tagging, Sensing, and Ranging for Industrial Applications *IEEE Frequency Control Symposium* **1** 2 263-272
[16] Koygerov A S and Dmitriev V F 2010 Radiomarker on surface acoustic waves with noise-proof frequency-manipulated code *Proc. Modeling of processes and systems Information management systems* 4
[17] Sorokin A V, Shepeta A P and Smirnov Y G Passive anticollision radiofrequency identification tag on surface acoustic waves with frequency–time difference Patent 2616342
[18] Malocha DC, Puccio D and Gallagher D 2004 Orthogonal Frequency Coding for SAW

Device Applications *In Proceedings of the IEEE Ultrasonics Symposium, Montreal, Cana-da* 1082–1085.

[19] Malocha D C, Gallagher D and Hines J 2004 SAW Sensors Using Orthogonal Frequency Cod-ing *In Proceedings of the 2004 IEEE International Frequency Control Symposium and Exposi-tion, Montreal, Canada* 307–310

Image encryption based on pixel bit modification

Kiswara Agung[1], Fatmawati[2], Herry Suprajitno[2]
[1]Mathematic Department, University of Jember, Jember, Indonesia
[2]Mathematic Department, AirlanggaUniversity, Surabaya, Indonesia

E-mail: kiswaras@gmail.com

Abstract. There are currently some publications about image encoding without changing pixel values. One of them encodes the image by dividing the image into blocks and then creating permutations and combinations into block positions. The disadvantage of this method is a grayscale histogram of an encoding image similar to a grayscale histogram of a plain image. This happens because the number of grayscale for the light intensity of the plain image is the same as the grayscale for the light intensity of the encoded image. In this article we propose image encoding by manipulating pixel bits. The result of this research is an image coding method where plain images and encoded images have big differences on grayscale histogram and visual.

1. Introduction

Nowadays, researchers about image encoding more focused on several aspects of which: [1] make permutations on some parts of the image, [5] transpose some parts of the image, [2] make combination on some parts of the image, [6] encode the image using affine transformations. [4] Some publications introduce an image encoding method by randomizing the position of pixels. [2] The position of pixels is randomized using permutation and combination processes. [5] divides the image into blocks, then the block position is scrambled to encode the image. [3] Scramble the pixel position on each layer of the colored image.

Image encoding performed by [2, 5, 10] produces a good visual encoding image but the weakness of this method is the greyscale histogram of plain and encoded images having the same shape because the number and pixel values are unchanged for the image before and after encoding. This weakness will be used by hackers as the entrance to destroy or manipulate the existing images. In statistical analysis, the difference between the plain image and the encrypt image can be seen from the mean sequare error (MSE) value, coeficient corelation, dan peak signal noise relation (PSNR).

In this article we offer an image encoding method by modifying its pixel value. We will manipulate the pixel value by redeeming between least significant bits (MSB) and least significant bits (LSB) in each pixel. The result of this method, the original image and the coded image have a significant difference both visually and from the histogram form. This happens because the number of grayscale of plain image and encrypt image is different, although the pixel position is no different.

2. Research Methods

2.1 Digital Image

Image data and text data is very different because an image contains very large data, and all data are interconnected one and the other. The image data also contains considerable data repetition [7]. The concept of the difference between text data and image data can be shown in the table below.

Table 1. Text and Image Encoding Differences

Type	Secret Data	Encrypted Data	Remarks
Text	"CSEMCKVIE"	"DTFNDLWJF"	Completely different

Image (RGB)	 Pixel (P_1)= (24,45, 233)	 Pixel (P_2)= (10, 65, 198)	P_1 and P_2 are completely different, but visually they look similiar
Image (Gray)	 Pixel (P_1)= (87)	 Pixel (P_1)= (114)	P_1 and P_2 are completely different, but visually they look similiar

A digital image consists of a number of elements, each element occupying a certain position and has a value. This element is usually called the picture element or pixel. [8]. An image is defined as a two-dimensional function *f*(*x, y*) where *x* and *y* are coordinate pairs. The value of *f*(*x, y*) is the light intensity of pixels in coordinates (*x,y*) and usually it's called grayscale. The value of a pixel can be converted to 8 binary digits (bits). 4 digits of the first are called the LSB (Least Significant Bit), where the change of value in this position will not give a significant change to the image. 4 digits of the second are called MSB (Most Significant Bit), where the change of value in this position will have a significant impact on the image [9]. The following figure shows the position of a value of bit.

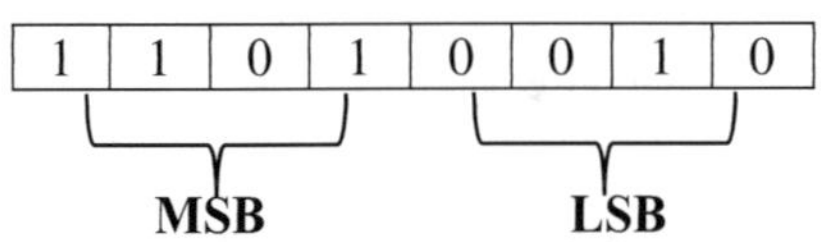

Figure1. MSB,LSB Interpretation

The maximum deviation of an image can be searched by creating a grayscale histogram and calculating the area. The smaller the deviation indicates that the coding results the better. To find the area of the histogram image can be searched by the formula: [3

$$L = \frac{h_0 + h_{255}}{2} + \sum_{i=1}^{254} h_i$$

L = deviation area

h_i = number of pixel that have different i

i = pixel value

The correlation coefficient of an image shows how the relationship between pixels that adjacent each other.The formula for determining the correlation coefficient of an image is [6]:

$E(x) = \frac{1}{N}\sum_{i=1}^{N} x_i \qquad E(y) = \frac{1}{N}\sum_{i=1}^{N} y_i$

$$CC = \frac{cov(x,y)}{\sigma_x \sigma_y}$$

$$= \frac{\sum_{i=1}^{N} \frac{x_i - E(x)}{y_i - E(y)}}{\sqrt{\sum_{i=1}^{N}(x_i - E(x))^2}\sqrt{\sum_{i=1}^{N}(y_i - E(y))^2}}$$

x_i = the pixel of plain image at i position
y_i = the pixel of encrypt image at i position
N = total of pixel

From the results of correlation coefficient analysis found that if the value of CC smaller then it is said that the results of coding is better. If $CC = 1$ it is said both identical image or encoding fail. If $CC = 0$ it's said the two images are perfectly different.The correlation of image is the relationship between pixel pairs that are adjacent to each other vertically, horizontally or diagonally [1].

2.2 Image Encryption Algorithm by Pixel Modification

In this article we propose an image coding method based on pixel modification. The modification of the pixels that we propose is the exchange between the most significant bits (MSB) and least significant bits (LSB). Because of the type of this image encoding is symmetry, so the encryption procedure as same as the decryption procedure. The following is the proposed image encoding algorithm:

1. Enter the initial image with the known pixel value
2. Convert the pixel value to 8 bits.
3. Exchange between MSB and LSB of each pixel

a. Exchange bits 1 and bit 5

b. Exchange bits 2 and bit 6

c. Exchange bits 3 and bit 7

d. Exchange bits 4 and bit 8

4. Apply step 2 to all pixels in the image
5. End

The above algorithm applies to encoding and decoding process. The algorithm of this method can be presented in the following flowchart form:

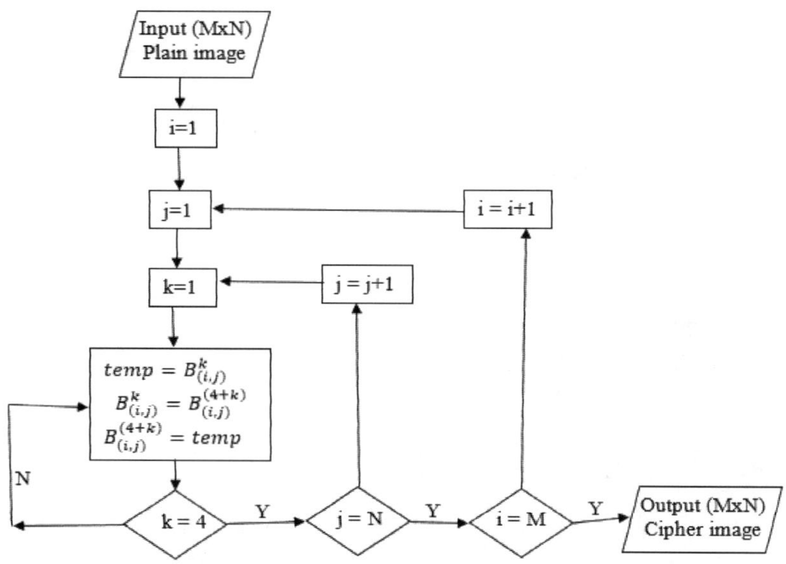

Figure 2. Flowchart Image Encoding

M	= width of image
N	= length of image
K	= bit position
(i,j)	= pixel position

$B^k_{(i,j)}$ – bit to k of pixel on the position (i,j)

For example, if known pixel data as follows:

Table 2. Pixel Data

Layer	Original Image				Encrypt Image			
	Value	**Bit**	**MSB**	**LSB**	**MSB**	**LSB**	**Bit**	**Value**
Red	69	01000101	0100	0101	0101	0100	01010100	84
Green	186	11011110	1101	1110	1110	1101	11101101	237
Blue	215	11010111	1101	0111	0111	1101	01111101	125

Based on the example above, we know that to encode an image only need to exchange between most significant bits (MSB) and least significant bits (LSB). This method is used for encryption or decryption of an image, so it is symmetry cryptography.

3. Results and Discussion

To test this algorithm, we do an experiment . This test is performed using a laptop with the I3 core microprocessor specification and Microsoft Windows 10 operating system .Program written using MATLAB R2016b and good quality image as the data. The results of this algorithm can be shown in the figure below. We use balloon images as data and histogram of grayscale for comparison

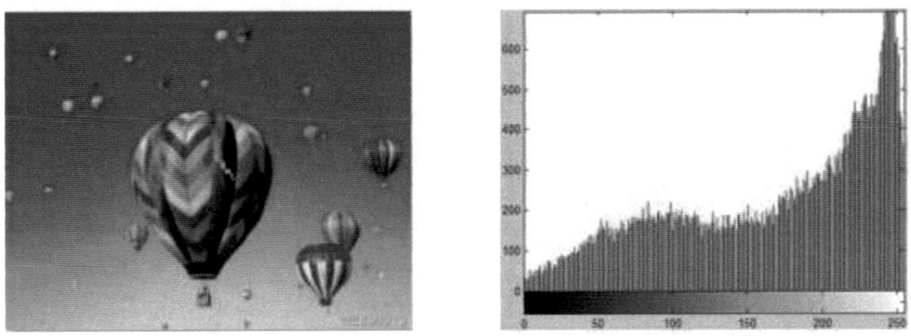

(a) (b)

Figure 3. a) Plain Image, b) Histogram Plain Image

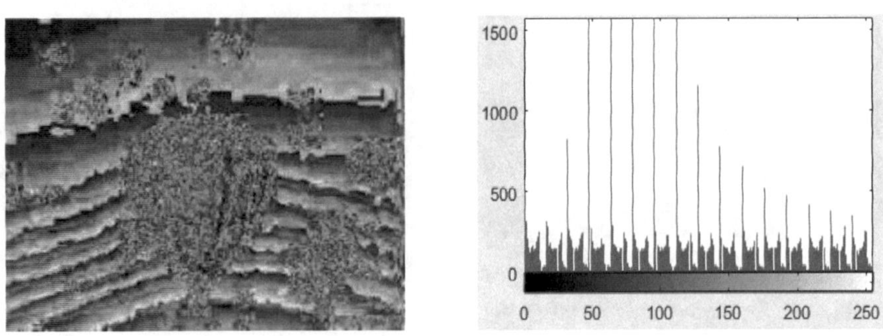

(a) (b)

Figure 4. a) Encrypt Image, b) Histogram Encrypt Image

From the visualization results, the image 3a and image 4a above show that the image encoding results are very difficult to interpretation. We can see that density of histogram plain image(figure.3b) is greater than the density of histogram encrypt image (figure.4b). It shows that there is a significant difference between the original image and image of the encoding. In the histogram of figure 3b and 4b above shows that the intensity of the encrypt image as not same as the plain image, it cause the image to be more bad. In addition to analyzing the histogram, we also analyze the correlation between adjacent pixels, by vertically,horizontally and diagonally. Here we choose 2000 pairs of adjacent pixels and we calculate the correlation coefficient.

Table 3. Koefisien korelasi antara plainimage/cipherimage

Proximity Pixel	Plain Image	Encrypt Image
Horizontal	0.82599	0.82589
Vertical	0.99076	0.44643
Diagonal	0.97199	0.31143

There is a difference of correlation coefficient between two adjacent pixels diagonally using our method is 0.97199 (plainimage) and 0.31143 (encrypt image). Similar results also occur in adjacent vertical and horizontal pixels, as shown in Table 3. It is apparent in Table 3. that the difference in coefficient of correlation between two diagonally adjacent pixels is significant enough to cause the two images to differ significant visually.

4. Conclusion

The proposed method produces significant grayscale histogram differences between the original image and the image after encoding so it can correct the weakness of some existing methods before. This method has been tested on many images and the result is very satisfactory because it has a high level of security and quite worthy of being called a good security system. This is demonstrated by the significant differences between plain images and both visual and histogram encoding images, in addition there are also differences in correlation coefficients between two adjacent pixels either vertically, horizontally or diagonally. We hope the next research not only analyze the histogram but also the MSE and PSNR so that the level of security to be better

References

[1] Dixit A, Drhuve P, Bhakwan D 2012Image Encryption Using Permutation *Computer Science & Information Technology* **6** 1-9
[2] Younes M A B and Jantan A 2008 An Image Encryption Approach Using a Combination of Permutation Technique Followed by Encryption *IJCSNS International Journal of Computer Science and Network Security* **8** 28-40
[3] Chang 2004 Gray-Level Image Encryption Scheme Using Full Phase Encryption and Phase Encoded Exclusive OR Operation *Optical Review* **11** 1 34 - 37
[4] Fhishawy 2007 Quality of Encryption Measurement of Bitmap Image with RC6, MRC6, and Rijndael Block Cipher Algorithm *International Journal of Network Security* **5** 3 241-251
[5] Kester Q A 2013 Image Encryption based on the RGB PIXEL Transposition and Shuffling *International Journal of Computer Network and Information Security* 43-50
[6] Kori P, Dubey P and Richhariya V 2015 Double Phase Image Encryption and Decryption Using Logistic Tent Map and Chaotic Logistic Map *IJSART International Journal and Research Technology* **1** 11 33-39
[7] Majumdar S 2014 An Analytical Survey on Different Secured Image Encryption Techniques *IJCAT International Journal of Computing and Technology* **1** 8 396-403
[8] Sakthidasan 2011 A New Algorithm for Image Encryption and Decryption of Digital Color Image *International Journal of Information and Education Technology* **1** 72-86

IOP Conf. Series: Journal of Physics: Conf. Series **1008** (2018) 012016 doi:10.1088/1742-6596/1008/1/012016

[9] Santoso K A, Fatmawati and Suprajitno H 2017 Image Encryption Technique Based on Pixel Exchange and XOR Operation *Proceeding of International Basic Science Conference* 286-288

[10] Zhou W and Alan B 2009 Mean Squared Error: A New Look At Signal Fidelity Measures *IEEE Signal Processing Magazine* 98-117

Stock price estimation using ensemble Kalman Filter square root method

D F Karya[1], P Katias[1], T Herlambang[2]
[1]Management Department, University of Nahdlatul Ulama Surabaya, Indonesia
[2]Information System Department, University of Nahdlatul Ulama Surabaya, Indonesia

E-mail: denisfk@unusa.ac.id

Abstract. Shares are securities as the possession or equity evidence of an individual or corporation over an enterprise, especially public companies whose activity is stock trading. Investment in stocks trading is most likely to be the option of investors as stocks trading offers attractive profits. In determining a choice of safe investment in the stocks, the investors require a way of assessing the stock prices to buy so as to help optimize their profits. An effective method of analysis which will reduce the risk the investors may bear is by predicting or estimating the stock price. Estimation is carried out as a problem sometimes can be solved by using previous information or data related or relevant to the problem. The contribution of this paper is that the estimates of stock prices in high, low, and close categorycan be utilized as investors' consideration for decision making in investment. In this paper, stock price estimation was made by using the Ensemble Kalman Filter Square Root method (EnKF-SR) and Ensemble Kalman Filter method (EnKF). The simulation results showed that the resulted estimation by applying EnKF method was more accurate than that by the EnKF-SR, with an estimation error of about 0.2 % by EnKF and an estimation error of 2.6 % by EnKF-SR.

1. Introduction

Shares are units of value or bookkeeping in various financial instruments refering to the ownership shares of a company [1]. The issuence of shares may enable companies that require long-term funding to sell their interests in business in return for cash. The fluctuation of stock value can be affected by various factors such as interest rate, investor psychology, political factor, and natural disaster [2]. The formation of stock prices occurs due to the demand and supply of the shares. The stock price index is an indicator that shows the value of the stock price. The value of the stock may fluctuate either in the form of increase or decrease. So investors should be able to predict whether stock prices are increasing or decreasing [3]. One method is the method of estimating the increase and decrease in stock prices. Estimation are made as a problem can very often be solved by using previous information or data related or relevant to the problem [3]. Kalman filter is a method of estimating state variables of a discrete linear dynamic system that minimizes the estimation error covariance [4].

Kalman *filter* was first introduced by Rudolph E. Kalman in 1960. It was about problem solving on linear discrete data filtering. Kalman filter methods are often used in robotics such as for estimation of missile trajectory [5], mobile robot trajectory [6] and Maglev ball position estimation [7]. A solution relies on the problem of linear discrete data filtering. However, in actual circumstances, there is often a continuous nonlinear dynamic system that requires another approach which is an extension of the Kalman filter called Ensemble Kalman Filter (EnKF). In the EnKF method, the algorithm is executed by generating a certain number of ensembles to calculate the mean value and the error covariance of the state variable [3]. The development of the EnKF method through modification of algorithm by adding a square root scheme at EnKF correction stage which can result in Ensemble Kalman Filter Square Root (EnKF-SR) method. This paper studied the implementation of the Squares Root Scheme on the Ensemble Kalman Filter (EnKF) method on the Stock Function obtained from Software Mathematica which is then applied to estimate the Stock Function simulated with Matlab software so as to generate error ratio of EnKF and EnKF-SR methods for close price, high price, and low price of stocks. The contribution of this paper is that the estimates of stock prices in high, low, and close category can be utilized as investors' consideration for decision making in investment.

2. Ensemble Kalman Filter Square Root (EnKF-SR)

This section presents EnKF-SR algorithm to estimated nonlinear or linear dynamic system and measurement model, the algorithm *Ensemble Kalman Filter Square Root* (EnKF-SR) can be seen [5]:

Model system and measurement model

$$x_{k+1} = f(x_k, u_k) + w_k \tag{1}$$

$$z_k = Hx_k + v_k \tag{2}$$

$$w_k \sim N(0, Q_k),\ v_k \sim N(0, R_k) \tag{3}$$

1. Initialization

Generate N *ensemble* as the first guess $\bar{x}_0$

$$x_{0,i} = [x_{0,1} \quad x_{0,2} \quad \cdots \quad x_{0,N}] \tag{4}$$

The first Mean *Ensemble*: $\bar{x}_{0,i} = x_{0,i}\, 1_N$ (5)

The first *Ensemble* error:

$$\tilde{x}_{0,i} = x_{0,i} - \bar{x}_{0,i} = x_{k,i}(I - 1_N) \tag{6}$$

2. Time Update

$$\hat{x}_{k,i}^{-} = f(\hat{x}_{k,-1,i}, u_{k-1,i}) + w_{k,i} \tag{7}$$

where $w_{k,i} = N(0, Q_k)$

Mean *Ensemble* : $\bar{x}_{k,i}^{-} = \hat{x}_{k,i}^{-}\, 1_N$ (8)

Error *Ensemble* :

$$\tilde{x}_{k,i}^{-} = \hat{x}_{k,i}^{-} - \bar{x}_{k,i}^{-} = \hat{x}_{k,i}^{-}\,(I - 1_N) \tag{9}$$

3. Measurement Update

$$z_{k,i} = Hx_{k,i} + v_{k,i} \tag{10}$$

where $v_{k,i} \sim N(0, R_k)$

$S_k = H\tilde{x}_{k,i}^{-}$, $E_k = (v_1, v_2, \ldots, V_N)$ and

$$C_k = S_k\, S_k^T + E_k\, E_k^T \tag{11}$$

Mean Ensemble :

$$\bar{x}_{k,i} = \bar{x}_{k,i}^{-} + \tilde{x}_{k,i}^{-} S_k^T C_k^{-}\left(\bar{z}_{k,i} - H\bar{x}_{k,i}^{-}\right) \tag{12}$$

Square Root Scheme:

- eigenvalue decompotition from $C_k = U_k\, \Lambda_k U_k^T$ (13)
- determine matrix $M_k = \Lambda_k^{\frac{1}{2}}\, U_k^T\, S_k^{-}$ (14)
- determine SVD from $M_k = Y_k\, \mathrm{L}_k\, V_k^T$ (15)

Ensemble Error: $\tilde{x}_{k,i} = \tilde{x}_{k,i}^{-}\, V_k\, (I - \mathrm{L}_k\, L_k^T)^{\frac{1}{2}}$ (16)

Ensemble Estimation: $\hat{x}_{k,i} = \tilde{x}_{k,i} + \bar{x}_{k,i}$ (17)

To evaluate of estimation result accuracy from EnKF and EnKF-SR algorithm, can be show with calculate Root Mean Square Error (RMSE) [8].

$$RMSE = \sqrt{\frac{\sum_i^n \left(X_{obs,i}(k) - X_{model,i}(k)\right)^2}{n}} \tag{18}$$

With

$X_{obs,i}(k)$ = observation data n = iteration

$X_{model,i}(k)$ = model data

IOP Conf. Series: Journal of Physics: Conf. Series **1008** (2018) 012017 doi:10.1088/1742-6596/1008/1/012017

3. Computational Result

This simulation made applying the EnKF and EnKF-SR algorithms to the stock functions obtained from Mathematical software simulation showed the stock data in Table 1. The simulation results were evaluated and compared to the established stock functions, and the stock functions for high, low and close prices in equation (19) - (21) are as follows:

$$f_{high}(x) = 11{,}856x^2 - 503{,}089x + 8010{,}62$$

$$f'_{high}(x) = 23{,}71x - 503{,}089 \tag{19}$$

$$f_{low}(x) = 53{,}51x^2 - 877{,}946x + 3875{,}59$$

$$f'_{low}(x) = 107{,}02x - 877{,}946 \tag{20}$$

$$f_{close}(x) = 44{,}2843x^2 - 627{,}876x + 7868{,}65$$

$$f'_{close}(x) = 88{,}5686x - 627{,}876 \tag{21}$$

Because the system requires discretitation, so the stock functions model in equation (19) – (21) must be discreted using the finite difference method. Equation (19) and (21), If f_{high_k} functional of high price stock and f_{low_k} functional of high price stock and f_{close_k} functional of close price stock

$$f_{high} = f_{high_k}, f_{low} = f_{low_k}, f_{close} = f_{close_k} \tag{22}$$

The change of state variables respect to the time are approximated by forward scheme of finite difference. Thus we will get

$$\dot{f_{high}} = \frac{df_{high}}{dt} \approx \frac{f_{high_{k+1}} - f_{high_k}}{\Delta t} \tag{23}$$

$$\dot{f_{low}} = \frac{df_{low}}{dt} \approx \frac{f_{low_{k+1}} - f_{low_k}}{\Delta t} \tag{24}$$

$$\dot{f_{close}} = \frac{df_{close}}{dt} \approx \frac{f_{close_{k+1}} - f_{close_k}}{\Delta t} \tag{25}$$

from equation (23) and (25) will be gotten the modified the stock functions model in (19)-(21) below

$$\begin{bmatrix} f_{high_{k+1}} \\ f_{low_{k+1}} \\ f_{close_{k+1}} \end{bmatrix} = \begin{bmatrix} (23{,}71x_k - 503{,}089)\Delta t \\ (107{,}02x_k - 877{,}946)\Delta t \\ (88{,}5686x_k - 627{,}876)\Delta t \end{bmatrix} \tag{26}$$

Stock data for high, low and close price in Table 1.

Table 1. Stock Data on for High, Low, Close Price

Month	High	Low	Close
1	7225	6875	6925
2	6950	6575	6625
3	6700	6150	6300
4	6425	5800	5850
5	6200	5700	5725
6	5700	4500	4900
7	4900	4495	4530

Month	High	Low	Close
8	4600	3870	3950
9	4245	3760	3760
10	4000	3725	3800
11	3900	3425	3425
12	3545	3400	3470
13	3500	3000	3125
14	3200	2885	3000
15	3400	2975	3235
16	4000	3150	3930
17	5075	3950	4290
18	4410	4030	4400
19	5500	4250	5275
20	6700	5275	6100
21	6700	5950	6150
22	6675	6025	6525
23	6550	5850	5950
24	6750	5850	6750
25	7100	6200	7000

This simulation consists of three simulations of 100, 200 and 300 ensembles applied for estimation of the water level and the steam temperature. The value of Δt used was $\Delta t = 0.1$ and the iteration was 150. Figure 1, 2 and 3 show the results of high, low and close stock price estimates by generating 300 ensembles. Figures 1, 2 and 3 show very accurate estimation results with an error of about 0.2 and accuracy of 98%. Figure 1 shows that the estimation result by the EnKF method was more accurate than that by the EnKF-SR method with a smaller error. Figure 2 and 3 also show that the EnKF method was more accurate than the EnKF-SR method.

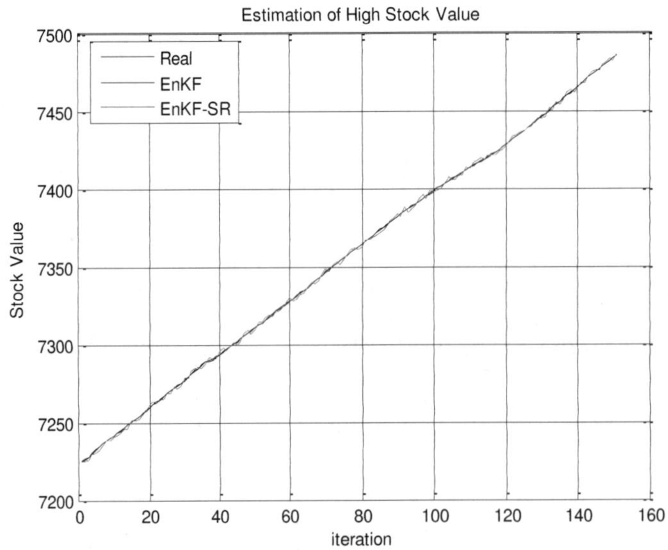

Figure 1. Estimation of High Stock Value using 300 Ensembles

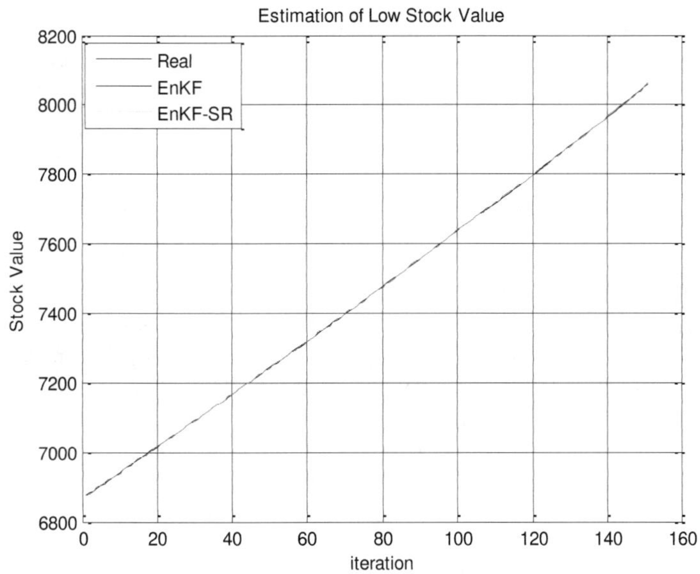

Figure 2. Estimation of Low Stock Value using 300 Ensemble

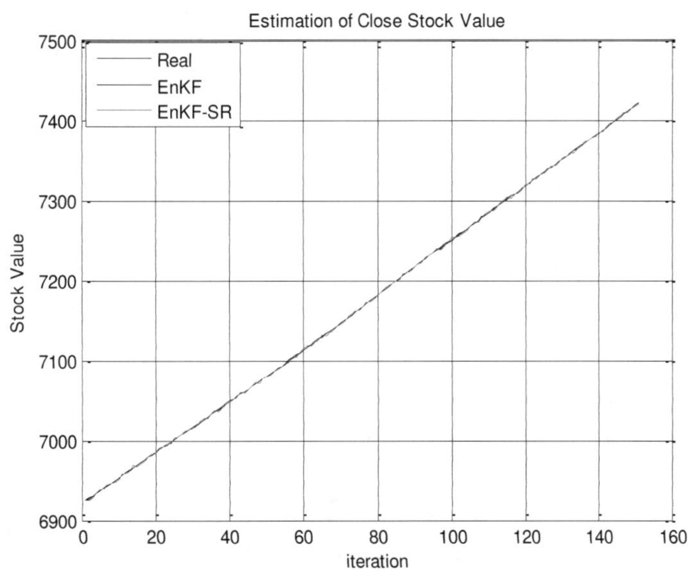

Figure 3. Estimation of Close Stock Value using 300 Ensemble

Then, comparison of estimation results with 100, 200 and 300 ensembles was done. Table 2 shows that with the 300 ensembles the most accurate results were in the estimates of high, low, and close price of the stocks by the EnKF method with errors of 0.258 for high price error, 0.251 for low price error, and 0.24 for close price error. The most accurate results by the EnKF-SR method by generating the 100 ensembles were in the estimstes of high, low, and close price of the stocks with errors of 0.273, 0.281, and 0.276 respectively. In general as seen in the table 2, the result of the three simulations were highly accurate with accuracy of 99%.

Table 2. Comparison of the values of RMSE by the EnKF and EnKF-SR method by generating 100, 200 dan 300 Ensembles

	100 Ensemble		200 Ensemble		300 Ensemble	
	EnKF	EnKF-SR	EnKF	EnKF-SR	EnKF	EnKF-SR
High	**0.2731**	1.5179	**0.26737**	2.1947	**0.25855**	2.634
Low	**0.28145**	1.5342	**0.27019**	2.1954	**0.25177**	2.7744
Close	**0.27693**	1.5794	**0.25515**	2.1805	**0.24098**	2.6757

4. Conclusion

Based on the results of the simulation analysis by generating 100, 200 and 300 ensembles, it could be concluded that the EnKF and EnKF-SR methods could be applied to estimate stock functions with high accuracy. The resulting errors were less than 2-3%. Regarding the accuracy result in the implementation of both methods, it was observed that the EnKF method gave more accurate stock price estimation result than the EnKF-SR method did.

Open problem. How to implemented Unscented Kalman Filter (UKF) and Unscented Kalman filter Square Root (UKF-SR) for estimation of stock price.

Acknowledgement

We gratefully acknowledge the support from LPPM - Nahdlatul Ulama Surabaya 0f University (UNUSA).

References

[1] Sia E R and Hansun S 2014 Rancang Bangun Aplikasi Peramalan Nilai Saham Menggunakan Algoritma Kalman Filter *Jurnal Ilmiah Komputer dan Informatika (Komputa)* **3** 2

[2] *http://forex.marketiva-id.com/tag/faktor-penggerak-pasar-saham* (Accessed on February 20, 2017, at 07.45 pm)

[3] Herlambang T, Djatmiko E B and Nurhadi H 2015 Navigation and Guidance Control System of AUV with Trajectory Estimation of Linear Modelling *Proc. of International Conference on Advance Mechatronics, Intelligent Manufactre, and Industrial Automation (IEEE ICAMIMIA)* 184-187

[4] Ermayanti E, Aprilini E, Nurhadi H and Herlambang T 2015 Estimate and Control Position Autonomous Underwater Vehicle Based on Determined Trajectory using Fuzzy Kalman Filter Method *International Conference on Advance Mechatronics, Intelligent Manufactre, and Industrial Automation (ICAMIMIA)-IEEE*

[5] Herlambang T 2017 Design of a Navigation and Guidance System of Missile with Trajectory Estimation Using Ensemble Kalman Filter Square Root (EnKF-SR) *The 4th International conference on Computer Application and Informations Processing Technology (CAIPT 2017)*

[6] Herlambang T, Rasyid R A, Hartatik S and Rahmalia D 2017 Estimasi Posisi *Mobile Robot* Menggunakan Metode Akar Kuadrat Unscented Kalman Filter (AK-UKF) *Technology Science and Engineering Journal* **1** 2

[7] Herlambang T 2017 Estimasi Posisi Magnetic Levitation Ball Menggunakan Metode Akar Kuadrat Ensemble Kalman Filter (AK-EnKF) Rekayasa Energi *Manufaktur Jurnal* **2** 1

[8] Herlambang T, Djatmiko E B and Nurhadi H 2015 Ensemble Kalman Filter with a Square Root Scheme (EnKF-SR) for Trajectory Estimation of AUV SEGOROGENI ITS *International Review of Mechanical Engineering IREME Journal* **9** 6 553-560

IOP Conf. Series: Journal of Physics: Conf. Series **1008** (2018) 012018 doi:10.1088/1742-6596/1008/1/012018

Statistical bias correction modelling for seasonal rainfall forecast for the case of Bali island

D Lealdi[1], S Nurdiati[1], A Sopaheluwakan[2]

[1]Department of Mathematics, Bogor Agricultural University, Dramaga Raya Street, West Java 16680, Indonesia
[2]Indonesian Agency for Meteorology, Climatology and Geophysics, 2 Angkasa 1 Street, Kemayoran, Jakarta 10720, Indonesia

Email : dedi_math07@yahoo.co.id, nurdiati@ipb.ac.id, ardhasena@bmkg.go.id

Abstract. Rainfall is an element of climate which is highly influential to the agricultural sector. Rain pattern and distribution highly determines the sustainability of agricultural activities. Therefore, information on rainfall is very useful for agriculture sector and farmers in anticipating the possibility of extreme events which often cause failures of agricultural production. This research aims to identify the biases from seasonal forecast products from ECMWF (European Centre for Medium-Range Weather Forecasts) rainfall forecast and to build a transfer function in order to correct the distribution biases as a new prediction model using quantile mapping approach. We apply this approach to the case of Bali Island, and as a result, the use of bias correction methods in correcting systematic biases from the model gives better results. The new prediction model obtained with this approach is better than ever. We found generally that during rainy season, the bias correction approach performs better than in dry season.

1. Introduction

The rainfall in the tropical area is one of the factors in climate with the highest variability. Indonesia is one of the countries with a tropical climate which is geographically located between 6° North Latitude and 11° South Latitude and between 95° West Longitude and 141° East Longitude crossing the equator. The characteristics of rainfall in several regions are definitely not the same. This condition is caused by several factors, namely position of region, condition of earth in a region, mountain and valley in a region, and even structure and orientation of island. As the result, the pattern of rainfall distribution tends to be uneven among the regions in the wide scope [1].

The rainfall is one of the factors in climate which is highly influential on the agricultural sector. The consistency in rain pattern and distribution highly determines the sustainability of agricultural activities. However, this condition will be disturbed during the phenomenon of extreme climates, such as the dry extreme (El Nino) and wet extreme (La Nina). The phenomenon of extreme climate can cause rainfall pattern which deviates from the normal condition.

According to the analysis of Indonesian Agency for Meteorological, Climatological and Geophysics, in Banten, Special Capital City District of Jakarta, and West Java, the possibility of extreme rain reached 500 mm/month in 1970-1999. This possibility increased to 13% in 1900-1929, the possibility of extreme rain in these three regions is only 3% [2]. Indonesian people also experienced the real climate change in 2010. In wet drought throughout 2010, this year was ended with the drought at last, it is the real description of how the climate has really changed [3].

Based on data released by BPS Bali Province in 2014, agricultural production decreased 2.74% to 857.944 tons of dry milled grain [8]. The decline is generally due to a decrease of paddy harvest area to 5.11% or 7,683 ha due to drought as a result of the dry season. From June to September in 2014, extreme climate change occurs, *ie* in the category of dry months (months with an average of less than 100 mm rainfall), resulting in drought or pneumonia area of 807 ha or higher than 54 ha in 2013 [9]. This dry season has an effect on the retreat of planting season and harvest season about one month. If you look at the total loss of the data, this amount is very large.

Therefore, the information regarding the pattern of rainfall is very useful for farmers in anticipating the possibility of extreme events which cause the failure of agricultural production. Many researchers studied the method for modelling the rainfall pattern. Piani et al. [4] planned and applied the bias correction on the output of daily climate model in Europe, so a distribution which approximates the observation model was obtained. This method is known as Statistical Bias Correction. The obtained result shows that the daily climate model is improved consistently with the method of bias correction. Bambang, et al. [5] identified the systematic error in TRMM data and built the transfer function to correct the bias of TRMM rainfall. From this research, it was found that the rainfall pattern from TRMM data which have been corrected and observed is similar, but the amount of rainfall is not similar.

In this research, Statistical Bias Correction method to see the relation between ECMWF (European Centre for Medium-Range Weather Forecasts) rainfall forecast and observation model from Indonesian Agency for Meteorological, Climatological, and Geophysics in 1996-2015 will be used. This research aims to identify the systematic error in ECMWF data and to build the transfer function in order to correct the up-to-date ECMWF rainfall bias as the prediction model. The systematic error can be identifiable through comparison of form and distribution of ECMWF hindcast data with observation data using the approaches of interpolation and the closest point. The resulted ensemble model in this research is the predicted rainfall approximation for the period of January - December 2016. Then will see how good the results based on observations of monthly rainfall in 2016.

2. Research Methodology

2.1. Data

The data used in this research are the rainfall data of ECMWF (*The European Centre for Medium-Range Weather Forecasts*) and observation in weather station from Central Indonesian Agency for Meteorological, Climatological and Geophysics in Jakarta from 1996 to 2016. The data taken is for observation in Bali Island.

2.2. Method of Data Analysis

2.2.1 Data Extraction. Data extraction was performed on the ECMWF rainfall data grid with MATLAB software to get the data output which is the same with the observation data. The data were extracted based on the location of latitude and longitude from every weather station. The process of data extraction used 2 methods, namely:

- *Method of the Closest Point.* It was performed by making an algorithm to determine the distance between 2 points and the closest distance. The determination of distance among the points $A(x_1, y_1)$ and $B(x_2, y_2)$ used the formula [6],

$$|AB| = ((x_2 - x_1)^2 - (y_2 - y_1)^2)^{1/2} \qquad (1)$$

- *Cubic Interpolation.* Interpolation is a way of finding value among some known data points. This method is a numerical approach that needs to be done if we need the value of a function $y = y(x)$ which is not known its formulation appropriately. In this research, cubic interpolation method will be used as one method for extracting ECMWF rainfall data. Cubic interpolation is done by determining the points between the four points by using the cubic polynomial approach as follows [7],

$$f_3(x) = b_0 + b_1(x - x_0) + b_2(x - x_0)b_1(x - x_1) + b_3(x - x_0)(x - x_1)(x - x_2) \quad (2)$$

with,

$$b_0 = f(x_0) \quad (3)$$

$$b_1 = \frac{f(x_1) - f(x_0)}{(x_1 - x_0)} \rightarrow f[x_1, x_0] \quad (4)$$

$$b_2 = \frac{f[x_2, x_1] - f[x_1, x_0]}{(x_2 - x_0)} = \frac{\frac{f(x_2) - f(x_1)}{(x_2 - x_1)} - \frac{f(x_1) - f(x_0)}{(x_1 - x_0)}}{(x_2 - x_0)} \rightarrow f[x_2, x_1, x_0] \quad (5)$$

$$b_3 = \frac{f[x_3, x_2, x_1] - f[x_2, x_1, x_0]}{(x_3 - x_0)} \rightarrow f[x_3, x_2, x_1, x_0] \quad (6)$$

2.2.2 Identification of Data Distribution. In this stage, the parameter estimation process and the distribution matching which was the closest to the data were performed. This process was required to determine the cdf for the bias correction process. This data distribution was checked with the support of MATLAB software to match data with one of the fittest distributions, namely beta distribution, birnbaum-saunders, exponential, extreme value, gamma, generalized extreme value, generalized pareto, inverse Gaussian, logistic, log-logistic, lognormal, nakagami, normal, rayleigh, rician, t location-scale and weibull. To sort the fittest distribution, it was characterized based on the parameter as follows: NLogL (Negative of the log likelihood), BIC (Bayesian information criterion), AIC (Akaike information criterion), and AICc (AIC with a correction for finite sample sizes).

2.2.3 Bias Correction. Bias correction was performed to see the relation between data of the observed rain and the predicted rainfall (ECMWF data) to make certain transfer function $[y = f(x)]$. The form of relation is cdfobserved= cdfpredicted. The relation between the observed and predicted data can be in form of regression equation, which is linear, exponential, or polynomial. The cdf can be formed with the formula as follows:

$$cdf(x) = \int_0^x \frac{e^{\left(-\frac{x}{\theta}\right)} x'^{(k-1)}}{\Gamma(k)\theta^k} dx' + cdf(0) \quad (7)$$

where x = average daily rainfall,
k = parameter of form,
θ = scale, and
$\Gamma(.)$ = Gamma function, and this function can be evaluated with the factorial function $(\Gamma(k) = (k-1)!)$.

The cumulative distribution function (*cdf*) for data with zero value means that there is no rain on that day (*dry spell*) [4]. The result of this bias correction procedure was then used to correct model of the predicted rainfall in 2016.

2.2.4 Evaluation of the Best Prediction and Determination Models. The prediction evaluation was performed by considering and comparing the mean values and standard deviations from distribution of PDF rainfall model. The best model was determined by the following metric:

$$x = \frac{\bar{x}_{cor} - x_{obs}}{|\bar{x}_{fut} - x_{obs}|} \quad \text{and} \quad y = \frac{\sigma_{cor}}{\sigma_{fut}} \tag{8}$$

with,

$\bar{x}_{cor}$ = mean from pdf of the corrected rainfall,
$\bar{x}_{fut}$ = mean from pdf of rainfall before correction,
x_{obs} = the observed monthly rainfall,
σ_{cor} = standard deviation from pdf of the corrected rainfall, and
σ_{cor} = standard deviation from pdf of rainfall before correction.

The statistical bias correction perform well when the values metrics above are within a unit interval: $-1 < x < 1$ and $y < 1$ [8–9]

3. Results

3.1. General Description of Data

General description of the pattern of ECMWF rainfall data and the observed rainfall data from Indonesian Agency for Meteorological, Climatological and Geophysics is shown in Figure 1.

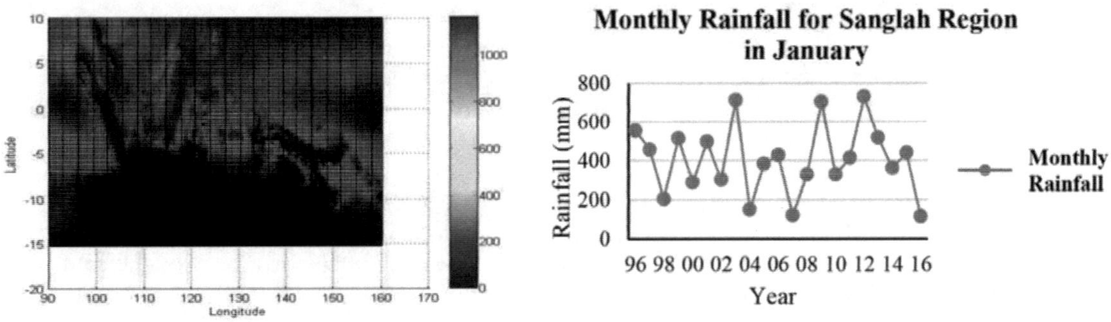

(a) (b)

Figure 1. Overview of Rainfall Data Year 1996-2016.
(a) Plot of Rainfall Data Prediction of EMWF for Indonesia Areas.
(b) Rainfall Observation Data Plot of BMKG Observation for Sanglah Region in January.

Based on Figure 1(a), it can be observed that rainfall data from ECMWF is grid-shaped. Data Grid is the data presentation with interpolation method that is the value estimation process in a region which is not sampled or measured, so map or distribution of region is formed. The rainfall data have 4 parameters, namely: average daily rainfall, latitude, longitude, and time.

Figure 1(b) is the monthly rainfall data from the observation of Indonesian Agency for Meteorological, Climatological, and Geophysics for Sanglah station in January. These two data were taken in 1996-2016, in which 1996-2015 data will be used to see the relation between these two types of data and 2016 data are the basis for verification of the predicted rainfall.

3.2. Data Extraction

After the data extraction on the ECMWF rainfall data grid with MATLAB *software*, it resulted in 11 data of ensemble members for every observation month in 1996-2015. Meanwhile, from the process of data extraction in 2016, it resulted in 51 ensemble members data.

IOP Conf. Series: Journal of Physics: Conf. Series **1008** (2018) 012018 doi:10.1088/1742-6596/1008/1/012018

3.3. Identification of Data Distribution

The distribution determination which is the most suitable with the data is consideration of values of *NLogL (Negative of the log likelihood), BIC (Bayesian information criterion), AIC (Akaike information criterion), and AICc (AIC with a correction for finite sample sizes)* which is the smallest in Table 1.

Table 1. Checking of ECMWF Rainfall Data Distribution for Sanglah Region in January 1996-2015.

VAR (p)	NLogL	BIC	AIC	AICc
Weibull	1344.238	2699.264	2692.477	2692.532
Details	1344.858	2700.503	2693.716	2693.771
Nakagami	1345.544	2701.876	2695.088	2695.144
Normal	1346.4	2703.588	2696.801	2696.856
Generalized Extreme Value	1345.807	2707.795	2697.614	2697.725
t-Location Scale	1346.399	2708.98	2698.799	2698.91
Logistic	1349.854	2710.496	2703.709	2703.764
Gamma	1350.809	2712.406	2705.618	2705.674
Loglogistic	1358.745	2728.278	2721.491	2721.546
Lognormal	1362.047	2734.881	2728.093	2728.149
Birnbaumsaunders	1363.756	2738.299	2731.512	2731.567
Rayleigh	1367.05	2739.494	2736.1	2736.119
Inverse Gaussian	1365.038	2740.863	2734.076	2734.132
Extreme Value	1369.441	2749.669	2742.882	2742.937
Exponential	1473.633	2952.659	2949.265	2949.284

Based on the result in Table 1, ECMWF rainfall data for Sanglah Region in January 1996-2015 have Weibull distribution.

3.4. Bias Correction

In this research, the process of rainfall bias correction, as a step in making a new predicted rainfall model, was performed. ECMW rainfall model in 1996-2015 was corrected based on the observed rainfall in the same period.

This process was performed for every month in different location, so every location will result in the rainfall model from January to December 2016. The process of statistical bias correction was simulated like in Figure 2 below.

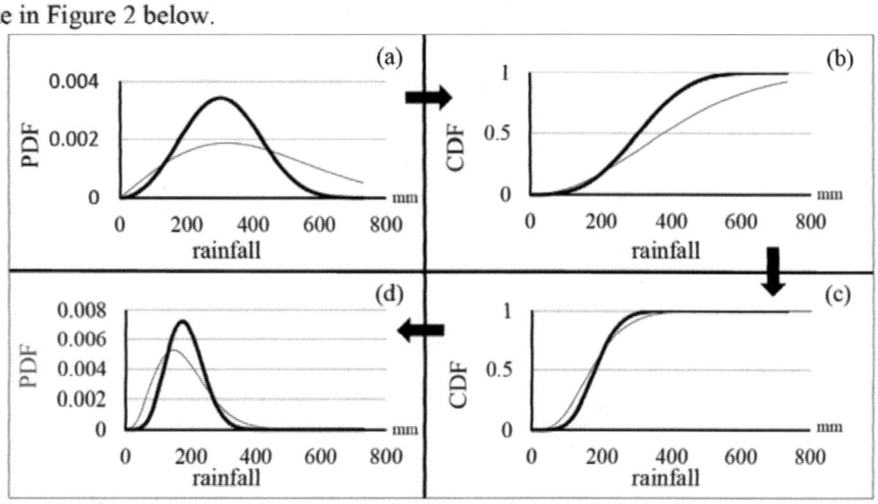

Figure 2. Process of statistical bias correction to compile rainfall model in January 2016 in Sanglah City.

IOP Conf. Series: Journal of Physics: Conf. Series **1008** (2018) 012018 doi:10.1088/1742-6596/1008/1/012018

(a) Figure of pdf of the predicted rain (thick line) and the observed rain (thin line),
(b) Figure of cdf value obtained from integral of pdf in Figure a,
(c) Figure of cdf of rain model before correction (thick line) and the corrected rain model (thin line) was obtained from relation of $cdf_{observed}(y)$ (thin line) = $cdf_{predicted}(x)$ (thick line) in Figure b, and
(d) Figure of pdf of rain model before correction (thick line) and rain prediction model after correction (thin line).

The result from bias correction process by considering the relation of $cdfobserved(y) = cdfpredicted(x)$ is then presented in PDF plot, so it is easier to be interpreted.

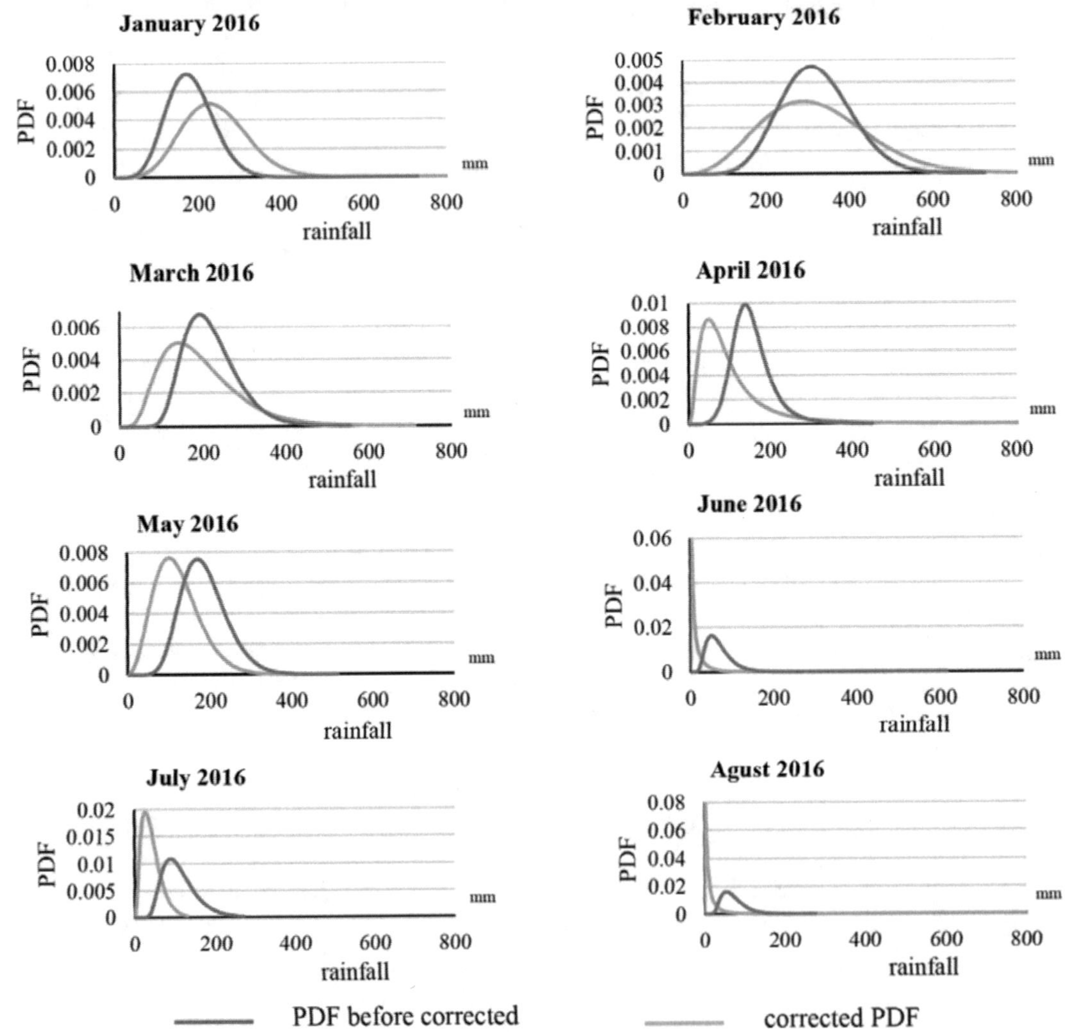

IOP Conf. Series: Journal of Physics: Conf. Series **1008** (2018) 012018 doi:10.1088/1742-6596/1008/1/012018

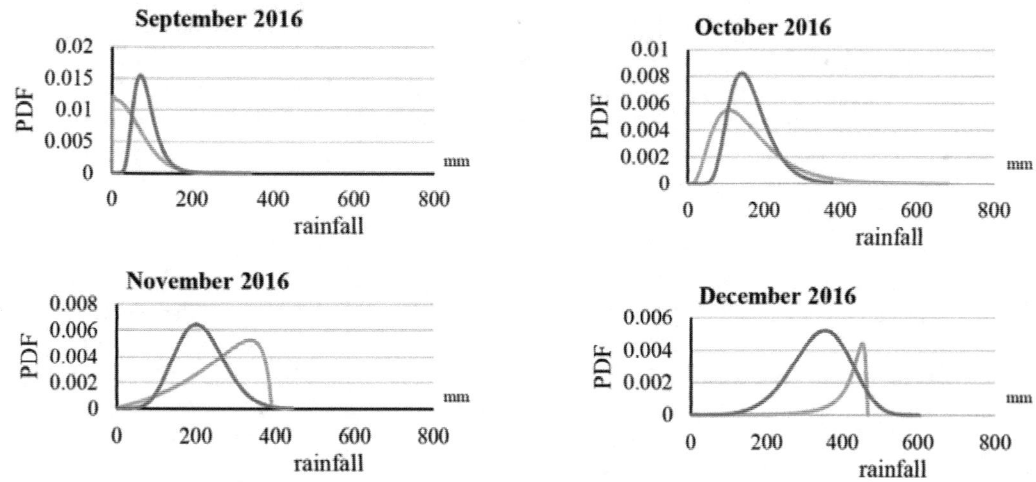

Figure 3. PDF of the Corrected Rainfall Model vs. Rainfall Model Before Correction in 2016 in Sanglah with Interpolation Approach.

Based on Figure 3, the bias correction method can identify well the bias of the historical rainfall forecast model. The value of the transfer function formed from the bias correction process can be used to improve the distribution of the 2016 rainfall prediction in Sanglah area. This can be seen from the rainfall prediction model in 2016 in Sanglah that has been improved based on information of rainfall bias from historical rainfall model.

3.5. The Best Prediction Model

After the precedures were applied, we can analyse the following result from the following table as follows:

Table 2. Evaluation of the Corrected Rainfall Model in 2016 in Sanglah with Interpolation Approach.

Month	$\overline{m}_{mod} - obs$	$\overline{m}_{cor} - obs$	σ_{mod}	σ_{cor}	$x = \frac{\bar{x}_{cor} - obs}{\lvert\bar{x}_{mod} - obs\rvert}$	$y = \frac{\sigma_{cor}}{\sigma_{mod}}$
January	113.9	67.56	189.11	213.62	0.86	1.13
February	-216.45	-323.49	332.24	279.22	-1.49	0.84
March	145.922	63.56	226.98	176.32	0.44	0.78
April	120.16	28.49	163.53	96.31	0.24	0.59
May	149.46	85.23	193.09	131.79	0.57	0.68
June	-57.304	-105.24	73.91	22.43	-1.84	0.30
July	33.92	-1.57	116.55	62.73	-0.05	0.54
August	33.375	-15.08	76.45	24.19	-0.45	0.32
September	-150.91	-196.36	89.65	58.71	-1.30	0.65
October	37.52	-32.12	173.05	137.25	-0.86	0.79
November	-111.68	6.87	219.56	329.54	0.06	1.50
December	-56.60	-53.20	349.17	390.07	-0.94	1.12

Based on Table 2, there is information that the predicted rainfall model in Sanglah with interpolation approach is good for prediction in March, April, May, July, and October. It can be seen from x and y values which have met the criteria. In the months, x value is on range of $-1 < x < 1$ which means that mean from the corrected rainfall distribution is not far from the real value of rainfall realization. This value also implies that the accuracy of the new prediction model obtained is better than the prediction model before the rainfall bias is corrected. Then, y value which is less than 1 indicates that the actual rainfall, is estimated by an improved distribution with reduced standard deviation. This value also shows that the new prediction model obtained has better precision than the prediction model before the rainfall bias is corrected. For the cases in other months, the predicted rainfall model after correction can be said quite good since it meets one of the desired criteria.

Based on Table 3, the predicted rainfall after correction which is modeled with the closest point approach also gets a good result for some cases, in February and May. However, it is not for performing well in some months, such as September and December, which do not meet the determined criteria. With the x value which does not meet this criterion, it indicates that the mean value of the corrected rainfall distribution in these months is far from the real value of rainfall. Besides that, having a larger y value, indicates larger uncertainties where obtained after the procedure was applied. This value also indicates that the new prediction model obtained has a precision that is not better than the prediction model before it is corrected with its rainfall bias. So using this method is still less good when compared with using interpolation method.

Table 3. Evaluation of the Corrected Rainfall Model in 2016 in Sanglah with the Closest Point Approach.

Month	$\overline{m}_{mod} - obs$	$\overline{m}_{cor} - obs$	σ_{mod}	σ_{cor}	$x = \frac{\bar{x}_{cor} - obs}{\lvert\bar{x}_{mod} - obs\rvert}$	$y = \frac{\sigma_{cor}}{\sigma_{mod}}$
January	50.63	51.12	173.21	208.33	1.01	1.20
February	167.89	107.83	294.19	279.69	0.64	0.95
March	78.63	19.86	203.36	176.91	0.25	0.87
April	24.11	-52.23	146.86	97.33	-2.17	0.66
May	68.35	-10.78	191.36	121.02	-0.16	0.63
June	20.87	-48.93	145.57	79.54	-2.35	0.55
July	-1.97	-45.81	120.38	62.34	-23.22	0.52
August	-42.62	-80.36	77.32	29.02	-1.89	0.38
September	-29.73	-77.12	89.54	55.42	-2.59	0.62
October	35.76	-48.29	159.67	133.76	-1.35	0.84
November	88.48	57.29	211.80	221.15	0.65	1.04
December	209.59	369.86	333.04	452.19	1.76	1.36

For modelling results in other areas of the island of Bali, generally we obtain improved results for rainfall data estimation using interpolation when compared to using the nearest point approach. The results of the evaluation of the bias correction process for all observation areas on the island of Bali can be seen in the graphs in Figure 4 and 5.

The average results obtained for all regions have met the criteria that have been determined. It is generally found that during the rainy season in February, March, November and December, the use of bias correction gives better results than other months for all areas of observation on the island of Bali. At least the corrected model for the season can meet one of the criteria in determining the best model that is seen from the mean value for the accuracy and standard deviation values for the precision level of the model [8-9]. Meanwhile, the results obtained by approaching the data using the nearest point cannot explain well if viewed the results based on the season that occurred in the territory of Indonesia.

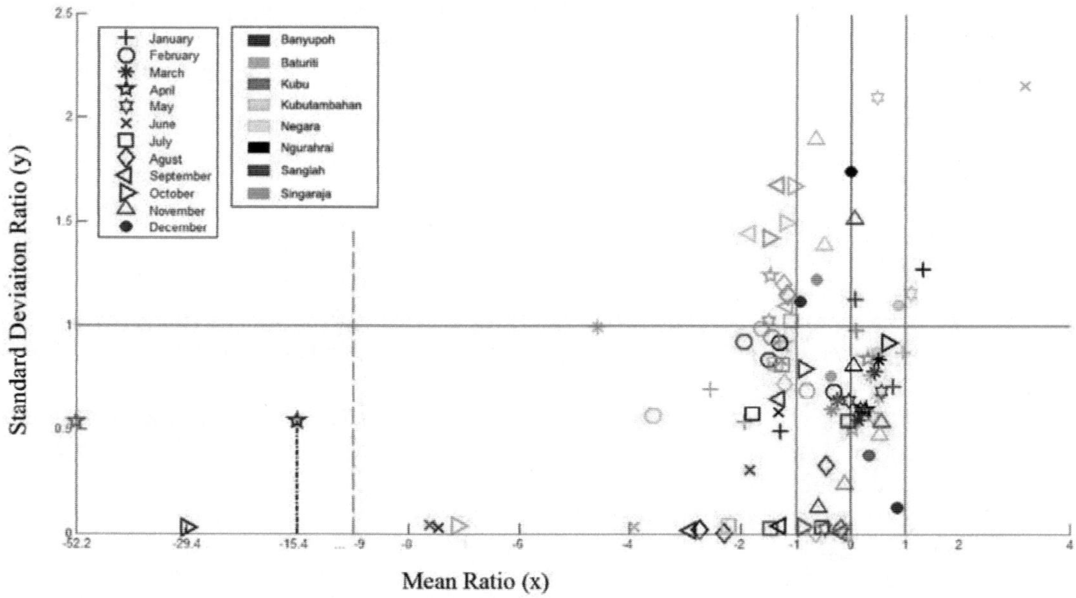

Figure 4. Graphic of Ratio Mean & Standard Deviation of Rainfall in Bali with Interpolation Approach in January – December 2016.

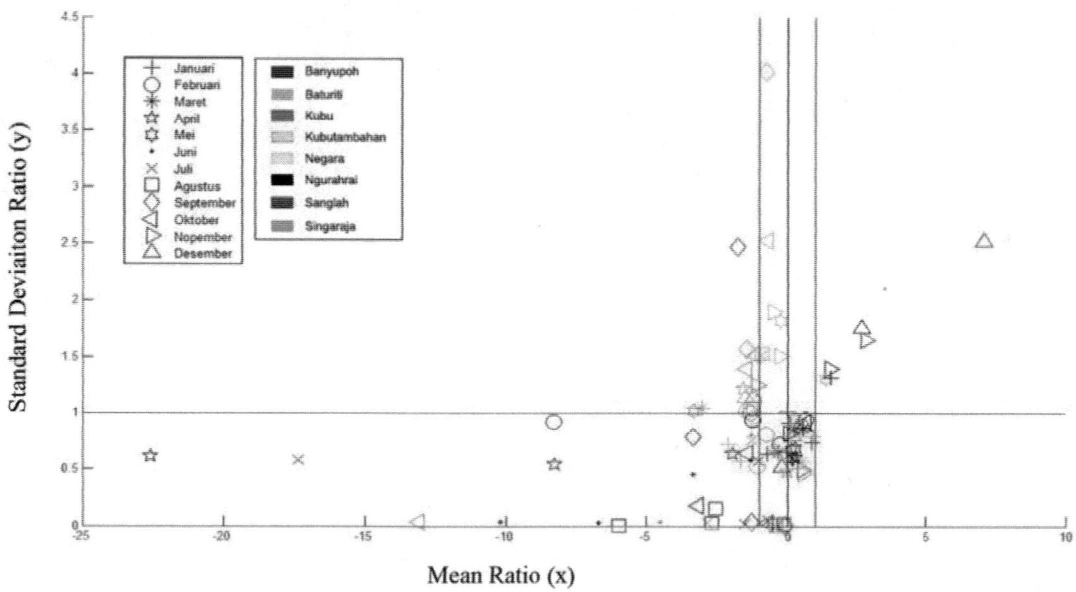

Figure 5. Graphic of Ratio Mean & Standard Deviation of Rainfall in Bali with the Closest Point Approach in January – December 2016.

4. Conclusion

The bias correction method can identify well the bias of the historical rainfall forecast model. The value of the transfer function formed from the bias correction process can be used to improve the distribution of the 2016 rainfall prediction on the island of Bali, to obtain better prediction. It is generally found that during the rainy season in February, March, November, and December, the use of bias correction provides better results than the dry season in modelling rainfall forecasts with data approach using interpolation. Whereas, the data approach using the nearest point in general has not provided good results for cases in the rainy or dry season. This can be caused of little information rainfall data used by researchers. For further research can use more rainfall data so that it can produce a better distribution.

References

[1] Swarinoto YS and Sugiyono 2011 Utilization of Air Temperature and Air Humidity in Regression Equations for Simulation Monthly Rain Total Prediction in Bandar Lampung. *Journal of Meteorology and Geophysics* **12** 271–281
[2] Aldrian E, M Karmini and Budiman 2011 Adaptation and Miti-gasi Climate Change in Indonesia *Climate Change and Air Quality Center of Indonesian Agency for Meteorological, Climatological and Geophysics* 67
[3] Aldrian, E 2014 Understanding Climate Dynamics in The Indonesian Archipelago Countries as The National Resilience Modality *Indonesian Agency for Meteorological, Climatological and Geophysics* 20–21
[4] Piani C J O, Haerter, and Coppola E 2010 Statistical Bias Correction for Daily Precipitation in Regional Climate Models Over Europe *J TheorApplClimatol* **99** 187–192
[5] Bambang DD, Rizaldi B, Bambang P, and Yuli S 2013 Evaluation of Rainfall TRMM Using Statistics Bias Correction Approach *Land and Climate Journal* **38** 20–34
[6] Purcell EJ, Varberg D, Rigdon SE 2003 *Calculus 8th Edition* (USA : Prentice Hall, Inc)
[7] Cheney W and Kincaid D 2008 *Numerical Mathematics and Computing 7th Edition* (USA : Thomson Brooks/Cole)
[8] JCGIM 2008 International Vocabulary of Metrology *Joint Committee for Guides in Metrology* **3** 21–22
[9] Taylor JR 1999 *An Introduction to Error Analysis: The Study of Uncertainties in Physical Measurements* (Colorado : University Science Books)

Ensemble averaging and stacking of ARIMA and GSTAR model for rainfall forecasting

D Anggraeni[1] , I F Kurnia[2] , A F Hadi[1,3]
[1] Department of Mathematics, University of Jember, Indonesia
[2] Graduated School of Mathematics, Department of Matematics, University of Jember, Indonesia

E-mail : dian_a.fmipa@unej.ac.id, ivafendikurnia@gmail.com, afhadi@unej.ac.id

Abstract. Unpredictable rainfall changes can affect human activities, such as in agriculture, aviation, shipping which depend on weather forecasts. Therefore, we need forecasting tools with high accuracy in predicting the rainfall in the future. This research focus on local forcasting of the rainfall at Jember in 2005 until 2016, from 77 rainfall stations. The rainfall here was not only related to the occurrence of the previous of its stations, but also related to others, it's called the spatial effect. The aim of this research is to apply the GSTAR model, to determine whether there are some correlations of spatial effect between one to another stations. The GSTAR model is an expansion of the space-time model that combines the time-related effects, the locations (stations) in a time series effects, and also the location it self. The GSTAR model will also be compared to the ARIMA model that completely ignores the independent variables. The forcested value of the ARIMA and of the GSTAR models then being combined using the ensemble forecasting technique. The averaging and stacking method of ensemble forecasting method here provide us the best model with higher acuracy model that has the smaller RMSE (Root Mean Square Error) value. Finally, with the best model we can offer a better local rainfall forecasting in Jember for the future.

1. Introduction

Extreme climate change is one of the impacts of global warming, the extreme impact of climate change, unconditinal temperature and season. In addition, the climate change leads to the changes in air temperature, rainfall, air pressure and wind speed. Of these factors, the rainfall is the most affect on human life, directly. Accurate weather forecasting of future rainfall data is one of the main foundations of a plan, effective and efficient in weather forecasting. The one of popular rainfall forecasting method is Autoregressive Integrade Moving Average (ARIMA) or often called the Box-Jenkins time series method. ARIMA matches the observations of the time series (series time) in conjunction with each other (depending on) or it may be said that the ARIMA model works with the relation of past storm events to the forecast in the future. The ARIMA model does not contain any element of location in the forecasting model so that another model containing time element, and location to predict rainfall data in 4 rainfall statiosn groups. The time-space model is one of the models that connects time-related elements, and location to a time series and location data. The time-space model developed by Pfeifer and Deutsch (1980a, 1980b) [1,2] has weaknesses on, which describes the locations, and the different times in a time series and location data. This weakness is chosen by Baravkova et al. (2002) [3] through a model known as the GSTAR model (Generelized Space Time

Autoregresive), which is a model that has an interrelationship of time and location where the location under study has vary characteristics (heterogeneous). Along with the development of the times, rainfall forecasting model is growing, One is the concept of combining several forecasting models. The Ensemble model is a forecasting method created, either with different models or different multiple initial conditions which are then combined into a single unit for the assessment of uncertainty in the formulation of the initial model and conditions. The data applied to the Ensemble model in this case study is the rainfall data of Jember, which has certain characteristics for each cluster of locations. Given Jember is one of the best produce of sugar cane plants. The productivity of sugar cane plants depends on climate change. Extreme climate changes affect the rainfall pattern that is uncertain and above nomal, can cause long drought that will also impact on the productivity of sugar cane plants. To minimize the impacts arising from the uncertain changes in rainfall by providing information on weather forecasts, especially opportunities through rainfall in an area within a certain period of time

2. Timeseries Model

2.1. ARIMA

ARIMA is often also called the Box-Jenkins time-series method. The Autoregressive Integrated Moving Average (ARIMA) model is a model that completely ignores independent variables in making forecasting. ARIMA uses past and present values of the dependent variable to produce accurate short-term forecasting. The basic assumption used in the ARIMA time series process discussion is a stationary process. To stationize the non-stationary data generation process is to do the first level difference, and so on. For example, the first or first difference is called $W_t = W_t - W_{t-1}$, the result of each difference is called an integrated process. While the order of the process that gets stationary time series is determined by the number of differences (differencing) is done.
The general model for the pure mixture of AR (p) and MA (q) pure processes, as ARIMA (1,0,1), also known as ARMA(p,q) is expressed as follows, where X_t regressed on itself lagged by some periods,

$$X_t = \mu + \phi_1 X_{t-1} + \phi_2 X_{t-2} + \cdots + \phi_p X_{t-p} + e_t - \theta_1 e_{t-1} - \theta_2 e_{t-2} - \theta_q e_{t-q} \quad (1)$$

with μ is mean, ϕ_i $(i = 1,2, \dots, p)$ are the autoregressive parameters, θ_j $(j = 1,2, \dots, q)$ are the moving average parameters, and e_t is lagged errors.
Furthermore, there are four steps in ARIMA modeling; identification, estimation and information criteria, diagnostic checking and model's use. Identification involves determining the order of the model required (p, d, and q) in order to capture the salient dynamic features of the data. This mainly leads to use graphical procedures (plotting the series, the ACF and PACF [4], etc). Then the second step, estimation and information criteria, involves estimation of the parameters of the different models (using step 1) and proceeds to a first selection of models (using information criteria). The next, diagnostic checking involves determining whether the model(s) specified and estimated is adequate. Notably, one uses residual diagnostics. Lastly, the best model will be used as a forecast model. To determine the best model can be used the following standard error estimate:

$$\mathrm{s} = \left[\frac{\mathrm{sse}}{\mathrm{n-n_p}}\right]^{1/2} = \left[\frac{\sum_{t-1}^{n}(Y_t - \hat{Y}_t)^2}{\mathrm{n-n_p}}\right]^{1/2} \quad (2)$$

2.2. GSTAR

Generalized Space Time is the development of the STAR model that fits the data heterogeneity location. Mathematically, the notation of GSTAR $(p; \lambda_1, \dots, \lambda_p)$ is the same as the model STAR (p_1)[3]. In the STAR model parameter values ϕ_{kl} assumed to be similar for all locations, while at GSTAR ϕ_{kl} parameter values in the same spatial lag allowed between different locations. GSTAR $(p; \lambda_1, \dots, \lambda_p)$ can be written as follows :

$$Z(t) = \sum_{k=1}^{p}[\phi_{k0} + \phi_{k1}\mathbf{W}]\,\mathbf{Z}\,(t-k) + \boldsymbol{e}\,(t) \tag{3}$$

Where ϕ_{k0} : diag $({\phi_{k0}}^{(1)}, \dots, {\phi_{k0}}^{(N)})$ and Φ_{k1} : diag $({\phi_{k1}}^{(1)}, \dots, {\phi_{k1}}^{(N)})$. The weights are chosen so that $W_{ii} = 0$ and $\sum_{i \neq j} W_{ij} = 1$

[5] said that the selection of spatial order of GSTAR model is generally limited to order 1, because higher order would be difficult to interpretation. While at the time order (autoregressive) can be determined with AIC (Akaike Information Criterian). The choice of best model order on GSTAR can be determined with the smallest AIC value. AIC calculations as according to Akaike (1973, 1974) [6], as:

$$AIC(p) = \ln\left|\widetilde{\sum}_u (p)\right| + \frac{2p}{T}K^2 \tag{4}$$

On the other hand, the selection of weight location GSTAR model is based on these three weights location:

2.2.1 *Weight location of uniform*

Uniform weight in $W_{ij} = \frac{1}{n_i}$ with $n_i =$ the number border location to location i. The weights this location give equal weight to each location. Therefore, the weight of this location often used on data that have a distance between the same location (homogeneous). [7]

2.2.2 *Weight location of inverse distance*

Weighting by the inverse distance refers to the distance between locations. Suppose the distance between the four locations are defined, r_1 = distance between location 1 to location 2; r_2 = distance between location 1 to location 3; r_3 = distance between location 1 to location 4; r_4 = distance between location 2 to location 3; r_5 = distance between location 2 to location 4; r_6 = distance between location 2 to location 4. Then, the matrix below are standardized in the form W_{ij} to get $\sum_{i \neq j} W_{ij} = 1$ [3]

$$W_{ij} = \begin{bmatrix} 0 & \frac{r_2+r_3}{r_1+r_2+r_3} & \frac{r_2+r_3}{r_1+r_2+r_3} & \frac{r_2+r_3}{r_1+r_2+r_3} \\ \frac{r_2+r_3}{r_1+r_2+r_3} & 0 & \frac{r_2+r_3}{r_1+r_2+r_3} & \frac{r_2+r_3}{r_1+r_2+r_3} \\ \frac{r_2+r_3}{r_1+r_2+r_3} & \frac{r_2+r_3}{r_1+r_2+r_3} & 0 & \frac{r_2+r_3}{r_1+r_2+r_3} \\ \frac{r_2+r_3}{r_1+r_2+r_3} & \frac{r_2+r_3}{r_1+r_2+r_3} & \frac{r_2+r_3}{r_1+r_2+r_3} & 0 \end{bmatrix} = \begin{bmatrix} 0 & W_{12} & W_{13} & W_{14} \\ W_{21} & 0 & W_{23} & W_{24} \\ W_{31} & W_{32} & 0 & W_{34} \\ W_{41} & W_{42} & W_{43} & 0 \end{bmatrix}$$

2.2.3 *Weight location of cross correlation*

Estimates of the cross-correlation of this data sample is:

$$r_{ij}(k) = \frac{\sum_{t=k+1}^{n}[Z_i(t) - \bar{Z}_i][Z_j(t-k) - \bar{Z}_j]}{\sqrt{(\sum_{t=1}^{n}[Z_i(t) - \bar{Z}_i]^2)((\sum_{t=1}^{n}[Z_j(t) - \bar{Z}_j]^2)}}$$

Furthermore, determination of the weight location can be done with the normalization from amount of the cross-correlation between locations at the corresponding time. This process generally produces weight following locations:

$$W_{ij} = \frac{r_{ij}(1)}{\sum_{k \neq 1}|r_{ik}(1)|}$$

This weight fulfilling $\sum_{i\neq j} W_{ij} = 1$ weights locations using normalization of the magnitude of the cross-correlation between locations at the corresponding time allows all forms of possible relationships between locations. This weights also gives flexibility on the magnitude and sign of the relationship between different locations: positive and negative. The weights of this location is the weight location includes uniform weights and binary locations.

Furthermore, the parameter estimation of GSTAR model by all location weights are using Least Square Method. Based on the GSTAR model with the order $p = 1$ and the spatial order 1 where we can write $\phi_{kl} = \phi_{1k}{}^{(i)}$ for $k = 0,1$ be derived as:

$$Z_i(t) = \phi_{10}{}^{(10)} Z_i(t-1) + \phi_{11}{}^{(i)} \sum_{j}^{N} W_{ij} Z_j(t-1) + e_i(t)$$

where $Z_i(t)$ observation on $t = 0,1,\ldots T$, for location $i = 1,2,\ldots,N$, so :

$$V_i(t) = \sum_{j=1}^{N} W_{ij}\, Z_j(t)$$

This applies for linear form, that is $\mathbf{Y}_i = \mathbf{X}_i \beta_i + \mathbf{e}_i$

$$Y_i = \begin{bmatrix} Z_i(1) \\ Z_i(2) \\ \vdots \\ Z_i(t) \end{bmatrix}, X = \begin{vmatrix} Z_i(0) & V_i(0) \\ Z_i(1) & V_i(1) \\ \vdots & \vdots \\ Z_i(t-1) & V_i(t-1) \end{vmatrix}, e_i = \begin{bmatrix} e_i(1) \\ e_i(2) \\ \vdots \\ e_i(t) \end{bmatrix},$$

$$\beta = \left(\phi_{10}{}^{(1)}, \phi_{10}{}^{(2)}, \ldots, \phi_{10}{}^{(N)};\ \phi_{11}{}^{(1)}, \phi_{11}{}^{(2)}, \ldots, \phi_{11}{}^{(N)}\right).$$

The modeling for all linear models can be write as $\phi_{10}{}^{(10)} = \mathbf{X}\beta + \mathbf{e}$, where

$$Y = (Y_1', \ldots, Y_N')', X = (X_1', \ldots, X_N')', \beta = (\beta_1', \ldots, \beta_N')', e = (e_1', \ldots, e_N')'$$

So the form of least squares estimates of $\widehat{\boldsymbol{\beta}}_{\mathbf{T}}$ is $\widehat{\boldsymbol{\beta}}_{\mathbf{T}} = (\mathbf{X}'\mathbf{X})^{-1}\mathbf{X}'\mathbf{y}$

The following step is to determine whether residuals meet white noise. The GSTAR model is considered feasible if it meets the constant variant (white noise). Assumption testing is necessary needed. The time-series of r_t is said to be white noise if r_t is an identically normal distributed sequence with a finite mean and variance. r_t is normally distributed with mean 0 and variance σ^2, it can be said that if the AIC value in lag AR (0) and MA (0), the residual meets the residual of white noise assumption [8]. Fulfillment of white noise assumption can be done by using portmanteu test with significance level of 5%.

At last, the purpose of a forecast is to produce an optimum prediction with the smallest possible error. This leads to the value of the MSE (Mean Square Error) forecast to use. In general, MSE can be formulated in the following form

$$RMSE = \sqrt{\frac{1}{n}\sum_t^n (y_t - \hat{y}_t)^2} \tag{5}$$

Where y_t = the observation of i time, $\hat{y}_t$ = the prediction of i time, n = sum of observation (*testing data*)

2.3. Ensemble

Forecasting ensemble is forecasting techniques combine the output of several methods of forecasting as a prediction value. In its development, in the last 15 years can be demonstrated that using metod ensemble forecasting has become one of the techniques that are widely used in weather forecasting [9]. There are two methods most commonly used in combining a different output of ensemble members [10], the average method (averaging) and merging (stacking).

2.3.1 Ensemble Averaging

Ensemble averaging using the average output of the ensemble obtained by calculating the average of the model output member ensemble models. Suppose N is the number of individual members models in an ensemble, a combination of function *f* that is

$$\hat{y}_i = (\hat{y}_i^k), i = 1, 2 \dots, m \tag{6}$$

with $\hat{y}_i$ is a predicted value of observation *i* obtained from the *k* models and forms of the function *f* is

$$f = (\hat{y}_i^k) = \frac{1}{N}\sum_{k=1}^{N}\hat{y}_i^k \tag{7}$$

2.3.2 Ensemble Stacking

Stacking is a common method that uses a combination of a model of high-level and lower level models to achieve a higher accuracy prediction.[10] suggest the minimization of a function *G* to improve the generalizability of the model, is

$$G = \sum_{i=1}^{m}\left[y_i - \sum_{k=1}^{N} c_k \hat{y}_i^k\right], \qquad c_k > 0$$

The coefficients $\hat{c}_1, \hat{c}_2, \hat{c}_3, \dots \hat{c}_n$ in the above equation are estimated to form or construct a predictive value of the end of an ensemble, is

$$\hat{y}_i = \sum_{k=1}^{N} c_k \hat{y}_i^k, \quad i = 1, 2, \dots, m \tag{8}$$

3. Research Methods

The data used in this research is rainfall dataset from 77 rainfall stations in Jember Regency from January 2005 to December 2016 in four clusters [11], which was kindly given by Central Bureau of Statistics (BPS) Jember. The dataset is divided into two kinds, namely in-sample data and out-sample data. In-sample data is used to form forecasting model. While the out-sample data is used to check the model's forecasting. Models used in this study are ARIMA, GSTAR and both combination models called Ensemble. Moreover, those forecasting models are applied to forecast another time periods (out-sample data). From this forecasting results, we will get the errors and compare each model's error. The smaller the error, the better the model. Generally, this research method is shown on Figure 1.

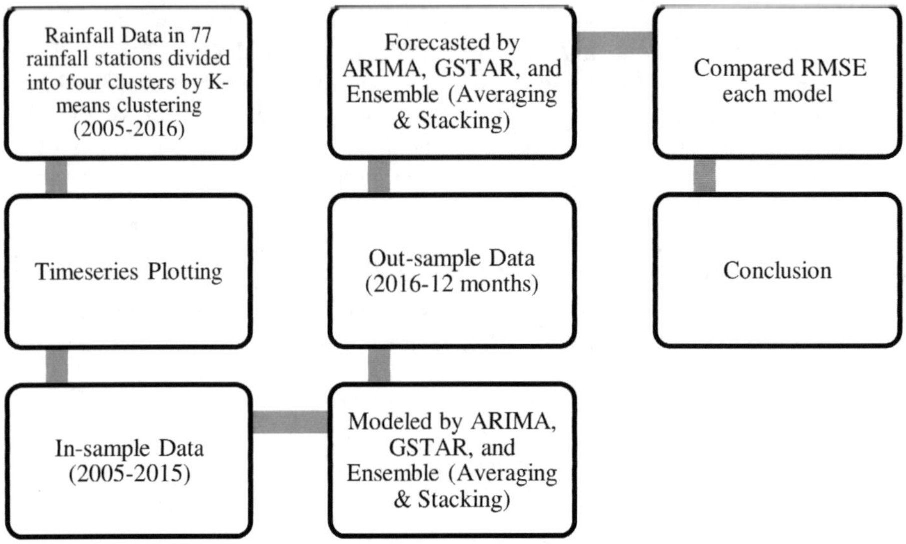

Figure 1. Research Flow Chart

4. Result adn Discussion

Time-series plotting as shown in Figure 2 says that the data plot are seasonal spread and relative sufficiently homogeneous for among four clusters. This paper concerns on non seasonal ARIMA model and seasonal GSTAR model due to expect to see combination of seasonal and non seasoanal two different models.

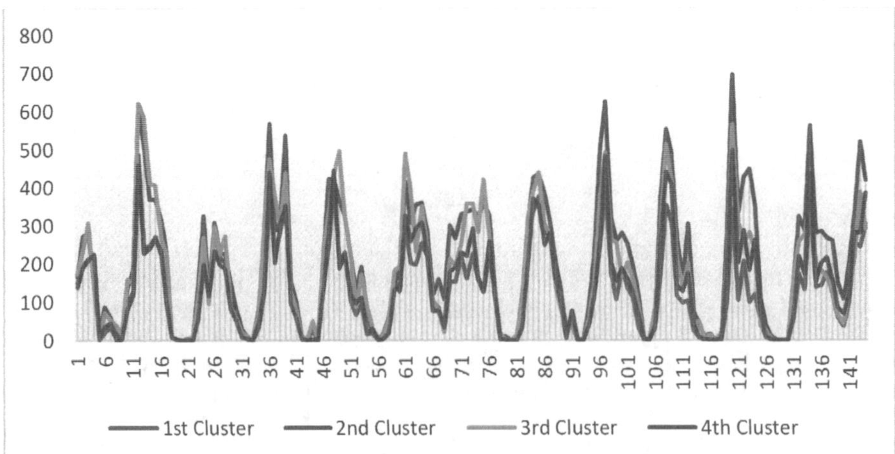

Figure 2. Timeseries Plotting of Rainfall Data from 77 Rainfall Stations in Jember (2005-2016)

4.1. ARIMA Models

ARIMA modeling starts with stationary checking. From Figure 3, we know that every single cluster is stationar, both in mean and variance. It also shows us the possible orde of ARIMA models from its lag. However, since ACF and PACF cut off after in certain time lags for each model,those order can be considered as space-time order candidates. ARIMA models also known as ARMA models as long as the timeseries data are already stationer in mean. In othe words, we no need to do differencing in this case, so the d orde is 0 (zero).

IOP Conf. Series: Journal of Physics: Conf. Series **1008** (2018) 012019 doi:10.1088/1742-6596/1008/1/012019

Table 1. ARMA Models In Every Single Cluster

Cluster	Model	Parameter	Parameter Value
		AR1	1.1899
		AR2	-0,332
1st Cluster	ARMA (4,1)	AR3	-0,0710
		AR4	-01878
		MA1	-07702
		AR1	1,5655
2nd Cluster	ARMA (2,1)	AR2	-0,7977
		MA1	-0,8296
		AR1	1,6773
3rd Cluster	ARMA (2,2)	AR2	-0,9239
		MA1	-1,1804
		MA2	0,3863
		AR1	1,0668
		AR2	-0,3781
4th Cluster	ARMA(5,2)	AR3	-0,0164
		AR4	-0,0414
		AR5	-0,2032
		MA1	-0,6123

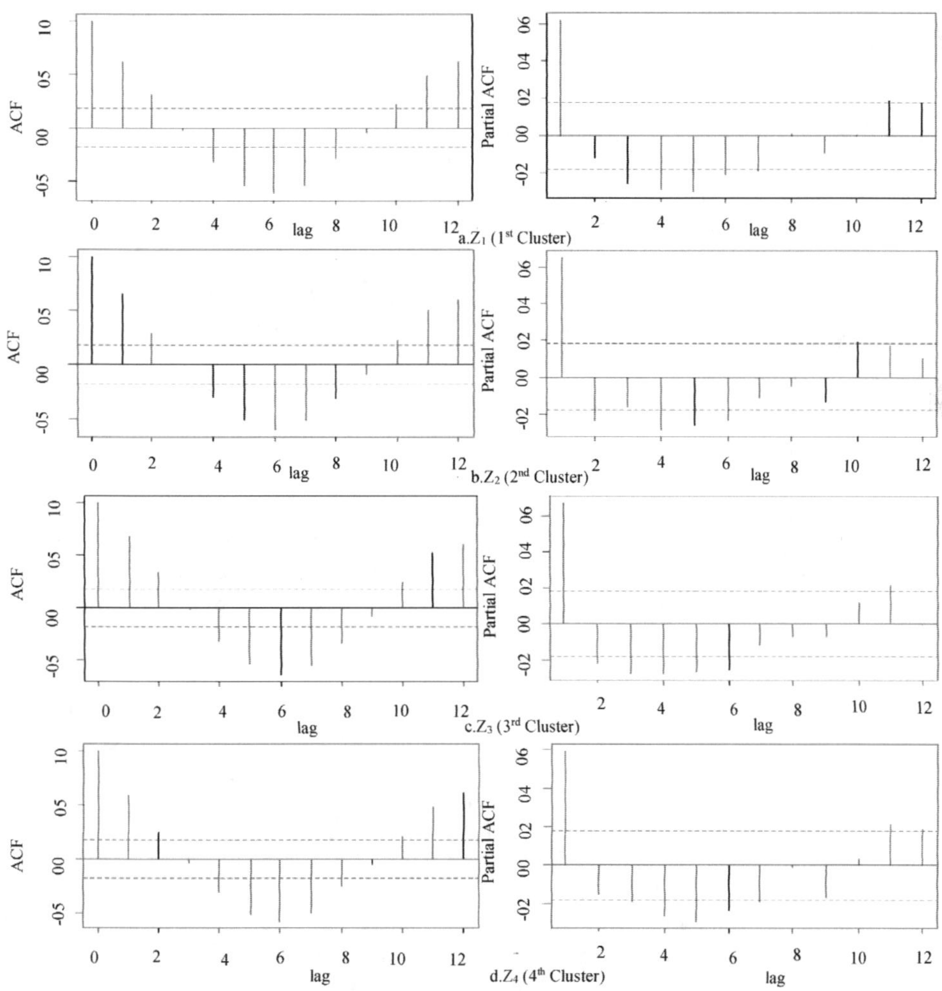

Figure 3. ACF and PACF Plot of ARIMA Models for four Clusters

Since there is no differencing in this case, then we could call the result models as ARMA (p,q). In addition, table 1 shows us the best ARMA model from all of space-time candidates in every single cluster based on Figure 3, with its parameter value. These models also have passed the diagnostic checking, so that it guarantees that they are white noise assumed.

4.2. GSTAR Models

In the other side, seasonal-GSTAR (1;1) is the simplest model of GSTAR $(p; \lambda_1, \dots, \lambda_p)$ model class defined in Eq. (3) because it is characterized by autoregressive terms lagged in time and spatial of order one. From Eq. (3), we can write seasonal-GSTAR (1;1) of four models for each cluster as follows:

$$Z_{K1}(t) = 0{,}234Z_{K1}(t-1) + 0{,}057Z_{K2}(t-1) + 0{,}039Z_{K3}(t-1) + 0{,}032Z_{K4}(t-1) + 0{,}129Z_{K1}(t-12) + 0{,}129Z_{K2}(t-12) + 0{,}088Z_{K3}(t-12) + 0{,}071Z_{K4}(t-12) + e_{K1}(t)$$

$$Z_{K2}(t) = -0{,}053Z_{K1}(t-1) + 0{,}579Z_{K2}(t-1) - 0{,}038Z_{K3}(t-1) - 0{,}035Z_{K4}(t-1) + 0{,}106Z_{K1}(t-12) + 0{,}275Z_{K2}(t-12) + 0{,}076Z_{K3}(t-12) + 0{,}071Z_{K4}(t-12) + e_{K2}(t)$$

$$Z_{K3}(t) = 0{,}072Z_{K1}(t-1) + 0{,}071Z_{K2}(t-1) + 0{,}311Z_{K3}(t-1) + 0{,}071Z_{K4}(t-1) + 0{,}098Z_{K1}(t-12) + 0{,}098Z_{K2}(t-12) + 0{,}203Z_{K3}(t-12) + 0{,}098Z_{K4}(t-12) + e_{K3}(t)$$

$$Z_{K4}(t) = 0{,}145Z_{K1}(t-1) + 0{,}176Z_{K2}(t-1) + 0{,}199Z_{K3}(t-1) - 0{,}119Z_{K4}(t-1) + 0{,}152Z_{K1}(t-12) + 0{,}185Z_{K2}(t-12) + 0{,}208Z_{K3}(t-12) - 0{,}057Z_{K4}(t-12) + e_{K4}(t)$$

4.3. Ensemble Models

4.3.1. Ensemble Averaging

In this section, we combine some forecasting models to cover possible factors than can influence rainfall data. First method is ensemble averaging. This method computes the average of two forecasting models used before, which are ARMA and GSTAR, based on Eq. (7) in every single cluster. Further, this forecasting result will help us to find the best performance of forecasting models by looking for each RMSE of the forecasting models before. In this case, each model brings its own error that possibly has large difference within both. Hence, it makes error of ensemble averaging would not be minimum, even larger than individual model's error. This could make ensemble averaging can not optimize forecasting result. Otherwise, ensemble averaging method will give better performance than indiviual forecasting model if one forecasting result negates the others so that the errors will be minimum. It will clearly stated in the next section.

4.3.2. Ensemble Stacking

In this section, we will forecast the rainfall data by Eq. (8), with estimasting the coefficient $c_k$by least square method. k here means the models involved in this ensemble, in this case, we used two models before, ARMA and GSTAR. In addition, to estimate coefficients in ensemble stacking, we involved the rainfall dataset in the year of 2016. From Eq. (8), we gives these four forecasting models of ensemble stacking:

$$Y_1 = 0.61\,ARMA + 0.22\,GSTAR$$
$$Y_2 = 0.58\,ARMA + 0.49\,GSTAR$$
$$Y_3 = 0.50\,ARMA + 0.03\,GSTAR$$
$$Y_4 = 0.66\,ARMA + 0.30\,GSTAR$$

4.4. The Best Performance of Forecasting Models

For the purpose of foracasting model, it would be useful if we also consider their forecast performance. Therefore, in this section we will examine one-step-ahead forecasting performance for each model candidate using out-sample data of rainfall dataset. RMSE value based on Eq. (5) determines the best performance of forecasting models. The smaller the value of RMSE model, it can be said the better the model. It can be seen on Table 2 that the model with the smallest RMSE is ensemble stacking of ARMA and GSTAR rather than individual models and ensemble averaging. Between individual models, ARMA and GSTAR, RMSE values show that ARMA gives the best performance (in average) based on the smaller velues. It means that seasonal models on ARIMA, also known as SARIMA, do not give much effects in rainfall forecasting due to extreme climate changes lately.

The RMSE value in Table 2 shows that weights given by each coefficient on each models can minimize the RMSE value in all clusters. In addition, combined forecasting models in this study may also reduce the RMSE model values, except for ensemble averaging. Further, ensemble averaging provides larger RMSE than GSTAR. It means that GSTAR shows better performance than ensemble averaging, unless the difference error of GSTAR and ARMA are smaller. So it can be concluded that the best model in this study is ensemble stacking of ARMA and GSTAR.

Table 2. RMSE Comparisons: Real data, ARIMA, GSTAR, and Ensemble

Cluster	ARMA	GSTAR	ENSEMBLE Averaging	Ensemble Stacking
1st Cluster	96,29834	100,05	90,88354	**87,49765**
2nd Cluster	145,4629	151,6928	132,3958	**131,0503**
3st Cluster	141,5264	129,6659	128,9629	**87,7086**
4th Cluster	101,9235	111,4439	100,1336	**98,98162**

Figure 4 also shows us that there is a large different performance between ARMA (in green) and GSTAR (in red) model like previous description. In the other side, we can see that ensemble stacking (in light blue) have the same pattern with real rainfall dataset (in dark blue) rather than three other forecasting models. It supports earlier statement that RMSE values of ensemble stacking will be smaller than the others. Ensemble averaging also gives better performance than individual models (ARMA and GSTAR), but not good enough rather than ensemble stacking.

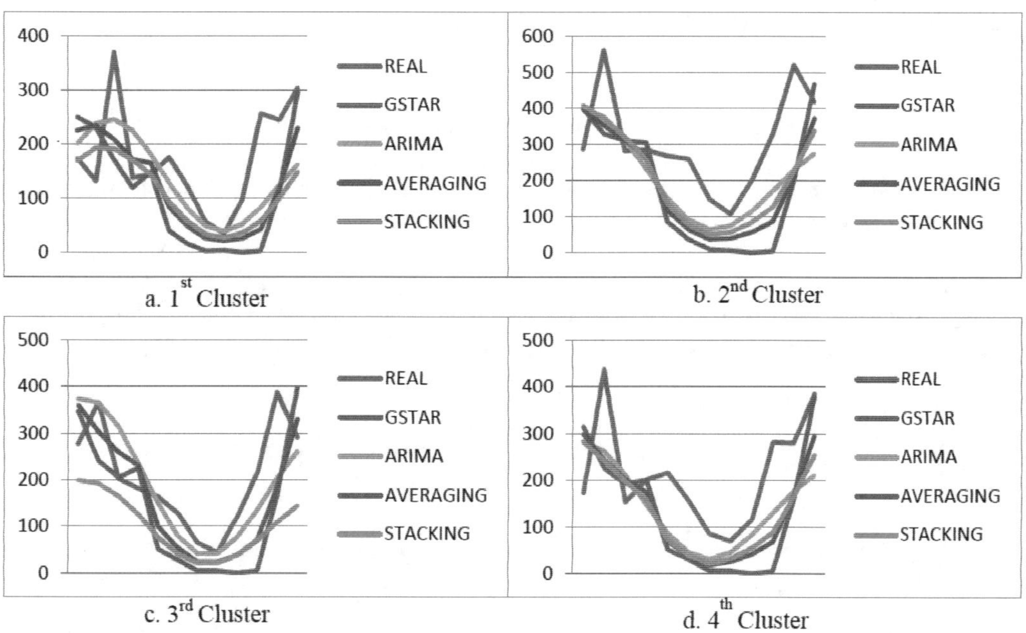

Figure 4. Plot of Forecasting Rainfall Dataset of 2015-2016 in Jember Regency

5. Conclusion

The conclusion from the above research is that each cluster has different RMSE values due to different topography of each region. The difference RMSE for each cluster is around 1,000 until 20,000. Ensemble stacking models of ARMA and GSTAR gives the best performance of forecasting models in all clusters rather than individual models and ensemble averaging of ARMA and GSTAR. While non seasonal ARMA model gives better performance in almost each cluster than GSTAR does. It tells us that in this case, seasonal models do not influence forecasting results since extreme climate changes occur lately.

Acknowledgement

This research was supported by DIPA Research Grant University of Jember no. 0715/UN25.3.1/LT.1/2017.

References

[1] Pfeifer P E and Deutsch S J A 1980a Three Stage Iterative Procedure for Space-Time Modeling, *Technometrics* **22** (1) 35-47

[2] Pfeifer P E and Deutsch S J 1980b Identification and Interpretation of First Order Space-Time ARMA Models *Technometrics* **22** (1) 397-408

[3] Borovkova S A Lopuhaa H P and Nurani B 2002 Generalized STAR Model With Experimental WEIGHTS In M Stasinopoulos & G Tou-louni (Eds.) *Proc. of the 17th International Workshop on Statictical Modeling* (Chania) 139-147

[4] Wei W W S 2006 *Time Series Analysis Univariate And Multivariate Methods* (Canada: Addison Wesley Publishing Company)

[5] Wutsqa D W Suhartono and Sujito B 2010 Generalized Space Time Autoregresive *Proc. Of The 6th IMT-GT Conf. On Mathematics, Statistics And Its Application* 120-131

[6] Nurcahyani F 2016 Pengelompokan Stasiun Curah Hujan Untuk Model Generalized Space Time Autoregresive (GSTAR) Pada Peramalan Curah Hujan Kabupaten Jember Dengan Tiga Pembobotan *Digital Repository University of Jember*

[7] Anggraeni D Prahutama A and Andari S 2013 Apliksi Generalized Space Time Autoregressive (GSTAR) Pada Pemodelan Volume Kendaraan Masuk Tol Semarang. *Media Statistika* **6 (2)** 71-80

[8] Tsay R S 2005 *Analysis of Financial Time Series* (New Jersey: John Wiley & Sons)

[9] Leutbecher M and Palmer T N 2008 Ensemble Forecasting *Journal of Computational Physics* 3515-39

[10] Zaier I, Shu C, Ouarda T, Seidou O and Chebana F 2010 Estimastion of Ice Thickness on Lakes Using Artificial Neural Network Ensembles *Journal of Hydrology* **383**(¾) 330-340

[11] Hadi A F Yudistira I Anggraeni D and Hasan M Geographical Clustering of The Rainfall Stations on Seasonal GSTAR Modeling for Rainfall Forecasting in Jember *preprint* Advanced Science Letters

IOP Conf. Series: Journal of Physics: Conf. Series **1008** (2018) 012020 doi:10.1088/1742-6596/1008/1/012020

A generalization of Cesàro sequence spaces in the Orlicz space

Haryadi[1], Supama[2], and A Zulijanto[3]

[1] Department of Mathematics, Universitas Muhammadiyah Palangkaraya, Indonesia
[2] Department of Mathematics, Universitas Gadjah Mada,Yogyakarta, Indonesia
[3] Department of Mathematics, Universitas Gadjah Mada,Yogyakarta, Indonesia

E-mail: haryadi_ump@yahoo.co.id

Abstract. In this paper, we generalize the Cesàro sequence spaces in the classic Banach space L_p to the generalized Orlicz space L_ϕ. We construct the space by replacing the norm $\|\cdot\|_p$ in L_p with modular ρ_ϕ in L_ϕ. This generalization has lead to the use of the Luxemburg norm to discuss some topological properties of the spaces. We prove results regarding to modular and norm convergence. We also describe some properties of the spaces and a closed subspaces of the space.

1. Introduction

Some new scalar-valued Cesàro sequence spaces were studied by many authors. Bala [1] constructs Cesàro sequence space using Orlicz function. The Cesàro-Orlicz space with the Δ_2-condition was discussed in [4] and [9]. In application, the sequence encountered also includes a vector-valued sequence, e.g. in the summation of fourier series and ergodic theorem. Hence, some authors are encouraged to concern their research on vector-valued sequence space ([6], [7] and [10]).

For the classical Banach space L_p, $1 \leq p < \infty$, the Cesàro sequence space W_p defined in [7], can be written as follows:

$$W_p = \left\{ (u_k) : \frac{1}{N} \sum_{k=1}^{N} \|u_k - u_0\|^p \to 0 \ (N \to \infty), \text{ for some } u_0 \in X \right\}. \tag{1}$$

The generalization of the L_p space to the Orlicz space, has led to the study of sequence space on the Orlicz space. In [2], Haryadi et.al. generalized the definition (1) by replacing the $\|\cdot\|^p$ in (1) with the $\int_E \phi(\cdot)$, where ϕ is the Orlicz function that satisfies Δ_2-condition. In this paper we generalize (1) to those sequence space with terms in the Orlicz space for arbitrary Orlicz function. Furthermore, we construct a modular on the space and describe some its topological properties.

Let X be a linear space over $\mathbb{R}$. The function $\rho : X \to [0, \infty]$ is called modular if the following conditions are satisfied:

(i) $\rho(x) = 0$ if and only if $x = \theta$, where θ is the zero element of X,

(ii) $\rho(-x) = \rho(x)$ for each $x \in X$.

(iii) $\rho(\alpha x + \beta y) \leq \rho(x) + \rho(y)$ for each $x, y \in X$ and $\alpha, \beta \geq 0$ with $\alpha + \beta = 1$.

If (iii) is replaced by
(iii') $\rho(\alpha x + \beta y) \leq \alpha\rho(x) + \beta\rho(y)$ for each $x, y \in X$ and $\alpha, \beta \geq 0$ with $\alpha + \beta = 1$, then ρ is called convex modular. If ρ is a modular on X, then

$$X_\rho = \left\{ x \in X : \lim_{\lambda \to 0} \rho(\lambda x) = 0 \right\} \tag{2}$$

is a linear space, and called a modular space. Let (x_k) be a sequence in X_ρ. The sequence (x_k) is called ρ-convergent in there exist $x_0 \in X_\rho$ such that $\rho(\lambda(x_k - x_0)) \to 0$ for some $\lambda > 0$ as $k \to \infty$ and it is called ρ-Cauchy if $\rho(\lambda(x_k - x_l)) \to 0$ for some $\lambda > 0$ as $k, l \to \infty$. The space X_ρ is called ρ-complete, if every ρ-Cauchy sequence id X_ρ is ρ-convergent. (see e.g. [8]).

The function $\phi : \mathbb{R} \to [0, \infty)$ is called Orlicz function, if it is continuous, even, convex, vanishing only at 0, and $\lim_{x\to\infty} \phi(x) = \infty$. An Orlicz function ϕ is said to satisfy the Δ_2-condition if there is a $K > 0$ such that $\phi(2x) \leq K\phi(x)$ for each $x \geq 0$. Give a Lebesgue measurable set $E \subset \mathbb{R}$. The generalized Orlicz space, denote by L_ϕ, is defined the space of all Lebesgue measurable real valued functions u on E such that $\int_E \phi(\lambda u(x))dx < \infty$ for some $\lambda > 0$. The function $\rho_\phi : L_\phi \to [0, \infty]$, where

$$\rho_\phi(u) = \int_E \phi(u(x))dx, \tag{3}$$

is a convex modular on L_ϕ. With respect to the modular ρ_ϕ, the space L_ϕ is ρ_ϕ-complete ([3]). Furthermore, with respect to the Luxemburg norm $\|\cdot\|_{(\phi)}$, i.e.

$$\|u\|_{(\phi)} = \inf\{t > 0 : \rho_\phi(u/t) \leq 1\}, \tag{4}$$

the Orlicz space is a Banach space (see e.g., [5] and [8]).

Let $\omega(L_\phi)$ denote the space of all sequence in L_ϕ. Now, we introduce the generalized Cesàro sequence space.

Definition 1. *For any Orlicz function ϕ, the Cesàro sequence spaces W_ϕ and $W_{\infty,\phi}$ are defined as*

$$W_\phi = \left\{ (u_k) \in \omega(L_\phi) : (\exists\lambda > 0, u_0 \in L_\phi) \lim_{N\to\infty} \frac{1}{N} \sum_{k=1}^{N} \rho_\phi(\lambda(u_k - u_0) = 0 \right\},$$

$$W_{\infty,\phi} = \left\{ (u_k) \in \omega(L_\phi) : \sup_N \frac{1}{N} \sum_{k=1}^{N} \rho_\phi(\lambda u_k) < \infty, \textit{ for some } \lambda > 0 \right\}.$$

2. The results

In this section we describe some properties of the Cesàro sequence space. We also give some topological properties of the space where the topology is generated by the Luxemburg norm.

Proposition 1. *If the Orlicz function satisfies Δ_2-condition, then*

(i) $W_\phi = \left\{ (u_k) \in \omega(L_\phi) : \lim_{N\to\infty} \frac{1}{N} \sum_{k=1}^{N} \rho_\phi(u_k - u_0) = 0, \textit{ for some } u_0 \in L_\phi \right\}$,

(ii) $W_{\infty,\phi} = \{(u_k) \in \omega(L_\phi) : \sup_N \frac{1}{N} \sum_{k=1}^{N} \rho_\phi(u_k) < \infty\}$.

IOP Conf. Series: Journal of Physics: Conf. Series **1008** (2018) 012020 doi:10.1088/1742-6596/1008/1/012020

Proof. We temporally write the right hand side of (i) by W_ϕ^0. It is clear that $W_\phi^0 \subseteq W_\phi$. Let $(u_k) \in W_\phi$, so there exist $\lambda > 0$ such that $\frac{1}{N}\sum_{k=1}^N \rho_\phi(\lambda(u_k - u_0) \to 0$ as $N \to \infty$. For $\lambda \geq 1$, by monotonicity of ϕ we have $(u_k) \in W_\phi^0$. For $\lambda < 1$, since ϕ satisfies the Δ_2-condition, then there exist $K(\lambda) > 0$ such that $\phi(\frac{x}{\lambda}) \leq K(\lambda)\phi(x)$ for each $x \geq 0$. Hence,

$$\begin{aligned}\frac{1}{N}\sum_{k=1}^N \rho_\phi(u_k - u_0) &= \frac{1}{N}\sum_{k=1}^N \int_E \phi\left(\frac{\lambda}{\lambda}(u_k(x) - u_0(x))\right) dx \\ &\leq \frac{K(\lambda)}{N}\sum_{k=1}^N \int_E \phi(\lambda(u_k(x) - u_0(x))dx \to 0\end{aligned}$$

as $N \to \infty$, i.e. $(u_k) \in W_\phi$. The proof of (ii) is similar. □

Note that from the above proposition, if $\phi(x) = |x|^p, 1 < p < \infty$, then $W_\phi = W_p$.

Using the fact that ρ_ϕ is a convex modular on L_ϕ, i.e. $\rho_\phi(\alpha u + \beta v) \leq \alpha\rho_\phi(u) + \beta\rho_\phi(v)$ for each $u, v \in L_\phi$ and $\alpha, \beta \geq 0$ with $1/\alpha + 1/\beta = 1$, it is easy to prove the following proposition.

Proposition 2. *For any Orlicz function ϕ, the function $\rho : \omega(L_\phi) \to [0, \infty]$ defined by*

$$\rho(u) = \sup_N \frac{1}{N}\sum_{k=1}^N \rho_\phi(u_k), \quad (u = (u_k)) \tag{5}$$

is a convex modular.

Proposition 3. *For any Orlicz function ϕ, the set*

$$W_{\infty,\phi} = \{u \in \omega(L_\phi) : \rho(\lambda u) < \infty, \ \textit{for some } \lambda > 0\} \tag{6}$$

is a modular space.

Proof. It is easy to see that $W_{\infty,\phi}$ is a linear space. Is clear that $(\omega(L_\phi))_\rho \subset W_{\infty,\phi}$. We are going to show that $W_{\infty,\phi} \subset (\omega(L_\phi))_\rho$. Let $u \in W_{\infty,\phi}$, i.e. $\rho(\lambda u) < \infty$ for some $\lambda > 0$. Let $\alpha \to 0$. By convexity of ρ, for $|\alpha/\lambda| < 1$ we have

$$\rho(\alpha u) = \frac{\alpha}{\lambda}\rho(\lambda u) \to 0$$

as $\alpha \to 0$. Hence, $u \in (\omega(L_\phi))_\rho$. □

Since ρ is a convex modular, we can define the Luxemburg norm $\|\cdot\|_\rho$ on $W_{\infty,\phi}$ as follows:

$$\|u\|_\rho = \inf\left\{t > 0 : \rho\left(\frac{u}{t}\right) \leq 1\right\}, \quad u \in W_{\infty,\phi}. \tag{7}$$

Given a sequence $(u^{(n)})$ in $W_{\infty,\phi}$. The sequence $(u^{(n)})$ is said ρ-convergent or modular convergent to $u \in W_{\infty,\phi}$, if there exists $\lambda > 0$ such that $\rho(\lambda(u^{(n)} - u)) \to 0$ as $n \to \infty$. It is said ρ-Cauchy if there exits $\lambda > 0$ such that $\rho(\lambda(u^{(n)} - u^{(m)})) \to 0$ as $n, m \to \infty$.

Proposition 4. *The space $W_{\infty,\phi}$ is ρ-complete, i.e. every ρ-Cauchy sequence is ρ-convergent.*

Proof. Given any ρ-Cauchy sequence $(u^{(n)})$ in $W_{\infty,\phi}$ and let $\varepsilon > 0$. There exist $\lambda_0 > 0$ and $n_0 \in \mathbb{N}$ such that

$$\sup_N \frac{1}{N}\sum_{i=1}^{N} \rho_\phi(\lambda_0(u_i^{(n)} - u_i^{(m)})) = \rho(\lambda_0(u^{(n)} - u^{(m)})) < \frac{\varepsilon}{3}, \quad \forall m, n \geq n_0. \tag{8}$$

Then for each $N \in \mathbb{N}$,

$$\frac{1}{N}\sum_{i=1}^{N} \rho_\phi(\lambda_0(u_i^{(n)} - u_i^{(m)})) < \epsilon, \quad \forall m, n \geq n_0. \tag{9}$$

Hence, for each $k \in \mathbb{N}$ and for each $m, n \geq n_0$,

$$\rho_\phi(\frac{\lambda_0}{k}(u_k^{(n)} - u_k^{(m)})) \leq \frac{1}{k}\rho_\phi(\lambda_0(u_k^{(n)} - u_k^{(m)})) \leq \frac{1}{k}\sum_{i=1}^{k} \rho_\phi(\lambda_0(u_i^{(n)} - u_i^{(m)})) < \varepsilon,$$

i.e. for each $k \in \mathbb{N}$, $(u_k^{(n)})_{n\in\mathbb{N}}$ is a ρ_ϕ-Cauchy in L_ϕ. Since L_ϕ is ρ_ϕ-complete, then there exist $u_k \in L_\phi$, $\lambda_k > 0$ and $n(k)$ such that

$$\rho_\phi(\lambda_k(u_k^{(n)} - u_k)) < \frac{\varepsilon}{3}, \quad \forall\, n \geq n(k).$$

Let $u = (u_k)$. For each $N \in \mathbb{N}$, let $\lambda = \min\{\lambda_k : 0 \leq k \leq N\}$ and $n' = \max\{n(k) : 1 \leq k \leq N\}$. Since ρ_ϕ monotonically increasing, then

$$\frac{1}{N}\sum_{k=1}^{N} \rho_\phi(\lambda(u_k^{(n)} - (u_k))) < \frac{\varepsilon}{3}, \quad \forall n \geq n'.$$

Hence, for $m, n \geq \max\{n_0, n'\}$,

$$\begin{aligned}\frac{1}{N}\sum_{k=1}^{N} \rho_\phi\left(\frac{\lambda}{2}(u_k^{(n)} - u_k)\right) &\leq \frac{1}{N}\sum_{k=1}^{N} \rho_\phi(\lambda(u_k^{(n)} - u_k^{(m)}) + \frac{1}{N}\sum_{k=1}^{N} \rho_\phi(\lambda(u_k^{(m)} - u_k))) \\ &< \frac{2\varepsilon}{3},\end{aligned}$$

which implies

$$\rho\left(\frac{\lambda}{2}(u^{(n)} - u)\right) < \varepsilon, \quad \forall n \geq \max\{n_0, n'\},$$

i.e. $(u^{(n)})$ is ρ-convergent to u.

We have to prove that $(u_k) \in W_{\infty,\phi}$. Let $n_1 = \max\{n_0, n'\}$. Since $u^{(n_1)} \in W_{\infty,\phi}$, then there exists $\alpha > 0$ such that

$$\sup_{N\in\mathbb{N}} \frac{1}{N}\sum_{k=1}^{N} \rho_\phi(\alpha u_k^{(n_1)}) < \infty.$$

Hence, for $\gamma = \min\{\alpha, \lambda/2\}$, we have

$$\begin{aligned}
\frac{1}{N}\sum_{k=1}^{N}\rho_\phi(\frac{\gamma}{2}u_k) &\le \frac{1}{N}\sum_{k=1}^{N}\rho_\phi(\gamma(u_k - u_k^{(n_1)})) + \frac{1}{N}\sum_{k=1}^{N}\rho_\phi(\gamma u_k^{(n_1)})\\
&\le \frac{1}{N}\sum_{k=1}^{N}\rho_\phi(\lambda/2(u_k - u_k^{(n_1)})) + \frac{1}{N}\sum_{k=1}^{N}\rho_\phi(\alpha u_k^{(n_1)})\\
&\le \rho\left(\frac{\lambda}{2}(u^{(n_1)} - u)\right) + \sup_{N\in\mathbb{N}}\frac{1}{N}\sum_{k=1}^{N}\rho_\phi(\alpha u_k^{(n_1)})\\
&\le M,
\end{aligned}$$

for some $M > 0$, and so $\sup_{N\in\mathbb{N}}\frac{1}{N}\sum_{k=1}^{N}\rho_\phi(\frac{\gamma}{2}u_k) < \infty$. □

Proposition 5. *With respect to the $\|\cdot\|_\rho$, the space $W_{\infty,\phi}$ is complete.*

Proof. Let $(u^{(n)})$ be any Cauchy sequence in $W_{\infty,\phi}$. Then for every $\lambda > 0$, $\rho(\lambda(u^{(n)} - u^{(m)}) \to 0$ as $m, n \to \infty$. Hence, there exists $u^0 \in W_{\infty,\phi}$ such that $\rho(\lambda(u^{(n)} - u^0)) \to 0$ as $n \to \infty$ for every $\lambda > 0$, for if no such u^0 exists, then there exists $\lambda_0 > 0$ such that $\rho(\lambda_0(u^{(n)} - u)) \not\to 0$ as $n \to \infty$ for every $u \in W_{\infty,\phi}$, and implies $\rho(\lambda_0(u^{(n)} - u^{(m)})) \not\to 0$ as $n \to \infty$ for every $m \in \mathbb{N}$, a contradiction. Let $n_0 \in \mathbb{N}$ such that

$$\rho\left(\frac{u^{(n)} - u^0}{1/\lambda}\right) \le 1,$$

for each $n \ge n_0$. Then $\|u^{n)} - u^0\|_\rho \le \frac{1}{\lambda}$ for each $n \ge n_0$. Since $\lambda > 0$ is arbitrary, then $(u^{(n)})$ is norm-convergent to u^0. □

Proposition 6. *If the Orlicz function ϕ satisfies the Δ_2-condition, then the norm convergent and the modular convergent are equivalent.*

Proof. It is enough to prove that $\rho(u^{(n)} - u) \to 0$ implies $\|u^{(n)} - u\|_\rho \to 0$ as $n \to \infty$. Given any $\lambda > 0$ and $\varepsilon > 0$. Since ϕ satisfies Δ_2-condition, then there exists $K(\lambda) > 0$ such that $\phi(\lambda x) \le K(\lambda)\phi(x)$ for each $x \ge 0$. Let $n_0 \in \mathbb{N}$ such that

$$\rho(u^{(n)} - u) < \frac{\varepsilon}{K^{l_0}}, \quad \forall n \ge n_0.$$

Then for each $n \ge n_0$,

$$\begin{aligned}
\rho(\lambda(u^{(n)} - u)) &= \sup_N \frac{1}{N}\sum_{k=1}^{N}\int_E \phi(\lambda(u_k^{(n)}(x) - u))\\
&\le K(\lambda)\sup_N \frac{1}{N}\sum_{k=1}^{N}\int_E \phi(u_k^{(n)}(x) - u_k(x))\\
&\le K^{l_0}\rho(u^{(n)} - u) < \varepsilon.
\end{aligned}$$

Since $\lambda > 0$ is arbitrary, then $\|u^{(n)} - u\|_\rho \to 0$ as $n \to \infty$. □

Proposition 7. *The space W_ϕ is a closed subspace of $W_{\infty,\phi}$.*

Proof. Obviously that W_ϕ is a linear subspace of $W_{\infty,\phi}$. Let $(u^{(n)})$ be any sequence in W_ϕ such that $\|u^{(n)} - u\|_\rho \to 0$ as $n \to \infty$ for some $u = (u_k) \in W_{\infty,\phi}$. Given any $\varepsilon > 0$ and $n \in \mathbb{N}$. Since $u^{(n)} = (u_k^{(n)})_{k\in\mathbb{N}} \in W_\phi$, then there exists $u_0^{(n)} \in L_\phi$ and $\lambda > 0$ such that $\frac{1}{N}\sum_{k=1}^{N} \rho_\phi(\lambda(u_k^{(n)} - u_0^{(n)})) \to 0$ as $N \to \infty$. Hence, there exists $N_0 \in \mathbb{N}$ such that

$$\frac{1}{N}\sum_{k=1}^{N} \rho_\phi(\lambda(u_k^{(n)} - u_0^{(n)})) < \frac{\varepsilon}{2}, \quad \forall N \geq N_0. \tag{10}$$

Since $\|u^{(n)} - u\|_\rho \to 0$, then $\rho(\lambda(u^{(n)} - u)) \to 0$, as $n \to \infty$. Therefore for any $N \in \mathbb{N}$, there exists $n_0 \in \mathbb{N}$ such that

$$\frac{1}{N}\sum_{k=1}^{N} \rho_\phi(\lambda(u_k^{(n)} - u_k)) \leq \rho(\lambda(u^{(n)} - u)) < \frac{\varepsilon}{2}, \quad \forall n \geq n_0. \tag{11}$$

Since

$$\frac{1}{N}\sum_{k=1}^{N} \rho_\phi(\frac{\lambda}{2}(u_k - u_0^{(n)})) = \frac{1}{N}\sum_{k=1}^{N} \rho_\phi(\lambda(u_k - u_k^{(n_0)})) + \frac{1}{N}\sum_{k=1}^{N} \rho_\phi(\lambda(u_k^{(n_0)} - u_0^{(n)})),$$

then by (10) and (11), for $n \geq n_0$,

$$\frac{1}{N}\sum_{k=1}^{N} \rho_\phi(\frac{\lambda}{2}(u_k - u_0^{(n)})) < \varepsilon, \quad \forall N \geq N_0$$

which mean $u = (u_k) \in W_\phi$. □

3. Conclusion and open problem

In this paper, we generalized the Cesàro sequence space in the generalized Orlicz space. We showed that the space is modular complete and norm complete. Furthermore, if the Orlicz function satisfies the Δ_2-condition, then the norm convergence and modular convergence on the space are equivalent. We also obtain that W_ϕ is a closed subspace of $W_{\infty,\phi}$.

Let (A_k) be a sequence of operators on the Orlicz space. From the topological properties we found above, we can further study conditions on (A_k) such that $\sum_{k=1}^{\infty} A_k u_k$ converges for each $(u_k) \in W_{\infty,\phi}$.

4. Acknowledgment

Authors gratefully acknowledge to the reviewer for making some useful corrections which improve the paper.

5. References

[1] Bala I 2012 *Communications in Mathematics and Applications* **3** (2) 197-204.
[2] Haryadi, Supama and Zulijanto 2017 *Far East Journal of Mathematical Sciences* **101** (9) 1897-1911.
[3] Khamsi W M and M A Koslowski 2015 *Fixed Point Theory in Modular Function Spaces* (Switzerland: Springer International Publishing)
[4] Kubiak D 2009 *J. Math. Anal. and App.* **349** 291-296.
[5] Krasnosel'skii M A and Rutickii Y B 1961 *Convex Functions and Orlicz Spaces* (Netherlands: P. Noordhoff Ltd)
[6] Maddox I J 1980 *Int. J. Math. Math. Sci.* **3** 423-432.
[7] Maddox I J 1974 *Composition Mathematica* **29** 35-42.
[8] Musielak J 1983 *Orlicz Spaces and Modular Space* (New York: Springer Verlag)
[9] Petrot N and Suantai S 2005 *Sci. Asia* **31** 173-177.
[10] Yilmaz Y and Ozdemir M K 2005 *Soochow J. Math.* **31** (3) 389-402.

IOP Conf. Series: Journal of Physics: Conf. Series **1008** (2018) 012021 doi:10.1088/1742-6596/1008/1/012021

An algorithm of Saxena-Easo on fuzzy time series forecasting

L C Ramadhani[1], **D Anggraeni**[1], **A Kamsyakawuni**[1], **A F Hadi**[1]
[1]Mathematics Department, University of Jember, Indonesia

E-mail: lutviacitraramadhani19@gmail.com

Abstract. This paper presents a forecast model of Saxena-Easo fuzzy time series prediction to study the prediction of Indonesia inflation rate in 1970-2016. We use MATLAB software to compute this method. The algorithm of Saxena-Easo fuzzy time series doesn't need stationarity like conventional forecasting method, capable of dealing with the value of time series which are linguistic and has the advantage of reducing the calculation, time and simplifying the calculation process. Generally it's focus on percentage change as the universe discourse, interval partition and defuzzification. The result indicate that between the actual data and the forecast data are close enough with Root Mean Square Error (RMSE)= 1.5289.

1. Introduction

One of soft computing methods that presented by Song and Chissom based on fuzzy set theory by Zadeh is known as Fuzzy time series. In recent years, it has been implemented in time series data such as enrollment data for the University of Alabama [1], [2], [3], [4], daily temperature [5], and car fatalities [6].

In 1998, Indonesia has experienced big economic crisis where inflation rate reach 77.63%. Various approaches have been developed by researchers to forecast Indonesia inflation rate so that goverment can anticipate impact from fluctuation of Indonesia inflation rate in future [7]. Now we try to implement fuzzy time series from Saxena and Easo on Indonesia inflation rate data. The Indonesia inflation rate is the change in the price increase of goods and services in Indonesia generally and continuously over a period of time which expressed as a percentage. This algorithm doesn't need stationarity like conventional forecasting method, capable of dealing with the value of time series which are linguistic, and has the advantage of reducing the calculation, time and simplifying the calculation process. It's also points out the forecast data of Alabama Universitu enrollments in 1978-1992 get the better result than the other algorithm which proposed before and produce better accuracy. The research is continued by calculating the accuracy from algorithm of Saxena-Easo which has been implemented on Indonesia inflation rate data. Here we used Root Mean Square Error (RMSE) and we compared the result of forecasting with actual data of Indonesia inflation rate.

The rest of this paper is organize as follow. In section 2, we give basic concept of fuzzy time series. Insection 3, we give explanation about algorithm of Saxena-Easo Fuzzy Time Series. The precision value of a forecasting method is given in section 4. The conclusions are discussed in section 5 where the result given value close to actual data.

2. The Basic Concept of Fuzzy Time Series

Definition 1 (Time Series): The sequence of data which is measured at successive time spaced at uniform time interval is called time series.

Definition 2 (Fuzzy set): The elements of sets with degrees of membership is called fuzzy set. Let $U = \{u1, u2, ..., u_n\}$ and let fuzzy set in U is A, A is defined below:

$$A = \frac{f_A(u_1)}{u_1} + \frac{f_A(u_2)}{u_2} + \cdots + \frac{f_A(u_n)}{u_n}$$

Where the membership function of the fuzzy set A is called f_A, $f_A: U \rightarrow [0,1]$. $f_A(u_i)$ denote the degree of membership of $u_i$in fuzzy set A, $f_A(u_i) \in [0,1]$ and $1 \leq i \leq n$.

Definition 3 (Fuzzy Time Series): Time series with fuzzy data. Let $Y(t)$ (t=...,0,1,2,...) is a subset of R and the universe discourse for $f_i(t)$ (i= 1, 2, ...). Then the collection of $f_i(t)$ is known as $F(t)$.$F(t)$ is fuzzy time series on Y(t) (t=...,0,1,2,...).

Definition 4 (Forecast Error): The difference between actual data and forecasted value of time series.

3. Algorithm of Saxena Easo Fuzzy Time Series

In this section, we will give explain about algorithm of Saxena-Easo Fuzzy Time Series that we use.

The procedure for this method is shown below:

Step 1.Change the actual data of Indonesia inflation rate into percentage of change using formula:

$$percChange = \left(\frac{x_t - x_{t-1}}{x_{t-1}}\right) \times 100 \quad (1)$$

rate for t-1.The percentage change of Indonesia inflation rate year to year from 1970 until 2016 is shown in Table 1. For example percentage change for 1970-1971:

$$\begin{aligned} percChange_{1970-1971} &= \left(\frac{x_{1971} - x_{1970}}{x_{1970}}\right) \times 100 \\ &= \left(\frac{2.62 - 8.94}{8.94}\right) \times 100 \\ &= -70.69 \end{aligned}$$

Table 1. The year to year percentage change of Indonesia inflation rate.

Years	**Percentage Change of Indonesia Inflation Rate**
1970-1971	-70.69
1971-1972	885.11
1972-1973	5.27
⋮	⋮
2013-2014	-0.24
2014-2015	-59.93
2015-2016	-9.85

Step 2.Define the universe of discourse U in U=[D_{min},D_{max}] and partition it into n equal intervals.D_{min} and D_{max} are the minimum and maximum from actual data of Indonesia inflation rate.For example, in this paper U = [-98, 886] is partitioned into seven equal intervals. Shown in Table 2.

Table 2. Intervals and frequency.

Interval	Frequency
[-98; 42.5714]	34
[42.5714; 183.1429]	8
[183.1429; 323.7143]	1
[323.7143; 464.2857]	1
[464.2857; 604.8571]	1
[604.8571; 745.4286]	0
[745.4286; 886]	1

Step 3.Divide the intervals before into several equal length based on the number of frequency from each interval in Table 2 and compute mid points from each subinterval. Next, give linguistic values for each subintervals. The result from this step is called fuzzy interval. For example, first interval in Table 2 has 34 frequency so that we divide first interval into 34 subinterval then we give linguistic value (X_1, X_2, …, X_{34}) and calculate mid points from each subinterval. The result shown in Table 3 where there's 46 subinterval from each interval in Table 2 then give linguistic value from X_1 until X_{46}.

Table 3. Linguistic value, interval and midpoint..

Linguistic	Interval	Mid Points
X1	[-98 ; -93.8655]	-95.9328
X2	[-93.8655; -89.7311]	-91.7983
X_3	[-89.7311; -85.5966]	-87.6639
⋮	⋮	⋮
X44	[323.7143 ;464.2857]	394
X45	[464.2857; 604.8571]	534.5714
X46	[745.4286 ; 886]	815.7143

Step 4. Compute prediction of percentage change with this formula (2):

$$t_j = \begin{cases} \frac{1+0.5}{\frac{1}{a_1}+\frac{0.5}{a_2}}, if\ j=1 \\ \frac{0.5+1+0.5}{\frac{0.5}{a_{j-1}}+\frac{1}{a_j}+\frac{0.5}{a_{j+1}}}, if 2 \leq j \leq n-1 \\ \frac{0.5+1}{\frac{0.5}{a_{n-1}}+\frac{1}{a_n}}, if\ j=n \end{cases} \quad (2)$$

Where a_{j-1}, a_j, a_{j+1} are mid points of fuzzy intervals X_{j-1}, X_j, X_{j+1}. For example prediction of percentage change for 1971, we have percentage change about -70.69 and it's in interval X_7. The result shown in Table 4.

$$t_{1971} = \frac{0.5+1+0.5}{\frac{0.5}{a_{j-1}}+\frac{1}{a_j}+\frac{0.5}{a_{j+1}}}$$
$$= \frac{0.5+1+0.5}{\frac{0.5}{-75.26}+\frac{1}{-71.13}+\frac{0.5}{-66.99}}$$
$$= -71.0057$$

Step 5. Defuzzify the fuzzy data with this formula (3).

$$F(t) = \left(\frac{t_j}{100} \times x_{t-1}\right) + x_{t-1} \quad (3)$$

Where x_{t-1} is actual data of Indonesia inflation rate for $t-1$, t_j is predicted percentage change. The result shown in Table 4 and for example below:

$$F(1971) = \left(\frac{-71.0057}{100} \times 8.94\right) + 8.94$$
$$F(1971) = 2.5921$$

Step 6. Count RMSE. The result from this step is shown in Table 4.

4. The Precision Value of A Forecasting Method

We use Root Mean Square Error (RMSE) to measure the accuracy of Saxena-Easo fuzzy time series for forecasting Indonesia inflation rate with formula (4):

$$RMSE = \sqrt{\frac{\sum_{t=1}^{n}(x_t - F_t)^2}{n}} \quad (4)$$

Where x_t is actual data of Indonesia inflation rate, *n* is number of data, F_t is forecasting data of Indonesia inflation rate.

Table 4. Comparison of actual data and forecast data

Years	Actual Inflation rate	Percentage Change	Fuzzy Set	Predicted Percentage	Forecast	SE
1970	8.94	-	-	-	-	-
1971	2.62	-70.6935	X_7	-71.0057	2.5921	0.0008
1972	25.81	885.1145	X_{46}	694.0434	20.8039	25.0607

1973	27.17	5.1145	X_{25}	-8.9238	23.5068	13.4192
1974	33.41	22.9665	X_{30}	23.6044	33.5833	0.03
⋮	⋮	⋮	⋮	⋮	⋮	⋮
2013	8.38	94.8837	X_{37}	84.6777	7.94111	0.1926
2014	8.36	-0.2387	X_{24}	-1.7564	8.2328	0.0162
2015	3.35	-59.9282	X_{10}	-58.5768	3.4630	0.0128
2016	3.02	-9.8507	X_{22}	-8.0632	3.0799	0.0036
					RMSE	1.5289

We implemented Saxena-Easo fuzzy time series using MATLAB software. Table 4 summarizes the result of Saxena Easo FTS method for Indonesia inflation rate in 1970 to 2016, where the universe of discourse is divided into 7 intervals and each intervals is divided based on frequency of each intervals with equal length until we get 46 of fuzzy set which is denoted (X_1, X_2, …, X_{46}). Every mid point of fuzzy set is used to compute predicted percentage. The predicted percentage change and actual data for *t*-1 are used to forecast Indonesia inflation rate for *t* period. The result shows between actual data and forecast data aren't too different. We can see in 1971, between actual data and forecast data have small square error (SE) about 0.0008. In the following, we compute the accuracy of this method with root mean square error (RMSE). The RMSE of the forecasting of this method is small enough about 1.5289. Figure 1 shows the black line for actual data and the grey line for forecast data with Saxena Easo fuzzy time series. We can see from this graph that the grey line is close enough to the black line, means that result of Saxena Easo fuzzy time series is quite accurate if it's compared with actual data.

Figure 1. The result of method for forecasting of Indonesia inflation rate

5. Conclusion

The study present algorithm of Saxena-Easo Fuzzy Time Series to be applied on Indonesia inflation rate data from 1970 – 2016. From table 4, we can see that this method provide RMSE = 1.5289 which is small enough so that this method gets a higher forecasting accuracy rate for forecasting. The smaller the RMSE is the more accurate are the forecast method. In future work, this method need to be developed to do forecasting for *t+1* period if there's no actual data for that period, provide smaller interval to make higher accuracy rate, and compare this method with conventional time series analysis.

Acknowledgement
This research was supported by DIPA Research Grant number 0715/UN25.3.1/LT.1/2017.

References:

[1] Jilani T A *et a.* 2007 Fuzzy metric approach for fuzzy time series forecasting based on frequency density based partitioning *Proc. of World Academy of Science, Engineering and Technology* **23** 33-38

[2] Saxena P and Easo S 2012 A New Method for forecasting enrollments based on fuzzy time series with higher accuracy rate *Int. J. Computer Technology and Applications* **3**(20) 33-37
[3] Stevenson M and Porter J E 2009 Fuzzy time series forecasting using percentage change as the universe of discourse *Proc. of World Academy of Science, Engineering and Technology* **55**(1) 54-57
[4] Song Q and Chissom B S 1993 Fuzzy time series and its models *Fuzzy Sets and Systems (North-Holland)* **54**(2) 69-77
[5] Lee L W *et al* 2006 Handling Forecasting Problems based on Two-Factors High-Orders Time Series *IEEE Transactions on Fuzzy Systems* **14**(4) 68-77

[6] Jilani T A *et al* 2007 Multivariate high order fuzzy time series forecasting for car road accidents *Int.Journal of Computational Intelligence* **4** 15-20
[7] Sari N R *et al* 2016 Backpropagation on Neural Network method for Inflation Rate Forecasting Indonesia *Int. J. Advance Soft Compu. Appl* **8**(7) 0-8

The modelling influence of water content to mechanical parameter of soil in analysis of slope stability

M Gusman[1], A Nazki[2] and R R Putra[1]

[1]Mining Engineering, Padang State University, Padang 25325, Indonesia
[2] Kementerian Energi dan Sumber Daya Mineral (KESDM)

Email : putra.roro21@gmail.com

Abstract. One of the parameters in slope stability analysis is the shear strength of the soil. Changes in soil shear strength characteristics lead to a decrease in safety factors on the slopes. This study aims to see the effect of increased moisture content on soil mechanical parameters. The case study study was conducted on the slopes of Sitinjau Lauik Kota Padang. The research method was done by laboratory analysis and simple liniear regression analysis and multiple. Based on the test soil results show that the increase in soil water content causes a decrease in cohesion values and internal shear angle. The relationship of moisture content to cohesion is described in equation Y = 55.713-0,6X with R^2 = 0.842. While the relationship of water content to shear angle in soil is described in the equation Y = 38.878-0.258X with R^2 = 0.915. From several simulations of soil water level improvement, calculation of safety factor (SF) of slope. The calculation results show that the increase of groundwater content is very significant affect the safety factor (SF) slope. SF slope values are in safe condition when moisture content is 50% and when it reaches maximum water content 73.74% slope safety factor value potentially for landslide.

1. Introduction

Landslide is one of the natural disasters that often occur in Indonesia including the city of Padang. Padang city is part of the Bukit Barisan mountain range area which geologically has the potential movement of land (landslide). An avalanche is a movement of slope constituents (soil, rocks, or rocks) that descend the slopes due to the disruption of the material stability of the slope [6]. According to Karyono [7] In general, the stability of the slope is controlled by several factors such as geometry, geological conditions (physical properties of slope constituents and geological structures), hydrogeological conditions, and engineering properties of slope constituents.

There are several areas prone to landslides in the city of Padang one of which is the slope along Sitinjau Lauik road. Disaster monitoring data mentions the number of avalanches during 2017 are as many as seven events, of which two of them impact the crisis. Avalanches often occur especially when high rainfall during the rainy season. High rainfall and long lasting play a role in triggering the movement of the land (landslide). Rainwater that seeps into the slopes can increase the saturation of the soil on the slopes, so the water pressure to stretch the soil bonds increases as well, and finally the ground mass is transported by the flow of water in the slope. Increased moisture content leads to a decrease in the value of mechanical properties (cohesion and internal shear angle) which affects the decrease in slope stability [5].

The objective of this research is to know the relationship of soil moisture content to mechanical properties (cohesion and internal shear angle) on the slope of Sitinjau Lauik road. Each value of mechanical properties simulated the value of Safety Factor (SF) so as to get correlation of water content to SF slope. Further data will be processed using multiple linear regression analysis to obtain the relationship between slope geometry (height and slope), water content, and SF grade of slope. The

IOP Conf. Series: Journal of Physics: Conf. Series **1008** (2018) 012022 doi:10.1088/1742-6596/1008/1/012022

outcome of this analysis will be obtained a safe slope design recommendation along the road of Sitinjau Lauik Kota Padang.

2. Theory

2.1 Water Content
The existence of water, especially groundwater (ground water) greatly affect the stability of a slope. Ground water has a pore water pressure which can lead to uplift force, thus lowering the shear strength and causing a landslide slope. Groundwater may affect the steepness of the slope in the following manner [7]:
a. Reduces rock or ground strength.
b. Changing mineral elements in rocks through chemical reactions and dissolution.
c. Change the density of rocks or soil.
d. Causes erosion.

The presence of ground water on the slope body will be a problem for the stability of the slope. This condition cannot be separated from outside influences, ie climate (represented by rainfall) that can increase ground water level, degree of saturation, or groundwater rise. The presence of water will decrease the physical properties and mechanical properties of the soil. The rising groundwater level increases the pore water pressure, which means it decreases the shear resistance of the slope masses, especially on soil material.

2.2 Parameter Mechanical Properties

2.2.1. Cohesion (c)
Cohesion is the tensile strength between the grains of soil expressed in units of weight per unit area. The greater the cohesion value the shear strength will be greater so that the slope will be more stable or safe [3].
The method of dealing with unstable slopes depends mainly on the nature of the soil encountered. The soil types and types of harmful soil formations above are layers composed of loose schist or soft clay in the form of fragments, cracked rigid clays, clays containing sand or silt, and cohesive soil masses containing layers or bags silt-tanned sands or sand [8].

2.2.2 Slide Angle (φ)
The inner sliding angle is an angle formed from the normal stress and shear stress relationships in the rock material. The inner sliding angle is the fracture angle formed when a rock is subjected to a voltage exceeding its shear stress. The larger the shear angle in a material, the material will be more resilient to accept the outside stress imposed so that the slope will be more stable or safe [2].

2.3 Safety Factor (SF)
The slope generally occurs through a particular plane called the slip surface. The stability of the slope depends on the driving force and the holding force acting on the slip. The resisting force is the force that holds out to avoid sliding, whereas the driving force is the force that causes the looms. The comparison between retaining forces against forces that move the soil is called the Safety Factor [2].
Thus, a minimum SF value with a certain value is recommended as the lowest FS boundary which is still safe so that the slope can be declared stable or not. In this study, the minimum safety factor used is SF ≥ (equal to or greater) of 1.25, in accordance with the procedures of Joseph E. Bowles [1], provided that:

SF ≥ 1.25: Safe Slope
SF <1.07: Unsafe Slope

SF> 1.07; <1.25: The Critical Slope

2.4 Simple and Multiple Linear Regression

Regression analysis is used to provide an explanation of the relationship between two types of variables, namely the dependent variable with the independent variable. The analysis of the relationship between two variables is called a simple regression analysis if it involves only one independent variable. The relationship between the dependent variable (Y) with the independent variable (X) is written in a general linear model on the next sheet.

$$y = i + xi \tag{1}$$

With i = 1,2, ... n is the regression coefficient which means the magnitude of the change in, if Xi increases one unit and the other variable is constant or called intercept. Multiple linear regression is almost the same as simple linear regression, only in multiple linear regression of independent variables more than one independent variable. The purpose of multiple linear regression analysis is to measure the intensity of the relationship between two or more variables and make predictions of the estimated values of x and y (Hastono, 2006). In general, multiple linear regression model as follows:

$$A = b_o + b_1.x_1.b_2.x_2 + \cdots + b_k.x_k \tag{2}$$
$$i = 1,2,\ldots,n$$
$$k = independent\ variable, ie\ x_1 \ldots x_k$$

From the regression equation obtained is known the interpretation of the relationship between the dependent variable with the independent through the value of the coefficient of determination (R^2). The greater the value of R^2, the stronger the relationship between the variables [4]. Interpretation of R^2 values can be seen in Table 1.

Tabel 1. Interpretation of coefficient of determination (R^2)

R^2	Interpretation
0,00 – 0,25	Low Correlation
0,26 – 0,50	Fair Correlation
0,51 – 0,75	High Correlatiion
0,76 – 1,00	Very High Correlation

3. Research Sites

The research sites are located on the slopes along Sitinjau Lauik Street in Padang City, with five points of data collection and laboratory testing sample (SL 1 until SL 5). The sampling map at the study site can be seen in Figure 1 (via Google earth).

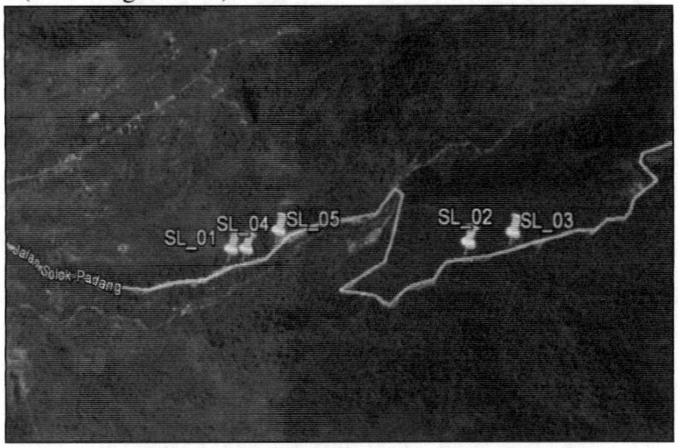

Figure 1. Sampling location map

Slope parameter data taken include the sampling coordinates (x, y, z), altitude, and slope. Sampling using a drill hand with an average depth of 0.9 to 1 meters. Slope parameter data in the study area can be seen in Table 2.

Table 2. Coordinate and geometry of laboratory test sampling slope

No	Coordinate (UTM)	Angle of Slope (^{0})	Height (meters)
1	East = 0666877 South = 9894886 Elevation = 353 m	51	23
2	East = 0667033 South = 9894960 Elevation = 390 m	54	14
3	East =0667215 Soth =9895055 Elevation = 400 m	60	11
4	East =0666807 South =9894884 Elevation = 405 m	43	20
5	East =0667937 South =9894866 Elevation = 385 m	65	15

it condition of one of the slopes of data collection and laboratory sample (SL2) can be seen in Figure 2. In Figure 2, the slopes have experienced landslide indicated by the decrease of soil material from the initial position. The type of soil material is brownish-brown clay with morphology at the top and sides of the slopes overgrown by trees and nail plants. The dimension of avalanche is 7.5 meters. In addition to the foot of the slope there are several points of water seepage.

Figure 2. Sampling location (SL 2)

IOP Conf. Series: Journal of Physics: Conf. Series **1008** (2018) 012022 doi:10.1088/1742-6596/1008/1/012022

4. Laboratory Test Result

Laboratory tests include the physical and mechanical properties of the slope samples. Physical properties tested were weight of content, specific gravity, and moisture content of soil samples. From the test results obtained the average value of each parameter as in Table 3.

Table 3. Physical properties testing results

No	Physical Parameters	Value	Unit
1	Natural Density	17.68	Kn/m^3
2	Saturated Density	21.93	Kn/m^3
3	Dry Density	10.80	Kn/m^3
4	Sfesicic Gravity	1.24	
5	Water Content	63.74	%

As for the mechanical properties of soil shear strength testing to obtain the value of normal load and shear load which is then processed into a normal voltage and shear stress. Furthermore, the normal voltage and shear stress relationship curves result in deep cohesion values and shear angles as in Table 4.

In this shear strength test, simulation of the influence of water content by gradually adding water to the soil sample. The simulation is carried out constantly with water addition of 2.5% of the sample weight each time the test is performed From the test results showed that the addition of water can only be done up to 10% of the sample weight, the addition of more than 10% causes the sample cannot be simulated again because the sample condition was destroyed by water added.

Table 4. Mechanical properties testing results

Water content (%)	Mechanical properties		information
	cohesion (Kn/m^3)	slide Angle (^{0})	
0.00	50.00	37.00	
10.00	46.71	35.33	
20.00	43.42	33.67	
30.00	40.12	32.00	simulation
40.00	36.83	30.33	
50.00	33.54	28.67	
60.00	30.25	27.00	
63.74	23.66	23.67	natural
66.24	16.16	20.68	
68.74	10.69	19.69	simulation
71.24	3.61	18.18	
73.74	1.42	17.40	
>73.74-100	-	-	Cannot be simulated again

Table 4 shows the simulated water content can be done in the range 0% to 73.74%. This means that the soil samples reach the maximum water content in the water content of 73.74%. The simulation cannot be done anymore exceeds the maximum water content.

4.1 Water Content Relation to Mechanical Properties

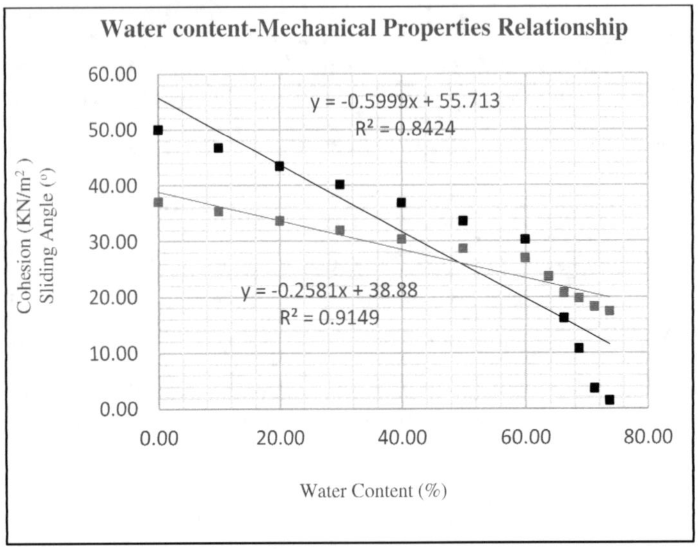

Figure 3. Graph of relation of water content to mechanical properties (cohesion and sliding angle)

The water content relationship with mechanical properties is obtained by plotting the water content test data and the mechanical properties (cohesion and internal shear angle) into the scatter plot. Figure 3 shows that linear regression is representative enough to describe the relationship between the variables because the graph tends to form a straight line (linear). From the results of this linear regression can be concluded that the higher the water content on the soil causes the cohesion parameters and sliding angle in the lower.

4.1.1 Water and Cohesion Relationship

From Figure 3, we get the equation of water and cohesion relationship as follows:

$$y = 55.713 - 0.6x_1 \quad (3)$$
$$With\ R^2 = 0.842$$

With Y is cohesion and X is the moisture content. The empirical equation shows a very strong (perfect) relationship, this is indicated by the value $R^2 = 0.842$. The relationship of moisture content to cohesion shows a correlation with inverse meaning that the increase in water content value leads to a decrease in the value of cohesion.

4.1.2 Water Content Relationships and Sliding Angle

From Figure 3 we get the equation of water content relationship and shear angle in the following:

$$y = 38.878 - 0.258x_1 \quad (4)$$
$$With\ R^2 = 0.915$$

With y is the internal sliding angle and x is the water content. The empirical equation shows a very strong (perfect) relationship, this is indicated by the value $R^2 = 0.915$. The association of water content with shear angle in showing correlation is inversely which means the increase of water content value causes decrease to deep shear angle value.

IOP Conf. Series: Journal of Physics: Conf. Series **1008** (2018) 012022 doi:10.1088/1742-6596/1008/1/012022

4.2 *Anaslisis Value of Slope Safety Factor*

From the simulation of the water content of 0%-73.74%, the slope value analysis of slope security actors based on the mechanical properties of each water content simulation result. Here is the result of processing the value of slope safety factor.

4.3 Water Content Relationship to Safety Factors (SF) of Slope

Table 5. Safety factor simulation values

Water content (%)	Safety factor (SF)
0.00	2.471
10.00	1.757
20.00	1.639
30.00	1.516
40.00	1.390
50.00	1.259
60.00	1.122
63.74	0.897
66.24	0.633
68.74	0.525
71.24	0.388
73.74	0.333

4.4 Water Content Relationship to Safety Factors (SF) Slope

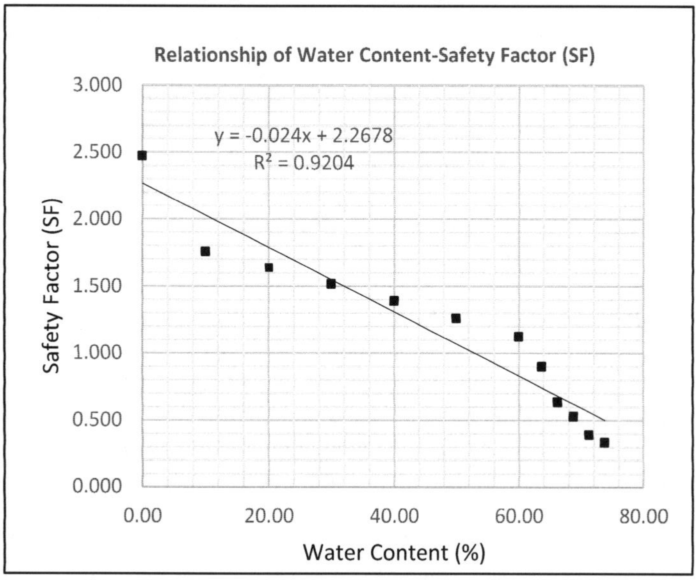

Figure 4. Water content relationship to safety factors (SF) of slope

The relationship of moisture content to SF slope is obtained by plotting the water content test and SF slope into scatter plot. Figure 4 shows that linear regression is representative enough to describe the relationship between the variables because the graph tends to form a straight line (linear).

From Figure 4 we get the equation of water content relationship to the value of SF slope as follows:

$$y = 2.267 - 0.024x \quad (5)$$
$$with\ R^2 = 0.92$$

With y is the Safety Factor (SF) and x is the water content. The empirical equation shows a very strong (perfect) relationship, this is indicated by the value of $R^2 = 0.92$. The relationship of moisture content to the SF slope showed a correlation between the water content and the decrease of the water content value causing the decrease of SF value.

From the graph shows that when the water content ranges between 0% to 50% so SF value> 1.25 which means slope conditions in safe category. When water content increases from 51 to 60% so SF value 1.0-1.25 which means slope conditions in the critical category. When the water content is ≥ 63.74% so SF value <1 which means this condition indicates the slope is unsafe (landslide).

5. Conclusion

From the data analysis done it can be concluded that:

1. Addition of water content on soil samples on the slopes of the Sitinjau Lauik causeway decrease to cohesion values and internal shear angle. This decrease in value causes the weakening of the strength of the slope so that the SF value on the slope also decreases
2. The slope of the road in Sitinjau Laut is in safe condition when the water content ranges between 0 untill 50%, critical when water content 51 to 60%, and unsafe (landslide) when water content ≥ 63.74%
3. High water content in the soil may cause a decrease in the level of stability (SF) on the slopes of the lava site so that it needs a way to reduce water content levels especially during rainfall such as slope or shotcrete installation.

6. Acknowledgments

The authors would like to express sincere appreciation for Soil Mechanics Laboratory of Mining Engineering Department Padang State University.

References

[1] Bowles J E 1971 *Foundation Analysis and Design* (Tokyo: McGraw- Hill Kogakusha Ltd)
[2] Hardiyatmo H C 2002 *Mekanika Tanah I ed III* (Yogyakarta: Gadjah Mada University Press)
[3] Hardiyatmo H C 2012 *Penanggulangan Tanah Longsor dan Erosi* (Yogyakarta: Gadjah Mada University Press)
[4] Hastono and Priyo S 2006 *Analisis Data SPSS* (Jakarta: Universitas Indonesia)
[5] Hustrulid W 2006 *Open Pit Mine Planning & Design 3rd ed* (London: CRC Press)
[6] Karnawati D 2005 *Bencana Alam dan Gerakan Massa Tanah di Indonesia dan Upaya Penanggulangannya* (Yogyakarta: Jurusan Teknik Geologi Universitas Gadjah Mada)
[7] Karyono 2004 *Kemantapan Lereng Batuan* (Hand Out Diklat Perencanaan Tambang UNISBA)
[8] Terzaghi K and Peck R B 1967 *Soil Mechanics in Engineering Practice* (New York: John Wiley & Sons, Inc)

Hybrid ARIMAX quantile regression method for forecasting short term electricity consumption in east java

M Prastuti[1], Suhartono[2], and NA Salehah[2]

[1] Business Statistics Department, Institut Teknologi Sepuluh Nopember, Surabaya, Indonesia 60111
[2] Statistics Department, Institut Teknologi Sepuluh Nopember, Surabaya, Indonesia 60111

Email: suhartono@statistika.its.ac.id

Abstract. The need for energy supply, especially for electricity in Indonesia has been increasing in the last past years. Furthermore, the high electricity usage by people at different times leads to the occurrence of heteroscedasticity issue. Estimate the electricity supply that could fulfilled the community's need is very important, but the heteroscedasticity issue often made electricity forecasting hard to be done. An accurate forecast of electricity consumptions is one of the key challenges for energy provider to make better resources and service planning and also take control actions in order to balance the electricity supply and demand for community. In this paper, hybrid ARIMAX Quantile Regression (ARIMAX-QR) approach was proposed to predict the short-term electricity consumption in East Java. This method will also be compared to time series regression using RMSE, MAPE, and MdAPE criteria. The data used in this research was the electricity consumption per half-an-hour data during the period of September 2015 to April 2016. The results show that the proposed approach can be a competitive alternative to forecast short-term electricity in East Java. ARIMAX-QR using lag values and dummy variables as predictors yield more accurate prediction in both in-sample and out-sample data. Moreover, both time series regression and ARIMAX-QR methods with addition of lag values as predictor could capture accurately the patterns in the data. Hence, it produces better predictions compared to the models that not use additional lag variables.

1. Introduction

Quantile regression is an estimation method that can overcome the data distribution that is not uniform. Data that has a non-uniform pattern is called heteroscedasticity, or a non-fulfillment of homoscedasticity assumption on regression analysis using Ordinary Least Square (OLS) method is also often called heteroscedasticity. This problem can be overcome by using the quantile regression method. The parameter estimation in the OLS method only provide a solution of the mean problem, so Koenker and Basset developed an alternative method of quantile regression [1]. Quantile regression extends the calculation of coefficient values across various quantiles, in order to provide more complete picture of the data conditions. This method is robust against outlier data, so it is highly recommended to be used to analyze a number of data that is not symmetrical and has a not homogeneous distribution. Forecasting using a quantile regression approach will produce better accuracy than the classical conditional mean method [2].

Combining two or more methods or hybrid method tend to improve the accuracy of prediction [3], so many researchers are using and developing it now. Hybrid method has been widely used in various field of science. For examples the development of combined method of quantile-based Regression Neural Network (QRNN) has been used to estimate the conditional density of multiperiod return value and compared with the results of the estimation using the GARCH-based Quantile Regression, and it show that the QRNN method is able to predict conditional density well [4]. Other studies apply hybrid ARIMA-LR (Linear Regression) method to predict the arrival of patients in an ER in China [5]. In 2015, Arunraj and Ahrens proposed hybrid SARIMAX-Quantile Regression model then applied it to forecast daily food sales and this study shows that hybrid SARIMAX-QR model provides better forecast accuracy for out-sample data than other individual method and provides better interval prediction [6].

This research will conduct a study about the development of hybrid models for forecasting the electricity energy consumption data in East Java. The data is suspected to have a calendar variation effect, resulting in the presence of certain patterns such as seasonal patterns, holiday effects, and effects of special days. This cause indicates that there were heteroscedasticity case in the electricity energy consumption data, so generating the forecast value of its data will not be easy. Based on this condition, this study aims to develop hybrid method using hybrid ARIMAX Quantile Regression (ARIMAX-QR) method for forecasting of electric energy consumption data in East Java. This method is expected to be able to yield forecasting model for data of electric energy consumption in East Java properly and minimize the cause of any forecasting errors. Forecasting of data on electricity consumption in East Java is very important, because it can be used as a material for distribution of electricity plan more efficient, so the distribution of electricity energy to consumers in East Java can be done optimally and fulfills the needs.

2. Literature Review

2.1. ARIMA

ARIMA is a time series analysis method composed of two models, namely autoregressive model (AR) and Moving Average (MA). ARMA model is composed of AR order p and MA order q, while ARIMA (p, d, q) is ARMA model (p, q) which get differencing as much as d. In the short-run electricity consumption data, double seasonal patterns are present in seasonal per-half hour and seasonal daily so it called has double seasonal multiplicative or Double Seasonal ARIMA patterns [7]. The Double Seasonal ARIMA model is written with ARIMA $(p,d,q)(P_1,D_1,Q_1)^{s1}(P_2,D_2,Q_2)^{s2}$ which has the following common form,

$$\phi_p(B)\Phi_{P_1}(B^{s_1})\Phi_{P_2}(B^{s_2})(1-B)^d(1-B^{s_1})^{D_1}(1-B^{s_2})^{D_2}Y_t=\theta_q(B)\Theta_{Q_1}(B^{s_1})\Theta_{Q_2}(B^{s_2})a_t. \quad (1)$$

In performing ARIMA model, there are several steps that must be done, such as model identification, parameter estimation and testing, model diagnostic, model selection, and forecasting [8].

2.2. Time Series Regression

The models in time series regression for the electricity forecasting are involved by trend, seasonal patterns, special days, and autocorrelated errors [7]. The effect of trend on time series regression is expressed by t. Weekly seasonal pattern consists of 7 variables i.e. Monday ($H_{1,t}$), Tuesday ($H_{2,t}$), Wednesday ($H_{3,t}$), Thursday ($H_{4,t}$), Friday ($H_{5,t}$), Saturday ($H_{6,t}$), and Sunday ($H_{7,t}$). The daily seasonal pattern consists of 48 variables because this study uses the electricity consumption data per half an hour ($D_{1,t}, D_{2,t}, ..., D_{48,t}$). Otherwise, the pattern of special day is indicated by dummy variable, $L_{1,t}$. The model of the time series regression used in this study is,

$$Y_t=\eta t+\gamma_1 H_{1,t}+\gamma_2 H_{2,t}+...+\gamma_7 H_{7,t}+\alpha_1 D_{1,t}+\alpha_2 D_{2,t}+...+\alpha_{48} D_{48,t}+\omega_1 L_{1,t}+\omega_2 L_{2,t}+...+\omega_8 L_{8,t}+\varepsilon_t \quad (2)$$

From the time series regression equation above there are 64 predictors of dummy variables. If the error does not fulfill the white noise assumption then lag of series will be used as an additional predictor variable. The selection of lag can be determined based on ACF and PACF plots [9]. Here are the steps of model building in the time series regression.

a. Parameter Estimation

Parameter estimation using Ordinary Least Square (OLS) minimizes the sum of squares of deviations or errors of observation values. The equation of a simple linear regression model is as follows [10]:

$$\mathbf{Y} = \mathbf{X}\boldsymbol{\beta} + \varepsilon \,, \tag{3}$$

with $\boldsymbol{\beta}$ being a vector containing the parameters that will be estimated from the model. Equation 3 then differentiated to each parameter and the estimated parameter that obtained from this process can be written as,

$$\hat{\boldsymbol{\beta}} = (\mathbf{X}'\mathbf{X})^{-1}\mathbf{X}'\mathbf{Y} \,, \tag{4}$$

where,

$$\mathbf{Y} = \begin{bmatrix} Y_1 \\ Y_2 \\ \vdots \\ Y_n \end{bmatrix}, \ \mathbf{X} = \begin{bmatrix} 1 & H_{1,1} & \cdots & L_{8,1} \\ 2 & H_{1,2} & \cdots & L_{8,2} \\ \vdots & \vdots & \ddots & \vdots \\ n & H_{1,n} & \cdots & L_{8,n} \end{bmatrix}, \ \boldsymbol{\beta} = \begin{bmatrix} \eta \\ \gamma_1 \\ \vdots \\ \omega_8 \end{bmatrix}.$$

b. Parameter Significance Test

Parameter significance test is done partially. Partial testing is a test that can be done as many parameters in the regression model [10]. The hypothesis testing parameters can be written as,

$H_0 : \beta_j = 0$

$H_1 : \beta_j \neq 0, j = 1, 2, ..., k$.

The t statistic will be used to test the significance of the parameters,

$$t = \frac{\hat{\beta}_j}{se(\hat{\beta}_j)} \,. \tag{5}$$

Reject H_0 if $|t| > t_{\alpha/2, n-k}$ or if p-value less than α , with n is the sample size and k is the number of regression coefficients.

2.3. *Quantile Regression*

Quantile is a technique of dividing a group of data into several equal parts, after the data is sorted from the smallest or the largest [11]. Quantile regression estimation is more robust to the outlier(s), and heteroscedasticity. It is also does not need any assumption testing. Heteroscedasticity or unequal variation implies that there is more than one slope representing the relationship between response and predictor measured on a subset of these factors. Quantile regression estimates the slope(s) from minimum to maximum response, giving a more complete explanation of the relationship between variables that are sometimes missed by other regressions [12].

The quantile regression approach assumes the various quantile functions of a Y distribution as a function of X. The use of this regression method is performed by divide or separate the data into several groups that suspected to have different alleged values on the quantile. The estimation of the response on the quantile regression uses the value of quantile-τ of Y with certain X. The model of a quantile regression equation can be written as,

$$\mathbf{y} = Q_\tau(Y \mid X) = \mathbf{X}\boldsymbol{\beta}(\tau) + \mathbf{u} \ \ , \tag{6}$$

where,

$$\mathbf{y}=\begin{bmatrix} y_1 \\ y_2 \\ \vdots \\ y_n \end{bmatrix},\ \mathbf{X}=\begin{bmatrix} x_{11} & x_{21} & \cdots & x_{p1} \\ x_{12} & x_{22} & \cdots & x_{p2} \\ \vdots & \vdots & \ddots & \vdots \\ x_{1n} & x_{2n} & \cdots & x_{pn} \end{bmatrix},\ \boldsymbol{\beta}(\tau)=\begin{bmatrix} \beta_1(\tau) \\ \beta_2(\tau) \\ \vdots \\ \beta_p(\tau) \end{bmatrix},\ \mathbf{u}=\begin{bmatrix} u_1 \\ u_2 \\ \vdots \\ u_n \end{bmatrix},$$

and $0<\tau<1$.

Quantile regression weighted τ for positive error and $(1-\tau)$ for negative error so the parameter estimation on quantile regression [11] can be expressed in the following form,

$$\hat{\boldsymbol{\beta}}(\tau) = \min_{\beta\in\mathbb{R}^p}\left[\sum_{\mathbf{y}\in\{t:\mathbf{y}\geq\mathbf{X}\boldsymbol{\beta}\}} \tau\left|\mathbf{y}-\mathbf{X}\boldsymbol{\beta}(\tau)\right| + \sum_{\mathbf{y}\in\{t:\mathbf{y}<\mathbf{X}\boldsymbol{\beta}\}} (1-\tau)\left|\mathbf{y}-\mathbf{X}\boldsymbol{\beta}(\tau)\right|\right], \quad (7)$$

where $\hat{\boldsymbol{\beta}}(\tau)$ is a regression coefficient on quantile-τ.

2.4. *ARIMAX Quantile Regression*

This method combined some univariate methods with quantile regression approach. The forecast values are done against quantile 0.50 or often called as median estimate. There are two models of hybrid ARIMAX Quantile Regression that used in this research.

2.4.1. *ARIMAX Quantile Regression Model 1*

In this ARIMAX-QR modelling its uses the dummy variable and lag of the data as predictor variable in quantile regression. ACF and PACF plot from stationary error can be used to determine the lag. Thus, ARIMAX-QR model with additional lag variables could be written as follows:

$$\hat{Y}_t(\tau) = Q_\tau(Y_t \mid t, H_{1,t},\ldots,H_{7,t},D_{1,t},\ldots,D_{48,t}L_{1,t},\ldots,L_{8,t},Y_{t-1},Y_{t-2},\ldots,Y_{t-p}), \quad (8)$$

where $t,H_{1,t},\ldots,H_{7,t},D_{1,t},\ldots,D_{48,t},L_{1,t},\ldots,L_{8,t}$ are exogenous variable(s).

2.4.2. *ARIMAX Quantile Regression Model 2*

This model is fitted by using its predicted values using the ARIMA method of data as well as other exogenous variable which are dummy variable(s) as predictor for quantile regression. This model can be written as follows:

$$\hat{Y}_t(\tau) = Q_\tau(Y_t \mid \hat{Z}_t, t, H_{1,t},\ldots,H_{7,t},D_{1,t},\ldots,D_{48,t},L_{1,t},\ldots,L_{8,t}), \quad (9)$$

where $\hat{Z}_t$ is prediction value of Y_t that developed using ARIMA model.

2.5. *Model Evaluation*

One of the purposes of forecasting is to improve the accuracy of forecast results and minimize the forecast error. This research using cross-validation approach to evaluate the model, using the model goodness criteria based on Root Mean Square Error (RMSE), Mean Absolute Percentage Error (MAPE) and Median Absolute Percentage Error (MdAPE), with the following formula [13],

$$RMSE = \sqrt{\frac{1}{M}\sum_{l=1}^{M} e_l^2}, \quad (10)$$

$$MAPE = \frac{1}{M}\sum_{l=1}^{M}\left|\frac{e_l}{Y_{n+l}}\right| \times 100, \quad (11)$$

$$MdAPE = \text{median}\left(\left|\frac{e_l}{Y_{n+l}}\right|\right) \times 100, \quad (12)$$

where $e_l = Y_{n+l} - \hat{Y}_n(l)$.

IOP Conf. Series: Journal of Physics: Conf. Series **1008** (2018) 012023 doi:10.1088/1742-6596/1008/1/012023

3. Methodology

3.1. Data and Variable

The data used in this research was electricity energy consumption per half hour data in East Java. The data was secondary data which obtained from the subsidiary of PT. PLN (Persero) is PLN P2B (Distribution and Control Load Center) in East Java region. Period of this data was from September 1, 2015 to April 30, 2016. In analyzing this research, the data was divided into two parts: in-sample data used for modeling, and out-sample data to validate the method used. In-sample data include data from 01 September 2015 to 29 February 2016 in period, and out-sample includes data from 1 March 2016 to 30 April 2016.

The dummy variables used in this study include trend dummy, per-half hour, days and special days (which occur other than on Sunday), and can be presented in Table 1.

Table 1 Dummy Variables

Dummy Variable	Explanation
Trend	$t = 1,2,3,\ldots,11664$
Days	$H_{1,t} = \begin{cases} 1, & \text{Monday} \\ 0, & \text{others} \end{cases}$
	$H_{2,t} = \begin{cases} 1, & \text{Tuesday} \\ 0, & \text{others} \end{cases}$
	M
	$H_{7,t} = \begin{cases} 1, & \text{Sunday} \\ 0, & \text{others} \end{cases}$
Per-half Hour	$D_{1,t} = \begin{cases} 1, & \text{at } 00.30 \\ 0, & \text{other} \end{cases}$
	$D_{2,t} = \begin{cases} 1, & \text{at } 01.00 \\ 0, & \text{other} \end{cases}$
	M
	$D_{48,t} = \begin{cases} 1, & \text{at } 24.00 \\ 0, & \text{other} \end{cases}$
Special Days	$L_{i,t} = \begin{cases} 1, & \text{special day } i \\ 0, & \text{other days} \end{cases}$

Special days used as dummy variables ($L_{i,t}$) are special days or national holidays occurring in the period of observation of September 2015 to April 2016 which occur other than Sunday. The list of special days used in this research is presented in Table 2.

Table 2 Special Days List

Dummy Variable	Special Day	Date
$L_{1,t}$	Feast of Sacrifice (Eid Adha)	24 September 2015
$L_{2,t}$	Islamic New Year	14 October 2015
$L_{3,t}$	Birth of Prophet (Mawlid)	24 December 2015
$L_{4,t}$	Christmas	25 December 2015
$L_{5,t}$	New Year's Day	01 January 2016
$L_{6,t}$	Chinese New Year	02 February 2016
$L_{7,t}$	Nyepi (Hindu's Holly Day)	09 March 2016
$L_{8,t}$	Good Friday	25 March 2016

4. Results and Analysis

4.1. Exploration of Electricity Consumption Characteristics

The amount of electricity consumption was influenced by the needs of each customer. The diverse customer activity that causes electricity consumption in East Java is fluctuating. Based on data on electricity consumption used from September 1, 2015 until September 30, 2016 consists of 19,008 observations. The average electricity consumption in East Java is 4,064.7 MW with standard deviation of 484,9 MW. Electricity consumption in East Java is the lowest at 07.00 with electricity consumption of 3,530.0 MW. This happens because at 7:00 am many people are just leaving for work so that not many people who use electronic equipment such as television, lights, and computers. This causes the consumption of electricity is still a little not too high, while the highest consumption occurred at 18.30 with a consumption of 4,743.2 MW. It was allegedly because the public had returned home after work and used various electronic equipment so that the use of electricity consumption reached the highest consumption.

The volatility of electricity consumption was not only influenced by hours of use but also affected by day. The highest average electricity consumption occurred on Tuesday at 4,029.8 MW. Generally, electricity consumption on weekdays is higher than on weekend (Saturday and Sunday). This happens because on Saturday and Sunday offices, factories or government agencies do not use electricity as much as on weekdays, such as turning on production machines in factories, turning on computers, lights, air conditioners and so on. The lowest electricity consumption occurred on Sunday with an average of 3,639.9 MW. This is supposedly happened because on weekend (Saturday and Sunday) most people often spend time for recreation outside home, and enjoying their break time, so the electricity consumption not as much as on weekdays.

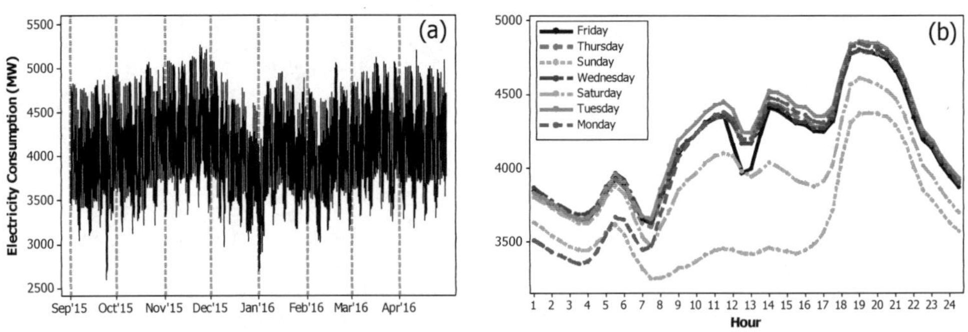

Figure 1. Time Series (a) and Mean Plot (b) of Electricity Consumption in East Java

Time series plot of electricity consumption in East Java Region shown in Figure 1. It is shown that from September 1, 2015 to April 30, 2016 the electricity consumption was increase. The electricity consumption on the weekdays from Monday to Thursday is relatively the same as shown in Figure 1b. But on Monday, from 0:30 to 05:00 and Friday at 12:00 to 12:30 shows the pattern of electricity consumption tend to be lower than the other weekdays in the same hour. On Monday at 0.30 to 05.00, electricity consumption is low, then it starts to increase significantly at 07.00 because most people started their activities. Otherwise, electricity consumption on Friday at 12.00 to 12.30 is lower than other working days because most people in East Java are doing the Friday prayer service.

4.2. Time Series Regression

Forecasting electricity consumption using time series regression was done by regressing the electricity consumption which is the response with the predictor such as dummy trend, hours, days, and special days (non-weekly national holidays). The model formed is shown in Table 3. From this model it is found that the residual still does not fulfill the white noise assumption, allegedly due to the effects of autocorrelated error. Therefore, modeling the data with adding lag of the series as a predictor was also done.

The ACF pattern of the data indicates a peak at lags 48, 96, 144, ... which indicates the daily seasonal period, and the peak in lag 336, 672, 1008, ... which denotes a seasonal period of the week. The PACF pattern shows a slow down to zero patterns. So it is concluded that the data has not been stationary. The process of differencing three times lag 1, 48, and 336 was done to make the data becomes stationary. From this result, lags used as predictors are lag 1, 48, and 336 ($Y_{t-1}, Y_{t-48}, Y_{t-336}$), where these lags show the seasonal patterns present in the data. In addition to these lags, also used its multiplicative lags such as lag 49, 337, 384 and 385 ($Y_{t-49}, Y_{t-337}, Y_{t-384}, Y_{t-385}$).

Table 3 Time Series Regression Model for Electricity Consumption in East Java

Predictor(s)	Model
Dummy	$\hat{Y}_t = -0{,}009t + 3909H_{1,t} + 3998H_{2,t} + 3972H_{3,t} + 3994H_{4,t} + 4010H_{5,t} + 3759H_{6,t} + 3428H_{7,t} - 65{,}2D_{1,t} - 104{,}7D_{2,t} - 149{,}0D_{3,t} - 185{,}4D_{4,t} - 221{,}6D_{5,t} - 245{,}1D_{6,t} - 236{,}4D_{7,t} - 167{,}6D_{8,t} - 54{,}5D_{9,t} + 17{,}8D_{10,t} - 46{,}6D_{11,t} - 177{,}0D_{12,t} - 281{,}4D_{13,t} - 305{,}6D_{14,t} - 148{,}5D_{15,t} - 15{,}9D_{16,t} + 134{,}9D_{17,t} + 195{,}5D_{18,t} + 252{,}2D_{19,t} + 315{,}4D_{20,t} + 358{,}5D_{21,t} + 371{,}9D_{22,t} + 314{,}7D_{23,t} + 188{,}9D_{24,t} + 185{,}4D_{25,t} + 302{,}9D_{26,t} + 415{,}2D_{27,t} + 403{,}1D_{28,t} + 371{,}1D_{29,t} + 325{,}2D_{30,t} + 315{,}6D_{31,t} + 284{,}8D_{32,t} + 294{,}9D_{33,t} + 376{,}8D_{34,t} + 648{,}7D_{35,t} + 855{,}3D_{36,t} + 893{,}6D_{37,t} + 882{,}7D_{38,t} + 881{,}4D_{39,t} + 849{,}7D_{40,t} + 777{,}4D_{41,t} + 655{,}8D_{42,t} + 477{,}7D_{43,t} + 323{,}0D_{44,t} + 254{,}1D_{45,t} + 150{,}6D_{46,t} + 73{,}6D_{47,t} - 882{,}8L_{1,t} - 414{,}2L_{2,t} - 609{,}6L_{3,t} - 764{,}1L_{4,t} - 1078L_{5,t} - 703{,}7L_{6,t}$
Dummy + Lag	$\hat{Y}_t = 0{,}00t + 58{,}9H_{1,t} + 52{,}5H_{2,t} + 51{,}0H_{3,t} + 53{,}9H_{4,t} + 52{,}9H_{5,t} + 44{,}1H_{6,t} + 37{,}9H_{7,t} + 1{,}9D_{1,t} + 11{,}5D_{2,t} + 8{,}1D_{3,t} + 9{,}3D_{4,t} + 10{,}1D_{5,t} + 13{,}5D_{6,t} + 25{,}3D_{7,t} + 49{,}0D_{8,t} + 66{,}8D_{9,t} + 52{,}7D_{10,t} + 2{,}8D_{11,t} - 23{,}1D_{12,t} - 15{,}7D_{13,t} + 13{,}4D_{14,t} + 80{,}7D_{15,t} + 74{,}9D_{16,t} + 84{,}8D_{17,t} + 52{,}8D_{18,t} + 52{,}3D_{19,t} + 56{,}6D_{20,t} + 49{,}8D_{21,t} + 39{,}2D_{22,t} + 12{,}6D_{23,t} - 15{,}7D_{24,t} + 29{,}5D_{25,t} + 75{,}6D_{26,t} + 75{,}9D_{27,t} + 30{,}1D_{28,t} + 22{,}2D_{29,t} + 17{,}9D_{30,t} + 28{,}8D_{31,t} + 21{,}3D_{32,t} + 37{,}6D_{33,t} + 63{,}7D_{34,t} + 133{,}6D_{35,t} + 116{,}6D_{36,t} + 59{,}8D_{37,t} + 40{,}4D_{38,t} + 43{,}7D_{39,t} + 31{,}6D_{40,t} + 16{,}7D_{41,t} - 4{,}91D_{42,t} - 28{,}6D_{43,t} - 23{,}3D_{44,t} + 6{,}7D_{45,t} - 7{,}1D_{46,t} + 0{,}7D_{47,t} - 39{,}4L_{1,t} - 19{,}7L_{2,t} - 29{,}1L_{3,t} - 22{,}6L_{4,t} - 30{,}7L_{5,t} - 32{,}8L_{6,t} + 0{,}9Y_{t-1} + 0{,}2Y_{t-48} - 0{,}2Y_{t-49} + 0{,}4Y_{t-336} - 0{,}4Y_{t-337} - 0{,}02Y_{t-384} - 0{,}03Y_{t-385}$

Both models were used to predict the out-sample data to check its performance in forecasting the several periods ahead data. These forecast values will be compared with its actual data (out-sample). The out-sample forecast plot of each model formed is shown in Figure 2. It is shown that the forecast values obtained by time series regression model were good enough to capture the electricity consumption data. After adding lag of the data as a predictor (Figure 2b), the forecast values tend to be better to capture the pattern of fluctuations than model that do not use lag as predictor (Figure 2a).

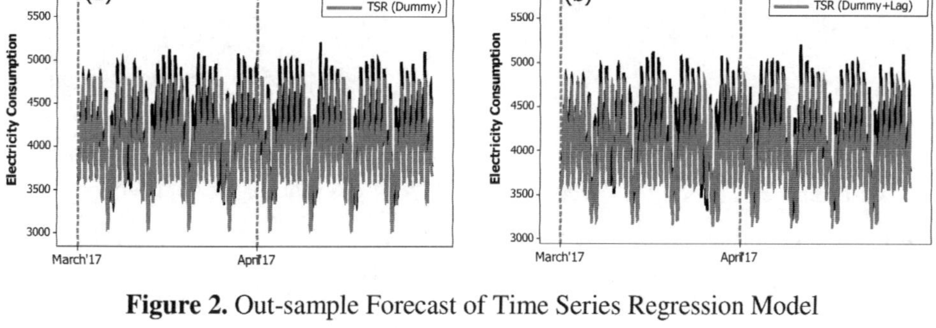

Figure 2. Out-sample Forecast of Time Series Regression Model

4.3. *ARIMAX Quantile Regression*

As we know, there are two models of ARIMAX Quantile Regression (ARIMAX-QR) applied in this research. ARIMAX-QR model 1 use lag and dummy variables as predictors. Determination of lag that used was same as the time series regression model using lag (subsection 4.2), lag 1, 48, and 336 ($Y_{t-1}, Y_{t-48}, Y_{t-336}$) and its multiplicative lags such as lag 49, 337, 384 and 385 ($Y_{t-49}, Y_{t-337}, Y_{t-384}, Y_{t-385}$) will be used as predictors. The last ARIMAX-QR model is model that used dummy variables and predicted values of ARIMA model for the data. Electricity consumption data contains two seasonal patterns, which are per-half hour and weekly patterns so Double Seasonal ARIMA need to be formed. For convenience, this research use $\text{ARIMA}(0,1,1)(0,1,1)^{48}(0,1,1)^{336}$ model as $\hat{Z}_t$. The model formed can be shown in Table 4.

Table 4 ARIMAX Quantile Regression Model for Electricity Consumption in East Java

ARIMAX-QR	Model
Model 1	$\hat{Y}_t = 0{,}00t + 34{,}9H_{1,t} + 28{,}9H_{2,t} + 29{,}9H_{3,t} + 31{,}0H_{4,t} + 29{,}7H_{5,t} + 23{,}3H_{6,t} + 19{,}6H_{7,t} + 3{,}5D_{1,t} + 12{,}4D_{2,t} + 13{,}7D_{3,t} + 12{,}9D_{4,t} + 15{,}4D_{5,t} + 19{,}5D_{6,t} + 34{,}2D_{7,t} + 54{,}2D_{8,t} + 76{,}4D_{9,t} + 53{,}2D_{10,t} + 2{,}1D_{11,t} - 20{,}7D_{12,t} - 16{,}3D_{13,t} + 20{,}0D_{14,t} + 99{,}2D_{15,t} + 88{,}1D_{16,t} + 98{,}1D_{17,t} + 58{,}9D_{18,t} + 55{,}7D_{19,t} + 64{,}4D_{20,t} + 53{,}8D_{21,t} + 40{,}4D_{22,t} + 19{,}7D_{23,t} - 18{,}0D_{24,t} + 34{,}1D_{25,t} + 83{,}1D_{26,t} + 88{,}8D_{27,t} + 29{,}9D_{28,t} + 23{,}3D_{29,t} + 21{,}3D_{30,t} + 30{,}1D_{31,t} + 25{,}9D_{32,t} + 36{,}5D_{33,t} + 66{,}5D_{34,t} + 139{,}2D_{35,t} + 130{,}3D_{36,t} + 52{,}1D_{37,t} + 40{,}5D_{38,t} + 45{,}53D_{39,t} + 30{,}6D_{40,t} + 14{,}2D_{41,t} - 7{,}3D_{42,t} - 29{,}2D_{43,t} - 26{,}4D_{44,t} + 7{,}6D_{45,t} - 6{,}5D_{46,t} - 1{,}9D_{47,t} - 28{,}2L_{1,t} - 24{,}4L_{2,t} - 25{,}4L_{3,t} - 15{,}7L_{4,t} - 25{,}4L_{5,t} - 24{,}9L_{6,t} + 0{,}9Y_{t-1} + 0{,}2Y_{t-48} - 0{,}2Y_{t-49} + 0{,}4Y_{t-336} - 0{,}4Y_{t-337} + 0{,}02Y_{t-384} - 0{,}03Y_{t-385}$
Model 2	$\hat{Y}_t = 0{,}00t + 83{,}6H_{1,t} + 85{,}3H_{2,t} + 88{,}3H_{3,t} + 86{,}5H_{4,t} + 86{,}0H_{5,t} + 80{,}8H_{6,t} + 72{,}4H_{7,t} + 3{,}0D_{1,t} + 4{,}5D_{2,t} - 1{,}9D_{3,t} - 3{,}1D_{4,t} - 1{,}2D_{5,t} + 0{,}5D_{6,t} - 3{,}5D_{7,t} + 3{,}2D_{8,t} + 2{,}3D_{9,t} + 3{,}1D_{10,t} + 5{,}7D_{11,t} - 0{,}6D_{12,t} - 8{,}7D_{13,t} + 3{,}1D_{14,t} + 1{,}9D_{15,t} + 4{,}7D_{16,t} + 4{,}0D_{17,t} + 9{,}6D_{18,t} + 11{,}6D_{19,t} + 7{,}9D_{20,t} + 15{,}1D_{21,t} + 8{,}8D_{22,t} + 12{,}8D_{23,t} + 8{,}7D_{24,t} + 1{,}5D_{25,t} + 11{,}5D_{26,t} + 7{,}6D_{27,t} + 14{,}1D_{28,t} + 20{,}6D_{29,t} + 4{,}5D_{30,t} + 11{,}2D_{31,t} + 7{,}7D_{32,t} + 0{,}9D_{33,t} + 13{,}8D_{34,t} + 9{,}7D_{35,t} + 25{,}6D_{36,t} + 11{,}9D_{37,t} + 24{,}2D_{38,t} + 19{,}7D_{39,t} + 23{,}1D_{40,t} + 11{,}7D_{41,t} + 21{,}2D_{42,t} + 14{,}2D_{43,t} + 12{,}1D_{44,t} + 12{,}0D_{45,t} + 4{,}7D_{46,t} + 1{,}3D_{47,t} - 16{,}4L_{1,t} - 19{,}9L_{2,t} - 16{,}9L_{3,t} - 16{,}3L_{4,t} - 18{,}7L_{5,t} - 14{,}2L_{6,t} + 0{,}98\hat{Z}_t$

From each ARIMAX-QR model formed, it will be used to forecast the out-sample data. The result of this process then compared to the actual as shown in Figure 3.

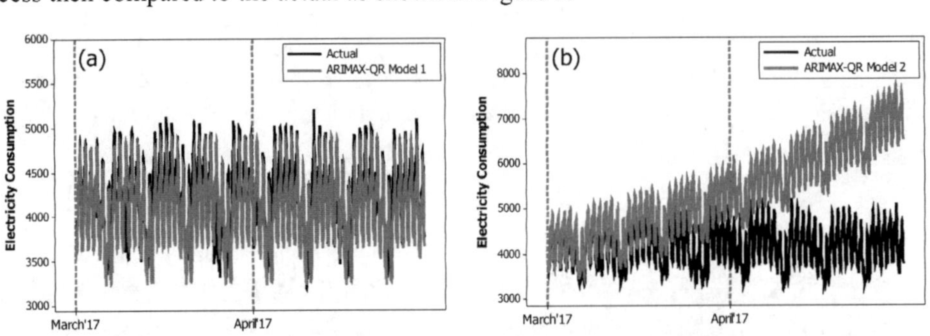

Figure 3. Out-sample Forecast of ARIMAX Quantile Regression Model

The predicted values generated by ARIMAX-QR model 1 (Figure 3a) show good results, where the forecast values tend to capture patterns from the data. The use of ARIMA prediction results in ARIMAX-QR model 2 aims to handle patterns that could not be captured using the decomposition model. The ARIMAX-QR model 2 prediction results as shown in Figure 3b show the disadvantages of

ARIMA method, where ARIMA method are suitable only for short-term forecasting. For long-term forecasting, ARIMA tends to produce a prediction that is getting wider or farther from actual data. Hence, this model shows that the longer the period to be predicted then the results will tend to farther from the actual data. It implies model 2 yields more accurate for short-term forecasting.

4.4. Comparison between Time Series Regression and ARIMAX Quantile Regression

Based on the results at the previous chapter about the forecasting of electricity consumption in East Java using Time Series Regression (TSR) and ARIMAX Quantile Regression (ARIMAX-QR) method, Table 5 shows goodness values at out-sample data, i.e. RMSE, MAPE and MdAPE for each model.

Table 5 Forecast Accuracy for Each Model

Method	RMSE	MAPE	MdAPE
TSR			
Dummy	220.787	4.573	4.397
Dummy + Lag	**189.935**	**4.474**	**4.256**
ARIMAX-QR			
Model 1	**177.982**	**3.374**	**2.761**
Model 2	1494.162	29.979	27.833

Note: The bold values show the best model in each method

The best model formed from modelling electricity consumption data using time series regression (TSR) method is a model involving lag and dummy variables as predictors compared to models involving only dummy variables (Table 5). For modeling using hybrid ARIMAX Quantile Regression (ARIMAX-QR), the best selected model is model 1 which use dummy and lag as predictors in the quantile regression model compared to the other model. When comparing the best TSR model with the best ARIMAX-QR model it was found that the ARIMAX-QR model yields a prediction value with a smaller error rate than other models. To find out the predicted performance of the best TSR model and the best ARIMAX-QR model, an adaptive approach for RMSE, MAPE and MdAPE values of each model was used. The adaptive RMSE, MAPE and MdAPE plots of each model for 7-days ahead forecast period are shown in Figure 4.

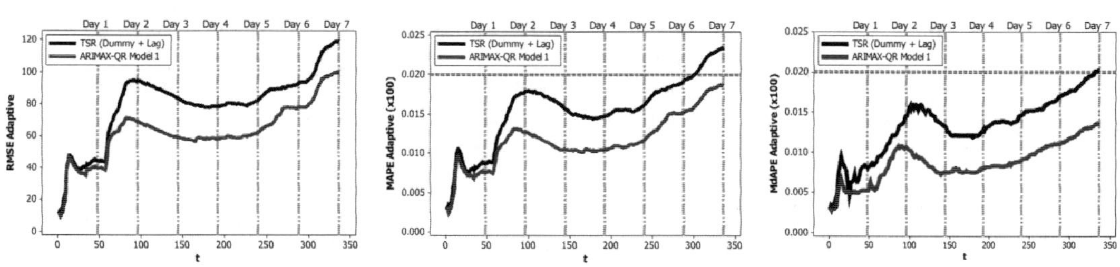

Figure 4. Adaptive Accuracy of Best Model

It is shown that the accuracy of the ARIMAX Quantile Regression model is better than Time Series Regression model (Figure 4). This is indicated by a lower forecast error value than Time Series Regression model, for forecasting the next 1 day, 2 days ahead, and still low until the next 7 days forecast. PT PLN (Persero) set the benchmark maximum forecast error of 2%. Both models produce an adaptive MAPE and MdAPE less than the benchmark set by the PLN, which means both models are well used to forecast data on short-term power consumption in East Java. Based on MAPE criteria, it is also known that Time Series Regression model is good to forecast electricity load in East Java up to 6 days ahead. Otherwise, it will cause prediction up to 7 days ahead become greater than benchmark set by PLN, i.e. 2%.

5. Conclusion

The results show that electricity consumption in East Java is depend on hours, days, and national holidays. Forecasting the electricity consumption in East Java using time series regression with the addition of lag as predictor produces the smallest forecast error values. Otherwise, ARIMAX Quantile Regression model that involves lag and dummy variables as predictors yield more accurate forecast than the model without lag and dummy variables as predictors. Generally, based on the evaluation criteria, it could be concluded that the ARIMAX Quantile Regression model yields better forecast performance than the time series regression model. Thus, the proposed method can be a potential alternative to forecast short-term electricity in East Java. Moreover, both time series regression and ARIMAX-QR methods with additional lag values as predictors could capture accurately the patterns in the data and produce better predictions than those which not using lag of variables.

Acknowledgements

This research was supported by LPPM ITS under scheme of “Penelitian Pemula 2017”, project No. 695/PKS/ITS/2017. The authors gratefully acknowledge the support from ITS for funding the research and anonymous for their useful suggestions.

References

[1] Koenker R and Bassett Jr. G 1978 *Econometrica* **46** 33-50
[2] Ma L and Pohlman L *The European Journal of Finance* **14** 409-425
[3] Zhang G P 2003 *Neurocomputing* **50** 159 – 175
[4] Taylor J W 2000 *Journal of Forecasting* **19** 299-311
[5] Xu Q, Tsui K-L, Jiang W and Guo H 2016 *Qual. Reliab. Engng. Int.* **32** 2751–2759
[6] Arunraj N S and Ahrens D 2015 *International Journal Production Economics* **170** 321-335
[7] Soares L J and Medeiros M C 2008 *International Journal of Forecasting* **24** 630-644
[8] Wei W W S 2006 *Time Series Analysis Univariate and Multivariate Methods* (New York : Pearson Education, Inc.)
[9] Suhartono, Lee M H and Hamzah N A 2010 *Proceedings of the Regional Conference on Statistical Sciences* 30-41
[10] Draper N R and Smith H 1998 *Applied Regression Analysis, 3rd Edition* (New York: John Wiley & Sons, Inc.)
[11] Koenker R 2005 *Quantile Regression* (New York: Cambridge University Press)
[12] Cade B S and Noon B R 2003 *Frontiers in Ecology and the Environment* **1** 8 412–420
[13] Hyndman R J and Koehler A 2006 *International Journal of Forecasting* **22** 679-688

Analysis of *Salmonella sp* bacterial contamination on Vannamei Shrimp using binary logit model approach

P P Oktaviana[1], **K Fithriasari**[1]
[1] Department of Statistics, Institut Teknologi Sepuluh Nopember
Surabaya, Indonesia

E-mail: paramitha.oktaviana@gmail.com

Abstract. Mostly Indonesian citizen consume vannamei shrimp as their food. Vannamei shrimp also is one of Indonesian exports comodities mainstay. Vannamei shrimp in the ponds and markets could be contaminated by *Salmonella sp* bacteria. This bacteria will endanger human health. *Salmonella sp* bacterial contamination on vannamei shrimp could be affected by many factors. This study is intended to identify what factors that supposedly influence the *Salmonella sp* bacterial contamination on vannamei shrimp. The researchers used the testing result of *Salmonella sp* bacterial contamination on vannamei shrimp as response variable. This response variable has two categories: 0 = if testing result indicate that there is no *Salmonella sp* on vannamei shrimp; 1 = if testing result indicate that there is *Salmonella sp* on vannamei shrimp. There are four factors that supposedly influence the *Salmonella sp* bacterial contamination on vannamei shrimp, which are the testing result of *Salmonella sp* bacterial contamination on farmer hand swab; the subdistrict of vannamei shrimp ponds; the fish processing unit supplied by; and the pond are in hectare. This four factors used as predictor variables. The analysis used is Binary Logit Model Approach according to the response variable that has two categories. The analysis result indicates that the factors or predictor variables which is significantly affect the *Salmonella sp* bacterial contamination on vannamei shrimp are the testing result of *Salmonella sp* bacterial contamination on farmer hand swab and the subdistrict of vannamei shrimp ponds.

1. Introduction

Indonesia has huge area which have benefits to the natural resources for people's welfare, for example aquaculture and mariculture. Ponds is one of the largest aquaculture industry in Indonesia which is indicated by the greatest contribution to the total value of aquaculture production, especially for shrimp and prawn. According to Directorate Generale of Aquaculture in Indonesia, there are two types of shrimps that become main export commodities, tiger shrimp (*Penaeaus Monodon*) and vannamei shrimp (*Litopenaeus Vannamei*) [4]. Both of them are highly contributed in export commodities, contains of high protein, and widely consumed by Indonesian society. In fact, vannamei shrimp has the highest increase in average production from 2010 to 2014, which is increase to 20.49% and according to Mahbubillah in [7], this shrimp has high demand both for national and international because of the affordable price and nutritional content. However, vannamei shrimp has high possibility to contaminated by the bacteria, such as *Salmonella sp* [5]. This kind of bacteria in vannamei shrimp can affect people's healthh and causing *salmonellosis* disease.

Based on previous description, the researchers are concern about *Salmonella sp* bacterial contamination on vannamei shrimp and using it as response varibale in this research. This study is intended to identify what factors that supposedly influence the *Salmonella sp* bacterial contamination on vannamei shrimp using binary logit model according to the response variable that has two categories.

2. Literature Review

2.1 Binary Logit Model

Binary Logit Model is a model that generated from Binary Logistic Regression analysis which is the analysis method used to find the relationship between the response variable Y that is binary or dichotomous and predictor variables X are polychotomus [3], where this variable X can be qualitative or quantitative. The data type on response variable Y is qualitative with two categories, for example; success ($Y=1$) and failed ($Y=0$). Binary logistic regression models are follows [1]:

$$\pi(x) = \frac{\exp(\beta_0 + \beta_1 x)}{1 + \exp(\beta_0 + \beta_1 x)} \tag{1}$$

$\pi(x) = P(Y = 1 \mid X = x) = 1 - P(Y = 0 \mid X = x)$; β_0 and β_1 are the parameters.

Equation (1) can be written *log odds* or called the *logit* model that have a relationship linear against x:

$$g(x) = \ln\left[\frac{\pi(\text{x})}{1 - \pi(\text{x})}\right] = \beta_0 + \beta_1 x \tag{2}$$

Where $\frac{\pi(\text{x})}{1-\pi(\text{x})}$ is odds ratio.If there are p as independent variables, the logit model is:

$$g(x) = \beta_0 + \beta_1 x_1 + \ldots + \beta_p x_p \tag{3}$$

$g(x)$ is a *binary logit model* in binary logistic regression.

a. Binary Logistic Regression Parameter Estimation

One method to estimate the parameters of binary logistic regression is *Maximum Likelihood Estimation* (MLE) [6,10]. The parameters of model being estimated from the vector $\boldsymbol{\beta}^{\mathbf{T}} = (\beta_0, \beta_1, \beta_2, \ldots, \beta_p)$. The value of the vector $\boldsymbol{\beta}^{\mathbf{T}}$ obtained by maximise function $\mathbf{L}(\boldsymbol{\beta})$ through differential of the parameters that will be estimated. Function $\mathbf{L}(\boldsymbol{\beta})$ is *log likelihood* function :

$$\mathbf{L}(\boldsymbol{\beta}) = \sum_{j=0}^{p}\left[\sum_{i=1}^{n} y_i x_{ij}\right]\beta_j - \sum_{i=1}^{n} n_i \ln\left[1 + \exp\left(\sum_{j=0}^{p} \beta_j x_{ij}\right)\right] \tag{4}$$

This function is a combination of ($Y1$, $Y2$, …, Yn) which independently, with a value of Y observations consist of success ($Y=1$) and failed ($Y=0$), with the binomial distribution and $E(Y_i) = n_i\pi(x)$; where $n_1 + n_2 + \ldots + n_i = N$. For the *log likelihood* equation in equation (4) is generated from each element of the $\boldsymbol{\beta}$ and equating with zero. So we obtain $\sum_{i=1}^{n} y_i x_{ij} - \sum_{i=1}^{n} x_{ij}\pi(x_i)$. The method to estimate variance and covariance from this parameter estimation is obtained from the second derivative of likelihood function. The second derivative as follows:

$$\frac{\partial^2 L(\beta)}{\partial \beta_j^2} = -\sum_{i=1}^{n} x_{ij}^2 \pi(x_i)(1 - \pi(x_i)) \tag{5}$$

To obtain maximum parameter $\boldsymbol{\beta}$ on logistic regression equation that could not be solved with MLE method then use Newton Raphson Method. This method is derivation of the *Maximum Likelihood* non linear method in order to be a form of linear parameters. Parameter $\boldsymbol{\beta}$ is estimated through iteration as below:

$$\boldsymbol{\beta}^{(t+1)} = \boldsymbol{\beta}^{(t)} - (\mathbf{H}^{(t)})^{-1}\mathbf{q}^{(t)} \tag{6}$$

where $t = 0, 1, 2, \ldots$, until convergence;

$$\mathbf{q}^{(t)} = \begin{bmatrix} \frac{\partial L(\beta)}{\partial \beta_0} & \frac{\partial L(\beta)}{\partial \beta_1} & \cdots & \frac{\partial L(\beta)}{\partial \beta_p} \end{bmatrix} \quad (7)$$

$$\mathbf{H} = \begin{bmatrix} h_{11} & h_{12} & \cdots & h_{1p} \\ h_{21} & h_{22} & \cdots & h_{2p} \\ \cdots & \cdots & \ddots & \vdots \\ h_{j1} & h_{j2} & \cdots & h_{jp} \end{bmatrix} \quad (8)$$

Elements of matrix **H** is $h_{jp} = \frac{\partial^2 L(\beta)}{\partial \beta_j \partial \beta_p}$.

Algorithm of Newton Raphson iteration process are :

1. Enter the initial value of $\beta^{(0)}$ to equation :
   $\boldsymbol{\beta}^{(t+1)} = \boldsymbol{\beta}^{(t)} + \{\mathbf{x}^T \boldsymbol{Diag}[\pi(x_i)^{(t)}(1-\pi(x_i)^{(t)})]\mathbf{x}\}^{-1}\mathbf{x}^T(y-m^{(t)})$ with value of $m^{(t)} = \pi(x_i)^{(t)}$ to obtain the value of $\pi(x_i)^{(0)}$
2. Looking for Hessian Matrix $\mathbf{H}^{(0)}$ and vector $\mathbf{q}^{(0)}$
3. For $t > 0$ then the iteration process followed by applying the equation:

$$\pi(x_i)^{(t)} = \frac{\exp\left(\sum_{j=0}^{p} \beta_j^{(t)} x_{ij}\right)}{1-\exp\left(\sum_{j=0}^{p} \beta_j^{(t)} x_{ij}\right)} \text{ and } \boldsymbol{\beta}^{(t+1)} = \boldsymbol{\beta}^{(t)} + \{\mathbf{x}^T \boldsymbol{Diag}[\pi(x_i)^{(t)}(1-\pi(x_i)^{(t)})]\mathbf{x}\}^{-1}\mathbf{x}^T(y-m^{(t)})$$

4. Iteration will be stopped if $\pi(x_i)^{(t)}$ and $\boldsymbol{\beta}^{(t)}$ have convergent when $|\beta_j^{(t+1)} - \hat{\beta}_j| \le c\,|\beta_j^{(t)} - \hat{\beta}_j|^2$, for $c > 0$.

b. Binary Logistic Regression Parameter Test

There are two kinds of parameters testing in binary logistic regression [6]:

i. Simultaneous Test

This test is used to check the significance of the coefficient β simultaneously against the dependent variables. The hypothesis are :

H_0 : $\beta_1 = \beta_2 = \ldots = \beta_l = 0$ (no effect coefficient significantly to model)

H_1 : at least there is one $\beta_j \neq 0$, with $j = 1, 2, \ldots, p$ (coefficient is significantly influence to model)

The statistical test is G^2 or *Likelihood Ratio Test* :

$$G^2 = -2\ln\left[\frac{\left(\frac{n_1}{n_0}\right)^{n_1}\left(\frac{n_0}{n}\right)^{n_0}}{\prod_{i=1}^{n} \hat{\pi}_i^{y_i}(1-\hat{\pi}_i)^{1-y_i}}\right] \quad (9)$$

where $n_0 = \sum_{i=1}^{n}(1-y_i)$ and $n_1 = \sum_{i=1}^{n} y_i$ and $n = n_0 + n_1$

Reject the null hypothesis if $G^2 > \chi^2_{(\alpha,p-1)}$ where p is the number of independent variables.

ii. Partial Test

This test is used to check the significance of the coefficient β partially by comparing alleged β and alleged of standard error. The hypothesis are :

H_0 : $\beta_j = 0$ (there is no effect of coefficient to model)

H_1 : $\beta_j \neq 0$, with $j = 1, 2, \ldots, p$ (coefficient is significantly influence to model)

The statistical test is W (*Wald*) :

$$W = \frac{\hat{\beta}_j}{SE(\hat{\beta}_j)} \quad (10)$$

where $SE(\hat{\beta}_j) = [Var(\hat{\beta}_j)]^{1/2}$.
Reject the null hypothesis if $|W| > Z_{\alpha/2}$.

c. Binary Logistic Regression of Goodness of Fit
Goodness of Fit test of binary logistic regression is used to check if there is significant difference between the observed and model predictions. The hypothesis are :
H_0 : model is fit (There is no significant difference between the observed and model predictions)
H_1 : model does not fitt (There is significant difference between the observed and model predictions)
The statistical test :

$$\hat{C} = \sum_{k=1}^{g} \frac{(o_k - n'_k \bar{\pi}_k)^2}{n'_k \bar{\pi}_k (1 - \bar{\pi}_k)} \tag{11}$$

where :
o_k = observation on group k ; $\bar{\pi}_k$ = average of probability estimation
g = number of group (combination of categories in simultaneous model)
n'_k = number of observations on group k
Reject the null hypothesis if $\hat{C} > \chi^2_{(\alpha,p-2)}$ where p is the number of independent variables.

2.2 Salmonella sp Bacteria and Vannamei Shrimp
According to D'aoust in [5], states that *Salmonella sp* is a short bacteria (1-2 μm), a stem which doesn't form spores. *Salmonella sp* is capable of fermenting glucose that produces acids and gases. This bacteria actively grows in the pH range from 3.6 to 9.5 and is optimal at a near-normal pH. *Salmonella sp* is one of the most dangerous bacteria in fishery products. These bacteria have a serious impact on human's health [8]. Vannamei shrimp (*Litopenaeus vannamei*) is a type of shrimp from South America that is widely cultivated in Indonesia since the late 90s to replace the tiger shrimp (*Penaeus monodon*) which has been difficult to cultivate. The advantages of cultivate the vannamei shrimp include: high adaptability to low temperature, relatively fast growth rate, wide market absorption, high survival rate [2]. *A salmonella sp bacterium is* a bacterium that often attacks the vannamei shrimp both in the waters and in the distribution.

2.3 Studies from Previous Research (State of the Art)
A study on the pollution of *Salmonella sp* bacteria in vannamei shrimp was done to examined the detection of *Salmonella sp* bacteria on fresh white shrimp (Penaeus Merguiensis) in the traditional market of Surabaya Municipality, with the result that fresh white shrimp which marketed in traditional markets in five areas of Surabaya polluted by *Salmonella sp* [9]. The highest pollution rate in South Surabaya (12%) and the lowest pollution in West and North Surabaya (4%). There is also a research examined the technique of *Salmonella sp* bacteria's identification on vannamei shrimp (Litopenaeus Vannamei) and its processing water in the development and quality testing laboratory of fishery products in South Sumatera Province with the result that *Salmonella sp* bacteria can cause Salmonellosis disease through food, especially foods that have errors in handling. This situation will provide an opportunity for the microorganisms that cause to grow and move to humans at the time of eating them [5].

3. Research Variables
Data that used in this research is the testing result of *Salmonella sp* bacterial contamination on vannamei shrimp. The research variables are shown in Table 1. *Y* is response variable and *X* is predictor variable.

Table 1. Research Variables

Variables	Explanation	Categories	Scale
Y	The testing result of *Salmonella sp* bacterial contamination on vannamei shrimp	0 : there is no *Salmonella sp* 1 : there is *Salmonella sp*	Nominal
X_1	The testing result of *Salmonella sp* bacterial contamination on farmer hand swab	0 : there is no *Salmonella sp* 1 : there is *Salmonella sp*	Nominal
X_2	The subdistrict of vannamei shrimp ponds	0 : Subdistrict A 1 : Subdistrict B 2 : Subdistrict C 3 : Subdistrict D 4 : Subdistrict E	Nominal
X_3	The fish processing unit supplied by	0 : PT. X 1 : PT. Y 2 : PT. Z	Nominal
X_4	The pond are in hectare	-	Ratio

Data structure of this research is shown in Table 2.

Table 2. Data Structure

Obs	Y	X_1	X_2	X_3	X_4
1	Y_1	X_{11}	X_{12}	X_{13}	X_{14}
2	Y_2	X_{21}	X_{22}	X_{23}	X_{24}
3	Y_3	X_{31}	X_{32}	X_{33}	X_{34}
4	Y_4	X_{41}	X_{42}	X_{43}	X_{44}
5	Y_5	X_{51}	X_{52}	X_{53}	X_{54}
⋮	⋮	⋮	⋮	⋮	⋮
N	Y_n	X_{n1}	X_{n2}	X_{n3}	X_{n4}

4. Result and Discussion

4.1 Descriptive Statistics

The cross tabulation analysis of fish processing unit and testing result of *Salmonella sp* bacterial contamination on vannamei shrimp that is indicated in Table 3; also the cross tabulation analysis of subdistrict of ponds location and testing result of *Salmonella sp* bacterial contamination on vannamei shrimp that is indicated in Table 4, show that there is *Salmonella sp* contamination on vanname shrimp.

Table 3. The Cross Tabulation Analysis Of Fish Processing Unit And Testing Result Of *Salmonella* Sp Bacterial Contamination On Vannamei Shrimp

Testing Result	PT. X	PT. Y	PT. Z
There is *Salmonella sp*	2	6	45
There is no *Salmonella sp*	3	14	260

Table 4. The Cross Tabulation Analysis Of Subdistrict Of Ponds Location And Testing Result Of *Salmonella sp* Bacterial Contamination On Vannamei Shrimp

Testing Result	Subdistrict A	Subdistrict B	Subdistrict C	Subdistrict D	Subdistrict E
There is *Salmonella sp*	4	10	8	7	24
There is no *Salmonella sp*	46	55	42	53	81

IOP Conf. Series: Journal of Physics: Conf. Series **1008** (2018) 012024 doi:10.1088/1742-6596/1008/1/012024

The results indicates that the problem of *Salmonella sp* contamination on vannamei shrimp is still found and need to cure. Thats why this research aim to identify the significant factors that influence *Salmonella sp* contamination on vannamei shrimp using binary logistic model approach.

4.2 Binary Logistic Regression

The first step of binary logistic regression is testing the parameters simultaneously and partially. Simultaneous test indicates to reject H_0; p-value = 0.013 (with H_0: $\beta_1 = \beta_2 = \beta_3 = \beta_4 = 0$ and H_1: at least one $\beta_j \neq 0, j = 1, 2, 3, 4$). It concludes that at least one of predictor variables affects response variable. Partial test concludes to reject H_0 for parameters of testing result of *Salmonella sp* bacterial contamination on farmer hand swab and subdistrict of vannamei shrimp ponds. It indicates that there are two significant factors or predictor variables, which are testing result of *Salmonella sp* bacterial contamination on farmer hand swab and subdistrict of vannamei shrimp ponds.

Table 5. The Results of Binary Logistic Regression

Variable	β	*Odds Ratio*
(Intercept)	-1.412	0.244
X_1 (1)	1.249	3.489
X_2 (1)	-0.482	0.618
X_2 (2)	-1.263	0.283
X_2 (3)	-0.420	0.657
X_2 (4)	-0.798	0.450

Based on binary logistic regression of response variables and those two significant predictor variables (the results is shown in Table 5), binary logit model could be developed as:

$$\hat{g}(x) = -1.412 + 1.249X_1(1) - 0.482X_2(1) - 1.263X_2(2) - 0.420X_2(3) - 0.798X_2(4)$$

The probability of there is *Salmonella sp* contamination on vannamei shrimp based on two significant predictor variables ($X_1(1)$: there is *Salmonella sp* on worker hand swab; $X_2(1)$: subdistrict B, $X_2(1)$: subdistrict C, $X_2(1)$: subdistrict D; $X_2(1)$: subdistrict E) is 0.0005. The calculation is shown as the following:

$$\hat{g}(x) = -1.412 + 1.249(1) - 0.482(1) - 1.263(2) - 0.420(3) - 0.798(4) = -7.623$$

$$\hat{\pi}(x) = \frac{\exp(-7.623)}{1 + \exp(-7.623)} = 0.0005$$

From the information, the probability of there is no *Salmonella sp* contamination on vannamei shrimp based on two significant predictor variables ($X_1(1)$: there is *Salmonella sp* on farmer hand swab; $X_2(1)$: subdistrict B, $X_2(1)$: subdistrict C, $X_2(1)$: subdistrict D; $X_2(1)$: subdistrict E) also can be calculated as 1 - 0.0005 = 0.9995.

The contribution of each predictor variable is explained by odds ratio. For example, based on odds ratio of variable $X_1(1)$ (that is predictor variable, category 1 : there is *Salmonella sp* on farmer hand swab) is 3.489. It concludes that the presence of *Salmonella sp* on farmer hand swab tend to cause the absence of *Salmonella sp* bacteria in vannamei shrimp 3.468 smaller than the absence of *Salmonella sp* bacteria on worker hand swab.

5. Conclusion

The result indicates that the predictor variables which is significanly affect the *Salmonella sp* bacterial contamination on Vannamei shrimp are testing result of *Salmonella sp* bacterial contamination on farmer hand swab and subdistrict of vannamei shrimp ponds. The binary logit model could be developed as $\hat{g}(x) = -1.412 + 1.249X_1(1) - 0.482X_2(1) - 1.263X_2(2) - 0.420X_2(3) - 0.798X_2(4)$.

Acknowledments
We are grateful to LPPM ITS who gives the chance to finish this research in part of the research for beginner (Penelitian Pemula) in Institut Teknologi Sepuluh Nopember, Surabaya, Indonesia.

References

[1] Agresti A 2002 *Categorical Data Analysis* (New Jersey : John Wiley & Sons, Inc.)
[2] http://www.agrotaninusantara.com /2014/11/apa-itu-udang-vannamei.html (Accessed on February 20, 2017, at 07.45 pm)
[3] Chen C, Zhang G, Tarefder R, Ma J, Wei H and Guan H 2015 A Multinomial Logit Model-Bayesian Network Hybrid Approach for Driver Injury Severity Analyses in Rear-End Crashes *Elsevier : Accident Analysis and Prevention* **80** 76–88
[4] http://www.djpb.kkp.go.id/arsip/c/246/Udang-Vannamei-dan-Udang-Windu-Masih-Andalan-Ekspor-Indonesia/?category_id=13 (Accessed on February 10, 2017, at 10.16 am)
[5] Hidayat H 2013 *Teknik Identifikasi Bakteri Salmonella sp pada Udang Vennamei (Liptopenaeus Vennamei) dan Air Pengolahannya di Laboratorium Pembinaan dan Pengujian Mutu Hasil Perikanan Provinsi Sumatera Selatan* (Universitas Sriwijaya: Ilmu Kelautan)
[6] Hosmer D W and Lemeshow S 2000 *Applied Logistic Regression, Second Edition* (Washingtong : John Wiley & Sons, Inc.)
[7] Kharisma A and Manan A 2012 Kelimpahan Bakteri Vibrio sp pada Air Pembesaran Udang Vannamei (Litopenaeus Vannamei) sebagai Deteksi Dini Serangan Penyakit Vibriosis *Media Jurnal Ilmiah Perikanan & Kelautan* **4**(2) 20-27
[8] http://mediapenyuluhanperkananpati.blogspot.co.id/2016/09/bahaya-samonella.html (Accessed on February 19, 2017, at 11.51 pm)
[9] Narumi H E, Zuhriansyah and Mustofa I 2009 Deteksi Pencemaran Bakteri Salmonella sp pada Udang Putih (Penaeus Merguiensis) Segar di Pasar Tradisional Kotamadya Surabaya *Jurnal Ilmiah Perikanan dan Kelautan* **1**(1) 57-63
[10] Nottorf F 2014 *Modeling The Clickstream Across Multiple Online Advertising Channels Using A Binary Logit with Bayesian Mixture of Normals* Elsevier : *Electronic Commerce Research and Applications* **13** 45–55

Copula-based model for rainfall and El- Niño in Banyuwangi Indonesia

R E Caraka[1,2], Supari [3], M Tahmid[3]

[1] Department of Statistics, Universitas Padjadjaran, Bandung, Indonesia
[2] School of Mathematical Sciences, Faculty of Science and Technology, The National University of Malaysia, Bangi, Malaysia
[3] Indonesia Agency for Meteorology Climatology and Geophysics (BMKG), Indonesia

E-mail : Rezzyekocaraka@gmail.com

Abstract. Modelling, describing and measuring the structure dependences between different random events is at the very heart of statistics. Therefore, a broad variety of varying dependence concepts has been developed in the past. Most often, practitioners rely only on the linear correlation to describe the degree of dependence between two or more variables; an approach that can lead to quite misleading conclusions as this measure is only capable of capturing linear relationships. Copulas go beyond dependence measures and provide a sound framework for general dependence modelling. This paper will introduce an application of Copula to estimate, understand, and interpret the dependence structure in a given set of data El-Niño in Banyuwangi, Indonesia. In a nutshell, we proved the flexibility of Copulas Archimedean in rainfall modelling and catching phenomena of El Niño in Banyuwangi, East Java, Indonesia. Also, it was found that SST of nino3, nino4, and nino3.4 are most appropriate ENSO indicators in identifying the relationship of El Nino and rainfall.

1. Introduction

Climate is a paradigm of a complex system. Analysing climate data is an exciting challenge. Analysis connects the two other fields climate measurements and climate models. Statistical methods can be used to learn about the evolution of climate See [9]. What's more, we have a limited amount of data, but we want to know the truth about the climate evolution influenced by noises. Similarly, we can not expect estimate based on data to equal the truth. Alternatively, we can justify the typical size of derivation. Related concepts are confidence intervals or bias. Error bars help to assess estimation results critically, they prevent us from making overstatements, they guide us on our way to enhance the knowledge about the climate. Estimates without error bars are useless. Copulas is a methods of formalizing dependence structures of random Vectors. Although they have been known about for a long time [11] they have been rediscovered relatively and recently in applied sciences (biostatistics, reliability, biology, etc.). Although the concept of copulas is well understood, it is now recognized that their empirical estimation is a harder and trickier task. Many traps and technical difficulties are present, and these are, most of the time, ignored or underestimated by practitioners. The problem is that the estimation of copulas usually implies that every marginal distribution of the underlying random vectors must be evaluated and plugged into an estimated multivariate distribution. Such a procedure produces unexpected and unusual effects concerning the usual statistical procedures: non-standard limiting behaviours, noisy estimations, etc.

Copulas is one of robust multivariate statistics methods. Researcher using copula function for modelling between random variables. The word copula appears for the first time (Sklar 1959), Introduced to financial applications [5], widely used in insurance, finance, energy, hydrology, survival analysis, etc. The copula is used to describe the relationship between random variables. Multivariate Copula is the probability distribution for the marginal probability distribution of each variable with a uniform distribution [10]. Theorem Sklar [11] stated that some joint multivariate distribution could be expressed in univariate marginal distribution functions and Copula can describe the structure of connection or relationship multivariate distribution function with marginal distributions. To estimate copula functions, the first issue consists in specifying how to estimate the margins and the common law separately. Moreover, some of these functions can be fully known. Depending on the assumptions made, some quantities have to be estimated parametrically, or semi or even non-parametrically. In the latter case, the practitioner has to choose between the usual statistical methodology of using "empirical counterparts" and invoking smoothing methods well-known in statistics: kernels, wavelets, orthogonal polynomials, nearest neighbors, etc.

Let F represent the joint distribution function (df) of a continuous random vector $X = (X_1, \dots, X_d) \in R^d$. let $F_1, \dots, F_d$ be their marginal dfs and let $X_i \to F_i(X_i), i = 1, \dots, d,$ be their probability integral transformation to the standard uniform distribution. Then the copula C pertaining to F is defined as

$$C(u_1, \dots, u_d) = F(F_1^{-1}(u_1), \dots, F_d^{-1}(u_d), \forall(u_1, \dots, u_d) \in (0,1)^d \quad (1)$$

where F_i^{-1} is the quantile function of $F_i, i = 1, \dots, d$. Thus, the copula is a multivariate distribution with uniform (0,1) margins. The copula C completely specified the distribution F as much as

$$\forall(x_1, \dots, x_d) = F(x_1, \dots, x_d) = C(F_1(x_1), \dots, F_n(x_d)) \quad (2)$$

When F is not continuous, there still exists the copula representation of F, but it may not be unique anymore. The copula of any F captures and summarizes the dependence between the X_i variables and is invariant under strictly increasing transformation of the X_i. By taking partial derivate one, obtains the multivariate density function

$$f(x_1, \dots, x_d) = c(F_1(x_1), \dots, F_d(x_d)) \prod_{i=1}^{d} f_i(x_i) \quad (3)$$

Where c denotes a d-dimensional copula density. Decomposition equation above simplifies the specification of the data underlying multivariate distribution and suggest estimating the marginal distributions f_i separated from the dependence structure given by the d-variate copula. Dependence measures defined only through $C(\cdot)$ or $c(\cdot)$ whose expressions do not involve the margins of F_i or f_i are knows as copula based dependence measures.

In this paper, we focus on the practical issues in modeling rainfall and El-Niño southern oscillation (ENSO). El Niño is defined by sea-surface temperature SST in the eastern tropical Pacific, while the Southern Oscillation Index is a measure of the atmospheric circulation response in the Pacific–Indian Ocean region.

2.Literature Preview

2.1 Sklar's Theorem

Sklar's theorem elucidates the role that copulas play in the relationship between multivariate distribution functions and their univariate margins.

A distribution function is a function F with domain $\overline{\mathbf{R}}$ such that

1. F is non-decreasing,
2. $F(-\infty) = 0$ and $F(\infty) = 1$

A joint distribution function is a function H with domain $\overline{\mathbf{R}}^2$ such that

1. H is 2-increasing

2. $H(x,-\infty) = H(-\infty,y) = 0$ and $H(\infty,-\infty) = 1$

Thus H is grounded, and because Dom $H = \overline{\mathbf{R}}^2$, H has margin F and G given by $F(x) = H(x,\infty)$ and $G(y) = H(\infty,y)$. F and G are distribution functions. Let H be a joint distribution function with margins F and G. Then there exists a copula C such that for all x, y in $\overline{\mathbf{R}}$

$$H(x,y) = C(F(x),G(y)) \tag{4}$$

If F and G are continuous, then C is unique. Let F be a distribution function. Then a quasi-inverse of F is any function $F^{(-1)}$ with domain $\boldsymbol{I}$ such that. If t is $RanF$, then $F^{(-1)}(t)$ is any number x in $\overline{\mathbf{R}}$ such that $F(x) = t$, for all t in $RanF$

$$F(F^{(-1)}(t)) = t; \tag{5}$$

$$F^{(-1)}(t) = \inf\{x|F(x) \geq t) = \sup\{x|F(x) \leq t\} \tag{6}$$

If F is strictly increasing, then it has but a single quasi-inverse, which is of course the ordinary inverse, for which we use the customary notation $F^{(-1)}$

and let $F^{(-1)}$ and $G^{(-1)}$ *be quasi-inverses of F and G, respectively. Then for any* (u,v) in Dom C'

$$C'(u,v) = H(F^{(-1)}(u),G^{(-1)}(v)) \tag{7}$$

$$C(u,v) = \frac{uv}{u+v-uv} \tag{8}$$

2.2 Families of Copula

There are many families Copula, some of which are family Copula Copula Archimedean, Copula Ellipse, Copula Bivariate Extreme Value, and Copula Marshal-Olkin. The univariate exponential distribution plays a central role in mathematical statistics because it is the distribution of waiting time in a standard Poisson process. The following bivariate exponential distribution, first described by Marshall and Olkin in [8] plays a similar role in a two-dimensional Poisson process the characteristics of copulas can be seen in Table 1 and Table 2.

The most popular Copula is Archimedean and Ellipse. Among Archimedean copulas, we are going to consider in particular the one parameter ones, choosing the generator, one obtains a subclass or family of Archimedean copulas. Table 1 describes some well-known families and their generators (for a more exhaustive list see [10]. The Gumbel family has been introduced by Gumbel in [7] it is also known as the Gumbel–Hougaard family. The Clayton family was first proposed by [4], and studied by [9]. To end up with, the Frank family, which appeared in [6]. We can see in Table 1 the explanation of each Copulas

Table 1. Summary characteristics of selected copulas. For the t-copula, $\lambda_L{=}\lambda_U = 2t_{v+1}(-\sqrt{v+1}\sqrt{1-\rho}\,/\sqrt{1+\rho}$. The Gumbel copula has $\lambda_U = 2 - 2^{1/\theta}$. Clayton copula possess $\lambda_L = 2^{-1/\theta}$. For the BB7 copula, $\lambda_U = 2 - 2^{1/\theta}$ may be different from $\lambda_L = 2^{1/\theta}$

Family	**Type**	**Parameters**	λ_L	λ_u	C^-	C^1	C^+
Gaussian	Elliptical	$-1 \leq \rho \leq 1$	0.0	0.0	$\rho = -1$	$\rho = 0$	$\rho = +1$
t-copula	Elliptical	$-1 \leq \rho \leq +1, v$	✓	✓	$\rho = -1$	-	$\rho = +1$
Frank	Archimedean	$-\infty < \theta < +\infty$	0.0	0.0	$\theta \to -\infty$	$\theta = 0$	$\theta \to +\infty$
Surv.Clayton	Archim./EV	$\theta \geq 0$	0.0	✓	-	$\theta = 1$	$\theta \to +\infty$
Gumbel	Archim./EV	$\theta \geq 1$	0.0	✓	-	$\theta = 1$	$\theta \to +\infty$
BB7	Archimedean	$\theta \geq 1, \delta > 0$	✓	✓	-	-	θ or $\delta \to \infty$

Table 2. Copula Archimedean

Copula	**Generator** $\phi(u)$	**Copula Bivariate** $C(u_1,u_2)$
Clayton	$\frac{u^{-\theta}-1}{\theta}, \theta \in (0,\infty)$	$\left(u_1^{-\theta}+u_2^{-\theta}-1\right)^{-\frac{1}{\theta}}$

Gumbel	$\left(-\log(u)\right)^{\theta}, \theta \in [1,\infty)$	$\exp\left\{-\left[\left(-\log(u_1)\right)^{\theta}+\left(-\log(u_2)\right)^{\theta}\right]^{\frac{1}{\theta}}\right\}$
Frank	$\log\left(\frac{e^{\theta u}-1}{e^{\theta}-1}\right), \theta \in R\backslash\{0\}$	$\frac{1}{\theta}\log\left(1+\frac{\left(e^{\theta u_1}-1\right)\left(e^{\theta u_2}-1\right)}{\left(e^{\theta}-1\right)}\right)$

Every copulas has a different tail-dependencies [1]. Clayton copulas bottom-tail dependencies, Copula Frank, does not have a tail dependencies, while Copula Gumbel has a top-tail dependencies [2]. The Copula each pattern is shown in Figure. 1.

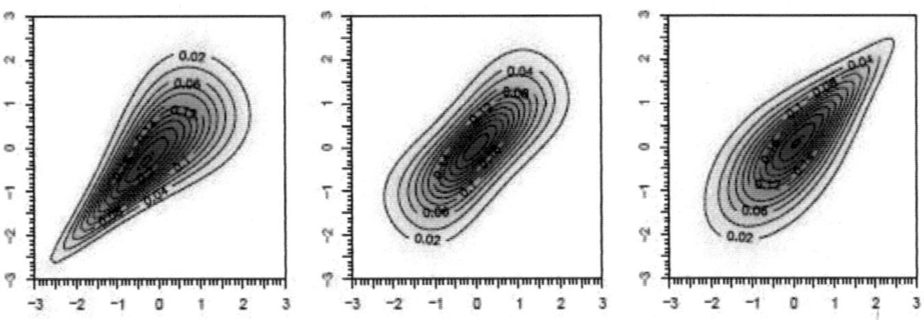

(a) (b) (c)

Figure. 1 Copula (a) Clayton, (b) Frank, and (c) Gumbel

From a statistical point of view, a copula function is a straightforward expression of a multivariate model and, as for most multivariate statistical models, much of the classical statistical inference theory is not applicable. The only theory that can be applied is the asymptotic maximum likelihood estimation (MLE) [3]

$$f(x_1,x_2,\ldots,x_n) = c(F_1(x_1),F_2(x_2),\ldots,F_n(x_n).\prod_{j=1}^{n} f_j(x_j) \qquad (9)$$

where

$$c(F_1(x_1),F_2(x_2),\ldots,F_n(x_n)) = \frac{\partial^n(C(F_1(x_1),F_2(x_2),\ldots F_n(x_n)))}{\partial(F_1(x_1)\partial(F_2(x_2),\ldots,\partial(F_n(x_n))} \qquad (10)$$

Is the n th mixed partial derivative of the copula C, c is the copula density and f is the standard univariate probability density function. This canonical representation for the multivariate density function permits us to say that, in general, a statistical modeling problem for copulas could be decomposed into two steps:

- identification of the marginal distributions;
- definition of the appropriate copula function.

This is an important point, to begin with, for estimation issues as we will see below.

Let$\aleph = \{x_{1t},x_{2t},\ldots,x_{nt}\}_{t=1}^{I}$ be the sample data matrix. Thus, the expression for the log-likelihood function is

$$l(\boldsymbol{\theta}) = \sum_{t=1}^{T}\ln c((F_1(x_{1t}),F_2(x_{2t}),\ldots,F_n(x_{nt})) + \sum_{t=1}^{T}\sum_{j=1}^{n}\ln f_j(x_{jt}) \qquad (11)$$

Where $\boldsymbol{\theta}$ is the set of all parameters of both the marginal and the copula. Hence, given a set of marginal probability discrete functions and a copula the previous log-likelihood may be written and by maximization I obtain the MLE estimator

$$\hat{\boldsymbol{\theta}}\,\mathrm{MLE} = \max_{\boldsymbol{\theta}\epsilon\Theta} l(\boldsymbol{\theta}) \qquad (12)$$

Throughout this section, we assume that the usual regularity conditions for asymptotic maximum likelihood theory hold for the multivariate model (i.e., the copula) as well as for all of its margins (i.e., the univariate p.d.f.s). After doing parameter estimation using MLE, then testing Copula parameter estimates to determine the significant parameters. Can be defined as

$H_0: C = C_\theta$
$H_1: C \neq C_\theta$
With H_0 C_θ follow Archimedean family and H_1 C_θ follow another model
moreover, the distribution becomes $\sqrt{n}Z_n \to N(0,1)$ standard normal distribution.
Hypotheses used can be defined as follows.
$H_0: \theta = \theta_0$
$H_1: \theta \neq \theta_0$
Statistical test:

$$z = \frac{\hat{\theta}-\theta_0}{SE\,(\hat{\theta})} \tag{13}$$

The null hypothesis is rejected if $|z| > z_{\frac{\alpha}{2}}$ or p-value<α

3. Simulation and Results

The data used in this research obtained from the Meteorology, Climatology and Geophysics (BMKG Indonesia) and the NOAA National Weather Service, which can be accessed http://www.cpc.ncep.noaa.gov/. Data obtained from BMKG is the rainfall data in Banyuwangi, East Java, Indonesia from 1982-2013. Data from NOAA's National Weather Service is the ENSO indicators data, SST Niño 1+2, SST Niño 3, SST Niño 4 dan SST Niño 3.4. El Niño refers to the periodic warming of the eastern equatorial Pacific Ocean that brings sea surface temperatures above average. Also, El Niño conditions, which can last for a year or two, develop concurrently with atmospheric changes leading to a variety of global effects. Including drier than normal weather in Indonesia and the Philippines, wetter than normal weather for Peru and Ecuador, a warmer than usual winter for much of the United States and above normal precipitation for the southern tier of the United States.

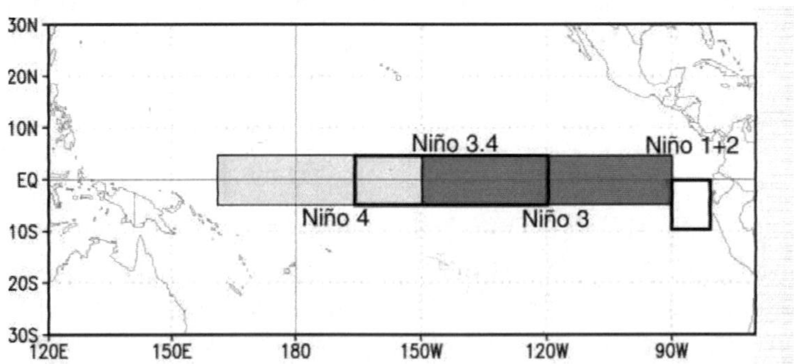

Figure. 2 El Niño Simulation (www.cpc.ncep.noaa.gov)

The formation of an El Niño is linked with the cycling of a Pacific Ocean circulation pattern known as the southern oscillation. In a typical year, a surface low pressure develops in the region of northern Australia and Indonesia and a high-pressure system over the coast of Peru. As a result, the trade winds over the Pacific Ocean move actively from east to west. The easterly flow of the trade winds carries warm surface waters westward, bringing convective storms to Indonesia and coastal Australia. Along the coast of Peru, cold bottom water wells up to the surface to replace the warm water that is pulled to the west. Before analysing must be known the pattern data of rainfall in Banyuwangi. In Figure 31 we can see that the rainfall in Banyuwangi experiencing fluctuating

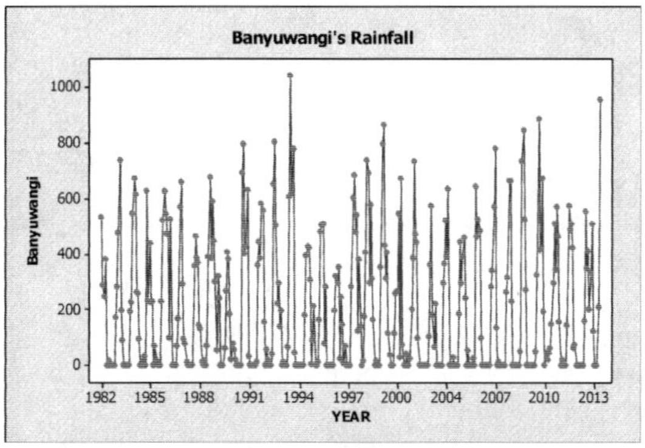

Figure. 3 Banyuwangi's Rainfall

Rainfall in Banyuwangi showed a high variation in each month and each year. The extreme climate often calls it, the climate is very low to cause flooding until the climate is very high and cause dryness. Based on descriptive statistics found that the maximum value of 1040.00 mm rain in Banyuwangi that occurred in January 1994 and obtained a mean value over 31 years amounted to 203.9 mm

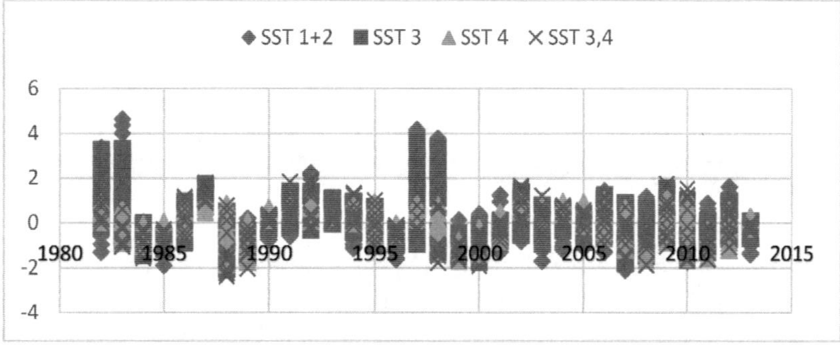

Figure. 4 El Niño from 1980 to 2013

Indonesia experiences ENSO influences from April on both El Niño (warm phase) and La Niña (cold phase). Although some indications of La Niña impacts exist as soon as March, the signal in March is too weak. The influences of ENSO reach their peaks in August and September by both types of events and decay afterward. By the end of the year, as the Asian monsoon comes, there is no clear signal that ENSO influences still exist. This is in agreement to what is found by Kirono (1999). The global climate during 1997 was affected by one of the most influential Pacific warm episodes on record. These warm episode conditions developed rapidly in March, with high ENSO conditions persisting from May through the end of the year (and subsequently well into 1998). Before the 1980s and 1990s, intense El Niño events occurred on average every 10 to 20 years. In the early 1980s, the first of a series of intense events developed. The El Niño of 1982-83 brought extreme warming to the equatorial Pacific. Surface sea temperatures in some regions of the Pacific Ocean rose 6° Celsius above average. The equatorial Walker circulation is characterized by ascending motion over Indonesia and the western tropical Pacific, and descending motion over the east-central equatorial Pacific, with upper-level westerly (low-level easterly), flow completing the "direct" circulation cell.

Table 3. Statistics Descriptive El Niño from 1980 to 2013

Variable	N	Mean	StDev	Variance	Minimum	Median	Maximum	Range
SST 1+2	384	0.0168	1.1838	1.4014	-2.1	-0.23	4.62	6.72
SST 3	384	-0.0137	0.9596	0.9207	-2.07	-0.155	3.62	5.69
SST 4	384	-0.0154	0.7057	0.498	-1.87	0.115	1.25	3.12
SST 3,4	384	-0.021	0.9439	0.8909	-2.38	-0.065	2.79	5.17

Based on the above table it can be seen that the maximum value of ENSO SST 1+2 was 4,62 which occurred in June 1983 and a minimum of -2,1 which occurred in November 2007. In the variable SST 3 maximum value of 3,62 that occurred in December 1997 and a minimum of -2.07 that occurred in December 1988. Thus, SST 4 maximum value of 1,25 that occurred in November 2009 and a minimum value of -1,87 in November 1988.

SST 3.4 variable maximum value amounted to 2,79 which occurred in January 1983 and a minimum of -2,38 that occurred in November 1988. Analyses were performed with Archimedean Copula approach, using Copula Frank, Clayton and Gumbel for variables that were analysed did not follow a normal distribution. However, for SST 3.4 also use the Gaussian Copula approach, because in testing the normal distribution cannot be concluded whether or not normal based on test Kolmogorov-Smirnov and Anderson-Darling.

Table 4. Parameter Estimation

Variables	Frank	Clayton	Gumbel	Gaussian
RF Banyuwangi and SST 1+2	-0,009	-0,019	0,991	-
P-value	0,723	0,721	-	-
RF Banyuwangi and SST 3	-0,104	-0,188	0,906	-
P-value	0,01	0,000	-	-
RF Banyuwangi and SST 4	-0,087	-0,160	0,920	-
p-value	0,010	0,005	-	-
RF Banyuwangi and SST 3.4	-0,115	-0,206	0,897	-0,116
p-value	0,000	0,000	-	0,000

Description: (-) = do not use the Gaussian Copula
***RF = Rain Fall**

Table 5 shows the structure of dependencies between rainfall and ENSO indicators in Banyuwangi. The significant parameters for rainfall and SST Niño 3, SST Niño in Bayuwangi is Frank and Clayton Copula because it has a *p-value*$<\alpha$. Parameter Copula Frank, Clayton, and Gaussian rainfall and SST Niño 3.4 in Banyuwangi significantly, the structure of dependencies of each variable follow Copula elected, whereas for rainfall and SST Niño 1 + 2 in Banyuwangi not follow Copula anything, because of *p-value*$>\alpha$. Copula Gumbel in Banyuwangi cannot be estimated because of the value of the parameter calculations in Table 3 for Copula Gumbel less than 1. Should the value θ of the Copula Gumbel is $\theta \in [1, \infty)$.

Table 5 Fitting Copula using MLE

Variabel	Copula	Estimation	Log-likelihood
RF Banyuwangi and SST 3	Frank	-0,935	4,262
	Clayton	-0,170	**5,042**
RF Banyuwangi and SST 4	Clayton	-0,323	*17,556*
RF Banyuwangi and SST 3.4	Frank	-0,913	4,071
	Clayton	-0,276	**14,381**
	Gaussian	-0,154	3,542

Description: Value in Bold indicates significant at α=0,05 and *log- likelihood italic* values showed significant on a α=0,10

Fitting Copula with the MLE (Maximum Likelihood Estimation) is performed for each Copula significant. We are using R software to anlaysis. Moreover, the best model for each pair of variables selected based on the fitting with the largest log-likelihood value and p-value is significant. Estimation shows that the parameter that indicates the close relationship between rainfalls in

IOP Conf. Series: Journal of Physics: Conf. Series **1008** (2018) 012025 doi:10.1088/1742-6596/1008/1/012025

Banyuwangi with 3.4 SST follow Clayton Copula. Some points indicate an outlier observation that is far from other observations. Scatterplot analysis in other districts is presented in more detail in Figure 5, Figure 6 and Figure 7.

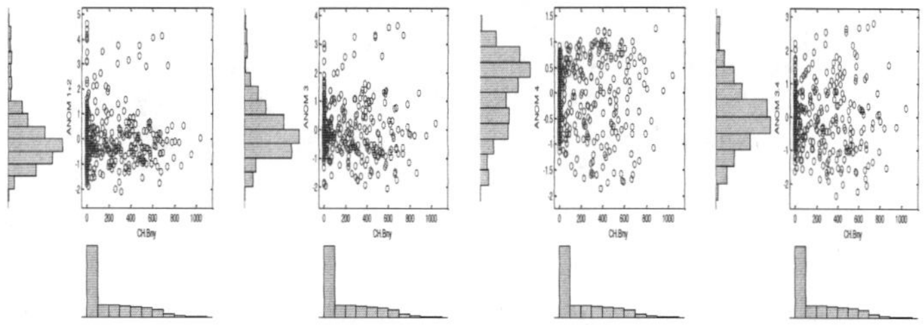

Figure. 5 Scatterplot and Histogram between Lag 1 Rainfall indicator and ENSO in Banyuwangi 1982-2013

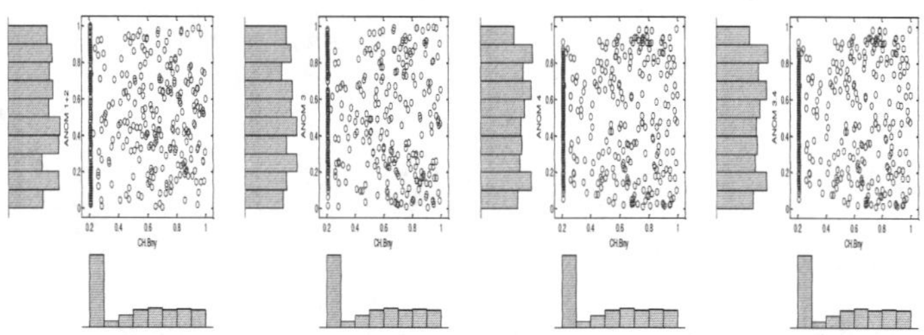

Figure. 6 Scatterplot and Histogram of Transformation to Uniform [0,1] between Lag 1 Rainfall indicator and ENSO in Banyuwangi 1982-2013

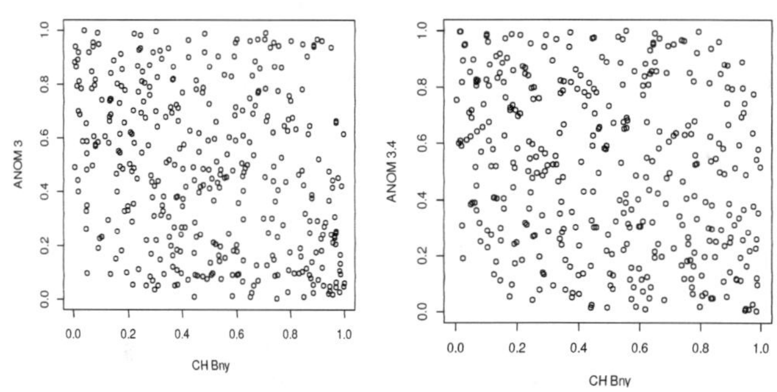

Figure. 7 Scatterplot Copula Rank between Lag 1 and Rainfall indicator and ENSO in Banyuwangi 1982-2013

In lag 1 does not form a symmetrical pattern, the distribution pattern has a skewness, especially in rainfall data. The rainfall distribution pattern has the right skewness, indicating that much data (mode) of rainfall is of little value.

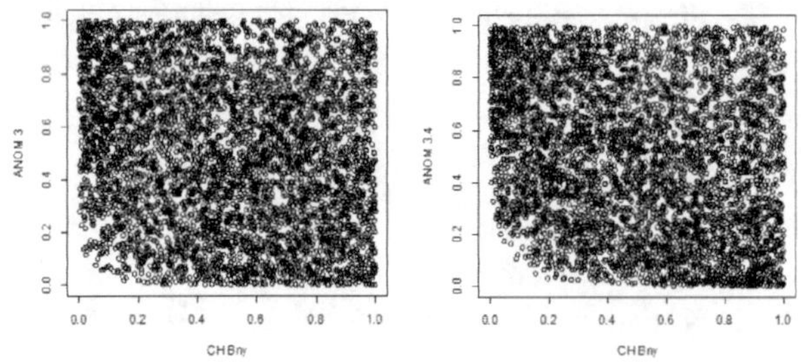

Figure. 8 Scatterplot Copula of Random Data (n=5000) between Lag 1 and Rainfall indicator and ENSO in Banyuwangi 1982-2013

Analysis of the relationship rainfall and ENSO indicators have been conducted on rainfall data and indicators of ENSO lag 0, lag 1 and lag 2. It aims to determine whether there are differences in the pattern of the relationship between rainfall and ENSO indicators of the time difference in the two variables. In addition to knowing there may be other factors that influence on the occurrence of rainfall in Banyuwangi, for example, there is the influence of wind, distance, and other factors that may affect the relationship of ENSO and rainfall in Banyuwangi. The results of the identification of patterns of relationships on 0 lag, lag 1 and lag two are shown in Table 4. Identification of relationship rainfall patterns with different time lags in Table 4 shows the difference Copula parameter and log - likelihood value in Banyuwangi. It shows the relationship between the two variables follows different Copula. Then relationship rainfall and SST El Niño 1 + 2 does not follow any Copula, which means SST El Niño 1 + 2 does not have a relationship with rainfall in Banyuwangi. However, for the relationship rainfall and Niño 3.4 SST based on the difference in a time lag, showed the same results on the relationship both follow Clayton Copula. Comparison of the three analyses is also to know at what time lag rainfall in Banyuwangi is closely related to ENSO indicators.

Table 6. Comparison of Parameters Copula on Lag 0, Lag 1 and Lag 2

Regional	**SST 1+2**	**SST 3**		**SST 4**		**SST 3.4**	
		Estimation	Log-likelihood	Estimation	Log-likelihood	Estimation	Log-likelihood
				Lag 0			
Banyuwangi	-	-0,915^{A}	4,183	-0,299^{B}	14,769	-0,247^{B}	10,547
				Lag 1			
Banyuwangi	-	-0,170^{B}	5,042	-0,323^{B}	17,556	-0,276^{B}	14,381
				Lag 2			
Banyuwangi	-	-0,189^{B}	6,297	-	10,387	-0,265^{B}	13,425

Description: A = Following Copula Frank (Not Having a Tail Dependency)
B = Following Clayton Copula (Having Tail Dependency)
- = Not Following Copula

Parameter estimation results with the difference in time lag which on lag 0, lag 1 and lags 2 between rainfall and ENSO indicators produces negative parameter values. It shows the relationship between rainfall and indicators of ENSO is reversed, i.e. in case of El Niño in the Pacific Ocean, the rainfall in Banyuwangi will decline could even lead to the occurrence of drought, and vice versa in case of La Nina, the rainfall will increase and could lead to flooding. Table 4 shows the relationship most strictly based on the lag time 1. This means that the ENSO events in the Pacific Ocean will affect rainfall next month. Also, scatterplot copula rank among lag 1 rainfall and ENSO indicators more amorphous than scatterplot Copula another rank. In addition, it can be concluded that the El Niño 3.4 SST is most

IOP Conf. Series: Journal of Physics: Conf. Series **1008** (2018) 012025 doi:10.1088/1742-6596/1008/1/012025

appropriately used in identifying the relationship of precipitation and ENSO indicators the lag 1 because it has the largest log-likelihood.

4. Conclusion

In this paper the possibility of analysing the copula family of the rainfall and El Niño has been demonstrated. Copulas Archimedean has already proven their flexibility in rainfall modeling and catching phenomena of El Niño in Banyuwangi, East Java, Indonesia. It can be concluded that 1997-98 El Niño is among the most reliable recorded and low rainfall in Indonesia. The relationship rainfall and SST El Niño 1 + 2 does not follow any Copula, which means SST El Niño 1 + 2 does not have a relationship with rainfall in Banyuwangi. ENSO indicators are most appropriate in identifying the relationship with rainfall is El Niño SST 3.4, SST 3 and SST 4. In general, the drought caused by climate anomalies and human activities. Prolonged drought that caused frequent droughts due to climate anomalies such as El Niño.

Acknowledgments

This research was supported by Department of Statistics, Universitas Padjadjaran, School of Mathematical Sciences, Faculty of Science and Technology, The National University of Malaysia, Bangi, Malaysia, Indonesia Agency for Meteorology Climatology and Geophysics (BMKG), Indonesia.

References

[1] Caraka R E, Haryono S, Hasbi Y, and Brilian W U 2018 *Exploratory MATLAB Software The Step Construction of Copula Gaussian Multivariate* (America: Institute of Physics)
[2] Caraka R E, Hasbi Y, Sugiyarto W, Sugiarto, and Ismail K M 2016 Time Series Analysis Using Copula Gauss And Ar(1)-N.Garch(1,1) *MEDIA STATISTIKA* 01–13
[3] Cherubini, Umberto, Luciano E, and Vecchiato W 2013 *Copula Methods in Finance* (America: Institute of Physics)
[4] Clayton D G 1978 A Model for Association in Bivariate Life Tables and Its Application in Epidemiological Studies of Familial Tendency in Chronic Disease Incidence *Biometrika* **65**(1) 141–51
[5] Embrechts, Paul, Lindskog F and Mcneil A 2003 *Modelling Dependence with Copulas and Applications to Risk Management* (Elsevier : Handbook of Heavy Tailed Distributions in Finance)
[6] Frank M J 1978 On the Simultaneous Associativity of F(x, Y) and X + Y - F(x, Y) *Aequationes Mathematicae* **18**(1–2) 266–67
[7] Heo J H and Salas J D 1996 Estimation of Quantiles and Confidence Intervals for the Log-Gumbel Distribution *Stochastic Hydrology and Hydraulics* **10**(3) 187–207
[8] Haijun L 2008 Tail Dependence Comparison of Survival Marshall-Olkin Copulas *Methodology and Computing in Applied Probability* **10**(1) 39–54
[9] David O 1982 A Model for Association in Bivariate Survival Data *Journal of the Royal Statistical Society. Series B (Methodological)* **44**(3) 414–22
[10] Bill R and Nelsen R B 2000 An Introduction to Copulas *Technometrics* **42**(3) 317
[11] Sklar A 1959 Fonctions de Répartition À N Dimensions et Leurs Marges *Publ. Inst. Statist. Univ. Paris* **8** 229–231

IOP Conf. Series: Journal of Physics: Conf. Series **1008** (2018) 012026 doi:10.1088/1742-6596/1008/1/012026

Estimation of water level and steam temperature using ensemble Kalman filter square root (EnKF-SR)

T Herlambang[1], Z Mufarrikoh[1], D F Karya[2], D Rahmalia[3]
[1]Information System Department, University of Nahdlatul Ulama Surabaya, Indonesia
[2]Magement Department, University of Nahdlatul Ulama Surabaya, Indonesia
[3]Mathematics Department, University of Islam Darul Ulum Lamongan, Indonesia

E-mail: teguh@unusa.ac.id

Abstract. The equipment unit which has the most vital role in the steam-powered electric power plant is boiler. Steam drum boiler is a tank functioning to separate fluida into has phase and liquid phase. The existence in boiler system has a vital role. The controlled variables in the steam drum boiler are water level and the steam temperature. If the water level is higher than the determined level, then the gas phase resulted will contain steam endangering the following process and making the resulted steam going to turbine get less, and the by causing damages to pipes in the boiler. On the contrary, if less than the height of determined water level, the resulted height will result in dry steam likely to endanger steam drum. Thus an error was observed between the determined. This paper studied the implementation of the Ensemble Kalman Filter Square Root (EnKF-SR) method in nonlinear model of the steam drum boiler equation. The computation to estimate the height of water level and the temperature of steam was by simulation using Matlab software. Thus an error was observed between the determined water level and the steam temperature, and that of estimated water level and steam temperature. The result of simulation by Ensemble Kalman Filter Square Root (EnKF-SR) on the nonlinear model of steam drum boiler showed that the error was less than 2%. The implementation of EnKF-SR on the steam drum boiler r model comprises of three simulations, each of which generates 200, 300 and 400 ensembles. The best simulation exhibited the error between the real condition and the estimated result, by generating 400 ensemble. The simulation in water level in order of 0.00002145 m, whereas in the steam temperature was some 0.00002121 kelvin.

1. Introduction

In Indonesia, electric power generation is carried out by various types of complex power plant installations. One of them is PLTU (Steam Power Plant). PLTU is one of the producers of electric power that provides electricity to meet the needs of all people who are expected to work optimally, therefore it is intended that the electricity distribution process is not hampered [1]. To support achieving that goal would require supporting components or equipments that always operate in good condition. Among the supporting equipments one unit that plays an important role is the boiler. Steam drum boiler is a tank that serves to separate the fluid into gas phase and liquid phase [2]. Its presence in the system plays a very important role. Measurement of water level and steam temperature in steam drum boilers is of great importance for the safety and operational efficiency of the boiler. It can be said that steam drum boilers are the heart of a boiler. This is where the steam used to turn turbines was first produced [3]. Than an estimator is required to predict and correct the water level and steam temperature of the steam drum boiler. Estimation is made for problem solving that requires previous information so that the next step can be determined in solving the problem. Estimation methods are often used in robotics such as for estimation of AUV trajectory [4,5,6] and Missile Trajectory [7], Maglev ball position estimation [8]. Kalman filter is a method of estimating state variables of a discrete linear dynamic system that minimizes the estimation error covariance.

Kalman filter was first introduced by Rudolph E. Kalman in 1960. It was about problem solving on linear discrete data filtering. In the nonlinear model another approach is required as an extension of the Kalman filter called Ensemble Kalman Filter (EnKF). In the EnKF method, the algorithm is executed by generating a certain number of ensembles to calculate the mean value and the error covariance of the state variable [5]. The development of EnKF method is through modification of algorithm by appending a square root scheme at EnKF correction stage which can produce Ensemble Kalman Filter Square Root (EnKF-SR) method.

The contribution of this paper is that estimation of both water level and steam temperature in the steam drum boiler can be an effective way to predict the increase and the decrease of water level and steam temperature in the steam boiler in future. In this paper, the implementation of square root scheme on Ensemble Kalman Filter (EnKF) method on Steam Drum Boiler equation was done, then it was applied to estimate the water level and the vapor temperature by simulation using Matlab software so that the error was observed in water level and vapor temperature determined with water level estimation and steam temperature.

2. Steam Drum Boiler

Steam drum is a container to store water in a large volume and to separate steam and water after heating takes place inside the boiler. The mathematical model of the boiler steam drum is formulatedwith two variables: the water level and the steam temperature

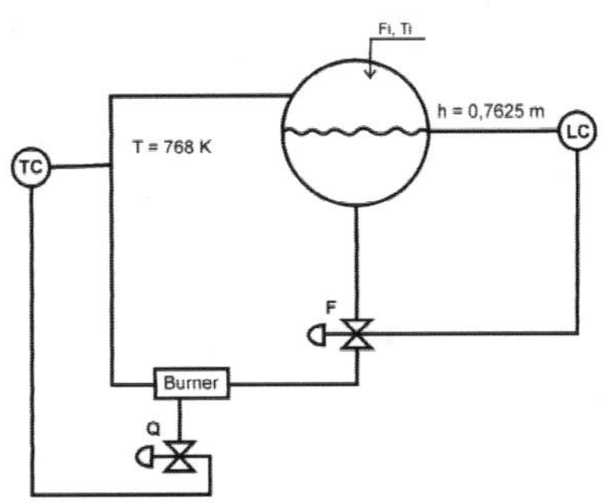

Figure 1 *steam drum boiler system* [2]

The mathematical model of the above figure is [2]:

$$A\frac{dh}{dt} = F_{in} - F_{out} \tag{1}$$

$$Ah\frac{dT}{dt} = F_{in}(T_{in} - T) + \frac{Q}{\rho C_p} \tag{2}$$

with $F_{out} = kw\sqrt{h}$

whereas :

F_{in} : Flow of in-coming water (kg/hour)
F_{out} : Flow of out-going water (kg/hour)
T : Temperature of vapor (K)
T_{in} : Temperatureof in-coming water (K)
Q : Flow of vapor (kg/hour)
V : Volume of water (m^3)
A : Area of *steam drum boiler* (m^2)
h : Height of water level (m)
ρ : Mass ofwater (kg/m^3)
C_p : Capacity of heat in steam drum (J/kg K)
w : *control valve*of water flow (m)
k: coeficiente of *control valve* ($m^{3/2}$/hour)

Because the system requires discretitation, so the steam drum boiler model in equation (1) – (2) must be discreted using the finite difference method.

Equation (1) and (2), If h_k water level and steam temperature T are

IOP Conf. Series: Journal of Physics: Conf. Series **1008** (2018) 012026 doi:10.1088/1742-6596/1008/1/012026

$h = \mathrm{h_k}, \mathrm{T} = \mathrm{T_k},$ (3)

The change of state variables respect to the time are approximated by forward scheme of finite difference. Thus we will get

$$\dot{\mathrm{h}} = \frac{\mathrm{dh}}{\mathrm{dt}} \approx \frac{\mathrm{h_{k+1}-h_k}}{\Delta \mathrm{t}} \quad (4)$$

$$\dot{\mathrm{T}} = \frac{\mathrm{dT}}{\mathrm{dt}} \approx \frac{\mathrm{T_{k+1}-T_k}}{\Delta \mathrm{t}} \quad (5)$$

from equation (4) and (5) will be gotten the modified steam drum boiler model in (6) below

$$\begin{bmatrix}\mathrm{h_{k+1}}\\ \mathrm{T_{k+1}}\end{bmatrix} = \begin{bmatrix}\mathrm{F_{in}} - \mathrm{kw}\sqrt{\mathrm{h_k}}\\ \mathrm{F_{in}(T_{in} - T_k)} + \frac{\mathrm{Q}}{\rho \mathrm{C_p}}\end{bmatrix} \quad (6)$$

3. Ensemble Kalman Filter Square Root (EnKF-SR)

This section present EnKF-SR algorithm to estimated nonlinear or linear dynamic system and measurement model, the algorithm *Ensemble Kalman Filter Square Root* (EnKF-SR) can be seen [7]:
Model system and measurement model

$$\mathrm{x_{k+1}} = \mathrm{f(x_k, u_k)} + \mathrm{w_k} \quad (7)$$

$$\mathrm{z_k} = \mathrm{Hx_k} + \mathrm{v_k} \quad (8)$$

$$\mathrm{w_k} \sim \mathrm{N(0, Q_k)},\ \mathrm{v_k} \sim \mathrm{N(0, R_k)} \quad (9)$$

1. Initialization
   Generate *N ensemble* as the first guess $\bar{x}_0$
   $$\mathrm{x_{0,i}} = [\mathrm{x_{0,1}} \quad \mathrm{x_{0,2}} \quad \cdots \quad \mathrm{x_{0,N}}] \quad (10)$$
   The first Mean Ensemble: $\bar{\mathrm{x}}_{0,\mathrm{i}} = \mathrm{x_{0,i}}\, 1_\mathrm{N}$ (11)
   The first Ensemble error: $\tilde{\mathrm{x}}_{0,\mathrm{i}} = \mathrm{x_{0,i}} - \bar{\mathrm{x}}_{0,\mathrm{i}} = \mathrm{x_{k,i}}(\mathrm{I} - 1_\mathrm{N})$ (12)
2. Time Update
   $$\hat{\mathrm{x}}^-_{\mathrm{k,i}} = \mathrm{f}(\hat{\mathrm{x}}_{\mathrm{k,-1,i}}, \mathrm{u_{k-1,i}}) + \mathrm{w_{k,i}} \quad (13)$$
   where $\mathrm{w_{k,i}} = \mathrm{N(0, Q_k)}$
   Mean Ensemble: $\bar{\mathrm{x}}^-_{\mathrm{k,i}} = \hat{\mathrm{x}}^-_{\mathrm{k,i}}\, 1_\mathrm{N}$ (14)
   Error Ensemble : $\tilde{\mathrm{x}}^-_{\mathrm{k,i}} = \hat{\mathrm{x}}^-_{\mathrm{k,i}} - \bar{\mathrm{x}}^-_{\mathrm{k,i}} = \hat{\mathrm{x}}^-_{\mathrm{k,i}}\,(\mathrm{I} - 1_\mathrm{N})$ (15)
3. Measurement Update
   $$z_{k,i} = Hx_{k,i} + v_{k,i} \quad (16)$$
   where $v_{k,i} \sim N(0, R_k)$
   $S_k = H\tilde{x}^-_{k,i}$, $E_k = (v_1, v_2, \ldots, V_N)$ and
   $$C_k = S_k\, S_k^T + E_k\, E_k^T \quad (17)$$
   Mean Ensemble :
   $$\bar{x}_{k,i} = \bar{x}^-_{k,i} + \tilde{x}^-_{k,i} S_k^T C_k^- \left(\bar{z}_{k,i} - H\bar{x}^-_{k,i}\right) \quad (18)$$
   Square Root Scheme:
   - eigenvalue decompotition from $C_k = U_k\, \Lambda_k U_k^T$ (19)
   - determine matrix $M_k = \Lambda_k^{\frac{1}{2}}\, U_k^T\, S_k^-$ (20)
   - determine SVD from $M_k = Y_k\, \mathrm{L}_k\, V_k^T$ (21)

   Ensemble Error: $\tilde{x}_{k,i} = \tilde{x}^-_{k,i}\, V_k \left(I - \mathrm{L}_k\, L_k^T\right)^{\frac{1}{2}}$ (22)
   Ensemble Estimation: $\hat{x}_{k,i} = \bar{x}_{k,i} + \bar{x}_{k,i}$ (23)

To evaluate of estimation result accuracy from EnKF-SR algorithm, can be show with calculate Root Mean Square Error (RMSE) [5].

$$RMSE = \sqrt{\frac{\sum_i^n \left(X_{obs,i}(k) - X_{model,i}(k)\right)^2}{n}} \tag{24}$$

With

$X_{obs,i}(k)$ = observation data

$X_{model,i}(k)$ = model data

n = iteration

4. Computational Result

This simulation was made by applying the EnKF-SR algorithm to the boiler steam drum model. The simulation results were evaluated and compared with real conditions. This simulation consisted of three simulations of 200, 300 and 400 ensembles applied for estimation of the water level and the steam temperature. The value of Δt used was Δt = 0.1, and the initial condition used was h = 0.765 m and T = 786 K.

Figures 2 and 3 showthe estimation of the water level and the vapor temperature of steam drum boilers by generating 400 ensembles, while figure 3 and 4 show the result of estimation by generating 300 ensembles. Figure 2 shows very accurate estimation results with an error of 0.0021 meters and accuracy of 98.2%. Figure 3 shows a very accurate estimation result with an error of 0.004 Kelvin and accuracy of 99.1%. Figures 2 and 3 illustrate that the estimation by EnKF-SR with 400 ensembles and iteration of 200 showed significantly accurate result which was very close to the real condition of the height of the water level.

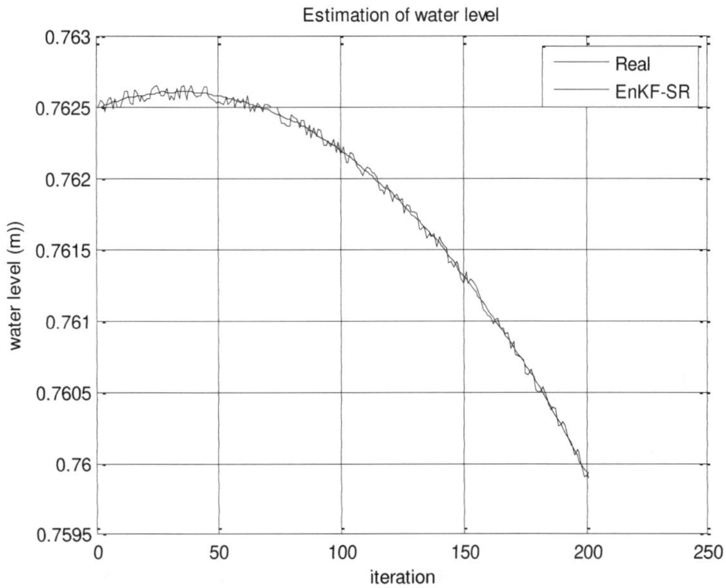

Figure 2. Estimation of water level using 400 Ensemble

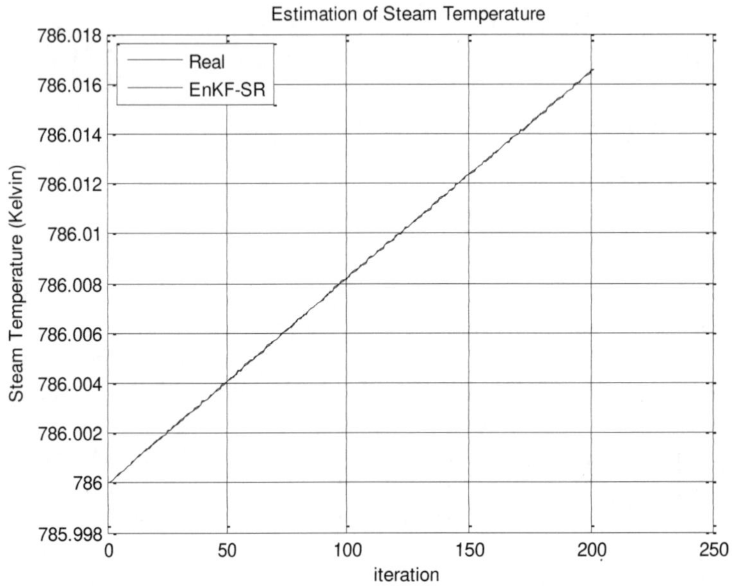

Figure 3. Estimation of Steam Temperature using 400 Ensemble

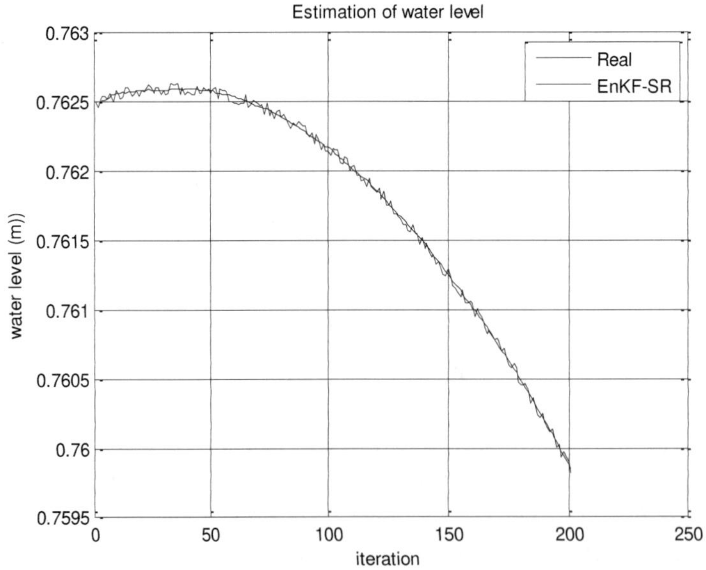

Figure 4. Estimation of water level using 300 Ensemble

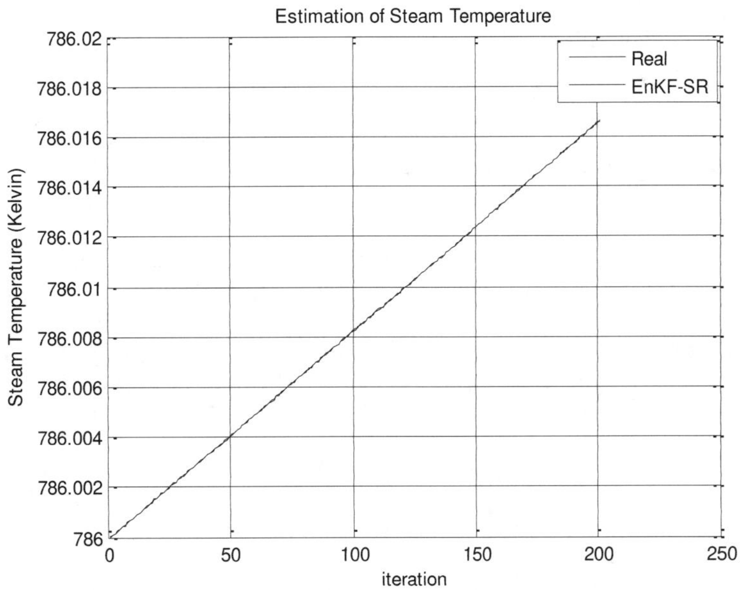

Figure 5. Estimation of Steam Temperature using 300 Ensemble

Further comparison of estimation results with 200, 300 and 400 ensembles in Table 1 shows that with the 400 ensembles the most accurate results were in water level and vapor estimates with errors of $2{,}145 \times 10^{-5}$ meter dan $2{,}121 \times 10^{-5}$ Kelvin. In terms of simulation time, with 400 ensembles it took a longer time than it did with 200 and 300 ensembles because the generation of numbers of ensemble greatly affected the computing time.

In general as seen in the table 1, the result of the three simulations were highly accurate. The first simulation by generate 200 ensemble with error of water level 0.0003704 meter or accuracy of 99% and error of steam temperature 0.00003655 kelvin or accuracy of 99.1%. the second simulation by generate 300 ensemble with error of water level 0.00003028 m or accuracy of 99.2% and accuracy of steam temperature 99.3%.

Table 1. Comparison of the values of RMSE by EnKF-SR method by generating 200, 300 and 400 ensembles

	200 Ensemble	**300 Ensemble**	**400 Ensemble**
Water Level	0.00003704 m	0.00003028 m	**0.00002145 m**
Steam Temperature	0.00003655 kelvin	0.00003015 kelvin	**0.00002121 kelvin**
Time Simulation	3.3438 s	5.2156 s	6.974 s

5. Conclusion

Based on the results of the simulation analysis by generating 200, 300 and 400 ensembles, the EnKF-SR method could be applied to estimate water levels and steam temperatures with a high degree of accuracy. Water level errors and vapor temperatures were of less than 2%, Square Root Scheme could affect estimation result because it could be implemented in the correction phase of the EnKF method.

Open problem. How to implemented Unscented Kalman Filter (UKF) and Unscented Kalman filter Square Root (UKF-SR) for estimation of watel level and steam temperature in steam drum boiler.

Acknowledgement
This research was supported by LPPM - Nahdlatul Ulama Surabaya 0f University (UNUSA).

References

[1] Herlambang T, Apriliani E, Cordova H and Mardlijah 2011 Desain Pengendalian Ketinggian Air dan Temperatur Uap pada Sistem *Steam Drum Boiler* dengan menggunakan Sliding Mode Control (SMC) *Seminar Nasional Penelitian, Pendidikan, dan Penerapan MIPA Universitas Negeri Yogyakarta*
[2] Herlambang T, Apriliani E, Cordova H, Mardlijah 2017 Dynamic Sliding Mode Control (DSMC) Untuk Sistem Kendali Water Level Pada Steam Drum Boiler *Technology Science and Engineering Journal* **1** (1)
[3] Moradi, Hamad 2009 *Robust Control of an Industrial Boiler Systema Comparison Between Two Approaches; Sliding Mode Control And* H_{∞} *Technique* (Tehran: Amirkabir University of Technology)
[4] Herlambang T, Djatmiko E B and Nurhadi H 2015 Navigation and Guidance Control System of AUV with Trajectory Estimation of Linear Modelling *Proc. of International Conference on Advance Mechatronics, Intelligent Manufactre, and Industrial Automation (ICAMIMIA)-IEEE* 184-187
[5] Herlambang T, Djatmiko E B and Nurhadi H 2015 Ensemble Kalman Filter with a Square Root Scheme (EnKF-SR) for Trajectory Estimation of AUV SEGOROGENI ITS *International Review of Mechanical Engineering IREME Journal* **9** (6) 553-560
[6] Ermayanti E, Aprilini E, Nurhadi H, and Herlambang T 2015 Estimate and Control Position Autonomous Underwater Vehicle Based on Determined Trajectory using Fuzzy Kalman Filter Method *International Conference on Advance Mechatronics, Intelligent Manufactre, and Industrial Automation (ICAMIMIA)-IEEE*
[7] Herlambang T 2017 Design of a Navigation and Guidance System of Missile with Trajectory Estimation Using Ensemble Kalman Filter Square Root (EnKF-SR) *The 4th International conference on Computer Application and Informations Processing Technology (CAIPT 2017)*
[8] Herlambang T 2017 Estimasi Posisi Magnetic Levitation Ball Menggunakan Metode Akar Kuadrat Ensemble Kalman Filter (AK-EnKF) *Rekayasa, Energi, Manufaktur Jurnal* **2** (1)

IOP Conf. Series: Journal of Physics: Conf. Series **1008** (2018) 012027 doi:10.1088/1742-6596/1008/1/012027

On the Total Edge Irregularity Strength of Generalized Butterfly Graph

Hafidhyah Dwi Wahyuna and Diari Indriati

Department of Mathematics, Faculty of Mathematics and Natural Sciences, Universitas Sebelas Maret, Surakarta, Indonesia

E-mail: dwhafidhyah@gmail.com, diari_indri@yahoo.co.id

Abstract. Let $G(V, E)$ be a connected, simple, and undirected graph with vertex set V and edge set E. A total k-labeling is a map that carries vertices and edges of a graph G into a set of positive integer labels $\{1, 2, ..., k\}$. An edge irregular total k-labeling $\lambda : V(G) \cup E(G) \to \{1, 2, ..., k\}$ of a graph G is a total k-labeling such that the weights calculated for all edges are distinct. The weight of an edge uv in G, denoted by $wt(uv)$, is defined as the sum of the label of u, the label of v, and the label of uv. The total edge irregularity strength of G, denoted by $tes(G)$, is the minimum value of the largest label k over all such edge irregular total k-labelings. A generalized butterfly graph, BF_n, obtained by inserting vertices to every wing with assumption that sum of inserting vertices to every wing are same then it has $2n+1$ vertices and $4n-2$ edges. In this paper, we investigate the total edge irregularity strength of generalized butterfly graph, BF_n, for $n \geq 2$. The result is $tes(BF_n) = \lceil \frac{4n}{3} \rceil$.

1. Introduction

Let $G(V, E)$ be a connected, simple, and undirected graph with vertex set V and edge set E. A labeling of a graph G is a mapping that carries a set of graph elements into a set of positive integers, called labels (Wallis [8]). If the domain of mapping is a vertex set, or an edge set, or a union of vertex and edge sets, then the labeling is called vertex labeling, edge labeling, or total labeling, respectively. In Gallian's survey [2], he showed that there were various kinds of labelings on graphs, and one of them was an irregular total labeling.

Bača *et al.* [1] introduced the notion of a total k-labeling, an edge irregular total k-labeling, and a vertex irregular total k-labeling. For a graph $G(V, E)$, they defined a labeling $f : V \cup E \to \{1, 2, ..., k\}$ to be a total k-labeling. An edge irregular total k-labeling with vertex set V and edge set E is a labeling $\lambda : V(G) \cup E(G) \to \{1, 2, ..., k\}$ such that for two different edges $e = u_i v_j$ and $f = u_k v_l$ then their weight $wt(e) \neq wt(f)$. The weight of an edge uv in G, denoted by $wt(uv)$, is defined as the sum of the label of u, the label of v, and the label of uv, that is

$$wt(uv) = \lambda(u) + \lambda(uv) + \lambda(v).$$

Furthermore, Bača *et al.* [1] also defined the total edge irregularity strength of G, denoted by $tes(G)$, as the minimum value of the largest label k over all such edge irregular total k-labelings. They also gave a lower bound and an upper bound on $tes(G)$ with vertex set V and a non-empty edge set E,

$$\left\lceil \frac{|E| + 2}{3} \right\rceil \leq tes(G) \leq |E|. \qquad (1)$$

Many researchers have investigated $tes(G)$ to some graph classes. By then, there are some result related to the $tes(G)$. Bača *et al.* [1] proved $tes(G)$ of paths, cycles, stars, wheels, and friendship graphs, that are, $tes(P_n) = tes(C_n) = \lceil \frac{n+2}{3} \rceil$, $tes(S_n) = \lceil \frac{n+1}{2} \rceil$, $tes(W_n) = \lceil \frac{2n+2}{3} \rceil$ for $n \geq 3$, and $tes(F_n) = \lceil \frac{3n+2}{3} \rceil$. Then, $tes(G)$ of tree G with edge set E and maximum degree $\triangle$ had been found by Ivančo and Jendrol' [6], that is $tes(G) = max\{\lceil \frac{\triangle+1}{2} \rceil, \lceil \frac{|E|+2}{3} \rceil\}$. In 2008, Nurdin *et al.* [7] proved $tes(G)$ of the corona product of paths with some graphs, namely paths P_n, cycles C_n, stars S_n, gears G_n, friendships F_n, and wheels W_n. Haque [3] in 2012 proved $tes(G)$ of generalized Petersen graphs $P(n,k)$ and Indriati *et al.* [4] found $tes(G)$ of Helm, H_n, and disjoint union of t isomorphic helms, tH_n. In 2013 Indriati *et al.* [5] found $tes(H_n^1) = \lceil \frac{4n+2}{3} \rceil$, $tes(H_n^2) = \lceil \frac{5n+2}{3} \rceil$, and $tes(H_n^m) = \lceil \frac{(m+3)n+2}{3} \rceil$ for $n \geq 3$ and $m \equiv 0 \pmod 3$.

Weisstein [9] defined the butterfly graph is a planar undirected graph with 5 vertices and 6 edges. In this paper, we investigate the total edge irregularity strength of generalized butterfly graph, BF_n, for $n \geq 2$.

2. Main Result

A generalized butterfly graph, BF_n, obtained by inserting vertices to every wing with assumption that sum of inserting vertices to every wing are same then it has $2n+1$ vertices and $4n-2$ edges. Let the vertex set of BF_n be $V(BF_n) = \{v_i \mid i = 0, 1, 2, ..., 2n\}$ and the edge set of BF_n be $E(BF_n) = \{(v_i, v_{i+1}) \mid i = 1, 2, ..., n-1, n+1, ..., 2n-1\} \cup \{(v_0, v_i) \mid i = 1, 2, ..., 2n\}$. Figure 1 illustrates the generalized butterfly graph BF_n. Based on Figure 1, BF_n has $\{v_0\}$ as an apex, $\{v_1, v_2, v_3, ..., v_{(n-1)}, v_n\}$ as vertices on right wing, and $\{v_{(n+1)}, v_{(n+2)}, v_{(n+3)}, ..., v_{(2n-1)}, v_{2n}\}$ as vertices on left wing.

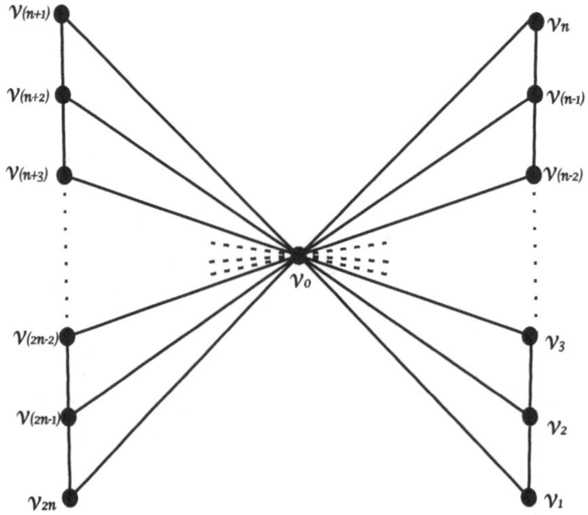

Figure 1. The generalized butterfly graph BF_n

In the next theorem, we present the total edge irregularity strength of generalized butterfly graph, BF_n, for $n \geq 2$ as follows:

Theorem 2.1 *For $n \geq 2$, $tes(BF_n) = \lceil \frac{4n}{3} \rceil$.*

Proof. From the lower bound of total edge irregularity strength we have that $tes(BF_n) \geq \lceil \frac{4n}{3} \rceil$, $n \geq 2$. To prove the equality, it is sufficient to show the existence of an edge irregular total k_1-labeling with $k_1 = \lceil \frac{4n}{3} \rceil$. Let $k_1 = \lceil \frac{4n}{3} \rceil$. Then from inequality (1) it follows that, $tes(BF_n) \geq \lceil \frac{|E(BF_n)|+2}{3} \rceil = \lceil \frac{(4n-2)+2}{3} \rceil = \lceil \frac{4n}{3} \rceil = k_1$, that is $tes(BF_n) \geq k_1$. To prove the reverse

IOP Conf. Series: Journal of Physics: Conf. Series **1008** (2018) 012027 doi:10.1088/1742-6596/1008/1/012027

inequality, we define a function f_1 as follows.
Case 1: For $n \equiv 2(mod\ 3)$, $n \geq 2$.

$$f_1(v_0) = \lceil \frac{4n}{3} \rceil.$$

$$f_1(v_i) = \begin{cases} 1, & \text{for } 1 \leq i \leq \frac{2n+5}{3}; \\ f_1(v_{i-1}) + (-1)^i(\frac{2n-1}{3} - i(mod\ 2)), & \text{for } \frac{2n+8}{3} \leq i \leq 2n. \end{cases}$$

$$f_1(v_iv_{i+1}) = \begin{cases} i, & \text{for } 1 \leq i \leq \frac{2n+2}{3}; \\ 2, & \text{for } \frac{2n+5}{3} \leq i \leq n-1; \\ 1, & \text{for } n+1 \leq i \leq 2n-1. \end{cases}$$

$$f_1(v_0v_i) = \begin{cases} \frac{2n-4}{3} + i, & \text{for } 1 \leq i \leq \frac{2n+5}{3}; \\ \lceil \frac{4n}{3} \rceil, & \text{for } \frac{2n+8}{3} \leq i \leq 2n. \end{cases}$$

Case 2: For $n \equiv 0(mod\ 3)$, $n \geq 3$ and n is odd.

$$f_1(v_0) = \lceil \frac{4n}{3} \rceil.$$

$$f_1(v_i) = \begin{cases} 1, & \text{for } 1 \leq i \leq \frac{2n}{3}; \\ \frac{2n}{3}, & \text{for } i = \frac{2n+3}{3}; \\ f_1(v_{i-1}) + (-1)^{i-1}(\frac{2n-3}{3} - i(mod\ 2)), & \text{for } \frac{2n+6}{3} \leq i \leq n; \\ f_1(v_{i-1}) + 1, & \text{for } i = n+1; \\ f_1(v_{i-1}) + (-1)^i(\frac{2n+3}{3} - i(mod\ 2)), & \text{for } n+2 \leq i \leq 2n. \end{cases}$$

$$f_1(v_iv_{i+1}) = \begin{cases} i, & \text{for } 1 \leq i \leq \frac{2n-3}{3}; \\ 1, & \text{for } i = \frac{2n}{3} \text{ and } n+1 \leq i \leq 2n-1; \\ 2, & \text{for } \frac{2n+3}{3} \leq i \leq n-1;. \end{cases}$$

$$f_1(v_0v_i) = \begin{cases} \frac{2n-3}{3} + i, & \text{for } 1 \leq i \leq \frac{2n+3}{3}; \\ \lceil \frac{4n}{3} \rceil, & \text{for } \frac{2n+6}{3} \leq i \leq 2n. \end{cases}$$

Case 3: For $n \equiv 1(mod\ 3)$, $n \geq 4$.

$$f_1(v_0) = \lceil \frac{4n}{3} \rceil.$$

$$f_1(v_i) = \begin{cases} 1, & \text{for } 1 \le i \le \frac{2n+7}{3}; \\ f_1(v_{i-1}) + (-1)^i(\frac{2n+1}{3} - i(mod\ 2)), & \text{for } 6 \le i \le 8 \text{ and } n = 4; \\ f_1(v_{i-1}) + 1, & \text{for } i = 8 \text{ and } n = 7; \\ f_1(v_{i-1}) + (-1)^{i-1}(\frac{2n-5}{3} + i(mod\ 2)), & \text{for } 9 \le i \le 14 \text{ and } n = 7; \\ f_1(v_{i-1}) + 1, & \text{for } \frac{2n+10}{3} \le i \le \frac{2n+13}{3} \text{ and } n \ge 10; \\ f_1(v_{i-1}) + (-1)^i(\frac{2n-5}{3} - i(mod\ 2)), & \text{for } \frac{2n+16}{3} \le i \le 2n \text{ and } n \ge 10. \end{cases}$$

$$f_1(v_iv_{i+1}) = \begin{cases} i, & \text{for } 1 \le i \le \frac{2n+4}{3}; \\ \frac{2n+4}{3}, & \text{for } i = \frac{2n+7}{3} \text{ and } n \ge 10; \\ f_1(v_{i-1}v_i) - 1, & \text{for } \frac{2n+10}{3} \le i \le n-2 \text{ and } n \ge 13; \\ 2, & \text{for } i = n-1 \text{ and } n \ge 16; \\ 1, & \text{for } n+1 \le i \le 2n-1. \end{cases}$$

$$f_1(v_0v_i) = \begin{cases} \frac{2n-5}{3} + i, & \text{for } 1 \le i \le \frac{2n+7}{3}; \\ \lceil \frac{4n}{3} \rceil, & \text{for } \frac{2n+10}{3} \le i \le 2n \text{ and } n \ge 7; \\ 5 + i(mod\ 2), & \text{for } \frac{2n+10}{3} \le i \le 2n \text{ and } n = 4; \end{cases}$$

It can be seen that the function f_1 is a map from $V(BF_n) \cup E(BF_n)$ into $\{1, 2, ..., \lceil \frac{4n}{3} \rceil\}$. Therefore, f_1 is a total k_1-labeling with $k_1 = \lceil \frac{4n}{3} \rceil$.

We observe that the weights of the edges are:

$$wt(v_iv_{i+1}) = \begin{cases} 2 + i, & \text{for } 1 \le i \le n-1; \\ 1 + i, & \text{for } n+1 \le i \le 2n-1. \end{cases}$$

Case 1: For $n \equiv 2(mod\ 3)$, $n \ge 2$.

$$wt(v_0v_i) = \begin{cases} 2n + i, & \text{for } 1 \le i \le \frac{2n+5}{3}; \\ wt(v_0v_{i-1}) + (-1)^i(\frac{2n-1}{3} - i(mod\ 2)), & \text{for } \frac{2n+8}{3} \le i \le 2n. \end{cases}$$

IOP Conf. Series: Journal of Physics: Conf. Series **1008** (2018) 012027 doi:10.1088/1742-6596/1008/1/012027

Case 2: For $n \equiv 0(mod\ 3)$, $n \geq 3$ and n is odd.

$$wt(v_0v_i) = \begin{cases} 2n+i, & \text{for } 1 \leq i \leq \frac{2n}{3}; \\ wt(v_0v_{i-1}) + (-1)^{i-1}(\frac{2n-3}{3} + i(mod\ 2)), & \text{for } \frac{2n+3}{3} \leq i \leq n; \\ wt(v_0v_{i-1}) + 1, & \text{for } i = n+1; \\ wt(v_0v_{i-1}) + (-1)^{i}(\frac{2n+3}{3} - i(mod\ 2)), & \text{for } n+2 \leq i \leq 2n. \end{cases}$$

Case 3: For $n \equiv 1(mod\ 3)$, $n \geq 4$.

$$wt(v_0v_i) = \begin{cases} 2n+i, & \text{for } 1 \leq i \leq \frac{2n+7}{3}; \\ wt(v_0v_{i-1}) + (-1)^{i}(\frac{2n-2}{3} - i(mod\ 2)), & \text{for } 6 \leq i \leq 8 \text{ and } n = 4; \\ wt(v_0v_{i-1}) + 1, & \text{for } i = 8 \text{ and } n = 7; \\ wt(v_0v_{i-1}) + (-1)^{i-1}(\frac{2n-5}{3} + i(mod\ 2)), & \text{for } 9 \leq i \leq 14 \text{ and } n = 7; \\ wt(v_0v_{i-1}) + 1, & \text{for } \frac{2n+10}{3} \leq i \leq \frac{2n+13}{3} \text{ and } n \geq 10; \\ wt(v_0v_{i-1}) + (-1)^{i}(\frac{2n-5}{3} - i(mod\ 2)), & \text{for } \frac{2n+16}{3} \leq i \leq 2n \text{ and } n \geq 10. \end{cases}$$

It can be seen that the weights of edges of BF_n under the total k_1-labeling, f_1, form consecutive integers from 3 to $4n$. It means that the weights of all edges are distinct, then we have $tes(BF_n) \leq k_1$. This completes the proof. So, the labeling is an edge irregular total k_1-labeling with $k_1 = \lceil \frac{4n}{3} \rceil$. Therefore, $tes(BF_n) = \lceil \frac{4n}{3} \rceil$, for $n \geq 2$. □

Figure 2 shows an edge irregular total labeling of BF_8.

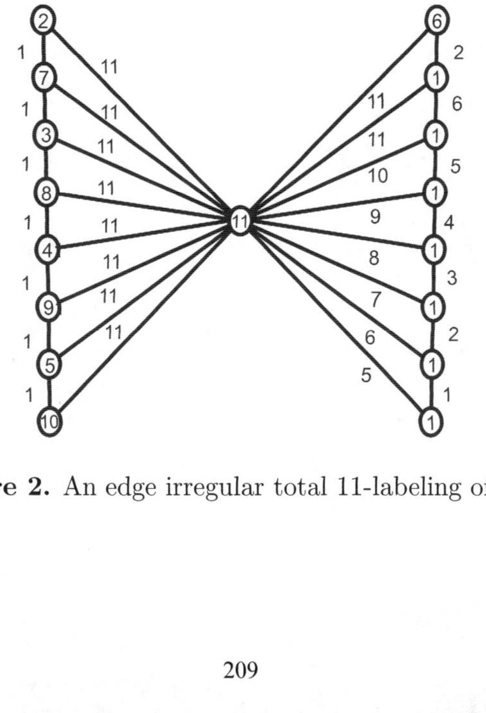

Figure 2. An edge irregular total 11-labeling of BF_8

3. Concluding Remark

Furthermore, we conclude this paper with the following open problem for the direction of further research which is still in progress.

Open Problem: Determine the total edge irregularity strength of generalized butterfly graph BF_n for $n \geq 3$ and n is odd. Then, generalized butterfly graph isomorphic with broken fan graph. Broken fan graph is fan graph which divide into n parts for $n = 2$. So, determine the total edge irregularity strength of broken fan graph which divide into n parts for $n \geq 3$.

Acknowledgments

We gratefully acknowledge the support from Department of Mathematics, Faculty of Mathematics and Natural Sciences, Universitas Sebelas Maret Surakarta. Then, we wish to thank the referees for their valuable suggestions and references, which helped to improve the paper.

References

[1] Bača M, Jendrol' S, Miller M and Ryan J. On Irregular Total Labelings 2007 *Discrete Math.* **307** 1378-1388
[2] Gallian J A. A Dynamic Survey of Graph Labeling 2016 *The Electronic Journal of Combinatorics.* **19** 1-284 #DS6
[3] Haque K M M. Irregular Total Labellings of Generalized Petersen Graphs 2012 *Theory Comput Syst.* **50** 537-544
[4] Indriati D, Widodo, Wijayanti I E and Sugeng K A. Kekuatan Tak Reguler Sisi Total pada Graf Helm dan Union Disjoinnya 2012 *Presented in Konferensi Nasional Matematika XVI*, Unpad, Jatinangor, Jawa Barat
[5] Indriati D, Widodo, Wijayanti I E and Sugeng K A. On the Total Edge Irregularity Strength of Generalized Helm 2012 *AKCE Int. J. Graphs Comb.* **10** (2) 147-155
[6] Ivančo J and Jendrol' S. Total Edge Irregularity Strength of Trees 2006 *Discuss. Math. Graph Theory.* **26** 449-456
[7] Nurdin, Salman A N M and Baskoro E T. The Total Edge Irregularity Strength of the Corona Product of Path with some Graphs 2013 *J. Combin. Math. Combin. Comput.* **65** 163-175
[8] Wallis W D. Magic Graphs 2001 Boston: Birkhäuser
[9] Weisstein, Eric W. Butterfly Graph *MathWorld*, From MathWorld-A Wolfram Web Resource *http://mathworld.wolfram.com/ButterflyGraph.html.*

The neighbourhood polynomial of some families of dendrimers

Mohamad Nazri Husin[1] and Roslan Hasni[1]

[1]School of Informatics and Applied Mathematics, Universiti Malaysia Terengganu, 21030 Kuala Terengganu, Terengganu, Malaysia.

Email: naz_reyhusin@yahoo.com; hroslan@umt.edu.my

Abstract. The neighbourhood polynomial $N(G,x)$ is generating function for the number of faces of each cardinality in the neighbourhood complex of a graph and it is defined as $(G,x) = \sum_{U\in N(G)} x^{|U|}$, where $N(G)$ is neighbourhood complex of a graph, whose vertices of the graph and faces are subsets of vertices that have a common neighbour. A dendrimers is an artificially manufactured or synthesized molecule built up from branched units called monomers. In this paper, we compute this polynomial for some families of dendrimer.

1. Introduction

A (simplicial) complex on a finite set X is a collection C of a subsets of X, closed under containment. Each set in C is called a face of the complex, and the maximal faces (with respect to containment) are called facets or bases. The dimension of a complex C is the maximum cardinality of a face. The *f*-vector (or face-vector) of a d-dimensional complex C is $(f_0, f_1, \dots, f_d)$, where f_i is the number of faces of cardinality i in C. The f – polynomial of a d-dimensional complex C is the generating function $f(C,x) = \sum ifix^i$ for the *f*-vector $(f_0, f_1, \dots, f_d)$ of the complex. For each graph polynomial, there is a complex for which the graph polynomial is a simple evaluation of the $f -$ polynomial. For instance, the independence complex $I(G)$ of graph G is the complex on the vertex set V of G whose faces are the independence set of G. The independence polynomial is merely the f – polynomial of the independent complex. One of the applications of simplicial complexes to graph theory is undoubtedly Lovasz's proof [4] of the chromatic number of Kneser graphs. His argument centers on the neighbourhood complex $N(G)$ of a graph, whose vertices are the vertices of the graph and whose faces are subsets of vertices that have common neighbour.

We consider a univariate polynomial, which called the *neighbourhood polynomial* of graph G, $N(G,x) = \sum_{U\in N(G)} x^{|U|}$ (see [3]), where $N(G)$ is neighbourhood complex of a graph, whose vertices are the vertices of the graph and faces are subsets of vertices that have a common neighbour.

The nanostars dendrimer is a part of a new group of macromolecules that seem photon funnels just like artificial antennas and also is a great resistant of photo bleaching. Recently some people investigated the mathematical properties of these nanostructures (see for example [1,2,5,7,8]).

Brown and Nowakowski [3] examined the neighbourhood polynomial for hypercube, for graphs not containing a four-cycle and for the graphs resulting from joins and Cartesian products.

IOP Conf. Series: Journal of Physics: Conf. Series **1008** (2018) 012028 doi:10.1088/1742-6596/1008/1/012028

Alikhani and Mahmoudi [6] studied some specific graphs and nanostructures and study their neighbourhood polynomial. So far not many results known on this polynomial. In this paper, we will consider some families of dendrimers and compute their neighbourhood polynomial.

2. Discussion and Main Results

In this section, we shall determine the neighbourhood polynomial of three families of dendrimers. We first state some properties of neighbourhood polynomial.

Theorem 1 [3] Let G be $C_4 - free$ with n vertices and m edges. Then

$$N(G,x) = \sum_{v \in V}(1+x)^{\deg v} - x(2m-n) - (n-1) \quad (1)$$

Am immediate observation is the following.

Corollary 1 [3] The neighbourhood polynomial for a $C_4 - free$ graph depends only on the degree sequence of the graph and can be calculated in polynomial time.

Theorem 1 gives us the neighbourhood polynomial for many graphs, including:

- If $G = C_n$ is a cycle of length $n > 4$, then $N(C_4, x) = 1 + nx + nx^2$;
- If G is an $r -$ regular graph of girth at least 5, then $N(G,x) = n(1+x)^r + n(r-1)x - (n-1)$. In particular, if G is Petersen graph, $N(G,x) = 1 + 10x + 30x^2 + 10x^3$;

3) If G is a tree, then $N(G,x) = \sum_{v \in V}(1+x)^{deg(v)} - x(n-1) - (n-1)$.
4) Let F_n be a friendship graph, then $N(F_n, x) = 2n(1+x)^2 + (1+x)^{2n} - (4n-1)x - 2n$.

We now determine the neighbourhood polynomial of some families of dendrimers. First, we compute the neighbourhood polynomial of PAMAM dendrimer which denoted by $PD_1[3]$ as depicted in Figure 1.

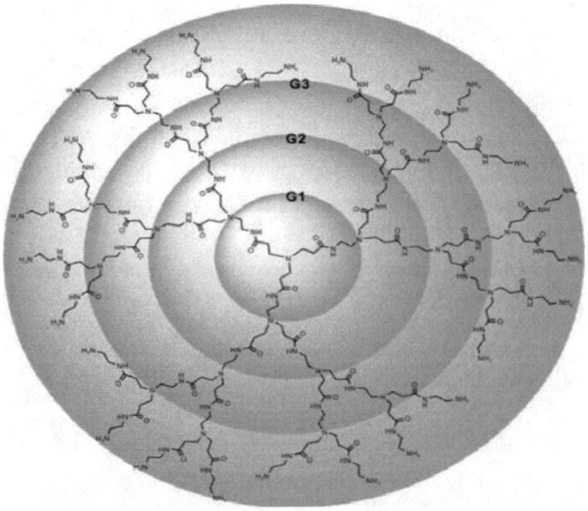

Figure 1. PAMAM dendrimer of generations G_n with growth stages, $PD_1[3]$

Table 1. Vertex partition of PAMAM dendrimer $PD_1[n]$ based on the number degree of vertex

Degree of Vertex	Number of Vertex
1	$9 \times 2^n - 3$
2	$15 \times 2^{n+1} - 15$

Degree of Vertex	Number of Vertex
3	$9 \times 2^n - 5$

Theorem 2 Let n be the number of growth stages in $PD_1[n]$ and $n \in N_0$. Then the neighbourhood polynomial of $PD_1[n]$ for $n = 0,1,2,3 \ldots$ is given by

$$N(PD_1[n], x) = (9 \times 2^n - 5)(1 + x)^3 + (30 \times 2^n - 15)(1 + x)^2 + (9 \times 2^n - 3)(1 + x) - x(12 \times 2^{n+2} - 25) - (12 \times 2^{n+2} - 24)$$

Proof. Let G be the PAMAM dendrimer $PD_1[n]$. The number vertices and edges in PAMAM dendrimer $PD_1[n]$ are $|V(PD_1[n])| = 12 \times 2^{n+2} - 23$ and $|E(PD_1[n])| = 12 \times 2^{n+2} - 24$. We find the vertex partition of the form 1, 2 and 3 for PAMAM dendrimer$PD_1[n]$ based on the number degree of vertex. Table 1 explain such partition for $PD_1[n]$. Now by using the partition given in Table 1, we can compute the neighbourhood polynomial for $PD_1[n]$. Since (1), this implies that

$$N(PD_1[n], x) = (9 \times 2^n - 3)(1 + x) + (15 \times 2^{n+1} - 15)(1 + x)^2 + (9 \times 2^n - 5)(1 + x)^3 - x(24 \times 2^{n+2} - 48 - 12 \times 2^{n+2} + 23) - (12 \times 2^{n+2} - 23 - 1)$$

By simple calculation, we get

$$N(PD_1[n], x) = (9 \times 2^n - 5)(1 + x)^3 + (30 \times 2^n - 15)(1 + x)^2 + (9 \times 2^n - 3)(1 + x) - x(12 \times 2^{n+2} - 25) - (12 \times 2^{n+2} - 24)$$

This completes the proof.□

The following theorem computes the neighbourhood polynomial for tetrathiafulvalene dendrimer $TD_2[n]$. The graph $TD_2[n]$ is depicted in Figure 2.

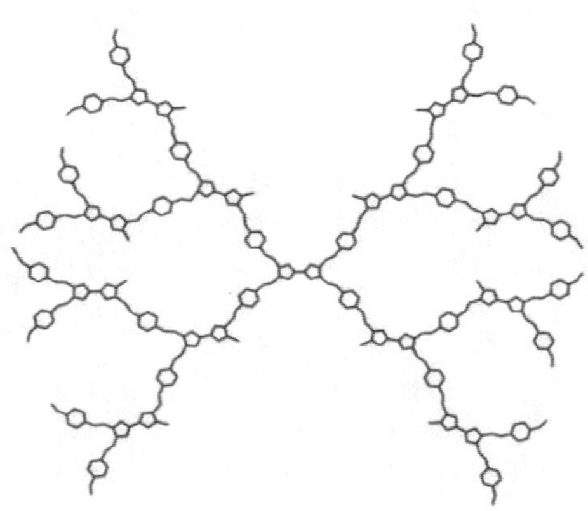

Figure 2. Tetrathiafulvalene dendrimer with 2-growth stages, $TD_2[2]$

Table 2. Vertex partition of Tetrathiafulvalene dendrimer $TD_2[n]$ based on the number degree of vertex

Degree of Vertex	Number of Vertex
1	$8 \times 2^n - 4$

IOP Conf. Series: Journal of Physics: Conf. Series **1008** (2018) 012028 doi:10.1088/1742-6596/1008/1/012028

2	$76 \times 2^n - 44$
3	$40 \times 2^n - 26$

Theorem 3 Let n be the number of growth stages in $TD_2[n]$ and $n \in N_0$. Then the neighbourhood polynomial of $TD_2[n]$ for $n = 0,1,2,3 \ldots$ is given by

$$N(TD_2[n], x) = (40 \times 2^n - 26)(1+x)^3 + (76 \times 2^n - 44)(1+x)^2 + (8 \times 2^n - 4)(1+x) - x(39 \times 2^{n+2} - 96) - (31 \times 2^{n+2} - 75)$$

Proof. Let G be the Tetrathiafulvalene dendrimer $TD_2[n]$. The number vertices and edges in tetrathiafulvalene dendrimer $TD_2[n]$ are $|V(TD_2[n])| = 31 \times 2^{n+2} - 74$ and $|E(TD_2[n])| = 12 \times 2^{n+2} - 24$. We find the vertex partition of the form 1, 2 and 3 for tetrathiafulvalene dendrimer $TD_2[n]$ based on the number degree of vertex. Table 2 explain such partition for $TD_2[n]$. Now by using the partition given in Table 2, we can compute the neighbourhood polynomial for $TD_2[n]$. Since (1), this implies that

$$N(TD_2[n], x) = (8 \times 2^n - 4)(1+x) + (76 \times 2^n - 44)(1+x)^2 + (40 \times 2^n - 26)(1+x)^3 - x(70 \times 2^{n+2} - 170 - 31 \times 2^{n+2} + 74) - (31 \times 2^{n+2} - 74 - 1)$$

By simple calculation, we get

$$N(TD_2[n], x) = (40 \times 2^n - 26)(1+x)^3 + (76 \times 2^n - 44)(1+x)^2 + (8 \times 2^n - 4)(1+x) - x(39 \times 2^{n+2} - 96) - (31 \times 2^{n+2} - 75)$$

This completes the proof.□

The following theorem computes the neighbourhood polynomial for POPAM dendrimer $POD_2[n]$. The graph $POD_2[n]$ is depicted in Figure 3.

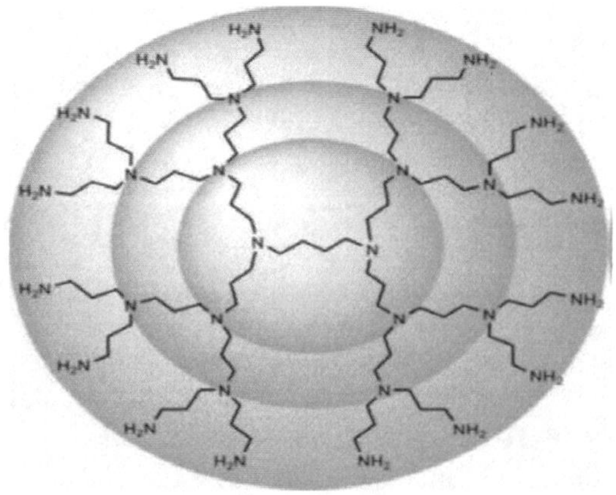

Figure 3. POPAM dendrimer of generations G_n with two growth stages, $POD_2[2]$

Table 3. Vertex partition of POPAM dendrimer $POD_2[n]$ based on the number degree of vertex

Degree of Vertex	Number of Vertex
1	4×2^n
2	$24 \times 2^n - 8$
3	$4 \times 2^n - 2$

IOP Conf. Series: Journal of Physics: Conf. Series **1008** (2018) 012028 doi:10.1088/1742-6596/1008/1/012028

Theorem 4 Let n be the number of growth stages in $POD_2[n]$ and $n \in N_0$. Then the neighbourhood polynomial of $POD_2[n]$ for $n = 0,1,2,3\ldots$ is given by

$$N(\ POD_2[n], x) = (4 \times 2^n - 2)(1+x)^3 + (24 \times 2^n - 8)(1+x)^2 + (4 \times 2^n)(1+x) - x(2^{n+5} - 12) - (2^{n+5} - 11)$$

Proof. Let G be the POPAM dendrimer $POD_2[n]$ with n growth stages. The number vertices and edges in POPAM dendrimer $POD_2[n]$ are $|V(POD_2[n]\ )| = 2^{n+5} - 10$ and $|E(POD_2[n]\ )| = 2^{n+5} - 11$. We find the vertex partition of the form 1, 2 and 3 for POPAM dendrimer $POD_2[n]$ based on the number degree of vertex. Table 5.3 explain such partition for $POD_2[n]$. Now by using the partition given in Table 5.3, we can compute the neighbourhood polynomial for $POD_1[n]$. Since (1), this implies that

$$N(\ POD_2[n], x) = (4 \times 2^n)\ (1+x) + (24 \times 2^n - 8)(1+x)^2 + (4 \times 2^n - 2)\ (1+x)^3 - x(2 \times 2^{n+5} - 22 - 2^{n+5} + 10) - (2^{n+5} - 10 - 1)$$

By simple calculation, we get

$$N(\ POD_2[n], x) = (4 \times 2^n - 2)(1+x)^3 + (24 \times 2^n - 8)(1+x)^2 + (4 \times 2^n)(1+x) - x(2^{n+5} - 12) - (2^{n+5} - 11)$$

This completes the proof.□

3. Conclusion and Open Problem

The concept of neighborhood polynomial in graph was introduced formally by Brown and Nowakowski [3]. Alikhani and Mahmoudi [6] later considered the neighborhood polynomial for some nanostructures. In this paper, we presented this polynomial for three families of dendrimer graphs namely, PAMAM $\boldsymbol{PD_1[n]}$, tetrathiafulvalene $\boldsymbol{TD_2[n]}$ and POPAM $\boldsymbol{POD_2[n]}$. One may consider the application of this polynomial in QSAR/QSPR researches for future investigation. Another research direction is study the neighborhood polynomial of various nano-structures such as nanotube, nanocones and fullerenes.

Acknowledgements
The authors would like to thank the referee for his/her suggestions that improve this paper.

References
[1] Alikhani S and Iranmanesh M A 2010 *J Comput Theor Nanosci* **7** 2314
[2] Alikhani S and Iranmanesh M A 2014 *MATCH Commun Math Comput Chem* **71** 39-55
[3] Brown J I and Nowakowski R J 2008 *Aust J Combin* **42** 55
[4] Lovasz L 1978 *J Combin Theory Ser* **25** 319
[5] Shabani H, Ashrafi A R and Diudea M V 2010 *Croat Chem Acta* **83** 439
[6] Alikhani S and Mahmoudi E 2014 *Iranian J Mathematics Chemistry* **5** (1) 21
[7] Arif N E and Hasni R 2011 *Digest J Nanomater Biostruct* **6(4)** 1551
[8] Arif N E, Hasni R and Kalaf A M 2013 *J Comput Theor Nanosci* **10** 285

IOP Conf. Series: Journal of Physics: Conf. Series **1008** (2018) 012029 doi:10.1088/1742-6596/1008/1/012029

On $P_2 \diamond P_n$-supermagic labeling of edge corona product of cycle and path graph

R Yulianto, Titin S Martini

Mathematics Department of Mathematics and Natural Sciences Faculty, Universitas Sebelas Maret, Surakarta, Indonesia

E-mail: yuliantoriza48@gmail.com, titinsmartini@gmail.com

Abstract. A simple graph $G = (V, E)$ admits a H-covering, where H is subgraph of G, if every edge in E belongs to a subgraph of G isomorphic to H. Graph G is H-magic if there is a total labeling $f : V(G) \bigcup E(G) \rightarrow 1, 2, ..., |V(G)| + |E(G)|$, such that each subgraph $H' = (V', E')$ of G isomorphic to H and satisfying $f(H') \overset{def}{=} \Sigma_{v\epsilon V'} f(v) + \Sigma_{e\epsilon E'} f(e) = m(f)$ where $m(f)$ is a constant magic sum. Additionaly, G admits H-supermagic if $f(V) = 1, 2, ..., |V|$. The edge corona $C_n \diamond P_n$ of C_n and P_n is defined as the graph obtained by taking one copy of C_n and n copies of P_n, and then joining two end-vertices of the i-th edge of C_n to every vertex in the i-th copy of P_n. This research aim is to find H-supermagic covering on an edge corona product of cycle and path graph $C_n \diamond P_n$ where H is $P_2 \diamond P_n$. We use k-balanced multiset to solve our reserarch. Here, we find that an edge corona product of cycle and path graph $C_n \diamond P_n$ is $P_2 \diamond P_n$ supermagic for $n \geq 3$.

1. Introduction

Let G be a simple graph $G = (V, E)$, where V is a set of vertices, and E is a set of edges. Chartrand and Lesniak [1] defined that cycle graph is a circuit with no repeated vertices, except the first and last vertices. The cycle graph with n vertices is denoted by C_n. They also defined path graph is a walk with no repeated vertices, path graph with n vertices is denoted by P_n.

Let G_1 and G_2 are two graphs on disjoint sets of n_1 and n_2 vertices, m_1 and m_2 edges, respectively. The edge corona $G_1 \diamond G_2$ is defined as the graph obtained by taking one copy of G_1 and m_1 copies of G_2, and then joining two end-vertices of the i-th edge of G_1 to every vertex in the i-th copy of G_2. Note that the edge corona $G_1 \diamond G_2$ of G_1 and G_2 has $n_1 + m_1n_2$ vertices and $m_1 + 2m_1n_2 + m_1m_2$ edges, for detail defnition of graph see [4].

Gallian [2] defined a graph labeling as an assignment of integers to the vertices or edges, or both, subject to certain condition. Magic labelings was first introduced in 1963 by Sedláčk [9]. The concept of H-magic graphs was introduced in [3]. An edge-covering of a graph G is a family of different subgraphs $H_1, H_2, ..., H_k$ such that each edge of E belongs to at least one of the subgraphs H_i, $1 \leq i \leq k$. Then, it is said that G admits an $(H_1, H_2, ..., H_k)$-edge covering. If every H_i is isomorphic to a given graph H, then we say that G admits an H-covering. Suppose that $G = (V(G), E(G))$ admits an H-covering. A bijective function $f : V(G) \cup E(G) \rightarrow \{1, 2, ..., |V(G)| + |E(G)|\}$ is an H-magic labeling of G if there exist a positive integer $m(f)$, which we call magic sum such that for each subgraph $H' = (V(G)', E(G)')$ of G isomorphic H, $f(H') = \sum_{v \in V(G)'} f(v) + \sum_{e \in E(G)'} f(e) = m(f)$. In this case we say that the graph G is H-magic. When $f(v) = \{1, 2, ..., |V(G)|\}$, then G is H-supermagic and we denote supermagic-sum is $s(f)$.

In [3], they proved that a complete bipartite graph $K_{n,n}$ could be covered by magic star covering $K_{1,n}$. Then Lladó and Moragas [5] proved in [3] the same graph containing a cycle cover, they also proved that C_3-supermagic labelings on a wheel graph W_n for $n \geq 5$ odd and C_4-supermagic labeling of a prism graph and a book graph. Marbun and Salman [6] then proved that W_n-supermagic labelings for a wheel W_n k-multilevel corona with a cycle C_n.

In this paper, we study an H-supermagic labeling of edge corona product of cycle and path graph. We prove that a edge corona product of cycle C_n and path P_n graph has a $P_2 \diamond P_n$-supermagic labeling for $n \geq 3$.

2. Main Result

A multiset is a set that allows the existence of same elements in it(Maryati et al. [7]). Let X be a set containing some positive integers. We use the notation $[a, b]$ to mean $\{x \in \mathbb{N} | a \leq x \leq b\}$ and ΣX to mean $\sum_{x \in X} x$. For any $k \in \mathbb{N}$, the notation $k + [a, b]$ means $k + x | x \in [a, b]$. According to Guitérrez and Llado [3], the set X is k-equipartion if there exist k subsets of X. say $X_1, X_2, \ldots, X_k$ such that $\bigcup_{i=1}^{k} X_i = X$ and $|X_i| = \frac{|X|}{k}$ for every $i \in [1, k]$.

2.1. *k-balanced multiset*

In this research, we used technique k-balancemultiset that introduced by Maryati et al. [7]. Let Y be a multiset of positive integers and $k \in \mathbb{N}$. A multiset Y is k-balanced if there are k subsets of Y where $Y_i = Y_1 = Y_2 = ... = Y_k$ then for each $i \in [1, k]$. We obtain $|Y_i| = |Y_k|, \sum Y_i = \frac{\sum Y}{k} \in N$ and $\biguplus_{i=1}^{k} Y_i = Y$.

Lemma 2.1 *[8] Let x, y, and k be integers, such that $1 \leq x \leq y$ and $k > 1$. If $X = [x, y]$ and $|X|$ is a multiple $2k$, then X is k-balanced.*

Here, we have several lemmas on k-balanced multiset to build theorem.

Lemma 2.2 *Let k and x be positive integers $k \geq 3$. Let $Y = [1, k] \uplus [1, k] \uplus [x + 1, x + k]$, then Y is k-balanced.*

Proof. For every $i \in [1, k]$ we define the multisets $Y_i = \{a_i, b_i, c_i\}$ with

$$\begin{aligned}
a_i &= \left\lfloor \frac{i+1}{2} \right\rfloor && \text{for } i \in [1, k] \\
b_i &= \left\lceil \frac{i+k}{2} \right\rceil && \text{for } i \in [1, k] \\
c_i &= x + k + 1 - i && \text{for } i \in [1, k].
\end{aligned}$$

Then, defined set

$$\begin{aligned}
A &= \{a_i | 1 \leq i \leq k\} = [1, k] \\
B &= \{b_i | 1 \leq i \leq k\} = [1, k] \\
C &= \{c_i | 1 \leq i \leq k\} = [x + 1, x + k].
\end{aligned}$$

Since $A \uplus B \uplus C = Y$ and $\biguplus_{i=1}^{k} Y_i = Y$, $|Y_i| = 3$ and $\sum Y_i = x + \frac{3k+3}{2}$ for every $i \in [1, k]$, so we have Y is k-balanced.

Lemma 2.3 *Let k and x be positive integers $k \geq 3$. If $Z = [x + 1, x + k^2]$ and $|Z|$ is k^2, then Z is k-balanced.*

Proof. For every $i \in [1, k]$ we define the multisets $Z_i = \{a^i_j | 1 \leq j \leq k\}$ where

$$a^i_j = \begin{cases} x + i, & \text{for } i \in [1, k] \text{ and } j = 1; \\ a^i_{j-1} + 1, & \text{for } j + i = k + 2; \\ a^i_{j-1} + x + 1, & \text{for } i \text{ and } j \text{ others}. \end{cases}$$

Since $|Z_i| = k$; $\biguplus_{i=1}^k Z_i = Z$ and $\sum Z_i = (x + k^2)\frac{k+1}{2}$ for every $i \in [1, k]$ then Z is k-balanced.

Lemma 2.4 *Let x, y and k be positive integers $k \geq 4$. If $W = [1, x] \uplus [1, x] \uplus [x + 1, x + k] \uplus [y + 1, y + k]$, then W is k-balanced.*

Proof. For every $i \in [1, k]$ we define the multisets $W_i = \{a_i, b_i, c_i, d_i\}$ with

$$\begin{aligned} a_i &= i \quad \text{for } i \in [1, k] \\ b_i &= \begin{cases} 1 + i, & \text{for } i \in [1, k-1]; \\ 1, & \text{for } i = k; \end{cases} \\ c_i &= \begin{cases} x + k - i, & \text{for } i \in [1, k-1]; \\ x + k, & \text{for } i = k; \end{cases} \\ d_i &= y + k + 1 - i \quad \text{for } i \in [1, k] \end{aligned}$$

Then, defined set

$$\begin{aligned} A &= \{a_i | 1 \leq i \leq k\} = [1, k] \\ B &= \{b_i | 1 \leq i \leq k\} = [1, k] \\ C &= \{c_i | 1 \leq i \leq k\} = [x + 1, x + k] \\ D &= \{d_i | 1 \leq i \leq k\} = [y + 1, y + k]. \end{aligned}$$

Since $A \uplus B \uplus C \uplus D = W$ and $\biguplus_{i=1}^k W_i = W$, $|W_i| = 4$ and $\sum W_i = 5k + 2$ for every $i \in [1, k]$ then W is k-balanced.

2.2. $P_2 \diamond P_n$-*Supermagic Labeling on A Cycle Graph Edge Corona with Path* $C_n \diamond P_n$

The edge corona product between C_n and P_n, denoted by $C_n \diamond P_n$ is a graph obtained by taking one copy of C_n and $|E(C_n)|$ copies of P_n and then joining two end-vertices of the i-th edge of C_n to every vertex in the i-th copy of P_n.

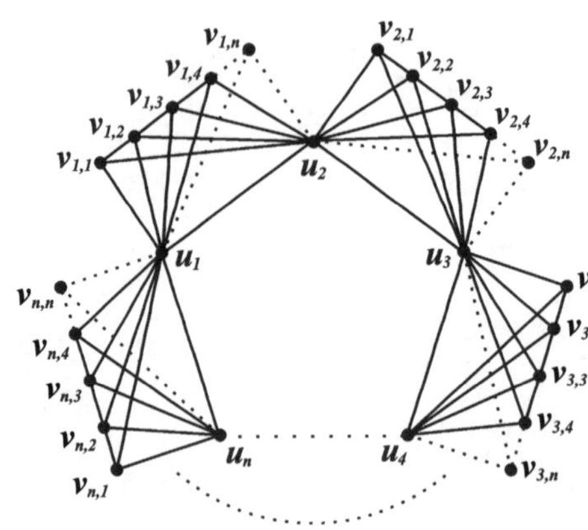

Figure 1. A Cycle Graph Edge Corona with Path $C_n \diamond P_n$

IOP Conf. Series: Journal of Physics: Conf. Series **1008** (2018) 012029 doi:10.1088/1742-6596/1008/1/012029

Theorem 2.1 *Let n be positive integers with $n \geq 3$. A graph $C_n \diamond P_n$ is $P_2 \diamond P_n$-supermagic.*

Proof. Let G be a $C_n \diamond P_n$ graph for any integer $n \geq 3$. Then $|V(G)| = n(n+1)$ and $|E(G)| = 3n^2$. Let $A = [1, 4n^2 + n]$. We define a bijective function $f : V(G) \cup E(G) \to \{1, 2, ..., 4n^2 + n\}$.

Here we have two cases to be considered.

Case 1. For n odd. Let $V(G) = \{v_i; 0 \leq i \leq n\} \uplus \{u^i_j; 0 \leq i \leq n, 0 \leq j \leq n\}$ and $E(G) = \{v_0v_1, v_1v_2, \dots v_nv_0\} \uplus \{e^i_j; 0 \leq i \leq n, 0 \leq j \leq n\}$. Given a set of labels for all vertices and edges of G denoted by A where $A = [1, 4n^2]$. Partition A into 3 sets, $A = X \uplus Y \uplus Z$, where $X = [1, n] \uplus [1, n] \uplus [n^2 + n + 1, n^2 + 2n]$, $Y = [n + 1, n^2 + n]$, and $Z = [n^2 + 2n + 1, 4(n^2) + n]$. Then we define the total labeling f on G as follows. First, partition X into 2 sets: $X_1 = [1, n]$ and $X_2 = [n^2 + n + 1, n^2 + 2n]$. The vertices v_i where $0 \leq i \leq n$ are labeled by set X_1 and edges $\{v_0v_1, v_1v_2, \dots v_nv_0\}$ are labeled by set X_2. According to Lemma 2.2, if $x = n^2 + n$ and $k = n$ we have n-balanced. Let $X_1 \uplus X_2$, then $\uplus^n_{i=1} X_i = X$ and we have $\sum X_i = \frac{2n^2+5n+3}{2}$. The vertices u^i_j where $0 \leq i \leq n$ and $0 \leq j \leq n$ are labeled by set Y. Define that u^i_j are vertices on path. According Lemma 2.3, if $x = n$, $k = n$, and $|Y| = n^2$ we have n-balanced where $\sum Y_i = \frac{n^3+2n^2+n}{2}$. Then, the edges e^i_j are labeled by set Z where e^i_j are edges on path and edge on product edge coronation. According Lemma 2.1, if $x = n^2 + 2n + 1$, $y = 4n^2 + n$, and $|Z| = 3n^2 - n$ we have n-balanced where $\sum Y_i = \frac{15n^3+4n^2-1}{2}$.

Case 2. For n even. Let $V(G) = \{v_i; 0 \leq i \leq n\} \uplus \{u_i; 0 \leq i \leq n\}$ and $E(G) = \{e^i_j; 0 \leq i \leq n, 0 \leq j \leq n\}$. Partition A into 3 sets, $A = P \uplus Q \uplus R$, where $P = [1, n] \uplus [1, n] \uplus [n+1, 2n] \uplus [2n+1, 3n]$, $Q = [3n+1, n^2+n)]$, dan $R = [n^2+n+1, 4n^2+n]$. Then we define the total labeling f on G as follows. First, partition P into 3 sets: $P_1 = [1, n]$, $P_2 = [n + 1, 2n]$, and $P_3 = [2n + 1, 3n]$. The vertices v_i where $0 \leq i \leq n$ are labeled by set P_1, P_2, and P_3. According to Lemma 2.4, if $x = n$ and $k = n$ we have n-balanced. Let $P_1 \uplus P_2 \uplus P_3$, then $\biguplus^n_{i=1} P_i = P$ and we have $\sum P_i = 5n+2$. The vertices u_i where $0 \leq i \leq n$ are labeled by set Q. Define that u_i are two vertices in path. According Lemma 2.1, if $x = 3n+1$, $y = n^2+n$, and $|Q| = 2n$ we have n-balanced where $\sum Q_i = n^2+4n+1$. Then, the edges e^i_j are labeled by set R where e^i_j are edges on graph G. According Lemma 2.1, if $x = n^2+n+1$, $y = 4n^2+n$, and $|R| = 3n^2$ we have n-balanced where $\sum R_i = \frac{15n^3+6n^2+3n}{2}$.

Furthermore, the constant supermagic sum of a subgraph $P_2 \diamond P_n$ are as follows

$$f(P_2 \diamond P_n) = \begin{cases} 8n^3 + 4n^2 + 3n + 1, & \text{for } n \text{ odd;} \\ \frac{15n^3+8n^2+21n+6}{2}, & \text{for n even.} \end{cases}$$

□

Figure 2 illustrates an example of $P_2 \diamond P_3$-supermagic labeling on $C_3 \diamond P_3$ graph.

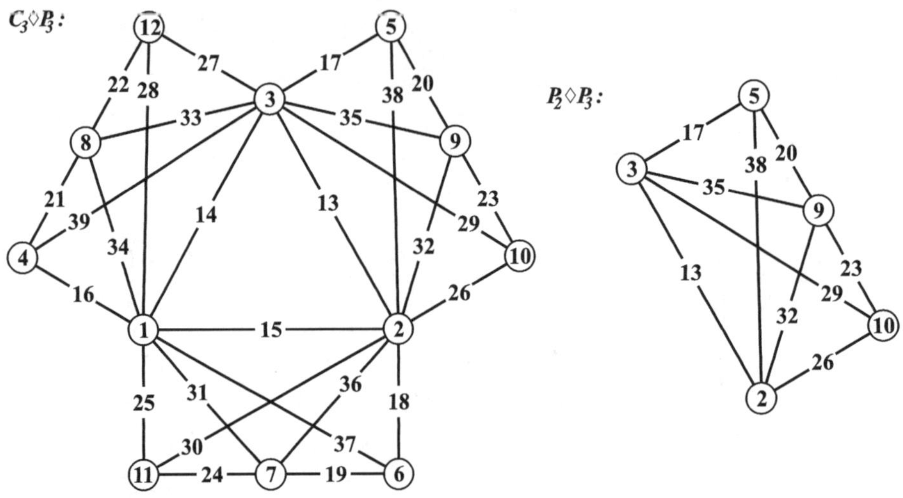

Figure 2. A $P_2 \diamond P_3$-supermagic labeling on $C_3 \diamond P_3$ graph

3. Conclusion

In this paper we have shown the $P_2 \diamond P_n$-supermagic labeling of edge corona product of cycle and path graph.

Open Problem: For further research we can studied $P_2 \diamond P_n$-supermagic labeling on $C_n \diamond P_m$ with $n \geq 3$ and $m \geq 2$.

Acknowledgments

We gratefully ackowledge the support from Mathematics Department of Mathematics and Natural Sciences Faculty, Universitas Sebelas Maret.

References

[1] Chartrand G and Lesniak L 1979 *Graphs and Digraphs* 2nd ed (California: Wadsworth Inc)
[2] Gallian J A 2016 A Dynamic Survey of Graph Labeling, The Electronic Journal of Combinatorics **19** DS6
[3] Guiterrez and Llado A 2005 Magic Coverings *J Combin Math Combin Computing* **55** 43-56
[4] Hou Y and Shiu W 2010 The Spectrum of The Edge Corona of Two Graphs *Electronic Journal of Linear Algebra* **20** 586-594
[5] Llado A and Moragas J 2008 Cycle-magic Graphs *Discrete Mathematics* **307** 2925-2933
[6] Marbun H T and Salman A N M 2013 wheel-Supermagic Labelings for A Wheel k-Multilevel Corona With A Cycle *J Graphs Cpmb* **10** 183-194
[7] Maryati T K, Salman A N M, Baskoro E T, Ryan J and Miller M 2010 On H-supermagic Labelings for Certain Shackles and Amalgamations of a Connected Graph *Utilitas Mathematica* **83** 333-342
[8] Maryati T K, Baskoro E T and Salman A N M 2008 Ph-supermagic Labelings of Some Trees *J Combin Math Combin Computing* **65** 182-189
[9] Sedlack J 1964 Theory of Graphs and Its Applications *House Czechoslovak Sci Prague* 163-164

Optimization of scheduling system for plant watering using electric cars in agro techno park

Nelly Oktavia Adiwijaya[1], Yudha Herlambang[1], Slamin[1]
[1]Faculty of Computer Science, University of Jember, Indonesia

Email: nelly.oa@unej.ac.id, yudhahimasif@gmail.com, slamin@unej.ac.id

Abstract. Agro Techno Park in University of Jember is a special area used for the development of agriculture, livestock and fishery. In this plantation, the process of watering the plants is according to the frequency of each plant needs. This research develops the optimization of plant watering scheduling system using edge coloring of graph. This research was conducted in 3 stages, namely, data collection phase, analysis phase, and system development stage. The collected data was analyzed and then converted into a graph by using bipartite adjacency matrix representation. The development phase is conducted to build a web-based watering schedule optimization system. The result of this research showed that the schedule system is optimal because it can maximize the use of all electric cars to water the plants and minimize the number of idle cars.

1. Introduction

Agro Techno Park in University of Jember is a special area used to develop agriculture, livestock and fisheries. Agro Techno Park area is about 14 hectares [1]. Processing and maintenance of plants in Agro Techno Park is still done manually, for example, in terms of plant watering schedule and plant management. The process of watering is done by the staff with a routine system and rely on instinct in the process of watering. Based on the interview, the plant watering is done twice a day. This leads to new problems because with less precise treatment resulted in the medicinal plants are not well developed. To handle these problems, Agro Techno Park University of Jember requires a technology that is able to perform watering automatically, can be controlled by scheduling system, and can be arranged according to the standard process. We assume the tools used here are electric cars as researched in University of Jember.

Methods often used to solve scheduling problems include graph coloring methods using edge coloring, or vertex coloring, or an addition with Artificial Intelligence. Welch-Powell algorithm is a graph coloring algorithm that performs staining based on the highest degree of its vertices or called Largest Degree Ordering (LDO) [2]. The algorithm can be used to color the edges of a graph efficiently. Another graph coloring algorithm was developed by Leighton using Recursive Largest First (RLF) [3].

The scheduling problem (university teaming) with more complicated variable also has been done. The specific case of lectures scheduling problem is a team teaching. One lecture is given by two or more teachers. The timetable is developed so that there is no clash of each two or more classes at the same time. The problem is solved by a vertex graph coloring [4]. The algorithm has also been developed into a website-based system [5]. More applications of graph coloring for scheduling system can be seen, for example, in [6,7,8,9].

The development of this automatic watering system begins with the development of the algorithm with the edge coloring graph technique. This technique can produce optimal number of colors as the option of watering schedule. The coloring result becomes the basis of the division of time on the watering schedule to avoid collisions on every itinerary. The smallest number of colors at each plant is chosen such that all the numbers result becomes the set of watering timetable. The algorithm is tested

IOP Conf. Series: Journal of Physics: Conf. Series **1008** (2018) 012030 doi:10.1088/1742-6596/1008/1/012030

randomly by giving data sets. This algorithm was designed in a common formula to be used both for the case scheduling watering plants with different frequencies and the different capacity of sprinklers cars [10]. This research develop watering scheduling system for Agro Techno Park using edge coloring of graph technique. This research was conducted in 3 stages: data collection phase, analysis phase and system development stage.

2. Data Collection Phase

At this stage, we collect data at Agro Techno Park in University of Jember. From the results of interviews conducted to one employee obtained data plants, data electric cars and employees. In the process of watering, it is done as much as 2 times a day. This treatment is carried out evenly to all existing plants although the frequency of watering each plant is different. Watering frequency and number of cars will be used as data parameters in determining the schedule of watering plants. The summary of collected data is presented in Table 1.

Table 1. Number of plants and electric cars

Number of plants	57
Number types of plants	2
number of plants with frequency of watering 1x	36
number of plants with frequency of watering 2x	17
number of plants with frequency of watering 3x	4
Number of electric cars	3

3. Analysis Phase

In this experiment, we use three cars and five plants, both of which will be linked to each other according to the needs of the plant. Figure 1 shows a bipartite graph that represent the relationship between cars and plants. The edges indicates the certain cars can water the certain plants. The electric cars consist of three types, each of which is assumed to have certain access to each plant in accordance to the size of the car in order to reach the plant.

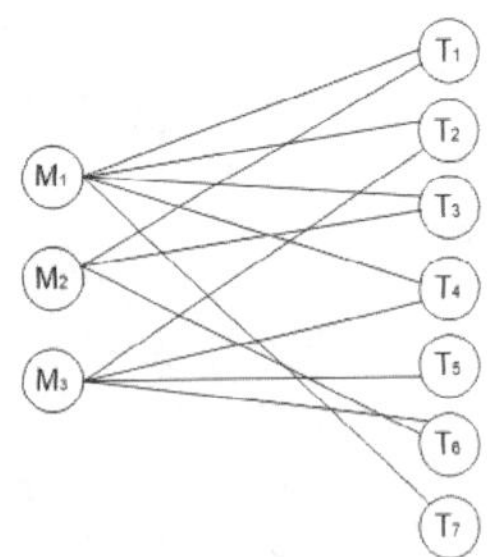

Figure 1. Graph representation of the relationship between cars and plants

The car is represented using variable *m,* while the plant is represented by varible *n*. The frequency of the watering is represented using *F[n]*. So the matrix representation of frequency and the degree as shown below. The adjacency matrix is represent as $MT_0[m][n]$. Using Welch Powell algorithm, the result of this experiment is matrix color *C[m][n]* and the graph representation as shown in Figure 2.

IOP Conf. Series: Journal of Physics: Conf. Series **1008** (2018) 012030 doi:10.1088/1742-6596/1008/1/012030

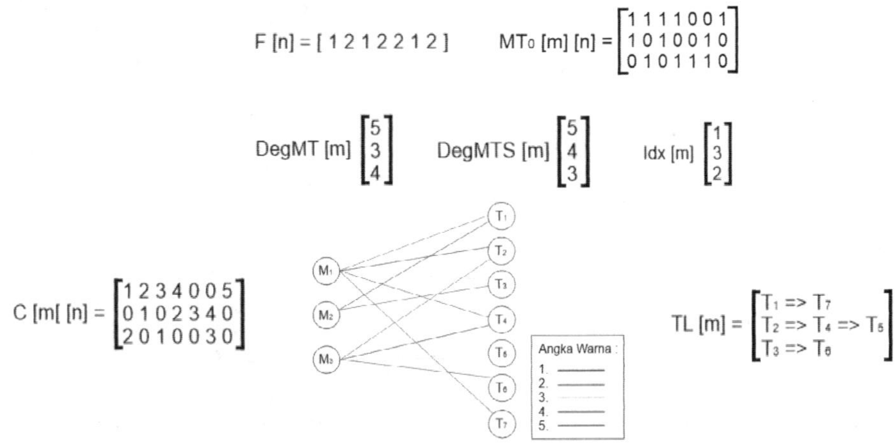

Figure 2. Graph coloring representation

The next process by looking for the smallest coloring number of cars. The car with the smallest coloring number will serve to water the plants. This matrix is denoted by *TL [m]*.

The results of the order of watering plants on the *TL[m]* means that the car 1 to water the plant 1 and plant 7. At the same time the car 2 watering plant 3 and plant 6, and at the same time the car 3 watering plant 2, plant 4 and plant 5. This coloring process is repeat until the frequency of watering the plant is 0. This means the plant has been finished watered to the maximum frequency the plant may be watered. The graph representation are as shown in Figure 3.

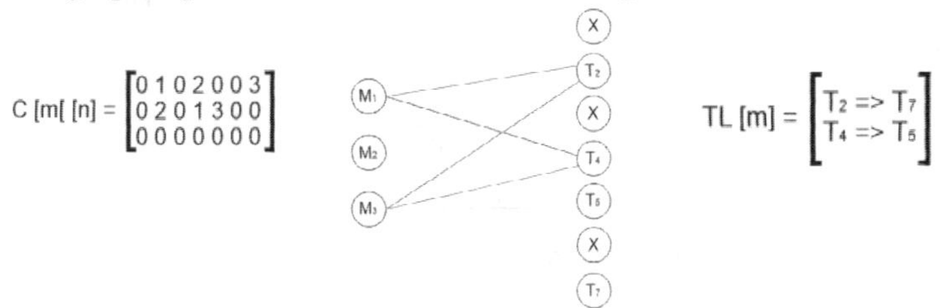

Figure 3. The second graph coloring representation

4. System development stage

The development stage begins with describing the business process system. In this system, it also developed the management of data features of plants, electric cars and employees. The complete system features are shown in use case images as illustrated in Figure 4.

IOP Conf. Series: Journal of Physics: Conf. Series **1008** (2018) 012030 doi:10.1088/1742-6596/1008/1/012030

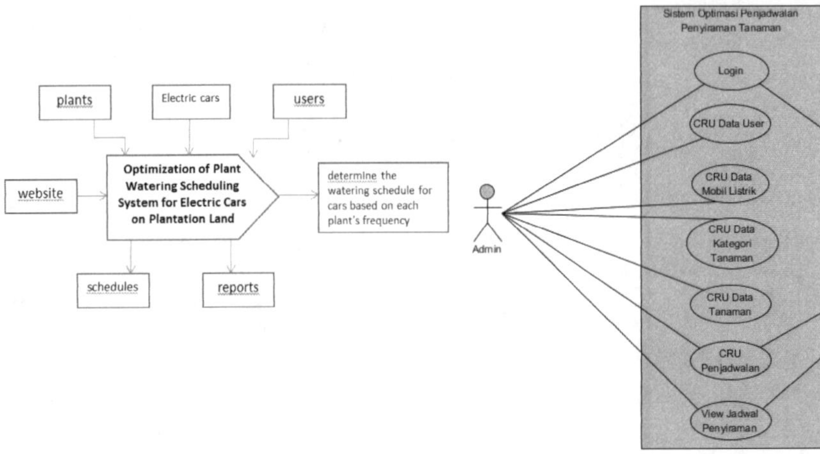

Figure 4. Business process system and Use Case diagram

Entity relationship diagram in this system presents the relationship between data in the database based on the basic objects of data that have interrelation relationship: plants, cars, user, watering. The entity relationshop diagram is shown in Figure 5.

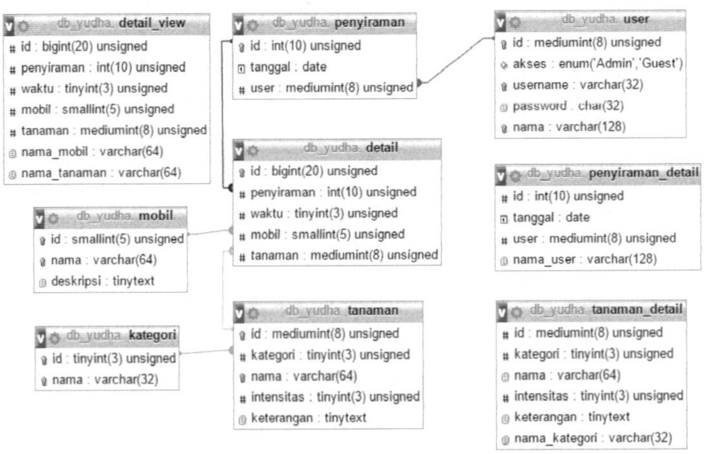

Figure 5. Entity reationship diagram

The scheduling page on this system can be accessed by the admin and users. In this page, data processing schedule watering can be done. This is the main feature of this system. The feature is presented the steps of making plant watering optimization schedule. The process starts from filling the adjacency table of the relationship between car and plant, line staining and watering sequence, and end up with the result of plant watering scheduling. The scheduling pages are illustrated in Figures 6.

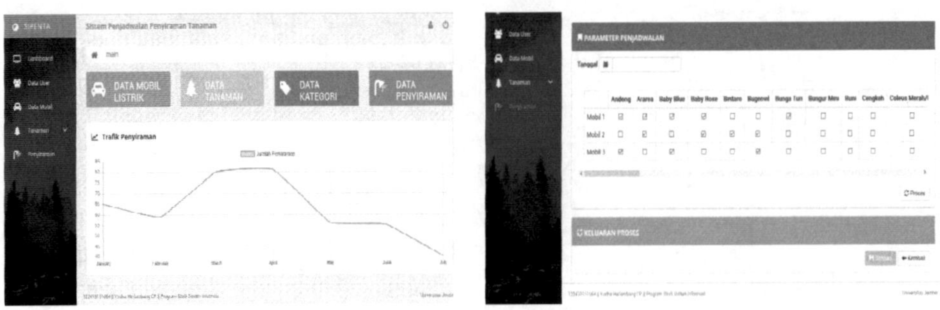

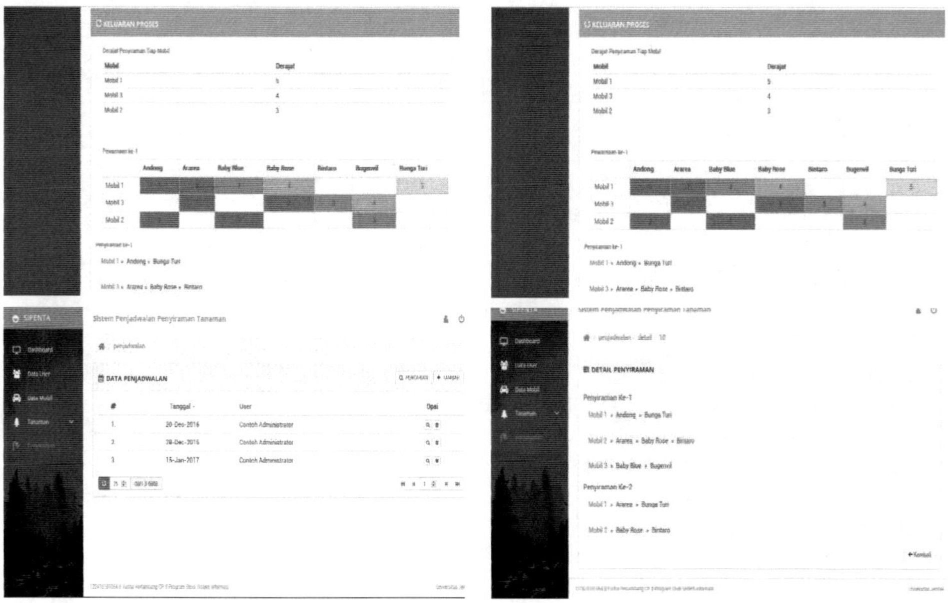

Figure 6. User interface of plants watering scheduling system

The optimization process of plant water scheduling system implements edge coloring of graph technique. Edge coloring techniques on the graph are used to illustrate the relationship between plants and electric cars. This relationship explains which electric car can water which plants. From this relationship, the optimal number of color will find through the edge coloring of graph technique. While adjacency matrix is used to convert the relationships on the graph, so that it becomes a table that produces the order of watering the plants. The relationship between the graph coloring and the adjacency matrix will provide optimal plant watering solutions.

Dashboard page with all the features of this system can be accessed by admin. These features are user data, car data, plants data and watering. The scheduling page on this system can be accessed by the admin and user. On this page, we can process data watering schedule. This feature presents the steps of making plant watering optimization schedule. The process starts from filling the adjacency table of the relationship between cars and plants, edge coloring and watering sequence which is the result of plant watering scheduling. The input data is sorted according to the highest degree then colored by representing the numbers and matrix adjacency. The next page contains the scheduling data stored in the database. On this page, it can be seen the detail of the watering process as a result of calculations that have been done. Then the scheduling detail page of the plant watering scheduling process contains the final calculation results of plant watering scheduling. This page contains a table showing which cars water which plants. With this calculation, the schedule of plant watering by electric car is optimised. In other words, this scheduling system can maximize the use of all electric cars to water the plants and minimize the number of idle cars.

5. Conclusion

From the discussion above, we can conclude that: (1) graph coloring is used to color the edge (relationship for watering) between plants with electric cars which is used the Welch Powell algorithm to determine the optimal color for edge coloring; (2) After coloring all edges, then the graph was represented into the matrix adjacency table; (3) the system of plant watering schedule is developed to optimize the watering schedule of plants in the Agro Technopark University of Jember; (4) the schedule is valid for daily schedule provided there is no addition of electric cars and plants; and (5) the acquisition of scheduling results depends on the plants, electric cars and the intensity of watering from each plant.

Acknowledgement

This research is supported by DRPM Ditjen Penguatan Risbang Kemenristekdikti under Penelitian Tim Pascasarjana 2017 research grant.

References

[1] UPT AgroTechnoPark U 2016 Kebun Agrotechnopark Jubung
[2] Astuti S 2011 Penyusunan Jadwal Ujian Mata Kuliah dengan Algoritma Pewarnaan Graf Welch Powell *J Dian* **11** 68–74
[3] Leighton F T 1979 A graph coloring algorithm for large scheduling problems *J Res Natl Bur Stand* **84** 489–506
[4] Adiwijaya N O and Slamin 2014 Algoritma Penjadwalan Perkuliahan pada Kasus Team Teaching dengan Metode Vertex Graph Coloring *Semin Nas Pendidik Mat Ahmad Dahlan (SENDIKMAD 2014)* 1458–65
[5] Darmah A N, Adiwijaya N O and Slamin 2015 Penjadwalan Perkuliahan dengan Kasus Team Teaching Menggunakan Teknik Pewarnaan Titik pada Graf dengan Optimasi Algoritma Welch-Powell *Repos. UNEJ*
[6] Marx D 2004 Graph coloring problems and their applications in scheduling *Proc John von Neumann PhD Students Conf* **48** 11–6
[7] Papaioannou E, Athanassopoulos S and Kaklamanis C 2017 Graph Coloring for Efficient Course and Exam Scheduling Using Matlab *4Th Int Conf Educ Soc Sci (Intcess 2017)* 401–10
[8] Malkawi M, Al-Haj Hassan M and Al-Haj Hassan O 2008 A New Exam Scheduling Algorithm Using Graph Coloring *Int Arab J Inf Techno* **5** 80–7
[9] Giaro K, Kubale M and Obszarski P 2009 A graph coloring approach to scheduling of multiprocessor tasks on dedicated machines with availability constraints *Discret Appl Math* **157** 3625–30
[10] Adiwijaya N O and Slamin 2017 Plants Watering Schedule Algorithm Using the Edge Coloring Graph Technique *Adv Sci Lett* **23** 2292–5

IOP Conf. Series: Journal of Physics: Conf. Series **1008** (2018) 012031 doi:10.1088/1742-6596/1008/1/012031

Alternative construction of graceful symmetric trees

I P Sandy[1], A Rizal[1], E N Manurung[1], K A Sugeng[1]
[1]Department of Mathematics, FMIPA Universitas Indonesia, Indonesia

E-mail: ikhlas.ps@gmail.com, ahmadrizal89@yahoo.com, nataliaelise15@gmail.com, kiki@sci.ui.ac.id

Abstract. Graceful labeling is one of the interesting topics in graph theory. Let $G = (V, E)$ be a tree. The injective mapping $f : V \to \{0, 1, ..., |E|\}$ is called graceful if the weight of edge $w(xy) = |f(x) - f(y)|$ are all different for every edge xy. The famous conjecture in this area is all trees are graceful. In this paper we give alternative construction of graceful labeling on symmetric tree using adjacency matrix.

1. Introduction

Graph labeling is an assignment of integers to the vertices or edges, or both, subject to certain conditions. Rosa [5] called a function f a β-valuation of a graph G with q edges if f is an injection from the vertices of G to the set $\{0, 1, ..., q\}$ such that, when each edge xy is assigned the label $|f(x) - f(y)|$, the resulting edge labels are all different. Golomb [3] subsequently called such labeling graceful and this term is used in this paper. Rosa [5] introduced β-valuations as well as a number of other labeling as tools for decomposing the complete graph into isomorphic subgraphs. In particular, β-valuations originated as a means of proving the conjecture of Ringel that $K_(2n + 1)$ can be decomposed into $2n + 1$ subgraphs that are all isomorphic to a given tree with n edges. Since the graceful labeling introduced, there are many results have been published. More results on graph labeling, including graceful labeling, can be found in Gallian's survey [2, 4].

Cavalier [1] has established several requirements that give an easy way to check the gracefulness of a labeling given the graph's adjacency matrix. The graph that we discussed in this paper is a symmetric tree, which is a root tree that every edge in the same level has the same degree.

Let G be a graph with $V(G) = \{v_1, v_2, \ldots, v_n\}$. Then the matrix $A_G = [a_{ij}]$ defined by

$$a_{ij} = \begin{cases} 1 & \text{if } v_i v_j \in E(G), \\ 0 & \text{otherwise.} \end{cases}$$

is called an adjacency matrix of G.

Let G be a graph with $m = |E(G)|$ and a mapping $f : V(G) \to \{0, 1, , m\}$. Then the $(m + 1) \times (m + 1)$ matrix $A_G = [a_i j]$ defined by:

$$a_{ij} = \begin{cases} 1 & \text{if } xy \in E(G), \text{and } f(x) = i \text{ and } f(y) = j \\ 0 & \text{otherwise,} \end{cases}$$

where where $i, j = 0, 1, , m$, is called the generalized adjacency matrix of G induced by the mapping f. The generalized adjacency matrix allows (all zero) rows/columns corresponding to missing labels; while in the adjacency matrix such rows correspond to vertices of degree 0. Since there is no orientation on the edges, the corresponding adjacency matrices will be symmetric and thus there is no distinction drawn between edges v_iv_j and v_jv_i. The property, that all the graphs considered are simple, guarantees that the main diagonal entries of the adjacency matrices will be zeros as a one on the main diagonal corresponds to a loop in the graph. Let A be an $n \times n$ matrix. Then the kth diagonal line of A is the collection of entries $D_k = a_{ij} : |j - i| = k$, counting multiplicity. Let $A = A_G$ be a (generalized) adjacency matrix, A is symmetric and $D_k = D_(- k)$. The entry 1 corresponding to an edge between vertices v_i and v_j lies in the $|j - i|^{th}$ diagonal line and the edge label for that edge is $|j - i|$, assuming $f(v_i) = i$ for each $i = 0, 1, , n - 1$ and f is a mapping on G.

Theorem 1.1 *[1] Let G be a labeled graph and let A_G be the generalized adjacency matrix for G. Then A_G has exactly one entry 1 in each diagonal line, except the main diagonal of zeros, if and only if the valuation f on G that induces A_G is graceful.*

Proof. We begin by noting that we need only consider the upper triangular part of A_G since it is symmetric. That is, for edges v_iv_j, we may assume $j > i$. Suppose A_G has exactly one entry of 1 in each diagonal line, other than the main diagonal of zeros. Suppose to the contrary that the labeling of G that induces A_G is not a graceful labeling. Then there are distinct edges v_gv_h and v_qv_l with edge labels $|h - g| = |l - q| = k > 0$. This implies

$$\sum_{i,j,|i-j|=k} a_{ij} \geq 2$$

contradicting the assumption that A_G has exactly one entry in each diagonal (not including main diagonal). Now suppose G is gracefully labeled by f and consider A_G. Then for all $k = 1, 2, , |E(G)|$, there is exactly one non-zero entry $a_ij = 1$, such that $j > i$ and $|j - i| = k$, contributing to $|D_k|$ since each edge has a unique label. That is, A_G has exactly one entry of 1 in each non-main diagonal. □

2. Alternative Construction of Graceful Symmetric Trees

Let G be a rooted tree with k level. We define this rooted tree as a symmetrical tree which every level contains vertices of the same degree. It has been shown in [2, 4] that all symmetrical trees are graceful. We are going to exhibit an alternative proof by constructing the graceful symmetrical tree from a graceful star.

Theorem 2.1 *All symmetric tree are graceful.*

Proof. We begin by considering a subgraph of the rooted tree, in which it has m vertices at level k connected to a vertex at level $k - 1$, as a star. It is clear that this star has $m + 1$ vertices.

This star is constructed by the generalized adjacency matrix $(m + 1) \times (m + 1)$ as follows.

$$A_{k-1} = \begin{bmatrix} 0 & \dots & 0 & 1 \\ \vdots & \ddots & \vdots & \vdots \\ 0 & \dots & 0 & 1 \\ 1 & \dots & 1 & 0 \end{bmatrix}$$

By Theorem 1.1, this star is graceful. Now consider a subgraph of the rooted tree in which it has every center vertices of p previous stars connected to a vertex at level $k - 2$.

This graph is constructed by the generalized adjacency matrix

IOP Conf. Series: Journal of Physics: Conf. Series **1008** (2018) 012031 doi:10.1088/1742-6596/1008/1/012031

$$A_{k-2} = \begin{pmatrix} 0 & \dots & 0 & A'_{k-1} & v \\ \vdots & \ddots & A'_{k-1} & 0 & v \\ 0 & \dots & 0 & \dots & \dots \\ A'_{k-1} & 0 & \dots & v & \\ v^t & v^t & \dots & v^t & 0 \end{pmatrix}.$$

The matrix has order $((m+1)p+1) \times ((m+1)p+1)$, while the p sub-matrices of the form

$$A'_{k-1} = \begin{bmatrix} 0 & 1 & \dots & 1 \\ 1 & 0 & \dots & 0 \\ \vdots & \vdots & \ddots & \vdots \\ 1 & 0 & \dots & 0 \end{bmatrix}$$

where O is zero matrix, and

$$v = \begin{bmatrix} 1 \\ 0 \\ \vdots \\ 0 \end{bmatrix}$$

with order $(m+1) \times 1$, and v^t is the transpose of v. Note that the sub-matrix A'_{k-1} is the generalized adjacency matrix A_{k-1} with entries displaced by rotating the matrix 180 degree clockwise. It means that if $A'_{k-1} = (a'_{ij})$ and $A_{k-1} = (a_{ij})$ then $a'_{ij} = a_{(n-1+i)(n+1-j)}$.

See the symmetric tree graph of level k in Figure 1.

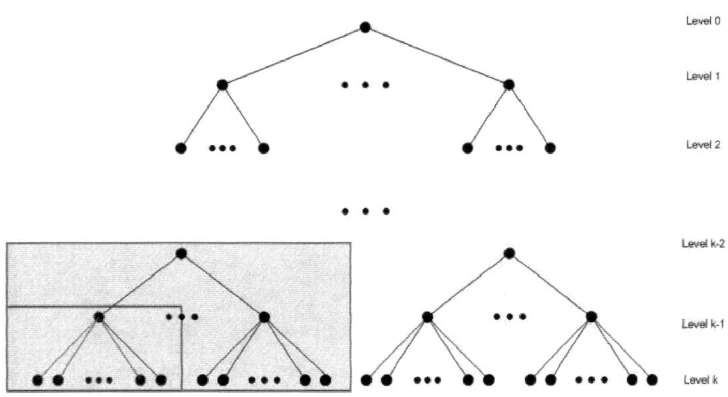

Figure 1. Symmetric Tree's construction.

By Theorem 1.1, this graph is graceful. Now consider a subgraph of the rooted tree in which it has every root vertices of q previous graphs connected to a vertex at level $k-3$.

This graph is constructed by the following generalized adjacency matrix

$$A_{k-3} = \begin{pmatrix} 0 & \dots & 0 & A'_{k-2} & v \\ \vdots & \ddots & A'_{k-2} & 0 & v \\ 0 & \ddots & \vdots & \vdots & v \\ A'_{k-2} & 0 & \dots & 0 & v \\ v^t & v^t & \dots & v^t & v \end{pmatrix}.$$

IOP Conf. Series: Journal of Physics: Conf. Series **1008** (2018) 012031 doi:10.1088/1742-6596/1008/1/012031

with order $(((m+1)p+1) \times (((m+1)p+1_q+1)))$ and the q sub-matrices has the form

$$A_{k-2} = \begin{pmatrix} 0 & v^{t'} & \dots & v^{t'} & v^{t'} \\ v' & 0 & \dots & 0 & A_{k-1} \\ \vdots & \vdots & \ddots & \vdots & \vdots \\ v' & 0 & A_{k-1} & \vdots & 0 \\ v' & A_{k-1} & 0 & \dots & 0 \end{pmatrix}.$$

where O is zero matrix, and

$$v = \begin{bmatrix} 1 \\ 0 \\ \vdots \\ 0 \end{bmatrix}$$

and v^t is the transpose of v. Note that the sub-matrix $A^{'}_{k-2}$ is the generalized adjacency matrix A_{k-2} with entries displaccd by rotating the matrix 180 degree clockwise.

By Theorem 1.1, this graph is graceful. This process can be repeated until we finally obtain a rooted tree with k level, which is constructed by the generalized matrix adjacency A_0.

$$A_0 = \begin{pmatrix} 0 & \dots & 0 & A^{'}_1 & v \\ \vdots & \ddots & A^{'}_1 & 0 & v \\ 0 & \dots & 0 & \dots & \dots \\ A^{'}_1 & 0 & \dots & v & \\ v^t & v^t & \dots & v^t & 0 \end{pmatrix}.$$

By the Theorem 1.1, this graph is graceful. Thus the symmetric tree is graceful. □

To make the construction clearer, we can see the following example.

Let we start from S_3, a star with three leaves with the label as in Figure 2.

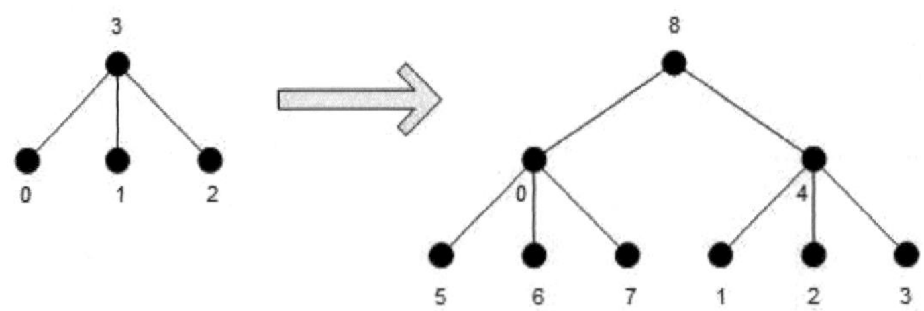

Figure 2. Tree's construction from star.

The adjacency matrix of the star is as follows.

$$A_2 = \begin{pmatrix} 0 & 0 & 0 & 1 \\ 0 & 0 & 0 & 1 \\ 0 & 0 & 0 & 1 \\ 1 & 1 & 1 & 0 \end{pmatrix}.$$

In the construction in Figure 2, we can have the adjacency matrix of the tree on the right side is as follows.

$$A_1 = \begin{pmatrix} 0&0&0&0&0&1&1&1&1\\ 0&0&0&0&1&0&0&0&0\\ 0&0&0&0&1&0&0&0&0\\ 0&0&0&0&1&0&0&0&0\\ 0&1&1&1&0&0&0&1&1\\ 1&0&0&0&0&0&0&0&0\\ 1&0&0&0&0&0&0&0&0\\ 1&0&0&0&0&1&1&1&1\\ 1&0&0&0&1&0&0&0&0 \end{pmatrix}.$$

If we continue the process to have A_0 then the resulting labeling of symmetric tree of three level can be seen in Figure 3.

$$A_0 = \begin{pmatrix} 0 & 0 & A_1' & v\\ 0 & A_1' & 0 & v\\ A_1' & 0 & 0 & v\\ v^t & v^t & v^t & 0 \end{pmatrix}.$$

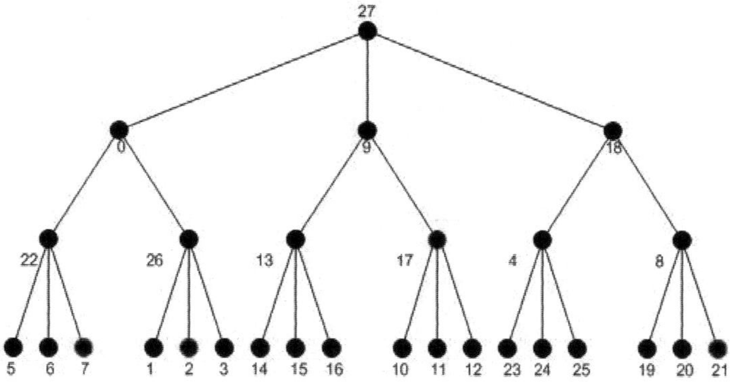

Figure 3. Example of graceful labeling.

3. Conclusion and Open Problem

The graceful construction using adjacency matrix can be use to prove that symmetric tree is graceful. The original conjecture on " all trees are graceful" is still open. The construction using adjacency matrix is promising to solve conjecture for the semi regular tree.

Acknowledgments

Part of this research is funded by Hibah PITTA UI 2018 research grant.

References

[1] Cavalier C M 2006 *Graceful Labelings* Thesis Lousiana State University
[2] Gallian J A 2016 Dynamic Survey of Graph Labeling it The Electronic Journal of Combinatorics 1-79
[3] Golomb S W 1972 *How to Number a Graph, Graph Theory and Computing* R C Read ed Academic Press New York 23-37
[4] Robeva E 2011 *An Extensive Survey of Graceful Trees* Thesis Stanford University
[5] Rosa A 1966 *On Certain Valuations of the Vertices of a Graph* Theory of Graphs (Internat. Symposium, Rome, July 1966) 349-355

IOP Conf. Series: Journal of Physics: Conf. Series **1008** (2018) 012032 doi:10.1088/1742-6596/1008/1/012032

On the strong metric dimension of sun graph, windmill graph, and möbius ladder graph

Mila Widyaningrum and Tri Atmojo Kusmayadi

Department of Mathematics, Faculty of Mathematics and Natural Sciences, Universitas Sebelas Maret, Surakarta, Indonesia

E-mail: milawidyaningrum4@gmail.com, tri.atmojo.kusmayadi@gmail.com

Abstract. Let G be a connected graph with vertex set $V(G)$ and edge set $E(G)$. The interval $I[u, v]$ between u and v is defined as the collection of all vertices that belong to some shortest u-v path. A vertex $s \in V(G)$ strongly resolves two vertices u and v if u belongs to a shortest v-s path or v belongs to a shortest u-s path. A set $S \subset V(G)$ is a strong resolving set if every two distinct vertices of G are strongly resolved by some vertex of S. The smallest cardinality of strong resolving set is called a strong metric basis. The strong metric dimension of G, denoted by $sdim(G)$, is defined as the cardinality of the strong metric basis. In this paper we determine the strong metric dimension of a sun graph S_n, a windmill graph K_n^m, and a Möbius ladder graph M_n. We obtain the strong metric dimension of sun graph S_n is n - 1 for $n \geq 3$. The strong metric dimension of windmill graph K_n^m is $(n$ - $1)m$ - 1 for $m \geq 2$ and $n \geq 3$. The strong metric dimension of Möbius ladder graph M_n with $n \geq 5$ is $2\lceil \frac{n+2}{4} \rceil$ for n even.

1. Introduction

Sebö and Tannier [8] introduced the concept of strong metric dimension in 2004. In 2007 the strong metric dimension of graph and digraph was introduced by Oellermann and Peters-Fransen [7]. Let G be a connected graph with vertex set $V(G)$ and edge set $E(G)$, Oellermann and Peters-Fransen [7] defined for two vertices $u, v \in V(G)$, the interval $I[u, v]$ between u and v to be the collection of all vertices that belong to some shortest u - v path. If u belongs to a shortest v-s path, denoted by $u \in I[v, s]$, or v belongs to a shortest u-s path, denoted by $v \in I[u, s]$ then a vertex s strongly resolves two vertices u and v. A set of vertices $S \subset V(G)$ is strong resolving set for G if every two distinct vertices of G are strongly resolved by some vertex of S. The smallest cardinality of strong resolving set is called strong metric basis. The cardinality of the strong metric basis is a strong metric dimension of G, denoted by $sdim(G)$.

Some authors have investigated the strong metric dimension to some graph classes. In 2004 Sebö and Tannier [8] observed the strong metric dimension of complete graph K_n, cycle graph C_n, and tree. Kratica [3] determined the strong metric dimension of hamming graph $H_{n,k}$ in 2012. At the same year, Kratica [4] already observed the strong metric dimension of some convex polytope graph. In 2013 Yi [9] determined that the metric dimension of G is 1 if and only if G is a path P_n and the strong metric dimension of G is n - 1 if and only if G is complete graph K_n. Kusmayadi et al. [5] determined the strong metric dimension of some related wheel graph such as sunflower graph, t-fold wheel graph, helm graph, and friendship graph in 2016. In this paper we determine the strong metric dimension of a sun graph S_n, a windmill graph K_n^m, and a Möbius Ladder graph M_n for n even.

IOP Conf. Series: Journal of Physics: Conf. Series **1008** (2018) 012032 doi:10.1088/1742-6596/1008/1/012032

2. Main Results

2.1. Strong Metric Dimension

Let G be a connected graph with a set of vertices $V(G)$ and a set of edges $E(G)$. Let S be a subset of $V(G)$. Oellermann and Peters-Fransen [7] defined for two vertices $u, v \in V(G)$, the interval $I[u, v]$ between u and v to be the collection of all vertices that belong to some shortest $u - v$ path. A vertex $s \in S$ strongly resolves two vertices u and v if $u \in I[v, s]$ or $v \in I[u, s]$. A vertex set S of G is a strong resolving set of G if every two distinct vertices of G are strongly resolved by some vertex of S. The smallest cardinality of strong resolving set is called strong metric basis of G. The strong metric dimension of a graph G is defined as the cardinality of strong metric basis denoted by $sdim(G)$. We often make use of the following lemma and properties about strong metric dimension given by Kratica et al. [4].

Lemma 2.1 *Let u, $v \in V(G)$, $u \neq v$,*

(i) $d(w,v) \leq d(u,v)$ for each w such that $uw \in E(G)$, and

(ii) $d(u,w) \leq d(u,v)$ for each w such that $vw \in E(G)$.

Then there does not exist vertex $a \in V(G)$, $a \neq u,v$ that strongly resolves vertices u and v.

Property 2.1 *If $S \subset V(G)$ is strong resolving set of graph G, then for every two vertices $u, v \in V(G)$ satisfying conditions* 1 *and* 2 *of Lemma* 2.1*, obtained $u \in S$ or $v \in S$.*

Property 2.2 *If $S \subset V(G)$ is strong resolving set of graph G, then for every two vertices $u, v \in V(G)$ satisfying $d(u,v) = diam(G)$, obtained $u \in S$ or $v \in S$.*

2.2. The Strong Metric Dimension of Sun Graph

Mirzakhah and Kiani [6] defined the sun graph S_n for $n \geq 3$ as a graph obtained by connecting each vertex of cycle C_n to a pendant vertex with an edge. The sun graph S_n can be depicted as in Figure 1.

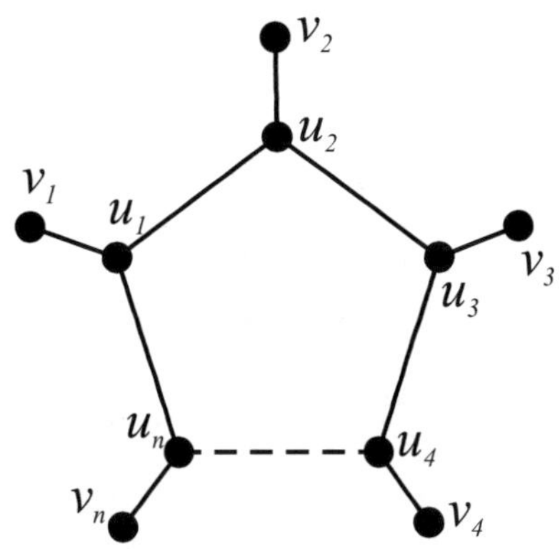

Figure 1. Sun graph S_n

Theorem 2.1 *Let S_n be the sun graph with $n \geq 3$. Then $sdim(S_n) = n - 1$.*

Proof. First, we prove that for every integer $n \geq 3$, if S is a strong resolving set of sun graph S_n then $| S | \geq n - 1$. We know that S is a strong resolving set of sun graph S_n. Suppose that S contains at most n - 2 vertices, then $| S | < n$ - 1. Let $V_1, V_2 \subset V(S_n)$, with $V_1 = \{v_1, v_2, \ldots, v_n\}$ and $V_2 = \{u_1, u_2, \ldots, u_n\}$. Now, we define $S_1 = V_1 \cap S$ and $S_2 = V_2 \cap S$. Without loss of generality, we may take $| S_1 | = a$, $a > 0$ and $| S_2 | = b$, $b \geq 0$. We obtain $a + b \leq n$ - 2, there are two distinct vertices v_i and $v_j \in V_1 \setminus S_1$, where $i \neq j$ such that for every $s \in S$, $v_i \notin I[v_j, s]$ and $v_j \notin I[v_i, s]$. This contradicts with the supposition that S is a strong resolving set. Thus $|S| \geq n - 1$.

Then, we prove that for every integer $n \geq 3$, if $S = \{v_1, v_2, \ldots, v_{n-1}\}$ then S is a strong resolving set of sun graph S_n. We prove that for every two distinct vertices $u, v \in V(S_n) \setminus S$ where $u \neq v$ is strongly resolved by a vertex $s \in S$. There are two possible pairs of vertices.

(i) A pair of vertices (u_i, v_n) with $i = 1, 2, \ldots, n$.
For every integer $1 \leq i \leq \lceil \frac{n-1}{2} \rceil$, we obtain the shortest path between $v_{\lceil \frac{n-1}{2} \rceil}$ and v_n is $v_{\lceil \frac{n-1}{2} \rceil}, u_{\lceil \frac{n-1}{2} \rceil}, u_{\lceil \frac{n-1}{2} \rceil - 1}, \ldots, u_1, u_n, v_n$. So that $u_i \in I[v_{\lceil \frac{n-1}{2} \rceil}, v_n]$. Then for every integer $\lfloor \frac{n+1}{2} \rfloor \leq i \leq n$, we obtain the shortest path between $v_{\lfloor \frac{n+1}{2} \rfloor}$ and v_n is $v_{\lfloor \frac{n+1}{2} \rfloor}, u_{\lfloor \frac{n+1}{2} \rfloor}, u_{\lfloor \frac{n+1}{2} \rfloor + 1}, \ldots, u_{n-1}, u_n, v_n$. So that $u_i \in I[v_{\lfloor \frac{n+1}{2} \rfloor}, v_n]$.

(ii) A pair of vertices (u_i, u_j) with $i, j = 1, 2, \ldots, n$ and $i \neq j$.
For every integer $i, j \in [1, n]$ without loss of generality we may assume $1 \leq i < j \leq n$, therefore u_i and u_j will be strongly resolved by v_i so that $u_i \in I[v_i, u_j]$.

From every possible pairs of vertices, there exists a vertex $v_i \in S$ with $i \in [1, n-1]$ which strongly resolves two distinct vertices of $V(S_n) \setminus S$. Thus S is a strong resolving set of sun graph S_n.

We obtain a set $S = \{v_1, v_2, \ldots, v_{n-1}\}$ is a strong resolving set of sun graph S_n with $n \geq 3$ and we have $|S| \geq n - 1$, so that S is a strong metric basis of sun graph S_n. Hence $sdim(S_n) = n - 1$. □

2.3. The Strong Metric Dimension of Windmill Graph

Gallian [1] defined the windmill graph K_n^m is a graph consists of m copies of K_n with a vertex in common. The windmill graph K_n^m can be depicted as in Figure 2.

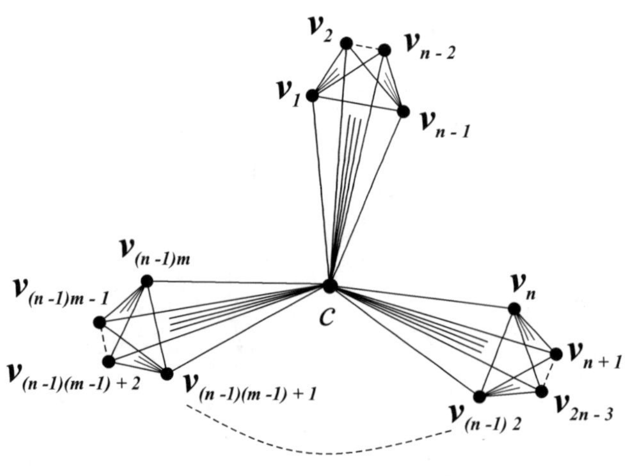

Figure 2. Windmill graph K_n^m

Theorem 2.2 *Let K_n^m be the windmill graph with $m \geq 2$ and $n \geq 3$. Then $sdim(K_n^m) = (n-1)m - 1$.*

Proof. First, we prove that for every integer $m \geq 2$ and $n \geq 3$, if S is a strong resolving set of windmill graph K_n^m then $|S| \geq (n - 1)m - 1$. Consider a pair of vertices (v_i, v_j) for $i, j = 1, 2, \ldots, (n-1)m$ and $i \neq j$ satisfying conditions 1 and 2 of Lemma 2.1. According to Property 2.1,

IOP Conf. Series: Journal of Physics: Conf. Series **1008** (2018) 012032 doi:10.1088/1742-6596/1008/1/012032

we obtain $v_i \in S$ or $v_j \in S$. It means S has at least one vertex from sets $X_{ij} = \{v_i, v_j\}$ with $i, j = 1, 2, \ldots, (n-1)m$ and $i \neq j$. The minimum number of vertices from sets X_{ij} is $(n$ - $1)m$ - 1. Therefore, $|\ S\ | \geq (n-1)m{-}1$.

After that, we prove that for every integer $m \geq 2$ and $n \geq 3$, if $S = \{v_1, v_2, \ldots, v_{(n-1)m-1}\}$ then S is a strong resolving set of K_n^m graph. For every two distinct vertices $u, v \in V(K_n^m)\backslash S$, there is a vertex $s \in S$ which strongly resolves u and v. Let us prove that a pair of vertices $(c, v_{(n-1)m})$ is strongly resolved by v_i for $i = 1, 2, \ldots, (n-1)(m-1)$. The shortest path between $v_{(n-1)m}$ and v_i is $v_{(n-1)m}$, c, v_i. Since vertex c belongs to this shortest path, v_i strongly resolves c and $v_{(n-1)m}$. Thus S is a strong resolving set of windmill graph K_n^m.

We have a set $S = \{v_1, v_2, \ldots, v_{(n-1)m-1}\}$ is a strong resolving set of windmill graph K_n^m and $|S| \geq (n$ - $1)m$ - 1, so that S is a strong metric basis of windmill graph K_n^m. Hence $sdim(K_n^m) = (n$ - $1)m$ - 1. □

2.4. The Strong Metric Dimension of Möbius Ladder Graph

Guy and Harary [2] defined the Möbius ladder graph M_n is a graph constructed by connecting vertices u and v in the cycle C_n if $d(u,v) = diam(C_n)$. The Möbius ladder graph M_n can be depicted as in Figure 3 and Figure 4.

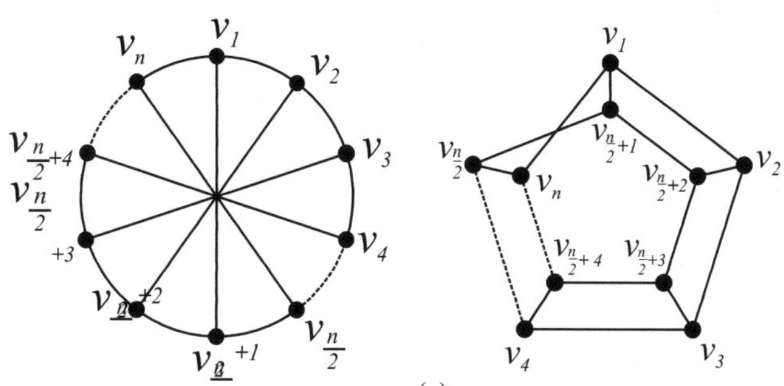

(a)

Figure 3. Möbius ladder graph M_n for n even

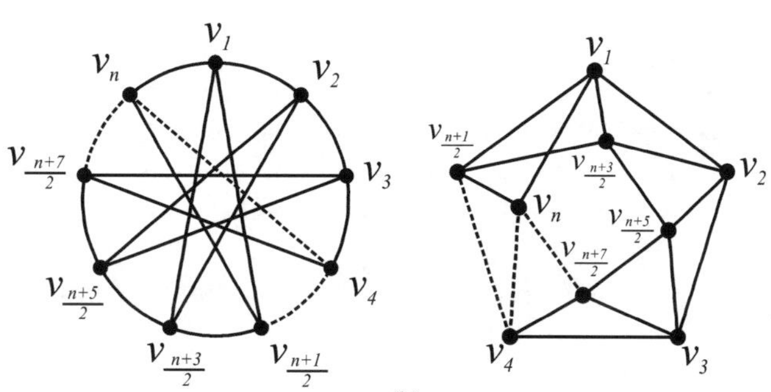

(b)

Figure 4. Möbius ladder graph M_n for n odd

Theorem 2.3 *Let M_n be the Möbius ladder graph with $n \geq 5$. Then $sdim(M_n) = 2\left\lceil\frac{n+2}{4}\right\rceil$ for n even.*

Proof. First, we prove that for every $n \geq 5$, if S is a strong resolving set of Möbius ladder graph M_n then $|\ S\ | \geq 2\left\lceil\frac{n+2}{4}\right\rceil$ for n even. We know that S is a strong resolving set of Möbius ladder graph M_n. Suppose that S contains at most $2 \left|\frac{n+2}{4}\right| - 1$ vertices, then $| S | < 2 \lceil\frac{n+2}{4}\rceil$. Let $V_1, V_2 \subset V(M_n)$, with $V_1 = \{v_1, v_2, \ldots, v_{\frac{n}{2}}\}$ and $V_2 = \{v_{\frac{n}{2}+1}, v_{\frac{n}{2}+2}, \ldots, v_n\}$. Now, we define $S_1 = V_1 \cap S$ and $S_2 = V_2 \cap S$. Without loss of generality, we may take $| S_1 | = p$, $p > 0$ and $| S_2 | = q$, $q > 0$. Clearly $p + q \geq 2 \lceil\frac{n+2}{4}\rceil$. Otherwise, there are two distinct vertices v_i and v_j, where $v_i \in V_1 \backslash S_1$ and $v_j \in V_2 \backslash S_2$ such that for every $s \in S$ we obtain $v_i \notin I[v_j, s]$ and $v_j \notin I[v_i, s]$. This contradicts with the supposition that S is a strong resolving set. Thus $| S | \geq 2 \lceil\frac{n+2}{4}\rceil$.

Then, we prove that for every even integer $n \geq 5$, a set $S = \{v_1, v_2, \ldots, v_{\lceil\frac{n+2}{4}\rceil}, v_{\frac{n}{2}+1}, v_{\frac{n}{2}+2}, \ldots, v_{\frac{n}{2}+\lceil\frac{n+2}{4}\rceil}\}$ is a strong resolving set of Möbius ladder graph M_n for n even. We prove that for every two distinct vertices $v_i, v_j \in V(M_n)\backslash S$, where $i, j = (\lceil\frac{n+2}{4}\rceil, \frac{n}{2}] \cup (\frac{n}{2}+\lceil\frac{n+2}{4}\rceil, n]$ is strongly resolved by a vertex $s \in S$. There are three possible pairs of vertices.

(i) A pair of vertices (v_i, v_j) where $\lceil\frac{n+2}{4}\rceil < i, j \leq \frac{n}{2}$, $i \neq j$.
For every integer $\lceil\frac{n+2}{4}\rceil < i, j \leq \frac{n}{2}$ with $\lceil\frac{n+2}{4}\rceil < i < j \leq \frac{n}{2}$, we obtain the shortest path between $v_{\lceil\frac{n+2}{4}\rceil}$ and v_j is $v_{\lceil\frac{n+2}{4}\rceil}, v_{\lceil\frac{n+2}{4}\rceil+1}, \ldots, v_i, \ldots, v_j$. So that $v_i \in I[v_{\lceil\frac{n+2}{4}\rceil}, v_j]$.

(ii) A pair of vertices (v_i, v_j) where $\frac{n}{2}+\lceil\frac{n+2}{4}\rceil < i, j \leq n$, $i \neq j$.
For every integer $\frac{n}{2}+\lceil\frac{n+2}{4}\rceil < i, j \leq n$ with $\frac{n}{2}+\lceil\frac{n+2}{4}\rceil < i < j \leq n$, we obtain the shortest path between $v_{\frac{n}{2}+\lceil\frac{n+2}{4}\rceil}$ and v_j is $v_{\frac{n}{2}+\lceil\frac{n+2}{4}\rceil}, v_{\frac{n}{2}+\lceil\frac{n+2}{4}\rceil+1}, \ldots, v_i, \ldots, v_j$. So that $v_i \in I[v_{\frac{n}{2}+\lceil\frac{n+2}{4}\rceil}, v_j]$.

(iii) A pair of vertices (v_i, v_j) where $\lceil\frac{n+2}{4}\rceil < i \leq \frac{n}{2}$ and $\frac{n}{2} + \lceil\frac{n+2}{4}\rceil < j \leq n$.
For every pair of vertices (v_i, v_j) with $\lceil\frac{n+2}{4}\rceil < i \leq \frac{n}{2}$ and $\frac{n}{2} + \lceil\frac{n+2}{4}\rceil < j \leq n$. Vertex v_i and v_j will be strongly resolved by a vertex $v_{\frac{n}{2}+\lceil\frac{n+2}{4}\rceil}$ if $\frac{n}{2} + \lceil\frac{n+2}{4}\rceil < j \leq i + \frac{n}{2}$, we obtain the shortest path between v_i and $v_{\frac{n}{2}+\lceil\frac{n+2}{4}\rceil}$ is $v_i, v_{i+\frac{n}{2}}, v_{i+\frac{n}{2}-1}, v_{i+\frac{n}{2}-2}, \ldots, v_{\frac{n}{2}+\lceil\frac{n+2}{4}\rceil}$. So that $v_j \in I[v_i, v_{\frac{n}{2}+\lceil\frac{n+2}{4}\rceil}]$. If $i + \frac{n}{2} \leq j \leq n$, then v_i and v_j will be strongly resolved by v_1 with the shortest path between v_i and v_1 is $v_i, v_{i+\frac{n}{2}}, v_{i+\frac{n}{2}+1}, \ldots, v_n, v_1$. So that $v_j \in I[v_i, v_1]$.

From every possible pairs of vertices, there exists a vertex $s \in S$ which strongly resolves two distinct vertices of $V(M_n) \setminus S$. Thus S is a strong resolving set of Möbius ladder graph M_n.

We obtain a set $S = \{v_1, v_2, \ldots, v_{\lceil\frac{n+2}{4}\rceil}, v_{\frac{n}{2}+1}, v_{\frac{n}{2}+2}, \ldots, v_{\frac{n}{2}+\lceil\frac{n+2}{4}\rceil}\}$ is a strong resolving set of Möbius ladder graph M_n and we have $| S | \geq 2 \lceil\frac{n+2}{4}\rceil$, so that S is a strong metric basis of M_n. Hence $sdim(M_n) = 2 \lceil\frac{n+2}{4}\rceil$ for n even. □

3. Conclusion

According to the discussion above, we have the strong metric dimension of a sun graph S_n, a windmill graph K_n^m, and a Möbius ladder graph M_n for n even. Furthermore, we conclude this paper with the following open problem for the direction of further research which is still in progress.

Open Problem: Determine the strong metric dimension of a Möbius ladder graph M_n for n odd.

Acknowledgments
The authors gratefully acknowledge the support from Department of Mathematics, Faculty of Mathematics and Natural Sciences, Universitas Sebelas Maret, Surakarta. Then, we wish to thank the referees for their suggestions which helped to improve the paper.

References

[1] Gallian J A *A dynamic survey of graph labeling* 2009 The electronic journal of combinatorics **16** 6 1-219
[2] Guy RK and Harary F 1967 *On The Möbius Ladders* University of Calgary Department of Mathematics **10** 4 493-496

IOP Conf. Series: Journal of Physics: Conf. Series **1008** (2018) 012032 doi:10.1088/1742-6596/1008/1/012032

[3] Kratica J, Kovacevic-Vujcic V, Cangalovic M and Stojanovic M 2012 Minimal Doubly Resolving Sets and The Strong Metric Dimension of Hamming Graphs *Applicable Analysis and Discrete Mathematics* **6** 63 - 71

[4] Kratica J, Kovacevic-Vujcic V and Cangalovic M 2012 Minimal Doubly Resolving Sets and The Strong Metric Dimension of Some Convex Polytype *Applicable Analysis and Discrete Mathematics* **218** 9790-9801

[5] Kusmayadi T A, Kuntari S, Rahmadi D and Lathifah F A 2016 On The Strong Metric Dimension of Some Related Wheel Graph **99** 9 1325-1334

[6] Mirzakhah M and Kiani D 2010 The sun graph is determined by its signless Laplacian spectrum *Electronic Journal of Linear Algebra* **20** 610 -620

[7] Oellermann OR and Peters-Fransen J 2007 The strong metric dimension of graphs and digraphs *Discrete Applied Mathematics* **155** 3 356-364

[8] Sebö A and Tannier E 2004 On metric generators of graphs *Mathematics of Operations Research* **29** 2 383-393

[9] Yi E 2013 On strong metric dimension of graphs and their complements *Acta Mathematica Sinica, English Series* **29** 8 1479-1492

IOP Conf. Series: Journal of Physics: Conf. Series **1008** (2018) 012033 doi:10.1088/1742-6596/1008/1/012033

On the r-dynamic chromatic number of the corronation by complete graph

Arika Indah Kristiana[1,2], M.Imam Utoyo[3], Dafik[1,2]
[1] CGANT University of Jember Indonesia
[2] Mathematics Edu. Depart. University of Jember Indonesia
[3] Mathematics Depart. University of Airlangga Surabaya Indonesia

E-mail: arikaindahkristiana@gmail.com, m.i.utoyo@fst.unair.ac.id, d.dafik@unej.ac.id

Abstract. In this paper we will study the r-dynamic chromatic number of the coronation by complete graph. A proper k-coloring of graph G such that the neighbors of any vertex v receive at least $\min\{r, d(v)\}$ different colors. The r-dynamic chromatic number, $\chi_r(G)$ is the minimum k such that graph G has an r-dynamic k-coloring. We will obtain lower bound of the r-dynamic chromatic number of $\chi_r(K_n \odot H)$, and $\chi_r(H \odot K_m)$ We also study the exact value of the r-dynamic chromatic number of $\chi_r(K_n \odot S_m), \chi_r(K_n \odot F_m), \chi_r(S_n \odot K_m), \chi_r(F_n \odot K_m)$ and $\chi_r(K_n \odot K_m)$ for $m, n \geq 3$.

1. Introduction

Let G be graph. Graphs in this paper are simple and finite. We denote the vertex set of G by $V(G)$ and the number of vertices of G is called the order of G. Thus for graph G, $\Delta(G)$ and $\delta(G)$ denote the maximum degree and minimum degree. For every $v \in V(G)$, $d(v), N(v), c(v)$ denote the degree of v, the neighbor set of v and the color of v. A proper vertex coloring of G by k colors is a function $c : V(G) \rightarrow \{1, 2, \cdots, k\}$ with this property: if $u, v \in V(G)$ are adjancent, then $c(u)$ and $c(v)$ are different. The r-dynamic coloring of a graph G, introduced by Montgomery [8] defined a proper k-coloring of graph G such that the neighbors of any vertex v receive at least $\min\{r, d(v)\}$ different colors. The r-dynamic chromatic number, $\chi_r(G)$ is the minimum k such that graph G has an r-dynamic k-coloring. The r-dynamic chromatic number has been studied by several authors, for instance in [1], [2], [3], [4], [6], [7], [9], [10], [11] and [12]. The following observations are useful for our study, proposed by Montgomery [8].

Observation 1.1 *Let $\Delta(G)$ be the maximum degree of graph G. It holds $\chi_r(G) \geq \min\{\Delta(G), r\} + 1$.*

Observation 1.2 *Let $G = K_n$ then $\chi_r(G) = n$.*

Observation 1.3 *Let $G = G_1 \odot G_2$ be corona graph, then we have $\delta(G) = \delta(G_2) + 1$ and $\Delta(G) = \Delta(G_1) + |V(G_2)|$, where $\Delta(G)$ is maximum degree of G and $\delta(G)$ is minimum degree of G.*

Furmanczyk [5], denoted by $G \odot H$, is a connected graph obtained by taking a number of vertices $|V(G)|$ copy of H, and making the ith of $V(G)$ adjacent to every vertex of the ith copy of $V(H)$. There have been many results already found, Pathinathan, et.al. in [11] find the b-chromatic number on corona graph of subdivision vertex path with path, corona graph of any graph with path, cycle, and complete graph. Ramya in [12] discuss about the acyclic coloring of corona of $C_n \odot K_{1,3}, P_n \odot K_2$ and star coloring of corona of $P_n \odot K_2$.

The Results

We will obtain lower bound of the r-dynamic chromatic number of $\chi_r(K_n \odot H)$, and $\chi_r(H \odot K_m)$. We alsa find the exact value of the r-dynamic chromatic number of $\chi_r(K_n \odot S_m), \chi_r(K_n \odot F_m), \chi_r(S_n \odot K_m), \chi_r(F_n \odot K_m)$ and $\chi_r(K_n \odot K_m)$ for $m, n \geq 3$.There are seven theorems found in this study. We will show the complete results in the followings theorem.

Lemma 1.1 *Let $G = K_n \odot H$ be a corona graph of K_n and graph $H \neq C_m, W_m$, for $n \geq 4$, the lower bound of the r-dynamic chromatic number is:*

$$\chi_r(G) \geq \begin{cases} \chi_r(K_n), & 1 \leq r \leq n-1 \\ r+1, & n \leq r \leq n+|V(H)|-1 \\ n+|V(H)|, & r \geq n+|V(H)| \end{cases}$$

Proof. Based defined of corona graph, $V(K_n \odot H) = V(K_n) \cup \bigcup_{i=1}^{n} V(H_i)$ and the maximum degree of $G = K_n \odot H$ is $\Delta(K_n \odot H) = \Delta(K_n) + |V(H)| = (n-1) + |V(H)|$. For $1 \leq r \leq n-1$, based Observation 1.1 and chosen $r = n-1$, we get $\chi_r(G = K_n \odot H) \geq \min\{\, r, \Delta(G)\} + 1 = \min\{\, n-1, n-1+|V(H)|\} + 1 = (n-1)+1 = n$. From Observation 1.2, $\chi(K_n) = n$ so proven $\chi_r(G = K_n \odot H) \geq \chi_r(K_n)$ for $1 \leq r \leq m$. For $r \geq n+|V(H)|$, based Observation 1.1 and chosen $r = n+|V(H)|$, we get $\chi_r(G = K_n \odot H) \geq \min\{\, r, \Delta(G)\} + 1 = \min\{\, n+|V(H)|, n+|V(H)|-1\} + 1 = (n+|V(H)|-1)+1 = n+|V(H)|$. It concludes the proof. □

Theorem 1.1 *Let $G = K_n \odot S_m$ be a corona graph of K_n and S_m, for $n \geq 4, m \geq 3$, the r-dynamic chromatic number is:*

$$\chi_r(G) = \begin{cases} n, & 1 \leq r \leq n-1 \\ r+1, & n \leq r \leq m+n \\ m+n+1, & r \geq m+n+1 \end{cases}$$

Proof. The graph $G = K_n \odot S_m$ is a connected graph with vertex set $V(K_n \odot S_m) = \{x_i; 1 \leq i \leq n\} \cup \{y_i; 1 \leq i \leq n\} \cup \{y_{ij}; 1 \leq i \leq n; 1 \leq j \leq m\}$. The order of this graph is $p = |V(K_n \odot S_m)| = mn + 2n$. Thus the maximum degree of graph $G = K_n \odot S_m$ is $\Delta(K_n \odot S_m) = m+n$ and the minimum degree of graph $G = K_n \odot S_m$ is $\delta(K_n \odot S_m) = 2$. To find the exact value of r-dynamic chromatic number of $G = K_n \odot S_m$, we define two cases, namely for $\chi_{1 \leq r \leq n-1}(K_n \odot S_m)$ and $\chi_{r \geq m+n+1}(K_n \odot S_m)$.

Case 1. For $1 \leq r \leq n-1$, based on Lemma 1.1 and Observation 1.2, $\chi_r(K_n) = n$ then we have the lower bound of the $r-$dynamic chromatic number of $G = K_n \odot S_m$ is $\chi_{1 \leq r \leq n-1}(K_n \odot S_m) \geq \chi_r(K_n) = n$. We will show that the upper bound of the $r-$dynamic chromatic number of $G = K_n \odot S_m$ is $\chi_{1 \leq r \leq n-1}(K_n \odot S_m) \leq \chi_r(K_n)$, define a map of the r-dynamic coloring of $G = K_n \odot S_m$ is $c_1 : V(K_n \odot S_m) \rightarrow \{1, 2, \ldots, k\}$ where $n \geq 4, m \geq 3$, by the following:

$$c_1(x_i) = i, \; 1 \leq i \leq n$$

$$c_1(y_i) = \begin{cases} 1, & \text{for } i = n \\ i+1, & \text{for } 1 \leq i \leq n-1 \end{cases}$$

$$c_1(y_{ij}) = \begin{cases} i+2, & \text{for } j \equiv 1(mod\ 3), 1 \le i \le n-2 \\ 1, & \text{for } j \equiv 1(mod\ 3),\ i = n-1 \\ 2, & \text{for } j \equiv 1(mod\ 3),\ i = n \end{cases}$$

$$c_1(y_{ij}) = \begin{cases} i+3, & \text{for } j \equiv 2(mod\ 3), 1 \le i \le n-3 \\ 1, & \text{for } j \equiv 2(mod\ 3), i = n-2 \\ 2, & \text{for } j \equiv 2(mod\ 3), i = n-1 \\ 3, & \text{for } j \equiv 2(mod\ 3), i = n \end{cases}$$

$$c_1(y_{ij}) = \begin{cases} i-1, & \text{for } j \equiv 0(mod\ 3), 2 \le i \le n \\ n, & \text{for } j \equiv 0(mod\ 3), i = 1 \end{cases}$$

It easy to see that c_1 is a map $c_1 : V(K_n \odot S_m) \to \{1, 2, \cdots n\}$. It is clearly that c_1 gives the upper bound of the $r-$dynamic chromatic number of $G = K_n \odot S_m$ is $\chi_{1 \le r \le n-1}(K_n \odot S_m) \le n$. It concludes that $r-$dynamic chromatic number of $G = K_n \odot S_m$ is $\chi_{1 \le r \le n-1}(K_n \odot S_m) = n$.
Case 2. For $r \ge m+n+1$, based on Lemma 1.1 with $|V(S_m)| = m+1$ then we have the lower bound of the $r-$dynamic chromatic number of $G = K_n \odot S_m$ is $\chi_{r \ge m+n+1}(K_n \odot S_m) \ge n + |V(S_m) = n + (m+1) = m+n+1$. We will show that the upper bound of the $r-$dynamic chromatic number of $G = K_n \odot S_m$ is $\chi_{r \ge m+n+1}(K_n \odot S_m) \le m+n+1$, define a map of the r-dynamic coloring of $G = K_n \odot S_m$ is $c_2 : V(K_n \odot S_m) \to \{1, 2, \ldots, k\}$ where $n \ge 4, m \ge 3$, by the following:

$$c_2(x_i) = i, 1 \le i \le n$$
$$c_2(y_i) = n+i, 1 \le i \le n$$
$$c_2(x_{ij}) = n+j+1, 1 \le i \le n, 1 \le j \le m$$

It easy to see that c_2 is a map $c_2 : V(K_n \odot S_m) \to \{1, 2, \cdots n+m+1\}$. It is clearly that c_2 gives the upper bound of the $r-$dynamic chromatic number of $G = K_n \odot S_m$ is $\chi_{r \ge m+n+1}(K_n \odot S_m) \le m+n+1$. It concludes that $r-$dynamic chromatic number of $G = K_n \odot S_m$ is $\chi_{r \ge m+n+1}(K_n \odot S_m) = m+n+1$. It concludes the proof. □

Theorem 1.2 *Let $G = K_n \odot F_m$ be a corona graph of K_n and F_m, for $n \ge 4, m \ge 3$, the r-dynamic chromatic number is:*

$$\chi_r(G) = \begin{cases} n, & \text{for } 1 \le r \le n-1 \\ r+1, & \text{for } n \le r \le m+n \\ m+n+1, & \text{for } r \ge\ m+n+1 \end{cases}$$

Proof. The graph $G = K_n \odot F_m$ is a connected graph with vertex set $V(K_n \odot F_m) = \{x_i; 1 \le i \le n\} \cup \{y_i; 1 \le i \le n\} \cup \{x_{ij}; 1 \le i \le n; 1 \le j \le m\}$. The order of this graph is $p = |V(K_n \odot F_m)| = mn + 2n$. Thus the maximum degree of graph $G = K_n \odot F_m$ is $\Delta(K_n \odot F_m) = m+n$ and the minimum degree of graph $G = K_n \odot F_m$ is $\delta(K_n \odot F_m) = 3$. To find the exact value of r-dynamic chromatic number of $G = K_n \odot F_m$, we define two cases, namely for $\chi_{1 \le r \le n-1}(K_n \odot F_m)$ and $\chi_{r \ge m+n+1}(K_n \odot F_m)$.
Case 1. For $1 \le r \le n-1$, based on Lemma 1.1 and Observation 1.2, $\chi_r(K_n) = n$ then we have the lower bound of the $r-$dynamic chromatic number of $G = K_n \odot F_m$ is $\chi_{1 \le r \le n-1}(K_n \odot F_m) \ge \chi_r(K_n) = n$. We will show that the upper bound of the $r-$dynamic chromatic number of $G = K_n \odot F_m$ is $\chi_{1 \le r \le n-1}(K_n \odot F_m) \le \chi_r(K_n)$, define a map of the r-dynamic coloring of $G = K_n \odot F_m$ is $c_3 : V(K_n \odot F_m) \to \{1, 2, \ldots, k\}$ where $n \ge 4, m \ge 3$, by the following:

$$c_3(x_i) = i, 1 \le i \le n$$

$$c_3(y_i) = \begin{cases} 1, & \text{for } i = n \\ i+1, & \text{for } 1 \le i \le n-1 \end{cases}$$

$$c_3(y_{ij}) = \begin{cases} i+2, & \text{for } j \text{ odd}, 1 \le j \le m, 1 \le i \le n-2 \\ 1, & \text{for } j \text{ odd}, 1 \le j \le m, i = n-1 \\ 2, & \text{for } j \text{ odd}, 1 \le j \le m, i = n \\ i-1, & \text{for } j \text{ even}, 1 \le j \le m, 2 \le i \le n \\ n, & \text{for } j \text{ even}, 1 \le j \le m, i = 1 \end{cases}$$

It easy to see that c_3 is a map $c_3 : V(K_n \odot F_m) \to \{1, 2, \cdots n\}$. It is clearly that c_3 gives the upper bound of the $r-$dynamic chromatic number of $G = K_n \odot F_m$ is $\chi_{1 \le r \le n-1}(K_n \odot F_m) \le n$. It concludes that $r-$dynamic chromatic number of $G = K_n \odot F_m$ is $\chi_{1 \le r \le n-1}(K_n \odot F_m) = n$.
Case 2. For $r \ge m + n + 1$, based on Lemma 1.1 with $|V(F_m)| = m + 1$ then we have the lower bound of the $r-$dynamic chromatic number of $G = K_n \odot F_m$ is $\chi_{r \ge m+n+1}(K_n \odot F_m) \ge n + |V(F_m)| = n + (m + 1)$. We will show that the upper bound of the $r-$dynamic chromatic number of $G = K_n \odot F_m$ is $\chi_{r \ge m+n+1}(K_n \odot F_m) \le m + n + 1$, define a map of the r-dynamic coloring of $G = K_n \odot F_m$ is $c_4 : V(K_n \odot F_m) \to \{1, 2, \ldots, k\}$ where $n \ge 4, m \ge 3$, by the following:

$$c_4(x_i) = i, 1 \le i \le n$$
$$c_4(y_i) = n + m + 1, 1 \le i \le n$$
$$c_4(x_{ij}) = j + i, 1 \le i \le n, 1 \le j \le m$$

It easy to see that c_4 is a map $c_4 : V(K_n \odot F_m) \to \{1, 2, \cdots m + n + 1\}$. It is clearly that c_4 gives the upper bound of the $r-$dynamic chromatic number of $G = K_n \odot F_m$ is $\chi_{r \ge m+n+1}(K_n \odot F_m) \le n+m+1$. It concludes that $r-$dynamic chromatic number of $G = K_n \odot F_m$ is $\chi_{r \ge m+n+1}(K_n \odot F_m) = m + n + 1$. It concludes the proof. □

Lemma 1.2 *Let $G = H \odot K_m$ be a corona graph of $H \ne C_n, W_n$ and graph K_m, for $m \ge 4$, the lower bound of the r-dynamic chromatic number is:*

$$\chi_r(G) \ge \begin{cases} \chi_r(K_m) + 1, & 1 \le r \le m \\ r + 1, & m + 1 \le r \le m + \Delta(H) \\ \Delta(H) + m + 1, & r \ge m + \Delta(H) + 1 \end{cases}$$

Proof. Based defined of corona graph, $V(H \odot K_m) = V(H) \cup \bigcup_{i=1}^{n} V((K_m)_i)$ and the maximum degree of $G = H \odot K_m$ is $\Delta(H \odot K_m) = \Delta(H) + |V(K_m)| = \Delta(H) + m$. For $1 \le r \le m$, based Observation 1.1 and choosen $r = m$, we get $\chi_r(G = H \odot K_m) \ge \min\{ r, \Delta(G)\} + 1 = \min\{ m, \Delta(H) + m\} + 1 = m + 1$. From Observation 1.2, $\chi(K_m) = m$ so proven $\chi_r(G = H \odot K_m) \ge \chi_r(K_m) + 1$ for $1 \le r \le m$. For $r \ge m + \Delta(H) + 1$, based Observation 1.1 and choosen $r = m+\Delta(H)+1$, we get $\chi_r(G = H \odot K_m) \ge \min\{ r, \Delta(G)\}+1 = \min\{ m + \Delta(H) + 1, \Delta(H) + m\} + 1 = m + \Delta(H) + 1$. It concludes the proof. □

Theorem 1.3 *Let $G = S_n \odot K_m$ be a corona graph of S_n and K_m, for $n \ge 3, m \ge 4$, the r-dynamic chromatic number is:*

$$\chi_r(G) = \begin{cases} r + 1, & n \le r \le m + n \\ m + n + 1, & r \ge m + n + 1 \end{cases}$$

Proof. The graph $g = S_n \odot K_m$ is a connected graph with vertex set $V(S_n \odot K_m) = \{x\} \cup \{x_i; 1 \le i \le n\} \cup \{y_j; 1 \le i \le m\} \cup \{x_{ij}; 1 \le i \le n; 1 \le j \le m\}$. The order of this graph is $p = |V(S_n \odot K_m)| = mn + m + n + 1$. Thus the maximum degree of graph $G = S_n \odot K_m$ is $\Delta(S_n \odot K_m) = m + n$ and the minimum degree of graph $G = S_n \odot K_m$ is $\delta(S_n \odot K_m) = m$.

To find the exact value of r-dynamic chromatic number of $G = S_n \odot K_m$, we define two cases, namely for $\chi_{1\le r\le m}(S_n \odot K_m)$ and $\chi_{r\ge m+n+1}(S_n \odot K_m)$.
Case 1. For $1 \le r \le n-1$, based on Lemma 1.2 and Observation 1.2, $\chi_r(K_m) = m$ then we have the lower bound of the $r-$dynamic chromatic number of $G = S_n \odot K_m$ is $\chi_{1\le r\le m}(S_n \odot K_m) \ge \chi_r(K_m)+1 = m+1$. We will show that the upper bound of the $r-$dynamic chromatic number of $G = S_n \odot K_m$ is $\chi_{1\le r\le m}(S_n \odot K_m) \le m+1$, define a map of the r-dynamic coloring of $G = S_n \odot K_m$ is $c_5 : V(S_n \odot K_m) \to \{1, 2, \ldots, k\}$ where $n \ge 3, m \ge 4$, by the following:

$$c_5(x) = 1$$
$$c_5(x_i) = 2, 1 \le i \le n$$

$$c_5(x_{ij}) = \begin{cases} 1, & \text{for } j = 1, 1 \le i \le n \\ 1+j, & \text{for } 2 \le j \le m, 1 \le i \le n \end{cases}$$

$$c_5(y_j) = \begin{cases} 2, & \text{for } j = 1 \\ 1+j, & \text{for } 2 \le j \le m \end{cases}$$

It easy to see that c_5 is a map $c_5 : V(S_n \odot K_m) \to \{1, 2, \cdots m+1\}$. It is clearly that c_5 gives the upper bound of the $r-$dynamic chromatic number of $G = S_n \odot K_m$ is $\chi_{1\le r\le m}(S_n \odot K_m) \le m+1$. It concludes that $r-$dynamic chromatic number of $G = S_n \odot K_m$ is $\chi_{1\le r\le m}(S_n \odot K_m) = m+1$.
Case 2. For $r \ge r \ge n+m+1$, based on Lemma 1.2 with $\Delta(S_n) = n$ then we have the lower bound of the $r-$dynamic chromatic number of $G = S_n \odot K_m$ is $\chi_{r\ge n+m+1}(S_n \odot K_m) \ge \Delta(S_n) + m + 1 = n + m + 1$. We will show that the upper bound of the $r-$dynamic chromatic number of $G = S_n \odot K_m$ is $\chi_{r\ge r\ge n+m+1}(S_n \odot K_m) \le m+n+1$, define a map of the r-dynamic coloring of $G = S_n \odot K_m$ is $c_6 : V(S_n \odot K_m) \to \{1, 2, \ldots, k\}$ where $n \ge 3, m \ge 4$, by the following:

$$c_6(x) = 1$$
$$c_6(x_i) = 1+i, 1 \le i \le n$$
$$c_6(x_{ij}) = 1+n+j, 1 \le i \le n, 1 \le j \le m$$
$$c_6(y_j) = 1+n+j, 1 \le j \le m$$

It easy to see that c_6 is a map $c_6 : V(S_n \odot K_m) \to \{1, 2, \cdots m+n+1\}$. It is clearly that c_6 gives the upper bound of the $r-$dynamic chromatic number of $G = S_n \odot K_m$ is $\chi_{r\ge m+n+1}(S_n \odot K_m) \le m+n+1$. It concludes that $r-$dynamic chromatic number of $G = S_n \odot K_m$ is $\chi_{r\ge m+n+1}(S_n \odot K_m) = m+n+1$. It concludes the proof. □

Theorem 1.4 *Let $G = F_n \odot K_m$ be a corona graph of F_n and K_m, for $n \ge 3, m \ge 4$, the r-dynamic chromatic number is:*

$$\chi_r(G) = \begin{cases} m+1, & 1 \le r \le m \\ r+1, & n \le r \le m+n \\ m+n+1, & r \ge m+n+1 \end{cases}$$

Proof. The graph $G = F_n \odot K_m$ is a connected graph with vertex set $V(S_n \odot K_m) = \{x\} \cup \{x_i; 1 \le i \le n\} \cup \{y_j; 1 \le i \le m\} \cup \{x_{ij}; 1 \le i \le n; 1 \le j \le m\}$. The order of this graph is $p = |V(F_n \odot K_m)| = mn+m+n+1$. Thus the maximum degree of graph $G = F_n \odot K_m$ is $\Delta(F_n \odot K_m) = m+n$ and the minimum degree of graph $G = F_n \odot K_m$ is $\delta(F_n \odot K_m) = m$. To find the exact value of r-dynamic chromatic number of $G = F_n \odot K_m$, we define two cases, namely for $\chi_{1\le r\le m}(F_n \odot K_m)$ and $\chi_{r\ge m+n+1}(F_n \odot K_m)$.
Case 1. For $1 \le r \le m$, based on Lemma 1.2 with $\chi_r(K_m) = m$ then we have the lower bound of the $r-$dynamic chromatic number of $F_n \odot K_m$ is $\chi_{1\le r\le m}(F_n \odot K_m) \ge \chi_r(K_m) + 1 = m+1$.

We will show that the upper bound of the $r-$dynamic chromatic number of $G = F_n \odot K_m$ is $\chi_{1\leq r\leq m}(F_n \odot K_m) \leq m+1$, define a map of the r-dynamic coloring of $G = F_n \odot K_m$ is $c_7 : V(F_n \odot K_m) \to \{1, 2, \ldots, k\}$ where $n \geq 3, m \geq 4$, by the following:

$$c_7(x) = 1$$

$$c_7(x_i) = \begin{cases} 2, & \text{for } i \text{ odd, } 1 \leq i \leq n \\ 3, & \text{for } i \text{ even, } 1 \leq i \leq n \end{cases}$$

$$c_7(x_i) = \begin{cases} 1, & \text{for } j = 1,\, 1 \leq i \leq n \\ 2, & \text{for } j = 2,\, i \text{ even, } 1 \leq i \leq n \\ 3, & \text{for } j = 2,\, i \text{ odd, } 1 \leq i \leq n \\ 1+j, & \text{for } 3 \leq j \leq m, 1 \leq i \leq n \end{cases}$$

$$c_7(y_j) = j+1, 2 \leq i \leq m$$

It easy to see that c_7 is a map $c_7 : V(F_n \odot K_m) \to \{1, 2, \cdots m+1\}$. It is clearly that c_7 gives the upper bound of the $r-$dynamic chromatic number of $G = F_n \odot K_m$ is $\chi_{1\leq r\leq m}(F_n \odot K_m) \leq m+1$. It concludes that $r-$dynamic chromatic number of $G = F_n \odot K_m$ is $\chi_{1\leq r\leq m}(F_n \odot K_m) = m+1$.

Case 2. For $r \geq m+n+1$, based on Lemma 1.2 with $\Delta(F_n) = n+1$ then we have the lower bound of the $r-$dynamic chromatic number of $G = F_n \odot K_m$ is $\chi_{r\geq m+n+1}(F_n \odot K_m) \geq \Delta(F_n) + m + 1 = n + m + 2$. We will show that the upper bound of the $r-$dynamic chromatic number of $G = F_n \odot K_m$ is $\chi_{r\geq m+n+1}(F_n \odot K_m) \leq m+n+1$, define a map of the r-dynamic coloring of $G = F_n \odot K_m$ is $c_8 : V(F_n \odot K_m) \to \{1, 2, \ldots, k\}$ where $n \geq 3, m \geq 4$, by the following:

$$c_8(x) = 1$$
$$c_8(x_i) = 1 + i, 1 \leq i \leq n$$
$$c_8(x_{ij}) = 1 + n + j, 1 \leq i \leq n, 1 \leq j \leq m$$
$$c_8(y_j) = 1 + n + j, 1 \leq j \leq m$$

It easy to see that c_8 is a map $c_8 : V(F_n \odot K_m) \to \{1, 2, \cdots m+n+1\}$. It is clearly that c_6 gives the upper bound of the $r-$dynamic chromatic number of $G = F_n \odot K_m$ is $\chi_{r\geq m+n+1}(F_n \odot K_m) \leq m+n+1$. It concludes that $r-$dynamic chromatic number of $G = F_n \odot K_m$ is $\chi_{r\geq m+n+1}(F_n \odot K_m) = m+n+1$. It concludes the proof. □

Theorem 1.5 *Let $G = K_n \odot K_m$ be a corona graph of K_n and K_m, for $m, n \geq 4$, the r-dynamic chromatic number is:*

i. For $n \leq m$

$$\chi_r(G) = \begin{cases} m+1, & 1 \leq r \leq m \\ r+1, & m+1 \leq r \leq m+n-1 \\ m+n, & r \geq m+n \end{cases}$$

ii. For $n > m$

$$\chi_r(G) = \begin{cases} n, & 1 \leq r \leq n-1 \\ r+1, & n \leq r \leq m+n-1 \\ m+n, & r \geq m+n \end{cases}$$

Proof. The graph $G = K_n \odot K_m$ is a connected graph with vertex set $V(K_n \odot K_m) = \{x_i; 1 \leq i \leq n\} \cup \{x_{ij}; 1 \leq i \leq n\ 1 \leq j \leq m\}$. The order of this graph is $p = |V(K_n \odot K_m)| = n(m+1)$. Thus the maximum degree of graph $G = K_n \odot K_m$ is $\Delta(K_n \odot K_m) = m+n-1$ and the minimum degree of graph $G = K_n \odot K_m$ is $\delta(K_n \odot K_m) = m$. To find the exact value of r-dynamic chromatic number of $K_n \odot K_m$, we define two cases, namely for $\chi_{1\leq r\leq m}(K_n \odot K_m)$ and $\chi_{r\geq m+n}(K_n \odot K_m)$.

Case 1. We take for $n \leq m$. For $1 \leq r \leq m$, based on Lemma 1.2 and Observation 1.2, $\chi_r(K_m) = m$ then we have the lower bound of the $r-$dynamic chromatic number of $G = K_n \odot K_m$ is $\chi_{1\leq r\leq m}(K_n \odot K_m) \geq \chi_r(K_m) + 1 = m + 1$. We will show that the upper bound of the $r-$dynamic chromatic number of $G = K_n \odot K_m$ is $\chi_{1\leq r\leq m}(K_n \odot K_m) \leq m + 1$, define a map of the r-dynamic coloring of $G = K_n \odot K_m$ is $c_9 : V(K_n \odot K_m) \to \{1, 2, \ldots, k\}$ where $m, n \geq 4$, by the following:

$$c_9(x_i) = i, 1 \leq i \leq n$$

$$c_1(x_{ij}) = \begin{cases} i+1, & \text{for } j = 1, 1 \leq i \leq n-1 \\ 1, & \text{for } j = 1, i = n \\ j+1, & \text{for } 1 \leq i \leq n, 2 \leq j \leq m \end{cases}$$

It easy to see that c_9 is a map $c_9 : V(K_n \odot K_m) \to \{1, 2, \cdots m+1\}$. It is clearly that c_9 gives the upper bound of the $r-$dynamic chromatic number of $G = K_n \odot K_m$ is $\chi_{1\leq r\leq m}(K_n \odot K_m) \leq m+1$. It concludes that $r-$dynamic chromatic number of $G = K_n \odot K_m$ is $\chi_{1\leq r\leq m}(K_n \odot K_m) = m+1$.
Case 2. For $r \geq m+n$, based on Lemma 1.1 with $\Delta(K_n) = n-1$ then we have the lower bound of the $r-$dynamic chromatic number of $G = K_n \odot K_m$ is $\chi_{r\geq m+n}(K_n \odot K_m) \geq \Delta(K_n) + m + 1 = (n-1) + m + 1 = m + n$. We will show that the upper bound of the $r-$dynamic chromatic number of $G = K_n \odot K_m$ is $\chi_{r\geq m+n}(K_n \odot K_m) \leq m + n$, define a map of the r-dynamic coloring of $G = K_n \odot K_m$ is $c_{10} : V(K_n \odot K_m) \to \{1, 2, \ldots, k\}$ where $m, n \geq 4$, by the following:

$$c_{10}(x_i) = i, 1 \leq i \leq n$$
$$c_{10}(x_{ij}) = i + j, 1 \leq i \leq n, 1 \leq j \leq m$$

It easy to see that c_{10} is a map $c_{10} : V(K_n \odot K_m) \to \{1, 2, \cdots n + m\}$. It is clearly that c_{10} gives the upper bound of the $r-$dynamic chromatic number of $G = K_n \odot K_m$ is $\chi_{r\geq m+n}(K_n \odot K_m) \leq m + n$. It concludes that $r-$dynamic chromatic number of $G = K_n \odot K_m$ is $\chi_{r\geq m+n}(K_n \odot K_m) = m + n$. It concludes the proof. □

Conclusion

We have found the lower bound of $\chi_r(K_n \odot H)$ and $\chi_r(H \odot K_m)$ where $H \neq C_m, W_m$. We also find the exact value some r-dynamic chromatic number of corronation by complete graphs, namely $\chi_r(K_n \odot S_m), \chi_r(K_n \odot F_m), \chi_r(S_n \odot K_m), \chi_r(F_n \odot K_m)$ and $\chi_r(K_n \odot K_m)$. We got $\chi_r(K_n \odot S_m) = \chi_r(K_n \odot F_m)$ and $\chi_r(S_n \odot K_m) = \chi_r(F_n \odot K_m)$. For the characterization of the lower bound of $\chi_r(G \odot H)$ for any connected graphs G and H, we have not found any result yet, thus we propose the following open problem.

Open Problem 1.1 *Given the any connected graphs G and H. Determine the sharp lower bound of $\chi_r(G \odot H)$.*

Acknowledgement

We gratefully acknowledge the support from CGANT - University of Jember of year 2017.

References

[1] Agustin I H, Dafik, Harsya A Y 2016 On r-Dynamic Coloring of Some Graph Operations. *International Journal of Combinatorics* **02(01)** 22-30

[2] S Akbari, M Ghanbari and S Jahanbekam 2014 On the dynamic coloring of cartesian product graphs, *Ars Combinatoria* **114** 161-167

[3] M Alishahi 2011 On the dynamic coloring of graphs *Discrete Appl. Math.* **159** 152-156.

[4] Dafik, Meganingtyas D E W, Purnomo K D, Tarmidzi M D, and Agustin I H 2017 Several classes of graphs and their r-dynamic chromatic numbers, *IOP Conf. Series: Journal of Physics: Conf. Series* 855 012011

[5] H Furmanczyk, M Kubale 2016 Equtable coloring of corona products of cubic graphs is harder than ordinary coloring, *Ars Mathematica Contemporanea* **10** 333-347

[6] S Jahanbekama, J Kimb, Suil O, Douglas B West 2016 On r-dynamic Coloring of Graphs *Discrete Applied Mathematics* **206** 65-72

[7] A I Kristiana 2017 On r-dynamic Chromatic Number of the corronation of path and several graphs, *International Journal of Advanced Engineering, Research and Science*, **04: 04** 96-101

[8] H J Lai, B Montgomery 2002 *Dynamic coloring of graph* Departement of Mathematics West Virginia University

[9] H J Lai, B Montgomery, H Poon 2003 Upper bounds of dynamic chromatic number, *Ars Combin.* **68** 193201

[10] B Montgomery 2001 Dynamic coloring of graphs (Ph.D Dissertation), West Virginia University

[11] Pathinahan T, Arokia Mary A, Bhuvaneswari D 2014 b-Chromatic Number of Subdivision Edge and Vertex Corona. *International Journal of Computing Algorithm.* **03** 516-520

[12] Ramya N 2014 On Coloring of Corona Graphs. *Indian Journal of Science and Technology.* **7** (3S) 9-11

Restricted Size Ramsey Number for $2K_2$ versus Dense Connected Graphs of Order Six

Denny Riama Silaban[1,2], **Edy Tri Baskoro**[1], **Saladin Uttunggadewa**[1]

[1]Combinatorial Mathematics Research Group, Faculty of Mathematics and Natural Sciences, Institut Teknologi Bandung, Jalan Ganesa 10 Bandung 40132 Indonesia

[2]Department of Mathematics, Faculty of Mathematics and Natural Sciences, Universitas Indonesia

E-mail: denny-rs@students.itb.ac.id, ebaskoro@math.itb.ac.id, s_uttunggadewa@math.itb.ac.id

Abstract. Let G and H be simple graphs. The Ramsey number $r(G,H)$ for a pair of graphs G and H is the smallest number r such that any red-blue coloring of the edges of K_r contains a red subgraph G or a blue subgraph H. The size Ramsey number $\hat{r}(G,H)$ for a pair of graphs G and H is the smallest number $\hat{r}$ such that there exists a graph F with size $\hat{r}$ satisfying the property that any red-blue coloring of the edges of F contains a red subgraph G or a blue subgraph H. Additionally, if the order of F in the size Ramsey number equals $r(G,H)$, then it is called the restricted size Ramsey number. In 1972, Chvátal and Harary gave the Ramsey number for $2K_2$ versus any graph H with no isolates. In 1983, Harary and Miller started the investigation of the (restricted) size Ramsey number for some pairs of small graphs with order at most four. In 1983, Faudree and Sheehan continued the investigation and summarized the complete results on the (restricted) size Ramsey number for all pairs of small graphs with order at most four. In 1998, Lortz and Mengenser gave both the size Ramsey number and the restricted size Ramsey number for all pairs of small forests with order at most five. Lately, we investigate the restricted size Ramsey number for $2K_2$ versus all connected graphs of order five. In this work, we continue the investigation on the restricted size Ramsey number for a pair of small graphs. In particularly, for $2K_2$ versus dense connected graph of order six.

1. Introduction

A graph G has the vertex set $V(G)$ and the edge set $E(G)$. The number of vertices and edges in G denoted by $v(G)$ and $e(G)$ respectively. Let G and H be graphs. If $H \subseteq G$, then $G - H$ is a graph with the vertex set $V(G)$ and the edge set $E(G) \setminus E(H)$ and if $H \subseteq \overline{G}$, then $G + H$ is a graph with the vertex set $V(G)$ and the edge set $E(G) \cup E(H)$. For further terminologies in graph, please see [3]. For a pair of graphs G and H, the *Ramsey number* $r(G,H)$ is the smallest number r such that any red-blue coloring of the edges of K_r contains a red subgraph G or a blue subgraph H. The *size Ramsey number* $\hat{r}(G,H)$ is the smallest size of F satisfying the property that any red-blue coloring of the edges of F contains a red subgraph G or a blue subgraph H. Furthermore, if the order of F is $r(G,H)$, then it is called the *restricted size Ramsey number*, $r^*(G,H)$. In addition, if any red-blue coloring of the edges of F contains a red subgraph G or a blue subgraph H, we say F *arrowing* G and H, denoted by $F \rightarrow (G,H)$. A red-blue coloring of F is called $(G,H)-$ *good* if F does not contain a red G or a blue H in this coloring. Then, $F \nrightarrow (G,H)$ if there is a red-blue coloring of F that $(G,H)-$good.

IOP Conf. Series: Journal of Physics: Conf. Series **1008** (2018) 012034 doi:10.1088/1742-6596/1008/1/012034

The concept of the size Ramsey number was introduced by Erdös *et al.* in 1978 [4], while the concept of the restricted size Ramsey number is a direct consequence of the concept of the size Ramsey number and the Ramsey number in graph. Some results on the (restricted) size Ramsey number of graph can be found in the survey of noncomplete Ramsey theory for graph given by Burr [1] and a survey of results on the size Ramsey number given by Faudree and Schelp [6].

In 1983, Harary and Miller [8] started to find the (restricted) size Ramsey number for any pair of small graphs with order at most four. Faudree and Sheehan [7] continued Harary and Miller's works and summarized the complete results on the (restricted) size Ramsey number for any pair of small graphs with order at most four. In 1998, Lortz and Mengenser [9] gave both the size Ramsey number and the restricted size Ramsey number for any pair of small forests with order at most five. Recently, we gave the restricted size Ramsey number for a P_3 versus any connected graph of order five, a P_3 versus a dense connected graph of order six, and a $2K_2$ versus any graph with no isolates of order five [11, 12, 14]. To proceed the research on the (restricted) size Ramsey number involving small graphs, we investigate the restricted size Ramsey number for $2K_2$ versus dense connected graphs of order six.

2. Peliminary Results

The are 112 non-isomorphic connected graph of order six. We generate the complete list of it using Mathematica software. The restricted size Ramsey number of $2K_2$ versus 12 of these graphs already known from (or as an implication of) previous results, as given in Figure 1. Most of these results are from our work in [13] except $r^*(P_3, H_1)$ is from [5].

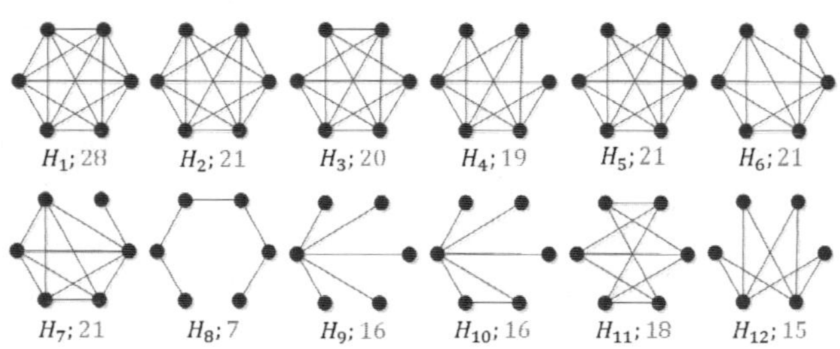

Figure 1: The known values of $r^*(2K_2, H)$, H connected graph of order six

In 1972, Chvátal and Harary [2] gave the Ramsey number for $2K_2$ and any graph with no isolates, as stated in Theorem 2.1. This theorem gives the order of the arrowing graph F in finding $r^*(2K_2, H)$.

Theorem 2.1. *[2] For any graph H with no isolates,*

$$r(2K_2, H) = \begin{cases} v(H) + 2, & H \text{ is complete}, \\ v(H) + 1, & \text{otherwise}. \end{cases}$$

From the definition of the (restricted) size Ramsey number of a pair of graphs we have a monotonicity property as follows. If $G' \subseteq G$ and $H' \subseteq H$, then

$$\hat{r}(G', H') \leq \hat{r}(G, H). \tag{1}$$

and

$$r^*(G', H') \leq r^*(G, H). \tag{2}$$

These inequalities are variants of the ones for Ramsey number of graphs given by Chvátal and Harary [2].

3. Main Results

In this section we present the restricted size Ramsey number of $2K_2$ versus a dense connected graph of order six. Lemmas 3.1 and 3.2 give the properties of graph F such that $F \to (2K_2, H)$ for any graph H with no isolates. Lemma 3.1 is a generalization of lemma given in [10] for $H = K_{1,n}$. Lemma 3.3 give the properties of F such that $F \to (2K_2, H)$ when graph H contains a clique of order four. We will use all these lemmas in proving our theorems.

Lemma 3.1. *[13] Let H be a graph.* $F \to (2K_2, H)$ *holds if and only if the following conditions are satisfied:*

(i) $H \subseteq F - v$ *for every* $v \in V(F)$ *and*

(ii) $H \subseteq F - C_3$ *for every* C_3 *in* F.

Lemma 3.2. *[14] Let H be a graph with no isolates. If* $F \to (2K_2, H)$ *and* $v(F) = r(2K_2, H)$, *then* $\delta(F) \geq 2$.

Lemma 3.3. *All graphs F with* $v(F) = 7$ *and* $e(F) = 18$ *except* $K_7 - K_{1,3}$ *satisfy* $F \nrightarrow (2K_2, K_4)$.

Proof. There are five graphs F with $v(F) = 7$ and $e(F) = 18$, namely $K_7 - K_{1,3}$, $K_7 - C_3$, $K_7 - P4$, $K_7 - (P_3 \cup K_2)$, and $K_7 - 3K_2$. In Figure 2 we give $K_7 - C_3$, $K_7 - P_4$, $K_7 - (P_3 \cup K_2)$, and $K_7 - 3K_2$ with respective $(2K_2, K_4)$−good coloring (red color is in dotted line). □

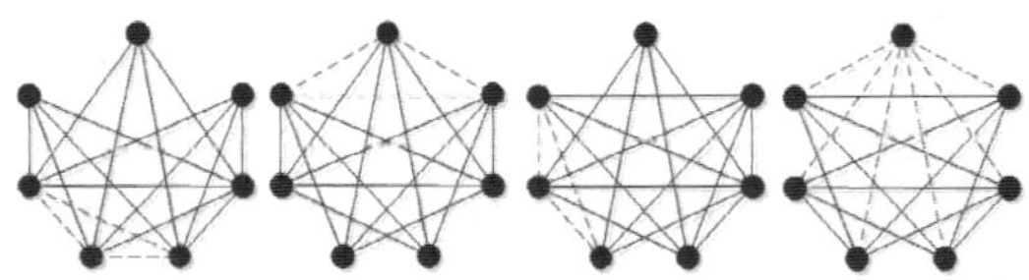

Figure 2: $(2K_2, K_4)$−good coloring of $K_7 - C_3$, $K_7 - P_4$, $K_7 - (P_3 \cup K_2)$, and $K_7 - 3K_2$

Theorem 3.4. *Let s and t be integers with* $2 \leq s \leq 5$ *and* $3 \leq t \leq 5$. $r^*(2K_2, K_6 - (K_{1,s} \cup K_{1,t})) = 18$.

Proof. From Theorem 2.1 we have $r(2K_2, K_6 - (K_{1,s} \cup K_{1,t})) = 7$ for $2 \leq s \leq 5$ and $3 \leq t \leq 5$. Note that $K_4 \subseteq K_6 - (K_{1,s} \cup K_{1,t})$ for $2 \leq s \leq 5$ and $3 \leq t \leq 5$. Now, consider $F = K_7 - K_{1,3}$. Let $u \in V(F)$ be a vertex of order six. It is clear that $K_6 - (K_{1,s} \cup K_{1,t}) \subseteq F - u \subseteq F - v$ for every vertex $v \in V(F)$. Furthermore, $K_6 - C_3 \subseteq F - C_3$ for every triangle C_3 in F. Since $K_6 - (K_{1,s} \cup K_{1,t}) \subseteq K_6 - C_3$ for $2 \leq s \leq 5$ and $3 \leq t \leq 5$, Lemma 3.1 implies $F \to (2K_2, K_6 - (K_{1,s} \cup K_{1,t}))$. As a consequence $r^*(2K_2, K_6 - (K_{1,s} \cup K_{1,t})) \leq 18$ for $2 \leq s \leq 5$ and $3 \leq t \leq 5$.

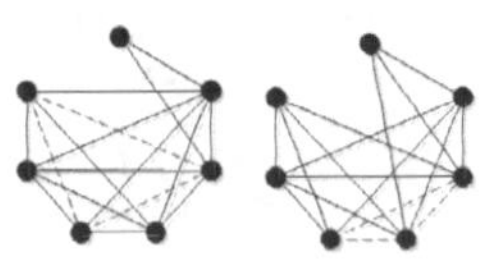

Figure 3: $K_7 - K_{1,4}$ and $K_7 - S_{3,1}$ with corresponding $(2K_2, K_4)$−good coloring

Now, we will consider all graphs F with $v(F) = 7$ and $e(F) = 17$. Using Lemma 3.3 and (2), we need to consider only two graphs, namely $K_7 - K_{1,4}$ and $K_7 - S_{3,1}$. In Figure 3 we give both graphs with corresponding $(2K_2, K_4)$−good coloring (red color is in dotted line). Since $K_4 \subseteq K_6 - (K_{1,s} \cup K_{1,t})$, by (2) we have $F \nrightarrow (2K_2, K_6 - (K_{1,s} \cup K_{1,t}))$. As a consequence $r^*(2K_2, K_6 - (K_{1,s} \cup K_{1,t})) \geq 18$ for $2 \leq s \leq 5$ and $3 \leq t \leq 5$. □

Theorem 3.5. *Let* $H_{a_1} = K_6 - (P_3 \cup K_2)$, $H_{a_2} = K_6 - 2P_3$, $H_{a_3} = K_6 - P_4$, *and* $H_{a_4} = K_6 - C_4$. $r^*(2K_2, H_{a_i}) = 20$ *for* $1 \leq i \leq 4$.

Proof. From Theorem 2.1 we have $r(2K_2, H_{a_i}) = 7$. Since $H_{a_i} \subseteq KH_3$ and $r^*(2K_2, H_3) = 20$ (see Figure 1), by (2) we have $r^*(2K_2, H_{a_i}) \leq 20$ for $1 \leq i \leq 4$.

Now, we will consider all graphs F with $v(F) = 7$ and $e(F) = 19$. There are only two graphs satisfying, namely $F = K_7 - P_3$ or $K_7 - 2K_2$. If $F = K_7 - P_3$, then the maximal graph of order six in $F - C_3'$ with C_3' is a triangle induced by a vertex of degree five and two vertices of degree six is $K_6 - C_3$. If $K_7 - 2K_2$, then the maximal graph of order six in $F - C_3''$ with C_3'' is a triangle induced by three vertices of degrees six is $K_6 - 3K_2$. Since $H_{a_i} \nsubseteq K_6 - C_3$ and $H_{a_i} \nsubseteq K_6 - 3K_2$, $F \nrightarrow (2K_2, H_{a_i})$ for both F. As a consequence $r^*(2K_2, H_{a_i}) \geq 20$ for $1 \leq i \leq 4$. □

Theorem 3.6. *Let* $H_{a_1} = K_{3,3} + K_2$ *and* $H_{a_2} = K_{3,3} + P_3$. $r^*(2K_2, H_{a_i}) = 18$ *for* $1 \leq i \leq 2$.

Proof. From Theorem 2.1 we have $r(2K_2, H_{a_i}) = 7$. Since $H_{11} \cong K_{3,3} \subseteq H_{a_1} \subseteq H_{a_2}$ and $r^*(2K_2, H_{11}) = 18$ (see Figure 1), by (2) we have $r^*(2K_2, H_{a_i}) \geq 18$ for $1 \leq i \leq 2$.

Now, let $F = K_7 - (P_3 \cup K_2)$. Let $u \in V(F)$ be a vertex of degree six. It is clear that $H_{a_i} \subseteq F - u \subseteq F - v$ for any $v \in V(F)$. Furthermore, in F there are four kind of triangles, namely a triangle induced by a vertex of degree four, five, and six respectively (denoted by C_3^1), a triangle induced by a vertex of degree four and two vertices of degrees six (denoted by C_3^2), a triangle induced by a vertex of degree six and two vertices of degrees five (denoted by C_3^3), a triangle induced by a vertex of degree five and two vertices of degrees six (denoted by C_3^4). It is clear that $K_6 - (P_4 + K_2) \cong F - C_3^2 \cong F - C_3^3 \subseteq F - C_3^1$, and $K_6 - (C_3 \cup K_2) \cong F - C_3^4$. Since $H_{a_2} \subseteq K_6 - (P_4 + K_2)$ and $H_{a_2} \subseteq K_6 - (C_3 \cup K_2)$, we have $F \rightarrow (2K_2, H_{a_2})$ and $r^*(2K_2, H_{a_2}) \leq 18$. Since $H_{a_1} \subseteq H_{a_2}$, by (1) we have $r^*(2K_2, H_{a_1}) \leq 18$. □

Theorem 3.7. *Let* $K_{1,5} \subseteq H \subseteq K_{1,5} + (K_4 - P_3)$. $r^*(2K_2, H) = 16$.

Proof. Since $H_9 \cong K_{1,5}$ and $r^*(2K_2, K_9) = 16$ (see Figure 1), by (2) we have $r^*(2K_2, H) \geq 16$. From Theorem 2.1 we have $r(2K_2, H) = 7$. Let $F = K_7 - K_{1,4} + K_2$ and let $u \in V(F)$ be a vertex of degree six. It is clear that $K_{1,5} + (K_4 - P_3) \subseteq F - u \subseteq F - v$ for any $v \in V(F)$. Furthermore, in F there are three kind of triangles, namely a triangle induced by a vertex of degree four and two vertices of degrees six (denoted by C_3^1), a triangle induced by three vertices of degrees six (denoted by C_3^2), and a triangle induced by a vertex of degree two and two vertices of degrees six (denoted by C_3^3). By observing that $K_{1,5} + (K_4 - P_3) \subseteq F - C_3^1 \subseteq F - C_3^2 \subseteq F - C_3^3$ we have $F \rightarrow (2K_2, K_{1,5} + (K_4 - P_3))$ and $r^*(2K_2, K_{1,5} + (K_4 - P_3)) \leq 16$. Since $H \subseteq K_{1,5} + (K_4 - P_3)$, by (2) we have $r^*(2K_2, H) \leq 16$. □

Theorem 3.8. $r^*(2K_2, K_{2,4} + K_2) = 15$.

Proof. Since $H_{12} \cong K_{2,4}$ and $r^*(2K_2, H_{12}) = 15$ (see Figure 1), by (2) we have $r^*(2K_2, K_{2,4} + K_2) \geq 15$. From Theorem 2.1 we have $r(2K_2, K_{2,4} + K_2) = 7$. Let $F = K_7 - C_4$. For a vertex u of degree six it is clear that $K_{2,4} + K_2 \subseteq F - u \subseteq F - v$ for any $v \in V(F)$. Furthermore, in F there are two kind of triangles, namely a triangle induced by a vertex of degree four and two vertices of degrees six (denoted by C_3^1) and a triangle induced by three vertices of degrees six (denoted by C_3^2). It is clear that $K_{2,4} + K_2 \subseteq F - C_3^2 \subseteq F - C_3^1$. Hence $F \rightarrow (2K_2, K_{2,4} + K_2)$. As a consequence $r^*(2K_2, K_{2,4} + K_2) \leq 15$. □

We summarize the new results for the restricted size Ramsey number of $2K_2$ versus dense connected graph of order six in Figure 4.

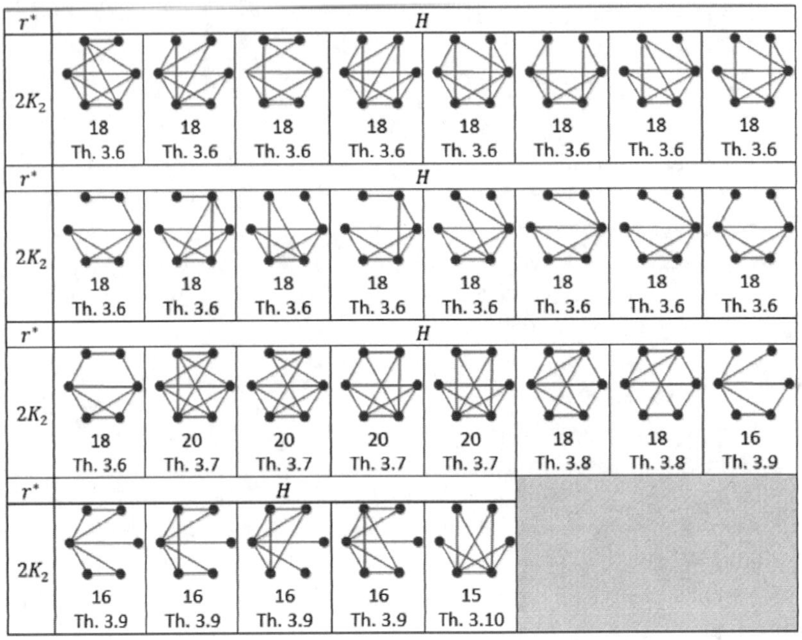

Figure 4: New results for $r^*(2K_2, H)$ with H is a dense graph of order six

4. Conclusion

The are 112 non-isomorphic connected graphs of order six. The restricted size Ramsey numbers of $2K_2$ versus each of twelve from these graphs already known. In this paper, the restricted size Ramsey numbers of $2K_2$ versus each of 29 other graphs are given. The restricted size Ramsey numbers of $2K_2$ versus the remaining 71 graphs are still open for further research.

Acknowledgement

This research was supported by the Research Grant "Program Penelitian Unggulan Perguruan Tinggi" and "Penelitian Disertasi Doktor 2017" Ministry of Research, Technology, and Higher Education, Indonesia.

References

[1] Burr S A 1979 A survey of noncomplete Ramsey theory for graphs *Ann. New york Acad. Sci.* **328** 58–75
[2] Chvátal V and Harary F 1972 Generalized Ramsey theory for graphs III: small off-diagonal number *Pacific Journal of Mathematics* **41** (2) 335–345
[3] Diestel R 2005 *Graph Theory* Springer-Verlag Heidelberg New York 4 edition
[4] Erdös P, Faudree R J, Rousseau C C and Schelp R H 1978 The size Ramsey number *Periodica Mathematica Hungarica* **9** (1-2) 145–161
[5] Erdös P and Faudree R J 1984 Size Ramsey number involving matchings, in: Sets, graphs, and numbers. Colloq. Math. Soc. Janos Bolyai **37** 247-264
[6] Faudree R J and Schelp R 2002 A survey of results on the size Ramsey numbers *Paul Erdös and His Mathematics* II **10** 291–309
[7] Faudree R J and Sheehan J 1983 Size Ramsey numbers for small-order graphs *J. Graph Theory* **7** 53–55
[8] Harary F and Miller Z 1983 Generalized Ramsey theory VIII: the size Ramsey number of small graphs *Studies in Pure Mathetematics - To the Memory of Paul Turán* Birkhaüser 271–283
[9] Lortz R and Mengersen I 1998 Size Ramsey results for paths versus stars *Australas. J. Combin.* **18** 3–12
[10] Mengenser I and Oeckermann J 1999 Matching-star Ramsey sets Discrete Applied Math. **95** 417-424
[11] Silaban D R, Baskoro E T and Uttunggadewa S 2017 On the restricted size Ramsey number for P_3 versus connected graph of order five Electronic Journal of Combinatoric and Applications **5** (2) 155-162

[12] Silaban D R, Baskoro E T and Uttunggadewa S 2017 Restricted size Ramsey number for P_3 versus small paths In AIP Conference Proceedings 1862
[13] Silaban D R, Baskoro E T and Uttunggadewa S 2018+ On the restricted size Ramsey number involving matchings Submitted
[14] Silaban D R, Baskoro E T and Uttunggadewa S 2018+ Restricted Size Ramsey Number Involving Matching and Graph of Order Five Submitted

IOP Conf. Series: Journal of Physics: Conf. Series **1008** (2018) 012035 doi:10.1088/1742-6596/1008/1/012035

On the local vertex antimagic total coloring of some families tree

Desi Febriani Putri[1,2], **Dafik**[1,3], **Ika Hesti Agustin**[1,2], **Ridho Alfarisi**[1,4]

[1]CGANT-University of Jember, Jember, Indonesia
[3]Department of Mathematics, University of Jember, Jember, Indonesia
[2]Department of Mathematics Education, University of Jember, Jember, Indonesia
[4]Department of Elementary School Teacher Education, University of Jember, Jember, Indonesia

E-mail: d.dafik@unej.ac.id, ikahestiagustin@gmail.com, desifebriani.putri@yahoo.com, alfarisi.fkip@unej.ac.id

Abstract. Let $G(V, E)$ be a graph of vertex set V and edge set E. Local vertex antimagic total coloring developed from local edge and local vertex antimagic coloring of graph. Local vertex antimagic total coloring is defined $f : V(G) \cup E(G) \longrightarrow \{1, 2, 3..., |V(G)| + |E(G)|\}$ if for any two adjacent vertices v_1 and v_2, $w(v_1) \neq w(v_2)$, where for $v \in G$, $w(v) = \sum_{e \in E(v)} f(e) + f(v)$, where $E(v)$ and $V(v)$ are respectively the set of edges incident to v and the set of vertices adjacent to v. Thus, any local vertex antimagic total coloring induces a proper vertex coloring of G if each vertex v is assigned the color $w(v)$. The chromatic number of local vertex antimagic total coloring denote $\chi_{lvat}(G)$ is the minimum number of colors taken over all colorings induced by local vertex antimagic total coloring of graph G. In this paper, we use some families tree graph. We also study the existence of local vertex antimagic total coloring chromatic number of some families tree namely star graph, double star graph, banana tree graph, centipede graph, and amalgamation of star graph.

Keywords: Local antimagic vertex total coloring, chromatic number, some families tree

1. Introduction

All graphs in this paper are finite, simple and undirected. A graph G has vertex set V and edge set E is denoted by $G = (V, E)$. The vertex set and edge set of G are often denoted by $V(G)$ and $E(G)$, respectively. The order of graph G is notated $|V(G)|$ and the size of graph G is notated $|E(G)|$. We refer the reader to [4], [9], [11] and [12] for all other terms and notation not provided in this paper. For graph theoretic terminology we refer to Chartrand and Lesniak [3].

Harsfield and Ringel [10] introduced the concept of an antimagic labeling. By a labeling of a graph, we mean any mapping that sends some set of graph elements to a set of positive integers. If the domain is a vertex set V(G) or an edge set E(G), the labelings are called respectively vertex labelings or edge labelings. If the domain is the set of all vertices and edges then the labeling is called a total labeling [5, 6, 7]. A general survey of graph labelings is found in [8].

Arumugam *et al.* [2] introduced the concept of local antimagic vertex coloring of graph G. Let $G = (V, E)$ be a graph of order n and size m having no isolated vertices. A bijection $f : E \longrightarrow \{1, 2, ..., m\}$ is called a local antimagic labeling it for all $uv \in E$ we have $w(u) \neq w(v)$ where $w(u) = \sum_{e \in E(u)} f(e)$. A graph G is local antimagic if G has a local antimagic labeling. The local antimagic chromatic number of graph G denoted by $\chi_{la}(G)$. [2].

Local edge antimagic coloring developed from local antimagic vertex coloring of graph G. Agustin *et al.* [1] studied the existence of local edge antimagic coloring of some special graphs. By local edge antimagic coloring, we mean a bijection $f : V(G) \longrightarrow \{1, 2, ..., |V(G)|\}$ is called a local edge antimagic coloring if for any two incident edges at the same vertices e_1 and e_2, $w(e_1) \neq w(e_2)$, where for $e = uv \in G, w(e) = f(u) + f(v)$. Thus, any local edge antimagic labeling induces a proper edge coloring of G if each edge e is assigned the color $w(e)$. The local edge antimagic chromatic number of graph G denoted by $\gamma_{lea}(G)$[1].

In this paper, we initiate to study a different type local antimagic coloring namely local vertex antimagic total coloring. A bijection of $f : V(G) \cup E(G) \longrightarrow \{1, 2, 3..., |V(G)| + |E(G)|\}$ is called local vertex antimagic total coloring if for any two adjacent vertices v_1 and v_2, $w(v_1) \neq w(v_2)$, where for $v \in G$, $w(v) = \sum_{e \in E(v)} f(e) + f(v)$, where $E(v)$ and $V(v)$ are respectively the set of edges incident to v and the set of vertices adjacent to v. Thus, any local vertex antimagic total coloring induces a proper vertex coloring of G if each vertex v is assigned the color $w(v)$. The chromatic number of local vertex antimagic total coloring denote $\chi_{lvat}(G)$ is the minimum number of colors taken over all colorings induced by local vertex antimagic total coloring of graph G.

2. Main Results

In this paper, we have studied the existance of local vertex antimagic total coloring of some families tree and amalgamation of path and star. We have found the local vertex antimagic total chromatic number of star graph S_n, double star graph $S_{n,m}$, banana tree graph $B_{m,n}$, centipede graph C_n, and amalgamation of star graph $Amal(S_n, v, m)$.

Observation 2.1. *If $\chi(G)$ is chromatic number vertex coloring then $\chi_{lvat}(G) \geq \chi(G)$.*

Proof: Based on the concept of local vertex antimagic total coloring that its coloring is proper coloring that is for every two vertex adjacent must have different colors or weights. Chromatic numbers are the number of different minimum color or weights variations. It depends not only on chromatic numbers. Thus the local vertex antimagic total chromatic number is obtained with more conditions than vertex chromatic number. Thus, $\chi_{lvat}(G) \geq \chi(G)$. □

Theorem 2.1. *For the graph star S_n with $n \geq 2$, the local vertex antimagic total chromatic number of S_n is $\chi_{lvat}(S_n) = 2$.*

IOP Conf. Series: Journal of Physics: Conf. Series **1008** (2018) 012035 doi:10.1088/1742-6596/1008/1/012035

Proof: The graph S_n is a connected graph with vertex set $V(S_n) = \{x\} \cup \{x_i, 1 \leq i \leq n\}$ and edge set $E(S_n) = \{xx_i, 1 \leq i \leq n\}$. Hence $|V(S_n)| = n+1$ and $|E(S_n)| = n$. Define a bijection $f : V(G) \cup E(G) \longrightarrow \{1, 2, ..., |V(G)| + |E(G)|\}$ by the following:

$$f(v) = \begin{cases} 2n+1, & v = x \\ 2n-i+1, & v = x_i, \ 1 \leq i \leq n \end{cases}$$

$$f(e) = i, \quad e = xx_i, \ 1 \leq i \leq n$$

Clearly f is a local vertex antimagic total labeling of S_n and we formulate the vertex weights as follows:

$$w_t(v) = \begin{cases} 2n+1, & v = x_i, \ 1 \leq i \leq n \\ \frac{n}{2}(1+n) + 2n + 1, & v = x \end{cases}$$

Hence, from the above the vertex weight, it is easy to see that f induces a proper local vertex antimagic total coloring of S_n and it gives $\chi_{lvat}(S_n) \leq 2$. Also it follow from Observation 2.1 that $\chi_{lvat}(S_n) \geq 2$. Hence $\chi_{lvat}(S_n) = 2$. □

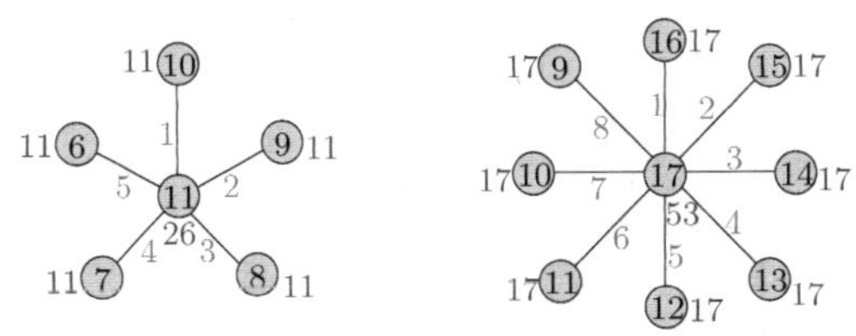

Figure 1. Example of local vertex antimagic total coloring for S_5 and S_8

Theorem 2.2. *For the double star graph $S_{n,m}$ with $n \geq 2$ and $m \geq 2$, the local vertex antimagic total chromatic number of $S_{n,m}$ is $\chi_{lvat}(S_{n,m}) \leq 3$.*

Proof: The graph S_n is a connected graph with vertex set $V(S_{n,m}) = \{x\} \cup \{y\} \cup \{x_i; 1 \leq i \leq n\} \cup \{y_j; 1 \leq j \leq m\}$ and edge set $E(S_{n,m}) = \{xy\} \cup \{xx_i; 1 \leq i \leq n\} \cup \{yy_j; 1 \leq j \leq m\}$. Hence $|V(S_{n,m})| = m+n+2$ and $|E(S_{n,m})| = m+n+1$. Define a bijection $f : V(S_{n,m}) \cup E(S_{n,m}) \longrightarrow \{1, 2, 3..., |V(S_{n,m})| + |E(S_{n,m})|\}$ by the following:

$$f(v) = \begin{cases} 2m+2n+2, & v = x \\ 2m+2n+3, & v = y \\ m+2n-i+2, & v = x_i, \ 1 \leq i \leq n \\ 2m+2n-j+2, & v = y_j, \ 1 \leq j \leq m \end{cases}$$

$$f(e) = \begin{cases} 1, & e = xy \\ m+i+1, & e = xx_i, \ 1 \leq i \leq n \\ j+1, & e = yy_j, \ 1 \leq j \leq m \end{cases}$$

Clearly f is a local vertex antimagic total labeling of $S_{n,m}$ and we formulate the vertex weights as follows:

$$w_t(v) = \begin{cases} 2m + 2n + 3, & v = x_i = y_j,\ 1 \le i \le n,\ 1 \le j \le m \\ 2m + 2n + \frac{n}{2}(2m + n + 3) + 3, & v = x \\ 2m + 2n + \frac{n}{2}(m + 3) + 4, & v = y \end{cases}$$

Hence, from the above the vertex weight, it is easy to see that f induces a proper local vertex antimagic total coloring of $S_{n,m}$ and it gives $\chi_{lvat}(S_{n,m}) \le 3$. □

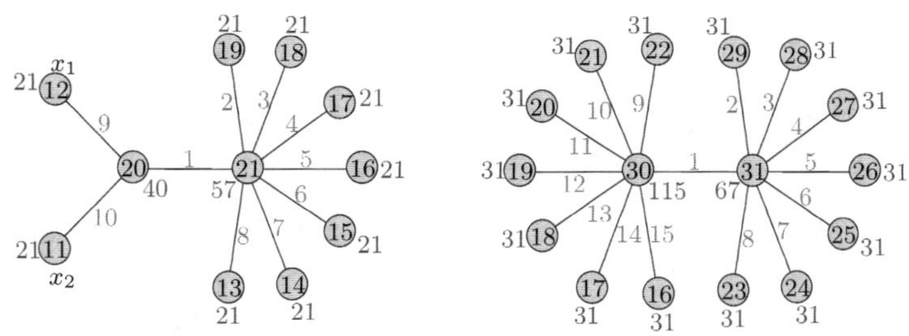

Figure 2. Example of local vertex antimagic total coloring for $S_{2,7}$ and $S_{7,7}$

Theorem 2.3. *For the banana tree graph $B_{m,n}$ with $n \ge 3$ and $m \ge 3$, the local vertex antimagic total chromatic number of $B_{m,n}$ is*

$$\chi_{lvat}(B_{m,n}) \le \begin{cases} 4 & \text{for, } n \text{ odd, } m \text{ odd or} \\ & \text{for, } n \text{ even, } m \text{ odd} \\ 5 & \text{for, } n \text{ odd, } m \text{ even} \\ 6 & \text{for, } n \text{ even, } m \text{ even} \end{cases}$$

.

Proof: The graph $B_{m,n}$ is a connected graph with vertex set $V(B_{m,n}) = \{x\} \cup \{x_i; y_i, 1 \le i \le n\} \cup \{y^i_j, 1 \le j \le m-1,\ \ 1 \le i \le n\}$ and edge set $E(B_{m,n}) = \{xx_i, 1 \le i \le n\} \cup \{x_iy_i, 1 \le i \le n\} \cup \{i_iy^i_j, 1 \le i \le n\ 1 \le j \le m-1\}$. Hence $|V(B_{m,n})| = mn + n + 1$ and $|E(B_{m,n})| = mn + n$.

Case 1. For n odd and m odd, define a bijection $f : V(B_{m,n}) \cup E(B_{m,n}) \longrightarrow \{1, 2, ..., |V(B_{m,n})| + |E(B_{m,n})|\}$ by the following:

$$f(v) = \begin{cases} 2mn + 2m + 1, & v = x \\ 2mn + m + \frac{j+1}{2}, & v = x_j,\ 1 \le j \le m,\ j \text{ odd} \\ 2mn + m + \frac{m+j+1}{2}, & v = x_j,\ 2 \le j \le m-1,\ j \text{ even} \\ 2mn + j, & v = y_j,\ 1 \le j \le m \\ mn + (n-i)m + j, & v = y^j_i,\ 1 \le j \le m,\ 2 \le i \le n-1,\ i \text{ even} \\ mn + (n-i+1)m - j + 1, & v = y^j_i,\ 1 \le j \le m,\ 1 \le i \le n-2,\ i \text{ odd} \end{cases}$$

$$f(e) = \begin{cases} \frac{j}{2}, & e = xx_j,\ 2 \leq j \leq m-1,\ j \text{ even} \\ \frac{m+j}{2}, & e = xx_j,\ 1 \leq j \leq m,\ j \text{ odd} \\ 2m-j+1, & e = x_jy_j,\ 1 \leq j \leq m \\ (i+1)m+j, & e = y_jy_i^j,\ 1 \leq j \leq m,\ 1 \leq i \leq n-1,\ i \text{ odd} \\ (i+2)m-j+1, & e = y_jy_i^j,\ 1 \leq j \leq m,\ 2 \leq i \leq n-2,\ i \text{ even} \end{cases}$$

Clearly f is a local vertex antimagic total labeling of $B_{m,n}$ and we formulate the vertex weights as follows:

$$w_t(x) = \tfrac{m}{2}(m+1) + 2mn + 2m + 1$$

$$w_t(x_j) = \tfrac{m+1}{2} + 2mn + 3m + 1,\ 1 \leq j \leq m$$

$$w_t(y_i^j) = 2mn + 2m + 1,\ 1 \leq i \leq n-1,\ 1 \leq j \leq m$$

$$w_t(y_j) = \tfrac{n-1}{4}(mn + m + 2) + \tfrac{n-1}{4}(mn + 5m) + 2mn + 2m + 1,\ 1 \leq j \leq m$$

Hence, from the above the vertex weight, it is easy to see that f induces a proper local vertex antimagic total coloring of $B_{m,n}$ and it gives $\chi_{lvat}(B_{m,n}) \leq 4$. □

Case 2. For n even and m odd, define a bijection $f : V(B_{m,n}) \cup E(B_{m,n}) \longrightarrow \{1, 2, ..., |V(B_{m,n})| + |E(B_{m,n})|\}$ by the following:

$$f(e) = \begin{cases} \frac{j}{2}, & e = xx_j,\ 2 \leq j \leq m-1,\ j \text{ even} \\ \frac{m+j}{2}, & e = xx_j,\ 1 \leq j \leq m,\ j \text{ odd} \\ 2m-j+1, & e = x_jy_j,\ 1 \leq j \leq m \\ 2m+\frac{j}{2}, & e = y_jy_i^j,\ i=1,\ \ 2 \leq j \leq m-1,\ j \text{ even} \\ 2m+\frac{m+j}{2}, & e = y_jy_i^j,\ i=1,\ \ 1 \leq j \leq m,\ j \text{ odd} \\ (i+1)m+j, & e = y_jy_i^j,\ 1 \leq j \leq m,\ 2 \leq i \leq n-2,\ i \text{ even} \\ (i+2)m-j+1, & e = y_jy_i^j,\ 1 \leq j \leq m,\ 3 \leq i \leq n-1,\ i \text{ odd} \end{cases}$$

$$f(v) = \begin{cases} 2mn+2m+1, & v = x \\ 2mn+m+\frac{j+1}{2}, & v = x_j,\ 1 \leq j \leq m,\ j \text{ odd} \\ 2mn+m+\frac{m+j+1}{2}, & v = x_j,\ 2 \leq j \leq m-1,\ j \text{ even} \\ 2mn+\frac{j+1}{2}, & v = y_j,\ 1 \leq j \leq m,\ j \text{ odd} \\ 2mn+\frac{m+j+1}{2}, & v = y_j,\ 2 \leq j \leq m-1,\ j \text{ even} \\ 2mn-m+\frac{m-j+2}{2}, & v = y_i^j,\ i=1,\ 1 \leq j \leq m,\ j \text{ odd} \\ 2mn-m+\frac{2m-j+2}{2}, & v = y_i^j,\ i=1,\ 2 \leq j \leq m-1,\ j \text{ even} \\ mn+(n-i)m+j, & v = y_i^j,\ 1 \leq j \leq m,\ 3 \leq i \leq n-1,\ i \text{ odd} \\ mn+(n-i+1)m-j+1, & v = y_i^j,\ 1 \leq j \leq m,\ 2 \leq i \leq n-2,\ i \text{ even} \end{cases}$$

Clearly f is a local vertex antimagic total labeling of $B_{m,n}$ and we formulate the vertex weights as follows:

$$w_t(x) = \tfrac{m}{2}(m+1) + 2mn + 2m + 1$$

$$w_t(x_j) = \tfrac{m+1}{2} + 2mn + 3m + 1,\ 1 \leq j \leq m$$

$$w_t(y_i^j) = 2mn + 2m + 1,\ 1 \le i \le n-1,\ 1 \le j \le m$$

$$w_t(y_j) = \tfrac{n-2}{4}(mn + 2m + 2) + \tfrac{n-2}{4}(mn + 6m) + \tfrac{m+1}{2} + 2mn + 4m + 1,\ 1 \le j \le m$$

Hence, from the above the vertex weight, it is easy to see that f induces a proper local vertex antimagic total coloring of $B_{m,n}$ and it gives $\chi_{lvat}(B_{m,n}) \le 4$. □

Case 3. For n odd and m even, define a bijection $f : V(B_{m,n}) \cup E(B_{m,n}) \longrightarrow \{1, 2, ..., |V(B_{m,n})| + |E(B_{m,n})|\}$ by the following:

$$f(v) = \begin{cases} 2mn + 2m + 1, & v = x \\ 2mn + m + \frac{j+1}{2}, & v = x_j,\ 1 \le j \le m-1,\ j \text{ odd} \\ 2mn + m + \frac{m+j}{2}, & v = x_j,\ 2 \le j \le m,\ j \text{ even} \\ 2mn + j, & v = y_j,\ 1 \le j \le m \\ mn + (n-i)m + j, & v = y_i^j,\ 1 \le j \le m,\ 2 \le i \le n-2,\ i \text{ odd} \\ mn + (n-i+1)m - j + 1, & v = y_i^j,\ 1 \le j \le m,\ 1 \le i \le n-1,\ i \text{ even} \end{cases}$$

$$f(e) = \begin{cases} \frac{j}{2}, & e = xx_j,\ 2 \le j \le m,\ j \text{ odd} \\ \frac{m+j+1}{2}, & e = xx_j,\ 1 \le j \le m-1,\ j \text{ even} \\ 2m - j + 1, & e = x_j y_j,\ 1 \le j \le m \\ (i+1)m + j, & e = y_j y_i^j,\ 1 \le j \le m,\ 1 \le i \le n-1,\ i \text{ odd} \\ (i+2)m - j + 1, & e = y_j y_i^j,\ 1 \le j \le m,\ 2 \le i \le n-2,\ i \text{ even} \end{cases}$$

Clearly f is a local vertex antimagic total labeling of $B_{m,n}$ and we formulate the vertex weights as follows:

$$w_t(x) = \tfrac{m}{2}(m+1) + 2mn + 2m + 1$$

$$w_t(x_j) = \tfrac{m+2}{2} + 2mn + 3m + 1,\ 1 \le j \le m,\ j \text{ odd}$$

$$w_t(x_j) = \tfrac{m+2}{2} + 2mn + 3m,\ 2 \le j \le m-1,\ j \text{ even}$$

$$w_t(y_i^j) = 2mn + 2m + 1,\ 1 \le i \le n-1,\ 1 \le j \le m$$

$$w_t(y_j) = \tfrac{n-1}{4}(mn + m + 2) + \tfrac{n-1}{4}(mn + 5m) + 2mn + 2m + 1,\ 1 \le j \le m$$

Hence, from the above the vertex weight, it is easy to see that f induces a proper local vertex antimagic total coloring of $B_{m,n}$ and it gives $\chi_{lvat}(B_{m,n}) \le 5$. □

Case 4. For n even and m even, define a bijection $f : V(B_{m,n}) \cup E(B_{m,n}) \longrightarrow \{1, 2, ..., |V(B_{m,n})| + |E(B_{m,n})|\}$ by the following:

$$f(v) = \begin{cases} 2mn + 2m + 1, & v = x \\ 2mn + m + \frac{j+1}{2}, & v = x_j,\ 1 \le j \le m-1,\ j \text{ odd} \\ 2mn + m + \frac{m+j}{2}, & v = x_j,\ 2 \le j \le m,\ j \text{ even} \\ 2mn + \frac{j+1}{2}, & v = y_j,\ 1 \le j \le m-1,\ j \text{ odd} \\ 2mn + \frac{m+j}{2}, & v = y_j,\ 2 \le j \le m,\ j \text{ even} \\ 2mn - m + \frac{m-j+1}{2}, & v = y_i^j,\ i = 1,\ 1 \le j \le m-1,\ j \text{ odd} \\ 2mn - m + \frac{2m-j+2}{2}, & v = y_i^j,\ i = 1,\ 2 \le j \le m,\ j \text{ even} \\ mn + (n-i)m + j, & v = y_i^j,\ 1 \le j \le m,\ 3 \le i \le n-1,\ i \text{ odd} \\ mn + (n-i+1)m - j + 1, & v = y_i^j,\ 1 \le j \le m,\ 2 \le i \le n-2,\ i \text{ even} \end{cases}$$

$$f(e)=\begin{cases}\frac{j}{2}, & e=xx_j,\ 2\le j\le m,\ j \text{ even}\\ \frac{m+j+1}{2}, & e=xx_j,\ 1\le j\le m-1,\ j \text{ odd}\\ 2m-j+1, & e=x_jy_j,\ 1\le j\le m\\ 2m+\frac{j}{2}, & e=y_jy_i^j,\ i=1,\ \ 2\le j\le m,\ j \text{ even}\\ 2m+\frac{m+j+1}{2}, & e=y_jy_i^j,\ i=1,\ \ 1\le j\le m-1,\ j \text{ odd}\\ (i+1)m+j, & e=y_jy_i^j,\ 1\le j\le m,\ 2\le i\le n-2,\ i \text{ even}\\ (i+2)m-j+1, & e=y_jy_i^j,\ 1\le j\le m,\ 3\le i\le n-1,\ i \text{ odd}\end{cases}$$

Clearly f is a local vertex antimagic total labeling of $B_{m,n}$ and we formulate the vertex weights as follows:

$$w_t(x)=\tfrac{m}{2}(m+1)+2mn+2m+1$$

$$w_t(x_j)=\tfrac{m+2}{2}+2mn+3m+1,\ 1\le j\le m,\ j \text{ odd}$$

$$w_t(x_j)=\tfrac{m+2}{2}+2mn+3m,\ 2\le j\le m-1,\ j \text{ even}$$

$$w_t(y_i^j)=2mn+2m+1,\ 1\le i\le n-1,\ 1\le j\le m$$

$$w_t(y_j)=\tfrac{n-2}{4}(mn+2m+2)+\tfrac{n-2}{4}(mn+6m)+\tfrac{m+2}{2}+2mn+4m+1,\ 1\le j\le m-1,\ j \text{ odd}$$

$$w_t(y_j)=\tfrac{n-2}{4}(mn+2m+2)+\tfrac{n-2}{4}(mn+6m)+\tfrac{m+2}{2}+2mn+4m,\ 2\le j\le m,\ j \text{ even}$$

Hence, from the above the vertex weight, it is easy to see that f induces a proper local vertex antimagic total coloring of $B_{m,n}$ and it gives $\chi_{lvat}(B_{m,n})\le 6$. □

Theorem 2.4. *For the centipede graph C_n with $n\ge 4$, the local vertex antimagic total chromatic number of C_n is $\chi_{lvat}(C_n)\le 4$.*

Proof: The graph C_n is a connected graph with vertex set $V(C_n)=\{x_i;\ y_i, 1\le i\le n\}$ and edge set $E(C_n)=\{x_ix_{1+1},\ 1\le 1\le n-1\}\cup\{x_iy_i,\ 1\le i\le n\}$. Hence $|V(C_n)|=2n$ and $|E(C_n)|=2n-1$.

Case 1. For n even, define a bijection $f: V(C_n)\cup E(C_n)\longrightarrow\{1,2,3...,|V(C_n)|+|E(C_n)|\}$ by the following:

$$f(v)=\begin{cases}\frac{3n-i}{2}, & v=x_i,\ 2\le i\le n,\ i \text{ even}\\ \frac{4n-i-1}{2}, & v=x_i,\ 1\le i\le n-1,\ i \text{ odd}\\ \frac{7n+i-1}{2}, & v=y_i,\ 1\le i\le n-1,\ i \text{ odd}\\ \frac{6n+i-2}{2}, & v=y_i,\ 2\le i\le n,\ i \text{ even}\end{cases}$$

$$f(e)=\begin{cases}\frac{i}{2}, & e=x_ix_{i+1},\ 2\le i\le n-2,\ i \text{ even}\\ \frac{n+i-1}{2}, & e=x_ix_{i+1},\ 1\le i\le n-1,\ i \text{ odd}\\ \frac{5n-i-1}{2}, & e=x_iy_i,\ 1\le i\le n-1,\ i \text{ odd}\\ \frac{6n-i}{2}, & e=x_iy_i,\ 2\le i\le n,\ i \text{ even}\end{cases}$$

Clearly f is a local vertex antimagic total labeling of C_n and we formulate the vertex weights as follows:

$$w_t(v) = \begin{cases} \frac{9n-2}{2}, & v = x_i, \quad i = n \\ 5n-1, & v = x_i, \quad 2 \le i \le n-2, \; i \text{ even} \\ 5n-2, & v = x_i, \quad 1 \le i \le n-1, \; i \text{ odd} \\ \frac{12n-2}{2}, & v = y_i, \quad 1 \le i \le n \end{cases}$$

Hence, from the above the vertex weight, it is easy to see that f induces a proper local vertex antimagic total coloring of C_n and it gives $\chi_{lvat}(C_n) \le 4$. □

Case 2. For n odd, define a bijection $f : V(C_n) \cup E(C_n) \longrightarrow \{1, 2, 3..., |V(C_n)| + |E(C_n)|\}$ by the following:

$$f(v) = \begin{cases} \frac{3n-i}{2}, & v = x_i, \; 1 \le i \le n, \; i \text{ odd} \\ \frac{4n-i}{2}, & v = x_i, \; 2 \le i \le n-1, \; i \text{ even} \\ \frac{6n+i-2}{2}, & v = y_i, \; 2 \le i \le n-1, \; i \text{ even} \\ \frac{7n+i-2}{2}, & v = y_i, \; 1 \le i \le n, \; i \text{ odd} \end{cases}$$

$$f(e) = \begin{cases} \frac{n+i}{2}, & e = x_i x_{i+1}, \; 1 \le i \le n-2, \; i \text{ odd} \\ \frac{i}{2}, & e = x_i x_{i+1}, \; 2 \le i \le n-1, \; i \text{ even} \\ \frac{5n-i}{2}, & e = x_i y_i, \; 1 \le i \le n, \; i \text{ odd} \\ \frac{6n-i}{2}, & e = x_i y_i, \; 2 \le i \le n-1, \; i \text{ even} \end{cases}$$

Clearly f is a local vertex antimagic total labeling of C_n and we formulate the vertex weights as follows:

$$w_t(v) = \begin{cases} \frac{7n-1}{2}, & v = x_i, \quad i = n \\ \frac{9n-1}{2}, & v = x_i, \quad 1 \le i \le n-2, \; i \text{ odd} \\ \frac{11n-1}{2}, & v = x_i, \quad 2 \le i \le n-1, \; i \text{ even} \\ \frac{12n-2}{2}, & v = y_i, \quad 1 \le i \le n \end{cases}$$

Hence, from the above the vertex weight, it is easy to see that f induces a proper local vertex antimagic total coloring of C_n for n odd and it gives $\chi_{lvat}(C_n) \le 4$. □

Theorem 2.5. *For $n, m \ge 3$, the local vertex antimagic total chromatic number of graph $G = Amal(S_n, v, m)$ is:*

$$\chi_{lvat}(G) \le \begin{cases} 3, & \text{for n odd or} \\ & \text{n even m odd} \\ 4, & \text{for n even m even} \end{cases}$$

Proof. The graph G is a amalgamation of star graph with vertex set $V(G) = \{x\} \cup \{x_j, 1 \le j \le m\} \cup \{y_i^j, 1 \le i \le n-1, \quad 1 \le j \le m\}$ and edge set $E(G) = \{xx_j, 1 \le j \le m\} \cup \{x_j y_i^j, 1 \le i \le n-1, \quad 1 \le j \le m\}$. Hence $|V(G)| = mn + 1$ and $|E(G)| = mn$.

IOP Conf. Series: Journal of Physics: Conf. Series **1008** (2018) 012035 doi:10.1088/1742-6596/1008/1/012035

Case 1. For n odd, define a bijection $f : V(G) \cup E(G) \longrightarrow \{1, 2, ..., |V(G)| + |E(G)|\}$ by the following:

$$f(v) = \begin{cases} 1, & v = x \\ 2m - j + 2, & v = x_j,\ 1 \le j \le m \\ (2n - i)m + j + 1, & v = y_i^j,\ 1 \le j \le m,\ 1 \le i \le n - 1,\ i\ odd \\ (2n + 1 - i)m - j + 2, & v = y_i^j,\ 1 \le j \le m,\ 1 \le i \le n - 2,\ i\ even \end{cases}$$

$$f(e) = \begin{cases} i + 1, & e = xx_j,\ 1 \le j \le m,\ j\ even \\ (i + 2)m - j + 2, & e = x_j y_i^j,\ 1 \le j \le m,\ 1 \le i \le n - 1,\ i\ odd \\ (i + 1)m + j + 1, & e = x_j y_i^j,\ 1 \le j \le m,\ 1 \le i \le n - 2,\ i\ even \end{cases}$$

Clearly f is a local vertex antimagic total labeling of graph G and we formulate the vertex weights as follows:

$$w_t(x) = \frac{(m-2)(m-3)}{2} + (m - 2)4 + 6$$

$$w_t(y_i^j) = 2mn + 2m + 3,\ 1 \le i \le n - 1,\ 1 \le j \le m$$

$$w_t(x_j) = \frac{n}{4}(mn + 4m + 2) + \frac{n-2}{4}(mn + 2m + 4) + \frac{3(m-1)}{2} + 5,\ 1 \le j \le m$$

Hence, from the above the vertex weight, it is easy to see that f induces a proper local vertex antimagic total coloring of graph $G = Amal(S_n, v, m)$ where n odd it gives $\chi_{lvat}(G) \le 3$. □

Case 2. For n even and m odd, define a bijection $f : V(G) \cup E(G) \longrightarrow \{1, 2, ..., |V(G)| + |E(G)|\}$ by the following:

$$f(v) = \begin{cases} 1, & v = x \\ \frac{3m+j+3}{2}, & v = x_j,\ 1 \le j \le m,\ j\ even \\ \frac{2m+j+3}{2}, & v = x_j,\ 1 \le j \le m,\ j\ odd \\ (2n - i)m + j + 1, & v = y_i^j,\ 1 \le j \le m,\ 1 \le i \le n - 1,\ i\ odd \\ (2n + 1 - i)m - j + 2, & v = y_i^j,\ 1 \le j \le m,\ 1 \le i \le n - 2,\ i\ even \end{cases}$$

$$f(e) = \begin{cases} \frac{j+2}{2}, & e = xx_j,\ 1 \le j \le m,\ j\ even \\ \frac{m+j+2}{2}, & e = xx_j,\ 1 \le j \le m,\ j\ odd \\ (i + 2)m - j + 2, & e = x_j y_i^j,\ 1 \le j \le m,\ 1 \le i \le n - 1,\ i\ odd \\ (i + 1)m + j + 1, & e = x_j y_i^j,\ 1 \le j \le m,\ 1 \le i \le n - 2,\ i\ even \end{cases}$$

Clearly f is a local vertex antimagic total labeling of graph $G = Amal(S_n, v = x_1, m)$ and we formulate the vertex weights as follows:

$$w_t(x) = \frac{(m-2)(m-3)}{2} + (m - 2)4 + 6$$

$$w_t(y_i^j) = 2mn + 2m + 3,\ 1 \le i \le n - 1,\ 1 \le j \le m$$

$$w_t(x_j) = \frac{n}{4}(mn + 4m + 2) + \frac{n-2}{4}(mn + 2m + 4) + \frac{3(m-1)}{2} + 5,\ 1 \le j \le m$$

IOP Conf. Series: Journal of Physics: Conf. Series **1008** (2018) 012035 doi:10.1088/1742-6596/1008/1/012035

Hence, from the above the vertex weight, it is easy to see that f induces a proper local vertex antimagic total coloring of graph $G = Amal(S_n, v, m)$ where n even and m odd and it gives $\gamma_{lvat}(G) \leq 3$. □

Case 3. For n even and m even, define a bijection $f : V(G) \cup E(G) \longrightarrow \{1, 2, ..., |V(G)| + |E(G)|\}$ by the following:

$$f(v) = \begin{cases} 1, & v = x \\ \frac{3m+j+3}{2}, & v = x_j,\ 1 \leq j \leq m,\ j\ odd \\ \frac{2m+j+2}{2}, & v = x_j,\ 1 \leq j \leq m,\ j\ even \\ (2n-i)m + j + 1, & v = y_i^j,\ 1 \leq j \leq m,\ 1 \leq i \leq n-1,\ i\ odd \\ (2n+1-i)m - j + 2, & v = y_i^j,\ 1 \leq j \leq m,\ 1 \leq i \leq n-2,\ i\ even \end{cases}$$

$$f(e) = \begin{cases} \frac{j+3}{2}, & e = xx_j,\ 1 \leq j \leq m\ j\ odd \\ \frac{m+j+2}{2}, & e = xx_j,\ 1 \leq j \leq m\ j\ even \\ (i+2)m - j + 2, & e = x_j y_i^j,\ 1 \leq j \leq m,\ 1 \leq i \leq n-1,\ i\ odd \\ (i+1)m + j + 1, & e = x_j y_i^j,\ 1 \leq j \leq m,\ 1 \leq i \leq n-2,\ i\ even \end{cases}$$

Clearly f is a local vertex antimagic total labeling of graph $G = Amal(S_n, v, m)$ and we formulate the vertex weights as follows:

$$w_t(x) = \frac{(m-2)(m-3)}{2} + (m-2)4 + 6$$

$$w_t(y_i^j) = 2mn + 2m + 3,\ 1 \leq i \leq n-1,\ 1 \leq j \leq m$$

$$w_t(x_j) = \frac{n}{4}(mn + 4m + 2) + \frac{n-2}{4}(mn + 2m + 4) + \frac{3(m-2)}{2} + 7,\ 1 \leq j \leq m,\ j\ odd$$

$$w_t(x_j) = \frac{n}{4}(mn + 4m + 2) + \frac{n-2}{4}(mn + 2m + 4) + \frac{3(m-2)}{2} + 6,\ 1 \leq j \leq m,\ j\ even$$

Hence, from the above the vertex weight, it is easy to see that f induces a proper local vertex antimagic total coloring of graph $G = Amal(S_n, v, m)$ where n even and m even and it gives $\chi_{lvat}(G) \leq 4$. □

3. Conclusion

In this paper we have to study the local vertex antimagic total coloring os some families tree. We have found chromatic number of star graph S_n, path graph P_n, double star graph $S_{n,m}$, banana tree graph $B_{m,n}$, centipede graph C_n, and amalgamation of star graph $Amal(S_n, v, m)$.

Open Problem 3.1. *Determine the lower bound of local vertex antimagic total coloring for any graph.*

Open Problem 3.2. *Determine the local vertex antimagic total chromatic number of other some families tree like caterpillar graph, firecracker graph and lobster graph.*

Acknowledgement
We gratefully acknowledge the support from CGANT - University of Jember of year 2018.

IOP Conf. Series: Journal of Physics: Conf. Series **1008** (2018) 012035 doi:10.1088/1742-6596/1008/1/012035

References

[1] Agustin I H, Hasan M, Dafik, Alfarisi R and Prihandini R M 2017 Local edge antimagic coloring of graphs *Far East Journal of Mathematical Sciences*

[2] Arumugam S, Premalatha K, Baca M and Semanicova-Fenovcikova A 2017 Local antimagic vertex coloring of a graph *Graphs and Combinatorics* **33** 275-285

[3] Chartrand G and Lesniak L 2005 *Graphs and digraphs 4th edn.* Chapman and Hall CRC

[4] Chartrand G and Zhang P 2011 *Discrete Mathematics* Waveland Press, Inc.

[5] Dafik 2015 *Antimagic Total Labeling of Disjoint Union Disconnected Graphs* CRC Jember

[6] Dafik, Hasan M, Azizah Y N and Agustin I H 2017 A generalized shackle of any graph H admits a super H-antimagic total labeling *Journal of Physics: Conf. Series* **893** 012042

[7] Dafik, Mirka M, Ryan J and Bâca M 2006 Super edge-antimagic total labelings of $mK_{n,n}$ *Ars Combinatoria* **101** 97-107

[8] Gallian J A 2016 A Dynamic survey of graph labeling *The Electronic Journal of Combinatorics* **18**

[9] Gross J L, Yellen J and Zhang P 2014 *Handbook of Graph Theory* Second Edition CRC Press Taylor and Francis Group

[10] Hartsfield N and Ringel G 1994 *Pearls in Graph Theory* Academic Press, INC. Boston

[11] Munir R 2009 *Matematika Diskrit Edisi 3* Informatika Bandung

[12] Rosen K H 2012 *Discrete Mathematics and Its Apllications Seventh Edition* McGraw-Hill New York

IOP Conf. Series: Journal of Physics: Conf. Series **1008** (2018) 012036 doi:10.1088/1742-6596/1008/1/012036

Super local edge antimagic total coloring of $P_n \triangleright H$

Elsa Yuli Kurniawati[1,2], Ika Hesti Agustin[1,2], Dafik[1,3], Ridho Alfarisi[1,4]

[1]CGANT-University of Jember, Jember, Indonesia
[2]Department of Mathematics, University of Jember, Jember, Indonesia
[3]Department of Mathematics Education, University of Jember, Jember, Indonesia
[4]Department of Elementary School Teacher Education, University of Jember, Jember, Indonesia

E-mail: elsakurniawati72@yahoo.co.id, ikahestiagustin@gmail.com, d.dafik@unej.ac.id

Abstract. In this paper, we consider that all graphs are finite, simple and connected. Let $G(V, E)$ be a graph of vertex set V and edge set E. A bijection $f : V(G) \longrightarrow \{1, 2, 3, ..., |V(G)|\}$ is called a local edge antimagic labeling if for any two adjacent edges e_1 and e_2, $w(e_1) \neq w(e_2)$, where for $e = uv \in G$, $w(e) = f(u) + f(v)$. Thus, any local edge antimagic labeling induces a proper edge coloring of G if each edge e is assigned the color $w(e)$. It is considered to be a super local edge antimagic total coloring, if the smallest labels appear in the vertices. The super local edge antimagic chromatic number, denoted by $\gamma_{leat}(G)$, is the minimum number of colors taken over all colorings induced by super local edge antimagic total labelings of G. In this paper we initiate to study the existence of super local edge antimagic total coloring of comb product of graphs. We also analyse the lower bound of its local edge antimagic chromatic number. It is proved that $\gamma_{leat}(P_n \triangleright G) \geq \gamma_{leat}(P_n) + \gamma_{leat}(G)$. Furthermore we have determine exact value local edge antimagic coloring of $P_n \triangleright P_m$, $P_n \triangleright C_m$ and $P_n \triangleright S_m$.

1. Introduction

We consider that all graphs in this paper are finite, simple and connected graph, for detail definition of graph see [1, 2].

A bijection mapping that assigns natural number to vertices of a graph is called a graph labeling. In this type of labeling, we consider all weights associated with each edge. The labelings will be said to be an antimagic if all edge weights value are all different. Hartsfield and Ringel [3] introduced the concept of antimagic labeling of a graph. There are a lot of results regarding to antimagic labeling, some of them can be found in Dafik *et. al* [5], [6]. They determined super edge-antimagic total labelings of $mK_{n,n}$ and super edge-antimagicness for a class of disconnected graphs, respectively.

IOP Conf. Series: Journal of Physics: Conf. Series **1008** (2018) 012036 doi:10.1088/1742-6596/1008/1/012036

In this study, we will examine the relation between antimagic labeling and coloring of graph, namely a local edge antimagic coloring. We know that a proper edge coloring of a graph G is a coloring of all edges of G by natural number such that any two adjacent edges receive different colors. While, by a local edge antimagic total labeling, we mean a bijection $f : V(G) \cup E(G) \longrightarrow \{1, 2, 3, \ldots, |V(G)| + |E(G)|\}$ such that for any two adjacent edges e_1 and e_2, $w_t(e_1) \neq w_t(e_2)$, for $e = uv \in G$ and $w_t(e) = f(u) + f(v) + f(uv)$. Thus, any local edge antimagic total labeling induces a proper edge coloring of G if each edge e is assigned by the color $w_t(e)$. It is considered to be a super local edge antimagic total coloring, if the smallest labels appear in the vertices. The super local edge antimagic chromatic number $\gamma_{leat}(G)$ is the minimum number of colors taken over all colorings induced by super local edge antimagic total labelings of G. This paper just initiate to study the super local edge antimagic total colorings.

Arumugam *et al.* [4] firstly introduced a local vertex antimagic coloring of a graph G. They gave a lower bound and an upper bound of local vertex antimagic coloring of joint graph and also exact value of local vertex antimagic coloring for path, cycle, complete graph, friendship, wheel, bipartite and complete bipartite. Ika, *et. al.* [7] has determined the lower bound of the local edge antimagic coloring, denoted by $\gamma_{lea}(G) \geq \Delta(G)$. If $\Delta(G)$ is maximum degrees of G.

Ika, *et. al.* [8] study super local edge antimagic total coloring of any graph using EAVL technique and has found the lower bound of super local edge antimagic chromatic number, denoted by $\gamma_{leat}(G) \geq \Delta(G)$, and determined exact value of ladder graph L_n, caterpillar graph $C_{n,m}$, and graph coronations $P_n \odot P_2$ and $C_n \odot P_2$. Ika, *et. al.* [9] has determined exact value of path graph $\gamma_{leat}(P_n) = 2$ if n is odd and $\gamma_{leat}(P_n) = 3$ if n is even, cycle graph $\gamma_{leat}(C_n) = 3$. Alfarisi, *et. al.* [10] also study the existence of local edge antimagic total coloring of some wheel related graphs, they found exact value of $\gamma_{leat}(F_n) = n + 2$ if n is odd and $\gamma_{leat}(F_n) = n + 3$ if n is even, wheel graph $\gamma_{leat}(W_n) = n + 4$ if $n = 4$ and $\gamma_{leat}(W_n) = n + 3$ if $n \geq 3$ and $n \neq 4$.

To show our results, we use following definition

Definition 1.1. *Let $G(V, E)$ be a graph of vertex set V and edge set E. A bijection $f : V(G) \cup E(G) \rightarrow \{1, 2, 3, \ldots, |V(G)| + |E(G)|\}$ is called a local edge antimagic total labeling if for any two adjacent edges e_1 and e_2, $w(e_1) \neq w(e_2)$, where for $e = uv \in G$, $w(e) = f(u) + f(uv) + f(v)$. It is considered to be a super local edge antimagic total labeling, if the smallest labels appear in the vertices.*

Noted that, any super local edge antimagic total labeling induces a proper edge coloring of G if each edge e is assigned by the color $w(e)$.

Furthermore, the following observation in Arumugam paper, see [4], is very important for our study.

Observation 1.1. [4] *For any graph G, the local vertex antimagic chromatic number $\chi_{la}(G) \geq \chi(G)$, where $\chi(G)$ is a chromatic number of vertex coloring of G.*

Let G and H be two connected graphs. Let o be a vertex of H. Comb product between G and H is a graph obtained by taking one copy of G and $|V(G)|$ copies of H and grafting the i-th copy of H at the vertex o to the i-th vertex of G, and denoted by $G \triangleright H$. By the definition of comb product, we can say that $V(G \triangleright H) = \{(a,u)|a \in V(G), u \in V(H)\}$ and $(a,u)(b,v) \in E(G \triangleright H)$ whenever $a = b$ and $uv \in E(H)$, or $ab \in E(G)$ and $u = v = o$.

In this paper we will determine super local edge antimagic total coloring of comb product of graphs and also analyse the lower bound of its local edge antimagic chromatic number. It is proved that $\gamma_{leat}(P_n \triangleright H) \geq \gamma_{leat}(P_n) + \gamma_{leat}(H)$. Furthermore we have determine chromatic number of local edge antimagic coloring of $P_n \triangleright H$

2. Main Result

Firstly, we present our result by showing the lower bound of super local edge antimagic total chromatic number of comb product graph.

Lemma 2.1. *Let P_n be path and H be any special graph. The super local edge antimagic total chromatic number of of comb product graph is $\gamma_{leat}(P_n \triangleright H) \geq \gamma_{leat}(P_n) + \gamma_{leat}(H)$.*

Proof. The comb product between P_n and H is a graph obtained by one copy of P_n and $| V(P_n) |$ copies of H, and grafting the i vertex of P_n to the u_i in i-th copy of G. Suppose P_n admits a local edge antimagic coloring with $\gamma_{leat}(P_n)$. Furthermore, the subgraph H_i admits a local edge antimagic total coloring with $\gamma_{leat}(H_i)$.The total edge weights of local edge antimagic total coloring from subgraph P_n and H_i are different.

To show that the total edge weights of local edge antimagic total coloring from subgraph P_n and H_i are different. If subgraph H_i; $1 \leq i \leq n$ are labeled by the function $f : V(H_i) \cup E(H_i) \longrightarrow \{1, 2, 3, ..., k\}$ with $k = |V(H_i)| + |E(H_i)|$. The largest total edge weight is $k + (k-1) + (k-2)$, thus we have $w_t(H_i) = 3k - 3$. Next, the label element of P_n by function $f : V(P_n) \cup E(P_n) \longrightarrow \{k+1, k+2, k+3, ..., |V(P_n)| + |E(P_n)|\}$, the smallest total edge weight is $(k+1) + (k+2) + (k+3)$ thus $w_t(P_n) = 3k + 6$. Since $w_t(H_i) < w_t(P_n)$, the total edge weight of *local edge antimagic total* coloring from subgraph P_n and H_i are different. The total edge weights in each subgraph H_i induce the local edge antimagic total coloring of the same weight, thus $\gamma_{leat}(H_1) = \gamma_{leat}(H_2) = \cdots = \gamma_{leat}(H_n) = \gamma_{leat}(H)$. We obtain that a local antimagic labeling of $P_n \triangleright H$.

$$\begin{aligned} \gamma_{leat}(P_n \triangleright H) &\geq |\{w_{P_n}(e), e \in V(P_n)\}| + |\{w_{H_i}(e), e \in V(H_i)\}| \\ &= \gamma_{leat}(P_n) + \gamma_{leat}(H_i) \\ &= \gamma_{lea}(P_n) + \gamma_{leat}(H) \end{aligned}$$

Hence, from the above edge weight it is easy to see that the lower bound of local antimagic total coloring of $P_n \triangleright H$ is $\gamma_{leat}(P_n \triangleright H) \geq \gamma_{leat}(P_n) + \gamma_{leat}(H)$. □

Theorem 2.1. *For $n, m \geq 3$, the super local edge antimagic total chromatic number of $P_n \triangleright P_m$ is*

$$\gamma_{leat}(P_n \triangleright P_m) = \begin{cases} 4, & \text{if } n \text{ is odd, } m \text{ is odd} \\ 5, & \text{if } n \text{ is odd, } m \text{ is even} \end{cases}$$

Proof. The graph $P_n \triangleright P_m$ is a connected graph with vertex set $V(P_n \triangleright P_m) = \{x_{i,j} : 1 \leq i \leq n, 1 \leq j \leq m\}$ and edge set $E(P_n \triangleright P_m) = \{x_{i,1}x_{i+1,1} : 1 \leq i \leq n-1\} \cup \{x_{i,j}x_{i,j+1}; 1 \leq i \leq n, 1 \leq j \leq m-1\}$. Hence $|V(P_n \triangleright P_m)| = mn$ and $|E(P_n \triangleright P_m)| = mn - 1$. We will

IOP Conf. Series: Journal of Physics: Conf. Series **1008** (2018) 012036 doi:10.1088/1742-6596/1008/1/012036

describe the proof into two cases.

Case 1: For n is odd and m is odd, we will show that the lower bound of the local edge antimagic total coloring of $P_n \triangleright P_m$ is $\gamma_{leat}(P_n \triangleright P_m) \geq 4$. Based on Lemma 2.1, the lower bound is $\gamma_{leat}(P_n \triangleright P_m) \geq \gamma_{leat}(P_n) + \gamma_{leat}(P_m) = 2 + 2 = 4$. It concludes that the lower bound of the local edge antimagic total coloring of $P_n \triangleright P_m$ is $\gamma_{leat}(P_n \triangleright P_m) \geq 4$. Furthermore, we will show that the upper bound of the local edge antimagic total coloring of $P_n \triangleright P_m$ is $\gamma_{leat}(P_n \triangleright P_m) \leq 4$.

Define a bijection $f : V(P_n \triangleright P_m) \cup E(P_n \triangleright P_m) \longrightarrow \{1, 2, 3, ..., |V(P_n \triangleright P_m)| + |E(P_n \triangleright P_m)|\}$ by the following.

$$f(v) = \begin{cases} \frac{i+(j-1)n+1}{2}, & \text{if } v = x_{i,j}, \text{ when } i \text{ is odd and } j \text{ is odd} \\ \frac{i+jn+1}{2}, & \text{if } v = x_{i,j}, \text{ when } i \text{ is even and } j \text{ is odd} \\ \frac{nm+i+nj}{2}, & \text{if } v = x_{i,j}, \text{ when } i \text{ is odd and } j \text{ is even} \\ \frac{i+(m+j-1)n}{2}, & \text{if } v = x_{i,j}, \text{ when } i \text{ is even and } j \text{ is even} \end{cases}$$

$$f(e) = \begin{cases} 2nm - n(1+j) - i + 1, & \text{if } e = x_{i,j}x_{i,j+1}, \text{ when } 1 \leq i \leq n \text{ and } j \text{ is odd} \\ 2nm + n(1+j) - i + 1, & \text{if } e = x_{i,j}x_{i,j+1}, \text{ when } 1 \leq i \leq n \text{ and } j \text{ is even} \\ 2nm - i - 1, & \text{if } e = x_{i,j}x_{i+1,j}, \text{ when } i \text{ is odd and } j = 1 \\ 2nm - i + 1, & \text{if } e = x_{i,j}x_{i+1,j}, \text{ when } i \text{ is even and } j = 1 \end{cases}$$

And the super total edge weight are as follows

$$w(e) = \begin{cases} \frac{5nm-2n+3}{2}, & \text{if } e = x_{i,j}x_{i,j+1}, \text{ when } 1 \leq i \leq n \text{ and } j \text{ is odd} \\ \frac{5nm+2n+3}{2}, & \text{if } e = x_{i,j}x_{i,j+1}, \text{ when } 1 \leq i \leq n \text{ and } j \text{ is even} \\ \frac{4nm+n+1}{2}, & \text{if } e = x_{i,j}x_{i+1,j}, \text{ when } i \text{ is odd and } j = 1 \\ \frac{4nm+n+5}{2}, & \text{if } e = x_{i,j}x_{i+1,j}, \text{ when } i \text{ is even and } j = 1 \end{cases}$$

Hence, from the above super total edge weights, it easy to see that f induces a proper edge coloring of $P_n \triangleright P_m$ and it gives $\gamma_{leat}(P_n \triangleright P_m) \leq 4$. Based on Lemma 2.1, the lower bound is $\gamma_{leat}(P_n \triangleright P_m) \geq \gamma_{leat}(P_n) + \gamma_{leat}(P_m) = 4$. It concludes that $\gamma_{leat}(P_n \triangleright P_m) = 4$ when n and m is odd.

Case 2: For n is odd and m is even, we will show that the lower bound of the local edge antimagic total coloring of $P_n \triangleright P_m$ is $\gamma_{leat}(P_n \triangleright P_m) \geq 5$. Based on Lemma 2.1, the lower bound is $\gamma_{leat}(P_n \triangleright P_m) \geq \gamma_{leat}(P_n) + \gamma_{leat}(P_m) = 2 + 3 = 5$. It concludes that the lower bound of the local edge antimagic total coloring of $P_n \triangleright P_m$ is $\gamma_{leat}(P_n \triangleright P_m) \geq 5$. Furthermore, we will show that the upper bound of the local edge antimagic total coloring of $P_n \triangleright P_m$ is $\gamma_{leat}(P_n \triangleright P_m) \leq 5$.

Define a bijection $f : V(P_n \triangleright P_m) \cup E(P_n \triangleright P_m) \longrightarrow \{1, 2, 3, ..., |V(P_n \triangleright P_m)| + |E(P_n \triangleright P_m)|\}$ by the following.

$$f(v) = \begin{cases} \frac{i+(j-1)n+1}{2}, & \text{if } v = x_{i,j}, \text{ when } i \text{ is odd and } j \text{ is odd} \\ \frac{i+jn+1}{2}, & \text{if } v = x_{i,j}, \text{ when } i \text{ is even and } j \text{ is odd} \\ \frac{n(m+j-1)+i}{2}, & \text{if } v = x_{i,j}, \text{ when } i \text{ is odd and } j \text{ is even} \\ \frac{n(m+j-2)+i}{2}, & \text{if } v = x_{i,j}, \text{ when } i \text{ is even and } j \text{ is even} \end{cases}$$

$$f(e) = \begin{cases} n(2m-1)-i+1, & \text{if } e = x_{i,j}x_{i,j+1}, \text{ when } 1 \le i \le n \text{ and } j = 1 \\ n(2m-1-j)-i+1, & \text{if } e = x_{i,j}x_{i,j+1}, \text{ when } 1 \le i \le n \text{ and } j \text{ is even} \\ n(2m+1-j)-i+1, & \text{if } e = x_{i,j}x_{i,j+1}, \text{ when } 1 \le i \le n \text{ and } j \text{ is odd} \\ 2nm-i-1, & \text{if } e = x_{i,j}x_{i+1,j}, \text{ when } i \text{ is odd and } j = 1 \\ 2nm-i+1, & \text{if } e = x_{i,j}x_{i+1,j}, \text{ when } i \text{ is even and } j = 1 \end{cases}$$

And the super total edge weight are as follows

$$w(e) = \begin{cases} \frac{5nm-n+3}{2}, & \text{if } e = x_{i,j}x_{i,j+1}, \text{ when } 1 \le i \le n \text{ and } j = 1 \\ \frac{5nm+n+3}{2}, & \text{if } e = x_{i,j}x_{i,j+1}, \text{ when } 1 \le i \le n \text{ and } j \text{ is odd} \\ \frac{5nm-3n+3}{2}, & \text{if } e = x_{i,j}x_{i,j+1}, \text{ when } 1 \le i \le n \text{ and } j \text{ is even} \\ 2nm+\frac{n+1}{2}, & \text{if } e = x_{i,j}x_{i+1,j}, \text{ when } i \text{ is odd and } j = 1 \\ 2nm+\frac{n+5}{2}, & \text{if } e = x_{i,j}x_{i+1,j}, \text{ when } i \text{ is even and } j = 1 \end{cases}$$

Hence, from the above super total edge weights, it easy to see that f induces a proper edge coloring of $P_n \triangleright P_m$ and it gives $\gamma_{leat}(P_n \triangleright P_m) \le 5$. Based on Lemma 2.1, the lower bound is $\gamma_{leat}(P_n \triangleright P_m) \ge \gamma_{leat}(P_n) + \gamma_{leat}(P_m) = 5$. It concludes that $\gamma_{leat}(P_n \triangleright P_m) = 5$ when n is odd and m is even. □

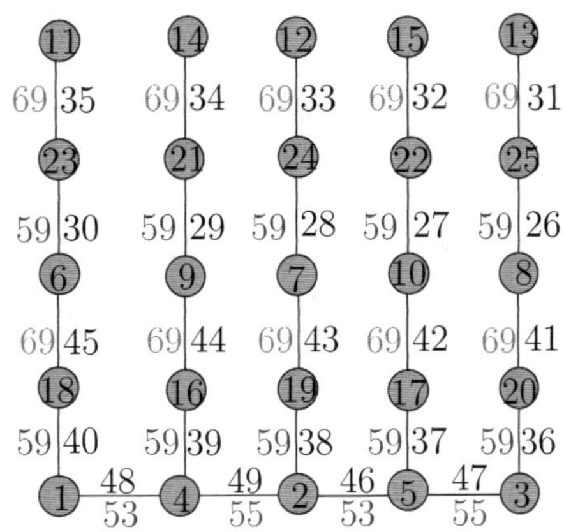

Figure 1. Illustration of local edge antimagic total coloring of $P_5 \triangleright P_5$

Theorem 2.2. *For $n, m \ge 3$, n is even and m is odd, the super local edge antimagic total chromatic number of $P_n \triangleright P_m$ is*

$$5 \le \gamma_{leat}(P_n \triangleright P_m) \le 7$$

Proof. For n is even and m is odd, we will show that the lower bound of the local edge antimagic total coloring of $P_n \triangleright P_m$. Based on Lemma 2.1, the lower bound is $\gamma_{leat}(P_n \triangleright P_m) \ge \gamma_{leat}(P_n) + \gamma_{leat}(P_m) = 3 + 2 = 5$. Furthermore, we will show the upper bound of the local edge antimagic total coloring of $P_n \triangleright P_m$ as follows.

Define a bijection $f : V(P_n \triangleright P_m) \cup E(P_n \triangleright P_m) \longrightarrow \{1, 2, 3, ..., |V(P_n \triangleright P_m)| + |E(P_n \triangleright P_m)|\}$ by the following.

$$f(v) = \begin{cases} \frac{i+(j-1)n+1}{2}, & \text{if } v = x_{i,j}, \text{ when } i \text{ is odd and } j \text{ is odd} \\ \frac{i+jn}{2}, & \text{if } v = x_{i,j}, \text{ when } i \text{ is even and } j \text{ is odd} \\ \frac{nm+i+nj+1}{2}, & \text{if } v = x_{i,j}, \text{ when } i \text{ is odd and } j \text{ is even} \\ \frac{n(m+j-1)+i}{2}, & \text{if } v = x_{i,j}, \text{ when } i \text{ is even and } j \text{ is even} \end{cases}$$

$$f(e) = \begin{cases} n(2m-1-j) - i + 1, & \text{if } e = x_{i,j}x_{i,j+1}, \text{ when } 1 \le i \le n \text{ and } j \text{ is odd} \\ n(2m+1-j) - i + 1, & \text{if } e = x_{i,j}x_{i,j+1}, \text{ when } 1 \le i \le n \text{ and } j \text{ is even} \\ 2nm - 1, & \text{if } e = x_{i,j}x_{i+1,j}, \text{ when } i = 1 \text{ and } j = 1 \\ 2nm - i + 1, & \text{if } e = x_{i,j}x_{i+1,j}, \text{ when } i \text{ is odd and } j = 1 \\ 2nm - i - 1, & \text{if } e = x_{i,j}x_{i+1,j}, \text{ when } i \text{ is even and } j = 1 \end{cases}$$

And the super total edge weight are as follows

$$w(e) = \begin{cases} \frac{5nm-2n+4}{2}, & \text{if } e = x_{i,j}x_{i,j+1}, \text{ when } i \text{ is odd and } j \text{ is odd} \\ \frac{5nm-2n+2}{2}, & \text{if } e = x_{i,j}x_{i,j+1}, \text{ when } i \text{ is even and } j \text{ is odd} \\ \frac{5nm+2n+4}{2}, & \text{if } e = x_{i,j}x_{i,j+1}, \text{ when } i \text{ is odd and } j \text{ is even} \\ \frac{5nm+2n+2}{2}, & \text{if } e = x_{i,j}x_{i,j+1}, \text{ when } i \text{ is even and } j \text{ is even} \\ \frac{4nm+n+2}{2}, & \text{if } e = x_{i,j}x_{i+1,j}, \text{ when } i = 1 \text{ and } j = 1 \\ \frac{4nm+n+4}{2}, & \text{if } e = x_{i,j}x_{i+1,j}, \text{ when } i \text{ is odd and } j = 1 \\ \frac{4nm+n}{2}, & \text{if } e = x_{i,j}x_{i+1,j}, \text{ when } i \text{ is even and } j = 1 \end{cases}$$

Hence, from the above super total edge weights, it gives $\gamma_{leat}(P_n \triangleright P_m) \le 7$. Based on Lemma 2.1, the lower bound is $\gamma_{leat}(P_n \triangleright P_m) \ge \gamma_{leat}(P_n) + \gamma_{leat}(P_m) = 5$. It concludes that $5 \le \gamma_{leat}(P_n \triangleright P_m) \le 7$ when n is even and m is odd. □

Theorem 2.3. *For $n, m > 3$, n and m is even, the super local edge antimagic total chromatic number of $P_n \triangleright P_m$ is*

$$6 \le \gamma_{leat}(P_n \triangleright P_m) \le 9$$

Proof. For n is even and m is odd, we will show that the lower bound of the local edge antimagic total coloring of $P_n \triangleright P_m$. Based on Lemma 2.1, the lower bound is $\gamma_{leat}(P_n \triangleright P_m) \ge \gamma_{leat}(P_n) + \gamma_{leat}(P_m) = 3 + 3 = 6$. Furthermore, we will show the upper bound of the local edge antimagic total coloring of $P_n \triangleright P_m$ as follows.

Define a bijection $f : V(P_n \triangleright P_m) \cup E(P_n \triangleright P_m) \longrightarrow \{1, 2, 3, ..., |V(P_n \triangleright P_m)| + |E(P_n \triangleright P_m)|\}$ by the following.

$$f(v) = \begin{cases} \frac{i+(j-1)n+1}{2}, & \text{if } v = x_{i,j}, \text{ when } i \text{ is odd and } j \text{ is odd} \\ \frac{i+jn}{2}, & \text{if } v = x_{i,j}, \text{ when } i \text{ is even and } j \text{ is odd} \\ \frac{n(m+j-1)+i+1}{2}, & \text{if } v = x_{i,j}, \text{ when } i \text{ is odd and } j \text{ is even} \\ \frac{n(m+j-2)+i}{2}, & \text{if } v = x_{i,j}, \text{ when } i \text{ is even and } j \text{ is even} \end{cases}$$

$$f(e)=\begin{cases} n(2m-1)-i+1, & \text{if } e=x_{i,j}x_{i,j+1}, \text{ when } 1\le i\le n \text{ and } j=1\\ n(2m+1-j)-i+1, & \text{if } e=x_{i,j}x_{i,j+1}, \text{ when } 1\le i\le n \text{ and } j \text{ is odd, } j\ne 1\\ n(2m-1-j)-i+1, & \text{if } e=x_{i,j}x_{i,j+1}, \text{ when } 1\le i\le n \text{ and } j \text{ is even}\\ 2nm-1, & \text{if } e=x_{i,j}x_{i+1,j}, \text{ when } i=1 \text{ and } j=1\\ 2nm-i+1, & \text{if } e=x_{i,j}x_{i+1,j}, \text{ when } i \text{ is odd, } i\ne 1 \text{ and } j=1\\ 2nm-i-1, & \text{if } e=x_{i,j}x_{i+1,j}, \text{ when } i \text{ is even and } j=1 \end{cases}$$

And the super total edge weight are as follows

$$w(e)=\begin{cases} \frac{5nm-n+4}{2}, & \text{if } e=x_{i,j}x_{i,j+1}, \text{ when } i \text{ is odd and } j=1\\ \frac{5nm-n+2}{2}, & \text{if } e=x_{i,j}x_{i,j+1}, \text{ when } i \text{ is even and } j=1\\ \frac{5nm+n+4}{2}, & \text{if } e=x_{i,j}x_{i,j+1}, \text{ when } i \text{ is odd and } j \text{ is odd, } j\ne 1\\ \frac{5nm+n+2}{2}, & \text{if } e=x_{i,j}x_{i,j+1}, \text{ when } i \text{ is even and } j \text{ is odd, } j\ne 1\\ \frac{5nm-3n+4}{2}, & \text{if } e=x_{i,j}x_{i,j+1}, \text{ when } i \text{ is odd and } j \text{ is even}\\ \frac{5nm-3n+2}{2}, & \text{if } e=x_{i,j}x_{i,j+1}, \text{ when } i \text{ is even and } j \text{ is even}\\ \frac{4nm+n+2}{2}, & \text{if } e=x_{i,j}x_{i+1,j}, \text{ when } i=1 \text{ and } j=1\\ \frac{4nm+n+4}{2}, & \text{if } e=x_{i,j}x_{i+1,j}, \text{ when } i \text{ is odd and } j=1\\ \frac{4nm+n}{2}, & \text{if } e=x_{i,j}x_{i+1,j}, \text{ when } i \text{ is even and } j=1 \end{cases}$$

Hence, from the above super total edge weights, it gives $\gamma_{leat}(P_n \triangleright P_m) \le 9$. Based on Lemma 2.1, the lower bound is $\gamma_{leat}(P_n \triangleright P_m) \ge \gamma_{leat}(P_n) + \gamma_{leat}(P_m) = 6$. It concludes that $6 \le \gamma_{leat}(P_n \triangleright P_m) \le 9$ when n and m is even. □

Theorem 2.4. *For $n, m \ge 3$ and n is odd, the local edge antimagic chromatic number of $P_n \triangleright C_m$ is*

$$\gamma_{leat}(P_n \triangleright C_m) = 5$$

Proof. The graph $P_n \triangleright C_m$ is a connected graph with vertex set $V(P_n \triangleright C_m) = \{x_{i,j} : 1 \le i \le n, 1 \le j \le m\}$ and edge set $E(P_n \triangleright C_m) = \{x_{i,1}x_{i+1,1} : 1 \le i \le n-1\} \cup \{x_{i,j}x_{i,j+1}; 1 \le i \le n, 1 \le j \le m-1\} \cup \{x_{i,1}x_{i,m} : 1 \le i \le n\}$. Hence $|V(P_n \triangleright C_m)| = mn$ and $|E(P_n \triangleright C_m)| = mn + n - 1$. We will describe the proof into two cases.

Case 1: For n is odd and m is odd, we will show that the lower bound of the local edge antimagic total coloring of $P_n \triangleright C_m$ is $\gamma_{leat}(P_n \triangleright C_m) \ge 5$. Based on Lemma 2.1, the lower bound is $\gamma_{leat}(P_n \triangleright C_m) \ge \gamma_{leat}(P_n) + \gamma_{leat}(C_m) = 2 + 3 = 5$. It concludes that the lower bound of the local edge antimagic total coloring of $P_n \triangleright C_m$ is $\gamma_{leat}(P_n \triangleright C_m) \ge 5$. Furthermore, we will show that the upper bound of the local edge antimagic total coloring of $P_n \triangleright C_m$ is $\gamma_{leat}(P_n \triangleright C_m) \le 5$.

Define a bijection $f : V(P_n \triangleright C_m) \cup E(P_n \triangleright C_m) \longrightarrow \{1, 2, 3, ..., |V(P_n \triangleright C_m)| + |E(P_n \triangleright C_m)|\}$ by the following.

$$f(v)=\begin{cases} \frac{i+1}{2}, & \text{if } v=x_{i,j}, \text{ when } i \text{ is odd and } j=1\\ \frac{n+i+1}{2}, & \text{if } v=x_{i,j}, \text{ when } i \text{ is even and } j=1\\ \frac{i+jn}{2}, & \text{if } v=x_{i,j}, \text{ when } i \text{ is odd and } j \text{ is odd, } j\ne 1\\ \frac{i+(j-1)n}{2}, & \text{if } v=x_{i,j}, \text{ when } i \text{ is even and } j \text{ is odd, } j\ne 1\\ 1-i+\frac{(m+j+1)n}{2}, & \text{if } v=x_{i,j}, \text{ when } 1\le i\le n \text{ and } j \text{ is even} \end{cases}$$

$$f(e)=\begin{cases}\frac{i}{2}+n(2m-3), & \text{if } e=x_{i,j}x_{i,j+1}, \text{ when } i \text{ is even and } j=1\\ \frac{n+i}{2}+n(2m-3), & \text{if } e=x_{i,j}x_{i,j+1}, \text{ when } i \text{ is odd and } j=1\\ \frac{i+1}{2}+n(2m-2), & \text{if } e=x_{i,j}x_{i,j+1}, \text{ when } i \text{ is odd and } j=2\\ \frac{i+1+n}{2}+n(2m-2), & \text{if } e=x_{i,j}x_{i,j+1}, \text{ when } i \text{ is even and } j=2\\ \frac{i+1}{2}+n(2m-2-j), & \text{if } e=x_{i,j}x_{i,j+1}, \text{ when } i \text{ is odd and } j \text{ is odd}\\ \frac{i+1+n}{2}+n(2m-2-j), & \text{if } e=x_{i,j}x_{i,j+1}, \text{ when } i \text{ is even and } j \text{ is odd}\\ \frac{i+1}{2}+2nm-nj, & \text{if } e=x_{i,j}x_{i,j+1}, \text{ when } i \text{ is odd and } j \text{ is even}\\ \frac{i+1+n}{2}+2nm-nj, & \text{if } e=x_{i,j}x_{i,j+1}, \text{ when } i \text{ is even and } j \text{ is even}\\ 2nm-i+1, & \text{if } e=x_{i,1}x_{i,m}, \text{ when } 1\le i\le n\\ 2nm+n-i-1, & \text{if } e=x_{i,j}x_{i+1,j}, \text{ when } i \text{ is odd, } j=1\\ 2nm+n-i+1, & \text{if } e=x_{i,j}x_{i+1,j}, \text{ when } i \text{ is even, } j=1\end{cases}$$

And the super total edge weight are as follows

$$w(e)=\begin{cases}\frac{5nm-2n+3}{2}, & \text{if } e=x_{i,j}x_{i,j+1}, \text{ when } 1\le i\le n \text{ and } j \text{ is odd}\\ \frac{5nm+2n+3}{2}, & \text{if } e=x_{i,j}x_{i,j+1}, \text{ when } 1\le i\le n \text{ and } j \text{ is even}\\ \frac{5nm+3}{2}, & \text{if } e=x_{i,1}x_{i,m}, \text{ when } 1\le i\le n\\ \frac{4nm+3n+1}{2}, & \text{if } e=x_{i,j}x_{i+1,j}, \text{ when } i \text{ is odd and } j=1\\ \frac{4nm+3n+5}{2}, & \text{if } e=x_{i,j}x_{i+1,j}, \text{ when } i \text{ is even and } j=1\end{cases}$$

Hence, from the super total above edge weights, it easy to see that f induces a proper edge coloring of $P_n \triangleright C_m$ and it gives $\gamma_{leat}(P_n \triangleright C_m) \le 5$. Based on Lemma 2.1, the lower bound is $\gamma_{leat}(P_n \triangleright C_m) \ge \gamma_{leat}(P_n) + \gamma_{leat}(C_m) = 5$. It concludes that $\gamma_{leat}(P_n \triangleright C_m) = 5$.

Case 2: For n is odd and m is even, we will show that the lower bound of the local edge antimagic total coloring of $P_n \triangleright C_m$ is $\gamma_{leat}(P_n \triangleright C_m) \ge 5$. Based on Lemma 2.1, the lower bound is $\gamma_{leat}(P_n \triangleright C_m) \ge \gamma_{leat}(P_n) + \gamma_{leat}(C_m) = 2 + 3 = 5$. It concludes that the lower bound of the local edge antimagic total coloring of $P_n \triangleright C_m$ is $\gamma_{leat}(P_n \triangleright C_m) \ge 5$. Furthermore, we will show that the upper bound of the local edge antimagic total coloring of $P_n \triangleright C_m$ is $\gamma_{leat}(P_n \triangleright C_m) \le 5$.

Define a bijection $f: V(P_n \triangleright S_m) \cup E(P_n \triangleright S_m) \longrightarrow \{1, 2, 3, ..., |V(P_n \triangleright P_m)| + |E(P_n \triangleright P_m)|\}$ by the following.

$$f(v)=\begin{cases}\frac{i+1}{2}, & \text{if } v=x_{i,j}, \text{ when } i \text{ is odd and } j=1\\ \frac{n+i+1}{2}, & \text{if } v=x_{i,j}, \text{ when } i \text{ is even and } j=1\\ \frac{i+jn}{2}, & \text{if } v=x_{i,j}, \text{ when } i \text{ is odd and } j \text{ is odd, } j\ne 1\\ \frac{i+(j-1)n}{2}, & \text{if } v=x_{i,j}, \text{ when } i \text{ is even and } j \text{ is odd, } j\ne 1\\ 1-i+\frac{(m+j+1)n}{2}, & \text{if } v=x_{i,j}, \text{ when } 1\le i\le n \text{ and } j \text{ is even}\end{cases}$$

$$f(e)=\begin{cases}\frac{i}{2}+n(2m-1), & \text{if } e=x_{i,j}x_{i,j+1}, \text{ when } i \text{ is even and } j=1\\ \frac{n+i}{2}+n(2m-1), & \text{if } e=x_{i,j}x_{i,j+1}, \text{ when } i \text{ is odd and } j=1\\ \frac{i+1}{2}+n(2m-j-2), & \text{if } e=x_{i,j}x_{i,j+1}, \text{ when } i \text{ is odd and } j \text{ is even}\\ \frac{i+1+n}{2}+n(2m-j-2), & \text{if } e=x_{i,j}x_{i,j+1}, \text{ when } i \text{ is even and } j \text{ is even}\\ \frac{i+1}{2}+n(2m-j), & \text{if } e=x_{i,j}x_{i,j+1}, \text{ when } i \text{ is odd and } j \text{ is odd}, j\neq 1\\ \frac{i+1+n}{2}+n(2m-j), & \text{if } e=x_{i,j}x_{i,j+1}, \text{ when } i \text{ is even and } j \text{ is odd}, j\neq 1\\ \frac{i}{2}+n(2m-2), & \text{if } e=x_{i,1}x_{i,m}, \text{ when } i \text{ is even}\\ \frac{i+n}{2}+n(2m-2), & \text{if } e=x_{i,1}x_{i,m}, \text{ when } i \text{ is odd}\\ 2nm+n-i-1, & \text{if } e=x_{i,j}x_{i+1,j}, \text{ when } i \text{ is odd}, j=1\\ 2nm+n-i+1, & \text{if } e=x_{i,j}x_{i+1,j}, \text{ when } i \text{ is even}, j=1\end{cases}$$

And the super total edge weight are as follows

$$w(e)=\begin{cases}\frac{5nm+n+3}{2}, & \text{if } e=x_{i,j}x_{i,j+1}, \text{ when } 1\le i\le n \text{ and } j \text{ is odd}\\ \frac{5nm-3n+3}{2}, & \text{if } e=x_{i,j}x_{i,j+1}, \text{ when } 1\le i\le n \text{ and } j \text{ is even}\\ \frac{6nm-3n+3}{2}, & \text{if } e=x_{i,1}x_{i,m}, \text{ when } 1\le i\le n\\ \frac{4nm+3n+1}{2}, & \text{if } e=x_{i,j}x_{i+1,j}, \text{ when } i \text{ is odd and } j=1\\ \frac{4nm+3n+5}{2}, & \text{if } e=x_{i,j}x_{i+1,j}, \text{ when } i \text{ is even and } j=1\end{cases}$$

Hence, from the above super total edge weights, it easy to see that f induces a proper edge coloring of $P_n \triangleright C_m$ and it gives $\gamma_{leat}(P_n \triangleright C_m) \le 5$. Based on Lemma 2.1, the lower bound is $\gamma_{leat}(P_n \triangleright C_m) \ge \gamma_{leat}(P_n)+\gamma_{leat}(C_m) = 5$. It concludes that $\gamma_{leat}(P_n \triangleright C_m) = 5$. Based on case 1 and case 2, so the super local edge antimagic total chromatic number of $P_n \triangleright C_m$, with n odd is $\gamma_{leat}(P_n \triangleright C_m) = 5$. □

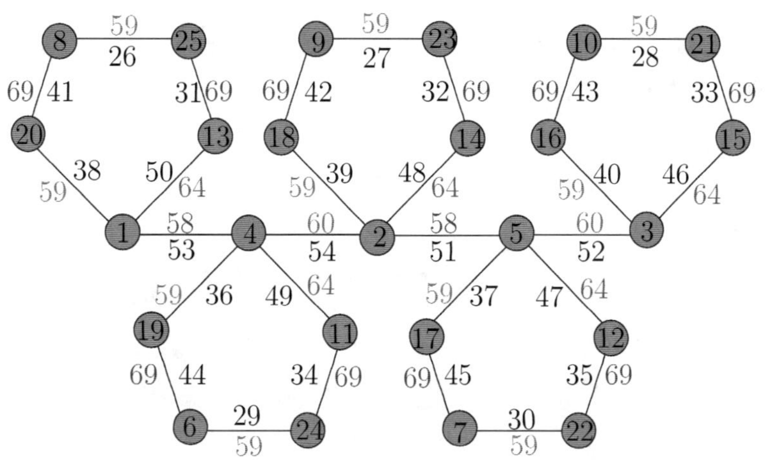

Figure 2. Illustration of local edge antimagic total coloring of $P_5 \triangleright C_5$

Theorem 2.5. *For $n, m \geq 3$ and n is odd, the local edge antimagic chromatic number of $P_n \triangleright S_m$ is*

$$\gamma_{leat}(P_n \triangleright S_m) = 2 + m$$

Proof. The graph $P_n \triangleright S_m$ is a connected graph with vertex set $V(P_n \triangleright S_m) = \{a_i, x_i; 1 \leq i \leq n\} \cup \{x_{i,j}; 1 \leq i \leq n, 1 \leq j \leq m-1\}$ and edge set $E(P_n \triangleright S_m) = \{a_i x_i; 1 \leq i \leq n\} \cup \{x_i x_{i+1}; 1 \leq i \leq n-1\} \cup \{a_i x_{i,j}; 1 \leq i \leq n, 1 \leq j \leq m-1\}$. Hence, $|V(P_n \triangleright S_m)| = mn + n$ and $|E(P_n \triangleright S_m)| = mn + n - 1$.

We will show that the lower bound of the local edge antimagic total coloring of $P_n \triangleright S_m$ is $\gamma_{leat}(P_n \triangleright S_m) \geq m + 2$. Based on Lemma 2.1, the lower bound is $\gamma_{leat}(P_n \triangleright S_m) \geq \gamma_{leat}(P_n) + \gamma_{leat}(S_m) = 2 + m$. It concludes that the lower bound of the local edge antimagic total coloring of $P_n \triangleright S_m$ is $\gamma_{leat}(P_n \triangleright S_m) \geq m + 2$. Furthermore, we will show that the upper bound of the local edge antimagic total coloring of $P_n \triangleright S_m$ is $\gamma_{leat}(P_n \triangleright S_m) \leq m + 2$.

Define a bijection $f : V(P_n \triangleright S_m) \cup E(P_n \triangleright S_m) \longrightarrow \{1, 2, 3, ..., |V(P_n \triangleright S_m)| + |E(P_n \triangleright S_m)|\}$ by the following.

$$f(v) = \begin{cases} \frac{i+1}{2}, & \text{if } v = x_i,\ i \text{ odd},\ 1 \leq i \leq n \\ \frac{i+n+1}{2}, & \text{if } v = x_i,\ i \text{ even},\ 1 \leq i \leq n \\ \frac{i}{2} + n, & \text{if } v = a_i,\ i \text{ even},\ 1 \leq i \leq n \\ \frac{i+n}{2} + n, & \text{if } v = a_i,\ i \text{ odd},\ 1 \leq i \leq n \\ \frac{i+1}{2} + n(j+1), & \text{if } v = x_{i,j},\ i \text{ odd},\ 1 \leq i \leq n,\ 1 \leq j \leq m-1 \\ \frac{i+n+1}{2} + n(j+1), & \text{if } v = x_{i,j},\ i \text{ even},\ 1 \leq i \leq n,\ 1 \leq j \leq m-1 \end{cases}$$

$$f(e) = \begin{cases} nm + 2n - i + 1, & \text{if } e = a_i x_i,\ 1 \leq i \leq n \\ n(m+j+2) - i + 1, & \text{if } e = a_i x_{i,j},\ 1 \leq i \leq n,\ 1 \leq j \leq m-1 \\ 2nm + 2n - i - 1, & \text{if } e = x_i x_{i+1},\ i \text{ odd},\ 1 \leq i \leq n-1 \\ 2nm + 2n - i + 1, & \text{if } e = x_i x_{i+1},\ i \text{ even},\ 2 \leq i \leq n-1 \end{cases}$$

And the super total edge weight are as follows

$$w_t(e) = \begin{cases} \frac{2nm+7n+3}{2}, & \text{if } e = a_i x_i,\ 1 \leq i \leq n \\ \frac{2nm+4nj+9n+3}{2}, & \text{if } e = a_i x_{i,j},\ 1 \leq i \leq n \text{ dan } 1 \leq j \leq m-1 \\ \frac{4nm+5n+1}{2}, & \text{if } e = x_i x_{i+1},\ i \text{ odd},\ 1 \leq i \leq n-1 \\ \frac{4nm+5n+5}{2}, & \text{if } e = x_i x_{i+1},\ i \text{ even},\ 1 \leq i \leq n-1 \end{cases}$$

Hence, from the above super total edge weights, it easy to see that f induces a proper edge coloring of $P_n \triangleright S_m$ and it gives $\gamma_{leat}(P_n \triangleright S_m) \leq 5$. Based on Lemma 2.1, the lower bound is $\gamma_{leat}(P_n \triangleright S_m) \geq \gamma_{leat}(P_n) + \gamma_{leat}(S_m) = 2 + m$. It concludes that $\gamma_{leat}(P_n \triangleright C_m) = 2 + m$. So the super local edge antimagic total chromatic number of $P_n \triangleright S_m$, with n odd is $\gamma_{leat}(P_n \triangleright S_m) = 2 + m$. □

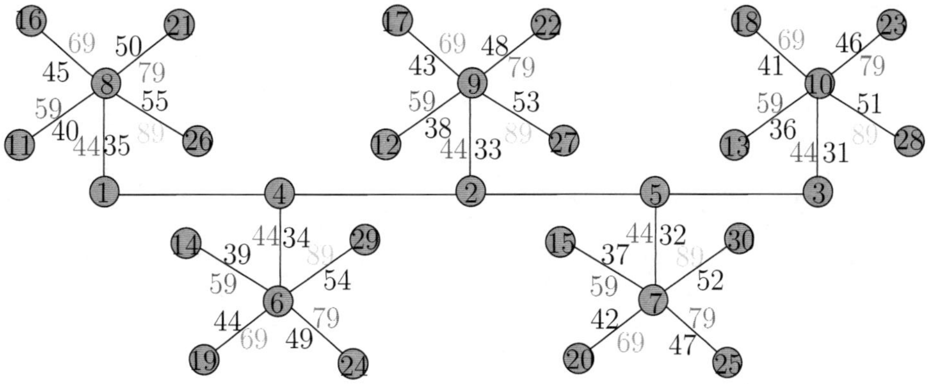

Figure 3. Illustration of local edge antimagic total coloring of $P_5 \rhd S_5$

3. Conclusion

In this paper we have given an asymptotically tight result on super local edge antimagic total chromatic number of comb products of path and any graphs. Hence the following problem aries naturally.

Open Problem 3.1. *Determine the lower bound and upper local edge antimagic coloring of* $G \rhd H$

Acknowledgement

We gratefully acknowledge the support from CGANT - University of Jember of year 2017.

References

[1] Gross J L, Yellen J and Zhang P 2014 *Handbook of graph Theory* Second Edition CRC Press Taylor and Francis Group

[2] Chartrand G and Lesniak L 2000 *Graphs and digraphs 3rd* ed (London: Chapman and Hall)

[3] Hartsfield N dan Ringel G 1994 *Pearls in Graph Theory* Academic Press. United Kingdom

[4] Arumugam S, Premalatha K, Baca M and Semanicova-Fenovcikova A 2017 Local Antimagic Vertex Coloring of a Graph, *Graphs and Combinatorics* **33** 275-285

[5] Dafik, Miller M, Ryan J, and Baca M 2011 Super edge-antimagic total labelings of $mK_{n,n}$ *Ars Combinatoria* **101** 97-107

[6] Dafik, M Mirka, Joe Ryan, and Martin Baca 2006 Super edge-antimagicness for a class of disconnected graphs

[7] Agustin I H, Dafik, Moh. Hasan, Alfarisi R and Prihandini R M 2017 Local edge antimagic coloring of graphs *Far East Journal of Mathematical Sciences.*

[8] Agustin I H, Dafik, Slamin, Ridho Alfarisi and Kurniawati E Y 2017 The construction of super local edge antimagic total coloring by using an EAVL Technique *accepted*

[9] Agustin I H, Dafik, Slamin, Ermita E R and Alfarisi R 2017 On the total local edge super antimagicness of special graph and graph with pendant edge *accepted*

[10] Alfarisi R, Dafik, Agustin I H and Kristiana A I 2017 The local edge antimagic total coloring of some wheel related graphs *Far East Journal of Mathematical Sciences*

IOP Conf. Series: Journal of Physics: Conf. Series **1008** (2018) 012037 doi:10.1088/1742-6596/1008/1/012037

On the modification Highly Connected Subgraphs (HCS) algorithm in graph clustering for weighted graph

E R Albirri[1], K A Sugeng[2], D Aldila[2]
[1]Mathematics Edu. Depart. University of Jember, Indonesia
[1]Mathematics Depart. Universitas Indonesia, Indonesia
E-mail: ermitara@unej.ac.id, kiki@sci.ui.ac.id, aldiladipo@sci.ui.ac.id

Abstract. Nowadays, in the modern world, since technology and human civilization start to progress, all city in the world is almost connected. The various places in this world are easier to visit. It is an impact of transportation technology and highway construction. The cities which have been connected can be represented by graph. Graph clustering is one of ways which is used to answer some problems represented by graph. There are some methods in graph clustering to solve the problem spesifically. One of them is Highly Connected Subgraphs (HCS) method. HCS is used to identify cluster based on the graph connectivity k for graph G. The connectivity in graph G is denoted by $k(G)$. If $k(G) > \frac{n}{2}$ that n is the total of vertices in G, then it is called as HCS or the cluster. This research used literature review and completed with simulation of program in a software. We modified HCS algorithm by using weighted graph. The modification is located in the Process Phase. Process Phase is used to cut the connected graph G into two subgraphs H and $\bar{H}$. We also made a program by using software Octave-401. Then we applied the data of Flight Routes Mapping of One of Airlines in Indonesia to our program.

1. Introduction

In this new world, technology, communication and human civilization are growing rapidly. Everyone can communicate and of course visits each other easily. All places in the world are almost connected. This connection can be represented by graph [3]. Graph Theory is a theory in discrite mathematics which discusses not only about theory but also the application. One of the graph theory application is graph clustering. There are many ways or methods to cluster the graph easily, one of them is Highly Connected Subgraphs (HCS) method.

Clustering is defined as a process of grouping element from a data into subset of its element of data (cluster) [6],[7]. *Graph Clustering* is a process of grouping elements in graph (edge or vertex) becoming subgraphs, it is called as *cluster* [9],[2]. *Highly Connected Subgraphs (HCS)*, is a method of graph clustering to find out the cluster based on its edge connectivity. This method can be used on two kinds of graph. They are classified as weighted graph and non-weighted graph. The determination of the graph is done to get the result of minimum cut. Next, we look for the subgraphs which result in clusters, identified as Highly Connected Subgraphs (HCS). The *edge connectivity* of G ($k(G)$) is minimum of the sum of edges in G which is cut such that G becomes disconnected [6],[10]. *Cut* in G is a set of edge S which if it is deleted, then it will make G be disconnected [5].

IOP Conf. Series: Journal of Physics: Conf. Series **1008** (2018) 012037 doi:10.1088/1742-6596/1008/1/012037

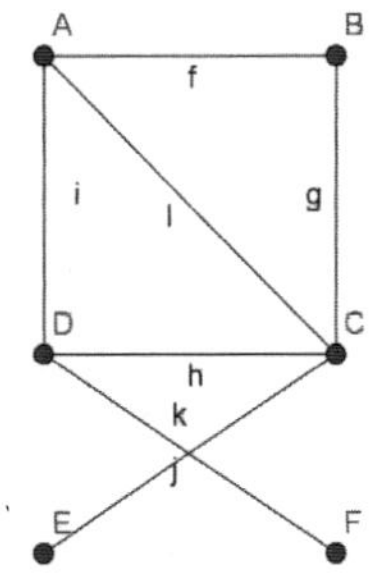

Figure 1. Simple Connected Weighted Graph

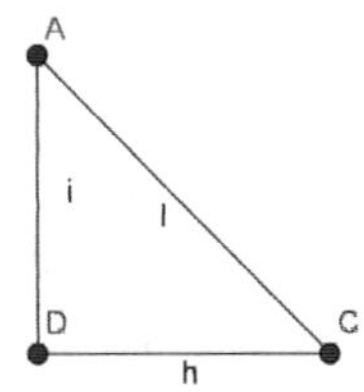

Figure 2. The Example of HCS Subgraph

Let G be a simple, connected, and weighted graph which can be seen in Figure 1. Based on the label, G can be labeled in three ways. It can be done in every vertex, edges, or both. Moreover, G which has been labeled in the edges is called *weighted graph* [11],[8]. Weighted graph consists of two, weighted vertex graph and weighted edge graph. In this research, we use weighted edge graph and call it as weighted graph.

One of matrix which represents a graph is adjacency matrix. Let G be a graph which has $V(G) = \{1, 2, \ldots, n\}$ and $E(G) = \{e_1, e_2, \ldots, e_n\}$. *Adjacency matrix* for non-weighted graph is denoted by $A(G) = [a_{ij}]$ [1]. $A(G)$ has size $n \times n$ with:

$$a_{i,j} = \begin{cases} 1 & \text{if } v_i \text{ adjacent } v_j; \\ 0 & \text{others.} \end{cases}$$

Next, adjacency matrix can contain edge weight of graph G as it is defined above. It changes entry 1 with its edge weight.

2. Research Method

The research methods that we use is the literature review and completed with simulation by using software Octave−401. The literature review has some steps. It consists of:

(i) study of general graph

(ii) study of general graph clustering

(iii) study of HCS algorithm for weighted graph

Then, we modify HCS algorithm of weighted graph in Octave language by using software Octave−401.

IOP Conf. Series: Journal of Physics: Conf. Series **1008** (2018) 012037 doi:10.1088/1742-6596/1008/1/012037

3. Result

In this section we partially discuss about Highly Connected Subgraphs, the algorithm of Highly Connected Subgraphs, the modification of algorithm for weighted graphs, and the application of Highly Connected Subgraphs for weighted graph in software Octave−401.

3.1. Highly Connected Subgraphs

Definition 1 *[6] Highly Connected Subgraphs (HCS) is a method of graph clustering which aims to look for cluster based on edge connectivity ($k(G)$) of graphs.*

Algorithm of this method has the same name as the name of the method. It is Highly Connected Subgraphs (HCS) algorithm. This algorithm is found by Hartuv and Shamir [6]. HCS algorithm is used to do graph clustering in a graph data which produces one or more clusters. The cluster is identified as HCS if the total of edge connectivity $k(G) > \frac{|V|}{2}$. For example, there is graph $G = (V(G), E(G))$ with $V(G) = \{A, B, C, D, E, F\}$ and $E(G) = \{f, g, h, i, j, k, l\}$. The graph in Figure 1 is not HCS because of $k(G) = 3$ and $\frac{|V|}{2} = 3$. Besides, subgraph $H = (V(H), E(H))$ with $V(H) = \{A, D, C\}$ and $E(H) = \{i, h, l\}$ in the Figure 1 is HCS because it has $k(H) = 2$ and $\frac{|V|}{2} = \frac{3}{3}$. Subgraph H can be seen in Figure 2. The Algorithm 1 is HCS algorithm owned by Hartuv and Shamir [6].

HCS($G(V, E)$)
begin
$(H, \bar{H}, C) \longleftarrow$ MINCUT (G)
if *G is Highly Connected Subgraph (HCS)* **then**
| return G
else
| HCS$\{H\}$ HCS$\{\bar{H}\}$
end

Algorithm 1: HCS Algorithm

3.2. Algorithm of Highly Connected Subgraphs

HCS algorithm for non-weighted graph is started by looking for the total of minimum cut in graph. *Minimum Cut* in this graph is defined as minimum number of cut edges such that it becomes disconnected graph. Next, we recheck whether G is HCS cluster. If it is not, then the plot of this algorithm recurs in subgraph H till we get the HCS cluster, as well as in subgraph $\bar{H}$. Figure 3 shows flowchart of using HCS in non-weighted graph [10].

3.3. Modification of HCS Algorithm for Weighted Graphs

It needs a modification to create a program by using HCS algorithm. It happens because the minimum cut used is the weighted graph. *Minimum Cut* is the minimum number of deleted graph edge which makes G be disconnected graph. In Figure 4 you can see the Process Phase (G, w) on a blue color. On weighted graph, the determination of *minimum cut* is different. It is minimum number of the weight on the graph which is lost to make graph G be disconnected. Because we use the weighted graph, we need an additional phase to determine the minimum cut which will be cut. This *phase* is called Process Phase. We can see the flowchart of Process Phase (G, w) which uses minimum cut algorithm for weighted graph in Figure 5. The implementation of this algorithm is problematic, that is why we do research on it.

The modification of HCS algorithm can be seen at Figure 4. Figure 4 is the center flowchart of HCS algorithm. There are other flowcharts in this Figure, those are Process Phase flowchart

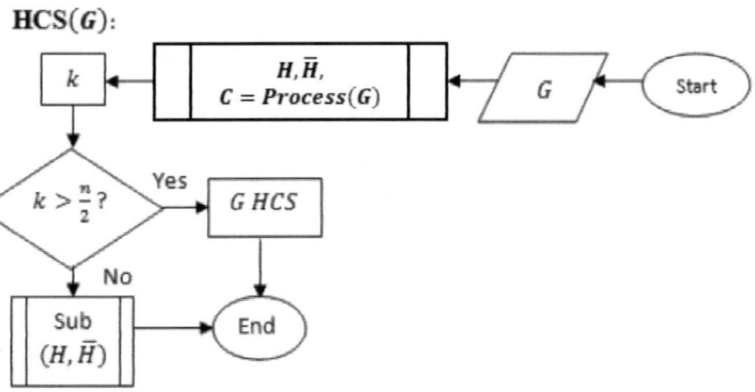

Figure 3. HCS Flowchart in Non-weighted Graph

in Figure 5 and Sub$(H, \bar{H})$ Phase flowchart in Figure 7. The algorithm modification for weighted graphs is in the Process Phase flowchart in Figure 5 and ofcourse we add another phase, that is the Minimum Cut Phase flowchart in Figure 6 to complete the algorithm. In the previous research [6] and [10], the previous researchers found how to get HCS from connected non-weighted graph. Since we have different object, weighted graph, so it can change the algorithm too. First, Process Phase (G, w) in Figure 5, there is another process, that is Minimum Cut Phase flowchart which can be seen in Figure 6. Process Phase is used to find out the weight of mincut $w(mincut)$ of G or we can say that how many minimum cuts there are such that we get two disconnected subgraphs H and $\bar{H}$. It starts by inputting graph G data as weighted adjacency matrix in program (software), then do initialization of weight w of mincut which is considered as ∞. It makes us be able to find minimum weight or minimum cut of graph G. Next, we make condition that the graph is not singleton (one-vertex-graph). If the graph is singleton, then the graph is not processed in graph clustering (clustering is done).

Next, if the graph is not singleton then it is continued to Minimum Cut Phase or $m = Mincutphase(G, w)$. The Minimum Cut Phase is used to find out where edges in G which will be cut based on the number of minimum cuts. Second, Sub$(H, \bar{H})$ Phase is used to repeat the program/algorithm for subgraph H and $\bar{H}$ till we find the HCS or singleton. The purpose of this phase is getting *phase-cut* so that program can cut some edges and finally it becomes 2-disconnected-subgraphs. First, we choose any vertex a in G. Then, a is saved in new vertex set A. Since vertex set $A \neq V$ then the process will continue to look for vertex v which is Most Tightly Connected Vertex (MCTV of vertex set $V - A$). Next, we get new set $A = A \cup \{v\}$. Its purpose is to look for two last vertices (s and t) in V. If it is fulfilled and there is a vertex $v = t$ left, then we get phase cut = cut $(A - \{t\}, \{t\})$.

After we get the value of Minimum Cut Phase (m), the Figure 5 shows that given a condition when weight w in Minimum Cut Phase is less than w in Minimum Cut (which has been initializated before in the first time), thus weight w in Minimum Cut which resulted is w in Minimum Cut Phase $(w(st))$. Next, Merge process in new graph G has new vertex set $V = V \setminus \{s, t\} \cup \{(st)\}$. Then, counting the total of weight $w(v, (st))$ in edges which is formed by Merge process. In this Phase, it will do looping till we have $|V| = 1$ or graph G is singleton and obtain Minimum Cut of G.

3.4. The Aplication of Highly Connected Subgraphs (HCS) Algorithm for Weighted Graphs in Software Octave–401

Using Highly Connected Subgraphs (HCS) algorithm can be applied in simple graph and real life problem, for example: the case of mapping of flight routes one of airlines in Indonesia. The

IOP Conf. Series: Journal of Physics: Conf. Series **1008** (2018) 012037 doi:10.1088/1742-6596/1008/1/012037

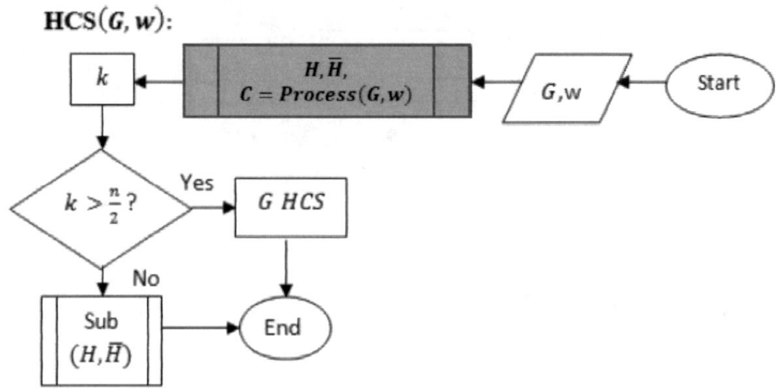

Figure 4. HCS Flowchart in Weighted Graph

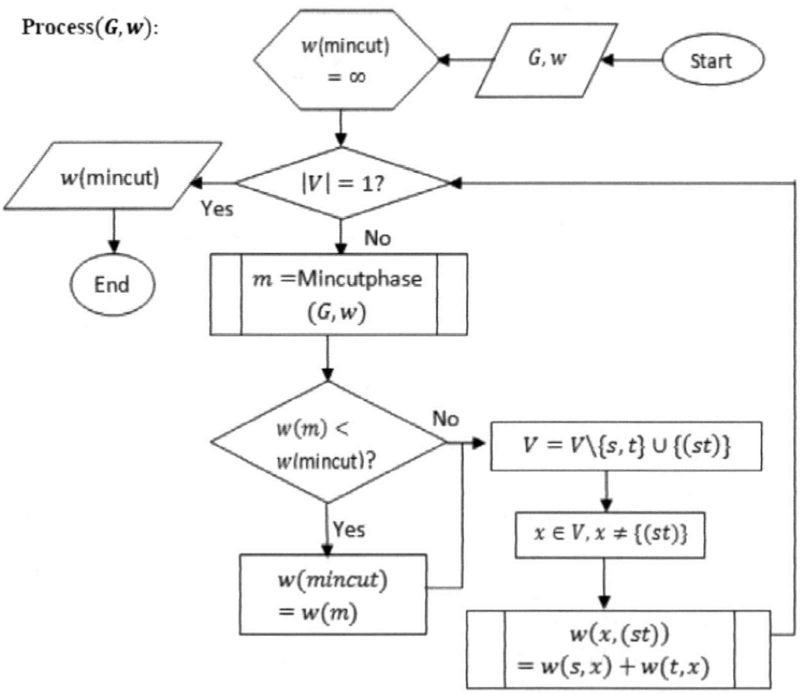

Figure 5. Process Phase

algorithm is written in Octave software language version 401 in computer with spec processor AMD Dual Core E-450 APU (1, 65 GHz), installed memory (RAM) 2 GB, and operating system Windows 7 Starter (32–bit).

3.4.1. The Aplication of HCS Algorithm of Simple Weighted Graph

To run the program, first we input graph data as weighted adjacency matrix and presume vertex h in editor of software Octave–401. Then "call" command to run syntax of Minimum Cut and HCS. Note the Figure 8, graph G has set of vertices $V = \{A, B, C, D, E, F, G\}$ which in syntax will define $A = 1$, $B = 2$, ..., $G = 7$. The following is an example to run the HCS

***Mincutphase*(*G*, *w*):**

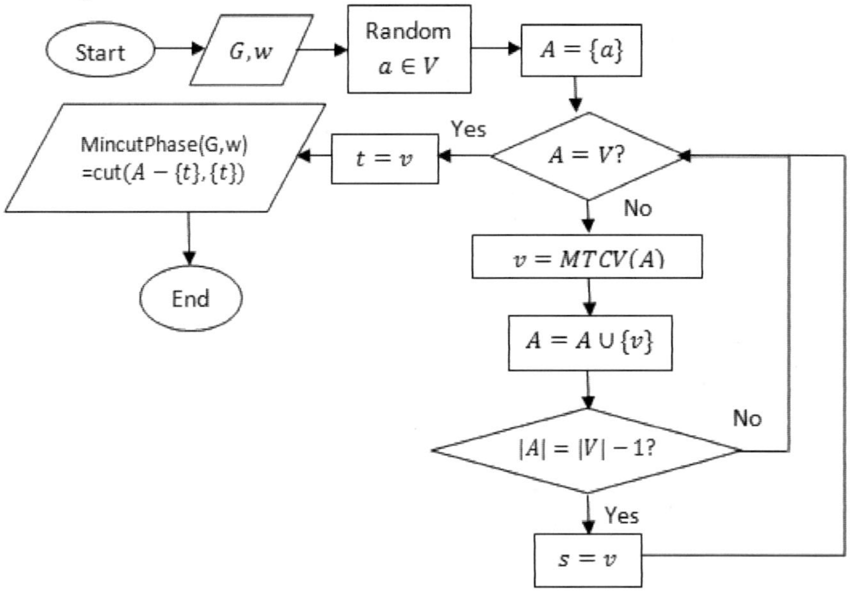

Figure 6. Minimum Cut Phase Flowchart

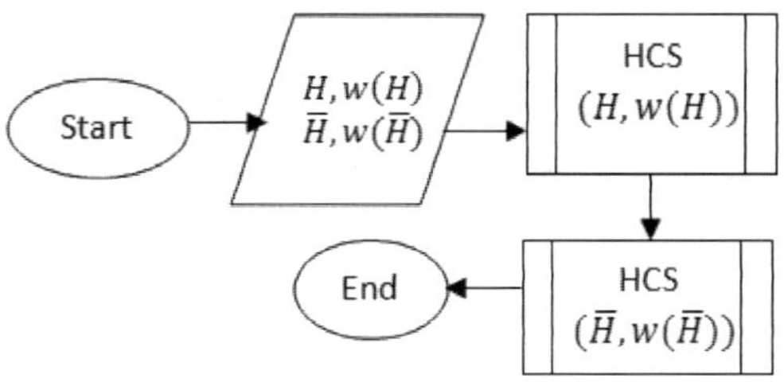

Figure 7. Flowchart of Phase Sub($H, \bar{H}$)

algorithm in Software Octave−401. Figure 9 is the result of running on the following Listing 1.

Listing 1. HCS Code of Graph G

```
n = 7 ;
g = zeros(n,n);

g(1,[2,4]) = [4 2];  g(2,3) = 4 ;
g(3,[4,6]) = [2 1];  g(4,[5,6]) = [2 1];
g(5,[6 7]) = [5 4];  g(6,7) = 3;

g = g + g';
(mv,mw) = mincut(g,[6 7])
(pb,qb) = HCS (g,mv,n)
```

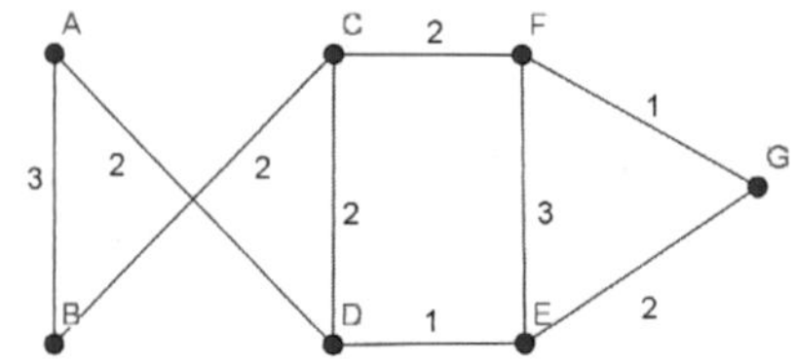

Figure 8. G Graph

```
mv =

   1
   1
   1
   1
   0
   0
   0


mw =

   4
```

Figure 9. Result of The (mv, mw) Running

Based on the result of the minimum-cut running (mv, mw) in Figure 9, it can be seen that there is Minimum Cut, symbolized mw, mw equals 4. Besides, mv has biner matrix which shows G has 7 vertices which is cut to separate edges in G into two subgraphs. They are pb subgraph which has 4 vertices and qb subgraph which has 3 vertices. Next, the command in HCS running (pb, qb) will be called as in Figure 10. There are two matrices which show two different subgraphs which is cut based on its Minimum Cut.

In this research, the program result which we created by using Octave−401 is a matrix. The matrix represented by subgraph is the Highly Connected Subgraph (HCS). The first running result in matrix which has binomial number. It explains the position of the "cut". The "cut" separates the graph G into two different subgraph H and $\bar{H}$. The subgraph H in this program is stated as pb and another one is stated as qb. Based on Listing 1 above, pb and qb are two subgraphs generated from cutting graph G. Since graph G is not HCS, it will check subgraph pb and all subgraphs generated from cutting by using minimum cut (mincut) in subgraph pb till it generates in subgraph (HCS cluster) or singleton. Next, through the same way, it checks subgraph qb. The command in this program uses mincut command and HCS command. So the result is that subgraph qb as HCS (cluster) of graph G with 7 vertices.

3.4.2. Aplication of Highly Connected Subgraphs (HCS) Algorithm of Flight Routes Mapping of One of Airlines in Indonesia

The aplication of HCS algorithm can be seen in a graph data in one of airlines in Indonesia. We use the distance of the vertices as the weight of edges in graph data. Based on Figure 11,

```
Not HCS

pb =

     0     3     0     2
     3     0     2     0
     0     2     0     2
     2     0     2     0


qb =

     0     3     2
     3     0     1
     2     1     0
```

Figure 10. Result of The (pb, qb) Running

there are 60 vertices connected. Next we input the weighted adjacency matrix of graph data. The Listing 2 for inputting data is like Listing 1 but we change the weight as the graph data of Flight Routes.

Listing 2. HCS Code of Flight Routes

```
clc;clear;
n = 60 ;
g = zeros(n,n);  % buat graf
p = zeros(n,n)

g(1,[7,22,29]) = [432.9 1769.83 2177.02];
g(2,[7,10,13,16,22,31]) = [635.59 307.94 681.16 406.31 838.77 1314.42];
g(3,[9,22]) = [300.67 607.89];
g(4,7) = 256.35;
g(5,[9,22]) = [170.46 661.62];
g(6,7) = 279.81;
g(7,[8,9,10,11,12,13,20,22]) = [385.87 877.12 454.36 223.7 431.68
1152.38 2204.96 1351.01];
g(8,[10,22]) = [114.6 976.83];
g(9,[13,14,16,20,22]) = [338.01 320.83 149.08 1334.36 500.54];
g(10,22) = 896.69;
g(11,22) = 1177.28;
g(13,22) = 207.12;
g(14,[16,22]) = [171.45 453.85];
g(15,22) = 717.53;
g(16,22) = 452.06;
g(17,[20,31]) = [867.02 577.88];
g(18,31) = 210.11;
g(19,[20,27,38]) = [393.89 271.41 341.06];
```

```
g(20,[21,22,24,25,27,28,29,31,32,33,38,40,55,59]) = [646 949.78 948.3
513.15 122.48  537.97 496.39 293.16 453.32 545.37 573.24 1463.33
2896.24 2437.42];
g(21,[24,25]) = [355.69 132.85];
g(22,[24,26,27,28,29,31,33,35,36,38,40,41,42,43,45,46,50,53,
54,55,57,59]) = [1897.97 648.94 1069.5 412.18 453.41 661.94 404.88
1918.99 1768.28 1397.57 2144.21 1592.92 1241.74 921.37 901.28 722.44
2389.2 2374.26 327.85 3777.06 3709.48 3334.98];
g(23,31) = 152.66;
g(24,[25,31,32]) = [466.51 1239.31 494.99];
g(25,32) = 130.83;
g(27,[28,30,31,38]) = [658.43 121.52 409.27 478.68];
g(28,[29,31,38]) = [42.62 255.38 1019.17];
g(31,[33,38,42,43,50,55]) = [262.96 771.49 807.36 489.34 1766.68
3143.02];
g(33,[38,42,43]) = [1026.58 980.73 645.57];
g(34,38) =350.59;
g(35,38) =686.36;
g(36,38) =407.66;
g(37,38) =669.59;
g(38,[39,40,41,42,50,53,54,55,56,58,59,60]) = [338.23 897.85 558.17
529.02 996.55 1042.66 1890.14 2380.13 1717.13 1429.91 1938.25 326.07];
g(40,[42,53,55,58,59]) = [915.73 302.83 1790.95 726.11  1449.59];
g(42,[43,44,45,46,48]) = [335.32 338.35 345.08 842.44 452.03];
g(45,46) = 554.75;
g(46,[47,49]) = [399.17 238.37];
g(50,[51,52,53,58]) = [550.62 648.79 345.68 479.95];
g(54,55) = 532.93;
g(55,[56,57,58,59]) = [767.44 667.81 1065.54 474.94];
g(56,58) =307.63;
g(58,59) =764.28;

g = g + g';
(mv,mw) = mincut(g,[...])
(pb,qb) = HCS (g,mv,n);
```

Based on Running Code in Listing 2 above, we get HCS cluster consisting 10 vertices (airports) in 10 cities. They are Denpasar, Jakarta, Surabaya, Makassar, Manado, Balikpapan, Palangkaraya, Pontianak, Jayapura, dan Timika. Figure 12 represents HCS cluster as a graph. There are some obstacles such that the graph data just gets a cluster. They are difference of degree in G and difference of edge weight in G. The difference of degree means that there is vertex which has a degree consisting one and there is also vertex which has degree consisting tens. It is like the difference of degree, the difference of edge weight also looks significant. It can be seen through Running Code in Listing 2, stating that weight of edge has a range from hundreds kilometer up to four a hundred and more kilometers up to four hundred kilometers. Due to the great difference between degree and edge weight, then it makes the graph G data get unbalance and it is not suitable for this HCS Algorithm. We can see that when the program do cutting on graph G, it results subgraph H and $\bar{H}$. The subgraph H is the biggest one. Meanwhile, the subgraph $\bar{H}$ is only singleton. Thus, the cluster result is only one-big-subgraph identified as HCS. Although the prediction results some subgraphs of HCS.

IOP Conf. Series: Journal of Physics: Conf. Series **1008** (2018) 012037 doi:10.1088/1742-6596/1008/1/012037

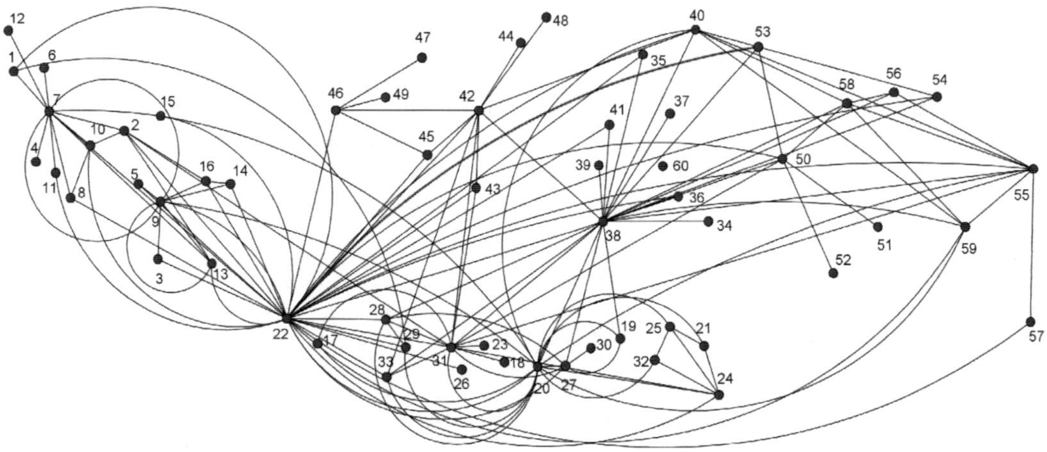

Figure 11. Flight Routes Mapping of One of Airlines in Indonesia as A Graph Data [4]

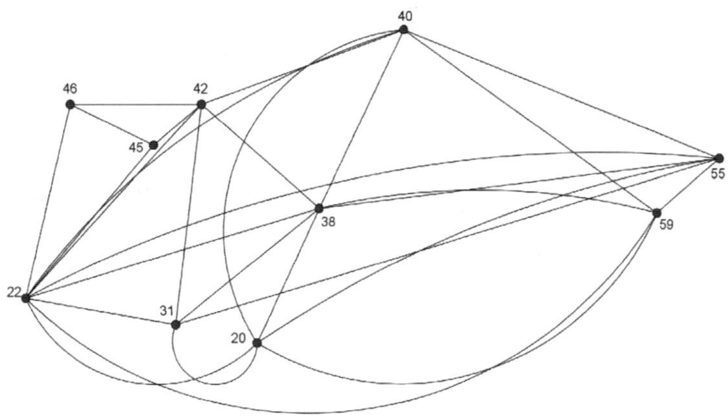

Figure 12. HCS Cluster from Running Program

4. Conclusion

We have discussed about modification of Highly Connected Subgraphs (HCS) algorithm for weighted graph and get conclusion such as:

(i) Modification of Highly Connected Subgraphs (HCS) algorithm for weighted graph can be done by adding variable h on algorithm to get two disconnected subgraphs by deleting some edges which have minimum weight (*minimum cut*).

(ii) Flight path data with edge weight as mileage is not recommended for implementation of Highly Connected Subgraphs (HCS) algorithm. The result is just has a cluster because there are significant differences. They are the difference of degree and the difference of edge weight.

Open Problem 1 *Highly Connected Subgraphs (HCS) algorithm for weighted graph by using another data, for example: protein data.*

Open Problem 2 *Modification of Highly Connected Subgraphs (HCS) algorithm for other kinds of graphs.*

Acknowledgments

This work was partially supported by the Mathematics Department, Universitas Indonesia 2018

References

[1] Bapat R B 2010 *Graphs and Matrices* New Delhi:Hindustan Book Agency
[2] Deo N 1990 *Graph Theory with Application to Engineering and Computer Science* New Delhi:Prentice Hall of India Private Limited
[3] Dafik, Miller M, Iliopoulos C, and Ryjacek Z 2009 On Diregularity of Digraphs of Defect at Most Two *Journal of Combinatorial Mathematics and Combinatorial Computing* **71** pp. 21-37
[4] *Flight Domestic Route of Garuda Indonesia* 2014 be quote April 15, 2016 from https://www.garuda-indonesia.com/id/id/destination/route-map/index-domestic.page
[5] Gross J L dan Yellen J 2006 *Graph Theory and Its Application* Boca Raton:Taylor & Francis Group
[6] Hartuv E and Shamir R 2000 *A clustering algorithm based on graph connectivity* Information Processing Letters **76(4-6)** pp. 175-181
[7] Przulj N 2005 *Knowledge Discovery in Proteomics:Graph Theory Analysis of Protein-Protein Interactions* Toronto:CRC Press
[8] Ruohonen K 2013 *Graph Theory* Kanada:University of British Columbia
[9] Schaeffer S E 2007 Graph Clustering *Journal of Computer Science* **Vol 1** pp. 27-64
[10] Stoer M F W 1997 A Simple Min Cut Algorithm *Journal of ACM* **44(4)** pp. 585-591
[11] Sugeng K A, Slamet S, and Silaban D R 2014 *Teori Graf dan Aplikasinya* Textbook:Mathematics Department FMIPA Universitas Indonesia

Local Edge Antimagic Coloring of Comb Product of Graphs

Ika Hesti Agustin[1,2], Moh. Hasan[1,2], Dafik[1,3], Ridho Alfarisi[1,4], A.I. Kristiana[1,3], R. M. Prihandini[1,4]

[1]CGANT-University of Jember, Jember, Indonesia
[3]Department of Mathematics Education, University of Jember, Jember, Indonesia
[2]Department of Mathematics, University of Jember, Jember, Indonesia
[4]Department of Elementary School Teacher Education, University of Jember, Jember, Indonesia

E-mail: d.dafik@unej.ac.id, ikahesti.fmipa@unej.ac.id, alfarisi.fkip@unej.ac.id

Abstract. All graph in this paper are finite, simple and connected graph. Let $G(V, E)$ be a graph of vertex set V and edge set E. A bijection $f : V(G) \longrightarrow \{1, 2, 3, ..., |V(G)|\}$ is called a local edge antimagic labeling if for any two adjacent edges e_1 and e_2, $w(e_1) \neq w(e_2)$, where for $e = uv \in G$, $w(e) = f(u) + f(v)$. Thus, any local edge antimagic labeling induces a proper edge coloring of G if each edge e is assigned the color $w(e)$. The local edge antimagic hromatic number $\gamma_{lea}(G)$ is the minimum number of colors taken over all colorings induced by local edge antimagic labelings of G. In this paper, we have found the lower bound of the local edge antimagic coloring of $G \rhd H$ and determine exact value local edge antimagic coloring of $G \rhd H$.

Keywords: Antimagic labeling, Local antimagic edge coloring, Local antimagic edge chromatic number, comb product.

1. Introduction

All graphs in this paper are finite, simple and connected graph, for detail definition of graph see [1, 2]. A bijection mapping that assigns natural number to vertices of a graph is called a graph labeling. In this type of labeling, we consider all weights associated with each edge. If all the edge weights have the different value then we call the labeling as an antimagic.

Hartsfield and Ringel [3] introduced the concept of antimagic labeling of a graph. A bijection $f : V(G) \longrightarrow \{1, 2, 3, ..., |V(G)|\}$ is called a local edge antimagic labeling if for any two adjacent edges e_1 and e_2, $w(e_1) \neq w(e_2)$, where for $e = uv \in G$, $w(e) = f(u) + f(v)$. Thus, any local edge antimagic labeling induces a proper edge coloring of G if each edge e is assigned the color $w(e)$. The local edge antimagic chromatic number $\gamma_{lea}(G)$ is the minimum number of colors taken over all colorings induced by local edge antimagic labelings of G.

There are some results related to local antimagic coloring. Arumugam *et al.* [4] firstly introduced a new notion of local antimagic vertex coloring of a graph G. They give an lower bound and upper bound of local antimagic vertex coloring of joint graph and also give the exact value of local antimagic vertex coloring for path, cycle, complete graph, friendship, wheel, bipartite and complete bipartite. Furthermore, we get the relation between local edge antimagic edge chromatic number and local antimagic vertex chromatic number studied in [4] . In their paper also showed the following observation.

Observation 1.1. [4] *For any graph G, $\chi_{lea}(G) \geq \chi(G)$, where $\chi(G)$ is a chromatic number of vertex coloring of G.*

Dafik, *et. al.* [5, 6] determined Super edge-antimagic total labelings of $mK_{n,n}$ and Super edge-antimagicness for a class of disconnected graphs, respectively. The research related to the coloring of a graph who has researched by ika, *et.al* [8] about On r-dynamic coloring of some graph operations. Marr *et, al* [11] stated the magic rectangles are a generalization of magic squares. Let local antimagic labeling with $\gamma_{lae}(G) = 2$ be the same as bimagic labeling. For more details on bimagic labelings we refer to Marr *et. al.* [10].

Agustin *et. al.* [9] studied a different type of local antimagic coloring, namely local edge antimagic coloring. Their research studied the existence of local edge antimagic coloring of some special graphs and also analyse the lower bound of its local edge antimagic chromatic number. On this research, they have found the local edge antimagic chromatic number of path graph P_n, cycle graph C_n, friendship graph $\mathcal{F}_n$, ladder graph L_n, star graph S_n, wheel graph W_n, complete graph K_n, prism graph Pr_n and the graphs $C_n \odot mK_1$ and $G \odot mK_1$. They also give a lower bound of local edge antimagic chromatic number its $\gamma_{lea} \geq \Delta(G)$. The chromatic number of local edge antimagic for path graph P_n, cycle graph C_n, friendship graph $\mathcal{F}_n$, ladder graph L_n, star graph S_n, wheel graph W_n, complete graph K_n, prism graph Pr_n by the following: $\gamma_{lea}(P_n) = 2$, $\gamma_{lea}(C_n) = 3$, $\gamma_{lea}(\mathcal{F}_n) = 2n + 1$, $\gamma_{lea}(L_n) = 3$, $\gamma_{lea}(S_n) = n$, $\gamma_{lea}(W_n) = n + 2$, $\gamma_{lea}(K_n) = 2n - 3$, and $\gamma_{lea}(Pr_n) = 5$.

In this paper, we investigate the local edge antimagic coloring of comb product graphs can be found in [7], [12]. Let G and H be two connected graphs. Let o be a vertex of H. The comb product between G and H, denoted by $G \triangleright H$, is a graph obtained by taking one copy of G and $|V(G)|$ copies of H and grafting the i-th copy of H at the vertex o to the i-th vertex of G. By the definition of comb product, we can say that $V(G \triangleright H) = \{(a, u) | a \in V(G), u \in V(H)\}$ and $(a, u)(b, v) \in E(G \triangleright H)$ whenever $a = b$ and $uv \in E(H)$, or $ab \in E(G)$ and $u = v = o$.

2. Main Results

In this paper, we have studied the existence of local edge antimagic coloring of comb product of graphs. We have found the local edge antimagic chromatic number of path comb path, path comb cycle, cycle and path, cycle and cycle, path and star, cycle and star. We also analyse the lower bound of its local edge antimagic coloring of $G \triangleright H$.

IOP Conf. Series: Journal of Physics: Conf. Series **1008** (2018) 012038 doi:10.1088/1742-6596/1008/1/012038

Theorem 2.1. *Let G and H be a connected graph, the local edge antimagic chromatic number of $G \triangleright H$ is $\gamma_{lae}(G \triangleright H) \geq \gamma_{lea}(G) + \gamma_{lea}(H)$.*

Proof. The comb product between G and H is a graph obtained by one copy of G and $|V(G)|$ copies of H, and grafting the i vertex of G to the u_i in i-th copy of H. Suppose G admits a local edge antimagic coloring with $\gamma_{lae}(G)$. Furthermore, the subgraph H_i admits a local edge antimagic coloring with $\gamma_{lae}(H_i)$. By claiming that edge weight in G and H_i are distict and edge weight in every subgraph H_i induces a local edge antimagic of H so that $\gamma_{lea}(H_1) = \gamma_{lea}(H_2) = \cdots = \gamma_{lea}(H_n) = \gamma_{lea}(H)$. Thus, we obtain that a local antimagic labeling of $G \triangleright H$.

$$\begin{aligned}\gamma_{lea}(G \triangleright H) &\geq |w_G(e), e \in V(G)| + |w_{H_i}(e), e \in V(H_i)| \\ &= \gamma_{lea}(G) + \gamma_{lea}(H_i) \\ &= \gamma_{lea}(G) + \gamma_{lea}(H)\end{aligned}$$

Hence, from the above edge weight it is easy to see that the lower bound local antimagic labeling of $G \triangleright H$ is $\gamma_{lea}(G \triangleright H) \geq \gamma_{lea}(G) + \gamma_{lea}(H)$. □

Theorem 2.2. *For $n, m \geq 3$, the local edge antimagic chromatic number of $P_n \triangleright P_m$ with grafting pendant vertex $x \in V(P_m)$ is $\gamma_{lea}(P_n \triangleright P_m) = 4$*

Proof. The graph $P_n \triangleright P_m$ is a connected graph with vertex set $V(P_n \triangleright P_m) = \{x_{i,j} : 1 \leq i \leq n, 1 \leq j \leq m\}$ and edge set $E(P_n \triangleright P_m) = \{x_{i,1}x_{i+1,1} : 1 \leq i \leq n-1\} \cup \{x_{i,j}x_{i,j+1}; 1 \leq i \leq n, 1 \leq j \leq m-1\}$. Hence $|V(P_n \triangleright P_m)| = mn$ and $|E(P_n \triangleright P_m)| = mn - 1$. Define a bijection $f : V(P_n \triangleright P_m) \longrightarrow \{1, 2, 3, ..., |V(P_n \triangleright P_m)|\}$ by the following

$$f(x_{i,j}) = \begin{cases} \frac{1-n+i+nj}{2}, & \text{if } i \equiv 1(mod\ 2) \text{ and } j \equiv 1(mod\ 2) \\ \frac{n+2+nj-i}{2}, & \text{if } i \equiv 0(mod\ 2) \text{ and } j \equiv 1(mod\ 2) \\ n(m+1) + \frac{1-i-nj}{2}, & \text{if } i \equiv 1(mod\ 2) \text{ and } j \equiv 0(mod\ 2) \\ mn + \frac{i-nj}{2}, & \text{if } i \equiv 0(mod\ 2) \text{ and } j \equiv 0(mod\ 2) \end{cases}$$

And the edge weight are as follows

$$w(x_{i,1}x_{i+1,1}) = \begin{cases} n+2 & i \equiv 0(mod\ 2); \\ n+1 & i \equiv 1(mod\ 2); \end{cases}$$

$$w(x_{i,j}x_{i,j+1}) = \begin{cases} nm+n+1 & j \equiv 0(mod\ 2); \\ nm+1 & j \equiv 1(mod\ 2); \end{cases}$$

Hence, from the above edge weights, it easy to see that f induces a proper edge colouring of $P_n \triangleright P_m$ and it gives $\gamma_{lea}(P_n \triangleright P_m) \leq 4$. Based on Theorem 2.1, the lower bound is $\gamma_{lea}(P_n \triangleright P_m) \geq \gamma_{lea}(P_n) + \gamma_{lea}(P_m) = 4$. It concludes that $\gamma_{lea}(P_n \triangleright P_m) = 4$. □

IOP Conf. Series: Journal of Physics: Conf. Series **1008** (2018) 012038 doi:10.1088/1742-6596/1008/1/012038

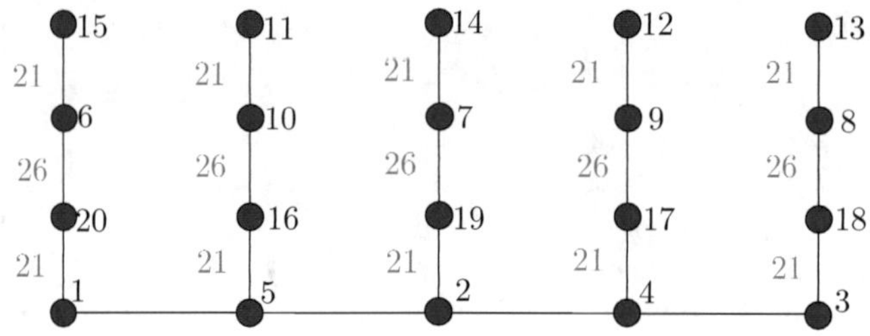

Figure 1. Example of Local antimagic edge coloring of $P_5 \triangleright P_4$

Theorem 2.3. *For $n \geq 3$ and m be even positive integers, the local edge antimagic chromatic number of $P_n \triangleright C_m$ is $\gamma_{lea}(P_n \triangleright C_m) = 5$*

Proof. The graph $P_n \triangleright C_m$ is a connected graph with vertex set $V(P_n \triangleright C_m) = \{x_{i,j} : 1 \leq i \leq n, 1 \leq j \leq m\}$ and edge set $E(P_n \triangleright C_m) = \{x_{i,1}x_{i+1,1} : 1 \leq i \leq n-1\} \cup \{x_{i,j}x_{i,j+1}, x_{i,m}x_{i,1}; 1 \leq i \leq n, 1 \leq j \leq m-1\}$. Hence $|V(P_n \triangleright C_m)| = mn$ and $|E(P_n \triangleright C_m)| = mn + n - 1$. For $n \geq 3$. Define a bijection $f : V(P_n \triangleright C_m) \longrightarrow \{1, 2, 3, ..., |V(P_n \triangleright C_m)|\}$ by the following

$$f(x_{i,j}) = \begin{cases} \frac{1-n+i+nj}{2}, & \text{if } i \equiv 1(mod\ 2) \text{ and } j \equiv 1(mod\ 2) \\ \frac{n+2+nj-i}{2}, & \text{if } i \equiv 0(mod\ 2) \text{ and } j \equiv 1(mod\ 2) \\ n(m+1) + \frac{1-i-nj}{2}, & \text{if } i \equiv 1(mod\ 2) \text{ and } j \equiv 0(mod\ 2) \\ mn + \frac{i-nj}{2}, & \text{if } i \equiv 0(mod\ 2) \text{ and } j \equiv 0(mod\ 2) \end{cases}$$

And the edge weight are as follows

$$w(x_{i,1}x_{i+1,1}) = \begin{cases} n+2 & i \equiv 0(mod\ 2); \\ n+1 & i \equiv 1(mod\ 2); \end{cases}$$

$$w(x_{i,j}x_{i,j+1}) = \begin{cases} nm+n+1 & j \equiv 0(mod\ 2); \\ nm+1 & j \equiv 1(mod\ 2); \end{cases}$$

$$w(x_{i,1}x_{i,m}) = \frac{mn}{2} + n + 1$$

Hence, from the above edge weights, it easy to see that f induces a proper edge colouring of $P_n \triangleright C_m$ and it gives $\gamma_{lea}(P_n \triangleright C_m) \leq 5$. Based on Theorem 2.1, the lower bound is $\gamma_{lea}(P_n \triangleright C_m) \geq \gamma_{lea}(P_n) + \gamma_{lea}(C_m) = 5$. It concludes that $\gamma_{lea}(P_n \triangleright C_m) = 5$. □

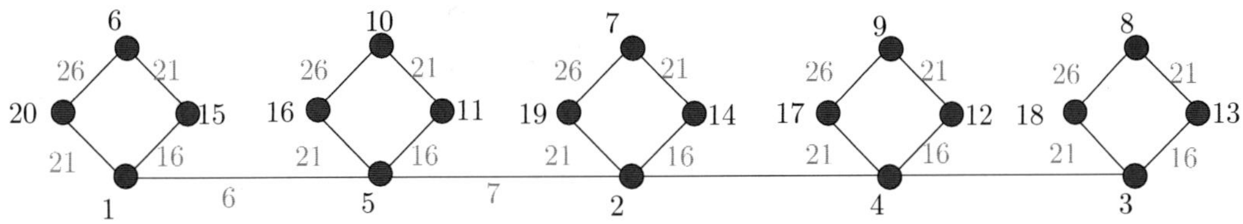

Figure 2. Example of Local antimagic edge coloring of $P_5 \triangleright C_4$

Theorem 2.4. *For $n, m \geq 3$, the local edge antimagic chromatic number of $C_n \triangleright P_m$ with grafting pendant vertex $x \in V(P_m)$ is $\gamma_{lea}(C_n \triangleright P_m) = 5$*

Proof. The graph $C_n \triangleright P_m$ is a connected graph with vertex set $V(C_n \triangleright P_m) = \{x_{i,j} : 1 \leq i \leq n, 1 \leq j \leq m\}$ and edge set $E(C_n \triangleright P_m) = \{x_{i,1}x_{i+1,1} : 1 \leq i \leq n-1\} \cup \{x_{i,j}x_{i,j+1}; 1 \leq i \leq n, 1 \leq j \leq m-1\} \cup \{x_{i,m}x_{i,1}\}$. Hence $|V(C_n \triangleright P_m)| = mn$ and $|E(C_n \triangleright P_m)| = mn$. For $n, m \geq 3$. Define a bijection $f : V(C_n \triangleright P_m) \longrightarrow \{1, 2, 3, ..., |V(C_n \triangleright P_m)|\}$ by the following

$$f(x_{i,j}) = \begin{cases} \frac{1-n+i+nj}{2}, & \text{if } i \equiv 1(mod\ 2) \text{ and } j \equiv 1(mod\ 2) \\ \frac{n+2+nj-i}{2}, & \text{if } i \equiv 0(mod\ 2) \text{ and } j \equiv 1(mod\ 2) \\ n(m+1) + \frac{1-i-nj}{2}, & \text{if } i \equiv 1(mod\ 2) \text{ and } j \equiv 0(mod\ 2) \\ mn + \frac{i-nj}{2}, & \text{if } i \equiv 0(mod\ 2) \text{ and } j \equiv 0(mod\ 2) \end{cases}$$

And the edge weight are as follows

$$w(x_{i,1}x_{i+1,1}) = \begin{cases} n+2 & i \equiv 0(mod\ 2); \\ n+1 & i \equiv 1(mod\ 2); \end{cases}$$

$$w(x_{1,1}x_{n,1}) = \begin{cases} \frac{n+4}{2} & \text{if } n \text{ is even;} \\ \frac{n+3}{2} & \text{if } n \text{ is odd;} \end{cases}$$

$$w(x_{i,j}x_{i,j+1}) = \begin{cases} nm+n+1 & j \equiv 0(mod\ 2); \\ nm+1 & j \equiv 1(mod\ 2); \end{cases}$$

Hence, from the above edge weights, it easy to see that f induces a proper edge colouring of $C_n \triangleright P_m$ and it gives $\gamma_{lea}(C_n \triangleright P_m) \leq 5$. Based on Theorem 2.1, the lower bound is $\gamma_{lea}(C_n \triangleright P_m) \geq \gamma_{lea}(C_n) + \gamma_{lea}(P_m) = 5$. It concludes that $\gamma_{lea}(C_n \triangleright P_m) = 5$. □

IOP Conf. Series: Journal of Physics: Conf. Series **1008** (2018) 012038 doi:10.1088/1742-6596/1008/1/012038

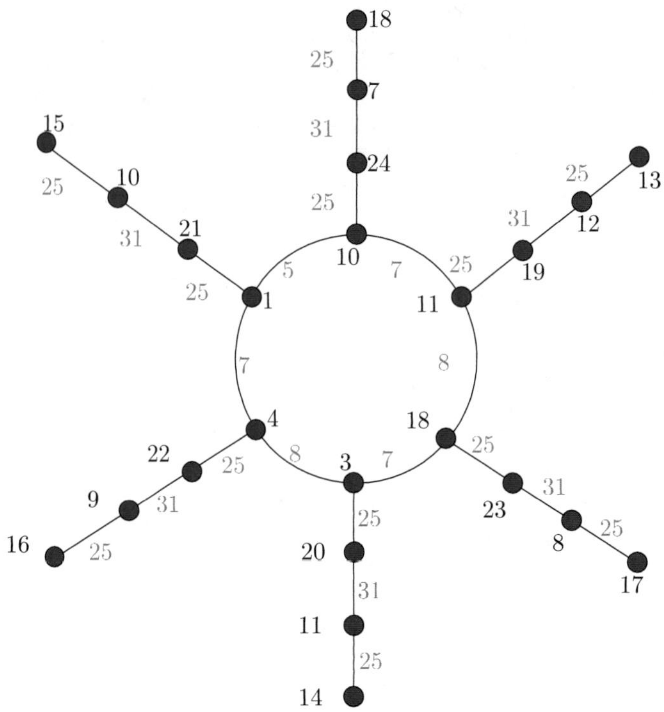

Figure 3. Example of Local antimagic edge coloring of $C_6 \triangleright P_4$

Theorem 2.5. *For $n \geq 3$ and m be even positive integers, the local edge antimagic chromatic number of $C_n \triangleright C_m$ is $\gamma_{lea}(C_n \triangleright C_m) = 6$*

Proof. The graph $C_n \triangleright C_m$ is a connected graph with vertex set $V(C_n \triangleright C_m) = \{x_{i,j} : 1 \leq i \leq n, 1 \leq j \leq m\}$ and edge set $E(C_n \triangleright C_m) = \{x_{i,1}x_{i+1,1} : 1 \leq i \leq n-1\} \cup \{x_{i,j}x_{i,j+1}; 1 \leq i \leq n, 1 \leq j \leq m-1\} \cup \{x_{1,1}x_{1,n}\} \cup \{x_{i,1}x_{i,m}; 1 \leq i \leq n\}$. Hence $|V(C_n \triangleright C_m)| = mn$ and $|E(C_n \triangleright C_m)| = mn + n$. For $n \geq 3$ and m be even positive integers. Define a bijection $f : V(C_n \triangleright C_m) \longrightarrow \{1, 2, 3, ..., |V(C_n \triangleright C_m)|\}$ by the following

$$f(x_{i,j}) = \begin{cases} \frac{1-n+i+nj}{2}, & \text{if } i \equiv 1(mod\ 2) \text{ and } j \equiv 1(mod\ 2) \\ \frac{n+2+nj-i}{2}, & \text{if } i \equiv 0(mod\ 2) \text{ and } j \equiv 1(mod\ 2) \\ n(m+1) + \frac{1-i-nj}{2}, & \text{if } i \equiv 1(mod\ 2) \text{ and } j \equiv 0(mod\ 2) \\ mn + \frac{i-nj}{2}, & \text{if } i \equiv 0(mod\ 2) \text{ and } j \equiv 0(mod\ 2) \end{cases}$$

And the edge weight are as follows

$$w(x_{i,1}x_{i+1,1}) = \begin{cases} n+2 & i \equiv 0(mod\ 2); \\ n+1 & i \equiv 1(mod\ 2); \end{cases}$$

$$w(x_{1,1}x_{n,1}) = \begin{cases} \frac{n+4}{2} & \text{if } n \text{ is even;} \\ \\ \frac{n+3}{2} & \text{if } n \text{ is odd;} \end{cases}$$

$$w(x_{i,j}x_{i,j+1}) = \begin{cases} nm+n+1 & j \equiv 0(mod\ 2); \\ nm+1 & j \equiv 1(mod\ 2); \end{cases}$$

$$w(x_{i,1}x_{i,m}) = \frac{mn}{2} + n + 1$$

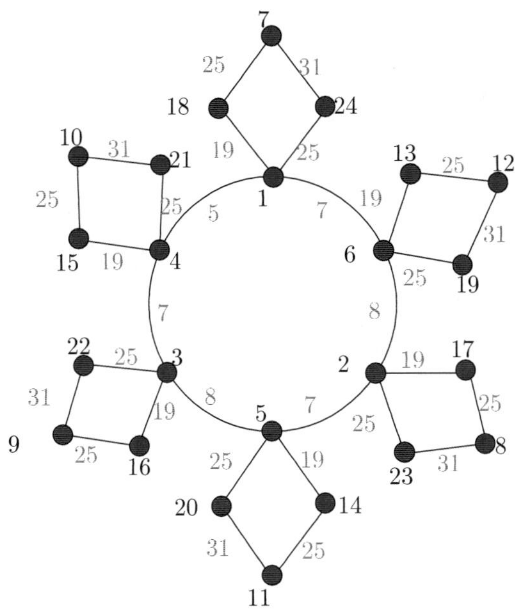

Figure 4. Example of Local antimagic edge coloring of $C_6 \triangleright C_4$

Hence, from the above edge weights, it easy to see that f induces a proper edge colouring of $C_n \triangleright C_m$ and it gives $\gamma_{lea}(C_n \triangleright C_m) \leq 6$. Based on Theorem 2.1, the lower bound is $\gamma_{lea}(C_n \triangleright C_m) \geq \gamma_{lea}(C_n) + \gamma_{lea}(C_m) = 6$. It concludes that $\gamma_{lea}(C_n \triangleright C_m) = 6$. □

Theorem 2.6. *For $n, m \geq 3$, the local edge antimagic chromatic number of $P_n \triangleright S_m$ with grafting central vertex $x \in V(S_m)$ is $\gamma_{lea}(P_n \triangleright S_m) = 2 + m$*

Proof. The graph $P_n \triangleright S_m$ is a connected graph with vertex set $V(P_n \triangleright S_m) = \{x_i : 1 \leq i \leq n\} \cup \{x_{i,j} : 1 \leq i \leq n, 1 \leq j \leq m\}$ and edge set $E(P_n \triangleright S_m) = \{x_{i,1}x_{i+1,1} : 1 \leq i \leq n-1\} \cup \{x_i x_{i,j}; 1 \leq i \leq n, 1 \leq j \leq m\}$. Hence $|V(P_n \triangleright S_m)| = n + mn$ and $|E(P_n \triangleright S_m)| = n + mn - 1$. For $n, m \geq 3$. Define a bijection $f : V(P_n \triangleright S_m) \longrightarrow \{1, 2, 3, ..., |V(P_n \triangleright S_m)|\}$ by the following

$$f(x_i) = \begin{cases} n+1-\frac{i}{2}, & \text{if } i \equiv 0(mod\ 2) \\ \frac{i+1}{2}, & \text{if } i \equiv 1(mod\ 2) \end{cases}$$

$$f(x_{i,j}) = \begin{cases} \frac{2nj+i}{2}, & \text{if } i \equiv 0(mod\ 2) \\ \frac{2nj+2n-i+3}{2}, & \text{if } i \equiv 1(mod\ 2) \end{cases}$$

And the edge weight are as follows

$$w(x_ix_{i+1}) = \begin{cases} n+2 & i \equiv 0(mod\ 2); \\ n+1 & i \equiv 1(mod\ 2); \end{cases}$$

$$w(x_ix_{i,j}) = nj+1$$

Hence, from the above edge weights, it easy to see that f induces a proper edge colouring of $P_n \triangleright S_m$ and it gives $\gamma_{lea}(P_n \triangleright S_m) \leq 2+m$. Based on Theorem 2.1, the lower bound is $\gamma_{lea}(P_n \triangleright S_m) \geq \gamma_{lea}(P_n) + \gamma_{lea}(S_m) = 2+m$. It concludes that $\gamma_{lea}(P_n \triangleright S_m) = 2+m$. □

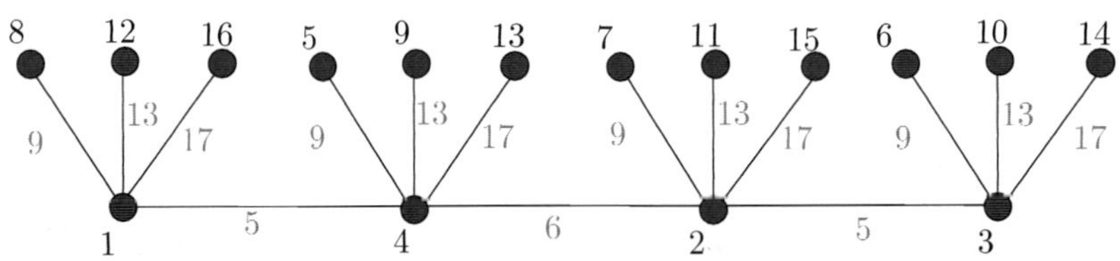

Figure 5. Example of Local antimagic edge coloring of $P_4 \triangleright S_3$

Theorem 2.7. *For $n, m \geq 3$, the local edge antimagic chromatic number of $C_n \triangleright S_m$ with grafting central vertex $x \in V(S_m)$ is $\gamma_{lea}(C_n \triangleright S_m) = 3+m$*

Proof. The graph $C_n \triangleright S_m$ is a connected graph with vertex set $V(C_n \triangleright S_m) = \{x_i : 1 \leq i \leq n\} \cup \{x_{i,j} : 1 \leq i \leq n, 1 \leq j \leq m\}$ and edge set $E(C_n \triangleright S_m) = \{x_{i,1}x_{i+1,1} : 1 \leq i \leq n-1\} \cup \{x_1x_n\} \cup \{x_ix_{i,j}; 1 \leq i \leq n, 1 \leq j \leq m\}$. Hence $|V(C_n \triangleright S_m)| = n+mn$ and $|E(C_n \triangleright S_m)| = n+mn$. For $n, m \geq 3$. Define a bijection $f : V(C_n \triangleright S_m) \longrightarrow \{1, 2, 3, ..., |V(C_n \triangleright S_m)|\}$ by the following

$$f(x_i) = \begin{cases} n+1-\frac{i}{2}, & \text{if } i \equiv 0(mod\ 2) \\ \frac{i+1}{2}, & \text{if } i \equiv 1(mod\ 2) \end{cases}$$

$$f(x_{i,j}) = \begin{cases} \frac{2nj+i}{2}, & \text{if } i \equiv 0(mod\ 2) \\ \frac{2nj+2n-i+3}{2}, & \text{if } i \equiv 1(mod\ 2) \end{cases}$$

And the edge weight are as follows

IOP Conf. Series: Journal of Physics: Conf. Series **1008** (2018) 012038 doi:10.1088/1742-6596/1008/1/012038

$$w(x_i x_{i+1}) = \begin{cases} n+2 & i \equiv 0(mod\ 2); \\ n+1 & i \equiv 1(mod\ 2); \end{cases}$$

$$w(x_{1,1}x_{n,1}) = \begin{cases} \frac{n+4}{2} & \text{if } n \text{ is even;} \\ \frac{n+3}{2} & \text{if } n \text{ is odd;} \end{cases}$$

$$w(x_i x_{i,j}) = nj + 1$$

Hence, from the above edge weights, it easy to see that f induces a proper edge colouring of $C_n \triangleright S_m$ and it gives $\gamma_{lea}(C_n \triangleright S_m) \leq 3 + m$. Based on Theorem 2.1, the lower bound is $\gamma_{lea}(C_n \triangleright S_m) \geq \gamma_{lea}(C_n) + \gamma_{lea}(S_m) = 3 + m$. It concludes that $\gamma_{lea}(C_n \triangleright S_m) = 3 + m$. □

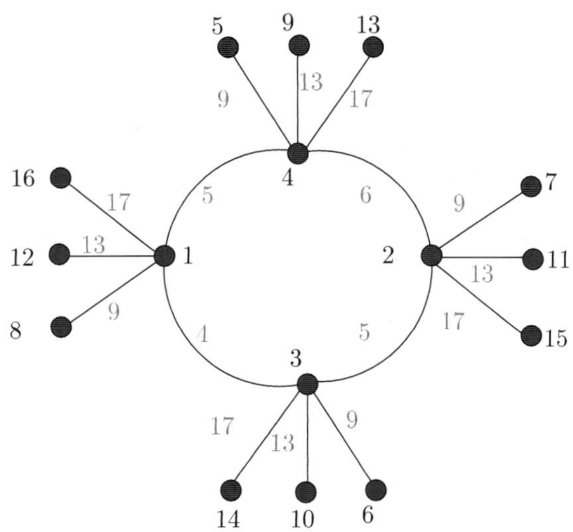

Figure 6. Example of Local antimagic edge coloring of $C_4 \triangleright S_3$

3. Conclusion

In this paper we have given an asymptotically tight result on local edge antimagic coloring of comb product of special graphs, namely path, cycle, and star. We also determine the lower bound of local edge antimagic coloring of comb product of any two graphs. Hence the following problem aries naturally.

Open Problem 3.1. *Determine exact value local edge antimagic coloring of comb product for another family graphs?*

Open Problem 3.2. *Determine exact value local edge antimagic total coloring of comb product $P_n \triangleright C_m$, if m is odd integer?*

Acknowledgement
We gratefully acknowledge the support from CGANT - University of Jember of year 2017.

References
[1] Gross J L, Yellen J and Zhang P 2014 *Handbook of graph Theory* Second Edition CRC Press Taylor and Francis Group
[2] Chartrand G and Lesniak L 2000 *Graphs and digraphs 3rd* ed (London: Chapman and Hall)
[3] Hartsfield N dan Ringel G 1994 *Pearls in Graph Theory* Academic Press. United Kingdom
[4] Arumugam S, Premalatha K, Baca M and Semanicova-Fenovcikova A 2017 Local Antimagic Vertex Coloring of a Graph, *Graphs and Combinatorics* **33** 275-285
[5] Dafik, Miller M, Ryan J, and Baca M 2011 Super edge-antimagic total labelings of $mK_{n,n}$ *Ars Combinatoria* 101 97-107.
[6] Dafik, Mirka M, Ryan J, and Baca M 2006 Super edge-antimagicness for a class of disconnected graphs.
[7] Agustin I H, Dafik, Prihandini R M 2018 $P_2 \rhd H$-super antimagic total labeling of comb product of graph *AKCE International Journal of Graphs and Combinatorics In Press*
[8] Agustin I H, Dafik and Harsya A Y 2016 On r-dynamic coloring of some graph operations *Indonesian Journal of Combinatorics* 1.1 22-30.
[9] Agustin I H, Dafik, Moh. Hasan, Alfarisi R, Prihandini R M 2017 Local Edge Antimagic Coloring of Graphs *Far East Journal of Mathematical Sciences.*
[10] Marr A, Phillips N C K, Wallis W D 2009 Bimagic labelings *AKCE Int. J. Graphs. Combin.* 6(1) 155-160.
[11] Marr A, Wallis W D 2013 *Magic Graphs 2nd* edn. Birkhauser, Switzerland
[12] Saputro S W, Mardiana N and Purwasi I A 2013 The metric dimension of comb product graph *In Graph Theory Conference in Honor of Egawas 60th Birthday* September **10**

IOP Conf. Series: Journal of Physics: Conf. Series **1008** (2018) 012039 doi:10.1088/1742-6596/1008/1/012039

The Construction of $P_2 \rhd H$-antimagic graph using smaller edge-antimagic vertex labeling

Rafiantika M. Prihandini[1,4], **I.H Agustin**[1,3], **Dafik**[1,2]
[1]CGANT University of Jember, Indonesia
[2]Mathematics Edu. Depart. University of Jember
[3]Mathematics Depart. University of Jember, Indonesia
[4]Elementary School Teacher Edu, University of Jember, Indonesia

E-mail: rafiantikap.fkip@unej.ac.id, ikahestiagustin@gmail.com, d.dafik@unej.ac.id

Abstract. In this paper we use simple and non trivial graph. If there exist a bijective function $g : V(G) \cup E(G) \rightarrow \{1, 2, \ldots, |V(G)| + |E(G)|\}$, such that for all subgraphs $P_2 \rhd H$ of G isomorphic to H, then graph G is called an (a, d)-$P_2 \rhd H$-antimagic total graph. Furthermore, we can consider the total $P_2 \rhd H$-weights $W(P_2 \rhd H) = \sum_{v \in V(P_2 \rhd H)} f(v) + \sum_{e \in E(P_2 \rhd H)} f(e)$ which should form an arithmetic sequence $\{a, a + d, a + 2d, ..., a + (n - 1)d\}$, where a and d are positive integers and n is the number of all subgraphs isomorphic to H. Our paper describes the existence of super (a, d)-$P_2 \rhd H$ antimagic total labeling for graph operation of comb product namely of $G = L \rhd H$, where L is a (b, d^*)-edge antimagic vertex labeling graph and H is a connected graph.

1. Introduction

In this paper we consider simple and nontrivial graphs. One of the graph operation is a comb product. Saputro, *et.al* in [16], defined a comb product of L and H, denoted by $L \rhd H$. Comb product is a graph obtained by taking one copy of L and $|V(L)|$ copies of H and grafting the i-th copy of H at the vertex o to the i-th vertex of L. Thus, we have $V(L \rhd H) = \{(a, v)|a \in V(L), v \in V(H)\}$ and $(a, v)(b, w) \in E(L \rhd H)$ whenever $a = b$ and $vw \in E(H)$, or $ab \in E(L)$ and $v = w = o$. Labeling is one to one mapping which maps the set of graph elements into a set of integer. Furthermore, an (a, d)-edge-antimagic vertex labeling is one to one mapping from $g : V(G) \rightarrow \{1, 2, \ldots, v\}$ which maps the set of vertices into a set of integer such that the set of edge weights of all edges in G is $\{a, a + d, \ldots, a + (e - 1)d\}$, where $a > 0$ and $d \geq 0$ are integer set [2]. In this paper we deal with labelings with domain either the set of all vertices and edges.

Suppose $G = L \rhd H$ and $H \subseteq G$, If there exist a bijective function $g : V(G) \cup E(G) \rightarrow \{1, 2, \ldots, |V(G)| + |E(G)|\}$, such that for all subgraphs $P_2 \rhd H$ of G isomorphic to H, then graph G is called an (a,d)-$P_2 \rhd H$-antimagic total graph. Furthermore, we can consider the total $P_2 \rhd H$-weights $W(P_2 \rhd H) = \sum_{v\in V(P_2\rhd H)} f(v) + \sum_{e\in E(P_2\rhd H)} f(e)$ which should form an arithmetic sequence $\{a, a+d, a+2d, ..., a+(n-1)d\}$, where a and d are positive integers and n is the number of all subgraphs isomorphic to H. Dafik *et al* in [7] proved cycle-super antimagicness of tensor product of two graphs, namely $C_r \otimes P_n$ for odd $r \geq 3$ any $n \geq 3$. They have showed the tensor product and its disjoint union for any $m \geq 2$, admit a super $(a,d)-C_{2r}$-antimagic total labeling for some feasible difference $d \in \{4r, 4r^2+2r, 6r^2\}$. Many paper have published, some other relevant results can be found in [1, 11, 12, 10] and [8, 14, 13, 15, 17, 18].

Recently, Dafik *et. al* in [5] showed the H-super antimagicness of a graph L when each edge of L is replaced by a graph H. They used a special technique, which is called *an integer set partition technique*, firstly introduced in [3]. They considered, for a connected version of graph and $k = 1, 2, \ldots, n-1$, a partition $\mathcal{P}^n_{c,d}(i,k)$ of the set $\{1, 2, \ldots, cn\}$ of n columns with $n \geq 2$, c-rows such that the sum of the numbers in the kth c-rows forms an arithmetic sequence of difference d.

Our paper investigate the existence of super (a,d)-$P_2 \rhd H$-antimagic total labeling of $G = L \rhd H$. We also show connection between $P_2 \rhd H$-antimagic total labeling and edge antimagic vertex labeling. Labels of vertices by following the EAVL pattern, furthermore we also found the general formula of feasible difference d of graph $G = L \rhd H$.

2. A Useful Lemma and Corollary

Let order of graph L, H be respectively $|V(L)|, |V(H)|$ and size $|E(L)|, |E(H)|$ respectively. The graph $G = L \rhd H$ is a connected graph with $|V(G)| = |V(L)||V(H)|$ and $|E(G)| = |V(L)||E(H)| + |E(L)|$.Thus $|V(G)| = np_H$ and $|E(G)| = nq_H + q_L$. The following lemma [4] is to define the upper bound of feasible d for $G = L \rhd H$ to be a super (a,d)-H-antimagic total labeling.

Lemma 1. [4] *Let G be a simple graph of order p and size q. If G is super (a,d)-H-antimagic total labeling then $d \leq \frac{(p_G-p_H)p_H+(q_G-q_H)q_H}{n-1}$, for $p_G = |V(G)|$, $q_G = |E(G)|$, $p_H = |V(H)|$, $q_H = |E(H)|$, and $n = |H|$.*

Corollary 1. *If the graph $G = L \rhd H$ admits super (a,d)-H-antimagic total labeling for integer $n \geq 3$, then $d \leq \frac{(p_H^2+q_H^2)(2p_L-4)+2q_Hq_L}{q_L-1}$*

Lemma 2. [6] *Let n and m be positive integers. The sum of $\mathcal{P}^n_{m,c_1}(i,k) = \{(k-1)n + k,\ \ 1 \leq i \leq c\}$ and $\mathcal{P}^n_{m,c_2}(i,k) = \{(k-1)c + i;\ \ 1 \leq i \leq c\}$ form an aritmatic sequence of difference $d_1 = c, d_2 = c^2\}$, respectively.*

3. The Results

Lemma 3. *Given that $G = L \rhd H$. If L admits an edge antimagic vertex labeling $(EAVL)$, then the sum of the corresponding partition label graph of H form an arithmetic sequence.*

Proof. By an (a,d)-edge-antimagic vertex labeling of a (p,q) graph L we mean a one to one mapping g from $V(L)$ onto $\{1, 2, \ldots, p\}$ such that the set of edge weights of all edges in G, $\{g(u)+g(v) : uv \in E(G)\}$, is $W = \{a, a+d, a+2d, \ldots, a+(q-1)d\}$, where $a > 0$ and $d \geq 0$. Let $\{h_i \in V(L); 1 \leq i \leq p\}$ indicates the order of the label, thus the arbitrary pairs, h_i, will form an arithmetic sequence. Let w be a set of an edge weight obtained by arbitrary sum of pairs h_i which $w = \{h_1 + h_2, h_2 + h_3, h_3 + h_4, \ldots, (h_1 + h_2) + (q_L - 1)d\} = \{a, a + d, a + 2d, \ldots, a + (q_L - 1)d\}$, thus we can have the following corresponding partition:

$$\begin{aligned}
\text{(i)} \sum_{\mathrm{i=1}}^{\mathrm{c}} \mathcal{P}^{\mathrm{n}}_{\mathrm{c,d^*}}(\mathrm{h_1}) + \sum_{\mathrm{i=1}}^{\mathrm{c}} \mathcal{P}^{\mathrm{n}}_{\mathrm{c,d^*}}(\mathrm{h_2}) &= \mathcal{C}^n_{c,d^*} + d^*h_1 + \mathcal{C}^n_{c,d^*} + d^*h_2 \\
&= \mathcal{C}^n_{c,d^*} + d^*h_1 + \mathcal{C}^n_{c,d^*} + d^*h_2 \\
&= 2\mathcal{C}^n_{c,d^*} + d^*(h_1 + h_2)
\end{aligned}$$

$$\begin{aligned}
\text{(ii)} \sum_{\mathrm{i=1}}^{\mathrm{c}} \mathcal{P}^{\mathrm{n}}_{\mathrm{c,d^*}}(\mathrm{h_2}) + \sum_{\mathrm{i=1}}^{\mathrm{c}} \mathcal{P}^{\mathrm{n}}_{\mathrm{c,d^*}}(\mathrm{h_3}) &= \mathcal{C}^n_{c,d^*} + d^*h_2 + \mathcal{C}^n_{c,d^*} + d^*h_3 \\
&= \mathcal{C}^n_{c,d^*} + d^*h_2 + \mathcal{C}^n_{c,d^*} + d^*h_3 \\
&= 2\mathcal{C}^n_{c,d^*} + d^*(h_2 + h_3)
\end{aligned}$$

$$\begin{aligned}
\text{(iii)} \sum_{\mathrm{i=1}}^{\mathrm{c}} \mathcal{P}^{\mathrm{n}}_{\mathrm{c,d^*}}(\mathrm{h_3}) + \sum_{\mathrm{i=1}}^{\mathrm{c}} \mathcal{P}^{\mathrm{n}}_{\mathrm{c,d^*}}(\mathrm{h_4}) &= \mathcal{C}^n_{c,d^*} + d^*h_3 + \mathcal{C}^n_{c,d^*} + d^*h_4 \\
&= \mathcal{C}^n_{c,d^*} + d^*h_3 + \mathcal{C}^n_{c,d^*} + d^*h_4 \\
&= 2\mathcal{C}^n_{c,d^*} + d^*(h_3 + h_4)
\end{aligned}$$

From the above, we can easily see that the sum of corresponding partition, $\mathcal{P} = \{2\mathcal{C}^n_{c,d^*} + d^*(a), 2\mathcal{C}^n_{c,d^*} + d^*(a + d), 2\mathcal{C}^n_{c,d^*} + d^*(a + 2d) \ldots, 2\mathcal{C}^n_{c,d^*} + d^*(a + (q_L - 1)d)\}$, form an arithmetic sequence with $b = dd^*$. □

Theorem 1. *Given the any graph H. If L admits $(b, d^*)-EAVL$, then the comb product of the connected graph $G = L \rhd H$ admits super $(a, d) - P_2 \rhd H$ antimagic total labeling with $d = d^* + d^*(d_v + d_e) + 1$.*

Proof. Let l be a $(b, d^*) - EAVL$ of graph L. The set of all edge weights of the edges of L under the labeling l is:
$\{w^l(e) : e \in E(L)\} = \{b, b + d^*, b + 2d^*, \ldots, b + (q_L - 1)d^*\}$ Denote the edges of graph L by the symbols $e_1, e_2, \ldots, e_{q_L}$ such that:
$\{w^l(e) = b + (k - 1)d^*$ with $1 \leq k \leq q_L\}$ Let H be a connected graph, and $G = L \rhd H$ contains p_L subgraphs isomorphic to H, say $H_1, H_2, \ldots, H_{p_L}$ where the subgraphs H_i replaces the vertex v_i in graph L, $i = 1, 2, \ldots, p_L$. Construct a total labeling g,$g : V(L \rhd H) \cup E(L \rhd H) \to \{1, 2, \ldots, p_L p_H + q_L + p_L q_H\}$ constitute the following set:

$$\begin{aligned}
g(V_{p_H}) &= \{\mathcal{P}^{p_L}_{p_H-1,d_v}(i, k) \oplus p_L\} \\
g(E_{q_L}) &= p_H p_L + j; 1 \leq k \leq q_L \\
g(E_{q_H}) &= \{\mathcal{P}^{p_L}_{q_H,d_e}(i, k) \oplus [p_L p_H + q_L]\}
\end{aligned}$$

where d_v depends on $p_H - 1$ and d_e depends on q_H. Furthermore the weight of the subgraphs H_i, $i = 1, 2, \ldots, p_L$ in the following way:

$$W = \sum_{v \in V(H_i)} f(v) + \sum_{e \in E(H_i)} f(e)$$

$$= \quad (b+(k-1)d^*) + (\sum_{i=1}^{p_H-1} (\mathcal{P}^{p_L}_{p_H-1,d_v}(k) \oplus p_L) + (p_H p_L + k) + (\sum_{i=1}^{q_H} (\mathcal{P}^{p_L}_{q_H,d_e}(k) \oplus p_H p_L + q_L)$$

Based on Lemma 3 we are obtained

$$\begin{aligned} &= [b+(k-1)d^*] + [2C^{p_L}_{p_H-1,d_v} + d_v(b+(k-1)d^*)] + [p_H p_L + k] + \\ &\quad [2C^{p_L}_{q_H,d_e} + d_e(b+(k-1)d^*)] \\ &= b - d^* + 2C^{p_L}_{p_H-1,d_v} + bd_v - d_v d_1 + p_H p_L + 2C^{p_L}_{q_H,d_e} + bd_e - d_e d^* + \\ &\quad (d^* + d^*(d_v+d_e)+1)k \end{aligned}$$

From the above, we can easily see that $W = \{[b - d^* + 2C^{p_L}_{p_H-1,d_v} + bd_v - d_v d_1 + p_H p_L + 2C^{p_L}_{q_H,d_e} + bd_e - d_e d^*] + d^* + d^*(d_v+d_e)+1, [b - d^* + 2C^{p_L}_{p_H-1,d_v} + bd_v - d_v d_1 + p_H p_L + 2C^{p_L}_{q_H,d_e} + bd_e - d_e d^*] + 2(d^* + d^*(d_v+d_e)+1), \dots, [b - d^* + 2C^{p_L}_{p_H-1,d_v} + bd_v - d_v d_1 + p_H p_L + 2C^{p_L}_{q_H,d_e} + bd_e - d_e d^*] + k(d^* + d^*(d_v+d_e)+1)\}$ form an arithmetic sequence. It completes the proof □

3.1. Special Families of Connected Graph

We have found a general formula for any graph. Now, in this section we describe the existence of super $(a,d) - P_2 \rhd H$ antimagicness of some special families namely $G = P_n \rhd H$ and $G = S_n \rhd H$.

Theorem 2. *For $n \geq 2$, the graph $G = P_n \rhd H$ admits a super (a,d)-$P_2 \rhd H$ antimagic total labeling with $a = n(c_1^2 - c_1 + t_1{}^2 - t_1) + 3(c_1 + c_2{}^2 + t_1 + t_2{}^2) + 2nc_1 + (c_2 - c_2{}^2) + 2c_2 n(c_1+1) + 2t_1(cn + 2n - 1) + (t_2 - t_2{}^2) + 2t_2(cn + nt_1 + 2n - 1) + 4$ and feasible $d = 2(c_1 + c_2{}^2 + t_1 + t_2{}^2) + 3$.*

Proof. Graph $G = P_n \rhd H$ has vertex set $V(G) = \{z_k; 1 \leq k \leq n\} \cup \{z_{i,k}; 1 \leq i \leq c; 1 \leq k \leq n\}$ and edge set $V(G) = \{e_k; 1 \leq k \leq n\} \cup \{e_{i,k}; 1 \leq i \leq c; 1 \leq k \leq n\}$. Suppose c and t are two fix positive integers, with $c = p_H - 1$ and $t = q_H$. By Lemma 2 for $i = 1, 2, \dots, c$ and $k = 1, 2, \dots, n$, we define the vertex and the edge labels as a linear combination of $\mathcal{P}^n_{c_1,c_1}(i,k); \mathcal{P}^n_{c_2,c_2^2}(i,k)$, written as follows:

$$\begin{aligned} g_1(z_k) &= \{k; 1 \leq k \leq n\} \\ g_1(z_{i,k}) &= \{\mathcal{P}^n_{c_1,c_1} \oplus n\} \cup \{\mathcal{P}^n_{c_2,c_2^2} \oplus n(c_1+1)\} \\ g_1(e_k) &= \{n(c+1) + k; 1 \leq k \leq n-1\} \\ g_1(e_{l,k}) &= \{\mathcal{P}^n_{t_1,t_1} \oplus [n(c+1) + (n-1)]\} \cup \{\mathcal{P}^n_{t_2,t_2^2} \oplus [n(c+1+t_1) + (n-1)]\} \end{aligned}$$

from the vertex and the edge labels, then it can be determined a function of the total vertex-weight and edge-weight, written as follows:

$$w^1_{g_1} = k + k + 1 = 2k + 1$$

$$\begin{aligned}
w^2_{g_1} &= [\sum_{i=1}^{c} \mathcal{P}^n_{c_1,c_1}(i,k) \oplus nc_1 + \sum_{i=1}^{c} \mathcal{P}^n_{c_1,c_1}(i,k+1) \oplus nc_1] + [\sum_{i=1}^{c} \mathcal{P}^n_{c_2,c_2^2}(i,k) \oplus \\
&\quad nc_2(c_1+1) + \sum_{i=1}^{c} \mathcal{P}^n_{c_2,c_2^2}(i,k+1) \oplus nc_2(c_1+1)] \\
&= [\mathcal{P}^n_{c_1,c_1}(k) \oplus nc_1 + \mathcal{P}^n_{c_1,c_1}(k+1) \oplus nc_1] + [\mathcal{P}^n_{c_2,c_2^2}(k) \oplus nc_2(c_1+1) + \\
&\quad \mathcal{P}^n_{c_2,c_2^2}(k+1) \oplus nc_2(c_1+1)] \\
&= \{[\frac{n}{2}(c_1^2 - c_1) + c_1k + nc_1] + [\frac{n}{2}(c_1^2 - c_1) + c_1k + c_1 + nc_1]\} + \\
&\quad \{[\frac{c_2 - c_2^2}{2} + c_2^2k + c_2n(c_1+1)] + [\frac{c_2 - c_2^2}{2} + c_2^2k + c_2^2 + c_2n(c_1+1)]\} \\
&= \{n(c_1^2 - c_1) + 2c_1k + c_1 + 2nc_1\} + \{(c_2 - {c_2}^2) + 2{c_2}^2k + {c_2}^2 + \\
&\quad 2c_2n(c_1+1)\}
\end{aligned}$$

$$\begin{aligned}
w^3_{g_1} &= [\sum_{l=1}^{t} \mathcal{P}^n_{t_1,t_1}(i,k) \oplus r_1(cn+2n-1) + \sum_{l=1}^{t} \mathcal{P}^n_{t_1,t_1}(i,k+1) \oplus t_1(cn+2n-1)] + \\
&\quad [\sum_{l=1}^{t} \mathcal{P}^n_{t_2,t_2^2}(i,k) \oplus t_2(cn+nt_1+2n-1) + \sum_{l=1}^{t} \mathcal{P}^n_{t_2,t_2^2}(i,k+1) \oplus t_2(cn+nt_1+ \\
&\quad 2n-1)] \\
&= [\mathcal{P}^n_{t_1,t_1}(k) \oplus t_1(cn+2n-1) + \mathcal{P}^n_{t_1,t_1}(k+1) \oplus t_1(cn+2n-1)] + [\mathcal{P}^n_{t_2,t_2^2}(k) \\
&\quad \oplus t_2(cn+nt_1+2n-1) + \mathcal{P}^n_{t_2,t_2^2}(k+1) \oplus t_2(cn+nt_1+2n-1)] \\
&= \{[\frac{n}{2}(t_1^2 - t_1) + t_1k + t_1(cn+2n-1)] + [\frac{n}{2}(t_1^2 - t_1) + t_1k + t_1 + t_1(cn+ \\
&\quad 2n-1)]\} + [\frac{t_2 - t_2^2}{2} + t_2^2k + t_2^2 + t_2(cn+nt_1+2n-1)]\}\{[\frac{t_2 - t_2^2}{2} + t_2^2k + \\
&\quad t_2(cn+nt_1+2n-1)] \\
&= \{\ n({t_1}^2 - t_1) + 2t_1k + t_1 + 2t_1(cn+2n-1)\} + \{(t_2 - {t_2}^2) + 2{t_2}^2k + {t_2}^2 + \\
&\quad 2t_2(cn+nt_1+2n-1)\}
\end{aligned}$$

The vertex and edge label under the labeling g_1 is a bijective function $g_1 : V(G) \cup E(G) \to \{1, 2, \ldots, p_G + q_G\}$. The total edge-weights of $G = P_n \rhd H$ under the labeling g_1, for $k = 1, 2, \ldots, n$, constitute the following sets:

$$\begin{aligned}
W^1_{g_1} &= w^1_{g_1} + w^2_{g_1} + n(c+1) + k + w^3_{g_1} \\
&= 2k + 1 + w^2_{g_1} + n(c+1) + k + w^3_{g_1} \\
&= C + [2(c_1 + c_2^2 + t_1 + t_2^2) + 3]k; 1 \le k \le n-1
\end{aligned}$$

with $C = n(c_1^2 - c_1) + c_1 + 2nc_1 + (c_2 - {c_2}^2) + {c_2}^2 + 2c_2n(c_1+1) + 1 + n({t_1}^2 - t_1) + t_1 + 2t_1(cn+2n-1) + (t_2 - {t_2}^2) + {t_2}^2 + 2t_2(cn+nt_1+2n-1)$. It is easy that the set of total edge-weights $W^1_{g_1}$ consists of an arithmetic sequence of the smallest value a when the total edge weights at $k = 1$ and the feasible difference $d = 2[c_1 + c_2^2 + t_1 + t_2^2] + 3$.

IOP Conf. Series: Journal of Physics: Conf. Series **1008** (2018) 012039 doi:10.1088/1742-6596/1008/1/012039

Since the biggest d is attained when $d = 2(c_2^2 + t_2^2)$ then, for $c = p_H$ and $t = q_H$, it gives $d \leq \frac{(p_H^2+q_H^2)(2n-4)+2(n-1)q_H}{n-2}$. It concludes the proof. □

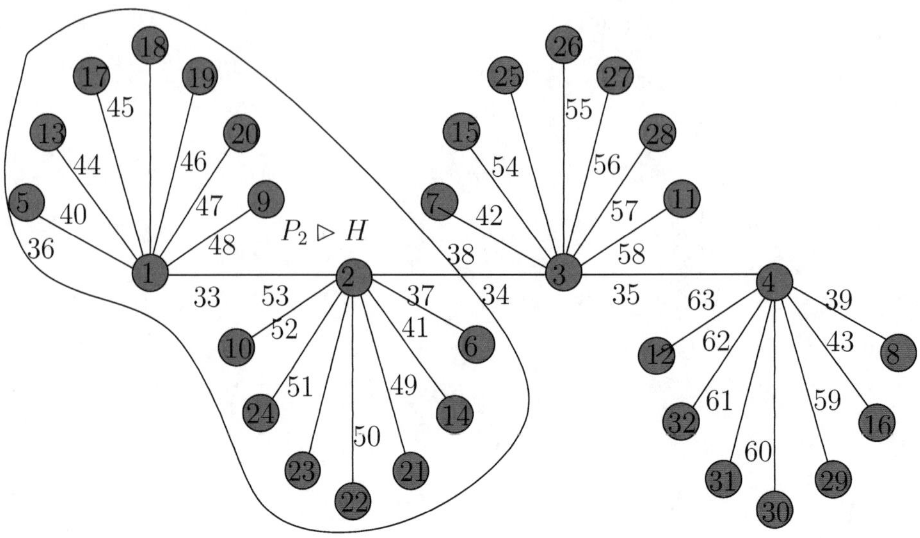

Figure 1. Illustration of graph $G = P_4 \rhd S_7$

Theorem 3. *For $n \geq 1$, the graph $G = S_n \rhd H$ admits a super (a,d)-$P_2 \rhd H$ antimagic total labeling with $a = n+2+(n+1)(c_1^2 - c_1) + c_1(n+2) + 2c_1(n+1) + c_2 + {c_2}^2(n+1) + 2c_2((n+1)(c_1+1)) + (n+1)({t_1}^2 - t_1) + t_1(n+2) + 2t_1((n+1)(c+1)+n) + t_2 + {t_2}^2(n+1) + 2{t_2}^2((n+1)(c+t_1+1)+n)$ feasible $d = c_1 + c_2^2 + t_1 + t_2^2 + 2$.*

Proof. Graph $G = S_n \rhd H$ has vertex set $V(G) = \{x\} \cup \{z_k; 1 \leq k \leq n\} \cup \{z_{i,k}; 1 \leq i \leq c; 1 \leq j \leq n\}$ and edge set $V(G) = \{e_k; 1 \leq k \leq n\} \cup \{e_{l,k}; 1 \leq l \leq t; 1 \leq k \leq n\}$. Let c and t be positive integers, with $c = p_H - 1$ and $t = q_H$. For $i = 1, 2, \ldots, c$ and $k = 1, 2, \ldots, n$, by Lemma 2, 3, 4 and 5 we define the vertex and the edge labels as a linear combination of $\mathcal{P}^n_{c_1,c_1}(i,k)$ and $\mathcal{P}^n_{c_2,c_2^2}(i,k)$ as follows:

$$\begin{aligned}
g_2(x) &= \{n+1\} \\
g_2(z_k) &= \{k; 1 \leq k \leq n\} \\
g_2(x_{i,k}) &= \{\mathcal{P}^n_{c_1,c_1} \oplus (n+1)\} \cup \{\mathcal{P}^n_{c_2,c_2^2} \oplus (n+1)(c_1+1)\} \\
g_2(e_k) &= \{(n+1)(c+1)+k; 1 \leq k \leq n\} \\
g_2(e_{l,k}) &= \{\mathcal{P}^n_{t_1,t_1} \oplus [(n+1)(c+1)+n]\} \cup \{\mathcal{P}^n_{t_2,t_2^2} \oplus [(n+1)(c+t_1+1)+n]\}
\end{aligned}$$

then can be determined a function of the total vertex-weight and edge-weight

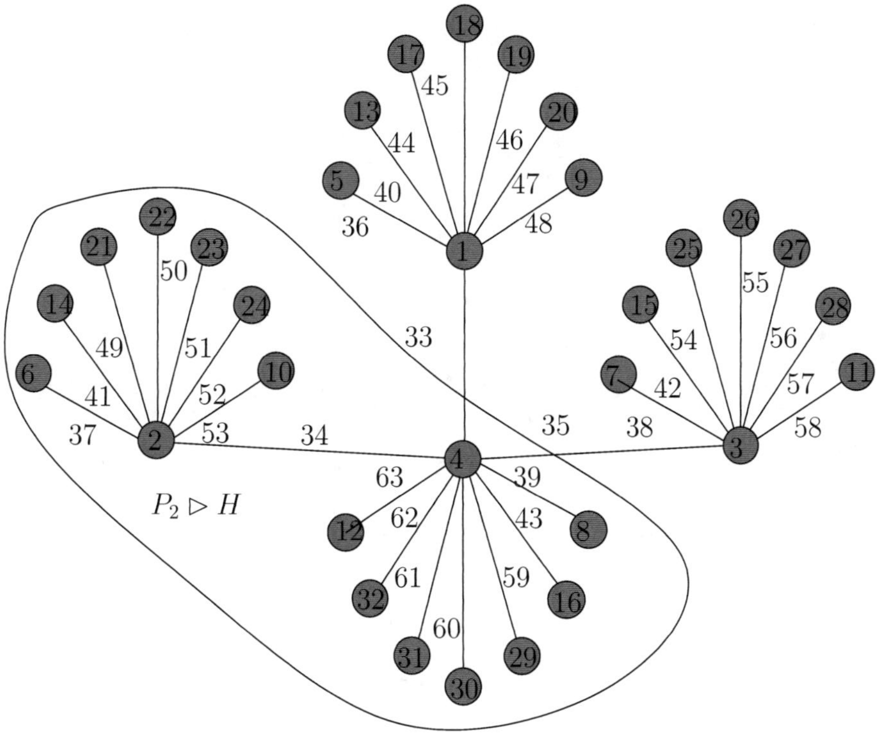

Figure 2. Illustration of graph $G = S_3 \rhd S_7$

$$
\begin{aligned}
w^1_{g_2} &= n+1+k \\
w^2_{g_2} &= [\sum_{i=1}^{c} \mathcal{P}^n_{c_1,c_1}(i,k) \oplus c_1(n+1) + \sum_{i=1}^{c} \mathcal{P}^n_{c_1,c_1}(i,n+1) \oplus c_1(n+1)] + \\
&\quad [\sum_{i=1}^{c} \mathcal{P}^n_{c_2,c_2^2}(i,k) \oplus c_2((n+1)(c_1+1)) + \sum_{i=1}^{c} \mathcal{P}^n_{c_2,c_2^2}(i,n+1) \\
&\quad \oplus c_2((n+1)(c_1+1))] \\
&= [\mathcal{P}^n_{c_1,c_1}(k) \oplus c_1(n+1) + \mathcal{P}^n_{c_1,c_1}(n+1) \oplus c_1(n+1)] + [\mathcal{P}^n_{c_2,c_2^2}(k) \\
&\quad \oplus c_2((n+1)(c_1+1)) + \mathcal{P}^n_{c_2,c_2^2}(n+1) \oplus c_2((n+1)(c_1+1))] \\
&= \{[\frac{n+1}{2}(c_1^2-c_1) + c_1k + c_1(n+1)] + [\frac{n+1}{2}(c_1^2-c_1) + 2c_1(n+1)] \\
&\quad +[\frac{c_2-c_2^2}{2} + c_2^2k + c_2^2 + c_2((n+1)(c_1+1))] + [\frac{c_2-c_2^2}{2} + c_2^2(n+1) \\
&\quad +c_2^2 + c_2((n+1)(c_1+1))]\} \\
&= [(n+1)(c_1^2-c_1) + c_1(n+k+1) + 2c_1(n+1)] + [c_2 + c_2^2(n+k) \\
&\quad +2c_2((n+1)(c_1+1))] \\
w^3_{g_2} &= [\sum_{l=1}^{t} \mathcal{P}^n_{t_1,t_1}(i,k) \oplus t_1((n+1)(c+1)+n) + \sum_{l=1}^{t} \mathcal{P}^n_{t_1,t_1}(i,n+1)
\end{aligned}
$$

IOP Conf. Series: Journal of Physics: Conf. Series **1008** (2018) 012039 doi:10.1088/1742-6596/1008/1/012039

$$
\begin{aligned}
& \oplus t_1((n+1)(c+1)+n)] + [\sum_{l=1}^{t} \mathcal{P}^n_{t_2,t_2^2}(i,k) \oplus t_2(n(c+t_1+1)+1+n) \\
& + \sum_{l=1}^{t} \mathcal{P}^n_{t_2,t_2^2}(i,n+1) \oplus t_2(n(c+t_1+1)+1+n)] \\
= & [\mathcal{P}^n_{t_1,t_1}(k) \oplus t_1((n+1)(c+1)+n) + \mathcal{P}^n_{t_1,t_1}(n+1) \oplus t_1((n+1)(c+1)+n)] \\
& +[\mathcal{P}^n_{t_2,t_2^2}(k) \oplus t_2((n+1)(c+t_1+1)+n) + \mathcal{P}^n_{t_2,t_2^2}(n+1) \\
& \oplus t_2((n+1)(c+t_1+1)+n)] \\
= & \{[\frac{n+1}{2}(t_1^2-t_1) + t_1k + t_1((n+1)(c+1)+n)] + [\frac{n+1}{2}(t_1^2-t_1) \\
& +t_1(n+1) + t_1((n+1)(c+1)+n)] + [\frac{t_2-t_2^2}{2} + t_2^2k + t_2^2 + \\
& t_2((n+1)(c+t_1+1)+n)] + [\frac{t_2-t_2^2}{2} + t_2^2(n+1) + t_2^2 + \\
& t_2((n+1)(c+t_1+1)+n)]\} \\
= & [(n+1)(t_1^2-t_1) + t_1(n+k+1) + 2t_1((n+1)(c+1)+n)] + [t_2 + t_2^2(n+k) \\
& +2t_2((n+1)(c+k_1+1)+n)]
\end{aligned}
$$

The vertex labeling g_2 is a bijective function $g_2 : V(G) \cup E(G) \to \{1, 2, \ldots, p_G + q_G\}$. The total edge-weights of $G = S_n \rhd H$ under the labeling g_2, for $k = 1, 2, \ldots, n$, constitute the following sets:

$$
\begin{aligned}
W^2_{g_2} &= w^1_{g_2} + w^2_{g_2} + (n+1)(c+1) + k + w^3_{g_2} \\
&= n+1+k+w^2_{k_2} + (n+1)(c+1) + k + w^3_{g_2} \\
&= C + [c_1 + c_2^2 + t_1 + t_2^2 + 2]k; 1 \le k \le n
\end{aligned}
$$

with $C = n+1+(n+1)(c_1^2-c_1)+c_1(n+1)+2c_1(n+1)+c_2+c_2{}^2n+2c_2((n+1)(c_1+1)) + (n+1)(t_1{}^2-t_1) + t_1(n+1) + 2t_1((n+1)(c+1)+n) + t_2 + t_2{}^2n + 2t_2{}^2((n+1)(c+t_1+1)+n)$. It is easy that the set of total edge-weights $W_{2_{g_2}}$ consists of an arithmetic sequence of the smallest value a when the edge weights at $k = 1$ and the difference $d = c_1 + c_2^2 + t_1 + t_2^2 + 2$. Since the biggest d is attained when $d = c_2^2 + t_2^2$ then, for $c = p_H$ and $t = q_H + n$, it gives $d \le \frac{(p_H^2+q_H^2)(2(n+1)-4)+2nq_H}{n-1}$ It concludes the proof. □

Concluding Remarks

We have shown the existence of super antimagic labeling for graph operation $G = L \rhd H$ where L is a $(b, d^*) - EAV$ labeling. We have found super(a, d)-$P_2 \rhd H$ antimagic labelings for all differences $d = d^* + d^*(d_v + d_e) + 1$ where d^* is the feasible value of difference in super edge antimagic graph L and d_v and d_e respectively are feasible values of differences in the partitions $\mathcal{P}^{p_L}_{p_H-1,d_v}$ and $\mathcal{P}^{p_L}_{q_H,d_e}$. We have not found the result for disconnected of graph G. Thus, we propose the following open problems.

Open Problem 1. *Let L be a subgraph of G and $G = s(L \rhd H)$. Does G admit a super $(a, d) - P_2 \rhd H$ antimagic total labeling for $n \geq 2$ and feasible d?*

Acknowledgement
We gratefully acknowledge the support from DP2M research grant HIKOM-DIKTI and CGANT - University of Jember of year 2017.

References

[1] Baca M, Lin Y and Semanicov-Fenovcikov A 2009 Note on super antimagicness of disconnected graphs *International J. Graphs and Comb* **6** 1 47-55
[2] Baca M, Lin Y, Miller M and Youssef M Z 2007 Edge-antimagic graphs *Discrete Mathematics* **307** 11 1232-1244
[3] Baca M, Brankovic L, Lascsakova M, Phanalasy O and Semanicova-Fenovciova A 2013 On d-antimagic labelings of plane graphs *Electronic Journal of Graph Theory and Applications (EJGTA)* **1** 1 28-39
[4] Dafik, Purnapraja A K and Hidayat R 2015 Cycle-Super Antimagicness of Connected and Disconnected Tensor Product of Graphs *Procedia Computer Science* **74** 93-99
[5] Dafik, Slamin, Tanna D, Semanicova-Fenovcikova A and Baca M 2017 Constructions of H-antimagic graphs using smaller edge-antimagic graphs *Ars Combinatoria* **133** 233-245
[6] Dafik, Hasan M, Azizah, Y N and Agustin I H 2017 A generalized shackle of any graph H admits a super H-antimagic total labeling *In Journal of Physics: Conference Series* IOP Publishing **893** 1 012042
[7] Dafik, Purnapraja A K, Hidayat R 2015 Cycle Super Antimagicness of Connected and Disconnected Tensor Product of Graphs *Procedia Computer Science* **74** 93-99
[8] Dafik and Agustin I H 2016 Super (a, d)- F_n-antimagic total labeling for a connected and disconnected amalgamation of fan graphs *In AIP Conference Proceedings* AIP Publishing **1707** 1 020003
[9] Gross J L and Yellen J eds 2004 *Handbook of graph theory* CRC press
[10] Agustin I H, Dafik, Prihandini R M 2018 $P_2 \rhd H$-super antimagic total labeling of comb product of graph *AKCE International Journal of Graphs and Combinatorics In Press*
[11] Inayah N, Salman A N M and Simanjuntak R 2009 On (a, d)-H-antimagic coverings of graphs *Journal of Combinatorial Mathematics and Combinatorial Computing* **71** 273
[12] Jeyanthi P and Selvagopal P 2010 More classes of H-supermagic Graphs *Intern. J. of Algorithms, Computing and Mathematics* **3** 1 93-108
[13] Maryati T K, Salman A N M, Baskoro E T, Ryan J and Miller M 2010 On H-supermagic labelings for certain shackles and amalgamations of a connected graph *Utilitas Mathematica* **83** 333
[14] Llad A and Moragas J 2007 Cycle-magic graphs *Discrete Mathematics* **307** 23 2925-2933
[15] Ngurah A A G, Salman A N M and Susilowati L 2010 H-supermagic labelings of graphs *Discrete Mathematics* **310** 8 1293-1300
[16] Saputro S W, Mardiana N and Purwasi I A 2013 The metric dimension of comb product graph *In Graph Theory Conference in Honor of Egawas 60th Birthday* September **10**
[17] Rizvi S T R, Ali K and Hussain M 2014 Cycle-supermagic labelings of the disjoint union of graphs *Utilitas Mathematica*
[18] Roswitha M and Baskoro ET 2012, May H-magic covering on some classes of graphs *In AIP Conference Proceedings* **1450** 1 135-138
[19] Simanjuntak R, Bertault F and Miller M 2000 Two new (a, d)-antimagic graph labelings *In Proc. of Eleventh Australasian Workshop on Combinatorial Algorithms* **11** 179-189

IOP Conf. Series: Journal of Physics: Conf. Series **1008** (2018) 012040 doi:10.1088/1742-6596/1008/1/012040

The non-isolated resolving number of k-corona product of graphs

Ridho Alfarisi[1,4], Dafik[1,2], Slamin[1,5], I. H. Agustin[1,3], A. I. Kristiana[1,2]
[1]CGANT-University of Jember, Jember, Indonesia
[2]Department of Mathematics Education, University of Jember, Jember, Indonesia
[3]Department of Mathematics, University of Jember, Jember, Indonesia
[4]Department of Elementary School Teacher Education, University of Jember, Jember, Indonesia
[5]Department of Information System, University of Jember, Jember, Indonesia

E-mail: alfarisi.fkip@unej.ac.id, d.dafik@unej.ac.id, s.slamin@unej.ac.id

Abstract. Let all graphs be a connected and simple graph. A set $W = \{w_1, w_2, w_3, \dots, w_k\}$ of veretx set of G, the $k-$vector ordered $r(v|W) = (d(x, w_1), d(x, w_2), \dots, d(x, w_k))$ of is a representation of v with respect to W, for $d(x, w)$ is the distance between the vertices x and w. The set W is called a resolving set for G if different vertices of G have distinct representation. The metric dimension is the minimum cardinality of resolving set W, denoted by $dim(G)$. Through analogue, the resolving set W of G is called non-isolated resolving set if there is no $\forall v \in W$ induced by non-isolated vertex. The non-isolated resolving number is the minimum cardinality of non-isolated resolving set W, denoted by $nr(G)$. In our paper, we determine the non isolated resolving number of k-corona product graph.

1. Introduction

In this paper, All graphs G is a nontrivial and connected graph, for more detail definition of graph see [1, 2]. The concept of metric dimension was independently introduced by Slater [3] and Harrary and Melter [4]. In his paper, Slater called this concept as a locating set. Let u, v be two vertices in G. The distance $d(u, v)$ is the length of a shortest path between two vertices u and v in connected graph G. For an ordered set $W = \{w_1, w_2, ..., w_k\}$ subset of vertex set $V(G)$. The representation $r(v|W)$ of v with respect to W is the ordered k-tuple $r(v|W) = (d(v, w_1), d(v, w_2), ..., d(v, w_k))$. The set W is called resolving set of G if every vertices of G have distinct representation respect to W, let u and v be two any vertices in G if $r(u|W) = r(v|W)$ implies that $u = v$. Hence if W is a resolving set of cardinality k for a graph G, then the representation set $r(v|W), v \in V(G)$ consists of $|V(G)|$ distinct k-vector. A resolving set of minimum cardinality for a graph G is called a minimum resolving set for G and this cardinality is the metric dimension of G, denoted by $dim(G)$. Saenpholphat and Zhang [7] introduced the concept of connected resolving set. A resolving set W of graph G is connected if each subgraph $\langle W \rangle$ induced by W has no isolated vertices in G. The minimum cardinality of a non-isolated set in a graph G is the non-isolated resolving number, denoted by $nr(G)$. For more detail notation of $nr(G)$ please see in Chitra and Arumugam [5].

Until today, Chartrand et.al.[6] determined the bounds of the metric dimensions for any connected graphs and determined the metric dimensions of some well known families of graphs such as tree, path, and complete graph. Citra and Arumugam [5] studied resolving set without isolated vertices of some family graphs. Furthermore, Baca, *et.al* [8] showed the metric dimension of regular bipartite graphs. In recent years, this metric dimension has been studied widely (see [7], [9], [10], [11], [12], [13]). In the following, we present some known results and Yunika [14] studied the metric dimension with non-isolated resolving number of corona product of graphs. Alfarisi et al [16] studied non-isolated resolving number of graphs with pendant edges

The known results on metric dimension and local metric dimension of some particular classes of graphs and graph operation have been discovered Chitra and Arumugam [5], Dafik [15], Yunika [14] as follows.

Proposition 1.1 *(Chitra and Arumugam [5]) Let G be a connected graph of order $n \geq 2$*

- *For P_n, $n \geq 2$, $nr(P_n) = 2$.*
- *For K_n, $n \geq 3$, $nr(K_n) = n - 1$.*
- *For $K_{m,n}$, $m, n \geq 2$, $nr(K_{m,n}) = m + n - 2$.*
- *For friendship G with k-triangles, $k \geq 2$, $nr(G) = k + 1$.*
- *For $P_n + K_1$, $n \geq 2$, $nr(P_n + K_1) = \lfloor n/2 \rfloor$.*

Proposition 1.2 *(Dafik [15]) Let G and H be a connected graph, then the metric dimensio with non-ioslated resolving number of $G + H$ is $\lfloor \frac{|V(G)|}{2} \rfloor + \lfloor \frac{|V(H)|}{2} \rfloor \leq nr(G + H) \leq \lceil \frac{|V(G)|}{2} \rceil + \lceil \frac{|V(H)|}{2} \rceil + max\{nr(G), nr(H)\}$*

Proposition 1.3 *(Yunika [14]) Let G be a connected graph of order $n \geq 2$*

- *For $G \cong K_1 \odot P_n$, $n \geq 2$, $nr(G) = \lfloor \frac{n}{2} \rfloor + 1$.*
- *For $G \cong K_1 \odot K_n$, $n \geq 2$, $nr(G) = n$.*
- *For $G \cong K_1 \odot C_n$, $n \geq 2$,*

$$nr(G) = \begin{cases} 3, & if\, 3 \leq n \leq 4 \\ \lfloor \frac{n}{2} \rfloor + 1, & if\, n \geq 5,\ n \text{ is odd} \\ \lfloor \frac{n}{2} \rfloor, & if\, n \geq 5,\ n \text{ is even} \end{cases}$$

Fruct and Harary [17] was introduced a type corona product of grpah. Let G be a connected graph of order n and H (not necessarily connected) be a graph of order m. A graph G corona product H, $G \odot H$, is defined as a graph obtained by taking one copy of G and $|V(G)|$ copies of graph $H_1, H_2, ..., H_n$ of H and connecting i-th vertex of G to the vertices of H_i, $1 \leq i \leq n$. By definition of corona product, we can say that

$$V(G \odot H) = V(G) \cup \bigcup_{i \in V(G)} V(H_i),$$

$$E(G \odot H) = E(G) \cup \bigcup_{i \in V(G)} (E(H_i) \cup \{iu_i | u_i \in V(H_i)\}),$$

For any integer $k \geq 2$, we define the corona product of graph $G \odot_k H$ recursively of $G \odot H$ as $G \odot_k H = (G \odot_{k-1} H) \odot H$. The graph $G \odot_k H$ is named as $k-$corona product or multicorona product of graph G and H for more detail definiton can be seen in [18]. Figure 1 is an ilustration of k-corona product graphs.

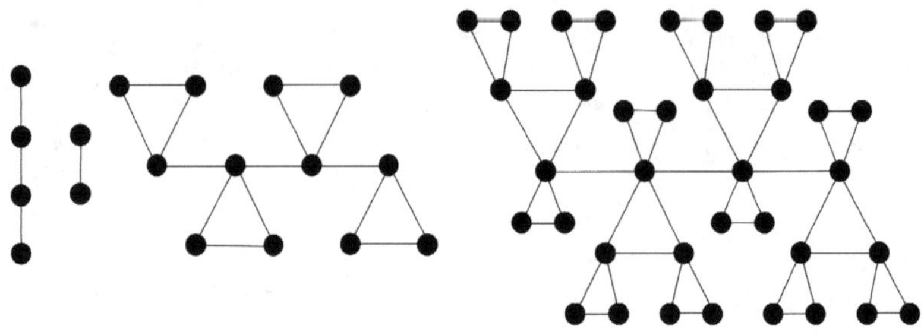

Figure 1. Example of $k-$corona product: (a) P_4; (b) P_2; (c) $P_4 \odot P_2$; (d) $P_4 \odot_2 P_2$

2. Main Results

In this section, we find the metric dimension with non-isolated resolving set of k-corona product graphs. We determine the sharp lower bound and the exact value of $P_n \odot_2 P_m$, $P_n \odot_2 K_m$, $K_m \odot_2 P_n$ and $K_n \odot_2 K_m$

Lemma 2.1 *Let G be a connected graph of order $|V(G)| \geq 2$ and H be a graph of order $|V(H)| \geq 2$, then non-isolated resolving number of $G \odot_k H$ is $nr(G \odot_k H) \geq |V(G \odot_{k-1} H)|nr(K_1 + H)$.*

Proof: The k-corona product of graph $G \odot_k H$ recursively of $G \odot H$ as $G \odot_k H = (G \odot_{k-1} H) \odot H$. Given a resolving set W' of $G \odot_k H$ with $W' \subset V(H) \neq \emptyset$ and resolving set $W"$ of $G \odot_k H$ with $W' \subset V(G \odot_{k-1} H) \neq \emptyset$. It can be shown with use Lemma 2.6.2 (i). Based on definition of non-isolated resolving set that we get $W_i = W_i' \cup W"_i$ for every $i \in V(G \odot_{k-1} H)$ with $W"_i = \{u_i\}$ so that $W_i = W_i' \cup \{u_i\}$. In this section, we can assume —$W_i' \cup \{u_i\}| = nr(K_1 + H)$. Thus, we get resolving set $W = W_1 \cup W_2 \cup W_3 \cup \ldots \cup W_{|V(G \odot_{k-1} H)|}$ or $W = \bigcup_{i=1}^{|V(G \odot_{k-1} H)|} W_i$, then we obtain the lower bound non-isolated resolving set W of $G \odot_{k-1} H$ is

$$
\begin{aligned}
|W| &\leq |\bigcup_{i=1}^{|V(G \odot_{k-1} H)|} (W_i)| \\
&= \Sigma_{i=1}^{|V(G \odot_{k-1} H)|} |W_i| \\
&= |W_1| + |W_2| + \ldots + |W_{V(G \odot_{k-1} H)|}| \\
&= \underbrace{nr(K_1 + H) + nr(K_1 + H) + \ldots + nr(K_1 + H)}_{|V(G \odot_{k-1} H)| times} \\
&= V(G \odot_{k-1} H)|nr(K_1 + H)
\end{aligned}
$$

Hence, it is clearly that the lower bound non-isolated resolving number of $G \odot_k H$ is $nr(G \odot_k H) \geq |V(G \odot_{k-1} H)|nr(K_1 + H)$. □

Theorem 2.1 *Let P_n and P_m be a connected graph of order $n, m \geq 2$, then non-isolated resolving number of $P_n \odot_2 P_m$ is $nr(P_n \odot_2 P_m) = (nm + n)(\lfloor \frac{m}{2} \rfloor + 1)$.*

Proof: Let $P_n \odot_2 P_m$ be a corona product of path P_n and P_m with vertex set $V(P_n \odot_2 P_m) = \{x_i; 1 \leq i \leq n\} \cup \{x_{i,j_1}; 1 \leq i \leq n, 1 \leq j_1 \leq m\} \cup \{x_{i,j_1,j_2}; 1 \leq i \leq n, 1 \leq j_1 \leq m, 1 \leq j_2 \leq m\} \cup \{x_{i,j_2}; 1 \leq i \leq n, 1 \leq j_2 \leq m\}$ and edge set $E(P_n \odot_2 P_m) = \{x_i x_{i+1}; 1 \leq i \leq n-1\} \cup \{x_i x_{i,j_1}, x_{i,j_1} x_{i,j_1,j_2}, x_i x_{i,j_2}; 1 \leq i \leq n, 1 \leq j_1 \leq m, 1 \leq j_2 \leq m\} \cup \{x_{i,j_1} x_{i,j_1+1}, x_{i,j_1,j_2} x_{i,j_1,j_2+1}, x_{i,j_2} x_{i,j_2+1}; 1 \leq i \leq n, 1 \leq j_1 \leq m-1, 1 \leq j_2 \leq m-1\}$

IOP Conf. Series: Journal of Physics: Conf. Series **1008** (2018) 012040 doi:10.1088/1742-6596/1008/1/012040

with the cardinality of vertex set $|V(P_n \odot_2 P_m)| = nm^2 + 2nm + n$ and the cardinality of edge set $|E(P_n \odot_2 P_m)| = 2nm^2 + 3mn - n - 1$.

For $n, m \geq 2$, based on Lemma 2.1 and Proposition 1.3 then we have $nr(P_n \odot_2 P_m) \geq |V(P_n \odot P_m)| nr(K_1 + P_m) = (nm+n)(\lfloor \frac{m}{2} \rfloor + 1)$. However, we can attain the sharpest lower bound. Furthermore, we prove that $nr(P_n \odot_2 P_m) \leq (nm+n)(\lfloor \frac{m}{2} \rfloor + 1)$. Choosing $W \subset V(P_n \odot_2 P_m)$ with $W = \{x_i, x_{i,j_1}; 1 \leq i \leq n, 1 \leq j_1 \leq m\} \cup \{x_{i,j_1,j_2}, x_{i,j_2}; 1 \leq i \leq n, 1 \leq j_1 \leq m-1, 1 \leq j_2 \leq m-1, j_1 \text{ and } j_2 \text{ is odd}\}$ is a non-isolated resolving set of $P_n \odot_2 P_m$ and the cardinality of non-isolated resolving set is $|W| = |\{x_i, x_{i,j_1}; 1 \leq i \leq n, 1 \leq j_1 \leq m\}| + |\{x_{i,j_1,j_2}, x_{i,j_2}; 1 \leq i \leq n, 1 \leq j_1 \leq m-1, 1 \leq j_2 \leq m-1, j_1 \text{ and } j_2 \text{ is odd}\}| = (nm+n)(\lfloor \frac{m}{2} \rfloor + 1)$. Thus, the representation of vertices $v \in V(P_n \odot_2 P_m) - W$ respecting to W are as follows.

The vertex representation of x_{i,j_1,j_2} for m even respect to W is

$$r(x_{i,j_1,j_2}|W) = (a, t_{j_1-1}, \underbrace{t'_{j_1-1}, \ldots, t'_{j_1-1}}_{\lfloor \frac{m}{2} \rfloor}, \ldots, t_1, \underbrace{t'_1, \ldots, t'_1}_{\lfloor \frac{m}{2} \rfloor}, 1, \underbrace{2, \ldots, 2}_{\frac{j_2}{2}-1}, 1, 1, \underbrace{2, \ldots, 2}_{\lfloor \frac{m}{2} \rfloor - \frac{j_2}{2} -}, t_1, \underbrace{t'_1, \ldots, t'_1}_{\lfloor \frac{m}{2} \rfloor}, \ldots,$$
$$t_{m-j_1}, \underbrace{t'_{m-j_1}, \ldots, t'_{m-j_1}}_{\lfloor \frac{m}{2} \rfloor} 2, \underbrace{3, \ldots, 3}_{\lfloor \frac{m}{2} \rfloor}, c); \; 1 \leq i \leq n, \; 1 \leq j_1 \leq m \; 1 \leq j_2 \leq m-1 \text{ and } j \text{ is even.}$$

with $t_l = l+1$, $1 \leq l \leq m - j_1$ and $t'_l = l+2$, $1 \leq l \leq m - j_1$, $1 \leq j_1 \leq m$; $t_k = k+1$, $1 \leq k \leq j_1 - 1$ and $t'_k = k+2$, $1 \leq k \leq j_1 - 1$, $1 \leq j_1 \leq m$.

The vertex representation of $x_{i,j_1,m}$ for m even respect to W is

$$r(x_{i,j_1,m}|W) = (a, t_{j_1-1}, \underbrace{t'_{j_1-1}, \ldots, t'_{j_1-1}}_{\lfloor \frac{m}{2} \rfloor}, \ldots, t_1, \underbrace{t'_1, \ldots, t'_1}_{\lfloor \frac{m}{2} \rfloor}, 1, \underbrace{2, \ldots, 2}_{\frac{m}{2}-1}, 1, t_1, \underbrace{t'_1, \ldots, t'_1}_{\lfloor \frac{m}{2} \rfloor}, \ldots, t_{m-j_1},$$
$$\underbrace{t'_{m-j_1}, \ldots, t'_{m-j_1}}_{\lfloor \frac{m}{2} \rfloor} 2, \underbrace{3, \ldots, 3}_{\lfloor \frac{m}{2} \rfloor}, c); \; 1 \leq i \leq n, \; 1 \leq j_1 \leq m, \; 1 \leq j_2 \leq m-1.$$

with $t_l = l+1$, $1 \leq l \leq m - j_1$ and $t'_l = l+2$, $1 \leq l \leq m - j_1$, $1 \leq j_1 \leq m$; $t_k = k+1$, $1 \leq k \leq j_1 - 1$ and $t'_k = k+2$, $1 \leq k \leq j_1 - 1$, $1 \leq j_1 \leq m$.

The vertex representation of x_{i,j_1,j_2} for m odd respect to W is

$$r(x_{i,j_1,j_2}|W) = (a, t_{j_1-1}, \underbrace{t'_{j_1-1}, \ldots, t'_{j_1-1}}_{\lfloor \frac{m}{2} \rfloor}, \ldots, t_1, \underbrace{t'_1, \ldots, t'_1}_{\lfloor \frac{m}{2} \rfloor}, 1, \underbrace{2, \ldots, 2}_{\frac{j_2}{2}-1}, 1, 1, \underbrace{2, \ldots, 2}_{\lfloor \frac{m}{2} \rfloor - \frac{j_2}{2} -}, t_1, \underbrace{t'_1, \ldots, t'_1}_{\lfloor \frac{m}{2} \rfloor}, \ldots,$$
$$t_{m-j_1}, \underbrace{t'_{m-j_1}, \ldots, t'_{m-j_1}}_{\lfloor \frac{m}{2} \rfloor} 2, \underbrace{3, \ldots, 3}_{\lfloor \frac{m}{2} \rfloor}, c); \; 1 \leq i \leq n, \; 1 \leq j_1 \leq m \; 1 \leq j_2 \leq m-2 \text{ and } j \text{ is even.}$$

with $t_l = l+1$, $1 \leq l \leq m - j_1$ and $t'_l = l+2$, $1 \leq l \leq m - j_1$, $1 \leq j_1 \leq m$; $t_k = k+1$, $1 \leq k \leq j_1 - 1$ and $t'_k = k+2$, $1 \leq k \leq j_1 - 1$, $1 \leq j_1 \leq m$.

The vertex representation of $x_{i,j_1,m-1}$ and $x_{i,j_1,m}$ for m odd respect to W is

$$r(x_{i,j_1,m-1}|W) = (a, t_{j_1-1}, \underbrace{t'_{j_1-1}, \ldots, t'_{j_1-1}}_{\lfloor \frac{m}{2} \rfloor}, \ldots, t_1, \underbrace{t'_1, \ldots, t'_1}_{\lfloor \frac{m}{2} \rfloor}, 1, \underbrace{2, \ldots, 2}_{\lfloor \frac{m}{2} \rfloor - 1}, 1, t_1, \underbrace{t'_1, \ldots, t'_1}_{\lfloor \frac{m}{2} \rfloor}, \ldots, t_{m-j_1},$$
$$\underbrace{t'_{m-j_1}, \ldots, t'_{m-j_1}}_{\lfloor \frac{m}{2} \rfloor} 2, \underbrace{3, \ldots, 3}_{\lfloor \frac{m}{2} \rfloor}, c); \; 1 \leq i \leq n, \; 1 \leq j_1 \leq m, \; 1 \leq j_2 \leq m-1.$$

$$r(x_{i,j_1,m}|W) = (a, t_{j_1-1}, \underbrace{t'_{j_1-1}, \ldots, t'_{j_1-1}}_{\lfloor \frac{m}{2} \rfloor}, \ldots, t_1, \underbrace{t'_1, \ldots, t'_1}_{\lfloor \frac{m}{2} \rfloor}, 1, \underbrace{2, \ldots, 2}_{\lfloor \frac{m}{2} \rfloor}, t_1, \underbrace{t'_1, \ldots, t'_1}_{\lfloor \frac{m}{2} \rfloor}, \ldots, t_{m-j_1},$$
$$\underbrace{t'_{m-j_1}, \ldots, t'_{m-j_1}}_{\lfloor \frac{m}{2} \rfloor} 2, \underbrace{3, \ldots, 3}_{\lfloor \frac{m}{2} \rfloor}, c); \; 1 \leq i \leq n, \; 1 \leq j_1 \leq m, \; 1 \leq j_2 \leq m-1.$$

with $t_l = l+1$, $1 \leq l \leq m - j_1$ and $t'_l = l+2$, $1 \leq l \leq m - j_1$, $1 \leq j_1 \leq m$; $t_k = k+1$, $1 \leq k \leq j_1 - 1$ and $t'_k = k+2$, $1 \leq k \leq j_1 - 1$, $1 \leq j_1 \leq m$.

where

$$a = (t^1_{i-1}, \underbrace{t^2_{i-1}, \dots, t^2_{i-1}}_{\lfloor \frac{m}{2} \rfloor}, \underbrace{t^3_{i-1}, \dots, t^3_{i-1}}_{m}, \underbrace{t^4_{i-1}, \dots, t^4_{i-1}}_{m\lfloor \frac{m}{2} \rfloor}, \dots, t^1_1, \underbrace{t^2_1, \dots, t^2_1}_{\lfloor \frac{m}{2} \rfloor}, \underbrace{t^3_1, \dots, t^3_1}_{m}, \underbrace{t^4_1, \dots, t^4_1}_{m\lfloor \frac{m}{2} \rfloor})$$

$$c = (t^1_1, \underbrace{t^2_1, \dots, t^2_1}_{\lfloor \frac{m}{2} \rfloor}, \underbrace{t^3_1, \dots, t^3_1}_{m}, \underbrace{t^4_1, \dots, t^4_1}_{m\lfloor \frac{m}{2} \rfloor}, \dots, t^1_{n-i}, \underbrace{t^2_{n-i}, \dots, t^2_{n-i}}_{\lfloor \frac{m}{2} \rfloor}, \underbrace{t^3_{n-i}, \dots, t^3_{n-i}}_{m}, \underbrace{t^4_{n-i}, \dots, t^4_{n-i}}_{m\lfloor \frac{m}{2} \rfloor})$$

with $t^1_l = l+2,\ 1 \le l \le n-i$; $t^2_l = l+3,\ 1 \le l \le n-i$; $t^3_l = l+3,\ 1 \le l \le n-i$; $t^4_l = l+4,\ 1 \le l \le n-i$; $t^1_k = k+2,\ 1 \le k \le i-1$; $t^2_k = k+3,\ 1 \le k \le i-1$; $t^3_k = k+3,\ 1 \le k \le i-1$; $t^4_k = k+4,\ 1 \le k \le i-1$

The vertex representation of x_{i,j_2} for m even respect to W is
$r(x_{i,j_2}|W) = (a, 1, \underbrace{2, \dots, 2}_{\frac{j_2}{2}-1}, 1, 1, \underbrace{2, \dots, 2}_{\lfloor \frac{m}{2} \rfloor - \frac{j_2}{2} -}, \underbrace{2, \dots, 2}_{m}, \underbrace{3, \dots, 3}_{m(\lfloor \frac{m}{2} \rfloor)}, c)$; $1 \le i \le n$, $1 \le j_2 \le m-1$ and j is even.
The vertex representation of $x_{i,m}$ for m even respect to W is
$r(x_{i,m}|W) = (a, 1, \underbrace{2, \dots, 2}_{\frac{m}{2}}, 1, c)$; $1 \le i \le n$, $1 \le j_1 \le m$, $1 \le j_2 \le m-1$.
The vertex representation of x_{i,j_2} for m odd respect to W is
$r(x_{i,j_2}|W) = (a, 1, \underbrace{2, \dots, 2}_{\frac{j_2}{2}-1}, 1, 1, \underbrace{2, \dots, 2}_{\lfloor \frac{m}{2} \rfloor - \frac{j_2}{2} -}, \underbrace{2, \dots, 2}_{m}, \underbrace{3, \dots, 3}_{m(\lfloor \frac{m}{2} \rfloor)}, c)$; $1 \le i \le n$, $1 \le j_2 \le m-2$ and j is even.
The vertex representation of $x_{i,m-1}$ and $x_{i,m}$ for m odd respect to W is
$r(x_{i,m-1}|W) = (a, 1, \underbrace{2, \dots, 2}_{\lfloor \frac{m}{2} \rfloor - 1}, 1, \underbrace{2, \dots, 2}_{m}, \underbrace{3, \dots, 3}_{m(\lfloor \frac{m}{2} \rfloor)}, c)$; $1 \le i \le n$.
$r(x_{i,m}|W) = (a, 1, \underbrace{2, \dots, 2}_{\lfloor \frac{m}{2} \rfloor}, \underbrace{2, \dots, 2}_{m}, \underbrace{3, \dots, 3}_{m(\lfloor \frac{m}{2} \rfloor)}, c)$; $1 \le i \le n$.
where

$$a = (t^1_{i-1}, \underbrace{t^2_{i-1}, \dots, t^2_{i-1}}_{\lfloor \frac{m}{2} \rfloor}, \underbrace{t^3_{i-1}, \dots, t^3_{i-1}}_{m}, \underbrace{t^4_{i-1}, \dots, t^4_{i-1}}_{m\lfloor \frac{m}{2} \rfloor}, \dots, t^1_1, \underbrace{t^2_1, \dots, t^2_1}_{\lfloor \frac{m}{2} \rfloor}, \underbrace{t^3_1, \dots, t^3_1}_{m}, \underbrace{t^4_1, \dots, t^4_1}_{m\lfloor \frac{m}{2} \rfloor})$$

$$c = (t^1_1, \underbrace{t^2_1, \dots, t^2_1}_{\lfloor \frac{m}{2} \rfloor}, \underbrace{t^3_1, \dots, t^3_1}_{m}, \underbrace{t^4_1, \dots, t^4_1}_{m\lfloor \frac{m}{2} \rfloor}, \dots, t^1_{n-i}, \underbrace{t^2_{n-i}, \dots, t^2_{n-i}}_{\lfloor \frac{m}{2} \rfloor}, \underbrace{t^3_{n-i}, \dots, t^3_{n-i}}_{m}, \underbrace{t^4_{n-i}, \dots, t^4_{n-i}}_{m\lfloor \frac{m}{2} \rfloor})$$

with $t^1_l = l+1,\ 1 \le l \le n-i$; $t^2_l = l+2,\ 1 \le l \le n-i$; $t^3_l = l+2,\ 1 \le l \le n-i$; $t^4_l = l+3,\ 1 \le l \le n-i$; $t^1_k = k+1,\ 1 \le k \le i-1$; $t^2_k = k+2,\ 1 \le k \le i-1$; $t^3_k = k+2,\ 1 \le k \le i-1$; $t^4_k = k+3,\ 1 \le k \le i-1$

It is clearly that every vertices $v \in V(P_n \odot_2 P_m) - W$ has the distinct representation respect to W. Furthermore, we need to shown that all vertices in non-isolated resolving set W without isolated vertex. All vertices in vertex set $W = \{x_i, x_{i,j_1}; 1 \le i \le n, 1 \le j_1 \le m\} \cup \{x_{i,j_1,j_2}, x_{i,j_2}; 1 \le i \le n, 1 \le j_1 \le m-1, 1 \le j_2 \le m-1, j_1 \text{ and } j_2 \text{ is odd}\}$ without isolated vertex by the edge set $\{x_i x_{i,j_1}, x_{i,j_1} x_{i,j_1,j_2}, x_i x_{i,j_2}; 1 \le i \le n, 1 \le j_1 \le m, 1 \le j_2 \le m\}$ which all vertices in W induces subgraph $P_n \odot_2 P_m$ with pendant edges. Hence, $\langle W \rangle$ has no isolated vertices. So, the upper bound non-isolated resolving number of $P_n \odot_2 P_m$ is $nr(P_n \odot_2 P_m) \le (nm+n)(\lfloor \frac{m}{2} \rfloor + 1)$.

Hence, the lower bound non-isolated resolving number of $P_n \odot_2 P_m$ is $nr(P_n \odot_2 P_m) \ge (nm+n)(\lfloor \frac{m}{2} \rfloor + 1)$. It concludes that $nr(P_n \odot_2 P_m) = (nm+n)(\lfloor \frac{m}{2} \rfloor + 1)$. □

Theorem 2.2 *Let P_n and K_m be a connected graph of order $n \ge 2$ and $m \ge 3$, then non-isolated resolving number of $P_n \odot_2 K_m$ is $nr(P_n \odot_2 K_m) = nm^2 + nm$.*

IOP Conf. Series: Journal of Physics: Conf. Series **1008** (2018) 012040 doi:10.1088/1742-6596/1008/1/012040

Proof: Let $P_n \odot_2 K_m$ be a be a corona product of path P_n and complete graph K_m with vertex set $V(P_n \odot_2 K_m) = \{x_i; 1 \leq i \leq n\} \cup \{x_{i,j_1}; 1 \leq i \leq n, 1 \leq j_1 \leq m\} \cup \{x_{i,j_1,j_2}; 1 \leq i \leq n, 1 \leq j_1 \leq m, 1 \leq j_2 \leq m\} \cup \{x_{i,j_2}; 1 \leq i \leq n, 1 \leq j_2 \leq m\}$ and edge set $E(P_n \odot_2 K_m) = \{x_i x_{i+1}; 1 \leq i \leq n-1\} \cup \{x_i x_{i,j_1}, x_{i,j_1} x_{i,j_1,j_2}, x_i x_{i,j_2}; 1 \leq i \leq n, 1 \leq j_1 \leq m, 1 \leq j_2 \leq m\} \cup \{x_{i,j_1} x_{i,j_1+r_1}, x_{i,j_1,j_2} x_{i,j_1,j_2+r_2}, x_{i,j_2} x_{i,j_2+r_2}; 1 \leq i \leq n, 1 \leq j_1 \leq m, 1 \leq j_2 \leq m, 1 \leq r_1 \leq m-j_1$ and $1 \leq r_2 \leq m-j_2\}$ with the cardinality of vertex set $|V(P_n \odot_2 K_m)| = nm^2 + 2nm + n$ and the cardinality of edge set $|E(P_n \odot_2 K_m)| = 2nm^2 + mn + n + mn(\frac{m^2-m}{2}) - 1$.

For $n \geq 2$ and $m \geq 3$, based on Lemma 2.1 and Proposition 1.3 then we have the lower bound non isolated resolving number of $P_n \odot_2 K_m$ is $nr(P_n \odot_2 K_m) \geq |V(P_n \odot K_m)| nr(K_1 + K_m) = (nm+n)m = nm^2 + nm$. However, we can attain the sharpest lower bound. Furthermore, we prove that $nr(P_n \odot_2 K_m) \leq nm^2 + nm$. Choosing $W \subset V(P_n \odot_2 K_m)$ with $W = \{x_i, x_{i,j_1}; 1 \leq i \leq n, 1 \leq j_1 \leq m\} \cup \{x_{i,j_1,j_2}, x_{i,j_2}; 1 \leq i \leq n, 1 \leq j_1 \leq m-1, 1 \leq j_2 \leq m-1\}$ is a non-isolated resolving set of $P_n \odot_2 K_m$ and the cardinality of non-isolated resolving set is $|W| = |\{x_i, x_{i,j_1}; 1 \leq i \leq n, 1 \leq j_1 \leq m\}| + |\{x_{i,j_1,j_2}, x_{i,j_2}; 1 \leq i \leq n, 1 \leq j_1 \leq m-1, 1 \leq j_2 \leq m-1\}| = nm^2 + nm$. Thus, the representation of vertices $v \in V(P_n \odot_2 K_m) - W$ respecting to W are as follows.

The vertex representation of x_{i,j_1,j_2} respect to W is

$$r(x_{i,j_1,j_2}|W) = (a, \underbrace{2, \underbrace{3, \ldots, 3}_{m-1}, \ldots, 2, \underbrace{3, \ldots, 3}_{m-1}}_{j_1-1}, \underbrace{1, \ldots, 1}_{m}, \underbrace{2, \underbrace{3, \ldots, 3}_{m-1}, \ldots, 2, \underbrace{3, \ldots, 3}_{m-1}}_{m-j_1}, 2, \underbrace{3, \ldots, 3}_{m-1}, c);\ 1 \leq i \leq n,\ 1 \leq j_1 \leq m,\ j_2 = m.$$

where

$$a = (t_{i-1}^1, \underbrace{t_{i-1}^2, \ldots, t_{i-1}^2}_{m-1}, \underbrace{t_{i-1}^3, \ldots, t_{i-1}^3}_{m}, \underbrace{t_{i-1}^4, \ldots, t_{i-1}^4}_{m(m-1)}, \ldots, t_1^1, \underbrace{t_1^2, \ldots, t_1^2}_{m-1}, \underbrace{t_1^3, \ldots, t_1^3}_{m}, \underbrace{t_1^4, \ldots, t_1^4}_{m(m-1})$$

$$c = (t_1^1, \underbrace{t_1^2, \ldots, t_1^2}_{m-1}, \underbrace{t_1^3, \ldots, t_1^3}_{m}, \underbrace{t_1^4, \ldots, t_1^4}_{m(m-1}, \ldots, t_{n-i}^1, \underbrace{t_{n-i}^2, \ldots, t_{n-i}^2}_{m-1}, \underbrace{t_{n-i}^3, \ldots, t_{n-i}^3}_{m}, \underbrace{t_{n-i}^4, \ldots, t_{n-i}^4}_{m(m-1)})$$

with $t_l^1 = l+2$, $1 \leq l \leq n-i$; $t_l^2 = l+3$, $1 \leq l \leq n-i$; $t_l^3 = l+3$, $1 \leq l \leq n-i$; $t_l^4 = l+4$, $1 \leq l \leq n-i$; $t_k^1 = k+2$, $1 \leq k \leq i-1$; $t_k^2 = k+3$, $1 \leq k \leq i-1$; $t_k^3 = k+3$, $1 \leq k \leq i-1$; $t_k^4 = k+4$, $1 \leq k \leq i-1$

The vertex representation of x_{i,j_2} respect to W is

$$r(x_{i,j_2}|W) = (a, \underbrace{1, \ldots, 1}_{m}, \underbrace{2, \ldots, 2}_{m}, \underbrace{3, \ldots, 3}_{m(m-1)}, c);\ 1 \leq i \leq n,\ j_2 = m.$$

where

$$a = (t_{i-1}^1, \underbrace{t_{i-1}^2, \ldots, t_{i-1}^2}_{m-1}, \underbrace{t_{i-1}^3, \ldots, t_{i-1}^3}_{m}, \underbrace{t_{i-1}^4, \ldots, t_{i-1}^4}_{m(m-1)}, \ldots, t_1^1, \underbrace{t_1^2, \ldots, t_1^2}_{m-1}, \underbrace{t_1^3, \ldots, t_1^3}_{m}, \underbrace{t_1^4, \ldots, t_1^4}_{m(m-1})$$

$$c = (t_1^1, \underbrace{t_1^2, \ldots, t_1^2}_{m-1}, \underbrace{t_1^3, \ldots, t_1^3}_{m}, \underbrace{t_1^4, \ldots, t_1^4}_{m(m-1}, \ldots, t_{n-i}^1, \underbrace{t_{n-i}^2, \ldots, t_{n-i}^2}_{m-1}, \underbrace{t_{n-i}^3, \ldots, t_{n-i}^3}_{m}, \underbrace{t_{n-i}^4, \ldots, t_{n-i}^4}_{m(m-1)})$$

with $t_l^1 = l+1$, $1 \leq l \leq n-i$; $t_l^2 = l+2$, $1 \leq l \leq n-i$; $t_l^3 = l+2$, $1 \leq l \leq n-i$; $t_l^4 = l+3$, $1 \leq l \leq n-i$; $t_k^1 = k+1$, $1 \leq k \leq i-1$; $t_k^2 = k+2$, $1 \leq k \leq i-1$; $t_k^3 = k+2$, $1 \leq k \leq i-1$; $t_k^4 = k+3$, $1 \leq k \leq i-1$

It is clearly that every vertices $v \in V(P_n \odot_2 K_m) - W$ has the distinct representation respect to W. Furthermore, we need to shown that all vertices in non-isolated resolving set W without isolated vertex. All vertices in vertex set $W = \{x_i, x_{i,j_1}; 1 \leq i \leq n, 1 \leq j_1 \leq m\} \cup \{x_{i,j_1,j_2}, x_{i,j_2}; 1 \leq i \leq n, 1 \leq j_1 \leq m-1, 1 \leq j_2 \leq m-1\}$ without isolated vertex by the edge set $\{x_i x_{i,j_1}, x_{i,j_1} x_{i,j_1,j_2}, x_i x_{i,j_2}; 1 \leq i \leq n, 1 \leq j_1 \leq m, 1 \leq j_2 \leq m\}$ which all vertices in W induces subgraph $P_n \odot_2 K_m$ with pendant edges. Hence, $\langle W \rangle$ has no isolated vertices. So, the upper bound non-isolated resolving number of $P_n \odot_2 K_m$ is $nr(P_n \odot_2 K_m) \leq nm^2 + nm$.

Hence, the lower bound non-isolated resolving number of $P_n \odot_2 K_m$ is $nr(P_n \odot_2 K_m) \geq nm^2 + nm$. It concludes that $nr(P_n \odot_2 K_m) = nm^2 + nm$

Theorem 2.3 *Let K_m and P_n be a connected graph of order $n \geq 2$ and $m \geq 3$, then non-isolated resolving number of $K_m \odot_2 P_n$ is $nr(K_m \odot_2 P_n) = (mn + m)(\lfloor \frac{n}{2} \rfloor + 1)$.*

Proof: Let $K_m \odot_2 P_n$ be a be a corona product of complete graph K_m and path P_n with vertex set $V(K_m \odot_2 P_n) = \{x_i; 1 \leq i \leq n\} \cup \{x_{i,j_1}; 1 \leq i \leq n, 1 \leq j_1 \leq m\} \cup \{x_{i,j_1,j_2}; 1 \leq i \leq n, 1 \leq j_1 \leq m, 1 \leq j_2 \leq m\} \cup \{x_{i,j_2}; 1 \leq i \leq n, 1 \leq j_2 \leq m\}$ and edge set $E(K_m \odot_2 P_n) = \{x_i x_{i+r}; 1 \leq i \leq n, 1 \leq r \leq n - i\} \cup \{x_i x_{i,j_1}, x_{i,j_1} x_{i,j_1,j_2}, x_i x_{i,j_2}; 1 \leq i \leq n, 1 \leq j_1 \leq m, 1 \leq j_2 \leq m\} \cup \{x_{i,j_1} x_{i,j_1+1}, x_{i,j_1,j_2} x_{i,j_1,j_2+1}, x_{i,j_2} x_{i,j_2+1}; 1 \leq i \leq n, 1 \leq j_1 \leq m-1, 1 \leq j_2 \leq m-1\}$ with the cardinality of vertex set $|V(K_m \odot_2 P_n)| = nm^2 + 2nm + n$ and the cardinality of edge set $|E(K_m \odot_2 P_n)| = 2nm^2 + 3mn - 2n + \frac{n^2-n}{2}$.

For $n \geq 2$ and $m \geq 3$, based on Lemma 2.1 and Proposition 1.3 then we have $nr(K_m \odot_2 P_n) \geq |V(K_m \odot P_n)| nr(K_1 + P_n) = (nm+m)(\lfloor \frac{n}{2} \rfloor + 1)$. However, we can attain the sharpest lower bound. Furthermore, we prove that $nr(K_m \odot_2 P_n) \leq (nm+m)(\lfloor \frac{n}{2} \rfloor + 1)$. Choosing $W \subset V(K_m \odot_2 P_n)$ with $W = \{x_i, x_{i,j_1}; 1 \leq i \leq n, 1 \leq j_1 \leq m\} \cup \{x_{i,j_1,j_2}, x_{i,j_2}; 1 \leq i \leq n, 1 \leq j_1 \leq m-1, 1 \leq j_2 \leq m-1, j_1 \text{ and } j_2 \text{ is odd}\}$ is a non-isolated resolving set of $K_m \odot_2 P_n$ and the cardinality of non-isolated resolving set is $|W| = |\{x_i, x_{i,j_1}; 1 \leq i \leq n, 1 \leq j_1 \leq m\}| + |\{x_{i,j_1,j_2}, x_{i,j_2}; 1 \leq i \leq n, 1 \leq j_1 \leq m-1, 1 \leq j_2 \leq m-1, j_1 \text{ and } j_2 \text{ is odd}\}| = (nm+m)(\lfloor \frac{n}{2} \rfloor + 1)$. Thus, the representation of vertices $v \in V(K_m \odot_2 P_n) - W$ respecting to W are as follows.

The vertex representation of x_{i,j_1,j_2} for n even respect to W is

$$r(x_{i,j_1,j_2}|W) = (a, t_{j_1-1}, \underbrace{t'_{j_1-1}, \dots, t'_{j_1-1}}_{\lfloor \frac{n}{2} \rfloor}, \dots, t_1, \underbrace{t'_1, \dots, t'_1}_{\lfloor \frac{n}{2} \rfloor}, 1, \underbrace{2, \dots, 2}_{\frac{j_2}{2}-1}, 1, 1, \underbrace{2, \dots, 2}_{\lfloor \frac{m}{2} \rfloor - \frac{j_2}{2} -}, t_1, \underbrace{t'_1, \dots, t'_1}_{\lfloor \frac{n}{2} \rfloor}, \dots,$$
$$t_{n-j_1}, \underbrace{t'_{n-j_1}, \dots, t'_{n-j_1}}_{\lfloor \frac{n}{2} \rfloor} 2, \underbrace{3, \dots, 3}_{\lfloor \frac{n}{2} \rfloor}, c);\ 1 \leq i \leq m,\ 1 \leq j_1 \leq n\ 1 \leq j_2 \leq n-1 \text{ and } j \text{ is even.}$$

with $t_l = l+1$, $1 \leq l \leq n-j_1$ and $t'_l = l+2$, $1 \leq l \leq n-j_1$, $1 \leq j_1 \leq n$; $t_k = k+1$, $1 \leq k \leq j_1 - 1$ and $t'_k = k+2$, $1 \leq k \leq j_1 - 1$, $1 \leq j_1 \leq n$.

The vertex representation of $x_{i,j_1,n}$ for n even respect to W is

$$r(x_{i,j_1,n}|W) = (a, t_{j_1-1}, \underbrace{t'_{j_1-1}, \dots, t'_{j_1-1}}_{\lfloor \frac{n}{2} \rfloor}, \dots, t_1, \underbrace{t'_1, \dots, t'_1}_{\lfloor \frac{n}{2} \rfloor}, 1, \underbrace{2, \dots, 2}_{\frac{n}{2}-1}, 1, t_1, \underbrace{t'_1, \dots, t'_1}_{\lfloor \frac{n}{2} \rfloor}, \dots, t_{n-j_1},$$
$$\underbrace{t'_{n-j_1}, \dots, t'_{n-j_1}}_{\lfloor \frac{n}{2} \rfloor} 2, \underbrace{3, \dots, 3}_{\lfloor \frac{n}{2} \rfloor}, c);\ 1 \leq i \leq m,\ 1 \leq j_1 \leq n,\ 1 \leq j_2 \leq n-1.$$

with $t_l = l+1$, $1 \leq l \leq n-j_1$ and $t'_l = l+2$, $1 \leq l \leq n-j_1$, $1 \leq j_1 \leq n$; $t_k = k+1$, $1 \leq k \leq j_1 - 1$ and $t'_k = k+2$, $1 \leq k \leq j_1 - 1$, $1 \leq j_1 \leq n$.

The vertex representation of x_{i,j_1,j_2} for n odd respect to W is

$$r(x_{i,j_1,j_2}|W) = (a, t_{j_1-1}, \underbrace{t'_{j_1-1}, \dots, t'_{j_1-1}}_{\lfloor \frac{n}{2} \rfloor}, \dots, t_1, \underbrace{t'_1, \dots, t'_1}_{\lfloor \frac{n}{2} \rfloor}, 1, \underbrace{2, \dots, 2}_{\frac{j_2}{2}-1}, 1, 1, \underbrace{2, \dots, 2}_{\lfloor \frac{n}{2} \rfloor - \frac{j_2}{2} -}, t_1, \underbrace{t'_1, \dots, t'_1}_{\lfloor \frac{n}{2} \rfloor}, \dots,$$
$$t_{n-j_1}, \underbrace{t'_{n-j_1}, \dots, t'_{n-j_1}}_{\lfloor \frac{n}{2} \rfloor} 2, \underbrace{3, \dots, 3}_{\lfloor \frac{n}{2} \rfloor}, c);\ 1 \leq i \leq m,\ 1 \leq j_1 \leq n\ 1 \leq j_2 \leq n-2 \text{ and } j \text{ is even.}$$

with $t_l = l+1$, $1 \leq l \leq n-j_1$ and $t'_l = l+2$, $1 \leq l \leq n-j_1$, $1 \leq j_1 \leq n$; $t_k = k+1$, $1 \leq k \leq j_1 - 1$ and $t'_k = k+2$, $1 \leq k \leq j_1 - 1$, $1 \leq j_1 \leq n$.

The vertex representation of $x_{i,j_1,n-1}$ and $x_{i,j_1,n}$ for n odd respect to W is

$$r(x_{i,j_1,m-1}|W) = (a, t_{j_1-1}, \underbrace{t'_{j_1-1}, \dots, t'_{j_1-1}}_{\lfloor \frac{n}{2} \rfloor}, \dots, t_1, \underbrace{t'_1, \dots, t'_1}_{\lfloor \frac{n}{2} \rfloor}, 1, \underbrace{2, \dots, 2}_{\lfloor \frac{n}{2} \rfloor - 1}, 1, t_1, \underbrace{t'_1, \dots, t'_1}_{\lfloor \frac{n}{2} \rfloor}, \dots, t_{n-j_1},$$
$$\underbrace{t'_{n-j_1}, \dots, t'_{n-j_1}}_{\lfloor \frac{n}{2} \rfloor} 2, \underbrace{3, \dots, 3}_{\lfloor \frac{n}{2} \rfloor}, c);\ 1 \leq i \leq m,\ 1 \leq j_1 \leq n,\ 1 \leq j_2 \leq n-1.$$

$r(x_{i,j_1,n}|W) = (a, t_{j_1-1}, \underbrace{t'_{j_1-1}, \ldots, t'_{j_1-1}}_{\lfloor\frac{n}{2}\rfloor}, \ldots, t_1, \underbrace{t'_1, \ldots, t'_1}_{\lfloor\frac{n}{2}\rfloor}, 1, \underbrace{2, \ldots, 2}_{\lfloor\frac{n}{2}\rfloor}, t_1, \underbrace{t'_1, \ldots, t'_1}_{\lfloor\frac{n}{2}\rfloor}, \ldots, t_{n-j_1},$
$\underbrace{t'_{n-j_1}, \ldots, t'_{n-j_1}}_{\lfloor\frac{n}{2}\rfloor} 2, \underbrace{3, \ldots, 3}_{\lfloor\frac{n}{2}\rfloor}, c);\ 1 \leq i \leq m,\ 1 \leq j_1 \leq n,\ 1 \leq j_2 \leq n-1.$

with $t_l = l+1,\ 1 \leq l \leq n-j_1$ and $t'_l = l+2,\ 1 \leq l \leq n-j_1,\ 1 \leq j_1 \leq n$; $t_k = k+1,\ 1 \leq k \leq j_1 - 1$ and $t'_k = k+2,\ 1 \leq k \leq j_1 - 1,\ 1 \leq j_1 \leq n.$
where

$$a = \underbrace{(3, \underbrace{4, \ldots, 4}_{\lfloor\frac{n}{2}\rfloor}, \underbrace{4, \ldots, 4}_{n}, \underbrace{5, \ldots, 5}_{n\lfloor\frac{n}{2}\rfloor}, \ldots, 3, \underbrace{4, \ldots, 4}_{\lfloor\frac{n}{2}\rfloor}, \underbrace{4, \ldots, 4}_{n}, \underbrace{5, \ldots, 5}_{n\lfloor\frac{n}{2}\rfloor})}_{i-1}$$

$$c = \underbrace{(3, \underbrace{4, \ldots, 4}_{\lfloor\frac{n}{2}\rfloor}, \underbrace{4, \ldots, 4}_{n}, \underbrace{5, \ldots, 5}_{n\lfloor\frac{n}{2}\rfloor}, \ldots, 3, \underbrace{4, \ldots, 4}_{\lfloor\frac{n}{2}\rfloor}, \underbrace{4, \ldots, 4}_{n}, \underbrace{5, \ldots, 5}_{n\lfloor\frac{n}{2}\rfloor})}_{m-i}$$

The vertex representation of x_{i,j_2} for n even respect to W is
$r(x_{i,j_2}|W) = (a, 1, \underbrace{2, \ldots, 2}_{\frac{j_2}{2}-1}, 1, 1, \underbrace{2, \ldots, 2}_{\lfloor\frac{n}{2}\rfloor-\frac{j_2}{2}-1}, \underbrace{2, \ldots, 2}_{n}, \underbrace{3, \ldots, 3}_{n(\lfloor\frac{n}{2}\rfloor)}, c);\ 1 \leq i \leq m,\ 1 \leq j_2 \leq n-1$ and j_2 is even.

The vertex representation of $x_{i,n}$ for n even respect to W is
$r(x_{i,n}|W) = (a, 1, \underbrace{2, \ldots, 2}_{\frac{n}{2}}, 1, c);\ 1 \leq i \leq m,\ 1 \leq j_1 \leq n,\ 1 \leq j_2 \leq n-1.$

The vertex representation of x_{i,j_2} for m odd respect to W is
$r(x_{i,j_2}|W) = (a, 1, \underbrace{2, \ldots, 2}_{\frac{j_2}{2}-1}, 1, 1, \underbrace{2, \ldots, 2}_{\lfloor\frac{n}{2}\rfloor-\frac{j_2}{2}-}, \underbrace{2, \ldots, 2}_{n}, \underbrace{3, \ldots, 3}_{n(\lfloor\frac{n}{2}\rfloor)}, c);\ 1 \leq i \leq m,\ 1 \leq j_2 \leq n-2$ and j_2 is even.

The vertex representation of $x_{i,n-1}$ and $x_{i,n}$ for n odd respect to W is
$r(x_{i,n-1}|W) = (a, 1, \underbrace{2, \ldots, 2}_{\lfloor\frac{n}{2}\rfloor-1}, 1, \underbrace{2, \ldots, 2}_{n}, \underbrace{3, \ldots, 3}_{n(\lfloor\frac{n}{2}\rfloor)}, c);\ 1 \leq i \leq m.$
$r(x_{i,n}|W) = (a, 1, \underbrace{2, \ldots, 2}_{\lfloor\frac{n}{2}\rfloor}, \underbrace{2, \ldots, 2}_{n}, \underbrace{3, \ldots, 3}_{n(\lfloor\frac{n}{2}\rfloor)}, c);\ 1 \leq i \leq m.$

where

$$a = \underbrace{(2, \underbrace{3, \ldots, 3}_{\lfloor\frac{n}{2}\rfloor}, \underbrace{3, \ldots, 3}_{n}, \underbrace{4, \ldots, 4}_{n\lfloor\frac{n}{2}\rfloor}, \ldots, 2, \underbrace{3, \ldots, 3}_{\lfloor\frac{n}{2}\rfloor}, \underbrace{3, \ldots, 3}_{n}, \underbrace{4, \ldots, 4}_{n\lfloor\frac{n}{2}\rfloor})}_{i-1}$$

$$c = \underbrace{(2, \underbrace{3, \ldots, 3}_{\lfloor\frac{n}{2}\rfloor}, \underbrace{3, \ldots, 3}_{n}, \underbrace{4, \ldots, 4}_{n\lfloor\frac{n}{2}\rfloor}, \ldots, 2, \underbrace{3, \ldots, 3}_{\lfloor\frac{n}{2}\rfloor}, \underbrace{3, \ldots, 3}_{n}, \underbrace{4, \ldots, 4}_{n\lfloor\frac{n}{2}\rfloor})}_{m-i}$$

It is clearly that every vertices $v \in V(K_m \odot_2 P_n) - W$ has the distinct representation respect to W. Furthermore, we need to shown that all vertices in non-isolated resolving set W without isolated vertex. All vertices in vertex set $W = \{x_i, x_{i,j_1}; 1 \leq i \leq n, 1 \leq j_1 \leq m\} \cup \{x_{i,j_1,j_2}, x_{i,j_2}; 1 \leq i \leq n, 1 \leq j_1 \leq m-1, 1 \leq j_2 \leq m-1, j_1$ and j_2 is odd$\}$ without isolated vertex by the edge set $\{x_i x_{i,j_1}, x_{i,j_1} x_{i,j_1,j_2}, x_i x_{i,j_2}; 1 \leq i \leq n, 1 \leq j_1 \leq m, 1 \leq j_2 \leq m\}$ which all vertices in W induces subgraph $K_m \odot_2 P_n$ with pendant edges. Hence, $\langle W \rangle$ has no isolated vertices. So, the upper bound non-isolated resolving number of $K_m \odot_2 P_n$ is $nr(K_m \odot_2 P_n) \leq (nm+m)(\lfloor\frac{n}{2}\rfloor + 1)$.

Hence, the lower bound non-isolated resolving number of $K_m \odot_2 P_n$ is $nr(K_m \odot_2$

$P_n) \geq (nm+m)(\lfloor \frac{n}{2} \rfloor + 1)$. It concludes that $nr(K_m \odot_2 P_n) = (nm+m)(\lfloor \frac{n}{2} \rfloor + 1)$ □

Theorem 2.4 *Let K_n and K_m be a connected graph of order $n, m \geq 3$, then non-isolated resolving number of $K_n \odot_2 K_m$ is $nr(K_n \odot_2 K_m) = nm^2 + nm$.*

Proof: Let $K_n \odot_2 K_m$ be a be a corona product of complete graph K_n and K_m with vertex set $V(K_n \odot_2 K_m) = \{x_i; 1 \leq i \leq n\} \cup \{x_{i,j_1}; 1 \leq i \leq n, 1 \leq j_1 \leq m\} \cup \{x_{i,j_1,j_2}; 1 \leq i \leq n, 1 \leq j_1 \leq m, 1 \leq j_2 \leq m\} \cup \{x_{i,j_2}; 1 \leq i \leq n, 1 \leq j_2 \leq m\}$ and edge set $E(K_n \odot_2 K_m) = \{x_i x_{i+r}; 1 \leq i \leq n, 1 \leq r \leq n-i\} \cup \{x_i x_{i,j_1}, x_{i,j_1} x_{i,j_1,j_2}, x_i x_{i,j_2}; 1 \leq i \leq n, 1 \leq j_1 \leq m, 1 \leq j_2 \leq m\} \cup \{x_{i,j_1} x_{i,j_1+r_1}, x_{i,j_1,j_2} x_{i,j_1,j_2+r_2}, x_{i,j_2} x_{i,j_2+r_2}; 1 \leq i \leq n, 1 \leq j_1 \leq m, 1 \leq j_2 \leq m, 1 \leq r_1 \leq m - j_1$ and $1 \leq r_2 \leq m - j_2\}$ with the cardinality of vertex set $|V(K_n \odot_2 K_m)| = nm^2 + 2nm + n$ and the cardinality of edge set $|E(K_n \odot_2 K_m)| = 2nm^2 + mn + mn(\frac{m^2-m}{2}) + \frac{n^2-n}{2}$.

For $n \geq 2$ and $m \geq 3$, based on Lemma 2.1 and Proposition 1.3 then we have $nr(K_n \odot_2 K_m) \geq |V(K_n \odot K_m)| nr(K_1 + K_m) = (nm+n)m = nm^2 + nm$. However, we can attain the sharpest lower bound. Furthermore, we prove that $nr(K_n \odot_2 K_m) \leq nm^2 + nm$. Choosing $W \subset V(K_n \odot_2 K_m)$ with $W = \{x_i, x_{i,j_1}; 1 \leq i \leq n, 1 \leq j_1 \leq m\} \cup \{x_{i,j_1,j_2}, x_{i,j_2}; 1 \leq i \leq n, 1 \leq j_1 \leq m-1, 1 \leq j_2 \leq m-1\}$ is a non-isolated resolving set of $K_n \odot_2 K_m$ and the cardinality of non-isolated resolving set is $|W| = |\{x_i, x_{i,j_1}; 1 \leq i \leq n, 1 \leq j_1 \leq m\}| + |\{x_{i,j_1,j_2}, x_{i,j_2}; 1 \leq i \leq n, 1 \leq j_1 \leq m-1, 1 \leq j_2 \leq m-1\}| = nm^2 + nm$. Thus, the representation of vertices $v \in V(K_n \odot_2 K_m) - W$ respecting to W are as follows.

The vertex representation of x_{i,j_1,j_2} respect to W is

$$r(x_{i,j_1,m}|W) = (a, \underbrace{2, \underbrace{3,\ldots,3}_{m-1}, \ldots, 2, \underbrace{3,\ldots,3}_{m-1}}_{j_1-1}, \underbrace{1,\ldots,1}_{m}, \underbrace{2, \underbrace{3,\ldots,3}_{m-1}, \ldots, 2, \underbrace{3,\ldots,3}_{m-1}}_{m-j_1}, 2, \underbrace{3,\ldots,3}_{m-1}, c);\ 1 \leq i \leq n,\ 1 \leq j_1 \leq m,\ j_2 = m.$$

where

$$a = (\underbrace{3, \underbrace{4,\ldots,4}_{m-1}, \underbrace{4,\ldots,4}_{m}, \underbrace{5,\ldots,5}_{m(m-1)}, \ldots, 3, \underbrace{4,\ldots,4}_{m-1}, \underbrace{4,\ldots,4}_{m}, \underbrace{5,\ldots,5}_{m(m-1)}}_{i-1})$$

$$c = (\underbrace{3, \underbrace{4,\ldots,4}_{m-1}, \underbrace{4,\ldots,4}_{m}, \underbrace{5,\ldots,5}_{m(m-1)}, \ldots, 3, \underbrace{4,\ldots,4}_{m-1}, \underbrace{4,\ldots,4}_{m}, \underbrace{5,\ldots,5}_{m(m-1)}}_{n-i})$$

The vertex representation of x_{i,j_2} respect to W is

$$r(x_{i,j_2}|W) = (a, \underbrace{1,\ldots,1}_{m}, \underbrace{2,\ldots,2}_{m}, \underbrace{3,\ldots,3}_{m(m-1)}, c);\ 1 \leq i \leq n,\ j_2 = m.$$

where

$$a = (\underbrace{2, \underbrace{3,\ldots,3}_{m-1}, \underbrace{3,\ldots,3}_{m}, \underbrace{4,\ldots,4}_{m(m-1)}, \ldots, 2, \underbrace{3,\ldots,3}_{m-1}, \underbrace{3,\ldots,3}_{m}, \underbrace{4,\ldots,4}_{m(m-1)}}_{i-1})$$

$$c = (\underbrace{2, \underbrace{3,\ldots,3}_{m-1}, \underbrace{3,\ldots,3}_{m}, \underbrace{4,\ldots,4}_{m(m-1)}, \ldots, 2, \underbrace{3,\ldots,3}_{m-1}, \underbrace{3,\ldots,3}_{m}, \underbrace{4,\ldots,4}_{m(m-1)}}_{n-i})$$

It is clearly that every vertices $v \in V(K_n \odot_2 K_m) - W$ has the distinct representation respect to W. Furthermore, we need to shown that all vertices in non-isolated resolving set W without isolated vertex. All vertices in vertex set $W = \{x_i, x_{i,j_1}; 1 \leq i \leq n, 1 \leq j_1 \leq m\} \cup \{x_{i,j_1,j_2}, x_{i,j_2}; 1 \leq i \leq n, 1 \leq j_1 \leq m-1, 1 \leq j_2 \leq m-1\}$ without isolated vertex by the edge set $\{x_i x_{i,j_1}, x_{i,j_1} x_{i,j_1,j_2}, x_i x_{i,j_2}; 1 \leq i \leq n, 1 \leq j_1 \leq m, 1 \leq j_2 \leq m\}$

which all vertices in W induces subgraph $K_n \odot_2 K_m$ with pendant edges. Hence, $\langle W \rangle$ has no isolated vertex. So, the upper bound non-isolated resolving number of $K_n \odot_2 K_m$ is $nr(K_n \odot_2 K_m) \leq nm^2 + nm$. Hence, the lower bound non-isolated resolving number of $K_n \odot_2 K_m$ is $nr(K_n \odot_2 K_m) \geq nm^2 + nm$. It concludes that $nr(K_n \odot_2 K_m) = nm^2 + nm$ □

3. Conclusion

The results show that the non-isolated resolving number attain the best lower bound. There are some open problem as follows

Open Problem 1 *Find the non-isolated resolving number of $G \odot_k H$ with $k \geq 3$ for G, H are connected graph except path P_n and complete graph K_n.*

Acknowledgement

We gratefully acknowledge the support from CGANT - University of Jember of year 2018.

References

[1] Gross J L, Yellen J and Zhang P 2014 *Handbook of graph Theory* Second Edition CRC Press Taylor and Francis Group
[2] Chartrand G and Lesniak L 2000 *Graphs and digraphs 3rd* ed (London: Chapman and Hall)
[3] Slater P J 1975 Leaves of trees. in:*Proc. 6th Southeast Conf. Comb., Graph Theory, Comput. Boca Rotan* No. 14 549-559
[4] Harary F and Melter R A 1976 On The metric dimension of a graph, *Ars Combin* No.2 191-195
[5] Chitra P J B and Arumugam S 2015 Resolving Sets without Isolated Vertices, *Procedia Computer Science* **74** 38-42
[6] Chartrand G, Eroh L, Johnson M A and Oellermann O R 2000 Resolvability in graphs and the metric dimension of a graph *Discrete Appl. Math* No.105 99-113
[7] Saenpholphat V and Zhang P 2002 Connected partition dimension of a graphs *Discussiones Mathematicae Graph Theory* **22** 305-323
[8] Baca M, Baskoro E T, Salman A N M, Saputro S W and Suprijanto D 2011 The Metric Dimension of Regular Bipartite Graphs *Bull. Math. Soc. Sci. Math. Roumanie, Tome* **54**(102) 15-28
[9] Iswadi H, Baskoro E T and Simanjuntak R 2011 On the Metric Dimension of Corona Product Graphs, *Far East Journal of Mathematical Sciences* **52** 155 170
[10] Rodriguez-Velazquez J A and Fernau H 2013 On the (Adjacency) Metric Dimension of Corona and Strong Product Graph and Their Local Variants *Combinatorial And Computational Results* Arxiv: 1309.2275.v1 (Math Co)
[11] Simanjuntak R, Uttunggadewa S and Saputro S W 2013 Metric Dimension of Amalgamation of Graphs arXiv:1312.0191v2 [math.CO]
[12] Saputro S W, Suprijanto D, Baskoro E T and Salman A N M 2012 The Metric Dimension of a Graph Composition Products With Star *J. Indones. Math. Soc.* **18** 85-92
[13] Yero I G, Kuziak D and Rodriguez-Velazquez J A 2010 On the Metric Dimension of Corona Product Graphs *Combinatorial and Computational Results* arXiv:1009.2586v2[math CO]
[14] Yunika S M 2017 *Analisa Dimensi Metrik dengan Himpunan Pembeda Tak-Terisolasi Graf Hasil Operasi Korona* Thesis
[15] Dafik, Agustin I H, Surahmat, Syafrizal Sy and Alfarisi R 2017 On non-isolated resolving number of some graph operations *Far East Journal of Mathematical Sciences* **102** (2) 2473 - 2492
[16] Alfarisi R, Dafik, Agustin I H, Kristiana A I 2017 On non-isolated resolving number of graphs with pendant edges *Journal Interconnection Network* Submitted
[17] Frucht R and Harary F 1970 On the corona of two graphs *Aequationes Math* **4** 322-325
[18] Furmanczyk H, Kubale M and Mkrtchyan V V 2012 Equitable Colorings of Corona Multiproducts of Graphs, arXiv:1210.6568v1 [cs.DM]

Locating domination number of m-shadowing of graphs

Dafik[1,2], Ika Hesti Agustin[1,3], Ermita Rizki Albirri[1,2], Ridho Alfarisi[1,4], R. M. Prihandini[1,4]

[1]CGANT-University of Jember, Jember, Indonesia
[3]Department of Mathematics, University of Jember, Jember, Indonesia
[2]Department of Mathematics Education, University of Jember, Jember, Indonesia
[4]Department of Elementary School Teacher Education, University of Jember, Jember, Indonesia

E-mail: d.dafik@unej.ac.id, ikahestiagustin@gmail.com, mitalby.ma@gmail.com

Abstract. Let $G = (V, E)$ be a connected, undirected and simple graph. We define a set D as a dominating set if for every vertex $u \in V - D$ is adjacent to some vertex $v \in D$. The domination number $\gamma(G)$ is the minimum cardinality of dominating set. A vertex set D in graph $G = (V, E)$ is called locating dominating set if for every pair of different vertex u and v in $V(G) - D$ which occupies $\emptyset \neq N(u) \cap D \neq N(v) \cap D$ where $N(u)$ is adjacency vertex set of u. The minimal cardinality of locating domination number is denoted by $\gamma_L(G)$. We present about the locating dominating set of m-shadowing graph. Let G be simple and connected graph. The m-shadow graph $D_m(G)$ of a connected graph G is constructed by taking m copies of G, say $G_1, G_2, \dots, G_m$, then join each vertex u in G_i to the neighbors of the corresponding vertex v in G_j, $1 \leq i$, $j \leq m$.

1. Introduction

Let G be a nontrivial, finite, simple, undirected and connected graphs, with vertex set $V(G)$, edge set $E(G)$ and with no isolated vertex. For more detail definition of graph, see [1, 2]. A set D of vertices of a graph $G = (V, E)$ is a dominating set if every vertex in $V(G) - D$ is adjacent to some vertex in D. The domination number, denoted by $\gamma(G)$, of G is the minimum cardinality of a dominating set in G. A locating dominating set is a dominating set D that locates all the vertices in the sense that every vertex not in D is uniquely determined by its neighborhood in D. The locating domination number of G, denoted $\gamma_L(G)$, is the minimum cardinality of a locating dominating set in G. A locating dominating set of order $\gamma_L(G)$ is called an $\gamma_L(G)$-set The concept of a locating dominating set was introduced and first studied by Slater [3, 4, 5, 6] and also there is some results about domination number (locating) of graphs, see in [8, 9, 10].

For definition and notation of locating dominating set in [7] explained that the open neighborhood of a vertex $v \in V(G)$ is $N_G(v) = \{u \in V(G); uv \in E(G)\}$ and its closed neighborhood is the set $N_G[v] = N_G(v) \cup \{v\}$. For a set D of vertex set of G, $N_G[D]$ is the union of all closed neighborhoods of vertices in D. The degree of v is $d_G(v) = |N_G(v)|$. If the graph G is a connected graph, we simply write $V(G), E(G), N(v), N[v], N[D]$ and $d(v)$ rather than $V(G), E(G), N_G(v), N_G[v], N_G[D]$ and $d_G(v)$, respectively.

A dominating set of G and denoted by D, if $N[v] \cap D \neq \emptyset$ for all vertex $v \in G$, or equivalently, $N[D] = V(G)$. Any two vertices u and $v \in V(G) \backslash D$ are located by D if they have distinct neighbors in D that is, $N(u) \cap D \neq N(v) \cap D$. If a vertex $u \in V(G) \backslash D$ is located from every other vertex in $V(G) \backslash D$, we simply say that u is located by D. A set D is a locating set of G if any two distinct vertices outside D are located by D. In particular, if S is both a dominating set and a locating set, then S is a locating dominating set. We remark that the only difference between a locating set and a locating dominating set in G is that a locating set might have a unique non-dominated vertex.

The m-shadow graph $D_m(G)$ of a connected graph G is constructed by taking m copies of G, say $G_1, G_2, \ldots, G_m$, then join each vertex u in G_i to the neighbors of the corresponding vertex v in G_j, $1 \leq i$, $j \leq m$.

2. The Results

In this section, we determine the exact values of locating dominating number of m-shadowing graphs. We obtain the exact value of m-shadowing of path P_n, m-shadowing of cycle C_n and m-shadowing of ladder L_n.

Theorem 1. *Let $D_m(P_n)$ be a connected graph where P_n is path graph. For any m and $n \geq 3$, the locating domination number of G is $\gamma_L(D_m(P_n)) = \lfloor \frac{mn}{2} \rfloor$.*

Proof. The graph $D_m(P_n)$ has vertex set $V(D_m(P_n)) = \{x_i^j; 1 \leq i \leq n; 1 \leq j \leq m\}$ and edge set $E(D_m(P_n)) = \{x_i^j x_{i+1}^j; 1 \leq i \leq n-1; 1 \leq j \leq m\} \cup \{x_i^j x_{i+1}^{j+1}; 1 \leq i \leq n-1; 1 \leq j \leq m-1\} \cup \{x_i^j x_{i-1}^{j+1}; 1 \leq i \leq n-1; 1 \leq j \leq m-1\}$.

The lower bound of $(D_m(P_n))$ is $\gamma_L(D_m(P_n)) \geq \lfloor \frac{mn}{2} \rfloor$. Suppose that $\gamma_L(D_m(P_n)) > \lfloor \frac{mn}{2} \rfloor$, we take $\gamma_L(D_m(P_n)) = \lfloor \frac{mn}{2} \rfloor - 1$. The m-shadow graph $D_m(P_n)$ of a connected graph P_n is constructed by taking m copies of P_n, say $(P_n)_1, (P_n)_2, \ldots, (P_n)_m$, then join each vertex u in $(P_n)_i$ to the neighbors of the corresponding vertex v in $(P_n)_j$, $1 \leq i$, $j \leq m$. We ilustrate for n even that for j even have $|D((P_n)_j)| = \frac{n}{2}$ and for j odd have $|D((P_n)_j)| = \frac{n}{2}$. If we remove one vertex in domintaor D for j even ($j = 2$), then there are one vertex which not dominated. If we remove one vertex in domintaor D for j odd ($j = 1$), then we have at least two vertices which have the same intersection $N(u) \cap D = N(v)$ for $u, v \in D((P_n)_j)$. We can conclude, $\gamma_L(D_m(P_n)) \geq \lfloor \frac{mn}{2} \rfloor$.

Furthermore, we will show that the upper bound of locating domination number of $D_m(P_n)$ is $\gamma_L(D_m(P_n)) \leq \lfloor \frac{mn}{2} \rfloor$. We choose a locating dominating set in $D_m(P_n)$ is $D = \{x_i^j; 1 \leq i \leq n; 1 \leq j \leq m; i \text{ even}; j \text{ odd}\} \cup \{x_i^j; 1 \leq i \leq n; 1 \leq j \leq m; i \text{ odd}; j \text{ even}\}$ The non-dominator set in $D_m(P_n)$ is $V(G) - D$, we get intersection between the neighborhood $N(V)$ with $V \in V(G) - D$ and dominator set D as follow: For $i = 1$ and $j = 1$, $\{N(x_i^j) = x_{i+1}^j, x_{i+1}^{j+1}\}$. So the intersection for $i = 1$ and $j = 1$ is $D \cap N(x_i^j) = \{x_{i+1}^{j+1}\}$. For $3 \leq i \leq n$, $1 \leq j \leq m$, and j odd, $N(x_i^j) = \{x_{i-1}^{j+1}, x_{i+1}^{j+1}, x_{i+1}^j, x_{i-1}^j\}$. Thus the

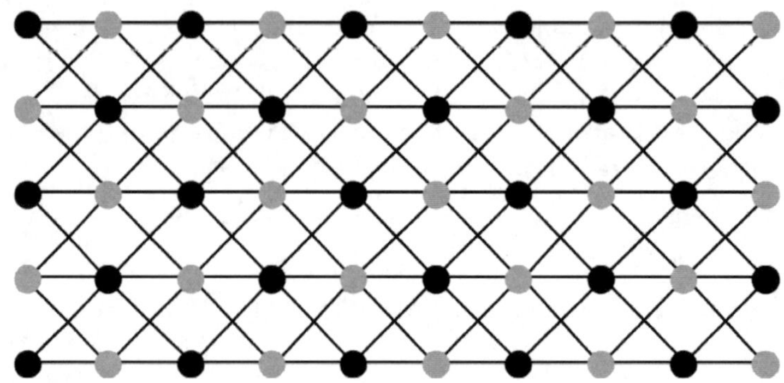

Figure 1. Example of locating domination number of $D_5(P_{10})$ and dominator with green color

intersection for $3 \leq i \leq n$, $1 \leq j \leq m$, and j odd is $D \cap N(x_i^j) = \{x_{i-1}^j, x_{i+1}^j\}$. For $1 \leq i \leq n$, i even, $1 \leq j \leq m$ and j even, $N(x_i^j) = \{x_{i-1}^{j-1}, x_{i+1}^{j-1}, x_{i-1}^j, x_{i+1}^j, x_{i-1}^{j+1}, x_{i+1}^{j+1}\}$. So the intersection for $1 \leq i \leq n$, i even, $1 \leq j \leq m$ and j even is $D \cap N(x_i^j) = \{x_{i-1}^j, x_{i+1}^j\}$

Based on the results above, it can be seen that the intersection between the neighborhood $N(V)$ with $V \in V(G) - D$ and dominator D is unique and not empty set, it mean the dominator set D dominates all edges in $D_m(P_n)$. Thus, it can be concluded that $\gamma_L(D_m(P_n)) \leq \lfloor \frac{mn}{2} \rfloor$. Hence, the lower bound and upper bound of locating domination number respectively are $\gamma_L(D_m(P_n)) \leq \lfloor \frac{mn}{2} \rfloor$ and $\gamma_L(D_m(P_n)) \geq \lfloor \frac{mn}{2} \rfloor$. It implies that the locating domination number of $D_m(P_n)$ is $\gamma_L(D_m(P_n)) = \lfloor \frac{mn}{2} \rfloor$. Figure 1 illustrate locating domination number of 5-shadow of path graph P_{10} with $\gamma_L(D_5(P_{10})) = 25$ □

Theorem 2. *Let $D_m(K_n)$ be a connected graph with $m \geq 1$ and $n \geq 3$, the locating domination number of $D_m(K_n)$ is $\gamma_L(D_m(K_n)) = \lfloor \frac{m}{2} \rfloor (n-2) + \lceil \frac{m}{2} \rceil$*

Proof. Let $D_m(K_n)$ be m-shadow graph of complete graph with $m \geq 1$ and $n \geq 3$. Vertex set $V(D_m(K_n)) = \{x_i^j; 1 \leq i \leq n, 1 \leq j \leq m\}$ and edge set $E(D_m(K_m)) = \{x_i^j x_{i+k}^j; 1 \leq i \leq n, 1 \leq k \leq n-i, 1 \leq j \leq m\} \cup \{x_i^j x_l^{j+1}; 1 \leq i, l \leq n, 1 \leq j \leq m-1, i \neq l\}$. The cardinality of vertex set and edge set, respectively are $|V(D_m(K_n))| = nm$ and $|E(D_m(K_n))| = m(\frac{n(n-1)}{2}) + (m-1)(n-1)n$

The lower bound of $(D_m(K_n))$ is $\gamma_L(D_m(K_n)) \geq \lfloor \frac{m}{2} \rfloor (n-2) + \lceil \frac{m}{2} \rceil$. Suppose that $\gamma_L(D_m(K_n)) > \lfloor \frac{m}{2} \rfloor (n-2) + \lceil \frac{m}{2} \rceil$, we take $\gamma_L(D_m(K_n)) = \lfloor \frac{m}{2} \rfloor (n-2) + \lceil \frac{m}{2} \rceil - 1$. The m-shadow graph $D_m(K_n)$ of a connected graph G is constructed by taking m copies of K_n, say $(K_n)_1, (K_n)_2, \ldots, (K_n)_m$, then join each vertex u in $(K_n)_i$ to the neighbors of the corresponding vertex v in $(K_n)_j$, $1 \leq i, j \leq m$. We ilustrate that for j odd have $|D((K_n)_j)| = 1$ and for j even have $|D((K_n)_j)| = n-2$. If we remove one vertex in domintaor D for j even ($j = 2$), then we have at least two vertices which have the same intersection $N(u) \cap D = N(v) \cap D = \{x_1^1, x_1^3, y\}$ for $u, v \in D((K_n)_j)$ and $y \in D((K_n)_2)$. We can conclude, $\gamma_L(D_m(K_n)) \geq \lfloor \frac{m}{2} \rfloor (n-2) + \lceil \frac{m}{2} \rceil$.

Furthermore, we will show that the upper bound of locating domination number of $D_m(K_n)$ is $\gamma_L(D_m(K_n)) \leq \lfloor \frac{m}{2} \rfloor (n-2) + \lceil \frac{m}{2} \rceil$. We choose a locating dominating set

IOP Conf. Series: Journal of Physics: Conf. Series **1008** (2018) 012041 doi:10.1088/1742-6596/1008/1/012041

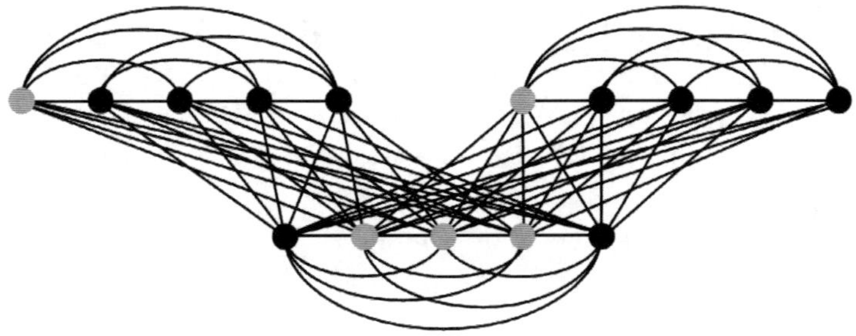

Figure 2. Example of locating domination number of $D_3(K_5)$ and dominator with green color

in $D_m(K_n)$ is $D = \{x_1^j; j \text{ is odd}\} \cup \{x_i^j; j \text{ is even}, 2 \leq i \leq n-1\}$. The non-dominator set in $D_m(K_n)$ is $V(G) - D$, we get intersection between the neighborhood $N(v)$ with $v \in V(G) - D$ and dominator set D as follow.

$N(x_i^1) \cap D = \{x_1^1, x_l^2; 2 \leq i \leq n, 2 \leq l \leq n-2, i \neq l\}$

For j is even, we have
$N(x_1^j) \cap D = \{x_i^{j-1}; 2 \leq i \leq n-1\}$
$N(x_n^j) \cap D = \{x_i^{j-1}, x_1^{j+1}; 2 \leq i \leq n-1\}$

For j is even and $j \geq 3$, we have
$N(x_i^j) \cap D = \{x_1^j, x_l^{j-1}, x_l^{j+1}; 2 \leq i \leq n, 2 \leq l \leq n-2, i \neq l\}$

Based on the results above, it can be seen that the intersection between the neighborhood $N(v)$ with $v \in V(G) - D$ and dominator D is unique and not empty set, it mean that dominator set D dominate all edges in $D_m(K_n)$.Thus, it can be concluded that $\gamma_L(D_m(K_n)) \leq \lceil mn \rceil$ attain the condition of locating dominating set. Hence, the lower bound and upper bound of locating domination number respectively, are $\gamma_L(D_m(K_n)) \leq \lfloor \frac{m}{2} \rfloor (n-2) + \lceil \frac{m}{2} \rceil$ and $\gamma_L(D_m(K_n)) \geq \lfloor \frac{m}{2} \rfloor (n-2) + \lceil \frac{m}{2} \rceil$. It can be concluded that the locating domination number of $D_m(K_n)$ is $\gamma_L(D_m(K_n)) = \lfloor \frac{m}{2} \rfloor (n-2) + \lceil \frac{m}{2} \rceil$. Figure 2 illustrate locating domination number of 3-shadow of complete graph K_5 with $\gamma_L(D_3(K_5)) = 5$ □

Theorem 3. *Let $D_m(C_n)$ be a connected graph where C_n is cycle graph. For any m and $n \geq 4$ then $\gamma(D_m(C_n)) \leq \lceil \frac{mn}{2} \rceil$*

Proof. The $D_m(C_n)$ graph has vertex set $V(G) = \{x_i^j; 1 \leq i \leq n; 1 \leq j \leq m\}$ and edge set $E(G) = \{x_i^j x_{i+1}^j; 1 \leq i \leq n-1; 1 \leq j \leq m\} \cup \{x_1^j x_n^j; 1 \leq j \leq m\} \cup \{x_i^j x_{i+1}^{j+1}; 1 \leq i \leq n-1; 1 \leq j \leq m-1\} \cup \{x_{i+1}^j x_i^{j+1}; 1 \leq i \leq n-1; 1 \leq j \leq m-1\} \cup \{x_1^j x_n^{j+1}; 1 \leq j \leq m-1\} \cup \{x_n^j x_1^{j+1}; 1 \leq j \leq m-1\}$.

Furthermore, we will show that the upper bound of locating domination number of $D_m(C_n)$ is $\gamma_L(D_m(C_n)) \leq \lceil \frac{mn}{2} \rceil$. We choose a locating dominating set in $D_m(C_n)$

is $D = \{x^j_{3i-2}; 1 \leq i \leq n; 1 \leq j \leq m; j \text{ odd}\} \cup \{x^j_{3i-1}; 1 \leq i \leq n; 1 \leq j \leq m; j \text{ even}\} \cup \{x^j_{3i}; 1 \leq i \leq n; 1 \leq j \leq m; j \text{ even}\}$. The non-dominator set in $D_m(C_n)$ is $V(G) - D$, we get intersection between the neighborhood $N(V)$ with $V \in V(G) - D$ and dominator set D as follow: For $1 \leq i \leq n$, $1 \leq j \leq m$ and j odd, $N(x^j_{3i-1}) = \{x^j_{3i-2}, x^{j+1}_{3i-2}, x^{j+1}_{3i}, x^j_{3i}\}$. So the intersection for $1 \leq i \leq n$, $1 \leq j \leq m$ and j odd is $D \cap N(x^j_{3i-1}) = \{x^j_{3i-2}, x^{j+1}_{3i}\}$. For $1 \leq i \leq n$, $1 \leq j \leq m$, and j odd, $N(x^j_{3i}) = \{x^j_{3i-1}, x^j_{3i-2}; 2 \leq i \leq n; x^{j+1}_{3i-1}; x^{j+1}_{3i-2}; 2 \leq i \leq n\}$. So the intersection for $3 \leq i \leq n$, $1 \leq j \leq m$, and j odd is $D \cap N(x^j_{3i}) = \{x^j_{3i-2}; 2 \leq i \leq n, x^{j+1}_{3i-1}\}$. For $i = 1$, $1 \leq j \leq m$ and j even, $N(x^j_i) = \{x^{j-1}_{i+1}, x^j_{i+1}, x^{j+1}_{i+1}\}$. So the intersection for $i = 1$, $1 \leq j \leq m$ and j even is $D \cap N(x^j_i) = \{x^j_{i+1}\}$. For $1 \leq i \leq n$, $1 \leq j \leq m$, and j even, $N(x^j_{3i+1}) = \{x^{j-1}_{3i}, x^j_{3i}, x^j_{3i+2}, x^{j-1}_{3i+2}, x^{j+1}_{3i}, x^{j+1}_{3i+2}\}$. So the intersection for $1 \leq i \leq n$, $1 \leq j \leq m$, and j even is $D \cap N(x^j_{3i+1}) = \{x^j_{3i}, \{x^j_{3i+2}\}$

Based on the results above, it can be seen that the intersection between the neighborhood $N(v)$ with $v \in V(G) - D$ and dominator D is unique and not empty set, it mean that dominator set D dominate all edges in $D_m(C_n)$. Thus, it can be concluded that $\gamma_L(D_m(C_n)) \leq \lceil \frac{mn}{2} \rceil$ attain the condition of locating dominating set. Hence, the upper bound of locating domination number of $D_m(C_n)$ is $\gamma_L(D_m(C_n)) \leq \lceil \frac{mn}{2} \rceil$. □

Theorem 4. *Let $D_m(L_n)$ be a connected graph where L_n is ladder graph. For any m and $n \geq 3$ then $\gamma(D_m(L_n)) \leq mn$*

Proof. The $D_m(L_n)$ graph has vertex set $V(G) = \{x^j_i; 1 \leq i \leq n; 1 \leq j \leq m\} \cup \{y^j_i; 1 \leq i \leq n; 1 \leq j \leq m\}$ and edge set $E(G) = \{x^j_i x^j_{i+1}; 1 \leq i \leq n-1; 1 \leq j \leq m\} \cup \{y^j_i y^j_{i+1}; 1 \leq i \leq n-1; 1 \leq j \leq m\} \cup \{x^j_i y^j_i; 1 \leq i \leq n; 1 \leq j \leq m\} \cup \{x^j_i x^{j+1}_{i+1}; 1 \leq i \leq n-1; 1 \leq j \leq m-1\} \cup \{x^j_i y^{j+1}_i; 1 \leq i \leq n; 1 \leq j \leq m-1\} \cup \{y^j_i y^{j+1}_{i+1}; 1 \leq i \leq n-1; 1 \leq j \leq m-1\} \cup \{y^j_i x^{j+1}_i; 1 \leq i \leq n; 1 \leq j \leq m-1\} \cup \{x^j_{i+1} x^{j+1}_i; 1 \leq i \leq n-1; 1 \leq j \leq m-1\} \cup \{y^j_{i+1} y^{j+1}_i; 1 \leq i \leq n-1; 1 \leq j \leq m-1\}$.

Furthermore, we will show that the upper bound of locating domination number of $D_m(L_n)$ is $\gamma_L(D_m(L_n)) \leq mn$. We choose a locating dominating set in $D_m(L_n)$ is $D = \{x^j_i; 1 \leq i \leq n; i \text{ even}; 1 \leq \mathrm{j} \leq \mathrm{m}\} \cup \{\mathrm{y}^\mathrm{j}_\mathrm{i}; 1 \leq \mathrm{i} \leq \mathrm{n}; \mathrm{i} \text{ odd}; 1 \leq \mathrm{j} \leq \mathrm{m}\}$. The non-dominator set in $D_m(L_n)$ is $V(G) - D$, we get intersection between the neighborhood $N(V)$ with $V \in V(G) - D$ and dominator set D as follow: For $1 \leq i \leq n$, i odd and $1 \leq j \leq m$, $N(x^j_{2i-1}) = \{x^j_{2i}, x^{j+1}_{2i}, x^{j-1}_{2i}, x^j_{2i-2}, x^{j+1}_{2i-2}, x^{j-1}_{2i-2}, y^j_{2i-1}\}$. So the intersection for $1 \leq i \leq n$, i odd and $1 \leq j \leq m$ is $D \cap N(x^j_{2i-1}) = \{x^j_{2i}, x^{j+1}_{2i}, x^{j-1}_{2i}, x^j_{2i-2}, x^{j+1}_{2i-2}, x^{j-1}_{2i-2}, y^j_{2i-1}\}$. For $1 \leq i \leq n$, i even and $1 \leq j \leq m$, $N(y^j_{2i}) = \{y^j_{2i-1}, y^{j+1}_{2i-1}, y^{j-1}_{2i-1}, y^j_{2i+1}, y^{j+1}_{2i+1}, y^{j-1}_{2i+1}, x^j_{2i}\}$. So the intersection for $1 \leq i \leq n$, i even and $1 \leq j \leq m$ is $D \cap N(y^j_{2i}) = \{y^j_{2i-1}, y^{j+1}_{2i-1}, y^{j-1}_{2i-1}, y^j_{2i+1}, y^{j+1}_{2i+1}, y^{j-1}_{2i+1}, x^j_{2i}\}$.

Based on the results above, it can be seen that the intersection between the neighborhood $N(v)$ with $v \in V(G) - D$ and dominator D is unique and not empty set, it mean that dominator set D dominate all edges in $D_m(L_n)$.Thus, it can be concluded that $\gamma_L(D_m(L_n)) \leq mn$ attain the condition of locating dominating set. Hence, the upper bound of locating domination number of $D_m(L_n)$ is $\gamma_L(D_m(L_n)) \leq mn$. □

3. Concluding Remarks

In this paper, we have determined the exact values of locating domination number on m-shadowing of graph, namely $D_m(P_n)$, $D_m(C_n)$ and $D_m(L_n)$. Thus, it still has many open problems.

Open Problem 1. *Let G be any connected graph, determine a sharp lower bound of* $\gamma_L D_m(G)$ *?*

Open Problem 2. *Let G and H be any connected graph, determine the locating domination number of the other operations in graph.*

Acknowledgement

We gratefully acknowledge the support from CGANT - University of Jember of year 2018.

References

[1] Gross J L, Yellen J and Zhang P 2014 *Handbook of graph Theory* Second Edition CRC Press Taylor and Francis Group
[2] Chartrand G and Lesniak L 2000 *Graphs and digraphs 3rd* ed (London: Chapman and Hall)
[3] Slater P J 1987 Dominating and location in acyclic graphs, *Networks* 17 5564.
[4] Slater P J 1988 Dominating and reference sets in graphs, *J. Math. Phys. Sci.* 22 445455
[5] Colbourn C J, Slater P J, Stewart L K 1987 Locating-dominating sets in series-parallel networks *Congr. Numer.* 56 135162.
[6] Finbow A, Hartnell B L 1988 On locating dominating sets and well-covered graphs *Congr. Numer.* 65 191200.
[7] Foucaud F, Henning M A, Lwenstein C and Sasse T 2016 Locating dominating sets in twin-free graphs *Discrete Applied Mathematics* 200 5258.
[8] Gembong A W, Slamin, Dafik and Agustin I H 2017 Bound of Distance Domination Number of Graph and Edge Comb Product Graph *Journal of Physics: Conference Series* **855** 012014
[9] Dafik, Agustin I H, Wardani D A R and Kurniawati E Y A 2018 study of local domination number of S n rtrie H graph *Journal of Physics: Conference Series* **943** 012003
[10] Wardani D A R, Dafik, Agustin I H and Kurniawati E Y 2018 On locating independent domination number of amalgamation graphs *Journal of Physics: Conference Series* **943** 012027

On the total irregularity strength of caterpillar with each internal vertex has degree three

Diari Indriati[1], Isnaini Rosyida[2] and Widodo[3]

[1]Faculty of Mathematics and Natural Sciences, Universitas Sebelas Maret, Surakarta, Indonesia
[2]Faculty of Mathematics and Natural Sciences, Universitas Negeri Semarang, Semarang, Indonesia
[3]Faculty of Mathematics and Natural Sciences, Universitas Gadjah Mada, Yogyakarta, Indonesia

E-mail: diari_indri@yahoo.co.id, diari_indri@staff.uns.ac.id, widodomathugm@gmail.com, iisisnaini@gmail.com

Abstract. Let G be a simple, connected and undirected graph with vertex set V and edge set E. A total k-labeling $f : V \cup E \to \{1, 2, \ldots, k\}$ is defined as totally irregular total k-labeling if the weights of any two different both vertices and edges are distinct. The weight of vertex x is defined as $wt(x) = f(x) + \sum_{xy \in E} f(xy)$, while the weight of edge xy is $wt(xy) = f(x) + f(xy) + f(y)$. A minimum k for which G has totally irregular total k-labeling is mentioned as total irregularity strength of G and denoted by ts(G). This paper contains investigation of totally irregular total k-labeling and determination of their total irregularity strengths for caterpillar graphs with each internal vertex between two stars has degree three. The results are $ts(S_{n,3,n}) = \lceil \frac{2n}{2} \rceil$, $ts(S_{n,3,3,n}) = \lceil \frac{2n+1}{2} \rceil$, and $ts(S_{n,3,3,3,n}) = \lceil \frac{2n+2}{2} \rceil$ for $n \geq 4$.

Keywords : total k-labeling, totally irregular total k-labeling, total irregularity strength, caterpillar.
Mathematics Subject Classification (2010): MSC 05C78

1. Introduction

Let us consider a connected, simple and undirected graph G with a vertex set $(V(G))$ and an edge set $(E(G))$. A labeling of a graph G is a mapping that carries a set of graph elements into a set of integers, called labels (see Wallis [10]). If the domain of mapping is a vertex set, or an edge set, or a union of vertex and edge sets, then the labeling is called *vertex labeling*, *edge labeling*, or *total labeling*, respectively. In his survey, Gallian [2] showed that there were various kinds of labelings on graphs, and one of them was an irregular total labeling.

Bača *et al.* [1] defined a labeling for a graph G, $f : V(G) \cup E(G) \to \{1, 2, \ldots, k\}$ to be a *vertex irregular total k-labeling* if for every two different vertices x and y the *vertex-weights* $wt_f(x) \neq wt_f(y)$, where the vertex-weight of vertex x is $wt_f(x) = f(x) + \sum_{xz \in E} f(xz)$. The *total vertex irregularity strength* of G, denoted by $tvs(G)$ is a minimum k for which G has a vertex irregular total k-labeling. They obtained the exact values of the total vertex irregularity strength for star, cycle, prisms and complete graphs. Moreover, Nurdin *et al.* [7] proved the exact value of the total vertex irregularity strength for any tree T with n pendant vertices and no vertex of degree two, that is

$$tvs(T) = \left\lceil \frac{n+1}{2} \right\rceil. \quad (1)$$

Bača *et al.* [1] also defined a labeling $g : V(G) \cup E(G) \to \{1, 2, \ldots, k\}$ to be an *edge irregular total k-labeling* of the graph G if for every two different edges xy and $x'y'$ of G, their *edge weights* are distinct. The *edge-weight* of xy defined as $wt_g(xy) = g(x) + g(xy) + g(y)$ and $wt_g(x'y') = g(x') + g(x'y') + g(y')$. The *total edge irregularity strength* denoted by $tes(G)$, is defined as the minimum k for which G has an edge irregular total k-labeling. They also obtained the exact values of the *tes* for path, cycle, star, wheel and friendship graphs. The *tes* of generalized web graphs have been determined by Indriati *et al.* [3]. Moreover, Ivančo and Jendrol [5] proved that for any tree T, satisfies

$$tes(T) = \max\left\{\lceil(|E(T)| + 2)/3\rceil, \lceil(\Delta(T) + 1)/2\rceil\right\}. \tag{2}$$

Marzuki *et al.* [6] combined the idea of vertex irregular total k-labeling and edge irregular total k-labeling, and introduced another irregular total k-labeling called the totally irregular total k-labeling. A labeling $h : V(G) \cup E(G) \to \{1, 2, \ldots, k\}$ to be a *totally irregular total k-labeling* of the graph G if for every two different vertices x and y, their *vertex-weights* $wt_h(x) \neq wt_h(y)$, and also for every two different edges xy and $x'y'$ of G, their *edge-weights* $wt_h(xy) \neq wt_h(x'y')$. The minimum k for which G has a totally irregular total k-labeling is defined as the *total irregularity strength*, and denote as $ts(G)$. For the total irregularity strength of a graph G, they observed that

$$ts(G) \geq \max\{tes(G), tvs(G)\}. \tag{3}$$

Moreover, Marzuki *et al.* [6] also determined the total irregularity strength of cycle and path. Ramdani and Salman [8] obtained the total irregularity strength of some cartesian product graphs, namely $K_{1,n}\Box P_2$, $P_n\Box P_2$, $(P_n + P_1)\Box P_2$, and $C_n\Box P_2$. In [9], Ramdani *et al.* determined the total irregularity strength of Gear graphs $G_n, n \geq 3$, fungus graphs $F_{g_n}, n \geq 3$ and disjoint union of stars $mS_n, n, m \geq 2$. In [4], the total irregularity strength of double stars $S_{n,m}$ and caterpillar $S_{n,2,n}$ has been determined. In this paper, we continue to investigate the total irregularity strength of caterpillar graphs $S_{n,3,n}$, $S_{n,3,3,n}$ and $S_{n,3,3,3,n}$ for $n \geq 4$.

2. Main Results

This part discussed about the total irregularity strength of three caterpillar graphs $S_{n,3,n}$, $S_{n,3,3,n}$ and $S_{n,3,3,3,n}$.

2.1. Caterpillar $S_{n,3,n}$

A caterpillar $S_{n,3,n}$ is a class of graph constructed from the double-star $S_{n,n}$ by inserting one vertex on the bridge connecting of the two centers of two stars, so that the inserted vertex has degree three. This inserted vertex called as the internal vertex. This caterpillar contains three stars with the center of the two end-stars have degree n respectively, while the center of the middle star has degree three. Therefore, there is no vertex with degree two. This graph is a tree with $2n + 2$ vertices, $2n + 1$ edges and $2n - 1$ pendant vertices. Maximal degree of the graph is $\Delta = n$.
According to (3), the lower bound of its total irregularity strength is the maximum value between its total edge irregularity strength and its total vertex irregularity strength. The total edge irregularity strength of graph $S_{n,3,n}$ can be found by (2), that is

$$tes(S_{n,3,n}) = \max\left\{\left\lceil\frac{\Delta+1}{2}\right\rceil, \left\lceil\frac{|E|+2}{3}\right\rceil\right\} = \max\left\{\left\lceil\frac{n+1}{2}\right\rceil, \left\lceil\frac{2n+3}{3}\right\rceil\right\} = \left\lceil\frac{2n+3}{3}\right\rceil. \tag{4}$$

Moreover, the total vertex irregularity strength of $S_{n,3,n}$ can be found by (1), that is

$$tvs(S_{n,3,n}) = \left\lceil\frac{(2n-1)+1}{2}\right\rceil = \left\lceil\frac{2n}{2}\right\rceil = n. \tag{5}$$

IOP Conf. Series: Journal of Physics: Conf. Series **1008** (2018) 012042 doi:10.1088/1742-6596/1008/1/012042

Furthermore, the lower bound of this *total irregularity strength* is obtained by (3), that is

$$ts(G) \geq \max\{tes(G), tvs(G)\} = \max\left\{\left\lceil\frac{2n+3}{2}\right\rceil, \left\lceil\frac{2n}{2}\right\rceil\right\} = \left\lceil\frac{2n}{2}\right\rceil = n. \tag{6}$$

In the Theorem 2.1, the exact value of its *total irregularity strength* is determined. The lower bound of this parameter can be seen in (6). We will show that the upper bound of this parameter is equal with the lower bound.

Theorem 2.1. *Let consider the caterpillar* $S_{n,3,n}$, $n \geq 4$. *Its total irregularity strength is* $ts(S_{n,3,n}) = \lceil\frac{2n}{2}\rceil = n$.

Proof. $S_{n,3,n}$ is a tree with $2n+2$ vertices, $2n+1$ edges and $2n-1$ pendant vertices. Figure 1 shows the illustration of this graph.

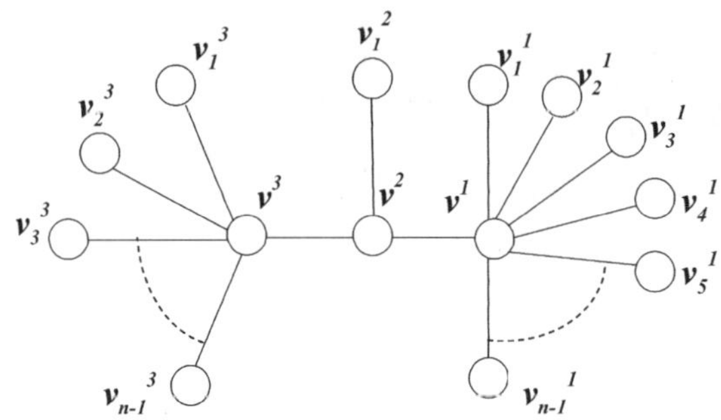

Figure 1. *The caterpillar* $S_{n,3,n}$

Let the vertex set of this graph be $V(S_{n,3,n}) = \{v_i^1 : 1 \leq i \leq n-1\} \cup \{v_i^3 : 1 \leq i \leq n-1\} \cup \{v^j : j = 1,2,3\} \cup \{v_1^2\}$ and the edge set be $E(S_{n,3,n}) = \{v^1v_i^1 : 1 \leq i \leq n-1\} \cup \{v^3v_i^3 : 1 \leq i \leq n-1\} \cup \{v^jv^{j+1} : j = 1,2\} \cup \{v^2v_1^2\}$. To determine the exact value of ts, construct the totally irregular total k-labeling h as follows. Label of vertices and edges:

$$h(v_i^1) = 1, \text{for } 1 \leq i \leq n-1.$$
$$h(v_i^3) = i, \text{for } 1 \leq i \leq n-1.$$
$$h(v^j) = \begin{cases} 1, & \text{for } j = 1, \\ n, & \text{for } j = 2, \\ n-1, & \text{for } j = 3. \end{cases}$$
$$h(v_1^2) = n.$$
$$h(v^1v_i^1) = i, \text{ for } 1 \leq i \leq n-1.$$
$$h(v^3v_i^3) = n, \text{for } 1 \leq i \leq n-1.$$
$$h(v^jv^{j+1}) = \begin{cases} n-2, & \text{for } j = 1, \\ n, & \text{for } j = 2. \end{cases}$$
$$h(v^2v_1^2) = n.$$

Under the total labeling h, it is shown that the greatest label for all vertices and edges is $k = n$. It means

IOP Conf. Series: Journal of Physics: Conf. Series **1008** (2018) 012042 doi:10.1088/1742-6596/1008/1/012042

that h is a total k-labeling with $k = \lceil \frac{2n}{2} \rceil = n$. The weight of vertices and edges are as follows.

$$wt_h(v_i^j) = \begin{cases} i+1, & \text{for } 1 \le i \le n-1,\ j=1, \\ n+i, & \text{for } 1 \le i \le n-1,\ j=3. \end{cases}$$

$$wt_h(v^j) = \begin{cases} \frac{n(n+1)}{2} - 1, & \text{for } j=1, \\ 4n-2, & \text{for } j=2, \\ (n+1)n-1, & \text{for } j=3. \end{cases}$$

$$wt_h(v_1^2) = 2n.$$

$$wt_h(v^j v_i^j) = \begin{cases} i+2, & \text{for } j=1,\ 1 \le i \le n-1, \\ 2n-1+i, & \text{for } j=3,\ 1 \le i \le n-1 \end{cases}$$

$$wt_h(v^j v^{(j+1)}) = \begin{cases} 2n-1, & \text{for } j=1, \\ 3n-1, & \text{for } j=2. \end{cases}$$

$$wt_h(v^2 v_1^2) = 3n.$$

The weight of vertices v_i^j for j=1, 3 and for v_1^2 form a consecutive integers from 2 up to n, for $n+1$ until $2n-1$ respectively and $2n$ for (v_1^2). The weights among vertices v^1, v^2, and v^3 are distinct. Then, it indicates that the weights of every pair of vertices are distinct. Furthermore, the weight of edges for every two different edges also distinct. There are $i+2$, $2n-1+i$, $2n-1$, $3n-1$ and $3n$ for the weight of all edges. Therefore, we conclude that h is a totally irregular total k-labeling and the total irregularity strength is $ts(S_{n,3,n}) = k = \lceil \frac{2n}{2} \rceil = n$. □

2.2. Caterpillar $S_{n,3,3,n}$

A caterpillar $S_{n,3,3,n}$ is a class of graph constructed from the double-star $S_{n,n}$ by inserting two vertices on the bridge connecting of the two centers of two stars, so that each inserted vertex has degree three. These inserted vertices called as the internal vertices. This caterpillar contains four stars with the center of the two end-stars have degree n respectively, while the center of each internal star has degree three. Therefore, there is no vertex with degree two. This graph is a tree with $2n+4$ vertices, $2n+3$ edges and $2n$ pendant vertices. Maximal degree of the graph is $\Delta = n$.

The lower bound of its total irregularity strength is the maximum value between its total edge irregularity strength and its total vertex irregularity strength (see (3)). According to (2), the total edge irregularity strength of graph $S_{n,3,3,n}$ is

$$tes(S_{n,3,3,n}) = \max\left\{ \left\lceil \frac{n+1}{2} \right\rceil, \left\lceil \frac{2n+5}{3} \right\rceil \right\} = \left\lceil \frac{2n+5}{3} \right\rceil. \tag{7}$$

While the total vertex irregularity strength of $S_{n,3,3,n}$ can be found by (1), that is

$$tvs(S_{n,3,3,n}) = \left\lceil \frac{2n+1}{2} \right\rceil. \tag{8}$$

Moreover, the lower bound of its *total irregularity strength* can be obtained by (3), that is

$$ts(G) \ge \max\{tes(G), tvs(G)\} = \max\left\{ \left\lceil \frac{2n+5}{3} \right\rceil, \left\lceil \frac{2n+1}{2} \right\rceil \right\} = \left\lceil \frac{2n+1}{2} \right\rceil. \tag{9}$$

The exact value of its *total irregularity strength* can be seen in the Theorem 2.2 below. It is enough to show that the upper bound is equal to the lower bound as in (9).

Theorem 2.2. *Let consider the caterpillar $S_{n,3,3,n}$, $n \ge 4$. Its total irregularity strength is $ts(S_{n,3,3,n}) = \lceil \frac{2n+1}{2} \rceil = n+1$.*

IOP Conf. Series: Journal of Physics: Conf. Series **1008** (2018) 012042 doi:10.1088/1742-6596/1008/1/012042

Proof. $S_{n,3,3,n}$ is a tree with $2n+4$ vertices, $2n+3$ edges and $2n$ pendant vertices. Let the vertex set of this graph be $V(S_{n,3,3,n}) = \{v_i^1 : 1 \le i \le n-1\} \cup \{v_i^4 : 1 \le i \le n-1\} \cup \{v^j : j = 1,2,3,4\} \cup \{v_1^j \text{for} j = 2,3\}$ and the edge set be $E(S_{n,3,3,n}) = \{v^1v_i^1 : 1 \le i \le n-1\} \cup \{v^4v_i^4 : 1 \le i \le n-1\} \cup \{v^jv^{j+1} : j = 1,2,3\} \cup \{v^jv_1^j \text{for} j = 2,3\}$. To determine the exact value of ts, construct the totally irregular total k-labeling f as follows.
Label of vertices and edges:

$$f(v_i^1) = 1, \text{for } 1 \le i \le n-1.$$

$$f(v_i^4) = \begin{cases} 1, & \text{for } i = 1, \\ i-1, & \text{for } 2 \le i \le n-1. \end{cases}$$

$$f(v^j) = \begin{cases} 1, & \text{for } j = 1, \\ n+1, & \text{for } j = 2,3, \\ 2, & \text{for } j = 4. \end{cases}$$

$$f(v_1^j) = \begin{cases} n-1, & \text{for } j = 2, \\ n, & \text{for } j = 3. \end{cases}$$

$$f(v^1v_i^1) = i, \text{ for } 1 \le i \le n-1.$$

$$f(v^4v_i^4) = \begin{cases} n, & \text{for } i = 1, \\ n+1, & \text{for } 2 \le i \le n-1. \end{cases}$$

$$f(v^jv^{j+1}) = \begin{cases} n, & \text{for } j = 1, \\ n+1, & \text{for } j = 2,3. \end{cases}$$

$$f(v^jv_1^j) = n+1, \text{ for } j = 2,3.$$

Let consider that under the total labeling f, the greatest label for all vertices and edges is $k = n+1$. Then, f is a total k-labeling with $k = \lceil \frac{2n+1}{2} \rceil = n+1$. The weight of vertices and edges are as follows.

$$wt_f(v_i^j) = \begin{cases} i+1, & \text{for } 1 \le i \le n-1,\ j = 1, \\ n+i, & \text{for } 1 \le i \le n-1,\ j = 4. \end{cases}$$

$$wt_f(v^j) = \begin{cases} \frac{n(n+1)}{2} + 1, & \text{for } j = 1, \\ 4n+3, & \text{for } j = 2, \\ 4n+4, & \text{for } j = 3, \\ n^2+n+1, & \text{for } j = 4. \end{cases}$$

$$wt_f(v_1^j) = \begin{cases} 2n, & \text{for } j = 2, \\ 2n+1, & \text{for } j = 3. \end{cases}$$

$$wt_f(v^jv_i^j) = \begin{cases} i+2, & \text{for } j = 1,\ 1 \le i \le n-1, \\ n+2+i, & \text{for } j = 4,\ 1 \le i \le n-1. \end{cases}$$

$$wt_f(v^jv^{(j+1)}) = \begin{cases} 2n+2, & \text{for } j = 1, \\ 3n+3, & \text{for } j = 2, \\ 2n+4, & \text{for } j = 3. \end{cases}$$

$$wt_f(v^jv_1^j) = \begin{cases} 3n+1, & \text{for } j = 2, \\ 3n+2, & \text{for } j = 3. \end{cases}$$

IOP Conf. Series: Journal of Physics: Conf. Series **1008** (2018) 012042 doi:10.1088/1742-6596/1008/1/012042

It is shown that the weight of vertices v_i^j for j=1, 4 form a consecutive integers from 2 up to n, for $n+1$ until $2n-1$ respectively and for v_1^2 and v_1^3 their weight are $2n$ for v_1^2, $2n+1$ for v_1^3. The weights among vertices v^1, v^2, v^3 and v^4 are distinct. Moreover, the weights of every pair of vertices are distinct. The weight of edges for every two different edges also distinct. There are $i+2$, $n+2+i$, $2n+2$, $2n+4$, $3n+1$, $3n+2$ and $3n+3$ for the weight of all edges. Moreover, it is can be concluded that f is a totally irregular total k-labeling and the total irregularity strength is $ts(S_{n,3,3,n}) = k = \lceil \frac{2n+1}{2} \rceil = n+1$. □

With the same reason as in Theorem 2.1 and 2.2, it can be proven the *total irregularity strength* of the caterpillar $S_{n,3,3,3,n}$ as in the Section 2.3.

2.3. Caterpillar $S_{n,3,3,3,n}$

Same as the previous statement, a caterpillar $S_{n,3,3,3,n}$ is a class of graph constructed from the double-stars $S_{n,n}$ by inserting three vertices on the bridge connecting of the two centers of two stars, so that each inserted vertex has degree three. This inserted vertices called as the internal vertices. This caterpillar contains five stars with the center of the two end-stars have degree n respectively, while the center of each internal star has degree three. There is also no vertex with degree two. This graph is a tree with $2n+6$ vertices, $2n+5$ edges and $2n+1$ pendant vertices. Maximal degree of the graph is $\Delta = n$.
The maximum value between its total edge irregularity strength and its total vertex irregularity strength become as its lower bound of its total irregularity strength. (see (3)). The total edge irregularity strength of graph $S_{n,3,3,3,n}$ can be found by (2).

$$tes(S_{n,3,3,3,n}) = \max\left\{\left\lceil\frac{n+1}{2}\right\rceil, \left\lceil\frac{2n+7}{3}\right\rceil\right\} = \left\lceil\frac{2n+7}{3}\right\rceil. \quad (10)$$

The total vertex irregularity strength of $S_{n,3,3,3,n}$ also can be found by (1), that is

$$tvs(S_{n,3,3,3,n}) = \left\lceil\frac{2n+2}{2}\right\rceil. \quad (11)$$

According to (3), the lower bound of its *total irregularity strength* can be obtained as follows.

$$ts(G) \geq \max\{tes(G), tvs(G)\} = \max\left\{\left\lceil\frac{2n+7}{3}\right\rceil, \left\lceil\frac{2n+2}{2}\right\rceil\right\} = \left\lceil\frac{2n+2}{2}\right\rceil. \quad (12)$$

Furthermore, in the Theorem 2.3, the exact value of its *total irregularity strength* can be found. The lower bound of this parameter can be seen in (12). Then, we will show that the upper bound of this parameter and the lower bound is equal.

Theorem 2.3. *The total irregularity strength of caterpillar* $S_{n,3,3,3,n}$, $n \geq 4$ *is* $ts(S_{n,3,3,3,n}) = \lceil \frac{2n+2}{2} \rceil = n+1$.

Proof. The definition of $S_{n,3,3,3,n}$ is similar with the previous graph, that is a tree with $2n+6$ vertices, $2n+5$ edges and $2n+1$ pendant vertices. The vertex set of this graph is $V(S_{n,3,3,3,n}) = \{v_i^1 : 1 \leq i \leq n-1\} \cup \{v_i^5 : 1 \leq i \leq n-1\} \cup \{v^j : j = 1,2,3,4,5\} \cup \{v_1^j \text{for} j = 2,3,4\}$ and the edge set is $E(S_{n,3,3,3,n}) = \{v^1v_i^1 : 1 \leq i \leq n-1\} \cup \{v^5v_i^5 : 1 \leq i \leq n-1\} \cup \{v^jv^{j+1} : j = 1,2,3,4\} \cup \{v^jv_1^j \text{for} j = 2,3,4\}$. In order to determine the exact value of ts, construct the totally irregular total k-labeling g as follows.

IOP Conf. Series: Journal of Physics: Conf. Series **1008** (2018) 012042 doi:10.1088/1742-6596/1008/1/012042

The Label of vertices and edges are:

$$g(v_i^1) = 1, \text{for } 1 \le i \le n-1.$$

$$g(v_i^5) = \begin{cases} 1, & \text{for } i = 1,2, \\ i-1, & \text{for } 3 \le i \le n-1. \end{cases}$$

$$g(v^1) = g(v^5) = 1.$$

$$g(v^2) = g(v^3) = g(v^4) = \begin{cases} n+1, & \text{for } n = 4, \\ n, & \text{for } n \le 5. \end{cases}$$

$$g(v_1^j) = \begin{cases} n-1, & \text{for } j = 2, \\ n, & \text{for } j = 3, \\ n+1, & \text{for } j = 4. \end{cases}$$

$$g(v^1 v_i^1) = i, \text{ for } 1 \le i \le n-1.$$

$$g(v^5 v_i^5) = \begin{cases} n, & \text{for } i = 1, \\ n+1, & \text{for } 2 \le i \le n-1. \end{cases}$$

$$g(v^1 v^2) = \begin{cases} n, & \text{for } n = 4, \\ n+1, & \text{for } n \le 5. \end{cases}$$

$$g(v^2 v^3) = \begin{cases} 2, & \text{for } n = 4, \\ 4, & \text{for } n \le 5. \end{cases}$$

$$g(v^3 v^4) = \begin{cases} 1, & \text{for } n = 4, \\ 3, & \text{for } n \le 5. \end{cases}$$

$$g(v^4 v^5) = \begin{cases} 3, & \text{for } n = 4, \\ n, & \text{for } n \le 5. \end{cases}$$

$$g(v^2 v_1^2) = g(v^3 v_1^3) = g(v^4 v_1^4) = n+1.$$

It is easy to see that under the total labeling g, the greatest label for all vertices and edges is $k = n+1$. Then, g is a totally total k-labeling with $k = \lceil \frac{2n+2}{2} \rceil = n+1$. The weight of vertices and edges are as follows.

$$wt_g(v_i^j) = \begin{cases} i+1, & \text{for } 1 \le i \le n-1,\ j = 1, \\ n+i, & \text{for } 1 \le i \le n-1,\ j = 5. \end{cases}$$

$$wt_g(v^j) = \begin{cases} \frac{n(n+1)}{2} + 2, & \text{for } j = 1, \\ 3n+6, & \text{for } j = 2, \\ 2n+8, & \text{for } j = 3, \\ 3n+4, & \text{for } j = 4, \\ n^2+n-1, & \text{for } j = 5. \end{cases}$$

$$wt_g(v_1^j) = \begin{cases} 2n, & \text{for } j = 2, \\ 2n+1, & \text{for } j = 3, \\ 2n+2, & \text{for } j = 4. \end{cases}$$

$$wt_g(v^j v_i^j) = \begin{cases} i+2, & \text{for } j=1,\ 1 \le i \le n-1, \\ n+1+i, & \text{for } j=5,\ 1 \le i \le n-1. \end{cases}$$

$$wt_g(v^j v^{(j+1)}) = \begin{cases} 2n+2, & \text{for } j=1, \\ 2n+4, & \text{for } j=2, \\ 2n+3, & \text{for } j=3, \\ 2n+1, & \text{for } j=4. \end{cases}$$

$$wt_g(v^j v_1^j) = \begin{cases} 3n, & \text{for } j=2, \\ 3n+1, & \text{for } j=3, \\ 3n+2, & \text{for } j=4. \end{cases}$$

It is easy to observe that the weight of vertices v_i^j for j=1, 5 and for v_1^2, v_1^3 and v_1^4 form a consecutive integers from 2 up to n, for $n+1$ until $2n-1$ and $2n$ for v_1^2, $2n+1$ for v_1^3 and $2n+2$ for v_1^4. Vertices v^1, v^2, v^3, v^4 and v^5 also have distinct weight. Therefore, every pair of vertices have a different weight. The weight of edges for every two different edges also distinct. There are $i+2$, $n+1+i$, $2n+1$, $2n+2$, $2n+3$, $2n+4$, $3n$, $3n+1$ and $3n+2$ for the weight of all edges. Therefore, it is can be shown that g is a totally irregular total k-labeling and the total irregularity strength is $ts(S_{n,3,3,3,n}) = k = \lceil \frac{2n+2}{2} \rceil = n+1$. □

3. Concluding Remark

Based on the above discussion, we conclude that the total irregularity strength of caterpillar graphs $S_{n,3,n}$, $S_{n,3,3,n}$ and $S_{n,3,3,3,n}$ are $\lceil \frac{2n}{2} \rceil$, $\lceil \frac{2n+1}{2} \rceil$, and $\lceil \frac{2n+2}{3} \rceil$ respectively. Moreover, we have the following open problem for the direction of further research.

Open Problem: What is the total irregularity strength of caterpillar $S_{n,3,3,\ldots,n}$ for t-times of 3's?

Acknowledgments

This research was supported by The Fundamental Research Project Grant 7186/UN1.P.III/DIT-LIT/LT/2017 from the Ministry of Research Technology and Higher Education of Indonesia. Thank you for the reviewer who giving some suggestions in order to improve this paper.

References

[1] Bača M,Jendrol S, Miller M, Ryan J On irregular total labeling 2007 *Discrete Math.* **307** 1378-1388.
[2] Gallian JA A Dynamic Survey of Graph Labeling 2016 *The Electronic Journal of Combinatorics* **19** #DS6.
[3] Indriati D, Widodo, Wijayanti IE, Sugeng KA and Bača M On total edge irregularity strength of generalized web graphs and related graphs 2015 *Mathematics in Computer Science* **9** (2) 161-167.
[4] Indriati D, Widodo, Wijayanti IE, Sugeng KA On total irregularity strength of double-stars and related graphs 2015 *Procedia-Computer Science* **74** 118-123.
[5] Ivančo J and Jendroľ S Total edge irregularity strength of trees 2006 *Discussiones Math. Graph Theory* **26** 449-456.
[6] Marzuki CC, Salman ANM and Miller M On the total irregularity strength of cycles and paths 2013 *Far East Journal of Mathematical Sciences* **82** 1-21.
[7] Nurdin, Baskoro ET, Salman ANM and Gaos NN On the total vertex irregularity strength of trees 2010 *Discrete Math.* **310** 3043-3048.
[8] Ramdani R and Salman ANM On the total irregularity strength of some cartesian product graphs 2013 *AKCE International J. Graphs Combin.* **10** (2) 199-209.
[9] Ramdani R, Salman ANM, Assyatun H, Semaničová - Feňovčíková A, Bača M Total irregularity strength of three families of graphs 2015 *Mathematics in Computer Sci.* **9** 229-237.
[10] Wallis WD 2001 *Magic Graphs* (Boston: Birkhauser).

On the locating domination number of $Pn \trianglerighteq H$ graph

Dwi Agustin Retno Wardani [1,4], **Ika Hesti Agustin**[1,3], **Dafik**[1,2], **Ridho Alfarisi** [1,5]

[1]CGANT University of Jember, Jember, Indonesia
[2]Mathematics Education Departement University of Jember, Jember, Indonesia
[3]Mathematics Departement University of Jember, Jember, Indonesia
[4]IKIP PGRI of Jember, Jember, Indonesia
[5]Department of Elementary School Teacher Education, University of Jember, Jember, Indonesia

E-mail: 2i.agustin@gmail.com; ikahestiagustin@gmail.com; d.dafik@unej.ac.id; alfarisi38@gmail.com

Abstract. A subset D of $V(G)$ is called a vertex dominating set of G if every vertex not in D is adjacent to some vertices in D. A graph $G = (V, E)$ is called a locating dominating set if for every two vertices $u, v\epsilon V(G) - D$, $N(v) \cap D \neq \emptyset$. Locating dominating number $\gamma_L(G)$ is the minimum cardinality of a locating dominating set. The value of locating domination number is $\gamma_L(G) \subseteq V(G)$. Edge comb product denoted by $G \triangleright H$ is a graph obtained by taking one copy of G and $|E(G)|$ copies of H and grafting the i-th copy of H at the edge e to the i-th edge of G. This paper studies about locating domination number in edge comb product graph $G \trianglerighteq H$ where G is path graph P_n and H is complete graph K_m, star graph S_m, triangular book graph BT_m, path graph P_m, friendship graph $\mathfrak{F}_m$ and fan graph F_m.

1. Introduction

Let G be a nontrivial, finite, simple, undirected and connected graphs, with vertex set $V(G)$, edge set $E(G)$ and with no isolated vertex, for more detail definition of graph see [6, 3]. A set D of vertices of a graph $G = (V, E)$ is dominating if every vertex in $V(G) - D$ is adjacent to some vertex in D. $\gamma(G)$ is the minimum cardinality of a dominating set in G.

Definition 1.1 *[7] A set $S \subseteq V$ of vertices in graph $G = (V, E)$ is called a dominating set if every vertex $v\epsilon V$ is either an element of S or is adjacent to an element of S*

Vertex set D in graph $G = (V, E)$ is a locating-dominating set if for each pair of distinct vertices u and v in $V(G) - D$ we have $N(u) \cap D \neq N(v) \cap D \neq \emptyset$, that is, each vertex outside of D is adjacent to a distinct, nonempty subset of the elements of D (Slater [8], [9]). By definition, V is trivially a locating-dominating set in $G = (V, E)$. The minimum cardinality locating-dominating number of graph G denoted by $\gamma_L(G)$. The value of locating dominating number is $\gamma_L(G) \subseteq V(G)$. The concept of a locating dominating set was introduced and first studied by Slater [8, 9, 2, 1] and also Waspodo *et. al.* [5] studied the bound of distance domination number of edge comb product, Wardani [12] studied about locating independent domination number of amalgamation graphs. In this paper, we will initiate to analyze locating domination number of edge comb product of two graphs, denoted by $G \trianglerighteq H$, where G is path graph and H is complete graph, star graph, triangular book graph, path graph, friendship graph and fan graph. Dafik *et.*

al. [4] studied locating domination number of edge comb product $S_n \trianglerighteq H$ with H are complete graph K_m, star graph S_m, triangular book graph Bt_m, and path graph P_m.

Firstly introduction a comb product of graph is Saputro [10]. Edge comb product is developed of comb product graph, let G and H be two connected graphs. Let e be an edge of H. $G \trianglerighteq H$ is symbol of edge comb product, Edge comb product is a graph obtained by taking one copy of G and $|E(G)|$ copies of H and grafting the i-th copy of H at the edge e to the i-th edge of G. By the definition of comb product. We can say that $p = |V(G \trianglerighteq H)| = q_1(p_2 - 2) + p_1$ and $q = |E(G \trianglerighteq H)| = q_1 q_2$ [11].

Figure 1 is example of edge comb product $P_3 \trianglerighteq H_3$ and Figure 2 is example of locating domination number of edge comb product $W_6 \trianglerighteq C_3$.

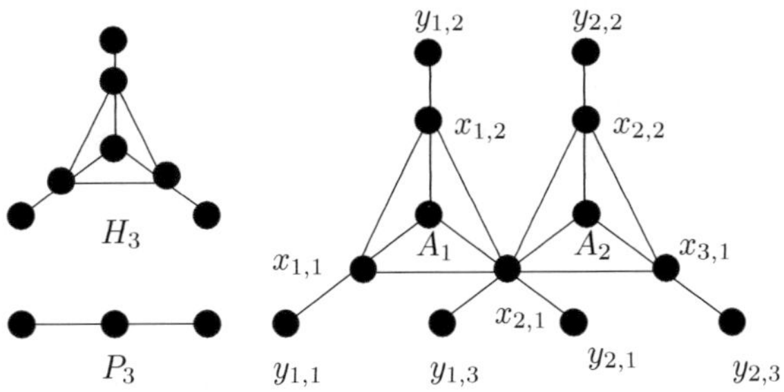

Figure 1. Example of Edge Comb Product $P_3 \trianglerighteq H_3$

$P_n \trianglerighteq H_m$ is connected graph with vertex set $V(P_n \trianglerighteq H_m) = \{A_i; 1 \le i \le n-1\} \bigcup \{x_i, j; 1 \le i \le n-1; 1 \le j \le m-1\} \bigcup \{x_{n,1}\} \bigcup \{y_{i,j}; 1 \le i \le n; 1 \le j \le m\}$ and edge set $E(P_n \trianglerighteq H_m) = \{A_i x_{i,j}; 1 \le i \le n-1; 1 \le j \le m-1\} \bigcup \{x_{i,j} x_{i,j+1}; 1 \le i \le n-1; 1 \le j \le m-2\} \bigcup \{x_{i,1} x_{i+1,1}; 1 \le i \le n\} \bigcup \{x_{i,m-1} x_{i+1,1}; 1 \le i \le n-1\} \bigcup \{x_{i,j} y_{i,j}; 1 \le i \le n; 1 \le j \le m\}$. The order and size of $(P_n \trianglerighteq H_m), n \ge 3, m \le 3$ are $|V(P_n \trianglerighteq H_m)| = 7n - 8$ and $|E(P_n \trianglerighteq H_m)| = 3nm - 3m$.

See Figure 2, the locating domination number of $W_6 \trianglerighteq C_3$ is 6 but we can expand the graph until $W_n \trianglerighteq C_m$ with the locating domination number of $W_n \trianglerighteq C_m$ is $\gamma_L(W_n \trianglerighteq C_m) = n$ we choose locating dominating set of $(P_n \trianglerighteq H_m)$ namely $D = \{A\} \bigcup \{x_i; 1 \le i \le n-1\}$ with cardinality of D is $|D| = n$. So the minimum cardinality of $\gamma_L(W_n \trianglerighteq C_m)$ is n.

2. The Result

In this section, we determine the exact values of locating dominating number of some edge comb product graphs $P_n \trianglerighteq H$. With H is Complete graph K_m $(P_n \trianglerighteq K_m)$, Star graph S_m $(P_n \trianglerighteq S_m)$, Triangular Book Bt_m $(P_n \trianglerighteq Bt_m)$, Path graph P_m $(P_n \trianglerighteq P_m)$, Friendship graph $\mathfrak{F}_m$ $(P_n \trianglerighteq \mathfrak{F}_m)$ and Fan graph F_m $(P_n \trianglerighteq F_m)$.

Theorem 2.1 *Let a comb product graph* $P_n \trianglerighteq K_m$ *and* $x_i x_{i+1}$ *is a graft edge of* K_m*, the locating domination number of* $\gamma_L\ (P_n \trianglerighteq K_m) = (n-1)(m-3) + 2$.

Proof. An edge comb product graph $P_n \trianglerighteq K_m$, $n \ge 3$ and $m \ge 4$, is a connected graph with vertex set $V(P_n \trianglerighteq K_m) = \{x_i; 1 \le i \le n\} \bigcup \{x_{i,j}; 1 \le i \le n-1; 1 \le j \le m-2\}$ and edge set $E(P_n \trianglerighteq K_m) = \{x_i x_{i+1}; 1 \le i \le n-1\} \bigcup \{x_i x_{i,j}; 1 \le i \le n; 1 \le j \le m-2\} \bigcup \{x_{i,j} x_{i,j}; 1 \le i \le n; 1 \le j \le m-2\}$. The order and size of $(P_n \trianglerighteq K_m), n \ge 3, m \ge 4$ are $|V(P_n \trianglerighteq K_m)| = nm - n - m + 2$ and $|E(P_n \trianglerighteq K_m)| = (n-1)(\frac{m(m-1)}{2})$.

The first step of proof we show that $\gamma_L(P_n \trianglerighteq K_m) \ge (n-1)(m-3) + 2$ with assume $\gamma_L(P_n \trianglerighteq K_m) < (n-1)(m-3) + 2$, we choose locating dominating set of $(P_n \trianglerighteq K_m)$

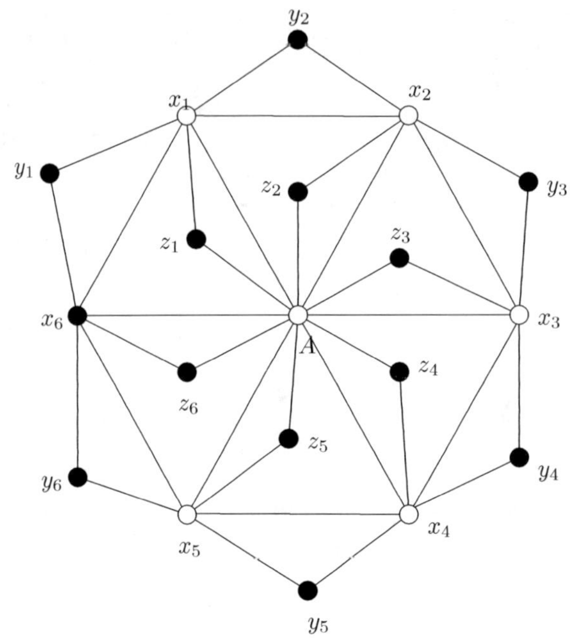

Figure 2. Example of Locating Domination Number of Edge Comb Product $W_6 \trianglerighteq C_3$

namely $D = \{x_{i,j}; 1 \leq i \leq n-1; 2 \leq j \leq m-2\} \cup \{x_n\}$ with cardinality of D is $|D| = (n-1)(m-3)+1$. The vertex set without locating dominating set of $(P_n \trianglerighteq K_m)$ is $V - D = \{x_{i_1}; 1 \leq i \leq n-1\} \cup \{x_i; 1 \leq i \leq n-1\}$ Intersection between vertex set $\forall v \epsilon (V-D)$ and D are as follows.

$$\begin{aligned} N(x_i) \cap D &= \{x_{i,j}, x_{i-1,j}; 1 \leq i \leq n-1; 2 \leq j \leq m-2\} \\ N(x_{i,1}) \cap D &= \{x_{i,j}, 1 \leq i \leq n-1; 2 \leq j \leq m-2\} \end{aligned}$$

For the vertices $x_1, x_{1,1} \in (V-D)$ can be seen that $N(x_1) \cap D = N(x_{1,1}) \cap D$, and D not satisfied properties of locating dominating set. So, we can conclude that $\gamma_L(P_n \trianglerighteq K_m) < (n-1)(m-3)+2$ not satisfied the definition of locating domination number, it is contradiction. Hence, the location domination number of $(P_n \trianglerighteq K_m)$ is $\gamma_L(P_n \trianglerighteq K_m) \geq (n-1)(m-3)+2$. Furthermore, we show that $\gamma_L(P_n \trianglerighteq K_m) \leq (n-1)(m-3)+2$, by choosing locating dominating set of $(P_n \trianglerighteq K_m)$ is $D = \{x_1\} \cup \{x_n\} \cup \{x_{i,j}; 1 \leq i \leq n-1; 2 \leq j \leq m-2\}$ with $|D| = nm - 3n - m + 5$. The vertex set without locating dominating set of $(P_n \trianglerighteq K_m)$ is $V - D = \{x_i; 2 \leq i \leq n-1\} \cup \{x_{i,1}; 1 \leq i \leq n-1\}$. Intersection between vertex set $\forall v \in (V-D)$ and D are as follows.

$$\begin{aligned} N(x_{i,1}) \cap D &= \{x_{i,j}; 1 \leq i \leq n-1; 2 \leq j \leq m-2\} \\ N(x_i) \cap D &= \{x_1, x_n, x_{i,j}, x_{i-1,j}; 1 \leq i \leq n-1; 2 \leq j \leq m-2\} \end{aligned}$$

For the vertices $x_{i,1}, x_i \in (V-D)$ can be seen that $N(x_{i,1}) \cap D \neq \emptyset$ and $N(x_i) \cap D \neq \emptyset$, D is locating dominating set of $(P_n \trianglerighteq K_m)$ because D satisfied the definition of domination number. In the otherhand $N(x_{i,1}) \cap D \neq N(x_i) \cap D$, it conclude that D satisfied to be locating dominating set, so D satisfied the definition of locating domination number. Hence the locating domination number of $(P_n \trianglerighteq K_m)$ is $\gamma_L(P_n \trianglerighteq K_m) \leq (n-1)(m-3)+2$. Because $\gamma_L(P_n \trianglerighteq K_m) \geq (n-1)(m-3)+2$ and $\gamma_L(P_n \trianglerighteq K_m) \leq (n-1)(m-3)+2$, then we can say that $\gamma_L(P_n \trianglerighteq K_m) = (n-1)(m-3)+2$. □

Theorem 2.2 *Let a comb product graph $P_n \trianglerighteq S_m$ and Ax_i is a graft edge of S_m, the locating domination number of γ_L $(P_n \trianglerighteq S_m) = (n-1)(m-1)+1$.*

Proof. An edge comb product graph $P_n \trianglerighteq S_m$ $n \geq 3$ and $m \geq 3$, is a connected graph with vertex set $V(P_n \trianglerighteq S_m) = \{x_i; 1 \leq i \leq n\} \bigcup \{x_{i,j}; 1 \leq i \leq n-1; 1 \leq j \leq m-1\}$ and edge set $E(P_n \trianglerighteq S_m) = \{x_i x_{i+1}; 1 \leq i \leq n-1\} \bigcup \{x_{i+1}x_{i,j}; 1 \leq i \leq n-1; 1 \leq j \leq m-1\}$. The order and size of $(P_n \trianglerighteq S_m), n \geq 3, m \geq 3$ are $|V(P_n \trianglerighteq S_m)| = nm - m + 1$ and $|E(P_n \trianglerighteq S_m)| = m(n-1)$.

The first step of proof we show that $\gamma_L(P_n \trianglerighteq S_m) \geq (n-1)(m-1)+1$ with assume $\gamma_L(P_n \trianglerighteq S_m) < (n-1)(m-1)+1$ We choose locating dominating set of $P_n \trianglerighteq S_m$ namely $D = \{x_{i,j}; 2 \leq i \leq n-1, 1 \leq j \leq m-1\}$ with cardinality of D is $|D| = nm - n - m + 1$. The vertex set without locating dominating set of $P_n \trianglerighteq S_m$ is $V - D = \{x_i; 1 \leq i \leq n\}$ Intersection between vertex set $\forall v \epsilon (V - D)$ and D are as follows.

$$\begin{aligned} N(x_i) \cap D &= \{x_{i,j}; 2 \leq i \leq n; 1 \leq j \leq m-1\} \\ N(x_1) \cap D &= \emptyset \end{aligned}$$

For the vertices $x_i \in (V - D)$ can be seen that $N(x_1) \cap D = \emptyset$, and D not satisfied properties of locating dominating set. So, we can conclude that $\gamma_L(P_n \trianglerighteq S_m) < (n-1)(m-1)+1$ not satisfied the definition of domination number, it is contradiction. Hence, the location domination number of $(C_m \trianglerighteq S_m)$ is $\gamma_L(P_n \trianglerighteq S_m) \geq (n-1)(m-1)+1$. Furthermore, we show that $\gamma_L(P_n \trianglerighteq S_m) \leq (n-1)(m-1)+1$, by choosing locating dominating set of $(P_m \trianglerighteq S_m)$ is $D = \{x_{i,j}; 2 \leq i \leq n; 1 \leq j \leq m-1\} \cup \{x_1\}$ with $|D| = nm - n - m + 2$. The vertex set without locating dominating set of $(P_m \trianglerighteq S_m)$ is $V - D = \{x_i; 2 \leq i \leq n\}$. Intersection between vertex set $\forall v \in (V - D)$ and D are as follows.

$$N(x_i) \cap D = \{x_{i,j}; 2 \leq i \leq n; 1 \leq j \leq m-1\}$$

For the vertices $x_{(i)} \in (V - D)$ can be seen that $N(x_i) \cap D \neq \emptyset$, D is locating dominating set of $P_n \trianglerighteq S_m$ because D satisfied the definition of domination number. In the otherhand $N(x_i) \cap D$ satisfied to be locating dominating set, so D satisfied the definition of locating domination number. Hence the locating domination number of $P_n \trianglerighteq S_m$ is $\gamma_L(P_n \trianglerighteq S_m) \leq (n-1)(m-1)+1$. Because $\gamma_L(P_n \trianglerighteq S_m) \geq (n-1)(m-1)+1$ and $\gamma_L(P_n \trianglerighteq S_m) \leq (n-1)(m-1)+1$, then we can say that $\gamma_L(P_n \trianglerighteq S_m) = (n-1)(m-1)+1$. □

Theorem 2.3 *Let a comb product graph $P_n \trianglerighteq Bt_m$ and x_1x_2 is a graft edge of Bt_m, the locating domination number of γ_L $(P_n \trianglerighteq Bt_m) = m(n-1)$.*

Proof. An edge comb product graph $P_n \trianglerighteq Bt_m$, $n \geq 3$ and $m \geq 3$, is a connected graph with vertex set $V(P_n \trianglerighteq Bt_m) = \{x_i; 1 \leq i \leq n\} \bigcup \{x_{i,j}; 1 \leq i \leq n-1; 1 \leq j \leq m\}$, and edge set $E(P_n \trianglerighteq Bt_m) = \{x_i x_{i+1}; 1 \leq i \leq n-1\} \bigcup \{x_i x_{i,j}; 1 \leq i \leq n-1; 1 \leq j \leq m\} \bigcup \{x_{i+1}x_{i,j}; 1 \leq i \leq n-1; 1 \leq j \leq m\}$. The order and size of $(P_n \trianglerighteq Bt_m), n \geq 3, m \geq 3$ are $|V(P_n \trianglerighteq Bt_m)| = n + nm - m$ and $|E(P_n \trianglerighteq P_m)| = 2nm - 2m + n - 1$.

The first step of proof we show that $\gamma_L(P_n \trianglerighteq Bt_m) \geq m(n-1)$ with assume $\gamma_L(P_n \trianglerighteq Bt_m) < m(n-1)$ We choose locating dominating set of $P_n \trianglerighteq Bt_m$ namely $D = \{x_{i,j}; 2 \leq i \leq n-1; 1 \leq j \leq m\} \bigcup \{x_{1,j}; 2 \leq m\}$ with cardinality of D is $|D| = m(n-1) - 1$. The vertex set without locating dominating set of $P_n \trianglerighteq Bt_m$ is $V - D = \{x_i; 1 \leq i \leq n-1\} \cup \{x_{1,1}\}$ Intersection between vertex set $\forall v \epsilon (V - D)$ and D are as follows.

$$\begin{aligned} N(x_i) \cap D &= \{x_{i,j}; 1 \leq i \leq n-1; 1 \leq j \leq m\} \\ N(x_{1,1}) \cap D &= \emptyset \end{aligned}$$

For the vertices $x_{1,1} \in (V - D)$ can be seen that $N(x_{1,1}) \cap D = \emptyset$, and D not satisfied properties of locating dominating set. So, we can conclude that $\gamma_L(P_n \trianglerighteq Bt_m) < m(n-1)$ not satisfied the definition of domination number, it is contradiction. Hence, the location domination number of

$(P_n \trianglerighteq Bt_m)$ is $\gamma_L(P_n \trianglerighteq Bt_m) \geq m(n-1)$. Furthermore, we show that $\gamma_L(P_n \trianglerighteq Bt_m) \leq m(n-1)$, by choosing locating dominating set of $(P_n \trianglerighteq Bl_m)$ is $D = \{x_{i,j}; 1 \leq i \leq n-1; 1 \leq j \leq m\}$ with $|D| = m(n-1)$. The vertex set without locating dominating set of $(P_n \trianglerighteq Bt_m)$ is $V - D = \{x_i; 1 \leq i \leq n-1\}$. Intersection between vertex set $\forall v \in (V-D)$ and D are as follows.

$$N(x_i) \cap D = \{x_{i,j}; 1 \leq i \leq n-1; 1 \leq j \leq m\}$$

For the vertices $x_{(i)} \in (V-D)$ can be seen that $N(x_i) \cap D \neq \emptyset$, D is locating dominating set of $P_n \trianglerighteq Bt_m$ because D satisfied the definition of domination number. In the otherhand $N(x_i) \cap D \neq \emptyset$ satisfied to be locating dominating set, so D satisfied the definition of locating domination number. Hence the locating domination number of $P_n \trianglerighteq Bt_m$ is $\gamma_L(P_n \trianglerighteq Bt_m) \leq m(n-1)$. Because $\gamma_L(P_n \trianglerighteq Bt_m) \geq m(n-1)$ and $\gamma_L(P_n \trianglerighteq Bt_m) \leq m(n-1)$, then we can say that $\gamma_L(P_n \trianglerighteq Bt_m) = m(n-1)$. □

Theorem 2.4 *Let a comb product graph $P_n \trianglerighteq P_m$ and x_1x_2 is a graft edge of P_m, the locating domination number of γ_L $(P_n \trianglerighteq P_m) = (n-1)(\lceil \frac{m}{2} \rceil - 1) + 1$.*

Proof. An edge comb product graph $P_n \trianglerighteq P_m$, $n \geq 3$ and $m \geq 5$, is a connected graph with vertex set $V(P_n \trianglerighteq P_m) = \{x_i; 1 \leq i \leq n\} \bigcup \{x_{i,j}; 1 \leq i \leq n-1; 1 \leq j \leq m-2\}$, and edge set $E(P_n \trianglerighteq P_m) = \{x_i x_{i+1}; 1 \leq i \leq n-1\} \bigcup \{x_{i+1}x_{i,1}; 1 \leq i \leq n-1\} \bigcup \{x_{i,j}x_{i,j+1}; 1 \leq i \leq n-1; 1 \leq j \leq m-3\}$. The order and size of $(P_n \trianglerighteq P_m), n \geq 3, m \geq 5$ are $|V(P_n \trianglerighteq P_m)| = nm - n - m + 2$ and $|E(P_n \trianglerighteq P_m)| = nm - n - m + 1$.

The first step of proof we show that $\gamma_L(P_n \trianglerighteq P_m) \geq (n-1)(\lceil \frac{m}{2} \rceil - 1) + 1$ with assume $\gamma_L(P_n \trianglerighteq P_m) < (n-1)(\lceil \frac{m}{2} \rceil - 1) + 1$ We choose locating dominating set of $P_n \trianglerighteq P_m$ namely $D = \{x_{i,j}; 2 \leq i \leq n; j \equiv 1 \bmod 2\}$ with cardinality of D is $|D| = (n-1)(\lceil \frac{m}{2} \rceil - 1)$. The vertex set without locating dominating set of $P_n \trianglerighteq P_m$ is $V - D = \{x_i; 1 \leq i \leq n\} \cup \{x_{i,j}; 2 \leq i \leq n; j \equiv 2+0 \bmod 2\}$ Intersection between vertex set $\forall v \epsilon (V-D)$ and D are as follows.

$$\begin{aligned} N(x_i) \cap D &= \{x_{i,1}; 2 \leq i \leq n\} \\ N(x_{i,j}; 2 \leq i \leq n; j \equiv 2+0 \bmod 2) \cap D &= x_{i,j}; 2 \leq i \leq n; j \equiv 1 \bmod 2 \\ N(x_1) \cap D &= \emptyset \end{aligned}$$

For the vertices $x_1 \in (V-D)$ can be seen that $N(x_1) \cap D = \emptyset$, and D not satisfied properties of locating dominating set. So, we can conclude that $\gamma_L(P_n \trianglerighteq P_m) < (n-1)(\lceil \frac{m}{2} \rceil - 1) + 1$ not satisfied the definition of domination number, it is contradiction. Hence, the location domination number of $(P_n \trianglerighteq P_m)$ is $\gamma_L(P_n \trianglerighteq P_m) \geq (n-1)(\lceil \frac{m}{2} \rceil - 1) + 1$. Furthermore, we show that $\gamma_L(P_n \trianglerighteq P_m) \leq (n-1)(\lceil \frac{m}{2} \rceil - 1) + 1$, by choosing locating dominating set of $(P_n \trianglerighteq P_m)$ is $D = \{x_1\} \cup \{x_{i,j}; 2 \leq i \leq n; j \equiv 1 \bmod 2\}$ with $|D| = (n-1)(\lceil \frac{m}{2} \rceil - 1) + 1$. The vertex set without locating dominating set of $(P_n \trianglerighteq P_m)$ is $V - D = \{x_i; 2 \leq i \leq n\} \cup \{x_{i,j}; 2 \leq i \leq n; j \equiv 2+0 \bmod 2\}$. Intersection between vertex set $\forall v \in (V-D)$ and D are as follows.

$$\begin{aligned} N(x_i) \cap D &= \{x_1\}, \{x_{i,1}; 2 \leq i \leq n\} \\ N(x_{i,j}; 2 \leq i \leq n; j \equiv 2+0 \bmod 2) \cap D &= \{x_{i,j}; 2 \leq i \leq n; j \equiv 1 \bmod 2\} \end{aligned}$$

For the vertices $x_{(i)}, x_{i,j} \in (V-D)$ can be seen that $N(x_i) \cap D \neq \emptyset$ and $N(x_{i,j}) \cap D \neq \emptyset$, D is locating dominating set of $P_n \trianglerighteq P_m$ because D satisfied the definition of domination number. In the otherhand $N(x_i) \cap D \neq N(x_{i,j}) \cap D$ satisfied to be locating dominating set, so D satisfied the definition of locating domination number. Hence the locating domination number of $P_n \trianglerighteq P_m$

IOP Conf. Series: Journal of Physics: Conf. Series **1008** (2018) 012043 doi:10.1088/1742-6596/1008/1/012043

is $\gamma_L(P_n \trianglerighteq P_m) \leq (n-1)(\lceil\frac{m}{2}\rceil - 1) + 1$. Because $\gamma_L(P_n \trianglerighteq P_m) \geq (n-1)(\lceil\frac{m}{2}\rceil - 1) + 1$ and $\gamma_L(P_n \trianglerighteq P_m) \leq (n-1)(\lceil\frac{m}{2}\rceil - 1) + 1$, then we can say that $\gamma_L(P_n \trianglerighteq P_m) = (n-1)(\lceil\frac{m}{2}\rceil - 1) + 1$.
□

Theorem 2.5 *Let a comb product graph $P_n \trianglerighteq \mathfrak{F}_m$ and Ax_1 is a graft edge of $\mathfrak{F}_m$, the locating domination number of γ_L $(P_n \trianglerighteq \mathfrak{F}_m) = nm - n$.*

Proof. An edge comb product graph $P_n \trianglerighteq \mathfrak{F}_m$, $n \geq 3$ and $m \geq 2$, is a connected graph with vertex set $V(P_n \trianglerighteq \mathfrak{F}_m) = \{x_i; 1 \leq i \leq n\} \bigcup \{x_{i,j}; 1 \leq i \leq n-1; 1 \leq j \leq 2m-1\}$, and edge set $E(P_n \trianglerighteq \mathfrak{F}_m) = \{x_i x_{i+1}; 1 \leq i \leq n-1\} \bigcup \{x_i x_{i,1}; 1 \leq i \leq n-1\} \bigcup \{x_{i+1} x_{i,j}; 1 \leq i \leq n-1; 1 \leq j \leq 2m-1\} \bigcup \{x_{i,j} x_{i,j+(m-1)}; 1 \leq i \leq n-1; 1 \leq j \leq m\}$. The order and size of $(P_n \trianglerighteq \mathfrak{F}_m), n \geq 3, m \geq 2$ are $|V(P_n \trianglerighteq \mathfrak{F}_m)| = 2nm - 2m + 1$ and $|E(P_n \trianglerighteq \mathfrak{F}_m)| = 3nm - 3m$.

The first step of proof we show that $\gamma_L(P_n \trianglerighteq \mathfrak{F}_m) \geq nm - n$ with assume $\gamma_L(P_n \trianglerighteq \mathfrak{F}_m) < nm - n$ We choose locating dominating set of $P_n \trianglerighteq \mathfrak{F}_m$ namely $D = \{x_{i,j}; 2 \leq i \leq n-1; 1 \leq j \leq m\} \bigcup \{x_{1,j}; 2 \leq j \leq m\}$ with cardinality of D is $|D| = nm - n - 1$. The vertex set without locating dominating set of $P_n \trianglerighteq \mathfrak{F}_m$ is $V - D = \{x_i; 1 \leq i \leq n-1\} \cup \{x_{i,j}; 2 \leq i \leq n-1; m+1 \leq j \leq 2m-1\} \bigcup \{x_{1,1}\} \bigcup \{x_1, j\}; m+1 \leq j \leq 2m-1\}$ Intersection between vertex set $\forall v \epsilon (V - D)$ and D are as follows.

$$\begin{array}{lcl} N(x_i) \cap D & = & \{x_{i,j}; 2 \leq i \leq n-1; m+1 \leq j \leq 2m-1\} \\ N(x_{i,j} \cap D & = & \{x_{i,j}; 2 \leq i \leq n-1; m+1 \leq j \leq 2m-1\} \\ N(x_{1,j}; m+1 \leq j \leq 2m-1 \cap D & = & \{x_{1,j}; 2 \leq j \leq m\} \\ N(x_{1,1} \cap D & = & \emptyset \\ N(x_1) \cap D & = & \emptyset \end{array}$$

For the vertices $x_{1,1} \in (V - D)$ and $x_1 \in (V - D)$ can be seen that $N(x_{1,1}) \cap D = \emptyset$ and $N(x_1) \cap D = \emptyset$, so D not satisfied properties of locating dominating set. We can conclude that $\gamma_L(P_n \trianglerighteq \mathfrak{F}_m) < nm - n$ not satisfied the definition of domination number, it is contradiction. Hence, the location domination number of $(P_n \trianglerighteq \mathfrak{F}_m)$ is $\gamma_L(P_n \trianglerighteq \mathfrak{F}_m) \geq nm - n$. Furthermore, we show that $\gamma_L(P_n \trianglerighteq \mathfrak{F}_m) \leq nm - n$, by choosing locating dominating set of $(P_n \trianglerighteq \mathfrak{F}_m)$ is $D = \{x_{i,j}; 1 \leq i \leq n-1; 1 \leq j \leq m\}$ with $|D| = nm - n$. The vertex set without locating dominating set of $(P_n \trianglerighteq \mathfrak{F}_m)$ is $V - D = \{x_i; 1 \leq i \leq n-1\} \cup \{x_{i,j}; 1 \leq i \leq n-1; m+1 \leq j \leq 2m-1\}$. Intersection between vertex set $\forall v \in (V - D)$ and D are as follows.

$$\begin{array}{lcl} N(x_1) \cap D & = & x_{1,1} \\ N(x_i; i \neq 1) \cap D & = & \{x_{i,j}; 1 \leq i \leq n-1; 1 \leq j \leq m\} \bigcup \{x_{i+1,1}\} \\ N(x_{i,j}; m+1 \leq j \leq 2m-1) \cap D & = & \{x_{i,j}; 1 \leq i \leq n-1; 1 \leq j \leq m\} \end{array}$$

For the vertices $x_{(i)}, x_{i,j} \in (V - D)$ can be seen that $N(x_i) \cap D \neq \emptyset$ and $N(x_{i,j}) \cap D \neq \emptyset$, D is locating dominating set of $P_n \trianglerighteq \mathfrak{F}_m$ because D satisfied the definition of domination number. In the otherhand $N(x_i) \cap D \neq N(x_{i,j}) \cap D$ satisfied to be locating dominating set, so D satisfied the definition of locating domination number. Hence the locating domination number of $P_n \trianglerighteq \mathfrak{F}_m$ is $\gamma_L(P_n \trianglerighteq \mathfrak{F}_m) \leq nm - n$. Because $\gamma_L(P_n \trianglerighteq \mathfrak{F}_m) \geq nm - n$ and $\gamma_L(P_n \trianglerighteq \mathfrak{F}_m) \leq nm - n$, then we can say that $\gamma_L(P_n \trianglerighteq \mathfrak{F}_m) = nm - n$. □

Theorem 2.6 *Let a comb product graph $P_n \trianglerighteq F_m$ and Ax_1 is a graft edge of F_m, the locating domination number of γ_L $(P_n \trianglerighteq F_m) = \lfloor\frac{m}{2}\rfloor(n-1)$.*

Proof. An edge comb product graph $P_n \trianglerighteq F_m$, $n \geq 3$ and $m \geq 5$, is a connected graph with vertex set $V(P_n \trianglerighteq F_m) = \{x_i; 1 \leq i \leq n\} \bigcup \{x_{i,j}; 1 \leq i \leq n-1; 1 \leq j \leq m-1\}$, and edge set $E(P_n \trianglerighteq F_m) = \{x_i x_{i+1}; 1 \leq i \leq n-1\} \bigcup \{x_{i+1} x_{i,j}; 1 \leq i \leq n-1; 1 \leq j \leq m-1\} \bigcup \{x_i x_{i,1}; 1 \leq i \leq$

$n-1\} \bigcup \{x_{i,j}x_{i,j+1}; 1 \le i \le n-1; 1 \le j \le m-2\}$. The order and size of $(P_n \trianglerighteq F_m), n \ge 3, m \ge 6$ are $|V(P_n \trianglerighteq F_m)| = nm - m + 1$ and $|E(P_n \trianglerighteq F_m)| = 2mn - 2m - n + 1$.

The first step of proof we show that $\gamma_L(P_n \trianglerighteq F_m) \ge \lfloor \frac{m}{2} \rfloor (n-1)$ with assume $\gamma_L(P_n \trianglerighteq F_m) < \lfloor \frac{m}{2} \rfloor (n-1)$ We choose locating dominating set of $P_n \trianglerighteq F_m$ namely $D = \{x_{i,j}; 1 \le i \le n-2; j \equiv 1 \mod 2\} \bigcup \{x_{n-1,j}; j \equiv 1 \mod 2 \ne 1\}$ with cardinality of D is $|D| = \lfloor \frac{m}{2} \rfloor (n-1) - 1$. The vertex set without locating dominating set of $P_n \trianglerighteq F_m$ is $V - D = \{x_i; 1 \le i \le n\} \cup \{x_{i,j}; 1 \le i \le n-1; j \equiv 2 \mod 2\} \bigcup \{x_{n-1,1}\}$ Intersection between vertex set $\forall v \epsilon (V - D)$ and D are as follows.

$$\begin{aligned} N(x_i) \cap D &= \{x_{i,j}; 1 \le i \le n-1; j \equiv 1 \mod 2\} \\ N(x_{i,j} \cap D &= \{x_{i,j}; 1 \le i \le n-1; j-1 \le j \le j+1\} \\ N(x_{n-1,1}) \cap D &= \emptyset \end{aligned}$$

For the vertices $x_{n-1,1} \in (V - D)$ can be seen that $N(x_{n-1,1}) \cap D = \emptyset$, so D not satisfied properties of locating dominating set. We can conclude that $\gamma_L(P_n \trianglerighteq F_m) < \lfloor \frac{m}{2} \rfloor (n-1)$ not satisfied the definition of domination number, it is contradiction. Hence, the location domination number of $(P_n \trianglerighteq F_m)$ is $\gamma_L(P_n \trianglerighteq F_m) \ge \lfloor \frac{m}{2} \rfloor (n-1)$. Furthermore, we show that $\gamma_L(P_n \trianglerighteq \mathfrak{F}_m) \le nm - n$, by choosing locating dominating set of $(P_n \trianglerighteq F_m)$ is $D = \{x_{i,j}; 1 \le i \le n-1; j \equiv 1 \mod 2\}$ with $|D| = \lfloor \frac{m}{2} \rfloor (n-1)$. The vertex set without locating dominating set of $(P_n \trianglerighteq F_m)$ is $V - D = \{x_i; 1 \le i \le n\} \cup \{x_{i,j}; 1 \le i \le n-1; j \equiv 2 \mod 2\}$. Intersection between vertex set $\forall v \in (V - D)$ and D are as follows.

$$\begin{aligned} N(x_i) \cap D &= \{x_{i,j}; 1 \le i \le n-1; j \equiv 1 \mod 2\} \\ N(x_{i,j}) \cap D &= \{x_{i,j}; 1 \le i \le n-1; j-1 \le j \le j+1\} \end{aligned}$$

For the vertices $x_{(i)}, x_{i,j} \in (V - D)$ can be seen that $N(x_i) \cap D \ne \emptyset$ and $N(x_{i,j}) \cap D \ne \emptyset$, D is locating dominating set of $P_n \trianglerighteq F_m$ because D satisfied the definition of domination number. In the otherhand $N(x_i) \cap D \ne N(x_{i,j}) \cap D$ satisfied to be locating dominating set, so D satisfied the definition of locating domination number. Hence the locating domination number of $P_n \trianglerighteq F_m$ is $\gamma_L(P_n \trianglerighteq F_m) \le \lfloor \frac{m}{2} \rfloor (n-1)$. Because $\gamma_L(P_n \trianglerighteq F_m) \ge \lfloor \frac{m}{2} \rfloor (n-1)$ and $\gamma_L(P_n \trianglerighteq F_m) \le \lfloor \frac{m}{2} \rfloor (n-1)$, then we can say that $\gamma_L(P_n \trianglerighteq F_m) = \lfloor \frac{m}{2} \rfloor (n-1)$. □

3. Concluding Remarks
In this paper, we have determined the exact values of locating dominating number of some edge comb product graph $P_n \trianglerighteq H$. With H is Complete graph K_m, Star graph S_m, Triangular Book Bt_m, Path graph P_n, Friendship graph $\mathfrak{F}_m$ and Fan graph F_m.

4. Open Problem
Let G be any connected graph, determine sharper lower bounds of $\gamma_L(G)$ in term of the degrees of the graph?

Acknowledgement
We gratefully acknowledge the support from CGANT - University of Jember of year 2017.

Reference
[1] Finbow A and Hartnell B L 1988 On locating dominating sets and well-covered graphs *Congr Numer* **65** 191200
[2] Colbourn C J, Slater P J and Stewart L K 1987 Locating-dominating sets in series-parallel networks *Congr. Numer* **56** 135162

[3] Chartrand G and Lesniak L 2000 *Graphs and digraphs 3rd* ed (London: Chapman and Hall).
[4] Dafik, Agustin I H, Wardani D A R and Kurniawati E Y 2017 On the local dominating set of $Sn \trianglerighteq H$ graph *Journal of Physics: Conference Series* **943**
[5] Waspodo G A, Slamin, Dafik and Agustin I H 2017 Bound of Distance Domination Number of Graph and Edge Comb Product Graph *IOP Conf Series: Journal of Physics: Conf Series* **855** 012014
[6] Gross J L, Yellen J and Zhang P 2014 *Handbook of graph Theory* Second Edition CRC Press Taylor and Francis Group
[7] Haynes T W, Hedetniemi S T and Slater P J 1998 *Fundamentals of Domination in Graphs.* CRC Press
[8] Slater P J 1987 Dominating and location in acyclic graphs *Networks* **17** 5564
[9] Slater P J 1988 Dominating and reference sets in graphs *J. Math. Phys. Sci.* **22** 445455
[10] Saputro S W 2013 The Metric Dimension of Comb Product Graphs Graph Theory Conference in Honor of Egawas 60th Birthday 1-2
[11] Wardani D A R, Dafik, Prihandoko A C and Kristiana A I 2016 On The Total r-Dynamic Coloring of Graph: A New Graph Coloring Study International Conference on Mathematics: Education Theory and Aplication Surakarta
[12] Wardani D A R, Dafik, Agustin I H, Sari M P, Kurniawati E Y 2017 On locating independent domination number of amalgamation graphs *Journal of Physics: Conference Series* **943**

On the local edge antimagicness of m-splitting graphs

E R Albirri[1,3], Dafik[1,3], Slamin[1,4], I H Agustin[1,2], R Alfarisi[1,5]
[1]CGANT-University of Jember, Jember, Indonesia
[2]Department of Mathematics, University of Jember, Jember, Indonesia
[3]Department of Mathematics Education, University of Jember, Jember, Indonesia
[4]Department of Information System, University of Jember, Jember, Indonesia
[5]Department of Elementary School Education, University of Jember, Jember, Indonesia

E-mail: d.dafik@unej.ac.id, ikahesti.fmipa@unej.ac.id, ermitara@unej.ac.id

Abstract. Let G be a connected and simple graph. A split graph is a graph derived by adding new vertex v' in every vertex v such that v' adjacent to v in graph G. An m-splitting graph is a graph which has m v'-vertices, denoted by ${}_mSpl(G)$. A local edge antimagic coloring in $G = (V, E)$ graph is a bijection $f : V(G) \longrightarrow \{1, 2, 3, ..., |V(G)|\}$ in which for any two adjacent edges e_1 and e_2 satisfies $w(e_1) \neq w(e_2)$, where $e = uv \in G$. The color of any edge $e = uv$ are assigned by $w(e)$ which is defined by sum of label both end vertices $f(u)$ and $f(v)$. The chromatic number of local edge antimagic labeling $\gamma_{lea}(G)$ is the minimal number of color of edge in G graph which has local antimagic coloring. We present the exact value of chromatic number γ_{lea} of m-splitting graph and some special graphs.

Keywords: *Local edge antimagic coloring, chromatic number of graph, m-splitting graph*

1. Introduction

This paper uses connected, simple and undirected graph. We denote by $V(G)$ and $E(G)$ the set of vertices and the set of edges of a graph. It can be seen at [4], [3] and [7]. Let G be a graph with p vertices and q edges. By *labeling* , we mean one-to-one mapping which carries a set of graph elements into a set of numbers (integers), called *labels.* The labeling on graph can be done on edge or vertex. In this research, we label the edge with its own uniqueness, it is *antimagic.* The *edge-antimagic labeling* is defined as labeling that all the weights of the edge have different values, which can be seen at [6] and [9]. Agustin I. H. et all [1] introduced a concept of an local edge antimagic of graph. For a graph G of a size q, *local edge antimagic* is defined as follows:

Definition 1.1 *A local edge antimagic coloring in* $G = (V, E)$ *graph is a bijection* $f : V(G) \longrightarrow \{1, 2, 3, ..., |V(G)|\}$ *in which for any two adjacent edges* e_1 *and* e_2 *satisfies* $w(e_1) \neq w(e_2)$*, where* $e = uv \in G$.

Lemma 1.1 *[1] If* $\delta(G)$ *is maximum degree of* G*, then we have* $\gamma_{lea}(G) \geq \delta(G)$

Hartsfield and Ringel [8] explain that the color of any edge $e = uv$ are assigned by $w(e)$ which is defined by sum of label both end vertices $f(u)$ and $f(v)$. The chromatic number of local edge antimagic labeling $\gamma_{lea}(G)$ is the minimal number of color of edge in G graph which has local antimagic coloring.

Some new result can be seen in Arumugam et al [2] about local antimagic vertex coloring which be a basic idea about local antimagic coloring. Then, there are another research about

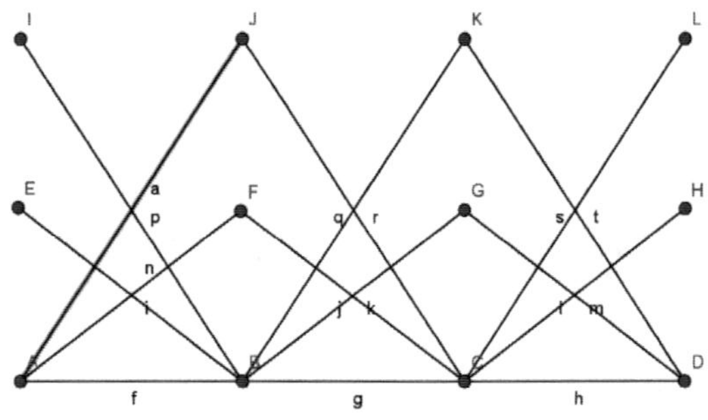

Figure 1. ${}_2Spl(P_4)$ Graph

super antimagic labeling which be found by Dafik et al [5]. The newest one is research about local edge antimagic of graph by Agustin et al [1]. Based on the previous research on it, we decide to do research in same topic such that local-edge-antimagic. But we choose different graph, it is split graph.

Definition 1.2 *Split graph is graph which be gotten by adding new vertex v' in every vertex v such that v' adjacent v in G graph. m-splitting graph is graph which has the number of vertex v' as m. m-splitting graph is denoted by ${}_mSpl(G)$.*

Based on Figure 1, we know that alphabet set $\{A, B, \ldots, L\}$ is the vertex set and $\{a, f, \ldots, t\}$ is the edge set. $\{A, B, C, D\}$ is the vertex set of basic graph or Path graph (P_4). $\{E, F, G, H\}$ is the vertex set of 1−splitting of P_4 and $\{I, J, K, L\}$ is the vertex set of 2−splitting of P_4.

2. Result

Some interesting theorem which be found by us such that,

Theorem 2.1 *For $n \geq 3$, the local edge antimagic of $m-$Splitting of P_n graph is*

$$\gamma_{lae}({}_mSpl(P_n)) = \gamma_{lae}(P_n) + 2m$$

Proof. The ${}_mSpl(P_n)$ graph has vertex set $V(G) = \{x_i : 1 \leq i \leq n\} \cup \{x_i^j : 1 \leq i \leq n, 1 \leq j \leq m\}$ and edge set $E(G) = \{x_i x_{i+1} : 1 \leq i \leq n-1\} \cup \{x_i x_{i+1}^j : 1 \leq i \leq n-1, 1 \leq j \leq m\} \cup \{x_i^j x_{i+1} : 1 \leq i \leq n-1, 1 \leq j \leq m\} \cup \{x_i^j x_{i+1}^{j+1} : 1 \leq i \leq n-1, 1 \leq j \leq m\} \cup \{x_i^j x_{i-1}^{j+1} : 1 \leq i \leq n-1, 1 \leq j \leq m\}$. The cardinality $|V(G)| = m(n+1)$ and $|E(G)| = 2m(n-1) + (n-1)$. Let split graph has set of vertices and edges. The set of vertices consists of set of inner and outter vertices. Inner vertex is the vertex set in center graph and outter vertex is the vertex set in split of graph. This also applies to inner and outer edge definitions. Define a bijection $f : V(G) \rightarrow \{1, 2, 3, \ldots, |V(G)|\}$ by

$$f(x_i) = \begin{cases} \frac{i+1}{2} & 1 \leq i \leq n,\ i \text{ odd}; \\ n - \frac{i-2}{2} & 1 \leq i \leq n,\ i \text{ even}. \end{cases}$$

$$f(x_i^j) = \begin{cases} \frac{2nj+i+1}{2} & 1 \leq j \leq m,\ i \text{ odd}; \\ \frac{2n(j+1)-(i-2)}{2} & 1 \leq j \leq m,\ i \text{ even}. \end{cases}$$

It is easy to see that f is a local antimagic labeling of ${}_mSpl(P_n)$ and the edge weights are as follows

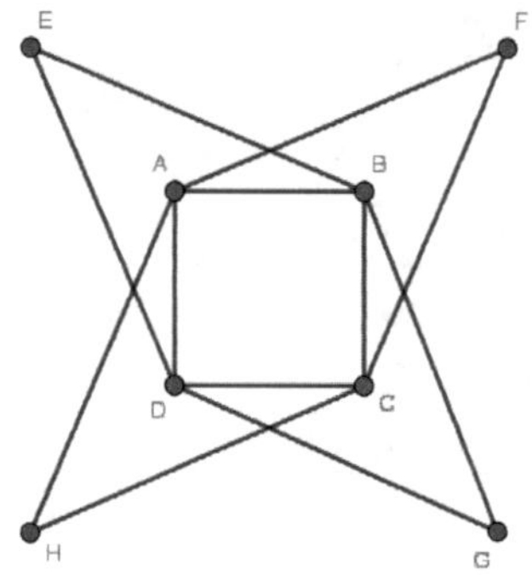

Figure 2. ${}_1Spl(C_4)$ Graph

$$w(x_i x_{i+1}) = \begin{cases} n+1 & 1 \leq j \leq m, \ i \text{ odd}; \\ n+2 & 1 \leq j \leq m, \ i \text{ even}. \end{cases}$$

$$w(x_i x^j_{i+1}) = \begin{cases} (j+1)n+1 & 1 \leq j \leq m, \ i \text{ odd}; \\ (j+1)n+2 & 1 \leq j \leq m, \ i \text{ even}. \end{cases}$$

$$w(x^j_i x_{i+1}) = \begin{cases} (j+1)n+1 & 1 \leq j \leq m, \ i \text{ odd}; \\ (j+1)n+2 & 1 \leq j \leq m, \ i \text{ even}. \end{cases}$$

Hence, from the above the edge weights, it is easy to see that f induces a proper edge colouring of ${}_mSpl(P_n)$ and it gives $\gamma_{lae}({}_mSpl(P_n)) \leq \gamma_{lae}(P_n) + 2m$. Then it will be showed that $\gamma_{lae}({}_mSpl(P_n)) \geq \gamma_{lae}(P_n) + 2m$.

Let $\gamma(G) \geq \delta$. We will show that $\gamma(G) \geq 2m+2$. Based on Lemma [1] that $\gamma(G) \geq \delta$, we know that $\delta(G) = 2m+2$. So, $\gamma(G) \geq \delta = 2m+2$, the lower bound of local edge antimagic of G. It conclude that $\gamma = 2m+2$. So $\gamma_{lae}({}_mSpl(P_n)) = \gamma_{lae}(P_n) + 2m$. □

Theorem 2.2 *For $n \geq 4$, the local edge antimagic of $m-$Splitting of C_n graph is*

$$\gamma_{lae}({}_mSpl(C_n)) = \gamma_{lae}(C_n) + 3m$$

Proof. The ${}_mSpl(C_n)$ graph has vertex set $V(G) = \{x_i : 1 \leq i \leq n\} \cup \{x^j_i : 1 \leq i \leq n, 1 \leq j \leq m\}$ and edge set $E(G) = \{x_i x_{i+1} : 1 \leq i \leq n-1\} \cup \{x_i x^j_{i+1} : 1 \leq i \leq n-1, 1 \leq j \leq m\} \cup \{x^j_i x_{i+1} : 1 \leq i \leq n-1, 1 \leq j \leq m\}$. The cardinality $|V(G)| = mn$ and $|E(G)| = 2mn$. Let split graph has set of vertices and edges. The set of vertices consists of set of inner and outter vertices. Inner vertex is the vertex set in center graph and outter vertex is the vertex set in split of graph. This also applies to inner and outer edge definitions. The illustration can be seen at Figure 2. Define a bijection $f : V(G) \rightarrow \{1, 2, 3, \ldots, |V(G)|\}$ by

$$f(x_i) = \begin{cases} \frac{i+1}{2} & 1 \leq i \leq n, \ i \text{ odd}; \\ n - \frac{i-2}{2} & 1 \leq i \leq n, \ i \text{ even}. \end{cases}$$

$$f(x^j_i) = \begin{cases} \frac{2nj+i+1}{2} & 1 \leq j \leq m, \ i \text{ odd}; \\ \frac{2n(j+1)-(i-2)}{2} & 1 \leq j \leq m, \ i \text{ even}. \end{cases}$$

It is easy to see f is a local antimagic coloring of ${}_mSpl(C_n)$ and the edge weights as follows. For n even,

$$w(e) = \begin{cases} n+1 & \text{for } e = x_ix_{i+1},\ 1 \le i \le n-1,\ i \text{ odd;} \\ n+2 & \text{for } e = x_ix_{i+1},\ 1 \le i \le n-1,\ i \text{ even;} \\ \frac{n+2}{2} & \text{for } e = x_nx_1; \end{cases}$$

For n odd,

$$w(e) = \begin{cases} n+1 & \text{for } e = x_ix_{i+1},\ 1 \le i \le n-1,\ i \text{ odd;} \\ n+2 & \text{for } e = x_ix_{i+1},\ 1 \le i \le n-1,\ i \text{ even;} \\ \frac{n+3}{2} & \text{for } e = x_nx_1; \end{cases}$$

$$w(x_ix_{i+1}^j) = \begin{cases} (j+1)n+1 & 1 \le j \le m,\ i \text{ odd;} \\ (j+1)n+2 & 1 \le j \le m,\ i \text{ even.} \end{cases}$$

$$w(x_i^jx_{i+1}) = \begin{cases} (j+1)n+1 & 1 \le j \le m,\ i \text{ odd;} \\ (j+1)n+2 & 1 \le j \le m,\ i \text{ even.} \end{cases}$$

Hence, from the above the edge weights, it easy to see that f induces a proper edge colouring of ${}_mSpl(C_n)$ and it gives $\gamma_{lae}({}_mSpl(C_n)) \le \gamma_{lae}(C_n) + 3m$. Then it will be showed that $\gamma_{lae}({}_mSpl(C_n)) \ge \gamma_{lae}(C_n) + 3m$.

We will show that $\gamma(G) \ge 3m+3$. Based on Lemma that $\gamma(G) \ge \delta = 2m+2$. *Case 1:* Assume that $\gamma(G) < 3m+3$, we take $\gamma = 3m+2$. Let c_1 and c_2 be the color of inner edge. Then $\frac{n(c_1+c_2)}{2} = \frac{n(2n+2)}{2}$ and hence $c_1+c_2 = 2n+2$. However, if x_i is the vertex with $f(x_i) = n$, then the colors received by inner edges are at least $n+1$ and one of them is at least $n+2$. Thus $c_1+c_2 = (n+1)+(n+2) = 2n+3$. *Case 2:* Assume that $\gamma(G) < 3m+3$, we take $\gamma = 3m+2$. Assume that each outer edge:

- for $m = 1$, then γ of outer edge is equal to 3
- for $m = 2$, then γ of outer edge is equal to 6

  ⋮
- for any m, then γ of outer edge is equal to $3m$

Based on *Case 1 and Case 2*, then it is contradiction. Hence $\gamma_{lae}({}_mSpl(C_n)) \ge 2+3m$. Since $\gamma_{lae}({}_mSpl(C_n)) \le 3+3m$ and $\gamma_{lae}({}_mSpl(C_n)) \ge \gamma_{lae}(C_n)+3m$. □

Theorem 2.3 *For $n \ge 4$, the local edge antimagic of $m-$Splitting of K_n graph is*

$$\gamma_{lae}({}_mSpl(K_n)) = \gamma_{lae}(K_n) + 4m$$

Proof. The ${}_mSpl(K_n)$ graph has vertex set $V(G) = \{x_i, x_i^j : 1 \le i \le n, 1 \le j \le m\}$ and edge set $E(G) = \{x_ix_{i+k} : 1 \le i \le n, 1 \le k \le n-i\} \cup \{x_ix_{i+k}^j : 1 \le i \le n, 1 \le j \le m, 1 \le k \le n-i\} \cup \{x_i^jx_{i+k} : 1 \le i \le n, 1 \le j \le m, 1 \le k \le n-i\}$. The cardinality $|V(G)| = mn$ and $|E(G)| = 3mn$. Let split graph has set of vertices and edges. The set of vertices consists of set of inner and outter vertices. Inner vertex is the vertex set in center graph and outter vertex is the vertex set in split of graph. This also applies to inner and outer edge definitions. The illustration can be seen at Figure 3. Define a bijection $f : V(G) \to \{1, 2, 3, \ldots, |V(G)|\}$ by

$$f(x_i) = \{\ i \quad 1 \le i \le n.$$

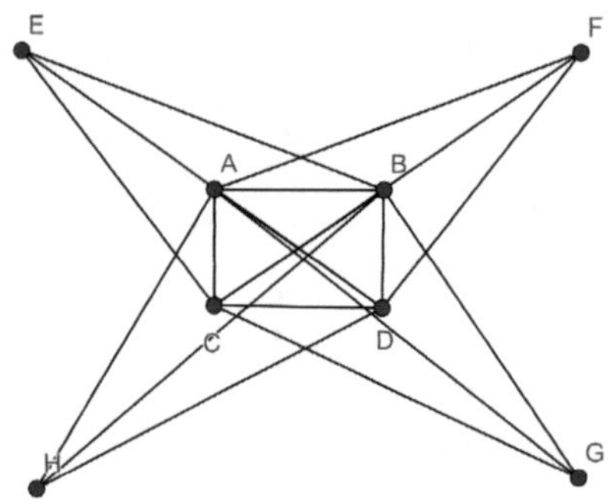

Figure 3. ${}_1Spl(K_4)$ Graph

$$f(x_i^j) = \begin{cases} \frac{2nj+i+1}{2} & 1 \le j \le m,\ i \text{ odd;} \\ nj+i & 1 \le j \le m,\ i \text{ even.} \end{cases}$$

It is easy to see f is a local antimagic coloring of ${}_mSpl(K_n)$ and the edge weights as follows.

$$w(e) = \begin{cases} i+k+1 & \text{for } e = x_i x_{i+k},\ 2 \le i \le n,\ 1 \le k \le n-1; \\ k+2 & \text{for } e = x_1 x_{1+k},\ 1 \le k \le n-2; \\ n+1 & \text{for } e = x_i x_n. \end{cases}$$

For $1 \le k \le n-i$,

$$w(x_i x_{i+k}^j) = \begin{cases} (j+1)n+1 & 1 \le j \le m,\ i \text{ odd;} \\ (j+1)n+2 & 1 \le j \le m,\ i \text{ even.} \end{cases}$$

$$w(x_i^j x_{i+k}) = \begin{cases} (j+1)n+1 & 1 \le j \le m,\ i \text{ odd;} \\ (j+1)n+2 & 1 \le j \le m,\ i \text{ even.} \end{cases}$$

Hence, from the above the edge weights, it is easy to see that f induces a proper edge colouring of ${}_mSpl(K_n)$ and it gives $\gamma_{lae}({}_mSpl(K_n)) \le \gamma_{lae}(K_n) + 4m$. Then it will be showed that $\gamma_{lae}({}_mSpl(K_n)) \ge \gamma_{lae}(K_n) + 4m$.

We will show that $\gamma(G) \ge 4m+4$. *Case 1:* Assume that $\gamma(G) < 4m+4$, we take $\gamma = 4m+3$. Let c_1 and c_2 be the color of inner edge. Then $\frac{n(c_1+c_2)}{2} = \frac{n(2k+4)}{2}$ and hence $c_1 + c_2 = 2k+4$. However, if x_i is the vertex with $f(x_i) = n$, then the colors received by inner edges are at least $k+2$ and one of them is at least $n+1$. Thus $c_1 + c_2 = (n+1) + (n+2) = k+n+3$.
Case 1: Assume that $\gamma(G) < 4m+4$, we take $\gamma = 4m+3$. Assume that each outer edge:

- for $m = 1$, then γ of outer edge is equal to 4
- for $m = 2$, then γ of outer edge is equal to 8

  ⋮

- for any m, then γ of outer edge is equal to $4m$

Based on *Case 1 and Case 2*, then it is contradiction. Hence $\gamma_{lae}({}_mSpl(K_n)) \ge 3+4m$. Since $\gamma_{lae}({}_mSpl(K_n)) \le 4+4m$ and $\gamma_{lae}({}_mSpl(K_n)) \ge \gamma_{lae}(K_n)+4m$. □

There are some theorem of special graph which be found as follows:

IOP Conf. Series: Journal of Physics: Conf. Series **1008** (2018) 012044 doi:10.1088/1742-6596/1008/1/012044

Theorem 2.4 *The local edge antimagic of* $_mSpl(P_4)$ *graph is* $\gamma_{lae}(_mSpl(P_4)) = 2(1+m)$.

Proof. The $_mSpl(P_4)$ graph has vertex set $V(G) = \{x_i : 1 \le i \le 4\} \cup \{x_i^j : 1 \le i \le 4, 1 \le j \le m\}$ and edge set $E(G) = \{x_i x_{i+1} : 1 \le i \le 3\} \cup \{x_i x_{i+1}^j : 1 \le i \le 3, 1 \le j \le m\} \cup \{x_i^j x_{i+1} : 1 \le i \le 3, 1 \le j \le m\} \cup \{x_i^j x_{i+1}^{j+1} : 1 \le i \le 3, 1 \le j \le m\} \cup \{x_i^j x_{i-1}^{j+1} : 1 \le i \le 3, 1 \le j \le m\}$. Let split graph has set of vertices and edges. The set of vertices consists of set of inner and outter vertices. Inner vertex is the vertex set in center graph and outter vertex is the vertex set in split of graph. This also applies to inner and outer edge definitions. Define a bijection $f : V(G) \to \{1, 2, 3, \ldots, |V(G)|\}$ by

$$f(x_i) = \begin{cases} \frac{i+1}{2} & 1 \le i \le 4,\ i \text{ odd;} \\ n - \frac{i-2}{2} & 1 \le i \le 4,\ i \text{ even.} \end{cases}$$

For $1 \le i \le 4$,

$$f(x_i^j) = \begin{cases} \frac{8j+i+1}{2} & 1 \le j \le m,\ i \text{ odd;} \\ \frac{8(j+1)-(i-2)}{2} & 1 \le j \le m,\ i \text{ even.} \end{cases}$$

It is easy to see that f is a local antimagic labeling of $_mSpl(P_4)$ and the edge weights are as follows. For $1 \le i \le 4$,

$$w(x_i x_{i+1}) = \begin{cases} 5 & 1 \le j \le m,\ i \text{ odd;} \\ 6 & 1 \le j \le m,\ i \text{ even.} \end{cases}$$

$$w(x_i x_{i+1}^j) = \begin{cases} (j+1)4+1 & 1 \le j \le m,\ i \text{ odd;} \\ (j+1)4+2 & 1 \le j \le m,\ i \text{ even.} \end{cases}$$

$$w(x_i^j x_{i+1}) = \begin{cases} (j+1)4+1 & 1 \le j \le m,\ i \text{ odd;} \\ (j+1)4+2 & 1 \le j \le m,\ i \text{ even.} \end{cases}$$

Hence, from the above the edge weights, it is easy to see that f induces a proper edge colouring of $_mSpl(P_4)$ and it gives $\gamma_{lae}(_mSpl(P_4)) \le 2(1+m)$. Then it will be showed that $\gamma_{lae}(_mSpl(P_4)) \ge 2(1+m)$.

Let $\gamma(G) \ge \delta$. We will show that $\gamma(G) \ge 10$. Based on Lemma [1] that $\gamma(G) \ge \delta$, we know that $\delta(G) = 10$. So, $\gamma(G) \ge \delta = 10$, the lower bound of local edge antimagic of G. It conclude that $\gamma = 10$. So $\gamma_{lae}(_mSpl(P_4)) = 2(1+m)$. □

Theorem 2.5 *The local edge antimagic of* $_2Spl(P_n)$ *graph is* $\gamma_{lae}(_2Spl(P_n)) = 6$.

Proof. The $_2Spl(P_n)$ graph has vertex set $V(G) = \{x_i : 1 \le i \le n\} \cup \{x_i^j : 1 \le i \le n, 1 \le j \le 2\}$ and edge set $E(G) = \{x_i x_{i+1} : 1 \le i \le n-1\} \cup \{x_i x_{i+1}^j : 1 \le i \le n-1, 1 \le j \le 2\} \cup \{x_i^j x_{i+1} : 1 \le i \le n-1, 1 \le j \le 2\} \cup \{x_i^j x_{i+1}^{j+1} : 1 \le i \le n-1, 1 \le j \le 2\} \cup \{x_i^j x_{i-1}^{j+1} : 1 \le i \le n-1, 1 \le j \le 2\}$. The cardinality $|V(G)| = 2(n+1)$ and $|E(G)| = 4(n-1) + (n-1)$. Let split graph has set of vertices and edges. The set of vertices consists of set of inner and outter vertices. Inner vertex is the vertex set in center graph and outter vertex is the vertex set in split of graph. This also applies to inner and outer edge definitions. Define a bijection $f : V(G) \to \{1, 2, 3, \ldots, |V(G)|\}$ by

$$f(x_i) = \begin{cases} \frac{i+1}{2} & 1 \le i \le n,\ i \text{ odd;} \\ n - \frac{i-2}{2} & 1 \le i \le n,\ i \text{ even.} \end{cases}$$

$$f(x_i^j) = \begin{cases} \frac{2nj+i+1}{2} & 1 \le j \le 2,\ i \text{ odd;} \\ \frac{2n(j+1)-(i-2)}{2} & 1 \le j \le 2,\ i \text{ even.} \end{cases}$$

It is easy to see that f is a local antimagic labeling of ${}_2Spl(P_n)$ and the edge weights are as follows

$$w(x_ix_{i+1}) = \begin{cases} n+1 & 1 \le j \le 2,\ i \text{ odd;} \\ n+2 & 1 \le j \le 2,\ i \text{ even.} \end{cases}$$

$$w(x_ix_{i+1}^j) = \begin{cases} (j+1)n+1 & 1 \le j \le 2,\ i \text{ odd;} \\ (j+1)n+2 & 1 \le j \le 2,\ i \text{ even.} \end{cases}$$

$$w(x_i^jx_{i+1}) = \begin{cases} (j+1)n+1 & 1 \le j \le 2,\ i \text{ odd;} \\ (j+1)n+2 & 1 \le j \le 2,\ i \text{ even.} \end{cases}$$

Hence, from the above the edge weights, it easy to see that f induces a proper edge colouring of ${}_mSpl(P_n)$ and it gives $\gamma_{lae}({}_2Spl(P_n)) \le 6$. Then it will be showed that $\gamma_{lae}({}_2Spl(P_n)) \ge 6$.

Assume that $\gamma_{lae}({}_2Spl(P_n)) = 6-1$. Since $\gamma_{lae}(P_n) = 2$, then $\gamma_{lae}({}_2Spl(P_n)) = 5$. Let c_1 till c_2 be the color of inner edge. Then $\frac{n(c_1+c_2)}{2} = \frac{n(2n+2)}{2}$ and hence $c_1 + c_2 = 2n+2$. However, if xi is the vertex with $f(xi) = n$, then the colors received by inner edges are at least $n+1$ and one of them is at least $n+2$. Thus $c_1 + c_2 = (n+1)+(n+2) = 2n+3$. It's contradiction. Hence $\gamma_{lae}({}_2Spl(P_n)) \ge 6$. Since $\gamma_{lae}({}_2Spl(P_n)) \le 6$ and $\gamma_{lae}({}_2Spl(P_n)) \ge 6$, it completes the proof. □

Theorem 2.6 *The local edge antimagic of* ${}_mSpl(C_4)$ *graph is* $\gamma_{lae}({}_mSpl(C_4)) = 3(1+m)$.

Proof. The ${}_mSpl(C_4)$ graph has vertex set $V(G) = \{x_i : 1 \le i \le 4\} \cup \{x_i^j : 1 \le i \le 4, 1 \le j \le m\}$ and edge set $E(G) = \{x_ix_{i+1} : 1 \le i \le 3\} \cup \{x_ix_{i+1}^j : 1 \le i \le 3, 1 \le j \le m\} \cup \{x_i^jx_{i+1} : 1 \le i \le 3, 1 \le j \le m\}$. The cardinality $|V(G)| = 4m$ and $|E(G)| = 8m$. Let split graph has set of vertices and edges. The set of vertices consists of set of inner and outter vertices. Inner vertex is the vertex set in center graph and outter vertex is the vertex set in split of graph. This also applies to inner and outer edge definitions. Define a bijection $f : V(G) \to \{1, 2, 3, \ldots, |V(G)|\}$ by

$$f(x_i) = \begin{cases} \frac{i+1}{2} & 1 \le i \le 4,\ i \text{ odd;} \\ n - \frac{i-2}{2} & 1 \le i \le 4,\ i \text{ even.} \end{cases}$$

$$f(x_i^j) = \begin{cases} \frac{8j+i+1}{2} & 1 \le j \le m,\ i \text{ odd;} \\ \frac{8(j+1)-(i-2)}{2} & 1 \le j \le m,\ i \text{ even.} \end{cases}$$

It is easy to see f is a local antimagic coloring of ${}_mSpl(C_4)$ and the edge weights as follows. For n even,

$$w(e) = \begin{cases} 5 & \text{for } e = x_ix_{i+1},\ 1 \le i \le 3,\ i \text{ odd;} \\ 6 & \text{for } e = x_ix_{i+1},\ 1 \le i \le 3,\ i \text{ even;} \\ 3 & \text{for } e = x_4x_1; \end{cases}$$

For n odd,

IOP Conf. Series: Journal of Physics: Conf. Series **1008** (2018) 012044 doi:10.1088/1742-6596/1008/1/012044

$$w(e) = \begin{cases} n+1 & \text{for } e = x_i x_{i+1},\ 1 \le i \le n-1,\ i \text{ odd;} \\ n+2 & \text{for } e = x_i x_{i+1},\ 1 \le i \le n-1,\ i \text{ even;} \\ \frac{n+3}{2} & \text{for } e = x_n x_1; \end{cases}$$

$$w(x_i x_{i+1}^j) = \begin{cases} (j+1)4+1 & 1 \le j \le m,\ i \text{ odd;} \\ (j+1)4+2 & 1 \le j \le m,\ i \text{ even.} \end{cases}$$

$$w(x_i^j x_{i+1}) = \begin{cases} (j+1)4+1 & 1 \le j \le m,\ i \text{ odd;} \\ (j+1)4+2 & 1 \le j \le m,\ i \text{ even.} \end{cases}$$

Hence, from the above the edge weights, it easy to see that f induces a proper edge colouring of ${}_mSpl(C_4)$ and it gives $\gamma_{lae}({}_mSpl(C_n)) \le 3(1+m)$. Then it will be showed that $\gamma_{lae}({}_mSpl(C_n)) \ge 3(1+m)$.

Assume that $\gamma_{lae}({}_mSpl(C_4)) = (\gamma_{lae}(C_4)-1)+3m$. Since $\gamma_{lae}(C_4) = 3$, then $\gamma_{lae}({}_mSpl(C_4)) = 3(1+m) - 1m$. Let c_1 till c_2 be the color of inner edge. Then $\frac{n(c_1+c_2)}{2} = \frac{n(2n+2)}{2}$ and hence $c_1 + c_2 = 2n+2$. However, if xi is the vertex with $f(xi) = n$, then the colors received by inner edges are at least $n+1$ and one of them is at least $n+2$. Thus $c_1+c_2 = (n+1)+(n+2) = 2n+3$. It's contradiction. Hence $\gamma_{lae}({}_mSpl(C_4)) \ge 3(1+m) - 1$. Since $\gamma_{lae}({}_mSpl(C_4)) \le 3(1+m)$ and $\gamma_{lae}({}_mSpl(C_4)) \ge 3(1+m)$, it completes the proof. □

Theorem 2.7 *The local edge antimagic of* ${}_3Spl(C_n)$ *graph is* $\gamma_{lae}({}_3Spl(C_n)) = 12$.

Proof. The ${}_3Spl(C_n)$ graph has vertex set $V(G) = \{x_i : 1 \le i \le n\} \cup \{x_i^j : 1 \le i \le n, 1 \le j \le 3\}$ and edge set $E(G) = \{x_i x_{i+1} : 1 \le i \le n-1\} \cup \{x_i x_{i+1}^j : 1 \le i \le n-1, 1 \le j \le 3\} \cup \{x_i^j x_{i+1} : 1 \le i \le n-1, 1 \le j \le 3\}$. The cardinality $|V(G)| = 3n$ and $|E(G)| = 6n$. Let split graph has set of vertices and edges. The set of vertices consists of set of inner and outter vertices. Inner vertex is the vertex set in center graph and outter vertex is the vertex set in split of graph. This also applies to inner and outer edge definitions. It also admits in edge set. Define a bijection $f : V(G) \to \{1, 2, 3, \dots, |V(G)|\}$ by

$$f(x_i) = \begin{cases} \frac{i+1}{2} & 1 \le i \le n,\ i \text{ odd;} \\ n - \frac{i-2}{2} & 1 \le i \le n,\ i \text{ even.} \end{cases}$$

$$f(x_i^j) = \begin{cases} \frac{2nj+i+1}{2} & 1 \le j \le 3,\ i \text{ odd;} \\ \frac{2n(j+1)-(i-2)}{2} & 1 \le j \le 3,\ i \text{ even.} \end{cases}$$

It is easy to see f is a local antimagic coloring of ${}_3Spl(C_n)$ and the edge weights as follows. For n even,

$$w(e) = \begin{cases} n+1 & \text{for } e = x_i x_{i+1},\ 1 \le i \le n-1,\ i \text{ odd;} \\ n+2 & \text{for } e = x_i x_{i+1},\ 1 \le i \le n-1,\ i \text{ even;} \\ \frac{n+2}{2} & \text{for } e = x_n x_1; \end{cases}$$

For n odd,

$$w(e) = \begin{cases} n+1 & \text{for } e = x_i x_{i+1},\ 1 \le i \le n-1,\ i \text{ odd;} \\ n+2 & \text{for } e = x_i x_{i+1},\ 1 \le i \le n-1,\ i \text{ even;} \\ \frac{n+3}{2} & \text{for } e = x_n x_1; \end{cases}$$

$$w(x_i x_{i+1}^j) = \begin{cases} (j+1)n+1 & 1 \le j \le 3,\ i \text{ odd}; \\ (j+1)n+2 & 1 \le j \le 3,\ i \text{ even}. \end{cases}$$

$$w(x_i^j x_{i+1}) = \begin{cases} (j+1)n+1 & 1 \le j \le 3,\ i \text{ odd}; \\ (j+1)n+2 & 1 \le j \le 3,\ i \text{ even}. \end{cases}$$

Hence, from the above the edge weights, it easy to see that f induces a proper edge colouring of ${}_mSpl(C_n)$ and it gives $\gamma_{lae}({}_mSpl(C_n)) \le 12$. Then it will be showed that $\gamma_{lae}({}_mSpl(C_n)) \ge 12$.

Assume that $\gamma_{lae}({}_3Spl(C_n)) = (\gamma_{lae}(C_n) - 1) + 9$. Since $\gamma_{lae}(C_n) = 3$, then $\gamma_{lae}({}_mSpl(C_n)) = 11$. Let c_1 till c_2 be the color of inner edge. Then $\frac{n(c_1+c_2)}{2} = \frac{n(2n+2)}{2}$ and hence $c_1 + c_2 = 2n + 2$. However, if xi is the vertex with $f(xi) = n$, then the colors received by inner edges are at least $n+1$ and one of them is at least $n+2$. Thus $c_1+c_2 = (n+1)+(n+2) = 2n+3$. It's contradiction. Hence $\gamma_{lae}({}_mSpl(C_n)) \ge 11$. Since $\gamma_{lae}({}_3Spl(C_n)) \le 12$ and $\gamma_{lae}({}_3Spl(C_n)) \ge 12$, it completes the proof. □

Theorem 2.8 *The local edge antimagic of* ${}_mSpl(K_4)$ *graph is* $\gamma_{lae}({}_mSpl(K_4)) = 4m + 5$.

Proof. The ${}_mSpl(K_4)$ graph has vertex set $V(G) = \{x_i, x_i^j : 1 \le i \le 4, 1 \le j \le m\}$ and edge set $E(G) = \{x_i x_{i+k} : 1 \le i \le 4, 1 \le k \le 4-i\} \cup \{x_i x_{i+k}^j : 1 \le i \le 4, 1 \le j \le m, 1 \le k \le 4-i\} \cup \{x_i^j x_{i+k} : 1 \le i \le 4, 1 \le j \le m, 1 \le k \le 4-i\}$. The cardinality $|V(G)| = 4m$ and $|E(G)| = 12m$. Let split graph has set of vertices and edges. The set of vertices consists of set of inner and outter vertices. Inner vertex is the vertex set in center graph and outter vertex is the vertex set in split of graph. This also applies to inner and outer edge definitions. Define a bijection $f : V(G) \to \{1, 2, 3, \dots, |V(G)|\}$ by

$$f(x_i) = \{\ i \quad 1 \le i \le 4.$$

$$f(x_i^j) = \begin{cases} \frac{8j+i+1}{2} & 1 \le j \le m,\ i \text{ odd}; \\ 4j+i & 1 \le j \le m,\ i \text{ even}. \end{cases}$$

It is easy to see f is a local antimagic coloring of ${}_mSpl(K_4)$ and the edge weights as follows.

$$w(e) = \begin{cases} i+k+1 & \text{for } e = x_i x_{i+k},\ 2 \le i \le n,\ 1 \le k \le 3; \\ k+2 & \text{for } e = x_1 x_{1+k},\ 1 \le k \le 2; \\ 4+1 & \text{for } e = x_i x_4. \end{cases}$$

For $1 \le k \le 4 - i$,

$$w(x_i x_{i+k}^j) = \begin{cases} (j+1)4+1 & 1 \le j \le m,\ i \text{ odd}; \\ (j+1)4+2 & 1 \le j \le m,\ i \text{ even}. \end{cases}$$

$$w(x_i^j x_{i+k}) = \begin{cases} (j+1)4+1 & 1 \le j \le m,\ i \text{ odd}; \\ (j+1)4+2 & 1 \le j \le m,\ i \text{ even}. \end{cases}$$

Hence, from the above the edge weights, it easy to see that f induces a proper edge colouring of ${}_mSpl(K_4)$ and it gives $\gamma_{lae}({}_mSpl(K_4)) \le 4m + 5$. Then it will be showed that $\gamma_{lae}({}_mSpl(K_4)) \ge 4m + 5$.

Assume that $\gamma_{lae}({}_mSpl(K_4)) = 4m+5-1$. Since $\gamma_{lae}(K_4) = 5$, then $\gamma_{lae}({}_mSpl(K_4)) = 4m+4$. Let c_1 till c_2 be the color of inner edge. Then $\frac{n(c_1+c_2)}{2} = \frac{n(2k+4)}{2}$ and hence $c_1 + c_2 = 2k + 4$.

However, if xi is the vertex with $f(xi) = n$, then the colors received by inner edges are at least $k+2$ and one of them is at least $n+1$. Thus $c_1+c_2 = (n+1)+(n+2) = k+n+3$. It's contradiction. Hence $\gamma_{lae}({}_mSpl(K_4)) \geq 4m+4$. Since $\gamma_{lae}({}_mSpl(K_n)) \leq 4m+5$ and $\gamma_{lae}({}_mSpl(K_n)) \geq 4m+5$, it completes the proof. □

Theorem 2.9 *The local edge antimagic of ${}_2Spl(K_n)$ graph is $\gamma_{lae}({}_2Spl(K_n)) = 2n+5$.*

Proof. The ${}_2Spl(K_n)$ graph has vertex set $V(G) = \{x_i, x_i^j : 1 \leq i \leq n, 1 \leq j \leq 2\}$ and edge set $E(G) = \{x_i x_{i+k} : 1 \leq i \leq n, 1 \leq k \leq n-i\} \cup \{x_i x_{i+k}^j : 1 \leq i \leq n, 1 \leq j \leq 2, 1 \leq k \leq n-i\} \cup \{x_i^j x_{i+k} : 1 \leq i \leq n, 1 \leq j \leq 2, 1 \leq k \leq n-i\}$. The cardinality $|V(G)| = 2n$ and $|E(G)| = 6n$. Let split graph has set of vertices and edges. The set of vertices consists of set of inner and outter vertices. Inner vertex is the vertex set in center graph and outter vertex is the vertex set in split of graph. This also applies to inner and outer edge definitions. Define a bijection $f : V(G) \to \{1, 2, 3, \ldots, |V(G)|\}$ by

$$f(x_i) = \{\ i \quad 1 \leq i \leq n.$$

$$f(x_i^j) = \begin{cases} \frac{2nj+i+1}{2} & 1 \leq j \leq 2,\ i \text{ odd;} \\ nj+i & 1 \leq j \leq 2,\ i \text{ even.} \end{cases}$$

It is easy to see f is a local antimagic coloring of ${}_2Spl(K_n)$ and the edge weights as follows.

$$w(e) = \begin{cases} i+k+1 & \text{for } e = x_i x_{i+k},\ 2 \leq i \leq n,\ 1 \leq k \leq n-1; \\ k+2 & \text{for } e = x_1 x_{1+k},\ 1 \leq k \leq n-2; \\ n+1 & \text{for } e = x_i x_n. \end{cases}$$

For $1 \leq k \leq n-i$,

$$w(x_i x_{i+k}^j) = \begin{cases} (j+1)n+1 & 1 \leq j \leq 2,\ i \text{ odd;} \\ (j+1)n+2 & 1 \leq j \leq 2,\ i \text{ even.} \end{cases}$$

$$w(x_i^j x_{i+k}) = \begin{cases} (j+1)n+1 & 1 \leq j \leq 2,\ i \text{ odd;} \\ (j+1)n+2 & 1 \leq j \leq 2,\ i \text{ even.} \end{cases}$$

Hence, from the above the edge weights, it easy to see that f induces a proper edge colouring of ${}_2Spl(K_n)$ and it gives $\gamma_{lae}({}_2Spl(K_n)) \leq 2n+5$. Then it will be showed that $\gamma_{lae}({}_2Spl(K_n)) \geq 2n+5$.

Assume that $\gamma_{lae}({}_2Spl(K_n)) = 2n+5-1$. Since $\gamma_{lae}(K_n) = 2n-3$, then $\gamma_{lae}({}_2Spl(K_n)) = 2n+4$. Let c_1 till c_2 be the color of inner edge. Then $\frac{n(c_1+c_2)}{2} = \frac{n(2k+4)}{2}$ and hence $c_1+c_2 = 2k+4$. However, if xi is the vertex with $f(xi) = n$, then the colors received by inner edges are at least $k+2$ and one of them is at least $n+1$. Thus $c_1+c_2 = (n+1)+(n+2) = k+n+3$. It's contradiction. Hence $\gamma_{lae}({}_2Spl(K_n)) \geq 2n+4$. Since $\gamma_{lae}({}_2Spl(K_n)) \leq 2n+5$ and $\gamma_{lae}({}_2Spl(K_n)) \geq 2n+5$. □

3. Concluding Remarks

We have discussed about the local antimagic edge coloring of some related $m-$splitting graphs for several sets of value (m, n) in this paper. Three basic theorem is about complete graph, circle graph, and path graph which has any solutions for being a basic graph of operation $m-$splitting.

IOP Conf. Series: Journal of Physics: Conf. Series **1008** (2018) 012044 doi:10.1088/1742-6596/1008/1/012044

Open Problem

Find local edge antimagic coloring of ${}_mSpl(H_n)$ graph for any n and m where H is any graph.

Acknowledgement

This work was partially supported by the CGANT University of Jember year 2018

References

[1] Agustin I H, Dafik, Moh. Hasan, Alfarisi R, Prihandini R M 2017 Local edge Antimagic Coloring of Graphs *Far East Journal of Mathematical Science* **102** 1925 - 1941
[2] Arumugam S, Premalatha K, Baca M and Semanicova-Fenovcikova A 2017 Local Antimagic Vertex Coloring of a Graph, *Graphs and Combinatorics* **33** 275-285
[3] Bondy J A and Murty U S R 1982 *Graph Theory and Application 3rd* ed (United States of America: North Holland)
[4] Chartrand G and Lesniak L 2000 *Graphs and digraphs 3rd* ed (London: Chapman and Hall)
[5] Dafik, Miller M, Ryan J, and Baca M 2011 Super edge-antimagic total labelings of ${}_mK_{n,n}$ *Ars Combinatoria,* **101** 97-107.
[6] Dafik, Baca M, Miller M, and Ryan J 2008 On super (a, d)-edge-antimagic total labeling of caterpillars *Journal of Combinatorial Mathematics and Combinatorial Computing (JCMCC)* **65** 61-70.
[7] Gross J L, Yellen J and Zhang P 2014 *Handbook of graph Theory* Second Edition CRC Press Taylor and Francis Group
[8] Hartsfield N dan Ringel G 1994 *Pearls in Graph Theory* Academic Press. United Kingdom
[9] Baca M, Miller M, Ryan J, and Dafik 2009 Antimagic labeling of disjoint union of s-crowns *Utilitas Mathematica* **78** 193-205

IOP Conf. Series: Journal of Physics: Conf. Series **1008** (2018) 012045 doi:10.1088/1742-6596/1008/1/012045

Non-isolated Resolving Sets of certain Graphs Cartesian Product with a Path

I M Hasibuan, A N M Salman and S W Saputro

Combinatorial Mathematics Reseach Group, Faculty of Mathematics and Natural Sciences, Institut Teknologi Bandung, Indonesia

E-mail: imhasibuan@uin-suska.ac.id

Abstract. Let G be a connected, simple, and finite graph. For an ordered subset $W = \{w_1, w_2, \cdots, w_k\}$ of vertices in a graph G and a vertex v of G, the metric representation of v with respect to W is the k-vector $r(v|W) = (d(v, w_1), d(v, w_2), \cdots, d(v, w_k))$. The set W is called a resolving set for G if every vertex of G has a distinct representation. The minimum cardinality of W is called the metric dimension of G, denoted by $dim(G)$. If the induced subgraph $\langle W \rangle$ has no isolated vertices, then W is called a non-isolated resolving set. The minimum cardinality of non-isolated resolving set of G is called the non-isolated resolving number of G, denoted by $nr(G)$. In this paper, we consider $H \square P_n$ that is a graph obtained from Cartesian product between a connected graph H and a path P_n. We determine $nr(H \square P_n)$, for some classes of H, including cycles, complete graphs, complete bipartite graphs, and friendship graphs.

1. Introduction

Let G be a connected, simple, and finite graph. For $u, v \in V(G)$, the *distance* betwen u and v in G, denoted by $d(u, v)$, is the lenght of the shortest path u to v in G. Let $W = \{w_1, w_2, \cdots, w_k\}$ be an ordered subset of $V(G)$. For a vertex $v \in V(G)$, a representation of v with respect to W is k-tuple $r(v|W) = (d(v, w_1), d(v, w_2), \cdots, d(v, w_k))$. The set W is called a *resolving set* of G if every distinct vertices u and v in G satisfies $r(u|W) \neq r(v|W)$. The minimum cardinality of a resolving set of G is called the metric dimension of G, denoted by $dim(G)$.

The metric dimension problem was firstly introduced independently by Harary-Melter [5] and Slater [12]. All connected graphs of order n which have metric dimension 1, $n-1$, or $n-2$ have been characterized [3]. Chartrand *et al.* [3] have proven that $dim(G) = 1$ if and only if G is a path, $dim(G) = n - 1$ if and only if G is a complete graph, and $dim(G) = n - 2$ if and only if $G = K_{s,t}$ for $s, t \geq 1$, $G = K_s + \bar{K}_t$ for $s \geq 1, t \geq 2$, or $G = K_s + (K_1 \cup K_t)$ for $s, t \geq 1$. Some authors also determined the metric dimension of certain classes of graph. Chartrand *et al.* [3] provided the metric dimension of cycles and paths. The metric dimension of some regular graphs have been proven by Baca *et al.* [1]. The metric dimension of amalgamation of a certain graph or a graph with some pendant edges have been studied by Iswadi *et al.* [6],[7],[9]. Meanwhile, some authors investigated the metric dimension of graphs obtained from certain operation of graphs [2],[8],[10],[11] and [13].

In this paper, we consider another version of metric dimension problem. Let W be a resolving set of a connected graph G. If an induced subgraph of G by W does not contain an isolated

IOP Conf. Series: Journal of Physics: Conf. Series **1008** (2018) 012045 doi:10.1088/1742-6596/1008/1/012045

vertex, then we called W as a non-isolated resolving set of G. The cardinality of the minimum non-isolated resolving set of G is called the non-isolated resolving number of G, denoted by $nr(G)$. Since a non-resolving set of G is also a resolving of G, it is clear that $dim(G) \leq nr(G)$.

The non-isolated resolving set problem was introduced by Chitra, and Arumugam [4]. They provided a sharp upper bound of $nr(G)$ for any graph G, namely $nr(G) \leq 2.dim(G)$. In the same paper, they characterized all connected graphs of order n with $nr(G) = n - 1$. They also gave some condition of a graph G such that $nr(G) = 2$.

The Cartesian product of graphs G and H, denoted by $G\Box H$, is the graph with vertex set $V(G\Box H) = \{(a,v)|a \in V(G), v \in V(H)\}$, where (a,v) is adjacent to (b,w) whenever $a = b$ and $vw \in E(H)$, or $v = w$ and $ab \in E(G)$. Chitra, and Arumugam [4] have initiated the research on non-isolated resolving set of certain Cartesian product graphs. They proved that $nr(P_n\Box P_n) = 4$ for $n \geq 2$, $nr(C_n\Box P_2) = 3$ for $n \geq 4$. They also provided the upper bound of $nr(G\Box P_2)$ for any non trivial connected graph G. In this paper, we determine the non-isolated resolving number of $H\Box P_n$ where H is either a cycle, a complete graph, a complete bipartite graph or a friendship graph.

2. Main Results

Let G and H be connected, simple, and finite graphs. For $a \in V(G)$ and $v \in V(H)$, we define $G(v) = \{(a,v)|a \in V(G)\}$ and $H(a) = \{(a,v)|v \in V(H)\}$. We can say that $G(v)$ and $H(a)$ as a column of v and a row of a, respectively. In the following theorem, we provide the non-isolated resolving number of $C_m\Box P_n$.

Theorem 2.1. *Let C_m be a cycle of order $m \geq 3$, and P_n be a path of order $n \geq 2$. Then*

$$nr(C_m\Box P_n) = \begin{cases} 2, & \text{if } m = 3, \\ 3, & \text{if } m = 4 \text{ or } n = 2, \\ 4, & \text{otherwise.} \end{cases}$$

Proof. Let $V(C_m) = \{a_1, a_2, \cdots, a_m\}$, $E(C_m) = \{a_1a_m, a_ia_{i+1}|1 \leq i \leq m-1\}$, $V(P_n) = \{v_1, v_2, \cdots, v_n\}$ and $E(P_n) = \{v_iv_{i+1}|1 \leq i \leq n-1\}$.
We define a vertex set

$$W = \begin{cases} \{(a_1,v_1),(a_2,v_1)\}, & \text{if } m = 3, \\ \{(a_1,v_1),(a_2,v_1),(a_3,v_1)\}, & \text{if } m = 4, \\ \{(a_1,v_1),(a_2,v_1),(a_1,v_2)\}, & \text{if } n = 2 \text{ and } m \neq 4, \\ \{(a_1,v_1),(a_2,v_1),(a_1,v_n),(a_2,v_n)\}, & \text{otherwise.} \end{cases}$$

It is easy to check that the induced subgraph of $C_m\Box P_n$ by W does not contain an isolated vertex.

Let $x = (a_i, v_p)$ and $y = (a_k, v_q)$ be a distinct vertices in $G\backslash W$. We distinguish four cases and for every cases, we show that $r(x|W) \neq r(y|W)$.

Case 1 : $m = 3$. We distinguish three subcases.
Subcase 1a : $p = q$. So, it is clear that $p = q \neq 1$. Note that $(a_i, v_1) \in W$ or $(a_k, v_1) \in W$. Without loss of generality, we assume that $(a_i, v_1) \in W$. Thus, we obtain $d(y,(a_i,v_1)) = d((a_k,v_1),(a_i,v_1)) + (p-1) > (p-1) = d(x,(a_i,v_1))$.
Subcase 1b : $i = k$. It is clear that $p \neq q$. We obtain that $d(x,(a_1,v_1)) = d((a_i,v_1),(a_1,v_1)) + (p-1) \neq d((a_k,v_1),(a_1,v_1)) + (q-1) = d(y,(a_1,v_1))$.

Subcase 1c : $i \neq k$, $p \neq q$. If $d(x,(a_1,v_1)) \neq d(y,(a_1,v_1))$, then we have $r(x|W) \neq r(y|W)$. Now, we assume that $d(x,(a_1,v_1)) = d(y,(a_1,v_1))$.
Therefore, we have $p = q+1$ or $q = p+1$. Without loss of generality, we assume $p = q+1$. So, we obtain

$$\begin{aligned} d(x,(a_2,v_1)) &= d((a_2,v_1),(a_1,v_1)) + (p-1) = d((a_2,v_1),(a_1,v_1)) + q \\ &\geq d((a_2,v_1),(a_k,v_1)) + q > d((a_2,v_1),(a_k,v_1)) + (q-1) = d(y,(a_2,v_1)) \end{aligned}$$

All the subcases imply $r(x|W) \neq r(y|W)$.

Case 2 : $m = 4$. We distinguish three subcases.
Subcase 2a : $p = q$. So, it is clear that $p = q \neq 1$. Note that $(a_i,v_1) \in W$ or $(a_k,v_1) \in W$. Without loss of generality, we assume that $(a_i,v_1) \in W$. Thus, we obtain $d(y,(a_i,v_1)) = d((a_k,v_1),(a_i,v_1)) + (p-1) > (p-1) = d(x,(a_i,v_1))$.
Subcase 2b : $i = k$. It is clear that $p \neq q$. We obtain that $d(x,(a_1,v_1)) = d((a_i,v_1),(a_1,v_1)) + (p-1) \neq d((a_k,v_1),(a_1,v_1)) + (q-1) = d(y,(a_1,v_1))$.
Subcase 2c : $i \neq k$, $p \neq q$. If $d(x,(a_1,v_1)) \neq d(y,(a_1,v_1))$, then we have $r(x|W) \neq r(y|W)$. Now, we assume that $d(x,(a_1,v_1)) = d(y,(a_1,v_1))$.
Therefore, we have $p = q+1$ or $q = p+1$. Without loss of generality, we assume $p = q+1$. So, we obtain

$$\begin{aligned} d(x,(a_2,v_1)) &= d((a_2,v_1),(a_1,v_1)) + (p-1) = d((a_2,v_1),(a_1,v_1)) + q \\ &\geq d((a_2,v_1),(a_k,v_1)) + q > d((a_k,v_1),(a_2,v_1)) + (q-1) = d(y,(a_2,v_1)) \end{aligned}$$

All the subcases imply $r(x|W) \neq r(y|W)$.

Case 3 : $n = 2, m \neq 4$. We distinguish three subcases.
subcase 3a : $p = q$. If $d(x,(a_1,v_1)) \neq d(y,(a_1,v_1))$, then $r(x|W) \neq r(y|W)$. Conversely, we assume that $d(x,(a_1,v_1)) = d(y,(a_1,v_1))$. We distinguish two subcases.
Subcase 3a(i) : $d(x,(a_2,v_1)) = d(x,(a_1,v_1))+1$. We obtain that $d(x,(a_2,v_1)) = d(x,(a_1,v_1)) + 1 = d(y,(a_1,v_1)) + 1 \neq d(y,(a_1,v_1)) - 1 = d(y,(a_2,v_1))$.
Subcase 3a(ii) : $d(x,(a_2,v_1)) = d(x,y) + d(y,(a_2,v_1))$. We obtain that $d(x,(a_2,v_1)) > d(y,(a_2,v_1))$ because $d(x,y) \geq 1$.
Subcase 3b : $i = k$. We obtain that $d(x,(a_1,v_1)) = d((a_i,v_1),(a_1,v_1)) \neq d((a_k,v_1),(a_1,v_1)) + 1 = d(y,(a_1,v_1))$.
Subcase 3c : $i \neq k$, $p \neq q$. If $d(x,(a_1,v_1)) \neq d(y,(a_1,v_1))$, then we have $r(x|W) \neq r(y|W)$. Now, we assume that $d(x,(a_1,v_1)) = d(y,(a_1,v_1))$.
Therefore, we have $p = 1$ or $p = 2$. Without loss of generality, we assume $p = 2$. So, we obtain

$$\begin{aligned} d(x,(a_1,v_2)) &= d((a_1,v_2),(a_i,v_2)) = d((a_i,v_2),(a_1,v_1)) - 1 \\ &\neq d((a_k,v_1),(a_1,v_1)) + 1 = d(y,(a_1,v_2)) \end{aligned}$$

All the subcases imply $r(x|W) \neq r(y|W)$.

Case 4 : $m \neq 5, n \geq 3$. We distinguish three subcases.
Subcase 4a : $p = q$. If $d(x,(a_1,v_1)) \neq d(y,(a_1,v_1))$, then $r(x|W) \neq r(y|W)$. Conversely, we assume that $d(x,(a_1,v_1)) = d(y,(a_1,v_1))$. We distinguish two cases.
Subcase 4a(i) : $d(x,(a_2,v_1)) = d(x,(a_1,v_1))+1$. We obtain that $d(x,(a_2,v_1)) = d(x,(a_1,v_1)) + 1 = d(y,(a_1,v_1)) + 1 \neq d(y,(a_1,v_1)) - 1 = d(y,(a_2,v_1))$.
Subcase 4a(ii) : $d(x,(a_2,v_1)) = d(x,y) + d(y,(a_2,v_1))$. We obtain that $d(x,(a_2,v_1)) > d(y,(a_2,v_1))$ since $d(x,y) \geq 1$.
Subcase 4b : $i = k$. We obtain that $d(x,(a_1,v_1)) = d((a_i,v_1),(a_1,v_1)) + (p-1) \neq$

$d((a_k, v_1), (a_1, v_1)) + (q-1)1 = d(y, (a_1, v_1))$.
Subcase 4c : $i \neq k$, $p \neq q$. If $d(x, (a_1, v_1)) \neq d(y, (a_1, v_1))$, then we have $r(x|W) \neq r(y|W)$. Now, we assume that $d(x, (a_1, v_1)) = d(y, (a_1, v_1))$.
Therefore, we have $p = q+1$ or $q = p+1$. Without loss of generality, we assume $p = q+1$. So, we obtain

$$\begin{aligned} d(x, (a_1, v_n)) &= d((a_i, v_n), (a_1, v_n)) + (n-p) = d((a_i, v_n), (a_1, v_n)) + (n-(q+1)) \\ &\neq d((a_i, v_n), (a_1, v_n)) + (n-q+2)) = (d((a_i, v_n), (a_1, v_n)) + 2) + (n-q)) \\ &= d((a_k, v_n), (a_1, v_n)) + (n-q)) = d(y, (a_1, v_n)) \end{aligned}$$

All the subcases imply $r(x|W) \neq r(y|W)$. Therefore, W is a non-isolated resolving set.

If $m = 3$, it is clear that $nr(G = C_m \Box P_n) \geq 2$ since G is not isomorphic to a path. Now, we assume that $m \geq 4$. We distinguish two cases.

Case 1 : $m = 4$ or $n = 2$
Suppose $nr(G) \leq 2$. Let $S \subseteq V(G)$ where $|S| = 2$ and the induced subgraph $\langle S \rangle$ does not contain an isolated vertex. Thus, $\langle S \rangle$ isomorphic with K_2. Therefore, S is one of the following forms :
$S_1 = \{(a_i, v_j), (a_i, v_{j+1})\}$ where $i \in \{1, 2, \cdots, m-1\}$ and $j \in \{1, 2, \cdots, n-1\}$, and
$S_2 = \{(a_i, v_j), (a_{i+1}, v_j)\}$ where $i \in \{1, 2, \cdots, m-1\}$ and $j \in \{1, 2, \cdots, n-1\}$.
If $S = S_1$, then $r((a_i, v_{j+2})|W) = (2, 1) = r((a_{i+1}, v_{j+1})|W)$. If $S = S_2$, then $r((a_{i+2}, v_j)|W) = (2, 1) = r((a_{i+1}, v_{j+1})|W)$.

Case 2 : $m \neq 4$ and $n \neq 2$.
Suppose $nr(G) \leq 3$. Let $S \subseteq V(G)$ where $|S| = 3$ and the induced subgraph $\langle S \rangle$ does not contain an isolated vertex. Thus $\langle S \rangle$ isomorphic with P_3 and S is one of the following forms :
$S_1 = \{(a_i, v_j), (a_i, v_{j+1}), (a_i, v_{j+2})\}$; $i \in \{1, 2, \cdots, m-1\}$ and $j \in \{1, 2, \cdots, n-2\}$,
$S_2 = \{(a_i, v_j), (a_i, v_{j+1}), (a_{i+1}, v_{j+1})\}$; $i \in \{1, 2, \cdots, m-1\}$ and $j \in \{1, 2, \cdots, n-1\}$,
$S_3 = \{(a_i, v_j), (a_{i+1}, v_j), (a_{i+2}, v)\}$; $i \in \{1, 2, \cdots, m-2\}$ and $j \in \{1, 2, \cdots, n-1\}$, and
$S_4 = \{(a_i, v_j), (a_{i+1}, v_j), (a_{i+1}, v_{j+1})\}$; $i \in \{1, 2, \cdots, m-2\}$ and $j \in \{1, 2, \cdots, n-2\}$
If $S = S_1$, then $r((a_{i-1}, v_{j+1})|W) = (2, 1, 2) = r((a_{i+1}, v_{j+1})|W)$. If $S = S_2$, then $r((a_{i+1}, v_{j+2})|W) = (3, 2, 1) = r((a_{i+2}, v_{j+1})|W)$. If $S = S_3$, then $r((a_{i+1}, v_{j-1})|W) = (2, 1, 2) = r((a_{i+1}, v_{j+1})|W)$. If $S = S_4$, then $r((a_{i+1}, v_{j+2})|W) = (3, 2, 1) = r((a_{i+2}, v_{j+1})|W)$.
All the cases lead to a contradiction. □

The following result, we determine the non-isolated resolving number of $K_m \Box P_n$.

Lemma 2.2. *Let W be a resolving set of $G = K_m \Box P_n$ for $m \geq 3$ and $n \geq 2$. Then there are no two different vertices x dan y of $V(K_m)$ such that $W \cap P_n(x) = \emptyset$ and $W \cap P_n(y) = \emptyset$*

Proof. We assume that there are two vertices x and y in K_m such that $W \cap P_n(x) \neq \emptyset$ and $W \cap P_n(y) \neq \emptyset$. Let $x = (a_i, v_p)$ and $y = (a_k, v_q)$ be a distinct vertices in $(K_m \Box P_n) \backslash W$ and $(a_l, v_r) \in W$. For $p = q$, we obtain

$$d(x, (a_l, v_r)) = d((a_i, v_r), (a_l, v_r) + (p-1) = d((a_k, v_r), (a_l, v_r) + (p-1) = d(y, (a_l, v_r)$$

We get a contradiction since W is a resolving set of G. □

Corollary 2.3. *Let m, and n be two natural numbers with $m \geq 3$, and $n \geq 2$. Let $G = K_m \Box P_n$. Then $nr(G) \geq m - 1$.*

Proof. This corollary is a direct consequence of Lemma 2.2. □

Theorem 2.4. *Let m, and n be two natural numbers with $m \geq 3$, $n \geq 2$. Let $G = K_m \Box P_n$. Then $nr(G) = m - 1$*

Proof. By Corollary 2.3, we only need to show that $nr(G) \leq m-1$. Let $V(K_m) = \{a_1, \cdots, a_m\}$, $E(K_m) = \{a_i a_j | i, j = 1, 2, \cdots, m\}$, $V(P_n) = \{v_1, \cdots, v_n\}$ and $E(P_n) = \{v_i v_{i+1} | i = 1, \cdots n-1\}$. We define a vertex set $W = \{(a_i, v_1) | i = 1, 2 \cdots m-1\}$.
It is easy to check that the induced subgraph of G by W does not contain an isolated vertex.

Let $x = (a_i, v_p)$ and $y = (a_k, v_q)$ be a distinct vertices in $G \backslash W$. We show that $r(x|W) \neq r(y|W)$. We distinguish three cases.

Case 1 : $p = q$. So, it is clear that $p = q \neq 1$. Note that $(a_i, v_1) \in W$ or $(a_k, v_1) \in W$. Without loss of generality, we assume that $(a_i, v_1) \in W$. Thus, we obtain $d(y, (a_i, v_1)) = d((a_k, v_1), (a_i, v_1)) + (p-1) = 1 + (p-1) = p > (p-1) = d(x, (a_i, v_1))$.

Case 2 : $i = k$. It is clear that $p = q$. We obtain that $d(x, (a_1, v_1)) = d((a_i, v_1), (a_1, v_1)) + (p-1) = 1 + (p-1) = p \neq q = 1 + (q-1) = d((a_k, v_1), (a_1, v_1)) + (q-1) = d(y, (a_1, v_1))$.

Case 3 : $i \neq k$, $p \neq q$. If $d(x, (a_1, v_1)) \neq d(y, (a_1, v_1))$, then we have $r(x|W) \neq r(y|W)$. Now, we assume that $d(x, (a_1, v_1)) = d(y, (a_1, v_1))$. Therefore, we have $p = q+1$ or $q = p+1$. Without loss of generality, we assume $p = q+1$. So, for some $i \in \{2, 3, \cdots, m-1\}$ we obtain

$$\begin{aligned} d(x, (a_i, v_1)) &= p - 1 = q \geq d((a_i, v_1), (a_k, v_1)) + q \\ &> d((a_i, v_1), (a_k, v_1)) + (q-1) = d(y, (a_i, v_1)) \end{aligned}$$

All the cases imply $r(x|W) \neq r(y|W)$. Thus, $nr(G) \leq m-1$. □

Let n, s, and t be three natural numbers with $n \geq 2$, $t \geq 2$, and $s \leq t$. We have the non-isolated resolving number of $G = K_{s,t} \Box P_n$ as follows

Lemma 2.5. *Let n, s, and t be three natural numbers with $n \geq 2$, $t \geq 2$, and $s \leq t$. Let $K_{s,t}$ be a complete bipartite graph and A be a partition of $K_{s,t}$ where $|A| \geq 2$, and P_n be a path graph. Let W be a resolving set of $G = K_{s,t} \Box P_n$. Then there are no two different vertices $x, y \in A$ such that $W \cap P_n(x) = \emptyset$ and $W \cap P_n(y) = \emptyset$.*

Proof. We assume that there are two vertices x and y in a partition A of $K_{s,t}$ such that $W \cap P_n(x) \neq \emptyset$ and $W \cap P_n(y) \neq \emptyset$. Let $x = (a_i, v_p)$, $y = (a_k, v_q)$ be a distinct vertices in $(K_{s,t} \Box P_n) \backslash W$ and $(a_l, v_r) \in W$. For $p = q$, we obtain

$$d(x, (a_l, v_r)) = d((a_i, v_r), (a_l, v_r) + (p-1) = d((a_k, v_r), (a_l, v_r) + (p-1) = d(y, (a_l, v_r)$$

Therefore, we have a contradiction since W is a resolving set of G. □

Corollary 2.6. *Let n, s, and t be three natural numbers with $n \geq 2$, $t \geq 3$, and $s \leq t$. Let $G = K_{s,t} \Box P_n$. Then $nr(G) \geq s + t - 2$*

Proof. This corollary is a direct consequence of Lemma 2.5. □

Theorem 2.7. *Let n, s, and t be three natural numbers with $n \geq 2$, $t \geq 3$, and $s \leq t$. Let $G = K_{s,t} \square P_n$. Then*

$$nr(G) = \begin{cases} t+1, & \text{if } s \in \{1,2\}, \\ s+t-2, & \text{if } s \geq 3. \end{cases}$$

Proof. Let $V(K_{s,t}) = \{a_1, a_2, \cdots, a_s, b_1, b_2, \cdots, b_t\}$, $E(K_{s,t}) = \{(a_i, b_j)|i = 1, \cdots, s; j = 1, \cdots, t\}$, $V(P_n) = \{v_1, v_2, \cdots, v_n\}$ and $E(P_n) = \{(v_i, v_{i+1})|i = 1, \cdots, n-1\}$.
We define a vertex set

$$W = \begin{cases} \{(a_1, v_1), (b_j, v_1)|j = 1, \cdots, t\}, & \text{if } s \in \{1,2\}, \\ \{(a_i, v_1), (b_j, v_1)|1 \leq i \leq s-1; 1 \leq j \leq t-1\}, & \text{if } s \geq 3. \end{cases}$$

It is easy to check that the induced subgraph of $K_{s,t} \square P_n$ by W does not contain an isolated vertex.

Let $x = (a_i, v_p)$, $y = (a_j, v_q)$, $z = (b_k, v_r)$ and $w = (b_l, v_s)$ be a distinct vertices in $G \backslash W$. We distinguish two cases, and for every cases, we show that $r(x|W) \neq r(y|W) \neq r(z|W) \neq r(w|W)$.

Case 1: $s = 1, 2$. We distinguish three subcases.
Subcase 1a : $p = q = r = s$. So, it is clear $p = q = r = s \neq 1$. For $s = 1$, it is clear. For $s = 2$, note that $(a_i, v_1) \in W$ or $(a_k, v_1) \in W$. Without loss of generality, we assume that $(a_i, v_1) \in W$. Thus, we obtain
$d(y, (a_i, v_1)) = d((a_j, v_1), (a_i, v_1)) + (p-1) > (p-1) = d(x, (a_i, v_1))$, $d(z, (a_i, v_1)) = d(x, (a_i, v_1)) + 1$, $d(w, (a_i, v_1)) = d(x, (a_i, v_1)) + 1$

$$\begin{aligned} d(z, (a_i, v_1)) &= d((b_k, v_1), (a_i, v_1)) + (p-1) = 1 + (p-1) = p \\ &< (p+1) = 2 + (p-1) = d((a_j, v_1), (a_i, v_1)) + (p-1) = d(y, (a_i, v_1)). \\ d(w, (a_i, v_1)) &= d((b_l, v_1), (a_i, v_1)) + (p-1) = 1 + (p-1) = p \\ &< (p+1) = 2 + (p-1) = d((a_j, v_1), (a_i, v_1)) + (p-1) = d(y, (a_i, v_1)). \end{aligned}$$

$d(w, (b_k, v_1)) = d(z, (b_k, v_1)) + 2$
Therefore, we have $r(x|W) \neq r(y|W) \neq r(z|W) \neq r(w|W)$.
Subcase 1b : $i = j$ or $k = l$. For $i = j$, we obtain that $d(x, (a_1, v_1)) = d((a_i, v_1), (a_1, v_1)) + (p-1) \neq d((a_j, v_1), (a_1, v_1)) + (q-1) = d(y, (a_1, v_1))$. Therefore, we have $r(x|W) \neq r(y|W)$. For $k = l$, we obtain that $d(z, (b_1, v_1)) = d((b_k, v_1), (b_1, v_1)) + (r-1) \neq d((b_l, v_1), (b_1, v_1)) + (s-1) = d(w, (b_1, v_1))$. Therefore, we have $r(z|W) \neq r(w|W)$.
Subcase 1c : $i \neq k$, $p \neq r$. If $d(x, (a_1, v_1)) \neq d(z, (a_1, v_1))$, then we have $r(x|W) \neq r(z|W)$. Now, we assume that $d(x, (a_1, v_1)) = d(z, (a_1, v_1))$.
Therefore, we have $p = r + 1$ or $r = p + 1$. Without loss of generality, we assume $r = p + 1$. So, we obtain

$$\begin{aligned} d(x, (b_1, v_1)) &= d((b_1, v_1), (a_i, v_1)) + (p-1) = d((b_1, v_1), (a_i, v_1)) + (r-2) \\ &\neq d((b_1, v_1), (b_k, v_1)) + r - 2 < d((b_1, v_1), (b_k, v_1)) + (r-1) = d(z, (b_1, v_1)) \end{aligned}$$

Thus, we have $r(x|W) \neq r(z|W)$. Therefore, $nr(G) \leq t + 1$.
Case 2 : $s \geq 3$. By using similar reason to proof case 1, we have $r(x|W) \neq r(y|W) \neq r(z|W) \neq r(w|W)$. Therefore, $nr(G) \leq s + t - 2$.

Let W be a resolving set that is not contain an isolated vertex of G. We distinguish two cases for the lower bound, :

Case 1 : $s = 1, 2$.
Suppose $nr(G) \leq t$. By using Lemma 2.5, we have $t - 1 \leq nr(G) \leq t$.

For $nr(G) = t - 1$, By using Lemma 2.5, each row of the partition A of $K_{s,t}$ minus one row, contribute a vertex for W. Therefore, the induced subgraph W is a null graph (graph without edge). We get a contradiction since W is a non-isolated set.
For $nr(G) = t$, we distinguish two subcases
Subcase 1: Each row of the partition A of $K_{s,t}$ contributes a vertex for W. Thus, the induced subgraph W of $K_{s,t}$ contains an isolated vertex.
Subcase 2: There is a row of the partition A of $K_{s,t}$ that does not contribute to W. Thus, there is a row of the other patition that contributes a vertek for W, e.g. (a_i, v_k). In order that W does not contain an isolated vertex, all vetices in W must be in the k-th column. Note that there are two vertices (b_t, v_k) and (a_i, v_{k+1}) that have the same representation $(1, 2, \cdots, 2)$. Therefore, we have a contradiction since W is a resolving set. Thus, $nr(G) \geq t + 1$.
Case 2 : $s \geq 3$. By using Corollary 2.6, we have $nr(G) \geq s + t - 2$. □

The following, we give the definition of a friendship graph. The friendship graph is the graph obtained by taking k copies of the cycle graph C_3 with a vertex in common, denoted by D_3^k. Let $V(D_3^k) = \{a_0, b_1, c_1, b_2, c_2, \cdots, b_k, c_k\}$, and $E(D_3^k) = \{a_0b_i, a_0c_i, b_ic_i | 1 \leq i \leq k\}$ be vertex set and edge set of D_3^k respectively. We define $T_i = \{b_i, c_i\}$ where $i \in \{1, 2, \cdots, k\}$.

Lemma 2.8. *Let D_3^k be a friendship graph with $k \geq 2$, and P_n be a path with $n \geq 2$. Let W be a resolving set of $G = D_3^k \Box P_n$. Then for $1 \leq i \leq k$, $W \cap P_n(b_i) \neq \emptyset$ or $W \cap P_n(c_i) \neq \emptyset$.*

Proof. We assume that there are two vertices x and y in a partition T_i of D_3^k such that $W \cap P_n(x) = \emptyset$ and $W \cap P_n(y) \neq \emptyset$. Let $x = (b_i, v_p)$, $y = (c_i, v_q)$ be a distinct vertices in $(D_3^k \Box P_n) \backslash W$ and $(b_1, v_r) \in W$. For $p = q$, we obtain

$$d(x, (b_l, v_r)) = d((b_i, v_r), (b_l, v_r) + (p - 1) = d((c_i, v_r), (b_l, v_r) + (p - 1) = d(y, (b_l, v_r)$$

We get a contradiction since W is a resolving set of G. □

Corollary 2.9. *Let k and n be natural numbers with $k, n \geq 2$. Let $G = D_3^k \Box P_n$. Then $nr(G) \geq k + 1$*

Proof. This corollary is a direct consequence of Lemma 2.8 □

Theorem 2.10. *Let k and n be natural numbers with $k, n \geq 2$. Let $G = D_3^k \Box P_n$. Then $nr(G) = k + 1$.*

Proof. By using Corollary 2.9, we only need to show that $nr(G) \leq m - 1$. Let $V(D_3^k) = \{a_0, b_1, c_1, b_2, c_2, \cdots, b_k, c_k\}$, $E(D_3^k) = \{a_0b_i, a_0c_i, b_ic_i | 1 \leq i \leq k\}$ and $V(P_n) = \{v_1, v_2, \cdots, v_n\}$ and $E(P_n) = \{(v_i, v_{i+1}) | i = 1, \cdots, n - 1\}$.
We define a vertex set $W = \{(a_0, v_1), (b_i, v_1) | i = 1, 2, \cdots, k\}$. It is easy to check that the induced subgraph of $C_m \Box P_n$ by W does not contain an isolated vertex.

Let $x_1 = (b_i, v_p)$, $x_2 = (b_j, v_q)$, $x_3 = (c_l, v_r)$, $x_4 = (c_m, v_s)$ and $x_5 = (a_0, v_t)$ be a distinct vertices in $G \backslash W$. We show that for every case, $r(x_1|W) \neq r(x_2|W) \neq r(x_3|W) \neq r(x_4|W) \neq r(x_5|W)$. We distinguish three cases.

Case 1 : $p = q = r = s = t$. Note that for $1 \leq i \leq k$, $(b_i, v_1) \in W$. We obtain $d(x_2, (b_i, v_1)) = d(x_1, (b_i, v_1)) + 2 = d(x_4, (b_i, v_1))$, $d(x_3, (b_i, v_1)) = d(x_1, (b_i, v_1)) + 1 = d(x_5, (b_i, v_1))$, $d(x_3, (b_j, v_1)) = d(x_2, (b_j, v_1)) + 2$, $d(x_4, (b_j, v_1)) = d(x_2, (b_j, v_1)) + 1 = d(x_5, (b_j, v_1))$, $d(x_4, (b_i, v_1)) = d(x_1, (b_i, v_1)) + 2 > d(x_1, (b_i, v_1)) + 1 = d(x_3, (b_i, v_1))$, $d(x_5, (b_j, v_1)) = d(x_1, (b_j, v_1)) + 1 < d(x_1, (b_j, v_1)) + 2 = d(x_3, (b_j, v_1))$, and $d(x_5, (b_i, v_1)) =$

$d(x_1,(b_i,v_1))+1 < d(x_1,(b_i,v_1))+2 = d(x_4,(b_i,v_1))$
Therefore, $r(x_1|W) \neq r(x_2|W) \neq r(x_3|W) \neq r(x_4|W) \neq r(x_5|W)$.

Case 2 : $i = j$ or $l = m$. For $i = j$, $d((x_2),(b_i,v_1)) = (q-1) \neq (p-1) = d(x_1,(b_i,v_1))$. For $l = m$, $d((x_4),(b_i,v_1)) = d((c_m,v_1),(b_i,v_1)) + (s-1) \neq d(c_l,v_1) + (l-1) = d(x_3,(b_i,v_1))$. Therefore, we have $r(x_1|W) \neq r(x_2|W)$ and $r(x_3|W) \neq r(x_4|W)$.

Case 3 : $i \neq l$, $p \neq r$. If $d(x_1,(b_1,v_1)) \neq d(x_3,(a_1,v_1))$, then we have $r(x_1|W) \neq r(x_3|W)$. Now, we assume that $d(x_1,(b_1,v_1)) = d(x_2,(b_1,v_1))$. Therefore, we have $p = r+1$ or $r = p+1$. Without loss of generality, we assume $p = r+1$. So, we obtain

$$\begin{aligned} d(x_1,(a_0,v_1)) &= d((b_1,v_1),(a_0,v_1)) + (p-1) \\ &= d((b_1,v_1),(a_0,v_1)) + r = d((c_l,v_1),(a_0,v_1)) + r \\ &> d((c_l,v_1),(a_0,v_1)) + (r-1) = d(x_3,(a_0,v_1)) \end{aligned}$$

Thus, we have $r(x_1|W) \neq r(x_3|W)$. All the cases imply that $nr(G) \leq m-1$. □

3. Conclusion

In this paper, we have studied non-isolated resolving set of the Cartesian product of a path and a certain graph, namely a cycle, a complete graph, a complete bipartite graph, and a frienship graph. We give an exact value of non isolated resolving number of them.

For future work, we provide an intresting question as follow.

Problem 3.1. *For $n \geq 2$. Let $H = G\Box P_n$ where G is a connected graph. Determine the relation between $nr(H)$ and $nr(G)$.*

Acknowlegment

The authors are also thankful to the anonymous referee for some comments that helped to improve the presentation of the manuscript.

References

[1] Bača M, Baskoro E T, Salman A N M, Saputro S W and Simanjuntak R 2011 The metric dimension of regular bipartite graphs *Bull. Math. Soc. Sci. Math. Roumanie* **54(102) No 1** 15-28
[2] Caceres J, Hernando C, Mora M, Puertas M L, Pelayo I M, and Wood D R 2005 On the metric dimension of some families graphs *Electronic Notes in Discrete Math.* **22** 129-133
[3] Chartrand G, Eroh L, Johnson M, and Oellermann O R 2000 Resolvability in graphs and the metric dimension of a graph *Discrete Appl. Math.* **105** 99-113
[4] Chitra P J B and Arumugam S 2015 Resolving sets without isolated vertices *Procedia Comput. Sci.* **74** 38-42.
[5] Harary F and Melter R A 1976 On the metric dimension of a graph *Ars. Combin.* **2** 191-195.
[6] Iswadi H, Baskoro E T, Salman A N M and Simanjuntak R 2010 The metric dimension of amalgamation of cycles, *Far East J. Math Sci.* **41 (1)** 19-31.
[7] Iswadi H, Baskoro E T, Salman A N M and Simanjuntak R 2010 The resolving graph of amalgamation of cycles, *Utilitas Mathematica* **83** 121-132.
[8] Iswadi H, Baskoro E T and Simanjuntak R 2011 On the metric dimension of corona product of graphs *Far East J. Math. Sci.* **52;2** 155-170.
[9] Iswadi H, Baskoro E T, Simanjuntak R and Salman A N M 2008 The metric dimension of graphs with pendant edges, *J. Combin. Math. Combin Comput.* **65** 139-145.
[10] Saputro S W, Baskoro E T, Salman A N M and Suprijanto D 2009 The metric dimension of a complete n-partite and its Cartesian product with a path *J. Combin. Math. Combin Comput.* **7** 283-293
[11] Saputro S W, Simanjuntak R, Uttunggadewa S, Assiyatun H, Baskoro E T, Salman A N M and Bača M 2013 The metric dimension of the lexicographic product of graphs *Discrete Math.* **213** 1045-1051
[12] Slater P J 1975 Leaves of trees *Congr. Numer.* **14** 549-559.
[13] Yero I G, Kuziak D and Rodrigues-Velazquez J A 2011 On the metric dimension of corona product graphs *Comput. Math. Appl.* **61** 2793-2799

IOP Conf. Series: Journal of Physics: Conf. Series **1008** (2018) 012046 doi:10.1088/1742-6596/1008/1/012046

On total irregularity strength of caterpillar graphs with two leaves on each internal vertex

I Rosyida[1,2], Widodo[2], D Indriati[3]

[1]Department of Mathematics, Universitas Negeri Semarang, Indonesia
[2]Department of Mathematics, Universitas Gadjah Mada, Yogyakarta, Indonesia
[3]Department of Mathematics, Universitas Sebelas Maret, Surakarta, Indonesia

E-mail: isnaini@mail.unnes.ac.id, widodo_mathugm@yahoo.com, diari_indri@yahoo.co.id

Abstract. Let $G(V,E)$ be a graph. A function f from $V(G) \cup E(G)$ to the set $\{1,2,\ldots,k\}$ is said to be a totally irregular total k-labeling of G if the weights of any two different vertices x and y in $V(G)$ satisfy $w_f(x) \neq w_f(y)$ where the weight $w_f(x)$ is the sum of label of x and labels of all edges incident to x, and the weights of any two different edges ux and vy in $E(G)$ satisfy $w_f(ux) \neq w_f(vy)$ where the weight $w_f(ux)$ is obtained from the sum of: label of x, label of u and label of edge ux. The total irregularity strength of the graph G, denoted by $ts(G)$, is the minimum number k for which G has a totally irregular total k-labeling. In this paper, we focus on a caterpillar graph with two leaves on each internal vertex $T_{2n+p,q}$, where n is the number of leaves on each end vertex of the central path, p is the number of leaves connected to internal vertices, q is the number of vertices of the central path, $p \geq 4$ and $q \geq 4$. Firstly, we do some experiments for constructing a formula for totally irregular total k-labeling of the caterpillar graph $T_{2n+p,q}$. Secondly, we determine the minimum number k which is ts of the caterpillar graph $T_{2n+p,q}$. We obtain that $ts(T_{2(q-1)+p,q}) = \left\lceil \frac{2(q-1)+p+1}{2} \right\rceil$ and $ts(T_{2n+p,q}) = \left\lceil \frac{2n+p+1}{2} \right\rceil$.

1. Introduction

Let $G = (V,E)$ be a simple graph. A labeling of G is a mapping from a set of graph elements into a set of integers, that are called labels. If the domain of the mapping is either a vertex set, or an edge set, or a union of vertex and edge sets, then the labeling is called *vertex labeling, edge labeling*, or *total labeling*, respectively (Wallis [19]; Marr and Wallis [10]).

Bača *et al.* [3] introduced a vertex irregular total k-labeling of G, that is a function $f : V \cup E \to \{1,2,\ldots,k\}$ such that the weights $wt_f(x) \neq wt_f(y)$ for every two different vertices x and y in $V(G)$, where $wt_f(x) = f(x) + \sum_{xy \in E} f(xy)$ and the minimum number k is called a vertex irregularity strength of the graph G, denoted by tvs(G). Meanwhile an edge irregular total k-labeling of G is defined as a function $f : V \cup E \to \{1,2,\ldots,k\}$ such that the weights $wt_f(uv) \neq wt_f(xy)$ for every two different edges uv and xy in $E(G)$, where $wt_f(uv) = f(u) + f(v) + \sum_{uv \in E} f(uv)$ and the minimum number k is called an edge irregularity strength of G, denoted by tes(G). Further, the function f is said to be a totally irregular total k-labeling of G if the weights of any two different vertices x and y in $V(G)$ satisfy $w_f(x) \neq w_f(y)$ and the weights of any two different edges uv and xy in $E(G)$ satisfy $w_f(uv) \neq w_f(xy)$. The total irregularity strength of the graph G, denoted by $ts(G)$, is the minimum number k for which G has a totally irregular total k-labeling.

IOP Conf. Series: Journal of Physics: Conf. Series **1008** (2018) 012046 doi:10.1088/1742-6596/1008/1/012046

Bača *et al.* [3] gave the lower and upper bound of the total vertex irregularity strength for any graph G with p vertices, q edges, the minimum degree δ, and the maximum degree $\triangle$ as follows:

$$\left\lceil \frac{p+\delta}{\triangle+1} \right\rceil \leq tvs(G) \leq p + \triangle - 2\delta + 1$$

Anholcer *et al.* [2] improved the bound given by Bača *et al.* as follows:

$$tvs(G) \leq 3\left\lceil \frac{p}{\delta} \right\rceil + 1.$$

Nurdin [14] obtained the total vertex irregularity strength of a tree T with n leaves and no vertex of degree 2 as follows:

$$tvs(T) = \left\lceil \frac{n+1}{2} \right\rceil.$$

Further, Nurdin [15] investigated the total vertex irregularity strength of a Caterpillar $T_{n,m}$ with n leaves and m vertices on the central path as follows: $tvs(T_{n,m}) = \left\lceil \frac{n+1}{2} \right\rceil$. Indriati et al. [5] found the total vertex irregularity strength of generalized helm graphs and prisms with outer pendant edges.

Ivanco and Jendrol et al. [9] investigated the total edge irregularity strength of a tree T, that is

$$tes(T) = \max\left\{ \left\lceil \frac{|E(T)|+2}{3} \right\rceil, \left\lceil \frac{\triangle(T)+1}{2} \right\rceil \right\}.$$

Hinding et al. [4] determined the total edge irregularity strength of subdivision of star. Miskuf and Jendrol [12] found a total edge irregularity strength (tes) of grid graph. Nurdin et al. [13] obtained the total edge-irregularity strengths of the corona product of paths with some graphs. Siddiqui [18] has obtained total edge irregularity strength of subdivision of star. Ahmad et al. [1] has done on total edge irregularity strength of strong product of two paths.

In the problem of determining the exact value of total irregularity strength of a graph, Marzuki et al. [11] gave an observation on the lower bound for the total irregularity strength of a graph G as follows:

$$ts(G) \geq \max\{tvs(G), tes(G)\}.$$

In the same paper, they determined a total irregularity strength of cycles and paths. After that, many researchers have invented the total irregularity strength of graphs. Such as, Indriati et al.[6] invented the total irregularity strength of double star $S_{n,n}$ and caterpillar $S_{n,2,n}$ that is a caterpillar with inserted vertex of degree 2. Ramdani et al. [16] investigated a total irregularity strength of some cartesian product graphs. Also, Ramdani et al.[17] obtained a total irregularity strength of gear, fungus and disjoint union of star. Further, Indriati et al.[7] found the total irregularity strength of star graph, double star $S_{n,n}$ and caterpillar $S_{n,2,2,m}$. Recently, Indriati et al.[8] has done on the total irregularity strength of $S_{n,2,m}$.

In this paper, we focus on caterpillar graphs that have two leaves on each internal vertex. We denote the caterpillar graphs as $T_{2n+p,q}$ where n is the number of leaves on each end vertex of the central path, p is the number of leaves connected to internal vertices, q is the number of vertices of the central path, $p \geq 2(q-2)$, and $q \geq 4$. We construct a totally irregular total k-labeling and investigate the total irregularity strength of the caterpillar graphs $T_{2(q-1)+p,q}$ and $T_{2n+p,q}$.

2. Main Results

In this paper, we have found the total irregularity strength of some caterpillar Graphs with two leaves on each internal vertex as stated in Theorem 1 and Theorem 2.

Theorem 1 *Let $T_{2(q-1)+p,q}$ be a caterpillar graph with $2(q-1)$ leaves connected to the two end vertices of the central path, $p = 2(q-2)$ is the number of leaves connected to internal vertices, q is the number of vertices of the central path, and $q \geq 4$. The total irregularity strength of $T_{2n+p,q}$ is*

$$ts(T_{2(q-1)+p,q}) = \left\lceil \frac{2(q-1)+p+1}{2} \right\rceil.$$

PROOF. It is obvious that the caterpillar graph $T_{2(q-1)+p,q}$ has $2(q-1)+p$ leaves. Assume that the vertex set of the caterpillar graph is $V = \{a_i^k, a_j^k, b_k\}$ where a_i^k, a_j^k are the leaves connected to vertices b_k for $i = 1, 2, \ldots, 2(q-1)+p, j = 1, 2, k = 1, 2, \ldots, q$ and b_k are the vertices in the central path of caterpillar. The edge set of the caterpillar is $E = \{a_i^k b_k, a_j^k b_k, b_k b_l\}$ for $k, l = 1, 2, \ldots, p$.

The total vertex irregularity strength of caterpillar $T_{2(q-1)+p,q}$ can be found by using a theorem given in [15], as follows:

$$tvs(T_{2(q-1)+p,q}) = \left\lceil \frac{2(q-1)+p+1}{2} \right\rceil.$$

Based on the total edge irregularity strength of a tree given by Ivanco and Jendrol [9], we have:

$$tes(T) = \max\left\{ \left\lceil \frac{|E(T_{2(q-1)+p,q})|+2}{3} \right\rceil, \left\lceil \frac{\triangle(T_{2(q-1)+p,q})+1}{2} \right\rceil \right\}$$

It follows from $|E(T_{2n+p,q})| = 2(q-1)+p+(q-1) = 3(q-1)+p$ and $\triangle(T_{2(q-1)+p,q}) = q-1+1 = q$, we obtain

$$tes(T) = \max\left\{ \left\lceil \frac{2(q-1)+p+(q-1)+2}{3} \right\rceil, \left\lceil \frac{q+1}{2} \right\rceil \right\} = \frac{3(q-1)+p+2}{3}$$

Further, by using an observation from Marzuki et al. [11], the lower bound for the total irregularity strength of caterpillar $T_{2n+p,q}$ as follows:

$$ts(T_{2n+p,q}) \geq \max\{tvs(T_{2n+p,q}), tes(T_{2n+p,q})\} = \left\lceil \frac{2(q-1)+p+1}{2} \right\rceil.$$

Further, we determine the exact value of the total irregularity strength of caterpillar $T_{2(q-1)+p,q}$ by constructing the totally irregular total k-labeling on vertex set and edge set: $f : V \cup E \to \{1, 2, \ldots, k\}$ where $k = \left\lceil \frac{2(q-1)+p+1}{2} \right\rceil$ as follows:

$$f(a_i^1) = \begin{cases} q, & \text{if } i = 1,2 \\ q+(i-2), & \text{if } 3 \leq i \leq q-2 \end{cases}$$

The label of vertex $f(a_{q-1}^1) = f(a_{q-2}^1) = 2q-4$.

$$f(a_i^k) = \left\lceil \frac{2(q-1)+p+1}{2} \right\rceil, i = 1,2 \text{ and } 2 \leq k \leq q-1.$$

$$f(a_i^q) = \begin{cases} 1, & \text{if } i = 1,2 \\ i-1, & \text{if } 3 \leq i \leq q-1 \end{cases}$$

If q is odd, then the labels of vertices b_k as follows:

$$f(b_k) = \begin{cases} 1, & \text{if } k = 1, q \\ 2k-2, & \text{if } 2 \le k \le q-1 \end{cases}$$

If q is even, then the labels of b_k as follows:

$$f(b_k) = \begin{cases} 1, & \text{if } k = 1, q \\ 2k-2, & \text{if } 2 \le k \le q-2 \\ 2k, & \text{if } k = q-1 \end{cases}$$

We define the labels of edges as follows:

$$f(a_i^1 b_1) = \begin{cases} 1, & \text{if } i = 1 \\ 2, & \text{if } 2 \le i \le (q-1)-1 \\ 3, & \text{if } i = q-1 \end{cases}$$

$$f(a_i^q b_q) = \begin{cases} 1, & \text{if } i = 1 \\ 2, & \text{if } 2 \le i \le (q-1) \end{cases}$$

$$f(a_i^k b_k) = i + 2k - 3, \text{ for } 2 \le k \le q-1 \text{ and } i = 1, 2.$$

Under the labeling f, the weights of vertices of the caterpillar $T_{2(q-1)+p,q}$ as follows:

$$w(a_i^k) = \begin{cases} \left\lceil \frac{2(q-1)+p+1}{2} \right\rceil + (i - q + 2), & \text{if } k = 1, 1 \le i \le (q-1) \\ i+1, & \text{if } k = q, 1 \le i \le (q-1) \\ (i + 2k - 3) + \left\lceil \frac{2(q-1)+p+1}{2} \right\rceil, & \text{if } 2 \le k \le q-1, i = 1, 2 \end{cases}$$

If q is odd, the weights of vertices b_k as follows:

$$w(b_k) = \begin{cases} (2q-1) + \left\lceil \frac{2(q-1)+p+1}{2} \right\rceil, & \text{if } k = 1 \\ 2\left\lceil \frac{2(q-1)+p+1}{2} \right\rceil, & \text{if } k = q \\ (6k-5) + 2\left\lceil \frac{2(q-1)+p+1}{2} \right\rceil, & \text{if } 2 \le k \le q-1 \end{cases}$$

If q is even, the weights of b_k as follows:

$$w(b_k) = \begin{cases} (2q-1) + \left\lceil \frac{2(q-1)+p+1}{2} \right\rceil, & \text{if } k = 1 \\ 2\left\lceil \frac{2(q-1)+p+1}{2} \right\rceil, & \text{if } k = q \\ (6k-5) + 2\left\lceil \frac{2(q-1)+p+1}{2} \right\rceil, & \text{if } 2 \le k \le q-2 \\ (6k-3) + 2\left\lceil \frac{2(q-1)+p+1}{2} \right\rceil, & \text{if } k = q-1 \end{cases}$$

The weights of edges $a_i^k b_k$ and $b_k b_{k+1}$ of the caterpillar $T_{2(q-1)+p,q}$ under the labeling f as follows:

Case 1: For q is odd

$$w(a_i^k b_k) = \begin{cases} \frac{1}{2}\left\lceil \frac{2(q-1)+p+1}{2} \right\rceil + 3, & \text{if } i = 1, k = 1 \\ \frac{1}{2}\left\lceil \frac{2(q-1)+p+1}{2} \right\rceil + (i+2), & \text{if } 2 \le i \le (q-1), k = 1 \\ \left\lceil \frac{2(q-1)+p+1}{2} \right\rceil + (4k+i-5), & \text{if } 2 \le k \le q-1,\ i = 1,2 \end{cases}$$

$$w(b_k b_{k+1}) = \begin{cases} \left\lceil \frac{2(q-1)+p+1}{2} \right\rceil + (4k-2), & \text{if } 2 \le k \le q-2 \\ 2\left\lceil \frac{2(q-1)+p+1}{2} \right\rceil - 1, & \text{if } k = q-1 \end{cases}$$

Case 2: For q is even

$$w(a_i^k b_k) = \begin{cases} \frac{1}{2}\left\lceil \frac{2(q-1)+p+1}{2} \right\rceil + 3, & \text{if } i = 1, k = 1 \\ \frac{1}{2}\left\lceil \frac{2(q-1)+p+1}{2} \right\rceil + (i+2), & \text{if } 2 \le i \le (q-1), k = 1 \\ \left\lceil \frac{2(q-1)+p+1}{2} \right\rceil + (4k+i-5), & \text{if } 2 \le k \le q-2,\ i = 1,2 \\ \left\lceil \frac{2(q-1)+p+1}{2} \right\rceil + (4k+i-3), & \text{if } k = q-1,\ i = 1,2 \end{cases}$$

$$w(b_k b_{k+1}) = \begin{cases} \left\lceil \frac{2(q-1)+p+1}{2} \right\rceil + (4k-2), & \text{if } 2 \le k \le q-2 \\ 2\left\lceil \frac{2(q-1)+p+1}{2} \right\rceil + 1, & \text{if } k = q-1 \end{cases}$$

It is proved that all labels of vertices and edges are at most $\left\lceil \frac{2(q-1)+p+1}{2} \right\rceil$ and the edge and vertex weights under the the labeling f are distinct. Thus, the total irregularity strength of the caterpillar is $ts(T_{2(q-1)+p,q}) = \left\lceil \frac{2(q-1)+p+1}{2} \right\rceil$. □

An example of the construction of the totally irregular total 14-labeling of $T_{2.7+12,8}$ is given in Figure 1. The red colored number represent the weight of a vertex or the weight of an edge.

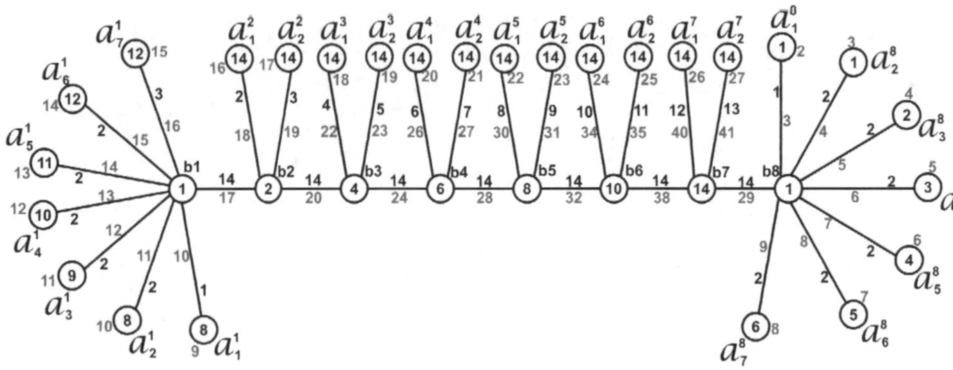

Figure 1. The totally irregular total 14-labeling of $T_{2.7+12,8}$.

Theorem 2 *Let $T_{2n+p,q}$ be a caterpillar graph as stated in Theorem 1, where $n \geq 3q-7$, and $q \geq 4$. The total irregularity strength of $T_{2n+p,q}$ is*

$$ts(T_{2n+p,q}) = \left\lceil \frac{2n+p+1}{2} \right\rceil .$$

PROOF. It is obvious that the caterpillar $T_{2n+p,q}$ has $2n+p$ leaves. Let V be the vertex set of the caterpillar $T_{2n+p,q}$ where $V = \{a_i^k, a_j^k, b_k\}$ where a_i^k, a_j^k are the leaves connected to vertices b_k for $i = 1, 2, \ldots, 2n, j = 1, 2, k = 1, 2, \ldots, p$ and b_k are the vertices in the central path of caterpillar. Let E be the edge set of the caterpillar where $E = \{a_i^k b_k, a_j^k b_k, b_k b_l\}$ for $k, l = 1, 2, \ldots, p$.

Based on observation from Marzuki et al. [11], the lower bound for the total irregularity strength of caterpillar $T_{2n+p,q}$ as follows:

$$ts(T_{2n+p,q}) \geq \max\{tvs(T_{2n+p,q}), tes(T_{2n+p,q})\} = \left\lceil \frac{2n+p+1}{2} \right\rceil .$$

Further, we determine the total irregularity strength of caterpillar $T_{2n+p,q}$ for $n \geq 3q-7$ by the totally irregular total k-labeling f on the vertex set and edge set $f : V \cup E \to \{1, 2, \ldots, k\}$ where $k = \left\lceil \frac{2n+p+1}{2} \right\rceil$ that is defined as follows:

$$f(a_i^k) = \begin{cases} \left\lceil \frac{2n+p+1}{2} \right\rceil + (i-(q-1)), & \text{if } 1 \leq i \leq q-1, k=1 \\ \left\lceil \frac{2n+p+1}{2} \right\rceil, & \text{if } q \leq i \leq n, k=1 \\ \left\lceil \frac{2n+p+1}{2} \right\rceil, & \text{if } i = 1, 2, \ \ 2 \leq k \leq q-1 \end{cases}$$

$$f(a_i^q) = \begin{cases} i, & \text{if } 1 \leq i \leq n-q+2 \\ i-1, & \text{if } n-q+3 \leq i \leq n, k=1 \end{cases}$$

$$f(b_k) = \begin{cases} 2k+(n-q)-1, & \text{if } n-q \ \text{ is odd, } 2 \leq k \leq q-1 \\ 2k+(n-q), & \text{if } n-q \ \text{ is even, } 2 \leq k \leq q-1 \\ 1, & \text{if } k=1 \text{ and } k=q \end{cases}$$

$$f(a_i^k b_k) = \begin{cases} 1, & \text{if } 1 \le i \le q-1, k=1 \\ i-q+2, & \text{if } q \le i \le n, k=1 \\ 1, & \text{if } 1 \le i \le n-q+2, k=q \\ 2, & \text{if } n-q+3 \le i \le n, k=q \\ (n-q)+(i+2(k-1)), & \text{if } i=1,2 \text{ and } 2 \le k \le q-2 \\ (n-q)+i+p, & \text{if } i=1,2 \text{ and } k=q-1 \end{cases}$$

$$f(b_k b_{k+1}) = \left\lceil \frac{2n+p+1}{2} \right\rceil \quad \text{for } 1 \le k \le q-1$$

We obtain the weights of vertices of the caterpillar $T_{2n+p,q}$ under the labeling f as follows:

$$w(a_i^k) = \begin{cases} \left\lceil \frac{2n+p+1}{2} \right\rceil + (i-q+2), & \text{if } k=1,\ 1 \le i \le n \\ i+1, & \text{if } k=q,\ 1 \le i \le n \\ (n-q)+i+2(k-1)+\left\lceil \frac{2n+p+1}{2} \right\rceil, & \text{if } 2 \le k \le q-2,\ i=1,2 \\ (n-q)+i+p+\left\lceil \frac{2n+p+1}{2} \right\rceil, & \text{if } k=q-1,\ i=1,2 \end{cases}$$

$$w(b_k) = \begin{cases} 2\left\lceil \frac{2n+p+1}{2} \right\rceil, & \text{if } k=q \\ 2\left\lceil \frac{2n+p+1}{2} \right\rceil + 3(n-q)+6k-1, & \text{if } 2 \le k \le q-2 \text{ and } n-q \text{ is even} \\ 2\left\lceil \frac{2n+p+1}{2} \right\rceil + 3(n-q)+6k-2, & \text{if } 2 \le k \le q-2 \text{ and } n-q \text{ is odd} \end{cases}$$

$$w(b_k) = \begin{cases} 2\left\lceil \frac{2n+p+1}{2} \right\rceil + 2k+3(n-q)+3+2p, & \text{if } k=q-1 \text{ and } n-q \text{ is even} \\ 2\left\lceil \frac{2n+p+1}{2} \right\rceil + 2k+3(n-q)+2+2p, & \text{if } k=q-1 \text{ and } n-q \text{ is odd} \end{cases}$$

Furthermore, the edge weights of the caterpillar under the labeling f as follows:

$$w(a_i^k b_k) = \begin{cases} i+2, & \text{if } 1 \le i \le n, k=q \\ n+i+2, & \text{if } 1 \le i \le n, k=1 \end{cases}$$

If $(n-q)$ is even, then the weights:

$$w(a_i^k b_k) = \begin{cases} \left\lceil \frac{2n+p+1}{2} \right\rceil + 2(n-q)+i+4k-2, & \text{if } i=1,2, 2 \le k \le q-2 \\ \left\lceil \frac{2n+p+1}{2} \right\rceil + 2(n-q)+i+2k+p, & \text{if } i=1,2, k=q-1 \end{cases}$$

If $(n-q)$ is odd, then we obtain the weights:

$$w(a_i^k b_k) = \begin{cases} \left\lceil \frac{2n+p+1}{2} \right\rceil + 2(n-q)+i+4k-3, & \text{if } i=1,2, 2 \le k \le q-2 \\ \left\lceil \frac{2n+p+1}{2} \right\rceil + 2(n-q)+i+2k-1+p, & \text{if } i=1,2, k=q-1 \end{cases}$$

IOP Conf. Series: Journal of Physics: Conf. Series **1008** (2018) 012046 doi:10.1088/1742-6596/1008/1/012046

$$w(b_1b_2) = \begin{cases} 5 + \left\lceil \frac{2n+p+1}{2} \right\rceil + (n-q), & \text{if } (n-q) \text{ is even} \\ 4 + \left\lceil \frac{2n+p+1}{2} \right\rceil + (n-q), & \text{if } (n-q) \text{ is odd} \end{cases}$$

$$w(b_kb_{k+1}) = \begin{cases} \left\lceil \frac{2n+p+1}{2} \right\rceil + 4k + 2(n-q), & \text{if } (n-q) \text{ is odd} \\ \left\lceil \frac{2n+p+1}{2} \right\rceil + 4k + 2(n-q) + 2, & \text{if } (n-q) \text{ is even} \end{cases}$$

$$w(b_kb_{k+1}) = \begin{cases} \left\lceil \frac{2n+p+1}{2} \right\rceil + 2k + (n-q), & \text{if } (n-q) \text{ is odd} \\ \left\lceil \frac{2n+p+1}{2} \right\rceil + 2k + (n-q) + 1, & \text{if } (n-q) \text{ is even} \end{cases}$$

It is clear that all labels of vertices and edges are at most $\left\lceil \frac{2n+p+1}{2} \right\rceil$ and the edge and vertex weights under the the labeling f are distinct. Thus, the total irregularity strength of the caterpillar is $ts(T_{2n+p,q}) = \left\lceil \frac{2n+p+1}{2} \right\rceil$. □

An illustration of the totally irregular total 16-labeling of $T_{2.11+8,6}$ can be seen on Figure 2. A number with the red color indicate the weight of a vertex or the weight of an edge.

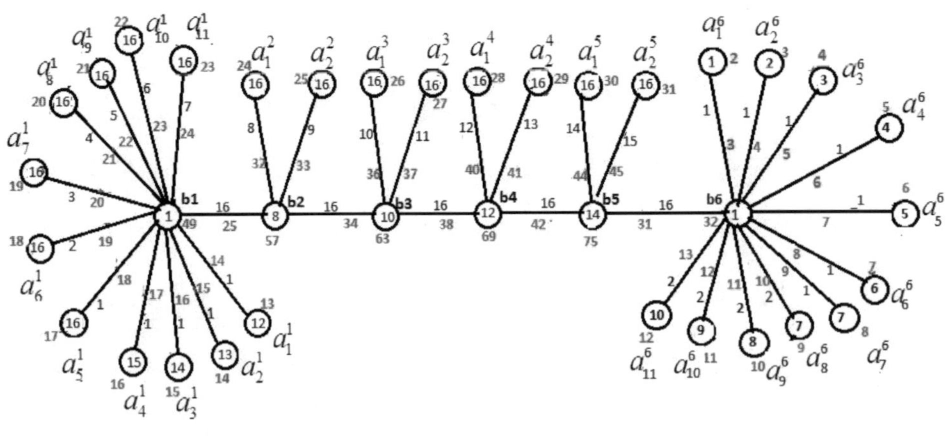

Figure 2. The totally irregular total 16-labeling of $T_{2.11+8,6}$.

3. Concluding Remarks

In this paper, we have found the exact value of total irregularity strength of caterpillar graphs that contain two leaves on each internal vertex and $2n$ leaves connected to the end vertices of central path in the caterpillar. We have obtained that $tvs(T_{2(q-1)+p,q}) = \left\lceil \frac{2(q-1)+p+1}{2} \right\rceil$ and $tvs(T_{2n+p,q}) = \left\lceil \frac{2n+p+1}{2} \right\rceil$ for $p \geq 4$ and $q \geq 4$.

An **Open problem** as the direction for further research is given as follows: The total irregularity strength of caterpillar graphs $T_{2n+p,q}$ where $q-1 < n \leq 2q-8$ and $n \geq 4$ and the total irregularity strength of caterpillar graphs $T_{2n+p,q}$ where $p = \alpha(q-2)$ is even for $\alpha = 2, 4, 6, \ldots$. Further, we can investigate the total irregularity strength of caterpillar graphs $T_{m+n+p,q}$, that is a caterpillar graph that have different number of leaves on the end vertices of central path in the caterpillar graph, $p = \alpha(q-2)$ for constant $\alpha = 2, 4, 6, \ldots$ is even. Another, the total irregularity strength of caterpillar graphs $T_{m+n+p,q}$ where $p = \alpha(q-2)$ is odd for constant $\alpha = 1, 3, 5, \ldots$.

4. Acknowledgments

This work was supported by the Fundamental Research Project from the Ministry of Research Technology and Higher Education of Indonesia, No.2238/UN1.P.III/DIT-LIT/LT/2017.

5. References

[1] Ahmad A, Baca M, Bashir Y and Siddiqui M K 2012 Total Edge Irregularity Strength of Strong Product of Two Paths *Ars Combin.* **106** 449–459.
[2] Anholcer M, Kalkowski M and Przybylo J 2009 A New Upper Bound for the Total Vertex Irregularity Strength of Graphs *Discrete Mathematics* **309** 6316–6317.
[3] Bača M, Jendroľ S, Miller M and Ryan J 2007 On Irregular Total Labelling *Discrete Math* **307** 1378–1388.
[4] Hinding N, Suardi N and Basir H 2015 Total Edge Irregularity Strength of Subdivision of Star *Journal of Discrete Mathematical Sciences and Cryptography* **18(6)** 869–875
[5] Indriati D, Widodo, Wijayanti I E and Sugeng K A 2016 The Total Vertex Irregularity Strength of Generalized Helm Graphs and Prisms with Outer Pendant Edges *Australasian Journal of Combinatorics* **65(1)** 14–26.
[6] Indriati D, Widodo, Wijayanti I E and Sugeng K A 2015 On Total Irregularity Strength of Double-Star and Related Graphs *Procedia Computer Science* **74** 118-123
[7] Indriati D, Widodo, Wijayanti I E and Sugeng K A 2016 On Total Irregularity Strength of Star Graphs, Double-Stars and Caterpillar *AIP: Conference Proceedings* **1707** 1-6
[8] Indriati D, Widodo, Wijayanti I E and Sugeng K A 2017 The Construction of Labeling and Total Irregularity Strength of Specified Caterpillar Graph, *Journal of Physics: Conf. Series* **855** 1-7
[9] Ivančo J and Jendroľ S 2006 Total Edge Irregularity Strength of Trees *Discussiones Math. Graph Theory* **26** 449–456.
[10] Marr A M and Wallis W D 2013 *Magic Graphs* (New York: Birkhauser-Springer).
[11] Marzuki C C, Salman, A N M and Miller M 2013 On the Total Irregularity Strength of Cycles and Paths *Far East Journal of Mathematical Sciences* **82(1)** 1–21.
[12] Miskuf J and Jendrol S 2007 On Total Edge Irregularity Strength of the Grids *Tatra Mt. Math. Publ.* **36** 147–151.
[13] Nurdin, Salman A N M and Baskoro E T 2008 The Total Edge-Irregular Strengths of the Corona Product of Paths with Some Graphs *J. Combin. Math. Combin. Comput.* **65** 163–175.
[14] Nurdin, Baskoro E T, Salman A N M and Gaos N N 2010 On the Total Vertex Irregularity Strength of Trees *Discrete Math* **310** 3043–3048.
[15] Nurdin, Zakir M, Salman and Firman 2013 Vertex Irregular Labeling and Vertex Irregular Total Labeling on Caterpillar Graph *International Journal of Applied Mathematics and Statistics* **40(10)** 99-105
[16] Ramdani R and Salman A N M 2013 On the Total Irregularity Strength of Some Cartesian Product Graphs *AKCE International J. Graphs Combin* **10(2)** 199–209.
[17] Ramdani R, Salman, A N M, Assiyatun H, Fenovckov A S and Baca M 2015 On the Total Irregularity Strength of Three Families of Graphs *Mathematics in Computer Science* **9(2)** 229–237.
[18] Siddiqui M K 2012 On Tes of Subdivision of Star *Int. J. of Math and Soft Comput.* **2(1)** 75–82.
[19] Wallis W D 2001 *Magic Graphs* (Boston: Birkhauser).

IOP Conf. Series: Journal of Physics: Conf. Series **1008** (2018) 012047 doi:10.1088/1742-6596/1008/1/012047

Super (a,d)-H-antimagic covering of möbius ladder graph

Novia Indriyani and Titin Sri Martini

Mathematics Department of Mathematics and Natural Sciences Faculty, Sebelas Maret University, Surakarta, Indonesia

E-mail: nov.indriyani8@gmail.com, titinsmartini@gmail.com

Abstract. Let $G = (V(G), E(G))$ be a simple graph. Let H-covering of G is a subgraph $H_1, H_2, \ldots, H_j$ with every edge in G is contained in at least one graph H_i for $1 \leq i \leq j$. If every H_i is isomorphic, then G admits an H-covering. Furthermore, an (a,d)-H-antimagic covering if there bijective function $\xi : V(G) \cup E(G) \rightarrow \{1, 2, 3, \ldots, \mid V(G) \mid + \mid E(G) \mid\}$. The H'-weights for all subgraphs H' isomorphic to H $\omega(H') = \sum_{v \in V(H')} \xi(v) + \sum_{e \in E(H')} \xi(e)$. The weights of subgraphs constitutes an arithmatic progression $\{a, a+d, \ldots, a+(t-1)d\}$ where a and d are positive integers and t is the number of subgraphs G isomorphic to H. If $\xi(V(G))$ $= \{1,2,\ldots,\mid V(G) \mid\}$ then ξ is called super (a,d)-H-antimagic covering. The research provides super (a,d)-H-antimagic covering with $d = \{1,3\}$ of Möbius ladder graph M_n for $n \geq 5$ and n is odd.

1. Introduction

Gallian [1] defined a graph labeling which is an assigment of integers on vertices $(V(G))$, or edges $(E(G))$, or both, subject to certain condition. Magic labeling was first introduced by Sedláčk in 1963 (Gallian [1]). In 1970, Kotzig and Rosa [5] was introduced edge-magic total labeling. Furthermore, Gutiérrez and Lladó [2] developed the study by defined a magic covering.

Let $G = (V, E)$ be a simple and finite graph. An edge-covering of G is a family of different subgraphs $H_1, H_2, \ldots, H_j$ such that every edge in G $(E(G))$ belongs to at least one of the subgraphs H_i for $1 \leq i \leq j$. If G admits an H-covering, then every H_j is isomorphic [1].

Suppose that G admits an H-covering. A bijective function $\xi : V \cup E \rightarrow \{1, 2, 3, \ldots, |V|+|E|\}$. Then ξ is an H-magic labeling of G if there is a positive integer m such that each subgraph H' isomorphic to H satisfies $f(H') = \sum_{v \in V(H')} \xi(v) + \sum_{e \in E(H')} \xi(e) = m(f)$, where $m(f)$ is a magic sum. In this case we say that the graph G is H-magic labeling. When $\xi(V(G)) = \{1, 2, 3, \ldots, |V(G)|\}$, then G is H-supermagic labeling. Lladó and Moragas [6] studied cycle-magic labeling of a wheel graph, a prism graph, a book graph, and a windmill graph. Ngurah et al. [8] contructed cycle-supermagic labeling on a book graph, a triangular ladder graph, and grids $P_m \times P_n$ for $m \geq 3$ and $n = 3, 4, 5$.

Then, an (a, d)-H-antimagic total labeling of G was introduced by Inayah et al. [3] is defined as a bijective function $\xi : V(G) \cup E(G) \rightarrow \{1, 2, 3, \ldots, |V(G)|+|E(G)|\}$ such that for all subgraphs H' isomorphic to H, the set of H'-weights $omega(H') = \sum_{v \in V(H')} \xi(v) + \sum_{e \in E(H')} \xi(e)$. The weights of the subgraphs constitute an arithmetic progression $\{a, a+d, a+2d, \ldots, a+(t-1)d\}$ where a and d are positive integers and t is the number of subgraphs of G isomorphic to H.

In this case, we say that G is (a, d)-H-antimagic labeling. Furthermore, ξ is called a super (a, d)-H-antimagic total labeling and G is called super (a, d)-H-antimagic when $\xi(V(G)) = \{1, 2, 3, \ldots, |V(G)|\}$. In [3], Inayah et al. studied some properties of (a, d)-H-antimagic total labelings of fans. In [1], Susilowati et al. proved that grid graph $P_n \times P_2$ is super (a, d)-C_n-antimagic total labeling for $n \in \{4, 6, 8\}$.

This research attempts to study super (a, d)-H-antimagic total labelings of Möbius ladder graph M_n for $n \geq 5$ odd.

2. Main Result

2.1. *k-balanced multiset*

Maryati et al. [7] defined k-balanced multiset as follows. Let $k \in N$ and Y be a multiset of positive integers. Y is said to be k-balanced if there k subsets from Y is $Y_1, Y_2, \ldots, Y_k$, such that for every $i \in [i, k]$. We obtain $|Y_i| = \frac{Y}{k}$, $\sum Y_i = \frac{\sum Y}{k} \in N$ and $\biguplus_{i=1}^{k} Y_i = Y$.

Here, we give several lemmas on k-balanced multiset.

Lemma 2.1 *Let x be positive integers, $x \geq 5$ and odd. Let $K = [3x + 1, 4x - 2] \uplus [4x, 5x - 3]$, then K is $(x - 2)$-balanced.*

Proof. For every $j \in [1, x - 2]$, we define the multisets $K_j = \{3x + j, 5x - j - 2\}$. It is obvious that for each $j \in [1, x-2]$, $|K_j| = 2$, $K_j \subset K$, and $\biguplus_{j=1}^{x-2} K_j = K$. Since $\sum K_j = 8x - 2$ for every $j \in [1, x-2]$, then K is $(x-2)$-balanced. □

2.2. *(k,δ)-anti balanced multiset*

Inayah [4] defined (k,δ)-anti balanced multiset as follows. Let k, $\delta \in N$ and X be a set containing the elements of positive integers. If there k subsets from X, say $X_1, X_2, \ldots, X_k$, such that for every $i \in [1, k]$ then X is (k,δ)-anti balanced. We obtain $|X_i| = \frac{|X|}{k}$, $\biguplus_{i=1}^{k} X_i = X$, and for every $i \in [1, k - 1]$ $\sum X_{i+1} - \sum X_i = \delta$.

Here, we give several lemmas on (k,δ)-anti balanced multiset.

Lemma 2.2 *Let x and y be positive integers, $x \geq 3$ and $y = 2x - 1$. Let $X = [1, x - 1] \uplus [x + 1, y] \uplus [y + 2, x + y] \uplus [x + y + 1, 2y]$ with $|X| = 4$ and $Y = [2y + 1, 2y + x - 1] \uplus [2y + x, 3y - 1]$ where $|Y| = 2$, then $K = X \biguplus Y$ is $(y - 2,3)$-anti balanced.*

Proof. For every $j \in [1, y - 2]$, we define the multisets $K_j = \{\lceil \frac{i+1}{2} \rceil, x + \lceil \frac{i}{2} \rceil, y + \lceil \frac{i+2}{2} \rceil, x + y + \lceil \frac{i+1}{2} \rceil, 2y + \lceil \frac{i}{2} \rceil, 2y + x + \lceil \frac{i-1}{2} \rceil\}$. Now we have for every $j \in [1, y - 2]$, $|K_j| = 6$, $K_j \subset K$, and $\biguplus_{j=1}^{y-2} K_j = K$. Since $\sum K_j = 15x + 3i - 3$ for every $j \in [1, y - 2]$, then $\sum K_{j+1} - \sum K_j = 3$, K is $(y - 2,3)$-anti balanced. □

Lemma 2.3 *Let x and y be positive integers, $x \geq 3$ and $y = 2x - 1$. Let $Y = [1, x] \setminus \{U_p | U_p = p + 1, p \in [1, x - 2]\} \uplus [x, 2x - 1] \setminus \{U_p | U_p = x + p, p \in [1, x - 2]\} \uplus \{2x\} \uplus [x + y, x + y + 1]$ with $|Y| = 4$ and $Z = [x + 2y - 1, x + 2y] \uplus \{3y\}$ where $|Z| = 2$, then $R = Y \biguplus Z$ is $(2,y - 3)$-anti balanced.*

Proof. For every $j \in [1, 2]$, we define the multisets $R_j = \{x(i - 1), i(x - 1) + 1, y + \lceil \frac{i}{2} \rceil, x + y + \lceil \frac{2-i}{2} \rceil, 2y + x - \lceil \frac{i-1}{2} \rceil, 3y - \lfloor \frac{2-i}{2} \rfloor\}$. Now we have for every $j \in [1, 2]$, $|R_j| = 6$, $R_j \subset R$, and $\biguplus_{j=1}^{2} R_j = R$. Since $\sum R_j = 15x + (y - 3)j$ for every $j \in [1, 2]$, then $\sum R_{j+1} - \sum R_j = y - 3$, R is $(2,y - 3)$-anti balanced. □

Lemma 2.4 *Let k be positive integers, $k \geq 5$ and odd. Let $X = [3k+1, 4k-2] \uplus [4k+2, 5k-1]$, then X is $(k - 2,2)$-anti balanced.*

IOP Conf. Series: Journal of Physics: Conf. Series **1008** (2018) 012047 doi:10.1088/1742-6596/1008/1/012047

Proof. For every $j \in [1, k-2]$, we define the multisets $X_j = \{3k + j, 4k + j + 1\}$. Now we have for every $j \in [1, k-2]$, $|X_j| = 2$, $X_j \subset X$, and $\biguplus_{j=1}^{k-2} X_j = X$. Since $\sum X_j = 7k + 2j + 1$ for every $j \in [1, k-2]$, then $\sum X_{j+1} - \sum X_j = 2$, X is $(k-2,2)$-anti balanced. □

Lemma 2.5 *Let k be positive integers, $k \geq 5$ and odd. Let $R = [4k-1, 4k] \uplus [4k+1, 5k] \setminus \{U_p | U_p = 4k + p + 1, p \in [1, k-2]\}$, then R is $(2, x-2)$-anti balanced.*

Proof. For every $j \in [1, 2]$, we define the multisets $R_j = \{4k - j + 1, 4k + (j-1)k + \lceil \frac{i}{2} \rceil - (j-1)\}$. Now we have for every $j \in [1, 2]$, $|R_j| = 2$, $R_j \subset R$, and $\biguplus_{j=1}^{2} R_j = R$. Since $\sum R_j = k(j+5) - 2j - 2$ for every $j \in [1, 2]$, then $\sum R_{j+1} - \sum R_j = x - 2$, R is $(2, x-2)$-anti balanced. □

Lemma 2.6 *Let y be positive integers, $y \geq 5$ and odd. Let $Y = [4y - 1, 5y - 2] \setminus \{U_p | U_p = 4y + p - 1, p \in [1, y-2]\} \uplus [5y - 1, 5y]$, then Y is $(2, y)$-anti balanced.*

Proof. For every $j \in [1, 2]$, we define the multisets $Y_j = \{4y + j(y-1), 5y + j - 2\}$. Now we have for every $j \in [1, 2]$, $|Y_j| = 2$, $Y_j \subset Y$, and $\biguplus_{j=1}^{2} Y_j = Y$. Since $\sum Y_j = y(j+8) - 2$ for every $j \in [1, 2]$, then $\sum Y_{j+1} - \sum Y_j = y$, Y is $(2, y)$-anti balanced. □

2.3. Super (a, d)-H-antimagic Labeling on Möbius Ladder Graphs

Gallian [1] defined that a Möbius ladder M_n is a graph obtained from the ladder $P_n \times P_2$ by joining the opposite end points of the two copies P_n. The Möbius ladder M_n can be depicted as in Figure 1

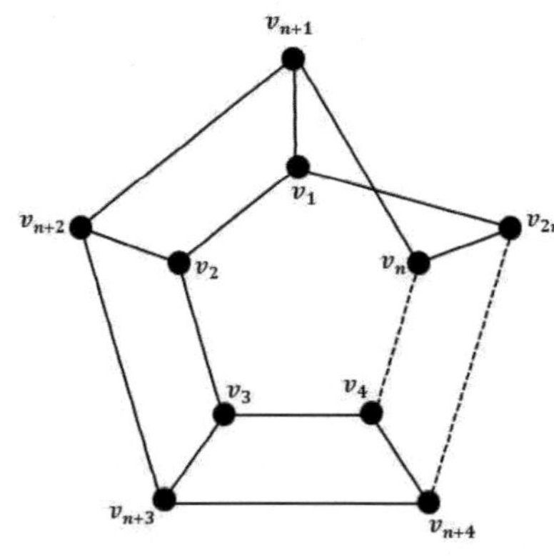

Figure 1. Möbius ladder graph M_n

Lemma 2.7 *If the graph M_n, $n \geq 5$ and odd, admits a super (a, d)-H-antimagic total labeling, then $d \leq \lfloor \frac{20n-32}{n-1} \rfloor$*

Proof. Let G and H be a Möbius ladder M_n for $n \geq 5$ odd and a cycle graph C_m for $m = 4$. Then $|V(G)| = v_G = 2n$, $|E(G)| = e_G = 3n$, $|V(H)| = v_H = 4$, and $|E(H)| = e_H = 4$. Suppose there exists a bijection $\xi : V(G) \cup E(G) \to \{1, 2, 3, \ldots, 5n\}$ which is a super (a, d)-H-antimagic total labeling of G. Let $\omega(H') = \Sigma_{v \in V(H')} \xi(v) + \Sigma_{e \in E(H')} \xi(e)$ be the weight of the subgraph H' isomorphic to H and let $W = \{\omega(H') : H' \cong H\} = \{a, a+d, a+2d, \ldots, a+(t-1)d\}$ be the set of H' weights and t be the number of subgraphs. Here $t = n$. Now, it is easy to see that the minimum possible weight of H' is at least $8n + 20$, i.e., $a \geq 8n + 20$. Also the maximum possible weight of H' is not more than $28n - 12$, i.e., $a + (t-1)d \leq 28n - 12$, $(n-1)d \leq 20n - 32$, thus $d \leq \lfloor \frac{20n-32}{n-1} \rfloor$. □

Theorem 2.1 *The graph G admits a super $(\frac{33n+9}{2}$,1)-H-antimagic total labeling for $n \geq 5$ and odd.*

Proof. Let G and H be a Möbius ladder M_n for $n \geq 5$ odd and a cycle graph C_m for $m = 4$. Let $V(G) = \{v_i; 1 \leq i \leq 2n\}$ and $E(G) = \{v_iv_{i+1}; 1 \leq i \leq n-1\} \uplus \{v_1v_{2n}\} \uplus \{v_iv_j; 1 \leq i \leq n, 1 \leq j \leq n\}$ where $|V(G)| = 2n$ and $|E(G)| = 3n$. Let $R = [1, 5n]$. Partition R into three sets such that $R = X \cup Y \cup Z$ where $X = [1, 2n]$, $Y = [2n+1, 3n]$, and $Z = [3n+1, 5n]$. Let $K = X \uplus Y$ where the vertices v_i are labeled by set X and edges v_iv_j are labeled by set Y. According to Lemma 2.2 and Lemma 2.3, if $x = \frac{n+1}{2}$, $y = n$ we have $(n-2, 3)$-anti balanced and $(2, n-2)$-anti balanced with $|K_i| = 6$ where $\Sigma K_i = \frac{17n+2i+5}{2})$ for $i \in [1, n]$. Then, the edges v_iv_{i+1} and edge v_1v_{2n} are labeled by set Z. According to Lemma 2.4 and Lemma 2.5 if k,$x = n$ we have $(n-2, 2)$-anti balanced and $(2, n-2)$-anti balanced with $|Y_i| = 2$ where $\Sigma Y_i = 8n+1$. Then we define the total labeling ξ on G as follows:

$$\xi(v_i) = \begin{cases} \frac{i+1}{2}, & \text{for } i \in [1, 2n-1]; \\ \frac{i+2}{2}, & \text{for } i = 2n; \\ \frac{2n+2+i}{2}, & \text{for } i = [1, 2n-2]. \end{cases}$$

$$\xi(v_iv_{n+(i)mod(n)}) = \begin{cases} \frac{5n+i}{2}, & \text{for } i \in \{1, 3, \ldots, n-2\}; \\ \frac{4n+i}{2}, & \text{for } i \in \{2, 4, \ldots, n-1\}; \\ 3n, & \text{for } i = n. \end{cases}$$

$$\xi_1(v_i\ v_{i+1}) = \begin{cases} 5n-i, & \text{for } i \in [1, n-1]; \\ 4n-i-4, & \text{for } i \in [n, 2n-2]; \\ 4n, & \text{for } i = 2n-1; \end{cases}$$

$$\xi_1(v_1\ v_{2n}) = 5n.$$

$$\begin{aligned} \omega(H) &= \sum_{v \in V(H')} \xi(v) + \sum_{e \in E(H')} \xi(e) \\ &= (4n+2) + (\frac{25n+5}{2}) \\ &= \frac{33n+9}{2}. \end{aligned}$$

Hence G has a super $(\frac{33n+9}{2}$,1)-H-antimagic total labeling. □

Figure 2 illustrates an example of super $(153, 1)$-C_4-antimagic labeling on M_9 graph.

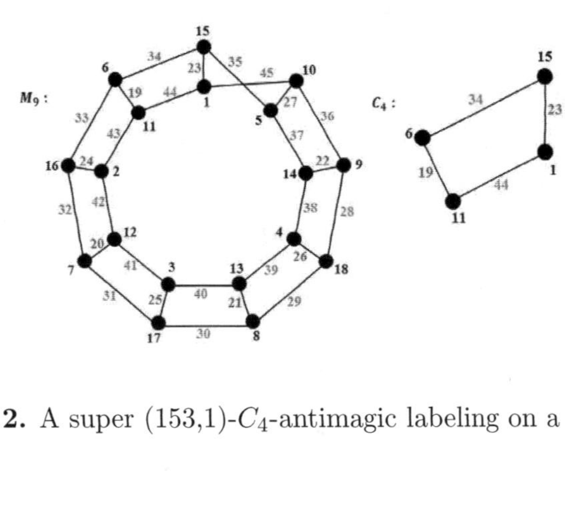

Figure 2. A super (153,1)-C_4-antimagic labeling on a M_9 graph

IOP Conf. Series: Journal of Physics: Conf. Series **1008** (2018) 012047 doi:10.1088/1742-6596/1008/1/012047

Theorem 2.2 *The graph G admits a super ($\frac{31n+11}{2}$,3)-H-antimagic total labeling for $n \geq 5$ and odd.*

Proof. Consider the partition of $[1, 5n]$ introduced in the proof of Theorem 2.1. According to Lemma 2.2 and Lemma 2.3, if $x = \frac{n+1}{2}$, $y = n$ we have $(n-2, 3)$-anti balanced and $(2, n-2)$-anti balanced with $|K_i| = 6$ where $\Sigma K_i = 8n + 2i + 2)$ for each $i \in [1, n]$. Then, the edges $v_i v_{i+1}$ and edge $v_1 v_{2n}$ are labeled by set Z. By Lemma 2.6 and Lemma 2.1, if y,$x = n$ we have $(2, n)$-anti balanced and $(n-2)$-balanced with $|Y_i| = 2$ where $\Sigma Y_i = \frac{15n+2i+1}{2}$ for each $i \in [1, n]$. Then we define the total labeling ξ on G as follows:

$$\xi(v_i) = \begin{cases} \frac{i+1}{2}, & \text{for } i \in [1, 2n-1]; \\ \frac{i+2}{2}, & \text{for } i = 2n; \\ \frac{2n+2+i}{2}, & \text{for } i = [1, 2n-2]. \end{cases}$$

$$\xi(v_i v_{n+(i)mod(n)}) = \begin{cases} \frac{5n+i}{2}, & \text{for } i \in \{1, 3, \ldots, n-2\}; \\ \frac{4n+i}{2}, & \text{for } i \in \{2, 4, \ldots, n-1\}; \\ 3n, & \text{for } i = n. \end{cases}$$

$$\xi_1(v_i\ v_{i+1}) = \begin{cases} 3n + i, & \text{for } i \in [1, n-1]; \\ 5n - i + 3, & \text{for } i \in [n, 2n-2]; \\ 5n - 1, & \text{for } i = 2n-1; \end{cases}$$

$$\xi_1(v_1\ v_{2n}) = 5n.$$

$$\begin{aligned} \omega(H) &= \sum_{v \in V(H')} \xi(v) + \sum_{e \in E(H')} \xi(e) \\ &= (\frac{7n+5}{2}) + (12n+3) \\ &= \frac{31n+11}{2}. \end{aligned}$$

Hence G has a super ($\frac{31n+11}{2}$,3)-H-antimagic total labeling. □

Figure 3 illustrates an example of super $(137, 3)$-C_4-antimagic labeling on M_9 graph.

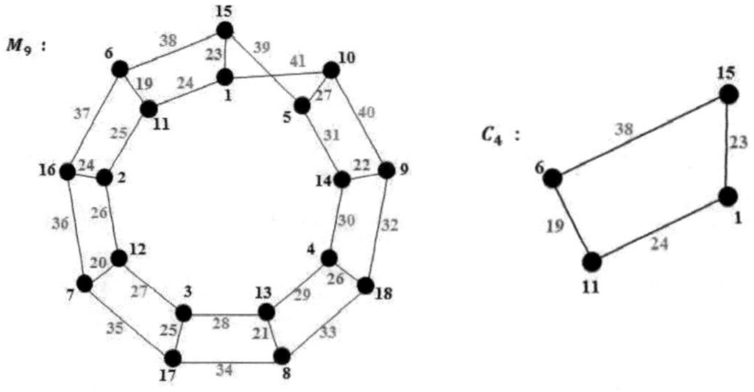

Figure 3. A super (137,3)-C_4-antimagic labeling on a M_9 graph

3. Conclusion

According to the discussion can be concluded that the super (a,d)-H-antimagic labeling of a Möbius ladder graph M_n for $n \geq 5$ odd with $d = \{1, 3\}$.

Open Problem: What is the super (a, d)-C_4-antimagic total labeling of Möbius ladder for all possible d.

Acknowledgments
The authors gratefully acknowledge the support from Department of Mathematics, Faculty of Mathematics and Natural Sciences, Universitas Sebelas Maret Surakarta.

References
[1] Gallian, J. A., *A Dynamic Survey of Graph Labeling*, The Electronic Journal of Combinatorics **18** (2016) #DS6.
[2] Gutiérrez, A. and A. Lladó, *Magic Covering*, J. Combin. Math. Combin. Comput. **55** (2005), 43-56.
[3] Inayah, N., A. N. M. Salman, and R. Simanjutak, *On (a,d)−H−Antimagic Coverings of Graphs*, J. Combin. Math. Combin. Comput. **71** (2009), 273-281.
[4] Inayah, N., *Pelabelan Selimut (a,d)−H−Anti Ajaib pada Beberapa Kelas Graf*, Disertasi, ITB Bandung (2013).
[5] Kotzig, A. and A. Rosa, *Magic Valuations of Finite Graphs*, Canad. Math. Bull (1970), 451-461.
[6] Lladó, A. and J. Moragas, *Cycle−magic Graphs*, Discrete Mathematics **307** (2007), 2925-2933.
[7] Maryati, T. K., A. N. M. Salman, E. T. Baskoro, J. Ryan, and M. Miller, *On H−Supermagic Labelings for Certain Shackles and Amalgamations of a Connected Graph*, Utilitas Mathematica **83** (2010), 333-342.
[8] Ngurah, A. A. G. and E. T. Baskoro, *On Magic and Antimagic Total labeling of Generalized Petersen Graph*, Utilitas Mathematica **63** (2003), 97-107.

IOP Conf. Series: Journal of Physics: Conf. Series **1008** (2018) 012048 doi:10.1088/1742-6596/1008/1/012048

On the strong metric dimension of generalized butterfly graph, starbarbell graph, and $C_m \odot P_n$ graph

Ratih Yunia Mayasari and Tri Atmojo Kusmayadi

Department of Mathematics, Faculty of Mathematics and Natural Sciences, Universitas Sebelas Maret, Surakarta, Indonesia

E-mail: ratihyuniamayasari@student.uns.ac.id, tri.atmojo.kusmayadi@gmail.com

Abstract. Let G be a connected graph with vertex set $V(G)$ and edge set $E(G)$. For every pair of vertices $u, v \in V(G)$, the interval $I[u, v]$ between u and v to be the collection of all vertices that belong to some shortest $u-v$ path. A vertex $s \in V(G)$ strongly resolves two vertices u and v if u belongs to a shortest $v-s$ path or v belongs to a shortest $u-s$ path. A vertex set S of G is a strong resolving set of G if every two distinct vertices of G are strongly resolved by some vertex of S. The strong metric basis of G is a strong resolving set with minimal cardinality. The strong metric dimension $sdim(G)$ of a graph G is defined as the cardinality of strong metric basis. In this paper we determine the strong metric dimension of a generalized butterfly graph, starbarbell graph, and $C_m \odot P_n$ graph. We obtain the strong metric dimension of generalized butterfly graph is $sdim(BF_n) = 2n - 2$. The strong metric dimension of starbarbell graph is $sdim(SB_{m_1,m_2,\ldots,m_n}) = \sum_{i=1}^{n}(m_i - 1) - 1$. The strong metric dimension of $C_m \odot P_n$ graph are $sdim(C_m \odot P_n) = 2m - 1$ for $m \geq 3$ and $n = 2$, and $sdim(C_m \odot P_n) = 2m - 2$ for $m \geq 3$ and $n \geq 2$.

1. Introduction

The concept of strong metric dimension was presented by Sebö and Tannier [6] in 2004. Oellermann and Peters-Fransen [5] defined for two vertices u and v in a connected graph G, the interval $I[u, v]$ between u and v to be collection of all vertices that belong to some shortest path. A vertex s strongly resolves two vertices u and v if $v \in I[u, s]$ or $u \in I[v, s]$. A set S of vertices in a connected graph G is a strong resolving set for G if every two vertices of G are strongly resolved by some vertex of S. The smallest cardinality of a strong resolving set of G is called its strong metric dimension and is denoted by sdim(G).

Some researchers have investigated the strong metric dimension to some graph classes. In 2004 Sebö and Tannier [6] observed that the strong metric dimension of complete graph K_n, cycle graph C_n, and tree. In 2012, Kuziak et al. [4] observed that the strong metric dimension of corona product graph. In the same year, Kratica et al. [1] determined that the metric dimension of hamming graph $H_{n,k}$. Kratica et al. [2] determined that the metric dimension of convex polytope D_n and T_n in 2012 too. In 2013 Yi [8] determined that $sdim(G) = 1$ if only if G is path graph and $sdim(G) = n - 1$ if only if G is complete graph. Kusmayadi et al. [3] determined the strong metric dimension of some related wheel graph such as sunflower graph, t-fold wheel graph, helm graph, and friendship graph. In this paper, we determine the strong metric dimension of a generalized butterfly graph, starbarbell graph, and $C_m \odot P_n$ graph.

2. Main Results

2.1. Strong Metric Dimension

Let G be a connected graph with vertex set $V(G)$, edge set $E(G)$, and $S = \{s_1, s_2, \ldots, s_k\} \in V(G)$. Oellermann and Peters-Fransen [5] defined the interval $I[u, v]$ between u and v to be the collection of all vertices that belong to some shortest $u - v$ path. A vertex $s \in S$ strongly resolves two vertices u and v if $u \in I[v, s]$ or $v \in I[u, v]$. A vertex set S of G is a strong resolving set of G if every two distinct vertices of G are strongly resolved by some vertex of S. The strong metric basis of G is a strong resolving set with minimal cardinality. The strong metric dimension of a graph G is defined as the cardinality of strong metric basis denoted by $sdim(G)$. We often make use of the following lemma and properties about strong metric dimension given by Kratica et al. [2].

Lemma 2.1 *Let* $u,\ v \in V(G)$, $u \neq v$,

(i) $d(w,v) \leq d(u,v)$ *for each* w *such that* $u\, w \in E(G)$, *and*

(ii) $d(u,w) \leq d(u,v)$ *for each* w *such that* $v\, w \in E(G)$.

Then there does not exist vertex $a \in V(G)$, $a \neq u,v$ *that strongly resolves vertices* u *and* v.

Property 2.1 *If* $S \subset V(G)$ *is strong resolving set of graph G, then for every two vertices* $u,v \in V(G)$ *satisfying conditions* 1 *and* 2 *of Lemma* 2.1, *obtained* $u \in S$ *or* $v \in S$.

Property 2.2 *If* $S \subset V(G)$ *is strong resolving set of graph G, then for every two vertices* $u, v \in V(G)$ *satisfying* $d(u,v) = diam(G)$, *obtained* $u \in S$ *or* $v \in S$.

2.2. The Strong Metric Dimension of Generalized Butterfly Graph

Weisstein [7] defined the butterfly graph as a planar undirected graph with 5 vertices and 6 edges. In this research, we obtain generalized butterfly graph. We define a generalized butterfly graph as a double shell in which each shell has any n order and for every shell have the same order. The generalized butterfly graph BF_n can be depicted as in Figure 1.

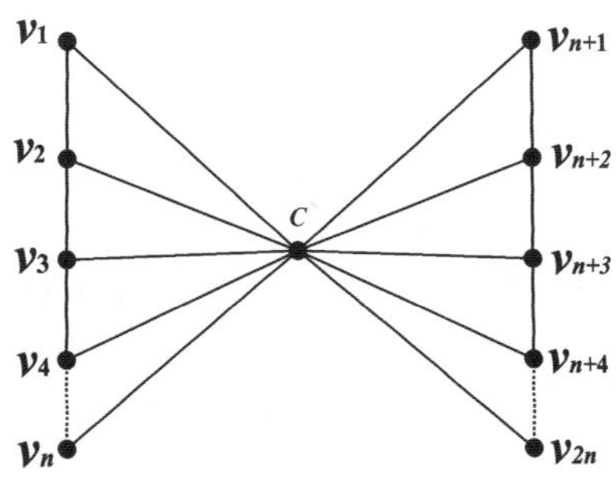

Figure 1. Generalized butterfly graph BF_n

Lemma 2.2 *For every integer* $n \geq 3$, *if S is a strong resolving set of generalized butterfly graph* BF_n *then* $\mid S \mid \geq 2n\text{-}2$.

Proof. Let us consider a pair of vertices (v_i, v_j) with i, j = 1,2,. . .,2n-1 satisfying both of the conditions of Lemma 2.1. According to Property 2.1, we obtain $v_i \in S$ or $v_j \in S$. It means that S contains one vertex from distinct sets $X_{ij}= \{v_i, v_j\}$ with i, j= 1,2, . . . ,2$n - 1$ and $i \neq j$.The minimum number of vertices from distinct sets X_{ij} is $2n-2$. Therefore, $\mid S \mid \geq 2n - 2$. □

Lemma 2.3 *For every integer $n \geq 3$, a set $S = \{v_1, v_2, \ldots, v_{2n-2}\}$ is a strong resolving set of generalized butterfly graph BF_n.*

Proof. We prove that every two distinct vertices $u, v \in (BF_n) \setminus S$, there exists a vertex $s \in S$ which strongly resolves u and v. There are two pairs of vertices from $V(BF_n) \setminus S$.

(i) A pair of vertices (c, v_i).
For every integer $i = 2n-1, 2n$ and $j = n+1, n+2, \ldots, 2n-2$, $d(v_i, v_j) = 2 = diam(BF_n)$, we obtain the shortest $v_i - v_j$ path : v_i, c, v_j. Thus, $c \in I[v_i, v_j]$.

(ii) A pair of vertices (v_{2n-1}, v_{2n})
For $i = 2n-2$, $d(v_i, v_j) = 2 = diam(BF_n)$, we obtain the shortest $v_{2n-2} - v_n$ path: $v_{2n-2}, v_{2n-1}, v_{2n}$. Thus, $v_{2n-1} \in I[v_{2n-2}, v_{2n}]$.

For every possible pairs of vertices, there exists a vertex $s \in S$ which strongly resolves every two distinct vertices $BF_n \setminus S$. Thus, S is a strong resolving set of BF_n. □

Theorem 2.1 *Let BF_n be the generalized butterfly graph with $n \geq 3$. Then $sdim(BF_n) = 2n-2$.*

Proof. By using Lemma 2.3 a set $S = \{v_1, v_2, \ldots, v_{2n-2}\}$ is a strong resolving set of generalized butterfly graph BF_n with $n \geq 3$. According to Lemma 2.2, $\mid S \mid \geq 2n$ - 2, S is a strong metric basis of generalized butterfly graph BF_n. Hence, $sdim(BF_n) = 2n$ - 2. □

2.3. The Strong Metric Dimension of Starbarbell Graph

Starbarbell graph $SB_{m_1,m_2,\ldots,m_n}$ is a graph obtained from a star graph S_n and n complete graph K_{m_i} by merging one vertex from each K_{m_i} and the ith-leaf of S_n, where $m_i \geq 3$, $1 \leq i \leq n$, and $n \geq 2$. In this paper, we obtain the strong metric dimension of starbarbell graph denoted by $SB_{m_1,m_2,\ldots,m_n}$. The starbarbell graph $SB_{m_1,m_2,\ldots,m_n}$ can be depicted as in Figure 2.

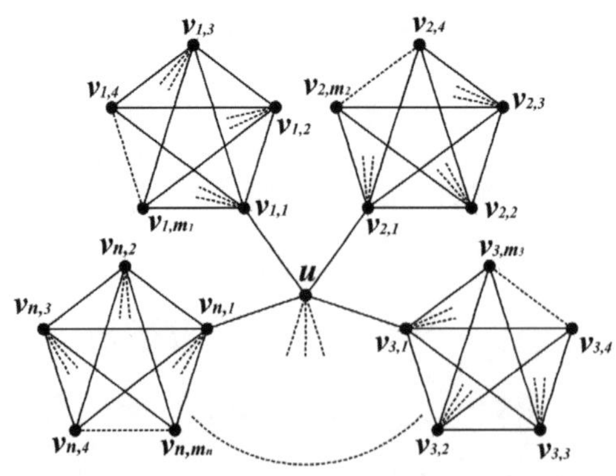

Figure 2. Starbarbell graph $SB_{m_1,m_2,\ldots,m_n}$

Lemma 2.4 *For every integer $m \geq 3$ and $n \geq 2$, if S is a strong resolving set of starbarbell graph $SB_{m_1,m_2,\ldots,m_n}$ then $\mid S \mid \geq \sum_{i=1}^{n}(m_i - 1) - 1$.*

Proof. We prove for every two distinct vertices $(v_{i,j}, v_{k,l}) \in V(SB_{m_1,m_2,\ldots,m_n})$ for every $i, k = 1, 2, \ldots, n$ with $i \neq k$ and $j, l = 2, \ldots, m_i$ so $d(v_{i,j}, v_{k,l}) = 4 = diam(SB_{m_1,m_2,\ldots,m_n})$ then by using Property 2.2, $v_{i,j} \in S$ or $v_{k,l} \in S$. Therefore, S contains at least one vertex from distinct set $X_{i,j,k,l} \in S = \{v_{i,j}, v_{k,l}\}$ for $i, k = 1, 2, \ldots, n$, $i \neq k$, and $j, l = 2, \ldots, m_i$. The minimum number of vertices from distinct set $X_{i,j,k,l}$ is $\sum_{i=1}^{n}(m_i - 1) - 1$. Therefore, $\mid S \mid \geq \sum_{i=1}^{n}(m_i - 1) - 1$. □

IOP Conf. Series: Journal of Physics: Conf. Series **1008** (2018) 012048 doi:10.1088/1742-6596/1008/1/012048

Lemma 2.5 *For every integer $m \geq 3$ and $n \geq 2$, a set $S = \{v_{1,2}, v_{1,3}, \ldots, v_{1,m_1}, v_{2,2}, v_{2,3}, \ldots, v_{2,m_2}, \ldots, v_{n,2}, v_{n,3}, \ldots, v_{n,m_n-1}\}$ is a strong resolving set of starbarbell graph $SB_{m_1,m_2,\ldots,m_n}$.*

Proof. We prove that for every two distinct vertices $u, v \in V(SB_{m_1,m_2,\ldots,m_n}) \backslash S$, $u \neq v$, there exists a vertex $s \in S$ which strongly resolves u and v. There are pairs four possible pairs of vertices.

(i) A pair of vertices (u, v_{n,m_n}).
For every integer $i \in \{1, 2, \ldots, n-1\}$ and $j \in \{2, 3, \ldots, m_i\}$, $d(v_{i,j}, v_{n,m_n}) = 4 = diam(SB_{m_1,m_2,\ldots,m_n})$, we obtain the shortest $v_{n,m_n} - v_{i,j}$ path: $v_{n,m_n}, v_{n,1}, u, v_{i,1}, v_{i,j}$. Thus, $u \in I[v_{n,m_n}, v_{i,j}]$.

(ii) A pair of vertices $(u, v_{i,1})$.
For every integer $i \in \{1, 2, \ldots, n-1\}$ and $j \in \{2, 3, \ldots, m_i\}$, $d(u, v_{i,j}) = 2$, we obtain the shortest $u - v_{i,j}$ path: $u, v_{i,1}, v_{i,j}$. Thus, $v_{i,1} \in I[u, v_{i,j}]$.
For $i = n$, $l \in \{2, 3, \ldots, m_i - 1\}$, $d(u, v_{n,l}) = 2$, we obtain the shortest $u - v_{n,l}$ path: $u, v_{n,1}, v_{n,l}$. Thus, $v_{n,1} \in I[u, v_{n,l}]$.

(iii) A pair of vertices $(v_{i,1}, v_{n,m_n})$.
For every integer $i \in \{1, 2, \ldots, n-1\}$ and $j \in \{2, 3, \ldots, m_i\}$, $d(v_{i,j}, v_{n,m_n}) = 4 = diam(SB_{m_1,m_2,\ldots,m_n})$, we obtain the shortest $v_{i,j} - v_{n,m_n}$ path: $v_{i,j}, v_{i,1}, u, v_{n,1}, v_{n,m_n}$. Thus, $v_{i,1} \in I[v_{i,1}, v_{n,m_n}]$.

(iv) A pair of vertices $(v_{i,1}, v_{k,1})$.
For every integer $i, k \in \{1, 2, \ldots, n-1\}$, $i \neq k$, and $l \in \{2, 3, \ldots, m_i\}$, $d(v_{i,1}, v_{k,l}) = 3$, we obtain the shortest $v_{i,1} - v_{k,l}$ path: $v_{i,1}, u, v_{k,1}, v_{k,l}$. Thus, $v_{k,1} \in I[v_{i,1}, v_{k,l}]$.
For $k = n$ and $l \in \{2, 3, \ldots, m_i - 1\}$, $d(v_{i,1}, v_{k,l}) = 3$, we obtain the shortest $v_{i,1} - v_{k,l}$ path: $v_{i,1}, u, v_{k,1}, v_{k,l}$. Thus, $v_{k,1} \in I[v_{i,1}, v_{k,l}]$.

From every possible pairs of vertices, there exists a vertex $s \in S$ which strongly resolves every two distinct vertices $SB_{m_1,m_2,\ldots,m_n} \setminus S$. Thus S is a strong resolving set of starbarbell graph $SB_{m_1,m_2,\ldots,m_n}$. □

Theorem 2.2 *Let $SBm_1, m_2, \ldots, m_n$ be the starbarbell graph with $m \geq 3$ and $n \geq 2$. Then $sdim(SB_{m_1,m_2,\ldots,m_n}) = \sum_{i=1}^{n}(m_i - 1) - 1$.*

Proof. By using Lemma 2.5, we have a set $S = \{v_{1,2}, v_{1,3}, \ldots, v_{1,m_1}, v_{2,2}, v_{2,3}, \ldots, v_{2,m_2}, \ldots, v_{n,2}, v_{n,3}, \ldots, v_{n,(m_n-1)}\}$ is a strong resolving set of $SB_{m_1,m_2,\ldots,m_n}$ graph with $m \geq 3$ and $n \geq 2$. According to Lemma 2.4, $\mid S \mid \geq \sum_{i=1}^{n}(m_i - 1) - 1$, S is a strong metric basis of $SB_{m_1,m_2,\ldots,m_n}$. Hence, $sdim(SB_{m_1,m_2,\ldots,m_n}) = \sum_{i=1}^{n}(m_i - 1) - 1$. □

2.4. The Strong Metric Dimension of $C_m \odot P_n$ Graph

The corona product $C_m \odot P_n$ graph is graph obtained from C_m and P_n by taking one copy of C_m and n copies of P_n and joining by an edge each vertex from i^{th}-copy of P_n with the ith-vertex of C_m. The $C_m \odot P_n$ can be depicted as in Figure 3.

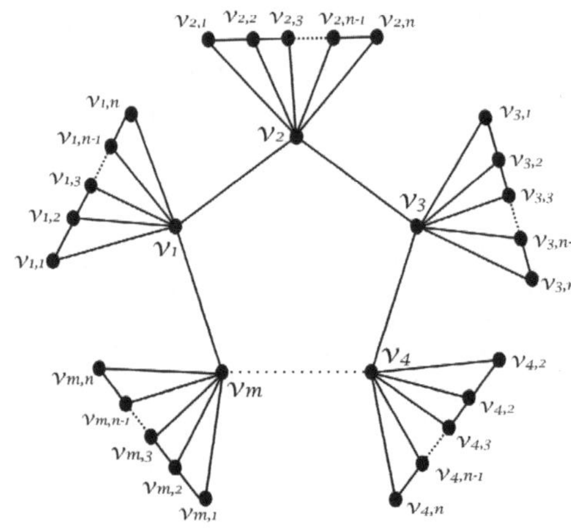

Figure 3. $C_m \odot P_n$ graph

Lemma 2.6 *For every integer $m \geq 3$ and $n \doteq 2$, if S is a strong resolving set of $C_m \odot P_n$ graph then $\mid S \mid \geq 2m - 1$.*

Proof. Let us consider a pair of vertices $(v_{i,j},_{k,l})$ with i, k = 1,2,. . .,m, $i \neq k$, and $j, l = 1, 2$, satisfying both of the conditions of Lemma 2.1. According to Property 2.1, we obtain $v_{i,j} \in S$ or $v_{k,l} \in S$. It means that S contains one vertex from distinct sets $X_{i,j,k,l}$= $\{v_{i,j}, v_{k,l}\}$ with i, k= 1,2,. . .,m, $i \neq k$, and $j, l = 1, 2$. The minimum number of vertices from distinct sets $X_{i,j,k,l}$ is $2m - 1$. Therefore, $\mid S \mid \geq 2m - 1$. □

Lemma 2.7 *For every integer $m \geq 3$ and $n \doteq 2$, a set $S = \{v_{1,1}, v_{1,2}, v_{2,1}, v_{2,2}, \ldots, v_{m,1}\}$ is a strong resolving set of $C_m \odot P_n$ graph.*

Proof. We prove that every two distinct vertices $u, v \in (C_m \odot P_n) \setminus S$, there exists a vertex $s \in S$ which strongly resolves u and v. There are two pairs of vertices from $V(C_m \odot P_n) \setminus S$.

(i) A pair of vertices $(v_i, v_{m,2})$.
For every integer $i = 1, 2, \ldots, m$ and $j = 1, 2$ with $1 \leq i \leq \lceil \frac{m-1}{2} \rceil$, $d(v_{\lceil \frac{m-1}{2} \rceil,j}, v_{m,2})$= $\lfloor \frac{m}{2} \rfloor +$ 2, we obtain the shortest $v_{\lceil \frac{m-1}{2} \rceil,j} - v_{m,2}$ path : $v_{\lceil \frac{m-1}{2} \rceil,j}, v_{\lceil \frac{m-1}{2} \rceil}, v_{\lceil \frac{m-1}{2} \rceil -1}, \ldots, v_1, v_m, v_{m,2}$. So that $v_i \in I[v_{\lceil \frac{m-1}{2} \rceil,j}, v_{m,2}]$.
For every integer $i = 1, 2, \ldots, m$ and $j = 1, 2$ with $\lfloor \frac{m-1}{2} \rfloor \leq i \leq n$, $d(v_{\lfloor \frac{m+1}{2} \rfloor,j}, v_{m2})$= $\lfloor \frac{m}{2} \rfloor$ + 2, we obtain the shortest $v_{\lfloor \frac{m+1}{2} \rfloor,j} - v_{m,2}$ path : $v_{\lfloor \frac{m+1}{2} \rfloor,j}, v_{\lfloor \frac{m+1}{2} \rfloor}, v_{\lfloor \frac{m+1}{2} \rfloor +1}, \ldots, v_{m-1}, v_m, v_{m,2}$. So that $v_i \in I[v_{\lfloor \frac{m+1}{2} \rfloor,j}, v_{m,2}]$.

(ii) A pair of vertices (v_i, v_k)
For every integer $i, k = 1, 2, \ldots, m - 1$, and $j, l = 1, 2$, without loss of generality, $1 \leq i < k \leq n$, v_i and v_k will be strongly resolved by $v_{i,j}$, so that $v_i \in I[v_{i,j}, v_k]$.

For every possible pairs of vertices, there exists a vertex $s \in S$ which strongly resolves every two distinct vertices $C_m \odot P_n \setminus S$. Thus S is a strong resolving set of $C_m \odot P_n$. □

Lemma 2.8 *For every integer $m \geq 3$ and $n \geq 3$, if S is a strong resolving set of $C_m \odot P_n$ graph then $\mid S \mid \geq mn - 2$.*

IOP Conf. Series: Journal of Physics: Conf. Series **1008** (2018) 012048 doi:10.1088/1742-6596/1008/1/012048

Proof. Let us consider a pair of vertices $(v_{i,j}, v_{k,l})$ with $i, k = 1,2,\ldots,m$, $i \neq k$, and $j, l = 1, 2, \ldots, n-1$, satisfying both of the conditions of Lemma 2.1. According to Property 2.1, we obtain $v_{i,j} \in S$ or $v_{k,l} \in S$. It means that S contains one vertex from distinct sets $X_{i,j,k,l} = \{v_{i,j}, v_{k,l}\}$ with $i, k = 1,2,\ldots,m$, $i \neq k$, and $j, l = 1, 2, \ldots, n-1$.The minimum number of vertices from distinct sets $X_{i,j,k,l}$ is $mn-2$. Therefore, $\mid S \mid \geq mn-2$. □

Lemma 2.9 *For every integer $m \geq 3$ and $n \geq 3$, a set $S = \{v_{1,1}, v_{1,2}, \ldots, v_{1,n}, v_{2,1}, v_{2,2}, \ldots, v_{2,n}, \ldots, v_{m,1}, v_{m,2}, \ldots, v_{m,(n-2)}\}$ is a strong resolving set of $C_m \odot P_n$ graph.*

Proof. We proof that every two distinct vertices $u, v \in (C_m \odot P_n) \setminus S$, there exists a vertex $s \in S$ which strongly resolves u and v. There are three pairs of vertices from $V(C_m \odot P_n) \setminus S$.

(i) A pair of vertices $(v_i, v_{m,l})$.
For every integer $i = 1, 2, \ldots, m$, $j = 1, 2, \ldots, n$, and $l \in \{n-1, n\}$ with $1 \leq i \leq \lceil \frac{m-1}{2} \rceil$, $d(v_{\lceil \frac{m-1}{2} \rceil, j}, v_{m,l}) = \lfloor \frac{m}{2} \rfloor + 2$, we obtain the shortest $v_{\lceil \frac{m-1}{2} \rceil, j} - v_{m,l}$ path : $v_{\lceil \frac{m-1}{2} \rceil, j}, v_{\lceil \frac{m-1}{2} \rceil}, v_{\lceil \frac{m-1}{2} \rceil - 1}, \ldots, v_1, v_m, v_{m,l}$. So that $v_i \in I[v_{\lceil \frac{m-1}{2} \rceil, j}, v_{m,l}]$.
For every integer $i = 1, 2, \ldots, m$, $j = 1, 2, \ldots, n$, and $l \in \{n-1, n\}$ with $\lfloor \frac{m-1}{2} \rfloor \leq i \leq n$, $d(v_{\lfloor \frac{m+1}{2} \rfloor, j}, v_{m,l}) = \lfloor \frac{m}{2} \rfloor + 2$, we obtain the shortest $v_{\lfloor \frac{m+1}{2} \rfloor, j} - v_{m,l}$ path : $v_{\lfloor \frac{m+1}{2} \rfloor, j}, v_{\lfloor \frac{m+1}{2} \rfloor}, v_{\lfloor \frac{m+1}{2} \rfloor + 1}, \ldots, v_{m-1}, v_m, v_{m,l}$. So that $v_i \in I[v_{\lfloor \frac{m+1}{2} \rfloor, j}, v_{m,l}]$.

(ii) A pair of vertices (v_i, v_k)
For every integer $i, k = 1, 2, \ldots, m-1$, and $j, l = 1, 2$, without loss of generality, $1 \leq i < k \leq n$, v_i and v_k will be strongly resolved by $v_{i,j}$, so that $v_i \in I[v_{i,j}, v_k]$.

(iii) A pair of vertices $(v_{m,(n-1)}, v_{m,n})$.
For every integer $i = n-2$, $d(v_{m,i}, v_{m,n}) = 2$, we obtain the shortest $v_{m,i} - v_{m,n}$ path: $v_{m,i}, v_{m,(n-1)}, v_{m,n}$. So that, $v_{m,(n-1)} \in I[v_{m,i}, v_{m,n}]$.

For every possible pairs of vertices, there exists a vertex $s \in S$ which strongly resolves every two distinct vertices $C_m \odot P_n \setminus S$. Thus S is a strong resolving set of $C_m \odot P_n$. □

Theorem 2.3 *Let $C_m \odot P_n$ be the corona product of cycle graph and path graph with $m \geq 3$ and $n \geq 2$, then*

$$sdim(C_m \odot P_n) = \begin{cases} 2m\text{-}1, & m \geq 3 \text{ and } n = 2; \\ mn\text{-}2, & m \geq 3 \text{ and } n \geq 3. \end{cases}$$

Proof. For $m \geq 3$ and $n \doteq 2$, it is easy to see that sdim$(C_m \odot P_n) = 2m-1$. By Lemma 2.8 and Lemma 2.9, we have sdim$C_m \odot P_n) = mn-2$ for $m \geq 3$ and $n \geq 3$. □

3. Concluding Remark

According to the discussion above it can be concluded that the strong metric dimension of a generalized butterfly graph BF_n, a starbarbell graph $SB_{m_1, m_2, \ldots, m_n}$, and $C_m \odot P_n$ graph are as stated in Theorem 2.1, Theorem 2.2, and Theorem 2.3 respectively.
Open Problem: Determine the strong metric dimension of Hanoi graph.

Acknowledgement
The authors gratefully acknowledge the support from Department of Mathematics, Faculty of Mathematics and Natural Sciences, Universitas Sebelas Maret Surakarta. Then, we wish to thank the referees for their suggestions, which helped to improve the paper.

IOP Conf. Series: Journal of Physics: Conf. Series **1008** (2018) 012048 doi:10.1088/1742-6596/1008/1/012048

References

[1] Kratica, J., V. Kovačević-Vujčić, and M. Čangalović, *Minimal Doubly Resolving Sets and The Strong Metric Dimension of Hamming Graphs*, Applicable Analysis and Discrete Mathematics **6** (2012), 63–71.

[2] Kratica, J., V. Kovačević-Vujčić, and M. Čangalović, *Minimal Doubly Resolving Sets and The Strong Metric Dimension of Some Complex Polytope*, Applied Mathematics and Computation **218** (2012), 9790–9801.

[3] Kusmayadi, T. A., S. Kuntari, D. Rahmadi, and F. A. Lathifah, *On The Strong Metric Dimension of Some Related Wheel Graph*, Far East Journal of Mathematical Sciences (FJMS) **99** (2016), no. 9, 1325–1334.

[4] Kuziak,D., I. G. Yero, and A. Rodríguez-Velázquez, *On The Strong Metric Dimension of Corona Product Graphs and Join Graph*, Discrete Applied Mathematics **161** (2013), 1022–1027.

[5] Oellermann, O. and J. Peters-Fransen, *The Strong Metric Dimension of Graphs and Digraphs*, Discrete Applied Mathematics **155** (2007), 356–364.

[6] Sebö, A. and E. Tannier, *On Metric Generators of Graph*, Mathematics and Operations Research **29(2)** (2004), 383–393.

[7] Weisstein, Eric. W., *Butterfly Graph*, MathWorld.

[8] Yi, E., *On Strong Metric Dimension Graphs and Their Complements*, Acta Mathematica Sinica **29(8)** (2013), 1479–1492.

IOP Conf. Series: Journal of Physics: Conf. Series **1008** (2018) 012049 doi:10.1088/1742-6596/1008/1/012049

Total edge irregularity strength of (n,t)-kite graph

Tri Winarsih and Diari Indriati

Department of Mathematics, Faculty of Mathematics and Natural Sciences, Universitas Sebelas Maret, Surakarta, Indonesia

E-mail: tri3.wiwin@gmail.com, diari_indri@yahoo.co.id

Abstract. Let $G(V,E)$ be a simple, connected, and undirected graph with vertex set V and edge set E. A total k-labeling is a map that carries vertices and edges of a graph G into a set of positive integer labels $\{1,2,\ldots,k\}$. An edge irregular total k-labeling $\lambda : V(G) \cup E(G) \rightarrow \{1,2,\ldots,k\}$ of a graph G is a labeling of vertices and edges of G in such a way that for any different edges e and f, weights $wt(e)$ and $wt(f)$ are distinct. The weight $wt(e)$ of an edge $e = xy$ is the sum of the labels of vertices x and y and the label of the edge e. The total edge irregularity strength of G, $tes(G)$, is defined as the minimum k for which a graph G has an edge irregular total k-labeling. An (n,t)-kite graph consist of a cycle of length n with a t-edge path (the tail) attached to one vertex of a cycle. In this paper, we investigate the total edge irregularity strength of the (n,t)-kite graph, with $n \geq 3$ and $t \geq 1$. We obtain the total edge irregularity strength of the (n,t)-kite graph is $tes((n,t)\text{-}kite) = \lceil \frac{n+t+2}{3} \rceil$.

Keywords : total k-labeling, edge irregular total k -labeling, total edge irregularity strength, $tes(G)$, (n,t)-kite graph.

Mathematics Subject Classification (2010) MSC 05C78

1. Introduction

Let $G(V,E)$ be a connected, simple, and undirected graph with vertex set V and edge set E. In [7], Wallis defined a labeling of a graph G is a mapping that carries a set of graph elements into a set of integers. The domain of mapping is a vertex set, or an edge set, or an union of vertex and edge sets. If the domain is vertex set or edge set, the labeling is called respectively vertex labeling or edge labeling. If the domain is union of vertex and edge set, then labeling is called total labeling. There are various kinds of labeling on graph and many ways to label graphs, one of them is an irregular total labeling (Gallian [2]).

A total k-labeling is a map that carries vertices and edges of a graph G into a set of positive integer labels $\{1,2,\ldots,k\}$. Bača et al. [1] defined an edge irregular total k-labeling $\lambda : V(G) \cup E(G) \rightarrow \{1,2,\ldots,k\}$ of a graph G is a labeling of vertices and edges of G in such a way that for any different edges e and f, weights e and f are distinct. The weight of an edge $e = xy$, denoted by $wt(e)$, is the sum of the labels of vertices x and y and the label of the edge e, that is $wt(e) = \lambda(x) + \lambda(e) + \lambda(y)$. Moreover, the minimum k for which the graph G has

IOP Conf. Series: Journal of Physics: Conf. Series **1008** (2018) 012049 doi:10.1088/1742-6596/1008/1/012049

an edge irregular total k-labeling is called the total edge irregularity strength of the graph G, denoted by $tes(G)$.

There are various classes of graph have been determined for the total edge irregularity strength. Bača et al. [1], put forward the lower bound and upper bound of total edge irregularity strength for any graph $G(V, E)$ in terms of $|V(G)|$ and $|E(G)|$, which may be state as the equation (1),

$$\left\lceil \frac{|E|+2}{3} \right\rceil \leq tes(G) \leq |E|. \tag{1}$$

In the same paper, Bača et al. [1] determined the total edge irregularity strength of path P_n, cycle C_n, star S_n, wheel W_n, and friendship graph F_n, that are $tes(P_n) = tes(C_n) = \lceil \frac{n+2}{3} \rceil$, $tes(S_n) = \lceil \frac{n+1}{2} \rceil$, $tes(W_n) = \lceil \frac{2n+2}{3} \rceil$ for $n \geq 3$, and $tes(F_n) = \lceil \frac{3n+2}{3} \rceil$, respectively. In 2012, Haque [3] determined $tes(G)$ of generalized Pertersen graphs $P(n, k)$. Then, $tes(G)$ of tree G with edge set E and maximum degree Δ had been found by Ivančo and Jendrol' [5], that is $tes(G) = max\{\lceil \frac{\Delta+1}{2} \rceil, \lceil \frac{|E|+2}{3} \rceil\}$.

Furthermore, Jendrol' et al. [6] have determined the total edge irregularity strength of complete graphs and complete bipartite graphs, that are $tes(K_n) = \lceil \frac{n^2-n-4}{6} \rceil$, $n \neq 5$ and $tes(K_{m,n}) = \lceil \frac{m \cdot n+2}{3} \rceil$ for $n, m \geq 2$. Indriati et al. [4] determined the total edge irregularity strength of generalized helm, that is $tes(H_n^1) = \lceil \frac{4n+2}{3} \rceil$, $tes(H_n^2) = \lceil \frac{5n+2}{3} \rceil$, and $tes(H_n^m) = \lceil \frac{(m+3)n+2}{3} \rceil$ for $n \geq 3$ and $m \equiv 0 \ (mod\ 3)$. In this paper, we investigate the total edge irregularity strength of (n, t)-kite graph, with $n \geq 3$ and $t \geq 1$.

2. Main Result

An (n, t)-kite graph consist of a cycle of length n with a t-edge path (the tail) attached to one vertex of a cycle (Wallis et al. [8]). An (n, t)-kite graph has $n+t$ vertices and $n+t$ edges. Let the vertex set of (n, t)-kite be $V((n, t)-kite) = \{v_i : 1 \leq i \leq n\} \cup \{u_j : 1 \leq j \leq t\}$ and the edge set of (n, t)-kite be $E((n, t)-kite) = \{v_i v_{i+1} : 1 \leq i \leq n-1\} \cup \{v_n v_1\} \cup \{u_j u_{j+1} : 1 \leq j \leq t-1\} \cup \{u_t v_1\}$. The (n, t)-kite graph ilustrated by Figure 1.

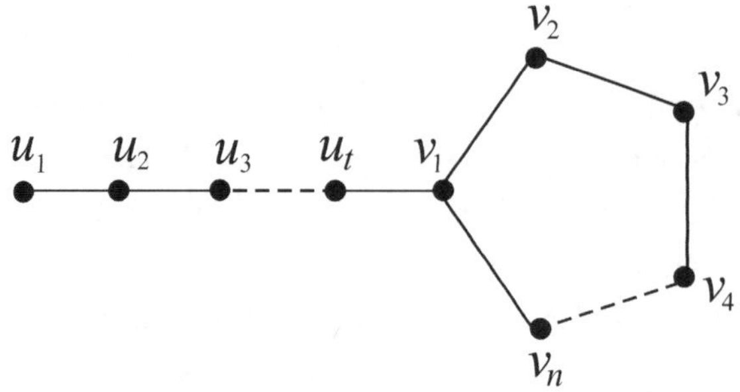

Figure 1. The (n, t)-kite graph

In the following theorem, we present the total edge irregularity strength of (n, t)-kite graph for $n \geq 3$ and $t \geq 1$.

Theorem 2.1 *Every (n,t)-kite with $n \geq 3$ and $t \geq 1$ satisfies tes((n,t)-kite)=$\lceil \frac{n+t+2}{3} \rceil$.*

Proof. From the lower bound of total edge irregularity strength we have that $tes((n,t)$-$kite) \geq \lceil \frac{n+t+2}{3} \rceil$, for $n \geq 3$ and $t \geq 1$. To prove the equality, it is sufficient to show the existence of an edge irregular total k-labeling, with $k = \lceil \frac{n+t+2}{3} \rceil$. We describe a total k-labeling $\lambda : V((n,t) - kite) \cup E((n,t) - kite) \to \{1, 2, \ldots, k\}$ as follows

$$\lambda(u_j) = \begin{cases} 1, & \text{if } 1 \leq j \leq 2; \\ \lceil \frac{j+1}{3} \rceil, & \text{if } 1 \leq j \leq t-1; \end{cases}$$

$$\lambda(u_j u_{j+1}) = \lceil \tfrac{j}{3} \rceil, \quad \text{for } 1 \leq j \leq t-1;$$

$$\lambda(u_t v_1) = \lceil \tfrac{t}{3} \rceil.$$

For $t \equiv 0 \ (mod\ 3)$

$$\lambda(v_i) = \begin{cases} \lceil \frac{t}{3} \rceil + \lceil \frac{i}{2} \rceil, & \text{if } 1 \leq i \leq n+1-\lceil \frac{n+t+1}{3} \rceil + \lceil \frac{t}{3} \rceil; \\ n+2+\lceil \frac{t}{3} \rceil - i, & \text{if } n+2-\lceil \frac{n+t+1}{3} \rceil + \lceil \frac{t}{3} \rceil \leq i \leq n; \end{cases}$$

$$\lambda(v_i v_{i+1}) = \begin{cases} \lceil \frac{t}{3} \rceil + \lceil \frac{i+1}{2} \rceil, & \text{if } 1 \leq i \leq n+1-\lceil \frac{n+t+1}{3} \rceil + \lceil \frac{t}{3} \rceil; \\ n+1+\lceil \frac{t}{3} \rceil - i, & \text{if } n+2-\lceil \frac{n+t+1}{3} \rceil + \lceil \frac{t}{3} \rceil \leq i \leq n-1; \end{cases}$$

$$\lambda(v_n v_1) = 1 + \lceil \tfrac{t}{3} \rceil.$$

For $t \equiv 1 \ (mod\ 3)$

$$\lambda(v_i) = \begin{cases} \lceil \frac{t}{3} \rceil - 1 + \lceil \frac{i+1}{2} \rceil, & \text{if } 1 \leq i \leq n-\lceil \frac{n+t+1}{3} \rceil + \lceil \frac{t}{3} \rceil; \\ n+1+\lceil \frac{t}{3} \rceil - i, & \text{if } n+1-\lceil \frac{n+t+1}{3} \rceil + \lceil \frac{t}{3} \rceil \leq i \leq n; \end{cases}$$

$$\lambda(v_i v_{i+1}) = \begin{cases} \lceil \frac{t}{3} \rceil - 1 + \lceil \frac{i+1}{2} \rceil, & \text{if } 1 \leq i \leq n-\lceil \frac{n+t+2}{3} \rceil + \lceil \frac{t}{3} \rceil; \\ n+1+\lceil \frac{t}{3} \rceil - i, & \text{if } n+1-\lceil \frac{n+t+2}{3} \rceil + \lceil \frac{t}{3} \rceil \leq i \leq n-1; \end{cases}$$

$$\lambda(v_n v_1) = 1 + \lceil \tfrac{t}{3} \rceil.$$

IOP Conf. Series: Journal of Physics: Conf. Series **1008** (2018) 012049 doi:10.1088/1742-6596/1008/1/012049

For $t \equiv 2 \ (mod\ 3)$

$$\lambda(v_i) = \begin{cases} \lceil \frac{t}{3} \rceil + \lceil \frac{i}{2} \rceil, & \text{if } 1 \le i \le n+1-\lceil \frac{n+t+2}{3} \rceil + \lceil \frac{t}{3} \rceil; \\ n+2+\lceil \frac{t}{3} \rceil - i, & \text{if } n+2-\lceil \frac{n+t+2}{3} \rceil + \lceil \frac{t}{3} \rceil \le i \le n; \end{cases}$$

$$\lambda(v_i v_{i+1}) = \begin{cases} \lceil \frac{t}{3} \rceil - 1 + \lceil \frac{i+1}{2} \rceil, & \text{if } 1 \le i \le n+1-\lceil \frac{n+t+2}{3} \rceil + \lceil \frac{t}{3} \rceil; \\ n+\lceil \frac{t}{3} \rceil - i, & \text{if } n+2-\lceil \frac{n+t+2}{3} \rceil + \lceil \frac{t}{3} \rceil \le i \le n-1; \end{cases}$$

$$\lambda(v_n v_1) = \lceil \tfrac{t}{3} \rceil.$$

Observe that under k-labeling λ the weights of the edges of (n,t)-kite graph are:

$$wt(u_j u_{j+1}) = j+2, \quad \text{for } 1 \le j \le t-1;$$

$$wt(u_t v_1) = t+2;$$

$$wt(v_i v_{i+1}) = \begin{cases} t + \lceil \frac{3(i+1)}{2} \rceil, & \text{if } 1 \le i \le \lceil \frac{2n-1}{3} \rceil; \\ t+3(n-i)+4, & \text{if } \lceil \frac{2n+2}{3} \rceil \le i \le n-1; \end{cases}$$

$$wt(v_n v_1) = t+4.$$

The weights of the edges of (n,t)-kite graph under the labeling λ constitute the set $\{3,4,5,\dots,n+t+2\}$, that is different for every edge irregular totak k-labeling. The function λ is mapping from $V((n,t)-kite) \cup E((n,t)-kite)$ into $\{1,2,\dots,\lceil \frac{n+t+2}{3} \rceil\}$.

The total k-labeling λ has the required properties on an edge irregular total labeling. We then have $tes((n,t)\text{-}kite) \le \lceil \frac{n+t+2}{3} \rceil$. However, by equation (1), $tes((n,t)\text{-}kite) \ge \lceil \frac{|E(G)|+2}{3} \rceil = \lceil \frac{n+t+2}{3} \rceil$. Then, we proved the theorem. □

We give an example of an edge irregular total labeling of (n,t)-kite graph for $n=5$ and $t=4$ in Figure 2.

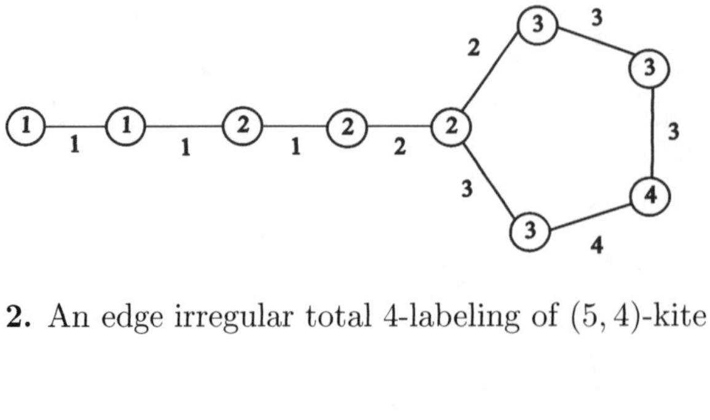

Figure 2. An edge irregular total 4-labeling of $(5,4)$-kite graph

3. Conclusion

According to the discussion above it can be concluded that the total edge irregularity strength of (n,t)-kite graph, $tes((n,t)\text{-}kite)=\lceil\frac{n+t+2}{3}\rceil$, for $n \geq 3$ and $t \geq 1$. Furthermore, we propose the following open problem.

Open Problem. Let G be a lollipop graph $L_{m,n}$ with $m \geq 3$ and $n \geq 1$ that obtained by joining a complete graph K_m to a path graph P_n with a bridge. Determine the total edge irregularity strength of graph G, $m \neq 5$.

Acknowledgments

We gratefully acknowledge the support from Department of Mathematics, Faculty of Mathematics and Natural Sciences, Universitas Sebelas Maret. Then, we wish to thank the referees for their valuable suggestions and references, which helped to improve the paper.

References

[1] Bača M, Jendrol' S, Miller M and Ryan J. On Irregular Total Labelings 2007 *Discrete Math.* **307** 1378-1388.

[2] Gallian J A. A Dynamic Survey of Graph Labeling 2016 *The Electronic Journal of Combinatorics.* **19** 1-284 #DS6.

[3] Haque K M M. Irregular Total Labelings of Generalized Petersen Graphs 2012 *Theory Comput Syst.* **50** 537-544.

[4] Indriati D, Widodo, Wijayanti I E and Sugeng K A. On the Total Edge Irregularity Strength of Generalized Helm 2013 *AKCE International Journal Graphs Combinatoric.* **10** (2) 147-155.

[5] Ivančo J and Jendrol' S. Total Edge Irregularity Strength of Trees 2006 *Discuss. Math. Graph Theory.* **26** 449-456.

[6] Jendrol' S, Miškuf J and Soták R. Total Edge Irregularity Strength of Complete Graphs and Complete Bipartite Graphs 2010 *Discrete Mathematics.* **310** 400-407.

[7] Wallis W D. Magic Graphs 2001 Boston: Birkhäuser.

[8] Wallis W D, Baskoro E T, Miller M and Slamin. Edge Magic Total Labeling 2000 *Australasian Journal of Combinatorics.* **22** 177-190.

The local metric dimension of starbarbell graph, $K_m \odot P_n$ graph, and Möbius ladder graph

Wahyu Tri Budianto and Tri Atmojo Kusmayadi

Department of Mathematics, Faculty of Mathematics and Natural Sciences, Universitas Sebelas Maret, Surakarta, Indonesia

E-mail: wahyutri25@gmail.com, tri.atmojo.kusmayadi@gmail.com

Abstract. For an ordered set $W = \{w_1, w_2, ..., w_n\}$ of n distinct vertices in a nontrivial connected graph G, the representation of a vertex v of G with respect to W is the n-vector $r(v|W) = (d(v, w_1), d(v, w_2), ..., d(v, w_n))$. W is a local metric set of G if $r(u|W) \neq r(v|W)$ for every pair of adjacent vertices u, v in G. Local metric set with minimum cardinality is called local metric basis of G and its cardinality is the local metric dimension of G and denoted by $lmd(G)$. Starbarbell graph $SB_{m_1,m_2,...,m_n}$ is a graph obtained from a star graph S_n and n complete graphs K_{m_i} by merging one vertex from each K_{m_i} and the ith leaf of S_n, where $m_i \geq 3$, $1 \leq i \leq n$, and $n \geq 2$. $K_m \odot P_n$ graph is a graph obtained from a complete graph K_m and m copies of path graph P_n, and then joining by an edge each vertex from the ith copy of P_n with the ith vertex of K_m. Möbius ladder graph M_n is a graph obtained from a cycle graph C_n by connecting every pair of vertices u, v in C_n if $d(u, v) = diam(C_n)$ for $n \geq 5$. In this paper, we determine the local metric dimension of starbarbell graph, $K_m \odot P_n$ graph, and Möbius ladder graph for even positive integers $n \geq 6$.

1. Introduction

In 1975, Slater [9] introduced the concept of metric dimension of a graph, where metric generator was called locating set. Then Harary and Melter [3] independently introduced the same concept in 1976, where metric generator was called resolving set. Let G be a connected graph with vertex set $V(G)$. A set $W \subset V(G)$ is called a metric generator of G if $d(u, x) \neq d(v, x)$ for any pair of vertices u and v in G and some vertices x in W, where $d(u, x)$ is the distance between u and x. Until now, there are several variations of metric generator, and one of them was introduced by Okamoto et al. [7] in 2010. For an ordered set $W = \{w_1, w_2, ..., w_n\}$ of n distinct vertices in a nontrivial connected graph G, the representation of a vertex v of G with respect to W is the n-vector $r(v|W) = (d(v, w_1), d(v, w_2), ..., d(v, w_n))$. The set W is a local metric set of G if $r(u|W) \neq r(v|W)$ for every pair of adjacent vertices u and v in G. The local metric set W with minimum cardinality is called local metric basis of G and the cardinality of local metric basis of G is the local metric dimension of G and denoted by $lmd(G)$.

Some authors have investigated the local metric dimension of some graph classes. In 2014 Kristina et al. [5] found the local metric dimension of comb product of cycle graph and star graph. In the same year, Ningsih et al. [6] found the local metric dimension of comb product of cycle graph and path graph. Then in 2015, Barragán-Ramírez et al. [1] observed the local metric dimension of subgraph amalgamation of graphs. In 2016 Rodríguez-Velázquez et al. [8] obtained the local metric dimension of corona product graphs. In this paper, we determine the local metric dimension of starbarbell graph $SB_{m_1,m_2,\ldots,m_n}$, $K_m \odot P_n$ graph, and Möbius ladder graph M_n for even positive integers n.

2. Main Results

2.1. The Local Metric Dimension of Starbarbell Graph

The starbarbell graph $SB_{m_1,m_2,\ldots,m_n}$ is a graph obtained from a star graph S_n and n complete graphs K_{m_i} by merging one vertex from each K_{m_i} and the ith leaf of S_n, where $m_i \geq 3$, $1 \leq i \leq n$, and $n \geq 2$. The starbarbell graph $SB_{m_1,m_2,\ldots,m_n}$ can be depicted as in Figure 1.

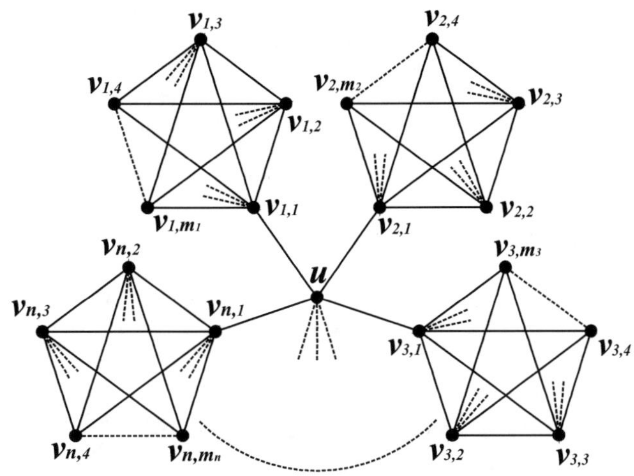

Figure 1. Starbarbell graph $SB_{m_1,m_2,\ldots,m_n}$

Theorem 2.1 *Let* $SB_{m_1,m_2,\ldots,m_n}$ *be the starbarbell graph, then* $lmd(SB_{m_1,m_2,\ldots,m_n}) = \sum_{i=1}^{n}(m_i-2)$.

Proof. Let $SB_{m_1,m_2,\ldots,m_n}$ be the starbarbell graph with $m_i \geq 3$, $1 \leq i \leq n$, and $n \geq 2$.

(i) We will show $lmd(SB_{m_1,m_2,\ldots,m_n}) \geq \sum_{i=1}^{n}(m_i-2)$.
To prove this case, the vertex set $V(SB_{m_1,m_2,\ldots,m_n})$ is partitioned into $n+1$ sets,

$$\begin{aligned} V_1 &= \{v_{1,1}, v_{1,2}, \ldots, v_{1,m_1}\}, \\ V_2 &= \{v_{2,1}, v_{2,2}, \ldots, v_{2,m_2}\}, \\ &\vdots \\ V_n &= \{v_{n,1}, v_{n,2}, \ldots, v_{n,m_n}\}, \text{ and} \\ V_{n+1} &= \{u\}. \end{aligned}$$

The cardinality of $V(SB_{m_1,m_2,\ldots,m_n})$ is $\sum_{i=1}^{n} m_i + 1$. Assume $W \subseteq V(SB_{m_1,m_2,\ldots,m_n})$ is chosen such that $|W| < \sum_{i=1}^{n}(m_i-2)$, then $|W^c| = |V(SB_{m_1,m_2,\ldots,m_n}) - W| > 2n+1$. Based on pigeonhole principle, there will be at least one partition V_i for $1 \leq i \leq n$ which at least three elements of V_i are elements of W^c too. Thus, there will be at least two distinct adjacent vertices $v_{i,j}$ and $v_{i,k}$, with $2 \leq j,k \leq m_i$ where

$$\begin{aligned} d(v_{i,j}, v_{i,1}) &= d(v_{i,k}, v_{i,1}) = 1, \\ d(v_{i,j}, v_{i,p}) &= d(v_{i,k}, v_{i,p}) = 1, \quad 2 \leq p \leq m_i, j \neq p \neq k, \\ d(v_{i,j}, u) &= d(v_{i,k}, u) = 2, \\ d(v_{i,j}, v_{q,1}) &= d(v_{i,k}, v_{q,1}) = 3, \quad 1 \leq q \leq n, i \neq q, \text{ and} \\ d(v_{i,j}, v_{q,r}) &= d(v_{i,k}, v_{q,r}) = 4, \quad 2 \leq r \leq m_q. \end{aligned}$$

IOP Conf. Series: Journal of Physics: Conf. Series **1008** (2018) 012050 doi:10.1088/1742-6596/1008/1/012050

It means we can not construct W as a local metric set if $|W| < \sum_{i=1}^{n}(m_i - 2)$. So $lmd(SB_{m_1,m_2,...,m_n}) \geq \sum_{i=1}^{n}(m_i - 2)$.

(ii) Assume $W = \{v_{i,j}\}$ where $1 \leq i \leq n$ and $3 \leq j \leq m_i$. The cardinality of W is $\sum_{i=1}^{n}(m_i - 2)$. Then representations of every vertex of $SB_{m_1,m_2,...,m_n}$ with respect to W are

$$\begin{aligned}
r(v_{1,1}|W) &= (1,1,...,1,3,3,...,3,...,3,3,...,3),\\
r(v_{1,2}|W) &= (1,1,...,1,4,4,...,4,...,4,4,...,4),\\
&\vdots\\
r(v_{1,m_1}|W) &= (1,1,...,0,4,4,...,4,...,4,4,...,4),\\
r(v_{2,1}|W) &= (3,3,...,3,1,1,...,1,...,3,3,...,3),\\
r(v_{2,2}|W) &= (4,4,...,4,1,1,...,1,...,4,4,...,4),\\
&\vdots\\
r(v_{2,m_2}|W) &= (4,4,...,4,1,1,...,0,...,4,4,...,4),\\
&\vdots\\
r(v_{n,1}|W) &= (3,3,...,3,3,3,...,3,...,1,1,...,1),\\
r(v_{n,2}|W) &= (4,4,...,4,4,4,...,4,...,1,1,...,1),\\
&\vdots\\
r(v_{n,m_n}|W) &= (4,4,...,4,4,4,...,4,...,1,1,...,0), \text{ and}\\
r(u|W) &= (2,2,...,2,2,2,...,2,...,2,2,...,2).
\end{aligned}$$

Every pair of adjacent vertices have distinct representations with respect to W, so W is a local metric basis for starbarbell graph $SB_{m_1,m_2,...,m_n}$. □

2.2. The Local Metric Dimension of $K_m \odot P_n$ Graph

By using the definition from Frucht and Harary [2], the corona product $K_m \odot P_n$ graph is a graph obtained from a complete graph K_m and m copies of path graph P_n, and then joining by an edge each vertex from the ith copy of P_n with the ith vertex of K_m. The $K_m \odot P_n$ graph can be depicted as in Figure 2.

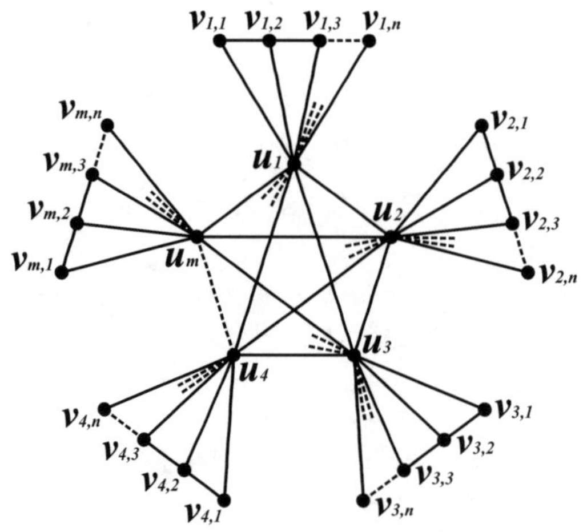

Figure 2. $K_m \odot P_n$ graph

Theorem 2.2 *For all positive integers m and n,*

$$lmd(K_m \odot P_n) = \begin{cases} 1, & m = n = 1; \\ 2, & m = 1,\ 2 \le n \le 5; \\ m-1, & m \ge 2,\ n = 1; \\ m\lfloor \frac{n+2}{4} \rfloor, & \begin{cases} m = 1, n \ge 6; \\ m \ge 2, n \ge 2. \end{cases} \end{cases}$$

Proof. We consider four cases based on the values of m and n.

Case 1. $m = n = 1$.
The corona product $K_1 \odot P_1$ graph is a path with two vertices, so it is easy to say that $lmd(K_m \odot P_n) = 1$ for $m = 1$ and $n = 1$.

Case 2. $m = 1$ and $2 \le n \le 5$.
The corona product $K_1 \odot P_n$ graph for $2 \le n \le 5$ is a graph which every single vertex in $K_1 \odot P_n$ belongs to some complete graphs K_3, so $lmd(K_1 \odot P_n) \neq 1$. Assume $W = \{u_1, v_k\}$ with $k = 2$ for $n = 2$ and $k = 3$ for $3 \le n \le 5$, then every pair of adjacent vertices have distinct representations with respect to W. Therefore, W is a local metric basis and $lmd(K_m \odot P_n) = 2$ for $m = 1$ and $2 \le n \le 5$.

Case 3. $m \ge 2$ and $n = 1$.
P_1 is an empty graph. Then by using the results of Rodríguez-Velázquez et al. [8] and Okamoto et al. [7] we have $lmd(K_m \odot P_1) = lmd(K_m) = m - 1$.

Case 4. $m = 1$, $n \ge 6$ and $m \ge 2$, $n \ge 2$.

(i) $m = 1$ and $n \ge 6$.
By the same reason with **Case 2**, we have $lmd(K_1 \odot P_n) \neq 1$ for $n \ge 6$. Assume $W \subseteq V(K_1 \odot P_n)$ is a non-empty set, then we consider three conditions below.

(a) If $v_{1,1}$, $v_{1,2}$, $v_{1,3} \notin W$ then $d(v_{1,1}, u_1) = (v_{1,2}, u_1) = 1$ and $d(v_{1,1}, v_{1,r}) = (v_{1,2}, v_{1,r}) = 2$, for $4 \le r \le n$, so $r(v_{1,1}|W) = r(v_{1,2}|W)$. We know that $v_{1,1}$ and $v_{1,2}$ are adjacent vertices, hence W is not a local metric set. In other words, if W is a local metric set then at least one of three vertices $v_{1,1}$, $v_{1,2}$ or $v_{1,3}$ belongs to W.

(b) By the same reason with (a), if W is a local metric set then at least one of three vertices $v_{1,(n-2)}$, $v_{1,(n-1)}$ or $v_{1,n}$ belongs to W.

(c) Assume a vertex $v_{1,t}$, for $1 \le t \le n-4$, belongs to W. If all of the vertices $v_{1,(t+1)}$, $v_{1,(t+2)}$, $v_{1,(t+3)}$, and $v_{1,(t+4)}$ do not belong to W, then $d(v_{1,(t+2)}, u_1) = (v_{1,(t+3)}, u_1) = 1$ and $d(v_{1,(t+2)}, v_{1,r}) = (v_{1,(t+3)}, v_{1,r}) = 2$, with $1 \le r \le t$ or $t+5 \le r \le n$, so $r(v_{1,(t+2)}|W) = r(v_{1,(t+3)}|W)$. We know that $v_{1,(t+2)}$ and $v_{1,(t+3)}$ are two adjacent vertices, then W is not a local metric set. In other words, if W is a local metric set and $v_{1,t} \in W$, then at least one of four vertices $v_{1,(t+1)}$, $v_{1,(t+2)}$, $v_{1,(t+3)}$, or $v_{1,(t+4)}$ belongs to W.

Based on three conditions above, the construction of W such that W is a local metric basis is by choosing every vertices $v_{1,r}$, for $r \equiv 3 \pmod 4$ and $1 \le r \le n$, as the elements of W. Then, if $n - r_{max} = 3$, we have to choose one of the vertices $v_{1,(n-2)}$, $v_{1,(n-1)}$, or $v_{1,n}$ as the element of W. Hence the cardinality of W is $\lfloor \frac{n+2}{4} \rfloor$.

(ii) $m \ge 2$ and $n \ge 2$.
By using the results of Rodríguez-Velázquez et al. [8], for $2 \le n \le 5$ we have $lmd(K_m \odot P_n) = m \cdot (lmd(K_1 + P_n) - 1) = m \cdot (lmd(K_1 \odot P_n) - 1) = m \cdot (2-1) = m$. For $2 \le n \le 5$ we have $m = m\lfloor \frac{n+2}{4} \rfloor$. Then for $n \ge 6$, again, by using the results of Rodríguez-Velázquez et al. [8], we have $lmd(K_m \odot P_n) = m \cdot lmd(K_1 + P_n) = m \cdot lmd(K_1 \odot P_n) = m\lfloor \frac{n+2}{4} \rfloor$. □

2.3. The Local Metric Dimension of Möbius Ladder Graph

Harary and Guy [4] defined the Möbius ladder graph M_n is a graph obtained from a cycle graph C_n by connecting every pair of vertices u, v in C_n if $d(u,v) = diam(C_n)$ for $n \geq 5$. The Möbius ladder graph M_n can be depicted as in Figure 3 and Figure 4.

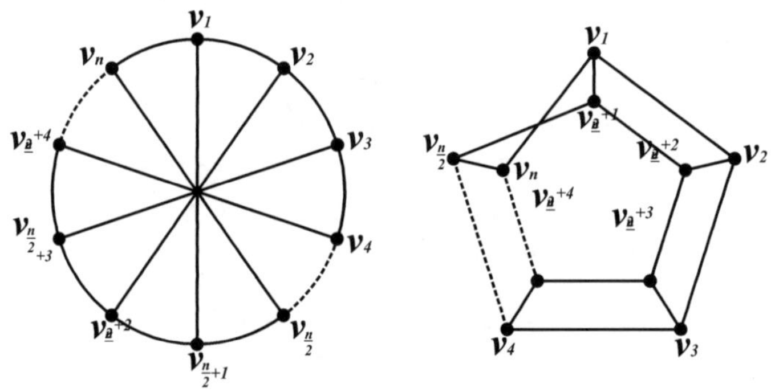

Figure 3. Möbius ladder graph M_n for even positive integers n

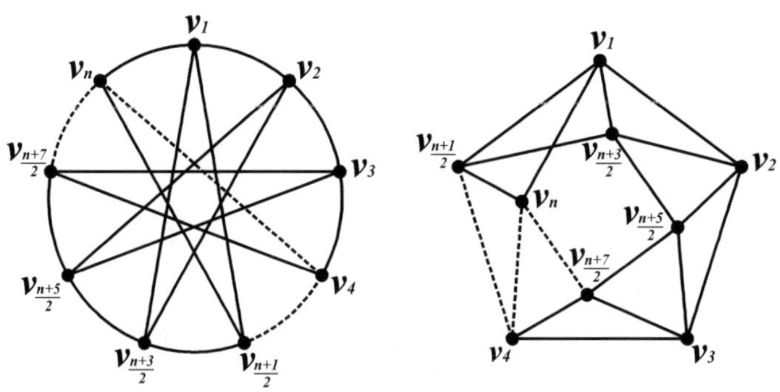

Figure 4. Möbius ladder graph M_n for odd positive integers n

Theorem 2.3 *Let M_n be the Möbius ladder graph. Then for even positive integers $n \geq 6$*

$$lmd(M_n) = \begin{cases} 1, & n \equiv 2(mod4); \\ 2, & n \equiv 0(mod4). \end{cases}$$

Proof. We consider two cases based on the values of n.

Case 1. $n \equiv 2(mod4)$.

The Möbius ladder graph M_n for $n \equiv 2(mod4)$ is a bipartite graph. Then by using the result of Okamoto et al. [7] we have $lmd(M_n) = 1$.

Case 2. $n \equiv 0(mod4)$.

(i) We will show that $lmd(M_n) \neq 1$.
Without loss of generality, assume $W = \{v_1\}$, then we obtain

$$r(v_{\frac{n+4}{4}}|W) = r(v_{\frac{n+8}{4}}|W) = r(v_{\frac{3n}{4}}|W) = r(v_{\frac{3n+4}{4}}|W) = (\tfrac{n}{4}).$$

Every pair of vertices $\{v_{\frac{n+4}{4}}, v_{\frac{n+8}{4}}\}$, $\{v_{\frac{3n}{4}}, v_{\frac{3n+4}{4}}\}$, and $\{v_{\frac{n+4}{4}}, v_{\frac{3n+4}{4}}\}$ are adjacent, hence W is not local metric set. So $lmd(M_n) \neq 1$, for $n \equiv 0(mod4)$.

(ii) Assume $W = \{v_1, v_2\}$, then representations of every vertex of M_n with respect to W are

$$
\begin{array}{lclcl}
r(v_1|W) & = & (0,1), & & \\
r(v_2|W) & = & (1,0), & & \\
r(v_3|W) & = & r(v_{\frac{n}{2}+2}|W) & = & (2,1), \\
r(v_4|W) & = & r(v_{\frac{n}{2}+3}|W) & = & (3,2), \\
 & \vdots & & & \\
r(v_{\frac{n+4}{4}}|W) & = & r(v_{\frac{3n}{4}}|W) & = & (\frac{n}{4}, \frac{n}{4}-1), \\
r(v_{\frac{n+8}{4}}|W) & = & r(v_{\frac{3n+4}{4}}|W) & = & (\frac{n}{4}, \frac{n}{4}), \\
r(v_{\frac{n+12}{4}}|W) & = & r(v_{\frac{3n+8}{4}}|W) & = & (\frac{n}{4}-1, \frac{n}{4}), \\
 & \vdots & & & \\
r(v_{\frac{n}{2}}|W) & = & r(v_{n-1}|W) & = & (2,3), \\
r(v_{\frac{n}{2}+1}|W) & = & r(v_n|W) & = & (1,2).
\end{array}
$$

Every pair of adjacent vertices have distinct representation, so W is a local metric basis and $lmd(M_n) = 2$, for $n \equiv 0(mod4)$. □

3. Conclusion

It can be concluded that the strong metric dimension of a starbarbell graph $SB_{m_1,m_2,\ldots,m_n}$, $K_m \odot P_n$ graph, and a Möbius ladder graph M_n for even positive integers $n \geq 6$ are as stated in Theorem 2.1, Theorem 2.2, and Theorem 2.3, respectively.

Open Problem : Determine the local metric dimension of Möbius ladder graph M_n for odd positive integers $n \geq 5$.

Acknowledgments

The authors gratefully acknowledge the support from Department of Mathematics, Faculty of Mathematics and Natural Sciences, Universitas Sebelas Maret Surakarta. Then, we wish to thank the referees for their suggestions, which helped to improve the paper.

References

[1] Barragán-Ramírez, G. A., R. Simanjuntak, S. W. Saputro, and S. Uttunggadewa, *The Local Metric Dimension of Subgraph-Amalgamation of Graphs*, Universitat Rovira i Virgili and Institut Teknologi Bandung (2015).
[2] Frucht, R. and F. Harary, *On The Corona of Two Graphs*, Aequationes Mathematicae **4** (1970), 322-325.
[3] Harary, F. and R. A. Melter, *On The Metric Dimension of A Graph*, Ars Combinatoria **2** (1976), 191-195.
[4] Harary, F. and R. K. Guy, *On The Möbius Ladders*, Canad. Math. Bull. **10** (1967), 493-496.
[5] Kristina, M., N. Estuningsih, and L. Susilowati, *Dimensi Metrik Lokal pada Graf Hasil Kali Comb dari Graf Siklus dan Graf Bintang*, Jurnal Matematika **1** (2014), 1-9.
[6] Ningsih, E. U. S., N. Estuningsih, and L. Susilowati, *Dimensi Metrik Lokal pada Graf Hasil Kali Comb dari Graf Siklus dan Graf Lintasan*, Jurnal Matematika **1** (2014), 24-33.
[7] Okamoto, F., B. Phinezy, and P. Zhang, *The Local Metric Dimension of A Graph*, Matematica Bohemica **135** (2010), 239-255.
[8] Rodríguez-Velázquez, J. A., G. A. Barragán-Ramírez, and C. G. Gómez, *On The Local Metric Dimension of Corona Product Graphs*, Bull. Malays. Math. Sci. Soc. **39** (2016), 157-173.
[9] Slater, P. J., *Leaves of Trees*, Congressus Numerantium **14** (1975), 549-559.

IOP Conf. Series: Journal of Physics: Conf. Series **1008** (2018) 012051 doi:10.1088/1742-6596/1008/1/012051

On the strong metric dimension of antiprism graph, king graph, and $K_m \odot K_n$ graph

Yuyun Mintarsih and Tri Atmojo Kusmayadi

Department of Mathematics, Faculty of Mathematics and Natural Sciences, Universitas Sebelas Maret, Surakarta, Indonesia

E-mail: yuyunaziz2@gmail.com, tri.atmojo.kusmayadi@gmail.com

Abstract. Let G be a connected graph with a set of vertices $V(G)$ and a set of edges $E(G)$. The interval $I[u, v]$ between u and v to be the collection of all vertices that belong to some shortest u-v path. A vertex $s \in V(G)$ is said to be strongly resolved for vertices u, $v \in V(G)$ if $v \in I[u, s]$ or $u \in I[v, s]$. A vertex set $S \subseteq V(G)$ is a strong resolving set for G if every two distinct vertices of G are strongly resolved by some vertices of S. The strong metric dimension of G, denoted by $sdim(G)$, is defined as the smallest cardinality of a strong resolving set. In this paper, we determine the strong metric dimension of an antiprism A_n graph, a king $K_{m,n}$ graph, and a $K_m \odot K_n$ graph. We obtain the strong metric dimension of an antiprim graph A_n are n for n odd and $n + 1$ for n even. The strong metric dimension of King graph $K_{m,n}$ is $m + n - 1$. The strong metric dimension of $K_m \odot K_n$ graph are n for $m = 1$, $n \geq 1$ and $mn - 1$ for $m \geq 2$, $n \geq 1$.

1. Introduction

The strong metric dimension was introduced by Sebö and Tannier [6] in 2004. Let G be a connected graph with a set of vertices $V(G)$ and a set of edges $E(G)$. Oelermann and Peters-Fransen [5] defined the interval $I[u, v]$ between u and v to be the collection of all vertices that belong to some shortest $u - v$ path. A vertex $s \in S$ is said to strongly resolve two vertices u and v if $u \in I[v, s]$ or $v \in I[u, v]$. A vertex set S of G is a strong resolving set for G if every two distinct vertices of G are strongly resolved by some vertices of S. The strong metric basis of G is a strong resolving set with minimal cardinality. The strong metric dimension of a graph G is defined as the cardinality of strong metric basis denoted by $sdim(G)$.

Some authors have investigated the strong metric dimension to some graph classes. Sebö and Tannier [6] observed the strong metric dimension of complete graph K_n, cycle graph C_n, and tree. Kratica et al. [2] observed the strong metric dimension of *hamming* graph $H_{n,k}$. At the same year, Kratica et al [3] determined the strong metric dimension of convex polytope D_n and T_n. Yi [8] determined that $sdim(G) = 1$ if only if G is path graph and $sdim(G) = n-1$ if only if G is complete graph. Kusmayadi et al. [4] determined the strong metric dimension of sunflower graph, t-fold wheel graph, helm graph, and friendship graph. In this paper, we determine the strong metric dimension of an antiprism graph, a king graph, and a $K_m \odot K_n$ graph.

2. Main Results

2.1. Strong Metric Dimension

Let G be a connected graph with a set of vertices $V(G)$, a set of edges $E(G)$, and $S=\{s_1, s_2, \ldots, s_k\} \in V(G)$. Oelermann and Peters-Fransen [5] defined the interval $I[u, v]$ between u and v to be the collection of all vertices that belong to some shortest $u - v$ path. A vertex $s \in S$ is said to strongly resolve two vertices u and v if $u \in I[v, s]$ or $v \in I[u, s]$. A vertex set S of G is a strong resolving set for G if every two distinct vertices of G are strongly resolved by some vertices of S. The strong metric basis of G is a strong resolving set with minimal cardinality. The strong metric dimension of a graph G is defined as the cardinality of strong metric basis denoted by $sdim(G)$. We often make use of the following lemma and properties about strong metric dimension given by Kratica et al. [3].

Lemma 2.1 *Let u, $v \in V(G)$, $u \neq v$,*

(i) $d(w,v) \leq d(u,v)$ for each w such that $u\ w \in E(G)$, and

(ii) $d(u,w) \leq d(u,v)$ for each w such that $v\ w \in E(G)$.

Then there does not exist vertex $a \in V(G)$, $a \neq u,v$ that strongly resolves vertices u and v.

Property 2.1 *If $S \subset V(G)$ is strong resolving set of graph G, then for every two vertices u, $v \in V(G)$ satisfying conditions* 1 *and* 2 *of Lemma* 2.1, *obtained $u \in S$ or $v \in S$.*

Property 2.2 *If $S \subset V(G)$ is strong resolving set of graph G, then for every two vertices u, $v \in V(G)$ satisfying $d(u, v) = diam(G)$, obtained $u \in S$ or $v \in S$.*

2.2. The Strong Metric Dimension of antiprism graph

Bača [1] defined the antiprism graph A_n for $n \geq 3$ is a 4-regular graph with $2n$ vertices and $4n$ edges. It consists of outer and inner C_n, while the two cycles connected by edges v_iu_i and $v_iu_{1+i(mod\ n)}$ for $i = 1, 2, 3, \ldots, n$. The antiprism graph A_n can be depicted as in Figure 1.

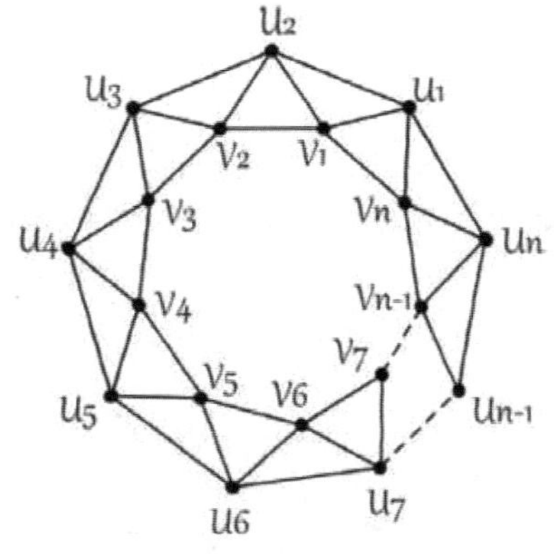

Figure 1. Antiprism graph A_n

Lemma 2.2 *For every integer $n \geq 3$ and n odd, if S is a strong resolving set of antiprism graph A_n then $|S| \geq n$.*

Proof. We know that S is a strong resolving set of antiprism graph A_n. Suppose that S contains at most n - 1 vertices, then $|S| < n$. Let $V_1, V_2 \subset V(A_n)$, with $V_1 = \{u_1, u_2, \ldots, u_n\}$ and $V_2 = \{v_1, v_2, \ldots, v_n\}$. Now, we define $S_1 = V_1 \cap S$ and $S_2 = V_2 \cap S$. Without loss of generality, we may take $\mid S_1 \mid = p$, $p > 0$ and $\mid S_2 \mid = q$, $q \geq 0$. Clearly $p + q \geq n$, if not then there are two distinct vertices v_a and v_b where $v_a \in V_1 \setminus S_1$ and $v_b \in V_2 \setminus S_2$ such that for every $s \in S$, we obtain $v_a \notin I[v_b, s]$ and $v_b \notin I[v_a, s]$. This contradicts with the supposition that S is a strong resolving set. Thus, $\mid S \mid \geq$ n. □

Lemma 2.3 *For every integer $n \geq 3$ and n odd, a set $S = \{u_1, u_2, \ldots, u_n\}$ is a strong resolving set of antiprism graph A_n.*

Proof. For every integer $i, j \in [1, n]$ with $1 \leq i < j \leq n$, a vertex $u_i \in S$ which strongly resolves v_i dan v_j so that $v_j \in I[v_i, u_j]$. Thus, $S = \{u_1, u_2, \ldots, u_n\}$ is a strong resolving set of antiprism graph A_n. □

Lemma 2.4 *For every integer $n \geq 3$ and n even, if S is a strong resolving set of antiprism graph A_n then $\mid S \mid \geq n+1$.*

Proof. We know that S is a strong resolving set of antiprism graph A_n. Suppose that S contains at most n vertices, then $\mid S \mid < n+1$. Let V_1, $V_2 \subset V(A_n)$, with $V_1 = \{u_1, u_2, \ldots, u_n\}$ and $V_2 = \{v_1, v_2, \ldots, v_n\}$. Now, we define $S_1 = V_1 \cap S$ and $S_2 = V_2 \cap S$. Without loss of generality, we may take $\mid S_1 \mid = p$, $p \geq 0$ and $\mid S_2 \mid = q$, $q \geq 0$. Clearly $p + q \geq n+1$, if not then there are two distinct vertices v_a and v_b where $v_a \in V_1 \setminus S_1$ and $v_b \in V_2 \setminus S_2$ such that for every $s \in S$, we obtain $v_a \notin I[v_b, s]$ and $v_b \notin I[v_a, s]$. This contradicts with the supposition that S is a strong resolving set. Thus, $\mid S \mid \geq$ n+1. □

Lemma 2.5 *For every integer $n \geq 3$ and n even, a set $S = \{u_1, u_2, \ldots, u_{\frac{n}{2}}, u_{\frac{n}{2}+1}, v_1, v_2, \ldots, v_{\frac{n}{2}}\}$ is a strong resolving set of antiprism graph A_n.*

Proof. We prove that for every two distinct vertices $u, v \in V(A_n) \setminus S$, $u \neq v$ there exists a vertex $s \in S$ which strongly resolves u and v. There are three possible pairs of vertices.

(i) A pair of vertices (u_i, u_j) with $i, j = \frac{n}{2}+2, \frac{n}{2}+3, \ldots, n$, $i \neq j$.
For every integer $i, j \in [\frac{n}{2}+2, n]$ with $i < j$, we obtain the shortest u_i - u_1 path: $u_i, u_{i+1}, \ldots, u_j, \ldots, u_n, u_1$. Thus, $u_j \in I[u_i, u_1]$.

(ii) A pair of vertices (v_i, v_j) with $i, j = \frac{n}{2}+1, \frac{n}{2}+2, \ldots, n$, $i \neq j$.
For every integer $i, j \in [\frac{n}{2}+1, n]$ with $i < j$, we obtain the shortest v_i - v_1 path: $v_i, v_{i+1}, \ldots, v_j, \ldots, v_n, v_1$. Thus, $v_j \in I[v_i, v_1]$.

(iii) A pair of vertices (u_i, v_j) with $i = \frac{n}{2}+2, \frac{n}{2}+3, \ldots, n$ dan $j = \frac{n}{2}+1, \frac{n}{2}+2, \ldots, n$.
For every integer $i \in [\frac{n}{2}+2, n]$ and $j \in [\frac{n}{2}+1, n]$ with $i \leq j$, we obtain the shortest u_i - v_1 path: $u_i, v_i, \ldots, v_j, \ldots, v_n, v_1$. Thus, $v_j \in I[u_i, v_1]$. Then, for every integer $i \in [\frac{n}{2}+2, n]$ and $j \in [\frac{n}{2}+1, n]$ with $i > j$, we obtain the shortest u_i - $v_{\frac{n}{2}}$ path: $u_i, v_{i-1}, \ldots, v_j, \ldots, v_{\frac{n}{2}+1}, v_{\frac{n}{2}}$. Thus, $v_j \in I[u_i, v_{\frac{n}{2}}]$.

From every possible pairs of vertices, there exists a vertex $s \in S$ which strongly resolves u, $v \in V(K_{m,n}) \setminus S$. Thus, S is a strong resolving set of antiprism graph A_n. □

Theorem 2.1 *Let A_n be the antiprism graph with $n \geq 3$, then*

$$sdim(A_n) = \begin{cases} n, & n \text{ odd;} \\ n+1, & n \text{ even.} \end{cases}$$

Proof. Let A_n be an antiprism graph with $n \geq 3$ and a set of vertices $V(A_n) = \{u_1, u_2, \ldots, u_n, v_1, v_2, \ldots, v_n\}$. We divide the proof into two cases according to the values of n.

(i) For n odd.
By using Lemma 2.3, a set $S = \{u_1, u_2, \ldots, u_n\}$ is a strong resolving set of antiprism graph A_n. According to Lemma 2.2, $\mid S \mid \geq n$ so that S is a strong metric basis of antiprism graph A_n. Hence, $sdim(A_n) = n$.

(ii) For n even.
By using Lemma 2.5, a set $S = \{u_1, u_2, \ldots, u_{\frac{n}{2}+1}, v_1, v_2, \ldots, v_{\frac{n}{2}}\}$ is a strong resolving set of antiprism graph A_n. According to Lemma 2.4, $\mid S \mid \geq n$ so that S is a strong metric basis of antiprism graph A_n. Hence, $sdim(A_n) = n+1$. □

2.3. The Strong Metric Dimension of King Graph
Weisstein [7] defined the king graph $K_{m,n}$ for $m, n \geq 2$ is the graph with mn vertices in which each vertex represents a square in an $m \times n$ chessboard, and each edge corresponds to a legal move by king.The king graph $K_{m,n}$ can be depicted as in Figure 2.

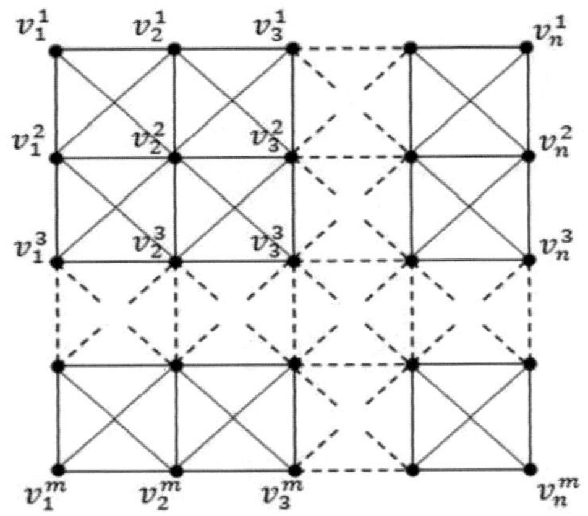

Figure 2. King graph $K_{m,n}$

Lemma 2.6 *For every integer m,n ≥ 2, if S is a strong resolving set of king graph $K_{m,n}$ then $\mid S \mid \geq$ m+n-1.*

Proof. We know that S is a strong resolving set of king graph $K_{m,n}$. Suppose that S contains at most $m + n - 2$ vertices, then $\mid S \mid < m + n - 1$. Let $V_1, V_2 \subset V(A_n)$, with $V_1 = \{v_1^1, v_2^1, \ldots, v_n^1, v_1^2, v_1^3, \ldots, v_1^m\}$ and $V_2 = \{v_2^2, v_3^2, \ldots, v_n^2, v_2^3, v_3^3, \ldots, v_n^3, \ldots, v_2^m, v_3^m, v_n^m\}$. Now, we define $S_1 = V_1 \cap S$ and $S_2 = V_2 \cap S$. Without loss of generality, we may take $\mid S_1 \mid = p$, $p > 0$ and $\mid S_2 \mid = q$, $q \geq 0$. Clearly $p + q \geq m+n-1$, if not then there are two distinct vertices v_a and v_b where $v_a \in V_1 \setminus S_1$ and $v_b \in V_2 \setminus S_2$ such that for every $s \in S$, we obtain $v_a \notin I[v_b, s]$ and $v_b \notin I[v_a, s]$. This contradicts with the supposition that S is a strong resolving set. Thus, $\mid S \mid \geq$ m+n-1. □

Lemma 2.7 *For every integer m,n ≥ 2, a set $S = \{v_1^1, v_2^1, \ldots, v_n^1, v_1^2, v_1^3, \ldots, v_1^m\}$ is a strong resolving set of king graph $K_{m,n}$.*

Proof. We prove that for every two distinct vertices $u, v \in V(K_{m,n}) \setminus S$, $u \neq v$ there exists a vertex $s \in S$ which strongly resolves u and v. There are three possible pairs of vertices.

(i) A pair of vertices (v_k^p, v_l^p) with $p = 2, 3, \ldots, m$ and $k, l = 2, 3, \ldots, n$.
For every integer $p \in [2, m]$ and $k, l \in [2, n]$ with $k < l$, we obtain the shortest v_l^p - v_1^p path: $v_l^p, v_{l-1}^p, \ldots, v_k^p, \ldots, v_2^p, v_1^p$. Thus, $v_k^p \in I[v_l^p, v_1^p]$.

(ii) A pair of vertices (v_q^i, v_q^j) with $i, j = 2, 3, \ldots, m$ and $q = 2, 3, \ldots, n$.
For every integer $i, j \in [2, m]$, $q \in [2, n]$ with $i < j$, we obtain the shortest v_q^j - v_q^1 path: $v_q^j, v_q^{j-1}, \ldots, v_q^i, \ldots, v_q^2, v_q^1$. Thus, $v_q^i \in I[v_q^j, v_q^1]$.

IOP Conf. Series: Journal of Physics: Conf. Series **1008** (2018) 012051 doi:10.1088/1742-6596/1008/1/012051

(iii) A pair of vertices (v_k^i, v_l^j) with $i, j = 2, 3, \ldots, m$ and $k, l = 2, 3, \ldots, n$.
For every integer $i, j \in [2, m]$, $k, l \in [1, n]$ with $i < j$ dan $k < l$, we obtain the shortest v_l^j - v_k^1 path: $v_l^j, v_{l-1}^{j-1}, v_{l-1}^{j-1}, \ldots, v_k^{j-1}, v_k^{j-2}, \ldots, v_k^i, v_k^{i-1}, \ldots, v_k^2, v_k^1$. Thus, $v_k^i \in I[v_l^j, v_k^1]$.

From every possible pairs of vertices, there exists a vertex $s \in S$ which strongly resolves $u, v \in V(K_{m,n}) \backslash S$. Thus S is a strong resolving set of king graph $K_{m,n}$. □

Theorem 2.2 *Let $K_{m,n}$ be the king graph with $m, n \geq 2$, then $sdim(K_{m,n}) = m + n - 1$.*

Proof. By using Lemma 2.7, a set $S = \{v_1^1, v_2^1, \ldots, v_n^1, v_1^2, v_1^3, \ldots, v_1^m\}$ is a strong resolving set of king graph $K_{m,n}$. According to Lemma 2.6, $|S| \geq m + n - 1$ so that S is a strong metric basis of king graph $K_{m,n}$. Hence, $sdim(K_{m,n}) = m + n - 1$. □

2.4. The Strong Metric Dimension of $K_m \odot K_n$ Graph

The corona product $K_m \odot K_n$ graph is graph obtained from K_m and K_n by taking one copy of K_m and n copies of K_n and joining by an edge each vertex from i^{th}-copy of K_n with the ith-vertex of K_m. The $K_m \odot K_n$ can be depicted as in Figure 3.

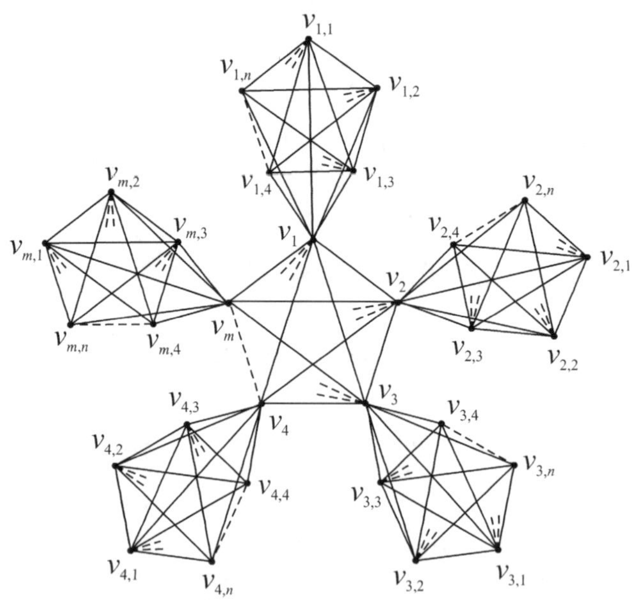

Figure 3. $K_m \odot K_n$ Graph

Lemma 2.8 *For every integer $m \geq 2$ and $n \geq 1$, if S is a strong resolving set of $K_m \odot K_n$ graph then $|S| \geq mn$-1.*

Proof. Let us consider a pair of vertices $(v_{i,k}, v_{j,l})$ for $i, j \in [1, m], k, l \in [1, n]$ satisfying both of the conditions of Lemma 2.1. According to Property 2.1, we obtain $v_{i,k} \in S$ or $v_{j,l} \in S$. It means that S contains one vertex from distinct sets $X = \{ v_{i,k}, v_{j,l} \}$ with $i, j = 1, 2, \ldots, m$ and $k, l = 1, 2, \ldots, n$. The minimum number of vertices from distinct sets X is $mn - 1$. Therefore, $|S| \geq mn - 1$. □

Lemma 2.9 *For every integer $m \geq 2$ and $n \geq 1$, a set $S = \{v_{1,1}, v_{1,2}, \ldots, v_{1,n-1}, v_{1,n}, v_{2,1}, v_{2,2}, \ldots, v_{2,n-1}, v_{2,n}, \ldots, v_{m,1}, v_{m,2}, \ldots, v_{m,n-1}\}$ is a strong resolving set of $K_m \odot K_n$ graph.*

Proof. For every integer $i \in [1, m-1]$, we obtain the shortest $v_{m,n}$ - $v_{i,k}$ path: $v_{m,n}, v_m, v_i, v_{i,k}$. So that $v_{i,k}$ strongly resolves a pair of vertices $(v_i, v_{m,n})$. Thus, $v_i \in I[v_{m,n}, v_{i,k}]$.

For a pair of vertices (v_i, v_j) with $i, j \in [1, m]$ and $i \neq j$, we obtain the shortest v_i - $v_{j,l}$ path: $v_i, v_j, v_{j,l}$. So that $v_{j,l}$ strongly resolves a pair of vertices (v_i, v_j). Thus, $v_j \in I[v_i, v_{j,l}]$.

Therefore $S = \{v_{1,1}, v_{1,2}, \ldots, v_{1,n-1}, v_{1,n}, v_{2,1}, v_{2,2}, \ldots, v_{2,n-1}, v_{2,n}, \ldots, v_{m,1}, v_{m,2}, \ldots, v_{m,n-1}\}$ is a strong resolving set of $K_m \odot K_n$ graph. □

Theorem 2.3 *Let $K_m \odot K_n$ graph with $m, n \geq 1$, then*

$$sdim(K_m \odot K_n) = \begin{cases} n, & m = 1,\ n \geq 1; \\ mn-1, & m \geq 2, n \geq 1. \end{cases}$$

Proof. Suppose $K_m \odot K_n$ graph with $m, n, k \geq 1$ and a set of vertices $V(K_m \odot K_n)=\{v_1, v_2, \ldots, v_m, v_{1,1}, v_{1,2}, \ldots, v_{1,n}, v_{2,1}, v_{2,2}, \ldots, v_{2,n}, \ldots, v_{m,1}, v_{m,2}, \ldots, v_{m,n}\}$. We divide the proof into two cases according to the values of m and n.

(i) For $m = 1$ and $n \geq 1$.
By using characterization Yi [8], $sdim(G) = n-1$ if only if $G \cong K_n$, so that $sdim(K_1 \odot K_n) = n$ because $K_1 \odot K_n \cong K_{n+1}$ then strong resolving set of $K_1 \odot K_n$ graph has n element. Hence $sdim(K_1 \odot K_n) = n$.

(ii) For $m \geq 2$ and $n \geq 1$.
By using Lemma 2.9 a set $S = \{v_{1,1}, v_{1,2}, \ldots, v_{1,n-1}, v_{1,n}, v_{2,1}, v_{2,2}, \ldots, v_{2,n-1}, v_{2,n}, \ldots, v_{m,1}, v_{m,2}, \ldots, v_{m,n-1}\}$ is a strong resolving set of $K_m \odot K_n$ graph. According to Lemma 2.8, $|\ S\ | \geq mn-1$ so that S is a strong metric basis of $K_m \odot K_n$ graph. Hence $sdim(K_m \odot K_n) = mn-1$. □

3. Conclusion

According to the discussion above, it can be concluded that the strong metric dimension of an antiprism graph, a king graph, and a $K_m \odot K_n$ graph are as stated in Theorem 2.1, Theorem 2.2, and Theorem 2.3 respectively.

Open Problem: Determine the strong metric dimension of a $K_m \odot^k K_n$ graph.

4. Acknowledgement

The authors gratefully acknowledge the support from Department of Mathematics, Faculty of Mathematics and Natural Sciences, Universitas Sebelas Maret, Surakarta. Then, we wish to thank the referees for their suggestions and references, which helped to improve the paper.

References

[1] Bača, M., F. Bashir, and A. Semaničová, *Face Antimagic Labeling of Antiprisms*, Util. Math **84** (2011), 209–224.

[2] Kratica, J., V. Kovačević-Vujčić, and M. Čangalović, *Minimal Doubly Resolving Sets and The The Strong Metric Dimension of Hamming Graphs*, Applicable Analysis and Discrete Mathematics **6** (2012), 63–71.

[3] Kratica, J., V. Kovačević-Vujčić, and M. Čangalović, *Minimal Doubly Resolving Sets and The The Strong Metric Dimension of Some Comvex Polytope*, Applied Mathematics and Computation **218** (2012), 9790–9801.

[4] Kusmayadi, T. A., S. Kuntari, D. Rahmadi, and F. A. Lathifah, *On The Strong Metric Dimension of Some Related Wheel Graph*, Far East Journal of Mathematical Sciences (FJMS) **99** (2016), no. 9, 1325–1334.

[5] Oelermann, O. and J. Peters-Fransen, *The Strong Metric Dimension of Graphs and Digraphs*, Discrete Applied Mathematics **155** (2007), 356–364.

[6] Sebö, A. and E. Tannier, *On Metric Generators of Graph*, Mathematics and Operations Research **29(2)** (2004), 383–393.

[7] Weisstein, E. W., *King Graph*, From MathWorld-A Wolfram Web Resource, http://mathworld.wolfram.com/KingGraph.html.

[8] Yi, E., *On Strong Metric Dimension Graphs and Their Complements*, Acta Mathematica Sinica **29(8)** (2013), 1479–1492.

On rainbow connection and strong rainbow connection number of amalgamation of prism graph $P_{3,2}$

C.D.R. Palupi[1], **W. Aribowo**[2], **Y. Irene**[3], **I. Hasanah**[4]
[1,2,3] Department of Mathematics, Faculty of Science and Technology, UIN Syarif Hidayatullah, Jakarta
[4] UIN Sultan Maulana Hasanuddin, Banten

E-mail: cynthia.dhevy13@mhs.uinjkt.ac.id

Abstract. Let G be a nontrivial connected graph. We follow Chartrand, et al in [2] for the definiton of rainbow connection and strong rainbow connection number. For $t \in N$ and $t \geq 2$, let $\{P_{(3,2)i} | i \in \{1, 2, \ldots, t\}\}$ is a finite collection of prism graph $P_{3,2}$ that has a fixed vertex v called a terminal. The amalgamation of prism graph, $Amal(P_{(3,2)i}, v)$, is a graph formed by taking all the element of finite collection of prism graph $P_{3,2}$ and identifying their terminal. This paper determines the rainbow connection and strong rainbow connection number from amalgamation of prism graph by a deductive reasoning method. The result are as follows: $rc(Amal(P_{(3,2)i}, v)) = 4$ and $src(Amal(P_{(3,2)i}, v)) = max\{4, t\}$ for $i \in \{1, 2, \ldots, t\}$.

1. Introduction

Let G be a nontrivial connected graph. We follow Chartrand, et al in [2] for the definition of rainbow connection and strong rainbow connection number. The following definition of an amalgamation of graphs is taken from [5]. For $t \in N$ and $i \in \{1, 2, \ldots, t\}$, let G_i be a simple connected graph and $|V(G_i)| = k_i \geq 2$ for some $k_i \in N$. For $t \geq 2$, let $\{G_1, G_2, \ldots, G_t\}$ be a finite collection of graphs and each G_i, $i \in \{1, 2, \ldots, t\}$, has a fixed vertex v called a terminal. The amalgamation $Amal(G_i, v)$ is a graph formed by taking all the G_is and identifying their terminals. Let $G \cong Amal(G_i, v)$. We define $V(G) = \{v\} \cup \{v_{i,j} | 1 \leq i \leq t, 1 \leq j \leq k_i - 1\}$.

Let $\{P_{(3,2)i}\}$, $i \in \{1, 2, \ldots, t\}$ be a finite set of prism graph, with $P_{(3,2)1}, P_{(3,2)2}, \ldots, P_{(3,2)t}$ as its element. Also, let $P_{(3,2)i}$ be the i-th prism graph component, with $V(P_{3,2}) = \{v_{a,b}; 1 \leq a \leq 3, 1 \leq b \leq 2\}$ and $E(P_{3,2}) = \{v_{a,b}v_{a+1,b}; 1 \leq a \leq 2, 1 \leq b \leq 2\} \cup \{v_{3,b}v_{1,b}; 1 \leq b \leq 2\} \cup \{v_{a,1}v_{a,2}; 1 \leq a \leq 3\}$. The component graph will have $|V(P_{3,2})| = 6$ and $|E(P_{3,2})| = 9$. An amalgamation from finite collection of prism graph $\{P_{(3,2)i}\}$ and has a fixed vertex v called terminal, denoted by $Amal(P_{(3,2)i}, v)$. Let $G \cong Amal(P_{(3,2)i}, v)$. Graph G have $V(G) = \{v\} \cup \{v_{i,j} | 1 \leq i \leq t, 1 \leq j \leq 5\}$ and $E(G) = \{v_{i,k}v_{i,k+1} | 1 \leq i \leq t, 1 \leq k \leq 4\} \cup \{vv_{i,1}\} \cup \{vv_{i,2}\} \cup \{vv_{i,5}\} \cup \{v_{i,1}v_{i,4}\} \cup \{v_{i,3}v_{i,5}\}$ with $|E(G)| = 9t$ and $|V(G)| = 5t + 1$. See Figure 1 for the illustration.

Chartrand, et al in [2] found out that if G is a nontrivial connected graph with size m which the diameter of G denoted by $diam(G)$, then:

$$diam(G) \leq rc(G) \leq src(G) \leq m \quad (1)$$

IOP Conf. Series: Journal of Physics: Conf. Series **1008** (2018) 012052 doi:10.1088/1742-6596/1008/1/012052

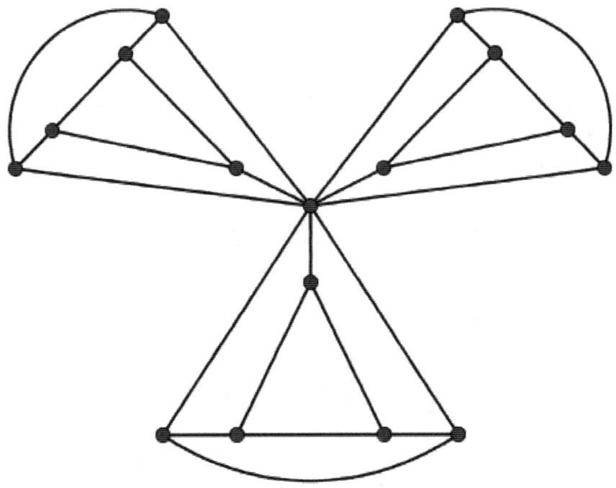

Figure 1. $Amal(G_i, v)$ where $G_i \cong P_{(3,2)i}$ with $t = 3$

Darmawan in [4] determined the rainbow connection number of special graphs and product of two graphs on his thesis in 2015. One of his result is the rainbow connection number and strong rainbow connection number of prism graph.

Theorem 1.1 *For every integer $m \geq 3$ and $n \geq 1$, the rainbow connection number and strong rainbow connection number of prism graph $P_{m,n}$ are*

$$rc(P_{m,n}) = src(P_{m,n}) = \begin{cases} n, & \text{for } m = 3 \\ \lceil \frac{m}{2} \rceil + (n-1), & \text{for } m \geq 4 \end{cases}$$

In 2016, Salman and Fitriani [5] provide a sharp lower and upper bound for the rainbow connection number of amalgamation of arbitrary graphs.

Theorem 1.2 *For $t \in N$, $t \geq 2$, let $\{G_i | i \in \{1, 2, \ldots, t\}\}$ be a finite collection of graphs and each G_i has a fixed vertex v called a terminal. If G is the amalgamation of $G_1, G_2, \ldots, G_t$, $Amal(G_i, v_{0i})$, then*

$$diam(G) \leq rc(G) \leq \sum_{i=1}^{t} rc(G_i)$$

2. The Results

This following section provides the theorem of the rainbow connection number and strong rainbow connection number of prism graph $P_{3,2}$. See Figure 3 and Figure 7 for the ilustration.

Theorem 2.1 *Let t be a positive integer with $t \geq 2$. If $G \cong Amal(G_i, v)$ where for each $i \in \{1, 2, \ldots, t\}$, G_i is a prism graph $P_{3,2}$, then $rc(G) = 4$.*

Proof. We can check that $diam(G) = 4$. From Theorem 1.2 we have

$$diam(G) \;\leq\; rc(G) \leq \sum_{i=1}^{t} rc(G_i)$$

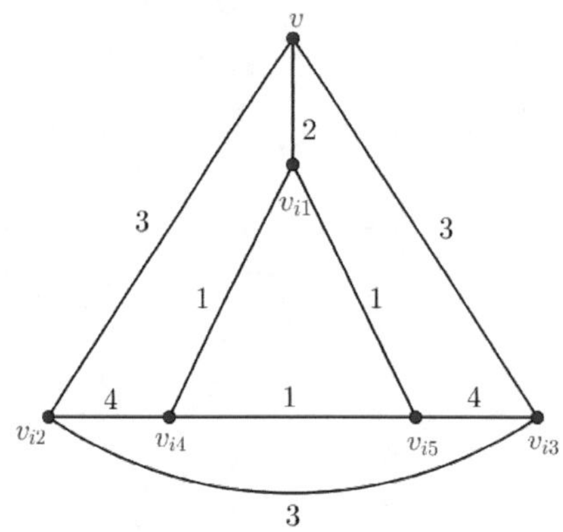

Figure 2. The rainbow coloring on the i-th component of the Amalgamation

$$\begin{aligned} 4 &\leq rc(G) \leq \sum_{i=1}^{t} rc(P_{(3,2)i}) \\ 4 &\leq rc(G) \leq 2t \end{aligned}$$

So the possible value of rainbow connection number from G are $4, 5, 6 \ldots, 2t$, such that the rainbow connection has a lower bound $diam(G) = 4$. We can show that $rc(G) = 4$ with the coloring c on the $i - th$ component graph is as shown in Figure 2. We define coloring $c : E(G) \rightarrow \{1, 2, 3, 4\}$ as follows.

$$c(e) = \begin{cases} 1, & e = v_{i1}v_{i4}; e = v_{i1}v_{i5}; e = v_{i4}v_{i5} \\ 2, & e = vv_{i1} \\ 3, & e = vv_{i2}; e = vv_{i3}\ e = v_{i2}v_{i3}, \\ 4, & e = v_{ix}v_{i(x+2)}, \text{with x} = 2, 3 \end{cases}$$

Now we need to proof for every two vertices $v_{ix}, v_{jy} \in V$ there exist a rainbow path.

For $i = j$ we have $d(v_{ix}, v_{jy}) \leq 2$, we can respectively choose rainbow path as follow. For v to v_{i1} we have path $v - v_{i1}$ with rainbow coloring $c(vv_{i1}) = 2$, v to v_{i2} we have path $v - v_{i2}$ with rainbow coloring $c(vv_{i2}) = 3$, v to v_{i3} we have path $v - v_{i3}$ with rainbow coloring $c(vv_{i3}) = 3$, v to v_{i4} we have path $v - v_{i1} - v_{i4}$ with rainbow coloring $c(vv_{i1}) = 2$ and $c(v_{i1}v_{i4}) = 1$, v to v_{i5} we have path $v - v_{i1} - v_{i5}$ with rainbow coloring $c(vv_{i1}) = 2$ and $c(v_{i1}v_{i5}) = 1$, v_{i1} to v_{i2} we have path $v_{i1} - v_{i4} - v_{i2}$ with rainbow coloring $c(v_{i1}v_{i4}) = 1$ and $c(v_{i4}v_{i2}) = 4$, v_{i1} to v_{i3} we have path $v_{i1} - v_{i5} - v_{i3}$ with rainbow coloring $c(v_{i1}v_{i5}) = 1$ and $c(v_{i5}v_{i3}) = 4$, v_{i1} to v_{i4} we have path $v_{i1} - v_{i4}$ with rainbow coloring $c(v_{i1}v_{i4}) = 1$, v_{i1} to v_{i5} we have path $v_{i1} - v_{i5}$ with rainbow coloring $c(v_{i1}v_{i5}) = 1$, v_{i2} to v_{i3} we have path $v_{i2} - v_{i3}$ with rainbow coloring $c(v_{i2}v_{i3}) = 3$, v_{i2} to v_{i4} we have path $v_{i2} - v_{i4}$ with rainbow coloring $c(v_{i2}v_{i4}) = 4$, v_{i2} to v_{i5} we have path $v_{i2} - v_{i4} - v_{i5}$ with rainbow coloring $c(v_{i2}v_{i4}) = 4$ and $c(v_{i4}v_{i5}) = 1$, v_{i3} to v_{i4} we have path $v_{i3} - v_{i5} - v_{i4}$ with rainbow coloring $c(v_{i3}v_{i5}) = 4$ and $c(v_{i5}v_{i4}) = 1$, v_{i3} to v_{i5} we have path $v_{i3} - v_{i5}$ with rainbow coloring $c(v_{i3}v_{i5}) = 4$, v_{i4} to v_{i5} we have path $v_{i4} - v_{i5}$ with rainbow coloring $c(v_{i4}v_{i5}) = 1$.

For $i \neq j$ we consider in 3 cases,
Case 1. $v_{ix} - v_{jy}$ with $x, y \in \{1, 2, 3\}$ (outer to outer vertices)

IOP Conf. Series: Journal of Physics: Conf. Series **1008** (2018) 012052 doi:10.1088/1742-6596/1008/1/012052

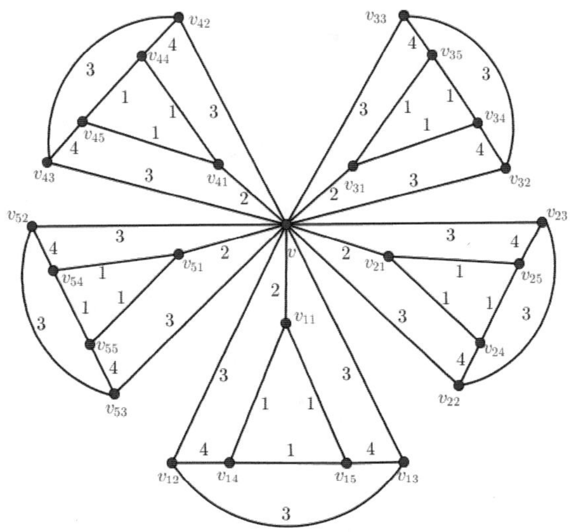

Figure 3. A rainbow 4-coloring of $Amal(G_i, v)$ where $G_i \cong P_{3,2}$ and $t = 5$

we can choose path $v_{ix} - v - v_{j1} - v_{j(y+2)} - v_{jy}$ with $c(v_{ix}v) = 3$, $c(vv_{j1}) = 2$, $c(v_{j1}v_{j(y+2)}) = 1$, $c(v_{j(y+2)}v_{jy}) = 4$

Case 2. $v_{ix} - v_{jy}$ with $x \leq 3$ and $y > 3$ (outer to inner vertices)
we can easily choose path $v_{ix} - v - v_{j1} - v_{jy}$, with $c(v_{ix}v) = 3$, $c(vv_{j1}) = 2$, $c(v_{j1}v_{jy}) = 1$.

Case 3. $v_{ix} - v_{jy}$ with $x, y > 3$ (inner to inner vertices)
we can choose path $v_{ix} - v_{i1} - v - v_{j(y-2)} - v_{jy}$, with $c(v_{ix}v_{i1}) = 1$, $c(v_{i1}v) = 2$, $c(vv_{j(y-2)}) = 3$, $c(v_{j(y-2)}v_{jy}) = 4$

Since 4 is the lower bound of rainbow connection number from amalgamation graf prism $P_{3,2}$ and there is a rainbow path for every two vertices $v_{ix}, v_{jy} \in V$, then $rc(G) = 4$. □
See Figure 3 for the illustration.

Theorem 3.1 *Let t be a positive integer with $t \geq 2$ and $G \cong Amal(G_i, v)$ where for each $i \in \{1, 2, \ldots, t\}$, G_i is a prism graph $P_{3,2}$. Then $src(G) = max\,\{4, t\}$*

Proof. Clearly $4 = diam(G)$ is a lower bound for $src(G)$. Now, if you want to find a geodesic path from any two vertices from different component graph, it needs to pass the terminal vertex (v). Hence, $src(G) = \max\{4, t\}$. For $t = 2$ and $t = 3$, we can find the coloring trivially as stated in Figure 4 and Figure 5 respectively.
As for the case when $t \geq 4$, let $c : E(G) \to \{1, 2, \ldots, t\}$ be the following coloring:

$$c(e) = \begin{cases} i, & \text{if e} = \text{vv}_{\text{ik}} \text{ with k} = 1,2,3 \text{ or e} = \text{v}_{\text{i2}}\text{v}_{\text{i3}} \\ i+1, & \text{if e} = \text{v}_{\text{i1}}\text{v}_{\text{ik}} \text{ with k} = 4,5 \text{ or e} = \text{v}_{\text{i4}}\text{v}_{\text{i5}} \\ i+2, & \text{if e} = \text{v}_{\text{ik}}\text{v}_{\text{i(k+2)}} \text{ with k} = 2,3 \end{cases}$$

See Figure 6 as illustration for the coloring. We need to show that for all $v_{ix}, v_{jy} \in V$, we can provide a strong rainbow path which connecting v_{ix} and v_{jy}.

If $i = j$, then we have $d(v_{ix}, v_{jy}) \leq 2$. We can provide geodesic paths as follows. For v to v_{i1} we have geodesic path $v - v_{i1}$ with strong rainbow coloring $c(vv_{i1}) = i$, v to v_{i2} we have geodesic

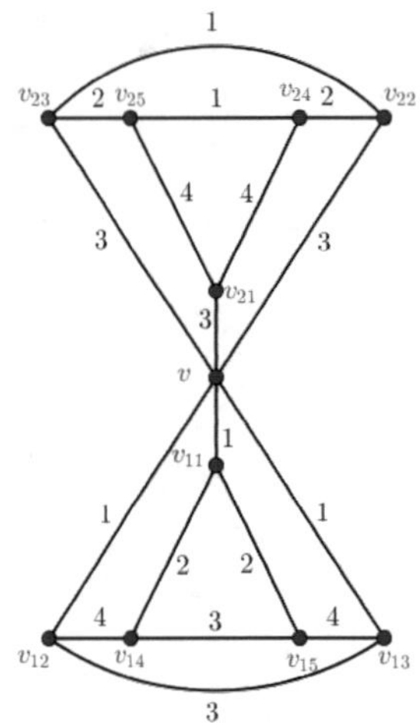

Figure 4. A strong rainbow 4-coloring of $Amal(G_i, v)$ where $G_i \cong P_{3,2}$ and $t = 2$

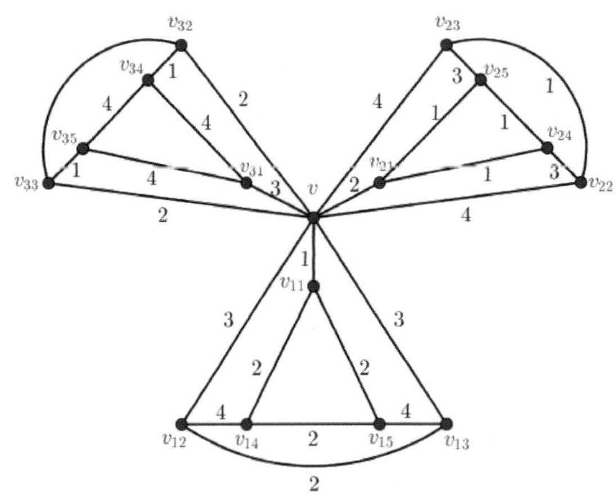

Figure 5. A strong rainbow 4-coloring of $Amal(G_i, v)$ where $G_i \cong P_{3,2}$ and $t = 3$

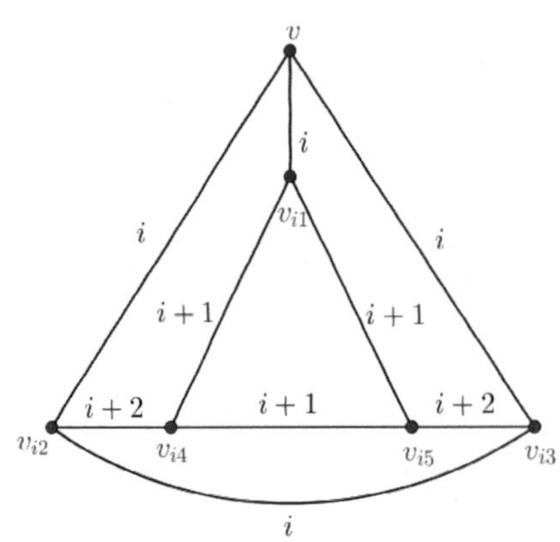

Figure 6. The strong rainbow coloring on the i-th component of the Amalgamation. All color are in *mod t*.

path $v - v_{i2}$ with strong rainbow coloring $c(vv_{i2}) = i$, v to v_{i3} we have geodesic path $v - v_{i3}$ with strong rainbow coloring $c(vv_{i3}) = i$, v to v_{i4} we have geodesic path $v - v_{i1} - v_{i4}$ with strong rainbow coloring $c(vv_{i1}) = i$ and $c(v_{i1}v_{i4}) = i+1$, v to v_{i5} we have geodesic path $v - v_{i1} - v_{i5}$ with strong rainbow coloring $c(vv_{i1}) = i$ and $c(v_{i1}v_{i5}) = i+1$, v_{i1} to v_{i2} we have geodesic path $v_{i1} - v_{i4} - v_{i2}$ with strong rainbow coloring $c(v_{i1}v_{i4}) = i+1$ and $c(v_{i4}v_{i2}) = i+2$, v_{i1} to v_{i3} we have geodesic path $v_{i1} - v_{i5} - v_{i3}$ with strong rainbow coloring $c(v_{i1}v_{i5}) = i+1$ and $c(v_{i5}v_{i3}) = i+2$, v_{i1} to v_{i4} we have geodesic path $v_{i1} - v_{i4}$ with strong rainbow coloring $c(v_{i1}v_{i4}) = i+1$, v_{i1} to v_{i5} we have geodesic path $v_{i1} - v_{i5}$ with strong rainbow coloring $c(v_{i1}v_{i5}) = i+1$, v_{i2} to v_{i3} we have geodesic path $v_{i2} - v_{i3}$ with strong rainbow coloring $c(v_{i2}v_{i3}) = i$, v_{i2} to v_{i4} we have geodesic path $v_{i2} - v_{i4}$ with strong rainbow coloring $c(v_{i2}v_{i4}) = i+2$, v_{i2} to v_{i5} we have geodesic path $v_{i2} - v_{i4} - v_{i5}$ with strong rainbow coloring $c(v_{i2}v_{i4}) = i+2$ and $c(v_{i4}v_{i5}) = i+1$, v_{i3} to v_{i4} we have geodesic path $v_{i3} - v_{i5} - v_{i4}$ with strong rainbow coloring $c(v_{i3}v_{i5}) = i+2$ and $c(v_{i5}v_{i4}) = i+1$, v_{i3} to v_{i5} we have geodesic path $v_{i3} - v_{i5}$ with strong rainbow coloring $c(v_{i3}v_{i5}) = i+2$, v_{i4} to v_{i5} we have geodesic path $v_{i4} - v_{i5}$ with strong rainbow coloring $c(v_{i4}v_{i5}) = i+1$.

For $i \neq j$, we consider 3 cases,

Case 1. $v_{ix} - v_{jy}$ with $x, y = 2$ and $x, y = 3$ (outer to outer)
Clearly $d(v_{ix}, v_{jy}) = 2$, then we can choose the geodesic path $v_{ix} - v - v_{jy}$ with color $c(v_{ix}v) = i$ and $c(vv_{yj}) = j$.

Case 2. $v_{ix} - v_{jy}$ with $x \leq 3$ and $y > 3$ (outer to inner)
We have $d(v_{ix}v_{jy}) \leq 3$

Subcase 2.1. If $(i - j) \equiv 1 \ mod \ t$ we obtain $i \equiv (j+1) \ mod$ then we can choose the geodesic path $v_{ix} - v - v_{j(y-2)} - v_{jy}$, with $c(v_{ix}v) = i$, $c(vv_{j(y-2)}) = j$, $c(v_{j(y-2)}v_{jy}) = j+2$

Subcase 2.2. If $(i - j) \neq 1 \ mod \ t$ we obtain $i \neq (j+1) \ mod$ then we can choose the geodesic path $v_{ix} - v - v_{j1} - v_{jy}$, with $c(v_{ix}v) = i$, $c(vv_{j1}) = j$, $c(v_{j1}v_{jy}) = j+1$

Case 3. $v_{ix} - v_{jy}$ with $x, y > 3$

Subcase 3.1. for $|i - j| \equiv 1 \ mod \ t$ we can choose the geodesic path $v_{ix} - v_{i(x-2)} - v - v_{j(y-2)} - v_{jy}$, with $c(v_{ix}v_{i(x-2)}) = i+2$, $c(v_{i(x-2)}v) = i$, $c(vv_{j(y-2)}) = j$, $c(v_{j(y-2)}v_{jy}) = j+2$

Subcase 3.2. for $|i - j| \neq 1 \ mod \ t$ we can choose the geodesic path $v_{ix} - v_{i1} - v - v_{j1} - v_{jy}$, with $c(v_{ix}v_{i1}) = i+1$, $c(v_{i1}v) = i$, $c(vv_{j1}) = j$, $c(v_{j1}v_{jy}) = j+1$

Since all paths chosen are geodesic and of different colors, then c is strong rainbow coloring. □
See Figure 7 for the illustration.

3. Conclusion

We have studied the rainbow connection number and strong rainbow connection number from amalgamation of prism graph $P_{3,2}$. The result show that all the rainbow connection number and strong rainbow connection number in this paper achieve the minimum value by using deductive reasoning methods, through the following theorem: Let t be a positive integer with $t \geq 2$ and $G \cong Amal(G_i, v)$ where for each $i \in \{1, 2, \ldots, t\}$, G_i is a prism graph $P_{3,2}$. Then $rc(G) = 4$ and $src(G) = max\{4, t\}$.

Open Problem

We have a few problems that we would like to investigate further in this topic such that finding the rainbow connection and strong rainbow connecton number of:
(i) amalgamation of prism graph $P_{m,n}$;

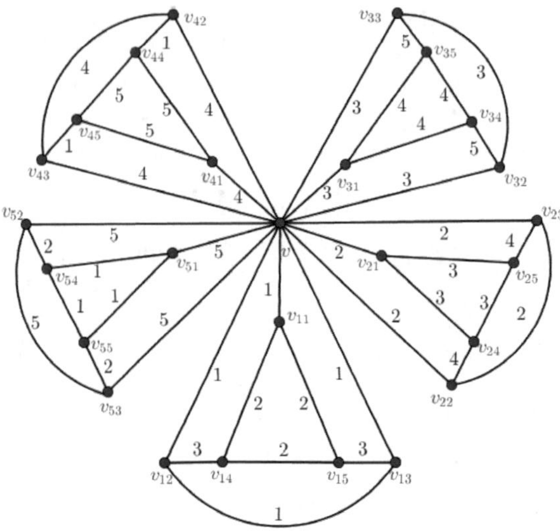

Figure 7. A strong rainbow 5-coloring of $Amal(G_i, v)$ where $G_i \cong P_{3,2}$ and $t = 5$

(ii) amalgamation of any graph G in general.

Acknowledgement

We gratefully acknowledge the support from Faculty of Sciences and Technology UIN Syarif Hidayatullah Jakarta.

References

[1] Chartrand, et all, Rainbow connection in graphs, *Math. Bohem.* **133**, 85-98, 2008
[2] Chartrand, et all, *Applied and Algoritmic Graph Theory*, New York: Mac Graw-Hill, inc, 1993
[3] Rosen, K, *Discrete Mathematics and Its Applications*, New York: Mac Graw-Hill, inc, 2012
[4] Darmawan, *Analisis Rainbow Connection Number pada Graf Khusus dan Hasil Operasinya, Graduate Thesis*, Pascasarjana Universitas Jember, 2015
[5] Salman and Fitriani, Rainbow Connection Number of Amalgamation of Some Graphs, *AKCE International J. Graph and Combinatorics* (13), 90-99, 2016
[6] Salman and Kumala, The Rainbow Connection Number of a Flower (C_m, K_n) Graph and a Flower (C_3, F_n) graph, *Procedia Computer Science* 74, 168-172, 2015
[7] Sugeng, et all, *Teori Graf dan Aplikasinya*, Indonesia: Departemen Matematika Universitas Indonesia, 2018
[8] Vilfred, A Theory of Cartesian Product and Factorization of Circulant Graphs, *Hindawi Publishing Corporation*, 1-10, 2013
[9] X. Li, Y. Sun, Characterization of graphs with large rainbow connection number and rainbow connection number of some graph operations, Discrete Math, 2016
[10] X. Li, Y. Shi, Y. Sun, Rainbow connection of graphs: a survey, Graph Combin, 29 (1) 1-38, 2013
[11] X. Li, Y. Sun, Rainbow connection of graphs, Springer-Verlag, New York, 2012

On the locating domination number of corona product

Risan Nur Santi[1,2], Ika Hesti Agustin[1,2], Dafik[1,3], Ridho Alfarisi[1,4]
[1]CGANT - University of Jember
[2]Department of Mathematics, University of Jember, Jember, Indonesia
[3]Department of Mathematics Education, University of Jember, Jember, Indonesia
[4]Department of Elementary School Teacher Education, University of Jember, Jember, Indonesia

E-mail: rissn.nursanti@gmail.com, ikahestiagustin@gmail.com, d.dafik@unej.ac.id, alfarisi38@gmail.com

Abstract. Let $G = (V(G), E(G)$ be a connected graph and $v\epsilon V(G)$. A dominating set for a graph $G = (V, E)$ is a subset D of V such that every vertex not in D is adjacent to at least one member of D. The domination number $\gamma(G)$ is the number of vertices in a smallest dominating set for G. Vertex set S in graph $G = (V, E)$ is a locating dominating set if for each pair of distinct vertices u and v in $V(G) - S$ we have $N(u) \cap S \neq \phi$, $N(v) \cap S \neq \phi$, and $N(u) \cap S \neq N(v) \cap S$, that is each vertex outside of S is adjacent to a distinct, nonempty subset of the elements of S. In this paper, we characterize the locating dominating sets in the corona product of graphs namely path, cycle, star, wheel, and fan graph.

Keywords : *Locating dominating sets, dominating sets, locating dominating number, corona product*

1. Introduction

Locating dominating set a natural expansion of dominating set. Historically, dominating set has been studied from the 1960 and developed in the 1970. Foucaud (2016) mention locating dominating set was first introduced and studied by Slater in 1987 [2,3]. Dominating set is a concept of determining a vertex set on a graph, where the vertex that has a condition can dominate the point around it, and the cardinality of members of set should be minimum. Minimum cardinality of the set dominance is called the domination number denoted by γ $[1, 4]$.

According to Haynes and Henning, the set D from vertex of a simple graph G is called dominating set if every vertex $u \in V(G) - D$ adjacent on some vertex $v \in D$ [9,10,11,12]. Let a directed graph not $G = (V, E)$, dominating set is a subset of $S \subseteq V$ of vertex at G, for all vertex $v \in V$ one of $v \in S$ or a neighbor from s ie u is at S $[7, 8]$.

A vertex set of graph $G = (V, E)$ is locating dominating set, if set of vertex dominator denoted by D qualifies that any vertex other than D,that is $V - D$ have different intersection with D. Let V be the vertex and E is the edge set of graph G so $\{u, v \in V \backslash D\}$ then $N(u) \cap D \neq \emptyset$, $N(v) \cap D \neq \emptyset$, and $N(u) \cap D \neq N(v) \cap D$, where $N(u)$ is vertex neighbors of u and $N(v)$ is vertex neighbors of v. Locating dominating number is the minimum cardinality of the locating dominating set. Locating dominating number is denoted by γ_L.

IOP Conf. Series: Journal of Physics: Conf. Series **1008** (2018) 012053 doi:10.1088/1742-6596/1008/1/012053

Some definitions used in this study are :

Definition 1.1 *Let G be a connected graph with $|V(G)| = p_1$ and $|E(G)| = q_1$ the graph H is a connected graph that has $|V(H)| = p_2$ dan $|E(H)| = q_2$. The cardinality of vertex set and edge set at $G \odot H$ respectly are $|V(G \odot H)| = p_1(p_2 + 1)$ and $|E(G \odot H)| = p_1(p_2 + q_2) + q_1$.*

2. Main Result

In this paper, we have studied locating domination number of some corona product. The results of this research, we found several lemma, theorem, and corollary about the locating domination number of corona products.

Lemma 2.1 *Let G and H be a connected graph, the locating domination number of $G \odot H$ is $\gamma_L(G \odot H) \geq |V(G)| \cdot \gamma_L(H)$.*

Proof. The corona products of G and H denoted by $G \odot H$ is a connected graph with the cardinality of vertex set and edge set respectively are $|V(G \odot H)| = |V(G)| \cdot (|V(H)| + 1)$ and $|E(G \odot H)| = |V(G)| \cdot (|V(H)| + |E(H)|) + |E(V)|$.

We prove that the lower bound of locating domination number is $\gamma_L(G \odot H) \geq |V(G)| \cdot \gamma_L(H)$. Based on definition of the corona graph, the graph $G \odot H$ has subgraph H as much as $|V(G)|$. If the vertex dominator on the subgraph H, that is $\gamma_L(H_1) = \gamma_L(H_2) = \gamma_L(H_3) = \ldots = \gamma_L(H_{|V(G)|})$, then $\gamma_L(G \odot H) \geq \gamma_L(H_1) + \gamma_L(H_2) + \gamma_L(H_3) + \ldots + \gamma_L(H_{|V(G)|})$. It concludes that $\gamma_L(G \odot H) \geq |V(G)| \cdot \gamma_L(H)$. □

Theorem 2.2 *For $n \geq 4$, the locating domination number of $G \odot C_n$ is $\gamma_L(G \odot C_n) = |V(G)| \cdot \lceil \frac{2n}{5} \rceil$.*

Proof. Corona products of G and cyle graph C_n denoted by $G \odot C_n$ is a connected graph with the cardinality of vertex set and edge set respectively are $|V(G \odot C_n)| = |V(G)|(n+1)$ and $|E(G \odot C_n)| = n(|V(G)| + |E(G)| + 1)$.

Based on Lemma 2.1, lower bound of locating domination number $G \odot C_n$ is $\gamma_L(G \odot C_n) \geq |V(G)| \cdot \lceil \frac{2n}{5} \rceil$. However we can attain sharpest lower bound. Forthemore, we prove the upper bound of locating domination number is $\gamma_L(G \odot C_n) \leq |V(G)| \cdot \lceil \frac{2n}{5} \rceil$. We determine vertex dominator of $G \odot C_n$ on the subgraph C_n, that is $D = \{x_{ij}; 1 \leq i \leq |V(G)|; 1 \leq j \leq \lceil \frac{2n}{5} \rceil\}$, so $|D| = |V(G)| \cdot \lceil \frac{2n}{5} \rceil$. The neighbors of $V - D = \{x_i, x_{ij}; 1 \leq i \leq |V(G)|; 1 \leq j \leq n - \lceil \frac{2n}{5} \rceil\}$ has different intersection sets with D, it is $N(x_{i(5k)}) \cap D = \{x_{i(5k-1)}, x_{i(5k+1)}\}, N(x_{i(5k-2)}) \cap D = \{x_{i(5k-1)}\}, N(x_{i(5k-3)}) \cap D = \{x_{i(5k-4)}\}, N(x_i) \cap D = \{x_{ij}; j \equiv 1 \ mod 5; j \equiv 4 \ mod 5\}$.

It is easy to see that $\gamma_(G \odot C_n)$ meet $\gamma_(G \odot C_n) \leq |V(G)| \cdot \lceil \frac{2n}{5} \rceil$. Therefore upper bound is $\gamma_(G \odot C_n) \leq |V(G)| \cdot \lceil \frac{2n}{5} \rceil$. It concludes that $\gamma_(G \odot C_n) = |V(G)| \cdot \lceil \frac{2n}{5} \rceil$. □

For the illustration of the locating domination number of $G \odot C_4$ can be seen in Figure 1, we mark the vertex is the vertex dominator of locating dominating set. $G \odot C_4$ has vertex dominator $D = \{x_{ij}; 1 \leq i \leq |V(G)|; 1 \leq j \leq n\}$ is that $D = \{x_{12}, x_{13}, x_{22}, x_{23}, \ldots, x_{|V(G)|2}, x_{|V(G)|3}\}$ and another vertex which is not vertex dominator $(V - D) = \{x_{11}, x_{14}, x_{21}, x_{24}, \ldots, x_{p_1 1}, x_{p_1 4}, x_1, x_2, x_3, \ldots, x_{|V(G)|}\}$ has different neighbors in D, that is $N(x_{11}) \cap D = \{x_{12}\}, N(x_{14}) \cap D = \{x_{13}\}, N(x_{21}) \cap D = \{x_{22}\}, N(x_{21}) \cap D = \{x_{22}\}, N(x_{24}) \cap D = \{x_{23}\}, \ldots, N(x_{p_1 1}) \cap D = \{x_{|V(G)|2}\}, N(x_{|V(G)|4}) \cap D = \{x_{|V(G)|3}\}, N(x_1) \cap D = \{x_{12}, x_{13}\}, N(x_2) \cap D = \{x_{22}, x_{23}\}, \ldots, N(x_{|V(G)|}) \cap D = \{x_{|V(G)|2}, x_{|V(G)|3}\}$. It is easy to see $N(u, v) \cap D \neq \emptyset$ and $N(u) \cap D \neq N(v) \cap D$. It is concludes that $\gamma_L(G \odot C_4) = 2|V(G)|$.

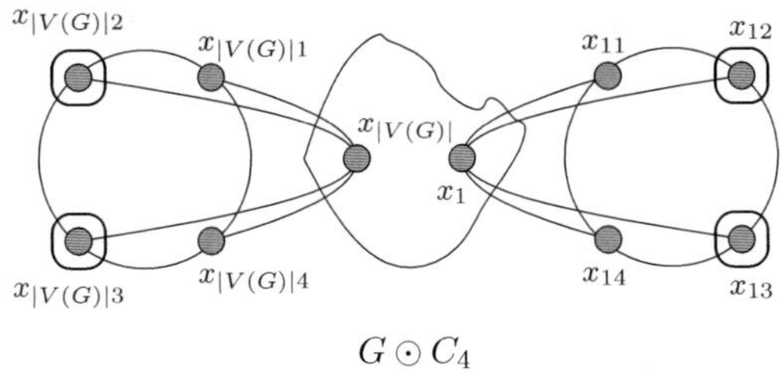

Figure 1. Locating domination number of $G \odot C_4$ is $2 \cdot |V(G)|$

The corollary of theorem :

Corollary 2.3 *For $n \geq 3$ and $m \geq 4$, the locating domination number of $P_n \odot C_m$ is* $\gamma_L(P_n \odot C_m) = n \cdot \lceil \frac{2m}{5} \rceil$.

Corollary 2.4 *For $n \geq 3$ and $m \geq 4$, the locating domination number of $C_n \odot C_m$ is* $\gamma_L(C_n \odot C_m) = n \cdot \lceil \frac{2m}{5} \rceil$.

Corollary 2.5 *For $n \geq 3$ and $m \geq 4$, the locating domination number of $S_n \odot C_m$ is* $\gamma_L(S_n \odot C_m) = (n+1) \cdot \lceil \frac{2m}{5} \rceil$.

Corollary 2.6 *For $n \geq 3$ and $m \geq 4$, the locating domination number of $W_n \odot C_m$ is* $\gamma_L(W_n \odot C_m) = (n+1) \cdot \lceil \frac{2m}{5} \rceil$.

Corollary 2.7 *For $n \geq 3$ and $m \geq 4$, the locating domination number of $F_n \odot C_m$ is* $\gamma_L(F_n \odot C_m) = (n+1) \cdot \lceil \frac{2m}{5} \rceil$.

Theorem 2.8 *For $n \geq 3$, the locating domination number of $G \odot P_n$ is $\gamma_L(G \odot P_n) = |V(G)| \cdot \lceil \frac{2n}{5} \rceil$.*

Proof. Corona products of G and path graph P_n denoted by $G \odot P_n$ is a connected graph with the cardinality of vertex set and edge set respectively are $|V(G \odot P_n)| = |V(G)|(n+1)$ and $|E(G \odot P_n)| = n(|V(G)| + |E(G)| + 1)$.

Based on Lemma 2.1, lower bound of locating domination number $G \odot P_n$ is $\gamma_L(G \odot P_n) \geq |V(G)| \cdot \lceil \frac{2n}{5} \rceil$. However we can attain sharpest lower bound. Forthemore, we prove the upper bound of locating domination number is $\gamma_L(G \odot P_n) \leq |V(G)| \cdot \lceil \frac{2n}{5} \rceil$. We determine vertex dominator of $G \odot P_n$ on the subgraph P_n, that is $D = \{x_{ij}; 1 \leq i \leq |V(G)|; 1 \leq j \leq \lceil \frac{2n}{5} \rceil\}$, so $|D| = |V(G)| \cdot \lceil \frac{2n}{5} \rceil$. The next, we will show that the neighbors of $V - D = \{x_i, x_{ij}; 1 \leq i \leq |V(G)|; 1 \leq j \leq n - \lceil \frac{2n}{5} \rceil\}$ has different intersection sets with D, it is $N(x_{i(5k)}) \cap D = \{x_{i(5k-1)}, x_{i(5k+1)}\}, N(x_{i(5k-2)}) \cap D = \{x_{i(5k-1)}\}, N(x_{i(5k-3)}) \cap D = \{x_{i(5k-4)}\}, N(x_i) \cap D = \{x_{ij}; j \equiv 1 \ mod 5; j \equiv 4 \ mod 5\}$.

It is easy to see that $\gamma_(G \odot P_n)$ meet $\gamma_(G \odot P_n) \leq |V(G)| \cdot \lceil \frac{2n}{5} \rceil$. Therefore upper bound is $\gamma_(G \odot P_n) \leq |V(G)| \cdot \lceil \frac{2n}{5} \rceil$. It concludes that $\gamma_(G \odot P_n) = |V(G)| \cdot \lceil \frac{2n}{5} \rceil$. □

For the illustration of the locating domination number of $G \odot P_4$ can be seen in Figure 2, we mark the vertex is the vertex dominator of locating dominating set. $G \odot P_4$ has vertex dominator $D = \{x_{ij}; 1 \leq i \leq |V(G)|; 1 \leq j \leq n\}$ is that $D = \{x_{12}, x_{13}, x_{22}, x_{23}, \ldots, x_{|V(G)|2}, x_{|V(G)|3}\}$ and another vertex which is not vertex dominator $(V - D) = \{x_{11}, x_{14}, x_{21}, x_{24}, \ldots, x_{p_11}, x_{p_14}, x_1, x_2, x_3, \ldots, x_{|V(G)|}\}$ has different neighbors in D, that is $N(x_{11}) \cap D = \{x_{12}\}, N(x_{14}) \cap D = \{x_{13}\}, N(x_{21}) \cap D = \{x_{22}\}, N(x_{21}) \cap D = \{x_{22}\}, N(x_{24}) \cap D = \{x_{23}\}, \ldots, N(x_{p_11}) \cap D = \{x_{|V(G)|2}\}, N(x_{|V(G)|4}) \cap D = \{x_{|V(G)|3}\}, N(x_1) \cap$

IOP Conf. Series: Journal of Physics: Conf. Series **1008** (2018) 012053 doi:10.1088/1742-6596/1008/1/012053

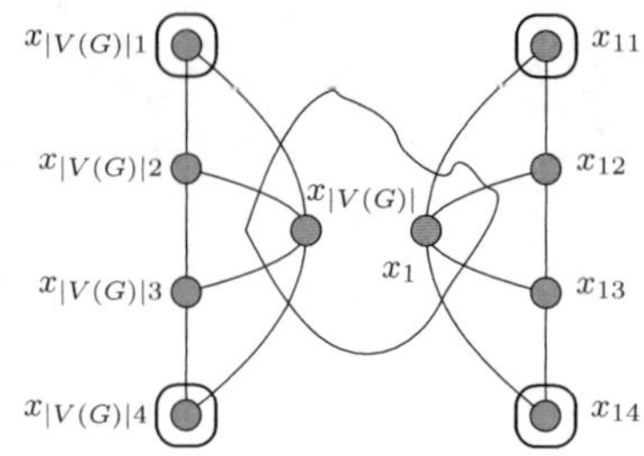

$G \odot P_4$

Figure 2. Locating domination number of $G \odot P_4$ is $2 \cdot |V(G)|$

$D = \{x_{12}, x_{13}\}, N(x_2) \cap D = \{x_{22}, x_{23}\}, \ldots, N(x_{|V(G)|}) \cap D = \{x_{|V(G)|2}, x_{|V(G)|3}\}$. It is easy to see $N(u,v) \cap D \neq \emptyset$ and $N(u) \cap D \neq N(v) \cap D$. It is concludes that $\gamma_L(G \odot P_4) = 2|V(G)|$.

The corollary of theorem :

Corollary 2.9 *For $n \geq 3$ and $m \geq 4$, the locating domination number of $P_n \odot P_m$ is $\gamma_L(P_n \odot P_m) = n \cdot \lceil \frac{2m}{5} \rceil$.*

Corollary 2.10 *For $n \geq 3$ and $m \geq 4$, the locating domination number of $C_n \odot P_m$ is $\gamma_L(C_n \odot P_m) = n \cdot \lceil \frac{2m}{5} \rceil$.*

Corollary 2.11 *For $n \geq 3$ and $m \geq 4$, the locating domination number of $S_n \odot P_m$ is $\gamma_L(S_n \odot P_m) - (n+1) \cdot \lceil \frac{2m}{5} \rceil$.*

Corollary 2.12 *For $n \geq 3$ and $m \geq 4$, the locating domination number of $W_n \odot P_m$. is $\gamma_L(W_n \odot P_m) = (n+1) \cdot \lceil \frac{2m}{5} \rceil$.*

Corollary 2.13 *For $n \geq 3$ and $m \geq 4$, the locating domination number of $F_n \odot P_m$ is $\gamma_L(F_n \odot P_m) = (n+1) \cdot \lceil \frac{2m}{5} \rceil$.*

Lemma 2.14 *Let G and H be a connected graph, the locating domination number of $G \odot H$ is $\gamma_L(G \odot H) \leq \gamma(G) + |V(G)| \cdot \gamma_L(H)$.*

Proof. The corona products of G and H denoted by $G \odot H$ is a connected graph with the cardinality of vertex set and edge set respectively are $|V(G \odot H)| = |V(G)| \cdot (|V(H)| + 1)$ and edge's cardinality $|E(G \odot H)| = |V(G)| \cdot (|V(H)| + |E(H)|) + |E(V)|$.

We prove that the upper bound of locating domination number is $\gamma_L(G \odot H) \leq \gamma(G) + |V(G)| \cdot \gamma_L(H)$. Based on definition of the corona graph, every $G \odot H$ has subgraph H as much as $|V(G)|$. If the vertex dominator on the subgraph G that is $\gamma(G)$ and on the subgraph H that is $\gamma_L(H_1) = \gamma_L(H_2) = \gamma_L(H_3) = \ldots = \gamma_L(H_{|V(G)|})$, then $\gamma_L(G \odot H) \geq \gamma_L(H_1) + \gamma_L(H_2) + \gamma_L(H_3) + \ldots + \gamma_L(H_{|V(G)|} + \gamma(G))$. It concludes that $\gamma_L(G \odot H) \leq \gamma(G) + |V(G)| \cdot \gamma_L(H)$. □

Theorem 2.15 *For $n \geq 4$, the locating domination number of $G \odot S_n$ is $\gamma_L(G \odot S_n) = \gamma(G) + |V(G)| \cdot n$.*

Proof. Corona products of G and path graph S_n denoted by $G \odot S_n$ is a connected graph with the cardinality of vertex set and edge set respectively are $|V(G \odot S_n)| = |V(G)| \cdot (n+2)$ and $|E(G \odot S_n)| = |V(G)| \cdot (2n+1) + |E(G)|$.

Based on Lemma 2.2, upper bound of locating domination number $G \odot S_n$ is $\gamma_L(G \odot S_n) \geq \gamma(G) + |V(G)| \cdot n$. However we can attain sharpest lower bound. Forthemore, we prove the lower bound of locating domination number is $\gamma_L(G \odot S_n) \geq \gamma(G) + |V(G)| \cdot n$. We determine set of vertex dominator of $G \odot S_n$ on the subgraph G, that is $D = \{x_i; 1 \leq i \leq \gamma(G)\}$ and on the subgraph S_n, that is $D = \{x_{ij}; 1 \leq i \leq |V(G)|; 1 \leq j \leq n\}$ so $|D| = \gamma(G) + |V(G)| \cdot n$. The next, we will show that the neighbors of $V - D = \{x_i, x_{ij}; 1 \leq i \leq |V(G)|; j = n+1\}$ has different intersection sets with D, it is $N(x_{i(n+1)}) \cap D = \{x_{i1}, x_{i2}, \ldots, x_{in}; 1 \leq i \leq n\}, N(x_i) \cap D = \{x_{i1}, x_{i2}, \ldots, x_{in}, \gamma(G); 1 \leq i \leq n\}$.

It is easy to see that $\gamma_(G \odot S_n)$ meet $\gamma_L(G \odot S_n) \geq \gamma(G) + |V(G)| \cdot n$. Therefore lower bound is $\gamma_L(G \odot S_n) \geq \gamma(G) + |V(G)| \cdot n$. It concludes that $\gamma_L(G \odot S_n) = \gamma(G) + |V(G)| \cdot n$. □

For the illustration of the locating domination number of $G \odot S_4$ can be seen in Figure 3, we mark the vertex is the vertex dominator of locating dominating set. $G \odot S_4$ has vertex dominator $D = \{x_i; x_{ij}; 1 \leq i \leq |V(G)|; 1 \leq j \leq n\}$ is that $D = \{x_{11}, x_{12}, x_{13}, x_{14}, x_{21}, x_{12}, x_{23}, x_{24}, \ldots, x_{p_11}, x_{p_12}, x_{p_13}, x_{p_14}\}$ and another vertex which is not vertex dominator $(V - D) = \{x_{15}, x_{25}, \ldots, x_{p_15}, \gamma(G)\}$ has different neighbors in D, that is $N(x_{15}) \cap D = \{x_{11}, x_{12}, x_{13}, x_{14}\}, N(x_{25}) \cap D = \{x_{21}, x_{22}, x_{23}, x_{24}\}, \ldots, N(x_{p_15}) \cap D = \{x_{p_11}, x_{p_12}, x_{p_13}, x_{p_14}\}, N(x_1) \cap D = \{x_{11}, x_{12}, x_{13}, x_{14}, \gamma(G)\}, N(x_2) \cap D = \{x_{21}, x_{22}, x_{23}, x_{24}, \gamma(G)\}, \ldots, N(x_{p_1}) \cap D = \{x_{p_11}, x_{p_12}, x_{p_13}, x_{p_14}, \gamma(G)\}$. It is easy to see $N(u,v) \cap D \neq \emptyset$ and $N(u) \cap D \neq N(v) \cap D$. It is concludes that $\gamma_L(G \odot S_4) = \gamma(G) + 4|V(G)|$.

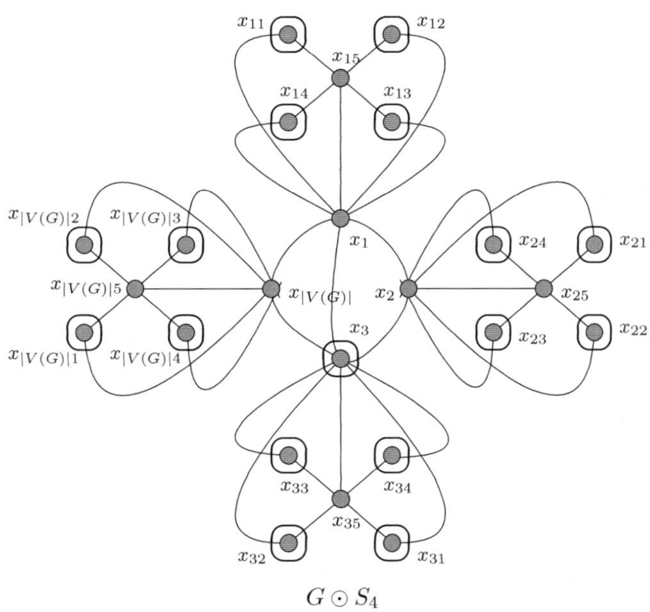

Figure 3. Locating domination number of $G \odot S_4$ is $\gamma(G) + 4 \cdot |V(G)|$

The corollary of theorem :

Corollary 2.16 *For $n \geq 3$ and $m \geq 4$, the locating domination number of $P_n \odot S_m$ is $\gamma_L(P_n \odot S_m) = \gamma(G) + n \cdot m$.*

Corollary 2.17 *For $n \geq 3$ and $m \geq 4$, the locating domination number of $C_n \odot S_m$ is $\gamma_L(C_n \odot S_m) = \gamma(G) + n \cdot m$.*

Corollary 2.18 *For $n \geq 3$ and $m \geq 4$, the locating domination number of $S_n \odot S_m$ is $\gamma_L(S_n \odot S_m) = \gamma(G) + (n+1) \cdot m$.*

Corollary 2.19 *For $n \geq 3$ and $m \geq 4$, the locating domination number of $W_n \odot S_m$ is $\gamma_L(W_n \odot S_m) = \gamma(G) + (n+1) \cdot m$.*

Corollary 2.20 *For $n \geq 3$ and $m \geq 4$, the locating domination number of $F_n \odot S_m$ is $\gamma_L(F_n \odot S_m) = \gamma(G) + (n+1) \cdot m$.*

Theorem 2.21 *For $n \geq 4$, the locating domination number of $G \odot W_n$ is $\gamma_L(G \odot W_n) = \gamma(G) + |V(G)| \cdot \lceil \frac{2n}{5} \rceil$.*

Proof. Corona products of G and path graph W_n denoted by $G \odot W_n$ is a connected graph with the cardinality of vertex set and edge set respectively are $|V(G \odot W_n)| = p_1(n+2)$ and $|E(G \odot W_n)| = 3np_1 + q_1$.

Based on Lemma 2.2, upper bound of locating domination number $G \odot W_n$ is $\gamma_L(G \odot W_n) \geq \gamma(G) + |V(G)| \cdot \lceil \frac{2n}{5} \rceil$. However we can attain sharpest lower bound. Forthemore, we prove the lower bound of locating domination number is $\gamma_L(G \odot W_n) \geq \gamma(G) + |V(G)| \cdot \lceil \frac{2n}{5} \rceil$. We determine set of vertex dominator of $G \odot S_n$ on the subgraph G, that is $D = \{x_i; 1 \leq i \leq \gamma(G)\}$ and on the subgraph S_n, that is $D = \{x_{ij}; 1 \leq i \leq |V(G)|; 1 \leq j \leq \lceil \frac{2n}{5} \rceil\}$ so $|D| = \gamma(G) + |V(G)| \cdot \lceil \frac{2n}{5} \rceil$. The next, we will show that the neighbors of $V - D = \{x_i, x_{ij}; 1 \leq i \leq |V(G)|; 1 \leq j \leq (n+1) - \lceil \frac{2n}{5} \rceil\}$ has different intersection sets with D, it is $N(x_{i(5k)}) \cap D = \{x_{i(5k-1)}, x_{i(5k+1)}\}, N(x_{i(5k-2)}) \cap D = \{x_{i(5k-1)}\}, N(x_{i(5k-3)}) \cap D = \{x_{i(5k-4)}\}, N(x_{i(n+1)}) \cap D = \{x_{i(5k-4)}, x_{i(5k-1)}; 1 \leq k \leq \lceil n \rceil\}, N(x_i) \cap D = \{x_{ij}, \gamma(G); j \equiv 1\ mod 5; j \equiv 4\ mod 5\}$.

It is easy to see that $\gamma_(G \odot W_n)$ meet $\gamma_L(G \odot W_n) \geq \gamma(G) + |V(G)| \cdot \lceil \frac{2n}{5} \rceil$. Therefore lower bound is $\gamma_L(G \odot W_n) \geq \gamma(G) + |V(G)| \cdot \lceil \frac{2n}{5} \rceil$. It concludes that $\gamma_L(G \odot W_n) = \gamma(G) + |V(G)| \cdot \lceil \frac{2n}{5} \rceil$.
□

For the illustration of the locating domination number of $G \odot W_4$ can be seen in Figure 4, we mark the vertex is the vertex dominator of locating dominating set. $G \odot W_4$ has vertex dominator $D = \{x_i; x_{ij}; 1 \leq i \leq |V(G)|; 1 \leq j \leq n\}$ is that $D = \{x_{11}, x_{14}, x_{21}, x_{24}, \ldots, x_{|V(G)|1}, x_{|V(G)|4}\}$ and another vertex which is not vertex dominator $(V - D) = \{x_{12}, x_{13}, x_{15}, x_{22}, x_{23}, x_{25}, \ldots, x_{|V(G)|2}, x_{|V(G)|3}, x_{|V(G)|5}, x_1, x_2, x_3, \ldots, x_{|V(G)|}\}$ has different neighbors in D, that is $N(x_{12}) \cap D = \{x_{11}\}, N(x_{13}) \cap D = \{x_{14}\}, N(x_{15}) \cap D = \{x_{11}, x_{14}\}, N(x_{22}) \cap D = \{x_{21}\}, N(x_{23}) \cap D = \{x_{24}\}, N(x_{25}) \cap D = \{x_{21}, x_{24}\}, \ldots, N(x_{|V(G)|2}) \cap D = \{x_{|V(G)|1}\}, N(x_{|V(G)|3}) \cap D = \{x_{|V(G)|4}\}, N(x_{|V(G)|5}) \cap D = \{x_{|V(G)|1}, x_{|V(G)|4}\}, N(x_1) \cap D = \{x_{11}, x_{14}, \gamma(G)\}, N(x_2) \cap D = \{x_{21}, x_{24}, \gamma(G)\}, \ldots, N(x_{|V(G)|}) \cap D = \{x_{|V(G)|1}, x_{|V(G)|4}, \gamma(G)\}$. It is easy to see $N(u, v) \cap D \neq \emptyset$ and $N(u) \cap D \neq N(v) \cap D$. It is concludes that $\gamma_L(G \odot W_4) = \gamma(G) + 2|V(G)|$.

The corollary of theorem :

Corollary 2.22 *For $n \geq 3$ and $m \geq 4$, the locating domination number of $P_n \odot W_m$ is $\gamma_L(P_n \odot W_m) = \gamma(G) + n \cdot \lceil \frac{2m}{5} \rceil$.*

Corollary 2.23 *For $n \geq 3$ and $m \geq 4$, the locating domination number of $C_n \odot W_m$ is $\gamma_L(C_n \odot W_m) = \gamma(G) + n \cdot \lceil \frac{2m}{5} \rceil$.*

Corollary 2.24 *For $n \geq 3$ and $m \geq 4$, the locating domination number of $S_n \odot W_m$ is $\gamma_L(S_n \odot W_m) = \gamma(G) + (n+1) \cdot \lceil \frac{2m}{5} \rceil$.*

Corollary 2.25 *For $n \geq 3$ and $m \geq 4$, the locating domination number of $W_n \odot W_m$. is $\gamma_L(W_n \odot W_m) = \gamma(G) + (n+1) \cdot \lceil \frac{2m}{5} \rceil$.*

Corollary 2.26 *For $n \geq 3$ and $m \geq 4$, the locating domination number of $F_n \odot W_m$ is $\gamma_L(F_n \odot W_m) = \gamma(G) + (n+1) \cdot \lceil \frac{2m}{5} \rceil$.*

IOP Conf. Series: Journal of Physics: Conf. Series **1008** (2018) 012053 doi:10.1088/1742-6596/1008/1/012053

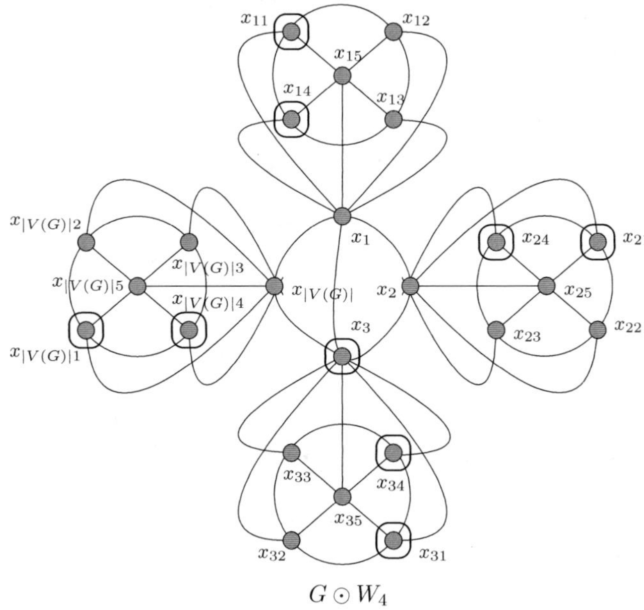

Figure 4. Locating domination number of $G \odot W_4$ is $\gamma(G) + 2 \cdot |V(G)|$

Theorem 2.27 *For $n \geq 4$, the locating domination number of $G \odot F_n$ is $\gamma_L(G \odot F_n) = \gamma(G) + |V(G)| \cdot \lceil \frac{2n}{5} \rceil$.*

Proof. Corona products of G and path graph F_n denoted by $G \odot F_n$ is a connected graph with the cardinality of vertex set and edge set respectively are $|V(G \odot F_n)| = p_1(n+2)$ and $|E(G \odot F_n)| = 2np_1 + q_1$.

Based on Lemma 2.2, upper bound of locating domination number $G \odot W_n$ is $\gamma_L(G \odot F_n) \geq \gamma(G) + |V(G)| \cdot \lceil \frac{2n}{5} \rceil$. However we can attain sharpest lower bound. Forthemore, we prove the lower bound of locating domination number is $\gamma_L(G \odot F_n) \geq \gamma(G) + |V(G)| \cdot \lceil \frac{2n}{5} \rceil$. We determine vertex dominator of $G \odot F_n$ on the subgraph G, that is $D = \{x_i; 1 \leq i \leq \gamma(G)\}$ and on the subgraph F_n, that is $D = \{x_{ij}; 1 \leq i \leq |V(G)|; 1 \leq j \leq \lceil \frac{2n}{5} \rceil\}$ so $|D| = \gamma(G) + |V(G)| \cdot \lceil \frac{2n}{5} \rceil$. The next, we will show that the neighbors of $V - D = \{x_i, x_{ij}; 1 \leq i \leq |V(G)|; 1 \leq j \leq (n+1) - \lceil \frac{2n}{5} \rceil\}$ has different intersection sets with D, it is $N(x_{i(5k)}) \cap D = \{x_{i(5k-1)}, x_{i(5k+1)}\}, N(x_{i(5k-2)}) \cap D = \{x_{i(5k-1)}\}, N(x_{i(5k-3)}) \cap D = \{x_{i(5k-4)}\}, N(x_{i(n+1)}) \cap D = \{x_{i(5k-4)}, x_{i(5k-1)}; 1 \leq k \leq \lceil n \rceil\}, N(x_i) \cap D = \{x_{ij}, \gamma(G); j \equiv 1\ mod 5; j \equiv 4\ mod 5\}$.

It is easy to see that $\gamma(G \odot F_n)$ meet $\gamma_L(G \odot F_n) \geq \gamma(G) + |V(G)| \cdot \lceil \frac{2n}{5} \rceil$. Therefore lower bound is $\gamma_L(G \odot F_n) \geq \gamma(G) + |V(G)| \cdot \lceil \frac{2n}{5} \rceil$. It concludes that $\gamma_L(G \odot F_n) = \gamma(G) + |V(G)| \cdot \lceil \frac{2n}{5} \rceil$. □

For the illustration of the locating domination number of $G \odot F_4$ can be seen in Figure 5, we mark the vertex is the vertex dominator of locating dominating set. $G \odot F_4$ has vertex dominator $D = \{x_i; x_{ij}; 1 \leq i \leq |V(G)|; 1 \leq j \leq n\}$ is that $D = \{x_{11}, x_{14}, x_{21}, x_{24}, \ldots, x_{|V(G)|1}, x_{|V(G)|4}\}$ and another vertex which is not vertex dominator $(V - D) = \{x_{12}, x_{13}, x_{15}, x_{22}, x_{23}, x_{25}, \ldots, x_{|V(G)|2}, x_{|V(G)|3}, x_{|V(G)|5}, x_1, x_2, x_3, \ldots, x_{|V(G)|}\}$ has different neighbors in D, that is $N(x_{12}) \cap D = \{x_{11}\}, N(x_{13}) \cap D = \{x_{14}\}, N(x_{15}) \cap D = \{x_{11}, x_{14}\}, N(x_{22}) \cap D = \{x_{21}\}, N(x_{23}) \cap D = \{x_{24}\}, N(x_{25}) \cap D = \{x_{21}, x_{24}\}, \ldots, N(x_{|V(G)|2}) \cap D = \{x_{|V(G)|1}\}, N(x_{|V(G)|3}) \cap D = \{x_{|V(G)|4}\}, N(x_{|V(G)|5}) \cap D = \{x_{|V(G)|1}, x_{|V(G)|4}\}, N(x_1) \cap D =$

$\{x_{11}, x_{14}, \gamma(G)\}, N(x_2) \cap D = \{x_{21}, x_{24}, \gamma(G)\}, \ldots, N(x_{|V(G)|}) \cap D = \{x_{|V(G)|1}, x_{|V(G)|4}, \gamma(G)\}$. It is easy to see $N(u, v) \cap D \neq \emptyset$ and $N(u) \cap D \neq N(v) \cap D$. It is concludes that $\gamma_L(G \odot F_4) = \gamma(G) + 2|V(G)|$.

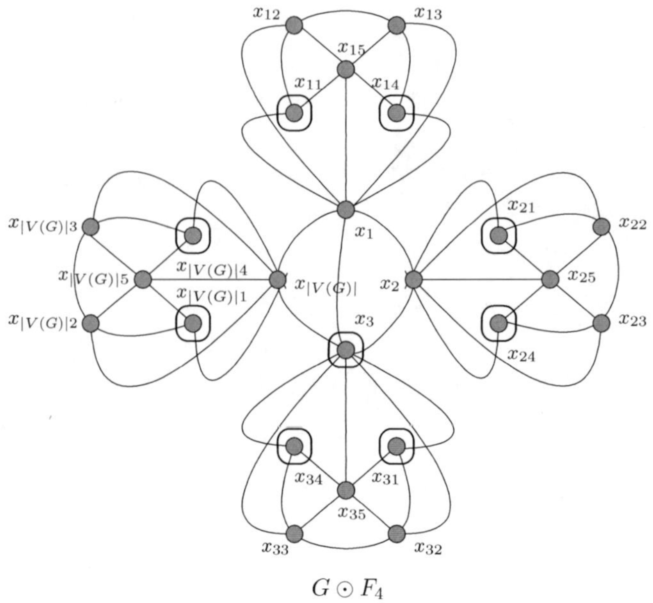

Figure 5. Locating domination number of $G \odot F_4$ is $\gamma(G) + 2 \cdot |V(G)|$

The corollary of theorem :

Corollary 2.28 *For $n \geq 3$ and $m \geq 4$, the locating domination number of $P_n \odot F_m$ is* $\gamma_L(P_n \odot F_m) = \gamma(G) + n \cdot \lceil \frac{2m}{5} \rceil$.

Corollary 2.29 *For $n \geq 3$ and $m \geq 4$, the locating domination number of $C_n \odot F_m$ is* $\gamma_L(C_n \odot F_m) = \gamma(G) + n \cdot \lceil \frac{2m}{5} \rceil$.

Corollary 2.30 *For $n \geq 3$ and $m \geq 4$, the locating domination number of $S_n \odot F_m$ is* $\gamma_L(S_n \odot F_m) = \gamma(G) + (n+1) \cdot \lceil \frac{2m}{5} \rceil$.

Corollary 2.31 *For $n \geq 3$ and $m \geq 4$, the locating domination number of $W_n \odot F_m$. is* $\gamma_L(W_n \odot F_m) = \gamma(G) + (n+1) \cdot \lceil \frac{2m}{5} \rceil$.

Corollary 2.32 *For $n \geq 3$ and $m \geq 4$, the locating domination number of $F_n \odot F_m$ is* $\gamma_L(F_n \odot F_m) = \gamma(G) + (n+1) \cdot \lceil \frac{2m}{5} \rceil$.

3. Conclusion

Based on the results of the above research, then we can conclude the locating domination number of $G \odot H$ is $\gamma_L(G \odot H) \geq |V(G)| \cdot \gamma_L(H)$ and $G \odot H$ is $\gamma_L(G \odot H) \leq \gamma(G) + |V(G)| \cdot \gamma_L(H)$.

Open Problem 3.1. *Define the locating domination number of the other operations graph*

Acknowledgement
We gratefully acknowledge the support from CGANT-University of Jember of year 2018.

References

[1] Dafik,Agustin I H, Fajariyato A, Alfarisi R 2016 *On the Rainbow Coloring for Some Graph Operations* AIP Conference Proceedings 1707 020004
[2] Darmaji, *et al* 2014 *Dominating Set Two Led on Jahangir Graph and Prism Graph* Paper Surabaya: ITS
[3] Chartrand G, Zhang P 2008 *Chromatic Graph Theory.* Chapman and Hall
[4] Gembong A W, Slamin, Dafik, Agustin I H 2017 *Bound of Distance Domination Number of Graph and Edge Comb Product Graph* Journal of Physics : Conf Series 855(2017)012014
[5] Goddard W, Henning M A 2006 *Independent Domination in Graphs: A Survey and Recent Results* South African National Research Foundation and The University Of Johannesburg.
[6] Harary F 1969 *Graph Teory* Wesley Publishing Company,Inc
[7] Hayness T W, Henning M A 2002 *Total Domination Good Vertices in Graphs* Australasian Journal Of Combinatorics 26 (2002):305-315.
[8] Henning M A, *et al* 2008 *On Matching and Total Domination in Graphs* Discrete Mathematics 308 (2008): 2313-2318
[9] Henning M A, Yeo A 2012 *Girth and Total Dominating in Graphs* Graphs Combin 28 (2012): 199-214
[10] Iswadi H 2011 *Lower Up of Matrix Locating Dominating Number on Crown Product Graphs* Proceedings of the National Seminar on Mathematics and Mathematics Education 1 vol(1)
[11] Murugesan N, *et al* 2004 *The Domination and Independence of Some Cubic Bipartite Graphs* Int J Contemp Math Sciences 6 13 (2009): 611-618.
[12] Ruan L, *et al.* 2004. *A greedy Approximation for Minimum Connected Dominating Set.* Theoretical Computer Science 329 (2004): 325-330.
[13] Snyder K. 2011. *C-Dominating Sets for Families of Graphs.* University of Mary Washington.
[14] Yannakakis M, Gavril F. 1980. *Edge Dominating Sets in Graphs.* SIAM Journal on Applied Mathematics, Vol 338, No.3 (Jun 1980) : 364-372.

IOP Conf. Series: Journal of Physics: Conf. Series **1008** (2018) 012054 doi:10.1088/1742-6596/1008/1/012054

On the total rainbow connection of the wheel related graphs

M. S. Hasan[1,2], Slamin[1,3], Dafik[1,4], I. H. Agustin[1,2] and R. Alfarisi[1,5]

[1]CGANT Research Group, University of Jember, Indonesia
[2]Department of Mathematics, University of Jember, Indonesia
[3]Department of Informatics, University of Jember, Indonesia
[4]Department of Mathematics Education, University of Jember, Indonesia
[5]Department of Elementary School Teacher Education, University of Jember, Indonesia

E-mail: saifulhasanmtk@gmail.com; slamin@unej.ac.id; d.dafik@unej.ac.id; hestyarin@gmail.com; alfarisi.fkip@unej.ac.id

Abstract. Let $G = (V(G), E(G))$ be a nontrivial connected graph with an edge coloring $c : E(G) \rightarrow \{1, 2, ..., l\}, l \in N$, with the condition that the adjacent edges may be colored by the same colors. A path P in G is called rainbow path if no two edges of P are colored the same. The smallest number of colors that are needed to make G rainbow edge-connected is called the rainbow edge-connection of G, denoted by $rc(G)$. A vertex-colored graph is rainbow vertex-connected if any two vertices are connected by a path whose internal vertices have distinct colors. The smallest number of colors that are needed to make G rainbow vertex-connected is called the rainbow vertex-connection of G, denoted by $rvc(G)$. A total-colored path is total-rainbow if edges and internal vertices have distinct colours. The minimum number of colour required to color the edges and vertices of G is called the total rainbow connection number of G, denoted by $trc(G)$. In this paper, we determine the total rainbow connection number of some wheel related graphs such as gear graph, antiweb-gear graph, infinite class of convex polytopes, sunflower graph, and closed-sunflower graph.

Keywords: rainbow edge-connected, rainbow vertex-connected, total rainbow connection, wheel related graphs.

1. Introduction

Let G be a nontrivial connected graph on which an edge-coloring $c : E(G) \rightarrow \{1, 2, ..., n\}, n \in N$ is defined, where the adjacent edges may be colored by the same color. A path is called a *rainbow path* if no two edges of the path have the same color. An edge-colored graph G is rainbow connected if any two vertices are connected by a rainbow path [2]. Chartrand et al. [1] defined the rainbow connection number of a connected graph G, denoted by $rc(G)$, as the smallest number of colors that are needed in order to make G rainbow connected.

A vertex-colored graph is rainbow vertex-connected if any two vertices are connected by a path whose internal vertices have distinct colors. The rainbow vertex-connection of a connected graph G, denoted by $rvc(G)$, is the smallest number of colors that are needed in order to make G rainbow vertex-connected. An easy observation is that if G is of order n, we always have $rvc(G) \leq n - 2$, and $rvc(G) = 0$ if and only if G is a complete graph. Notice that $rvc(G) = diam(G) - 1$ with equality if the diameter is 1 or 2 [4]. For rainbow connection and rainbow vertex-connection, some examples are given to show that there is no upper bound for one of parameters in terms of the other in [4].

A total-coloured path is total-rainbow if its edges and internal vertices have distinct colours. A total colouring of a k-connected graph G, not necessarily proper, is total-rainbow k-connected if any two vertices of G are connected by k disjoint total-rainbow paths. The total rainbow k-connection number of G, denoted by $trc(G)$, is the minimum integer t such that, there exists a total-rainbow k-connected colouring of G using t colours. We have $trc_k(G)$ is well-defined if and only if G is k-connected. We write $trc(G)$ for $trc_1(G)$, and similarly for $rc(G)$ and $rvc(G)$ [7]. Li *et al.* [6] proved that for any fixed $k \geq 1$ it is NP-complete to decide whether $trc_k(G) = 3$. The following theorem due to Liu *et al.* [7].

Theorem 1.1 *[7] Let G be a nontrivial connected graph, the total rainbow connection number of G is $trc(G) \geq 2diam(G) - 1$.*

In this paper, we investigate the total rainbow connection number $trc_k(G)$ of some wheel related graphs such as gear graph [10], antiweb-gear graph [9], infinite class of convex polytopes [9], sunflower graph [8], and closed-sunflower graph.

2. Main Results

We start this section with the total rainbow connection of a gear graph as follows.

Theorem 2.1 *If J_n is a gear graph with $n \geq 3$, then the total rainbow connection of J_n is*

$$trc(J_n) = \begin{cases} 3 & for\ n = 2 \\ 5 & for\ n = 3 \\ 7 & for\ n \geq 4\ and\ n\ even \\ 8 & for\ n \geq 5\ and\ n\ odd. \end{cases}$$

Proof. Let J_n, for $n \geq 3$, be a gear graph. Then J_n is a connected graph with vertex set $V(J_n) = \{x, y_i, z_i; 1 \leq i \leq n\}$ and $E(J_n) = \{xy_i; 1 \leq i \leq n\} \cup \{y_i z_i; 1 \leq i \leq n\} \cup \{z_i y_{i+1}, z_n y_1; 1 \leq i \leq n-1\}$. The cardinality of vertex set and edge set, respectively are $|V(J_n)| = 2n + 1$ and $|E(J_n)| = 3n$, and $diam(J_n)$ is

$$diam(J_n) = \begin{cases} 2 & \text{for } n = 2 \\ 3 & \text{for } n = 3 \\ 4 & \text{otherwise.} \end{cases}$$

For $n = 2$, based on Theorem 1.1 that $trc(J_2) \geq 2diam(J_2) - 1 = 2(2) - 1 = 3$. However, we can attain the lower bound. Furthermore, we prove that $trc(J_2) \leq 3$, by defining the edge coloring $f_1(e)$ with

$$f_1(e) = \begin{cases} 1 & \text{for } e \in \{y_i z_i | 1 \leq i \leq n\} \cup \{xy_1\} \\ 2 & \text{for } e \in \{z_1 y_2, z_2 y_1, xy_2\} \end{cases}$$

and by defining the vertex coloring $f_1(v)$ with

$$f_1(v) = 3 \text{ for } v \in \{x\} \cup \{y_i | 1 \leq i \leq n\} \cup \{z_i | 1 \leq i \leq n\}.$$

Hence, it can be seen that it $trc(J_2) \leq 3$. Thus, the total rainbow connection of J_2 is $trc(J_2) = 3$.

For $n = 3$, based on Theorem 1.1 that $trc(J_3) \geq 2diam(J_3) - 1 = 2(3) - 1 = 5$. However, we can attain the lower bound. Furthermore, we prove that $trc(J_3) \leq 5$, by defining the edge coloring $f_2(e)$ with

$$f_2(e) = \begin{cases} i & \text{for } e \in \{xy_i | 1 \le i \le n\} \\ 1 & \text{for } e \in \{y_1z_1, z_2y_3\} \\ 2 & \text{for } e \in \{z_1y_2, y_3z_3\} \\ 3 & \text{for } e \in \{y_2z_2, z_3y_1\} \end{cases}$$

and by defining the vertex coloring $f_2(v)$ with

$$f_2(v) = \begin{cases} 4 & \text{for } v \in \{x\} \\ 5 & \text{for } v \in \{y_i | 1 \le i \le n\} \cup \{z_i | 1 \le i \le n\} \end{cases}$$

Hence, it can be seen that it $trc(J_3) \le 5$. Thus, the total rainbow connection of J_3 is $trc(J_3) = 5$.

For $n \ge 4$ and n even, based on Theorem 1.1 that $trc(J_n) \ge 2diam(J_n) - 1 = 2(4) - 1 = 7$. However, we can attain the lower bound. Furthermore, we prove that $trc(J_n) \le 7$, by defining the edge coloring $f_3(e)$ with

$$f_3(e) = \begin{cases} 1 & \text{for } e \in \{xy_i | 1 \le i \le n \text{ and } i \text{ is odd}\} \\ 2 & \text{for } e \in \{xy_i | 1 \le i \le n \text{ and } i \text{ is even}\} \\ 3 & \text{for } e \in \{y_iz_i | 1 \le i \le n \text{ and } i \text{ is odd}\} \cup \\ & \quad \{z_iy_{i+1} | 1 \le i \le n-1 \text{ and } i \text{ is even}\} \cup \{z_ny_1\} \\ 4 & \text{for } e \in \{y_iz_i | 1 \le i \le n \text{ and } i \text{ is even}\} \cup \\ & \quad \{z_iy_{i+1} | 1 \le i \le n-1 \text{ and } i \text{ is odd}\} \end{cases}$$

and by defining the vertex coloring $f_3(v)$ with

$$f_3(v) = \begin{cases} 5 & \text{for } v \in \{y_i | 1 \le i \le n \text{ and } i \text{ is odd}\} \\ 6 & \text{for } v \in \{y_i | 1 \le i \le n \text{ and } i \text{ is even}\} \\ 7 & \text{for } v \in \{x\} \cup \{z_i | 1 \le i \le n\}. \end{cases}$$

Hence, it can be seen that it $trc(J_n) \le 7$. Thus, the total rainbow connection of J_n is $trc(J_n) = 7$.

Now, for $n \ge 5$ and n is odd, we get $f(e)$ and $f(v)$ above such that it show $trc(J_n) \le 7$. But if n is odd, there is a path that has the same color. A path y_i to z_n *(i odd)* has two choices of *path* that can be passed, that is $y_i - x - y_1 - z_n$ *(i odd)* or $y_i - x - y_n - z_n$ *(i odd)*, but of the two options the *path* has the same color, so that when n odd $f(e)$ is not enough if only colored with four colors. Then there is a color change on the edge $x - y_n$, colored with 5. So that $trc(J_n) \ge 8$, when n is odd. □

Figure 1 shows the example of the total rainbow connection of the gear graph J_6 with $trc(J_6) = 7$.

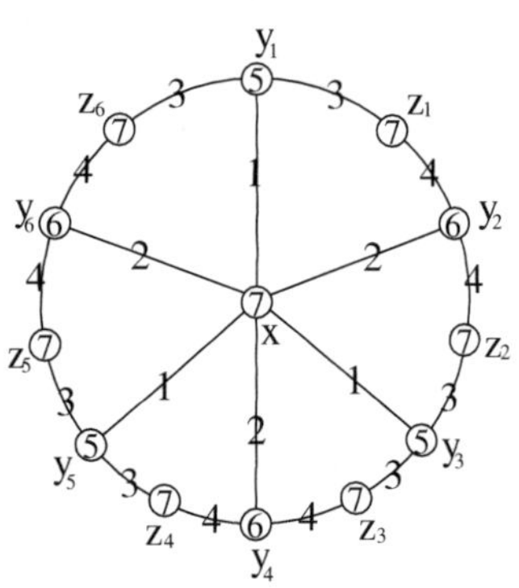

Figure 1. The total rainbow connection of the gear graph J_6 with $trc(J_6) = 7$.

The following theorem presents the total rainbow connection of an antiweb-gear graph.

IOP Conf. Series: Journal of Physics: Conf. Series **1008** (2018) 012054 doi:10.1088/1742-6596/1008/1/012054

Theorem 2.2 *If AWJ_n for $n \geq 3$ is an antiweb-gear graph, then the total rainbow connection of AWJ_n is*

$$trc(AWJ_n) = \begin{cases} 3 & \text{for } 3 \leq n \leq 4 \\ 5 & \text{for } 5 \leq n \leq 6 \\ 7 & \text{for } n \geq 7 \end{cases}$$

Proof. Let AWJ_n for $n \geq 3$ be an antiweb-gear graph. Then AWJ_n is a connected graph with vertex set $V(AWJ_n) = \{x, y_i, z_i; 1 \leq i \leq n\}$ and $E(AWJ_n) = \{xy_i; 1 \leq i \leq n\} \cup \{y_iz_i; 1 \leq i \leq n\} \cup \{z_iy_{i+1}, z_ny_1; 1 \leq i \leq n-1\} \cup \{y_iy_{i+1}, y_ny_1; 1 \leq i \leq n-1\} \cup \{z_iz_{i+1}, z_nz_1; 1 \leq i \leq n-1\}$. The cardinality of vertex set and edge set, respectively are $|V(AWJ_n)| = 2n+1$ and $|E(AWJ_n)| = 5n$, and $diam(AWJ_n)$ is

$$diam(AWJ_n) = \begin{cases} 2 & \text{for } 3 \leq n \leq 4 \\ 3 & \text{for } 5 \leq n \leq 7 \\ 4 & \text{otherwise.} \end{cases}$$

For $3 \leq n \leq 4$, based on Theorem 1.1 that $trc(AWJ_n) \geq 2diam(AWJ_n) - 1 = 2(2) - 1 = 3$. However, we can attain the lower bound. Furthermore, we prove that $trc(AWJ_n) \leq 3$, by defining the edge coloring $f_4(e)$ with

$$f_4(e) = \begin{cases} 1 & \text{for } e \in \{xy_1, xy_2, y_1y_2, y_4y_1, z_1y_2, z_2y_3, y_3z_3, y_4z_4\} \cup \{z_iz_{i+1} | 1 \leq i \leq n \text{ and } i \text{ is odd}\} \\ 2 & \text{for } e \in \{xy_3, xy_4, z_3y_4, z_4y_1, y_2y_3, y_3y_4, z_2z_3, z_4z_1, y_1z_1, y_2z_2\} \end{cases}$$

and the vertex coloring $f_4(v)$ with

$$f_4(v) = 3 \text{ for } v \in \{x\} \cup \{y_i | 1 \leq i \leq n\} \cup \{z_i | 1 \leq i \leq n\}.$$

Hence, it can be seen that it $trc(AWJ_n) \leq 3$. Thus, the total rainbow connection of AWJ_n is $trc(AWJ_n) = 3$.

For $5 \leq n \leq 6$, based on Theorem 1.1 that $trc(AWJ_n) \geq 2diam(AWJ_n) - 1 = 2(3) - 1 = 5$. However, we can attain the lower bound. Furthermore, we prove that $trc(AWJ_n) \leq 5$, by defining the edge coloring $f_5(e)$ with

$$f_5(e) = \begin{cases} 1 & \text{for } e \in \{xy_1, xy_2, xy_3\} \\ 2 & \text{for } e \in \{xy_4, xy_5, xy_6\} \\ 3 & \text{for } e \in \{y_iz_i | 1 \leq i \leq n\} \cup \{z_iy_{i+1} | 1 \leq i \leq n-1\} \cup \{z_ny_1\} \\ i & \text{for } e \in \{y_1y_2, y_2y_3, y_3y_4, z_1z_2, z_2z_3, z_3z_4\} \\ i-3 & \text{for } e \in \{y_iy_{i+1} | 4 \leq i \leq n-1\} \cup \{z_iz_{i+1} | 4 \leq i \leq n-1\} \\ n-3 & \text{for } e \in \{y_ny_1, z_nz_1\} \end{cases}$$

and the vertex coloring $f_5(v)$ with

$$f_5(v) = \begin{cases} 4 & \text{for } v \in \{y_i | 1 \leq i \leq n\} \cup \{z_i | 1 \leq i \leq n \text{ and } i \text{ is odd}\} \\ 5 & \text{for } v \in \{x\} \cup \{z_i | 1 \leq i \leq n \text{ and } i \text{ is even}\}. \end{cases}$$

Hence, it can be seen that it $trc(AWJ_n) \leq 5$. Thus, the total rainbow connection of AWJ_n is $trc(AWJ_n) = 5$.

Now, we prove the lower bound for $n = 7$. For $n = 7$ when $diam(AWJ_7) = 3$, based on Theorem 1.1 that $trc(AWJ_7) \geq 5$. But AWJ_7 have a path $z_1 - z_2 - z_3 - \cdots - z_n$, when a path is *cycle graph* with $n = 7$. By [7] $trc(C_7) \geq 6$. So that $trc(AWJ_7) \geq 6$.

For $n \geq 8$, based on Theorem 1.1 that $trc(AWJ_n) \geq 2diam(AWJ_n) - 1 = 2(4) - 1 = 7$. However, we can attain the lower bound. Furthermore, we prove that $trc(AWJ_n) \leq 7$, by defining the edge coloring $f_6(e)$ with

$$f_6(e) = \begin{cases} 1 & \text{for } e \in \{xy_i | 1 \le i \le n \text{ and } i \text{ is odd}\} \\ 2 & \text{for } e \in \{xy_i | 1 \le i \le n \text{ and } i \text{ is even}\} \\ 3 & \text{for } e \in \{y_i y_{i+1} | 1 \le i \le n-1\} \cup \{y_n y_1\} \cup \{z_i z_{i+1} | 1 \le i \le n-1\} \cup \{z_n y_1\} \cup \\ & \quad \{y_i z_{i+1} | 1 \le i \le n-1 \text{ and } i \text{ is odd}\} \cup \{z_i y_{i+1} | 1 \le i \le n-1 \text{ and } i \text{ is even}\} \\ 4 & \text{for } e \in \{y_i z_i | 1 \le i \le n \text{ and } i \text{ is even}\} \cup \{z_i y_{i+1} | 1 \le i \le n-1\} \cup \{z_n y_1\} \end{cases}$$

and the vertex coloring $f_6(v)$ with

$$f_6(v) = \begin{cases} 5 & \text{for } v \in \{y_i | 1 \le i \le n \text{ and } i \text{ is odd}\} \\ 6 & \text{for } v \in \{y_i | 1 \le i \le n \text{ and } i \text{ is even}\} \\ 7 & \text{for } v \in \{x\} \cup \{z_i; 1 \le i \le n\}. \end{cases}$$

Hence, it can be seen that it $trc(AWJ_n) \le 7$. Thus, the total rainbow connection of AWJ_n is $trc(AWJ_n) = 7$ for $n \ge 7$. □

Figure 2 shows the example of the total rainbow connection of the antiweb-gear graph AWJ_5 with $trc(AWJ_5) = 5$.

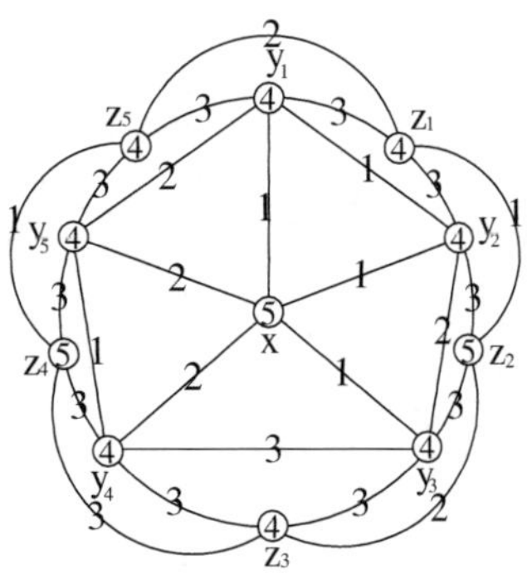

Figure 2. The total rainbow connection of the antiweb-gear graph AWJ_5 with $trc(AWJ_5) = 5$

The total rainbow connection of an infinite class of convex polytopes is presented below.

Theorem 2.3 *If $\mathbf{Q}_n$ for $n \ge 3$ is an infinite class of convex polytopes, then the total rainbow connection of $\mathbf{Q}_n$ is $trc(\mathbf{Q}_n) = 5$.*

Proof. Let $\mathbf{Q}_n$ for $n \ge 3$ be an infinite class of convex polytopes. Then $\mathbf{Q}_n$ is a connected graph with vertex set $V(\mathbf{Q}_n) = \{P, x, y_i, z_i; 1 \le i \le n\}$ and $E(\mathbf{Q}_n) = \{xy_i; 1 \le i \le n\} \cup \{y_i y_{i+1}, y_n y_1; 1 \le i \le n-1\} \cup \{y_i z_i; 1 \le i \le n\} \cup \{z_i y_{i+1}, z_n y_1; 1 \le i \le n-1\} \cup \{z_i z_{i+1}, z_n z_1; 1 \le i \le n-1\} \cup \{Pz_i; 1 \le i \le n\}$. The cardinality of vertex set and edge set, respectively are $|V(\mathbf{Q}_n)| = 2n+1$ and $|E(\mathbf{Q}_n)| = 5n$, and $diam(\mathbf{Q}_n) = 3$. Based on Theorem 1.1 that $trc(\mathbf{Q}_n) \ge 2diam(\mathbf{Q}_n) - 1 = 2(3) - 1 = 5$. However, we can attain the lower bound. Furthermore, we prove that $trc(\mathbf{Q}_n) \le 5$, by defining the edge coloring $f_7(e)$ with

$$f_7(e) = \begin{cases} 1 & \text{for } e \in \{xy_i | 1 \le i \le n \text{ and } i \text{ is odd}\} \cup \{Pz_i | 1 \le i \le n \text{ and } i \text{ is odd}\} \\ 2 & \text{for } e \in \{xy_i | 1 \le i \le n \text{ and } i \text{ is even}\} \cup \{Pz_i | 1 \le i \le n \text{ and } i \text{ is even}\} \\ 3 & \text{for } e \in \{y_i y_{i+1} | 1 \le i \le n-1\} \cup \{y_n y_1\} \cup \{z_i y_{i+1} | 1 \le i \le n-1\} \cup \\ & \quad \{z_i z_{i+1} | 1 \le i \le n-1\} \cup \{z_n z_{i+1} | 1 \le i \le n-1\} \cup \{y_i z_i | 1 \le i \le n\} \end{cases}$$

and the vertex coloring $f_7(v)$ with

IOP Conf. Series: Journal of Physics: Conf. Series **1008** (2018) 012054 doi:10.1088/1742-6596/1008/1/012054

$$f_7(v) = \begin{cases} 4 & \text{for } v \in \{y_i|1 \leq i \leq n\} \cup P \\ 5 & \text{for } v \in \{x\} \cup \{z_i|1 \leq i \leq n\}. \end{cases}$$

Hence, it can be seen that it $trc(\mathbf{Q}_n) \leq 5$. Thus, the total rainbow connection of $\mathbf{Q}_n$ is $trc(\mathbf{Q}_n) = 5$. □

Figure 3 shows the example of the total rainbow connection of the infinite class of convex polytopes $\mathbf{Q}_5$ with $trc(\mathbf{Q}_5) = 5$.

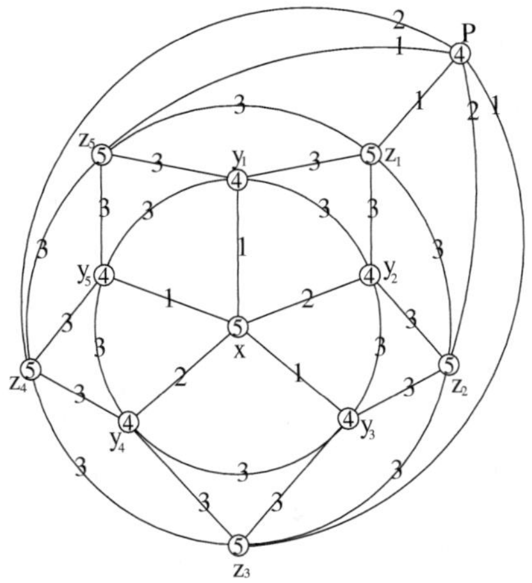

Figure 3. The total rainbow connection of the infinite class of convex polytopes $\mathbf{Q}_5$ with $trc(\mathbf{Q}_5) = 5$

The following theorem provide the total rainbow connection of a sunflower graph.

Theorem 2.4 *If SF_n for $n \geq 3$ is a sunflower graph, then the total rainbow connection of SF_n is*

$$trc(SF_n) = \begin{cases} 3 & \text{for } n = 3 \\ 5 & \text{for } n = 4 \\ 7 & \text{for } n \geq 5. \end{cases}$$

Proof: Let SF_n for $n \geq 3$ be a sunflower graph. Then SF_n is a connected graph with vertex set $V(SF_n) = \{x, y_i, z_i; 1 \leq i \leq n\}$ and $E(SF_n) = \{xy_i; 1 \leq i \leq n\} \cup \{y_iy_{i+1}, y_ny_1; 1 \leq i \leq n-1\} \cup \{y_iz_i; 1 \leq i \leq n\} \cup \{z_iy_{i+1}, z_ny_1; 1 \leq i \leq n-1\}$. The cardinality of vertex set and edge set, respectively are $|V(SF_n)| = 2n+1$ and $|E(SF_n)| = 4n$, and $diam(SF_n)$ is

$$diam(SF_n) = \begin{cases} 2 & \text{for } n = 3 \\ 3 & \text{for } 4 \leq n \leq 5 \\ 4 & \text{otherwise.} \end{cases}$$

For $n = 3$, based on Theorem 1.1 that $trc(SF_3) \geq 2diam(SF_3) - 1 = 2(2) - 1 = 3$. However, we can attain the lower bound. Furthermore, we prove that $trc(SF_3) \leq 3$, by defining the edge coloring $f_8(e)$ with

$$f_8(e) = \begin{cases} 1 & \text{for } e \in \{xy_i|1 \leq i \leq n\} \cup \{y_iz_i|1 \leq i \leq n\} \\ 2 & \text{for } e \in \{y_iy_{i+1}|1 \leq i \leq n-1\} \cup \{y_ny_1\} \cup \{z_iy_{i+1}|1 \leq i \leq n-1\} \cup \{z_ny_1\} \end{cases}$$

and the vertex coloring $f_8(v)$ with

$$f_8(v) = 3 \text{ for } v \in \{x\} \cup \{y_i|1 \leq i \leq n\} \cup \{z_i|1 \leq i \leq n\}.$$

Hence, it can be seen that it $trc(SF_3) \leq 3$. Thus, the total rainbow connection of SF_3 is $trc(SF_3) = 3$.

For $n = 4$, based on Theorem 1.1 that $trc(SF_4) \geq 2diam(SF_4) - 1 = 2(3) - 1 = 5$. However, we can attain the lower bound. Furthermore, we prove that $trc(SF_4) \leq 5$, by defining the edge coloring $f_9(e)$ with

$$f_9(e) = \begin{cases} 1 & \text{for } e \in \{xy_i|1 \leq i \leq 2\} \cup \{y_iz_i|1 \leq i \leq n\} \\ 2 & \text{for } e \in \{z_iy_{i+1}|1 \leq i \leq n-1\} \cup \{z_nz_1\} \\ 3 & \text{for } e \in \{xy_i|3 \leq i \leq 4\} \cup \{y_ny_1\} \cup \{y_iy_{i+1}|1 \leq i \leq n-1\} \end{cases}$$

and the vertex coloring $f_9(v)$ with

$$f_9(v) = \begin{cases} 4 & \text{for } v \in \{x\} \cup \{y_i|1 \leq i \leq n \text{ and } i \text{ is odd}\} \\ 5 & \text{for } v \in \{y_i|1 \leq i \leq n \text{ and } i \text{ is even}\} \cup \{z_i|1 \leq i \leq n\}. \end{cases}$$

Hence, it can be seen that it $trc(SF_4) \leq 5$. Thus, the total rainbow connection of SF_4 is $trc(SF_4) = 5$.

Now, we prove the lower bound for $n = 5$. For $n = 5$ when $diam(SF_5) = 3$, based on Theorem 1.1 that $trc(SF_5) \geq 5$. But SF_5 have a path $y_1 - y_2 - y_3 - y_4 - y_5 - y_1$, when a path is *cycle graph* with $n = 5$. By [7] $trc(C_5) \geq 3$. Then, by adding colors to the other vertices and other edges, we have $trc(SF_5) \geq 7$.

For $n \geq 6$, based on Theorem 1.1 that $trc(SF_n) \geq 2diam(SF_n) - 1 = 2(4) - 1 = 7$. However, we can attain the lower bound. Furthermore, we prove that $trc(SF_n) \leq 7$, by defining the edge coloring $f_{10}(e)$ with

$$f_{10}(e) = \begin{cases} 1 & \text{for } e \in \{xy_i|1 \leq i \leq n \text{ and } i \text{ is odd}\} \\ 2 & \text{for } e \in \{xy_i|1 \leq i \leq n \text{ and } i \text{ is even}\} \\ 3 & \text{for } e \in \{y_iy_{i+1}|1 \leq i \leq n-1\} \cup \{y_ny_1\} \cup \{y_iz_i|1 \leq i \leq n \text{ and } i \text{ is odd}\} \cup \\ & \quad \{z_iy_{i+1}|1 \leq i \leq n-1 \text{ and } i \text{ is even}\} \cup \{z_ny_1|n \text{ even}\} \\ 4 & \text{for } e \in \{z_iy_{i+1}|1 \leq i \leq n-1 \text{ and } i \text{ is odd}\} \cup \{z_ny_1|n \text{ odd}\} \cup \\ & \quad \{y_iz_i|1 \leq i \leq n \text{ and } i \text{ is even}\} \end{cases}$$

and the vertex coloring $f_{10}(v)$ with

$$f_{10}(v) = \begin{cases} 5 & \text{for } v \in \{y_i|1 \leq i \leq n \text{ and } i \text{ is odd}\} \cup \{z_i|1 \leq i \leq n\} \\ 6 & \text{for } v \in \{y_i|1 \leq i \leq n \text{ and } i \text{ is even}\} \\ 7 & \text{for } v \in \{x\}. \end{cases}$$

Hence, it can be seen that it $trc(SF_n) \leq 7$. Thus, the total rainbow connection of SF_n is $trc(SF_n) = 7$ for $n \geq 6$. □

Figure 4 shows the example of the total rainbow connection of the sunflower graph SF_6 with $trc(SF_6) = 7$.

We close this section with the total rainbow connection of a closed-sunflower graph as presented in the following theorem.

Theorem 2.5 *If CSF_n for $n \geq 3$ is a closed-sunflower graph, then the total rainbow connection of CSF_n is*

$$trc(CSF_n) = \begin{cases} 3 & \textit{for } 3 \leq n \leq 4 \\ 5 & \textit{for } 5 \leq n \leq 6 \\ 7 & \textit{for } n \geq 7. \end{cases}$$

IOP Conf. Series: Journal of Physics: Conf. Series **1008** (2018) 012054 doi:10.1088/1742-6596/1008/1/012054

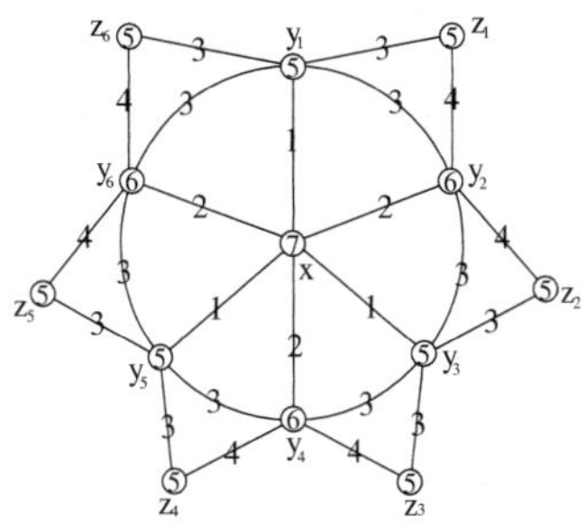

Figure 4. The total rainbow connection of the sunflower graph SF_6 with $trc(SF_6) = 7$

Proof: Let CSF_n for $n \geq 3$ be a closed-sunflower graph. Then CSF_n is a connected graph with vertex set $V(CSF_n) = \{x, y_i, z_i;\ 1 \leq i \leq n\}$ and $E(CSF_n) = \{xy_i; 1 \leq i \leq n\} \cup \{y_iy_{i+1}, y_ny_1; 1 \leq i \leq n-1\} \cup \{y_iz_i; 1 \leq i \leq n\} \cup \{z_iy_{i+1}, z_ny_1; 1 \leq i \leq n-1\} \cup \{z_iz_{i+1}, z_nz_1; 1 \leq i \leq n-1\}$. The cardinality of vertex set and edge set, respectively are $|V(CSF_n)| = 2n+1$ and $|(CSF_n)E| = 5n$, and $diam(CSF_n)$ is

$$diam(CSF_n) = \begin{cases} 2 & \text{for } 3 \leq n \leq 4 \\ 3 & \text{for } 5 \leq n \leq 7 \\ 4 & \text{for } n \geq 8. \end{cases}$$

For $3 \leq n \leq 4$, based on Theorem 1.1 that $trc(CSF_n) \geq 2diam(CSF_n) - 1 = 2(2) - 1 = 3$. However, we can attain the lower bound. Furthermore, we prove that $trc(CSF_n) \leq 3$, by defining the edge coloring $f_{11}(e)$ with

$$f_{11}(e) = \begin{cases} 1 & \text{for } e \in \{xy_i|1 \leq i \leq n \text{ and } i \text{ is even}\} \cup \{y_iy_{i+1}|1 \leq i \leq n-1 \text{ and } i \text{ is odd}\} \cup \\ & \{z_iy_{i+1}|1 \leq i \leq n-1 \text{ and } i \text{ is odd}\} \cup \{y_iz_i|1 \leq i \leq n \text{ and } i \text{ is odd}\} \cup \\ & \{y_ny_1\} \cup \{z_ny_1\} \cup \{z_nz_1|n \text{ odd}\} \cup \{z_iz_{i+1}|1 \leq i \leq n-1 \text{ and } i \text{ is odd}\} \\ 2 & \text{for } e \in \{xy_i|1 \leq i \leq n \text{ and } i \text{ is odd}\} \cup \{y_iy_{i+1}|1 \leq i \leq n-1 \text{ and } i \text{ is even}\} \cup \\ & \{z_iy_{i+1}|1 \leq i \leq n-1 \text{ and } i \text{ is even}\} \cup \{y_iz_i|1 \leq i \leq n \text{ and } i \text{ is even} \cup \\ & \{y_ny_1\} \cup \{z_ny_1\} \cup \{z_nz_1|n \text{ even}\} \cup \{z_iz_{i+1}|1 \leq i \leq n-1 \text{ and } i \text{ is even}\} \end{cases}$$

and the vertex coloring $f_{11}(v)$ with

$$f_{11}(v) = 3 \text{ for } v \in \{x\} \cup \{y_i|1 \leq i \leq n\} \cup \{z_i|1 \leq i \leq n\}.$$

Hence, it can be seen that it $trc(CSF_n) \leq 3$. Thus, the total rainbow connection of CSF_n is $trc(CSF_n) = 3$.

For $5 \leq n \leq 6$, based on Theorem 1.1 that $trc(CSF_n) \geq 2diam(CSF_n) - 1 = 2(3) - 1 = 5$. However, we can attain the lower bound. Furthermore, we prove that $trc(CSF_n) \leq 5$, by defining the edge coloring $f_{12}(e)$ with

$$f_{12}(e) = \begin{cases} 1 & \text{for } e \in \{z_4z_5\} \cup \{xy_i|1 \leq i \leq n \text{ and } i \text{ is odd}\} \\ 2 & \text{for } e \in \{xy_i|1 \leq i \leq n \text{ and } i \text{ even}\} \cup \{z_nz_1|n \text{ odd}\} \cup \{z_{n-1}z_n|n \text{ even}\} \\ 3 & \text{for } e \in \{y_iy_{i+1}|1 \leq i \leq n-1\} \cup \{z_iy_{i+1}|1 \leq i \leq n-1\} \cup \{y_ny_1\} \cup \{z_ny_1\} \cup \\ & \{y_iz_i|1 \leq i \leq n\} \cup \{z_nz_1|n \text{ even}\} \\ i & \text{for } e \in \{z_iz_{i+1}|1 \leq i \leq 3\} \end{cases}$$

and by defining the vertex coloring $f_{12}(v)$ in the following:

$$f_{12}(v) = \begin{cases} 4 & \text{for } v \in \{x\} \cup \{z_i|1 \leq i \leq n \text{ and } i \text{ is odd}\} \\ 5 & \text{for } v \in \{y_i|1 \leq i \leq n\} \cup \{z_i|1 \leq i \leq n \text{ and } i \text{ is even}\} \end{cases}$$

IOP Conf. Series: Journal of Physics: Conf. Series **1008** (2018) 012054 doi:10.1088/1742-6596/1008/1/012054

Hence, it can be seen that it $trc(CSF_n) \leq 5$. Thus, the total rainbow connection of CSF_n is $trc(CSF_n) = 5$.

Now, we prove the lower bound for $n = 7$. For $n = 7$ when $diam(CSF_7) = 3$, based on Theorem 1.1 that $trc(CSF_7) \geq 5$. But CSF_7 have a path $z_1 - z_2 - z_3 - z_4 - z_5 - z_6 - z_7 - z_1$, when a path is *cycle graph* with $n = 7$. By [7] $trc(C_7) \geq 6$. Then, by adding colors to the other vertices and other edges, we have $trc(CSF_7) \geq 7$.

For $n \geq 8$, based on Theorem 1.1 that $trc(CSF_n) \geq 2diam(CSF_n) - 1 = 2(4) - 1 = 7$. However, we can attain the lower bound. Furthermore, we prove that $trc(CSF_n) \leq 7$, by defining the edge coloring $f_{13}(e)$ with

$$f_{13}(e) = \begin{cases} 1 & \text{for } e \in \{xy_i | 1 \leq i \leq n \text{ and } i \text{ is odd}\} \\ 2 & \text{for } e \in \{xy_i | 1 \leq i \leq n \text{ and } i \text{ is even}\} \\ 3 & \text{for } e \in \{y_iy_{i+1} | 1 \leq i \leq n-1\} \cup \{z_iz_{i+1} | 1 \leq i \leq n-1\} \cup \{y_ny_1\} \cup \\ & \quad \{y_iz_i | 1 \leq i \leq n \text{ and } i \text{ is odd}\} \cup \{e = z_nz_1\} \cup \{z_nz_1 | n \text{ even}\} \cup \\ & \quad \{z_iy_{i+1} | 1 \leq i \leq n-1 \text{ and } i \text{ is even}\} \\ 4 & \text{for } e \in \{y_iz_i | 1 \leq i \leq n \text{ and } i \text{ is even}\} \cup \{z_iy_{i+1} | 1 \leq i \leq n-1 \text{ and } i \text{ is odd}\} \cup \\ & \quad \{z_ny_1 | n \text{ odd}\} \end{cases}$$

and the vertex coloring $f_{13}(v)$ with

$$f_{13}(v) = \begin{cases} 5 & \text{for } v \in \{y_i | 1 \leq i \leq n \text{ and } i \text{ is odd}\} \cup \{z_i | 1 \leq i \leq n\} \\ 6 & \text{for } v \in \{y_i | 1 \leq i \leq n \text{ and } i \text{ is even}\} \\ 7 & \text{for } v \in \{x\}. \end{cases}$$

Hence, it can be seen that it $trc(CSF_n) \leq 7$. Thus, the total rainbow connection of CSF_n is $trc(CSF_n) = 7$ for $n \geq 8$. □

Figure 5 shows the example of the total rainbow connection of the closed-sunflower graph CSF_6 with $trc(CSF_6) = 5$.

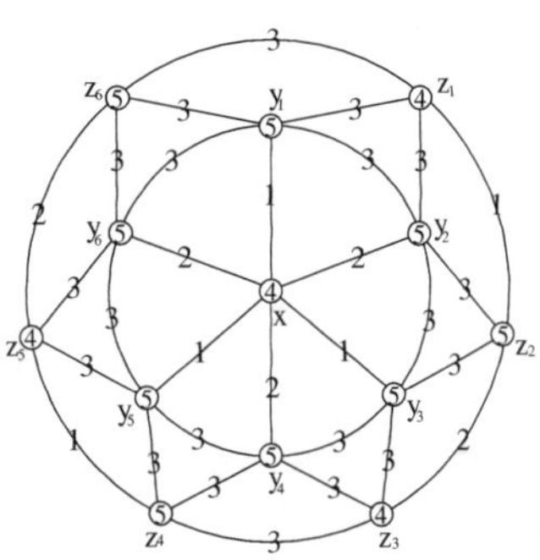

Figure 5. The total rainbow connection of the closed-sunflower graph CSF_6 with $trc(CSF_6) = 5$

3. Conclusion

In this paper we have given an total rainbow connection of some wheel related graphs namely gear graph, antiweb-gear graph, infinite class of convex polytopes, sunflower graph, and closed-sunflower graph. We give the following open problem for future work.

Open Problem 3.1 *Determine the total rainbow connection number of other wheel related graphs that have not been discovered.*

Acknowledgement

We gratefully acknowledge the support from CGANT Research Group - University of Jember.

IOP Conf. Series: Journal of Physics: Conf. Series **1008** (2018) 012054 doi:10.1088/1742-6596/1008/1/012054

References

[1] Dafik, Agustin I H, Fajariyato A and Alfarisi R 2016 On the rainbow coloring for some graph operations *AIP Conference Proceedings* **1707** 020004

[2] Sun Y 2015 On rainbow total-coloring of a graph *Discrate Applied Mathematics. Department of Mathematics, PR China: Shaoxing University, Zhejiang* 312000

[3] Agustin I H, Dafik, Gembong A W and Alfarisi R 2017 On rainbow k-connection number of special graphs and it's sharp lower bound *International Conference on Mathematics: Education, Theory and Application. Journal of Physics: Conf. Series* **855** 012003

[4] Krivelevich M and Yuster R 2010 The rainbow connection of a graph is (at most) reciprocal to its minimum degree *J. Graph Theory* 63 185-191 doi:/10.1002/jgt.20418

[5] Li X and Shi Y 2013 On the rainbow vertex-connection *Discussiones Mathematicae Graph Theory* 33 307-313

[6] Li W, Li X and Wu D 2017 Hardness result for the totoal rainbow k-connection of graphs *Applied Mathematics and Computation* 305 27-31

[7] Liu H, Mestre A and Sousa T 2014 Total rainbow k-connection in graphs *Discrate Applied Mathematics.*

[8] Jeyanthi P and Sudha A 2016 Total edge irregularity strength of wheel related graphs *J. Graph Label* **2**(1) 45-57

[9] Siddiqui H M A and Imran M 2014 Computing the metric dimension of wheel related graphs *Applied Mathematics and Computation* 242 624632

[10] Rahim M T, Farooq M, Ali M and Jan S 2012 Multi-level distance labelings for generalized gear graphs *International Journal of Mathematics and Soft Computing* Vol. **2**, No. 1 57-63

IOP Conf. Series: Journal of Physics: Conf. Series **1008** (2018) 012055 doi:10.1088/1742-6596/1008/1/012055

On the (Strong) Rainbow Vertex Connection of Graphs Resulting from Edge Comb Product

Dafik[1,2], **Slamin**[1,3], **Agustina Muharromah**[1,4,*]

[1] CGANT Research Group, University of Jember, Indonesia
[2] Mathematics Education Study Program, University of Jember, Indonesia
[3] Information System Study Program, University of Jember, Indonesia
[4] Mathematics Department, University of Jember, Indonesia

E-mail: d.dafik@unej.ac.id, slamin@unej.ac.id, mahagustina@yahoo.co.id

Abstract. The vertex-colored graph $G = (V, E)$ is said rainbow vertex-connected, if for every two vertices u and v in V, there is a $u - v$ path with all internal vertices have distinct color. The rainbow vertex connection number of G, denoted by $rvc(G)$, is the smallest number of colors required in order to make graph G to be rainbow vertex-connected. If every two vertices u and v in V are connected by at least one shortest rainbow path, then G is strongly rainbow vertex-connected. The strong rainbow vertex- connection number, denoted by $srvc(G)$, is the minimum number of colors required in order make graph G to be strongly rainbow vertex-connected. In this paper, we determine the rainbow vertex connection number and the strong rainbow vertex connection number of graphs which are resulted from edge comb product.

Keywords: rainbow connection, rainbow vertex connection, strong rainbow vertex connection, coloring, comb product

1. Introduction

Let $G = (V, E)$ be a non trivial, finite, simple, connected and undirected graph. The edge colored graph G is said *rainbow connected*, if no two edges in the path are colored the same. The *rainbow connection number* of G, denoted by $rc(G)$, is the smallest number of colors required in order to make graph G to be rainbow connected. This concept was introduced by Chartrand *et al.* [2] in 2008.

One year later, Krivelevich and Yuster [6] extended the concept into rainbow edge-connection and rainbow vertex-connection. The rainbow edge-connection is actually the rainbow connection, while the rainbow vertex-connection is defined as follows. The vertex-colored graph G is said *rainbow vertex-connected*, if for every two vertices u and v in V, there is a $u - v$ path with all internal vertices have distinct color. The *rainbow vertex connection number* of G, denoted by $rvc(G)$, is the smallest number of colors required in order to make graph G to be rainbow vertex-connected. In the same paper [6], they provided the lower bound on the rainbow vertex connection number of any connected graph G, that is $rvc(G) \geq diam(G) - 1$, with equality if the diameter is 1 or 2.

There have been several results on rainbow vertex-connection number of some families of graphs such as:

- Pencil graph Pc_n for $n \geq 2$, $rvc(Pc_n) = \lceil \frac{n}{2} \rceil$ if $n \leq 7$ and $rvc(Pc_n) = \lceil \frac{n}{2} \rceil + 1$ otherwise [4].

- Cycle C_n of order $n \geq 3$, $rvc(C_n) = 0$ if $n = 3$, $rvc(C_n) = 1$ if $n = 4, 5$, $rvc(C_n) = 3$ if $n = 9$, $rvc(C_n) = \lceil \frac{n}{2} \rceil - 1$ if $n = 6, 7, 8, 10, 11, 12, 13$, or 15, and $rvc(C_n) = \lceil \frac{n}{2} \rceil$ if $n \geq 16$ or $n = 14$ [3].
- Any 2-connected graph G, $rvc(G) \leq rvc(C_n)$ [3].
- Any 3-connected graph G, $rvc(G) \leq \frac{3(n+2)}{5}$ [9].

The vertex-colored graph G is said *strong rainbow vertex-connected*, if every two vertices of G are connected by at least one shortest rainbow path. The *strong rainbow vertex-connection number*, denoted by $srvc(G)$, is the minimum number of colors required to make a strongly rainbow vertex-connected graph G [8].

Similar to the rainbow vertex-connection, there have been several results on strong rainbow vertex-connection number of some families of graphs such as:

- Complete cubic Halin graph H_n for $n \geq 4$, $srvc(H_n) = 3(2^{n-1} - 1 + (3.2^{n-4})) + 1$ [1].
- Necklace graph Ne_n for $n \geq 1$, $srvc(Ne_n) = diam(Ne_n) - 1$ if $1 \leq n \leq 7$ and $srvc(Ne_n) = diam(Ne_n)$ if $n \geq 8$ [1].
- Complete bipartite graph, $K_{m,n}$ for $m \geq 2$ and $n \geq 1$, $srvc(K_{m,n}) = 1$ [8].
- Complete multipartite graph $K_{n_1,n_2,\ldots,n_k}$ for $n_i \geq 2$ for each $i = 1, 2, \ldots, k$ and $k \geq 3$, $srvc(K_{n_1,n_2,\ldots,n_k}) = 1$ [8].
- Wheel W_n for $n \geq 3$, $srvc(W_n) = 1$ [8].
- Path P_n for $n \geq 3$, $srvc(P_n) = n - 2$ [8].

Furthermore, X. Li and Y. Shi [7] observed the following theorem.

Theorem 1. *Let G be a nontrivial connected graph of order n. Then*

(a) $diam(G) - 1 \leq rvc(G) \leq srvc(G)$;

(b) $srvc(G) = 0$ if and only if G is a complete graph;

(c) $srvc(G) = 1$ if and only if $diam(G) = 2$

In this paper, we determine the rainbow vertex connection number and the strong rainbow vertex connection number of graphs which are resulted from edge comb product. The edge comb product between two graphs G and H, denoted by $G \trianglerighteq H$, is a graph obtained by taking one copy of G and $|E(G)|$ copies of H and grafting the i-th copy of H at the i-th edges of G [10].

2. Main Results

We start this section with a new result on (strong) rainbow vertex connection number of edge comp product of a path on n vertices P_n and a triangular book graph. The triangular book graph, denoted by Bt_m, for $m \geq 2$, is defined as the graph cartesian product $S_{m+1} \times P_2$, where S_{m+1} is a star graph on $m + 1$ vertices and P_2 is the path graph on two nodes.

Theorem 2. *If P_n is a path on $n \geq 2$ vertices and Bt_m is a triangular book graph of $m \geq 2$ sheets, then the rainbow vertex connection number and strong rainbow vertex connection number of $P_n \trianglerighteq Bt_m$ with grafting edge $x_1x_2 \in V(Bt_m)$ is $rvc(P_n \trianglerighteq Bt_m) = srvc(P_n \trianglerighteq Bt_m) = n - 2$*

Proof. Let G be the graph obtained from edge comb product of P_n and Bt_m with grafting edge $x_1x_2 \in V(Bt_m)$. Then $V(P_n \trianglerighteq Bt_m) = \{x_i; 1 \leq i \leq n\} \cup \{x_{i,j}; 1 \leq i \leq n-1, 1 \leq j \leq m\}$ and $E(P_n \trianglerighteq Bt_m) = \{x_ix_{i+1}; 1 \leq i \leq n-1\} \cup \{x_ix_{i,j}; 1 \leq i \leq n-1, 1 \leq j \leq m\} \cup \{x_{i+1}x_{i,j}; 1 \leq i \leq n-1, 1 \leq j \leq m\}$. Thus $|V(P_n \trianglerighteq Bt_m)| = n + mn - m$ and $|E(P_n \trianglerighteq Bt_m)| = n - 2m + 2mn - 1$.

We first show that $rvc(P_n \trianglerighteq Bt_m) = n-2$. Since $diam(P_n \trianglerighteq Bt_m) = n-1$ if the grafting edge $x_1x_2 \in V(Bt_m)$, by Theorem 1, we have $rvc(P_n \trianglerighteq Bt_m) \geq n-2$. To show that $rvc(P_n \trianglerighteq Bt_m) \leq n-2$, color the vertex v according to the following formula:

$$c(v) = \begin{cases} 1, & \text{for } v = x_i;\ i = 1, n \\ i-1, & \text{for } v = x_i;\ 2 \leq i \leq n-1 \\ 1, & \text{for } v = x_{i,j};\ 1 \leq i \leq n,\ 1 \leq j \leq m \end{cases}$$

From the above formula, it is easy to see that the color of the vertices are $i-1$ for $2 \leq i \leq n-1$, that is, $c : V(P_n \trianglerighteq Bt_m) \to \{1, 2, 3, \ldots, n-2\}$. Thus $rvc(P_n \trianglerighteq Bt_m) \leq n-2$. Combining with the lower bound, we have $rvc(P_n \trianglerighteq Bt_m) = n-2$.

We now show that $srvc(P_n \trianglerighteq Bt_m) = n-2$. By Theorem 1, we have $n-2 \leq rvc(P_n \trianglerighteq Bt_m) \leq srvc(P_n \trianglerighteq Bt_m)$. To show that $srvc(P_n \trianglerighteq Bt_m) \leq n-2$, color the vertex v according to the following formula:

$$c(v) = \begin{cases} 1, & \text{for } v = x_i;\ i = 1, n \\ i-1, & \text{for } v = x_i;\ 2 \leq i \leq n-1 \\ 1, & \text{for } v = x_{i,j};\ 1 \leq i \leq n,\ 1 \leq j \leq m \end{cases}$$

Again from the above formula, it is easy to see that the color of the vertices are $i-1$ for $2 \leq i \leq n-1$, that is, $c : V(P_n \trianglerighteq Bt_m) \to \{1, 2, 3, \ldots, n-2\}$. Every pair of vertices are connected by geodesic path. Thus $srvc(P_n \trianglerighteq Bt_m) \leq n-2$. Combining with the lower bound, we also have $srvc(P_n \trianglerighteq Bt_m) = n-2$. This concludes that if the grafting edge $x_1x_2 \in V(Bt_m)$ $rvc(P_n \trianglerighteq Bt_m) = srvc(P_n \trianglerighteq Bt_m) = n-2$. □

The following theorem presents the rainbow vertex connection number and the strong rainbow vertex connection number of edge comp product of a path on n vertices P_n and a triangular book graph Bt_m with grafting edge $x_iu_j \in V(Bt_m)$.

Theorem 3. *If P_n is a path on $n \geq 2$ vertices and Bt_m is a triangular book graph of $m \geq 2$ sheets, then the rainbow vertex connection number and strong rainbow vertex connection number of $P_n \trianglerighteq Bt_m$ with grafting edge $x_iu_j \in V(Bt_m)$ is $rvc(P_n \trianglerighteq Bt_m) = srvc(P_n \trianglerighteq Bt_m) = n-1$*

Proof. Let G be the graph obtained from edge comb product of P_n and Bt_m with grafting edge $x_1u_j \in V(Bt_m)$. Then $V(P_n \trianglerighteq Bt_m) = \{y_i; 1 \leq i \leq n-1\} \cup \{x_i; 1 \leq i \leq n\} \cup \{y_{i,j}; 1 \leq i \leq n-1, 1 \leq j \leq m-1\}$ and $E(P_n \trianglerighteq Bt_m) = \{x_1x_{i+1}; 1 \leq i \leq n-1\} \cup \{x_iy_j; 1 \leq i \leq n-1\} \{y_ix_{i+1}; 1 \leq i \leq n-1\} \cup \{x_iy_{i,j}, y_iy_{i,j}; 1 \leq i \leq n-1, 1 \leq j \leq m-1\}$. Thus $|V(P_n \trianglerighteq Bt_m)| = n + mn - m$ and $|E(P_n \trianglerighteq Bt_m)| = n - 2m + 2mn - 1$.

We first show that $rvc(P_n \trianglerighteq Bt_m) = n-1$. Since $diam(P_n \trianglerighteq Bt_m) = n$ if the grafting edge $x_iu_j \in V(Bt_m)$, by Theorem 1, we have $rvc(P_n \trianglerighteq Bt_m) \geq n-1$. To show that $rvc(P_n \trianglerighteq Bt_m) \leq n-1$, color the vertex v according to the following formula:

$$c(v) = \begin{cases} 1, & \text{for } v = y_i, y_{i,j};\ 1 \leq i \leq n-1,\ 1 \leq j \leq m-1 \\ 1, & \text{for } v = x_n \\ i, & \text{for } v = x_i;\ 1 \leq i \leq n-1 \end{cases}$$

From the above formula, it is easy to see that the color of the vertices are $i-1$ for $1 \leq i \leq n-1$, that is, $c : V(P_n \trianglerighteq Bt_m) \to \{1, 2, 3, \ldots, n-1\}$. Thus $rvc(P_n \trianglerighteq Bt_m) \leq n-1$. Combining with the lower bound, we have $rvc(P_n \trianglerighteq Bt_m) = n-1$.

We now show that $srvc(P_n \trianglerighteq Bt_m) = n-1$. By Theorem 1, we have $n-1 \leq rvc(P_n \trianglerighteq Bt_m) \leq srvc(P_n \trianglerighteq Bt_m)$. To show that $srvc(P_n \trianglerighteq Bt_m) \leq n-1$, color the vertex v according to the following formula:

$$c(v) = \begin{cases} 1, & \text{for } v = y_i, y_{i,j};\ 1 \leq i \leq n-1,\ 1 \leq j \leq m-1 \\ 1, & \text{for } v = x_n \\ i, & \text{for } v = x_i;\ 1 \leq i \leq n-1 \end{cases}$$

IOP Conf. Series: Journal of Physics: Conf. Series **1008** (2018) 012055 doi:10.1088/1742-6596/1008/1/012055

Again from the above formula, it is easy to see that the color of the vertices are $i-1$ for $1 \leq i \leq n-1$, that is, $c : V(P_n \trianglerighteq Bt_m) \rightarrow \{1,2,3,\ldots,n-1\}$. We see that every pair of vertices are connected by geodesic path. Thus $srvc(P_n \trianglerighteq Bt_m) \leq n-1$. Combining with the lower bound, we also have $srvc(P_n \trianglerighteq Bt_m) = n-1$. This concludes that if the grafting edge $x_i u_j \in V(Bt_m)$, then $rvc(P_n \trianglerighteq Bt_m) = srvc(P_n \trianglerighteq Bt_m) = n-1$. □

We now presents the rainbow vertex connection number and the strong rainbow vertex connection number of edge comp product of a star S_n and a triangular book graph Bt_m with grafting edge $x_1x_2 \in V(Bt_m)$.

Theorem 4. *If S_n is a star on $n+1$ vertices for $n \geq 4$ and Bt_m is a triangular book graph of $m \geq 2$ sheets, then the rainbow vertex connection number and strong rainbow vertex connection number of $S_n \trianglerighteq Bt_m$ with grafting edge $x_1x_2 \in V(Bt_m)$ is $rvc(S_n \trianglerighteq Bt_m) = srvc(S_n \trianglerighteq Bt_m) = 1$*

Proof. Let G be the graph obtained from edge comb product of S_n and Bt_m with grafting edge $x_1x_2 \in V(Bt_m)$. Then $V(S_n \trianglerighteq Bt_m) = \{A\} \cup \{x_i; 1 \leq i \leq n-1\} \cup \{x_{i,j}; 1 \leq i \leq n-1; 1 \leq j \leq m\}$ and $E(S_n \trianglerighteq Bt_m) = \{Ax_i; 1 \leq i \leq n-1\} \cup \{Ax_{i,j}; 1 \leq i \leq n-1; 1 \leq j \leq m\} \cup \{x_i x_{i,j}; 1 \leq i \leq n-1; 1 \leq j \leq m\}$. Thus $|V(S_n \trianglerighteq Bt_m)| = nm+n-m-1$ and $|E(S_n \trianglerighteq Bt_m)| = 2mn+n-2m-1$.

We first show that $rvc(S_n \trianglerighteq Bt_m) = 1$. Since $diam(S_n \trianglerighteq Bt_m) = 2$ if the grafting edge $x_1x_2 \in V(Bt_m)$, by Theorem 1, we have $rvc(S_n \trianglerighteq Bt_m) \geq 1$. To show that $rvc(S_n \trianglerighteq Bt_m) \leq 1$, color the vertex v according to the following formula:

$$c(v) = \begin{cases} 1, & \text{for } v = A \\ 1, & \text{for } v = x_i;\ 1 \leq i \leq n-1 \\ 1, & \text{for } v = x_{i,j};\ 1 \leq i \leq n-1,\ 1 \leq j \leq m \end{cases}$$

From the above formula, it is easy to see that the color of the vertices is 1 for $1 \leq i \leq n-1$, that is, $c : V(S_n \trianglerighteq Bt_m) \rightarrow \{1\}$. Thus $rvc(S_n \trianglerighteq Bt_m) \leq 1$. Combining with the lower bound, we have $rvc(S_n \trianglerighteq Bt_m) = 1$.

We now show that $srvc(S_n \trianglerighteq Bt_m) = 1$. By Theorem 1, we have $1 \leq rvc(S_n \trianglerighteq Bt_m) \leq srvc(S_n \trianglerighteq Bt_m)$. To show that $srvc(P_n \trianglerighteq Bt_m) \leq 1$, color the vertex v according to the following formula:

$$c(v) = \begin{cases} 1, & \text{for } v = A \\ 1, & \text{for } v = x_i;\ 1 \leq i \leq n-1 \\ 1, & \text{for } v = x_{i,j};\ 1 \leq i \leq n-1,\ 1 \leq j \leq m \end{cases}$$

Again from the above formula, it is easy to see that the color of the vertices is 1 for $1 \leq i \leq n-1$, that is, $c : V(S_n \trianglerighteq Bt_m) \rightarrow \{1\}$. We see that every pair of vertices are connected by geodesic path. Thus $srvc(S_n \trianglerighteq Bt_m) \leq 1$. Combining with the lower bound, we also have $srvc(S_n \trianglerighteq Bt_m) = 1$. This concludes that if the grafting edge $x_1x_2 \in V(Bt_m)$ $rvc(S_n \trianglerighteq Bt_m) = srvc(P_n \trianglerighteq Bt_m) = 1$. □

The following theorem presents the rainbow vertex connection number and the strong rainbow vertex connection number of edge comp product of a star on $n+1$ vertices S_n and a triangular book graph Bt_m with grafting edge $x_i u_j \in V(Bt_m)$.

Theorem 5. *If S_n is a star on $n+1$ vertices for $n \geq 4$ and Bt_m is a triangular book graph of $m \geq 2$ sheets, then the rainbow vertex connection number and strong rainbow vertex connection number of $S_n \trianglerighteq Bt_m$ with grafting edge $x_i u_j \in V(Bt_m)$ is $rvc(S_n \trianglerighteq Bt_m) = srvc(S_n \trianglerighteq Bt_m) = 3$*

Proof. Let G be the graph obtained from edge comb product of S_n and Bt_m with grafting edge $x_1 u_j \in V(Bt_m)$. Then $V(S_n \trianglerighteq Bt_m) = \{a\} \cup \{x_i; 1 \leq i \leq n-1\} \cup \{x_{i,j}; 1 \leq i \leq n-1; 1 \leq j \leq m\}$ and $E(S_n \trianglerighteq Bt_m) = \{ax_i; 1 \leq i \leq n-1\} \cup \{ax_{i,j}; 1 \leq i \leq n-1; 1 \leq j \leq m\} \cup \{x_i x_{i,j}; 1 \leq i \leq n-1; 1 \leq j \leq m\}$. Thus $|V(S_n \trianglerighteq Bt_m)| = nm+n-m-1$ and $|E(S_n \trianglerighteq Bt_m)| = 2mn+n-2m-1$.

We first show that $rvc(S_n \trianglerighteq Bt_m) = 3$. Since $diam(S_n \trianglerighteq Bt_m) = 4$ if the grafting edge $x_iu_j \in V(Bt_m)$, by Theorem 1, we have $rvc(S_n \trianglerighteq Bt_m) \geq 3$. To show that $rvc(S_n \trianglerighteq Bt_m) \leq 3$, color the vertex v according to the following formula:

$$c(v) = \begin{cases} 1, & \text{for } v = a_{i,j}, x_i; 1 \leq i \leq n-1, 1 \leq j \leq m-1 \\ 2, & \text{for } v = y_i;\ 1 \leq i \leq n-1 \\ 3, & \text{for } v = a \end{cases}$$

From the above formula, it is easy to see that the color of the vertices are 3, that is, $c : V(S_n \trianglerighteq Bt_m) \rightarrow \{1, 2, 3\}$. Thus $rvc(S_n \trianglerighteq Bt_m) \leq 3$. Combining with the lower bound, we have $rvc(S_n \trianglerighteq Bt_m) = 3$.

We now show that $srvc(S_n \trianglerighteq Bt_m) = 3$. By Theorem 1, we have $3 \leq rvc(S_n \trianglerighteq Bt_m) \leq srvc(S_n \trianglerighteq Bt_m)$. To show that $srvc(S_n \trianglerighteq Bt_m) \leq 3$, color the vertex v according to the following formula:

$$c(v) = \begin{cases} 1, & \text{for } v = a_{i,j}, x_i; 1 \leq i \leq n-1, 1 \leq j \leq m-1 \\ 2, & \text{for } v = y_i;\ 1 \leq i \leq n-1 \\ 3, & \text{for } v = a \end{cases}$$

Again from the above formula, it is easy to see that the color of the vertices are 3, that is, $c : V(S_n \trianglerighteq Bt_m) \rightarrow \{1, 2, 3\}$. We see that every pair of vertices are connected by geodesic path. Thus $srvc(S_n \trianglerighteq Bt_m) \leq 3$. Combining with the lower bound, we also have $srvc(S_n \trianglerighteq Bt_m) = 3$. This concludes that if the grafting edge $x_iu_j \in V(Bt_m)$, then $rvc(S_n \trianglerighteq Bt_m) = srvc(S_n \trianglerighteq Bt_m) = 3$. □

3. Conclusion

From the results discussed in the previous section, we can see that rainbow vertex connection number and strong rainbow vertex connection number the graphs obtained from edge comb product of a path and a triangular book graph have the same value. So do the graphs obtained from edge comb product of a star and a triangular book graph. For future work, it is interesting to characterize the edge comb product of two families of graphs G and H such that $rvc(G \trianglerighteq H) = srvc(G \trianglerighteq H)$.

Acknowledgement

This research is supported by DRPM Ditjen Penguatan Risbang Kemenristekdikti under Penelitian Tim Pascasarjana 2017 research grant.

Bibliography

[1] Arputhamary I A and Mercy M H 2017 Strong rainbow vertex-coloring of cubic Halin graphs *International Journal of Pure and Applied Mathematics (2017).* **114** (6) 11–20.
[2] Chartrand G, Johns G L, McKeon K A and Zhang P 2008 Rainbow connection in graphs *Mathematica Bohemica (2008).* **133** (1) 85–98.
[3] Li X and Liu S 2011 Rainbow vertex-connection number of 2-connected graphs *arxiv:1110.5770v1[math.CO].*
[4] Simamora D N S and Salman A N M 2015 The rainbow (vertex) connection number of pencil graphs *Procedia Computer Science (2015).* **74** 138–42.
[5] Nuriyeva F, Ugurlu O and Kutucu H 2013 A Mathematical Model for the Rainbow Vertex-Connection Number *ACSIJ Advance in Computer Science: an International Journal (2013).* **2** (4-5).
[6] Krivelevich M and Yuster R 2009 The rainbow connection of a graph is (at most) reciprocal to its minimum degree three *J. Graph Theory* (2009) **63** (3) 185–91.
[7] Li X and Shi Y 2013 On the rainbow vertex-connection *Discussiones Mathematicae Graph Theory (2013).* **33** 307–13.
[8] Li X, Mao Y and Shi Y 2014 The strong rainbow vertex-connection of graphs, *Util Math (2014).* **93** 213–23.
[9] Wang Z, Xu X and Liu Y 2017 Rainbow vertex-connection number of 3-connected graph *Applied Mathematical Sciences (2017).* **11** (16) 751–7.
[10] Waspodo G A, Slamin, Dafik and Agustin I H 2017 Bound of Distance Domination Number of Graph and Edge Comb Product Graph *IOP Conf. Series: Journal of Physics: Conf. Series (2017).* **855**.

IOP Conf. Series: Journal of Physics: Conf. Series **1008** (2018) 012056 doi:10.1088/1742-6596/1008/1/012056

Comparison of learning models based on mathematics logical intelligence in affective domain

Arif Widayanto[1], Hasih Pratiwi[1], Mardiyana[1]

[1]Mathematics Education Departement, Universitas Sebelas Maret, Surakarta, Indonesia
Email : arifwidayanto23@gmail.com

Abstract. The purpose of this study was to examine the presence or absence of different effects of multiple treatments (used learning models and logical-mathematical intelligence) on the dependent variable (affective domain of mathematics). This research was quasi experimental using 3x3 of factorial design. The population of this research was VIII grade students of junior high school in Karanganyar under the academic year 2017/2018. Data collected in this research was analyzed by two ways analysis of variance with unequal cells using 5% of significance level. The result of the research were as follows: (1) Teaching and learning with model TS lead to better achievement in affective domain than QSH, teaching and learning with model QSH lead to better achievement in affective domain than using DI; (2) Students with high mathematics logical intelligence have better achievement in affective domain than students with low mathematics logical intelligence have; (3) In teaching and learning mathematics using learning model TS, students with moderate mathematics logical intelligence have better achievement in affective domain than using DI; and (4) In teaching and learning mathematics using learning model TS, students with low mathematics logical intelligence have better achievement in affective domain than using QSH and DI.

1. Introduction

Mathematics is one science that has an important role in the development of logical thinking ability, analytical, systematic, critical, and creative, and can be used to solve the problems faced in everyday life. However, the fact that mathematics is considered as a difficult subject to grasp, causing mathematics achievement is still low, including junior high school students. The low mathematics achievement experienced by junior high school students in the district of Karanganyar.

The low in learning outcomes of mathematics can be seen from the value of the National Examination of junior high school under academic year 2015/2016. The average score of Indonesian language is 76.53, the average score of English is 51.88, the average score of mathematics is 45.71, and the average score of natural science is 56.71. From these values, it can be clearly seen that the value of national examination for mathematics is the lowest when being compared with other subjects. The government has made various efforts to improve learning outcomes in learning mathematics, but it has not be able to achieve the expected results.

Some of the factors that are the most common cause of low mathematics values are ineffective teaching methods, unavailability of teaching materials, foreign language in teaching materials, and the symbolism used in the subject [1]. The learning model is one of the factors that need to be considered in the learning process because it can affect student learning outcomes. The use of learning models tailored to the material and learning objectives will enable learners to understand the existing material so that their learning outcomes can be improved.

1.1. Learning Outcomes

According to Bloom, learning outcomes can be divided into three aspects: cognitive, affective, and psychomotor [2]. The cognitive domain contains intellectual-centered behavior, such as knowledge and thinking skills [3]. As time gone and the development of the era, cognitive domain of Bloom's Taxonomy is evolved [4]. The cognitive domain dimension of the Revised Bloom's Taxonomy is remembering, understanding, applying, analyzing, evaluating, and creating [5]. Bloom's Taxonomy of Educational Objectives on affective domain includes objectives on interests, values, appreciation and attitudes [6]. The first level of the affective taxonomy is referred to as receiving. At this level, the learner is aware of the topic, stimuli, event or issue and is willing and ready to learn about it or respond to it. The Second level, responding ranges from compliance by voluntary response to having a sense of satisfaction in doing what is required. The third level is referred to as valuing. At this level the learner voluntarily manifests behaviors that are consistent with certain beliefs. At the fourth level of the taxonomy; organization the learner organizes a set of values into a value system that are used to respond to diverse situations. The fifth level and which is the highest level in the hierarchy of affective taxonomy is characterization by a value or value set and this occurs when a student's behavior is consistent and predictable as if it has been adopted as a life style [7]. Psychomotor domain contain behavior that centers on the function of manipulation and physical abilities. It was established to address skills development relating to the physical dimensions of accomplishing a task [8]. In this paper, the author will discuss about the affective domain.

1.2. Learning Models

One of the obstacles to improving learning outcomes is teacher-centered learning or textbooks. Several studies in the area of mathematics have shown that instruction, especially at the secondary school level remains overwhelmingly teacher-centered, with greater emphasis being placed on lecturing and textbook than on helping student to think critical across subject area and applying their knowledge to read-worlds situation [9].

Teaching-centered learning or often we call the Direct Instruction model does not require students to be active in digging information in learning. It is an instructional model that focuses on the interaction between teachers and students. Key components of direct instruction include modeling, reinforcement, feedback, and successive approximations [10]. During the use of active learning, students move from being passive recipients of knowledge to being participants in activities. Active learning, through which students become active participants in the learning process, is an important means for development of higher level thinking [11]. The one of the active learning that can improve the affective domain is the cooperative learning model. Cooperative learning is concerned with framing student interaction in ways that are likely to raise positive interdependence and promotive interaction [12]. The result came up with five additional main components that a cooperative learning approach should have to be instrumental and these were: (1) positive interdependence; (2) individual accountability; (3) face-to-face interaction; (4) interpersonal and small group skills; and (5) group processing [13]. To be effective, cooperative learning must be well planned and structured with suitable learning materials and guidelines given to all participants [14]. It is explained that cooperative learning model Talking Stick type has usability such as making the students more focused when following the learning activities because at the time of the scepter the students must be involved in the learning activities [15]. The step of the Talking Stick learning model is the formation of heterogeneous groups, the students are given daily problems to solve, the rods are rolled out, the students who get the sticks are assigned to the presentation of the results of the discussion, and the group awards. Question is a sign of the complexity and fascination of life, a sign of cognitive vitality, of freedom of the mind [16]. The step of the learning model Question Student Have is the formation of heterogeneous groups, each student writing everyday problems on a piece of paper, the best issues will be selected and exchanged for completion by other groups, the results of the discussions presented, and the group awards.

1.3. Logical Mathematical Intelligence
Another factor that affects the affective domain is multiple intelligence. According [17], Gardner argues for nine relatively autonomous intelligences, there are linguistic intelligence, logical mathematical intelligence, musical intelligence, bodily kinesthetic intelligence, spatial intelligence, interpersonal intelligence, intrapersonal intelligence, naturalist intelligence, and existential intelligence.

Gardner described logical/mathematical intelligence as the ability to study problems, to carry out mathematical operations logically and analytically, and to conduct scientific investigations. Gardner identified mathematicians, logicians, and scientists as persons who would possess high levels of this hypothesized intelligence. The kinds of processes used in the service of logical-mathematical intelligence include categorization, classification, inference, generalization, calculation, and hypothesis testing. People who prefer to use their logical-mathematical intelligence usually do well on standardized comprehension/written language tests. They like to solve abstract problems and often do so by trial and error [18].

2. Research Method
This research is quasi experimental using 3x3 of factorial design. The dependent variable is affective domain and the independent variables are learning models (Talking Stick, Qustion Student Have, and Direct Instruction) and mathematics logical intelligence. The population of this research was VIII grade students of junior high school in Karanganyar under the academic year 2017/2018. Two schools were taken as samples using stratified cluster random sampling technique which are SMP Negeri 2 Karanganyar and SMP Negeri 2 Gondangrejo. Three classes from each school are taken as experimental classes.

3. Result and Discussion
3.1. Normality Test
The normality test is performed to determine whether the selected sample is from a normally distributed population. In this study used the Lilliefors method for normality test.

Table 1. Normality Test Data

Sample	$L_{observation}$	$L_{0.05;183}$	Decision	Conclusion
TS	0.0732	0.1109	H_0 is not rejected	Normal
QSH	0.0962	0.1099	H_0 is not rejected	Normal
DL	0.0896	0.1109	H_0 is not rejected	Normal

Based on Table 1, it appears that L_{obs} for each sample does not exceed L_{table} so that H_0 is not rejected, it means that each sample comes from a normally distributed population.

3.2. Homogeneity Test
The homogeneity test is conducted to determine whether the variance of a number of populations is equal or not. To test it used Barlett test.

Table 2. Homogeneity Test Data

Sample	$\mathcal{X}^2{}_{observation}$	$\mathcal{X}^2{}_{0.05;2}$	Decision	Conclusion
TS, QSH, DI	3.418	5.991	H_0 is not rejected	Homogen

Based on Table 2, it appears that $\mathcal{X}^2{}_{obs}$ does not exceed $\mathcal{X}^2{}_{tabel}$ so that H_0 is not rejected, it means that the variance of some populations is the same or homogeneous.

3.3. Two Ways Analysis of Variance with Unequal Cells

Two ways analysis of variance with unequal cells was performed to test whether or not there were significant differences in effect on some treatments on the affective domain.

Table 3. Marginal Average Data

Learning Model (A)	Logical Mathematical Intelligence (B)			Marginal Average
	High	Moderate	Low	
TS	126,59	123,59	128.87	125,72
QSH	122,92	120,50	113.00	119,42
DI	117,65	110,81	107.69	112,60
Marginal Average	122,01	118,87	116,19	-

Table 4. Two Ways Analysis of Variance with Unequal Cells Data

Source	Sum of Square	df	Mean of Square	$F_{observation}$	$F_{0,05}$	Decision
(A)	5973,99	2	2987,00	28,4388	3,00	H_{0A} is rejected
(B)	1057,40	2	528,70	5,0337	3,00	H_{0B} is rejected
(AB)	1268,30	4	317,08	3,0188	2,37	H_{0AB} is rejected
Error	18275,62	182	105,03	-	-	-
Total	26575,31	190	-	-	-	-

Based on Table 4 can be summarized as follows: (a) H_{0A} is rejected, hence there are difference of affective domain to model of learning, (b) H_{0A} is rejected, then there are different of affective domain to students mathematical logical intelligence, and (c) H_{0AB} is rejected, then there is the interaction between the learning model and the intelligence of mathematical logic to the affective domain students. Due to H_{0A}, H_{0B}, and H_{0AB} are rejected, it is necessary to test further anava with Scheffe method to know which have significantly different mean. In Table 5, the results of the calculation of further tests of intercross rows and the results of the mean further test calculations between the columns in Table 6.

3.4. Double Comparison Test

Double comparison test is a follow up of the analysis of variance if the result of variance analysis shows that H_0 is rejected. This test is used Scheffe method to see which gives significantly different mean.

Table 5. Inter-Line Comparison Test Results

H_0	$F_{observation}$	$2F_{0,05;2,183}$	$Critical\ Value$	Decision
$\mu_{1.} = \mu_{2.}$	11,6264	6,00	$\{F\|F > 6,00\}$	H_0 is rejected
$\mu_{1.} = \mu_{3.}$	49,5824	6,00	$\{F\|F > 6,00\}$	H_0 is rejected
$\mu_{2.} = \mu_{3.}$	13,5004	6,00	$\{F\|F > 6,00\}$	H_0 is rejected

Based on Table 5, it was clear that all of the hypothesis is rejected. This means that there is a significant effect between learning models and affective domain. In the current research, teaching and learning with model Talking Stick lead to better achievement in the affective domain than Question

Student Have, teaching and learning with model Question Student Have to lead to better achievement in the affective domain than using Direct Instruction.

Table 6. Inter-Columns Comparison Test Results

H_0	$F_{observation}$	$2F_{0,05;2,183}$	*Critical Value*	Decision
$\mu_{\cdot 1} = \mu_{\cdot 2}$	3,1718	6,00	$\{F \mid F > 6,00\}$	H_0 is not rejected
$\mu_{\cdot 1} = \mu_{\cdot 3}$	8,9284	6,00	$\{F \mid F > 6,00\}$	H_0 is rejected
$\mu_{\cdot 2} = \mu_{\cdot 3}$	1,9529	6,00	$\{F \mid F > 6,00\}$	H_0 is not rejected

Subsequently, we looked into overall results for Table 6. If H_0 is not rejected, then there is no significant effect between the intelligence of mathematical logic and affective domain. The only exception is hypothesis $\mu_{\cdot 1} = \mu_{\cdot 3}$, there is a significant effect between the intelligence of mathematical logic and affective domain. In this case, students with high mathematics logical intelligence have better achievement in the affective domain than students with low mathematics logical intelligence.

Table 7. Inter-Cell Comparison Test Results on the Same Line

H_0	$F_{observation}$	$8.F_{(0,05;8;183)}$	*Critical Value*	Decision
$\mu_{11} = \mu_{12}$	0,9196	15,52	$\{F \mid F > 15,52\}$	H_0 is not rejected
$\mu_{11} = \mu_{13}$	0,3939	15,52	$\{F \mid F > 15,52\}$	H_0 is not rejected
$\mu_{12} = \mu_{13}$	2,6246	15,52	$\{F \mid F > 15,52\}$	H_0 is not rejected
$\mu_{21} = \mu_{22}$	0,6195	15,52	$\{F \mid F > 15,52\}$	H_0 is not rejected
$\mu_{21} = \mu_{23}$	9,4807	15,52	$\{F \mid F > 15,52\}$	H_0 is not rejected
$\mu_{22} = \mu_{23}$	4,9213	15,52	$\{F \mid F > 15,52\}$	H_0 is not rejected
$\mu_{31} = \mu_{32}$	4,8935	15,52	$\{F \mid F > 15,52\}$	H_0 is not rejected
$\mu_{31} = \mu_{33}$	8,9204	15,52	$\{F \mid F > 15,52\}$	H_0 is not rejected
$\mu_{32} = \mu_{33}$	0,8427	15,52	$\{F \mid F > 15,52\}$	H_0 is not rejected

Table 7 presented all of the hypothesis is not rejected. In this regard, there is no interaction between affective domain and mathematic logical intelligence viewed from each learning models.

Table 8. Inter-Cell Comparison Test Results on the Same Columns

H_0	$F_{observation}$	$8F_{(0,05;8;183)}$	*Critical Value*	Decision
$\mu_{11} = \mu_{21}$	6,8060	15,52	$\{F \mid F > 15,52\}$	H_0 is not rejected
$\mu_{11} = \mu_{31}$	7,4317	15,52	$\{F \mid F > 15,52\}$	H_0 is not rejected
$\mu_{21} = \mu_{31}$	3,1649	15,52	$\{F \mid F > 15,52\}$	H_0 is not rejected
$\mu_{12} = \mu_{22}$	1,0734	15,52	$\{F \mid F > 15,52\}$	H_0 is not rejected
$\mu_{12} = \mu_{32}$	18,9304	15,52	$\{F \mid F > 15,52\}$	H_0 is rejected
$\mu_{22} = \mu_{32}$	9,1587	15,52	$\{F \mid F > 15,52\}$	H_0 is not rejected
$\mu_{13} = \mu_{23}$	19,1002	15,52	$\{F \mid F > 15,52\}$	H_0 is rejected
$\mu_{13} = \mu_{33}$	33,0632	15,52	$\{F \mid F > 15,52\}$	H_0 is rejected
$\mu_{23} = \mu_{33}$	2,2148	15,52	$\{F \mid F > 15,52\}$	H_0 is not rejected

After analyzing Table 8, the result showed that hypothesis is not rejected expect hypothesis for $\mu_{12} = \mu_{32}$, $\mu_{13} = \mu_{23}$, and $\mu_{13} = \mu_{33}$. In the hypothesis $\mu_{12} = \mu_{32}$, H_0 is rejected. It means that when viewed by students who have the intelligence of middle-class mathematical logic, there is a significant influence on the learning model of Talking Stick and Direct Instruction against the affective domain of mathematics. Be based on Table 3, in teaching and learning mathematics using learning model Talking Stick, students with moderate mathematics logical intelligence have better achievement in the affective domain than using Direct Instruction.

Hypothesis $\mu_{13} = \mu_{23}$ is rejected. It signifies that the interaction between the learning models (Talking Stick and Question Student Have) and the intelligence of mathematical logic that affects the affective domain mathematics. By looking at the average in Table 3, it can be concluded that the Talking Stick model is more effective in the affective domain than the Question Student Have model if given to students who have low mathematical logic intelligence.

Finally, Scheffe method also reported hypothesis $\mu_{13} = \mu_{33}$ is rejected. It declared the affective domain of mathematics is influenced by the model of learning and the intelligence of mathematical logic. In teaching and learning mathematics using learning model Talking Stick, students with low mathematics logical intelligence have better achievement in the affective domain than using Direct Instruction.

4. Conclusion

Based on the results of the research and the discussion that has been done, it can be concluded that : (1) Teaching and learning with model Talking Stick lead to better achievement in affective domain than Question Student Have, teaching and learning with model Question Student Have lead to better achievement in affective domain than using Direct Instruction; (2) Students with high mathematics logical intelligence have better achievement in affective domain than students with low mathematics logical intelligence have; (3) In teaching and learning mathematics using learning model Talking Stick, students with moderate mathematics logical intelligence have better achievement in affective domain than using Direct Instruction; and (4) In teaching and learning mathematics using learning model Talking Stick, students with low mathematics logical intelligence have better achievement in affective domain than using Question Student Have and Direct Instruction.

Acknowlegment

Praise be to Allah SWT and Rasulullah SAW who has given grace so that can be finished this paper. Do not forget to say thank you very much to both parents, my family, a big family of Post Graduate Mathematics Education, Universitas Sebelas Maret, and Government of Karanganyar Regency which has provided facilities.

References

[1] Njoroge J N and Githua, B N 2013 Effects of Cooperative Learning/ Teaching Strategy on Learners Mathematics Achievement by Gender Asian Journal of Social Sciences & Humanities **2** 2 567-576

[2] Budiyono 2015 Pengantar Penilaian Hasil Belajar *UNS Press* **1** 69

[3] Bloom B S, Engelhart M D, Furst E J, Hill W H and Krathwohl D R (Eds) 1956 *Taxonomy of Educational Objectives; The Classification of Educational Goals.* Handbook I; The cognitive domain. New York: David McKay Co Inc

[4] Anderson L W, Krathwohl D R, Airasian P W, Cruikshank K A, Mayer, R E, Pintrich P R, Raths J and Wittrock MC 2001 A Taxonomy for Learning, Teaching, and Assessing: A revision of Bloom's Taxonomy of Educational Objectives

[5] Shmelev V, Karpova M and Dukhanov A 2015 An Approach of Learning Path Sequencing based on Revised Bloom's Taxonomy and Domain Ontologies with the use of Genetic Algorithms *Procedia Computer Science* **66** 711-719

[6] Adetayo JO 2014 Assessing the Affective Behaviours in Learners *Journal of Education and Practice* **5** 16

[7] Olatunji M O 2013 Teaching And Assessing of Affective Characteristics: A Critical Missing Link In Online Education *International Journal on New Trends in Education and Their Implications* **4** 9

[8] Dave R H 1975 Developing and Writing Behavioural Objectives *Educational Innovators Press*

[9] Akinsola M K & Olowojaiye F B 2008 Teacher Instructional Methods and Student Attitudes Towards Mathematics *International Electronic Journal of Mathematics Education* **3** 1 60-73

[10] Magliaro S G, Lockee B B and Burton J K 2005 Direct instruction revisited: A key model for instructional technology *Journal of Educational Research Technology and Development* **53** 41-55

[11] Aydede M N and Kesercioğlu T 2010 The Effect of Active Learning Applications on Students Views About Scientific Knowledge *Procedia -Social and Behavioral Sciences* **2** 2 3783–3786

[12] Herrmann K J 2013 The impact of cooperative learning on student engagement: Results from an intervention *Active Learning in Higher Education* **14** 3 175–187

[13] Johnson D W & Johnson R 1994 Leading the cooperative school (2nd ed.) *Edina, MN: Interaction Book Company*

[14] Azmin N H 2015 Effect of the Jigsaw-Based Cooperative Learning Method on Student Performance in the General Certificate of Education Advanced-Level Psychology: An Exploratory Brunei Case Study *Canadian Center of Science and Education* **9** 1 2016

[15] Ryan T G 2013 The Scholarship of Teaching and Learning within Action Researh : Promise and Possibilities *Inquiry in Education* **4** 2 3

[16] Albu G 2014 The teacher and the questions of students. Case study *Procedia - Social and Behavioral Sciences* **187** 34 – 39

[17] Nicolini P, Alessandri G & Bilancioni G 2010 Web-Ob for multiple intelligences observation *Procedia Social and Behavioral Sciences* **2** 728-732

[18] Maftoon P 2012 The Realization of Gardner's Multiple Intelligences (MI) Theory in Second Leanguage Acquisition (SLA) *Journal of Language Teaching and Research Academy Publisher Manufactured in Finland* **3** 6 1233-241

Remembering the hindu festivities mathematically by the balinese using integer operations and least common multiple

Jero Budi Darmayasa[1,2], Wahyudin[2], Tatang Mulyana[2], Muchamad Subali Noto[3]
[1]Department of Mathematics Education, Universitas Borneo Tarakan, Tarakan, Kalimantan Utara, Indonesia.
[2]Department of Mathematics Education, Universitas Pendidikan Indonesia, Indonesia.
[3]Study Program of Mathematics Education, Universitas Swadaya Gunung Djati, Cirebon, Indonesia.

Email : jeromath@student.upi.edu.

Abstract. Ethnomathematicsis considered as a new study in mathematic education. As a study, numerous regions in this world starts to explore through ethnomathematics, including Indonesia. As the intersection between mathematics and mathematical modelling and culture, ethnomathematics exists in various society's cultural elements, including in the Balinese Hindus' festivities. To find the mathematical concept used in determining the festivity days, the researcher(s) conducted ethnographic research in *Bali Mula* society in Kintamani District, Bali. Participation observation, in-depth interview, and literature and documentation were used in collecting the data. As the result, the researcher(s) revealed that the mathematical concept used is integer operations, least common multiple, mixed fraction, and number sequences. Since it contains mathematical concept used in junior high, thus ethnomathematics of "4-hindu's festivities" may be used as context in mathematics learning. By using ethnomathematics as the context, the researcher(s) expect that it will help teachers in motivation their students to learn mathematics.

1. Introduction

Bali is the name of an island in Indonesia. The majority of people in this island are Hindus Balinese tribes. There, religion and customs unite themselves and transform into the unique characteristics of Balinese people. Also, supported by its extraordinary nature, Bali becomes both national and international tourism destination. Its abundant cultures at least affects Balinese people's mindset, including the ability in mathematics. However, someone's ability in mathematics is affected by his cultural background, and hung on what he sees and feels or experiences [1].

Mathematics, as one of disciplines, is definitely correlated with culture. There are people's daily activities with mathematical applications within. For example, people in *Kudus* has used the concept of circle, triangle, trapezium, rectangle, block, square pyramid and triangular prism in their daily life [2].

The correlation between mathematics and culture is called as Ethnomathematics. It combines mathematics practiced, adopted and combined within cultural practice by numerous communities ordinarily [3]. Mathematical concept can also be learnt in each stage that allows the learner to attach it in the cultures of Balinese society or others.

To date, there are a lot of researches on Ethnomathematics in mathematics and mathematics education. As the research area, Ethnomatematics may refer to a research to describe how mathematics is created, taught, emphasized and specialized in cultural system [4]. Thus, such researches may be in the form of exploration or its use.

As the attempt to use ethnomathematics, the thing we can learn from mathematics teacher is how to use ethnomathematics in mathematical learning. It is possible since children who go to school basically do not establish themselves in "tabula rasa". They have abundant collection of ideas and experiences according to cultural aground and daily activities of the integral part of cognitive function, and both of them are visible and popping up [5].

Ehtnomathematics may also be advantageous in the development of mathematics curriculum. This is in line with a thought mentioning that it very makes sense to integrate ethnomathematics an mathematics curriculum, because such integration may result in a comprehended and respected noble value within a culture [4]. Besides, the use of facts in students' daily life makes mathematical learning be more meaningful and exciting [6].

The above mentioned experts' perspective implies the significance of the investigation of the noble value contained in society's culture to improve the education quality, therefore, exploration by ethnomathematics is the initial step to figure out the concepts of mathematics taught at school in the cultural activities of Bali Mula Society. As conducted by the previous researchers, the research(s) observed each cultural activity to find out the types of applied mathematical concepts.

The Ethnomathematics of Balinese Sculptors applied similarity, shifting, and rotation; meanwhile, the Ethnomathematics of Balinese Calendar Compiler adopted repetition and confluence [7] On the other hand, ethnomathematics applied in Balinese webbing may be found in the use of tessellation in webbing pattern. The existing pattern uses the concept of rectangle, square, and regular tessellation [8].

According to the above notion, the research(s) conducted an ethnographic research to explore the ethnomathematics deeper, especially in the society of Bali Mula. The religious system in determining festivity days is the main focus in this research. Furthermore, this research aimed to figure out the types of mathematical concepts used by Balinese society in determining their festivities. After the researcher(s) found the mathematical concepts in the cultural elements, it was mapped with the mathematics materials used in the corresponding education levels. It is because Ethnomathematics will be relevant to mathematics learning used at schools [9].

2. Method

This research is conducted in qualitative way. A qualitative research may refer to as approach to explore and comprehend the meaning of social issues within the society, either individually or in group [10]. Specifically, this research is an ethnographic research with *Bali Mula* society in an area as a study. From March to June 2017, the researchers collected the data by using participant observation technique, in depth interviews, and documentation. In detail, the researchers described the data collection as follow:

Table 1 . Data Collection Technique

Data collection technique	Situation
Participant observation	Interacted and discussed with Bali Mula society in the area of Kintamani; participated in festivities.
In dept interview	Asked several questions relating to festivities or other sacred dats. The researchers asked such question to illiterate and literate elderly people, village figures, youth, and junior high students.
Literature study and Documentation	Looked for books related to Hindus festivities, learnt then matched them with the result of observation and interview.

The data resources in thi research are *Bali Mula* society living in the area of Batur Global Geopark, especially in *Songan* Village A and B, *Terunyan*, and *Belandingan*.

3. Result and Discussion

There are many festivities in Hindu. There are daily festivities, which are based on *wewaran*, *pananggal* and *panglong*, *pawukon*, and *sasih*. Daily festivities are in the form of *Yadnya Sesa* and *Tri Sandya*. *Yadnya Sesa* refers to festivities of food offerings (in small amount) before eating or after cooking.

Generally, *wewaran*-based festivities are divided into two categories: based on 3-days week (*triwara*) and 5-days week (*pancawara*); and 5-days week (*pancawara*) and 7-days week (*saptawara*). *Wewaran* refers to the rhythm of days based on the rhythm of universe. In *wewaran*, there are 10 types of week that has been preserved by Balinese society. The naming is according to the number of days in each week, they are 1-day week (*ekawara*), 2-days week (*dwiwara*), 3-days week (*triwara*), and so on until 10-days week (*dasawara*). Meanwhile, for society living in several areas in *Java* island and West *Java*, they commonly knows two of those week types: 5-days week and 7-days week.

Purnama (full moon) and *Tilem* (new moon) are the festivities based on *pananggal* and *panglong*. Furthermore, *Kajeng Kliwon* is festivities determined based on the confluence of *triwara* and *pancawara*. One *Kajeng* is among the three days in triwara (*Pasah, Beteng, Kajeng*). *Kliwon* is a day in *pancawara* (*Umanis, Paing, Pon, Wage,Kliwon*).

Based on the confluence of *pancawara* and *saptawara*, there are four festivities: *Buda Kliwon, Saniscara Kliwon, Buda Wage (Buda Cemeng),* and *Anggara Kliwon (Anggara Kasih)*. *Redite* = Sunday, *Coma* = Monday, *Anggara* = Tuesday, *Buda* = Wednesday, *Wraspati* = Thursday, *Sukra* = Friday, and *Saniscara* = Saturday are name of days in *saptawara*. The Cigugur society also used the combination of *pancawara* dan *saptawara* to count their rice harvesting period [11].

Each one cycle of *saptawara* has its own name. There are 30 names for *saptawara* popularly called *pawukuan* or *pawukon*. So, one *wuku* occurs in 7 days started from Sunday to Saturday. The name of 30 *wukus* are as follow: *1. Sinta, 2. Landep, 3. Ukir, 4. Kulantir, 5. Tolu, 6. Gumbreg, 7. Wariga, 8. Warigadian, 9. Julungwangi, 10. Sungsang, 11. Dungulan, 12. Kuningan, 13. Langkir, 14. Medangsia, 15. Pujut, 16. Pahang, 17. Klurut, 18. Merakih, 19. Tambir, 20. Medangkungan, 21. Matal, 22. Uye, 23. Menail, 24. Prangbakat, 25. Bala, 26. Ugu, 27. Wayang, 28. Klau, 29. Dukut, dan 30. Watugunung.*

Galungan and *Kuningan* are festiviries conducted based on *pawukon*. Meanwhile, festivities of *Siwalatri* dan *Nyepi* are commemorated based on *sasih* (month) in *pranatamangsa*, in which the former occurs in *sasih kepitu* (7th month) while the latter in *sasih Kesanga* (9th month).

Balinese Hindus depends such festivities on *Desa-Kala-Patra*. Such terms mean Place-Time-Situation. The term verily affects the festivities, mainly for Balinese Hindus living outside Bali Island. Other than the influence of *Desa* (place), *Kala* (time) also plays critical role. While time, in its largest scope, may refer to the exact time to conduct such festivities.

Considering numerous festivities, a mean for the purpose of remembering or calculating them are surely needed. Self calculation, moon's position observation, wall calendar check, internet access, or android-based application are the means to determine the festivities. In this section, the researchers present how *Bali Mula* society in *Kintamani* figure out the 4-Hindus festivities (*Buda Kliwon, Tumpek, Buda Cemeng,* and *Anggara Kasih*).

a. Illiterate elderly people

Illiterate *Bali Mula* society in *Kintamani* memorize the 4-Hindus festivities by using *wuku*. They memorize the days of festivities in particular *wukus*. Of 4 festivities, 3 of them are more dominant in their memory. Those festivities are *Buda Kliwon, Tumpek, and Anggara Kasih*.

Table 2. Matrix of 4-Hindus festivities within 210 days

Saptawara / Wuku	Redite	Coma	Anggara	Buda	Wrespati	Sukra	Saniscara
1. Sinta	Paing	Pon	Wage	Kliwon	Umanis	Paing	Pon
2. Landep	Wage	Kliwon	Umanis	Paing	Pon	Wage	Kliwon
3. Ukir	Umanis	Paing	Pon	Wage	Kliwon	Umanis	Paing
4. Kulantir	Pon	Wage	Kliwon	Umanis	Paing	Pon	Wage
5. Tolu	Kliwon	Umanis	Paing	Pon	Wage	Kliwon	Umanis
6. Gumbreg	Paing	Pon	Wage	Kliwon	Umanis	Paing	Pon
7. Wariga	Wage	Kliwon	Umanis	Paing	Pon	Wage	Kliwon
8. Warigadian	Umanis	Paing	Pon	Wage	Kliwon	Umanis	Paing
9. Julungwangi	Pon	Wage	Kliwon	Umanis	Paing	Pon	Wage
10. Sungsang	Kliwon	Umanis	Paing	Pon	Wage	Kliwon	Umanis
11. Dungulan	Paing	Pon	Wage	Kliwon	Umanis	Paing	Pon
12. Kuningan	Wage	Kliwon	Umanis	Paing	Pon	Wage	Kliwon
13. Langkir	Umanis	Paing	Pon	Wage	Kliwon	Umanis	Paing
14. Medangsia	Pon	Wage	Kliwon	Umanis	Paing	Pon	Wage
15. Pujut	Kliwon	Umanis	Paing	Pon	Wage	Kliwon	Umanis
16. Pahang	Paing	Pon	Wage	Kliwon	Umanis	Paing	Pon
17. Krulur	Wage	Kliwon	Umanis	Paing	Pon	Wage	Kliwon
18. Merakih	Umanis	Paing	Pon	Wage	Kliwon	Umanis	Paing
19. Tambir	Pon	Wage	Kliwon	Umanis	Paing	Pon	Wage
20. Medangkungan	Kliwon	Umanis	Paing	Pon	Wage	Kliwon	Umanis
21. Matal	Paing	Pon	Wage	Kliwon	Umanis	Paing	Pon
22. Uye	Wage	Kliwon	Umanis	Paing	Pon	Wage	Kliwon
23. Menail	Umanis	Paing	Pon	Wage	Kliwon	Umanis	Paing
24. Prangbakat	Pon	Wage	Kliwon	Umanis	Paing	Pon	Wage
25. Bala	Kliwon	Umanis	Paing	Pon	Wage	Kliwon	Umanis
26. Ugu	Paing	Pon	Wage	Kliwon	Umanis	Paing	Pon
27. Wayang	Wage	Kliwon	Umanis	Paing	Pon	Wage	Kliwon
28. Klau	Umanis	Paing	Pon	Wage	Kliwon	Umanis	Paing
29. Dukut	Pon	Wage	Kliwon	Umanis	Paing	Pon	Wage
30. Watugunung	Kliwon	Umanis	Paing	Pon	Wage	Kliwon	Umanis

Somehow, there is interesting thing behind the correlation among such festivities. *Buda Kliwon* occurs 10 days before *Tumpek*, and so does *Tumpek* to *Anggara Kasih*. Such fact then becomes the guideline of the illiterate elderly people, for example, if today is *Buda Kliwon*, then *Tumpek* and *Anggara Kasih* will respectively occur in the next 10 and 20 days.

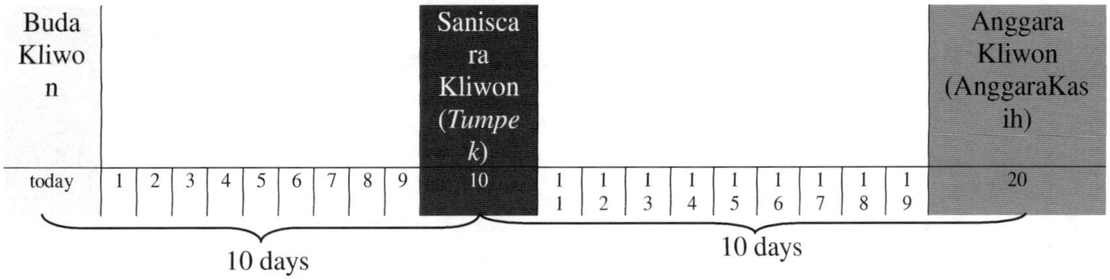

Since such festivities occurs in sequence, those people may easily memorize them. For instance, if today is *Buda KliwonSinta,Tumpek Landep* and *Anggara Kasih Kulantir* will respectively occur in the next 10 and 20 days. With such guideline, the illiterate elderly people may easily determine the days for 4-Hindus festivities

Besides, it is known that *Buda Kliwon* occurs in every 35 days and so do the other festivities. In other words, if today is *Anggara Kasih*, then the nearest *Buda Kliwon* and *Tumpek* will respectively occur in the next 15 and 25 days.

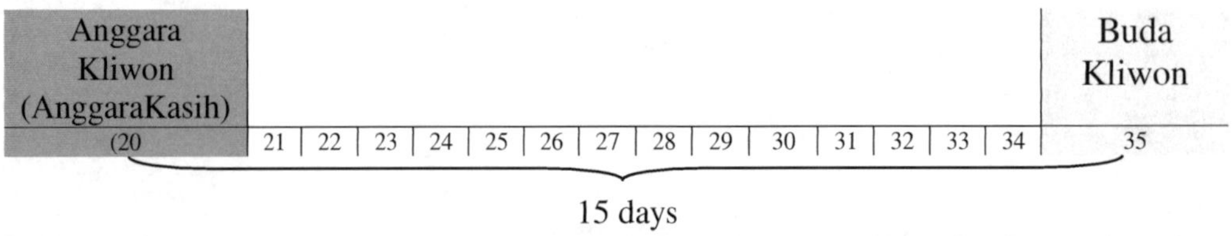

Similar with the other society, several important events related to family tend to be more memorable for *Bali Mula* society. Day of birth, marriage and other events are more likely quite memorable. The most common event is birth day, which is well known as "*Otonan*". *Otonan* will occur every 210 days or 6 months according to Balinese month.

To memorize his *otonan*, one usually uses the guideline of the 4-Hindues festivities. The elderly peope will find out the nearest festivities. For example, for a child born in *Anggara Umanis Landep*, his *otonan* will occur 4 days before *Tumpek Landep,* not 6 days after *Buda Kliwon Sinta.*

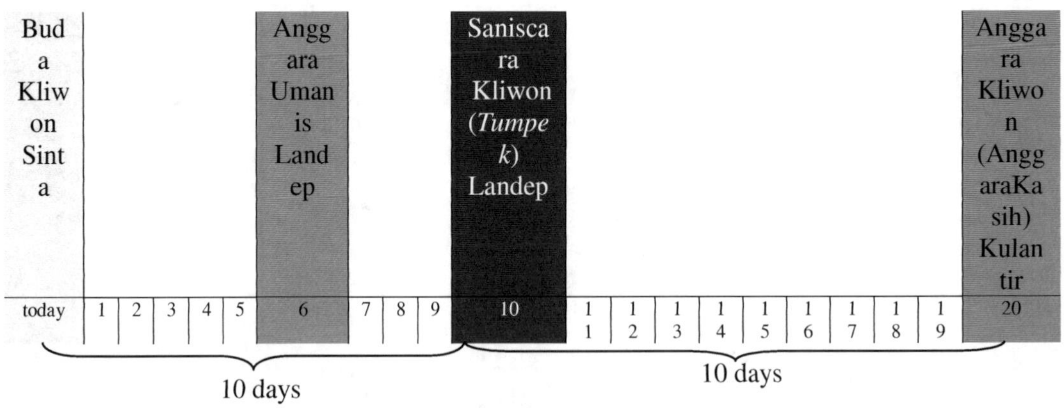

The general question that new Hindus parents mostly find is "how old is your child?." To properly answer such question, the parents will start to count it. The degree of difficulty in calculating such age can be categorized as follow: between the age of 1-210 days (0-6 months) and more than 6 months. For example, people who ask about the child's age in following days:

1) ≤ 6 months
   - On the day *Anggara Kasih Kulantir* of
     Anggara kasih Kulantir occurs 10 days after *Tumpek Landep*(+10). The child is born 4 days before *Tumpek Landep*(+4). So, on the day of *Anggara Kasih Kulantir*, the child is 14 days. The mathematical operations used in this case is $4 + 10 = 14$.
   - On the day of *Coma Paing Warigadian*
     Warigadian occur in the 8th wuku. When the question comes on the day of *Coma Paing Warigadian*, then the guideline used should be *Tumpek Wariga* that has occurred two days before (+2).*Tumpek Landep* occurs 35 days (+35) before *Tumpek Wariga*. The child was born 4 days before *Tumpek Landep*(+4). So, on the day of *Coma Paing Warigadian*, the child is 41 days. The mathematical operations used in this case is $4 + 35 + 2 = 41$.
   - On the day of *Anggara Wage Ugu*

Ugu occurs on the 26th wuku. When the question comes on the day of *Anggara Wage Ugu,*then the guideline used should be the next *Buda Kliwon Ugu* (-1).*Buda Kliwon Sinta* occurs 5 *x* 35 days before *Buda Kliwon Ugu*. The child was born 6 days after *Buda Kliwon Sinta* (-6). So, on the day of *Coma Paing Warigadian*, the child is 168 days. The mathematical operations used in this case is $-1+5x35-6=(5x35)-7=168$.

In this case, parents tend to tell their kid's age in this month unit: "7 days to five months"

2) > 6 months
   - On *Redite Umanis Ukir*
     Ukir exactly occurs after *Landep*. It is quite easy to answer question related to child's age when such question comes on the day of *Redite Umanis Ukir* or the other day in *Ukir* wuku. Parents will celebrate the child's birth day (*otonan*) by creating *upakara* according to their respective capability. Therefore, question related to age comes on the day of *Redite Umanis ukir* for a children who was born on the day *Anggara Umanis Landep* may be easily answered. The distance between *Anggara Umanis Landep* to *Anggara Umanis Landep* is 6 months (1 *oton*) or its multiple. *Anggara Umanis Landep* occurs 5 days before *Redite Umanis Ukir*. Thus, the child is 6 months 5 days or 12 months 5 days or 18 months 5 days on the day of corresponding*Redite Umanis Ukir*.

b. Literate society
   1) Self-calculation.
   People who are literate and have books as their reference may count the festivities days by themselves. There are standard rules specified in a printed book on how to count the 4-Hindus Festivities. Specifically, the method to determine *Buda Keliwon*, *Tumpek, Buda Cemeng, and Anggara Kasih* is by using the following formula [12]:

The number of *Wuku* should be divided by 5, if
The remainder is 1, thus it means the festivities day of *Buda Keliwon*
The remainder is 2, thus it means the festivities day of *Tumpek*
The remainder is 3, thus it means the festivities day of *BudaCemeng*
The remainder is 4, thus it means the festivities day of *Anggara Kasih*
The remainder is 5, thus it means of *Pengembang*

For example, *Wuku Sinta* containing 1 as the *wuku* number, then 1 : 5 = 0 with 1 as the remainder. So, in *Wuku Sinta* the festivities to celebrate is *Buda Kliwon.*
The calculation above may be simplified by ordering all of *wukus* in sequence by using fingers. It starts with folding the thumb for *Wuku Sinta,* index finger for *Wuku Landep* so on until the five *wukus* calculated.

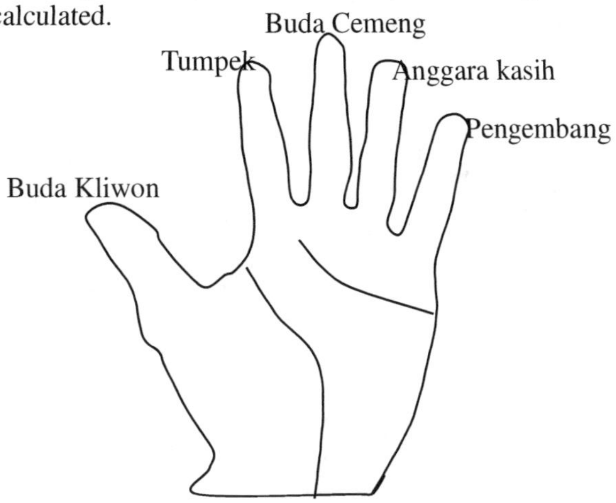

Figure 1. Calculation 4-Hindus Festivities using fingers

IOP Conf. Series: Journal of Physics: Conf. Series **1008** (2018) 012057 doi:10.1088/1742-6596/1008/1/012057

Since there are 30 *Wukus,* it means that within six months (210 days), there are 6 *Buda Kliwon* days. Similar pattern also occurs to *Tumpek, Buda Cemeng,* dan *Anggara Kasih. Buda Kliwon* that are celebrated in *Wuku (1) Sinta,(6)Gumbreg, (11) Dungulan, (16) Pahang, (21)Matal,* and *(26) Ugu*. Meanwhile, the festivities day of *Tumpek* occurs in *Wuku (2) Landep, (7) Wariga, (12) Kuningan, (17) Klurut, (22) Uye,* and *(27) Wayang*. The festivities of *Buda Cemeng* occurs in *Wuku (3) Ukir, (8) Warigadian, (13) Langkir, (18) Merakih, (23) Menail,* and *(28) Klau*. Furthermore, the day of *Anggara Kasih* occurs in *Wuku (4) Kulantir, (9) Julungwangi, (14) Medangsia, (19) Tambir, (24) Prangbakat,*and *(29) Dukut*. Moreover, wuku *(5) Tolu, (10) Sungsang, (15) Pujut, (20) Medangkungan, (25) Bala,* and *(30) Watugunung* are called *Pengembang.*
There are mathematical concepts for junior high that is in line with the above mentioned method. The concepts are:

a) Sequences
In the existing 30 wukus, the sequence containing *Buda Kliwon, Saniscara Kliwon, Buda Wage,* and *Anggara Kliwon* forms the following patterns:
a) 1st, 6th, 11th, 16th, 21st and 26th wuku refers to the festivities day of *Buda Kliwon.*
b) 2nd, 7th, 12th, 17th, 22nd and 27th wuku refers to the festivities day of *Tumpek.*
c) 3rd, 8th, 13th, 18th, 23rd, and 28th wuku refers to the festivities day of *Buda Cemeng.*
d) 4th, 9th, 14th, 24th, and 29th refers to the festivities day of *Anggara Kasih.*
The patterns above are the real examples of pattern or sequence.

b) Fraction and mixed fraction
There are 30 *wukus* in Balinese six months. If it is put in sequence from 1 to 30, there will be 30 *wuku* numbers. In reference to the condition of *wuku* number division by 5 and the remainders, then the mixed fraction is:

$$\left.\begin{array}{l} 1:5=\frac{1}{5} \\ 6:5=1\frac{1}{5} \\ 11:5=2\frac{1}{5} \\ 16:5=3\frac{1}{5} \\ 21:5=4\frac{1}{5} \\ 26:5=5\frac{1}{5} \end{array}\right\} \text{Represents } \textit{Buda Kliwon}$$

$$\left.\begin{array}{l} 2:5=\frac{2}{5} \\ 7:5=1\frac{2}{5} \\ 12:5=2\frac{2}{5} \\ 17:5=3\frac{2}{5} \\ 22:5=4\frac{2}{5} \\ 27:5=5\frac{2}{5} \end{array}\right\} \text{Represents } \textit{Tumpek}$$

$$\left.\begin{aligned} 3:5&=\frac{3}{5}\\ 8:5&=1\frac{3}{5}\\ 13:5&=2\frac{3}{5}\\ 18:5&=3\frac{3}{5}\\ 23:5&=4\frac{3}{5}\\ 28:5&=5\frac{3}{5}\end{aligned}\right\}\text{Represents } \textit{Buda Cemeng}$$

$$4:5=\frac{4}{5}$$

$$\left.\begin{aligned} 9:5&=1\frac{4}{5}\\ 14:5&=2\frac{4}{5}\\ 19:5&=3\frac{4}{5}\\ 24:5&=4\frac{4}{5}\\ 29:5&=5\frac{4}{5}\end{aligned}\right\}\text{Represents } \textit{Anggara Kasih}$$

c) Least Common Multiple

The days of 4 festivities are based on *Pancawara* and *Saptawara*. *Buda Kliwon* is celebrated in the day of *Buda* (*Saptawara*) and *Kliwon* (*Pancawara*), while *Tumpek* in the day of *Saniscara* (*Saptawara*) and *Kliwon* (*Pancawara*), and *Anggara Kasih* in *Anggara* (*Saptawara*) and *Kliwon* (*Pancawara*). The determination of the sequence of the 4 festivities days can be calculated by using Least Common Multiple. Since the days in *Saptawara* re-occur in each 7 days and *Pancawara* in each 5 days, then the determination should use the least common multiple of 7 and 5, which results in 35 days.

2) Wall calendar

Most of Balinese society use wall calendar and so do others living outside Bali Island. Such calendar is called Balinese calendar. Basically, some of the contents in such calendar are not originally inherited from Balinese culture. Combination and modernization exist within. The pattern of Balinese calendar is the same with the common Gregorian one [13].

One of the most popular printed Balinese calendars is made by the late I Ketut Bambang Geder Rawi and his sons.

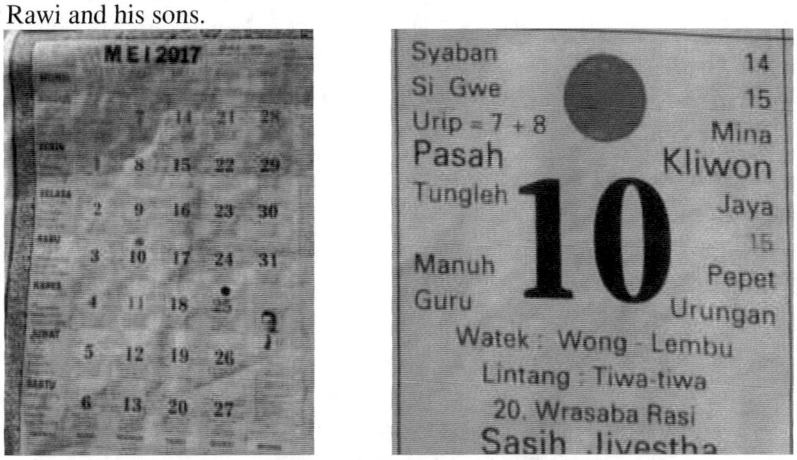

Figure 2. Balinese wall calendar

In order to be able to read such calendar, the competence of table reading is required. Similar with the common wall calendar, there are 5 columns and 7 rows in each month. The rows show *saptawara* while the columns indicate *wuku*. The information about *pancawara* is

IOP Conf. Series: Journal of Physics: Conf. Series **1008** (2018) 012057 doi:10.1088/1742-6596/1008/1/012057

presented in the square as the confluence of row and column. For example, in Wednesday row (*Buda*), the second column showing the date of 10 May 2017 refers to the day of *Kliwon* (*Pancawara*). Therefore, the festivities day of *Buda kliwon* occurs in such day.

3) Modern application
The information may be accessed on a website. Other than database-based on website, this information can also be accessed through application. There are various choices of android-based application containing the information of Hindus festivities available on playstore.

4. Conclusions

Balinese Hindus society have many methods to figure out their festivities days. The differences between background and literacy also determine their distinctive methods. For illiterate society, they use their environmental observation as their guideline in determining the festivities days. Meanwhile, the literate society have more methods in determining such events, such as self-calculation, wall calendar check, moon's position observation, or internet access and android-based application. The calculation of the 4-Hindus Festivities (*Buda Kliwon, Tumpek, Buda Cemeng,* dan *Anggara Kasih*) may be done in mathematical concepts, they are: integers operations, least common multiple, mixed fractions and sequence. Since it fully contains mathematical concepts, the determination of the 4-Hindus Festivities may be used as the context in mathematical learning at junior high school. The researchers hope that such utilization may motivate the students in learning mathematics as mandated in the regulation of minister of education.

References

[1] Zusmelia 2016 Matematika dalam Perspektif Indegenous People dan Indegeneous Knowledge (Kasus pada Masyarakat Matrilineal Minangkabau Sebuah Tinjauan Sosiologis) Seminar Nasional Matematika dan Pendidikan Matematika Padang: STKIP PGRI Sumatera Barat 1 – 12

[2] Zaenuri, Teguh A W and Dwidayanti N 2017 Ethnomathematics Exploration on Culture of Kudus City and Its Relation to Junior High School Geometri Concept. International Journal of Education and Research 161-167

[3] Mosimege M 2004 Ethnomathematical Studies on Indegenous Games: Examples from Southern Africa Pisa: Tipografia Editrice Pisana Ethnomathematics and Mathematics Education 119-137

[4] Zhang W and Zhang Q 2010 Ethnomathematics and Its Integration within the Mathematics Curriculum *Journal of Mathematics Education* **3** 1 151 - 157

[5] Kinard J T and Kozulin A (2008) *Rigorus Mathematical Thinking (Conceptual Formation in Mathematical Classroom* New York: Cambridge

[6] Arisetyawan A, Suryadi D, Herman T and Rahmat C 2014 Study of Ethnomathematics: A lesson from the Baduy Culture *International Journal of Education and Research* 681-688

[7] Suharta I G 2016 Akrab dengan Matematika, Tanpa Belajar Matematika. *Seminar Nasional MIPA* 8-13 Singaraja: Undiksha Press

[8] Puspadewi K and Putra I 2014 Etnomatematika di Balik Kerajinan Anyaman Bali *Jurnal Matematika* 80-89

[9[Marsigit 2016 Pengembangan Pembelajaran Matematika Berbasis Etnomatematika. Seminar Nasional Matematika dan Pendidikan Matematika Padang: STKIP PGRI Sumatera Barat 1-32

[10] Creswell J W 2014 *Research Design: Qualitative, Quantitative, and Mixed Methods Approaches* Los Angeles: SAGE

IOP Conf. Series: Journal of Physics: Conf. Series **1008** (2018) 012057 doi:10.1088/1742-6596/1008/1/012057

[11] Dewanto S P, Kusuma D A, Ruchjana B N and Abdullah A S 2017 October. Intangible heritage for sustainable future: mathematics in the paddy field In *Journal of Physics: Conference Series* **893** 1 012003 IOP Publishing
[12] Guweng I K 1991 Sarining Wariga Denpasar: C2
[13] Chaterjee S 1997 Balinese Traditional Calender. Indian Journal of History of Science 325 - 346

Students' misconception on equal sign

N F Kusuma[1], S Subanti[1], B Usodo[1]

[1]Mathematics Education [1]Departement, Universitas Sebelas Maret, Surakarta, Indonesia

Email: niyankusuma@gmail.com

Abstract. Equivalence is a very general relation in mathematics. The focus of this article is narrowed specifically to an equal sign in the context of equations. The equal sign is a symbol of mathematical equivalence. Studies have found that many students do not have a deep understanding of equivalence. Students often misinterpret the equal sign as an operational rather than a symbol of mathematical equivalence. This misinterpretation of the equal sign will be label as a misconception. It is important to discuss and must resolve immediately because it can lead to the problems in students' understanding. The purpose of this research is to describe students' misconception about the meaning of equal sign on equal matrices. Descriptive method was used in this study involving five students of Senior High School in Boyolali who were taking Equal Matrices course. The result of this study shows that all of the students had the misconception about the meaning of the equal sign. They interpret the equal sign as an operational symbol rather than a symbol of mathematical equivalence. Students merely solve the problem only single way, which is a computational method, so that students stuck in a monotonous way of thinking and unable to develop their creativity.

1. Introduction

Mathematical equivalence concept has been taught to students since elementary school, specifically in material arithmetic. Learning mathematics in junior high school is a transition from learning arithmetic to algebra. An example of the current discontinuity between elementary school arithmetic and the algebra learnt in upper grades, relates to understanding the concept of an equal sign. The equal sign is a symbol of mathematical equivalence. The equal sign should be understood as a relational symbol, indicating that a balanced relationship exists between numbers on the two sides of the equivalence symbol [6].

Students who defined the equal sign as a symbol of mathematical equivalence were more likely to recognize not only that the equations $2 \times \square + 15 = 31$ and $2 \times \square + 15 - 9 = 31 - 9$ had the same solutions, but this also could be determined without actually solving the two equations [7]. Students who already hold a relational view of the equal sign, successfully solve $84 + 104 = \square + 103$ by discourage computation and encourage looking for relationships across the equal sign. In this case, a student who is thinking relationally might notice that 103 is one less than 104 and that the number to be placed in the box must therefore be one more than 84.

Studies have found that many students do not have a deep understanding of equivalence. Based on a research of [8], in the exploratory study of the mathematical thinking of a selection of Year 7 and Year 8 students in Brazil, they found that when students were asked to solve numerical expressions using four arithmetic operations, most students opted for computational methods. Students often misinterpret the equal sign as an operational rather than a symbol of mathematical equivalence [1, 4, 6, 7]. This misinterpretation of the equal sign will be label as a misconception. Misconception is mistaken idea or view resulting from a misunderstanding of something. It arises from the fact that before knowing the correct concept they already have their own concept which is formed from reasoning, intuition, culture, or the other [5].

This misconception can lead to the problems of students' understanding if not resolve immediately, for example students who define the equal sign as an operational found difficulty in solving algebraic related problems. When students were asked to solve the equation $5 + 3x = 14$, students tend to move 5 to the right side into $3x = 14 - 5$, then students moved to the right side into $x = \frac{9}{3}$ without knowing the reasons why students do this. Students who view the right-hand side of an equation as the answer and who prefer to solve equations by transposing lack an understanding of the balance between the right and left-hand sides of the equation [2].

There are five factors that causes misconceptions in students namely students, teachers, textbooks, context, and way of teaching [9]. Arithmetic algorithms are teachable as procedures for solving algebraic equations, by focusing on relations rather than calculating the answer [3]. In the Indonesian curriculum, there is a serious discontinuity between the arithmetic taught in elementary school and the algebra taught in upper grades. The elementary school arithmetic syllabus focuses upon computational performance, without attending to relations and fundamental properties of arithmetic operations. As a result, students hold the misconception about mathematical equivalence and experience difficulty when they begin to study algebra in upper grades. It also influences the student who is finding an answer without understanding the equation concept and merely find it only single way, which is a computational method. The consequence is students stuck in a monotonous way of thinking and unable to develop their creativity. Creativity is one of important skills required for 21st century education besides critical thinking, communication, and collaboration.

The concept of equivalence is also taught to students in High School, specifically in material equal matrices. A matrix is a group of numbers arranged in a square-shaped range or a rectangle made up of rows and columns. Two matrices A and B are equal $(A = B)$ if and only if they are of the same size and their corresponding entries are equal [10]. The equal matrices can often be use to determine the value of the variable or variables on the entries of a matrix.

2. Method

This research uses a qualitative method with descriptive research type. The strategy used in this research is a case study. This research was conducted at one of senior high school in Boyolali Central Java Province, Indonesia. There are 5 students as participants. The subjects of the study are students of grade 12 with relatively the same scores and with good communication skills. They are SR, SD, SW, SA, and SL. This qualitative study was carried out in the following procedures: Firstly, researchers collected data from an individual written test. Secondly, data collected from an individual interview. Thirdly, data analyzed by comparing data that were collected from the individual written test with data from the individual interview. Fourthly, sum up for conclusion.

3. Result and Discussion

According to data analyzing, the results classified into two sections. The sections are seen from students' test result and students' interview data results. Three questions are given by researcher to identify students' misconception on the equal sign. Students' misconception will be described in the following section.

Question 1:

$$“\begin{pmatrix} 0 & -1 \\ 4 & 5 \end{pmatrix} \underset{\uparrow}{=} \begin{pmatrix} 0 & -\frac{1}{1} \\ \frac{16}{4} & \frac{10}{2} \end{pmatrix}$$

(a) The arrow points to a symbol. What is the name of this symbol?
(b) What does this symbol mean?"

Figure 1 below is the student SR answer of question 1.

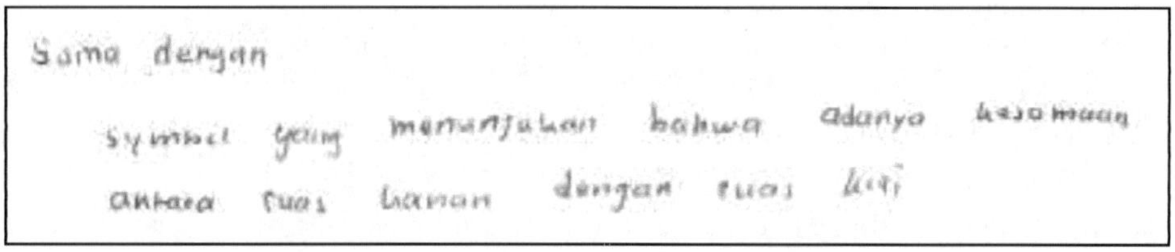

Figure 1. The student SR answer of question 1

According to the student's test result, SR defined the equal sign as a symbol that indicates the existence of an equivalence between the right side and the left side. On the other hand, student SD, SW, SA, and SL defined the equal sign as a symbol that indicates corresponding entries on both matrices are equal.

Question 2:
"Given the matrices as mentioned below:

$$A = \begin{pmatrix} 74 & p \\ s & 3{,}8 \end{pmatrix} \quad B = \begin{pmatrix} q & -900 \\ -78 & r \end{pmatrix} \quad C = \begin{pmatrix} 75 + 23 & 1000 - 86 \\ 92{,}3 - 77{,}8 & 3{,}7 + 1{,}4 \end{pmatrix}$$

If$A + B = C$, determine the value of $p, q, r,$ and s!"

According to student's test result, all of the students solve the problem in the same way. Figure 2 below is an example of student SW answer to question 2.

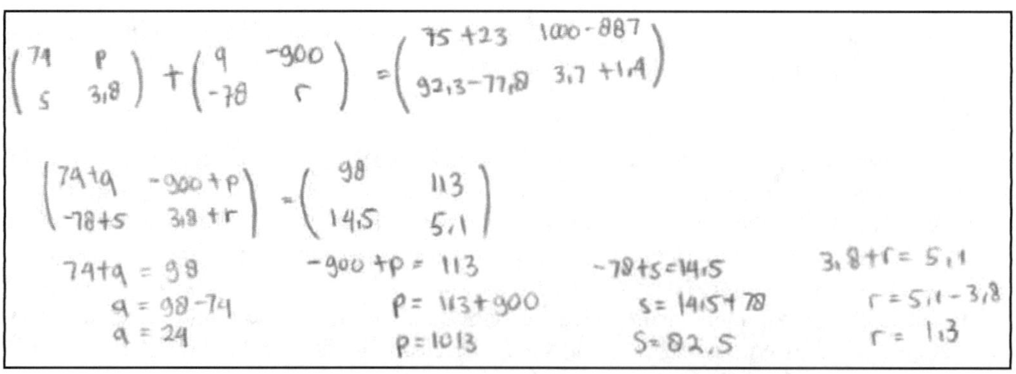

Figure 2. The student SW answer of question 2

First, students SW wrote the matrix in the form of known sum operation which was $A + B = C$. On the right side, student SW calculated the sum or subtraction of numbers on matrix C. On the left side, student SW summed every corresponding entries on matrices A and B. According to the calculation result on both sides, student SW use the definition of equal matrices to identify the unknown value. Based on the definition, matrix $A + B$ and matrix C are equal $(A + B = C)$ if and only if they are of the same size and their corresponding entries are equal [10].

Furthermore, student SW given four new equations, which were $74 + p = 98$; $-900 + q = 113$; $-78 + s = 14{,}5$; and $3{,}8 + r = 5{,}1$. Student SW moves the numbers on both sides in every equation to determine the value of p, q, r, and s. The result of the researcher's interview with the student SR on the first question listed in Table 1 below:

Table 1. Result of interview with student SW of question 2

Code	Interview Result
R	*"Just look at the sheet on you, what is the value of p?"*
SW1	*"This ismom. The value of p is 1013."*
R	*"How do you get the value of p?"*
SW2	*"Move the numbers -900 on left side to right side."*
R	*"Why are you moving the numbers -900? How does the negative sign may change into pofitive sign?"*
SW3	*"Because we will find the value of p, then the numbers -900 is moved to the right side, so the only remaining variable p on the left side. A sign of negative numbers 900 can be turned into a positive because if a number of the moved side, then the mark will change."*

The interview result with student SW indicates that student SW experienced misconception concerning on equal sign. It was proved by code SW3, in which student SW moves the segment on equation without knowing the right reason for doing so. Student SW did not use the concept of equivalence at all thus students SW could not explain why number 900 appeared on the right side and how number -900 which located on the left side at first, then left only with variable p. This also happens to identify the value of q, r, and s. In the same way, student SW got the right value of q, r, s, although the concept that being used was not exact.

Question 3:
"Given matrices as mentioned below:

$$P = \begin{pmatrix} a \times 10 & b \div 6 \\ c \times 4 & d \div 8 \end{pmatrix}$$

$$Q = \begin{pmatrix} 27 \times 30 & 90 \div 18 \\ 2400 \times 0{,}4 & 12 \div 0{,}8 \end{pmatrix}$$

If $P = Q$, determine the value$a, b, c,$ and d!"

The student's test result shows that all of the students solve the problem in the same way. Figure 3 below is an example of student SA answer to question 3 on first activity.

Figure 3. The student SA answer of question 3 on first activity

The question above indicates $P = Q$. Student SA uses the definition of equal matrices to determine the unknown value. Matrix P and matrix Q are equal $(P = Q)$ if and only if they are of the same size and their corresponding entries are equal [10]. The figure 3 shows that student SA given four new equations, consist of $a \times 10 = 27 \times 30$ and $c \times 4 = 2400 \times 0{,}4$. In order to determine the value of a and c, student SA divides the numbers on the right side with the numbers on the left side in every equation. The result of the researcher's interview with the student SA on the first question listed in Table 2 below:

IOP Conf. Series: Journal of Physics: Conf. Series **1008** (2018) 012058 doi:10.1088/1742-6596/1008/1/012058

Table 2. Result of interview with student SA on question 3 at first activity

Code	Interview Result
R	*"Just look at the sheet on you, what is the value of a?"*
SA1	*"This is mom. The value of a is81."*
R	*"How do you get the value of a?"*
SA2	*"Yes like that. The numbers*27×30*divided by 10."*
R	*"Oke. Then, where the number 10 on the left side?"*
SA3	*"The number 10 on the left side does not exist because it has been used to divide the numbers*27×30*on the right side."*
R	*"Is there another way to get the value of a?"*
SA4	*"hmmm...I think there is no, mom. I only know the way."*

The interview result with student SA explains that student SA experienced misconception concerning equal sign. It was proved by code SA2 and SA3, in which student SA divides number 27×30 on the right side with number 10 on the left side. Student SA did not use the concept of equivalence. Student SA unable to understand that number on the left side also should be divided with number 10. The same case also carried out in finding the value of c as student SA obtain the right value of c, even though the concept that used was not accurate. Defining the value of b and d, all of the students also solve the problem in the same way. Figure 4 below is an example of student SR answer to question 3 on second activity.

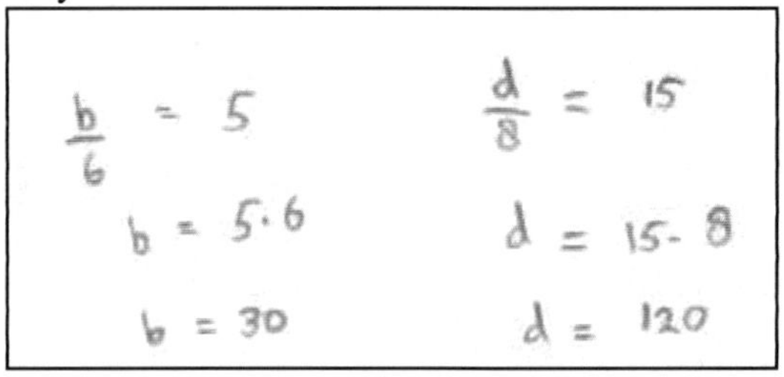

Figure 4. The student SR answer of question 3 on second activity

Figure 4 informs that student SR got two new equations, which were $\frac{b}{6} = 5$ and $\frac{d}{8} = 15$. In finding the value of b and d, student SR uses cross multiplication on that equation. The result of the researcher's interview with the student SR on the second question listed in Table 3 below:

Table 3. Result of interview with student SR on question 3 at second activity

Code	Interview Result
R	*"Just look at the sheet on you, what is the value of b?"*
SR1	*"This is mom. The value of a is 30."*
R	*"How do you get the value of b?"*
SR2	*"The number 6 on left side multiplied by the number 5 on the right side."*
R	*"Do you mean the cross product?"*
SR3	*"Yes mom. That I mean."*
R	*"Why do you do multiplication? In other words, why are you not doing the addition, substraction or division against the number?"*
SR4	*"No mom, because on the left side that is forms division. So, if the form division, how to find the unknown number is multiply the denominator on the left side with the number on the right side. If the form multiplication, how to find the unknown number is divide the number that is in the right side with coefficients on the left side, such as finding the value of a and c."*
R	*"Ok. Then, where is the number 6 in the left side?"*

SR5	*"The number 6 are already multiplied by the number 5, so there is no number 6 on the left side."*
R	*"Is there another way to get the value of a?"*
SR6	*"No mom. From I used to work on the problem in a way like it."*

According to interview result, student SR also experienced misconception regarding the equal sign. It was proved by code SR2-SR5, as student SR multiplies number 6 on the left side with number 5 on the right side. Student SR did not use the concept of equivalence so that student SR unable to understand that on the left side should also multiply with number 6. In other words, both sides multiplied by number 6. The same case applied in finding the value of d. In the same way, students SR got the right value of d, even though the concept that being used was not appropriate.

4. Conclusion

Based on the analysis of the result, the researcher concluded that all of the students hold the misconception about the meaning of the equal sign. They recognize that the equal sign as a symbol of mathematical equivalence, but did not have a deep understanding of equivalence. The students move the numbers on both sides when faced with forms of the equation with an addition or subtraction as the operation. On the other hand, the students conducted cross multiplication on that equation with multiplication or division as the operation. Students merely solve the problem only single way, which is a computational method , so that students stuck in a monotonous way of thinking and unable to develop their creativity. The solution offered to reduce that misconception, teacher can develop learning method that demands the student to more focus on creative thinking and often several kinds of question in order significantly increase the students' relational thinking.

Acknowledgments

We thank students for their help and participation

References

[1] Andrews P and Sayers J 2012 Teaching Linear Equations: Case Studies from Finland, Flanders and Hungary *Journal of Mathematical Behavior* **31** 476-488

[2] Aydin S, Çelikkol E S, Of M, and Mutlu C 2013 On the Two Connected Concepts in Secondary School Mathematics: Algebraic Expressions and Linear Equations *Proc. Sosial and Behavioral Sciences* **106** 1966-1970

[3] Carpenter 2005 Algebra in Elementary School Developing Relational Thinking *ZDM* **32**

[4] Napaphun V 2012 Relational Thinking: Learning Arithmetic in order to Promote Algebraic Thinking *Journal of Science and Mathematics Education in Southeast Asia* **35** 84-101

[5] Ningrum R W, Yulianti M, Hellingo D D Z and Budiarto M T 2018 Students' Misconceptions on Properties of Rectangles *Journal of Physics (IOP Publishing Ltd)* **947**

[6] Powell S R and Fuchs L S 2010 Contribution of Equal-Sign Instruction Beyond Word-Problem Tutoring for Third-Grade Students with Mathematics Difficulty *Journal of Educational Psychology* **102** 381-394

[7] Stephens A C 2006 Equivalence and Relational Thinking: Preservice Elementary Teachers' Awareness of Opportunities and Misconceptions *Journal of Mathematics Teacher Education* **9** 249-278

[8] Stephens M and Ribeiro A 2012 Working Towards Algebra the Importance of Relational Thinking *Revista Latinoamericana de Investigatión en Matemática Educativa* **15** 373-402

[9] Suparno P 2005 *Analisis Miskonsepsi Fisika dan Remidiasi* (Salatiga: Universitas Kristen Satya Wacana)

[10] Wirodikromo S 2006 *Matematika Jilid 3 IPA untuk Kelas XII* (Jakarta: Erlangga) 81

IOP Conf. Series: Journal of Physics: Conf. Series **1008** (2018) 012059 doi:10.1088/1742-6596/1008/1/012059

The 21st century skills with model eliciting activities on linear program

Septriana Handajani[1], Hasih Pratiwi[1], Mardiyana[1]

[1]Mathematics Education Departement of Graduate Program, Universitas Sebelas Maret, Surakarta, Indonesia

E-mail: septrianahandajani@gmail.com

Abstract. Human resources in the 21st century are required to master various forms of skills, including critical thinking skills and problem solving. The teaching of the 21st century is a teaching that integrates literacy skills, knowledge, skills, attitudes, and mastery of ICT. This study aims to determine whether there are differences in the effect of applying Model Elliciting Activities (MEAs) that integrates 21st century skills, namely 4C and conventional learning to learning outcomes. This research was conducted at Vocational High School in the odd semester of 2017 and uses the experimental method. The experimental class is treated MEAs that integrates 4C skills and the control class is given conventional learning. Methods of data collection in this study using the method of documentation and test methods. The data analysis uses Z-test. Data obtained from experiment class and control class. The result of this study showed there are differences in the effect of applying MEAs that integrates 4C skills and conventional learning to learning outcomes. Classes with MEAs that integrates 4C skills give better learning outcomes than the ones in conventional learning classes. This happens because MEAs that integrates 4C skills can improved creativity skills, communication skills, collaboration skills, and problem-solving skills.

1. Introduction

The 21st century is a century marked by the rapid development of science and technology. This condition should be followed by learning that can meet the demands of the times. The teaching of the 21st century is expected to produce human resources that can master various forms of 21st century skills. The Educational Testing Service (ETS) in its publication, *Digital Transformation: A Literacy Framework for ICT Literacy* (2007), defies 21st century learning skills as the ability to a) collect and/or retrieve information, b) organize and manage information, c) evaluate the quality, relevance, and usefulness of information, and d) generate accurate information through the use of existing resources [1]. The skills have been grouped into three main areas: 1) Learning and innovation skills: critical thinking and problem solving, communications and collaboration, creativity and innovation, 2) Digital literacy skills: information literacy, media literacy, Information and communication technologies (ICT) literacy, 3) Career and life skills: flexibility and adaptability, initiative and self-direction, social and cross-cultural interaction, productivity and accountability [2].

Linear program (also called linear optimization) is a method to achieve the best outcome (such as maximum profit or lowest cost) which in its completion using mathematical model. Linear program is one of the material that is considered difficult by students in Vocational High School grade XI. Basic competence to be achieved on linear program material is that students can solve contextual problems related to linear program. The ability of students to solve problems related to daily life such as problems of linear program is relatively low. This condition indicates that students have difficulty in using mathematical concepts to solve real problems. White state that one of the students' major difficulties is to express mathematically inequlity terms [3]. The result of Ogan's study also showed on Linear Program, student experience difficulties in constructing mathematical meanings of symbols [4].

Along with the development of the era, the ability to solve problems is needed in daily life. Problems in linear program are related to daily life, so it is important to implement a learning that can improve students' ability to solve problems and make students active in learning. According to Chamberlin and

Moon, Model Eliciting Activities (MEAs) are written specifically for instruction in mathematics. This model was developed after the Problem-Based Learning approach. MEAs is one of the recommended instructional strategis for teacher in mathematics [5]. Lesh states that MEAs would encourage students to creative mathematical models to solve complex problems [6]. MEAs are a model of mathematical learning to understand, explain, and communicate mathematical concepts contained in a problematic presentation through mathematical modeling. There are six principles of learning with MEAs : 1) *Model Construction Principle,* 2) *The Reality Principle,* 3) *The self-Assessment Principle*, 4) *The Construct Documentation Principle*, 5) *The Construct Shareability and Reusability Principle*, 6) *The Effective Prototype Principle* [7]. MEAs have five learning characteristics when viewed in relation to student learning: collaboration, multiple processes, self-directed learning and self-assessment, fostering of ownership, and model development. Student collaboration is fostered when doing MEAs because students work in groups of three or four [7]. A MEAs is implemented in several steps : 1) The teacher reads a simulated newspaper article that develops a context for students, 2) The students respond to readiness questions that are based on the article, 3) The teacher reads the problem statement with the students and makes sure each group understands what is being asked and students subsequently attempt to solve the problem, 4) After creating multiple iterations of the solution and revising when necessary, students present their models to the class [5]. The result of Altay research is that mathematics learning with MEAs is enjoyable and meaningful as these activities requaire students to make relation between mathematical concept and real life [8].

The Partnership for 21st Century Skills identifies 21st century skills as critical thinking and problem solving, communication, collaboration, and creativity and innovation- more commonly known as the 4C skills [9].

1.1. Critical Thinking and Problem Solving Skills
Critical thinking Skill is the ability to understand a complex problem, connecting information one with other information, so that eventually emerged various perspectives in solving the problem. Critical thinking skills required students in daily life. Students who are able to think critically can solve problems effectively in social, scientific, and practical problems. Merely having knowledge and information is not enough to be able to solve problems and make decisions. To be effective in the workplace (and in their personal lives), students must be able to solve problems and make effective decisions; they must be able to think critically [10]. Characteristics of Critical Thinking Skills: a) reason effectively, b) ask pointed questions and solve problems, c) analyze and evaluate alternative points of view, and d) reflect critically on decisions and processes [1] .

1.2 Communication Skills
Learning is a basic social activity - whether in school, workplace, or other environment. Communication refers to an individual's ability to communicate clearly, using spoken, written, and non-verbal language. Characteristics Communication skills : a) Articulate thoughts and ideas effctively using oral, written and nonverbal communication skills in a variety of forms and contexts, b) Listening effctively to decipher meaning, including knowledge, values, attitudes and intentions, c) Using communication for a variety of purposes, d) Utilize multiple media and technologies, and know how to judge their effctiveness apriori as well as assess their impact, and e) Communicate effctively in diverse environments (including multi-lingual)[1].

1.3 Collaboration Skills
Collaboration has become a twenty-first-century trend. The need in society to think and work together on issues of critical concern has increased shifting the emphasis from individual efforts to group work, from independence to community [11]. Collaboration is the act of working together to achieve common goals.The Partnership for 21st Century Expertise says that mastering collaborative skills requires the ability to work effectively with diverse teams. Collaboration is not simply a synonym for students

IOP Conf. Series: Journal of Physics: Conf. Series **1008** (2018) 012059 doi:10.1088/1742-6596/1008/1/012059

working in groups but also aims to instill the ability to socialize and control the ego and emotions. Through collaboration will create togetherness, ownership, responsibility, and caring among members.

Characteristics Collaboration Skills : a) Demonstrate the ability to work effctively and respectfully with diverse teams, b) Exercise flxibility and willingness to be helpful in making necessary compromises to accomplish a common goal, c) Assume shared responsibility for collaborative work, and value individual contributions made by each team member [1].

1.4 Creativity and Innovation Skills

Creativity is one's ability to generate new ideas from existing ideas. Creativity depends on one's creative thinking, the process of thinking to create new ideas that lead to new discoveries and often called innovation. Characteristics Creativity and Innovation Skills : a) Develop, apply and convey new ideas to others new invention and often called innovation, b) Open and responsive in viewing new and different views, c) Using technology to make decisions, solve problems, and create new ideas [2].

Based on the characteristics of MEAs and the characteristics of each 4C skills, MEAs that integrates 4C skills in classroom learning is very relevant because it has similarities that are collaboration, communication, critical thinking and Problem Solving, and creativity and innovation skills. The results of Wahyu's research is the mathematical problem solving skills possesed by the students who use MEAs learning are better than those who taught in conventional learning [12]. Smith L S in his research stated that there is a potential for developing 4C skills with problem-based mat learning. The results show that 4C skills are required in problem based learning [13].

2. Methods

Type of research used in this study is the experimental method. This research was conducted at Vocational High School 2 Surakarta and Vocational High School 4 Surakarta. In this study, four classes were selected as experimental classes and control classes. Sample of this research is 60 students of Vocational High School 2 in Surakarta and Vocational High School 4 Surakarta chosen by random sampling as experimental classes and 60 students of Vocational High School 2 Surakarta and Vocational High School 4 Surakarta chosen by random sampling as control classes . Experimental classes were treated with MEAs learning integrated 4C skills and control classes were given conventional learning treatment. Methods of data collection in this study using documentation method and test method. Documentation method used as initial data. The test method is used as the final test. The test method with linear program material is used as the final test. The instrument consisting 20 items is validated by the head of Mathematics Teachers Association Vocational High School Surakarta and the head of Mathematics Teachers Association Senior High School Central Java. This is then tested in Vocational High School 6 in Surakarta and analyzed. Instrument analysis consist of discrimination power (DP), difficulty level (D), and reliability test with Alpha Cronbach formula. An item is said to be good if it has a discrimination power of more than or equal to 0.30 ($DP \geq 0.30$), difficulty level $0.30 \leq D \leq 0.70$ and reliability index greater than or equal to 0.70 ($r_{11} \geq 0.70$). The results of the instrument analysis are 15 items were obtained to qualify for discrimination power and difficulty. The reliability value of the 15 items, $r_{11} = 0.76$. Reliability value used describe the usefulness of an item in measurement. It means that the result of measurement having a reliability index 0.70 or more then instrument can be used to measurement. The data analysis used in this research is Z-test by taking significance level $\alpha = 5\%$ which is to compare two models of learning. Alpha (α) level is the probability of rejecting the null hypothesis given that it is true. As a matter of good scientific practice, a significance level is chosen before data collection and is usually set to 0.05 (5%). Other significance levels (e.g., 0.01) may be used, depending on the field of study [14].

3. Results and Discussion

3.1. Preliminary data analysis

This section, we present the initial data analysis. Preliminary data is taken from final exam score of the previous semester. Table 1. shows the result of normality test with Liliefors method on experimental and control classes. Test were tested with Excell.

Table 1. Normality test in the experimental and control class preliminary data

Class	L-value	N	$L_{0,05;60}$
Experimental	0.1049	60	0.1143
Control	0.0997	60	0.1143

Based on table 1, in the experimental class obtained L-value = 0.1049 and in the control class L-value = 0.0997. L-value is obtained from final exam score of the previous semester each class is sorted from the smallest value to the largest value, then tested with Liliefors method. In experimental classes obtained the maximum value of L-value = 0.1049 and control classes obtained the maximum value of L-value = 0.0997. $L_{0,05;\,60}$ = 0.1143; C = {L | L > 0.1143}, because the L-value in each class is not included in C then data of two classes are said to be normally distributed.

The Z test can be used to test the data whose sample is large. The number of samples of 30 or more is considered a large sample. In addition, the Z test is used to analyze data whose population variance is known. However, if the population variance is unknown, then the variance of the sample can be used instead. Criteria use of Z test 1) Normally distributed data, 2) Variance ($\sigma 2$) known, 3) Large sample size (n) $\geq$ 30, and 4) Used only to compare 2 pieces of observation. Tabel 2. Describe the Z test result with Excell to find uot whether there are differences in the experimental class and control class.

Table 2. Z test on initial ability of experimental and control class

Class	N	Mean	SD	Z-value	$Z_{0,05}$
Experimental	60	62.58	11.06	1.515	1.65
Control	60	59.58	10.65		

The result of Z test, obtained Z-value = 1,515, it is smaller than $Z_{0,05}$ = 1,65. It can be said there is no differences in both classes. Both classes have the same initial capability. Furthermore, for experimental classes using MEAs learning integrated with 4C skills and for control classes with conventional learning.

3.2 Analysis of final data

Table 3 present normality test of experimental and control classes final test. Data obtained from the final test scores of linear program material in the experimental classes and control classes.

Table 3. Normality test in the experimental and control classes final test

Class	L-value	N	$L_{0,05;60}$
Experimental	0.1076	60	0.1143
Control	0.0948	60	0.1143

Result in table 3 shows L-value in experimental classes and control classes less than $L_{0.05;60}$, then it is said the final test data is normally distributed.

Table 4 shows the result of Z test on the final test of experimental and control classes with Excell. Data obtained from the final test scores of linear program material in the experimental classes and control classes.

Table 4. Z test after experimental

Class	N	Mean	SD	z-value	$z_{0,05}$
Experimental	60	78,37	13,17	2,93	1,65
Control	60	71,07	14,12		

It was obtained from the data analysis that Z-value = 2.93 and $Z_{0.05}$ = 1.65. Z-value is higher than $Z_{0,05}$ so it is said there was a difference in the final test of experimental and control classes. On initial ability of experimental classes, mean = 62.58 and modus = 70. After experimental classes were treated with MEAs learning integrated 4C skills, mean = 78.37 and modus = 80. It means a significant change in learning outcomes. Based on the result of the analysis, it can be concluded that there are differences in the effect of applying Model Elliciting Activities (MEAs) that integrates 21st century skills, namely 4C and conventional learning to learning outcomes. Learning using MEAs by integrating 4C skills in learning activities provides better learning outcomes than conventional learning. This condition can occur because there are differences in characteristics in conventional learning and MEAs learning that integrated 4C skills.

Table 5 shows the differences characteristic of the two lessons.

Table 5. Table of conventional learning characteristics and MEAs learning that integrate 4C skills.

Conventional learning	MEAs by integrated 4C skills
Learning Center	
teacher centered and teachers is a source of learning.	Student centered, students can find learning resources not only from the teacher. Teacher as a facilitator.
Classroom environment	
Students work individually	Students work in groups
Learning Activities	
Teachers perform the learning steps of explaining material, examples, and exercises.	Teachers begin learning by presenting an article or video about daily life related to problems in linear programming. Students are also active to look for problems with the linear program and solve the problem in the group and then present the group results.

The character difference in both learning influences students' reponse during classroom learning. In conventional learning, students wait for teacher and passive explanation, students do not want to ask when they do not understand, that it can cause boredom. Students do not motivate each other, students easily give up when experiencing difficulties because there is no discussion friend. Students' creativity is not built and students do not built communication skills. On the contrary, in learning with MEAs integrated with 4c skills students can develop their creativity and ideas so they are not passive. Students are accustomed to collaborate, respect each other, accept the advantages and disadvantages of others and students can motivate each other. Students' creativity is awakened and students can develop communication skills.

Table 6 describes activities with MEAs that are integrated with 4C skills.

Table 6. Activities MEAs learning that are integrated with 4C skills

MEAs	4C skills	Activities
The teacher reads a simulated newspaper article that develops a context for students The students respond to readiness questions that are based on the article	*Critical thinking and Problem solving*	Observe an article given by a teacher or observe a video containing daily activities related to linear programming issues. This activity trains students to find out and apply mathematical concepts in daily life.
	Creativity and Innovation skills	Students search for articles that contain daily activities related to linear program then create problems or problems related to the article and students complete the group. This activity can build students' creativity when student search and solving problems
The teacher reads the problem statement with the students and makes sure each group understands what is being asked and students subsequently attempt to solve the problem	*Collaboration skills*	Work together in groups, help each other and motivate when completing tasks. This activity can develop the attitude of cooperation and mutual respect
After creating multiple iterations of the solution and revising when necessary, students present their models to the class.	*Communication skills*	Discussion to process the given problem and solve the problem, write down the results of the discussion and then present the results. This activity encourages students to dare to communicate opinions

Students' difficulties to solving problems in linear program can be helped by students being invited to see daily events related to the problems in the linear program that determines maximum results or minimum expenditure. Students are invited to critical thinking how to get maximum profit and how to spend the minimum cost. Students learn creatively to construct a mathematical model in the form of inequalities that can be used to solve problems. Students work together in groups so that communication occurs. Students who have difficulty can ask in groups. This condition causes students not bored so that it can successfully solve the problem. When students in groups are looking for problems related to linear programming then completing the group and presenting the results, making students more confident. Vocational high schools are expected to produce a person who is able to think critically in the marketplace, creative in work and able to collaborate and communicate in the markerplace.

4. Conclusion

The 21st century, 4C skills integrated in MEAs learning can provide better learning outcomes than conventional learning. MEAs and 4C skills have the same characteristics, that is *Critical thinking and Problem solving, Creativity and Innovation skills, Collaboration skills, and Communication skills.* This learning causes students feel not bored because can be actively involved and happen mutual cooperation. Students also become more courageous to express opinions because there is mutual respect. The 21st century, 4C skills are needed in today's development. The ability of problem solve is also very necessary in daily life. Based on the research findings, the author suggested that learning using 4C skills are integrated in the model of learning MEAs can applied in the learning of linear program. Suggestion for further research is the researcher using technology to integrate 21st century skills in MEAs

Acknowledgments

The authors thank Universitas Sebelas Maret and Vocational High School 2 Surakarta and Vocational High School 4 Surakarta due to give permission for this study.

References

[1] Pasific Policy Research Center 2010 *21st Century Skills for Students and Teachers* (Honolulu: Honolulu Kamehameha Schools Research & Evaluation Division)

[2] Bernie T and Charles F 2009 *21st Century Skills* Printed in the United States of America

[3] White K M, Kilda CBC St 1996 Secondary School Students' Understanding Of Inequalities In Linear Programming

[4] Ogan G C and George N R 2015 Investigating Difficult Concepts In Senior Secondary School Mathematics Curriculum As Perceived By Students *International Journal of Academic Research and Reflection* **3** 67 – 74

[5] Chamberlin S A and Moon S M 2008 How Does the Problem Based Learning Approach Compare to The Model Eliciting Activity Approach in Mathematics? *International Journal for Matthematics Teaching and Learning*

[6] Lesh R and Doerr H M 2003 Foundation of model and modeling perspective on mathematics teaching and learning In R. Lesh & H.M Doerr (Eds) *Beyond Constructivism: Model and modeling perspectives on mathematics problem solving, learning, and teaching* 3-34 (Hillsdale, NJ: Lawrence Erlbaum and Associates)

[7] Chamberlin S A and Moon S M 2005 Model Eliciting Activities as a Tool to Develop and Identify Creatively Gifted Mathematicians *The Journal of Secondary Gifted Education* **17** 37 – 47

[8] Altay M K, Özdemir E Y and Akar S S 2014 Pre-service Elemetary Teachers' Views on Model Eliciting Activities *Procedia – Social and Behavioral Science* **116** 345-349

[9] Levin J 2012 Teaching Generation TechX with the 4Cs: Using Technology to Integrate 21st Century Skills *Journal of Instructional Research* **1** 59 – 64

[10] Snyder L G and Snyder M J 2008 Teaching Critical Thinking and Problem Solving Skills *The Delta Pi Epsilon Journal* **1**(2) 90 – 99

[11] Laal M and Laal M 2012 Collaborative Learning: what is it? *Procedia – Social and Behavioral Science* **31** 491 – 495

[12] Wahyu H 2014 The Implementation of MEAs Instruction to Students' Mathematics Problem Solving and Connecting Ability *Proc. Int. Conf. On Research, Implementation and Education of Mathematics and Science* 607 – 614

[13] Smit L S 2016 A Better Understanding of 21st Century Skills in Mathematics Education and a View on these Skills in Current Practice (Netherlands:Utrecht University Netherlands)

[14] Craparo, Robert M 2007 "Significance level". In Salkind, Neil J. *Encyclopedia of Measurement and Statistics* **3** (Thousand Oaks, CA: SAGE Publications) 889–891

Global conjecturing process in pattern generalization problem

Sutarto[1], Toto Nusantara[2], Subanji[2], Intan Dwi Hastuti[3], Dafik[4]

[1]Mathematics Education Departement, IKIP Mataram, Mataram, Indonesia
[2]Department of Mathematics, UM Malang, Malang, Indonesia
[3]Mathematics Education Departement, Universitas Muhammadiyah Mataram, Mataram, Indonesia
[4]Mathematics Education Departement, University of Jember, Jember, Indonesia

Email: sutarto@ikipmataram.ac.id

Abstract. The aim of this global conjecturing process based on the theory of APOS. The subjects used in study were 15 of 8th grade students of Junior High School. The data were collected using Pattern Generalization Problem (PGP) and interviews. After students had already completed PGP; moreover, they were interviewed using students work-based to understand the conjecturing process. These interviews were video taped. The result of study reveals that the global conjecturing process occurs at the phase of action in which subjects build a conjecture by observing and counting the number of squares completely without distinguishing between black or white squares, finaly at the phase of process, the object and scheme were perfectly performed.

1. Introduction

One of the standard assessments used in this study begins from preschool until secondary school level [19] is making and examining the mathematic conjecture. Furthermore, it is explained that making conjecture is important work because it functions as the basic to develop and increase new perception for further study. Making and examining conjecture is a step in mathematical study [12], reasoning [21], and mathematical thinking [16].

Conjecture is the logical statement where its truth is not yet certain [3, 8, 16, 21]. Along with this idea, conjecture is a statement concerning all the possible cases, based on the empirical facts, but with the doubtful elements [4]. Based on these arguments, it can be argued that conjecture is the statement based on the reasoning process where its truth is not yet certain.

Conjecture and problem solving are linked activity. Conjecture and problem solving are the important parts and interrelated in the mathematic activity [4, 2] Moreover, it is said that the problem solving involves the finding whiler conjecture is the main road for the finding [19]. In problem solving, conjecture helps the problem solver to find the solution for the problems faced. As a result, conjecture does not just show up, but there is a process and the process is the conjecturing process.

The conjecturing process is the process of constructing the conjecture [4,16]. Assumes that the conjecturing process is the mental activity expressed in problem solving based on the knowledge which has been owned and the trust is necessary to be proven [8]. Based on the argument, it can be concluded that the conjecturing process is the mental activity in constructing conjecture based on the possessed knowledge. The mental activity is the process in the mind which can be seen ins the students' behavior in problem solving [11, 30, 31].

There will be the conjecturing process if the students face any problems. In the conjecturing process, the students construct the conjecture based on their knowledge. There is no validation in the conjecture built by the students. It depends on the students' involment. If there is a validation for the conjecture they built, the conjecture is considered as the correct one. Only after the conjecture is validated then one can consider it as correct or incorrect. If the conjecture does not have incorrect value the process of conjecturing will be processed again until the it has the correct value. The conjecture with the correct value is the solution for the problems faced by the students.

In relation to conjecturing process, the familiar one used is mathematic problem solving as it is the conjecturing type of empirical induction from a finite number of discrete cases [4]. This type of conjecturing process consists of seven steps namely observing the case, organizing the case, finding and predicting the pattern, formulating the conjecture, validating the conjecture, generalizing the conjecture, and validating the generalization. The conjecturing process of this type—by inducting from a finite number of discrete cases—is mostly found in the problem related to the numbers, where the pattern under observation is consistent. In the problem solving involving the numbers with consistent pattern, the seven conjecturing processes do not always take place; there are many factors affecting process such as the type of task or the students' characteristics involved [3].

The pattern described as the regularity which can be predicted above commonly involves the numerical, spatial or logical relations [17]. Many mathematicians state that the mathematics is a 'science about pattern' [22, 25]. They highlight the pattern existence in all mathematic fields [25]. In particular the pattern is considered by some researchers as the strategy applied in algebraic field because the pattern is the basic measure to construct the generalization which is the mathematic essence [29].

Generalization of pattern is an important aspect in school mathematical activities [5, 18, 26, 29]. Along with this idea, the generalization must be the core of the school mathematical activities [10]. The generalization of pattern itself is the activity making the pattern common rule based on some special examples. The common rule obtained is the conjecture, and the generalization is the specific type of conjecture obtained from common reasoning [28].

The significant contribution to the conjecture or conjecturing has been studied by researchers who consider the conjecture as the intuition of expression [8]. It shows the important of conjecturing atmosphere [15]. Research has been carried out on the production of conjectures within a dynamic geometry environment [9] by analyzing how the students verify the conjecture and how the teachers' trust related to this process [1]. Researchers developed an open classical analogy in geometry construction [13] by designing mathematic conjecturing activities to foster thinking and constructing actively [14]. Then, they analyze various familiar types and steps of conjecturing process in problem solving [4]. Among the studies, it is not revealed yet how the students—through conjecturing process—generalizing the pattern of problem solving.

The conjecturing process described above corelates to some theories [3, 4, 20, 21, 23]. These theories are the basic of conjecturing process of empirical induction from a finite number of discrete cases. They consist of four inductive reasoning processes in problem solving, namely (1) observing specific cases, (2) formulating the conjecture based on previous case, (3) generalization, and (4) conjecture verification with specific new cases [20]. Reid uses the inductive reasoning process in the context of empirical induction from a finite number of discrete cases by the steps: (1) observing specific cases, (2) observing the pattern, (3) formulating the conjecture for common cases (with doubtfulness), (4) generalization, and (5) using generalization to prove [21]. there are seven steps in describing the inductive reasoning process, namely (1) Observing cases, (2) Organizing cases, (3) Searching for and predicting patterns, (4) Formulating a conjecture, (5) Validating the conjecture, (6) Generalizing the conjecture, (7) Justifying the generalization [3].

There are seven steps in describing the inductive reasoning process drive from Canada's, as such a types of conjecturing process namely empirical induction from finite number of discrete cases. The conjecturing process in this study is empirical induction from a finite number of discrete cases. The explanation of those seven steps take place in conjecturing process [24]. Based on its explanation and indicator adapted from the students. the conjecturing process in generalization of pattern problem solving is grouped into two types, i.e. global conjecturing, and local conjecturing. The global conjecturing is the mental activity in constructing the conjecture by observing the problems intact, and The local conjecturing is the mental activity of constructing the conjecture by observing the problems separately. While, The global conjecturing process is often done by the students in generalization of pattern problem solving. Thus, this study will describe the global conjecturing process based on APOS theory.

The mental activity in conjecture analyzes using APOS theory, because APOS theory is a theory that can be used as the analytical tool to describe one's developmental scheme in a mathematic topic--as it is the totality of the related knowledge (aware or unaware) to the topic [6]. This theory is based on the hypothesis of mathematic knowledge that may use to solve the situation as the mathematical problem by constructing the action, process and object as well as regulating the scheme to comprehend the situation and solve the problem [7]. This theory is called APOS and use to describe an action at the interiorization as the Process. The Process is encapsulated in an object. Then, it is related to other knowledge in a schema. A schema can also be encapsulated as an object. Problem of Research: How is the global conjecturing process in the solving of pattern generalization problem based on APOS theory?

2. Methodology of Research

2.1. Subjects

The subjects in this study are 15 students of class VIII derived from 9 students of VIII State Junior High School 1 Malang, and 6 students of State Junior High School 3 Malang.

2.2. Instrument

There are two types of instruments use. The main instrument is the researchers themselves who act as planners, data collectors, data analysts, interpreters, and reporters of research results. The auxiliary instrument used in this study is a Pattern Generalization Problem (PGP) and interviews. The problem given aims to obtain a description of the process of conjecturing of the students, while the interview used was unstructured interview. The PGP is presented in Figure 1.

Pay attention to the square pattern forms by the 1st figure, 2nd figure and 3rd figure below!

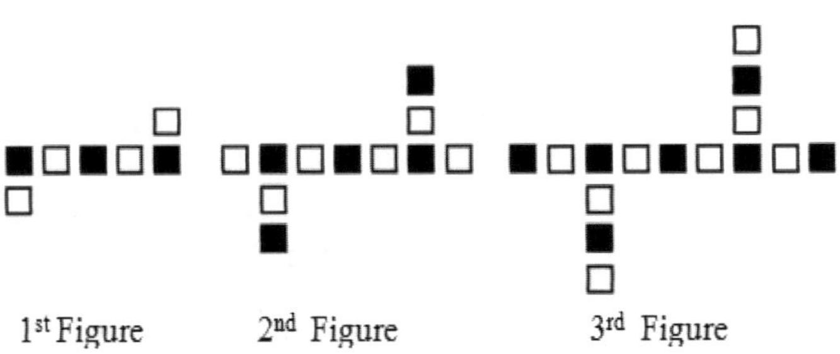

Find out the common formula to set the number of square at the nth figure from the formed pattern! Explain how do you find the common formula!

Figure 1. The Pattern Generalization Problem (PGP)

2.3. Data Analysis

This study is a qualitative research with descriptive exploratory approach. At the data analysis step, the activities conducted by researchers were (1) transcribing the data obtained from interviews, (2) data condensation, including explaining, choosing principal matters, focusing on important things, removing the unnecessary ones, and organizing raw data obtained from the field (3) encoding the data from PGP answer sheet and interviews refer based on indicators of local conjecturing process are presented in Table 1, (4) describing the global conjecturing process in the solving of pattern generalization problem based on APOS theory, and (5) conclusion.

3. Results of Research

Based on the analysis results of the answer sheets and the interview results, it is obtained the data on the global conjecturing process conducted by the students in the generalization of pattern problem solving based on the APOS theory. After getting bored for the subject taking process, it is obtained 6 subjects who conduct the global conjecturing process, 5 subjects who conduct the contrast conjecturing and 3 subjects who conduct the local conjecturing and generalizing symbolic. Out of 6 subjects, it will be described two subjects making the global conjecturing process of generalization of pattern problem-solving which is the S1 subject and S2 subject. The data presented is obtained by the procedures (1) the subjects complete the PGP, and (2) after the subjects complete the PGP, then they are interviewed to explore about the global conjecturing process which has been conducted. The data presentation and analysis of the global conjecturing processes in the generalization of pattern problem solving is as the following.

3.1. S1 subject data presentation

In generalizing the S_1 patterns, it has realized that 1st figure, 2nd figure, and 3rd figure form a pattern. To find a common formula on the number of square at the nth figure, S_1 observes and counts the number of square regardless the black square and white ones at the 1st figure, 2nd figure, and 3rdFigure Here are the interview quotation and the S_1 work results in completing the following PGP.

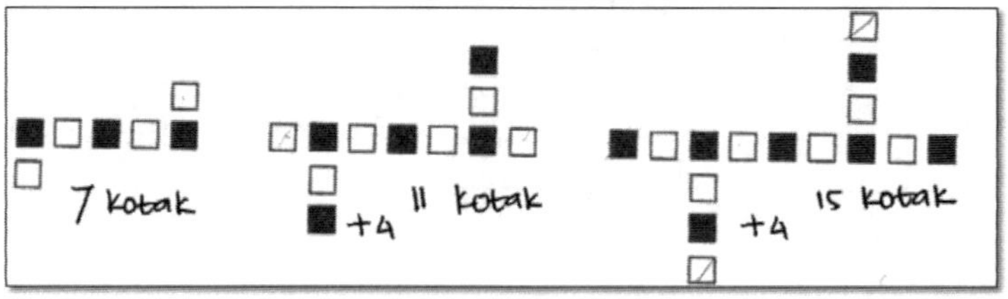

Figure 2.S_1 subject work result

$S_1$04 : this is the different of the figure, Sir. This is the first, 7 and the second one is 11, and the third one, the number is 15 (while pointing to the square figure). The different is 4, so the following figure is plus 4, plus 4, plus 4.

Based on the number of square in the 1st figure, 2nd figure, and 3rd figure, S_1 organizes the cases by ordering the number row pattern. Then, S_1 finds and predicts the pattern by seeing the different between the 2nd figure and the 2nd figure, the 3rd figure and the 2nd figure and thinking how the following figure is plus 4, plus 4, plus 4. This is confirmed by the $S_1$04 interview quotation and the students' work results in completing the following PGP.

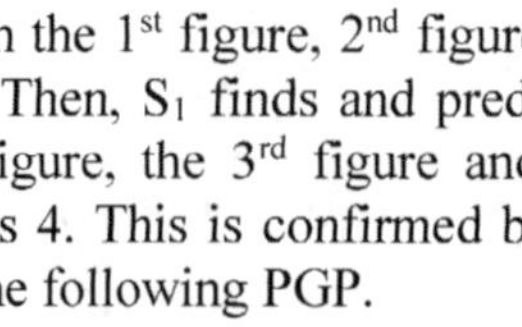

Figure 3. S_1 subject work result

To formulate the conjecture, S_1 subject sees the addition of 1st figure into the 2nd figure is 2, and the 2nd figure into the 3rd figure is also 4, by seeing this addition, S_1 formulates the nth formula conjecture in setting the number of square at the figure is $n = n + 4$. After that, S_1 validates the conjecture by seeing the appropriateness with the number of square at the 4th figure and 3rd figure, then saying that the nth formula is incorrect. The following is the S_1 interview quotation.

IOP Conf. Series: Journal of Physics: Conf. Series **1008** (2018) 012060 doi:10.1088/1742-6596/1008/1/012060

$S_1$08 : this adds by 4, then adds by 4, then adds by 4. But, thinking it continuously, it may be incorrect. It is why, when 4 adds by 4, it is 8, then when n is 3, it adds by 4 the result is only 7. So it is incorrect.

After realizing that the conjecture formulated is incorrect, S_1 tries new strategy to formulate the conjecture namely by finding the initial number before it is plus 4 because the pattern always adds by 4. S1 finds the initial number by looking for the different of the number of the square at the 1st figure, 2nd figure and 3rd figure by 4. The subtraction results consecutively are 3, 7, 11. S1 realizes that the initial number searched is not yet correct because its initial number is still different. This is shown by the interview quotation and the S_1 work results as the following.

$S_1$11 : this is initially still different (3, 7, 11) meaning that it is incorrect (while pointing out the work result)

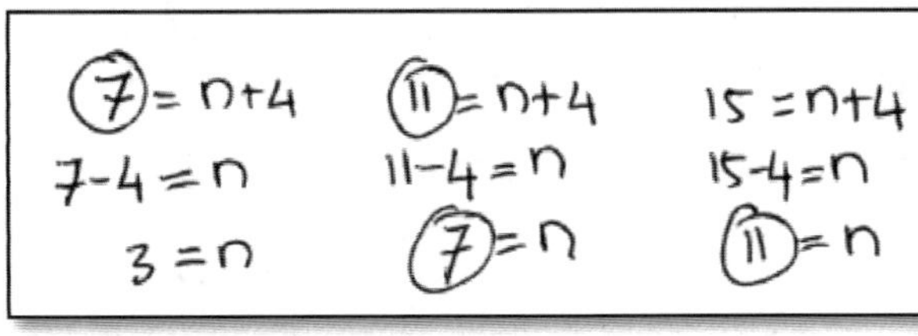

Figure 4. S_1 subject work result

S_1 subject then uses the new strategy which is to look for the initial numbers before adding by 4 times n. S1 writes down the initial number symbols with x, for the 2nd figure $x + (4 \times 2) = 11$, then $x = 3$, for the 3rd figure$x + (4 \times 3) = 15$ then $x = 3$ so the initial number before plus 4 times n is 3. After finding the initial figure, S_1 formulates the conjecture namely the common formula which is $3 + (4 \times n)$and validates the conjecture based on the number of squares which are already known. After validating S_1 generalizes the conjecture as to believe that the common formula is $3 + (4 \times n)$It is also shown from the quotation interview and S_1 work result as the following.

$S_1$11 : I look for the initial number before it is plus 4 times n. the second figure is the same to $x + (4 \times 2) = 11$ so $x = 3$. Then the 3rd figure is similar to $x + (4 \times 3) = 15$ so $x = 3$. So, the intial number before it is plus 4 times n is 3.
P 13 : Okay, then are you sure by the common formula you obtain?
$S_1$13 : yes, Sir I am...

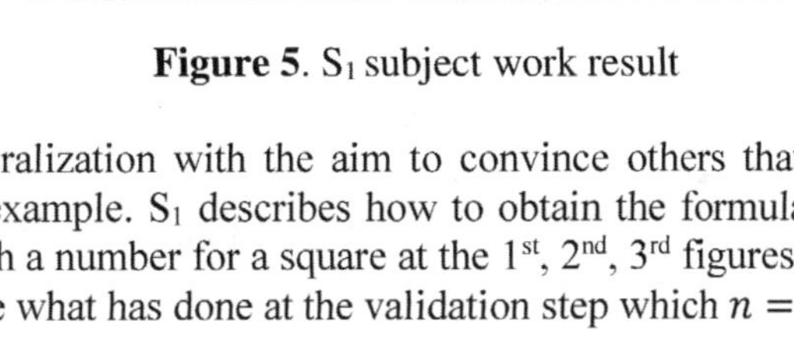

Figure 5. S_1 subject work result

S_1justifying the generalization with the aim to convince others that the conjecture obtained is correct with a particular example. S_1 describes how to obtain the formula, and shows an example of the formula suitability with a number for a square at the 1st, 2nd, 3rd figures and calculate the number of square at the 4th figure like what has done at the validation step which $n = 3 + (4 \times 4) = 19$ and 19 is

IOP Conf. Series: Journal of Physics: Conf. Series **1008** (2018) 012060 doi:10.1088/1742-6596/1008/1/012060

also obtained from the 3rd figure plus 4 is 19, from this example S_1 justifying the resulting generalizations. This is shown by the interview quotation as the following.

P 19 : Okay, then how do you explain that the resulting *formula is correct?*
$S_1$19: I will explain how I get the formula and show the example for the 1st, 2nd, 3rd and 4th figures. For example, for the 4th figure, $n = 3 + (4 \times 4) = 19$. *19 is also obtained from the 3rd fig.,* $15 + 4 = 19$. *(while pointing put the work). s*

From the data described based on the conjecturing process steps, it can be described the S_1 subject thinking structure analyzed based on the APOS step. The S1 conjecturing process in the generalization of pattern problem solving begins with the action steps namely observing case, and organize the case, then S_1 internalizes the action into prose by finding and predicting the pattern. Once internalized the action into the process, S_1 encapsulates the process into the object by formulating the conjecture and validating the conjecture. At the following scheme step, S_1 generalizes the conjecture and justifying the resulting conjecture. S_1 thinking structure is presented in Figure 6.

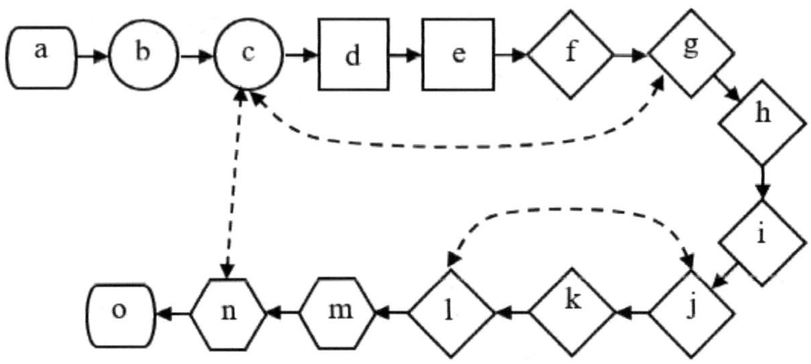

Figure 6. S_1 subject thinking structure

Notes:

a :	The problem proposed is to find the common formula to set the number of square at nth figure	l :	The nth formula is $3 + (4 \times n)$
b :	Counting and observing the number of square at the 1st, 2nd, and 3rd figures.	m :	Believing in that The nth formula is $4n + 3$
c :	Counting the number of square at the 1st, 2nd, and 3rd figures	n :	Validating the nth formula by pointing out at the example at the 1st, 2nd, 3rd, and 4th figures, supposed the 4th figure $n = 3 + (4 \times 4) = 19$. 19 is also obtained from the 3rd figure, $15 + 4 = 19$.
d :	Writing down the row pattern of 7, 11, 15	o :	Done
e :	Counting the square different at the 1st, 2nd, and 3rd figures and thinking of the following object	→	Activity sequence
f :	Stating the row different is 4	◄- -►	Validation activity, for example from g to c, then go back to g; from l to i, then go back to l, etc
g :	The addition is 4 so $n = n + 4$	○	Action
h :	Finding the initial number before being	□	Process

	added by 4.		
i :	Counting the different of number of square at the 1st, 2nd, 3rd and 4th figures, the results are 3, 7, 9.	◇	Object
j :	$x + (4 \times 2) = 11, x = 3$ $x + (4 \times 3) = 15, x = 3$	⬡	Schema
k :	The initial number before 4 times n is 3	▢	Initial and final activities

3.2 S_2 subject data presentation

In generalizing the patterns, S_2 subject has been aware that the 1st figure, 2nd figure and 3rd figure form a pattern. To find a common formula of the number of square at the nth figure, S_2 observes and counts the number of square regardless the black square and white one at 1st figure, 2nd figure and 3^{rd}Figure Here is the interview quotation of S_2.

P 04 : what do you think first when reading this problem?
$S_2$04 : at first I look at the figure, then from this figure, I look at another figure continuously, then it compare both (while pointing out at the PGP)
P 05 : Then you compare, what does it mean?
$S_2$05 : ...When comparing both, I find if in each figure there is 4 addition, four square addition. I still can not see the white and black. I don't see it. Then at first, I think of this continuously, the pattern is always like this(while pointing out at the PGP)

Based on the number of square at the that the 1st figure, 2nd figure and 3rd figure, S_2 subject organizes cases by signing up to number one with the 1st figure, number two with the 2nd figure, number three with the 3rd figure, and so on. Furthermore, S_2 locates and predicts the patterns by comparing the number of squares at the 1st figure, 2nd figure and 3rd figure and finds that the number of additional figure is always 4 and thinks of that the pattern always continues. This is confirmed by $S_2$05 interview quotation and the student's work results in completing the following generalization of pattern problem solving.

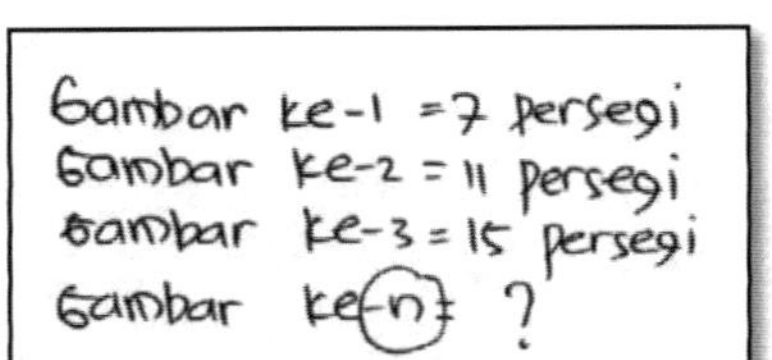

1st figure = 7 squares
2nd figure = 11 squares
3rd figure = 15 squares
N^{th} figure = ?

Figure 7. S_2 subject work result

To formulate a conjecture S_2 tries to determine suitable n based on the figure sequence. For example, 7 squares and 11 squares, this means that it has to plus 4, if n then n + 4 it can not be. So it must determine a suitable n based on the figure sequence. After S_2 tries to enter 1st figure (one) into the formula because one is also n, by trying one by one starting from $(1 \times 1) + 6$, $(1 \times 2) + 5$, and $(1 \times 4) + 3$. Then, S_2 formulates a common formula of conjecture to determine the number of square at the nth figure $= (n \times 4) + 3$ and validates the conjecture based on the number of squares at the 4th figure and 5^{th}Figure After validating, S1 generalizes the conjecture as to believe that the common formula is $= (n \times 4) + 3$. It is also shown from the interview quotation and the students' work in completing the following generalization of pattern problem solving.

$S_2$08 : so, I try the first one, supposed 1×1, 1×1 must be added with what number to be 7, eee.. then it is plus 6, but if it is supposed $2 \times 1 + 6$ then the results is not 11, so I keep trying the

closest ine which is the most appropriate answer for this square ((while pointing out at the square figure at PGP). I keep trying $(1 \times 2) + 5 = 7$ *is correct, then this one* $(2 \times 2) + 5$ *the result is 9 and not 11, so I keep trying four* $(1 \times 4) + 3 =$ *this 7, I try if it is* $(2 \times 4) + 3 = 11$*, I try again the* $(3 \times 4) + 3 = 15$*, because Iam still doubt I try this one* $(4 \times 4) + 3 = 19$*,* $15 + 4 = 19$*. So that is my thinking pattern.*

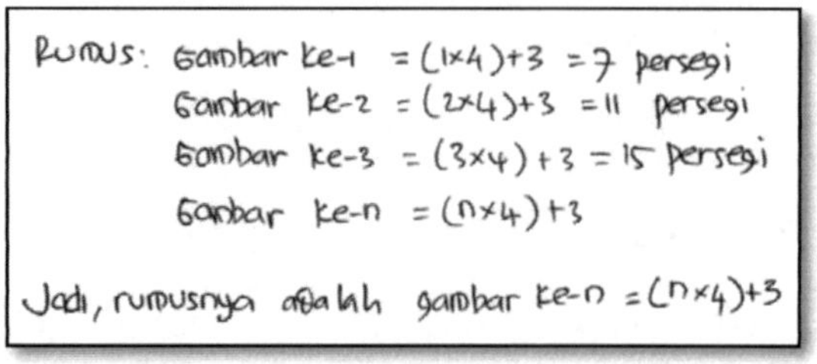

Formula : 1st figure = (1 x 4) + 3 = 7 squares
2nd figure = (2 x 4) + 3 = 11 squares
3rd figure = (3 x 4) + 3 = 15 squares
nth figure = (n x 4) + 3
So, the formula is the nth figure is (n x 4) + 3

Figure 8. S_2 subject work result

S_2 subject justifying the generalization with the aim of convincing others that the resulting conjecture is correct with a particular example. S_2 points out, from this example, S_2 justifying the generalization results in. This is shown by the following interview quotation.

P 14 : Okay, then how do you explain to others that the resulting formula is correct?
$S_2$14: I will show the results at the 1st, 2nd, 3rd figures and so on. This is the proof (while pointing out the work result). I have tried it continuously, then it is correct. So I believe in the formula.

From the data described based on the conjecturing process steps, it describes S_2 subject thinking structure which is analyzed based on the APOS step. S_2 Conjecturing process in the generalization of pattern problem solving begins by observing case the action step, and organizing the case, then S_2 internalizes the action into prose by finding and predicting the pattern. Once internalized into the action, S_2 encapsulates the action into the by formulating the conjecture and validating the conjecture. Then at the following schema step, S_2 Generalizing the conjecture and justifying the resulting conjecture. S_2 thinking structure is presented in Figure 9.

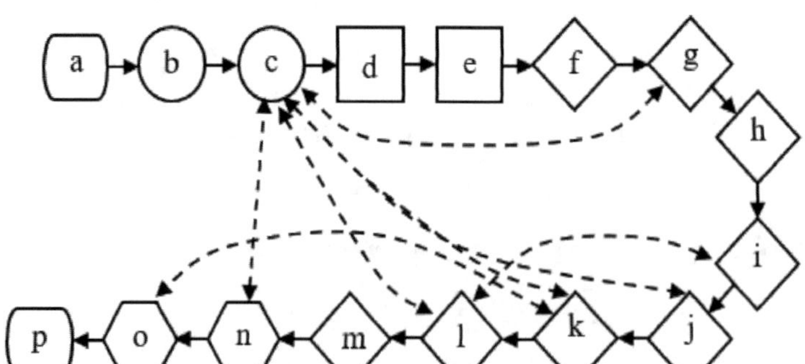

Figure 9. S_2 subject thinking structure

Notes:

a :	The problem proposed is to find the common formula to set the number of square at nth figure	m :	The nth formula is $(n \times 4) + 3$
b :	Counting and observing the number of square at the 1st, 2nd, and 3rd figures.	n :	Believing in that The nth formula is $(n \times 4) + 3$
c :	Counting the number of square at the 1st, 2nd,	o :	Validating the nth formula by specific case

	and 3rd figures		
d :	Making list of table to sort the pattern of 7, 11, 15	p :	Done
e :	Counting the square different at the 1st, 2nd, and 3rd figures and thinking of the following object	→	Activity sequence
f :	Stating the row different is 4	←-→	Validation activity, for example from g to c, then go back to g; from l to i, then go back to l, etc
g :	Looking for suitable n to sort the 1st, 2nd, and 3rd figures	○	Action
h :	Supposed, $7 + 4 = 11$, if n + 4 it can not be	□	Process
i :	Trying to input 1st (one) figure into the formula because one is n	◇	Object
j :	Trying (1×1) adds with what number to be 7. $(1 \times 1) + 6 = 7$ correct, $(2 \times 1) + 6 = 8$ incorrect	⬡	Schema
k :	Trying (1×2) adds with what number to be 7. $(1 \times 2) + 5 = 7$ correct, $(2 \times 2) + 5 = 9$ incorrect	▢	Initial and final activities
l :	Trying (1×4) adds with what number to be 7. $(1 \times 4) + 3 = 7$ correct, $(2 \times 4) + 3 = 11$ correct $(3 \times 4) + 3 = 15$ correct, $(4 \times 4) + 3 = 19$ correct		

3.3 Global Conjecturing Process Schema of S1 subject and S2 subject in Generalization of Pattern Problem Solving based on APOS

In generalizing the patterns, S_1 and S_2 have been aware that the 1st figure, 2nd figure and 3rd figure form a pattern. To find a common formula of the number of square at the nth Figure At this action step, S_1 and S_2 observe and count the number of square regardless the black square and white one, at the 1st figure, 2nd figure and 3rdFigure Based on the number of square at the 1st figure, 2nd figure and 3rd figure, S_1 organizes cases by sorting the number row patterns and S_2 registers to relate number one with the 1st figure, number 2 with 2nd figure, number 3 with the 3rd figure, and so on. Then, at the process step, S_1 and S_2 are searching for and predicting the pattern by looking at the difference between the 2nd figure and the 1st figure, 3rd figure and the 2nd figure and think that the following figure increases by 4.

The *object* step, to formulate conjecture, S_1 sees the 1st figure addition to the 2nd figure is 4, and the 2nd figure to the 3rd figure is also 4, by looking at the addition, S_1 looks for the initial number before adding by 4 times n. For the 2nd figure $x + (4 \times 2) = 11$, then $x = 3$, for the 3rd figure $x + (4 \times 3) = 15$ then $x = 3$ so the initial number before added 4 times n is 3. S_2 Subject tries to determine the suitable n based on the figure sequence, then S_2 tries to enter the 1st (one) figure into the formula because one is n, by trying one by one starting from $(1 \times 1) + 6$, $(1 \times 2) + 5$, and $(1 \times 4) + 3$. The conjecture produced by S_1 and S_2 is $3 + (4 \times n)$ and validate the conjecture based on specific examples obtained at the action or process step.

The scheme step, S_1 and S_2 subjects generalize the conjecture to believe that the conjecture resulted is correct after validating the conjecture in the previous steps. In justifying the generalization with the aim of convincing others that the resulting conjecture is correct, S_1 and S_2 use specific examples obtained at the action step or process step. Justifying the generalizations made by the subject

S_1 and S_2 is the same as what has done at the validation step namely using the specific examples. The Global conjecturing process scheme of S_1 subject and S_2 subject in the solving of pattern generalization problem is presented in Figure 10.

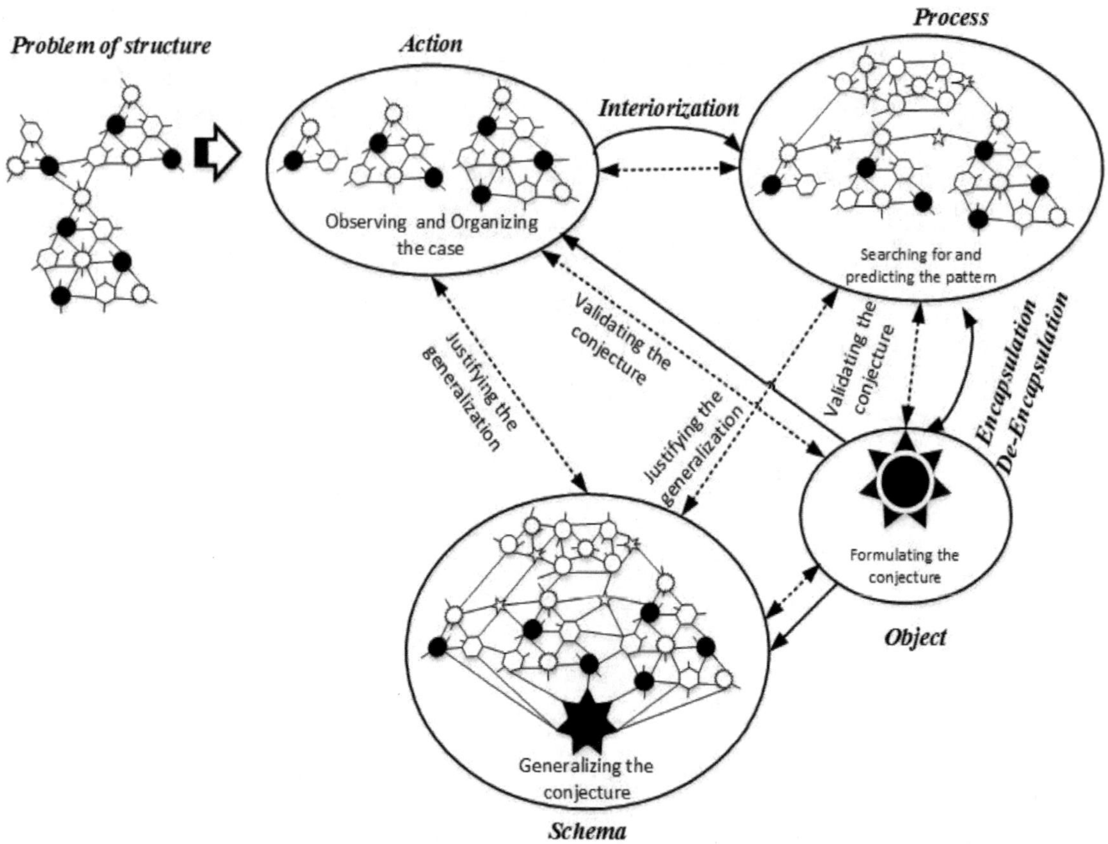

Figure 10. The schema of global conjecturing process

4 Discussion

This section will discuss the research findings related to global conjecturing process to solve problem of pattern generalization. In generalizing the pattern, S_1 and S_2 on the "action" step have realized that the 1st figure, 2nd figure, and 3rd figure form a pattern. To find a common formula number of square in the nth , S_1 and S_2, , it observed by counting number of square regardless the black and white squares on the 1st figure, 2nd figure, and 3rd figures. Based on the number of square at the 1st figure, 2nd figure, and 3rd figure, S_1 organizes the case by sorting the row pattern of 7, 11, 15 and so on while S_2 makes a list or a table to relate number one with the 1st figures, number 2 with the 2nd figure, number 3 with the 3rd figure. This shows that at the "action" step, S_1 subject and S_2 subject observe and organizes the cases regardless the black and white squares, therefore the conjecturing process conducted by the subjects is referred to as the global conjecturing process. Observing cases and organizing cases regardless the black and white squares are based on the Gestlat laws in observation, called similarity Law which is a person tends to perceive the same holistic stimulus [27].

This "process" step, the subjects internalize the action to find and predict the pattern by inverstigating the distinguish between number of square at the 2nd , 1st , 3rd , and 2nd figures. and think that the following figure has the same pattern, namely obtaining the increased 4. In formulating the conjecture step, S_1 conducts the encapsulation to generate the object which is to see the 1st figure addition to the 2nd figure is 4, and the 2nd figure to 3rd figure is also 4. S_1 seeks the initial number before adding 4 times n. for the 2nd figure, -2 $x + (4 \times 2) = 11$ so $x = 3$, for the 3rd figure -3 $x + (4 \times 3) = 15$ so $x = 3$ so the initial number before being added 4 times n is 3. After finding the initial number, S_1 formulate the common formula of the conjecture namely $3 + (4 \times n)$ and validating the conjecture based on the number of square known. S_2 tries to det the appropriate n based on the figure sequence, after that S_2 tries to enter the 1st figure into the formula because one is also n, by trying one by trying one by one starting from $(1 \times 1) + 6$, $(1 \times 2) + 5$, and $(1 \times 4) + 3$ then S_2formulates the common formula of the conjecture to set the number of square at the nth figure $n = (n \times 4) + 3$ and validate the conjecture based on the number of squares on the 4th figure and 5th figure. The way done by S_1 is looking for the initial number using the x symbol and S_2 seeks the appropriate based on the figure sequence. Both ways are different but meaningful for itself to find the common formula of the conjecture, it describes the knowledge possessed. This is consistent to what expressed by[28]hat the mathematical symbol is a tool for coding and describing the knowledge as well as communicating the mathematical knowledge. At this process step and object step, the subjects conduct it perfectly.

At this scheme step, S_1 subject and S_2 subject generalize the conjecture to believe that the resulting conjecture is correct. In justifying the generalization with the aim of convincing others that the resulting conjecture is correct, S_1 and S_2 use the specific examples obtained at the action step and object step. S_1 justifying the generalization with the aim of convincing others that the resulting conjecture is correct with the specific example. S_2 counts the number of square at the 4th figure namely $n = 3 + (4 \times 4) = 19$ and 19 is also obtained from 3rd figure plus 4 is 19, from this example, S_1 validates the resulting generalization. S_2 validates the generalization with aim of convincing others that the resulting conjecture is correct with specific example obtained at the object step by pointing out $(1 \times 4) + 3 = 7$, $(2 \times 4) + 3 = 11$, $(3 \times 4) + 3 = 15$, and $(4 \times 4) + 3 = 19$. In justifying the generalizations, S_1 subject and S_2 subject conduct their own way, this is based on the [3] that the students do not just simply use the notation or symbols but also their presentation and give a reason mathematically, make conclusions and generalizations in their way. At this scheme step, it is also conducted perfectly.

5 Conclusions

Based on the findings in the global conjecturing process conducted by the students in the generalization of pattern problem solving, it increases the conjecturing process theories [4] about the type of empirical induction from a finite number of discrete cases which has seven steps and not study the students' thinking process in constructing the conjecture generalization. The results show that the global conjecturing process occurs at the step of action in which subjects build a conjecture by observing and counting the number of squares complete, at the step of process, the object and scheme were perfectly performed.

Acknowledgements
This research is a grant from the Direktorat Riset dan Pengabdian Kepada Masyarakat (DRPM) Ristekdikti. The author would like to express sincere appreciation for all the support provided

References

[1] Bergqvist T 2005 How students verify conjectures: teachers' expectations. *Journal of Mathematics Teacher Education* **8** 171–191
[2] Sutarto and Hastuti I D 2015 Conjecturing Dalam Pemecahan Masalah Generalisasi Pola. *Jurnal Ilmiah Mandala Education* **1** 2 172-178
[3] Cañadas M C and Castro E 2005 A proposal of categorisation for analysing inductive reasoning *In M. Bosch (Ed.), Proceedings of the CERME 4 International Conference* 401-408
[4] Cañadas M C, Deulofeu J, Figueiras L, Reid D A and Yevdokimov O 2007 The conjecturing process: Perspectives in theory and implications in practice *Journal of Teaching and Learning*, 1 55–72

[5] Dindyal J 2007 High school students' use of patterns and generalisations *In J. Watson & K.Beswick (Eds), Proceedings of the 30th annual conference of the Mathematics Education Research Group of Australasia* 236-245

[6] Dubinsky E 2001 Using a theory of learning in college *TaLUM* **12** 10 – 15

[7] Dubinsky E and Mc Donald M 2011 APOS: A constructivist theory of learning in undergraduate mathematics education research *In D. Holton et al. (Eds), The teaching and learning of mathematics at university level: An ICMI Study* 273-280

[8] Fischbein E 2002 *Intuition in science and mathematics* An Educational Approach NewYork: Kluwer Academic Publisher.

[9] Furinghetti F and Paola D 2003 To produce conjectures and to prove them within a dynamic geometry environment: A case study *In N. A. Pateman, B. J. Doherty, & J. Zilliox (Eds.), Proceedings of the Twenty-Seventh Annual Conference of the International Group for the Psychology of Mathematics Education* 397-404

[10] Küchemann D 2010 Using patterns generically to see structure *Pedagogies: An International Journal* **5** 3 233–250

[11] Hastuti I D, Nusantara T and Susanto H 2016 Constructive metacognitive activity shift in mathematical problem solving *Educational Research and Reviews* **11** 8 656

[12] Lakatos I 2015 *Proofs and refutations: The logic of mathematical discovery Cambridge* university press

[13] Lee K H and Sriraman B 2010 Conjecturing via reconceived classical analogy *Educational Studies in Mathematics* **76** 2 123–140

[14] Lin F L 2006 Designing mathematics conjecturing activities to foster thinking and constructing actively. *APEC-TSUKUBA International Conference*, Tsukuba, Japan

[15] Mason J 2002 Generalization and algebra: Exploiting Children's Powers *In L. Haggerty (Ed.), Aspects of Teaching Secondary Mathematics: Perspectives on Practice* 105-120

[16] Mason J, Burton L and Stacey K 2010 *Thingking mathematically second edition*, England: Pearson Education Limited

[17] Mulligan J and Mitchelmore M 2009 Awareness of pattern and structure in early mathematical development *Mathematics Education Research Journal* **21** 2 33-49

[18] Mullingan J T, Mitchelmore MC, English LD and Robertson G 2011 Implementing a Pattern and Structure Mathematics Awareness Program (PASMAP) In Kindegarden *In L. Sparrow, B. Kissane, & C. Hurst (Eds.) Shaping the Future of Mathematics Education. Proceedings of the 33rd Annual Conference of the Mathematics Education Research Group of Australasia* 797–804

[19] Nasional Council of Teacher of Mathematics 2000 *Principles and standards for school mathematics* Reston, VA: NCTM

[20] Pólya G 1967 *Le découverte des mathématiques* París: DUNOD

[21] Reid D 2002 Conjectures and refutations in grade 5 matematics *Journal of Research Mathematics and Education* **33** 1 5-29

[22] Resnik M D 2005 *Mathematics as a science of mathematics* Oxford: University Press

[23] Sutarto Toto N and Subanji S 2015 Indicator of conjecturing process in a problem solving of the pattern generalization *In Proceding International Conference on Educational Research and Development (ICERD), UNESA Surabaya* 32-45

[24] Sutarto T N and Subanji S 2016 Local conjecturing process in the solving of pattern generalization problem *Educational Research and Reviews* **11** 8 732

[25] Tikekar V G 2009 Deceptive patterns in mathematics *International Journal of Mathematical Science Education* **2** 1 13-21

[26] Vogel R 2005 Patterns: A fundamental idea of mathematical thinking and learning *ZDM* **37** 5 445-449

[27] King D B and Wertheimer M 2009 *Max Wertheimer & Gestalt Theory* New Brunswik, New Jersey: Transaction Publisher

[28] Schwartz J L, Yerushalmy M and Wilson B *The geometric supposer: what is it a case of*? Hillsdale, NJ Lawrence Erlbaum Associates
[29] Zazkis R and Liljedahl P 2002 Generalization of patterns: the tension between algebraic thinking and algebraic notation *Educational Studies in Mathematics* **49** 379–402
[30] Hastuti I D 2016 Pergeseran aktivitas metakognitif siswa dalam pemecahan masalah matematika Disertasi tidak diterbitkan Malang: PPs UM
[31] Sutarto T N and Subanji S 2016 Local conjecturing process in the solving of pattern generalization problem *Educational Research and Reviews* **11** 8 732

IOP Conf. Series: Journal of Physics: Conf. Series **1008** (2018) 012061 doi:10.1088/1742-6596/1008/1/012061

The characteristics of failure among students who experienced pseudo thinking

D Anggraini[1], T A Kusmayadi[1] and I Pramudya[1]

[1]Mathematics Education, Sebelas Maret University
Jl. Ir. Sutami 36A Kentingan Jebres Surakarta 57126, Indonesia

Email: dellaramandha@gmail.com

Abstract. The purpose of this research is to describe the thinking process of students who experienced pseudo thinking when solving the straight line equation. The result of this study shows the characteristics of error that caused students to experience pseudo thinking when solving the problem and their relation with students' metacognition skill. This qualitative research was conducted in State 16 Junior High School in Surakarta, Indonesia during the odd semester of 2017/2018 academic year. The subjects of the study were students Junior High School students of 8th grade chosen using purposive sampling technique. Data were collected through the administration of think aloud method. The result showed that the characteristics of errors among the subjects are: 1) the answers resulted from pseudo thinking when solving the problem were obtained from the spontaneous, fast, unconscious and uncontrolled thinking process; 2) students had misconception; 3) students had tendency to memorize the formula and imitate the completion procedure; 4) students experienced fuzzy memory when solving the problem. From the mistakes among students who experienced pseudo thinking, their metacognition ability could be inferred.

1. Introduction

When solving a math problems, there are two possible results students that can be given by the students, correct and incorrect answers. However, the correct answers may not be merely resulted from the right thinking process as the incorrect ones are not merely resulted from the wrong thinking process. In other words, when solving math problems, students can experience pseudo thinking. The pseudo thinking term in Indonesian educational field was firstly introduced by Subanji in 2007 on his dissertation entitled "The Thinking Process of Pseudo's Reasonable Reasoning in Constructing the Dynamic Genesis Functions Chart".

The pseudo thinking has been studied by many experts using different terms and contexts: Pseudo analytic versus Pseudo conceptual in the context of routine mathematical problem solving [1]; Established Experience (EE) versus Plausible Reasoning (PR) in the context of problem solving non routine [2]; Direct Translation Approach (DTA) versus Meaning Based Approach (MBA) in the context of solving word problems [3]; Dual Process Theory of Kahneman (Process System 1 versus Process System 2) in the context of solving algebra problems [4]; The true pseudo construction versus the false pseudo construction in constructing mathematical concepts and solving mathematical problems [5].

The given answers of the pseudo thinking are divided into two; the true pseudo thinking and the false pseudo thinking [6]. The true pseudo thinking happens when the students get the correct answers but in fact students' reasoning is wrong while the false pseudo thinking happens when the students' answers are incorrect, but after reflection, the students are able to give reasons correctly [7].

The pseudo thinking is one of students' errors in constructing mathematical concepts [8]. The results of Subanji and Nusantara's research [9] show that students often seem to understand the concept well based on the final results they obtained when solving the problem, but after deeper tracing, it turns out that the process of thinking is wrong. These kinds of thinking process are known as the pseudo thinking process.

Based on the preliminary study conducted in State 16 Junior High School in Surakarta, Indonesia among 8th grade students when the students were asked to determine the gradient line $6x - 3y + 10 = 0$, most of the students determined the line gradient by using the formula $m = -\frac{a}{b}$ so obtained $m = 2$. However, when students were asked why they used the formula and how they got it, and what a and b meant, students were unable to explain. Furthermore, when students were given similar problem as the previous to determine the gradient line $8y - 4x + 16 = 0$, most of the students replied that the line gradient was the same as the line gradient in the previous problem that is $m = 2$, despite their knowledge that in determining the gradient of the line students being able to use the formula $m = -\frac{a}{b}$. However, students' thinking was dominated by thinking fast without the reflection process so that they chose the incorrect answer. These conditions show that students experienced pseudo thinking when determining the line gradient.

The pseudo thinking is defined as the visible result of the problem solving process, and is not the result of a genuine thinking process, but rather comes from a pseudo thinking process that occurs when students learn or when solving a problem [10]. The mistakes among students experiencing pseudo thinking in constructing concepts and solving mathematical problems need to get attention as if they are not resolved soon, those errors will have a consequential impact on subsequent math problems [11].

Pseudo thinking process is related to student's metacognition skill. Metacognition skill is a conscious knowledge and someone's control into thinking process and result [12]. Students with pseudo thinking showed that they experienced difficulty in using metacognition skill in solving math problem. For instance, students with true pseudo thinking, even if they produced the correct answer, their thinking process in solving the answer was still wrong and they did not understand what their fault was. Those findings showed inability among students who experienced pseudo thinking process in identify the difficult metacognition.

Metacognition in this research is someone's consciousness towards thinking process and result. Metacognition includes three aspects, which are the knowledge of strategy, the knowledge of cognitive tasks and the self-knowledge [13]. Not all students were able to use their metacognition well as well as with students who experience pseudo thinking. The inability to identify the difficulties metacognition experienced by students.

Student who experiences difficulty in using metacognition skill are those who were unable or having difficulties to use metacognitive knowledge. Metacognitive knowledge is a knowledge about consciousness in general is the same thing as vigilance and knowledge about self-knowledge [12]. Metacognitive knowledge consists of the knowledge of strategy, the knowledge of cognitive tasks and the self-knowledge [13].

Based on those problems, it is important to examine the thinking process and errors of students who experienced pseudo thinking when solving math problem in straight line equation in order to find out students' metacognition skills. This case was influenced by students' metacognition skill in solving the problem, especially in math problem influencing learning process and students' performance. For those reasons, this research examines the thinking process of students who experienced pseudo thinking and the errors of students who experienced pseudo thinking in solving math problem in the straight line equation. In addition, this research was also aimed to find out metacognition skill of students who experienced pseudo thinking.

IOP Conf. Series: Journal of Physics: Conf. Series **1008** (2018) 012061 doi:10.1088/1742-6596/1008/1/012061

2. Research Method

This research is a qualitative research that aims to describe the thinking process of students who experienced the pseudo thinking when solving the problems of straight line equation in order to reveal what mistakes were made by students who experienced pseudo thinking in solving the problems of straight line equation.

The instruments of data collection in this study was a sheet of straight line equation and interview guidelines. Interviews were used to find out whether students experienced pseudo thinking. Based on their work actually experience pseudo thinking or not and explore more about the thinking process delivered by the students. In this research, to increase the validity of the data, the triangulation method was used in the steps of verification in which the results of test data obtained were compared with the results of interviews.

This research was conducted in State 16 Junior High School in Surakarta, Indonesia on 8th grade. The research subjects were chosen with purposive sampling method. Students chosen as the subjects were those who met the following criteria: the subjects experienced pseudo thinking when solving the straight line equation problem and was able to communicate their thought well through speaking or writing so the exploration of student learning process could be done optimally.

The data collection in this research was done by think aloud method where students were given the sheet of straight line equation to solve. In the process of solving the problems the students were asked to express what students was thinking. After the solving the problems, the researcher then grouped the students' answers into correct and incorrect answers. Students with correct answers were asked to justify the answer. Students who were not able to justify the answer were classified as subjects experiencing the true pseudo thinking. Students with incorrect answer were given the opportunity to do the reflection. Students who were able to correct the answers after the reflection and explain the answers correctly were classified as subjects experiencing the false pseudo thinking.

3. Result and Discussion

Based on the data collected from the sheet of straight line equation given to 26 students of 8th grade in State 16 Junior High School in Surakarta, Indonesia, the student answers can be categorized as follows.

Table 1. Categories of Students' Job Results

Category	Students
True pseudo thinking	4
False pseudo thinking	8
Totally right	2
Totally wrong	12

Based on Table 1, it can be inferred that almost half of the students experienced pseudo thinking when solving the straight line equation. However, in this article, the researcher describe the pseudo thinking process of the students with 2 samples: a student who experienced true pseudo thinking (subject 1) and a student who experienced false pseudo thinking (subject 2). The description of those 2 subjects were expected to represent those who could describe the actual conditions in the field. The following image describes the results of the work of subject 1 and subject 2 when solving the sheet of straight line equation.

Find the line equation through $(-2,4)$ and
a. Parallel to line $4x - 2y + 12 = 0$
b. Perpendicular to line $9y - 3x + 18 = 0$

Figure 1. The question of straight line

IOP Conf. Series: Journal of Physics: Conf. Series **1008** (2018) 012061 doi:10.1088/1742-6596/1008/1/012061

3.1 Subject 1 (S1)

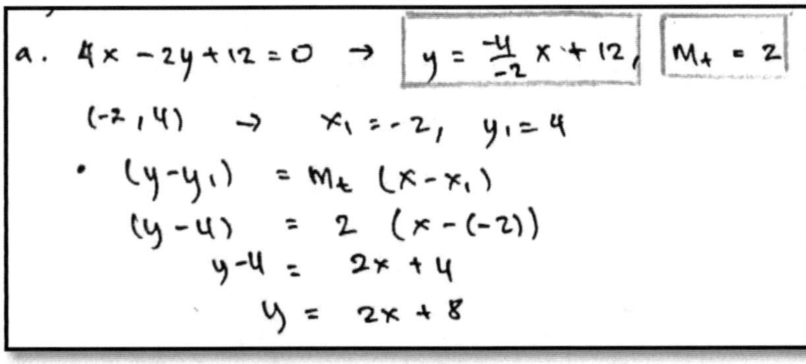

Figure 2. The S1's answer for the (*a*) question

Based on S1's answer for the (*a*) question, S1 was wrong in changing $4x - 2y + 12 = 0$ to $y = mx + c$ although S1 was correct in determining gradient line $4x - 2y + 12 = 0$. Other than that, S1 mentioned m_t as $4x - 2y + 12 = 0$ gradient line. Whereas on the question, not mentioned t line, S1 did not put the line as a t line. Furthermore, S1 also did not write a parallel two line gradient relationship, so it was unknown whether S1 really knew the gradient relationship of two parallel lines or not.

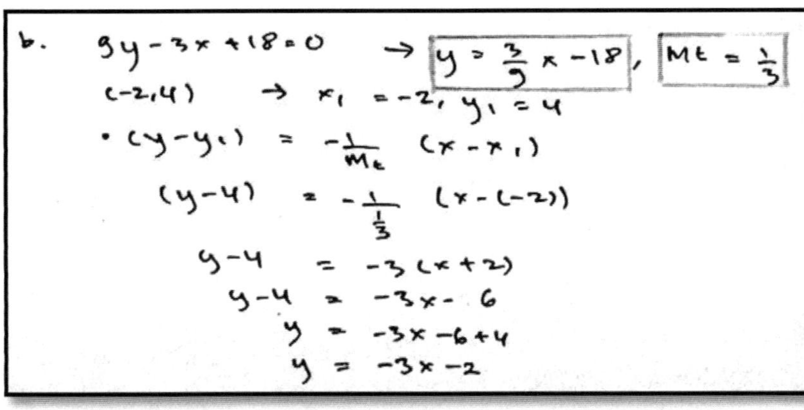

Figure 3. The S1's answer for the (*b*) question

When solving the (*b*) question, S1 was also wrong in changing the $9y - 3x + 18 = 0$ to $y = mx + c$ even though S1 was correct in determining the gradient of line $9y - 3x + 18 = 0$. Other than, that S1 also still wrote the initial m_t as the grid of the line $9y - 3x + 18 = 0$ and did not write a two line perpendicular gradient relationship, so it was unknown whether S1 really knew the two line perpendicular gradient relationship or not.

Based on the steps given by S1 when solving the problem, it seemed S1 experienced the true pseudo thinking, to convince researcher whether S1 experienced the true pseudo thinking or not based on the results of his work when completing the problem, the researcher conducted interviews with S1. There are "R" for Researcher and "S1" for Subject 1.

Excerpt 1

R : Are you sure that you change line equation of $4x - 2y + 12 = 0$ and $9y - 3x + 18 = 0$ into the general form of line equation is already correct?

S1: The general form of line equation is $y = mx + c$, isn't it?
R : Yes, now. Can you explain to me how you change those line equation?
S : Yes, I just have to change that $y = \frac{-4}{-2}x + 12$ and $y = \frac{3}{9}x - 18$.
R : So, the 12 isn't divided with negative 2 and the negative 18 isn't divided with 9, why is it?
S1: We just need to find the gradient, so only the number in front of x will be divided with the number in front of y
R : So, what is the number in front of x and the number in front is y? Is it variable or coefficient?
S1: It is coefficient
R : Are you sure 12 and 18 should not be divided with y coefficient?
S1: I'm sure. Because we just need to find the gradient from those line

Based on the interview results, it is known that S1 has not understood how to change the form of the equation line $ax + by + c = 0$ into $y = mx + c$ form correctly. Although the result was only coincidence of S1 results in determining the correct gradient of the line, the construction process was not appropriate (having an error) in changing the form of the equation line. This is because S1's process of thinking in changing the form of the equation line was done spontaneously, quickly, automatically, unconsciously and without any control.

Excerpt 2

R : Do you remember the requirement of two parallel gradient lines?
S1: Same
R :Here you wrote requirement of two parallel gradient lines, but directly input the gradient line known into the line that will be looked for. Why is it?
S1: If the question is like this, I will directly input it into the formula, miss. The gradient is equal, I did not need to write the requirement.

From the excerpt of the interview, it is known that S1 had understood about the relationship of two parallel lines gradient. Although the result of S1's completion does not write the relationship of two parallel lines, it is because the habit of S1 does not write the concepts, what matter is the result.

Excerpt 3

R : So, in the (*b*) question. It will find the perpendicular line to $9y - 3x + 18 = 0$, what is the requirement of a two line perpendicular gradient?
S1: (the student was silent)
R : Here when you find the line equation that through point $(-2,4)$ and perpendicular with $9y - 3x + 18 = 0$, you used $(y - y_1) = -\frac{1}{m_t}(x - x_1)$. Why did you use those formula?
S1: Because in my notebook the formula to determine the perpendicular line equation is like that
R : So, Do you know what is the requirement of two perpendicular line?
S1: I forget, miss. I remembered, but like those the formula and based on the exercise given by you also for determine the line that through points (x_1, y_1) and perpendicular with the other line that using those formula

The result of the interview shows that S1 does not understand the concept of a two line perpendicular gradient relationship. This is because, S1 only memorizes the formula and imitate the completion procedure that causes the reasoning S1 in constructing the concept of lines that are perpendicular to each other is not developing optimally.

Excerpt 4

R : I want to ask about the gradient that you already wrote in the question (*a*) and (*b*) question, you wrote m_t, what is it about?

S1: m_t is the gradient line. So, I presuppose the gradient for the line known in the question with m_t
R : In that case, can I presuppose the gradient for the line equation known with m_x or m_y initials?
S1:Yes, you can
R :What is your concept to write m_t?
S1: I don't know. I'm just following what I already wrote in my notebook

Based on the results of the interviews, S1 misunderstood the writing of gradients, as several times S1 wrote m_t, m_x *or* m_y. In other words, S1 had misconception of gradient writing. Misconception is defined as an interpretation of concepts in an unacceptable statement. Therefore, the results of the work and interview results shows that S1 experienced true pseudo thinking. Although S1 was by chance able to find the correct answer, but the reason given by S1 is not right. This can be seen from the S1's wrong process of thinking in changing the form of the line equation $ax + by + c = 0$ into the form $y = mx + c$, wrong in understanding the gradient writing, and not understanding the concept of the two line perpendicular gradient relationship.

The mistakes made by S1 were related to metacognition skill of S1 in understanding the concept of straight line equation. Based on the work and interview result, it was found that the metacognition of S1 in self-knowledge aspect was that S1 felt confident that problem completion steps was right, despite the wrong implementation. This case was taken from S1's result in which the mistake in changing the line form $ax + by + c = 0$ to general form of straight line equation $y = mx + c$ to find out the known gradient was found. The case with metacognition of S1 in the knowledge of cognitive tasks was that S1 knew what formula that should be applied to solving the problem, however S1 did not understand the concept applied appropriately. This is in accordance with the interview result in which the fact that S1 did not understand the concept of gradient relationship of two lines perpendicular to each other was found. The case with metacognition of S1 in aspect of strategy was that S1 could not mention the reasons why S1 used $(y - y_1) = -\frac{1}{m_t}(x - x_1)$ formula in determining straight line equation appropriately and S1 also failed to reason why S1 failed to writing m_t initial appropriately.

This was caused by S1's thinking process in solving the problem was done in spontaneous, quick, automatic, unconscious and uncontrolled manner in changing the straight line equation form. S1 also has a habit to only memorize the formula and imitate the solving procedure that caused S1's reasoning in constructing the concept of straight line equation to be wrong even if the final result was correct.

3.2 Subject 2 (S2).

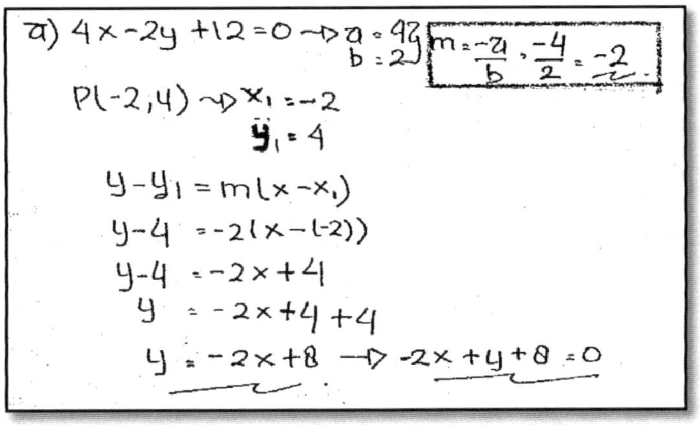

Figure 4. The S2's answer for the (*a*) question

IOP Conf. Series: Journal of Physics: Conf. Series **1008** (2018) 012061 doi:10.1088/1742-6596/1008/1/012061

Based on Figure 4. S2's answer for the (*a*) question it can be inferred that S2 knew that to determine the line gradient in the form of $ax + by + c = 0$ is $m = \frac{-a}{b}$, with $b \neq 0$. However, S2 was incorrect in determining the *y* coefficient on the line $4x - 2y + 12 = 0$. Whereas the coefficient of *y* should be negative 2. In addition, S2 did not write a parallel two line gradient relationship so it cannot be known whether S2 really knows the relationship of two parallel or parallel gradient lines.

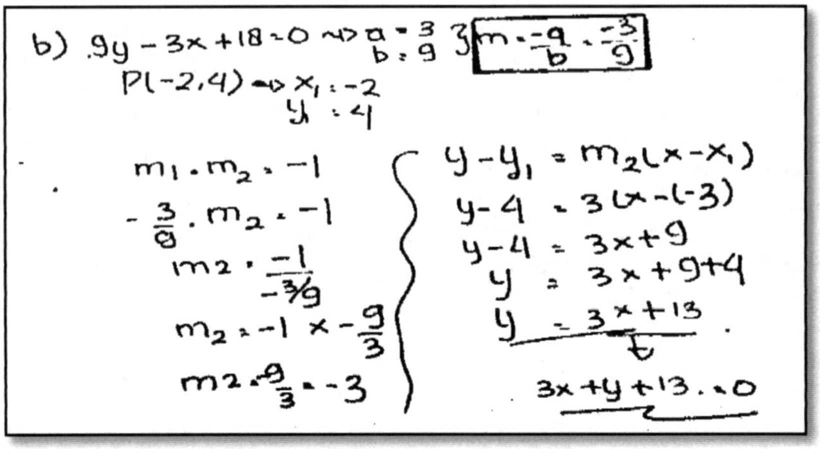

Figure 5. The S2's answer for the (*b*) question

From S2's answer when solving the (*b*) question, S2 made a mistake which is the same as the mistake in (*a*) question. S2 made mistake in determining the x coefficient on the line $9y - 3x + 18 = 0$. When solving (*b*) question, the error made by S2 was equal to the error made in the (*a*) question in which S2 wrongly determined the coefficient x on the line $9y - 3x + 18 = 0$. However, in the (*b*) question S2 wrote the two perpendicular gradient relation, S2 gave the line known as the first line having a gradient m_1 and the line sought as the second line having a gradient m_2. However, since S2 was wrong in determining the first line gradient (m_1) then in determining the second line gradient (m_2) S2 still got the incorrect answer.

Based on the steps given by S2 when solving the problem, it seems that S2 is having false pseudo thinking, to convince the researcher whether S2 had false pseudo thinking or not based on the result of his work when solving the problem, the researcher conducted interview with S2. As for "R" for Researcher and "S2" for Subject 2.

Excerpt 1

R : Okay, in the (*a*) question and (*b*)question. Do you know your mistake?
S2 : Is it when I was determining the line gradient? Isn't it?
R : True. What you actually should do?
S2: *b* value should be negative 2 because the *y* coefficient is negative 2, so the line gradient $4x - 2y + 12 = 0$ has $m = 2$ gradient and the line gradient $9y - 3x + 18 = 0$ is $m = \frac{-(-3)}{9}$ because the *x* coefficient is negative 3, so $a = -3$. Is it true, miss?
R : That's true. How did you get the wrong answer when you do the test?
S2: I actually forget about the sign in front of the *x* coefficient or *y* coefficient can be impacting or not in determining *a* and *b* value

Based on the interview result above, it can be seen that after doing the reflection, S2 was able to improve the answer to be the correct answer in determining the line gradient. However, because the thinking process of S2 in constructing the concept of determining the gradient is generated from fuzzy memory, it caused the incorrect answer.

Excerpt 2

R : Do you the the requirement of two parallel line?
S2: They have the same gradient
R : Who has the gradient?
S2: The parallel line
R : Why did you put the gradient directly into the formula?
S2: Because, it is the same. So, I just put it into the formula. I don't have to write down the requirement.

From the excerpts, it is known that S2 could understand the concept of the relation of two parallel line. Based on the result of S2's answer when solving the question and the interview result, it can be seen that S2 experienced false pseudo thinking.

S2's metacognition in aspect of cognitive task knowledge was shown by the finding that S2 knew what formula that should be applied when solving the problem, but had an error in deciding determine y coefficient in the (a) question and x coefficient in the (b) question that caused S2's answer to be incorrect in determining the gradient. However, after the reflection activity, S2 showed metacognition in self-knowledge aspect which are S2 aware of the mistakes made resulting in the final result answer and on strategy aspect, S2 was able to manage or control his knowledge to fix the answer become the correct one by rechecking the answer and fixing the mistake when determines the line gradient.

It is because S2 was actually able to reason correctly when solving the question, but as thinking process in constructing the concept of determining the gradient was generated from fuzzy memory, the final result was impacted. However, actually S2 was able to understand the concept of two parallel lines and the concept of two lines that are perpendicular to each other.

4. Conclusion

The results showed that there were some error characteristics students tend to have in solving straight line equations that can cause them to experience pseudo thinking. First, the thinking process that students do when solving problems results from spontaneous, fast, unconscious and unchecked (control or reflection) thinking process. Second, students tend to experience misconceptions. Third, students tend to memorize the formula and imitate the completion procedure that has been given by the teacher. Fourth, students experienced fuzzy memory when solving the problem.

Based on the conclusion of the resut of this research, the suggestions given to math teachers in Junior High School level is that they are expected to know how the process of students' thinking when getting answer in resolving the problem whether the answer produced is in accordance with the right thinking process or not to minimize the occurrence of pseudo thinking. The results of this study need to be explored deeper that it is advisable for advanced researcher to conduct research on other focus like picking up mathematical topics other than straight line equation. Advanced researcher should be able to use certain knowledge construction theory to find the characteristics of student mistakes among students who experienced pseudo thinking. In an instance, the theory of Action, Process, Object, and Schema (APOS) can be used as a knife of analysis so that the description of students' misconception of students' pseudo thinking becomes more complete.

Acknowledgments

The authors would like to thank the referees for their suggestions, State 16 Junior High School in Surakarta, Indonesia who supported this research implementation and Sebelas Maret University, Surakarta, Indonesia for funding this research in the academic year of 2017.

IOP Conf. Series: Journal of Physics: Conf. Series **1008** (2018) 012061 doi:10.1088/1742-6596/1008/1/012061

References

[1] Vinner S 1997 The Pseudo-Conceptual and The Pseudo-Analytical Thought Processes in Mathematics Learning *Educational Studies in Mathematics* **34** pp 97-129
[2] Lithner J 2000 Mathematical Reasoning in Task Solving *Educational Studies in Mathematics* **41** pp 165-190
[3] Pape S J 2004 Middle School Children's Problem-Solving Behaviour: A Cognitive Analysis from A Reading Comprehension Perspective *Journal of Research in Mathematics Education* **35** pp 187-219
[4] Leron U and Hazzan O 2009 Intuitive vs Analytical Thinking Four Perspective *Educational Studies in Mathematics* **71** pp 263-278
[5] Subanji 2015 *Teori Kesalahan Konstruksi Konsep dan Pemecahan Masalah Matematika* (Malang: Universitas Negeri Malang)
[6] Subanji 2011 *Teori Berpikir Pseudo Penalaran Kovariasional* (Malang: Universitas Negeri Malang)
[7] Herna, Nusantara T, Subanji and Mulyati S 2016 The Characterization of True Pseudo Construction in Understanding Concept of Limit Function *Journal of Research & Method in Educations* **6** pp 77-87
[8] Subanji and Nusantara 2013 Karakteristik Kesalahan Berpikir Siswa dalam Mengonstruksi Konsep Matematika *Jurnal Ilmu Pendidikan (JIP)* **19** pp 208-217
[9] Subanji and Nusantara 2016 Thinking Process of Pseudo Construction in Mathematics Concept *International Education Studies* **9** 1913-9039
[10] Hayati I N 2015 *Berpikir Pseudo Siswa dalam Memecahkan Masalah PISA* (Malang: Tesis Universitas Negeri Malang)
[11] Subanji 2016 *Teori Defragmentasi Struktur Berpikir dalam Mengonstruksi Konsep dan Pemecahan Masalah Matematika* (Malang: Universitas Negeri Malang)
[12] Fitria C 2016 *Analisis Kesulitan Metakognisi Siswa dalam Memecahkan Masalah Sistem Pertidaksamaan Linear Dua Variabel Ditinjau dari Tipe Kepribadian Guardian, Artisan, Rational, dan Idealist Kelas X SMKN 1 Jombang* (Surakarta: Tesis Universitas Sebelas Maret)
[13] Anderson L W and Krathwohl D R 2001 *A Taxonomy For Learning, Teaching, and Assessing (A Revision of Bloom's Taxonomy of Educational Objectives* (New York: Addision Wesley Longman, Inc)

IOP Conf. Series: Journal of Physics: Conf. Series **1008** (2018) 012062 doi:10.1088/1742-6596/1008/1/012062

Metacognitive experience of mathematics education students in open start problem solving based on intrapersonal intelligence

D P Sari[1], B Usodo[1], S Subanti[1]

[1]Master of Mathematics Education Program, Universitas Sebelas Maret

E-mail: diana.witasari94@gmail.com

Abstract. This research aims to describe metacognitive experience of mathematics education students with strong, average, and weak intrapersonal intelligence in open start problem solving. Type of this research was qualitative research. The research subject was mathematics education students in Muhammadiyah University of Surakarta in academic year 2017/2018. The selected students consisted of 6 students with details of two students in each intrapersonal intelligence category. The research instruments were questionnaire, open start problem solving task, and interview guidelines. Data validity used time triangulation. Data analyses were done through data collection, data reduction, data presentation, and drawing conclusion. Based on findings, subjects with strong intrapersonal intelligence had high self confidence that they were able to solve problem correctly, able to do planning steps and able to solve the problem appropriately. Subjects with average intrapersonal intelligence had high self-assessment that they were able to solve the problem, able to do planning steps appropriately but they had not maximized in carrying out the plan so that it resulted incorrectness answer. Subjects with weak intrapersonal intelligence had high self confidence in capability of solving math problem, lack of precision in taking plans so their task results incorrectness answer.

1. Introduction

Mathematical problem solving is an important thing that should get teacher's attention, especially to develop students' ability in their problem solving, either in the real problems or mathematical problems. Mathematical problem solving is a base which determines students' success in mathematics learning. Problem solving in mathematics learning is important thing and should be carried out by students [9]. That statement has a reason that mathematical problem solving is a medium for students to develop their ability, to build new mathematical knowledge, and to reflect mathematical problem solving process [2].

To get optimal result in problem solving, there are several problem solving steps that should be well-organized used. According to Polya [17] problem solving consists of four steps, (1) understanding the problem; (2) devising the plan; (3) carrying out the plan; and (4) looking back. This systematic problem solving steps or procedures will shape well-structured mindset of individuals in solving problem.

Problem solving is individuals' activity or effort for finding the right solution of a problem [20]. Problem solving has a connection to thinking process. The process shows the sequence of events that must be passed. This means that problem solving is carried out by process sequence, such as individuals should think about problem solving strategies and control every action in obtaining the right problem solving. The individuals' awareness to get the right answer through their way of thinking is the role of metacognitive experience.

Flavel [8] defines that metacognitive experience is individuals' awareness on strategies related to solve the problem. Metacognitive experience is the important aspect in problem solving. The reason is metacognitive experience helps individuals to devise problem solving strategies until they achieve the right result [6]. The same statement is also explained by Kuzle [13], metacognitive experience assists individuals in identifying the problem that should be solved, in looking back the real problem, and in achieving the goal or solution. Thus, metacognitive experience is individuals' awareness to do useful things rather than do negative things in solving problems.

Jacob & Paris [12] explain that metacognitive experience includes planning, monitoring and evaluation process. Planning is individuals' ability to plan their learning activities. The planning activities are choosing the right strategies that can affect the result and the availability of accurate information sources. Monitoring is individuals' ability to monitor the learning process and matters related to the learning process, such as monitoring the plan steps to get the right answer. Evaluation is individuals' ability to evaluate the effectiveness of learning strategies, such as modifying the learning strategies or discontinuing their learning activities.

Metacognitive experience in problem solving is important for students. Students who involve their metacognitive experience are more effective solving their problem; the reason is metacognitive experience is a skill that is able to make one's thinking clearly [15]. Therefore, students should have good understanding about metacognitive experience to use in their problem solving. Anggo [2] explains that there are some factors that can stimulate metacognitive experience in problem solving, i.e the use of challenging and non-routine mathematical problems. Open start problem is one of the problems that has challenging and non-routine characteristics because it has various ways to solve problems in which the first step in solving the problem shows its vagueness. The characteristic of mathematical problem in open start, is students do not know how to solve the problem directly, there will be hesitation about the questions, time for accomplishing problem solving steps, and time to find out the answers [14].

Problem in analytic geometry of space has the same characteristic with open start problem. The reason is problem in analytic geometry space has a single answer with various ways of solving the problem although it is not automatically known how to accomplish it. Thus, one of the problems that can stimulate metacognitive experience for mathematic education students is open start problem in analytical geometry of space.

The researcher conducted the document analysis on semester learning plan of analytic geometry of space subject on Mathematics Education Study Program, Faculty of Teacher Training and Education, at University of Muhammadiyah Surakarta. The analysis result of semester learning plan showed that teachers were more focused on the use of active learning model in carrying out the learning rather than on emphasis of students' metacognitive experience. The analysis result was strengthened with interview result from lecturer of analytic geometry of space subject who explained that the varied learning model is needed in teaching the material of space analytic geometry. The researcher did further interview about the students' metacognitive experience in solving the problem of analytic geometry of space, but the lecturer could not answer about it.

The researcher also conducted the document analysis of score on subject of analytic geometry of space in academic year 2016/2017. The result of document analysis showed that 34% of students get score C, D, and E (final score less than 56). These results indicate that many mathematics teacher candidates have not good understanding about metacognitive experience which makes them lack of capability in planning, monitoring, evaluation their own teaching process so that it causes students get low score on the subject of analytic geometry of space.

Furthermore, the researcher conducted a pre-survey activity to know Mathematics Education students' metacognitive experience at Muhammadiyah University of Surakarta in solving the problem of analytic geometry of space. The researcher conducted a task-based interview to a student who pursued analytic geometry of space subject. The selected students are those who have quite achievement and activeness, and are able to argue well related to lecturer of analytic geometry of space explanation. Pre-survey result showed that the students had not maximized the evaluation process on carrying out the plan and looking back step, so that their solving tasks result the wrong answer. This matter shows that the evaluation of individuals' performance is not only due to adequate knowledge, but also the individual's awareness to organize, carry out and evaluate their knowledge [19]. This awareness arises when individuals can understand themselves well so they can solve the problems effectively and efficiently. The ability of individuals to understand themselves is called intrapersonal intelligence.

Armstrong [3] explains that intrapersonal intelligence is self-knowledge to act adaptively based on the knowledge. Intrapersonal intelligence involves the ability to create accurate and realistic overview of personal strengths and weaknesses, awareness of thinking processes, and the ability to have self-discipline and self-understanding. Hoer, Bogeman, & Wallach [11] states that through strong intrapersonal intelligence individuals will be able to reflect on their own shortcomings to set useful life goals. The same statement is explained by Gangadevi & Ravi [9] that the learning experience of individuals who have strong intrapersonal intelligence tend to be independent, have self-confidence, and able to make important decisions for themselves.

Strengths and weaknesses are important aspects of intrapersonal intelligence. The strengths and weaknesses information of individuals can correct the mistakes [4]. Individuals with average intrapersonal intelligence know their strengths and weaknesses but they have not maximized the strengths and minimize the weaknesses. Habeeb & Fatema [10] explain that individuals with weak intrapersonal intelligence have not been able to reflect themselves from their shortcomings in solving a problem.

Based on the previous explanation, researcher needs to conduct the research related to metacognitive experience in open start problem solving. The aim of this research is to describe metacognitive experience of mathematics education students with strong, average, and weak intrapersonal intelligence in open start problem solving.

2. Research Methods

This research was a qualitative research. The research was conducted in Mathematics Education Study Program of Muhammadiyah University of Surakarta in academic year 2017/2018. The selected subjects were done by purposive sampling with snowball sampling. Creswell [7] explains that the purpose of selection of subjects with purposive sampling is the researchers can choose individuals as subjects of research and understand the phenomenon on their research focus. Snowball sampling is subject recruitment in which the subject can increase as time progressed depending on the research need [5]. Firstly, the selected subjects consisted of 6 students with details of two students in each intrapersonal intelligence category, and then the subjects could be increased depending on the research need. The category of intrapersonal intelligence was obtained based on questionnaire developed by researcher based on aspects and indicators written by Alder [1].

Other instruments in this research were open start problem solving tasks on analytic geometry of space and interview guidelines. The open start problem solving tasks given to the subjects during the interview are as follows.

Task of *open start* I problem solving

Given points $P(0,6,4)$, $Q(0,-1,4)$, $R(-5,0,0)$, and S lies on the x-axis. Determine the point S so that QS intercrosses perpendicular to PR, and then find the distance between QS to PR!

Task of *open start* II problem solving

Given points $P(0,6,4)$, $Q(0,-1,4)$, $R(-5,0,0)$, and S lies on the x-axis. Determine the point S so that QS intercrosses perpendicular to PR, and then find the distance between PS to QR!

Task of *open start* III problem solving

Determined points $A(1,2,3)$, $B0,0,1)$, and $C(x_C, y_C, 7)$ lies on a line g_1. Given a line g_2 that that passes through the point $P(2,0,0)$ and perpendicular plane $V: 5x - 2y + 3z - 3 = 0$. Determine the distance point C to line g_2!

All instruments in this research fulfilled the criteria since they were validated by experts of their fields. Research data were obtained through task-based interview and documentation of students' work. Data validity used time triangulation. Testing credibility with time triangulation is carried out by collecting data at different times [18]. The researcher conducted two task-based interviews that held at different times. The researcher would add third task-based interviews, if the first task-based interview result found the difference from the second result. Data analyses were done through data collection, data reduction, data presentation, and drawing conclusion. Analysis model of this research used interactive analysis. This means that data reduction, data presentation, and conclusion drawing are reciprocal [5]. The research methods which described above is illustrated by Figure 1 below.

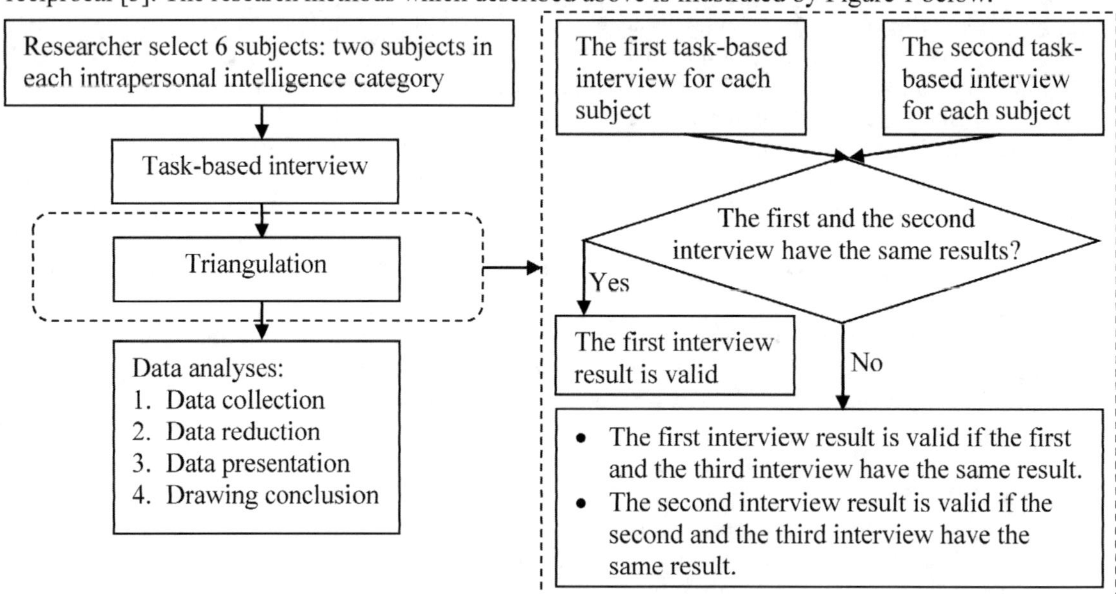

Figure 1. Flowchart of the research methods

3. Research Results

Researcher conducted task-based interviews on subjects with their intrapersonal intelligence tendency. Task-based interview activities were conducted to determine the metacognitive experience of math education students in solving open start problems. The result of metacognitive experience description of each subject is shown as follows.

3.1. Metacognitive Experience of Mathematics Education Students with Strong Intrapersonal Intelligence in Open Start Problem Solving

a. Understanding the problem

The subjects understood the problem by reading the question, writing down coordinate geometry information and the question, and finding the keywords to help them in solving the problem. The interview results with the subjects at understanding the problem are presented as follows (P = researcher; T = subjects with strong intrapersonal intelligence).

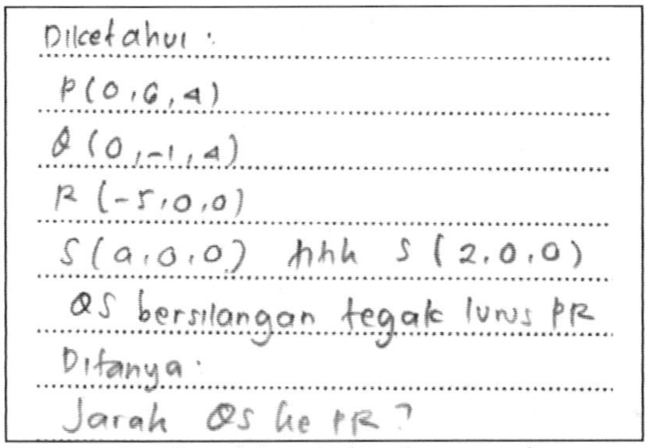

$P_{1.1}$: Have you understood the given problem?

$T_{1.1}$: Yes I have. The problem that should be solved is calculating the distance of line *QS* to *PR*.

$P_{1.2}$: What do you do to understand the problem?

$T_{1.2}$: Reading the question at first, understand the coordinate geometry information and the question, searching for keyword related to the lesson material.

Figure 2. The task result of strong intrapersonal intelligence subjects on understanding the problem

Data $T_{1\cdot1}$ indicates that the subject understands the problem. How subject understand the problem is shown by data $T_{1\cdot2}$. The subject understands the problem by writing down coordinate geometry information and the question as seen in Figure 2.

b. Devising a Plan

The subjects devised a plan by thinking the steps through imagination and thinking of formulas to solve the problem. Excerpts of interview with subjects on devising a plan are presented as follows.

$P_{2.1}$: What are your plans for solving the given problem?

$T_{2.1}$: Determine the point *S* at first and then make the plane that passes through *QS* and crosses perpendicular *PR*.

$P_{2.2}$: Why do you plan on it?

$T_{2.2}$: Because *QS* and crossing perpendicular *PR* so it can be found the distance with point space concept to plane.

$P_{2.3}$: What is on your mind when you devise a plan?

$T_{2.3}$: Imagining the distance of two perpendicular lines crossing then thinking a usable formula.

Data $T_{2\cdot1}$ shows the subject knows the steps taken to solve the problem. Why subject choose plan $T_{2\cdot1}$ is explained by data $T_{2\cdot2}$. How subject devise the plan is shown by data $T_{2\cdot3}$.

c. Carrying Out The Plan

Subjects solved the problems which were appropriate with the plan. Subjects carry out a step-by-step plan for solving. Subjects were able to accomplish the tasks correctly, although the subjects did not use effective and efficient strategies as shown in Figure 3.

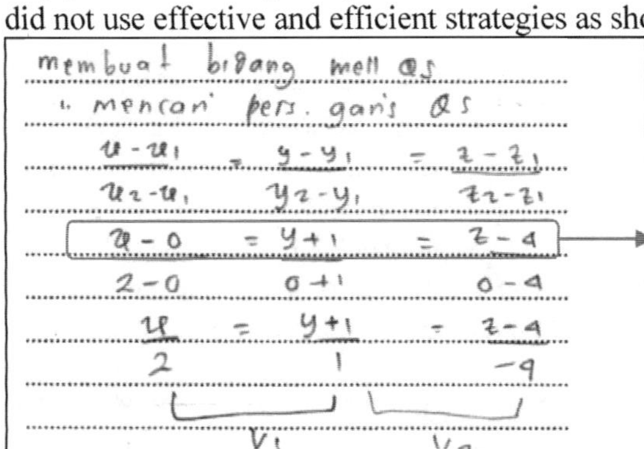

Subject substitutes the point *Q* to replace the value of x_1, y_1, and z_1 (actually it is easier to substitute point *S*). This indicates that subject have not yet monitored the task results toward effective and efficient strategies.

Figure 3. Task result of strong intrapersonal intelligence subject on carrying out the plan

d. Looking Back

The subjects checked the answer by looking back the task results. This step was taken to convince the correctness of the answers. Excerpts of interviews with the subjects on looking back are presented as follows.

$P_{4.1}$: What are you doing? (This question was asked when the researcher saw the subject moving the pencil from top to bottom of the answer sheet)

$T_{4.1}$: Looking back the steps and the calculations in order to make sure the correctness of my answer.
$P_{4.2}$: Are you sure that the answer that you write is correct?
$T_{4.2}$: Yes I'm sure that my answer is correct.
$P_{4.3}$: How do you know your answer is correct?
$T_{4.3}$: Because the steps are correct, the formula is correct, and the calculation is correct.

The subjects are looking back the task results. Why subject looks back are shown by data $T_{4.1}$. Data $T_{4.2}$ indicates that the subject believes the correctness of the answer. Why the subject believes the correctness of the answer is shown by data $T_{4.3}$.

3.2. Metacognitive Experience of Mathematic Education Students with Average Intrapersonal Intelligence in Open Start Solving Problem

a. Understanding the problem

The subjects understood the problem by reading the question, writing down coordinate geometry information and the question. The interview results with the subjects on understanding problem are presented as follows (P = researcher; S = subjects with average intrapersonal intelligence).

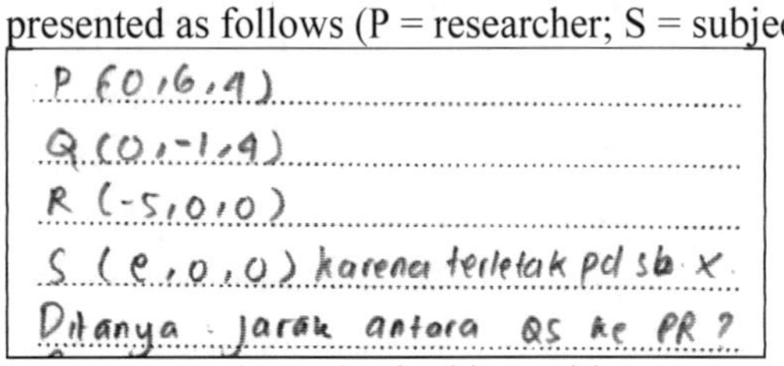

Figure 4. Task result of subject with average intrapersonal intelligence on understanding the problem

$P_{1.1}$: Have you understood the given problem?
$S_{1.1}$: Yes I have. The problem that should be solved is calculating the distance of the *QS* line to *PR*.
$P_{1.2}$: What do you do to understand the problem given?
$S_{1.2}$: Reading the question at first and then writing down coordinate geometry information and the question.

Data $S_{1.1}$ shows that subject understands the given problem. How subject understands the problem is shown by data $S_{1.2}$. When subjects understand the problem they write coordinate geometry information and the question as seen in Figure 4.

b. Devising a plan

Subjects devised a plan by thinking the steps through imagination and thinking of formulas that could be used to solve the problem. Excerpts of interview with subjects on devising a plan are presented as follows.

$P_{2.1}$: What are your plans to solve the given problem?
$S_{2.1}$: Determine point S and then make a plane that passes through *PR* and find the distance of *QS* to the plane.
$P_{2.2}$: Why do you plan it ?
$S_{2.2}$: Because *QS* crosses perpendicular to *PR*.
$P_{2.3}$: What is on your mind when you devise a plan?
$S_{2.3}$: Finding ways and thinking about the steps that will be used to solve the problem.

Data $S_{2.1}$ shows that subject knows the steps taken to solve the problem. Why subject choose plan $S_{2.1}$ is explained by data $S_{2.2}$. How subject devises a plan is shown by data $S_{2.3}$.

c. Carrying out the plan

Subject solved the problems which were appropriate with the plan. Subjects carried out a step by step plan, it was proved that subjects seemed changing their answers several times because they doubted. Subject believed that their answers were correct although the results were incorrect. The results of subjects' task on carrying the plan are shown in Figure 5.

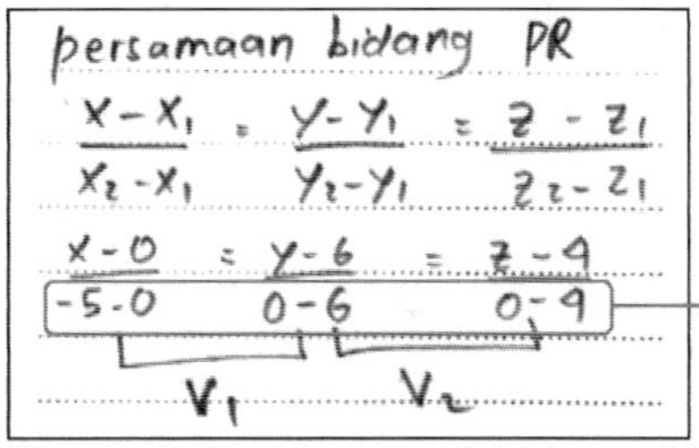

Subject substitutes point Q to replace value of x_2, y_2, dan z_2 (the correct answer is to substitute point R). It is shown that subject has not been maximized on monitoring and evaluation process toward the completion so that he does not improve the mistake.

Figure 5. Task result of subject with average intrapersonal intelligence on carrying out the plan

d. Looking back

Subjects believed the correctness of the answer by looking back the steps of work results. Excerpts of the interview with subjects on looking back step are presented as follows.

$P_{4.1}$: Are you sure that your given answer is correct?
$S_{4.1}$: Yes, I am sure.
$P_{4.2}$: How do you know that your answer is correct?
$S_{4.2}$: Because the steps are correct.

The subjects are looking back the work results. Data $S_{4.1}$ shows that subject believe his answer was correct. Why subject believes the correctness of his answer is shown by data $S_{4.2}$.

3.3. Metacognitive Experience of Mathematics Education Students with Weak Intrapersonal Intelligence in Solving Open Start Problem

a. Understanding the problem

Subjects understood the problem by reading the question, writing down coordinate geometry information and the question. The interview results with the subjects on understanding the problem are presented as follows (P = researcher; R = weak intrapersonal intelligence subjects).

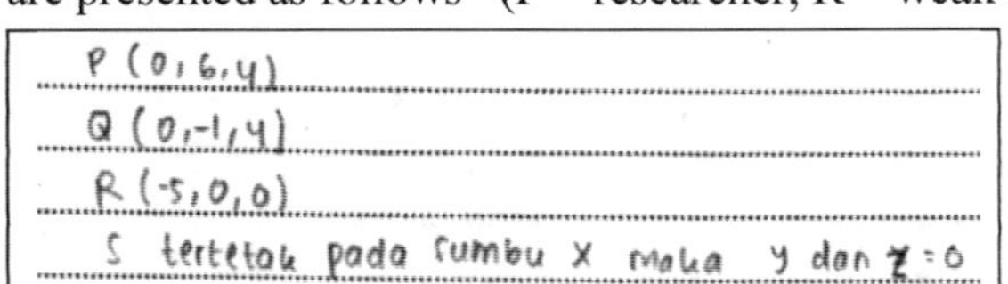

Figure 6. Task result of subject with weak intrapersonal intelligence on understanding the problem

$P_{1.1}$: Have you understood the given problem?
$R_{1.1}$: Yes, I have. The problem that should be solved is calculating the distance of *PS* line to *QR*.
$P_{1.2}$: What do you do to understand the problem?
$R_{1.2}$: Reading the question at first, understand the coordinate geometry information and the question.

Data $R_{1.1}$ shows that the subject understand the given problem. How subject understands the problem is shown by data $R_{1.2}$. The subject understands the problem by writing coordinate geometry information and the question as seen in Figure 6.

b. Devising a plan

The subjects devised a plan by thinking the steps through imagination and thinking of formulas to solve the problem. Excerpts of interview with subjects on devising a plan are presented as follows.

$P_{2.1}$: What are your plans to solve the given problem?
$R_{2.1}$: Determine the point *S* then find the distance *PS* to *QR*.
$P_{2.2}$: Why do you plan it?
$R_{2.2}$: Because finding distance of *PS* to *QR* the coordinate of point *S* should have been known.
$P_{2.3}$: What is on your mind when you devise a plan?
$R_{2.3}$: Thinking about the steps that will be used to solve the problem.

Based on data $R_{2.1}$, subject knows the steps to solve the problem. Why subject choose plan $R_{2.11}$ is explained by data $R_{2.2}$. How the subject devise a plan is shown in data $R_{2.3}$.

IOP Conf. Series: Journal of Physics: Conf. Series **1008** (2018) 012062 doi:10.1088/1742-6596/1008/1/012062

c. Carrying out the plan
Subjects solved the problems which were appropriate with the plan. Subjects carried out a step-by-step plan for the solving. Subjects believed the correctness of his answer, although the answers were still not correct. The subjects' task results on carrying out the plan are shown in Figure 7.

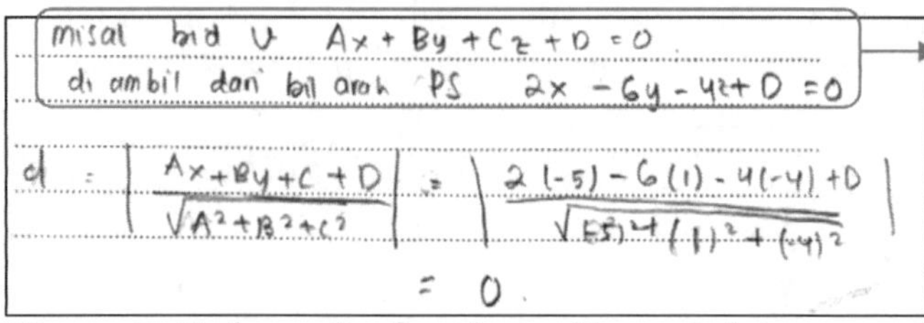

Figure 7. Task result of subject with weak intrapersonal intelligence on carrying out the plan

Subject make the plane passing through *PS*. The task result of the subject is shown in Figure 7 which shows use of formula and step to make the plane that passes through *PR* are not correct. This result shows that subject has not carried out the solving steps correctly.

d. Looking back
The subjects checked the answer by looking back the work results. Excerpts of interviews with the subjects on looking back step are presented as follows.
$P_{4.1}$: Are you sure that your answer is correct?
$R_{4.1}$: Yes, I am sure my answer is correct, although the distance is zero
$P_{4.2}$: How do you know that your answer is correct?
$R_{4.2}$: Because the steps and the formula are correct.
Subjects look back the result of their works. Data $R_{4.1}$ shows that subject believes his answer is correct. Why subject believes his answer is shown by data $R_{4.2}$.

4. Discussion

The research result toward the subjects with strong intrapersonal intelligence showed that subjects carried out planning, monitoring, and evaluation process on understanding problem step. This was proved that the result of data aimed to know the correctness of core question and the statement sentence related to the question. In devising a plan step, subjects did appropriately planning, monitoring, and evaluation process. It was shown that students were able to plan the steps, able to choose the formulas and their initial knowledge to solve the problem. In carrying out the plan, subjects did planning process appropriately but they had not yet maximize the process of monitoring and evaluation. This was proved that the result of subjects' tasks used less effective and efficient strategy, although the final answers of the task were correct. In looking back step, subjects did planning, monitoring, and evaluation process well. It was applied by subjects to make sure their answers were correct.

Based on the result, subjects with strong intrapersonal intelligence had high self confidence that they were able to solve problem correctly, able to do planning steps and able to solve the problem appropriately. This is appropriate with Gangadevi & Ravi [9] statement. They state that individual learning experience who have strong intrapersonal intelligence tend to be independent, have self-confidence, and able to make important decisions for themselves.

The result on subjects with average intrapersonal intelligence showed that subjects did planning, monitoring, and evaluation process appropriately on understanding the problem step. It could be seen from the data obtained showing that the coordinate geometry information and the question were correct, as well as the statement sentence was in accordance with the question. In devising a plan step, subjects carried out the process of planning, monitoring, and evaluation appropriately. This was proved that subjects were able to decide the formula and initial knowledge used to solve the problem. In carrying out the plan, subjects had not carried out the planning process correctly and maximized the monitoring and evaluation process. It was shown that subjects were not aware of an error so they did not make improvements. The subjects had not carried out well the planning, monitoring and evaluation process on looking back step. This was shown that the subjects only looked back the steps to do without looking back the calculation and the formula used so that their final answers were incorrect.

Based on the result, subjects with average intrapersonal intelligence had high self-assessment that they were able to solve the problem, able to do planning steps appropriately but they had not maximized in carrying out the plan so that it resulted incorrectness answer. It was proved that subjects with average intrapersonal intelligence had not yet maximized their strength and minimized their weakness so they could not maximized their solving plan. Strength and weakness are important aspects of intrapersonal intelligence; through those aspects person can correct the mistake [4]. The reason is intrapersonal intelligence is the ability to understand and apply self-discipline in order to develop the problem-solving strategies based on their limitations [16].

The result of the subjects on weak intrapersonal intelligence showed that subjects did planning, monitoring, and evaluation process appropriately on understanding problem step. It was shown that data obtained contained about coordinate geometry information and statement sentence which were appropriate with the question. In devising a plan, subjects had not carried out planning process correctly. In addition, subjects had not maximized yet monitoring and evaluation process. It was shown that subjects were incapable to give reason why those planning were chosen. In carrying out the plan step, subjects solved the problem based on the previous plan. In looking back step, subjects checked the answer by looking back the task results. Subjects believed the correctness of their answer, although their answers were incorrect and very out of appropriateness with the question.

Based on result above, subjects with weak intrapersonal intelligence had high self confidence in capability of solving math problem, lack of precision in taking plans so their task results incorrectness answer. The result has in accordance with Habeeb & Fatema's statements [10], individuals who have weak intrapersonal intelligence have not been able to reflect themselves from their weaknesses in problem solving.

5. Conclusion

Students with strong intrapersonal intelligence had high self confidence that they were able to solve problem correctly, able to do planning steps and able to solve the problem appropriately. Students with average intrapersonal intelligence had high self-assessment that they were able to solve the problem, able to do planning steps appropriately but they had not maximized in carrying out the plan so that it resulted incorrectness answer. Students with weak intrapersonal intelligence had high self confidence in capability of solving math problem, lack of precision in taking plans so their task results incorrectness answer.

Educators are expected to have a good understanding of metacognitive experience because metacognitive experience is important for students. Students who involved their metacognitive experience are more effective solving their problem and able to solve the problem appropriately.

6. Open Problem

Mathematics teacher are expected to create learning that can enhance metacognitive experience of students. Can Android media enhance students metacognition?

7. Acknowledgement

The authors would like to thank to the leader of Universitas Sebelas Maret for giving permission to the researcher to carry out the research. The authors are very grateful to the Mathematics Education department, Muhammadiyah University of Surakarta for giving the opportunity to do retrieval research data.

References

[1] Alder H 2001 *Boost Your Intelligence* (Jakarta: Erlangga)
[2] Anggo M 2011 Pelibatan Metakognisi dalam Pemecahan Masalah Matematika *Edumatica* **1** 25-32

[3] Armstrong T 2009 *Multiple Intelligences in the Classroom* (USA: Alexandria)
[4] Azid N H & Yaacob A 2016 Enriching Orphans' Potentials through Interpersonal and Intrapersonal Intelligence Enrichment Activities *International Journal of Instruction* **9** 17-32
[5] Budiyono 2017 *Pengantar Metodologi Penelitian Pendidikan* (Surakarta: UNS Press)
[6] Chimuma L L & Johnson I. D 2016 Assesing Students' Use of Metacognition during Mathematical Problem Solving Using Smartpens *Educational Research: Theory & Practice* **28** 22-36
[7] Creswell J W 2012 *Educational Research: Planning, Conducting, and Evaluating Quantitative and Qualitative Research* (Upper Saddle River: Pearson Education, Inc)
[8] Flavell J H 1979 Metacognition and Cognitive Monitoring: A new Area of Cognitive E-Developmental Inquiry *American Pschycological* **31** 906-911
[9] Gangadevi S & Ravi 2014 Multiple Intelligence Based Curriculum to Enhance Inclusive Education to Bring Out Human Potential *International Journal of Advanced Research* **2** 619-626
[10] Habeeb K T & Fatema M 2016 Affect of Intrapersonal and Interpersonal Awareness Dimensions of Emotional Intelligence on Stress Management of Adolescents *International Journal of Applied Research* **2** 589-59
[11] Hoer T R Bogeman S & Wallach C 2010 *Celebrating Every Learner* (San Francisco: Jossey Bass)
[12] Jacob J E & Paris S G 1987 Children's Metacognition About Reading: Issues in Definition, Measurement, and Instruction *Educational Psychologist* **22** 255-278
[13] Kuzle A 2013 Patterns of Metacognitive Behavior during Mathematics Problem Solving in a Dynamic Geometry Envoronment *International Electronic Journal of Mathematics Education* **8** 20-40
[14] Monaghan J, Pool P, Roper T and Threlfall J 2009 Open-Start Mathematics Problems: An Approach to Assesing Problem Solving *Oxford University Press on behalf of The Institute of Mathematics and Its Application* **28** 21-31
[15] Ozsoy G and Ataman A 2009 The Effect of Metacognitive Strategy Training on Mathematical Problem Solving Achievement *International Electronic Journal of Elementary Education* **1** 68-83
[16] Perez M M P and Ruz N R 2014 Intrapersonal Intelligence and Motivation in Foreign Language Learning *European Scientific Journal* **10** 142-150
[17] Polya G 1973 *How to Solve It* (Princeton, New Jersey : Princeton University Press)
[18] Satori D and Komariah A 2013 *Metode Penelitian Kualitatif* (Bandung: Alfabeta)
[19] Telaumbanua Y N, Sinaga B, Mukhtar and Surya E 2017 Development of Mathematics Module Based on Metacognitive Strategy in Improving Students' Mathematical Problem Solving Ability at High School *Journal of Educational and Practice* **8** 73-80
[20] Wismath S L 2015 Collaborative Learning in Problem Solving: A Case Study in Metacognitive Learning *The Canadian Journal for the Scholarship of Teaching and Learning* **6** 1-17

Analysis of difficulties in mathematics problem solving based on revised Bloom's Taxonomy viewed from high self-efficacy

R D E Prismana [1], T A Kusmayadi [1], I Pramudya[1]

[1]Postgraduate School of Mathematics Education Sebelas Maret University. Jl. Ir. Sutami 36A, Kentingan, Surakarta 57126, Indonesia.

E-mail: rosanadianedyprismana@yahoo.co.id

Abstract. The ability of solving problem is a part of the mathematic curriculum that is very important. Problem solving prefers the process and strategy that is done by students in solving a problem rather than the result. This learning concept in accordance with the stages on the revised bloom's taxonomy. The revised Bloom's Taxonomy has two dimensions, namely the dimension of cognitive process and the dimension of knowledge. Dimension of knowledge has four categories, but this study only restricted on two knowledge, conceptual knowledge and procedural knowledge. Dimensions of cognitive processes are categorized into six kinds, namely remembering, understanding, applying, analyzing, evaluating, and creating. Implementation of learning more emphasis on the role of students. Students must have their own belief in completing tasks called self-efficacy. This research is a qualitative research. This research aims to know the site of the students' difficulty based on revised Bloom's Taxonomy viewed from high self-efficacy. The results of the study stated the students with high self efficacy have difficulties site. They are evaluating conceptual knowledge, evaluating procedural knowledge, creating conceptual knowledge, and creating procedural knowledge. It could be the consideration of teachers in the teaching, so as to reduce the difficulties of learning in students.

1. Introduction

Mathematics has a very important role in developing the learners' ability to think and to be logical. Besides that, math is a tool and servant of sciences, not only for itself but also for the other sciences, and it's good for the importance of theoretical and practical as an application of mathematics. The purpose of mathematics education is actualizing the student learning at the highest level [7]. But most of students get difficulty. When studying mathematics students have varying views; "*students may not have the same conceptions of understanding in mathematics learning when they are studying primary, secondary, or tertiary mathematics*" [8].

The National Council for Excellence in Critical Thinking defines critical thinking as the "intellectually disciplined process of actively and skillfully conceptualizing, applying, analyzing, synthesizing, and/or evaluating information gathered from, or generated by, observation, experience, reflection, reasoning, or communication, as a guide to belief and action. The most famous conceptualization of critical thinking comes from Bloom's Taxonomy [5]. It has been revised for the revised Bloom's taxonomy. The revised bloom's taxonomy has two dimensions, namely the dimension of knowledge and the dimension of cognitive processes. Dimension of cognitive process contains six categories. Those are remembering, understanding, applying, analyzing, evaluating and creating.

Dimensions of knowledge contains four categories. They are factual, conceptual, procedural, and metacognitif [1].

The Common Core State Standards and Partnership for 21st Century Skills emphasize that education should focus on both core academic subject mastery and 21st century skills development [10]. These skills help students to be ready for college and career life. There are various strategies by which to achieve 21st century skills; problem solving, critical thinking, collaborative learning, integration environment, and digital tools in teaching [2][12]. One of the strategies used in 21st century skills is problem solving.

The ability of solving problem is a part of the mathematic curriculum that is very important, because in the process of learning as well as the solution, the students might be possible to get experience using knowledge and skills that already owned to apply in solving problem which is not in routine [8]. Problem solving prefers the process and strategy that is done by students in solving a problem rather than the result [9]. Problem solving that gives priority to the process related to the methods in problem solving (procedural knowledge). Problem solving that gives priority to the strategy related to the basic concepts in completing it (conceptual knowledge).

The implementation of learning mathematics is also inseparable from the role of the students to understand, finish, and solve mathematic problems. Students must have confidence of their ability in completing at ask given to them. The belief in their own ability in completing a certain task is known as Self-efficacy. Self-efficacy refers to personal beliefs about one's capabilities to learn or perform action at designated levels [3][6]. Self-efficacy is also meant as "Belief refer to the specific expectations that we hold about our abilities to accomplish specific task" [3]. Self-efficacy has a positive impact on person's life especially in learning [14]. The opinions related to self-efficacy indicate that self-efficacy has anenough important role to achieve one's goal. This is due to the student's beliefs about the subjects being studied thatguessed to have an effect on the students' achievement towards learning [9]. Self-efficacy produces the difference on how someone thinks, acts, and feels [4][13]. Students with high self-efficacy also have difficulty in finishing the problem solving task. It can be seen in the results of research that has ever been done that students with an average high self-efficacy have difficulty in solving mathematic problem in story task with a percentage more than 50% [9].

One of the materials taught in fulfilling the aspect of algebra is linear equation system with two variables (SPLDV). The material presented in class VIII. Choosing is linear equation system with two variables in this studyis caused by a variety of problem in the form of problem solving so can help the researcher in analyzing the students' ability of mathematics problem solving. This material can be made in various types of non-routine task based on the difficulty level on revised Bloom's Taxonomy so can be known the site of difficulty in finishing the problem solving task in terms of self-efficacy.

2. Research Methods

This research is a qualitative research. This research is a qualitative research. This research aims to know the site of the students' difficulty based on revised Bloom's Taxonomy viewed from high self-efficacy. Categories self efficacy can be seen in table 1.

Table 1. The Result of Self-efficacy category

High self-efficacy	$X > 76.62$	S4, S5, S10.
Moderate self-efficacy	$62.57 < X < 76.62$	-
Low self-efficacy	$X < 62.57$	-

The question form Self efficacy have 22 grain statement. The subject of the research taken by using purposive sampling technique. The subjects were students with high self efficacy category. They are subject A and subject B. Subjects were given a matter of mathematics problem solving. The answers are analyzed based on revised Bloom's Taxonomy. Sites of difficulty based on the revised Bloom's Taxonomy are; remember the concept, remember the procedure, understand the concept, understand the procedure, apply the concept, apply the procedure, analyze the concept, analyze the procedure, evaluate the concept, evaluate the procedure, create the concept, and create the procedure. They can be seen in table 2. To check the validation of analysis results, researcher uses the triangulation methods. That is by conducting interviews of each subject after answering the problem solving sheet.

Table 2. The revised Bloom's Taxonomy limited on 12 stages of difficulty

Cognitive Process Dimension (DPK)	Knowledge Dimensions (DP)	
	Conceptual	Procedural
Remember	Remember Conceptual	Remember Procedural
Understanding	Understanding Conceptual	Understanding Procedural
Applying	Applying Conceptual	Applying Procedural
Analyzing	Analyzing Conceptual	Analyzing Procedural
Evaluating	Evaluating Conceptual	Evaluating Procedural
Creating	Creating Conceptual	Creating Procedural

3. Result and Discussion

The instrument used to determine the selection of the subject based on the student self-efficacy is the questionnaire of self-efficacy (SE). The results of the questionnaire analysis is used to classify the subject SE with the high, moderate, and low category. This research takes 2 subjects with high SE category. The result of the analysis based on the revised Bloom's Taxonomy points out that the subject with a high SE category has the difficulty site in evaluating conceptual knowledge, evaluating procedural knowledge, creating conceptual knowledge, and creating procedural knowledge. For more details, the result of the analysis can be seen below:

3.1. Analysis of Problem Solving

3.1.1. Question No. 1 for Subject A and Subject B who have a high SE,

Given the system of equations:

$$\begin{cases} 5^2 p - 9^{\frac{1}{2}} q = 2.6^2 - 31 \\ 3^2 p + 10^2 q = 2.10^2 + 118 \end{cases}$$

There are values p and q satisfying the equation system. Is this equations system a linear equation system with two variables? Justify your answer.

Figure 1. Question 1

Yes, because p and q satisfy the system of linear equations with two variables

Figure 1.1 The answers of subject A with high SE on question No. 1

In Figure 1.1 it seems that Subject A can write what is meant; rememberingthe concept and remembering the procedural. The subject can mention that SPLDV has 2 variables and has a completion of p and q that fulfill both equations. So it can be concluded that the students do not have difficulty in remembering conceptually and procedurally.

Yes, because there are two variables in linear system are p and q

Figure 1.2 The answer of subject B with high SE on question No. 1

In Figure 1.2 it seems that Subject B only mentions that the equation has two variables p and q. But the mean of answer from subject B was correct, yet Subject B did not answer that p and q is the completion of the second equation.

3.1.2. Question No. 2 for Subject A and Subject B who have a high SE,

Given two triangles are ΔABC and ΔPQR. where $\angle A = 2y$, $\angle B = 3x + 4y$, $\angle C = x + 2y$, $\angle P = 3x - y$, $\angle Q = 2x + 3y$, and $\angle R = x + y$.

a. Based on the information existed on $\Delta ABC$and ΔPQR, please make the similarity related to the information above?

b. Determine x and y the values of satisfying the obtained equations?

Figure 2. Question 2

a. $\angle A + \angle B + \angle C = 180^0$

$3x + 2x + 3y - 20^0 + 2y - 4x - 20^0 = 180^0$

$x + 5y = 180^0$

$\angle P + \angle Q + \angle R = 180^0$

$4x - 20^0 + 5x - 2y + 3x + y = 180^0$

$12x + 2y = 180^0$

b. $x + 5y = 180^0$ x 12 $\quad 12x + 60y = 180^0$

$12x + 2y = 180^0$x 1 $\quad 12x + 2y = 180^0$ -

$58y = 0^0$

$y = 0^0$

$x + 5y = 180^0$

$x = 180^0$

Figure 2.1 The answer of subject A with high SE with on question No. 2

In Figure 2.1 it seems that Subject A understands the information given by the question 2a. So subject A can make the equation of task information. This means subjects do not have difficulty in understanding the problem conceptually. It's just in the calculation of algebra subject A get miscalculations. On a equation in ΔPQR is meant $4x - 20^0 + 5x - 2y + 3x + y = 180^0$, the correct $12x - y = 200^0$, but the subject A writing $12x + 2y = 180^0$. This means that subject A is less careful in working on it. So, it can be concluded that subject Adoes not get difficulty in working on question No. 2a, it's just less careful in doing the calculation of algebra.

In Figure 2.1 it seems that subject A can understand that equation of 2a can be used to find the value of x and y by using the method of elimination. So that the subject does not have difficulty in understanding procedurally.But it's just because the initial equation gets wrong in calculating so that the value of x and y is also getting wrong.

IOP Conf. Series: Journal of Physics: Conf. Series **1008** (2018) 012063 doi:10.1088/1742-6596/1008/1/012063

a. $\angle A + \angle B + \angle C = 180^0$
$3x + 2x + 3y - 20^0 + 2y - 4x - 20^0 = 180^0$
$x + 5y - 40^0 = 180^0$
$x + 5y = 220^0$
$\angle P + \angle Q + \angle R = 180^0$
$4x - 20^0 + 5x - 2y + 3x + y = 180^0$
$12x - y = 200^0$

b. $x \quad + 5y = 220^0$ x 1 $\quad x + 5y = \ 220^0$
$12x - y = 200^0$x 5 $\quad 60x - 5y = 1000^0$ +
$\qquad\qquad 61x \quad = 1220^0$
$x = \dfrac{1220^0}{61} = 20^0$
$x + 5y = 220^0$
$20 + 5y = 220^0$
$5y = 220^0 - 20^0$
$y = \dfrac{200^0}{5} = 40^0$

Figure 2.2 The answer of subject B with high SE on question no. 2

In Figure 2.2subject B can write the equation of 2a correctly so subject B do not get any difficulties in applying it conceptually. Subject B can also apply both equations to look for the values of x and y using elimination method and substitution methods related to the command no. 2b. So subject B does not have any difficulty in applying procedurally.

3.1.3. Question No. 3 for Subject A and Subject B who have a high SE,

Given an angle of the triangle is 20^0. The second angle is 3 times bigger than the third angle. What is the third angle?

Figure 3. Question 3

$\angle A = 20^0$
$\angle B = 3\angle$ 恒
$\angle A + \angle B + \angle C = 180^0$
$20^0 + 3\angle C + \angle C = 180^0$
$3\angle C + \angle C = 180^0 - 20^0$
$3\angle C + \angle C = 160^0$
$5\angle C = 160^0$
$\angle C = 32^0$

Figure 3.1 The answer of subject A with high SE on question No. 3

In Figure 3.1 itseems that subject A can solve the problem. SubjectA can analyze question No. 3 so subject A can find the known elements. So subject A has no difficulty in analyzing conceptually. Subject A can link from what is known to find the the third angle by changing the third angle with a symbol. So subject Adoes not have difficulty in procedural analysis. But when doing algebra calculation Subject A gets wrong that should produce subject A writes down. $20^0 + 3\angle C + \angle C = 180^0$ the correct $4\angle C = 160^0$, but the subject A writing $5\angle C = 160^0$. This means that the subject A are less thorough in working on it.

$$\angle I + \angle II + \angle III = 180^0$$
$$20^0 + 3\angle III + \angle III = 180^0$$
$$20^0 + 4\angle III = 180^0$$
$$4\angle III = 180^0 - 20^0$$
$$4\angle III = 160^0$$
$$\angle III = \frac{160^0}{4} = 40^0$$

Figure 3.2 The answer of subject B with high SE on question No. 3

In Figure 3.2 it seems that Subject B can analyze the question by knowing the known elements by changing the first, second, third anglewith $\angle I, \angle II, and\ \angle III$. It can be seen when the subject B writes$\angle I + \angle II + \angle II$ 犎$= 180^0$ be $20^0 + 3\angle III + \angle III = 180^0$. This means that Subject B knows$\angle I = 20^0$and $\angle II = 3\angle III$. So subject B does not experience any difficulties in analyzing conceptually. Then it can be seen that subject B can use what is known to find the third angleusing substitution method. Subject B can dothe question by analyzingon what elements that can be used to find the 3 angles. This means that Subject B does not have difficulty in analyzing procedurally.

3.1.4. Question No. 4 for subject A and subject B who have a high SE,

Given the sum of two positive integers are 55 and the different of two positive integers are 25. Find the two positive integers? Given the reason why do you use that method?

Figure 4. Question 4

Because $30^0 + 25^0 = 55^0$
$55^0 - 25^0 = 30^0$

Figure 4.1 The answersof subject A with a high SE on question No. 4

In Figure 4.1 subject A does not write the requested answer. Subject A just writes $30^0 + 25^0 = 55^0$and $55^0 - 25^0 = 30^0$. SubjectA shows a degree of the settlement suddenly. It's means the answer of subject A is no direction and unwarranted. From theanswer written by subjectA,subjectA getsdifficulty on question No. 4. So it can be said that subjectAfaces difficulty in evaluating the question conceptually and evaluating question procedurally.

$$A + B = 55$$
$$C + D = 25$$
$$\underline{\hspace{3cm}} -$$
$$30$$
reasons = because it's easier.

Figure 4.2 The answer of subject B with high SE on question No. 4

In Figure 4.2 Subject B can write the mathematic model from the task "the sum of 2 integers of positive is 55" that is written by separating the 2 integers of positive with A and B. In the second statement, both students have difficulty in writing down the mathematic models. In the statement "the difference of 2 integers is 25". Subject B writes it by changing the 2 integers of cacah with a symbol which is different with the first symbol that's C and Dbe $C + D = 25$ the correct$A - B = 25$. Then Subject B only deducts without seeing that there are no variables that can be eliminated in the equation. Subject B writes "the reasons using the method is because easier to understand." Obviously, that's not a logical reason. It's means that subject B gets difficulty in determining the mathematic models and difficulty in deciding what method will be used. Seen from the answer that's not directional, subject B is not able to give the right reason in the use of method which written. So it can be concluded that subject B finds difficulty in evaluating conceptually and procedurally.

3.1.5. Question No. 5 for subject A and subject B who have a high SE,

After a representative of the group determines the bus Agency to rent a bus, now they weigh a package price for the cost of lodging and entrance ticket to the museum. The second package is shown as below. For example, there is a quote from the tour study tour agents offering the package for other costs of hotels and tourist attractions.
For the package A, 3 nights lodging accommodations and attractions 2 tickets are 415 000 USD,-per person.
For the package B. 4 nights lodging accommodations and attractions 4 tickets are 620 000 USD,-per person. Create an problem that can be solved by using systems of linear equations two variables and solve the problems that you have created.

Figure 5. Question 5

$$\begin{array}{l} \sqrt{4} + \sqrt{8} \;\;= 80\,000 \\ \sqrt{3} + \sqrt{10} = 70\,000 \\ \hline \qquad 10\,000 \end{array} \; -$$

Figure 5.1 The answer of subject B with high SE on question No. 5

In Figure 5.1 subject A is asked to identify the questions related to SPLDV in accordance with the problems given. Written on the answer sheets of subject A $(\sqrt{4} + \sqrt{8} = 80000)$, subject A can not write any variables that can be written to be linear equations of two variables. Subject A writestheanswers that is not directedtothe answer sheet No. 5. The subject A has difficulty on question No. 5 in conceptually even procedurally.So it can be procedural knowledge.

Rina *Rp* 80 000
4 Boards
8 Pencils
Nawa *Rp* 70 000
3 Boards
10 Pencils

Figure 5.2 The answer of subject B with high SE on question No. 5

In Figure 5.2 Subject B tries to write what is known without aiming to linear equations of two variables. Subject B only writes 4 boards = ... and 8 pencils= .... That should be made into equation 4 boards + 8 pencils = 80 000. Subject B is allegedly still having problem in making SPLDV problem conceptually. Subject B is not using the method of SPLDV at all and do not solve the question no 5. It can be concluded that subject B is also still having difficulty in creating procedurally. So subject B gets difficulty in creating the conceptual knowledge and creating procedural knowledge.

3.2. Analysis of the Interview

3.2.1. Partial results of the Interviews with subject A:

Researcher	:	Do you have any difficulties on the questions given?
Subject A	:	Yes, there is a question that i could not do.
Researcher	:	What number?
Subject A	:	Question number 5.
Researcher	:	Do you have difficulty in making a matterrelated to question No.5?
Subject A	:	Yes. I can't make a question related to question no. 5. I was confused the way how to make it.
Researcher	:	For the question number 4, can you do it?
Subject A	:	I have answered it but i am still hesitant.
Researcher	:	Have you made the mathematic model for question number 4correctly?
Subject A	:	I am still confused when i should write the reason using the samemethod.
Researcher	:	For question number 2, is your answer correct?
Subject A	:	No, i was wrong in calculating. It should be negative y.
Researcher	:	Do you know the mistake of question number 3?
Subject A	:	Yes, i should write it.

3.2.2. Partial results of the Interviews with subject B:

Researcher	:	From question number 1 to 5, any difficult questions?
Subject B	:	Question number 4 and 5
Researcher	:	Question number 4, in which part do you get difficulty?
Subject B	:	I was confused to find cacah integer.
Researcher	:	In the first statement "the sum of positive integer is 55" it can be amathematic model. Why isn't the difference?
Subject B	:	Yes, it should use minus.
Researcher	:	Which is the easiest method to use from the mathematic model? What is the reason?
Subject B	:	I was confused how to determine the easiest method.
Researcher	:	For question number 5, which part do you get difficult?
Subject B	:	I could not make a question related to SPLDV.
Researcher	:	For question number 1,2,3, do you have difficulty?
Subject B	:	I can do the question number 1,2,3 as well.
Researcher	:	For question number 1. Can you mention the other reason?
Subject B	:	P and q must satisfy both equations.

Based on the results of the analysis and interviews, it can be concluded that based on the revised Bloom's Taxonomy, subject with high SE generally still got difficulties in the process of solving problem. In general subject with high SE had difficulty in evaluating conceptually, evaluating procedurally, creating conceptually, and creating procedurally.

Based on the research that has been conducted for an average high self-efficacy, the difficulties experienced by junior high school students in solving problem of mathematics is the story questionsof understanding the concept 66.5%, applying the concept 72.6%, applying the procedure 72.6%, analyzing the procedure 96.7%, evaluating the concept 70.7%, and evaluating the procedure 98.8% [11].

4. Conclusion

Based on the purpose of the research and data analysis that has been conducted by researchers about the site of the students' difficulty in solving material problem of SPLDV based on the revised Bloom's Taxonomy, then it can be concluded that students with high SE are generally able to remember conceptually, remember procedurally, understand conceptually, understand procedurally, apply conceptually, apply procedurally, analyze conceptually, and analyze procedurally in problem solving questions. But in general, students with high SE are having difficulties at the level of evaluating and creating the problem solving questions both conceptually and procedurally. Factors that may cause difficulty in solving the problem besides cognitive factors and students' knowledge is that the studentsare less thorough in working on the question and they are in a hurry.

This can be a concern for educators, that students who have high SE does not mean they have no difficulty level in the process of finishing the problem solving question. Thus, educators can consider SE category owned by students and expected to be a consideration of teachers in doing the learning process in the classroom, so as to reduce the difficulties of learning in students.

Acknowledgment

The authors would like to thank the referees for their suggestions and thank to Sebelas Maret University, Surakarta, Indonesia for funding this research in the academic year of 2017.

References

[1] Anderson L W and Krathwohl D R 2010 *Kerangka landasan untuk pembelajaran pengajaran, dan asesmen: revisi taksonomi BloomTerjemahan Agung Prihantoro* (New York: Pearson Addison-Wesley)
[2] Alismail H A and McGuire P 2005 21st Century Standards and Curriculum: Current Researchand Practice *Journal of Education and Practice* **6** 150-154
[3] Bandura A 1997 *Self-efficacy: The exercise of control* (New York: Freeman)
[4] Bandura A and Adams N E 1997 Analysis of Self Efficacy Theory of Behavioral Change. *Cognitive Theory and Research* **1** 287-310
[5] Bialik M and Fadel C 2015 *Skills for the 21st Century: What Should Students Learn?* (Montes Alti: Education Foundation)
[6] Schunk D H 2012 *Learning theories* (Boston MA: Pearson Education)
[7] Ciltas A and Tatar E 2011 *Teaching mathematics: developing as a reflective secondary teacher* (London UK: Sage Publication)
[8] Khiat H 2010 A grounded theory approach: conceptions of understanding in engineering mathematics learning. *The Quantum Report* **15** 1459-1488
[9] Novferman N 2016 Analisis Kesulitan dan Self-Efficacy Siswa SMP dalam Pemecahan Masalah Matematika Berbentuk Soal Cerita *Jurnal Edukasi* **3** 76-87
[10] Suherman E 2003 *Strategi pembelajaran matematika kontemporer* (Bandung: JICA-Universitas Pendidikan Indonesia)
[11] Tarigan D E 2012 *Analisis Kemampuan Pemecahan Masalah Matematika Berdasarkan Langkah-*

Langkah Polya pada Materi Sistem Persamaan Linear Dua Variabel Bagi Siswa Kelas VIII SMP Negeri 9 Surakarta Ditinjau Dari Kemampuan Penalaran Siswa (Surakarta: UNS)
[12] Tyack D 1974 *The one best system: A history of American urban education* (Cambridge, MA: Harvard University Press)
[13] Zarch M K and Kadivar P 2006 The Role of Mathematics Self-Efficacy and Mathematics Ability in the Structural Model of Mathematics Performance. *Proceedings of the 9th WSEAS International Conference on Applied Mathematics* **9** 242-249
[14] Zimmermann B J, Sebastian B, and Robert K 1996 *Developing self-regulated learners beyond achievement to self efficacy* (Washington DC: American Psychological Association)

Investigating students' failure in fractional concept construction

Henry Kurniawan[1,2], Akbar Sutawidjaja[2], Abdur Rahman As'ari[2], Makbul Muksar[2], Iwan Setiawan[3]
[1]Politeknik Negeri Lampung, Indonesia
[2]Universitas Negeri Malang, Indonesia
[3]SMA Negeri 2 Maratapura OKU Timur, Sum-Sel, Indonesia

Email: henry_stk@polinela.ac.id

Abstract. Failure is a failure to achieve goals. This failure occurs because a larger scheme integrates the schemes in mind that are related to the problem at hand. These schemes are integrated so that they are interconnected to form new structures. This new scheme structure is used to interpret the problems at hand. This research is a qualitative research done to trace student's failure which happened in fractional concept construction. Subjects in this study as many as 2 students selected from 15 students with the consideration of these students meet the criteria that have been set into two groups that fail in solving the problem. Both groups, namely group 1 is a search group for the failure of students of S1 subject and group 2 is a search group for the failure of students of S2 subject.

1. Introduction

Fractions studied by students in elementary school are part of a rational number that can be written in the form of a/b with a and b being integers and b not equal to zero [3]. Fractions can be given various meanings according to the context in which the concept is used. While the concept is defined as an abstract idea that can be used to classify a set of objects [6]. [12] states in order to understand the meaning of fractions one must "pay attention to the mathematical theory of embedded fractions, applied, and referential mapping between their theory and situation" (p 54). According to [17] the fraction model assigned to elementary students is the area or area model, the measurement model or the length and the set model. Freudenthal [7] sees fractions as a phenomenological source of rational numbers. Fractions are interpreted into five major subconstructs: part of the whole or set, division, ratio, operator, and measurement [2, 4, 11, 16].

Meanwhile, according to Kennedy [8] fractions have 3 meanings that will be used in this study, as follows:

(1) Fractions show a unit partitioned into parts of the same size.
Fractions can be interpreted as part of something intact. Unity broken into smaller parts produces fractions of the same size [2].

(2) Fractions show a set that is partitioned into sub-subsets of equal size.
If a set of objects are grouped into equal parts, then the situation is clearly related to the division.

(3) Fractions show the ratio (ratio).
The relationship between a pair of numbers is often expressed as a comparison.

IOP Conf. Series: Journal of Physics: Conf. Series **1008** (2018) 012064 doi:10.1088/1742-6596/1008/1/012064

According to [13], elementary students are in concrete operational period only able to recognize concrete facts logically, but still can not think abstractly. In the concrete operation phase, this experience must be concrete [14,15]. This means that the object of action to be applied must be concrete, ie it can be perceived by the student's senses [14,15]. This concrete experience gives students constructive meaning of knowledge.

The success of constructing fractional concepts by students is evident from the work they do in the absence of errors in the process and perform the steps in correctly solving the problem on the outcome. However, not all the fractional concept constructs performed by students succeed, there are times when the fractional concept construction is done fail. The failure may be wrong to use the strategy, ignore the error in the process of obtaining the solution, assessing the wrong solution is right, wrong in doing so causing deadlock or wrong in changing the problem so that it is not in accordance with the concept structure.

Failure according to Big Indonesian Dictionary is the failure to reach the goal. So it can be interpreted in the failure of student construction on fractional concept is the student's failure to do a series of thinking process to solve the problem of fractions in writing, because the concept of the student is different from the scientific concept.

Examples of student construction failures in Arnon's research, *et al* [1] which illustrates the interiorization of false action in constructing fractions as a set that is partitioned into sub-subsets of equal size.

In Avi interviews of images created representing –, are as follows:

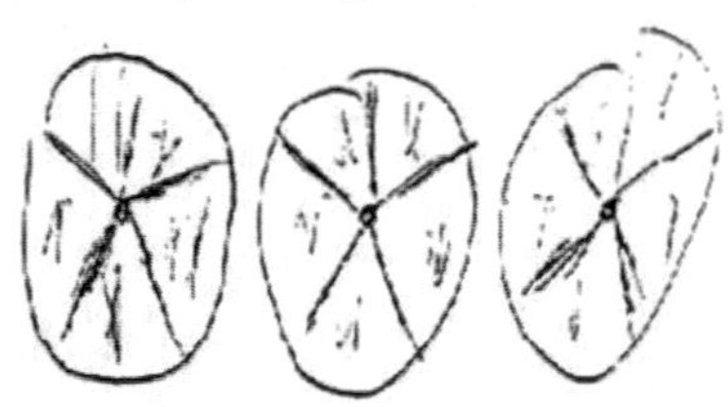

Figure 1. The work of Avi

Researcher (Arnon)	: How many fifths do we see in your picture?
Avi	: Ah, so in every circle there are five, and if we combine all fifteen.
Researcher	: So three fifths or five fifteen?
Avi	: this (pointed pictures), five more, er, three, lim ... .. three more.
Researcher	: Explain again, why are you doing here three circles?
Avi	: because each has, e because it is written in a numerator three.

Avi's action (action) is wrong in two ways: first it misrepresents the numerator by thinking it can be represented by three circles, the second of which divide each of the three circles into five equal parts of 15 parts. None of these steps of action are true.

If we see from the mistakes Avi made above it can be felt Avi uses the wrong strategy in working on the problem, ignoring the error in the process of obtaining the solution, assessing the wrong solution is correct and changing the problem so that it is not in accordance with the concept structure.

Another example of the failure of the students in the first article researcher Kurniawan, *et al* [9] so that the need for limited intervention provided by teachers to students.

The form of failure can be seen from the meaning of fractions by kennedy are:

a. Fractions show a unit partitioned into parts of the same size

determine the fraction as a congruent part of the whole based on blue or pink

student answers

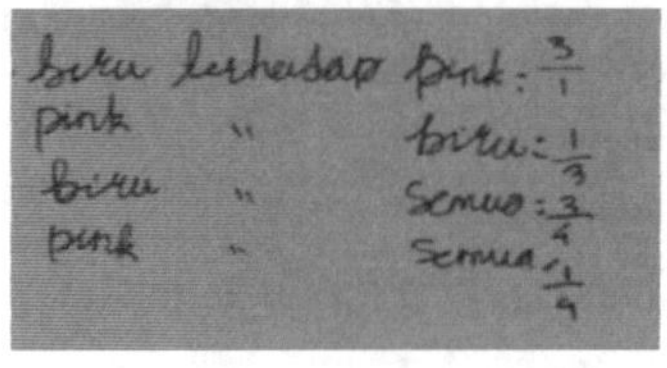

b. Fractions show a set that is partitioned into sub-subsets of the same size.

specify fractions as part of a set that is congruent based on orange, black or blue student answer

c. Fraction show the ratio (ratio).

determine the fraction as the ratio of the whole based on orange color with chocolate student answers

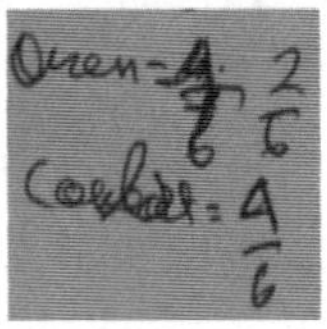

The three answers of 3 students in determining the fractions of the given problem fail in obtaining a solution that is inconsistent with the fractional concept structure, so it needs to be traced more thoroughly, why it fails, what causes failure, and describes the failure process to find the solution of the solution. From this preliminary research, the researcher tried to trace the failure of students in constructing the concept of fractions by using different problem from the research.

2. Research Methods

This research reveals how the process of fractional concept construction by students in problem solving failure. So as to get the picture of the process of failure of fractional concept construction. According to Creswel [5], this kind of research belongs to qualitative research. The subjects in this study are 2 students of SD N Blimbing 3 Malang class V and VI selected from 15 students with consideration of two students meet the criteria that have been set into two groups that fail in solving the problem. The problem used in this study is "Nadia cuts a square ABCD into small squares (like a red square) there are 4 pieces, large squares (like a blue square) there are 2 pieces and there are two rectangles. If the number 1 is expressed by the square area of the ABCD and the red square is the square unit that makes up the square ABCD. While the rectangle is a cut of two square units. Write down the pieces on the red, blue and red and blue combined parts. Serve fragments of red, blue and red and blue combinations of rectangles, circles and equilateral triangles? "
The given problem is done with think alout. Then the students are interviewed according to the flow of the students. Then the two students are grouped into two groups, namely: group 1 (the search for the failure of the students of S1) and group 2 (the search for the failure of the students of S2)

3. Results and Discussion

Search for student failure of subject S1

Subject S1 is a student who has high ability in his class, visible from the test scores, the value of raport and the researchers' communication with classroom teachers and the results of observations and research conducted by researchers at school.
From the results of observations of researchers on the results of work and subject S1 interviews indicate that the subject S1 experiencing confusion that resulted in not able to connect the fractional scheme on the square ABCD and rectangle to the fractional scheme on the circle [10]. Here are the work and interviews taken to show the fractions on the red part.

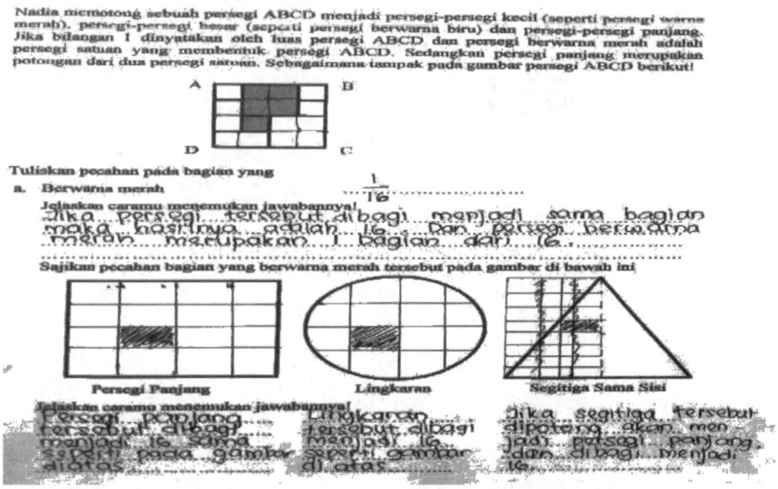

Figure 2. The work of subject S1 work in the presentation of red denominations

Figure 2 provides information on failures made by subject S1. After confirming the fraction on the red part in the square ABCD, subject S1 succeeded in presenting a fraction of the worth of square ABCD to rectangle and equilateral triangle. While the representation of fractions from square ABCD to circle images is not well constructed. The subject S1 knows that the number of pieces in the circle is as much as the one on the square ABCD, rectangles and equilateral triangles, but is unable to give justfication to the cut shape on the circle. (the integrated problem scheme is greater than the initial scheme)

When observing and starting cutting parts of the circle the subject S1 realizes there is something wrong with the pieces he did on the circle image or not the same shape of the cuts. S1 subject tries to think, move the ruler, move the pen and measure the circle. Then he pours it into a circle image of the results that are in his mind but still wrong. This is because the subject S1 can not connect the scheme that belongs to the new scheme on the fraction in the circle. Footage of interview result with subject S1.

R : What do you do for fractions on the red part in square ABCD

S1 : I am trying to equate part of this ABCD square. (pointing to a red square)

R : Then can you explain where the fraction of 1/16 can be found?

S1 : I divide this square ABCD into the same part as the red part.
And the result is 16, the red one is a part of 16 parts.
(while pointing and cutting into square ABCD)

then

R : You serve the fragment (pointing to 1/16) here (pointing to the rectangle) and you shaft one part here. Right?

S1 : yes ......

R : You also do on the circle. Why do you hesitate and are looking for a form for presentation, can you explain!

S1 : I am trying to make this (pointing around the circle), like this (pointing rectangle) or this (pointing square ABCD).

R : Owh so ..... Why are pieces like this that you do?
S1 : Confused me how to share (with a smile)

This dialog illustrates the confusion experienced by the subject S1 when deciding which piece will be done in the circle image, whereas in rectangular and equilateral triangles can be done correctly. This confusion occurs because the subject S1 tries to equate (shape) the shape of the piece on a circle with a square ABCD and a rectangle. Subject S1 realized there are 4 ways to cut the circle into equal parts. How to cut on the circle, namely: pieces in a simple way, pieces using percent, pieces using degrees and pieces with similarity of form (*similarity*). Of the four ways that the subject S1 using pieces with the same shape, because it is considered easier.

Subject S1 tries to continue to solve the next problem. This is evident from the work and interview footage conducted to show fractions on the blue.

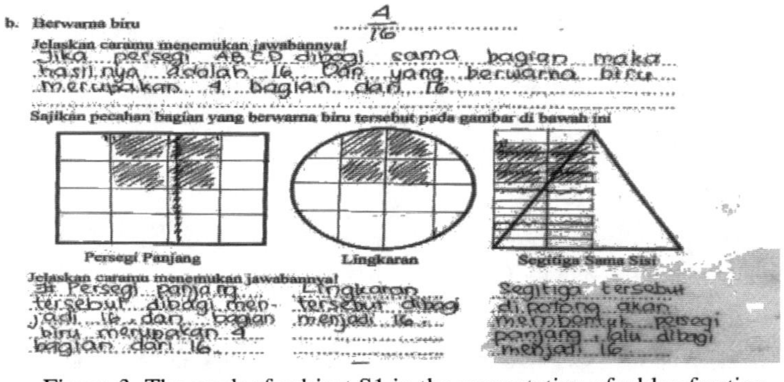

Figure 3. The work of subject S1 in the presentation of a blue fraction

Figure 3 still seen subject S1 failure. Although already trying to find other ways. In the fractional representation of the square ABCD to the same triangular image of the subject S1 performs the cut on the equilateral triangle into two equal parts, then the result of the piece of the right part is moved to the left side so that the shape becomes rectangle. The rectangle is cut into parts of the same shape so that the number is the same as the square ABCD. While in the presentation of fractions from square ABCD to circle images are not constructed again properly [10].

Subject S1 can not take advantage of the schemes that are owned (construction is not intact) on the fractions in the circle. When observing and starting cutting parts of a circle the subject S1 is still aware there is something wrong with the pieces he did on the circle's image or not the same shape of the cuts.

Here's a snippet of the interview.

R : You present the fraction (pointing to 4/16) here (pointing to the rectangle) and you shove one here.
S1 : yes (heeeh)
R : Well, this one (pointing to the circle) also you still hesitate ......?
S1 : Hehehe
R : But still you keep doing like this, why not change?
S1 : Still confused how to share it (hehehe)

Subject S1 still fails in constructing the similarity (similarity) on the shape of a loop cut. This happens because the subject has not maximized the scheme it has.

Then subject S1 continues to resolve the last problem. This is evident from the work and interview footage conducted to show the fractions on the combined parts of red and blue.

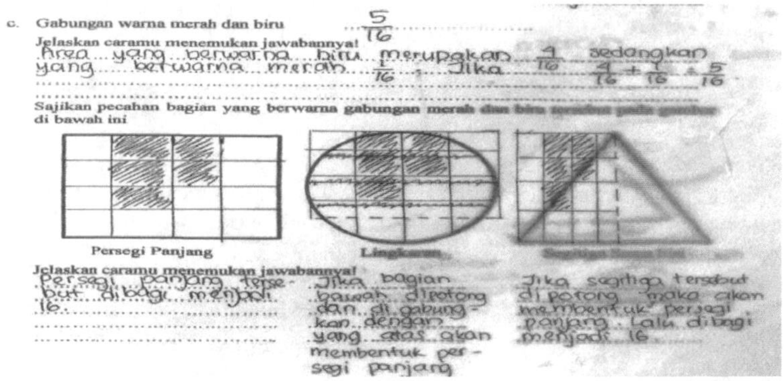

Figure 4. The work of subject S1 in the presentation of red and blue mixed fractions

Figure 4 still shows the subject S1 has failed because the scheme that has been in can not be utilized (construction is not intact). This is seen in the presentation of the combined fractions of the square ABCD to the circle image by the subject S1 by trying to change the shape of the cut on the circle image. The piece is formed into 3/4 of the top and 1/4 of the bottom. The bottom of the circle is placed on the right and left side above 3/4 of the circle.

Here's a snippet of the interview.

R : You present a fraction (pointing to 5/16) here (pointing to the rectangle). Well here you start to change it. Can you explain, why you change.

S1 : I just realized, if this (pointed circle) can be cut and placed here (the bottom piece of the circle)

R : Is this piece (pointing to the bottom piece of the circle) the same as the part you put this on?

S1 : Yes (hehe) .... I confuse this one (pointing to the bottom piece of the circle) not the same as this one (pointing to the combined pieces of the top of the circle)

R : Well just keep on doing it, why?

S1 : Yes, how do I change it like this (his expression still shows confusion)

Behavior of the stages of the subject S1 on the fractions of red, blue and combined red and blue, so the occurrence of failure in constructing the concept of fractions (fractions as equal parts of the whole) can be seen in the following figure.

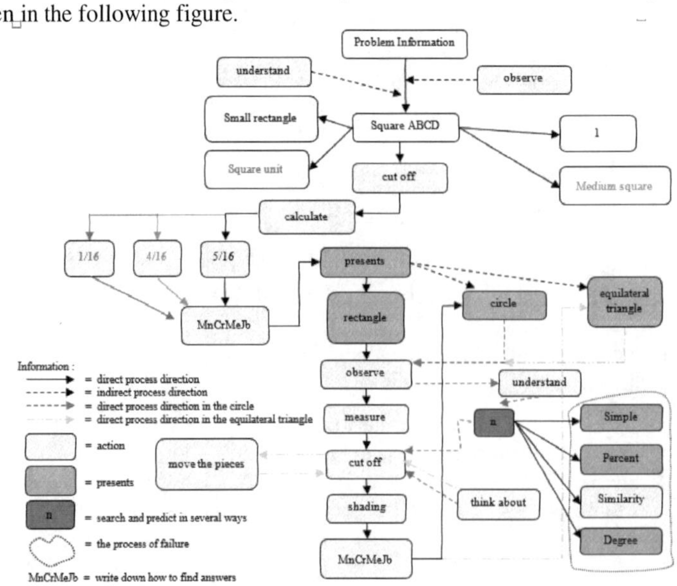

Figure 5. Failure on fractional concept observed on the subject of S1

Figure 5 gives an overview of the stages of the subject S1, beginning from the stage of understanding, observing, measuring, cutting, moving pieces, shading to belief (by writing how to find answers) of the given problem.

At the stage of the process of determining how to cut the circle, subject S1 knows there are 4 ways (simple, percent, *similarity* and degree). The subject then uses the similarity of the square ABCD as the ultimate image which is most easily done by cutting resulting in another image to be the same shape as the square ABCD. So the subject S1 is not able to prune the cut shape on the circle that has the same fraction as the fraction in square ABCD.

Search for student failure of subject S2

The subject of S2 is a student who has a moderate ability in his class and netted based on observations and communication of researchers with classroom teachers. Based on the observation of the researcher at the time of the subject S2 of study failed in presenting the concept of fracture (fraction as equal parts of whole) and the concept of fractions worth of two different images.

Here are the work and interviews taken to show the fractions on the red part.

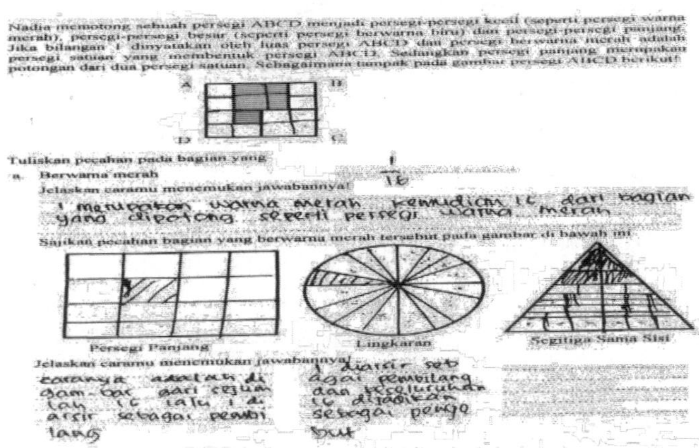

Figure 6. The work of the subject of S2 in the presentation of red denominations

The failure performed by the subject S2 of Figure 5 is seen in the unsuccessful S2 in presenting the equivalent fractions in equilateral triangle images. The representation of a fraction of the value of the square ABCD to an equilateral triangle image is done by cutting the sides of the equilateral triangle into triangular, square and rectangular shapes. This happens because the subject of S2 has an existing schema integrated greater than the problem scheme that is being encountered [10]. So the construction that happened failed, while the presentation of a fraction of the value of square ABCD to rectangle and circle successfully constructed well. The subject of S2 can not connect the scheme owned by the new scheme to fractions in equilateral triangles. When observing and initiating the cutting of equilateral triangle parts the subject S2 realizes that the cuts are not of the same shape. Here's a snippet of interviews with the subject of S2.

R : What do you do for fractions on the red part in square ABCD

S2 : I cut this part of square ABCD into the same part as the red part. (pointing to a red square)

R : can you explain where fractions of 1/16 can be found?

S2 : I divide this square ABCD into the same part as the red part. the result is 16, its red 1. So 1/16 (while pointing and cutting into square ABCD)

then

R : You present the shards (pointing to 1/16) here (pointing to rectangles and circles) and you shove one here. Right?

S2 : right sir.

R : how to draw this equilateral triangle?

S2 : I tried to cut it out so like this (pointed around the first equilateral triangle) not the same shape, then I draw again like this (pointing to the equilateral equilateral triangle), keep me shading one part. (thinking)

R : Owh so ..... Are you sure of this piece?

S2 : Confused (hehe), not sure sir.

This dialogue illustrates the confusion and uncertainty of the subject of S2 in determining the shape of the cut made on an equilateral triangle image. This confusion occurs because the subject S2 can not cut out the same cuts on an equilateral triangle as it does on the square ABCD, rectangles and circles. The subject of S2 goes on to the next issue. This is evident from the work and interview footage conducted to show fractions on the blue.

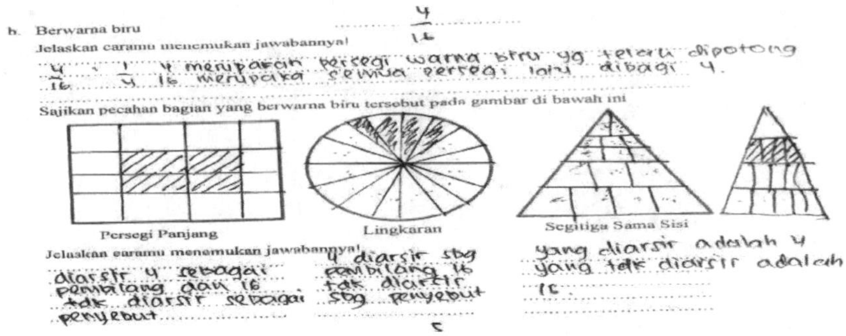

Figure 7. The work of the subject S2 in the presentation of a blue fraction

Figure 7 still shows the subject S2 to failure. Although it has tried to bypass the other way on an equilateral triangle image, either an equilateral triangle image is provided as well as illustrated by the subject S2. It is seen that the subject S2 did not succeed in cutting the same triangular image with the same shape. The subject of S2 can not take advantage of the schemes they have (old schemes not yet activated) on fractions in equilateral triangles.

Here's a snippet of the interview.

R : You present a denomination (pointing to 4/16) here (pointing to the rectangle and circle) and you shove one here.

S2 : Yes, Sir.

R : how is this (pointing to an equilateral triangle) are you still confused?

S2 : right sir. I cut it here (pointing to the first equilateral triangle image) not sure yet, keep me drawing again its triangle. Then I cut also still not sure ..... hehe

R : But still you keep doing like this, why not change?

S2 : Still confused how to share it sir (hehehe)

The subject of S2 still fails to construct similarity on the equilateral triangle pieces. This happens because the subject has not maximized the scheme it has.

S2 subject tries to continue to solve the next problem. This is evident from the work and interview footage conducted to show the fractions on the combined parts of red and blue.

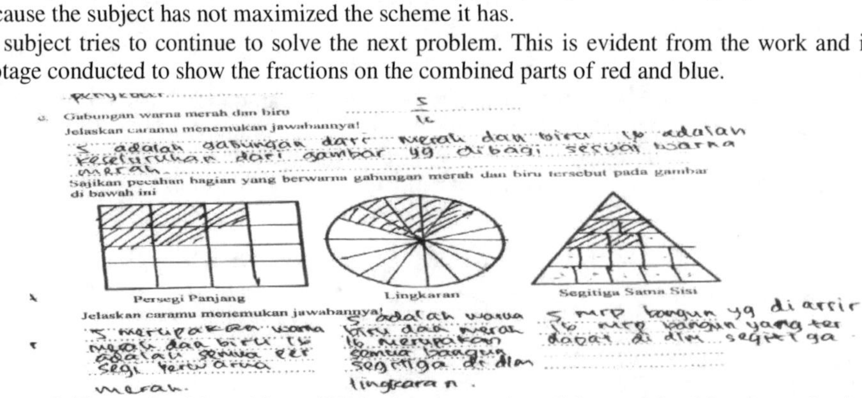

Figure 8. The work of the subject of S2 in the presentation of the combined fractions of red and blue

Figure 8 still visible subject of S2 failure. The subject of S2 does not have a scheme to solve the problem because it has not been in the activation of schemes owned on fractions in equilateral triangles.
Here's a snippet of the interview.

R : You present a fraction (pointing to 5/16) here (pointing to the rectangle and circle). Well here you start to change it (pointing to an equilateral triangle). Can you explain, why you change.
S2 : I believe it looks like this (pointing an equilateral triangle)
R : Does this snippet (pointing to the same piece of triangle) is appropriate?
S2 : if the number is the same Mr. .... but the shape is not the same
R : So what?
S2 : Yes, just like this sir (his expression still shows confusion)

The behavior of the stages performed by the subject S2 on the combined fractions of red and blue so that the failure to construct the concept of fractions (fractions as equal parts of the whole) can be seen in the following figure.

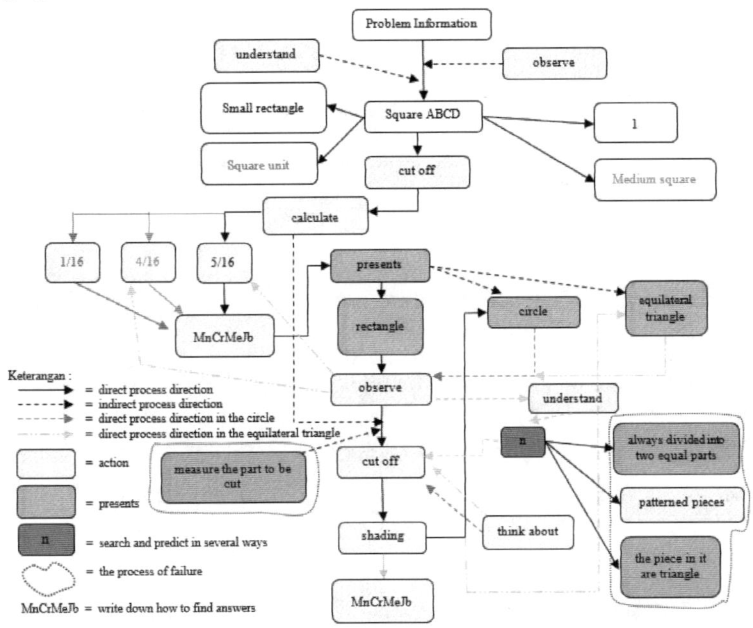

Figure 9. Failure on fractional concepts observed on the subject of S2

Figure 9 gives an overview of the stages of the subject of S2, beginning with the stage of understanding, observing, cutting, thinking, counting, shading to belief (by writing how to find answers) of the given problem.
At the stage of the process of determining how to cut an equilateral triangle, the subject of S2 knows there are 3 ways (divided into two equal parts, patterned pieces, and pieces inside there are triangles). Then the subject uses similarity in the previous image (eg in the square ABCD and the equilateral triangle (on the issue)) as part of the resulting image on the cutting of the equilateral triangle which is cut into two equal parts and the piece inside is the triangle.
The subject of S2 fails in constructing the similarity of the equilateral triangle pieces. This happens because the Subject S2 can not connect the schemes that belong to the new scheme on fractions in equilateral triangles and the subject S2 has not yet maximized the scheme it has.
So the subject of S2 is incapable of imaging the cut shape on the equilateral triangle having the same fraction as the fraction in the square ABCD.

4. Conclusion

Search of student failure in fractional concept construction divided into two groups, that is group 1 is group of student failure search of subject S1 and group 2 is group of student failure search subject S2.

Group Search 1 is marked with subject of S1
a) integrated problem scheme greater than the initial scheme (owned scheme)
b) can not connect the scheme that belongs to the new scheme in the fractional scheme on the square ABCD to the fractional scheme on the circle.
c) realize there are 4 ways to cut the circle into equal parts, namely: cut in a simple way, cut with percent, cut with degree and cut with similarity. Of the four ways that the subject S1 using pieces with the same shape, because it is considered easier.
d) the schemes that have been owned can not be utilized (construction is not intact)

Group Search 2 is marked with subject of S2
a) the initial schema integrated is greater than the problem scheme
b) Unable to maximize the scheme, so as not to be able to imitate the cut shape on the equilateral triangle having the same fraction as the fraction in square ABCD.
c) know there are 3 ways to cut the triangle equally, ie a. divided into two equal parts, b. patterned pieces, and c. pieces inside there is a triangle.
d) has no scheme to solve the problem because it has not been in activation of schemes owned on fractions in equilateral triangles.

Acknowledgments

We are thankful to the Head of SDN Blimbing 3 Malang who permit us to do this research, Mrs. Sumiatun as a teacher at SDN Blimbing 3 for her cooperation and direction, Head & Friends in EKBIS Polinela who support us in work, and Head & Supervisor of our dissertation at Postgraduate of State University of Malang who have entrusted us in doing research and make this article. As well as families who support us in prayers so this article can be quickly resolved. Our special gratitude to Kemenristekdikti who have funded this research through the Doctoral Dissertation Grant in 2017.

References

[1] Arnon I, Nesher P, & Nirenburg R 2001 Where do fractions encounter their equivalents? Can this encounter take place in elementary school? *International Journal of Computers for Mathematical Learning* **6** 167–214
[2] Behr M J, Harel G, Post T & Lesh R 1992 Rational Number, Ratio, and Proportion. In D. A. Grouws (Ed.), *Handbook of Research on Mathematics Teaching and Learning* pp. 296-333 (New York, Macmillan)
[3] Bennett Jr A B Burton, L J & Nelson L T 2012 *Mathematics for elementary teachers a conceptual approach* (New York, McGraw-Hill)
[4] Chapin S H & Johnson A 2006 *Math Matters: Understanding the Math You Teach, Grades K-8 (2nd edition)* Sausalito, CA : Math Solutions Publications
[5] Creswel J W 2012 *Educational Research : planning, conducting, and evaluating quantitative and qualitative research Fourth Edition* (Boston, USA: Pearson)
[6] Depdiknas 2003 *Pedoman Khusus Pengembangan Sistem Penilaian Berbasis Kompetensi SMP* Jakarta, Depdiknas.
[7] Freudenthal H 1983 *Didactical Phenomenology of Mathematical Structures* Dordrecht D Reidel.

[8] Kennedy L 1994 *Guiding children's learning of mathematics* Belmont CA Wadsworth, Kennedy LM Tipps, S 1994 *Guiding children's learning of mathematics,* 7th ed. Belmont CA, Wadsworth.

[9] Kurniawan H, Nusantara T, Subanji, Susiswo, Setiawan I, Sutawidjaja A, As'ari AR and Muksar M 2016 *Int. Educ. Stud.* **9** 145

[10] Kurniawan H, Sutawidjaja A, As'ari A R, and Muksar M 2018 *J. Phys.: Conf. Ser.* **983** 012173

[11] Lamon S J 1999 *Teaching Fractions and Ratios for Understanding: Essential Content Knowledge and Instructional Strategies for Teachers* Mahwah, NJ: Lawrence Erlbaum Associates, Inc., Publishers.

[12] Ohlsson S 1988 *Mathematical meaning and applicational meaning in the semantics of fractions and related concepts.* In J. Hiebert & M. Behr (Eds.), Number concepts and operations in the middle grades pp. 53 – 92 Reston, VA: Lawrence Erlbaum Associates.

[13] Piaget J 1965 *The child's conception of number* (C. Gattegno & F. M. Hodgson, Trans.). New York: W. W. Norton. (Original work published 1941).

[14] Piaget J 1975 Piaget's theory (G. Cellerier & J. Langer, Trans.). in P. B. Neubauer (Ed.), *The process of child development* (pp. 164-212). New York:Jason Aronson.

[15] Piaget J 1976 *The grasp of consciousness* (S. Wedgwood, Trans.) Cambridge, MA: Harvard University Press. (Original work published 1974).

[16] Van De Walle A.J, Karp S K and Bay-Williams M J 2010 *Elementary and Middle School Mathematics Teaching Developmentally* 7th Pearson Education Inc United States of America

[17] Van de Walle J 2003 *Elementary and middle school mathematics: Teaching developmentally 5th edition* Boston MA: Allyn & Bacon p.123

IOP Conf. Series: Journal of Physics: Conf. Series **1008** (2018) 012065 doi:10.1088/1742-6596/1008/1/012065

Analysis of students' creative thinking level in problem solving based on national council of teachers of mathematics

Hobri[1], Suharto[1] and Ahmad Rifqi Naja[1]
[1]Department of Mathematics Education, University of Jember, Indonesia

Email: hobri.fkip@unej.ac.id

Abstract. This research aims to determine students' creative thinking level in problem solving based on NCTM in function subject. The research type is descriptive with qualitative approach. Data collection methods which were used are test and interview. Creative thinking level in problem solving based on NCTM indicators consists of (1) Make mathematical model from a contextual problem and solve the problem, (2) Solve problem using various possible alternatives, (3) Find new alternative(s) to solve the problem, (4) Determine the most efficient and effective alternative for that problem, (5) Review and correct mistake(s) on the process of problem solving. Result of the research showed that 10 students categorized in very satisfying level, 23 students categorized in satisfying level and 1 students categorized in less satisfying level. Students in very satisfying level meet all indicators, students in satisfying level meet first, second, fourth, and fifth indicator, while students in less satisfying level only meet first and fifth indicator.

1. Introduction

Mathematics is one of the most important subjects that is so instrumental in advancing the civilization of a nation, mathematics can be interpreted as a study of patterns and relations, a path or thinking pattern, an art, a language, and a tool, therefore mathematics is not a solitary knowledge, but its existence to assist human beings in understanding and mastering various other problems by Jihad [3]. Problem solving is a process that students take to solve a problem so that the problem will not be a problem for him anymore by Hudojo [2]. Problem solving in NCTM naturally develops students' ability in solving problems through their exploration of mathematical knowledge by Xie [10]. Therefore, problem solving has a very important role and is the core to mathematics learning by Ninik *et. al.* [7] and also there is some results about creative thinking skills in [11, 12, 13].

The expected problem solving is problem solving which is involving and exercising creativity. The statement is supported by one of the ideals of mathematics learning formulated by Ministry of Education and Culture's regulation number 59 of year 2014 which reads "students are able to have creative attitude and behavior in solving problems" by Ministry of Education and Culture [4]. So, it can be concluded that creativity becomes an important discussion because it has a share in achieving education's goals by Febriyanti et. al. [1]. However, creativity in problem solving is rarely noticed in mathematics learning by Prianggono [8]. Sometimes teachers emphasize more to memorize formulas in order to solve problems such as math. Although sometimes this method is considered to be more instantaneous but it actually inhibits the development of students' reasoning and creativity. Whereas it has been emphasized that one of the objectives of learning mathematics by the government is that students are able to have creative attitude and behavior in solving problems. The goal of mathematics learning, especially problem solving, is to build new mathematical knowledge through problem solving, solving problems that arise in mathematics and in other contexts, able to apply and adapt various appropriate strategies to solve a problem and to monitor and reflect on the process of solving mathematical problems by NCTM [6]. Therefore, we should start to abandon those bad learning habits and turn to the learning ways that can develop and enhance the creativity of students.

Senior High School is one level of education which is equivalent to SMK or MA in Indonesia. Learning in senior high schools equip their students with science knowledge that will benefit them to study in college. According to National Ministry of Education's regulation, several mathematical subjects for high school skills include conceptual comprehension, problem solving, reasoning, mathematical communication, and mathematical utility by National Education Standards Agency [5]. This capability is aligned with the five standard mathematical learning processes formulated by NCTM namely: problem solving, reasoning, communicating, making connections, and presenting. To be able to realize the learning process that uses this standard, qualified teachers are required to be able to improve students' skills in these five aspects.

In this research, problem which was used is open middle contextual type problem. The school chosen as the research location is SMA Negeri 1 Situbondo, this school is one of the best senior high schools in Situbondo. Subjects in this research is the class XI MIPA 1 SMA Negeri 1 Situbondo. The purpose of this research is to determine the level of creative thinking ability of students in solving mathematical problems based on NCTM (National Councils of Teachers of Mathematics)

2. Research Methods

The type of research used is descriptive research with qualitative approach. The subjects of the study were the students of class XI MIPA 1 SMA Negeri 1 Situbondo, which consists of 34 students. To achieve the objectives of this research, research procedure is conducted. Research procedure consists of some steps that need to be done sequentially. The first step is making research instruments which are test and interview guidelines. The test instrument contains a mathematical functions subject problem based on NCTM that serves to measure the creative thinking level to solve mathematical problems of students. Problem which was used is an open middle contextual type problem which was arranged for students to find or use a variety of strategies to solve the given problem. The next step is validity testing. Validation was performed on both instruments by two lecturers of FKIP Mathematics University of Jember and one mathematics teacher of SMAN 1 Situbondo. If it meets the criteria specified then the instrument can be said to be valid and can be used. However, if it is not yet valid, it will be revised and re-validated until the instrument is valid. The third step is determining location and subject of research and data collecting. In this step location and subject of research is determined and then datas which were needed for the research were collected. The data were collected at SMAN 1 Situbondo with the subjects of research were 34 students of class XI MIPA 1 SMAN 1 Situbondo. The datas ware collected by conducting test to 34 students of class XI MIPA 1 with an open middle contextual mathematics function subject problem which has been compiled based on NCTM. The students answer then analysed in order to determine their creative thinking level in problem solving and then some students are interviewed as sample to strengthen and enrich the analysis result. As we know, the earlier commonly used methods to identify and measure this are CAMT, TTCT and Silver's statement about the most suitable approach to identfy students creative thinking. The latest research about creative thinking levelling in problem solving in Indonesia used three components of creative thinking (fluency, flexibility, and novelty) as theoritical basis and develop it as indicators to measure students' creative thinking level in solving and posing mathematical problem which is devided into 5 level by Siswono *et. al.* [9].

In this research, in order to analyse and determine the level of students' creative thinking in problem solving, researcher combine theories and indicators by earlier researcher with NCTM's mathematics problem solving's goals. This action results indicators of problem solving's goals based on NCTM and creative thinking level in problem solving based on NCTM which are presented in Table 1 and Table 2.

Table 1: Indicators of problem solving's goals based on NCTM's achievements

No	Goals of Problem Solving Based on NCTM	Indicator(s)
1.	Solve problems that arise in mathematics and in other contexts.	Create mathematical model of a contextual problem and solve it correctly.
2	i. Apply and adapt various appropriate strategies to solve the problem and ii. Build new mathematical knowledge through problem solving	a) Solve problem using various possible alternatives b) Find new alternative(s) to solve the problem c) Determine the most efficient and effective alternative for that problem.
3	Monitor and reflect on the process of solving mathematical problems	Review and correct mistake(s) on the process of problem solving.

In the Table 1, the meaning of new alternative solution(s) in this research is another alternative that teachers have never taught to students or not present in their mathematics students' book.

Table 2: Indicators of creative thinking level in problem solving based on NCTM

Creative Thinking Level in Problem Solving Based On NCTM	Indicator(s)
Very satisfying	Students meet all indicators of achievement of problem-solving goals according to NCTM.
Creative Thinking Level in Problem Solving Based On NCTM	**Indicator(s)**
Satisfying	Students meet all indicators of: 1) Solve problems that arise in mathematics and in other contexts 2) Monitor and reflect on the process of solving mathematical problems But only meet some indicator(s) of: i. Apply and adapt various appropriate strategies to solve the problem and ii. Build new mathematical knowledge through problem solving.
Less satisfying	Students meet all indicators of: 1) Solve problems that arise in mathematics and in other contexts and monitor and reflect on the process of solving mathematical problems 2) Monitor and reflect on the process of solving mathematical problems.
Not satisfying	Students do not meet any indicators of achievement of problem-solving goals according to NCTM.

The final step is drawing conclusions. In this step researcher draw conclusions from data analysis' result which has been done before

3. The Results

The analysis result of students' answer using indicators of problem solving's goals based on NCTM's achievements showed that the percentage of students who meet the first indicator is 100 percent. All of the students are able to create mathematical model of a contextual problem and solve it correctly. The following is an example figure of the student's work that relates to the first indicator.

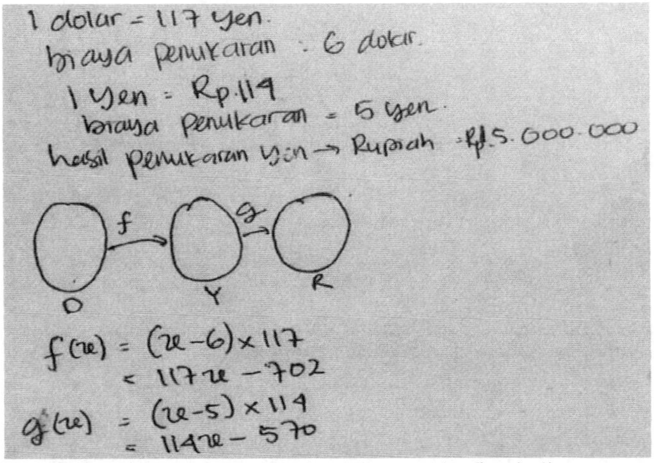

Figure 1: Student's work that relates to the first indicator

From this Figure 1, it is known that this student wrote down all of the known things in the given problem and what is asked there and made mathematical model of it in the form of mathematical functions. Some students also made composite function's figure of the given problem.

The second indicator, solve problem using various possible alternatives, can only be fulfilled by 97,1% of the class which means only one student who couldn't use more than one way to solve the given problem. The following figures are some alternatives which were used by the other 33 students to solve the given problem.

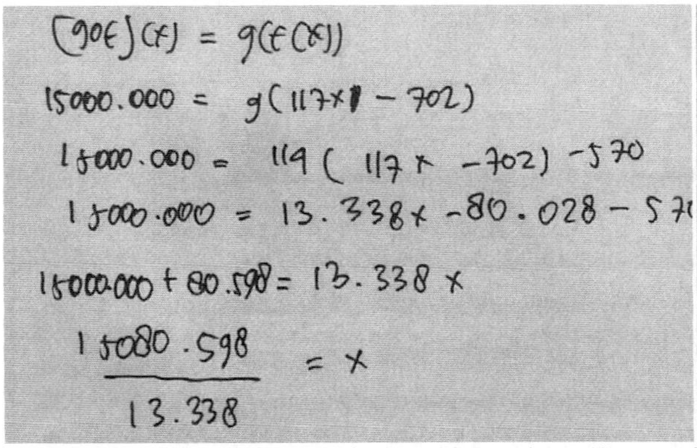

Figure 2: Alternative solution used by students (1)

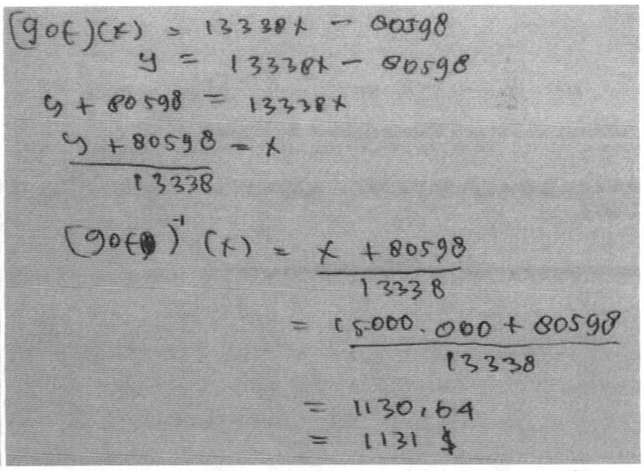

Figure 3: Alternative solution used by students (2)

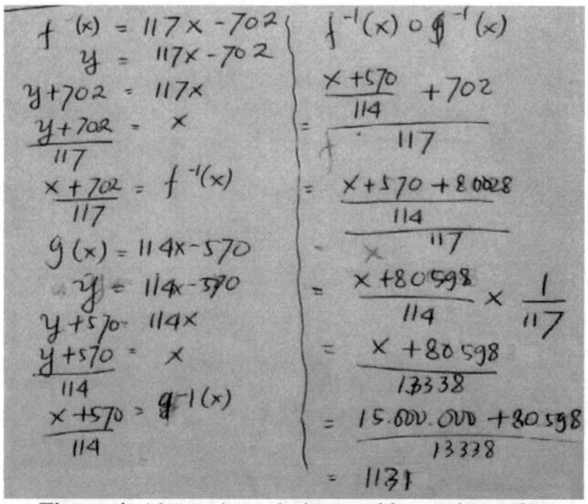

Figure 4: Alternative solution used by students (3)

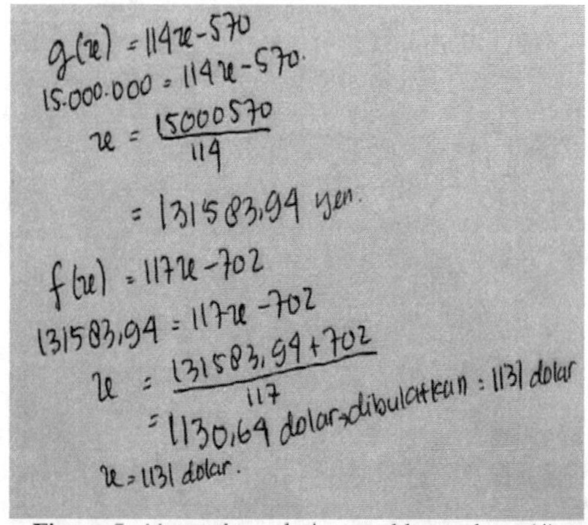

Figure 5: Alternative solution used by students (4)

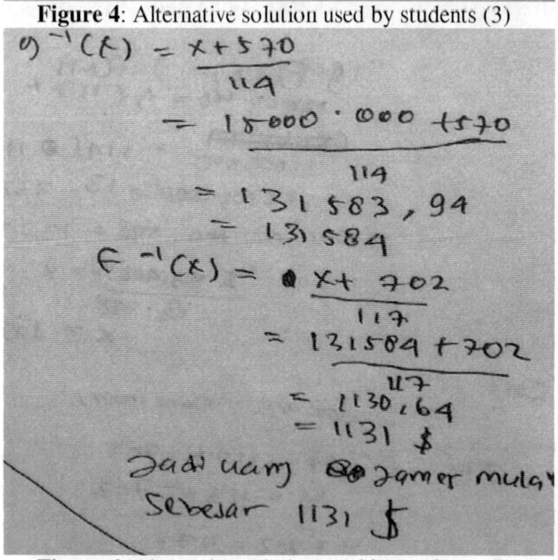

Figure 6: Alternative solution used by students (5)

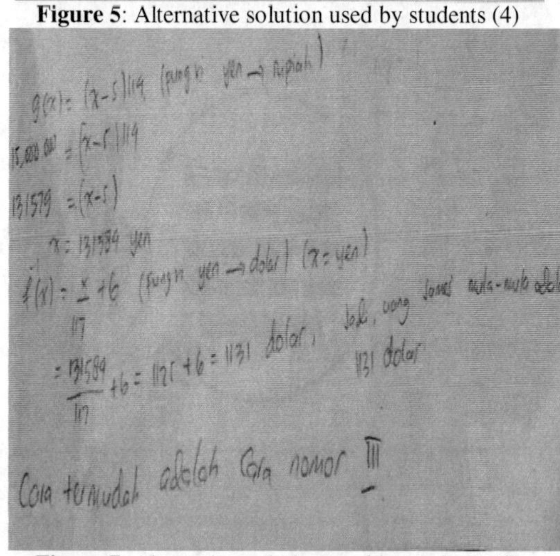

Figure 7: Alternative solution used by students (6)

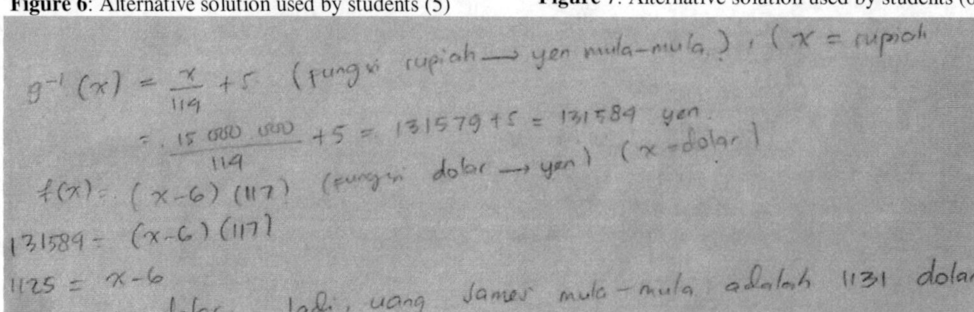

Figure 8: Alternative solution used by students (7)

The number of ways students used to solve the given problem in the test is 7. In figure 2 student substituted value of the final money which is one of known things in problem to $g \circ f(x)$. In figure 3 student tried to find the inverse of $g \circ f(x)$ by defining it as a function y then arranging the terms there so it becomes form of the equation of $x = y + \frac{80.598}{13.338}$ with x as the invers of $g \circ f(x)$ and y as the value of final money in Rupiah. In figure 4 student used the inverse properties of the composite function. In figure 5 student manually substituted the value of final money in Rupiah to $g(x)$ to find the value of its x then substituted that x value to $f(x)$ to find the value of initial money in Dollar. In figure 6 student manually substituted the value of final money in Rupiah to $g^{-1}(x)$ and the substituted its value to $f^{-1}(x)$ to find the value of initial money in Dollar. In figure 7 student manually substituted the value of final money in Rupiah to $g(x)$ to find the value of its x and substituted that x value to $f^{-1}(x)$ to find the value of initial money in Dollar In figure 8 student manually substituted the value of final money in Rupiah to $g^{-1}(x)$ to find the value of $g^{-1}(x)$ and then substituted its value to $f(x)$ to find the value of initial money in Dollar.

The third indicator, find new alternative(s) to solve the problem, can only be fulfilled by 29,4% of the class. This indicator sure is the hardest for this class' students to fulfill. Finding a new way other than ways that found in students' book or teachers taught is not an easy task because it requires a deep understanding of the theory which is used to solve the given mathematical problem and other knowledge that relates to it. For this class, ways that categorized as old way are: substituting value of the final money which is one of known things in problem to $g \circ f(x)$, finding the inverse of $g \circ f(x)$ by defining it as a function y then arranging the terms there so it becomes form of the equation of $x = y + \frac{80.598}{13.338}$ with x as the invers of $g \circ f(x)$ and y as the value of final money in Rupiah, and using the inverse properties of the composite function. While the other four such as shown in Figure 5, Figure 6, Figure 7, and Figure 8 are categorized as the new way.

The forth indicator, determine the most efficient and effective alternative for that problem, is fullfilled by 100% of the class. Most of them selected substituting value of the final money which is one of known things in problem to $g \circ f(x)$ as the most efficent and effective alternative for the given problem. The fifth indicator, review and correct mistake(s) on the process of problem solving, is fullfilled by 100% of the class. Making mistakes or errors while solving mathematical problems is common. Almost all of the students in this class made mistakes in the process of solving the problem given. The students' effort to fix can be seen from the stationaries' scribbles like pen and correction pen marks. They also admitted to having made mistakes while doing given math problem and have fixed them.

From the result of determining students' creative thinking level in problem solving of mathematics function problem which done at class XI MIPA 1 with 34 subject students hence got 10 student which categorized in very satisfying level, 23 students categorized in satisfying level and 1 person categorized in less satisfying level. If it is observed from the achievement of the indicator, it is found that 34 students meet the first indikator, 33 students meet the second indicator, 10 students meet third indicator, 34 students meet forth indicator, and 34 students meet fifth indicator. The following table will describe the summary of XI MIPA 1 students' achievements of problem solving's goals based on NCTM achievement indicators and the result of determining creative thinking level in problem solving based on NCTM.

Table 3: Creative Thinking Level Determination in Problem Solving Result of XI MIPA Students

No	Name	Indicator					Info	Level
		1st	2nd	3rd	4th	5th		
1	A A	√	√	×	√	√	2W,2O,0N	Satisfying
2	A A F	√	√	×	√	√	3W,3O,0N	Satisfying
3	A Am	√	√	×	√	√	3W,3O,0N	Satisfying
4	A R F	√	√	√	√	√	4W,3O,1N	V. Satisfying
5	A A M	√	√	×	√	√	3W,3O,0N	Satisfying
6	A H	√	√	×	√	√	3W,3O,0N	Satisfying
7	A R P	√	√	√	√	√	5W,3O,2N	V. Satisfying
8	A E A	√	√	√	√	√	4W,2O,2N	V. Satisfying

9	A S I	√	√	×	√	√	3W,3O,0N	Satisfying
10	A N	√	√	×	√	√	3W,3O,0N	Satisfying
11	A W	√	×	×	√	√	1W,1O,0N	L. Satisfying
12	D B H	√	√	×	√	√	2W,2O,0N	Satisfying
13	D C M S	√	√	×	√	√	3W,3O,0N	Satisfying
14	D S P	√	√	×	√	√	3W,3O,0N	Satisfying
15	D P	√	√	√	√	√	5W,3O,2N	V. Satisfying
16	F S U	√	√	×	√	√	2W,2O,0N	Satisfying
17	F T M	√	√	√	√	√	5W,3O,2N	V. Satisfying
18	H R	√	√	×	√	√	3W,2O,1N	V. Satisfying
19	I A	√	√	×	√	√	2W,2O,0N	Satisfying
20	K R P	√	√	×	√	√	3W,3O,0N	Satisfying
21	M N A	√	√	×	√	√	3W,3O,0N	Satisfying
22	M. K K D	√	√	×	√	√	2W,2O,0N	Satisfying
23	M N R	√	√	√	√	√	6W,2O,4N	V. Satisfying
24	M R P	√	√	×	√	√	2W,2O,0N	Satisfying
25	N S T W	√	√	×	√	√	3W,3O,0N	Satisfying
26	N N D N	√	√	√	√	√	5W,2O,3N	V. Satisfying
27	R A P	√	√	×	√	√	2W,2O,0N	Satisfying
28	S N	√	√	×	√	√	3W,3O,0N	Satisfying
29	S M. R D	√	√	×	√	√	2W,2O,0N	Satisfying
30	Y A A	√	√	√	√	√	6W,2O,4N	V. Satisfying
31	Y K P	√	√	×	√	√	3W,3O,0N	Satisfying
32	Z N	√	√	×	√	√	2W,2O,0N	Satisfying
33	Z R	√	√	√	√	√	4W,3O,1N	V. Satisfying
34	F I B	√	√	×	√	√	2W,2O,0N	Satisfying

Information:
Info : Additional info
W : Way(s)/ alternative solution(s)
O : Old way(s)/old alternative solution(s)
N : New way(s)/new alternative solution(s)

From the analysis result can be summarized information as follows:
1) Percentage of class XI MIPA 1 students who categorized in very satisfying level is 29.4%
2) Percentage of class XI MIPA 1 students who categorized in satisfying level is 67.6%
3) Presentage of class XI MIPA 1 students who categorized in less satisfying level is 2.9%
4) Percentage of class XI MIPA 1 students who meet the first indicator 100%
5) Percentage of class XI MIPA 1 students who meet the second indicator 97.1%
6) Percentage of class XI MIPA 1 students who meet the third indicator 29.4%
7) Percentage of class XI MIPA 1 students who meet the forth indicator 100%
8) Percentage of class XI MIPA 1 students who meet the fifth indicators 100%

4. Conclusion

Based on the result of the analysis and discussion, the determination of students' creative thinking level in problem solving based on NCTM in function subject can be concluded. Indicators of problem solving's goals based on NCTM achievements which are most fulfilled by students of class XI MIPA 1 SMAN 1 Situbondo are first, forth and fifth indicators that is 100% of students of this class are able to meet these indicators. While indicators of problem solving's goals based on NCTM achievements which are least fulfilled by the students of class XI MIPA 1 SMAN 1 Situbondo is third indicator that is only 29.4% of students of this class are able to meet this indicator. In first indicator, all students are able to fulfill it. In second indicator, there are five students who are able to fulfill it. In third indicator, all students are able to fulfill it. In forth indicator, all students are able to fulfill it. And in fifth indikator all studets are able to fulfill it. Students are not able to meet second indicator because they are unable to use more than one alternative solution to solve the given problem. While third indicator can not be fulfilled because they are unable to find new problem solving solution that can be used to solve the given function subject problem. The least amounted of alternative solutions used by students is 1 alternative solution, whereas the most amounted of alternatives solutions used by students are 6 alternative solutions. The number of most amounted new problem solving alternatives found by students to solve the given function subject problem are 4 new alternatives.

The number of students of class XI MIPA 1 who categorized in very satisfying level is as many as 10 students or 29.4% of the total students in the class, the number of students of class XI MIPA 1 who categorized in satisfying level is as many as 23 students or 67.6% of the total students in the class, the number of students of class XI MIPA 1 who categorized in less satisfying level is as much as 1 person or 2.9% of the total students in the class and no students of class XI MIPA 1 who is categorized in not satisfying level. Suggestions from the result of this research that can be described by researcher are (1) To the next researcher, it is suggested to find literature as much as possible to strengthen the theory, (2) Selecting classes for research's subjects with more heterogeneous math skills, (3) Solidify creative thinking level in problem solving of mathematical problems indicators in order to be able to analyse the level of creative thinking in problem solving better.

Acknowledgments
We gratefully acknowledge the support from FKIP-University of Jember of year 2018.

References

[1] Febriyanti U A, Hobri, Setiawani S 2016 *Seventh Grader Creative Thinking Level in Solving Open-Ended Problems on Rectangular and Square Subject.*

[2] Hudojo H 1988 *Teaching and Learning Mathematics* Jakarta: Ministry of Education P2LPTK Project.

[3] Jihad A 2008 *Mathematics Curriculum* Development Yogyakarta: Multi Pressindo.

[4] Ministry of Education and Culture 2014 *The Curriculum of 2013 for Senior High School/Madrasah Aliyah.*

[5] National Education Standards Agency 2006 *Content Standards for Basic and Secondary Education Units* Jakarta: National Education Standards Agency.

[6] NCTM 2000 *Principles and Standards for School Mathematics* Reston Virginia: NCTM.

[7] Ninik, Hobri, Suharto 2014 *Analysis of Problem Solving Ability for Each Stage of Polya's Model from Ibu Pakusari Vocational High School Multimedia Department's Students on Linear Program Subject.*

[8] Prianggono A 2012 *Analysis of Vocational High School Students' Creative Thinking Process in Solving and Submitting Mathematical Problems on Quadratic Equationt Subject.*

[9] Siswono, Tatag Y E *2010 Leveling Students' Creative Thinking in Solving and Posing Mathematical Problem*

[10] Xie X 2005 *The Cultivation of Problem-Solving and Reason in NCTM and Chinese National Standards. School of Education.* Nanjing Normal University

[11] Çetinkay C 2014 The effect of gifted students' creative problem solving program on creative thinking *Procedia - Social and Behavioral Sciences* **116** 3722 – 3726

[12] Sitorus J, Masrayati 2016 Students' creative thinking process stages: Implementation of realistic mathematics education *Thinking Skills and Creativity* **22** 111-120

[13] Montag-Smit T, Maertz C P 2017 Searching outside the box in creative problem solving: The Role of creative thinking skills and domain knowledge *Journal of Business Research* **81** 1-10

Discover the pythagorean theorem using interactive multimedia learning

I Adhitama[1], I Sujadi[1], and I Pramudya[1]
[1]Department of Mathematics Education, Sebelas Maret University, Jl. Ir. Sutami No.36A, Jebres, Kota Surakarta, Jawa Tengah 57126, Indonesia

Email : indranisa69@gmail.com

Abstract. In learning process students are required to play an active role in learning. They do not just accept the concept directly from teachers, but also build their own knowledge so that the learning process becomes more meaningful. Based on the observation, when learning Pythagorean theorem, students got difficulty on determining hypotenuse. One of the solution to solve this problem is using an interactive multimedia learning. This article aims to discuss the interactive multimedia as learning media for students. This was a Research and Development (R&D) by using ADDIE model of development. The results obtained was multimedia which was developed proper for students as learning media. Besides, on Phytagorian theorem learning activity we also compare Discovery Learning (DL) model with interactive multimedia and DL without interactive multimedia, and obtained that DL with interactive gave positive effect better than DL without interactive multimedia. It was also obtainde that interactive multimedia can attract and increase the interest ot the students on learning math. Therefore, the use of interactive multimedia on DL procees can improve student learning achievement.

1. Introduction

Students' mathematic achievement can be seen in the report of Program for International Student Assessment (PISA). Based on the results of PISA in 2015, Indonesia was ranked to 62 from 70 countries [1, 3]. Apart from the results of the PISA mathematics learning achievements can also be seen from the results of the Trends in International Mathematics and Science Study (TIMMS) [2, 3], Indonesia was scored 386 in average and ranked 38 of 42 countries. This indicated that the mathematics achievement of Indonesian students was still low in international level. While at the national level, the achievements of learning mathematics can also be seen on the results of the National Examination (UN). In the 2015/2016 academic year , students' achievement of geometry and measurement material was in the lowest level compared to other material [4].

The difficulty of students in learning the Pythagorean theorem was in determining the hypotenuse, specifically in the algebra operation, and stated an idea [5]. Based on the previous statements, obtained several problem to solve: (1) in mathematics there are many abstract elements of algebra, especially for Junior High School students is hard to understand abstract things. According to Piaget's learning theory that students in junior level (ages 12-15) had not yet fully able to understand the abstract material [6]. (2)

the student's got difficulties in determining the hypotenuse because of the media applied in classroom was less effective. (3) students' also got difficulty in conveying an idea or opinion. This was also caused of students accepted direct material without engaging in learning. The selection of the appropriate learning model is important in the learning process. (4) Students tend to be lazy and not interested in learning mathematics. So students achievements did not show the good results. This is in line with the opinion of House, that students who have serious or interested in learning mathematics has a higher learning achievement of students who are bored or not interested in learning mathematics [7].

Based on these problems, some suggestions are likely to be running well is: first, the selection of the model of Discovery Learning (DL). DL is a learning model that focuses on the discovery process of a concept or principle, so learning will be more meaningful and can last longer stored in memory [8, 9, 10, 11]. The learning model DL has advantages in training the students to be active in the learning process, improve students' curiosity, and train the skills of students in solving their own problems with how to collect, process, and analyze data by themselves [8, 9]. The application of DL was appropriate to discover the Pythagorean theorem included in the geometry. It is because most students learn geometry based on his memory [11]. Based on the results of research and Oyebayin Martins [12], indicates that DL is better than traditional learning. The steps of DL is as follows: 1) Giving Stimulus, 2) Identifying Problems, 3) Collecting Data, 4) Processing Data, 5) Verifying, and 6) Making Conclusion [13, 14].

The second suggestion is the use of media of instruction, i.e. interactive multimedia. Interactive multimedia is merging/blend two or more media such as text, images, animations, graphics, sounds, video, and audio which is used for the presentation of which the target can control their own media [15, 16, 17, 18, 19, 20]. In addition to the realization of the principle of Mathematic Curriculum, interactive multimedia can also help in attracting and facilitate students in learning process. This is in accordance with the results of the study [21, 22], "Multimedia lessons certainly make learning easier, especially in the fields which are more abstract and which would be better understood with the help of pictures and animations".

Based on both the suggestions of the previous problems, obtained one solution that is using interactive multimedia in application of DL to learn the Pythagorean theorem. For that we developed interactive multimedia with DL on the Pythagorean theorem material to improve learning achievement.

2. Methods

This research was research development (R&D) which focuses on product development. Development of the research method is research in generating new products, and further test the effectiveness of such products [23]. This research aims to produce mathematics multimedia learning on the Pythagorean theorem material by using DL in curriculum 2013 and packaged in the form of CD-assisted learning software Adobe Flash Pro CS6.

The development model used in this study is a model of ADDIE (Analysis, Design, Development, Implementation, and Evaluation). ADDIE model of development was selected because it is one of the most widely used models to develop effective products [24, 25]. Through the stages the ADDIE gained a decent media used in learning. The media is said to be valid if it fulfill the criteria of feasible, practical, and effective [26]. Valid criteria obtained from the results of the assessment by media experts, practical criteria obtained from the assessment of the media by the students, and criteria for effective analysis of data obtained from a test-t. If the test result indicates that the null hypothesis is rejected, then the DL with interactive multimedia learning proved to be more effective than learning DL without using interactive multimedia [23, 27].

The population in this research is the student of MTs Al Furqon in academic year 2017/2018 and obtained students of two class as chosen sample by using random sampling techniques. Each class is used as an experimental class and the class of the control. Experimental class would be applied to DL learning

IOP Conf. Series: Journal of Physics: Conf. Series **1008** (2018) 012066 doi:10.1088/1742-6596/1008/1/012066

model with interactive multimedia while the class control applied DL that doesn't use interactive multimedia. Research instrument used are interviews, observation, and test.

3. Result and Discussion

The following below are the results of the research development by following the flow of the ADDIE model of development:

3.1. Analysis

On the stage there is a field of study with Analysis do observations conducted at MTs Al Furqon. Results found less computer lab is underutilized. Activities conducted in the laboratory of computer learning, practice only computer course. Based on the results of the interviews one of the teachers that students less interested by mathematics, and tend to get bored when lessons take place.

The current curriculum is a curriculum of 2013. The material was selected as research materials is the Pythagorean theorem with a low absorbance [4]. The model of learning tailored to the needs of students, namely model DL. Based on the study of the literature of the difficulty students in learning the Pythagorean theorem, then needed an interactive multimedia learning DL model on the material of the Pythagorean theorem to solve problems that have been discussed before.

Interactive multimedia will be presented in the form of a learning CD. Minimal specification your computer in order to run interactive multimedia are: (1) computer with Windows XP/7/8/10 or Linux, (2) with Computer-equipped flash Player, (3) Intel Pentium IV or higher, (4) CD-ROM (Compact Disc Read-Only 52x dive speed memory), (6) of RAM (Random Access Memory) minimum of 512 megabytes, VGA (Video Grapics Array) 32 megabytes, and (7) active or there is Headphone Speaker.

3.2. Design

The design phase is carried out after the stage of analysis. In such activities is obtained early in the framework of creating interactive multimedia such as Flowchart.

IOP Conf. Series: Journal of Physics: Conf. Series **1008** (2018) 012066 doi:10.1088/1742-6596/1008/1/012066

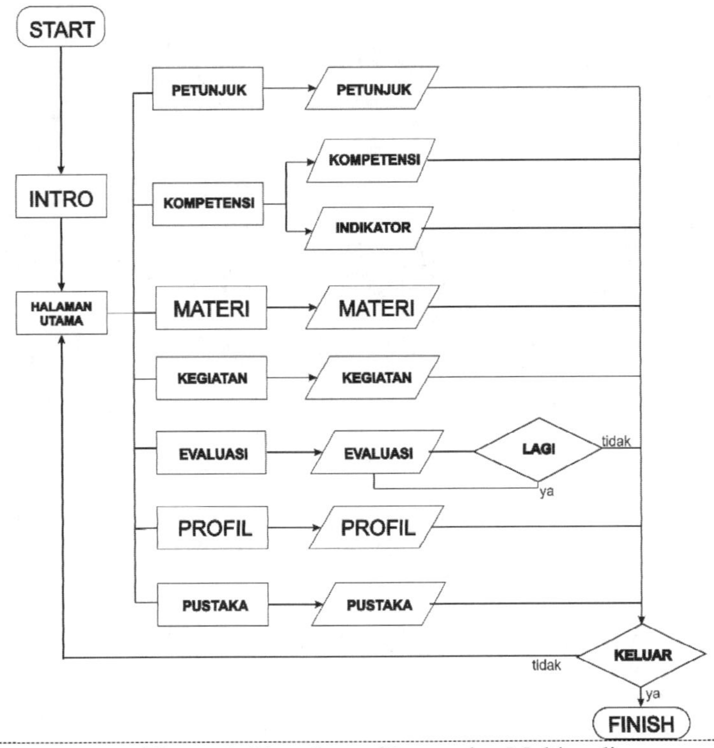

Figure 1. Flowchart of Interactive Multimedia

This interactive multimedia in learning-centered "activities" which contained student learning materials model DL. In addition there is an "evaluation" that are embedded in interactive multimedia this as material students after learning exercise done.

3.3. Development

In this stage of development, there is the process of making media, after it's done the validation by an expert (expert media and expert material). Experts assess media now by completing the assessment of the media. If in the judgment of the media by experts there is input, then the corresponding revised media expert input.

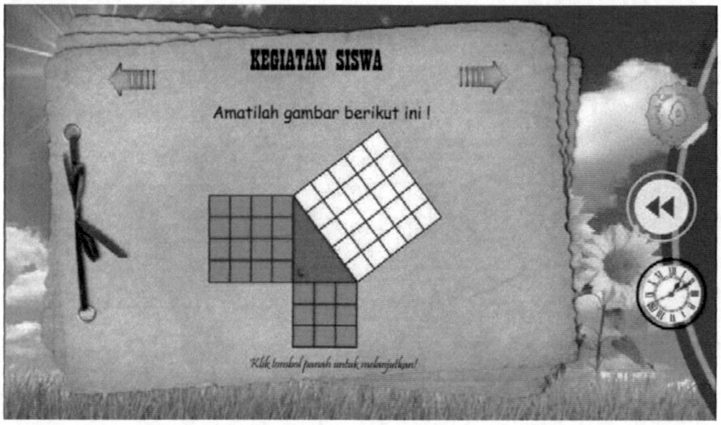

Figure 2. One of the display of student activities in interactive multimedia

Figure 2 shows one of the display of student learning activities in finding Pythagorean theorem based on DL model stage. In that stage students will discover the concept of Pyhagoran theorem by themselves, so learning will be stored long in memory.

Table 1. Summary of Development Stage Result

Stage	Test Subject	Average Score	Percentage	Classification	Information
Validation	Media Expert	4,2	83,00	Very high	Valid
	Materi Expert	4,0	80,67	Very high	
Alpha test	Students of MTs Al-furqon	4,3	86,00	Very high	Practical

.

Based on table 1 in the validation stage, it was found that the average score of media assessment from media experts was 4.2 with percentage of 83.00%, and it was found that the average score of media valuation from the material expert was 4.0 with the percentage of 80.67% obtained very classification high. It can be concluded that interactive multimedia has been valid based on the assessment of media experts and material experts.

Prior to the implementation phase, the first phase of trials conducted in small classes (alpha test), in which there are 10 students. Students are asked to use multimedia and then given the now assessment of media for users (students). If there are entries from students, then the media will be revised again before testing in the larger class (beta test).

Based on table 1 in the alpha test stage that interactive multimedia is considered practical with the average score of 4.3 and the percentage of 86.00%, so multimedia is ready to be implemented to a larger class.

3.4. Implementation

The implementation was done in experiments class which in there are 23 students. Students were asked to use the multimedia learning. After the study is completed students prompted the now assessment of media.

Table 2. Summary of the results of the assessment by media students

Aspect	Indicator	Average	Average	Total

		per indicator	per aspect	Average
The suitability of the media with a purpose or competence	The clarity of the learning objectives	4,2	4,2	4,1
	The clarity of the instructions of use	4,3		
	The suitability of the material	3,9		
	The interactivity of the media	4,4		
Technical quality	The clarity of the multimedia display	4,3	3,9	
	The readability of the text	3,9		
Media Ability	The availability of example and practice	3,9	4,2	
	Simulation quality	3,8		
The suitability of media targeting	Increase to the interest in learning the Pythagorean theorem	4,1	4,0	
	Provide learning aid	3,7		
	Interest in learning other mathematical material	4,4		

Based on table 2 is obtained that for the aspect of conformity in a study on indicators of interest in learning the Pythagorean theorem with an average score of 4.1 with very high criteria. This means that 82% of 23 students feel interactive multimedia learning interest adds to their Pythagorean theorem. Moreover, in the interest of learning indicators in other mathematical material has an average score of 4.4 that includes high criteria. This means that 87% of the 23 students want to learn using interactive multimedia on mathematical material to another. In line with the opinion that multimedia can enhance Arsyad interest student learning [28].

Interactive multimedia that is developed also provide learning aid for students. As shown in table 1 with an average score of 3.7 which means it belongs to the category. So obtained 74% of students find it helpful in learning the Pythagorean theorem with the interactive multimedia. In line with the opinions and research results Milavanovic and Hardiyanto that multimedia helps the learning process easier especially abstract elements are more easily understood by the existence of images and animation [21, 27].

Other results obtained is the percentage of students scoring against interactive multimedia model DL developed amounted to 82%, this includes very high. So, interactive multimedia model DL assessed practical use by users (students).

3.5. Evaluation

At this stage of the evaluation of the students do a written test in order to obtain data on student learning achievement. Proven data obtained with Gaussian test and Lilliefors test proven to be homogeneous by Bartlett. After all the prerequisites are met that test done testing hypotheses with t-test.

Table 3. Summary of t-test Achievement of Math Learning

Class	N	Average	t_{obs}	t_{α}	Decision
Experiment	23	81,65	2,66	1,68	H_0 is rejected
Control	24	77,0			

Based on table 3 looks that t_{obs}=2,66>t_{table}=1,68 so that H_0 is rejected. The conclusion is a model learning DL by using interactive multimedia better than on learning model DL without using interactive multimedia. It contents in line with the results of the study [27] Hardiyanto and Adrian [29] that learning with interactive multimedia gives better results than without use of interactive multimedia learning.

Interactive multimedia that is developed more effective because in the interactive multimedia embedded learning model DL so that students learning more meaningful and long-stored in memory [8, 9, 10, 11]. In addition, the interactive multimedia enhances the interest of students who have been described previously in the implementation phase outcome that good may increase interest in the achievements of learning math. The last was developed interactive multimedia can help instructional criteria either. In accordance with the opinion of the [21, 22], "Multimedia lessons certainly make learning easier, especially in the fields which are more abstract and which are better understood with the help of pictures and animations".

4. Conclusion and Suggestion

Based on the result and discussion can be inferred that the interactive multimedia developed viable for use in learning the Pythagorean theorem. This is because the interactive multimedia has met the criteria of a valid, practical, and effective.

Interactive multimedia judged can raise the interest of students in learning mathematics and help learning model DL. If student learning increased interest, then it can spur student learning achievement because the increasing of interest is one of the deciding factor whether or not learning achievements. It is therefore Expected to develop interactive multimedia content on other particularly in learning mathematics.

5. Acknowledgements

Researchers say thanks to the head of the MTs Al Furqon Sanden who have given permission to do research at the place. Do not forget thanks to Anisa Oktarina Dwi and Rangga Mahendra for the help in writing this article.

6. References

[1] PISA 2016 *Results Excellence and Equity in Education Volume I* (Paris: OECD Publishing)
[2] Mullis I V S, Martin M O, Foy P and Arora A 2012 *TIMMS 2011 International Result in Mathematics* (Chestnut Hill: TIMSS & PIRLS International Study Center)
[3] Udil P, Tri A K, and Riyadi 2017 Profile of Mathematics Anxiety of 7th Graders *AIP Confrence Proceedings* **1868**
[4] BSNP *Laporan Hasil Ujian Nasional Tahun Pelajaran* 2015/2016
[5] Anggraini G R 2017 Analisis Kesalahan Pemahaman Konsep pada Materi teorema Pythagoras di Kelas VIII SMP Negeri 3 Kartasura (UMS : Naskah Publikasi)
[6] Syahbana A 2012 Peningkatan Kemampuan Berpikir Kritis Matematis Siswa SMP Melalui Pendekatan Contextual Teaching and Learning *Edumatica* **2** 45-57
[7] House J D 2006 Mathematics Beliefs and Achievement of Elementary School Students in Japan and the United States : Result From the Third International Mathematics and Science Study *The Journal of Genetic Psychology*. **167** 31-45
[8] Balm A G 2009 The Effect of Discovery Learning on Students' succsess and Inquiry Learning Skills *Eurasian Journal of Education Research* **35** 1-20
[9] In'am A, and Siti H 2016 Learning Geometry through Discovery Learning using a Scientific *Approach Int Jour of Instruction* **10** 55-70
[10] Tampo B, Ahmad A, and Muris M 2016 The Development of Discovery Learning-Inquiry Learning Model to Reduce the Science Misconceptions of Junior High School Student *Int Jour of Environtmental & Science Education* **11** 5676-5686
[11] Ramdhani M R, B Usodo, S Subanti 2017 Discovery Learning with Scientific Approach on Geometry *IOP Conf. Series : Jurnal of Physics* **895**

[12] Martins O and Oyebanji R K 2000 *The effects of inquiry and lecture teaching approaches on the cognitive achievement of integrated science students* Journal of Science Teachers' Association of Nigeria **35** 25-30
[13] Jaw S H 2008 Scaffolding Discovery Learning Spaces *MERLOT Journal of Online Learning and Teaching* **4** 533-548
[14] Klahr D and Nigam M 2004 The equivalence of learning paths in early science instruction: Effects of direct instruction and discovery learning *Psychological Science* **15** 661–667
[15] Mayer R E 2009 *Multimedia Learning Prinsip-prinsip dan Aplikasinya* (Yogyakarta : Pustaka Pelajar)
[16] Nusir S, Izzat A, Mohammed A K, and Fatima S 2012 Studying the Impact of using Multimedia Interactive Programs at Children Ability to Learn Basic Math Akills *Acta Didactica Bapocensia* **5** 17-32
[17] Amine A T, Chellali B, and Boutkhil G 2012 Using Multimedia to Motivate Students in EFL Classrooms : A Case Study of English Master's Students at Jijel University Algeria *Malaysian Journal of Distance Education* **14** 63-81
[18] Ampa A T 2015 The Implementation of Interctive Multimedia Learning Materials in Teaching Listening Skills *English Language Teaching* **8** 56-62
[19] Daryanto 2016 *Media Pembelajaran, Peranannya Sangat Penting dalam Mencapai Tujuan Pembelajaran* (Yogyakarta : Gava Media)
[20] Mishra S, & Ramesh C S 2004 Interactive Multimedia in Education and Training *India Idea Group Publising* **6** 115-122
[21] Milavanovic M, Durdica T, and Aleksander M 2012. Multimedia Approach in Teaching Mathematics – Examples of Interactive Lesson From Mathematical Analisis and Geometry *Interactive Multimedia* **6** 112-116
[22] Babiker M E A 2015 For Effective Use of Multimedia in Education, Teachers Must Develop their Own Educational Multimedia Applications TOJET **14** 62-68
[23] Sugiyono 2013 Metode Penelitian Pendidikan (Pendekatan Kuantitatif, Kualitatif, dan R&D) (Bandung : Alfabeta)
[24] Widyastuti P D, M Mardiyana, and D R S Saputro 2017 An Instructional Media Using Comics on the System of Linear Equation *IOP Conf. Series: Journal of Physics* **985**
[25] Aldoobie N 2015 ADDIE Model *American International Journal of Contemporary Research* **5** 69-72
[26] Nieveen N 1999 Prototyping to reach product quality. In Jan Van den Akker, R.M Branch, K. Gustafson, N. Nieveen & Tj. Plomp, Dordrecht, Netherland : *Kluwer Academic Pub.* 125-135.
[27] Hardiyanto E L, Budiyono, and Budi S 2015 Pengembangan Multimedia Interaktif untuk Pembelajaran Materi Pokok Balok Siswa SMP Kelas VIII *Jurnal Elektronik Pembelajaran Matematika* **3** 215-221
[28] Arsyad A 2013 *Media Pembelajaran* (Jakarta : PT Raja Grafindo)
[29] Adrianus I W, I Yuda S, I Made C, and I Made C 2013 Pengembangan Multimedia Pembelajaran Matematika Berpendekatan Kontekstual untuk Siswa Kelas VIII di SMP Negeri 4 Singaraja *e-Jour Program Pascasarjana Universitas Ganesha Prodi Teknologi Pembelajaran* **3** 12-21

Technological pedagogical content knowledge of junior high school mathematics teachers in teaching linear equation

S Wati[1], L Fitriana[1], and Mardiyana[1]

[1]Department of Mathematics Education, Universitas Sebelas Maret, Jl. Ir. Sutami No. 36 A, Kentingan, 57126, Jebres, Surakarta, Indonesia

Email: susilowati.maths@gmail.com

Abstract. Linear equation is one of the topics in mathematics that are considered difficult. Student difficulties of understanding linear equation can be caused by lack of understanding this concept and the way of teachers teach. TPACK is a way to understand the complex relationships between teaching and content taught through the use of specific teaching approaches and supported by the right technology tools. This study aims to identify TPACK of junior high school mathematics teachers in teaching linear equation. The method used in the study was descriptive. In the first phase, a survey using a questionnaire was carried out on 45 junior high school mathematics teachers in teaching linear equation. While in the second phase, the interview involved three teachers. The analysis of data used were quantitative and qualitative technique. The result PCK revealed teachers emphasized developing procedural and conceptual knowledge through reliance on traditional in teaching linear equation. The result of TPK revealed teachers' lower capacity to deal with the general information and communications technologies goals across the curriculum in teaching linear equation. The result indicated that PowerPoint constitutes TCK modal technological capability in teaching linear equation. The result of TPACK seems to suggest a low standard in teachers' technological skills across a variety of mathematics education goals in teaching linear equation. This means that the ability of teachers' TPACK in teaching linear equation still needs to be improved.

1. Introduction

Algebra has been widely recognized as one of the most difficult topics within secondary school mathematics curricula [1]. One of algebra materials is the linear equation. A linear equation is an algebra material that has existed in junior high school to university. A linear equation is critical to the study of more advanced mathematical topics such as derivative in calculus [2], the line of best fit in statistics [3], and to describe nonlinear functions in advanced algebra [4]. Therefore, linear equation material is essential to be mastered. The linear equation is difficult for students to understand and becomes a problem in learning for teachers [5]. Studies have found students experienced various difficulties with a linear equation. Students got difficulties and misunderstood linear equations, graphs, slopes and could not comprehend the connection between slope and the x- and y-intercepts [6]. Research has documented procedural and conceptual students' difficulties with slope [7]. Students also have difficulty in interpreting linear equations and their graphs [8]. The misconception and deficiencies of the students in the linear equation also may lead to severe learning difficulties in the subjects of functions, limit, derivation, and integral through high school and the university years.

Student difficulties of understanding the concept of the linear equation can be caused by lack of understanding of the concept and the way of teachers teach. Teachers quality was identified one of the crucial factors attributed to the situation.

The Indonesia goverment launched a program called teacher sertification. The program was designed as a tool to improve teacher quality through three strategies such as attraction channel (increasing the enrolments of preservice teachers), upgrading (improving teacher qualifications), and the behavioral (changing teachers' teaching practice). To obtain the certificate a teacher as to pass a certification qualification test conducted at the end of a training program especially designed for the purpose. However, a study suggest that teacher certification does not lead to improvement of teachers' teaching performance, teachers' behavior and teachers' participation in professional development [9]. The study also finds that there is no performance difference between certified and uncertified teachers.

Teachers' perspective and competency with the regard to teachers' attitudes, beliefs, subject matter knowledge, pedagogical knowledge, pedagogical content knowledge, and technological pedagogical content knowledge can influence what and how students learn mathematics [10]. Effective teachers not only know their subject matter but they also able to create a stimulating learning environment and apply pedagogical strategies, including technology-rich strategies that engage students while helping them improve their achievement [11]. Departmen of National Education of Indonesia also requires that every teacher in Indonesia must have standard competencies namely the mastery of knowledge (content), technology, pedagogy, culture, humanity, and civilization [12].

Technology is considered important to be integrated into learning to respond to the challenges of the 21st century. Information Communication Technology (ICT) has been developed rapidly and has been used in all areas of knowledge, including in the field of education. Realizing the importance of information technology in education, mathematics teachers should utilize the technology to assist the learners achieve their goals and to ease them in understanding and using the concepts of mathematics. Content-specific mathematics technologies support students in exploring and identifying mathematical concepts and relationships. Content-neutral technologies increase students' access to information and ideas and enhance student-student and student-teacher interactions to support and enrich sense making. Strategic use of technological tools can support learning of mathematical procedures and skills as well as the development of advanced mathematical proficiencies, such as problem solving, reasoning, and justifying [13, 14, 15, 16].

Many of researches result from mathematics education illustrate that the integrating of ICT may change the environment of teaching and learning mathematics [17, 18]. ICT seems to provide a focal point that encourage interaction between learners and the technology itself. This implies that ICT used in instruction will support constructivist pedagogy, where learners use technology to explore and reach an understanding of mathematical concepts [19]. Thus, ICT can make students more active and not solely dependent on their teacher so they can study independently. Furthermore, for ICT to be used effectively in everyday teaching, radical changes are advocated in approaches to the teaching of mathematics. For this reason, it is important to promote research and practice that are able to provide teachers the opportunities to adequately utilize and integrate the technology into mathematics classrooms.

Currently, one of the most important ways of providing technological support is to use a framework for integrating complex problems of knowledge from pedagogy, content, technology and different forms of interactions among these elements in classroom. Teachers are not only required have the ability of Pedagogical Content Knowledge (PCK) but also apply technology in learning, so technology, pedagogy, and content have become a part of teachers. The Technological Pedagogical Content Knowledge (TPACK) framework was developed by Mishra and Koehler based on Lee Shulman's concept of PCK by adding technology. Ability to use appropriate strategies supported by the right technology tools on specific topics is a knowledge model called the Technological Pedagogical Content Knowledge (TPACK) [20].

TPACK in any discipline is the perfect union of three knowledge domains (content, pedagogy, and technology) to develop a knowledge base from which a teacher can view a lesson and understand how

IOP Conf. Series: Journal of Physics: Conf. Series **1008** (2018) 012067 doi:10.1088/1742-6596/1008/1/012067

technology can enhance the learning opportunities and experiences for students while also knowing the correct pedagogy to enhance the learning content. In mathematics education, a teacher with a TPACK perspective is a teacher that understand the correct pedagogy with this technology. By having a proper TPACK, he will be able to engage and motivate students as they explore the content of mathematics to a greater degree. The TPACK framework suggests that integrated knowledge of technology, pedagogy, and content is an essential condition to effective and innovative classroom teaching using technology.

TPACK comprises seven constructs which are Technological Knowledge (TK), Pedagogical Knowledge (PK), Content Knowledge (CK), Technological Pedagogical Knowledge (TPK), Technological Content Knowledge (TCK), Pedagogical Content Knowledge (PCK), and Technological Pedagogical Content Knowledge (TPACK). TK refers to knowledge about various technology that can be implemented in everyday life as well as in teaching and learning activities. PK is the knowledge about teaching and learning strategies. CK is knowledge about the actual subject matter to be learned. TPK, TCK, and PCK refer to knowledge about the interplay between the original constructs. Thus, TPK is knowledge relates to how technology can be used appropriately for particular teaching and learning strategy. TCK relates to how specific technology play importance role in delivering certain content so as to it is being conveyed successfully. Meanwhile, PCK relates to how certain teaching and learning strategy fits subject matter delivery [20]. This research was conducted in order to analyze TPACK of junior high school mathematics teachers in teaching linear equation.

2. Method

This study combines both quantitative and qualitative approaches in two phases. In the first phase, a survey method using a questionnaire was carried out on 45 teachers in Karanganyar. The instrument used was a questionnaire that measures the level of Pedagogical Content Knowledge (PCK), Technological Content Knowledge (TCK), Technological Pedagogical Knowledge (TPK), and Technological Pedagogical Content Knowledge (TPACK) in teaching linear equation. Items were drawn from the current literature and from previous TPACK-related instrumens and in some cases modified to reflect the mathematical focus of this study [21, 22]. The instrument constructed contained 19 items for measuring teachers' self-assessments of the four TPACK domains: 6 PCK items, 4 TPK items, 4 TCK items, and 5 TPACK items. For these 19 items, participants answered each question using the five-level Likert scale: strongly disagree, disagree, neither agree nor disagree, agree, and strongly agree. The measure TPACK domains were used in this study to represent participants' self-assessment of their knowledge. Internal reability of the four construct was established through Cronbach Alphas for all constructs: PCK(0.801), TPK (0.762), TCK (0.765), and TPACK (0.920). Data were analyzed using descriptive statistics, namely the mean. While in the second phase, the interview method involved three teachers were performed. They were selected using purposive sampling technique. The instrument used was the semi-structured interview protocol to asses the teachers' TPACK integration in their teaching practice.

3. Results and Discussion

3.1. Pedagogical Content Knowledge (PCK)

Pedagogical Content Knowledge refers to the content knowledge that deals with the teaching process. Pedagogical Content Knowledge is different for various content areas, as it blends both content and pedagogy with the goal being to develop better teaching practices in the content areas. Based on the analysis, the profile of PCK junior high school of mathematics teachers is presented in table 1.

Table 1. Pedagogical Content Knowledge Score Junior High School Mathematics Teachers

No	Statement	M	SD	Criteria
1	I can help students keep understanding the material of the linear equation using various strategies without using technology.	3.29	0.61	Good

No	Statement	M	SD	Criteria
2	I can engage students actively in meaningful discussions on the material of linear equation even without the use of technology.	3.63	0.77	Good
3	I can make students actively solve real problems on the topics of the linear equation without using technology.	3.79	0.70	Good
4	I can help change the thinking process of students in order to master a difficult topic on the material of the linear equation without using technology.	3.51	0.52	Good
5	I can evoke meaningful reflection on the students on the material of the linear equation without using technology.	3.65	0.75	Good
	Mean	3.57		Good

The average score PCK of junior high school mathematics teachers is 3.57, it is in the good criteria this indicates the ability of Pedagogical Content Knowledge teachers has been good average that the highest on the item statement "I can make students actively solve real problems on the topics of the linear equation without using technology". However, the result of the interview show that teachers normally prefer teacher-centered approach namely the 'chalk-and-talk" in the explaining and elaboration of the principles, theories, definitions, and concepts. It also demonstrated that they adopt a passive lecture teaching style with textbooks as the primary source in their teaching practices. Students too still make the teacher as a dominant learning center.

3.2. Technological Pedagogical Knowledge (TPK)

Technological Pedagogical Knowledge refers to the knowledge of how various technologies can be used in teaching and to understanding that using technology may change the way teachers teach. Based on the analysis, the profile of TPK junior high school of mathematics teachers is presented in table 2.

Table 2. Technological Pedagogical Knowledge Score Junior High School Mathematics Teachers

No	Statement	M	SD	Criteria
1	I use computer apps in the teaching linear equation.	3.43	0.75	Adequate
2	I choose technology that suits the approach and strategy of learning in the classroom.	3.79	0.70	Good
3	I use internet facilities (like social networks, emails, and blogs) to communicate with students.	3.83	0.64	Good
4	I use online quiz (such as Quipper School, Edmodo, Socrative) to do the tests.	3.36	0.50	Adequate
	Mean	3.60		Good

The average score TPK of junior high school mathematics teachers is 3.60, it is in the good criteria this indicates the ability of technological Pedagogical Knowledge Teachers has been good average that the highest on the item statement "I use internet facilities (like social networks, emails, and blogs) to communicate with students". However, the result of interview show that teachers were quite familiar with the various types of techniques or methods of teaching, but the dominant methods among them were expository and discussion. With the disclosure of knowledge about the constructivist approached, teacher have the option to try the method of demonstration, cooperative, module-based activities and game-based learning. This technology integration should encourage teachers to conduct student-centered activities, where students actively involved in many instructional activities. Technology should be a tool that helps students to experiment and investigate and to help teachers create contests and accelerate the process of problem-solving. In this study, the use of technology synonymous with the method of lecture, in which teachers deliver the lessons and students just listen. Although there were bilateral discussions, teachers still clearly shown as a single authority 'knowledge' providers. It reflects that the teachers have yet to be able to select appropriate teaching approaches with appropriate technology how to teach a subject. If teachers could handle the demonstration method, cooperative, games and module-based methods, it means that the teacher can also customize any appropriate technology that can be used in conjunction with the teaching methods.

3.3. Technological Content Knowledge (TCK)

Technological Content Knowledge refers to the knowledge of how technology can create new representations for specifics content. It suggests that teachers understand that by using a specific technology, they can change the way leaners practice and understand concepts in a specific content area. Based on the analysis, the profile of TCK junior high school mathematics teachers is presented in table 3.

Table 3. Technological Content Knowledge Score Junior High School Mathematics Teachers

No	Statement	M	SD	Criteria
1	I use technology to help understand the concept of the linear equation.	3.29	0.61	Adequate
2	I can download the application of computer and animation related to a linear equation.	3.79	0.70	Good
3	I can develop student activities and tasks that involve the use of technology	3.59	0.51	Good
4	I can create self-learning activities for students using ICT devices (eg web quest or flip)	3.43	0.65	Adequate
	Mean	3.52		Good

The average score TCK of junior high school mathematics teachers is 3.52, it is in the good criteria this indicates the ability of Technological Content Knowledge Teachers has been good average that the highest on the item statement "I can download the application of computer and animation related to a linear equation". However, the result of interview show that the mathematics contents were flatly referred without modification and only referring to the textbooks or workbooks provided. Not just confined to the requirements of the syllabus, the process of teaching also limited to the scope teaching the basic concepts, and the procedural aspects which emphasized only on writing the right steps in solving questions. The use of technology only in the introduction phase of the lesson to introduce the basic concepts and also for finding questions sources. PowerPoint constitutes TCK modal technological capability in teaching linear equation. The teachers were able to name same technology software such as the Geogebra instead of PowerPoint presentations which were considered suitable for teaching topic the linear equation. This means that teachers can choose the appropriate technology to teach a particular subject, but they have not been able to explore specifically how the software can be integrated into teaching and learning process which is very helpful in learning activities and accelerate the process of understanding the concepts. In the learning process carried out, almost all teachers use learning resources in the form of the same textbooks used by students, in addition to using the internet as other supporting material. Research conducted by the Son, Robb and Charismiaji regarding the Computer Literacy and Competency revealed that teachers in Indonesia generally have basic computing skills are quite high, but the frequency of utilization of these capabilities are limited only to a few applications only (one quite dominant is a word processing application) [23].

3.4. Technological Pedagogical Content Knowledge (TPACK)

Technological Pedagogical Content Knowledge refers to the knowledge required by teachers for integrating technology into their teaching in any content area. Teachers have an intuitive understanding of the complex interplay between the three basic components of knowledge (content, pedadogy, and technology) by teaching content using appropriate pedagogical methods and technologies. Based on the analysis, the profile of TPACK junior high school of mathematics teachers is presented in table 4.

Table 4. Technological Pedagogical Content Knowledge Score Junior High School Mathematics Teachers

No	Statement	M	SD	Criteria
1	I understand how to integrate mathematical knowledge (especially the linear equation), pedagogical knowledge and technological knowledge.	3.71	0.61	Good
2	I can choose a learning strategy and technology that match the material of the linear equation that will be delivered in class.	3.71	0.73	Good
3	I can integrate knowledge of the content, pedagogical knowledge and knowledge of technology to realize meaningful learning.	3.50	0.85	Adequate
4	I can apply different learning strategies and varying technologies in the implementation of mathematics learning	3.50	0.65	Adequate
5	I can enable students to construct deep and meaningful insights into the material of the linear equation using various technologies.	3.50	0.52	Adequate
	Mean	3.58		Good

The average score of TPACK junior high school mathematics teachers is 3.58 are in good criteria this indicates TPACK junior high school mathematics teachers are good. Although the means to integrate the three content knowledge, pedagogy and technology are still sufficient criteria, the selection and application of technology in learning strategies is also sufficient, but teachers are good enough in constructing meaningful knowledge in learning the linear equation. The result of interview show that the teachers said that the limitations of the technology or media used in learning are the main problem in the learning process. Teachers also have difficulty in choosing learning strategy and model, allocation of a limited time, while very much a concept that should be taught. The procedure teaching is tailored to the teacher's lesson plan, but the teaching procedures used to focus more on the order of the material in the textbook. The method of teaching often utilized by teachers is the conventional method. Sometimes adds the discussion method in learning. But overall learning implementation that teachers use still conventional methods. When asked in an interview, the reason teachers are still using conventional methods because of limited facilities in schools so that there is no other choice than conventional methods. The lack of supporting facilities such technology in learning make teachers tend to retain the conventional method. This indicates that the pedagogy of teachers is still less because it can not create meaningful learning on the grounds no technology can be utilized.

4. Conclusion

This study aims to identify TPCK of junior high school mathematics teachers in teaching linear equation. The result PCK revealed teachers emphasized developing procedural and conceptual knowledge through reliance on traditional in teaching linear equation. The result of TPK revealed teachers' lower capacity to deal with the general information and communications technologies goals across the curriculum in teaching linear equation. The result indicated that PowerPoint constitutes TCK modal technological capability in teaching linear equation. The result of TPACK seems to suggest a low standard in teachers' technological skills across a variety of mathematics education goals in teaching linear equation. This means that the ability of teachers' TPACK in teaching linear equation still needs to be improved. Although TPACK may not directly relate to teaching practice, however, TPACK shapes teachers their practice. Deliberate and systematics efforts are needed to facilitate the improvement of teachers TPACK.

References

[1] Jupri A, Drijvers P and Van den Heuvel-Panhuizen M 2014 *Mathematics Education Research*

Journal **26(4)** pp 683-710
[2] Stanton M and Moore-Russo D 2012 *School Science and Mathematics* **112(5)** pp 270-277
[3] Casey S and Nagle C 2016 *Educational Studies in Mathematics* **92(2)** pp 163-177
[4] Hattikudur S, Prather R, Asquith P, Knuth E, Nathan M and Alibali M 2011 School Science and Mathematics **112(4)** pp 230-240
[5] Dhoruri A and Markaban 2011 *Pembelajaran persamaan garis lurus* (Yogyakarta: PPPPTK Matematika) pp: 2
[6] Birgin O 2012 *Bolema, Rio Claro (SP)* **26(42A)** pp 139-162
[7] Cho P and Nagle C 2017 International Journal Research in Education and Science **3(1)** pp 135-150
[8] Nathan M J and Kim S 2007 *Mathematical Thinking and Learning* **9(3)** pp 193-219
[9] Chang M C, Shaeffer S, Al-Samarrai S, Ragatz A B, de Ree J and Stevenson R 2014 *Teacher reform in Indonesia: The reles of politics and evidence in policy making* (Washington DC: The World Bank)
[10] Bingolbali E, Akkoc H, Ozmantar M F, and Demir S 2011 *International Electronic Journal of Mathematics Education* **6(1)** pp 40-59
[11] NCATE *What make a teacher effective?* National Council for the Accreditation of Teacher Education Retieved 22.11.17 from http://www.ncate.org/LINKCLIK.aspx?fileticket=JFRrmWqa1jU%3d&tabid=361
[12] Kementrian Pendidikan dan Kebudayaan Indonesia 2013 *Kurikulum 2013* Retieved 22.11.17 from http://www.kemdiknas.go.id/
[13] Gadanidis G and Geiger V 2010 *ZDM* **42(1)** pp 91-104
[14] Nelson J, Christopher A and Mims C 2009 *Tech Trends* **53(5)** pp 80-85
[15] Pierce R and Stacy K 2010 *International Journal of Computers for Mathematical Learning* **15(1)** pp 1-20
[16] Roschelle J, Shechtman N, Tatar D, Hegedus S, Hopkins B, Empson S, Knudsen J and Gallagher L 2010 *American Educational Research Journal* **47(4)** pp 833-878
[17] Chandra V and Briskey J 2012 *International Journal of Pedagogies and Learning* **7(1)** pp 73-83
[18] Tay L Y, Lim S K, Lim C P and Koh L 2012 *Australasian Journal of Educational Technology* **28(4)** pp 740-754
[19] Sang G, Valcke M, Braak J, Tondeur J and Zhu C 2011 *Journal of Computer Assisted Learning* **27(2)** pp 160-172
[20] Mishra M L and Koehler M J 2006 *Teachers College Record* **108(6)** pp 1017-54
[21] Widodo A 2017 *21QL Integration (TPACK) Survey*
[22] Pamuk S, Ergun M, Cakir R, Yilmaz H B and Ayas 2013 *Education and Information Technologies* **20(2)** pp 241-263
[23] Son J B, Robb T and Charismiadji I 2011 *Computer-Assisted Language Learning - Electronic Journal* **12** 26-42

IOP Conf. Series: Journal of Physics: Conf. Series **1008** (2018) 012068 doi:10.1088/1742-6596/1008/1/012068

Problem solving of student with visual impairment related to mathematical literacy problem

A R Pratama, D R S Saputro, and Riyadi

Department of Mathematics Education, Faculty of Teacher Training and Education, Sebelas Maret University, Indonesia

E-mail: ade.restu@student.uns.ac.id

Abstract. The student with visual impairment, total blind category depends on the sense of touch and hearing in obtaining information. In fact, the two senses can receive information less than 20%. Thus, students with visual impairment of the total blind categories in the learning process must have difficulty, including learning mathematics. This study aims to describe the problem-solving process of the student with visual impairment, total blind category on mathematical literacy issues based on Polya phase. This research using test method similar problems mathematical literacy in PISA and in-depth interviews. The subject of this study was a student with visual impairment, total blind category. Based on the result of the research, problem-solving related to mathematical literacy based on Polya phase is quite good. In the phase of understanding the problem, the student read about twice by brushing the text and assisted with information through hearing three times. The student with visual impairment in problem-solving based on the Polya phase, devising a plan by summoning knowledge and experience gained previously. At the phase of carrying out the plan, students with visual impairment implement the plan in accordance with pre-made. In the looking back phase, students with visual impairment need to check the answers three times but have not been able to find a way.

1. Introduction

In recent decades, problem-solving skills have been considered a critical aspect of mathematics education [1]. In fact, problem-solving is a basic competency in 21st-century skills [2] [3]. In the mathematics education, problem-solving is the heart of mathematics which means that all aspects of mathematics originate in problem-solving [4] [5]. Mathematical learning that focuses on problem-solving can make students understand the essence of the math material under study [6] [7] [8]. Based on this, problem-solving in mathematics learning has three important points, namely problem solving as the main objective of mathematics learning, problem-solving as a process in mathematics learning, and problem-solving as basic skills that must be possessed when one learns mathematics [9].

The importance of problem-solving in mathematics learning requires that students have that ability. Problem-solving is an important activity in mathematics learning since problem-solving abilities gained in math lessons can be used in solving other problems in the real world [10]. Students' ability to solve problems can be seen based on how students solve problems. One of the problem-solving phases is the Polya phases. The Polya phases provide a useful framework for problem-solving that has a heuristic conceptual problem-solving feature. Meanwhile, the mathematical problem is quite good in training problem-solving abilities in the real world that is the problem of similar mathematical literacy

PISA [11]. Mathematical literacy is an individual's capacity to formulate, employ, and interpret mathematics in a variety of contexts. It includes reasoning mathematically and using mathematical concepts, procedures, facts, and tools to describe, explain, and predict phenomena. It assists individuals to recognize the role that mathematics plays in the world and to make the well-founded judgments and decisions needed by constructive, engaged and reflective citizens [12]. The statement also corresponds to the expert's disclosure that the mathematical literacy as the ability to use mathematical knowledge and experience in solving real-life problems [11] [13] [14].

Problem-solving and mathematical literacy problems are two important things in mathematics learning. They have a relationship in scoring students for the challenges of life in the future. Supply students in developing problem-solving abilities in real life. Therefore, The Organization for Economic Co-operation and Development (OECD) through the Project for International Student Assessment (PISA) examines students 15-year-old related to literacy content, mathematical literacy one of them. The selection of normal children studied by the OECD indicated that normal children are a top priority in education. In fact, based on the Salamanca Statement and the Framework for Action that education is for all, which means education is appropriate for all children, whether the child is normal or who has special needs [15]. The Salamanca statement has been followed up by various countries including Indonesia by holding an inclusive school.

The existence of inclusive education is an alternative to equate the rights of normal children and children with special needs in education. In such a way as to be able to provide the provision of competence of children with special needs in the future when the plunge in the community. One of the children with special needs that need attention is children with visual impairment. Children with visual impairment are children who have total blindness and low vision. Children with visual impairment of the total blind categories in obtaining information utilize the sense of hearing and the sense of touch. Meanwhile, children with low vision categories are still able to use the sense of sight to obtain information, but they must use visual aids. In the learning process, 80% of information is obtained through sight [16]. For visually impaired students this manner of receiving information is a challenge since they both experience the world differently than people with full vision [17].

Students with visual impairment, the total blind category cannot use the sense of sight as a means of obtaining information to make them need special attention. Whereas, almost all mathematical material is presented in visual conceptions such as statistics, geometry, arithmetic [18] [19]. Thus, students with visual impairment of the total blind category will have difficulty in learning mathematics. However, students with a visual impairment conceive the estimation of volume, surface area, and length better, in concrete concepts and concepts that are familiar to them, when compared to sighted students [20]. This makes the process of solving the problem of students with visual impairment are totally blind category interesting to study.

There are other interesting things in the testing conducted by OECD, the subject of testing ie 15-year-old students. 15-year-olds can follow the trajectory to be constructive, reflective and able to make good decisions and opinions [12]. However, views on the development of children, students age 12 or at the Junior High School level VII entered the early stages of the development of logical thinking, abstract and rational [21]. This means that at this stage the child can be directed to abstract things like mathematical concepts. Children can make decisions based on the information they have received. Children are able to make rational excuses based on decisions taken. Thus, it is very suitable to know how the process of solving problems at the first grade level of schooling grade VII.

1.1 Problem Solving

Problem-solving can be interpreted as a process starting from the minute students are faced with the problem until the end when the problem is solved [22]. This means when the student is faced with a problem and the student is trying to solve the problem to find the solution to the problem then the party can be said in the process of solving the problem. Another definition states that problem-solving is the cognitive process focusing on changing the given state to the final state where the solution procedure is not clear [23]. Problem-solving is an important thing students must have in the 21st century. Problem-solving in the 21st century requires individuals to find solutions to complex problems [24]. The problem-solving process in mathematics can help students be a good problem solver. This means problem-solving in mathematics can help students solve more complex problems. There is no specific method of solving in Mathematics, but the process of problem-solving is a systematic thing that can be seen in general. Thus, Polya formulates four basic phases in the problem-solving process:

- Understanding the problem, in this phase students should be able to answer questions, such as "What is unknown", "What is asked" and "how the solution" to the problem.
- Devising a plan, in this phase developing a plan and establishing the worker's steps to get a solution based on what is already known: previous student experiences are important.
- Carrying out the plan, in this phase algebraic manipulation activities, simplifying and using other math skills are indispensable.
- Looking back, students review the results have been obtained, check the results that have been obtained as requested and apply the solution to the problem under other circumstances [22].

1.2 Mathematical Literacy Problem
Mathematical literacy is an individual's capacity to formulate, employ, and interpret mathematics in a variety of contexts. It includes mathematical concepts, procedures, facts, and tools to describe, explain, and predict phenomena. It assists individuals to recognize the role of constructive, engaging and reflective citizens [12]. Important points contained in the definition of mathematical literacy is how the concepts, procedures, and facts of mathematics are used as a tool to solve problems that occur. Simply literacy mathematics is the knowledge to know and apply basic math in our daily lives [13]. That sense identifies that students who have good math literacy will be able to solve real-world problems using mathematical principles well too. Problem-solving on mathematical literacy problems begins with realistic problems. The process of solving mathematical literacy problems ranging from identifying realistic problems and formulating problems mathematically, employing certain mathematical procedures to obtain mathematical results, which then reinterpret them as the initial problem [25]. Based on this description, problems related to mathematical literacy are real life problems similar to PISA and solutions that use concepts, faculties, procedures and mathematical principles.

The ability of mathematical literacy is indispensable to students in facing 21st-century competition. In the 21st-century mathematics skills are treated as the ability to use math in answering real-world problems [26]. 21st-century mathematics is no longer just interpreted as arithmetic, but as a tool to solve problems in the real world. There is a common thread between 21st-century mathematics and mathematics literacy, both of which focus on mathematics as a tool for solving real-world problems. This means students who will be able to survive in the 21st-century global competition that is students who have good mathematical literacy skills. Problem-solving processes related to mathematical literacy need to be studied in order to capture student literacy skills.

2. Method
This research used the qualitative method with descriptive research type. The strategy used is case study strategy. The subject of the study was one student from a junior secondary school of inclusion. The student is in grade VII who has a visual impairment that is categorized as total blindness. The data from this study comes from the results of related tests of mathematical literacy problem and the results of in-depth interviews. The test assigned to the study subjects amounted to one mathematical literacy problem similar to that of PISA. Data collection techniques in this study by providing written tests of mathematical literacy problems on the subject of research. The question of written test first has been changed into the form of writing braille. Furthermore, an in-depth interview was conducted with the research subjects. The results of written test and interview transcripts were analyzed, then described the problem solving based on the Polya phase relating to mathematical literacy issues.

3. Result and Discussion

Problem-solving of a student with visual impairment related to mathematical literacy is described based on Polya phase. Problems related to mathematics literacy similar to PISA that corresponds to the 7th grade of Junior High School. The problem similar to PISA as follows:
Theme: Telephone Communication
Mr. Robert is the father of Anita who works as an Indonesian ambassador in Spain. While Anita remains in Indonesia, precisely in the city of Surakarta. They often communicate by phone. However, there is a time difference between Indonesia and Spain. For example, in Indonesia showed at 09.00 WITA (Indonesia Central Time/ UTC + 8) then In Spain showed at 03.00 local times. If Mr. Robert called Anita at 23:00 local time, how much time was shown at Anita's house?
Based on the problem, students will solve the problem until the discovery of the solution, then the process of solving student problems is described by Polya's phase. Students' answers are displayed in part according to the phase of the Polya.

3.1 Understanding the Problem

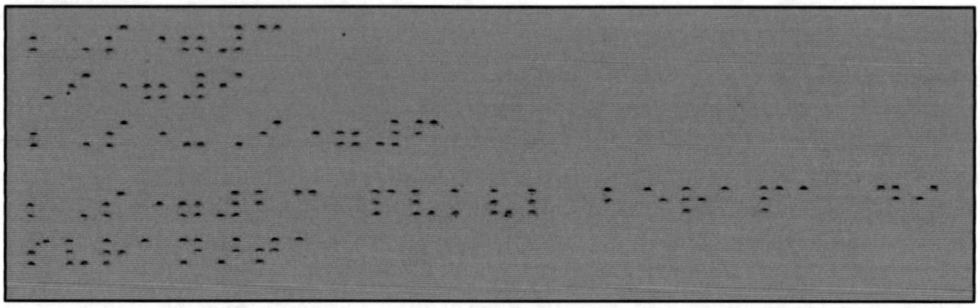

Figure 1. Photograph of Answer Writing Test in The First Activity

Based on Figure 1 shows the student's answer about what is known and asked. The known thing in the matter is written in the coded braille. Spain is given code "S" and "I" for Indonesia. Writing time was shortened without writing completely. At 09.00 written 9 and 03.00 written 3. Questions asked on the matter also written with the code S = 23 at what time in Surakarta. Writing answers in braille still have some errors. The errors are spelling errors and punctuation marks on braille. Write a symbol in braille there must be a mistake let alone by writing conventionally from right to left [27]. The errors found in the student's answers are not fatal. Based on interviews with teacher companion students said the students are not yet proficient in reading and writing braille. But for simple writing still able to follow. So in the practice of this study, researcher help student by reading the problem three times. Without the help of information through hearing, student seems confused. Regardless of student writing errors, students can understand the problem with a well-proven student is able to know what is at the core of the problem. Based on the picture 1 the student has gone through the process of solving the problem in Polya phase, that is understanding the problem.

Table 1. The Result of First Activity Interview

Code	Interview Result
R	*"How many times have you read the question?*
A1	*"Twice Sir*
	(after that the researcher read out the questions three times)
R	*"Do you already understand about the problem?*
A2	*"Yes Sir*
R	*"What do you understand from the problem?"*
A3	*"Time in Spain at nine am, in Indonesia at three am, if in Spain 11 at night, ask what time in Surakarta?"*

Table 1. The Result of First Activity Interview

Code	Interview Result
R	*"Now, let me explain what you wrote"*
A4	*"S = 9, I = 3, S - I = 6, S = 23 then in Surakarta?, its S is Spain Sir, I same as Indonesia, so In Spain at 9 o'clock then in Indonesia at 3 o'clock, I keep looking for it is the time difference spanyol same indonesia, 6 hours."*
R	*"Well if that's what you mean s = 3 and so on?"*
A5	*"If in Spain at 11 o'clock then in Surakarta what time, so that's what asked "*
R	*"Why did you immediately write down the time difference between Spain and Indonesia?"*
A6	*"That's to find the answer sir. It used to be told Mr. X (the name of his teacher) there are students here who participated in swimming championships abroad. It's also different. That's why I wrote it right now"*
R	*"Does it mean to have problems like this?"*
A7	*"Yes Sir"*
R	*"Then, after knowing the time difference?"*
A8	*"Because the clock is first in Indonesia, so stay on add the difference earlier with the hour in Spain"*

Interview Activity 1 in Sections A2, A3, A4, and A5 indicates that the student has done the first process in Polya phase that is understanding the problem. Understanding the problem that a student does is to read twice by fingering problems in braille. The student understanding of the problem is reflected in the students understand what is known and to ask and how to find the solution [22]. Based on the result of the student' test and interview, a student with visual impairment in solving problems based on Polya phase, understanding the problem can be said well. Understanding this problem, students read twice by touching the text and through hearing three times. Information gleaned through the sense of touch and touching of the hearing can make students understand the problem. The first thing a student should do is understand the real-world problem and then formulate it into a math problem. Making connections between real-world problems and math problems is the hardest part. This activity requires critical thinking. This makes solving problems related to mathematical literacy problems useful in developing real-world problem-solving. So students can have basic skills of the 21st century

3.2 Devising a Plan

Visible to the students' work through Figure 1 that in addition to the known and asked questions, the student looks for and write the time difference between Spain and Indonesia. This indicates that student start planning problem solving and the students' initial plan is to find the time difference between the two countries. This is also reinforced by the results of interviews on A6, A7, and A8 that students have done the devising a plan. Devising a plan, student summons the knowledge gained from experience with similar problems [22]. The experienced student received was conveyed by previous teachers on issues based on reality. Problems conveyed by teachers by taking samples from the student to make the student understand the core problem. So that the essence of math material taught can be well understood by students [6] [7] [8]. Based on the results of a written test of student and interviews, the student with visual impairment in problem-solving based on the Polya phase, devising a plan by summoning knowledge and experience gained previously. The knowledge and experience of the student are gained through the process of learning in the process. Mathematics learning in the classroom has included examples of real-world problems to solve using mathematical concepts, facts, procedures, and principles. Collaborating information from previously acquired knowledge and experience with new information from problems to find solutions is critical to the 21st-century problem-solving.

3.3 Carrying Out the Plan

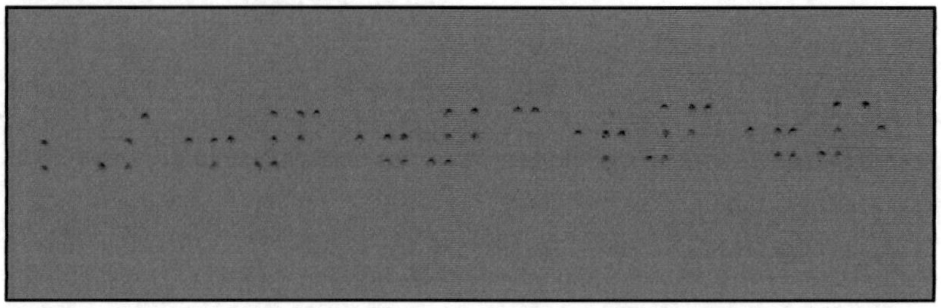

Figure 2. Photograph of Answer Writing Test in The Second Activity

Seen in Figure 2 that the student finds the solution to the proposed problem. The solution has been planned then student writes in braille. Writing answers using the code, S + 6 = 23 + 6 = 5. S is a code from Spain then added with six. 6 is the time difference between Spain and Indonesia. Then S is substituted with twenty-three and added by six to five. The workmanship of the student has succeeded in doing according to plan and in accordance with mathematics principle.

Table 2. The Result of Second Activity Interview

Code	Interview Result
R	*"Try to explain what you write?*
A9	*"S + 6 = 23 + 6 = 5, so like last sir, because in Indonesia first 6 hours and in know in Spain at 11 pm. I counted 11 nights plus 6 hours, he answered at five in the morning* (pointing fingers, counting from twelve, one, two, three, four, five)"
R	*"Why do you write down your answer briefly, like at 23:00, you write 23 only?"*
A10	*"I am still learning braille sir"*

In the interview section, A9 illustrated that the previously prepared plan is well implemented by the student. The student explains in detail the purpose of the written answer. In fact, student reinforces the argument by giving a calculation mark by using a finger. Based on the result of the students' written test and interview, the student with visual impairment in solving the problem based on Polya phase, carry out the plan, the student carries out the plan according to the pre-made [22]. The plan is structured systematically based on mathematical principles. The use of mathematical principles in solving real-world problems well is one indicator of 21st-century skills. This means that in carrying out the plan the student has been able to perform of problem-solving leading to 21st-century skills.

3.4 Looking Back

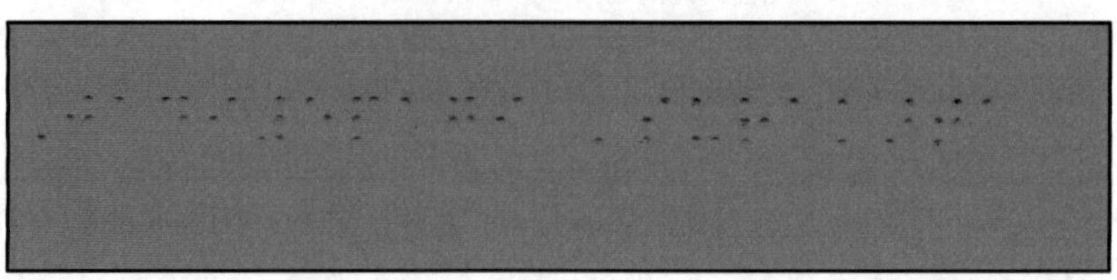

Figure 3. Photograph of Answer Writing Test in The Third Activity

In Figure 3 shows the student's last answer. It appears that in Figure 3 the student concludes what he has done. The conclusion says at five o'clock in the morning. The student's conclusions do not seem so precise. Five o'clock in the morning is not the final answer to the question. Still, need one more phase to solve the problem.

Table 3. The Result of Third Activity Interview

Code	Interview Result
R	*"Is your work done?*
A11	*"It's finished Sir*
R	*"How many results?"*
A12	*"Five o'clock in the morning at Anita's house, when in Spain at eleven o'clock"*
R	*"Are you sure the answer is at five in the morning?"*
A13	*"Sure sir"*
R	*"Have you checked your answer?"*
A14	*"Already sir, I check twice* (then students re-read their work)"
R	*"Try reading again?* (researchers tried to assert*)"*
A15	*"Just a moment sir.* (students start reading from scratch, from the given problem)*"*
R	*"It's done?*
A16	"(Silent. Looks doubt on students, students scratching their head*)"*
R	*"(researchers read out the problem)"*
A17	*"Just a moment sir* (visible students feel the problem and answer again. The student face looks serious, there are wrinkles on the forehead Students write back.)"

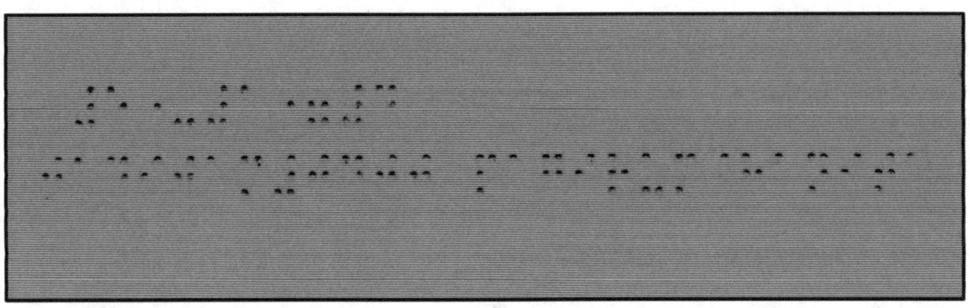

Figure 4. Photograph of Answer Writing Test in The Fourth Activity

In Figure 4 there is an additional answer to the previous answer. The previous answer is added with 1. Visible with the addition of the answer the student actually has found the solution. The solution requires a variety of other knowledge related to the problem posed.

Table 4. The Result of Fourth Activity Interview

Code	Interview Result
A18	*"It's finished Sir"*
R	*"The answer remains at five in the morning?"*
A19	*"Not Sir. The answer is 4 am. Earlier I forgot that still in WITA (Indonesia Middle Time). Surakarta entered in WIB (Indonesia West Time)."*
R	*"Sure this time the answer at 4 am?"*
A20	*"Sure sir. I thought long ago, recall the difference between time and WITA. The answer I checked 3 times sir".*
R	*"What do you think is the time difference between WIB and WITA?"*
A21	*"Difference time one hour, first an WITA Sir"*

Table 4. The Result of Fourth Activity Interview

Code	Interview Result
R	*"You are sure? Not faster WIB?"*
A22	*"Sure sir. I remember the sun rising from the east, so it was faster WIT, then WITA, last WIB. Each one hour difference"*
R	*"Four o'clock in the morning the solution?"*
A23	*"Yes Sir"*
R	*"Is there any other way to solve it?"*
A24	*"(*silent a few minutes) *Do not know sir. I know it's just that way"*

Based on interviews on the A14, the student makes the process of looking back the answer twice. Examination by fingering the answer in braille. Then, in the interview in section A15 the student solidifies the answer by reviewing the answers that have been written. The researcher tried to dig deeper information on the student because the solution given by the student is not yet a final solution. At first, the student remained confident about the solutions found (A14 and A15). However, after the student re-read the problems and answers and the researcher read about the faces of the doubts and gestures of the body that signal doubts (A15 and A16). Afterwards, the student appears seriously continuing to work on (A17). There is an indication that the student forgot to change the time at the time of western Indonesia first (A19). The forgetfulness that occurs may be the inaccuracy of the students themselves and because student obtains information only from the sense of touch and hearing without involving the sense of sight then something is missing from the information. As has been disclosed that in learning 80% of information is obtained from sight [15]. This can result in the loss of some information that will lead to the brain. Regardless of what the pupil resumes work and finds the final solution to the problem (A19)(A20)(A22)(A23). Even student check the answer back three times. However, the student has not been able to find another way to solve the problem. So it can be concluded that at the stage of problem-solving phase fourth Polya that is looking back, students check again by groping the answer as much as 2-3 times [22]. However, the student has not been able to find other ways to solve the problem. Checking answers repeatedly identifies that the student is thinking about how they are being used accordingly or not. This means students in the stage of looking back students experience the process of metacognition.

4. Conclusion

Based on the objectives and data analysis that has been done by the researcher can be concluded that solving problem of literacy based on Polya's phase can be passed by the students with visual impairment well, although there are some deficiencies. In the phase of understanding the problem, the student read about twice by brushing the text and assisted with information through hearing three times. The student with visual impairment in problem-solving based on the Polya's phase, devising a plan by summoning knowledge and experience gained previously. At the phase of carrying out the plan, students with visual impairment implement the plan in accordance with pre-made. In the looking back phase, students with visual impairment need to check the answers three times but have not been able to find a way. At the solving of problems based on the Polya's phase undertaken by the student with visual impairment, it is found that solving the problem leads to the solving of 21st-century skills problems. In the phase of understanding the problem and looking back, it is identified that there are 21st-century skills of critical thinking and metacognition.

Some things seem to need to be addressed for the advancement of mathematics education for the visually impaired. The problem solving for the blind is good enough because of the problem solving oriented learning by taking the real world problem. For that reason, teachers should provide more such learning. By giving a problem in a real context it will develop student problem-solving abilities. The material essence in mathematics learning can be well received by students. Second, students with visual impairment who obtain information from other than the senses of vision should develop abilities in braille literacy. Basically when writing and reading braille or good braille literacy skills then problem solving related to math literacy will also increase.

Based on the results and findings of the study there are several suggestions that researchers recommend. The phase of understanding the problem, students should read the problem or listen to the problem repeatedly so that the students really understand their problems. Write down all the information that is on the issue in braille to minimize the forgotten information. Imagine the problems facing real as happened thus forming an internal representation of the problem. The phase of devising a plan, students should read the back issues to examine plans that have been made before carrying out his plans. Students must plan more than one, when the first plan fails then the next plan can be used. The phase of carrying out the plan, the plan that has been prepared is applied in detail and thorough. This is so that errors in calculations can be minimized. Students must understand what is written, meaning that students know the reason for each step of work in solving the problem. The phase looking back, students should re-understand their problem, then read the information that has been written. Contemplating the plan made whether it has been appropriate or not to solve the problem at hand. Check each calculation that has been made. Check each calculation that has been made. Checking the final answer whether in accordance with the solution of the issue at hand. In addition, there are indications of critical thinking and metacognition on the blind side in solving problems can be used as a study material for further research.

References

[1] Tai W C and Lin S W 2015 Relationship Between Problem Solving Style and Mathematical Literacy *Educational Research and Reviews* **10** 1480-6

[2] P21 2007*a The Intellectual and Policy Foundations of the 21st Century Skills Framework.* (Washington DC: Partnership for 21st Century Skills)

[3] P21 2007*b 21st Century Curriculum and Instruction (*Washington DC: Partnership for 21st Century Skills)

[4] Wilson J W, Fernandez M L, and Hadaway N 1993 Mathematical Problem Solving In Problem Solving *Research Ideas for the Classroom: High School Mathematics*, ed Wilson (New York: Macmillan)

[5] Halmos P 1980 The Heart of Mathematics *American Mathematical Monthly* **87** p 519-24

[6] Olkun S and Toluk S 2002 Textbooks,Word Problems, and Student Success on Addition and Subtraction *International Journal for Mathematics Teaching and Learning* **11** p 162-70

[7] Rizvi N F 2004 Perspective Teacher's Ability to Pose Word Problem *International Journal for Mathematics Teaching & Learning* **10** 79 -88

[8] Carlan V G, Rubin S, and Morgan B M 2005 Cooperative Learning, Mathematical Problem Solving and Latinos *International Journal for Mathematics Teaching and Learning* **6** p 47-56

[9] Branca N A 1980 *Problem Solving as a Goal, Process, and Basic Skill In Problem Solving in School Mathematics 1980* Yearbook of the National Council of Teachers of Mathematics, ed S Krulik and R. E. Reys (Reston, VA: National Council of Teachers of Mathematics)

[10] Bell F H 1978 *Teaching and Learning Mathematics in Scondary School* (New York : Wm C. Brown Company Publiser)

[11] Stacey K 2011 The PISA View of Mathematical Literacy in Indonesia *Journal on Mathematics Education (IndoMS-JME)* **2** 95-126

[12] OECD 2013 *PISA 2015 Draft Mathematics Framework* (Paris: OECD publising)

[13] Ojose B 2011 Mathematics Literacy: Are We Able to Put the Mathematics We Learn Into Everyday Use? *Journal of Mathematics Education* **4** 89-100

[14] Tokada D, Herman T, and Suhendra 2017 Discovery Learning for Mathematical Literacy Ability *in Journal of Physics: Conference Series* **895** p 1-5

[15] UNESCO 1994 *The Salamanca Statement and Framework for Action on Special Needs Education Adopted by the World Conference on Special Needs Education: Access and*

Equity (Paris: UNESCO)
[16] Hill E W and Blasch B 1980 *Concept Development;Foundations of Orientation and Mobility* (New York: American Foundation for the Blind)
[17] Toenders F, De Putter L, Sanders W and Brok P D 2017 Analysing the Physics Learning Environment of Visually Impaired Students in High Schools *Phys. Educ* **52** p 1-8
[18] Quek F and *McNeill* D 2006 Embodiment Awareness, Mathematics Discourse, and the Blind *Annals of the New York Academy of Sciences* **1093** 266–79
[19] Sahin M and Yorek N 2009 Teaching Science to Visually Impaired Students: A Smallscale Qualitative Study *US-China Education Review* **6** 19–26
[20] Andreou Y and Kotsis K 2005 The Estimation Of Length, Surface Area, and Volume by Blind and Sighted Children *International Congress Series* **1282** 780-4
[21] Piaget J 1952 *The Origins of Intelligence In Children* (New York: International Universities Press)
[22] Polya G 1973 *How to Solve It Second Edition* (New Jersey: Princeton University Press)
[23] Mayer R E and Wittrock M C 1996 Problem Solving Transfer *Handbook of Educational Psychology* ed R. Calfee, R. Berliner (New York: Macmillan)
[24] Scott L C 2015 *The Futures Of Learning 2: What Kind Of Learning For The 21st Century?* Education Research and Foresight Working Papers (Paris: UNESCO)
[25] Oktiningrum W, Zulkardi, and Hartono Y 2016 Developing PISA-Like Mathematics Task with Indonesia Natural and Cultural Heritage as Context to Assess Students' Mathematical Literacy *Journal on Mathematics Education* **7** 1-8
[26] Mahajan S 2014 *Mathematics for the 21st Century: What Should Students Learn?* (Boston, Massachusetts: Center for Curriculum Redesign)
[27] Kway E H, Salleh N M, and Majid R A 2010 Slate and Stylus: An Alternative Tool for Braille Writing *Procedia Social and Behaviour Sciences* **7** p 326-35

IOP Conf. Series: Journal of Physics: Conf. Series **1008** (2018) 012069 doi:10.1088/1742-6596/1008/1/012069

Interference thinking in constructing students' knowledge to solve mathematical problems

W E Jayanti[1], B Usodo[1] and S Subanti[1]

[1]Department of Mathematics Education, Universitas Sebelas Maret, Indonesia

E-mail: wandaekajayantiHZ@yahoo.co.id

Abstract. This research aims to describe interference thinking in constructing students' knowledge to solve mathematical problems. Interference thinking in solving problems occurs when students have two concepts that interfere with each other's concept. Construction of problem-solving can be traced using Piaget's assimilation and accommodation framework, helping to know the students' thinking structures in solving the problems. The method of this research was a qualitative method with case research strategy. The data in this research involving problem-solving result and transcripts of interviews about students' errors in solving the problem. The results of this research focus only on the student who experience proactive interference, where student in solving a problem using old information to interfere with the ability to recall new information. The student who experience interference thinking in constructing their knowledge occurs when the students' thinking structures in the assimilation and accommodation process are incomplete. However, after being given reflection to the student, then the students' thinking process has reached equilibrium condition even though the result obtained remains wrong.

1. Introduction

Basically, the process of mathematical learning is closely related to the formation and using of thinking ability. This is due to the influence of 21st-century skills that require students to have the ability in understanding the concepts and improve high-quality thinking skills in analyzing the concepts, evaluating and finding alternative solutions to solving the problems [1]. In the same direction with the 21st-century skills, where each students must have metacognition that is used to refer to the awareness individuals have their own thinking, their evaluation that thinking, and their regulation of that thinking [2]. When students have been able to form their thinking, it allows students to have the ability to understand the concept to solve a problem.

While students in understanding a concept then they receive new information obtained from their brain works. The information will be processed to find a solution in justification that will be stored in the memory of brain. Furthermore, according to information processing theory which states that when information comes from the environment, then the information is processed temporarily (perceived), and puts into short-term memory, continued into long-term memory [3]. During the process of recalling information contained in the long-term memory, which is used to solve the problem in the short-term memory may occur interference or exchange of information. This form of interference is called interference thinking.

Interference thinking in solving problems occurs when student has a construction of two or more different concepts in which the two or more concepts are interrelated [4]. Meanwhile, interference refers to the lack of ability to recall information when it is similar to the one that stored in memory [5].

Interference thinking is an error that occurs because of the conception that interferes each other, so one concept disturbs or interferes another concept [6].

Rodieger et. al (2010) in their book discusses Mc Geoch's statement that forgetting comes from the results of memory error accessing information is divided into two forms, they are due to proactive interference and retroactive interference [7]. Proactive interference is old information interfering with the ability to remember new information. Meanwhile, retroactive interference is new information interfering with the ability to remember old information [8].

Problems that can cause interference thinking in mathematics quite a lot, among others: the comparison of value and reversed values, Least Common Multiple (LCM) and Greatest Common Divisor (GCD), angluar and contradictory, permutations and combinations, and so forth [4,6]. However, in this research, the problem that can cause interference thinking is the sequence and series.

Based on research, interference output occurs when two categories are tested separately rather than tested together [9]. This is in accordance with the case of students in this research, ie students are not only difficult to distinguish the problems done using sequences and arithmetic series but also sequences and geometry series when displayed together. Whereas for solving sequence and arithmetic series problems with sequences and geometry series are different, arithmetic sequences have a distinction denoted by b and the difference between fixed rates, whereas the geometric sequences have a ratio denoted by r and the value of comparison between indefinite tribes [10].

Less optimal problem-solving in learning mathematical and instructional emphasis on procedures, resulting in errors in the construction of mathematical problem-solving [11]. Problem-solving is a major part of learning math. Problem-solving skills can be transferred to solve other problems in life [12]. The importance of developing problem-solving in mathematical learning is expressed by NCTM which states that "solving the problems is not only a goal of learning mathematical but also a major means of doing so... By learning to problem-solving in mathematical, the student should acquire ways of thinking, habits of persistence and curiosity, and confidence in an unfamiliar situation" [13]. Meanwhile, according to Polya problem-solving is interpreted as an attempt to find a way out from a difficulty in order to achieve a goal that is not immediately achievable [6].

Construction of problem-solving depends on the construction of mathematical concepts experienced. Good concept construction on students will help to simplify the process of problem-solving construction [4]. It can be traced using Piaget's assimilation and accommodation framework, helping to know the student thinking structures in solving the problems.

The assimilation process is the process of integrating new stimulus into the scheme that has been formed. The accommodation process is a process of integrating the new stimulus through the modification of the old scheme or through the formation of a new scheme to adapt to the newly acquired stimulus [14]. The balance between assimilation and accommodation are called state of equilibrium and disequilibrium occurs when the child is in a new environmental phenomenon that is incompatible with the child's mental scheme [15].

The analytical process of assimilation and accommodation has been illustrated by Subanji and Supratman [14]. So it can help to understand the two forms of assimilation and accommodation processes.

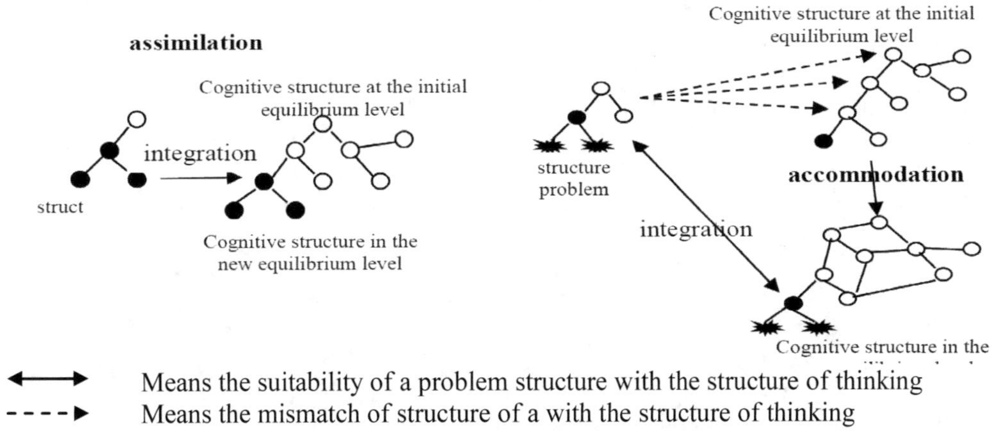

Figure 1. Assimilation and Accommodation Process (Adapted Subanji and Supratman, 2015)

In the process of solving the problem, when students are faced with a construction of two concepts in which the structure of thinking the student has no scheme in accordance with the problems it faces the interference will occur thinking. This is due to the similarity between the two concepts that interfere with each other. Therefore, this research will describe the process of interference thinking in solving the problem of sequence and arithmetic series with sequences and geometry series based on assimilation and accommodation framework.

2. Research Methods

The method of this research was a qualitative method with case research strategy. The qualitative research is aimed to uncover a problem and develop it in detail to understand the central phenomenon of a problem [16]. Subjects in this research involving 28 students of 9th grade in Junior High School Al-Islam 1 Surakarta that were selected by purposive sampling. Those data were analyzed descriptively using Miles and Huberman steps.

The data in this research involves the result of solving the problem of students' interference test and interviews transcripts about students' errors in solving the problem. This interview was conducted to trace the causes of students' interference thinking in solving the problem. The interview was a semi-structured interview. The focus of the research was to analyze the interference of students' thinking in solving the sequence and series problem based on assimilation and accommodation framework from Piaget's. However, this study will only describe students who have only proactive interference.

3. Results and Discussion

Results and discussion obtained by researcher can trace the process of interference thinking in solving the problem of sequence and series. The tests were conducted on 28 of 9th-grade junior high school students and the results of classification of interference thinking will be presented in Table 1 as follow.

Table 1. Classification of Thinking Interference

Type of Interference	Students
Proactive	7
Retroactive	3

Based on Table 1, it was found that the classification of students' thinking interference has two types of interference, namely proactive interference and retroactive interference. But, this research focused on students who have only proactive interference. Therefore, Subject (S) was chosen to be a

research subject that represented proactive interference. Moreover, the problem related to geometric sequences for proactive thinking interference are as follows:
A car has been purchased for Rp. 80.000.000,00. The selling value becomes ¾ of the previous price every year. What is the sale value of the car after 3 years of use?

The following will be presented on the answers of students who experience proactive interference on the matter of sequence and series. When the student has given a matter of geometry sequence, but at the time of answering the question, the student responded by using an arithmetic sequence formula. The result of the student work is shown in Figure 2 below.

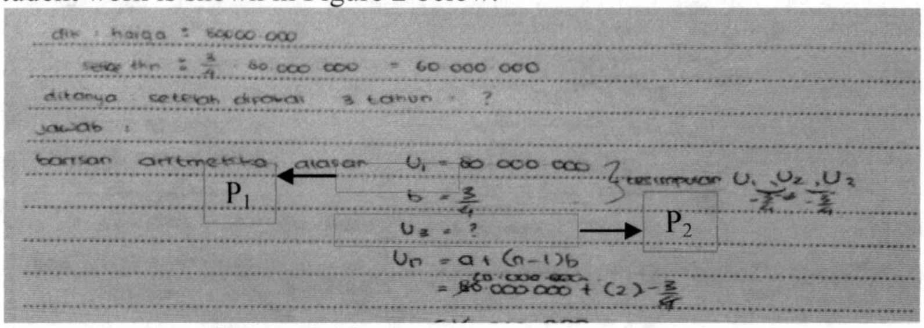

Figure 2. The result of student work

Information :
P_1 = Problem 1
P_2 = Problem 2

Based on Figure 2 above, interference test results show that the student solves the geometry sequence with the completion of the arithmetic sequence. At the beginning of the construction process of her knowledge, S writes the known and asked questions. After that, S states that the question is a matter of arithmetic sequence with reasons U_1 = 80.000.000 and difference (b) = $\frac{3}{4}$. For the first problem (P_1), the student assumed that $\frac{3}{4}$ is the difference. Furthermore, for the second problem (P_2), the student assumed that formula $Un = a + (n - 1)b$ is the right solution to solve the problem. Therefore, S searched for the 3rd term by using the formula $Un = a + (n - 1)b$. Having obtained the calculation of the 3rd term, S determined the value of U_3 for 14.000.000.

The researcher conducted interviews to find out more in the process of work S. Interviews show that S considered important information on the matter is the selling value to 3/4 of the previous price every year. According to her, 3/4 is the difference so that the selling price per year is fixed. In addition, students considered other important information is the asking price after 3 years of use. From the information known and asked, S could determine the solution to the problem by using arithmetic sequence. Below will show interview conversation as follows.

Researcher : From that question, what information is obtained?
Subject : Hmm.. That, miss (*Students pointed toward the question*)
Researcher : Please, pay attention back to the question.
Subject : This is, miss. A car has been purchased for Rp. 80.000.000,00. It means, the purchase price of his car is Rp. 80.000.000,00 (*Student mumbling*)
Researcher : What's next?
Subject : The selling value becomes ¾ of the previous price every year.
Researcher : Next?
Subject : Continue.. What is the sale value of the car after 3 years use?.

Based on the dialogue above, after the student obtains the information data and information questions, it makes it easier for the student to understand the problem. Furthermore, the interview was continued to find out how the students drafted the settlement plan on the question as follows.

Researcher : After you got the information, what's steps do you take?
Subject : From the question, I can find out that a = 80.000.000.
Researcher : What is called a?
Subject : a is the beginning, the first term, miss.
Researcher : After that, what should you do?
Subject : Looking for the difference, miss. Because the sale value becomes ¾ from the previous price. So.. ¾ it is the difference, miss ..
Researcher : Why do you call ¾ the difference?
Subject : Yes, It is because each year the selling value is constant, miss. As I recall, if it is constant, it calls difference. But, I also forget miss. (*Students smile and look confused*)
Researcher : How should it be?
Subject : Hmm.. Is it the ratio, miss? (*Students look confused between the difference and the ratio*)
Researcher : What is the ratio?
Subject : The ratio is not fixed value, miss.
Researcher : Then, what is the different between the ratio and the difference (b)?
Subject : As I far as I know, the ratio is multiplication and the difference is a sum, miss.
Researcher : What does it mean?
Subject : Yaa .. The ratio is the value of each tribe multiplied, but if the difference is the value of each tribe summed, miss. (*Students are hesitant in answering and looking up while recalling*)
Researcher : So, ¾ what is it?
Subject : The difference is, miss.

From the dialogue, it appears that the subject is very confident with the answer that the answer is correct. In the conversation above, the student is indicated to experience interference thinking. As Subanji has pointed out that interference thinking occurs when the student has a construction of two or more different concepts in which the two or more concepts are interrelated [4]. When S preparing a problem-solving plan, student seems to forget in determining ¾. The existence of error about the construction of the concept experienced by student causes the student could not distinguish between the ratio with a difference, so that arithmetic sequence concept interferes geometry sequence concept. Researchers continued the dialogue with the student by providing a trick question that leads to conflict cognitive.

Researcher : After that, what else should you do?
Subject : Hmm, the value of a same b already existed, after that continued to U_3, miss. So I was using arithmetic.
Researcher : What is your reason for using an arithmetic sequence? Do you know the formula for arithmetic sequences?
Subject : As I recall that the difference must be using arithmetic, miss. (*Student smiling*). Then, I was looking for the answer with the formula $Un = a + (n - 1) b$. Next, continued to include $a = 80.000.000$ and $b = ¾$. After I counted, the result was 14.000.000.
Researcher : Are you sure about your answer?
Subject : Hmm.. Not really, miss (*Students shook her heads and looked down*)
Researcher : Why are you unsure of the answers you get?
Subject : Actually I am still confused must use difference or ratio, miss.
Researcher : Can you use information to rework it in a different way?
Subject : (*pause for a moment*) May I try again, miss?
Researcher : Yes, please.

Prior to the interview, the results of the analysis of interference thinking experienced by S based on the assimilation and accommodation framework show that when S faced a problem, S actually had a scheme related to the sequence and series problem. This shows the students' thinking structure occurs assimilation process. However S used the arithmetic sequence scheme into the geometry sequence, so the answer is not appropriate. The student's thinking structure before being given a reflection in solving the geometry sequence problem and assuming the settlement using the arithmetic sequence is presented in the diagram flow as follows.

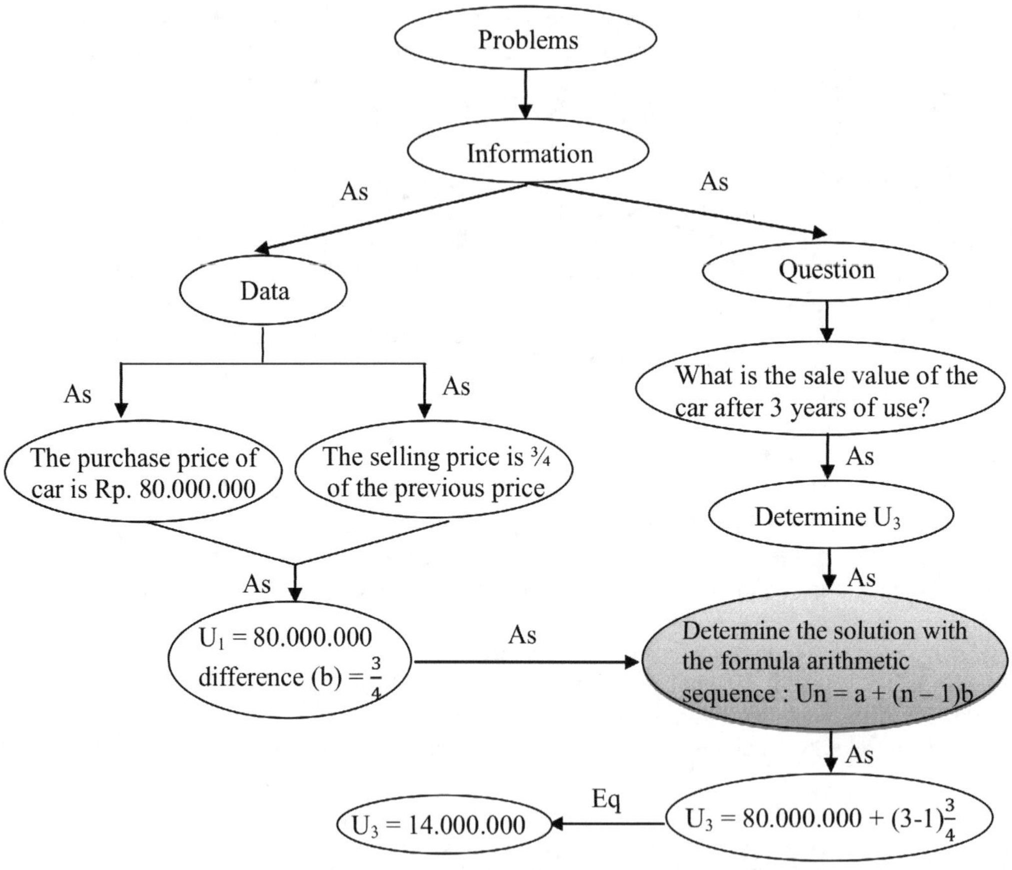

Diagram 1. Students' Construction Process Who Experience Thinking Interference Before Reflection

Information:
As : Assimilation Process
Eq : Equilibrium Process

Based on Diagram 1 above, it appears that when students are faced with a problem about the geometry sequence, the student obtains the data information and the question information. However, as the student construction her knowledge to solving the problem, the assimilation process dominates her thinking space. Assimilation occurs when S knows that "The selling value becomes ¾ of the previous price every year". Meanwhile, S considers that if every year the sale value of the car becomes ¾ from the previous price then ¾ is a difference value always fixed. Regardless of the prior information and questions were given, S determines that if the values of a and b are known in solving the problem, the next step is to use the arithmetic sequence formula to solving the problem. After the

IOP Conf. Series: Journal of Physics: Conf. Series **1008** (2018) 012069 doi:10.1088/1742-6596/1008/1/012069

calculation by S then obtained the result U_3 = 14.000.000 and this signifies when doing the test results obtained to reach disequilibrium conditions. This means S uses assimilation as the process of thinking.

After the interview, the student thinking process in solving the problem was still at the assimilation stage. However, when S was asked again "Why is ¾ called difference?" and S responded with doubt whether it is a ratio. The question was asked, "What is different of ratio with a difference (b)?". This suggests that when S thinks back to different in ratios and differences (b), it means that in the students' thinking process, S experiences accommodation process and occur the interference of thinking. Thus, S interprets the arithmetic sequence scheme into a geometry sequence. Once given a new stimulus, then the thinking process S arises the state wants to try to resolve the problem.

In the process of problem-solving, students are faced with challenges that arouse curiosity to solve them. The curiosity indicates an imbalance between assimilation and accommodation process called disequilibration [4]. Therefore, the student thinking process continued on accommodation process, where student modified the old scheme to form a new stimulus. The following will show the results of the work S when doing reflection on the problems encountered.

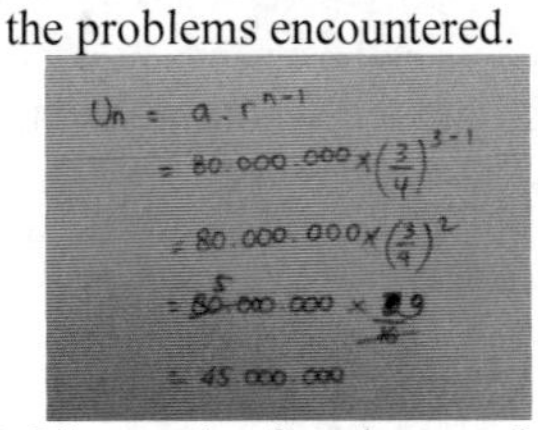

Figure 3. The result of student work reflection

Based on Figure 3 the results of student work reflection can be analyzed further using student's thinking structure in solving the problem of geometry sequence. The structure of students' thinking interference after being given a reflection as follows.

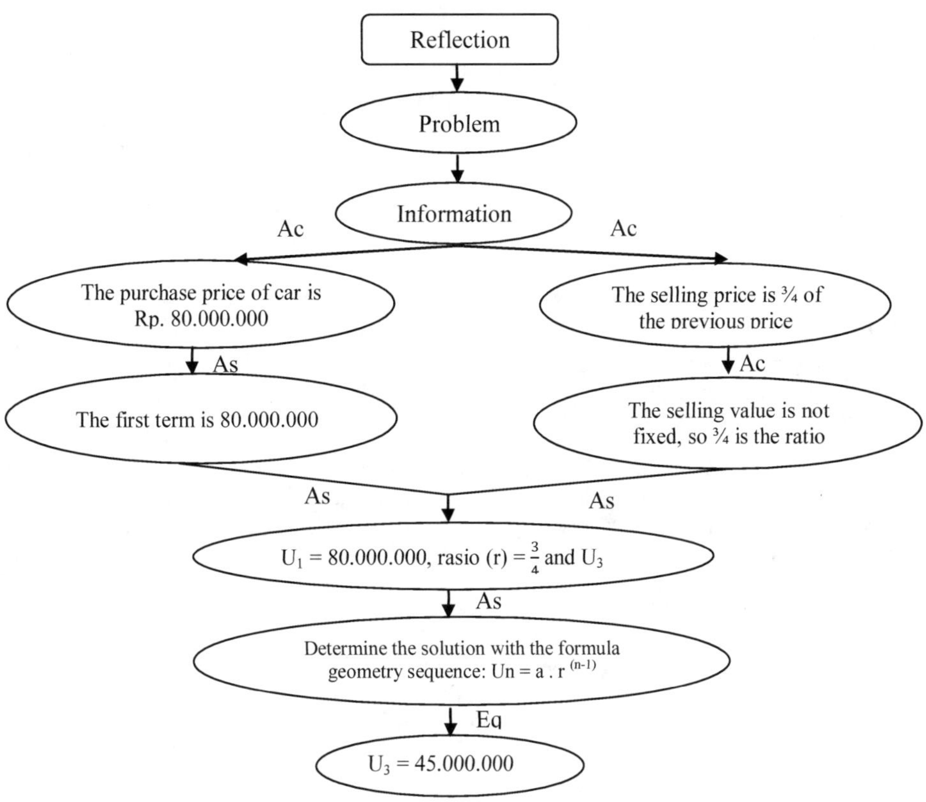

Diagram 2. Interference t ink ng Structure After Reflection

Information:
As : Assimilation Process
Ac : Acommodation Process
Eq : Equilibrium Process

Based on Figure 3, when student rewrites her work, the thinking process of student begins again through the assimilation process with the view that if each year the car's sale value becomes ¾ from the previous price then ¾ is a ratio whose the value is not fixed. After knowing the first tribe (a) and the ratio, the student uses a geometry sequence formula to solve the problem it faces. It indicates that the accommodation process is in progress. Furthermore, students' thinking interference after reflection shows that the error occurs when the student determines the value of n of the tribe so that at the time of calculating the students' answer had not the correct answer, ie U_3 is 45.000.000.

After reflection of the students' thinking process have reached equilibrium because of the balance between the assimilation and accommodation process, but the result obtained still wrong after the reflection. This is in line with Black and Pope (2008) who stated that the balance between assimilation and accommodation are called state of equilibrium [15]. Then, research conducted by Sukoriyanto, et. al. (2016) also stated that the real global interference is global interference if it continued with reflection, the answer is still wrong [6]. Therefore, it can be said that the student who experiences proactive interference is caused by interpreting the arithmetic sequence as a geometry sequence and while the reflection results obtained is still wrong even though it has reached an equilibrium condition.

4. Conclusion

In this research, the occurrence of interference thinking in the student is characterized by the student interpret the concept of sequence and arithmetic series as a sequence and geometry series. The interference experienced by the student in sequence and series is proactive interference because the students' memory of the sequence and the arithmetic series dominates the thinking space. Moreover, when student constructs her knowledge, the student also constructs what problems faced her. Construction of problem-solving has been using Piaget's assimilation and accommodation framework, thus it can help to know the student thinking structures in solving the problems. Construction of problem-solving depends on the construction of mathematical concepts experienced. Before reflecting on the answer, the student experiences a thinking process of assimilation and accommodation are incomplete. In other words, the concept of arithmetic sequences interferes with the concept of geometry sequence. However, when reflected on the answer, the student reaches equilibrium condition, where there is a balance between the assimilation and the accommodation process even though the final result obtained still wrong after the reflection.

As an implication of these findings, once the teacher knows what the possible interference occurrences happen to the student, the teacher can design the lesson according to the sequence and series material. Therefore, teachers should emphasize an understanding of the context of the problem that has been given about the differences between the two types of sequences and series with the aim that students can make clearer memories of both materials and no matter more dominates the students' thinking space. In addition, teachers should direct students to have metacognition that is used to shape student self-awareness, so that students have the ability to understand concepts, improve thinking skills that trigger to develop critical thinking and creative thinking skills in analyzing and looking for alternative solutions to problems encountered.

Acknowledgments

The authors would like to thank to the referees for their suggestions and this research is part of the thesis of magister study at Sebelas Maret University, Surakarta, Indonesia in the academic year of 2018.

References

[1] Etistika Y W, Dwi A S, and Amat N 2016 Transformasi pendidikan abad 21 sebagai tuntutan pengembangan sumber daya manusia di era global *Proceedings of the National Seminar on Mathematics Education* **1** 263-278

[2] Wilson J and Clarke D 2004 Towards the Modelling of Mathematical Metacognition *Math. Educ. Res. J.* **16** 25–48

[3] Kandarakis A G and Poulus M S 2008 Teaching Implications of Information Processing Theory and Evaluation Approach of Learning Strategies Using LVQ Neural Network *Journal Wseas Transaction on Advances in Engineering Education* **5** 111-119

[4] Subanji 2015 *Teori Kesalahan Konstruksi Konsep dan Pemecahan Masalah Matematika* (Malang: Penerbit Universitas Negeri Malang)

[5] Anderson M C and Neely J H 1996 *Interference Inhibition in Memory Retrieval* in Bjork E L and Bjork R A (Eds) Handbook of Perception and Cognition 2nd Edition pp 237-313 (California: Academic Press)

[6] Sukoriyanto, Nusantara T and Subanji 2016 Student Interference thinking of Real Global Type in Solving Permutation and Combination Problems **6** 47-50

[7] Roediger H L, Weinstein Y and Agarwal P K 2010 *Forgetting: Preliminary Considerations* in Sala S D (Ed) Forgetting 1-22 (New York: Psychology Press)

[8] Wade C and Tavris C 2010 *Psychology* 10th (Ed) (Boston: Pearson Education)

[9] Malmberg K J, Criss A H, Gangwani T H, and Shiffrin R M 2012 Overcoming the Negative Consequences of Interference From Recognition Memory Testing. *Psychological Science Journal* **20** 1-5

[10] Dalal A and Atri A 2014 An Introduction to Sequence and Series *International Journal of Research (IJR)* **1** 1286-1292

[11] Subanji 2011 *Teori Berpikir Pseudo Penalaran Kovariasional* (Malang: Universitas Negeri Malang)

[12] Subanji and Nusantara T 2013 Karakterisasi Kesalahan Berpikir Siswa dalam Mengonstruksi Konsep Matematika. *Jurnal Ilmu Pendidikan (JIP)* **19** 208-217

[13] NCTM 2000 *Principles and Standards for School Mathematical* NCTM

[14] Subanji and Supratman AM 2015 The Pseudo-Covariational Reasoning Thought Processes in Constructing Graph Function of Reversible Event Dynamics Based on Assimilation and Accommodation Frameworks *J. Korean Soc. Math. Educ., Ser. D, Res. Math. Educ.* **19** 61-79

[15] Blake B and Pope T 2008 Developmental Psychology: Incorporating Piaget's and Vygotsky's Theories in Classroom *J. Cross Disciplinary Perspective Educ.* **1** 59-67

[16] Creswell J W 2012 *Research Design: Planing, Conducting and Evaluating Quantitative and Qualitative* (Research United States of America: Person)

IOP Conf. Series: Journal of Physics: Conf. Series **1008** (2018) 012070 doi:10.1088/1742-6596/1008/1/012070

High profile students' growth of mathematical understanding in solving linier programing problems

Utomo[1], TA Kusmayadi[1], I Pramudya[1]

[1]Department of Mathematics Education, Sebelas Maret University, Surakarta, Indonesia

E-mail: utomo@student.uns.ac.id

Abstract. Linear program has an important role in human's life. This linear program is learned in senior high school and college levels. This material is applied in economy, transportation, military and others. Therefore, mastering linear program is useful for provision of life. This research describes a growth of mathematical understanding in solving linear programming problems based on the growth of understanding by the Piere-Kieren model. Thus, this research used qualitative approach. The subjects were students of grade XI in Salatiga city. The subjects of this study were two students who had high profiles. The researcher generally chose the subjects based on the growth of understanding from a test result in the classroom; the mark from the prerequisite material was ≥ 75. Both of the subjects were interviewed by the researcher to know the students' growth of mathematical understanding in solving linear programming problems. The finding of this research showed that the subjects often folding back to the primitive knowing level to go forward to the next level. It happened because the subjects' primitive understanding was not comprehensive.

1. Introduction

Program linear is one of the obligatory materials for competency tests in mathematics class of senior high school students grade XI [5]. Mathematically, the program linear is a part of applied mathematics, with the model of mathematics which is made up of equation or inequality linear, regional work level unit containing a program development for solving various problems [14]. Linear program is widely applied to help solve economic problem, industry, military, and social [16]. Therefore, it is important for the students to comprehend the material well that it will be used in workplaces.

Understanding is an important aspect in learning mathematics [12].Understanding is also the center of developing curriculum in every country [10]. The importance of understanding in learning makes many curricula suggest to create mathematics learning by understanding [12]. In addition, understanding is the point of learning mathematics in school [17]. The importance of understanding makes education experts try to define it. One of the experts is Pirie-Kieren.

Understanding is a level that is not linear and it will take place continuously. Understanding will never stop as long as the learning level takes place and has eight levels of growth such as primitive knowing (PK), image making (IM), image having (IH), property noticing (PN), formalising (F), observing (Ob), structuring (S), and inventising (I) [12,15].

Piere-Kieren defined the description of each level of growth of mathematical understanding; the first level is primitive knowing. The growth of mathematical understanding process starts from this level. This level also becomes the background knowledge of mathematical understanding which is used to construct an understanding about a certain concept. In this research, the students could write the information of the problem. The second level is image making. On this level, the students can make a difference from the previous knowledge and use it in a new way. In this research, the students could write the relation and arrange the information from the problem. The third level is image having. On this level, the students can use the mental illustration about a certain topic without doing any activity that triggers the illustration. In this research, the students could create examples, write constraint and goal functions from the information. The fourth level is property noticing. On this level, the students can manipulate or combine the aspects from an image to construct a characteristic in a context. In this research, the students could develop that the maximum and minimum values can be determined from the goal function. The fifth level is formalising. On this level, the students can create an abstract of mathematic concept based on the characteristic that appears. This research shows that the students can give true meaning to all values that have been obtained, and formalize or confirm how the optimum value is true. The sixth level is observing. On this level, the students can reflect and coordinate formal activities so they are able to be used in some problems that they face as well as express them as theorems. The students also ask about how a question about concepts can be related to each other and how to find patterns in order to define ideas as algorithm or theorem. In this research, the students could say that a linear programming problem has an optimum value at a critical point. The seventh level is structuring. This level happens when the students think about a formal observation as the theory and say a logical argument in form of evidence. In this research, the students could say that a linear programming problem has a limited area of set completion so the optimum value is at the critical point. The eighth level is inventising. On this level, the students can think about a formal observation as the theory and they can get an understanding about a concept as well as ask questions, and then construct a new concept.

There are much research, because of the importance in learning process, in order to improve students' mathematical understanding in the last period [7]. Research that are related to understanding are as followed: Investigating the use of initial knowledge about how to teach logarithm function [4], Illustrating the theory of Pirie-Kieren in the context of language and exam contest [6], The use of Pirie-Kieren's understanding model to investigate the role of definition in proving the theorem [11], and Investigating the increase of students' mathematical understanding that uses learning model of problem posing [3].

In Indonesia, students' understanding about the linear program material is still low. This statement is supported by data based on the results of Senior High School (Science Stream) National Examination in 2015/2016. The results of the examination show that students' absorption in solving daily problems which are related to the linear program is only 52.70%. This shows that the ability of students in solving mathematics problems is still low.

The description above shows that the importance of mathematical understanding in learning process is no exception for linear program material. For that reason, the researcher is interested to investigate the students' growth of mathematical understanding in solving the linear programming problems based on the growth of the mathematical understanding by Pirie-Kieren.

IOP Conf. Series: Journal of Physics: Conf. Series **1008** (2018) 012070 doi:10.1088/1742-6596/1008/1/012070

2. Method

This research aims to find out the students growth of mathematical understanding in solving the linear programming problems based on the theory of Pirie-Kieeren. This research used the descriptive qualitative method. This research was conducted in SMA Negeri 1 Salatiga (Grade XI) whereas the subjects of the research were 2 students. The researcher generally chose the subjects based on the growth of understanding in the classroom from the test result, the mark obtained from the prerequisite material (≥ 75) and teachers' feedback.

The instruments of this research were test worksheets. The students' answers are strengthened by interviews that were used to know the growth of subjects' understanding. The data was obtained from the students' test answers and interviews. The data analysis technique was reducing the data, presenting the data and conclusion. Reducing data is selecting, focusing and simplifying the data obtained. Presenting the data arranges the information in narration and the reducing result is used to get the conclusion.

3. Result and Discussions

At the beginning, I gave a test to the students. It was used to find out students' growth of mathematical understanding toward the problem given by the researcher. The subjects of this research were a male student and a female student. In choosing the subject for this research, the researcher took a look at the prerequisite material which was about inequality, by drawing on an inequality graphic whose value was ≥ 75.

In subject S1 the value of prerequisite material was 75 and for the S2 was 87. I also generally chose the subject based on the students' growth of mathematical understanding in the classroom in solving the problem that was given by the teacher and also feedback from the teacher.

3.1. Result

3.1.1. Subject 1

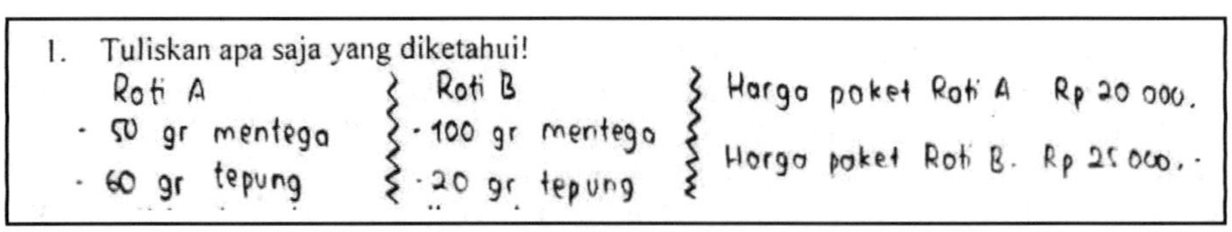

Figure 1. S1's answer for No.1

The initial process done by S1 was writing the information that could be found in the problem. S1 wrote the information about bread A, bread B and the price of the package of each bread. S1 had not written the availability of the butter and the flour (Figure 1). According to Pirrie- Kieren, this level means that the student is in primitive knowing level. S1 also had the prerequisite material to learn the linear program. The prerequisite material was linear inequalities and drawing the linear inequalities' graphics. The student's answer was strengthened by an interview which was done by the researcher with S1. Below is the interview result:

IOP Conf. Series: Journal of Physics: Conf. Series **1008** (2018) 012070 doi:10.1088/1742-6596/1008/1/012070

R : "What can you find from the problem given?"

S1 : "There are bread A and bread B. For bread A, there are 50 gr of butter and 60 gr of flour and there are 100 gr of butter and 20 gr of flour for bread B. Then, the price of package of bread A is Rp20. 000,00 and the price of package of bread B is Rp25.000,00"

R : "Do you think it is enough to know that information only?"

S1 : "(nodding) yes, it is enough"

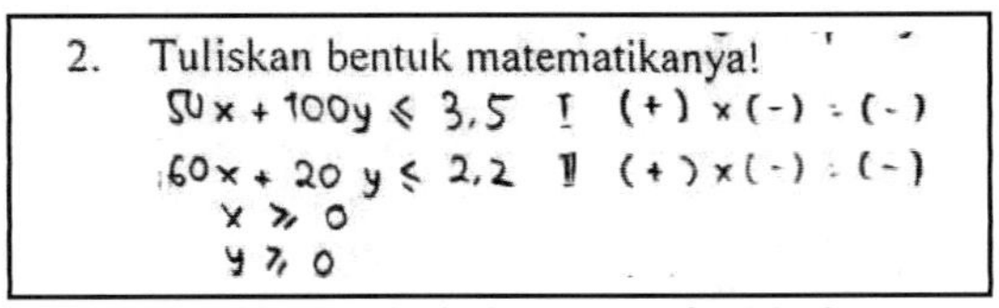
2. Tuliskan bentuk matematikanya!
50x + 100y ≤ 3,5 I (+) x (-) = (-)
60x + 20 y ≤ 2,2 II (+) x (-) = (-)
x ≥ 0
y ≥ 0

Figure 2. S1's answer for No.2

The next step was finding out the relation of the information that was known from the problem. From the written constraints, they show that the student can write the example of the information (Figure 2). When S1 was able write the example, it gave a strong clue that S1 moved from the primitive knowing level to the image making level. It was also strengthened by the result of the interview below:

P : "Okay. I want to ask. So, what is x?"

S : "The butter of bread A"

P : "Is it correct if x is the butter of bread A? Please try to understand carefully, you wrote x in making a package of bread A, it needs 50 gr of butter and 60 gr of flour. So, what is x? Is it butter or what?"

S : "(the student was still reading and smiling)"

The following step was writing the constraints. The constraints were $50 + 100y \leq 3{,}5$ with $60x + 20y \leq 2{,}2$ and $x \geq 0$ with $y \geq 0$. The S1's growth of mathematical understanding moved from the image making level to the image having. S1 faced difficulties when S1 wrote $50x + 100y \leq 3{,}5$ and $60x + 20y \leq 2{,}2$. Therefore, she needed to go back from the image having level to the primitive knowing level. It was strengthened by the interview result of the researcher with S1 as follows:

P : "According to you, what is the mathematical form of that kind of problem?"

S : "The mathematical form is determined by the type, the first is 50x+100y≤3,5. It means that 50 gr of bread A and 100 gr of butter. Then, the butter means the butter."

Figure 3 shows S1's flow of thought. S1 also had a concept that will be used to solve the problem given even though it was not correct. According to Piere-Kieren, S1's answer shows the level from image having level to property noticing level. S1 was able to determine the intersection of the line with the x and y axes. S1 did not change the inequality of $50x + 100y \leq 3{,}5$ into the equation at the beginning. That constraint still needs to be fixed, so the result was $y \leq 350$ which is not correct. In the second constraint, S1 still did the same thing and the result was $x \leq 44$. S1 also did the same thing while determining 17 intersections of two lines, so the intersections of two lines were $x \leq \frac{17}{500}$ and $y \leq \frac{18}{100} \times \frac{1}{100}$. S1 also made the linear inequality graphics and determined 100 areas of set completion even though they were incorrect. The constraints faced by S1 made S1 went back from the property noticing level to the primitive knowing level again.

Based on the data obtained by S1, I can make a scheme from stages of the student's growth of mathematical understanding based on Pirie-Kieren model. The scheme of S1's student growth of mathematical understanding is shown in Figure 4.

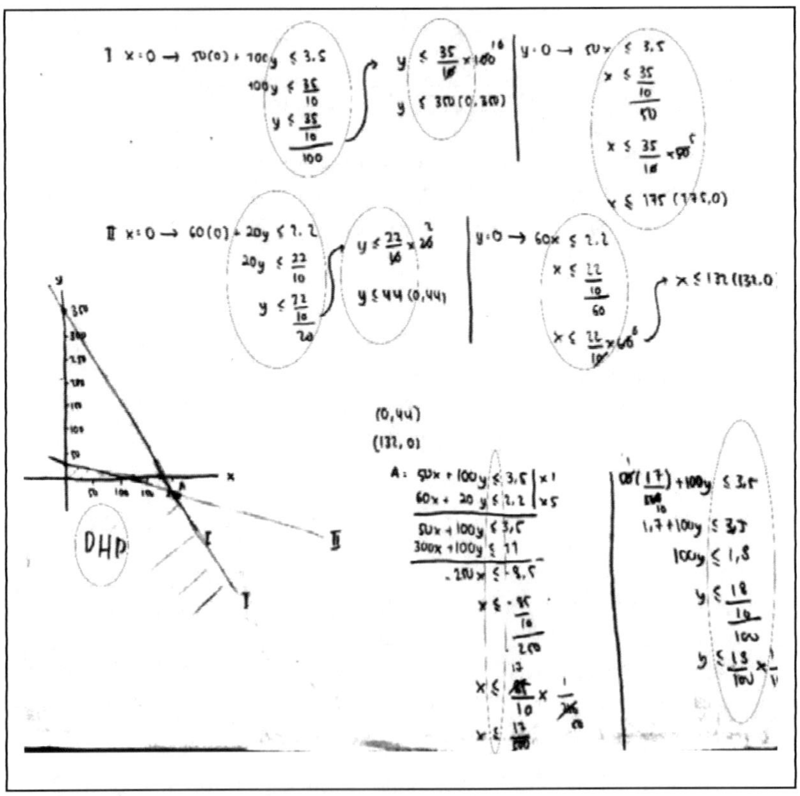

Figure 3. S1's answer for No. 3

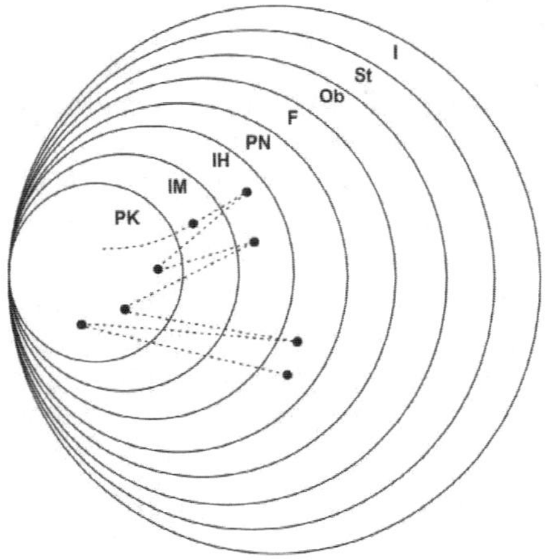

Figure 4. Pirie-Kieren's model of S1's growth of mathematical understanding

Details of Figure 4:

PN : *Primitive Knowing*
IM : *Image Making*
IH : *Image Having*
PN : *Property Noticing*
F : *Formalising*
Ob : *Observing*
St : *Structuring In: Inventising*

-------------------- : Student's growth of mathematical understanding which is incorrect

———————— : Student's growth of mathematical understanding which is correct

3.1.2 Subject 2

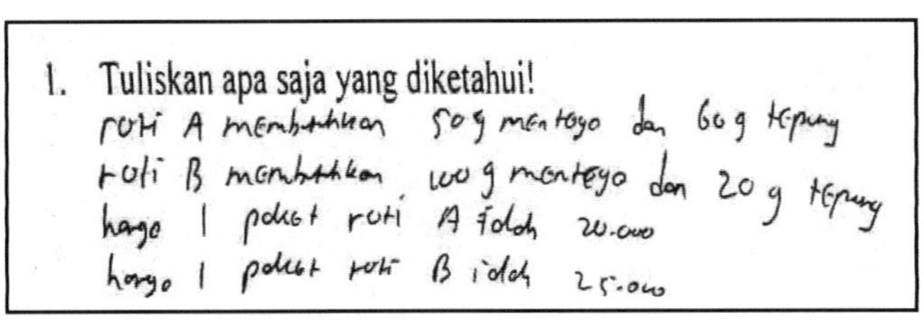

Figure 5. S2's answer for No.1

The initial level done by S2 was writing the information that it could be found in the problem. S2 wrote the information about bread A, bread B and the prices of the package of each bread. S2 had not written the availability of the butter and the flour. According to Pirrie- Kieren, this level means that student is in primitive knowing level. S2 also had the prerequisite material to learn the linear

IOP Conf. Series: Journal of Physics: Conf. Series **1008** (2018) 012070 doi:10.1088/1742-6596/1008/1/012070

program. The prerequisite material was linear inequalities and drawing the linear inequalities' graphics.

The next process was finding out the relation of the information from the problem. It was showed by S2 in Figure 5. S2 could make a good example of this. According to Pirie-Kierin, what was done by S2, there is a movement from the primitive knowledge level to the image making level. Then, S2 wrote the constraints that S1 found in the problem. The constraints were $100m + 20t \geq 0$ and $50m + 60t \geq 0$. The level done by S2 was a movement from the image making level to the image having level. While writing the constraints, S2 faced difficulty in writing so S2 needed to be back from the image having level to the primitive knowing level. S1's student growth of mathematical understanding suits with Pirie – Kierin model which moves into the higher level. The level is property noticing. S2 was able to write the representative of the functions of the goal from the problem given correctly. S2 could determine the intersection from the line even it was not correct. The S2's answer were $100m + 20(0) = (100, 0)$ and $100(0) + 20t = (0, 20)$. The other intersections were $50m + 60(0) = (50, 0)$ and $50(0) + 60t = (0, 60)$. The student had not known how to draw the graphic function from the constraints. Based on the data obtained by S2, the researcher can make a scheme from the stages of student's growth of mathematical understanding based on Pirie-Kierin model. The scheme of student's growth of mathematical understanding is shown in the Figure 6.

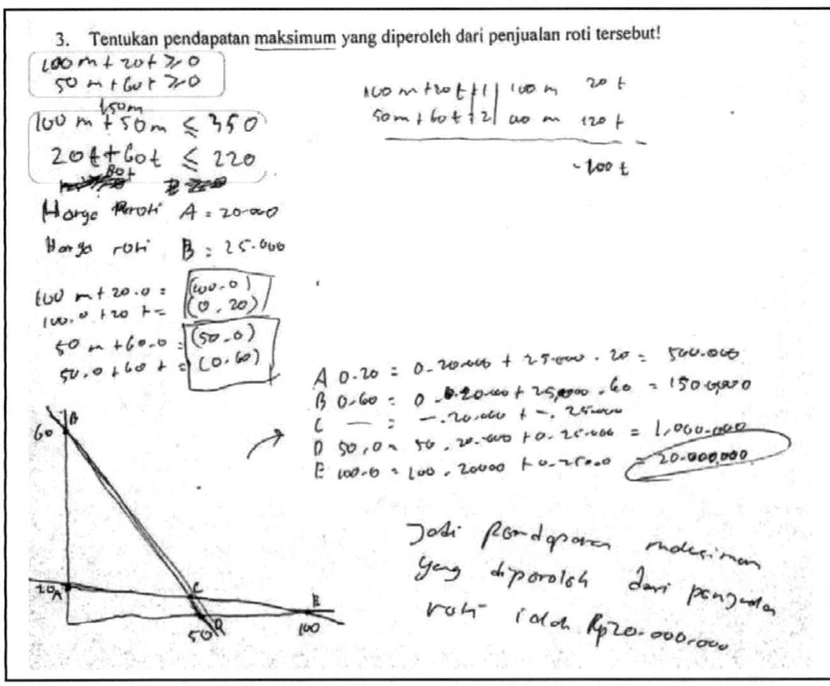

Figure 6. S2's answer for No.3

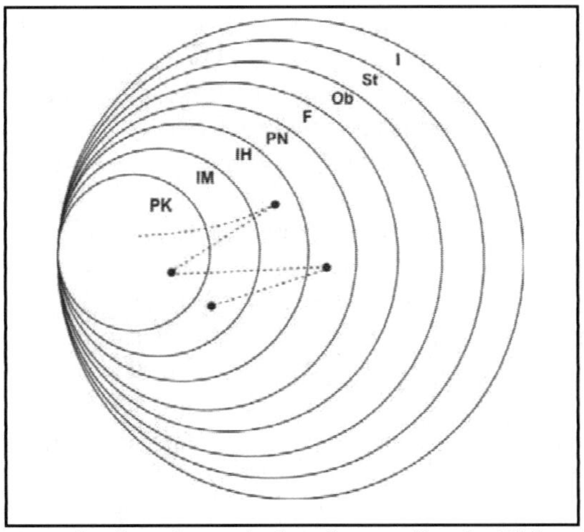

Figure 7. S2's growth of mathematical understanding by Pirie – Kieren model

Details of Figure 7:

PN : *Primitive Knowing*
IM : *Image Making*
IH : *Image Having*
PN : *Property Noticing*
F : *Formalising*
Ob : *Observing*
St : *Structuring*
In : *Inventising*
------------------- : Student's growth of mathematical understanding which is incorrect
——————— : Student's growth of mathematical understanding which is correct

The subject S1 and S2's growth of mathematical understanding were started from the primitive knowing level until the property noticing level and the process had not run well. Both of the subjects wrote the information from the problem incompletely which means that the primitive knowing level had not yet been fulfilled. It was based on the answers and strengthened by interview results.

3.2. Discussions

I will discuss further about subject S1. S1's growth of mathematical understanding occurred in several stages. Those levels will be elaborated further from the subject's answers. S1' answer shows that when S1 wrote the information from the problem (primitive knowing level), S1 had not written the information completely therefore researcher should ask again. One of the student's mistakes was the student did not understanding the problem well [2]. The next level done by S1 was seeing the relation between the information in the problem (image making level), so S1 could determine the variables (image having level). It could be done by the student well. The next level (property noticing level) was writing the S1's constraints. S1 still faced difficulties because S1 forgot to adjust the unit of the problem at the beginning. S1 needed to come back to the primitive knowing level in order to continue to the higher level. Furthermore, one of the student's mistakes was modeling the problem into mathematical form incorrectly [2]. S1 had also determined the intersection point of the x and y axes. However, S1 had not changed the constraints (inequality) into an equation. Most students are still confused with the operation that they are going to use [13]. S1 also drew the graphics from the constraints, but S1 faced difficulties again when S1 determined areas of set completion. This is the same with a research conducted by Cahyani [1]. She stated that students' mistake is the mistake when the students draw the areas worthy of linear programming problems.

S1's growth of mathematical understanding apparently must have often been returning to the lower level (primitive knowing) to expand S1's understanding. It is suitable with the Pieri – Kieren's theory that is when someone faces the condition where he/she cannot answer directly related to the problem, he/she can return to the lower level. It is similar to what expressed by [9]. She stated that when someone finds a mistake that can be solved immediately, he/she must fold back to a deeper level.

Growth of mathematical understanding of the subjects toward the mistakes given is suitable with Pirie – Kieren theory. The growth of mathematical understanding started with primitive knowing, image making, image having and property noticing even though it was not correct. It means that the subjects must master the prerequisite material before learning new materials. The previous researchers stated that college students still often return to the previous understanding level in order to understand the concept of limit [17] and also folding back is an important activity to the growth of understanding [8].

4. Conclusion

Growth of mathematical understanding of high profile students occurred in some steps. The first step (primitive knowing step) was started since the subjects have initial knowledge about inequality, drawing graphics of linear inequality, after reading the problem, the subjects wrote the information again from the problem. The second step (image making step) happened when the subjects wrote examples (image having step). The fourth (property noticing step) happened when then subjects determined the constraints, and the maximum value of the goal's function.

Growth of mathematical understanding in solving the problem of linear programming problems is based on Pirie – Kieren theory on the primitive knowing level until property noticing level. When the students are on the higher level, the students should still return to primitive knowing level. It is because the students' primitive knowing level had not been fulfilled. In other words, the students had not mastered the prerequisite material to solve the programming linear problems. Therefore, the students should be given questions to strengthen their understanding toward a growth of mathematical understanding level.

Hence, in teaching mathematics materials in classroom, it is very important for the teachers to give a deep reinforcement of the prerequisite material before starting new material. Furthermore, the process of learning mathematics should pay attention to each layer of understanding by Pirie-Kieren so that the process of students' understanding runs well.

Acknowledgments

The authors would like to thank the references for their suggestions and Sebelas Maret University, Surakarta Indonesia for finding this research in the academic years of 2017.

References

[1] Cahyani D, Ismail Y, and Yahya, L 2015 *Identifikasi Kesalahan Siswa dalam Menyelesaikan Soal Cerita Matematika pada Materi Program Linier* (Gorontalo: Prodi Matematika Universitas Negeri Gorontalo)

[2] Fitria T N 2013 Analisis Kesalahan Siswa dalam Menyelesaikan Soal Cerita Berbahasa Inggris pada Materi Persamaan dan Pertidaksamaan Linier Satu Variabel. *MATHEdunesa* **2** (1)

[3] Ghanny and Ferdinanto 2014 Meningkatkan Kemampuan Pemahaman Matematis Siswa melalui Problem Posing. *Journal Euclid* **1** 1-59

[4] Kastberg 2002 *Understanding mathematical concepts: The case of The Logarithmic Function* Georgia (Disertasi of The University of Georgia in Partial)

[5] Kemedikbud 2016 *Lampiran Permedikbud Nomor 24 Tahun 2016 Tentang KI KD Kurikulum 2013* (Jakarta: Kementerian Pendidikan dan Kebudayaan RI)

[6] Manu S 2005 Growth of Mathematical Understanding in a Bilingual Context: Analysis and Implications. *Proceedings of the 29th Conference of the International Group for the Psychology of Mathematics Education* **3** 289-296

[7] Martin L, LaCroix L, and Fownes L 2005 Folding back and Growth of Mathematical Understanding in Workplace Training. *ALM International Journal* **1** 19-35

[8] Martin L 2008 Folding Back and Dynamical Growth of Mathematical Understanding: Elaborating The Pirie-Kieren Theory. *The Journal of Mathematical Behavior* **27** 64 – 85

[9] Mell D E 2003 Models and theories of mathematical understanding: comparing pirie and kieren's model of the growth of mathematical understanding and APOS theory *USA: CBMS Issues in Mathematical Education* **12**

[10] Mousley J 2005 What Does mathematics Understanding Look Like?. *Paper presented at the the Annual Conference held at RMIT, Melbourne 7 – 9 Juli.*

[11] Paremeswaran R 2010 Expert Mathemticians' Approach to Understanding Definitions. *The mathematics Educator 2010* **20**(1) 43 – 51

[12] Pirie S and Kieren T 1994 Growth in Mathematical Understanding: How We Can Characterize it and How We can Represent it. *Educational Studies in Mathematics* **9** *160–190*

[13] Priawan I M, Abbas N, and Nurwan 2015 Pemecahan Masalah Matematis pada Materi Persamaan dan Pertidaksamaan Linier Satu Variabel di Kelas VII SMP Negeri 1 Batuda (Gorotalo: Prodi Pendidikan Matematika Universitas Gorontalo)

[14] Rahmi and Mulia S 2016 *Buku Ajar Program linier* (Yogyakarta: Deepublish).

[15] Siswono T Y E 2009 Pengembangan Model Pembelajaran Matematika Berbasis Pengajuan dan Pemecahan Masalah untuk Meningkatkan Kemampuan Berpikir Kreatif Siswa (*Departemen Pendidikan Nasional: Makalah Simposium Pusat Penelitian)*

[16] Sri Mulyono 2005 *Riset operasi* (Jakarta: Fakultas Ekonomi UI)

[17] Susiswo 2014 Folding Back Mahasiswa dalam Menyelesaikan Masalah Limit berdasarkan Pengetahuan Konseptual dan Pengetahuan Prosedural. *Prosiding Seminar Nasional TEQIP (Teachers Quality Improvement Program) dengan tema "Membangun Karakter Bangsa melalui Pembelajaran Bermakna TEQIP", 1st December 2014 in Malang of University*

IOP Conf. Series: Journal of Physics: Conf. Series **1008** (2018) 012071 doi:10.1088/1742-6596/1008/1/012071

Students' logical-mathematical intelligence profile

D P Arum[1], T A Kusmayadi[1], and I Pramudya[1]

[1]Department of Mathematics Education, Sebelas Maret University, Surakarta, Indonesia

E-mail: depi.arum@gmail.com

Abstract. One of students' characteristics which play an important role in learning mathematics is logical-mathematical intelligence. This present study aims to identify profile of students' logical-mathematical intelligence in general and specifically in each indicator. It is also analyzed and described based on students' sex. This research used qualitative method with case study strategy. The subjects involve 29 students of 9th grade that were selected by purposive sampling. Data in this research involve students' logical-mathematical intelligence result and interview. The results show that students' logical-mathematical intelligence was identified in the moderate level with the average score is 11.17 and 51.7% students in the range of the level. In addition, the level of both male and female students are also mostly in the moderate level. On the other hand, both male and female students' logical-mathematical intelligence is strongly influenced by the indicator of ability to classify and understand patterns and relationships. Furthermore, the ability of comparison is the weakest indicator. It seems that students' logical-mathematical intelligence is still not optimal because more than 50% students are identified in moderate and low level. Therefore, teachers need to design a lesson that can improve students' logical-mathematical intelligence level, both in general and on each indicator.

1. Introduction

Gardner thinks that intelligence as a biological factor that depends on the environment, the culture, the community around, and with whom he/she interacts [1]. Gardner also defined intelligence as "*a bio-psychological potential to process information that can be activated in a cultural setting to solve problems or create products that are of value in a culture*" [2]. Gardner & Hatch stated that intelligence is traditionally defined in terms of Intelligence Quotient (IQ), which measures the range of simple Verbal/Linguistic and Logic/Mathematical abilities [3]. Furthermore, Gardner's statement regarding the theory of multiple intelligences gives the dimension where each individual is different from the others. It means that in the learning process every student has different abilities and characteristic, including their intelligence. This intelligence profile can influence the learning process and learning outcomes [1]. Therefore, it is necessary for the teachers to know the characteristics of multiple intelligences of students to choose strategies that can improve the effectiveness of learning process in the classroom, include mathematics learning.

The purpose of teaching and learning of mathematics is not only the students have good knowledge of numerical facts, but also problem-solving skills that can be adapted according to the strengths and weaknesses of each student [4]. In Indonesia, one of the benchmarks to monitor, to encourage and to improve the quality of learning in the classroom is national exam results. Based on the results of mathematics national examination in SMP Al-Irsyad Surakarta shows that student achievement has not optimal yet. Based on Table 1 shows that the average value of mathematics subjects is still below 50 points despite there was an increase of the average value from 2016 to 2017 [5, 6, 7].

Table 1. National Exam Results in Mathematics Subject

Year	Scope		
	SMP Al-Irsyad (%)	Central Java (%)	National (%)
2015	42.71	47.43	56.28
2016	39.75	43.39	50.24
2017	47.16	48.65	50.31

Source: Badan Standar Nasional Pendidikan (BSNP)

The low result of mathematics learning is influenced by several factors. Multiple intelligence is one of the factors that influence the capability of students in mathematics. It cannot be separated from the various theoretical and empirical studies that found a positive relationship between the multiple intelligences and students' mathematics learning achievement [8, 9]. Learning achievement is a strong indicator that explains the students who succeed in learning process in the classroom, it is very important to pay attention to the student's characteristics of intelligence and so that we can improve the achievement of students with moderate intelligence level and low intelligence level. It can make the students know about their strengths and weaknesses [10].

One of the multiple intelligences that a role in mathematics learning is logical-mathematical intelligence. A student with well-developed logical-mathematical intelligence will be able to observe and recognize the patterns and relationships [11]. It means that students' logical-mathematical intelligence plays a huge role in problem-solving activities. Someone who is able to solve math problems is the person that has good logical-mathematical intelligence [12]. In another case, to solve the problems of mathematics is required the component of logical-mathematical intelligence (ability to finding and reasoning) [13]. The results of Rahbarnia, et al. [14] also showed that there is a positive correlation between logical-mathematical intelligence and mathematical problem solving.

Gardner described logical-mathematical intelligence as the ability to study the problems, complete the logical and analytical, mathematical operations, and conduct scientific investigations [15]. Logical-mathematical intelligence can also be interpreted as "*an individual's sensitivity to, and capacity to discern, logical or numerical patterns, and ability to handle long chains of reasoning. These individuals like to experiment, solve puzzles, and ask cosmic questions*" [16]. Based on those opinions above, it seems that students' logical-mathematical intelligence includes several individual elements of ability such as the ability to classify, study and solve problems, make logical reasoning, and analytical thinking.

Saban explained that students with high level of logical-mathematical intelligence are more able to classify objects in several categories, identify logical relationships between events, and perform certain quantitative calculations to identify real relationships between events [17]. Onay argued, "*logical-mathematical intelligence is a skill about thinking with numbers, calculation, deriving conclusions, constituting logical relationships, problem solving, critical thinking, introducing with abstract symbols like numbers, geometrical shapes, relating knowledge pieces*" [18]. Meanwhile, Willis and Johnson revealed that logical-mathematical intelligence encompasses five core areas: (1) classifying; (2) comparing; (3) basic numerical operations; (4) inductive and deductive reasoning; and (5) making and testing hypothesis [19]. Thus, logical-mathematical intelligence is the ability of students to identify and classify objects, perform mathematical calculations, solve the problems, think logically and critically, and make conclusions. Based on the definition of logical-mathematical intelligence above it appears that students' logical-mathematical intelligence can be identified based on several indicators. Moreover, the level of logical-mathematical intelligence of students in this study can be identified and measured using a logical-mathematical intelligence test was developed based on the indicators presented in Table 2.

Table 2. Logical-Mathematical Intelligence Indicators

Indicator	Descriptions of Characteristics
Ability to understand patterns and relationship	Students are able to identify the patterns and relationships between several numbers or objects.
Ability to classify	Students are able to group the objects into several categories
Ability to compare	Students are able to find the comparison between two object group
Ability to do the mathematical calculation	Students are able to finish the calculation well
Ability to inductive and deductive reasoning	Students are able to make a conclusion from the information logically

Another important issue is sex influences on educational outcomes. Several studies have been conducted to review the concepts of multiple intelligences and sex [20, 21]. Therefore, this study will find how the student's logical-mathematical characteristics specifically based on students' sex.

Based on the explanation above, it seems that the characteristics of students' multiple intelligences play an important role in the success of learning process. Specifically, this study focuses on the profile of students' logical-mathematical intelligence in mathematics learning. Several studies have been conducted to find the correlation between students' multiple intelligences and student achievement [22, 23, 24, 25]. However, there are a few studies that pay attention to students' logical-mathematical intelligence profiles generally and specifically on each indicator. Most of the previous research just to look at the profile of multiple intelligences generally [1, 26, 27], or just to develop a valid instrument that can identify multiple intelligences [3, 28, 29] and also just to see how huge the impact of theoretical-based learning of multiple intelligences for student learning achievement was [10, 22, 30]. It is important for teachers to recognize the characteristics of students' logical-mathematical intelligence to improve and develop the process of mathematics learning in the classroom. This is in accordance with the opinion of Emmiyatti, et al. [1] that teachers need to understand multiple profiles of students' multiple intelligences in order to provide appropriate learning activities in the classroom that accommodate multiple students' intelligence. Therefore, this study aims to identify students' logical-mathematical intelligence profile. In particular, this study focuses on the analysis and description of logical-mathematical intelligence generally and specifically on each indicator, namely the ability to understand the patterns and relationships, classification, comparison, mathematical calculations, and deductive-inductive reasoning. In addition, logical-mathematical intelligence is analyzed and described based on students' sex.

2. Method

This research is a type of descriptive analysis research using qualitative method and case study strategy. Qualitative research is a study to understand the phenomenon experienced by research subjects holistically through descriptions in the form of words [31]. Merriam stated the definition of a case study as "*intensive and holistic description and analysis of limited phenomena such as a program, an institution, a person, a process, or a social unit*" [32]. Thus, qualitative research with case study strategies is a study to explore and understand the subject and phenomenon intensively and holistically through the oral description. In this research, the phenomenon refers to students' logical-mathematical intelligence in general and specifically in each indicator.

This research was conducted at SMP Al Irsyad Surakarta, Indonesia from September until October 2017. Subjects in this research involved 29 students 9th grade selected by purposive sampling technique. The data in this research were the students' logical-mathematical intelligence test results and interview recordings. The instrument of this research includes a logical-mathematical intelligence test and interview guidelines. Logical-mathematical intelligence test consists of 20 multiple choice questions developed by researchers based on several indicators, namely ability to understand patterns and relationships, classification, comparison, mathematical calculations, and deductive-inductive reasoning. Before using logical-mathematical intelligence test, the instrument was validated by four validation expert. In addition, researchers conducted a 30-point test instrument to measure grain difficulty, grain distinguishing and the reliability. The test results of the logical-mathematical intelligence test were imposed on 92 students showed the grain difficulty level of $0.3 \leq p \leq 0.7$, the divergent power index $D \geq 0.3$ and the reliability coefficient $r_{11} = 0.74$.

Subjects were asked to complete a logical-mathematical intelligence test before being interviewed. The results of the logical-mathematical intelligence test and interview were analyzed qualitatively by using Miles and Huberman's steps to describe the students' logical-mathematical intelligence level [33]. Naderi studied Male Versus Female Intelligence Among Undergraduate Students: Does Gender Matter? [34]. Activities in data analysis include data reduction, data display, and verification. Firstly, researchers analyzed the results of logical-mathematical intelligence test consisting of 29 students to know the average and level of logical-mathematical intelligence of students. After that, some students were interviewed to investigate and explore students' logical-mathematical intelligence. In this study, it focuses on the analysis and description of students' logical-mathematical intelligence in general and specifically in each indicator. In addition, logical-mathematical intelligence is analyzed and described based on students' sex.

3. Result and Discussion

The result of this study represents the capability of student from logical-mathematical intelligence test and interview recorded. Logical-mathematical intelligence test result was described by Table 3 and classified into high level, moderate level, and low level. Table 3 also describes data frequency and percentage of students based on level of intelligence and students' sex. On the other hand, Table 4 explains the average score of students and the percentage of each logical-mathematical intelligence indicator of male and female students.

Table 3 described that 15 students (51.7%) have moderate level of logical-mathematical intelligence with average score 11.17. Eight students (27.6%) have high level of logical-mathematical intelligence and only 6 students (20.7%) have low level of logical-mathematical intelligence. Based on students' sex, both of male and female student have moderate level of logical-mathematical intelligence, 5 students (41.67%) for male and 10 students (58.82%) for female with average score of male students is higher than female students (11.42:11.00).

Table 3. Description of Students' Logical-mathematical Intelligence Test Result

Level LMI	Group					
	Male		Female		TOTAL	
	f	%	*f*	%	*f*	%
High	4	33.33	4	23.53	8	27.6
Moderate	5	41.67	10	58.82	15	51.7
Low	3	25	3	17.65	6	20.7

Based on the results, it appears that male students have higher logical-mathematical intelligence than female students. These results were supported by several other researchers who identified intelligence based on sex differences [34, 35]. Andrian & Buchanan's research on sex differences in 20 countries shows that males are significantly higher in outcomes than female for intelligence in general. Adrian and Buchanan state that these differences of intelligence are consistent across countries and populations, although there is little difference [20]. More specifically, in research conducted by Rammstedt and Rammsayer show that male students have higher logical-mathematical and spatial intelligence than female students [35].

IOP Conf. Series: Journal of Physics: Conf. Series **1008** (2018) 012071 doi:10.1088/1742-6596/1008/1/012071

Based on each indicator, Table 4 shows that logical-mathematical intelligence is strongly influenced by the student's capability to classify and understand the patterns and the relationship. It can be seen that 25% of the total score of students in this study were influenced by the students' classification ability with the average score of 2.79. In the second place, indicator of ability to understand the pattern and the relationship was 24% of the total score of students with average score of 2.72. The weakest indicator is ability to comparison (only 8.6% of the total student score with an average score of 0.97). Meanwhile, for indicator of ability to deductive-inductive reasoning was 22.5% of the total score of students with average score of 2.52 and indicator of ability to computing was 19.4% from the total score of students with the average score 2.17.

Table 4. Students' Average Score and Percentage of Each Logical-mathematical Intelligence Indicator

Indicator LMI	Group					
	Male		Female		TOTAL	
	$\bar{x}$	%	$\bar{x}$	%	$\bar{x}$	%
Understand patterns and relationships	2.75	24.1	2.71	24.6	2.72	24.4
Classification	2.75	24.1	2.82	25.7	2.79	25.0
Comparison	1.25	10.9	0.76	7.0	0.97	8.6
Calculation	2.5	21.9	1.94	17.6	2.17	19.4
Deductive-inductive reasoning	2.17	19.0	2.76	25.1	2.52	22.5
TOTAL	11.42		11.00		11.17	

Based on the students' sex, the most influencing indicator of the logical-mathematical intelligence of male students from strongest to weakest is the ability to understand the patterns and the relationships, classification, computation, inductive-deductive reasoning and comparison. On the other side, for female students, indicators that influence the logical-mathematical intelligence from strongest to weakest are the ability of classification, inductive-deductive reasoning, understanding the patterns and the relationships, computation and comparison. Male students are stronger than female students in indicators of ability to understand the patterns and relationships, comparisons, and computations. While female students are stronger than male students in indicators of classification ability and inductive-deductive reasoning.

Based on the results of interviews with several students, it appears that major students did not experience significant difficulties in any points of problem questions from classification capability indicators and ability to understand patterns and relationships. Students look easy to understand the problem and find the right answer. This indicates that students' logical-mathematical intelligence in general is strongly influenced by the ability of classification and the ability to understand patterns and relationships. Furthermore, for the item number of the comparison indicator, some students appear to be confused between the ratio of values with the reversed ratio. In addition, students look less thorough when performing mathematical calculations. The same thing happened to some point number about indicator of computation ability. Students said they were not interested in questions involving numbers and calculations. Some students said that the item number of the computational ability indicator is the last part done when performing a logical-mathematical intelligence test. As for the item number indicator of the ability to make inductive-deductive reasoning, some students said that confused determine the exact conclusion of the premises of the question.

These findings are supported by previous researchers who stated that students in the high, medium, low category can classify the information on the problem, be able to compare the relationship between the existing information on the problem with the knowledge they possess, and capable of performing mathematical calculation operations. Meanwhile, for the ability to use inductive or deductive reasoning can only be done by students from high and moderate category [36]. Research conducted by Nofianti, et al. [37] also found that classification characteristics are characteristics with the highest percentage compared to other logical-mathematical intelligence characteristics.

4. Conclusion

Based on the results and discussion above, it can be concluded that most of students were moderate level in logical-mathematical intelligence (51.7% of students have an average score of 11.17). In addition, both of logical-mathematical intelligence level of male and female students was a moderate. The intelligence level of male and female student was strongly influenced by the capability to classify and understand the patterns and the relationships. Furthermore, the comparative ability was the weakest indicator. It seems that students' logical-mathematical intelligence had not been optimal yet because more than 50% of them had moderate-low level. Students' logical-mathematical intelligence point on certain indicators was still low. Therefore, teachers need to design a learning method that can improve students' logical-mathematical intelligence level generally and also in every single indicator. This can be solved by facilitating students in the process of learning mathematics by providing problem-solving questions in accordance with students' logical-mathematical intelligence indicators. In addition, there is a need for further efforts to develop valid and reliable instances for measuring logical-mathematical intelligence particularly. This will be useful for teachers to know the students' logical-mathematical intelligence profile and use it for evaluation of the learning method.

Acknowledgments

The authors would like to thank to the referees for their suggestions and also to the mentors for helping authors in preparation of this paper.

References

[1] Emmiyatti N, Rasyid, M A, Rahman, M A, Arsyad, A, and Dirawan, G D 2014 Multiple Intelligences Profiles of Junior Secondary School Students in Indonesia *Int Ed Studies* **7** 103-110
[2] Gangadevi S and Ravi 2014 Multiple Intelligence Based Curriculum to Enhance Inclusive Education to Bring Out Human Potential *Int Jof Advanced Research* **2** 619-626
[3] McClellan J A and Conti G J 2008 Identifying the Multiple Intelligences of Your Students *J of Adult Ed* **37** 13-18
[4] Jones S and Tanner H 2002 Teachers' Interpretations of Effective Whole-Class Interactive Teaching in Secondary Mathematics Classrooms *Ed Studies* **28** 265-274
[5] Badan Standar Nasional Pendidikan (BSNP) 2015 Laporan Hasil Ujian Nasional Tahun Pelajaran 2014/2015
[6] Badan Standar Nasional Pendidikan (BSNP) 2016 Laporan Hasil Ujian Nasional Tahun Pelajaran 2015/2016
[7] Badan Standar Nasional Pendidikan (BSNP) 2017 Laporan Hasil Ujian Nasional Tahun Pelajaran 2016/2017
[8] Agustin M 2011 *Permasalahan Belajar dan Inovasi Pembelajaran* (Bandung: PT Rafika Aditama)
[9] Xie J and Lin R 2009 Research on Multiple Intelligences Teaching and Assessment *Asian J of Management and Humanity Sc* **4** 106-124
[10] Ibnian S S K and Hadban A D 2013 Implications of Multiple Intelligences Theory in ELT Field *Int J of Humanities and Soc Sc* **3** 292-7
[11] Gunawan A W 2011 *Born To Be A Genius* (Jakarta: Gramedia Pustaka Utama)
[12] Yaumi 2012 *Pembelajaran Berbasis Multiple Intelligences* (Jakarta: Dian Rakyat)

[13] Karamikabir N 2012 Gardner's Multiple Intelligence and Mathematics Education *Proc Soc and Behavioral Sc* **31** 778-781
[14] Rahbarnia F, Hamedian S, and Radmehr F 2014 A Study on The Relationship Between Multiple Intelligences and Mathematical Problem Solving Based on Revised Bloom Taxonomy *J of Interdisciplinary Math* **17** 109-134
[15] Maftoon P and Sarem S N 2012 The Realization of Gardner's Multiple Intelligences (MI) Theory in Second Language Acquisition (SLA) *J of Language Teaching and Research* **3** 1233-41
[16] Lunenburg F C and Lunenburg M R 2014 Applying Multiple Intelligences in The Classroom: A Fresh Look at Teaching Writing. *Int J of Scholarly Academic Intellectual Diversity* **16** 1-13
[17] Korkmaz O 2016 The Effect of Scratch and Lego Mindstorms Ev3-Based Programming Activities on Academic Achievement, Problem-Solving Skills and Logical-Mathematical Thinking Skills of Students *Malaysian Online J of Ed Sc* **4** 73-88
[18] Kaplan A, Calp U F and Ozdemir E 2011 The Determination of Visually Impaired Students' Multiple Intelligences Dimensions and Their Relationships with Success in Mathematics *Scientific Research and Essays* **6** 2841-2849
[19] Aini N A, Sunardi, Slamin and Hobri 2017 Student's Mathematics Creative Thinking Skills in Terms of Logical Mathematical Intelligence *Int J of Sc Research and Management (IJSRM)* **5** 6930-34
[20] Naderi H, Abdullah R, Aizan H T and Sharir J 2010 Intelligence and Academic Achievement: An Investigation of Gender Differences *Life Sc J* **7** 83-87
[21] Chandra R and Azimmudin S 2013 Influence of Intelligence and Gender on Academic Achievement of Secondary School Students of Lucknow City *IOSR J of Humanities and Soc Sc (IOSR-JHSS)* **17** 09-14
[22] Niroo M, Nejhad G H H and Haghani M 2012 The Effect of Gardner Theory Application on Mathematical/Logical Intelligence and Student's Mathematical Functioning Relationship *Proc Soc and Behavioral Sc* **47** 2169-75
[23] Lipi I 2013 Student Intelligence and Academic Achievement in Albanian Universities. Case of Vlora University *Acta Universitatis Danubius (AUDA)* **5** 48-58
[24] Shahzada G, Khan U A, Islam F U, Faqir K 2014 Interrelation of Multiple Intelligences and their Correlation with Students' Academic Achievements: A Case Study of Southern Region, Khyber Pakhtunkhwa *FWU J of Soc Sc* **8** 59-64
[25] Ahvan Y R and Pour H Z 2016 The Correlation of Multiple Intelligences For The Achievements of Secondary Students *Ed Research and Reviews* **11** 141-5
[26] Tafti M A, Heidarzadeh M, and Khademi M 2014 A Comparison of Multiple Intelligences Profile of Students with and without Learning Disabilities *Int J of Applied Psychology* **4** 121-5
[27] Dolu G and Urek H 2014 A Study Related to the Investigation of Multiple Intelligence Profiles of Gifted and Talented Children across Turkey *J for the Ed of the Young Scientist and Giftedness* **2** 58-66
[28] Al-Kabani M S and Al-Wahaibi S 2015 Testing the Multiple Intelligences Theory in Oman *Proc Soc and Behavioral Sc*190 106-12
[29] Saban A I, Shearer B, Kayiran B K and Isik D 2012 The Validity and Reliability Study of Turkish Version of The Multiple Intelligences Developmental Assessment Scales *Int J of Human Sc* **9** 651-66
[30] Mousavi S S and Ahmadi F 2013 Education Effect Based on Gardner Multiple Intelligence Hypotheses in Students Mathematics' Education Progress of High School Second Grade in Garmsar City *Int J of Soc Sc (IJSS)* **3** 25-31
[31] Moleong L. J 2007 *Metodologi Penelitian Kualitatif* (Bandung: Remaja Rosdakarya)
[32] Yazan B 2015 Three Approaches to Case Study Methods in Education: Yin, Merriam,and Stake *The Qualitative Report* **20** 134-152

IOP Conf. Series: Journal of Physics: Conf. Series **1008** (2018) 012071 doi:10.1088/1742-6596/1008/1/012071

[33] Sugiyono 2008 *Metode Penelitian Kuantitatif, Kualitatif dan R&D* (Bandung: Alfabeta)

[34] Naderi H, Abdullah R and Aizan H. T 2008 Male Versus Female Intelligence Among Undergraduate Students: Does Gender Matter? *Asian J of Scientific Research* **1** 539-543

[35] Rammstedt B and Rammsayer T. H 2002 Self-Estimated Intelligence: Gender Differences, Relationship to Psychometric Intelligence and Moderating Effects of Level of Education *European Psychologist* **7** 275-284

[36] Hasanah W and Siswono T. Y. E 2013 Kecerdasan Logis-Matematis Siswa Dalam Memecahkan Masalah Matematika Pada Materi Komposisi Fungsi *MathEdunesa* **2** 1-6

[37] Nofianti L, Sugiarti T and Susanto 2015 Kecerdasan Visual-Spasial dan Logika Matematika dalam Menyelesaikan Soal Geometri Siswa Kelas XI IPA 8 SMA Negeri 2 Jember (*Visual Spatial and Logical-mathematical Intelligence in Solving Geometry Problems Class XI IPA 8 SMA Negeri 2 Jember*) *Artikel Ilmiah Mahasiswa* **1** 1-7

IOP Conf. Series: Journal of Physics: Conf. Series **1008** (2018) 012072 doi:10.1088/1742-6596/1008/1/012072

Students creative thinking skills in solving two dimensional arithmetic series through research-based learning

M Tohir[1,2], Z Abidin[1,3], Dafik[1], Hobri[1]

[1]Department of Mathematics Education, University of Jember, Indonesia
[2]State Junior High School 2 of Jember, Indonesia
[3]Pelita Hati School of Jember, Indonesia

E-mail: tohir.unej@gmail.com

Abstract: Arithmetics is one of the topics in Mathematics, which deals with logic and detailed process upon generalizing formula. Creativity and flexibility are needed in generalizing formula of arithmetics series. This research aimed at analyzing students creative thinking skills in generalizing arithmetic series. The triangulation method and research-based learning was used in this research. The subjects were students of the Master Program of Mathematics Education in Faculty of Teacher Training and Education at Jember University. The data was collected by giving assignments to the students. The data collection was done by giving open problem-solving task and documentation study to the students to arrange generalization pattern based on the dependent function formula i and the function depend on i and j. Then, the students finished the next problem-solving task to construct arithmetic generalization patterns based on the function formula which depends on i and $i + n$ and the sum formula of functions dependent on i and j of the arithmetic compiled. The data analysis techniques operative in this study was Miles and Huberman analysis model. Based on the result of data analysis on task 1, the levels of students creative thinking skill were classified as follows; 22,22% of the students categorized as "*not creative*"; 38.89% of the students categorized as "*less creative*" category; 22.22% of the students categorized as "*sufficiently creative*"; and 16.67% of the students categorized as "*creative*". By contrast, the results of data analysis on task 2 found that the levels of students creative thinking skills were classified as follows; 22.22% of the students categorized as "*sufficiently creative*", 44.44% of the students categorized as "*creative*"; and 33.33% of the students categorized as "*very creative*". This analysis result can set the basis for teaching references and actualizing a better teaching model in order to increase students creative thinking skills.

1. Introduction

Mathematics education at higher education tends to direct learners to understand mathematical formula by applying it in a lesson involving problem solving. This is the case particularly when the material contains certain formulas, such as arithmetic sequence and series. In fact, the ability to think in identifying and constructing formulas in Mathematics is required to foster students understanding on the material and accrue meaningful learning. One of thinking abilities essential for the students to experience more meaningful learning and improve their thinking skill in solving everyday problems is creative thinking.

Munandar suggests that creative thinking can be defined as the ability to reflect the aspects of fluency, flexibility, originality, and elaboration [1]. In creative thinking, learner will go a series of stages, *inter alia,* synthesizing ideas, building ideas, planning the implementation of ideas, and applying these ideas to produce something [2].

Based on the aforementioned discussion, it is imperative to develop an examination focusing on the graduate students creative thinking skills concerne with combinatorics course. Knowing the graduate students creative thinking skills will shed light on what is essential to increase their mastery level, especially in the course of Combinatorics.

Learning method is also very important in measuring students creativity. As stated in the 21st century skills framework, creativity is one of the important things for students. For achieving and increasing their creativity, a student-centered learning method needs to be applied. One of student-centered learning methods is research-based learning, a method recently applied by several universities in several countries. The learning method has one of the ambitions of many universities, that is to bring the two core of educational endeavour and research closer together and to integrate research into teaching more than what has been done so far [3].

2. Research Methods

The research method applied in this research was triangulation with qualitative and quantitative approach. Qualitative research adheres to the natural background in which a study takes place (it views the context as a whole). In this regard, human serves as an instrument. Qualitative method requires inductive data analysis and theory development, which are all based on the data, which is descriptive in nature and more concerned with the process under investigation. The research is limited by its focus, the specific criteria for ensuring data validity, and its temporary design. Also, qualitative data are generated through joint decision [4]. However, this qualitative research was more concerned with results, rather than process, and applied by analyzing the students performance on task 1 and task 2 in order to gain details regarding their creative thinking skills and metacognition levels. Afterwards, the effectiveness of learning method operationalized can be proved by the result of data analysis.

The research subjects were 18 students of the Master Program of Mathematics Education in Faculty of Teacher Training and Education Jember University in the 2016/2017 academic year. The students were expected to give a comprehensive illustration of the level of their creative thinking skills.

The data collection technique was the most important step in the research as it determined the success of gaining accurate data, which were then analyzed the students answers on the problem given. The data were collected by giving the assignment. The task was given to gather data pertinent to the level of students creative thinking skills in Combinatorics. The data collected were reduced, served, concluded, and verified. Data validity was ensured by triangulation, peer checking, and extended observation.

The researchers applied research-based learning when implementing the teaching and learning on Arithmetic series. Based on research-based learning conducted by Mr. Dafik (the lecturer), which adhered to Tremp's model and focused on Syntax, seven steps were applied systematically [5]. The steps of research-based learning operative in the study included formulating a general question, overview of research-literature, defining the question, planning research activities, undertaking investigation, interpretation and consideration of results, and report as well as presentation.

The researchers implemented the research-based learning as a part of steps for getting the data related to the creative thinking analysis. Escalating students creative thinking skills can be done in numerous ways, one of which is by analyzing and predicting it. According to Siswono, the distribution of creative thinking levels is useful in predicting students ability to think creatively, especially in the field of Mathematics [2].

The data presentation in qualitative research can be made in a brief description and relations between categories. In Miles and Huberman's model, as quoted by Sugiyono, the most common form of the data presentation in qualitative research, to date, has been narrative text [6]. Commonly, the data presentation in qualitative research is narrative text, but in this research, data presentation included the classification and identification of the data, followed by the collection of data organized and categorized for drawing conclusion.

3. Research Finding

The data collection was commenced by providing learning material first and then distributing the task to the students, which focused on arithmetic progression. The arithmetic progression formula was $U_n = a + (n-1)b$, where $U_n = f(x_i)$ and a series of to-function i,a is the first term of the series i, n is the number of rows i, b is different from the series *to-i,* and i own a tribe *to-n*. While, formula of arithmetic sequence is $S_n = \frac{n}{2}(a + U_n)$.

3.1 Formula of Arithmetic Sequence Function $f(x_i)$ Based on Term i

In this step, the students were required to determine the formula of arithmetic sequence. The formula might consist of 1 or more formulas, based on the sequence pattern. For example, if there are two kinds of formula on the sequence pattern, so the formula should be written as the following:

i	**1**	**2**	**3**	**4**	**5**	**6**
f(x)	4	1	5	2	6	3

$$\Rightarrow \qquad f(x_i) = \begin{cases} \dfrac{i+n+2}{2} & ,i\,odd \\ \dfrac{i}{2} & ,i\,even \end{cases}$$

The material above was discussed by the lecturer and students. Most of the student came up with the following results.

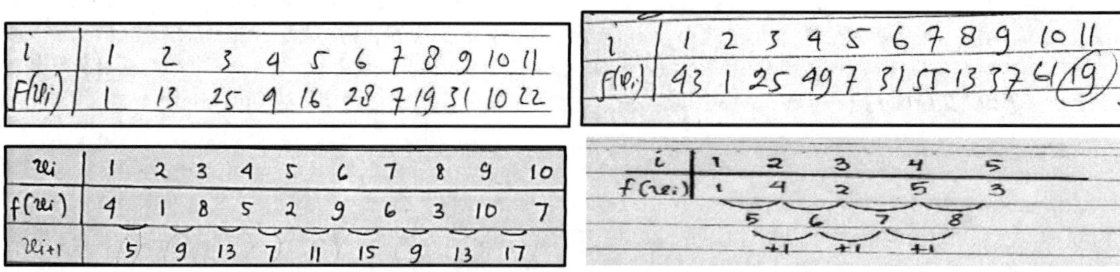

Figure 1. Many Arithmetic sequences found by the students

3.2 Formula of Functions Which Depends on i and j or $P^n_{m,d}(i, j)$

In the next step, arithmetic sequence had to be generalized. The formula could consist of many formulas based on the sequence pattern at each *i*-row or *j*-column. There were many examples about generalized arithmetic sequence, depending on variables *i* and *j* :

	j \ i	1	2	3	4	5
	1	1	2	3	4	5
m	2	6	7	8	9	10
	3	11	12	13	14	15
		18	21	24	27	30
	d	3	3	3	3	

j \ i	1	2	3	4	5
1	1	4	7	10	13
2	2	5	8	11	14
3	3	6	9	12	15
	6	15	24	33	42
d	9	9	9	9	

Based on the example, students tried to find the new arithmetic sequence and they came up with the following results:

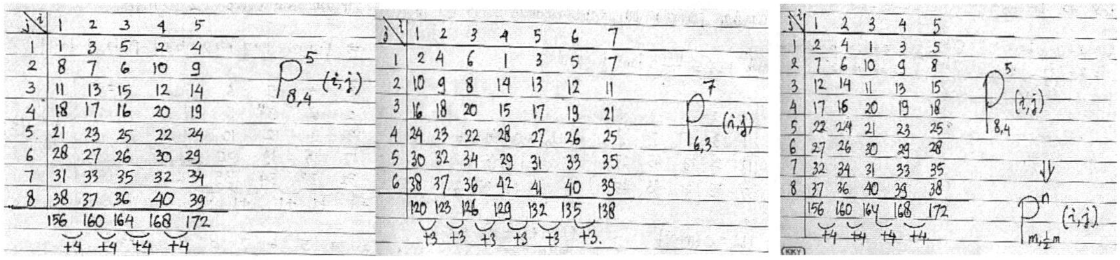

Figure 2. Many Arithmetic sequences depending on *i* and *j* foun by the students

3.3 Finding Formulas of Total Function $\sum_{j=1}^{m} P_{m,d}^{n}(i,j)$

If the formula $P_{m,d}^{n}(i,j)$ can be determined correctly, the formula amount $\sum_{j=1}^{m} P_{m,d}^{n}(i,j)$ exactly can be determined as well. Therefore, the formula of $\sum_{j=1}^{m} P_{m,d}^{n}(i,j)$ can be generalized properly. The following example portrays how the students found the function formula:

j \ i	1	2	3	4	5
1	1	2	3	4	5
2	6	7	8	9	10
	7	9	11	13	15

for $U_m = a + (n-1)b = 1 + (j-1)n$
Then for $U_n = a + (n-1)b = 1 + (j-1)n + (i-1)1 = (j-1)n + i$
Thus, the formula $P_{m,m}^{n} = (j-1)n + i$

Then formula Number $\sum_{j=1}^{m} P_{m,d}^{n}(i,j)$

$$\sum_{j=1}^{m} P_{m,d}^{n}(i,j) = \sum_{j=1}^{m}(j-1)n + i = \sum_{j=1}^{m} jn - n + i = \sum_{j=1}^{m} jn + \sum_{j=1}^{m}(-n+i) = (1+2+3+\ldots+m)n + (-n+i)m$$

$$= \frac{m}{2}(1+m)n - mn + mi = \frac{mn}{2}(m-1) + mi$$

3.4 Achievement of Test Result Based on Student Creative Thinking Skill's Indicators

The data concerning the students creative thinking skills classified them into high, medium, and low group, the complete of which details are presented in Figure 3.

IOP Conf. Series: Journal of Physics: Conf. Series **1008** (2018) 012072 doi:10.1088/1742-6596/1008/1/012072

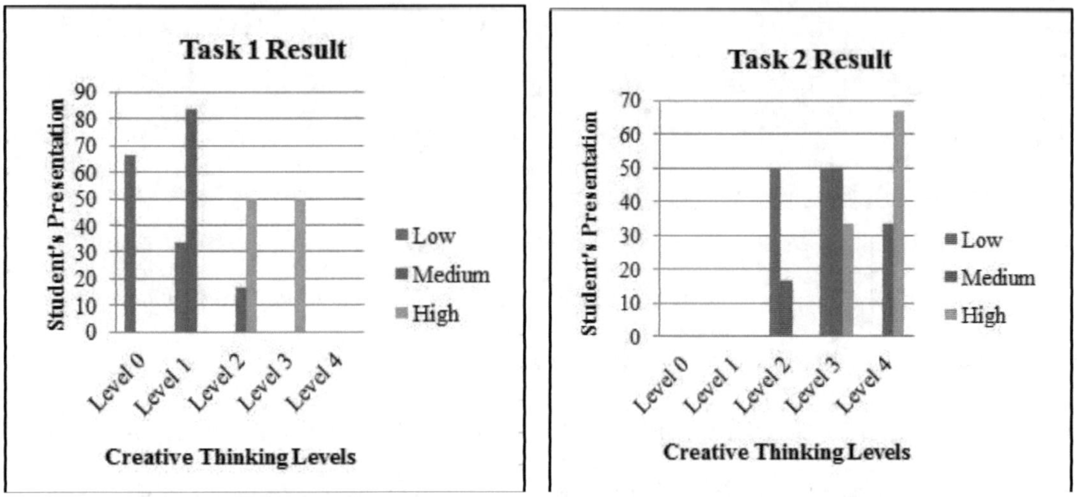

Figure 3. Data Achievement of Test Result Based on The Students Creative Thinking Skill

Figure 3 shows that the students performance on task 1 and task 2 results were significantly different at all levels. The low-level students improved their performance from level 0 to level 2 or 3. Similar increase was also evident among medium-level students, who successfully increased their performance from level 1 to level 3. The other improvement, among high-level students, was evinced by the improvement from level 2 or 3 to level 4. The improvement in all indicators of students creative thinking skills could not be separated from the guidance of research-based learning process. This finding was in line with the results of research conducted by Bani, the results of which indicated that guided discovery methods can improve students mathematical reasoning [8]. This teaching method not only encourages the students to be more active, but also guides them to be more active in learning. Hudojo states that if students are actively involved in finding concepts, then they will understand the concept better, and their memory will retain the learnt concept longer, allowing it to be used in other contexts [9].

Based on the research results, the researchers classified the students critical thinking skills into the following categories, as stipulated by the indicators of critical thinking skills.

Table 1. The Students Creative Thinking Skills on Task 1 and Task 2

Task 1		Category	Task 2	
Sum	Percentage (%)		Sum	Percentage (%)
4	22,22	Level 0 (Not Creative)	-	-
7	38,89	Level 1 (Less Creative)	-	-
4	22,22	Level 2 (Sufficiently Creative)	4	22,22
3	16,67	Level 3 (Creative)	8	44,44

Task 1		Category	Task 2	
Sum	Percentage (%)		Sum	Percentage (%)
-	-	Level 4 (Very Creative)	6	33,33

Based on the data presented in Table 1, it is obvious that majority of the students creative thinking skill levels, based on task 1, are in the "*less creative*" category, and the level of students creative thinking skill, upon accomplishing task 2, is in "*creative*" category. Therefore, the levels of students creative thinking skill were significantly different between the two tasks. However, based on the results of the interview, there were some interesting obstacles experienced by students. These findings revealed (1) that the students arranged non-arithmetic sequence; (2) that they tried to arrange a multilevel arithmetic sequence; (3) that when constructing the formula, many students assumed the parameters as constants, upon operating them; (4) that the students generally did not think that the formula obtained had to be applied in a general fashion to any extent; (5) that when finding the function of formula from the existing data, they found it difficult modifying the first term of sequence, this was the case because the first term always moved when generalized, be the first term start from the front or back or even the middle in arithmetic sequence; and (6) that when generalizing arithmetic pattern, there was a new arithmetic sequence with more n-terms and the initial value lied on the same position as previous one. Thus, research-based learning was proven effective in aiding the students in improving their creative thinking skills, although there were some obstacles experienced by the students.

4. Discussion

The following figure is an example of students work when finding the arithmetic sequence, along with its analysis result.

IOP Conf. Series: Journal of Physics: Conf. Series **1008** (2018) 012072 doi:10.1088/1742-6596/1008/1/012072

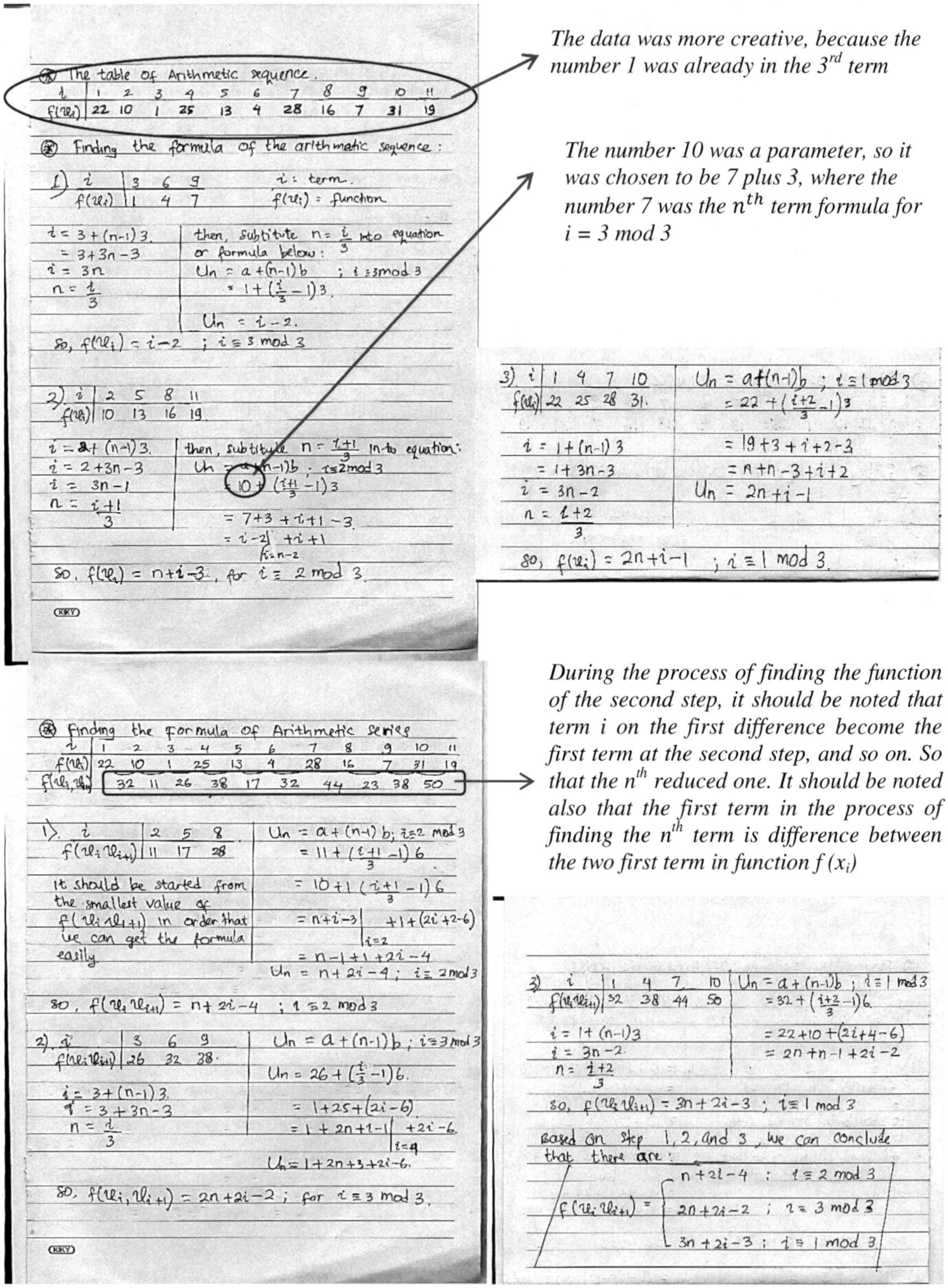

Figure 4. Students work upon finding the new arithmetic sequence and its formula

IOP Conf. Series: Journal of Physics: Conf. Series **1008** (2018) 012072 doi:10.1088/1742-6596/1008/1/012072

The following interview result portrays the response of very-creative students, describing the process he went through upon finding new arithmetic sequence up until getting the generalization of formula.

Lecturer : Did you face a problem when finding the arithmetic sequence which could be generalized?

Student A : Yes, I did. I ran into a problem when finding it because I tried to find an unusual arithmetic sequence. It was the first time I tried it.

Lecturer : Can you explain more about your problems?

Student A : I think it was because I did not do it this way before. We should tried to find an unusual arithmetic sequence. In the other courses, there was no task like this. In this course, we should find the unique arithmetic sequence which could be generalized and then expand it very carefully. When expanding the sequence, I experienced many problems. For example, sometimes it could not be expanded for *m* or *n* even. When it happened, I had to find another arithmetic sequence or use it and then tried to generalize it with many cases of *m* and *n*.

Lecturer : From all of the learning steps you went through, what was your most difficult problem when finishing it?

Student A : This learning method was new for me. I got many finishing steps clearly but the most difficult problem was when it came to predicting of arithmetic sequence which I already found. I should think deeply about the possibility of expanding the sequence and generalizing it. Otherwise, if I did not think about it deeply about it then I got stucked or got many risks. For example, I got an arithmetic sequence going through careful thoughts, and, if I could not expand it, then I had to repeat the whole process all the way from the beginning. Unluckily, sometimes, I got an arithmetic sequence and already managed to expand but I could not find the formula, then I started all steps over again.

Lecturer : How can you handle your problems in finishing the test?

Student A : I should understand the task very carefully, think deeply before going onto the next step and being more creative when finding unique sequence. Once, I tried to never give up.

Lecturer : What is your opinion about the learning method already applied in your class?

Student A : It was good for me. Even though, I got difficulties but I enjoyed the learning process. It was meaningful for me.

Based on figure 4 and interview result above, the study has concluded that the students work signified creative thinking ability at "very creative" category because the first term already started from in the third position and all of the numbers were not in ordered. According to Wulantina's research, student who has high ability tries to get more detailed information and can indentify the problems given very well when choosing the information during problem solving steps [10]. On the other hand, the results of Saefudin's research concludes that when applying ideas, students with high Mathematics ability does not make a mistake in solving problems, and they feel so much encouraged to solve problems in different ways using various answers [11]. The accuracy to apply a chosen strategy is based on student's previous experiences. This is in line with the opinion Cropley and Urban, which corroborates that the incubation stage is the stage at which students put their relation completion of the ideas they have had previously [7]. Based on the above findings, the formulas of second step or $f(x_i x_{i+1})$ were created. At this stage, only 61.11% of graduate students managed to find a formula function properly.

Based on the analysis results of data obtained from the students, some students were found unable to relate the first function to another function at the next level. Therefore, the research shifted to the formula functions that depend on *i* and *j* or $P^{n}_{m,d}(i, j)$ formulas. The following description shown in Figure 5 provides more details on that finding. What follows is student B's work upon generalizing formula, followed by the result of interview conducted to him.

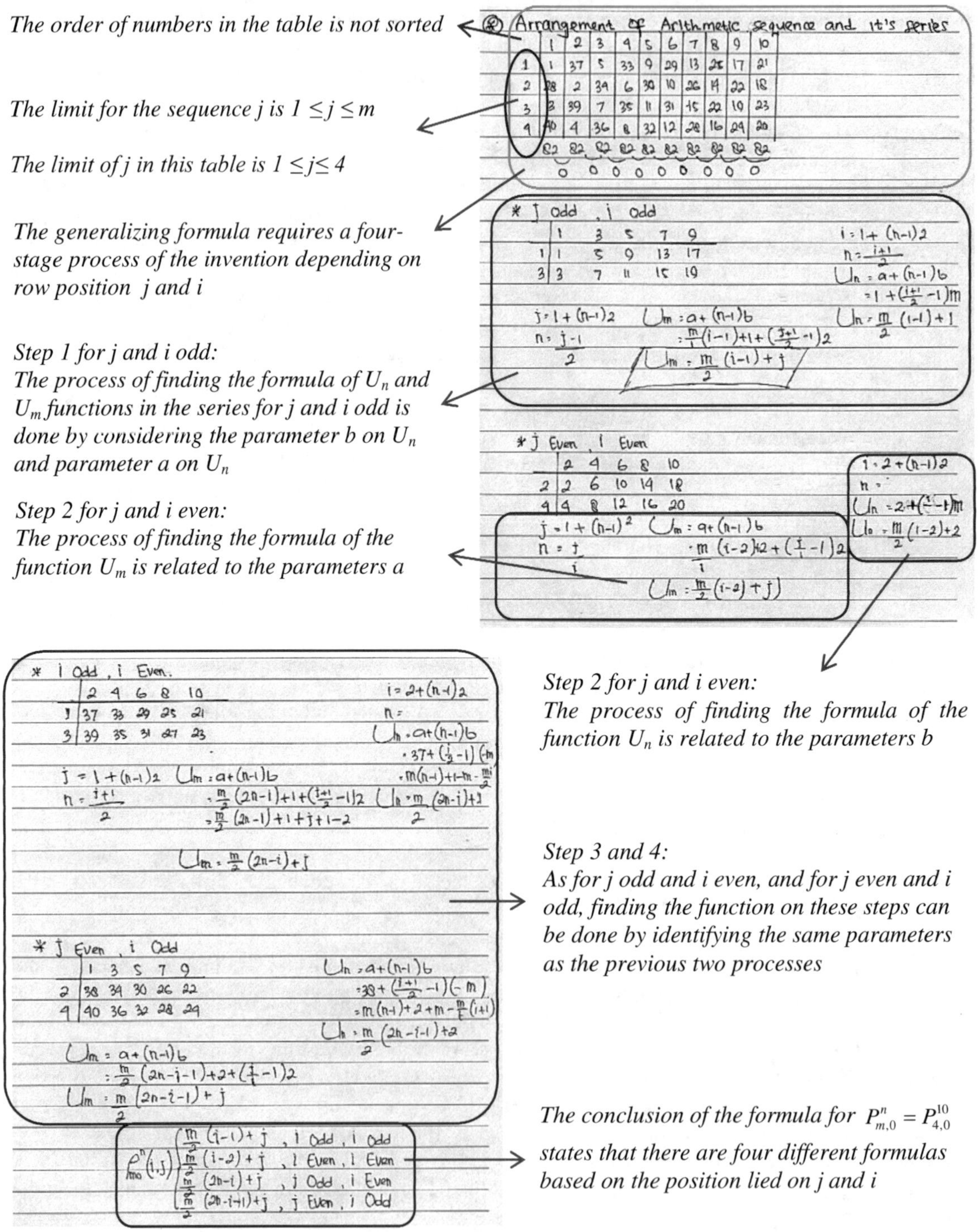

Figure 5. Student's work of finding the new arithmetic sequence and it's formula

Lecturer : Did you face any problem when finding the arithmetic sequence which could be generalized?

Student B : Yes, I did. I faced many problems about it.

Lecturer : Can you explain more about your problems?
Student B : I got many problems. For example, I found difficulties finding a new arithmetic sequence at first, so I had to try it again and again using multiple ways. When generalizing formulas, sometimes, I forgot to assume the current number as parameters so, I got stucked in the last step
Lecturer : From all of the steps of the learning which you went through, what was your most difficult problem when finishing it?
Student B : Finding the new arithmetic sequence at first was tough, and also formulation part was hard.
Lecturer : How did you handle your problems in finishing the test?
Student B : I had to find a new arithmetic sequence through many different ways and then tried to finish the formulation very carefully.
Lecturer : What was your opinion about the learning method already applied in your class?
Student B : I faced many problems as I explained before but I think that learning method was unusual and different from others. As long as I was capable of finishing the task on every step, I had to think seriously and very carefully.

Based on the students performance portrayed on figure 5 and interview to student B, the step of generalizing formula required a four-stage invention process, depending on the row position j and i. There was also one part in the work done by student B that indicated "creative" level because the arrangement of the numbers was already scrambled but the formulas could not be found. The student possessing creative ability tried to get more details information yet he could not indentify the problems very well when choosing the right information during problems solving steps. This premise supported the research undertaken by Wulantina, claiming that students with fair ability to gain information can precisely analyse question being asked, but he is not so much consistent in selecting the information and the information needed to solve the problem [12]. Another research result by Defitriani contends that students are not so creative to solve problems at hand, although they try to think of a solution to the problems they face [10]. While based on a hypothetical theory developed by Siswono, the less creative students tend to end up with uncheckd answers once they complete a task [12].

The students who have low-level creative thinking skills just answered the given problem without thinking deeply whether the answer was correct or false. It happened because students had not understood the main problems and some of them believed that their answer was correct, but they got stuck in the middle of finishing the steps. Referring to Deftriani, student categorized at *not creative* level tries to understand problems given but he misunderstands it [12]. Another research conducted by Siswono states that student categorized at *not creative* is convinced about his ideas for problem solution,but he makes mistakes when operationalizing the ideas [12].

5. Conclusion

Based on the abovementioned results and discussion, the study has drawn the following conclusions. Firstly, based on test results, the students who possess high, middle, and low creative thinking skills make a significant improvement between task 1 and task 2. All of the results are related to the learning process implemented, i.e. guided research-based learning. Then, based on the results of interviews to the students, several obstacles are present upon task accomplishment. These occur because the students are not familiar with guided research-based learning. There are only small number of students who can adapt to the learning quickly and well in this research-based learning. Secondly, the levels of students creative thinking skill based on their performance on task 1 are in the "*less creative*" category, and those based on their performance on task 2 are in "*creative*" category. Thirdly, based on the results of analysis and discussion, factors affecting the levels of students creative thinking skills in Combinatorics include the accuracy in expanding the existing data to find the function, the misunderstanding about the parameters when generalizing formulas, the inclination of graduate students to rely on rote (learning), and motivation. Lastly, developing students creative thinking skills can be carried out by increasing their motivation in developing the concepts taught by lecturers, continuously doing exercises in problem solving, and understanding problems given more carefully. So and so, the guided research-based learning method, when integrated into Combinatorics, can successfully lead to improved level of students critical thinking skills.

Acknowledgements
We gratefully acknowledge the support from Faculty of Teacher Training and Education - University of Jember 2017.

References

[1] Munandar U 2012 *Development of Talented Student Creativity* (Jakarta: Rineka Cipta)
[2] Siswono 2008 *Based Mathematics Instructional Model and Problem Solving to Improve The Ability of Creative Thinking* (Surabaya: Unesa University Press)
[3] Dekker H and Wolff S W 2016 Re-inventing Research-Based Teaching and Learning, *Paper prepared for presentation at the meeting of the European Forum for Enhanced Collaboration in Teaching of the European University Association in Brussels* on 5 December 2016 Belgium p 2
[4] Tohir Mohammad 2016 Analysis of Prospective Mathematics Teachers Ability in Applying Scientific Approach Based The Curriculum 2013, *Proceedings of The National Seminar on Mathematics Education 2016 at The State University of Surabaya* 431-446. ISBN 978-602-449-023-2
[5] Tremp Peter 2010 *Research-Based Teaching and Learning* (Munich: University of Zurich)
[6] Sugiyono 2017 *The Research Method of Quantitative, Qualitative, and R&D* (Bandung: Alfabeta)
[7] Bani A 2011 Improving Mathematical Understanding and Mathematical Reasoning of Junior High School Students Through Guided Discovery, *Journal of Research and Educational Thought* Special Edition (1) p 12-20 ISSN 1412-565X
[8] Hudojo H 2005 *Development of Curriculum and Mathematics Learning* (Malang: Malang University Publisher)
[9] Wulantina E Kusmayadi T A & Riyadi 2015 Process of Student 's Creative Thinking In Mathematics Problem Solving Based on Mathematics Ability of Student Grade 10 MIA SMAN 6 Surakarta, *Mathematics Learning of Electronic Journal* **3** (6) p 671-682
[10] Saefudin, A.A 2011 Creative Thinking Process of Elementary School Students (ESS) With High Mathematics Skills in Open Mathematics Problem-Solving, *The University of PGRI Yogyakarta: Mathematics and Character Education in Learning* p 1-9 ISBN 978-979-16353-6-3.
[11] Sari Agus P 2017 Creative Thinking Process of Student in Solving Mathematical Problems Based on Wallas Model, *Beta Tadris Mathematics Journal* **10** (1) p 18-32

IOP Conf. Series: Journal of Physics: Conf. Series **1008** (2018) 012073 doi:10.1088/1742-6596/1008/1/012073

The errors of metacognitive evaluation on metacognitive failure of students in mathematical problem solving

Nizlel Huda[1], Akbar Sutawidjaja[2], Subanji[2] dan Swasono Rahardjo[2]
[1]Mathematics Department, Jambi University, Jambi, Indonesia,
[2]Mathematics Department, State University of Malang, Indonesia

E-mail: nizlel@yahoo.com

Abstract: Metacognitive activity is very important in mathematical problems solving. Metacognitive activity consists of metacognitive awareness, metacognitive evaluation and metacognitive regulation. This study aimed to reveal the errors of metacognitive evaluation in students' metacognitive failure in solving mathematical problems. 20 students taken as research subjects were grouped into three groups: the first group was students who experienced one metacognitive failure, the second group was students who experienced two metacognitive failures and the third group was students who experienced three metacognitive failures. One person was taken from each group as the reasearch subject. The research data was collected from worksheets done using think aload then followed by interviewing the research subjects based on the results' of subject work. The findings in this study were students who experienced metacognitive failure in solving mathematical problems tends to miscalculate metacognitive evaluation in considering the effectiveness and limitations of their thinking and the effectiveness of their chosen strategy of completion.

1. Introduction

Problem solving is very important in mathematics learning because it can improve students' ability to choose the right resolution strategy and apply it accurately to get the right solution [3]. Realizing the importance of problem solving, Mathematics Education Department, faculty of teacher training and education, Jambi university, created a curriculum that included problem solving and set forth in the subject of Kapita Selekta I [6]. Problems of mathematical problem solving are not routine issues. Therefore, students need more complex thinking by using high-level thinking processes on this . High-level thinking processes are in dire need of students' metacognitive processes.

The metacognitive aspect is a very important aspect in mathematical problem solving because metacognitive ability is important in decision making [20]. When a person has a high metacognitive ability and knows how to apply it, there is a higher probability of problem solving to be successful [2]. Some researchers have concluded that metacognitive processes can improve the results of problem solving [1, 11, 23,4] and students who use metacognitive in problem solving often produce the correct response than those who do not use it [4].

There is a link between metacognitive and metacognition. Metacognition is thinking about thinking [16, 19, 8, 9, 15] and the metacognition component consists of metacognitive knowledge and metacognitive regulation [21, 17, 9, 5]. Metacognition refers to students' awareness of their own cognitive processes and the arrangement of those processes to achieve a particular purposes [7]. Metacognition is a person's ability to understand and monitor their own learning and how to use specific learning strategies in problem solving [13]) or decision-making [18].

Although metacognitive is an important aspect of problem solving but many students who have metacognitive activities in mathematical problem solving but they still fail to do metacognition. According to [10], there are 3 metacognitive failures (*metacognitive mirages*, *metacognitive blindness* and *metacognitive vandalism*). Metacognitive mirage is when the student does not see any difficulties, mistakenly abandon the useful strategy, changes the wrong calculation and refuses the true answer.

IOP Conf. Series: Journal of Physics: Conf. Series **1008** (2018) 012073 doi:10.1088/1742-6596/1008/1/012073

Metacognitive vandalism is a metacognitive failure that occurs when a student overcomes the deadlock by taking destructive actions by means of which students can change the problem by improperly implementing conceptual structures. *Metacognitive mirage* is a metacognitive failure that occurs when a student fails to notice something is wrong, for example, the student survives the wrong strategy or ignores the miscalculation. As a result of all three metacognitive failures, a student experiences errors in problem solving.

Metacognitive failure can be observed from 3 metacognitive activities such as *metacognitive awareness, metacognitive evaluation and metacognitive regulation* [22, 14]. According to [14], *metacognitive awareness* relates to self-made statements or other people's mathematical thinking and shows thoughts about: (1) what one knows, (2) where the presence of someone in the process of problem solving, and (3) what needs to be done, what has been done, or what someone can do. *Metacognitive evaluation* deals with self-made thoughts or others' mathematical thinking that show about: (1) the effectiveness and limitations of one's thinking, (2) the effectiveness of the chosen strategy, (3) the assessment of results, (4) (5) assessment of progress, ability and understanding of a person. Metacognitive regulation deals with self-made statements or other people's mathematical thoughts that suggest (1) problem-solving planning strategies, (2) goal setting, and (3) choosing problem-solving strategies.

Several studies related to metacognitive failure have been done. [10] found the identification of various forms of metacognitive failure by highlighting the ways in which verbal interactions could affect the thinking of each student. Furthermore, [12] states that students experience metacognitive failure in mathematical proof based on assimilation and accommodation framework.

Based on the researchers' observation results at Mathematics Education study of teaching faculty in Jambi University on March 15, 2017, researchers found that 77% has done metacognitive activity in problem solving but still fails in metacognitifnya activities. Based on the above explanation, this study examines how students' error in metacognitive evaluation occurs in students who have: (1) one metacognitive failure, (2) two metacognitive failures and (3) three metacognitive failures.

2. Research Methods

This research was conducted on April 2017 with the students of Mathematics Education Department, faculty of teacher training and education, Jambi university , who have studied *Kapita Selekta Sekolah Menengah 1.* The type of study was qualitative. There were 3 subjects in this study, one subject experiencing one metacognitive failure, one subject experiencing 2 metacognitive failures and 1 subject experiencing 3 metacognitive failures. The research data was collected by using the main instument of the researcher himself and the supporting instrument in the form of duty worksheets and interview guidelines which had been validated by the validator. Research data was analyzed by the following steps: (1) transcribing result of think aloud and result of student's interview, (2) conducting data reduction by making abstraction, (3) encoding every thinking process done by students, (4) describing structure of students' thinking process , (5) analyzing what happened during the study, (6) drawing conclusions.

3. The Results and Discussion

3.1. Result

This section describes the errors of metacognitive evaluation on the first subject (S1), second subject (S2) and the third subject (S3).

3.1.1. Exposure Data of first subject's (S1)

S1 experienced metacognitive blindness in the process of thinking. It can be seen in Diagram 1.

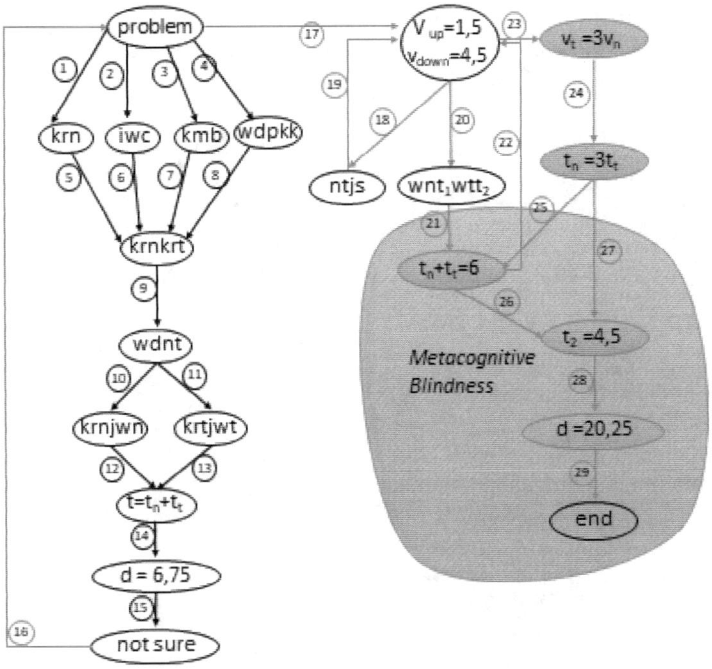

Diagram 1. S1's thinking process

Table 1. The terms of code on S2's thinking process

Code	Terms
krn	The average velocity of climbing up to hill
iwc	Having Enough time for taking a rest
kmb	The average velocity of downing the hill
wdpkk	The time needed to walk from the foothills to the top of the hill and climbing back to the foothills
krnkrt	The average velocity of climbing plus the average velocity of going down the hill
wdnt	Time needed to climb up and go down the hill
krnjwn	The average velocity of climbing is the distance per time of climbing
krtjnt	The average velocity of climbing is the distance per time of climbing
ntjs	Climbing and going down through the same path
wnt_1wnt_2	The time of climbing is t1 and the time of going down time is t2
tttlc	t going down three times faster
d	distance

It can be seen from diagram 1 that S1 rethought that the average velocity of going down was three times of the average velocity to climb the hill (on process 23), so the time to climb was three times going down the hill (on process 24). S1 also thought again that the time to climb the hill and went down from the hill was 6 hours (on process 25). This can be seen from the reseracher's interview results (R) with S1.

S1: I think again because the average velocity to climb a hill is greater than the average velocity to go down thehill then I determine that the time to go down was 4.5 hours. The time needed to climb the hill is longer than the time to go down the hill. *(Metacognition evaluation: Consider the effectiveness and limitations of thinking).*

R: Next, what do yu think by setting the time of going down the hill for 4.5 hours?

S1: By setting the time to go down is 4.5 hours then I get the distance from the foothills to the top of the hill is 20.25 km. I think my answer is true again *(Metacognition evaluation: Considering the results)*

3.1.2. Exposure data of second subject (S2)

S2 experienced two metacognitive failures in mathematical problem solving namely *metacognitive mirage* and *metacognitive blindness* in the process of thinking. The thinking process of S2 can be seen in Diagram 2.

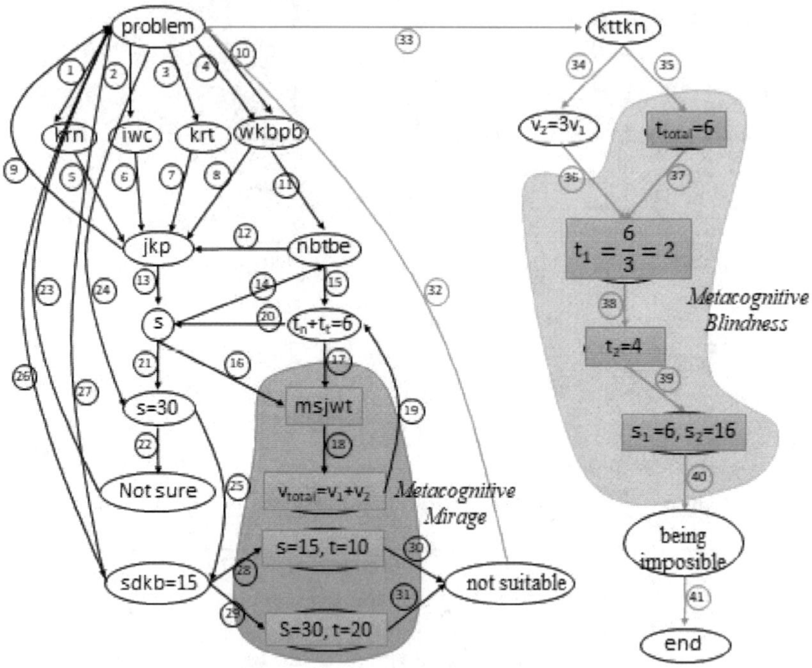

Diagram 2. S2's thinking process

Table 2. The terms of code on S2's thinking process

Code	Terms
krn	The average velocity of climbing the hill
iwc	Having Enough Time for taking a rest
krt	The average velocity of going down from the hill
wkbpb	Time from the foothills to the top of the hill coming back to the foothills
jkp	The distance from the foothills to the top of the hill
nbtbe	Climbing the hill and going down from the hill are 6 hours
msjwt	The recent problem is the total distance and time
Sdkb	Distance from the foothills
kttkn	The velocity goes down three times from the velocity of climbing the hill

3.1.2.1. S2 experienced metacognitive mirage

In diagram 2, S2 thought again about the relationship between the distance travelled from the hill to the top of the hill (at process 17). S2 also recalled the link between average velocity to climb and the average velocity to go down the hill at a total average of velocity. It is obtained from the average velocity of climbing the hill plus the average velocity of going down the hill (at process 18). S2 thought again that the average velocity of climbing and going down the hill with total time. The total time was the time of climbing plus time of going down the hill (at process 19). This can be seen from researcher's interview results (R) with S2.

S2: I think again that the average total of velocity is same to the average velocity of climbing the hill plus going down the hill. I think that it is multiplied by the total time which will result in the total of distance which is the distance from the foothills to the top hill and back to the foothills. The average total of velocity is equal to the average velocity of climbing the hill plus the average velocity of going down from the hill. I think I can get a total average speed. (*Metacognitive evaluation: considering the effectiveness and limitations of his thinking*)

R: Next ?

S2: I think if the distance from the foothills to the top of the hill is 15 km then the time needed to climb from the foothills to the top of the hill is 10 hours. Then if the distance is 30 km then the time needed to climb is 20 hours. I think again that this way does not help me to solve this problem. (*Metacognitive evaluation: considering the effectiveness of the selected strategy*).

3.1.2.2. S2 experienced metacognitive blindness

In diagram 2, S2 thought again about the time needed to climb the hill and go down from the hill was 6 hours. S2 also thought again if the time to climb the hill was 2 hours and to go down from the hill was 4 hours, then S2 got that the distance to climb and go down the hill were not same. This can be seen from the researcher's (R) interview results with S2.

S2: I think again that I have time to go down that hill. (*Metacognitive evaluation: considering the effectiveness and limitations of thinking*).

R: Then, what do you think for the next?

S2: I think again where I can specify that the time to climb is 2 hours and the time to go downhill is 4 hours. I think I can do it this way. *(Metacognitive evaluation: considering the effectiveness of the selected strategy).* I checked back the distance of climbing the hill is not the same as the distance of going down the hill. I think the information is not enough to solve this problem. *(Metacognitive evaluation: considering the difficulty of the problem).*

3.1.3. Data Exposure of third subject (S3)

S3 experienced *metacognitive mirage, metacognitive vandalism and metacognitive blindness* in the process of thinking. S3's thinking process can be seen in Diagram 3.

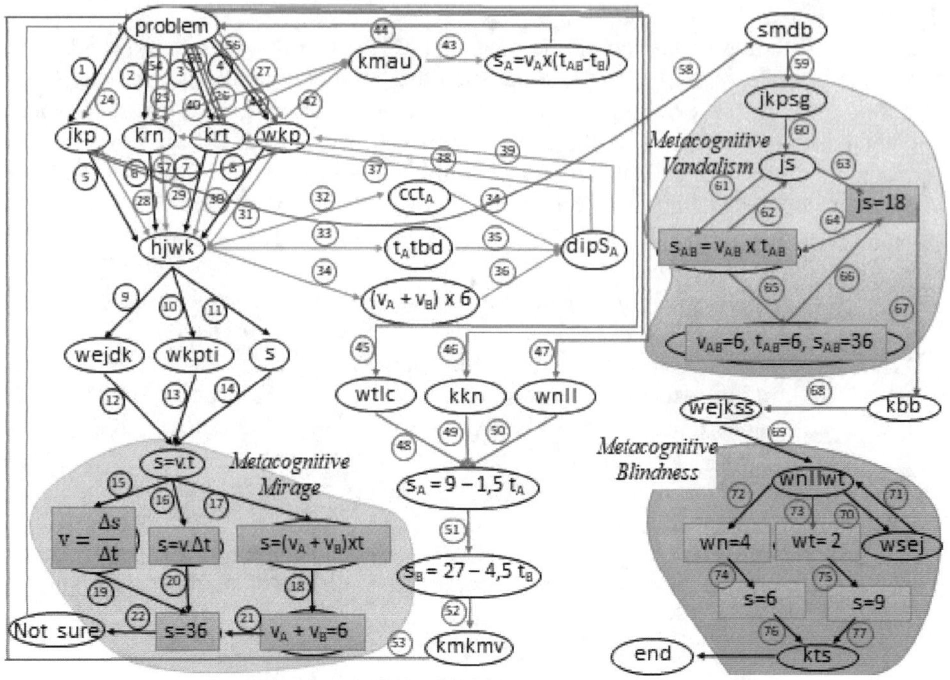

Diagram 3. S3's thinking process

Table 3. Terms of code on S3's thinking process

Code	Terms
Jkp	Distance from foothills to the top of the hill
Krn	The average velocity of climbing the hill
Krt	The average of going down from the hill
Wkp	Time from foothills to the top of the hill
Hjwk	The relation between distance, time and velocity
Wejdk	The time is 6 hours from the foothills
Wkpti	The time from foothills to the top of the hill
S	Being questioned S
Jvt	The same distance with v equals t
ccv_A	Try from v_A
$v_A tbd$	V_A can not be determined
$dipS_A$	Distance determined from the foothills to the top of the hill
Kmav	Why there are still variables
Kkn	The velocity goes up
Wnll	The time of going up is much longer
Wtlc	The time of going down is faster
Wmltj	The time of climbing is 3 hours
Kmkmv	How could that there are still the variables

3.1.3.1. S3 Experienced the metacognitive mirage

In diagram 3 it can be seen that S3 thought again that the average velocity could be calculated by Δv, distance was calculated by ΔS and time was calculated by Δt. (at process 15). The distance was the average velocity multiplied by Δt (at process 16) and the distance was also same with the sum of the average velocity of climbing the hill and going down from the hill multiplied by the time (at process 17). Then S3 determined that the distance from the foothills to the top of the hill was 36 km (at process 18, 19, 20 and 21). This can be seen from the researcher's (R) interview results with S3.

IOP Conf. Series: Journal of Physics: Conf. Series **1008** (2018) 012073 doi:10.1088/1742-6596/1008/1/012073

S3: I remember when I was learning physic, mam. I think again that the average velocity is the distance change divided by the time change. Distance is the average velocity multiplied by the time change then I can get the distance from the foothills to the top of the hill. (*Metacognitive evaluation: considering the effectiveness and limitations of their own thoughts or thoughts of others)*

R: Then, what do you think for the next ?

S3: I think again that the distance can be obtained from the average total of velocity multiplied by the time. It is obtained from the average velocity of climbing the hill plus the average velocity going down the hill. I can do this way and my answer is still correct (*Metacognitive evaluation: considering the results)*.

3.1.3.2. S3 experienced metacognitive vandalism.

It can be seen from diagram 3, S3 has obtained the distance's formula of climbing the hill and going down the hill. S3 thought again that the time to climb could be substituted with the time to go down the hill. S3 could not continue the completion and got stuck. S3 thought again that the distance from the foothills to the top of the hill and went back to the foothills was obtained from combined average velocity multiplied by the total time (process 61). S3 thought again that the distance from the foothills to the top of the hill was the overall distance divided by two (at process 62, 63, 64 and 65). This can be seen from the researcher's (R) interview results with S3.

S3: I think that the distance from the foothills to the top of the hill is 9 minus 1.5 multiplied by the time of climbing the hill. The distance from the top of the hill to the foothills is 27 minus 4.5 multiplied by the time of going down the hill. Why is there still a variable? I think this way does not help me to solve the problem *(Metacognitive evaluation: considering the effectiveness of the selected strategy)* and I also think that there is no information to solve this problem *(Metacognitive evaluation: considering the difficulty of the problem)*

R: What do you think?

S3: S3: I think about it for a long time, but I do not find the solution. I think again and change the way of the solution. It comes to my mind that the distance from the foothills to the top of the hill is same with th distance from teh top of the hill to the foothills. Therefore, I divide two of the overall distance and I get the distance from foothills to the top of the hill which is 18 km. I also think again that my answer is correct *(Metacognitive evaluation: considering results)*.

3.1.3.3. S3 experienced metacognitive blindness

It can be seen from diagram 3, S3 has already understood that the time needed to climb the hill is longer than the time to go down the hill. It needed 6 hours to climb and go down the hill (at process 70 and 71). S3 thought again that 4 hours was the time to climb (at process 72) and 2 hours was the time to go down the hill (at process 73). S3 does not think about the relationship among the specified time, velocity and distance, so that S3 obtained the distance of climbing the hill was different with going down the hill (at process 74 and 75). This can be seen from researcher's interview results (R) with S3.

S3: I think again that the time of climbing the hill is longer than the time of going down the hill which is 6 hours. .... (S3 kept silent for a moment)

R: What are you thinking when you are silent?

S3: I think again that I set the time to climb is 4 hours and the time to go down is 2 hours because the time of climbing is much longer than going down the hill and the specified time needed to do both is 6 hours. I think I can solve it *(Metacognitive evaluation: considering the strategy chosen)* and I think my answer is correct. *(Metacognitive evaluation: considering results)*

3.2. Discussions

The activity of *metacognitive evaluation* occurs to subjects who experienced the metacognitive failure in problem solving. It is explored through the results of think aloud, the researcher's interview results with the subjects based on the duty worksheets.

Subjects who experienced metacognitive blindness at first has thought correctly. The subject thought if the average velocity of climbing the hill was 3 times of going down the hill, so that the time to climb the hill was same which was 3 times of going down the hill. The overall time required to climb and go down the hill was 6 hours, then the subject thought that the time needed to go down the hill was only 4.5 hours. The subject has a failure of considering the effectiveness and limitations of his thinking. Furthermore, by setting the time to go down the hill, 4.5 hours, the subject got 20.25 km as the distance from the foothills to the top of the hill. In this case, the subject experienced an error of *metacognitive evaluation's* in considering the outcome.

Subjects who experienced *metacognitive mirage* and *metacognitive blindness*, at first, thought that the problem was the distance and time needed to climb and go down the hill. The subject thought that that the average total of velocity can be obtained from the average velocity of climbing the hill plus the average velocity of going down the hill. In this case the subject has been failure in considering the effectiveness and limitations of his thinking. Next, the subject thought again if the distance from foothills and the top of the hill was 15 km, then the time to climb the hill was 10 hours, likewise if the distance from foothills to the top of the hill was 30 km then time needed to climb the hill was 20 hours. This does not help the subject to solve the problem , so that the subject experienced the error of *metacognitive evaluation* in considering the effectiveness of the selected strategy.

The subject thought that the average velocity of going down the hill was 3 times of the average velocity of climbing the hill and the time to climb and go down the hill wass 6 hours, then the time to climb the hill was 6 divided by 3 which was 2 hours and 4 hours to go down. In this case, the subject failed to consider the effectiveness of the chosen strategy. Furthermore, the subject thought that there was not enough information to solve these problems. Thus, it can be said that the subject experienced an error of metacognitive evaluation in considering the problems' difficulties

Subjects experienced *the metacognitive mirage, metacognitive vandalism* and *metacognitive blindness*. At first, the subject thought that distance was the average velocity multiplied by time. Then he thought again that the distance was the distance multiplied by the Δt. Next, the subject also thought again that the distance was the sum of the average velocity of climbing and going down the hill multiplied by the time. The subject thought again that he could get the distance from foothills to the top of the hill. In this case, he experienced the error of *metacognitive evaluation* in considering the effectiveness and limitations of his thinking.

The subject also thought that he could get distance from the time and the average velocity of climbing and going down the hill. In this case, he experienced an error of *metacognitive evaluation* in considering the effectiveness of the new strategy and the error of *metacognitive failure* in considering the outcome.

The subject thought again that there were still variables if the subject calculated the distance from the foothills to the top and this way did not help him to solve his problem. In this case, he failed in considering the effectiveness of the chosen strategy. Next the subject thought again that there was no information to solve this problem. This could be said that the subject experienced the error of *metacognitive evaluation* in considering the difficulty of the problem.

The subject thought again that the distance from the foothills to the top was same with the distance from the top to the foothills, then the overall distance was divided by 2. The subject thought again that the answer was true. In this case, the subject experienced an error of *metacognitive evaluation* in considering the outcome.

The subject thought that the time needed to climb and go down the hill was 6 hours. He also thought that the time to climb the hill was longer than go down the hill. Next, he thought that the time to climb the hill was 4 hours and the time to go down the hill was 2 hours. Then he also thought that he could solve the problem and the subject also thought that his answer was correct. In this case the subject experienced an error of *metacognitive evaluation* in considering the chosen strategy and the subject also experienced an error of *metacognitive evaluation* in considering the outcome.

4. Conclusion and Suggestions

4.1. Conclusion

The conclusions of this study are:

1. Subjects experiencing a *metacognitive failure* have an error of *metacognitive evaluation* in considering: the effectiveness and limitations of their thinking and the results obtained.
2. Subjects experiencing two *metacognitive failures* have the errors of *metacognitive evaluation* in considering: the effectiveness and limitations of their thinking, the effectiveness of the chosen strategy and the difficulty of the problems encountered.
3. Subjects experiencing three *metacognitive failures* have the errors *of metacognitive evaluation i*n considering: the effectiveness and limitations of their thinking, the effectiveness of new strategies and strategies chosen, the difficulty of the problems encountered and the results obtained.

4.2. Suggestions

Based on the results of the study about the error of *metacognitive evaluation* towards the *metacognitive failure* of students in solving mathematical problems, researchers give suggestion to other researchers to investigate the error of metacogninive evaluation at every step of problem solving.

References

[1] Artz A F and Armour-Thomas E 1992 Development of Cognitive-metacognitive framework for Protocol Analysis of Mathematical Problem Solving in Small Groups *Cognition and Instruction* **9**(2) 137-175

[2] Aurah C M, Koloi-Keaikitse S, Isaacs C and Finch H 2011 The Role of Metaconitionin Everyday Problem Solving among Primary Students in Kenya. Problems of education in the 21st century **30**

[3] Aydoğdu M and Ayaz M F 2009 The Important of Problem Solving in Mathematics Curriculum *e-Journal of New World Sciences Academy Natural and Applied Sciences* **3** (4) A0095 538-545

[4] Barbacena L B and Sy N R 2015 *Metacognitive Model in Mathematical Problem Solving. Intersection* **12** (1)

[5] Brown A 1987 Metacognition, Executive Control, Self-Regulation and Other More Mysterious Mechanisms In: Weinert F E and Kluwe R H Eds *Metacognition, Motivation and Understanding* Hillsdale 65-116

[6] Curriculum Team 2015 The *Curriculum of Mathematics Education Program Department of PMIPA Faculty of Teacher Training and Education University of Jambi* Jambi

[7] Efklides A 2006 Metacognition and effect: What can Metacognitive Experience Tell Us about the Learning Process? *Educational Research Review* 3-14

[8] Flavel J H 1979 Metacognition and Cognitive Monitoring. *American Psychologist* **34** 906-911

[9] Garofalo J and Lester F K 1985 Metacognition Cognitive Monitoring and Mathematical Performance. *Journal for Research in Mathematics Education* **16** (3) 163-176

[10] Goos M 2002 Understanding, Metacognitive Failure. *http://espace.library.uq.edu.aw/.../JMB.Metafailure,pdf*

[11] Goos M, Galbraith P and Renshaw P 2000 A Money Problem : A Source of Insight into Problem Solving. *Journal of Research Mathematics Education* **25** 660-675

[12] Huda N, Subanji, Nusantara T, Susiswo, Sutawidjaja A and Rahardjo S 2016 University students' metacognitive failures in mathematical proving investigated based on the

framework of assimilation and accommodation. Educational Research and Reviews **11**(12) 1119-1128

[13] Karran E P and Irrizarry J 2014 Effects of Metacognitive Strategies on Problem Solving Ability in Construction Education *50th ASC Annual International Conference Proceedings.*

[14] Magiera M T and Zawojewski J S 2011 Characterization of Social-Based and Self-Based Context Associated with Students' Awareness, Evaluation, and Regulation of Their Thinking Small-Group Mathematical Modelling. *Journal for Reserch in Mathematics* Education **42**(3) 334-370

[15] Mahdavi M 2014 An Overview: Metacognition in Education. *International Journal of Multidisciplinary and Current Research* **2**

[16] Schoenfeld A H 1992 Learning to Think Mathematically : Problem Solving, Metacognition and Sense Making in Mathematics. In Douglas A. Grouws (ed) *Handbook of Research on Mathematics Teaching and Learning.* Micmillan Publishing Company New York 355-358

[17] Scoot B M and Leviy M G 2013 Examining the Component of Fuzzy *Educational Research,* **2** (2) 1-12

[18] Sengul S, Katranci Y 2012 Metacognitive aspects of solving function problems. *Procedia-Social and Behavioral Sciences* **4** 2178–2182

[19] Toit S D and Kotze G 2009 Metacognitive Strategies in the Teaching and Learning of Mathematics *Pythagoras* 70 55-67

[20] Toit S D and Toit G F 2013 Learner metacognition and mathematics achievement during problem-solving in a mathematics classroom. *The Journal for Transdisciplinary Research in Southern Africa* **9**(3) 505-518

[21] Veenman M V J, Wolters V H and Afflerbach P 2006 Metacognition and Learning : Conceptual and Methodological Considerations. *Metacognition and Learning* **1** 3-14.

[22] Wilson J and Clarke D 2004 Toward Modelling of Mathematical Metacognition *Mathematics Education Research Journal* **16** (2) 25-48

[23] Yimer A and Ellerton N F 1994 *Cognitive and Metacognitive Aspects of Mathematical Problem Solving*: An Emerging Model

Gender differences in prospective teachers' mathematical literacy: problem solving of occupational context on shipping company

N D S Lestari[1,2], D Juniati[3], St. Suwarsono[4]

[1] Mathematics Education Department, University of Jember, Indonesia
[2] Department of Mathematics Education, Universitas Negeri Surabaya, Indonesia
[3] Department of Mathematics, Universitas Negeri Surabaya, Indonesia
[4] Mathematics Education Department, Faculty of Teacher Training and Education, Universitas Sanatha Dharma, Indonesia

E-mail: nurcholiflestari@mhs.unesa.ac.id

Abstract. The purpose of this paper is to describe to what extent the prospective teachers can be considered as mathematically literate and how they communicate their reasoning in solving the problem based on the sex differences. Data were collected through mathematics literacy test on occupational context by 157 of prospective teachers from three universities in East Java, Indonesia. Their written responses were collected, organized based on the sex differences, analyzed and categorized to one of three levels of mathematical literacy. The examples of interesting students' response altogether with the scoring are discussed to describe their characteristic on mathematical literacy and their communication. The result showed that in general the mathematical literacy of female prospective teachers tend to be better than male prospective math teachers. Female prospective teachers are more capable of logical reasoning, using concepts, facts and procedures and algebraic operations to draw conclusions; make an interpretations and evaluations. This study has an implication that gender differences in mathematical literacy of prospective math teachers do exist, therefore this issue should be given a serious concern from the development programs of the faculty.

1. Introduction

Mathematical literacy differs from mathematics in general. Mathematical literacy is the knowledge to know and apply basic mathematics to everyday life [1]. Mathematical literacy is not about studying higher levels of formal mathematics, but about making mathematics relevant and empowering for everyone [2]. Mathematical literacy is an individual's capacity to formulate, employ, and interpret mathematics in a variety of contexts. It includes reasoning mathematically and using mathematical concepts, procedures, facts, and tools to describe, explain, and predict phenomena. It assists a person to recognize the role that mathematics play in the world and to make the well-founded judgement and decision needed [3].

The problems in mathematical literacy are categorized by the nature of the situation (the context category) and the major domain of mathematics involved (the content category). In solving the problem, three components of the nature of mathematical thought and action are needed [4]. The first component is content mathematics classified as shape and space, change and relationships, quantity, and uncertainty and data. The second component is the process of solving problems classified as formulate, employ and interpret. The last component is fundamental mathematics capability classified as mathematisation, devising strategies, representation, reasoning, symbols and communication. While Personal, societal, occupational and scientific are labelled for the mathematical literacy context categories. Every problem are categorized based on to the one closest type of every components [3].

Prospective math teachers are specially prepared to become a qualified mathematics teachers. Teachers has an important role played in helping students to learn and affecting the educational outcomes-achievement, future participation, and attitudes of their students [5]. They should have to know deeply about what they teach, how they teach it [6], [7] and how the curriculum supported the teaching [7]. All of this knowledge are known as pedagogical content knowledge (PCK) which is stated as important element to be an effective mathematics teacher [8].

Prospective math teachers, especially in Indonesia, should have mathematical literacy and knowledge about mathematical literacy for at least of four reasons. *Firstly*, mathematical literacy become a hidden educational objectives of mathematics teaching and learning in Indonesia. The competence related to mathematical literacy has not yet stated clearly but implicitly seen from the standard of students graduated from elementary and secondary schools. It stated that graduated students should have the factual, conceptual, procedural and metacognitive knowledge relates to sciences, technology, art, and culture; and able to use it for their self, and society, as a citizen [9]. This standard is in line with the definition of mathematical literacy that to be mathematically literate one must have the ability to use the mathematical knowledge. However, based on previous study [10] many mathematics teachers are misinterpreted mathematical literacy and did not attempt to teach the lessons that led to the teaching of mathematical literacy mathematics teachers It implies that mathematics education students as prospective teachers have to know deeply about what they teach, i.e. mathematics and literacy mathematics. *Secondly*, to teach mathematical knowledge (conceptual and procedural knowledge), teachers should ensure that the concept of mathematics is firstly developed, and it requires an appropriate action of teachers to use correct mathematical language, symbols and notation in their teaching [11]. *Thirdly,* to teach students such that they able to use their mathematical knowledge and enhance their problem solving skills, then mathematics should be thought in context and bring a real-life situation into the classroom [11]. Since the teaching and learning should be meaningful then the usage of everyday context in the teaching mathematics should be used as a means to access mathematical knowledge. Therefore mathematical literacy can be an alternative to present problem with daily life context. *Fourthly,* the teacher's ability to communicate their reasoning become important to teach mathematical knowledge [12]. This ability is one of mathematical ability needed for mathematical literacy [3]. Therefore, it also can be concluded that mathematical literacy is important for prospective math teachers as it is suitable to use in the teaching and learning mathematics.

The research about gender differences in mathematical literacy has investigated in many studies. A program assessment, PISA, aimed to evaluate the education system of countries in the world by testing 15-years-old students of their reading, mathematical, and science literacy, it also concerned to gender difference of it [3]. The result of PISA showed that in general, males performance in mathematical literacy are better than females[13] but it is contrary with Indonesian students' performance [14]. Furthermore the study of Arora and Pawlowski [13] tried to find the why factors of this results. They claimed that females' lack of self-confidence in their mathematics and science ability and their anxiety toward mathematics contribute to their under presentation among high performers in mathematics and science. Individuals with more self-confidence are more likely to allow themselves the freedom to engage in the trial-and-error processes that are essential to build these types of knowledge [15]. Another program assessment that study about literacy is PIAAC. Compared to PISA, PIAAC subject research were adult, around 16-65 years old and the focus were about proficiency in key information-processing skills-literacy, numeracy and problem solving in technology-rich environments and gathering information and data on how adults use their skills at home, at work and in the wider community [16]. The survey result of PIAAC in 2012 showed that most countries that participated in PISA 2003, males were more engaged in and had more positive attitudes toward learning mathematics [13]. Both, PISA and PIAAC claimed that males were tends to be better than females in mathematical literacy, except for Indonesia (by PISA test). In other words in Indonesia females' performance in mathematical literacy is better than males. However, Lailiyah [17] research about mathematical literacy skills of students' in term of gender differences give a different results. In her research, she involved four prospective teachers to take a mathematical literacy test and used six indicators of mathematical literacy. Her research found that male students have well of each indicators in mathematical literacy. While, female students are categorized as good in almost each indicator, except for fifth indicator that is scored enough. This means that male students were slightly better than female students.

The mathematical literacy problem used in this paper are focused on occupational context especially on "shipping company". The popularity of online shop in young people, including students, has made the shipping company much needed to facilitate the delivery of purchased goods. Based on the preliminary study, the users of shipping companies usually want the security of goods is guaranteed with a minimum cost. The calculation of the shipping cost depends on special terms of each shipping company. Commonly, the cost of sending goods depends on the volume, weight and choice of insurance used. Therefore, the ability of a person to perform spatial calculations, algebraic calculations, arguments and decision-making is necessary. They have gained these knowledge from their formal education before. However, many of them do not care how the costs are set and they will be so surprised to know the shipping cost from the shipping company officer. Therefore, the problem in occupational context about shipping company is interesting to use in this mathematical literacy test.

Table 1. Description of the questions on occupational context of shipping company

No	Domain category of each questions	Description
1	Content: Shape and Space Context: Occupational Process: Using concepts, facts, procedures and mathematical reasoning	Determining the weight of goods package to be used as a basis for determining the cost of delivery by first determining the type of package weight if the weight and volume of package of goods are known
2	Content: change and relationships Context: Occupational Process: Interpreting, applying and evaluating	Deciding whether it is important or not a delivery service officer advises the use of package insurance that is sent in terms of the amount of compensation money if the goods are lost
3	Content: Quantity Context: Occupational Process: Using concepts, facts, procedures and mathematical reasoning	Counting the number of transactions that can be served by the delivery company every day if the hours of operation, the number of officers and the average service time are known.

The purpose of this study is to describe to what extent male and female prospective teachers do are called mathematically literate and how they communicate their reasoning in solving the problem based on the gender differences.

2. Method.

In order to answer our research questions above, the mathematical literacy test with "shipping company" problem case are needed. On the test, they were asked to response/answer the question by giving brief explanation of reasoning and to erase the wrong response/answer is forbidden. They are only allowed to strike off the wrong answer and rewrite the correct ones at above or below it such that their way of thinking and reasoning can be traced. The test was conducted in three different time as there were three different university at East Java participated. Mathematical literacy test were followed by 157 students of mathematics education at least on 5th semester. At this grade they had obtained more than 75% pure mathematics courses, and 80% mathematics education courses.

Furthermore, the prospective teachers written responses were collected, and organized based on the sex differences. A scoring rubric was used to assess their response on each questions. The score represented their skill and characteristics of mathematical literacy on that question. The score on each questions were varied depending on their difficulties level (see Table 3, 4 and 5). Then, their mathematical literacy was categorized into three categories of mathematical literacy (low, moderate, or good). Analysis of the data was done using the frequency of student on each categories of mathematical literacy and frequency of student performance on each question. The analysis were continued by the qualitative methods by presenting the examples of interesting prospective teachers responses altogether with the scoring.

3. Result and Discussion

Mathematical literacy test followed by 157 prospective teachers, consisted of 24 male and 133 female prospective math teachers. It is a custom condition in teacher training faculty at Indonesia that the number of male students are fewer than female students. Their responses on the mathematical literacy were collected and analyzed. The result of the data analysis were used to answer the question to what extent male and female prospective teachers do are called mathematically literate and how they communicate their reasoning in solving the problem based on the sex differences, Table 2 bellow shows the frequency of mathematical literacy category of prospective teachers on the theme of "shipping company".

Table 2. The mathematical literacy of prospective teachers

Category	Male		Female	
	n	%	N	(%)
Low	8	33.33	26	19,55
Moderate	13	54.17	77	57,89
Good	3	12.5	30	22,56
Total	**24**	**100**	**133**	**100**

Table 2 above shows that the percentage of male prospective teachers in the category "low" mathematical literacy was 13.78% more than female prospective math teachers. While, for both "good" and “moderate” category the percentage of female prospective teachers was more than male prospective math teachers. It shows that in general the mathematical literacy of female prospective teachers tended to be better than male prospective math teachers. Again, this result are in line with the PISA results that Indonesian performance in mathematical literacy in PISA for female and male participants is not in line to the general trends of other country performance [12] [13]

The next following result will describe how the mathematical literacy performance of male and female prospective teachers on each question.

3.1. Performance of mathematical literacy and communication of reasoning for question 1.

Context for Question 1.

A shipping company charges the cost of package (goods to be shipped and the packaging) based on the weight of the package. There are two types of package weights that are actual weight and volumetric weight. Actual weight is weight by volume of package. The package should be weighed first to know the actual weight, then proceed to the volume. The following rules apply:

i. If the packet volume is less than $18{,}000 cm^3$, the package weight is the actual weight and the cost of shipping is done by actual weight.

ii. When the package volume of goods $\geq$ 18,000 cm^3, then the weight of the package is the volumetric weight calculated by following formula, and the calculation of shipping cost is done by volumetric weight of package to be shipped (Bv)

$$Bv = \frac{p \times l \times t}{6000} \times 1kg$$

Bv is volumetric weight, p is length in cm, $l$is width in cm, t is height in cm

Question 1

If Budi want to send a package measuring 30cm × 30cm × 25cm with actual weight 1.50 kg. Determine the weight of the parcels and the calculation basis used to calculate the cost of sending the package!

As mentioned before that every question have a unique scoring system. There are 0-4 scale used as score for the first question. Students respond were compared to the description of the score and then determine the most corresponding one to describe the prospective math teacher performance. The highest score (4) is given to them that able to select and apply the rules of package weight calculation with the right reasoning. It mean that they should have some fundamental mathematics ability [9]. They able to *devising a strategy* to solve the problem, working with given information *representation* from the text and formula, develop and *communicate* their *argument and reasoning* to select appropriates rule, use a *mathematisation* and *symbols* to modify the given representation and calculate the weight of goods package. Question 1 only involve two terms of conditions. Table 3 bellow shows the frequency of Prospective Teachers' Performance of mathematical literacy on question 1

Table 3. Prospective Teachers' Performance of mathematical literacy on question 1

Performance descriptions (score)	Percentage (%)	
	Male	Female
Does not respond or respond without explanation (0)	4.17	0
Responding with reasoning explanation but does not lead to the right answer (1)	16.67	12.78
Choosing the right rules of package weight calculation with correct reasoning but do some misunderstanding to identify the statement or do some miscalculation(2)	25.00	14.29
Selecting and applying the rules for package weight calculation correctly but giving incomplete reasoning explanation (3)	16.67	10.53
Selecting and applying the rules of package weight calculation with the right reasoning (4)	37.5	62.41

Table 3 above shows that the performance of male prospective teachers in question 1 was much better than female prospective math teachers. There was no female prospective teachers who did not respond or responded without explanation, while there were a small number of male prospective teachers who did not respond to anything or respond without any explanation. Moreover, 87.22% female prospective teachers were able to choose the rules for calculating the weight of package with correct reasoning, 72.94% female prospective teachers were able to select and applied the rules for package weight calculation and 62.41% of female prospective teachers who did it perfectly. In answering this question male prospective math often failed in communicate their arguments and reasoning. They tend to directly do the calculation without any argument or reason of choosing the rules. This characters is contrary to female prospective math teacher. They often show a brief explanation about their argument or reasoning before they took an action. It implies that for first question, female prospective teachers are more mathematically literate than male prospective math teachers.

Fig. 1 and Fig 2 bellow show about how the characteristics of mathematical literacy of male and female prospective math teachers, both are in low category of mathematical literacy, in reasoning and communicating.

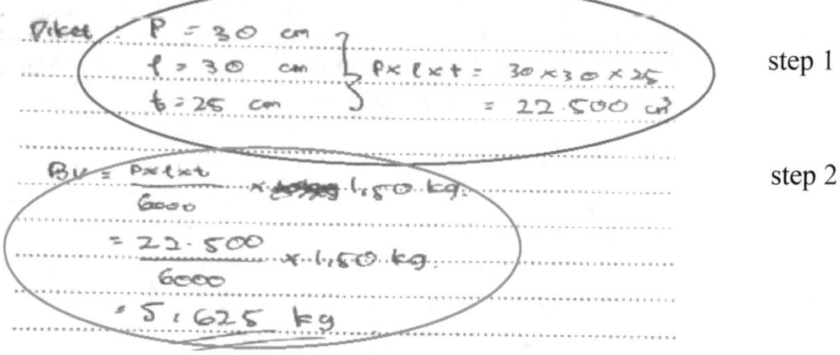

Figure 1. Example of a male prospective teachers' work in question 1

Figure 1 shows that he had poor written communication skill since there is only mentioned the calculation without explanation of his logical reasoning. There was a leap of reasoning in step 1 to step 2. He did not understand the meaning of the symbols used in the given formula. In the formula problem it was written $Bv = \frac{p \times l \times t}{6000} \times 1kg$ with p was the package length, l was the package width, t was the package height whereas 1 was the constants given. However, he replaced the constant value 1 with the actual weight of the goods namely 1.5. Surprisingly, this mistake was not only occurred to some male prospective teachers but also for female perspective math teachers. Based on this analysis, it can be concluded that he was tended to only do computations according to the order of rules given to the problem. He did not present any reason or explanation at each step of understanding symbols contained within the given formula. Therefore the prospective teachers' works shown in Figure 1 were given a score of 2.

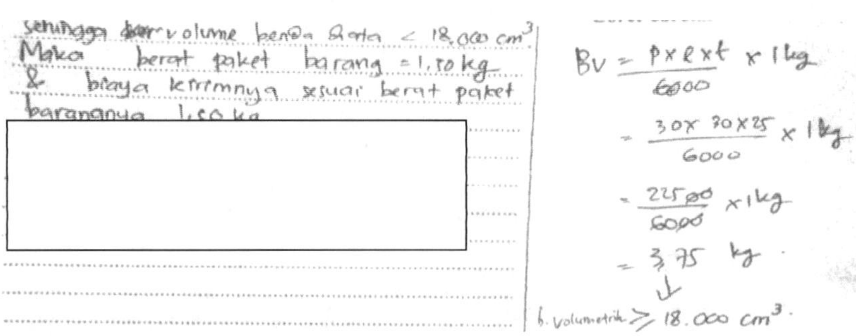

Figure 2. Example of female prospective teacher's respond for question 1

Figure 2 above is an example of female prospective teacher's respond. It shows that the female student tried to use a written communication to explain her flow of thinking. She showed two part of independent solution (on the right and on the left) with a contradict statement. On the left side she stated that the volume is less than 18.000cm^3 but on the other side are the contrary. It indicated that she did not understand the meaning of the given rules, she only made a calculation and unable to draw a connection between their work. She made a mistake in understanding the introduced concepts such as volumetric weight and actual weight. She compared two different magnitudes, volumetric weight and package volume as shown in Figure 2 on the right sides. This incorrect identification lead to a mistake in making conclusion that only found in female prospective math teachers. The prospective teachers' performance as shown in Figure 2 was given a score of 1.

IOP Conf. Series: Journal of Physics: Conf. Series **1008** (2018) 012074 doi:10.1088/1742-6596/1008/1/012074

3.2 Performance of mathematical literacy and communication of reasoning for question 2.

Context for question 2

A shipping company has stipulated the provisions of the packaged delivery cost based on the weight and use of package insurance as follows:

i. If the weight of the package is not more than 1 kg then the delivery cost is a minimal cost of 25,000
ii. If the weight of the package is more than 1 kg then the excess weight of the goods package will be charged additional delivery costs of 5.000/Ounces.
iii. Electronic goods must be packed in a wooden box, and the packaging will be done by the shipping service provider with a packed fee charged as an adder of the cost of delivery the package.
iv. If there is an incident of package to be delivered missing then the shipping company will replace the lost item with a maximum of 10 times the delivery cost. If the shipped item is more than 10 times the delivery cost, it is advisable to insure the package at a cost of 0.2% of the item price and administrative fee of 5.000

Question 2

A customer intends to send electronic goods for 3,500,000 using this shipping company. The dispatcher informs us that the packing fee is 200,000 and the item weight is 3.2 kg. If you were the dispatcher, would you recommend using insurance or not? Explain your answer!

There are 0-5 scale used as score for the second question. The highest score (5) is given to them who able to select, apply the rules to calculate the delivery cost and drawn right reasoning. It mean that they should have some fundamental mathematics ability. They able to devise a strategy to solve the problem, working with given information *representation* from the text and formula, develop and *communicate* their *argument and reasoning* to select appropriates rule, use a *mathematisation* by constructing mathematical models for complex situation and use an appropriates *symbols* to modify the given representation and calculate the weight of goods package. Differs from question 1, question 2 is containing more complex term. It contains of five terms that need to be comprehend by the subjects.

Table 4. Percentage of prospective teachers' performance in question 2

Performance description [score]	Percentage (%)	
	Male	Female
Does not respond or respond without explanation (0)	4.17	6.77
Responding with reasoning explanation but it does not lead to the right answer (1)	33.33	30.08
Only choosing the calculation rule of delivery cost but incorrect in applying the rule because she is wrong to understand the statement. (2)	45.83	38.35
Being able to select and apply rules to calculate delivery cost with correct reasoning but she does not conclude or explain them (3)	0	3.76
Being able to select and apply rules to calculate delivery cost with correct reasoning but incorrect in drawing conclusion (4)	4.17	5.26
Being able to select and apply rules to calculate delivery cost, as well as draw conclusion by providing correct reasoning (5)	12.5	15.79

Table 4 above shows that in question 2 the performance of female prospective teachers was better than male prospective math teacher even though the difference is small. More than 82% for both male and female mathematics prospective teachers could not draw a conclusion correctly, most of them were failed in understanding the statement rule and reasoning thinking. It also shows that no more than 1/6 of the prospective teachers of both male and female were able to solve question 2 correctly and logically. This result also showed that question 2 was a more difficult question than question 1

IOP Conf. Series: Journal of Physics: Conf. Series **1008** (2018) 012074 doi:10.1088/1742-6596/1008/1/012074

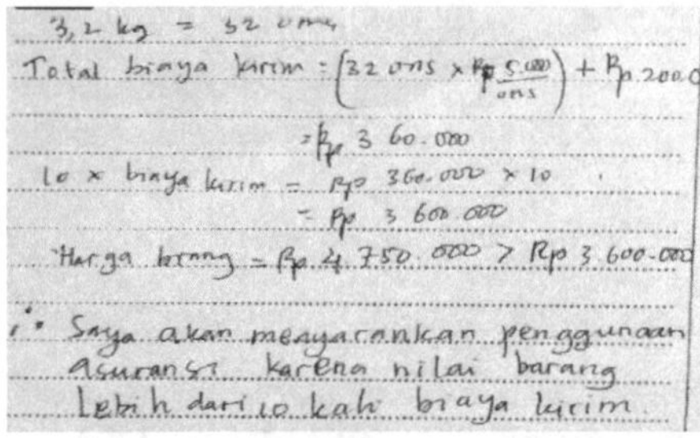

Translation:
3.2kg=3.2 ons
Total of delivery cost=$\left(32ons \times Rp\frac{5000}{ons}\right) + Rp200.000$ = Rp 360.000
10×delivery cost = Rp 360.000 ×10 = Rp 3.600.000
The things price=Rp 4.750.000>Rp 3.600.000
∴ I will suggest to use assurance because the things value is more than 10 times of the delivery cost

Figure 3. Example of a prospective teachers' answer in question 2

Figure 3 shows the student's mistake in understanding the statements of the calculation rule of the delivery cost causing a response error. On that problem, it was stated that: (i) If the weight of the package of goods was not more than 1 kg then the delivery cost was the minimal cost namely Rp. 25.000,00 and (ii) If the weight of the goods package was more than 1 kg then the excess weight of the goods package would be charged additional delivery fee of Rp5.000,00/Ounces. However, the student assumed that if the weight of the goods was more than 1 kg then rule 2 was applied regardless of rule 1. Therefore, the student's performance was given a score of 2 even though the student showed algebraic ability to make mathematics and mathematical representations

3.3 Performance of mathematical literacy and communication of reasoning for question 3
Question 3 and the context:
A shipping company operates from 08.00 to 17.00 every day, closes on break from 5 pm to 6 pm and reopens at 6 pm to 9 pm. The working hours are divided into 2 shifts of the same length. The time required for a shift change is about 20 minutes. Each shift consists of 3 shipping operational officers who perform the package delivery transaction service. Standard time of service delivery officer is 7 minutes per transaction. How many transactions can this shipping company serviced on a daily basis? Write down your reasons

As question 1 and 2, In question 3 prospective teacher also should show a good reasoning. They should have a good ability to represent the given text to their own representation using numbers or diagrams. They also should make a correct mathematisation to modelling the situation and in the same time they have to use their reasoning to make their model to be logic and acceptable. At this question, different ways of thinking will lead to a different correct answer. It is mean that this question is an open ended question. Finally, they have finished the answer by correct calculation.

Table 5. Percentage of frequency of prospective teacher responses to question 3

Performance description [score]	Percentage	
	male	Female
Not responding or responding without explanation [0]	0.00	8.27
Responding with explanation of reasoning but it does not lead to the right answer [1]	20.83	16.54
Calculating operational time correctly with logical reasoning but making a reasonable error to calculate the number of transactions [2]	54.17	48.87
Calculating the operational time correctly with logical reasoning but making an operation mistake to calculate the number of transactions [3]	4.17	2.26
Calculating operational time and the number of transactions correctly with logical reasoning [4]	20.83	22.56

Table 5 above shows that female prospective teachers responded perfectly by 2.17% more than male students of mathematics education, but 8.27% of female students chose not to respond or responded without giving explanations/reasoning. In addition, Table 5 also shows that about half of the prospective teachers participated on this test, both male and female, made a mistake in calculating the number of transactions. Figure 4 below shows an example of a correct but different way response.

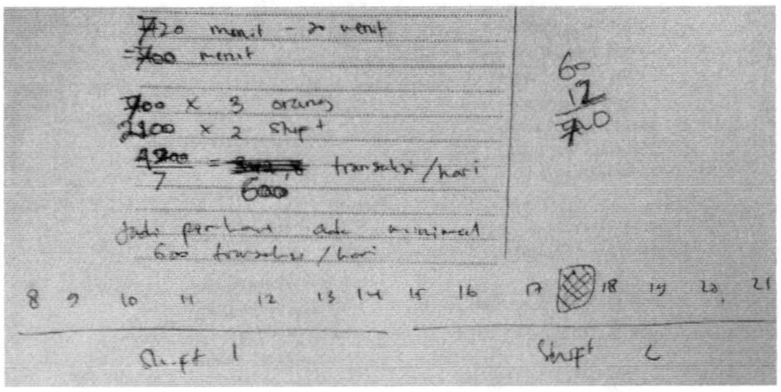

Translation
720 minutes – 20 minutes= 700 minutes
700 x 3 people
2100 x shift
4200/7=600 transactions/day
So in each day, there are 600 transaction/day
60
12
700

Figure 4. The answer of male prospective teachers in calculating operational time

Figure 4 above shows they started to answer by calculating the total operational time. He was using a list of numbers from 8 to 21 to describe a time and shaded a space between number 17 and 18 to indicate the rest time. He also draw two segment under the list of number to identify the operational time in every shift for watch person. It brought him to a conclusion that the total time for each shift were 6 hours. Unfortunately, on his work there was no notes about this total time. It may lead him to a mistake in reasoning. He had used 12 hours or 720 minutes as total operational time as one shift total time, as instead of the total time for two shift. So he doubled the total time again and brought him to the incorrect answer. The figure also indicated that he did not use a good communication as there was no hand writing about his reasoning. Again he only wrote down the important point for him to make himself easier to answer but not to communicate to other. Therefore, the performance of this student was given a score of 2. This kind of performance was only done by 1 male prospective math teacher. In addition to the responses in Figure 4, some male prospective teachers also showed different creativity to solve questions such as not dividing the shift into 2 equal lengths but the shift replacement time remained a reduction at the end of the operational hours. Then the prospective teachers should make a rounds to count the number of transactions.

4. Conclusion.

Based on the research finding and discussion that has been described above it can be concluded that in general the mathematical literacy of female prospective teachers tend to be better than male prospective math teachers. Female prospective teachers are more capable of logical reasoning, using concepts, facts and procedures and algebraic operations to draw conclusions, interpretations and evaluations. In addition, female prospective teachers also show a more communicative attitude in writing their way of thinking, while male students tend to only write important points for them. However, male prospective teachers have better creativity than female students by using unique methods. This study has an implication that gender differences in mathematical literacy of prospective math teachers do exist therefore this issue should be given a serious concern from the development programs of the faculty.

Acknowledgments
We gratefully akcnowledge the support from Kemenristek Dikti Grant of Indonesia of year 2017

References

[1] Ojose B 2011 Mathematics literacy: are we able to put the mathematics we learn into everyday use? *Journal of Mathematics Education* **4** 89-100

[2] McCrone S S and Dossey J A 2007 Mathematical literacy--it's become fundamental *Principal Leadership* **7** 32-37.

[3] OECD 2015 PISA 2015 *Draft mathematics framework* (Paris: OECD Publishing)

[4] Stacey K and Turnerr R 2015 *Assessing mathematical literacy: The PISA experiences* ed K stacey and R Tunner (Switzerland: Springer International Publishing) chapter 1 pp 5-33

[5] Leder G C, Forgasz H J, Jackson G 2014 Mathematics, english and gender issues: do Teachers count? *Australian Journal of Teacher Education* **39**

[6] Ma'rufi, I Ketut B and Dwi J 2017 The analysis of mathematics teachers' leraning on algebra function limitmaterial based on teaching experience difference *The 1st International Conference on Mathematics: Pure, Applied and Computation 2016* (AIP Publishing) p. 020056-1

[7] Shulman L S 1986 Those who understand: Knowledge growth in teaching. American Educational Research Association **15** p 4-14

[8] Ma'rufi, I Ketut B and Dwi J 2016 Pedagogical content knowledge: knowledge of pedagogy novice teachers in mathematics learning on limit algebraic function *The 1st International Conference on Mathematics, Science, and Computer Science (ICMSC) 2016* (AIP Publishing) p 050003-1

[9] Salinan Lampiran Permendikbud no 20 tahun 2016 tentang standart kompetensi lulusan pendidikan dasar dan menengah retrieved [20/01/2018] from [http://bsnp-indonesia.org/wp-content/uploads/2009/04/Permendikbud_Tahun2016_Nomor020_Lampiran.pdf]

[10] Lestari N D S, Juniati D and Suwarsono St 2017 Preliminary study: mathematics' teacher conception in supporting the integration of mathematics' literacy and mathematics teaching and learning *International Conference on Mathematics Education, Theory and Application (ICMETA) 2016 (Solo)* (Solo: UNS) pp 76-83

[11] Machaba, F M 2017 Pedagogical demands in mathematics and mathematical literacy teachers and facilitators. *Open Access Eurasia Journal of Mathematics, Science and Technology Education* **4** 95-108

[12] Pantaleon K V, Juniati D, Lukito A and Mandur K 2018 *J. Phys.: Conf. Ser* **947** 012070

[13] Arora A and Pawlowksi E 2017 Examining gender differences in the mathematical literacy of 15-year-olds and the numeracy skills of the age cohorts as adults Retrieved [05/08/2017] from [https://static1.squarespace.com/static/51bb74b8e4b0139570ddf020/t/588a18c13a04118a0c68116e/1485445313820/Arora_Pawlowski_PIAAC_2017.pdf]

[14] Stacey K 2011 The PISA view of mathematical literacy in Indonesia *Journal of Indonesian Mathematics Society B (Journal on Mathematics Education)* **2** pp 95-126

[15] OECD 2015 *The ABC of gender equality in education: aptitude, behaviour, and confidence* (Paris: OECD Publishing)

[16] OECD 2016 *Survey of adult skills: reader's companion, second edition, OECD skill studies* (Paris: OECD Publishing)

[17] Lailiyah S 2017 Mathematical literacy skills of students' in term of gender differences *The 4th International conference on Research, Implementation, and education of Mathematics science (4th ICRIEMS) 2017* (AIP Publishing) p 050019-1

IOP Conf. Series: Journal of Physics: Conf. Series **1008** (2018) 012075 doi:10.1088/1742-6596/1008/1/012075

The Use of Interactive Media *Ispring Suite 8* Supported by *Google SketchUp* to Improve Students' Geometry Skills Based on Hoffer's Theory

A Nurwijayanti [1*], **Budiyono**[2], **L Fitriana**[3]

[1,2,3]Department of Mathematics Education, Sebelas Maret University
Jl. Ir. Sutami 36.A, Jebres, Surakarta, Jawa Tengah 57126, Indonesia

*Corresponding author: aniinurwijayanti@gmail.com

Abstract. The basic Geometry skills are needed by the students to solve the geometrical tasks in daily life. There are five aspects of the Geometry ability based on the Hoffer's theory. They are visual, verbal, drawing, logical, and application. These are the abilities that the students in junior high school level need to master. The purpose of this study is to find out and describe the effectiveness of the interactive media supported by *Google SketchUp* to improve the students' basic Geometry skills based on Hoffer's theory. The subject in this research is 30 students from class 9E in Junior High School of Mojogedang 1, Karanganyar regency. This study uses a pre-test and post-test experiment and analyzed with the t-test hypothesis with the significant level of 5%. The result of this study can be seen from the diffeence average score between the pre-test and post-test, which shows a significance difference. It means that through the interactive media supported by *Google SketchUp*, the students' five basic abilities are improved. Therefore, it can be concluded that the interactive media supported by *Google SketchUp* is potential and can be used to help the students in improving their basic Geometry skills based on Hoffer's theory.

1. Introduction

Geometry is a branch of mathematics concerned with a point, straight line, plane figures, space, spatial figures, and the relation between them [4]. In the teaching and learning process, Geometry has a distinctive spot in the mathematics curriculum since it has a lot of variety of concepts [5]. These varieties such as shape and size, or its system related to visual thinking are the basis for Geometry thinking. They make the teaching and learning of Geometry as a complex mathematics lesson. On the geometrical problem solving, it needs a clear concept for the students to be able to apply the Geometry skills that they have. These skills are such as visualizing, recognizing various types of plane and Geometry, describing a picture, sketching, labeling particular spots, and also knowing the difference and the similarity among Geometry shapes. Therefore, Geometry becomes one of the most difficult and most disliked lessons by the students [9]. Besides, it is also one of the mathematics materials which the students also find it hard to understand [2]. However, it will still become the part of mathematics that has to be learned by the students.

There are several reasons why it is essential to learn Geometry [21]. First, it helps the students to have a complete appreciation of their world. Second, the Geometry exploration can help them to develop their problem-solving skill. Third, it has the main role in the other mathematics fields. Fourth, Geometry is used by people on the daily basis. Fifth, it is challenging and interesting. It also covers every aspect of life. These are the reasons why the students have to explore about shape, line, angle,

IOP Conf. Series: Journal of Physics: Conf. Series **1008** (2018) 012075 doi:10.1088/1742-6596/1008/1/012075

and Geometry, which links to their life [12]. It is supported by some of the statements that Geometry is a part of mathematics which related to student's life because almost all of the objects around them are Geometry [13], and the study of Geometry contributes to helping students develop the skills of visualization, critical thinking, intuition, problem-solving, deductive reasoning, logical argument and proof [17]. It makes the Geometry as an important part of mathematics.

Considering the importance of learning Geometry, especially for the students, therefore they need to have several basic skills. Generally, based on the Hoffer's theory, these are the skills that they have to master: 1) able to analyze the characteristics of geometrical shape in two or three dimensions, and able to make a statement about the relation between them; 2) able to specifically determine the position of a point and describe the spatial relationship with the other systems; 3) apply the transformation and use it simmetrically to analyze mathematical situation; 4) use the visual skill, spatial reasoning, and Geometry modelling to solve the problem [16].

These Geometry skills can be broken down into five types. They are visual, verbal, drawing, logical, and application [10]. These skills are linked to each other. For example, students will be able to apply the concept of Geometry into the daily life tasks if they are able to develop their logical thinking and able to explain it verbally. This is supported by the statement that says the verbal and visual skills has to develop the basic logical reasoning [9]. Furthermore, it is also important to balance these skills with proving and problem-solving which needs an integration it [10]. This is supported by a statement which says a broad and structured Geometry knowledge has a positive correlation with a better geometrical problem-solving skill [15]. So, these five basic skills are an intact unit that the students on junior high school level have to master to solve geometrical problems.

To improve it, it is necessary to have an innovation in the teaching and learning process. One of them is by using a software-based media. Since these days, the technology has a rapid development, especially in the software development. On the education itself, the use of the software also grows along with this development to support the teaching and learning process. This is supported by the statement that the use of technology in school has broadly known as one of the ways to elevate the teaching and learning process [3]. The term Technology-Enhanced Learning has also increasingly been used in England, Europe, and other countries in the world to shows that the use of technology can be used on the lessons in school [19]. So, the teaching and learning process with the computer-based program is getting more attention to develop an innovation on it.

One of the software that is continuously changing is the interactive presentation flash-based *Ispring Suite 8* software. It is a flash-based program which is integrated with the presentation program of *power point*. It presents the learning materials in the form of a flash slide that contains pictures, animation, video, or audio. Basically, it can be used as the alternative learning media for the teacher to deliver the learning materials in an interesting way. It is also more practical and optimal since it also can engage the students more in the lesson.

In mathematics, there are also several available software to be used in the lesson. This makes many educators conduct some research about the teaching and learning process which utilizes that software. For example, the software Cabri 3D is used in the learning of three-dimensional shapes for junior high school students [19], and the Geogebra program as the media which is used in teaching calculus for the university level students [11]. Further, *Google SketchUp* presents as a 3D graphics software which rapidly developed these days. It is able to help the students in their visualizing, constructing, associating, and also thinking process. Moreover, it has the orbit tool that can be used to rotate the plane or the Geometry in all directions easily. Other than that, the *Google SketchUp* also has a distinctive feature on its push/pull tool to draw the object, so it does not take a long time for the students to form it. The students who use *Google SketchUp* with this feature has a better performance compared to the students who do not use it [8]. The students who use it also do not have a lot of difficulties to draw in it [6]. This supports the statement that says *Google SketchUp* is easy to be used and learned. It also has a variety of objects in its library [7]. The diversity of its feature and the simplicity to use it makes the *Google SketchUp* is used in many fields especially as the interactive media in learning, in which beliefs can compete or even exceeds the other 3D graphics software [18].

To find further about the effect of using interactive media supported by *Google SketchUp* in teaching and learning process, it needs to have a research on it. The research is also to find out the students' Geometry skills based on the Hoffer's theory. Moreover, it is also important to conduct this research to be the teachers' consideration in choosing the right media on their lesson. The right choice on the media can help the students in learning the lesson or specifically in this study, is to improve the students' Geometry skills.

2. Methods of the Study

This study is an experimental study with the design of one group pre-test and post-test. The study is conducted in Junior High School of Mojogedang 1, Karanganyar regency. The population of this study is all of the students in grade 9. This study uses the cluster random-sampling technique, then chooses one group/class which will be given the treatment on the teaching and learning process of mathematics especially in Geometry using the interactive media supported by *Google SketchUp*. The students' Geometry skills are measured before and after the experiment. In this study, the pre-test is given to the students to find out their skill first. Then, it implements the interactive media to the students. After that, the students are given the post-test to measure their skills. The result of the pre-test and the post-test then is compared to find the conclusion. Whether there is an improvement in their Geometry skills or not.

3. Result and Discussion

Before the treatment, (using the interactive media supported by the *Google SketchUp* on the teaching and learning process) the students are given the pre-test to find out their initial skills on Geometry material especially on the curved-face-three dimensional objects. The result of the pre-test then is analyzed by the researcher and gets the $\mu = 2.84$ as the mean score and $\sigma = 1.67$ as the standard deviation score. This score is used as the reference to categorize the students pre-skills. This categorization is based on their pre-test result which can be seen from the Table 1 below.

Table 1. The Students' Pre-Skills

Interval	Category	Frecuency
$X \geq 4.78$	High	3
$1.44 \leq X < 4.78$	Medium	23
$X < 1.44$	Low	4

Based on the analysis of the students' pre-test score, the students are divided into three categories. They are the high, moderate, and low category. The data shows that there are 3 students (10%) on the high category, 23 students (76.6%) on the moderate category, and 4 students (13.3%) on the low category. Then, the researcher conducts the treatment in the form of the curved-face-three dimensional objects teaching and learning process using the interactive media *Ispring Suite 8* supported by *Google SketchUp*. The experiment is broken down into 5 meetings. The 4 meetings are for cylinder and cone materials. The last meeting is for conducting a test. In the beginning of the experiment, the students are introduced to the media that will be used, the function of the *Google SketchUp* tools, and how to use it. After that, the students are given the post-test to find the effectiveness of the media that has been used on the teaching and learning process, also to get the result of the students' basic Geometry skills. The post-test is related to the five basic Geometry skills based on the Hoffer's theory. The result of the pre-test and post-test then are analyzed to find the improvement of the students' basic Geometry skills based on the Hoffer's theory. Before it is analyzed to test the hypothesis, these data are tested using the normality test to find whether these two data has a normal distribution or not. The normality test is used the SPSS software with the significant level of 5%.

Table 2. The Normality Test from the Pre-Test and Post-Test of Basic Geometry Skills

	Kolmogorov-Smirnov[a]			Shapiro-Wilk		
	Statistic	Df	Sig.	Statistic	Df	Sig.
Pre-Test	0.127	30	0.200*	0.946	30	0.135
Post-Test	0.145	30	0.109*	0.942	30	0.104

a. Lilliefors Significance Correction

*. This is a lower bound of the true significance

Table 2 is the result of the normality test from the pre-test and post-test. The result shows that the observed significance level from the pre-test is 0.135 > 0.05 and on the post-test is 0.104 > 0.05. It indicates that from these results the sample for both data drawn from population whic has a normal distribution. After that, these output is analyzed by the hypothesis test. It is used is the Paired Sample Test, because the data has the normal distribution.

Table 3. Paired Sample Test of Geometry Ability

	Paired Differences					T	df	Sig. (2-tailed)
				95% Confidence Interval of the Difference				
	Mean	Std. Deviation	Std. Error Mean	Lower	Upper			
PreTest-PostTest	-2.90000	1.07198	.19572	-3.30028	-2.49972	-14.817	29	0.000

Table 3 shows that the observed significance level that reaches 0.000 which less than 0.05, so the H_0 is rejected. Therefore, the result of this test is that both pre-test and post-test average scores are not identical. It means that there is a significant difference between the pre-test and post-test score of the students' basic Geometry skills. Furthermore, it shows that the teaching and learning process using the interactive media *Ispring Suite 8* supported by *Google SketchUp* is proved to improve the students' basic Geometry skills based on the Hoffer's theory significantly. Based on the analysis result table, the average score on the pre-test is 3.11 and 6.01 on the post-test. The use of both media at the same time or separately is considered advantageous on the teaching and learning process. It also has the benefit on the result of the learning on the prior research [8] [1] [20].

3.1 The Basic Geometry Skills based on the Post-Test Result

Before the hypothesis test on the students' basic skills of Geometry, the data of the post-test has to be tested with the normality test first. This is to determine whether the result of the post-test has a normal distribution or not.

Table 4. The Normality Test Based on the Pre-test Result.

	Kolmogorov-Smirnov[a]			Shapiro-Wilk		
	Statistic	Df	Sig.	Statistic	Df	Sig.
High	0.385	3	.	0.750	3	0.000
Medium	0.134	23	0.200*	0.958	23	0.427
Low	0.333	4	.	0.763	4	0.051

a. Lilliefors Significance Correction

*. This is a lower bound of the true significance

Based on the table above, the score of the students' basic Geometry skills based on the Hoffer's theory on the high category has a observed significance level of 0.000 < 0.05.This result makes the H_0 is rejected, in the other words, the population in this category do not have a normal distribution. On the moderate and low categories, the observed significance level of 0.427 > 0.05 is on the moderate level and 0.051 > 0.05 on the low level so the H_0 is accepted. It means that the population on these categories has a normal distribution. Next, the hypothesis test from the average of the post-test result is conducted to find the differences on the students' basic Geometry skills based on the Hoffer's theory. This test is calculated using the Kruskal-Wallis H, since it has an nonnormal distribution. It is also

IOP Conf. Series: Journal of Physics: Conf. Series **1008** (2018) 012075 doi:10.1088/1742-6596/1008/1/012075

used three categories in it. Table 5 below shows the result of the hypothesis result using the Kruskal-Wallis H test.

Table 5. Test Statistic[a,b]

	Score
Chi-Square	7.010
Df	2
Asymp. Sig.	0.030

Based on the result of Table 5 above, it shows the result of 0.030 which is less than 0.05, so the H_0 is rejected. It indicates that there will be at least one category which has the difference average. To find which category that has the difference average, it has to be calculated using the Mann-Whitney test.

Table 6. The Diffence of the Students' Basic Geometry Skills Based on the Pre-Categories

	Mann-Whitney U	Wilcoxon W	Z	Asymp. Sig. (2-tailed)	Exact Sig. [2*(1-tailed Sig.)]
High-Moderate	11.000	287.000	-1.894	0.058	0.064[b]
High-Low	0.000	10.000	-2.160	0.031	0.057[b]
Moderate-Low	20.500	30.500	-1.749	0.080	0.082[b]

Table 6 shows the difference between the high and moderate categories. This difference has the observed significance level of 0.058, where it is more than 0.05. It means the H_0 is accepted. Therefore, there is no significance average difference in the students' basic Geometry skills. It also occurs in the moderate and low categories. The result shows the observed significance level of 0.080, where this is more than 0.05. Since this result is less than 0.05, it means that the H_0 is accepted. On the other words, there is a significance average difference between the high and low categories.

Those results show that the implementation of the interactive media *Ispring Suite 8* supported by *Google SketchUp* significantly improves the students' basic Geometry skills. Some of the best answers from the students on the pre-test and post-test are chosen to find out the improvement of the five basic Geometry skills based on the Hoffer's theory. The explanation can be seen as follows.

Table 7. The Example of Students Answers Improvement

Pre-Test

1. Take a look at the following picture.

a.What are the shapes which form this figure?
b. How many sides of those shapes? Mention and explain it.

Answer:

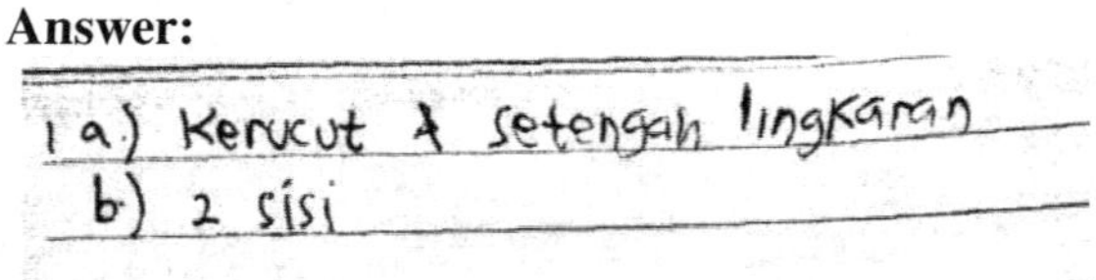

Figure 1. The Visual and Drawing Skills:
The students were still mistakenly mentioned the elements which formed that structure. Therefore, it could be said that they were not able to recognize the elements of it. The students were also still unable to show and draw the sides of the figure yet.

Post-Test

1. Take a look at the following picture.

a.What are the shapes which form this figure?
b.How many sides of those shapes? Mention and explain it.

Answer:

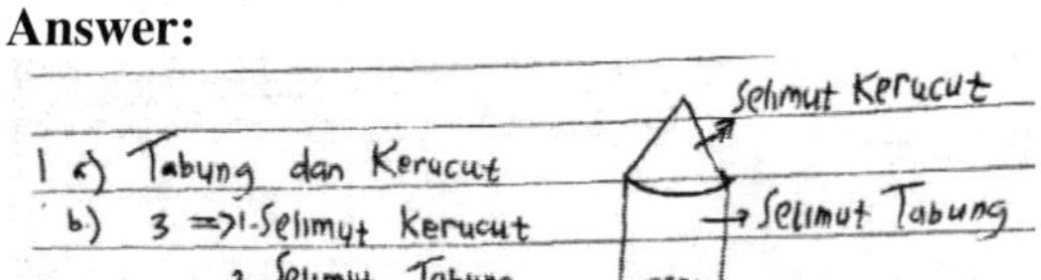

Figure 2. The Visual and Drawing Skills:
The development of these skills is improved. Students are able to answer it correctly and able to show the sides of the figure by drawing it.

2. Take a look at the following picture.

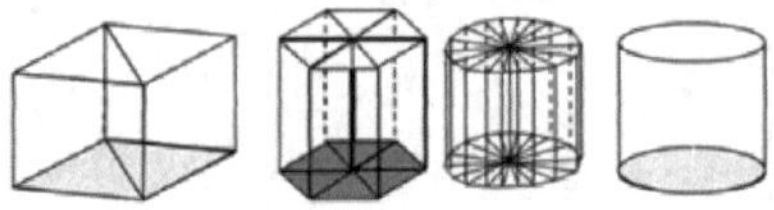

a. Based on the base side from the pictures above, write the definition of picture 4!
b. What are the shapes which form the figure in picture 4? Try to figure it out and write the area formula on each of it!
c. Based on your answer to the point b, find the area formula for the picture 4!

Answer:

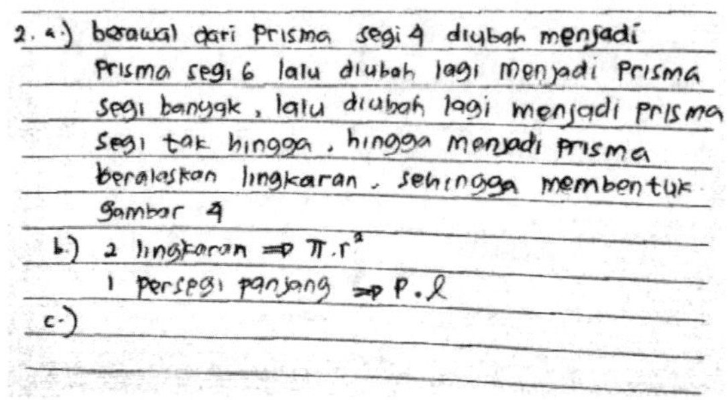

Figure 3. The Visual, Verbal, and Logical Skills: Based on the answers above, it could be said that the students were able to describe the figure verbally. However, they still could not show their logical analysis skill yet. They were not able to write the formula of the figure.

2. Take a look at the following picture.

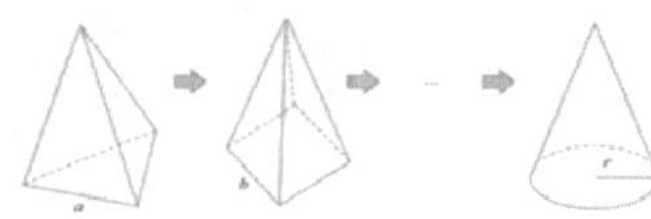

a. Based on the base side from the picture above, write the definition of picture 3!
b. What are the shapes which form the figure in picture 3? Try to figure it out and write the area formula on each of it!
c. Based on your answer to the point b, find the area formula for the picture 3!

Answer:

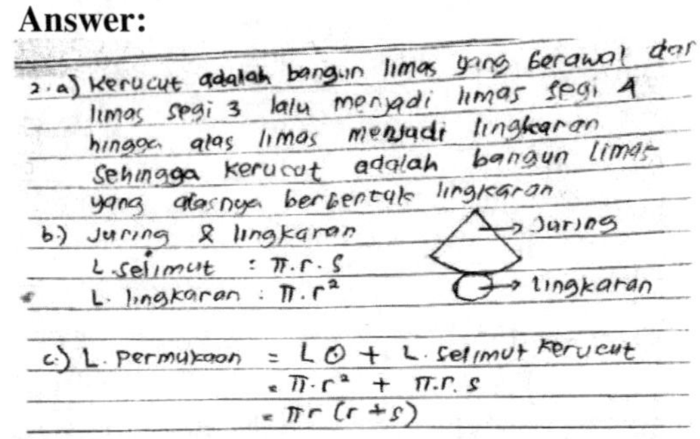

Figure 4. Visual, Verbal, and Logical Skills: The development of these skills is improved. The students are able to describe the figure verbally. They are also able to show their logical analysis, by writing down the formula of the figure. Therefore, they can get the conclusion of the formula.

3. A company will make a water storage container in a cone shape with the radius 2meter. This container is expected able to store water for about 37.68 m^3. Then, how much is the minimum material area needed to make it?

Answer:

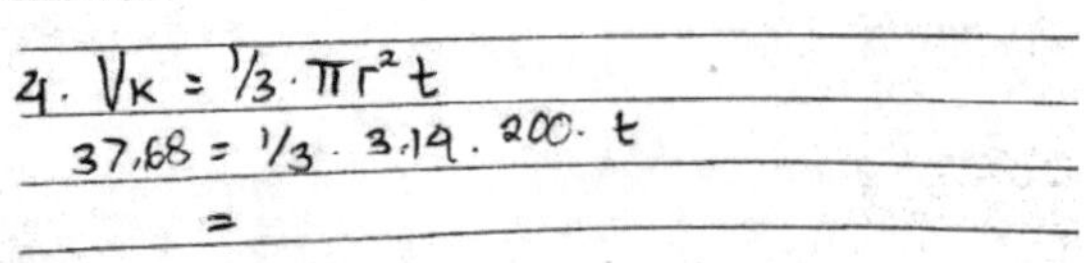

Figure 5. Logical and Application Skills: Based on the answer above, the students were only able to write the formula correctly. However, they could not apply it to its calculating. Therefore, the students were indicated to not fully capable yet to apply the formula that they have learned on the daily basis events.

3. A company will make a water storage container in a cone shape with the radius 2 meter. This container is expected able to store water for about 37.68 m^3. Then, how much is the minimum material area needed to make it?

Answer:

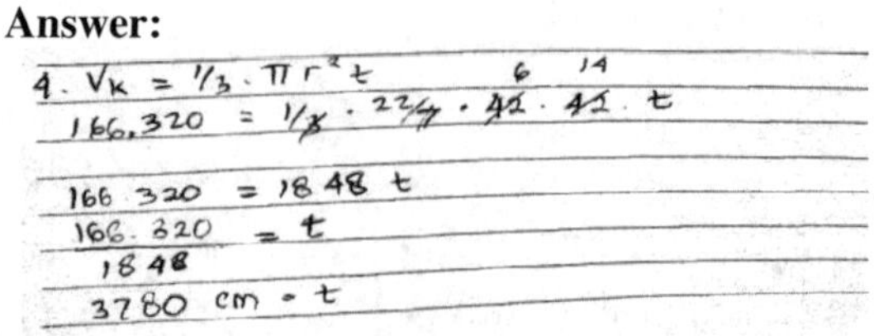

Figure 6. Logical and Application Skills: There is an improvement or development of these skills. The students are able to write the formula correctly. They are also able to apply it to its calculating process, even they still have not get the correct answer yet. It indicates that the students are already able to apply the formula that they have learned on the daily basis events.

4. Conclusion
Based on the test results and its analysis above, it can be concluded that the interactive media *Ispring Suite 8* supported by *Google SketchUp* is potential can help to improve the students' basic Geometry skills (visual, verbal, drawing, logical, and application) based on the Hoffer's theory. The improvement itself has a significance difference. This can be seen from the comparison of the pre-test and post-test average results. Further, it also can be concluded from the analysis of the students' answers. Based on this study, interactive media supported by *Google SketchUp* can be an alternative teachers to enhance the students' basic Geometry skills and to create learning activity that more variation. This study can be developed further in other geometries and materials at a higher level, in order to sharpen the assumption that the use of this media can be widely recognized.

Acknowledgments
The authors thank to the Department of Magister Mathematics Education, Faculty of Teacher Training and Education, Sebelas Maret University.

References
[1] Abu M S, Ali M B and Hock T T 2012 Assisting primary school children to progress through their Van Hiele's levels of Geometry thinking using Google Sketchup *Procedia-Social and Behavioral Sciences,* **64**(2012), 75-84
[2] Adolphus T 2011 Problems of teaching and learning of Geometry in secondary schools in rivers state, Nigeria *Int. J. of Emerg. Sci.,* **1**(2), 143-152
[3] Bennison A and Goos M 2010 Learning to teach mathematics with technology: A survey of professional development needs, experiences and impacts *Mathematics Education Research Journal,* **22**(1), 31-56
[4] Biber C, Tuna A and Korkmaz S 2013 The mistakes and the misconceptions of the eighth grade students on the subject of angles *European. J. of science and mathematics education,* **1**(2), 50-59
[5] Burger W F and Culpepper B 1993 *Restructuring Geometry in Research Ideas for the Classroom: High School Mathematics* ed P. S. Wilson (New York: MacMillan) chapter 8
[6] Burghardt M D, Hecht D, Russo M, Lauckhardt J and Hacker M 2010 A study of mathematics infusion in middle school technology education classes *J. of Technology Education,* **22**(1), 58-74
[7] Chandra H 2013 *SketchUp 8 untuk Orang Awam* (Palembang: Maxikom)
[8] Chou P N and Wu Y C 2014 Integrating 3D visualization tools into teaching surface area in elementary school classrooms: An example of Google SketchUp *Taiwan. J. of Mathematics Education,* **1**(1), 1-18
[9] Connolly S 2010 The immpact of Van Hiele-based Geometry instruction on student understanding *Mathematical and Computing Sciences Masters,* p 97
[10] Hoffer A 1981 Geometry is more than proof. *NCTM J.,* **74**(1), 11-18
[11] Hohenwarter M, Hohenwarter J, Kreis Y and Lavicza Z 2008 Teaching and Learning Calculus with Free Dynamic Mathematics Software GeoGebra *Int. Cong. on Mathematics Education* 11 (Monterrey Mexico: International Council of Scientific Unions) pp 1-9
[12] Jane M S 2006 *Developing Geometric Reasoning* (Washington DC: GED Mathematics Training Institute)
[13] Jones K 2002 Issues in the teaching and learning of Geometry in Linda Haggarty (Ed), aspects of teaching secondary mathematics: perspectives on practice (London: Routledge Falmer) chapter 8 pp 121-139. ISBN: 0-415-26641-6
[14] Kirkwood A and Price L 2014 Technology-Enhanced learning and teaching in higher education: what is enhanced and how do we know? A critical literature review *Learning, Media, and Technology,* **39**(1), 6-36

[15] Lawson M J and Chinnappan M 2000 Knowledge connectedness in Geometry problem-solving *J. for Research in Mathematics Education,* **31**(1), 26-43

[16] NCTM 2000 *Principles and Standards for School Mathematics* (Reston, VA: National Council of Teachers of Mathematics)

[17] Safrina K, Ikhsan M and Ahmad A 2014 Peningkatan kemampuan pemecahan masalah geometri melalui pembelajaran kooperatif berbasis teori Van Hiele *J. Didaktik Matematika,* **1**(1), 2355-4185

[18] Setiawan S I A 2011 Google SketchUp: Perangkat alternatif dalam pemodelan 3D *Ultimatics,* **III**(2) 6-10

[19] Subroto T 2011 The use of cabri 3d software as virtual manipulation tool in 3-dimension Geometry learning to improve junior high school students' spatial ability *Int. Conf. and the Fourth National Conference Mathematics Education* (Yogyakarta: Department of Mathematics Education Yogyakarta State University) pp 609-618

[20] Uygan C and Kurtulus A 2016 Effect of teaching activities via google sketchup and concrete models on spatial skills of preservice mathematics teachers *Turkish. J. of Computer and Mathematics Education*, **7**(3), 510-535

[21] Van de Walle J A 2001 *Geometric Thinking and Geometric Concepts In Elementary and Middle School Mathematics: Teaching Developmentally,* 4th ed (Boston: Allyn and Bacon)

IOP Conf. Series: Journal of Physics: Conf. Series **1008** (2018) 012076 doi:10.1088/1742-6596/1008/1/012076

Analysis of difficulties in mathematics learning on students with guardian personality type in problem-solving HOTS geometry test

R K N Karimah[1], T A Kusmayadi[1] and I Pramudya[1]

[1]Department of Mathematics Education, Sebelas Maret University, Indonesia.

E-mail: rudhita.kislamiyanti@gmail.com

Abstract. Learning in the current 2013 curriculum is based on contextual issues based on questions that can encourage students to think broadly. HOTS is a real-life based assessment of everyday life, but in practice, the students are having trouble completing the HOTS issue. Learning difficulty is also influenced by personality type Based on the fact that the real difference one can see from a person is behavior. Kersey classifies the personality into 4 types, namely Idealist, Rational, Artisan ,and Guardian. The researcher focuses on the type of guardian personality that is the type of personality that does not like the picture. This study aims to describe the difficulty of learning mathematics in students with a type of guardian personality in the completion of Geometry materials especially in solving HOTS. This research type is descriptive qualitative research. Instruments used in this study were the researchers themselves, personality class test sheets, learning difficulty test sheets in the form of HOTS Geometry test, and interview guides. The results showed that students with guardian personality it was found that a total of 3.37 % difficulties of number fact skill, 4.49 % difficulties of arithmetics skill, 37.08 % difficulties of information skill, 31.46% difficulties of language skill, 23.60 % difficulties of visual-spatial skill.

1. Introduction

One of the teaching strategies that promote HOTS is problem-solving. Problem solving is an active that involves various actions in the mind of thought including accesing and using knowledge and experience [17]. Students need to equip themselves with various knowladge and high skills in problem-solving, wich involves higher order thinking. Assessment of the 2013 Curriculum adapts international standard assessment models, hopefully helps students in increasing their level of thinking ability, because high-level thinking of learners can be encouraged to think broadly and deeply about the subject matter [9].The Bloom Revision Taxonomy is used to measure high-level thinking skills (HOTS), Anderson & Krathwohl classified the dimensions of the cognitive process is used to measure high-level thinking skills by category of analyze, evaluate and create [3]. Assessment of the 2013 Curriculum adapts international standard assessment models, hopefully helps students in increasing their level of thinking ability, because high-level thinking of learners can be encouraged to think broadly and deeply about the subject matter [9]. The high-level thinking ability, it is shown that based on the Trends in International Mathematics and Science Study (TIMSS) survey in 2011 the achievement of Indonesian mathematics students ranked 38th out of 42 countries with an average score of 386. By the Organization for Economic Co-operation and Development (OECD) in 2015 using the Program for International Student Assessment (PISA) test that Indonesia's mathematical achievement ranks 69 out of 76 countries that follow PISA [11]. Thus, it appears that there is still a low level of mathematics achievement in the ability to think high levels of Indonesian high school students in the International level. The instilment of high-level thinking ability should have been begun by schools in Indonesia order to meet the demands of the 21st century. This is in line with the characteristics of the 21st century community's skills according to the partnership of 21st century skills that identifies students in the 21st century should be able to develop the competitive skills required in the 21st century that are focused on the development of HOTS [6].

The results of National Examination Junior High School data in 2015/2016 in Kediri obtained from the BSNP PAMER program show that the absorptive capacity of the students in mathematics was low, AL-HUDA Junior High School was one of the schools that had low absorption, Geometry polyhedron was below 50%, Geometry material had the lowest absorption power, so it is concluded that students had learning difficulties in Geometry materials. This was supported by the results of research conducted by Ozeram stated that the results from the study revealed that seventh year secondary school students have a number of misconceptions and lack of knowledge related to Geometry subject [20]. The difficulty of this learning is the lack of understanding of Geometry concepts that ultimately inhibit the further learning process. Kennedy & Tipps argued that learning Geometry was capable of developing problem-solving abilities and being the supporters of many other topics in mathematics [15]. As in Jones stated that "The study of Geometry contributes to helping students develop the skills of visualization, critical thinking, intuition, perspective, problem-solving, conjecturing, deductive reasoning, logical argument and proof" [13]. With regard to this, it seem that students are experiencing difficulty in mathematical learning in solving mathematical issues that are in the form of HOTS questions that are questioned with high-level thinking, especially in Geometry material.

The difficulty of learning mathematical students is not only caused by ineffective learning but may also be caused by differences in personality types possessed by each individual. Dacey & Travers stated that psychosocial factor could influence the difficulties faced by students. In addition, differences in student personality become one of the causes of varied difficulties faced by students [7].One of these personality types was done by Keirsey. Keirsey classified the personality into 4 types, namely Idealist, Rational, Artisan and Guardian [14]. The classification by Keirsey was based on the fact that the real difference one can see from a person is behavior (behave). Referring to research conducted by Dewiyani stated that the guardian personality type did not like the image, but rather the preferred word and type of test was the objective of the test [8]. Based on the opinion of Dewiyani it is possible to prevent students' understanding of the geometric material. Based on this researcher analyzed the difficulty of learning mathematics at a student with guardian personality in problem-solving HOTS Geometry. Many mathematics skills were involved in problem-solving. However, large numbers of students have not acquired the basic skills they need in mathematics [2]. Ineffectiveness in the learning process will adversely affect the difficulty of acquiring mathematical skills. Understanding the difficulties students experience in mathematics skills will help students in the learning process. Lack of student math skills in the problem-solving process will have an impact on students' difficulties.

The researcher will analyze students' with guardian personality type of difficulties on the HOTS Geometry test to know the extent of students' understanding and what difficulties are often encountered in the material. This study refers to the results of previous studies which conducted by Tarzimah, which also discussed the issue of students' learning difficulties. This study looked difficulties in mathematics learning of into five types of mathematics skills i.e. number fact skill, arithmetic skill, information skill, language skill, visual-spatial skill [23]. Abdullah et al [1] studied Analysis of Students' Errors in Solving Higher Order Thinking Skills (HOTS) Problems for The Topic of Fraction and also Jonhson et al [12] studied Learning Disabilities.

2. Research Methods

The research design used was the qualitative method with descriptive approach. In this research, it formed a description of difficulty learning student mathematics with guardian personality at solving HOTS Geometry issue. This method was carried out by analyzing oral tests and interviews with students. Subjects in this study were students with guardian personality based on high achievement, medium and low achievement in AL-Huda Kediri Junior High School. Selection of subject in this research used purposive sample technique. The analysis was conducted on the student with a guardian personality with high ability, medium ability, and low ability. The classification was based on previous semester grades and consultations with teachers while the guardian personality was taken from the Kersey personality test.

The data in this study was based on the results of the written exam on HOTS Geometry and interviews were processed in such a way that it could be concluded that the learning difficulties of students with the guardian personality of each student's ability. The research procedures undertaken in this study referred to the procedure of the study according to Bogdan modified by Moleong. The research stages included: (1) pre-field stage; (2) fieldwork; and (3) data analysis stage [18]. The data credibility test was completed by using techniques triangulation that finding the suitability of data that was sourced from the results sheet of problem-solving tasks and interviews.

3. Results and Discussion

The data collection was conducted by giving personality type test to Junior High School at AL-Huda Kediri which amounted to 40 students. Personality type tests were done as a determination in the categorization of student personality types. The following student personality test results are shown in Table 1. The result of personality type test is 13 students with guardian personality type.

Table 1. Personality Type Test Result

Personality	Frequency	Percentage
Artisan	9	22.5 %
Guardian	13	32.5 %
Idealist	7	17.5 %
Rationalist	11	27.5 %
Total	40	100 %

The following results group difficulties in problem solving based on students' mathematical skills with guardian personality type on HOTS solution completion.

Table 2. Total Number of Difficulties Mathematics Skill

Hots Question	Types of difficulties mathematics skill					Total
	Number fact skill	Arithmetic skill	Information skill	Language skill	Visual-spatial skill	
1	0	0	13 (14.60 %)	9 (10.11%)	6 (6.75 %)	28 (31.46 %)
2	3 (3.37 %)	4 (4.49 %)	12 (13.48 %)	10 (11.24 %)	8 (8.99 %)	37 (41.57 %)
3	0	0	8 (8.99 %)	9 (10.11 %)	7 (7.87 %)	24 (26.97 %)
Total	3 (3.37 %)	4 (4.49 %)	33 (37.08 %)	28 (31.46 %)	21 (23.60 %)	89 (100 %)

IOP Conf. Series: Journal of Physics: Conf. Series **1008** (2018) 012076 doi:10.1088/1742-6596/1008/1/012076

Table 2 shows the number of difficulties according to mathematics skill from the sample of the study in solving HOTS mathematics problem-solving for the topic of Geometry. The highest number of difficulties by the students was in HOTS question 2, wich was 41.57% in various types of difficulties compared to other items HOTS question. Meanwhile, the lowest number of difficulties by students was in HOTS question 3, wich was 26.97% in various types of difficulties. In addition, information skill was the highest type of difficulties in mathematics skill by students, wich was 37.08%.

In this study, the selection of this subject interview was conducted with the consideration of mathematics teachers at AL-HUDA Kediri Junior High School. The consideration of the subject selection was based on the student's ability in mathematics. The subjects determined by the researcher were students with high, medium and low ability. Selected student as subject of researcher were students S1 with low ability, S2 with moderate ability, and S3 with high ability.

3.1 Analysis of problem-solving HOTS Question 1

Cognitive Levels: Analysis and Evaluation

A carpenter has 32 meters of timber and wants to make a border around a garden bed. He is considering the following designs for the garden bed. Circle either "Yes" or "No" for each design to indicate whether the garden bed can be made with 32 meters of timber.

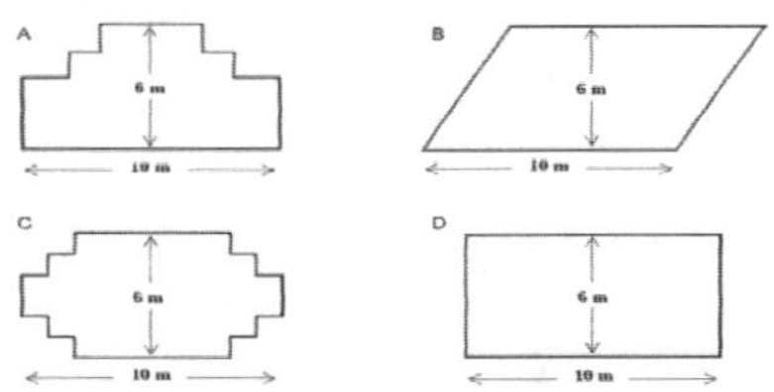

Figure 1. Garden Design Option

Garden bed using this design, can the garden bed be made with 32 meters of timber? Explain!

Design A Yes / No
Design B Yes / No
Design C Yes / No
Design D Yes / No

The question of HOTS 1 on the cognitive levels evaluation and analysis of Subject S1 encountered difficulties in solving. Below is the answer to Subject S1 with low ability with guardian personality types answer to the completion of the HOTS question 1.

Design A	:	No, because timber has different size
Design B	:	No, because it will be less for design B
Design C	:	No, because this design curvy shape and it is hard to become fence
Design D	:	Yes, because design D has the same size and circumference of 32 m

Figure 2. The answer of subject S1with high ability on HOTS question 1

Figure 2 it seems that subject S1 on the solution of the HOTS 1 test based on aspects category cognitive level of analysis and evaluation. It appears that students do not answer correctly on design A and design C, while for design B and design D students answer correctly but the arguments given are not correct, so it appears that students S1 has difficulties in the work of HOTS 1.

The following are interviews with the subject of S1 research:

R : "What information is in the question?"
S1 : "timber with length of 32 meters"
R : "Ok, then what is the question, and How is to find?"
S1 : "Perfect match design for 32 meters timber, by calculating the park design circumference".

R : "Are you sure of your choice?"
S1 : "Design 1, 2 and 3 I am not sure about it, but I'm sure Design 4 is right."
R : "Ok, give an explanation!"
S1 : "Design 4 is a rectangular and a formula to find the circumference of rectangular is = 2 × (p + l) = 2 × (10 + 6) = 32 m".
R : "Ok, right, then how to design 1,2 and 3?"
S1 : "I do not know why I just imagined it from the picture it looks like 32 m wood is not enough for the three designs."

Based on interviews and answers analysis on the solution of the HOTS questions 1 on aspects of analysis and evaluation questions with the type of guardian personality is started by analyzing the questions appropriately but at the evaluation stage, the student could not complete the answer correctly. Subject S1 experienced difficulties mathematics skills i.e. *information skill* that are difficulties in problem-solving planning, students could not see what was known and what was being asked the question but at the time of interview the student knew what was asked and what was known, but did not write down into the answer, enabling students to experience difficulties, this was due to the laziness factor and would take a long time to solve, and students merely memorized the formula alone without understanding the concepts, as the student's answer correctly on D design, where the student memorized the formula in looking around from the rectangle, while for the other design the student was difficult in the determination of the circumference, this means the students just memorize the formula without knowing the concept of looking around, *language skill* that are difficulties in understanding question, and *visual spatial skill* that are subject S1 can not afford in manipulate geometrical shapes. According to Polya, understanding problem is the first phase of problem-solving process [21]. During the phase of understanding problem, first of all, student must know the story from the information given and the objective of the problem.

Design A (Yes) = 10 + 5 + 5 + 6 + 2 + 1 + 2 + 1 = 32
Design B (No) = -
Design C (Yes) = 8 + 2 + 2 + 6 + 2 + 2 + 4 + 4 +1 +1 = 32
Design D (Yes) = 10 + 10 + 6 + 6 = 32

Figure 3. The Answer of Subject S2 with Moderate Ability on HOTS Question 1

Design A (Yes)
Design B (No)
Design C (Yes)
Design D (Yes)

Figure 4. The Answer of Subject S3 with High Ability on HOTS Question 1

Figure 3 shows subject S2 answers all question correctly but the arguments given in each answer are less precise. This means that in the category of analysis has difficulties in the information skills of the student's difficulties in problem-solving planning. It appears that subject S2 is less precise in writing information about questions and arguments written on each answer option. In Figure 4 subject S3 looks at all the answers correctly but does not provide an argument in every answer. This means that on the category of cognitive and analysis has difficulties in the information skills of students has difficulties in planning problem-solving, it appears that the absence of information about the questions written on the answer and not given the right reasons on each answer option.

3.2 Analysis of problem-solving HOTS Question 2

Cognitive Level: Evaluation
Look at Figure 5 below!

IOP Conf. Series: Journal of Physics: Conf. Series **1008** (2018) 012076 doi:10.1088/1742-6596/1008/1/012076

Mrs. Ana has a cupboard under the stairs. She will measure the size of the cupboard, and measured height and width for each stair in a row of 9 cm and 12 cm. Identify the size cupboard for Mrs. Ana!

Figure 5. Cupboard

The question of HOTS 2 on the cognitive levels evaluation to Subject S2 encountered difficulties in solving. Below is the answer of Subject S2 with moderate ability with guardian personality types answer to the completion of the HOTS question

$=\sqrt{12+9}$ $=144+81$ $=\sqrt{225}$ $=15$ $=15\times 10 = 150 =$ sisi miring $=\sqrt{150-90}$ $=22500-8100$ $=\sqrt{14400}=120$	$=\frac{120\times 12}{2} = 5400\ cm^2$ $= 90\times 12 = \underline{1080}$ hypotenusa $0 + 1080$ $00\ cm^2$

Figure 6. The Answer of Subject S2 with Moderate Ability on HOTS Question 2

Figure 6 it seems that the solution given by subject S2 of the HOTS question 2 based on aspects cognitive category evaluation it seems that subject S2 has difficulties in problem-solving planning, so it was not written what was known and asked the answer of the student became unstructured and allowed many errors in solving.

The following are interviews with the subject of S2:

R : "Build what's in the cupboard?"
S2 : "Area of a rectangle and a triangle."
R : "Then how to find a cupboard?"
S2 : "Triangle area plus rectangular area"
R : "ok, how to find wide of a triangle?"
S2 : "The base height is divided by 2, looking for hypotenuse and then searched the area.
R : "try to look at your answer, where is the answer saying that"
S2 : "this it is" (answers)
R : "You answered no information about it, should be given what statement is known and what is being asked?"
S2 : "it was too long."
R : "now look at your answer looking for hypotenuse, please explain!"
S2 : "square root for 12 squares plus 9 squares"
R : "why in your answer there is no squares?"
S2 : "I am forget it, because I'm doing it fast"

Based on interviews and answers analysis on the completion of the HOTS questions on aspects of S2 with moderate ability on the type of guardian personality evaluation questions to start by analyzing the questions appropriately but at the evaluation, stage students could not complete the answer correctly. S2 knew what was being asked and knew this meant that S2 was able to analyze the matter, but S2 was lazy in writing a long answer. This is the cause of the student's answer to be unstructured so there were many mistakes, in the answer it meant the student was in trouble with *information skills*. The subject S2 also has difficulties in arithmetics skill, it seem that the answer is not accurate in the planning procedure of the calculation experienced by subject S2. Furthermore, the difficulties in recognizing and understanding number fact skill, students wrote: $\sqrt{13-12} = 169 - 144 = \sqrt{25} = 5$, meaning student made incorrect calculations, although the end result was in line with the correct answers. Incompetency in conceptual knowladge of fraction caused error in procedural skill of fraction. According to Narayanan, students were lacking in aritcmetic and procedure knowladge as a result from weak conceptual understanding [19].

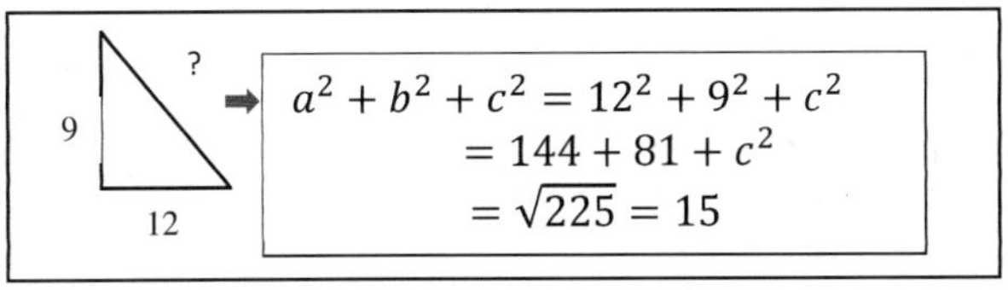

Figure 7. The Answer of Subject S1 With Low Ability on HOTS Question 2

In Figure 7 it seem that subject S1 responded incorrectly, also did not finish the work until the end. This is because the subject S1 has difficulties in understanding the category of HOTS question 2. It seem in accuracy and logarithm in computational and mathematical working procedure subject S1 responded incorrectly, not writing the information received on the question, and subject S1 also difficulty in manipulate geometrical shape. This means that subject S1 has difficulties on arithmetic skill, information skill, language skill, visual-spatial skill. As a result student has failed to carry out problem-solving correctly. Refers to Tambychik, students faced with difficulties in problem-solving due to their lacking in the ability to visualize mathematics problem and concept [23]. the affect of mathematics difficulties is cumulative because the development of mathematics skills is based on the previous skills [5]. While subject S3 answered with less precise that is at the final step, subject S3 is not having difficulties in the process, but the lack of accuracy in the final step.

3.3 *Analysis of problem-solving HOTS Question 3*

Cognitive Level: Analysis

Bobi cuts a cake with rectangular surfaces into four parts as shown in Figure 8, according to Bobi's cookie is cut as wide. Susi does not agree with Bobi's opinion. Among the statements of Bobi and Susi, which statements are true? Explain!

Figure 8. Bobi's Cake

The question of HOTS 3 on the cognitive levels analysis of Subject S3 encountered difficulties in solving. Below is the answer to Subject S3 with high ability of guardian personality types answer to the completion of the HOTS question

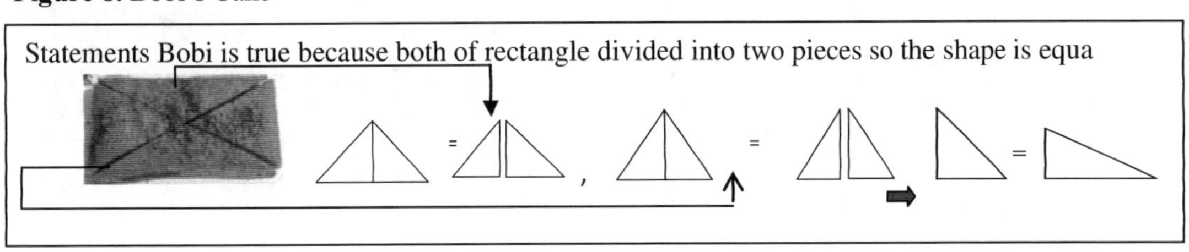

Figure 9. The Answer of Subject S3 with High Ability on HOTS Question 3

Figure 9 it seems that the solution given by subject S3 of the HOTS question 3 based on aspects category cognitive level analysis seems that subject S3 has difficulties in visual-spatial skill, subject S3 has difficulties in manipulate geometrical shape.

The following are interviews with S3 research subjects:

R : "what's the matter?"
S3 : "cake"
R : "Only cake? Try to look at the base of the flat-shaped cake? "
S3 : "rectangle."
R : "Ok, then what's the question?"
S3 : "Cake is cut as wide or not"
R : "then, how to cut the cake?"
S3 : "Cut with hypotenuse"
R : "Yes, the cake is cut tilted according to each diagonal, this is what is known in the matter. What's your answer?"
S3 : "the same piece of cake"
R : "why so ?"
S3 : "because if viewed from its shape is equally big and both types of triangle are equal if divided into two then the same size"
R : "take note of the illustration of the answer you are drawing, then see the pieces of the triangle you cut into two then you compare the pieces of the triangle. Is it true?"
S3 : "Yes it's true"
R : "Well, now take a look at the question on the question!" Is it about sending a cut of the triangle?
S3 : "no miss, my answer is wrong"
R : "You've tried to answer well, but it's just not right in understanding the meaning of the matter,
S3 : "Yes, I will be more careful again in understanding the question"

Based on interviews and answers on HOTS solving questions on aspects analysis of subject S3 with high ability of guardian personality types started by analyzing questions incorrectly. The subject S3 was difficulties in the visual-spatial skill that is difficulty in understanding the forms of Geometry. According to Lerner's being a visual abnormality child who is often unable to distinguish Geometry form [16]. Furthermore, S3 was difficulties in language skill, the conclusion of S3 that is the triangles could be divided into two equals. Whereas in the matter there is no command of the triangle divided into two. Factors that cause students to experience difficulties are lack of understanding in reading and understanding of the problem. Meanwhile, the answer subject S1 is correct, but the argument given is not correct so the answer wrong becaused between the answer and the question are not consisten. This means that subject S1 has difficulties in mathematics skills such as information skill and language skill. The subject S1 has difficulties in manipulate geometrical shape particular in visual-spatial skill. According to Garderen, deficiency in visual-spatial skill might cause difficulty in differentiating and relating information meaningfully [10]. In another situation, the answer of subject S2 is wrong, it seems that S2 has difficulties in understanding the problem so that the answer is not correct, it means subject S2 has difficulties in manipulate geometrical shape especially in subject S2 was difficulties in visual-spatial skill.

Based on the results of the analysis, this study found some difficulties for students. These show evidence of earlier studies that show that students have much difficulty in understanding of Geometry. Geometry it self is a crucial component of mathematics education [4]. The results of this research also strengthen previous research conducted by Utami stating that the mistakes of students in settlement of problems of Geometry as they are hasty and not conscientious in conducting process calculation [24]. Further, this research also shows difficulties of students who have the guardian personality type for answering HOTS with dominant material of Geometry. The most difficult parts of this are language skills, information skills, and visual spatial skill with various causes. One of the causes is students only have limited understanding of any information given. So that, students are hardly solving questions of the HOTS. These findings strengthen findings Tambychik & Meerah's study which resulting that the weaknesses of many student are at number fact skill, visual-spatial skills, and information skill [22].

4. Conclusion

This study concludes that students with guardian personality type have difficulties in mathematics problem-solving HOTS Geometry due to incompetency in acquiring many mathematics skills. It was found that a total of 89 type of difficulties based mathematics skill have been identified in the problem solving process involving HOTS, also found a total of 3.37% difficulties of number fact skill, 4.49% difficulties of arithmetics skill, 37.08 % difficulties of information skill, 31.46% difficulties of language skill, 23.60% difficulties of visual-spatial skill. Information skill was found to be the most critical mathematics skill.

The HOTS problem with cognitive level analysis and evaluation category, students with high and moderate abilities are experiencing difficulties in information skill, while students with low ability in information skill, language skill, and visual-spatial skill. The HOTS problem with cognitive level evaluation category, students with high ability did not have difficulties, students with moderate abilities difficulty in information skill, number fact skill, and arithmetics skill, while students with low ability in information skill, language skill, visual-spatial skill, and arithmetics skill. The HOTS problem-solving with cognitive analysis category of students with high ability were difficulties in language skill and visual-spatial skill, moderate ability in visual-spatial skill, while students with low ability were difficulties in information skill, language skill and visual-spatial skill. Therefore, teachers should be able to recognize the difficulties faced by students in solving problems related to HOTS. Furthermore, the researcher suggest to look for each other type of personality type, the research to a compare between each type of personality type David Kersey and also be able develop problems related to the problem of HOTS, especially on the Geometry material.

Acknowledgments
The authors would like to thank the referees for their suggestions and Sebelas Maret University, Surakarta, Indonesia for funding this research in the academic year of 2017.

References

[1] Abdullah A H, Abidin N L Y and Ali M 2015 Analysis of Students' Errors in Solving Higher Order Thinking Skills (HOTS) Problems for The Topic of Fraction *Asian Sosial Sciences* **9** (21) 133-142

[2] Ali M N and Desa R 2004 Meningkatkan Kemahiran Mengingat Sifir 3,6 dan 9 di Kalangan Murid Pemulihan Tahap 2 *Jurnal Kajian Tindakan Pelajar PSPK. Kementerian Pendidikan Malaysi* 102 – 112

[3] Anderson L W and Krathwohl D R 2001 *A Taxonomy for Learning, Teaching, and Assessing: A Revision of Bloom's Taxonomy of Educational Objectives* (Allyn & Bacon Boston: MA Pearson Education Group)

[4] Biber C, Tuna A and Korkmaz S 2013 The mistakes and The Misconseptions of The Eight Grade Students on The Subject of Angles *European Journal of Science and Mathematics Education* I (2) 50-59

[5] Carnine M 1997 Instructional design in Mathematics for student with learning disabilities *Journal of Learning Disabilities* 30:130-141
[6] Conklin W 2012 *Higher-order thinking skills to develop 21st century learners* (Huntingon Beach: Shell Education Publishing, Inc)
[7] Darcey J S and Travers J F 2006 *Human Development Across the Lifespan* 6th ed (Boston: Mc Graw Hill)
[8] Dewiyani M J 2009 Karakteristik Proses Berpikir Siswa dalam Mempelajari Matematika Berbasis Tipe Kepribadian *Prosiding Seminar Nasional Penelitian, Pendidikan dan Penerapan MIPA Fakultas MIPA, Universitas Negeri Yogyakarta 16 Mei 2009* M-481
[9] Direktorat Pembinaan SMA 2015 *Penyusunan Soal Higher Order Thinking Skill's Sekolah* Menengah *Atas (*Jakarta: Direktorat Jenderal Pendidikan Menengah Kementerian Pendidikan dan Kebudayaan)
[10] Garderen D V 2006 Spatial Visualization, visual imagenary and mathematical problem solving of students with varying abilities. *Journal of Learning Disabilities* **39** (6): 496-506
[11] TIMSS *and TIMSS Advanced 2015* 2016 *Highlights from TIMSS and TIMSS Advanced 2015* (Washington: NCES, IES, U.S. Department of Education)
[12] Johnson D J and Myklebust H R 1967 *Learning Disabilities* (New York: Grume and Stratton)
[13] Jones K 2002 Aspects of Teaching Secondary Mathematics: perspectives on practice *Issues in the Teaching and Learning of Geometry* In: Linda Haggarty (Ed) (London: Routledge Falmer) 8:121-139. ISBN: 0-415-26641-6
[14] Keirsey D and Bates M 1984 *Please Understand Me* (Promotheus Nemesis Book)
[15] Kennedy L M and Tipps S 1994 *Guiding's Learning of Mathematics (7th ed)*. (California: Wadsworth)
[16] Lerner J W 1988 *Learning Disabilities: Theories, Diagnosis and Theacing Strategies* (London: Houghton Mifflin Company)
[17] Lester F K and Kehle P E 2003 From Problem Solving to Modeling: The Evolution of Thinking About Research on Complex Mathematical Activity. In R. Lesh, & H. M. Doerr (Eds.), *Beyond Constructivism –Models and Modeling Perspectives on Mathematical Problem Solving, Learning, and Teaching* pp 501-517 (Mahwah, NJ: Lawrence Erlbaum Associates)
[18] Moleong L J 2014 *Metodologi Penelitian Kualitatif* (Bandung: PT Remaja Rosdakarya)
[19] Narayanan L M 2007 *Analysis of error in addition and subtraction of fraction among form 2* (Universiti Malaya: Kertas Projek Sarjana Pendidikan)
[20] Ozerem A 2012 Misconceptions in Geometry and Suggested Solutions for Seventh Grade Students *International Journal of New Trends in Arts, Sports & Science Education* **1**(4):23-35
[21] Polya G 1981 *Mathematical Discovery on Understanding Learning and Theacing Problem Solving* (New York: John Wiley and Sons)
[22] Tambychik T, Meerah T S M and Aziz Z 2010 Mathematics Skills Difficulties: A Mixture of Intricacies. Procedia *Social and Behavioral Science* **7**(C):142-151
[23] Tambychik T and Meerah T S M 2010 Students' Difficulties in Mathematics Problem-Solving: What do they Say? *Procedia Social and Behavioral Science* **8**:142-151
[24] Utami A D 2016 Tipe Kesalahan Mahasiswa dalam Menyelesaikan Soal-soal Geometri Berdasar Newman's error analysis (NEA) *Jurnal Ilmiah Pendidikan Matematika* **4**(2): 85-92

Geometry in flipbook multimedia, a role of technology to improve mathematics learning quality: the case in madiun, east java

S Andini[1]**, L Fitriana**[1]**, Budiyono** [1]

[1]Mathematics Education Department of Graduate Program, Sebelas Maret University, Surakarta, Indonesia

E-mail: andini08@yahoo.com

Abstract. Abstract. This research is aimed to describe the process and to get product development of learning material using flipbook. The learning material is developed in geometry, especially quadrilateral. This research belongs to Research and Development (R&D). The procedure includes the steps of Budiyono Model such as conducting preliminary research, planning and developing a theoretical and prototype product, and determining product quality (validity, practicality, and effectiveness). The average assessment result of the theoretical product by the experts gets 4,54, while validity result of prototype product by the experts is 4,62. Practicability is obtained by the implementation of flipbook prototype in each meeting of limited-scale try out based on learning observation, with the average score of 4,10 and increasing of 4,50 in wide-scale try out. The effectiveness of the prototype product is obtained by the result from pretest and posttest on a limited-scale and a wide-scale try out. The limited-scale pre-test result showed a significant increase in average score of wide-scale pre-test of 25,2, and there is an increase in the average score of posttest on limited-scale try out and wide-scale try out is 8,16. The result of product quality can be concluded that flipbook media can be used in the geometry learning in elementary school which implemented curriculum 2013.

1. Introduction

Geometry is an important branch of mathematics and it has been identified as a basic mathematical skill [8, 9, 17]. Geometry is important for students as it is also applied in other branches of mathematics [1]. Geometry is also seen as one of the most important topics in mathematics [13, 14, 18]. The tendency in the study of geometry can enhance the ability to solve mathematics problems. Many people enjoy the challenge of solving riddles and other types of puzzles, studied geometry offers similar intriguing challenges [11]. But, many studies [3, 15, 16, 21, 22, 26] have shown that students have difficulties in learning geometry, and in thinking geometrically. Moreover, although geometry has been introduced since the beginning age of school. However, some students still assume that learning geometry is something difficult. Most researchers focus on the geometrical curricula of secondary school; however, discovering the implications of the van Hiele theory for elementary school students is also very important. Thus, the focus of this study is at the elementary level [10].

IOP Conf. Series: Journal of Physics: Conf. Series **1008** (2018) 012077 doi:10.1088/1742-6596/1008/1/012077

The result of field research shows that the average thinking level of elementary school students is still on the lowest level. Almost all students have not been able yet to apply geometry learning skills comprehensively which consist of visual, verbal, drawing, logic, and application skill. It causes low ability in a problem-solving. The observation result of quadrilateral plane geometry learning process in grade 6th shows there is no interesting prop and media for supporting student's learning activity. Questionnaire result of need assessment gives data that respondents state that they are interested towards audiovisual learning media (graphic and animation), video, also the facilitation in applying mathematics learning.

The multimedia which will be used must unite among the participant, learning material given, also the way in solving student's boredom [19, 27]. Research on technology integration in education has demonstrated that using computers in the classroom benefits students learning in mathematics [12, 25]. Given the importance of mathematics, geometry, and spatial thinking on educational success, it is important to find better ways to teach them. There is ample evidence that real and virtual manipulatives can help elementary school children learn mathematics better [5]. Yet computers are still rarely used in the classroom to teach mathematics even in so-called developed countries, such as the USA [7].

Based on previous researches and the result of field research about the application of technology in mathematics, that the right multimedia development can improve the quality of learning mathematics. Flipbook (e-book) multimedia which has characteristic hypermedia has more interesting features than a printed book. Based on the literature study, flipbook learning media can raise student's geometry learning interest [2]. Geometry material in flipbook multimedia has been constructed, and students are able to reach an informal deductive level and have geometry basic skill [23, 24]. Determining flipbook multimedia quality is based on quality criteria of development study result. We have been referring to the quality of educational products from the perspective of developing learning materials. However, we consider the three quality aspects (validity, practicality, and effectiveness) and also to be applicable to a much wider array of the educational product [20,19].

2. Methods

This research uses research and development method. This research modifies the Budiyono model of research and development [3]. Education research and development is a process that is used to improve and validate an education product such as teaching-learning model, instrument, textbook, module, and media [4]. Thus, the aim of this research is to develop a flipbook in the quadrilateral plane which is valid, practical, and effective. The media validity is assessed by the experts, they are the material substantial expert, learning and curriculum expert, also the multimedia expert. That expert's validation is done in the early of design development (theoretical product) until prototype media construction. Assessment consideration and theoretical product validation refers to 1) material substantial, 2) Van Hiele theory characteristic analysis with material substance, 3) substantial discrepancy between the product and learning material of Curriculum 2013-Book of Integrative Thematic, 4) media display design, 5) exercise questions, and 6) evaluation in the form of test question. After theoretical product gets experts' advice and revision, prototype media construction can be done soon. During this phase, the experts also give assessment and suggestion in order to make the media more perfect and be ready to use.

Practicability assessment and media effectiveness can be seen from the limited scale try out result and wide-scale try out. Media practicability level is measured through a questionnaire, interview, and observation paper of learning activity, while media effectiveness uses pre-post test design.

The limited-scale try out involves sample with 1 school from 13 elementary schools population which implement curriculum 2013. The sample is chosen by simple random sampling. For sample election result, it gets a school with 28 students. This try out is done by applying flipbook as learning media with the same teacher for four times meeting. Each meeting consists of 3 x 35 minutes. Pre-test and post-test instruments consist of 6 essay questions. Interview activity is done for the teacher and 6 students who have high, medium, and low academic level. Practicability questionnaire instrument in 5 scales is only given to the teacher, while observation activity in the classroom is done by other teacher and the researcher. Practicability experiment observer gives notes and suggestions for completing material of flipbook multimedia.

The wide-scale try out is done with the same technique as in limited-scale try out. This ty out involves samples from two elementary schools outside limited scale try out samples. The number of respondent in interview activity is 12 students in high, medium, and low ability.

3. Result and Discussion

3.1 Validity of Flipbook Multimedia

The expert's validation for a theoretical product is conducted by the scale of Likert (1-4) with interval score and category are shown in **Table 1**.

Table 1. The Interval Score of Theoretical Product Assesment

Interval Score	Scale	Category
X > 4,21	5	Very good
3,40 < X ≤ 4,21	4	Good
2,60 < X ≤ 3, 40	3	Good Enough
1,79 < X ≤ 2,60	2	Less Good
X ≤ 1,79	1	Very Unfavourable

Based on **Table 1**, a minimum score of the theoretical product is "good" results obtained from the assessment of experts. If the overall aspects of the obtained results are "good", then the theoretical product can be made into a product prototype. Theoretical product assessment result by the experts is shown in **Table 2**.

Table 2. Theoretical Product Assessment Results by The Experts

Criteria	Skala	Material Expert		Curriculum and Learning Expert		Multimedia Expert	
		Frequency	The Total of Score	Frequency	The Total of Score	Frequency	The Total of Score
Very Good	5	38	190	13	65	7	35
Good	4	22	88	7	28	13	52
Good Enough	3	0	0	0	0	0	0
Less Good	2	0	0	0	0	0	0
Very Unfavourable	1	0	0	0	0	0	0
The Amount of Frequency		**60**		**20**		**20**	
Average Value		**4,63**		**4,65**		**4,35**	

Based on **Table 2** above, it is stated by material experts that flipbook theoretical product gets an average assessment of 4,63 with criteria 68% belongs to extremely good and 32% belongs to good. The result of assessment average score from curriculum and learning experts is 4,65 with criteria 70% belongs to extremely good and 30% belongs to good, while validity result of multimedia experts state assessment average score is 4,35 with criteria 40% belongs to extremely good and 60% belongs to good. From all validity results, it can be concluded that flipbook theoretical product is proper to be developed as a prototype product.

As in the assessment of the theoretical product, the expert's validation for prototype product is conducted by the scale of Likert with interval score and category are shown in **Table 3**.

Table 3. The Interval Score of Prototype Product Assesment

Interval Score	Scale	Category
X > 4,21	5	Very good
3,40 < X $\leq$ 4,21	4	Good
2,60 < X $\leq$ 3, 40	3	Good Enough
1,79 < X $\leq$ 2,60	2	Less Good
X $\leq$ 1,79	1	Very Unfavourable

Based on **Table 3**, a minimum score of prototype product is "good" results obtained from the assessment of experts. If the overall aspects of the obtained results are "good", then the prototype product can be applied to a limited-scale try out and a wide-scale try out. The result of prototype product by the expert is shown in **Table 4**.

Table 4. Prototype Assessment Results by The Experts

Criteria	Skala	Material Expert		Curriculum and Learning Expert		Multimedia Expert	
		Frequency	The Total of Score	Frequency	The Total of Score	Frequency	The Total of Score
Very Good	5	40	200	15	75	9	45
Good	4	20	80	5	20	11	44
Good Enough	3	0	0	0	0	0	0
Less Good	2	0	0	0	0	0	0
Very Unfavourable	1	0	0	0	0	0	0
The Amount of Frequency		**60**		**20**		**20**	
Average Value		**4,67**		**4,75**		**4,45**	

Based on **Table 4** above, it is stated by material experts that flipbook prototype gets an average assessment of 4,67 with criteria 71% belongs to extremely good and 29% belongs to good. The result of average assessment score from the curriculum and learning experts is 4,65 with criteria 79% belongs to extremely good and 21% belongs to good, while validity result of multimedia experts state the assessment average score is 4,35 with criteria 51% belongs to extremely good and 49% belongs to good. From all validity results, it can be concluded that flipbook prototype is proper to use in the limited-scale try out and the wide-scale try out. The Figure 1 is a display of a flipbook multimedia have been validated.

Figure 1. The Display of Flipbook Multimedia

3.2 Effectiveness of Flipbook Multimedia

The achievement result of the average pre-post test on a limited-scale and a wide-scale try out is shown in **Figure 2** below.

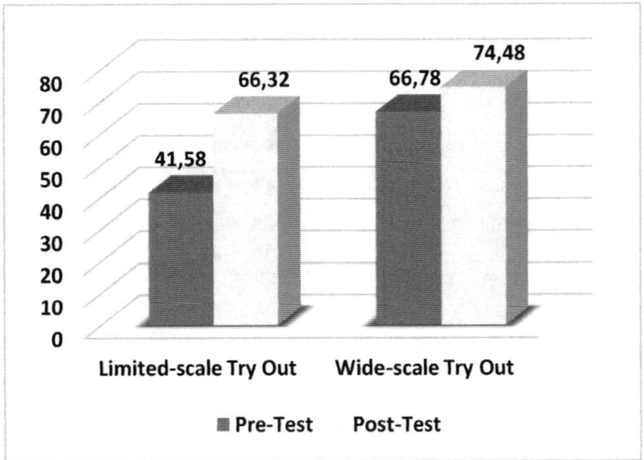

Figure 2. Graph of Average Score of Pre-Post Test on Limited-scale and Wide-scale Try Out

Six questions are given to students consist of polygon material, the properties of the quadrilateral, the relationship between quadrilateral, and quadrilateral problem solving involving measurement and calculation.

The results of limited-scale tryout pretest analysis found that the low tendency of the average score of students is influenced by several aspects, those are skills in solving problems, weaknesses in analyzing relationships between structures such as "is the rectangle a parallelogram?", and mention the types of quadrilateral that have the same properties.

The limited-scale post-test results showed a significant increase in average score of 24,74. In general, the students are able to solve the problem of determining the types of polygons (concave-convex, regular-and irregular). It shows that students' drawing skills are improving. Students can also identify the necessary mathematical properties of a quadrilateral, in which case students apply logical skills with good and correct language/verbal skill. Questions which are related to problem-solving and constructing a quadrilateral have been done in the right way.

Based on **Figure 2**, there is an increase in the average score of wide-scale test on the pre-post test is 7,7. In this pre-test, it is known that students have been able to draw polygons and/or it is not polygon correctly, students can also solve problem-solving problems and get the right results even though the verbal skills are still limited to simply write out the wide-formula of a quadrilateral. Most students are able to give a name the quadrilateral based on visualization, but they have not been able to identify the necessary mathematical properties of the quadrilateral. Viewed from the level of logic skills, most students are only able to express the similarities and differences of quadrilateral properties. As in the limited-scale try out, students' ability to link information by developing geometric models tends to be low. The average post-test results in wide-scale try out show an increase in learning outcomes. Students do not experience obstacles when working on problem-solving problems with the writing of language and how to work properly, the ability to measure and calculate the area is also getting better. Students do not experience difficulty when identifying equations and differences in a square with rectangular views from angles, sides, and diagonals. Drawing skills and declaring the concept of the circumference of a quadrilateral is also increases.Results the average post-test on the wide-scale try out indicate improved learning result. Students are not hard working on problem-solving. They can write with the precise language and in the right way. The ability to measure and calculate the area of also getting better. Students do not find it difficult to identify similarities and differences between a rectangular and a parallelogram in terms of square angle, side, and diagonal. The skills of drawing and stated the concept of the circumference of a quadrilateral is also increasing.

3.3 Practicality of Flipbook Multimedia

Table of Prototype Practicality Distribution on the Limited-scale Try Out and Wide-scale Try Out are shown in **Table 5** below.

Table 5. Distribution Tabel of Practicality Frequency Based of Observation

Kriteria	Skala	Limited-scale Try Out		Wide-scale Try Out	
		Frequency	The Total of Score	Frequency	The Total of Score
Very Good	5	2	10	20	100
Good	4	18	72	20	80
Good Enough	3	0	0	0	0
Less Good	2	0	0	0	0
Very Unfavourable	1	0	0	0	0
The Total of Frequency		**20**		**40**	
Average Score		**4,10**		**4,50**	

Based on **Table 5** above, it can be said that the observer gives a positive response to the prototype product with the average score of learning observation of 4.10 and increasing of 4.50. In this try out, the observer stated that the intensity of the use and attention of students to the prototype is very high so that the learning objectives are achieved optimally. From an interview with can find that, the teachers feel comfortable in using prototype product because the product is user-friendly, the buttons work well, and the supporting apps run seamlessly. In terms of material substance, the prototype has been developed based on van Hiele levels of geometry thinking, there are prerequisite materials that meet the element of adequacy, and do not have a dependence on other teaching materials.

4. Conclusion

Conclusions of this study are: 1) the result of field research shows that the average thinking level of elementary school students is still on the lowest level. Almost all students have not been able yet to apply geometry learning skills comprehensively which consist of visual, verbal, drawing, logic, and application skill, 2) the result of average assessment from the expert on the validation stage of theoretical and prototype product is good category, 3) effectiveness test on limited-scale try out and wide-scale try out show a significant improvement and students are able to reach an informal deductive level and have geometry basic skill, 4) the average score of the learning observation of practicality aspects is increasing.

References

[1] A Halim, Abdullah A H and Zakaria E 2013 *Procedia-Social & Behavioral Sciences Journal.* **102** 251-266
[2] Arsyad A 2011 *Learning Media* Fifth edition (Jakarta: Rajawali Pers)
[3] Budiyono 2017 *Introduction To The Methodology Of Educational Research* (Surakarta: UNS Press)
[4] Borg W R and Gall M D 1979 *Educational Research An Introduction Third Edition Longman* (New York)
[5] Clements D H 1999 *Contemporary Issues in Early Childhood* **1**(1) 45–60

[6] Gecü Z and Özdener N 2010 *Procedia Soscial and Behavioral Science* **2** 2824-2828

[7] Flores A, Knaupp J E, Middleton J A and Staley F A 2002 *Contemporary Issues in Technology and Teacher Education* **2** (1)

[8] Hoffer A R and Hoffer S A K 1992 *Geometry and visual thinking In T R Post (Eds) Teaching mathematics in grades K-8*: Research based mathematics 2nd ed (Boston: Allyn and Bacon) 249-227

[9] Hong L T 2005 *Van Hiele Levels and Achievement in Writing Geometry Proofs Among Form 6 Students* (Kuala Lumpur : Universiti of Malaya)

[10] Hsiu-Lan Ma, De-Chih Lee , Szu-Hsing Lin and Der-Bang Wu 2015 *Eurasia Journal of Mathematics, Science & Technology Education* **11**(5) 1181-1196

[11] Littel and Houghton 1991 *Geometry-based Constructivist Learning Realistic* (Pekalongan: A Paper of MGMP Mathematics Conference)

[12] Işiksal M and Aşkar P 2005 *Educational Research Journal* **47** (3) 333-350

[13] Ministry of Education of Taiwan (MET) 1993 *Curriculum Standards for National Elementary School in Taiwan*

[14] Ministry of Education of Taiwan (MET) 2008 *Grade 1-9 Curriculum Provisional Guidelines Mathematics*

[15] Mitchelmore M C 1997 *Children's informal knowledge of physical angle situations. Learning and Instruction* **7** 1-19

[16] Mullis I V S, Martin M O, Gonzalez E J, Gregory K D, Garden R A, O'Connor K M, Chrostowski S J and Smith T A TIMSS 1999 *International Mathematics Report: findings from IEA's repeat of the third International mathematics and science study at the eighth grade* (Chestnut Hill MA: Boston College)

[17] National Council of Teachers of Mathematics 2000 *Principles and standards for school mathematics* (Reston: VA)

[18] National Council of Teachers of Mathematics (NCTM) 2000 *Curriculum and evaluation standards for school mathematics*

[19] Nazeri 2003 Proceedings Of The Seminar On The Investigation Of IPG East Zon **1** (1) 3-4

[20] Zulkifli 2013 *IOSR Journal of Research & Method in Education (IOSR-JRME)* **3** (1) 24-27

[21] Prescott A, Mitchelmore M and White P 2002 *Proceedings of the Annual Conference of the Mathematics Education Research Group of Australia Incorporated ED 472950*

[22] Purnomo A 1999 *Mastery of the concepts of Geometry in conjunction with the theory of progression of Thought Van Hiele at Grade 8th SMP Negeri 6 Malang* (Malang: Pascasarjana FKIP IKIP Malang)

[23] Purwanto N 2004 *Psychology Education* (Bandung: PT Remaja Rosdakarya)

[24] Purwanto N 2004 *Theoretical and Practical Education* (Bandung: PT Remaja Rosdakarya)

[25] Olkun S, Altun A and Smith G 2005 *British Journal of Educational* **36** (2) 317-326

[26] Sudarman 2000 *Development of Computer-aided Learning Package Material is Broad and Round The Triangle for Grade 5th in Elementary School* (Malang: Pascasarjana Universitas Negeri Malang)

[27] Vaughan T 2008 *Multimedia: Making it Work Seventh Edition* (United States of America: McGraw-Hill)

IOP Conf. Series: Journal of Physics: Conf. Series **1008** (2018) 012078 doi:10.1088/1742-6596/1008/1/012078

Profile of mathematical reasoning ability of 8th grade students seen from communicational ability, basic skills, connection, and logical thinking

Sumarsih[1], Budiyono[1] and D Indriati[1]
[1]Department of Mathematics Education, Universitas Sebelas Maret, Surakarta, Indonesia

Email: smarsih74@yahoo.com

Abstract. This research aims to understand the students' weaknesses in mathematical reasoning ability in junior secondary school. A set of multiple choice tests were used to measure this ability involve components mathematical communication, basic skills, connection, and logical thinking. A total of 259 respondents were determined by stratified cluster random sampling. Data were analyzed using one-way Anova test with $F_{obs} = 109.5760$ and F = 3.0000. The results show that students' ability from schools with high National Exam in mathematics category was the best and followed by medium and low category. Mathematical connection is the most difficult component performed by students. In addition, most students also have difficulty in expressing ideas and developing logical arguments.

1. Introduction

Reasoning is needed in solving mathematical questions [1]. It is a basic skill of mathematics and is necessary for a number of purposes – to understand mathematical concepts, to use mathematical ideas and procedures flexibility, and to recontruct [2]. Kaur and Lam describe mathematical reasoning as components consistsing of communication, basic skills, connection, and logical thinking [3]. It is clear that the "reasoning of mathematics" need to be fully integrated into classrooms, schools, and districts around the country to produce citizens and employees adequately prepared for the 21st century [4]. Based on the above-mentioned research, it can be seen that reasoning skills needs to be applied in learning mathematics in the classroom so that students are able to become proficient communicators, creators, critical thinker, and collaborators.

Based on interviews with some Junior Secondary School students, it is found that the students' process of reasoning was not yet developed well. The implication is that the students' critical thinking ability and communication skills were not well-developed. There are some examples showing the lack of students' mathematical reasoning. First, the students had studied the basic concepts of numerical operations and they had been able to solve arithmetic operations in positive integers without problem. However, in the reality, the students faced difficulties in determining the arithmetic operation results on negative integers and fractions. Related to these, their answers were various and incorrect. For example, in solving $3x = 12$, the students answered correctly in a little amount of time. However, in solving basic questions like $5x = 1$ which should be answered as $x = \frac{1}{5}$, some students answered $x = 1$, $x = -5$, or $x = -4$. Therefore, the students' ability to make simple examples into complex ones was not yet well-developed [5].

Second, in learning mathematics, the students were experiencing difficulties in writing a statement into mathematical sentence related to operations of addition, subtraction, multiplication, and division. Problem solving-type questions are full of statements and/or sentences, so that in order to solve it, students are expected to rewrite it into mathematical sentences correctly. A hindering factor, which was factor coming from within the students, was found which causes them to have trouble in expressing their ideas [6]. Students in general have troubles in expressing opinion with their own words, mostly because they are afraid to be wrong. The other hindering factors vary: laziness, afraid to be laughed about, afraid to be named as "the smart one", afraid of teachers, and afraid of teachers' wrath.

The other difficulties relate to two-dimensional figure discourse matter. Two-dimensional figures have two main elements: sides and corners. Figures like trapezoids, kites, parallelograms, and rhombi can be divided into several two-dimensional figures, which may take forms of squares, rectangles, and triangles. Students were having troubles in determining the side length, even more if stated in variables. The low ability of the students in finding facts from pictures was caused by the students' inability to free themselves from their mindset. Students often thought about an elaborate solution, while in fact the solution itself is merely a simple one [7].

Based on the description above, it can be concluded that the students' low ability of mathematical reasoning was indicated by the ability of students in making simple examples into complex ones, which was not yet developed well, the students' difficulties in expressing their ideas in their own words because of fear of doing wrong, and their low ability in finding facts from pictures because the students were hindered by their own faraway thoughts.

Tajudin and Chinnappan said that: "Further research should be conducted to generate higher level of clarity about the roles of reasoning in mathematics learning. This study indicates that reasoning skills are important at least for the Low-Achievement group" [8]. Ayele argued that to improve students' reasoning skills, teachers should value their creativity in the classroom as well as help and appreciate students' creative work and ideas [9]. Tonelsdeveloped Wind Phoenix game with problem solving process through creation of conjectures. Tonels concluded that, with role assumption in the game, the players form logical problem uttering process – puzzles [10].

The research by Poon & Leung shows that students in schools with higher academic achievements had higher scores in geometry test. Besides, strong correlations are found between the students' achievement in geometry and their fundamental logical reasoning ability. This finding offers a reason for reflecting present geometry curriculum in junior secondary schools in Hong Kong [11]. Based on the research of Ayele, Tonels, and Poon & Leung, it can be concluded that mathematical reasoning can be developed by appreciating students' creative works and ideas through games, or by reflecting curriculum which implements learning process that emphasizes on students gaining their own experience. Asyari et al [22] studied Improving critical thinking skills through the integration of problem based learning and group investigation.

This research analyzed more deeply about the profile of students' mathematical reasoning ability in terms of mathematical communication, basic mathematical skills, mathematical connections, and logical think. The components of mathematical reasoning ability in this study are part of "4C", which is included in the critical thinking and communication skills component. This research further experiment with Creative Problem Solving and Group Investigation learning model through Mind Mapping technique. As the result of this research, the teachers and/or educators are expected to identify the characteristics of students' ability and to determine which components of mathematical reasoning is weak, in order to find an alternative way to improve it.

2. Theoretical basis

Ball and Bass suggested that mathematical reasoning assumes mathematical communication. Communication is an integral part of the process of reasoning [2]. According to Kaur and Lam, communication refers to the ability to use mathematical language to express mathematical ideas and arguments precisely, concisely and logically, to use model with mathematics, and use tools strategically. The tools are table, chart, diagram, or list [3]. Ball and Bass explained that reasoning is a basic skill of mathematics and is necessary for a number of purposes –to understand mathematical concepts, to use mathematical ideas and procedures flexibility, and to reconstruct [2]. As stated by Kaur and Lam in their quotations about syllabus document of the Ministry of Education of Singapore, mathematical reasoning refers to the ability to analyse mathematical situations and construct logical arguments [3].

Russell stated that mathematical reasoning is essentially about the development, justification and use of mathematical generalizations. The generalizations create an interconnected web of mathematical knowledge–conceptual understanding. Seeing mathematics as a web of interrelated ideas is both a result of an emphasis on mathematical reasoning and a foundation for reasoning further [12]. In the words Kaur and Lam explained that connections refer to the ability to see and make linkage among mathematical ideas, between mathematics and other subjects, and between mathematics and everyday life [3]. Logic is the study of the methods and principles used to distinguish correct from incorrect reasoning. Reasoning is a special kind of thinking in which inference take place, in which conclusions are drawn from premises [13]. It can be concluded that mathematical reasoning is further described into four components, which are communication, basic skills, connection, and the way of mathematical logical thinking.

Higher Ability Selection Test (HAST) is a test package written by the Australian Council Educational Research (ACER), consisting of mathematical reasoning, reading comprehension, abstract reasoning, and written expression [14]. The test are used for selection, or to better differentiate student intakes, and to distinguish between stronger and weaker students. The mathematical reasoning test developed by ACER includes tests to understand, interpret, and analyze mathematical information at the junior level as well as math and science at the middle and senior level. Test participants should apply logical and strategic thinking to work through questions in the exam. Tasks are presented as numbers, texts, diagrams, graphs, and tables including number, measurement, space, time, logical relationships and problems, which must be solved to obtain mathematical components. Two important processes in reasoning are connections and further changes to form an argument or to solve a problem [2]. The process of reasoning consists of deductive reasoning and inductive reasoning [15]. Deduction is defined as the reasoning process from general to specific, whereas induction is defined as the process of reasoning from specific to general. Deductive reasoning includes: ponens mode, tollens mode, and syllogism [13]. Premises, in the argument the propositions upon which inference is bassed, the propositions that are claimed to provide grounds or reason for the conclusion. Conclusion in any argument, the proposition to which the other propositions in the argument are claimed to give support, or for which they are given as reasons. Conclusions with ponens mode, tollens mode, and syllogism can be distinguished as seen in Table 1.

Table 1. Deductive conclusion

	Ponens mode	Tollens mode	Syllogism
Premise I	$a \rightarrow b$	$a \rightarrow b$	$a \rightarrow b$
Premise II	a	$\sim b$	$b \rightarrow c$
Conclusion	b	$\sim a$	$a \rightarrow c$

Inductive reasoning includes: analogies, generalizations, and causal relationships. Analogy is a process of reasoning that departs from two or more specific events that have similarities with each other. The greater the number of respects in which the entity in the conclusion is similar to the entities in thr premise, the more probable in that the conclusion. Generalization is a process of reasoning that departs from a number of facts or the particular phenomenon observed, then drawing general conclusions about the observed phenomenon. Causal relationships are reasoning derived from interconnected phenomena. Descriptive causal laws asserting a necessary connection between event of two kinds, of which one is the cause and other the effect [13].

Referring to the opinions of Ball & Bass, Russell, Kaur & Lam, and Copi as well as the adaptation of ACER, the ability of mathematical reasoning is the ability or skills in the form of test results about mathematical communication, basic math skills, mathematical connection, and logical thinking. Indicators of mathematical reasoning ability as measured in this research are stated in Table 2.

Table 2. Indicators of mathematical reasoning ability

Reasoning Component	Indicators
Mathematical communication	Using mathematical expressions to express mathematical ideas. Preparing arguments appropriately, concisely, and logically. Determining images/sketch
Mathematical basic skills	Understanding the concept and interpret the data Using flexible mathematical ideas and procedures. Building a logical argument.
Mathematical connection	The ability to build inter-discourse material or interdisciplinary connection. Stating relations Making generalizations
Logical thinking	Using analogies, implications, and syllogism Using reasoning in solving problems

3. Methods

The design of this research is mix-methods research with quantitative and qualitative descriptive approach. This research was conducted on Junior Secondary School students in Sragen, Indonesia. The samples were taken by stratified cluster random sampling [16]. Sampling was done by dividing 49 state SMPs into three categories namely high, medium, and low. Categorization was determined based on the result of National Exam in mathematics of Junior Secondary School in 2016 with mean of 44.35, standard deviation of 18.27, and $\alpha = 0.5$. The categorization and the number of samples are shown in Table 3.

Table 3. The number of research samples

No	Category	Number of SMPs	Number of Sample
1	High	5	96
2	Medium	41	96
3	Low	3	67
	Total	49	259

Data were obtained by interview and test methods. Interview method was used to know students' difficulties in understanding mathematics related to the process of reasoning. A set of mathematical reasoning ability tests consist of four components and are outlined in 11 indicators. Distribution of questions item can be seen in Table 4.

Tabel 4. Distribution of question item

Components	Sum		
	Indicator	Tryout item	Test item
Mathematical communication	3	6	3
Mathematical basic skills	3	10	7

Components	Sum		
	Indicator	Tryout item	Test item
Mathematical connection	3	7	4
Logical thinking	2	17	11
Total	11	40	25

Content validation was conducted through expert judgment by three experts, namely two lecturers on psychology major and a counseling tutor. Furthermore, tryouts of the mathematical reasoning test instrument consisted of 40 items of question, conducted on 60 8th grade students. A total of 25 question items were selected, having fulfilled power distinct criteria of $D \geq 0.30$ and difficulty level of $0.3 < P \leq 0.7$ [17] and an index of reliability of 0.84 was obtained.

4. Result and discussion

4.1. Requirements test of one-way anova with unbalanced cells

A set of mathematical reasoning ability tests was conducted to students of schools with National Examination result in three categories namely high, medium, and low. These three categories of data have met the requirements of one-way Anova test with unbalanced cells, namely normality and homogeneity [18] as shown in Table 5 and Table 6.

Table 5. Requirements normality of variance analysis

Requirements	NE Results Category		
	High	Medium	Low
n_j	96	96	67
$T_j = \Sigma X$	7220	6020	2924
Mean	75.2083	62.7083	43.6418
Sd Dev	13.9124	13.4930	12.4702
Var (S_j^2)	193.5561	182.0614	155.5061
L_{Max}	0.0675	0.0759	0.1021
L_{Table}	0.0904	0.0904	0.1082

Table 6. Requirements homogeinity of variance analysis

S_p^2	179.480741
b_{obs}	0.99635599
$b_k(\alpha;n_1;n_2;n_3)$	0.97643359

Based on Table 5, the normality test using the Lilliefors method shows that $L_{max} < L_{table}$. This means that each group comes from a normally distributed population. Based on Table 6, the homogeneity test using Bartlett method, it is obtained that $b_{obs} > b_k$ (0.05;96; 96; 67). This means that all three populations have the same variance.

4.2. One-way variance analysis test with unbalanced cells

ANOVA test results with unbalanced cells as shown in Table 7. Based on $F_\alpha = 3.0000$ and $F_{obs} = 109.5760 > F\alpha$, the decision is H_0 is rejected [18]. It was concluded that not true that all categories have the same mean. Because of the decision that H_0 is rejected, then it conducted the multiple comparison test by the *Scheffe'* method. The results are as shown in Table 8.

Table 7. Summary of one-way anova test with unbalanced cells

Source	SS	Df	MS	F_{obs}	F_{α}
Mathematical reasoning ability (A)	39333.548	2	19666.774	109.5760	3.0000
Error (G)	45947.070	256	179.481		
Total (T)	85280.618	258			

Table 8. Summary of multiple comparison test

Comparison	F_{obs}	$F\alpha$
F_{1-2}	41.787213	6.0000
F_{2-3}	79.925334	
F_{1-3}	219.0759	

The analysis result shows that student's mathematical reasoning ability in schools with high National Exam (NE) in mathematics category are better than students in schools with medium and low category. As well as student's mathematical reasoning ability in schools with medium categories are better than low category.

4.3. Qualitative descriptive analysis of student's mathematical reasoning ability

Descriptive qualitative analysis was carried out based on the percentage of score on each component and indicator of mathematical reasoning ability. Table 9 demonstrates score result on each component of mathematical reasoning ability from schools with low, medium, and high NE results category in 2016.

Table 9. Score result of mathematical reasoning ability components

Mathematical Reasoning Components	NE Results Category		
	High	Medium	Low
a. Mathematical communication	82%	67%	47%
b. Mathematical basic skills	71%	64%	33%
c. Mathematical connection	59%	31%	32%
d. Logical thinking	82%	72%	40%

Table 9 illustrates that in the schools that fall into the high NE result category, students have low levels of basic mathematical skills and mathematical connection abilities. Meanwhile, students from schools with medium and low category have low levels of abilities on all components. In addition, Table 9 demonstrates that students have low levels of mathematical connection abilities even though they are from the schools with high NE result category. Connection relate with how people can organize the ideas. In this global era, the ability to analyze an information will determine someone's success to take advantage of opportunities and win the competition [4]. To overcome this case, mathematics teachers need to apply learning method that can improve connections ability.

Further, the analysis of each indicator from reasoning components in Table 9 are then shown in Table 10, 11, 12, and 13. Table 10 illustrates the score results of the mathematical communication component.

Table 10. Scores results of mathematical communication component

Component / Indicators	NE Results Category		
	High	Medium	Low
a. Mathematical communication	**82%**	**67%**	**47%**
1) Using the language of mathematics to express mathematical ideas.	100%	88%	71%
2) Develop an argument with precise, concise, and logical.	86%	72%	45%
3) Determining picture/ sketches	60%	43%	27%

Table 10 illustates that schools in all NE result categories, students have low levels of determining picture/sketches. Communication is related to how students are able to transform a problem into a symbolic mathematical problem, to reorganize the mathematical system, and to interact and dialogue with existing media. The importance of mathematization as a crucial process in the learning and teaching of mathematics [6]. There is a need for a better understanding of teachers' learning and the quality of their development of know-how and skills about language in mathematics [20]. Communication is so crucial in the 21st Century that the cooperative and collaborative approaches to learning depicted [19]. Based on interview result with some students, it is found that the cause of the lack of ability to identify pictures is that students have difficulty in analysing information from images into appropriate mathematical models. To overcome this problem, it is necessary for teachers to apply mind mapping learning technique.

Table 11. Scores result of mathematical basic skills component

Component / Indicators	NE Results Category		
	High	Medium	Low
b. Mathematical Basic Skills	**71%**	**64%**	**33%**
1) Understanding the concepts and interpret data	83%	80%	41%
2) Using mathematical ideas and procedures with flexible.	67%	47%	28%
3) Building a logical argument	47%	65%	27%

Based on the results shown in Table 11, it can be seen that most students already understand concepts and interpret information. However, all students have not succeeded in applying mathematical ideas and building logical arguments. In this component, teachers need to improve students' competence in applying mathematical ideas and building logical arguments. This basic skills indicator is really important in mathematics learning [12]. Because to solve every mathematics questions, students must understand the concept mathematically and be able to give a logical argumentation. Based on interview with students, it is found that the cause of the lack of ability in applying mathematical ideas and building logical arguments is that students are not confident to express their ideas. Self-confidence is necessary in mathematics learning. This is in line with Warner and Kaur's opinion, they stated that without confidence/self-efficacy in one's ability, students cannot perform to their potential or at their highest standard. It is even possible that learners with lesser abilities, but with confidence, can outperform higher ability students because belief in oneself can be a powerful influence. Bandura refers to situation-specific, self-confidence as self-efficacy, which is the strength of an individual's belief that they can successfully perform a given activity or task [19].

Table 12. Score result of mathematical connection component

Component / Indicators	NE Results Category		
	High	Medium	Low
c. Mathematical connection	**59%**	**31%**	**32%**
1) Connection capabilities between subjects or materials	70%	42%	29%
2) Stating relations	51%	21%	42%
3) Make generalizations	47%	20%	27%

Table 12 illustrates that schools in all NE result categories, students have low levels on all indicators of the mathematical connection ability component. Based on the interview, the weakness of mathematical connection ability is caused by the low creativity in thinking and organizing the ideas. They understand the lessons when it described, but students still difficulty making connection between subjects or materials. Based on this analysis, teachers are important to develop innovative learning models to produce students who ready within the competitive world of work in the 21st century [1]. Bertoncelli and Lynass state that employers in the 21st century stimulate that employees be critical thinkers, effective collaborators, innovators, and excellent communicators [21]. Hence, there is a challenge for educators to transform how they prepare learners for the inevitable and impending workforce.

Table 13. Scores acquisition of logical thinking component

Component / Indicators	NE Results Category		
	High	Medium	Low
d. Logical thinking	**82%**	**72%**	**40%**
1) Using the analogy, the implications, the syllogism	81%	73%	39%
2) Using the reasoning in problem solving	85%	66%	40%

In the logical thinking component (Table 13), the schools with high category obtained that the highest percentage score followed by medium and low category. The results of the interviews obtained the information that students can solve the problems which were described, but they can not use analogy to solve more complicated problems. Analogy is parallel drawn between two (or more) entities by indicating one or more respects in which they are similar [21]. In fact, if the students can develop analogy and logical thinking then the student will be easily to solve the problems. Based on this description teachers are be advised to improve the frequency in learning to develop analogical and logical thinking.

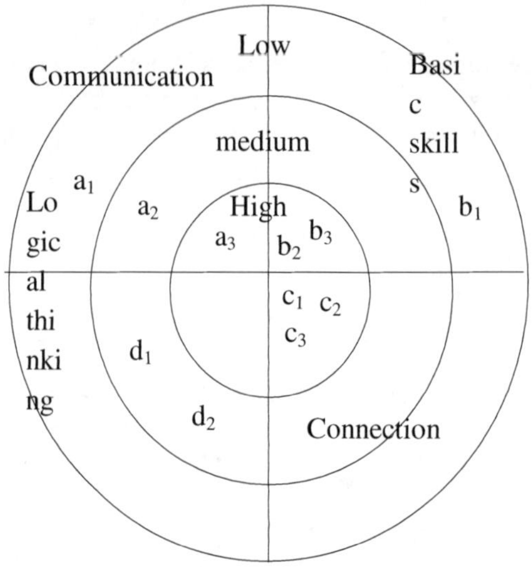

Figure 1. Map the low of student's mathematical reasoning ability

Based on illustrations in Table 10, 11, 12, and 13, the low of student's mathematical reasoning ability from schools with high, medium, and low category of NE results can be mapped out as shown in Figure 1. Based on Figure 1, it is concluded that students in high category have no problem on logical thinking component. Students in medium category have no problem on indicators using the language of mathematics to express mathematical ideas (a_1) and understanding the concepts and interpret data (b_1). Students in low category have low percentage on all indicators.

5. Conclusions

Based on the results of the research and discussion, it was concluded that the students's mathematical reasoning ability from schools with high category in mathematics National Examination result is better than medium and low category. The medium is better than the low category. The mathematical connection component is the most difficult component for students in all categories. The weakness of mathematical connection is due to the low creativity of thinking and organizating ability between ideas and concepts in mathematics. In addition to the components connection, in the parts of the mathematical communication component students also have difficulty in pouring ideas and developing logical arguments. This is because, by their lack of confidence in the opinion. This condition needs to be addressed with teachers to apply learning models which can facilitate the students to develop creative thinking and good communication. Looking at the 21st century, schools should prepare their students in order to be able to be critically thinker and be able to organize the ideas. Therefore, there is a challenge for educators to transform how to prepare the students for the inevitable and impending workforce.

Acknowledgments

The authors would like to thank to the referees for their suggestions and this research is part of Sebelas Maret University, Surakarta, Indonesia in the academic year of 2018

References

[1] Zivkovic S 2016 A model of critical thinking as an important attribute for success in the 21st century *Procedia–Social and Behavioral Sciences* **232** 102 (Turkey: Elsevier)

[2] Ball D L and Bass H 2003 *Making mathematics reasonable in school* (Reston, VA: National Council of Teachers of Mathematics)

[3] Kaur B and Lam T 2012 *Reasoning, Communication and Connections in Mathematics.* (Singapore: World Scientific Publishing Co. Pte. Ltd.)
[4] Roekel D V 2011 *Preparing 21 st century students for a global society an educator's guide to the "Four Cs"* (USA: National Education Assosiation)
[5] Forgues H L, Tian J and Siegler R S 2016 Why is learning fraction and decimal arithmetic so difficult? *Developmental Review ScienceDirect* **38** 201
[6] Jupri A and Drijvers P 2016 Student difficulties in mathematizing word problems in algebra *Eurasia Journal of Mathematics, Science & Technology Education* **12** 2481
[7] Pantziara M, Gagatsis A and Elia I 2009 Using diagrams as tools for the solution of non-routine mathematical problems Published online: 10 February 2009 *Springer Science + Business Media*
[8] Tajudin N M and Chinnappan M 2016 Relationship between scientific reasoning skills and mathematics achievement among malaysian student *Geografia online TM Malaysian Journal of Society* **46** 96
[9] Ayele M 2016 Mathematics teachers' perceptions on enhancing students' creativity in mathematics. *IEJME Mathematics Education* **11** 3521
[10] Tonels C N 2016 The act of playing and the logical and mathematical reasoning in digital games. *Elsevier B.V. All rights reserved.* http://dx.doi.org/10.1016/j.entcom.2016.10.001
[11] Poon, K. K. & Leung, C. K. (2016) A study of geometric understanding via logical reasoning in Hong Kong *International Journal for Mathematics Teaching and Learning (IJMT).*17 3
[12] Brodie K 2010 *Teaching Mathematical Reasoning in Secondary School Classrooms (*New York: Springer Science+Business Media)
[13] Copi I M, Cohen c and McMahon K 2014 *Person New International Edition Introduction to Logic (*USA: Person Education Limited)
[14] ACER 2017 The Tests Higher Ability Selection from http:// www.acer.org/hast-secondary/the-tests
[15] Solso R L 1991 *Cognitive Psychology* (Boston: Alliyn and Bacon)
[16] Phrasisombath K 2009 *Sample Size and Sampling Methods* (Vientiane: University of Health Sciences)
[17] Hale C D 2014 *Measuring Learning & Performance: A Primer* (Florida: Charles Dennishale.org)
[18] Heron E 2009 *Analysis of Variance-ANOVA* from https://www.tcd.ie
[19] Warner S and Kaur A 2017 The perceptions of teachers and students on a 21st century mathematics instructional model *IEJME - Mathematics Education* **12** 193
[20] Hajer M 2017 Teachers' knowledge about language in mathematics professional development courses: from an intended curriculum to a curriculum in action *EURASIA Journal of Mathematics Science and Technology Education* **13** 4087
[21] Bertoncelli T, Mayer O and Lynass M 2016 Creativity, learning technique and TRIZ *Elsevier Procedia CIRP* **39** 191
[22] Asyari M, Al Muhdhar M, Susilo H and Ibrohim 2016 Improving critical thinking skills through the integration of problem based learning and group investigation *IJLLS* **10** 42

IOP Conf. Series: Journal of Physics: Conf. Series **1008** (2018) 012079 doi:10.1088/1742-6596/1008/1/012079

Students' thinking preferences in solving mathematics problems based on learning styles: a comparison of paper-pencil and geogebra

Umi Farihah

Department of Mathematics Education, State Islamic Institute of Jember, Indonesia

E-mail: u_farihah@yahoo.com

Abstract. The purpose of this study was to analyze students' thinking preferences in solving mathematics problems using paper pencil comparing to geogebra based on their learning styles. This research employed a qualitative descriptive study. The subjects of this research was six of eighth grade students of Madrasah Tsanawiyah Negeri 2 Trenggalek, East Java Indonesia academic year 2015-2016 with their difference learning styles; two visual students, two auditory students, and two kinesthetic students.. During the interview, the students presented the Paper and Pencil-based Task (PBTs) and the Geogebra-based Task (GBTs). By investigating students' solution methods and the representation in solving the problems, the researcher compared their visual and non-visual thinking preferences in solving mathematics problems while they were using Geogebra and without Geogebra. Based on the result of research analysis, it was shown that the comparison between students' PBTs and GBTs solution either visual, auditory, or kinesthetic represented how Geogebra can influence their solution method. By using Geogebra, they prefer using visual method while presenting GBTs to using non-visual method.

1. Introduction

Research studies showed that the students use visual (such as graphic) and non-visual (such as Algebraic, verbal, and numeric) methods while solving the mathematics problems [1, 2]. Visual methods involve visual imagery and non-visual methods do not involve visual imagery [3]. The recent technology devices support the assessable and affordable visualization from the idea of mathematics to the visual imagery. It also supports the non-visual method by organizing the data and computing it effectively and accurately [4]. By investigating the student's solution methods in solving mathematics problems, researchers contend that the students have a cognitive preference for visual or non-visual solution methods [5]. This cognitive preferences determine the methodological type to which students belong.

By using an innovative technology, visualization has become an important part of mathematics education. Curriculum reformation efforts in the mathematics education proved the importance of the process of visual and the use of visual representations [4]. Besides, the connecting of the visual and non-visual representations, it is expected to be able to produce a better understanding about mathematics [6, 7, and 8]. It is proved that technology build the relationship with the difference computer program and offer a wide range of students' perspective with the visual support [9].

The importance of visualization in the problem solving is also stated by Rif'at. He argued that to solve the mathematics problems, there is a need a visual presentations beside analytical presentations.

Despite the presentation has been exploited in the instruction, it is taken as a device to assist the instruction, so the problem solving analytically done. Whereas, the visual presentations is not only exploited as a tool to assist but also exploited as a strategy as well as a thinker in solving a problem specifically a visual-characterized and visualized problem [10]

Technology devices are not only support visual representation but also show the relationship among different representations. NCTM proved the importance of students' ability in choosing, implementing, and interpreting mathematics representation [4]. The students had got a deep understanding about the concept after they investigate and synthesize the relationship between graphic and analytic representation [11]. The effective mathematics product might be exist through the coordination for both graphic and numeric representation as well as algebra [9].

Technology assists coordinating the difference representation. It consists of a set of computer which specifically designed to serve this coordination like Geogebra [12]. Geogebra software supports several representations by offering several devices to make a graph function, show the expression of algebra, and performing numeric counting. It also support the data from numeric into visual representation. Some researchers state that the ability to employ every representative and interpret among representations will reveal a deep understanding about mathematics [13].

The effect of thinking preferences in the students' learning and working during the use of technology has been the most important subject in the late decade. The study of the topic shows the different result. Study the effect of instruction of dynamic geometry environment of students' work toward the geometry task such as determining the width of triangle and parallelograms [14]. They also investigate whether or not using a dynamic geometry can reduce students' cognitive burden by accommodating their different thinking preferences. The research result shows that there is no relationship between thinking preference and the students' work toward the geometry tasks. In his research, verbalizer utilizes visual instruction more than a dynamic geometry software. However, the result come into a contrary with the previous research result in the mathematics education [15]. It states that optimal learning is happened since the students are thought with the appropriate way of their thinking preference. Riding and Douglas also found out that verbalizer can accomplish the assignment better than those visualizer (analytical thinking) in the text-based environment while visualizer can do better than those verbalizers in every visual and graphic presentation [16].

Krutetskii's research concluded that there is a correlation between the ability of visualizing abstract relationship and the ability of geometry concept of space. Nevertheless, both of them are not the important component of the mathematics ability. Furthermore, he found out that the strength or weaknesses of visual or analytical thinking does not determine how far the students' mathematics ability is but determines the kind of it. It means that someone who have a different correlation mathematics ability between two components mentioned in the previous sessions (visual-pictorial, verbal-logical). This correlation determine thinking domain (analytic, geometric, and harmonic) where the people comes from [2].

Yerushalmy's research report states that the students experience the positive effect from the use of technology in the Algebra instruction, specifically in the field of symbol, equality, and the problematic context [17]. In addition, Yerushalmy, Shternberg and Gilead conducted a research focused on the software that support visualization [18]. Sirin Coskun conducted a multi-cased research about the effect of technology of the students' visual and non-visual thinking preference by comparing the strategy to solve the Algebra paper-based test (PBTs) of Senior High School students in Florida, USA. In his research, Coskun found out the comparison between their PBTs and GBTs shows is this dynamic software can influence their solution method [2].

2. Method

This current research has employed a qualitative descriptive method. The subject of this research was eight graders of the second semester academic year 2015-2016. This current research was conducted in MTsN 2 Trenggalek, East Java Indonesia which consisted of six students who had different learning style: two visual students, two auditory students, and two kinesthetic students. Students who were selected as participants in this study were those who have a good communication, good in operating Geogebra program, and have one of dominant learning styles among others which was shown by a higher score obtained from learning styles test for all learning styles consisting of auditory, kinesthetic, and visual learning style.

Table 1. Research Subject Based on Their Learning Style

No	Code	Visual Score	Auditory Score	Kinesthetic Score	Learning Style
1	LN	36	27	24	Visual
2	ESP	37	32	29	Visual
3	ATR	36	38	34	Auditory
4	NDP	33	36	24	Auditory
5	FAM	30	32	36	Kinesthetic
6	YE	28	31	36	Kinesthetic

The data collection was obtained from questionnaires, test, observation, and interview. Questionnaires of learning style were used to get the data about the students' learning style. Meanwhile, a test in this study was in the form of paper and pencil-based test (PBTs) and Geogebra-based test (GBTs) which were used to get the data about the students' visual and non-visual thinking preferences in solving mathematics problems. The questionnaires used in this study had been employed by Sagitasari which consisted of 30 items [19], while PBTs and GBTs tests were adopted from Sirin Coskun that consisted of 10 mathematic problem [2] which many adaptation compatible with the students of MTs or secondary level as a subject of this study, they were number 6, 7, and 8 .

The data analysis used in this study was obtained from two sources; first, students' learning style analysis by adding all scores obtained by the students based on their learning style, they were visual, auditory, and kinesthetic learning style. Those data were then compared by those three sources obtained by the students. The highest score obtained by the students indicates their learning style; second, the students' visual and non-visual thinking preferences in solving mathematics problems employed a model of Miles & Huberman. The activity of data analysis were reduction, data display, and drawing conclusion or verification [20]. Choi-Koh [28] studied Effect of a graphing calculator on a 10th-grade student's study of trigonometry and also Huntly et al [29] studied Effects of Standards-based mathematics education: A study of the Core-Plus Mathematics Project algebra and functions strand.

3. Research Result

The result of the research showed that all of the students who have either visual learning style, auditory learning style, or kinesthetic learning style used non-visual method to solve PBTs which consists of 10 items except number 1 and 2. Both of them were solved by kinesthetic students using visual method. The students who have visual learning style used a procedural way, while the students who have auditory learning style prefer using verbal logic. This was shown from the result of the students' presentation in solving PBTs as shown in this following question:

PBTs 1	*Altogether there are 8 tables in a house. Some of them have four legs, and the others have three legs. Altogether they have 27 legs. How many tables are there with four legs?*

The students with visual learning style solved PBTs 1 by using non-visual solution method, that was separated the table with four legs as x and a table with three legs as y, they used elimination mixed method and substitution. Subject LN eliminate x earlier so that she found the value of y, then, she found the value of x by substituting it. Meanwhile, subject ESP eliminated y earlier so that she found the value of x, then, she found out the value of y by substituting it. However, both of them found out the same result.

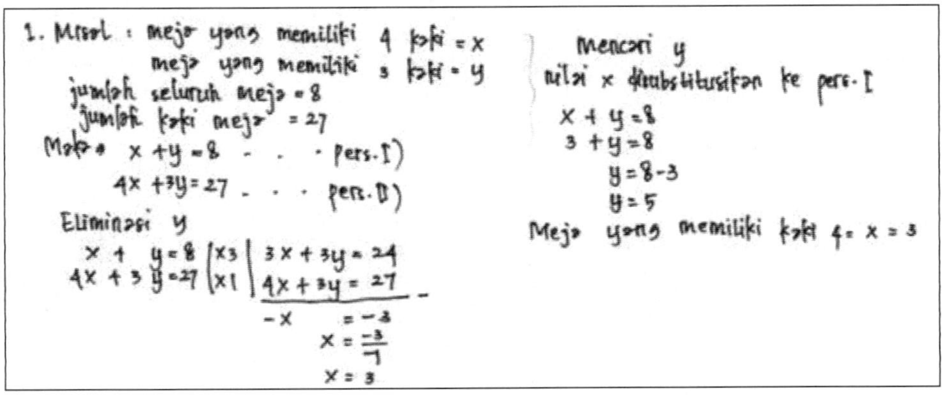

Figure 1. ESP's Solution in Solving PBTs 1

All of the students who have an auditory learning style solved PBTs 1 using non-visual solution method by guessing the number. ATR guessed the number if there were three tables that consisted of four legs so $4 \times 3 = 12$, legs left = 15, then, it was divided by 3, it became 5. She found out that there were three tables that has four legs. Meanwhile, NDP guessed that there were five tables that has three legs so it became 27 if they add them. Even though they used different way to count but they have the same way of thinking, they prefer using their logical way of thinking.

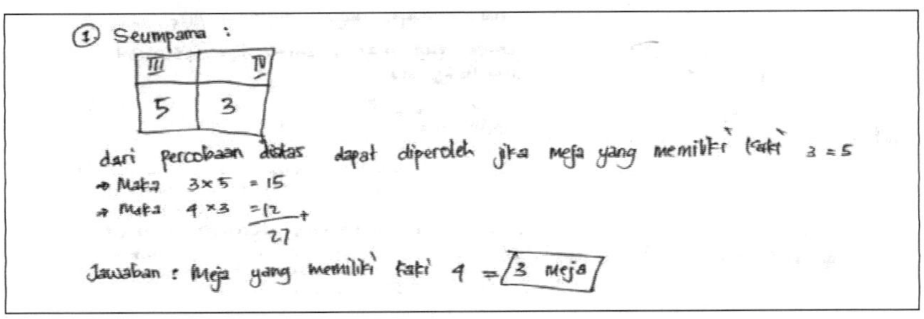

Figure 2. NDP's Solution in Solving PBTs 1

The students who have kinesthetic learning style, they tend to solve PBTs 1 using visual solution method. Subject FAM solved this problem by drawing a table. Firstly, she pulled the tables as if they have only three legs, then she added the legs to the table and it reaches 27. On the other hands, subject YE drew a table that has four legs and three legs so she found out the numbers of legs are 27. Secondly, subject FAM and YE found out the same final result they were the tables that has four legs and three legs.

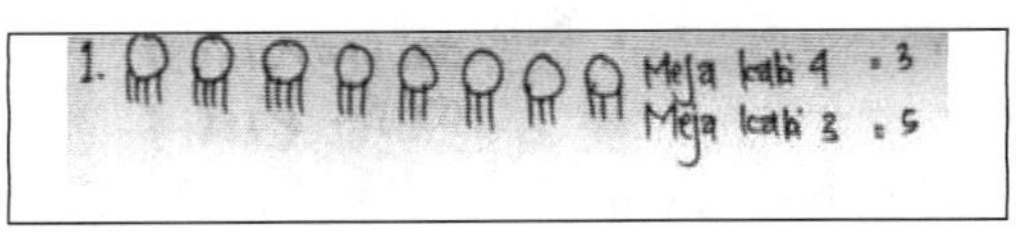

Figure 3. FAM's Solution in Solving PBTs 1

PBTs 3, PBTs 5, PBTs 6, and PBTs 7 were solved by all students who have visual, auditory, and kinesthetic learning style using elimination and substitution mix method. Meanwhile, PBTs 4, PBTs 8, PBTs 9, and PBTs 10 were solved using verbal logic and logical way of thinking. The following figure is the result of students' presentation towards PBTs 4 and PBTs 9.

PBTs 4	*Budi rode his bike to his friend Eko's house, which was 18 miles away. After he had been riding for half an hour, he got a flat tire. He walked his bike the rest of the way. The total trip took him 3 hours. If his walking rate was one-fourth as fast as his riding rate, how fast did he ride?*

All of the research subjects either visual, auditory, or kinesthetic learning style used the same way in solving PBTs 4 by using the speed formula and reducing into the formula of $distance\ (s) = speed\ (v) \times time\ (t)$. After they find $v_2 = 4$, then they findv_1 by multiply v_2 with 4, therefore they found v_1 equal to 16. So, they found that the speed is 16 mil/hour.

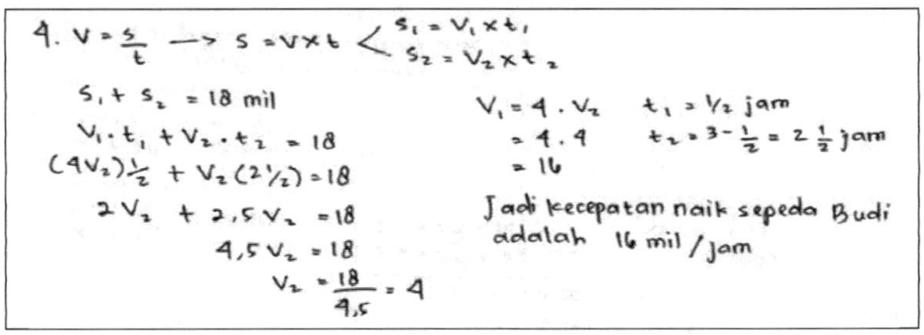

Figure 4. LN and ESP's Solution in Solving PBTs 4

PBTs 9	*There are 9 boys to every 10 girls in a particular high school. There are 2622 students at the school. How many girls are there?*

All the subjects of the research solved PBTs 9 by using non-visual method, it is a guessing numbers. If the number of the girls were 10, it would be 9 for boys, but their numbers reach = 19 (wrong), then they guessed other numbers. If the number of the girls were 100, it would be 90 for the boys, but their numbers reach = 190 (wrong). Finally, they found the number of the girls were 1380 girls and 1242 boys. Therefore, the number of them were 2622 (correct).

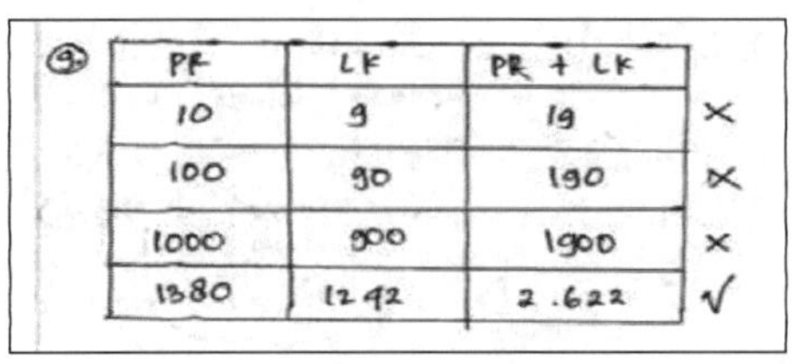

Figure 5. NDP's Solution in Solving PBTs 9

After accomplishing all Paper-based Tasks, they turned to the following steps, it was delivering Geogebra-based Tasks (GBTs) that consisted of 10 items followed by conducting an interview. While the students work and solve the problem using GBTs, it had been provided by an installed dynamic software called Geogebra where the students can select to use visual solution method in the form of graph as well as non-visual solution method in the form of numeric by opening the window *spreadsheets* as well as Algebra by opening the window *Computer Algebra System* (CAS).

The result of the research shows that all the students who have visual, auditory, and kinesthetic learning style employed visual solution method to solve all the Geogebra-based Tasks, except GBTs 4 and GBTs 9. They would solve the problem by numeric or algebra representation. The comparisons which were could not made by them are the comparisons which were could not found by them as they learn in the class. The comparison which is not in the form of $ax + by = c$ as stated in the example of GBTs 9.The following is the solution method used by the students in solving GBTs 1, GBTs 4, and GBTs 9.

GBTs 1	*Bill had Rp 10.500 coin money, in one thousand coins and five hundred coins. If he had 17 coins, how many coins he had?*

All the subjects used the same solution method to solve GBTs 1, it is a graphic representation assisted by the window of Algebra. They told that *x* as a coin of five hundreds while *y* as a coin of one thousand. They made equations $x + y = 17$ and $1000x + 500y = 10500$, then they drew a graph of both equations, then they found out intersection point of graphs. They interpreted the intersection point and stated that there had to be 4 coins of one thousand and 13 coins of five hundreds.

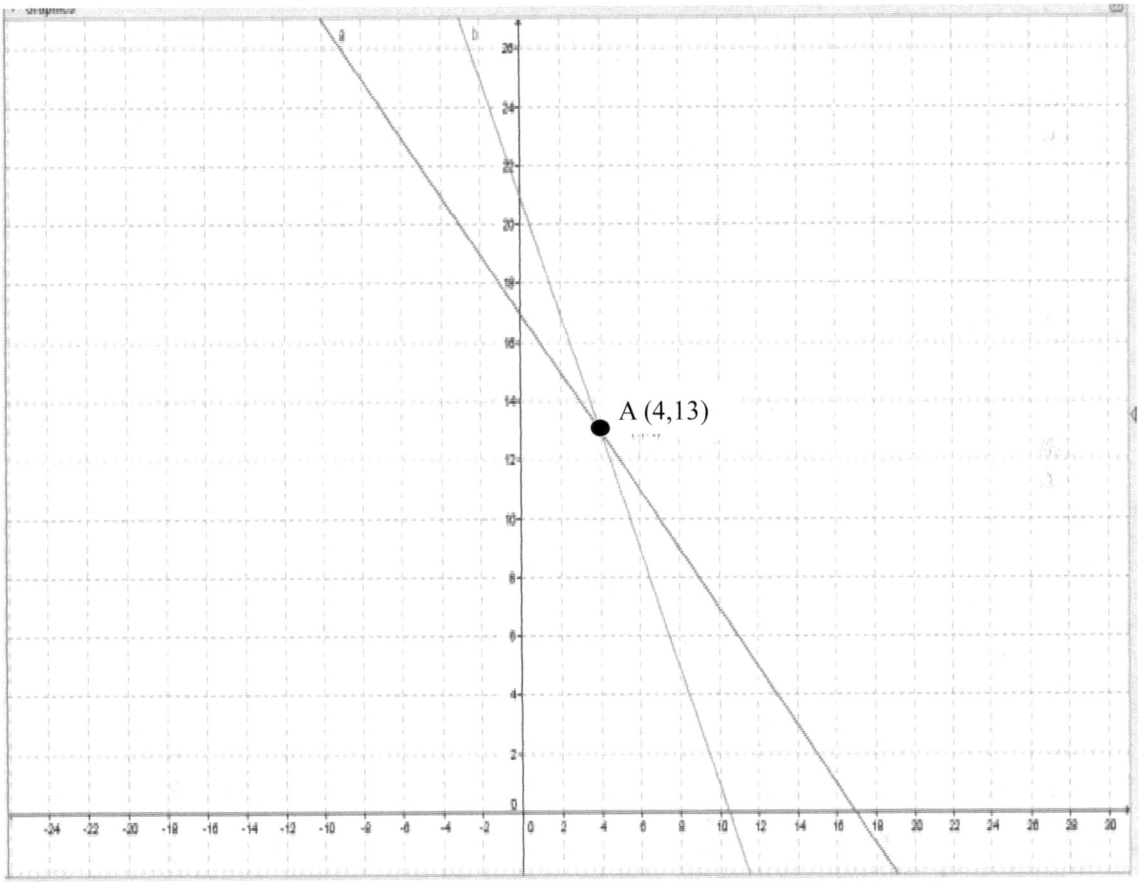

Figure 6. All subject's solution in Solving GBTs 1

GBTs 4	*An airplane trip for five hours is 640 mil. It is flight by different speed. After flying for two hours, the pilot flies the airplane twice faster that the first flight. How fast the speed for the first and second flight in that trip?*

Subject LN, ATR, and FAM used solution method which were different from the subject ESP, NDP, and YE which are still categorized as non-visual. Subject LN, ATR, and FAM use numeric method assisted by spreadsheet display till they found the same value in cell C4 and D4, it was 2560.

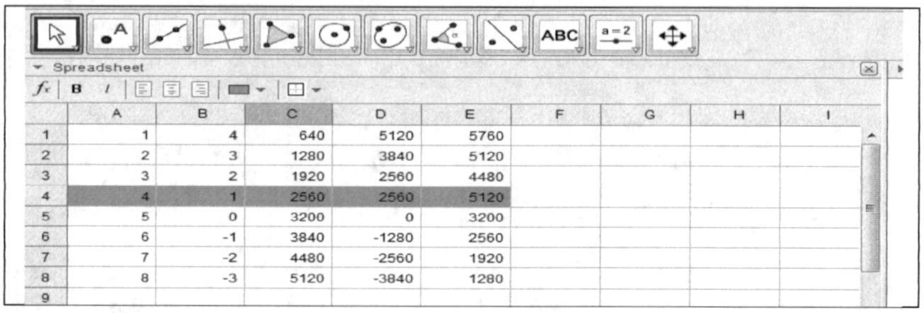

	A	B	C	D	E	F	G	H	I
1	1	4	640	5120	5760				
2	2	3	1280	3840	5120				
3	3	2	1920	2560	4480				
4	4	1	2560	2560	5120				
5	5	0	3200	0	3200				
6	6	-1	3840	-1280	2560				
7	7	-2	4480	-2560	1920				
8	8	-3	5120	-3840	1280				
9									

Figure 7. LN, ATR, and FAM's Solution in Solving GBTs 4

Meanwhile, ESP, NDP, and YE used Algebra representation by opening the window CAS, they assumed the first speed = x and the second flight speed = y until they found the value of $x = 80$ and $y = 160$. They interpreted the first flight speed was 80 mil/hour and the second flight speed was 160 mil/hour.

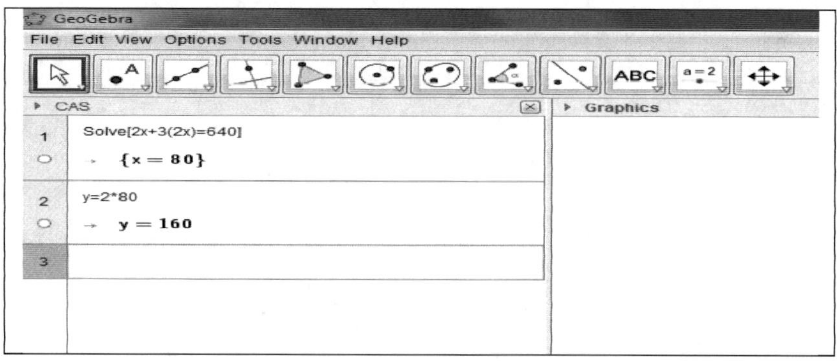

Figure 8. ESP, NDP, and YE's Solution in Solving GBTs 4

GBT 9	*There are 2 students wearing glasses to every 7 students not wearing glasses in a particular classroom. There are 63 students in that classroom. How many students are wearing glasses?*

All subjects of visual, auditory, and kinesthetic learning style solved GBTs 9 using numeric method. They inserted 2 into cell B1 and 7 in cell C1 to represent the number of students who wear glasses and those who don't wear glasses. In the second row, they inserted 4 to cell B2 and 14 in cell C2. They selected cells to create the remnant from the pattern. They drag the pattern till the fifteenth row. After they investigated the seventh row number till it was found 63 in cell D7, they stated that there were 14 persons who wore glasses

.

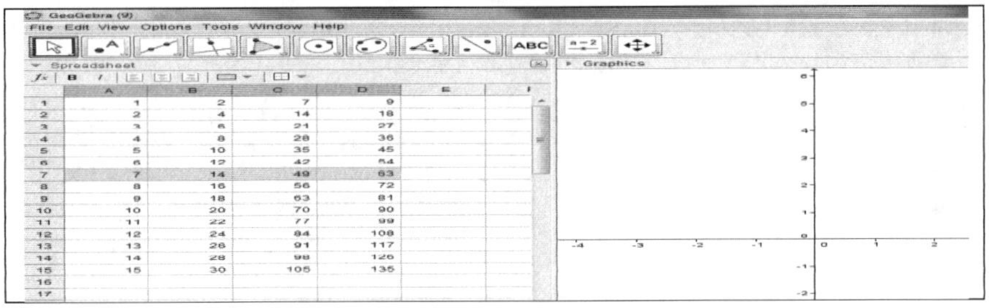

Figure 9. Solution of All Subject in Solving GBTs 9

At the first step, they would like to use graph representation but they couldn't solve the problem using a graph because they couldn't make equations.

4. Discussion

Technology devices give an opportunity to the students to use some representations. However, all the students who have either visual, auditory, or kinesthetic learning style state that they prefer solving GBTs problem using graphic representation because it is accessible easily. The Geogebra program has provided the Algebra window to draw a graph even the smallest scale of that graph.

This result supported the finding of Harskamp et. al. He reported that the function of automatic graphic picture of Geogebra and the availability of coordinate system are the important factor that caused this result. Whereas, the Geogebra did not support the visualizers. The automatic creation from the toolbar in *spreadsheet* page supported the non-visualizer. However, the students tend to use visual solution more often in solving GBTs than those in PBTs [21]. Slavit's research result shows that the students who use a graph is only when they found a problem that should be solved with a graph. This research finding was not support Slavit's finding because the students often use a graph in this study in solving mathematics problem although they did not found problems must be solved using a graph [22].

These overall research findings shows that there were different solution methods used by the students as they accomplish the paper and pencil-based task and Geogebra-based task. All the subjects of this research either students of auditory, kinesthetic, or visual have a powerful preference to think verbal logic as they solve paper and pencil-based task. The most selected main method is a numeric and an algebra, so it can be stated that most of them have non-visual preference thinking as they solve mathematics problem without any Geogebra assistant. Nevertheless, while they are provided by Geogebra, their thinking preference turn to change into a powerful preference to think visually. As they solve Geogebra-based task using a main method, they prefer to use a graph. This evidence proved that Geogebra influences students' visual and non-visual thinking preferences in solving mathematics problems. This research support the previous finding performed by Coskun who found an effect of Geogebra on students' non-visual and visual thinking preferences.

This current study also support the finding of Ruthven [23], Harskamp et al. [21], and Yerushalmy [1]. They reported that the students use different solutions as they solve the problem using technology. Harskamp et al. stated that the group of experiment use a strategy of graph more than a controlled group. However, there is no significance differences between the use of Heuristic (Guess and Check) and Algorithmic strategy [21].

Due to the use of Geogebra affects students preference to use visual representation compared with the paper pencil-based solution, the second usage of this technology might support the second usage of visual and non-visual representation which involve their visual and non-visual thinking. As has been stated in the previous session, researchers argue that the ability to use visual and non-visual representation as well as interpret both of them will affect a deep understanding about mathematics [13, 2].

The National Research Council identifies the significant indicator of conceptual understanding as the ability to use a different representation in the mathematic situation as well as to know how the form of different representation could be beneficial for different purposes [24]. For this reason, there is a close relationship between the representation fluency and conceptual understanding as well as improve the result of representation fluency in developing a conceptual understanding [2].

Some representation theories argue that the most important factor to understand the idea of mathematics is the ability to interrelate the representation to obtain the representational fluency and interpret the ideas of mathematics in different representation [25]. The result of this research shows that the use of a dynamic software facilitate the process of students interpretation to support previous findings performed by Choi-Koh [26] and Huntley et al. [27]. Therefore, the use of a dynamic software might contribute the students' conceptual understanding by supporting their representation fluency.

5. Conclusion and Suggestion

Based on the result of questionnaires, tests, observation, and interview to the students, the researcher found out that Geogebra has influenced students' visual and non-visual thinking preferences in solving mathematics problem for the students who have visual, auditory, and kinesthetic learning style. The students' responses toward PBTs and GBTs reveal the important information about their thinking. They use various type of representations as they solve the mathematics problems. Auditory and kinesthetics' students whose basically prefer non-visual but when they used geogebra, they changed their preference into visual. All the subjects have a difference solution in solving PBTs and GBTs, this difference indicate the changing of their preference to think visually or non-visually as they solve the problem within two different instructional media.

Realizing that this research limits the scope only on mathematics problems that can be solved by the linier equality and two dimension graph, further research can be broaden by changing the students accomplishment in the form of interview, for example; mathematics involvement task that can be solved by using polynomial and three dimension graph. Subjects of the research also can be changed into the higher level; senior high school and university level.

Aknowledgements

The author is thankful to the anonymous referee for valuable comments and kind suggestions

References

[1] Yerushalmy M 2006 Slower algebra students meet faster tools: Solving algebra word problems with graphing software. *Journal for Research in Mathematics Education* **37**(5) 356-387

[2] Coskun S 2011 A Multiple Case Study Investigating The Effect of Technology on Students' Visual and Nonvisual Thinking Preferences: Comparing Paper-Pencil and Dynamic Software Based Strategies of Algebra Word Problem. Unpublished Ph.D Dissertation. University of Central Florida.

[3] Presmeg N C 1985 *The role of visually mediated processes in high school mathematics: A classroom investigation*. Unpublished Ph.D Dissertation. University of CambridgeAnother reference

[4] NCTM 2000 *Principles and Standards for School Mathematics* USA

[5] Presmeg N C 1986 Visualization and Mathematical Giftedness. *Educational Studies in Mathematics* **17** 297-311

[6] Ainsworth S, Bibby P A and Wood D J 1997 Evaluating principles for multi representational learning environments. Paper presented at the 7th European Conference for Research on Learning and Instruction Athens

[7] Seufert T 2003 Supporting coherence formation in learning from multiple representations. *Learning and Instruction* **13** 227-237

[8] Labeke N V and Ainsworth S 2001 *Applying the DeFT framework to the design of multi-*

representational instructional simulations. Paper presented at the AIED'2001 - 10th International Conference on Artificial Intelligence in Education, San Antonio, Texas.

[9] Borba M C and Villarreal M E 2005 Visualization, mathematics education and computer environments. In B. S. Jones, & R. Z. Smith (Eds.), *Humans-with-Media and the Reorganization of Mathematical Thinking* (pp 89) (New York: Springer)

[10] Rif'at M 2001 Pengaruh Pola-pola Pembelajaran Visual dalam Rangka Meningkatkan Kemampuan Menyelesaikan Masalah-Masalah Matematika. UnpublishedDisertation (UPI Bandung)

[11] Aspinwall L, Shaw K L and Presmeg N 1997 Uncontrollable mental imagery: Graphical connections between a function and its derivative. *Educational Studies in Mathematics,* **33**, 301-317

[12] Hohenwarter M 2002 Geogebra - Ein Software system für dynamische Geometrie und Algebra der Ebene Master thesis University of Salzburg

[13] de Jong T and Joolingen W R V 1998 Scientific discovery learning with computer simulations of conceptual domains. *Review of Educational Research* **68** 179-201

[14] Pitta-Pantazi D and Christou C 2009 Cognitive styles, task presentation mode and mathematical performance. *Research in Mathematics Education* **11**(2) 131-148

[15] Bishop A J 1989 Review of the research on visualization in mathematics education. *Focus On Learning Problems in Mathematics* **11** (1) 7-16

[16] Riding R and Douglas G 1993 The effect of cognitive style and mode of presentation on learning performance. *British Journal of Educational Psychology* **63** 297-307

[17] Yerushalmy M 1997 Emergence for new schemes for solving algebra word problems: The impact of technology and the function approach. In E. Pehkonen (Ed.), *Proceedings of the 21st PME International Conference* **1** 165-172

[18] Yerushalmy M, Shternberg G and Gilead S 1999 Visualization as a vehicle for meaningful problem solving in algebra. In O Zaslavsky (Ed) *Proceedings of the 23rd PME International Conference* **1** 197-211

[19] Sagitasari Dewi A 2010 Hubungan antara Kreativitas dan Gaya Belajar dengan Prestasi Belajar Matematika Siswa SMP Unpublished Essay Universitas Negeri Yogyakarta

[20] Miles M B and Huberman A M 1994 *Qualitative Data Analysis*. New Delhi: SAGE Publication.

[21] Harskamp E, Suhre C and Streun A V 2000 The graphics calculator and students' solution strategies *Mathematics Education Research Journal* **12**(1) 37-52

[22] Slavit D 1998 Three women's understandings of algebra in a Precalculus course integrated with the graphing calculator. *Journal of Mathematical Behavior* **17**(3) 355-372

[23] Ruthven K 1990 The influence of graphic calculator use on translation from graphic to symbolic forms *Educational Studies in Mathematics* **21** 431-450

[24] National Research Council 2001 *Adding it up: Helping children learn mathematics*

[25] Kilpatrick J, Swafford J and Findell B (Eds.) Mathematics Learning Study Committee, Center for Education, Division of Behavioral and Social Sciences and Education. Washington, DC: National Academy Press.

[26] Nemirovsky R and Noble T 1997 On mathematical visualization and the place where we live *Educational Studies in Mathematics* **33** 99-131

[27] Pierce R and Stacey K 2010 Mapping pedagogical opportunities provided by mathematics analysis software. *International Journal of Computers for Mathematical Learning*.

[28] Choi-Koh S S 2003 Effect of a graphing calculator on a 10th-grade student's study of trigonometry *The Journal of Educational Research* **96**(6) 359-369

[29] Huntley M A, Rasmussen C L, Villarubi R S, Sangtong J and Fey J T 2000 Effects of Standards-based mathematics education: A study of the Core-Plus Mathematics Project algebra and functions strand. *Journal for Research in Mathematics Education* **31**(3) 328-361